Graduate Texts in Mathematics 242

Graduate Texts in Mathematics

(continued after index)

Pierre Antoine Grillet

Abstract Algebra

Second Edition

Pierre Antoine Grillet
Dept. Mathematics
Tulane University
New Orleans, LA 70118
USA
grillet@math.tulane.edu

Mathematics Subject Classification (2000): 20-01 16-01

ISBN-13: 978-1-4419-2450-6   eISBN-13: 978-0-387-71568-1

Printed on acid-free paper.

springer.com

Dedicated in gratitude to

Anthony Haney
Jeff and Peggy Sue Gillis
Bob and Carol Hartt
Nancy Heath
Brandi Williams
H.L. Shirrey
Bill and Jeri Phillips

and all the other angels of the Katrina aftermath, with special thanks to

Ruth and Don Harris

Preface

This book is a basic algebra text for first-year graduate students, with some additions for those who survive into a second year. It assumes that readers know some linear algebra, and can do simple proofs with sets, elements, mappings, and equivalence relations. Otherwise, the material is self-contained. A previous semester of abstract algebra is, however, highly recommended.

Algebra today is a diverse and expanding field of which the standard contents of a first-year course no longer give a faithful picture. Perhaps no single book can; but enough additional topics are included here to give students a fairer idea. Instructors will have some flexibility in devising syllabi or additional courses; students may read or peek at topics not covered in class.

Diagrams and universal properties appear early to assist the transition from proofs with elements to proofs with arrows; but categories and universal algebras, which provide conceptual understanding of algebra in general, but require more maturity, have been placed last. The appendix has rather more set theory than usual; this puts Zorn's lemma and cardinalities on a reasonably firm footing.

The author is fond of saying (some say, overly fond) that algebra is like French pastry: wonderful, but cannot be learned without putting one's hands to the dough. Over 1400 exercises will encourage readers to do just that. A few are simple proofs from the text, placed there in the belief that useful facts make good exercises. Starred problems are more difficult or have more extensive solutions.

Algebra owes its name, and its existence as a separate branch of mathematics, to a ninth-century treatise on quadratic equations, *Al-jabr wa'l muqabala*, "the balancing of related quantities", written by the Persian mathematician al-Khowarizmi. (The author is indebted to Professor Boumedienne Belkhouche for this translation.) Algebra retained its emphasis on polynomial equations until well into the nineteenth century, then began to diversify. Around 1900, it headed the revolution that made mathematics abstract and axiomatic. William Burnside and the great German algebraists of the 1920s, most notably Emil Artin, Wolfgang Krull, and Emmy Noether, used the clarity and generality of the new mathematics to reach unprecedented depth and to assemble what was then called modern algebra. The next generation, Garrett Birkhoff, Saunders MacLane, and others, expanded its scope and depth but did not change its character. This history is documented by brief notes and references to the original papers. Time pressures, sundry events, and the state of the local libraries have kept these references a bit short of optimal completeness, but they should suffice to place results in their historical context, and may encourage some readers to read the old masters.

This book is a second edition of *Algebra*, published by the good folks at Wiley in 1999. I meant to add a few topics and incorporate a number of useful comments, particularly from Professor Garibaldi, of Emory University. I ended up rewriting the whole book from end to end. I am very grateful for this chance to polish a major work, made possible by Springer, by the patience and understanding of my editor, Mark Spencer, by the inspired thoroughness of my copy editor, David Kramer, and by the hospitality of the people of Marshall and Scottsville.

Readers who are familiar with the first version will find many differences, some of them major. The first chapters have been streamlined for rapid access to solvability of equations by radicals. Some topics are gone: groups with operators, Lüroth's theorem, Sturm's theorem on ordered fields. More have been added: separability of transcendental extensions, Hensel's lemma, Gröbner bases, primitive rings, hereditary rings, Ext and Tor and some of their applications, subdirect products. There are some 450 more exercises. I apologize in advance for the new errors introduced by this process, and hope that readers will be kind enough to point them out.

New Orleans, Louisiana, and Marshall, Texas, 2006.

Contents

I
Groups

Group theory arose from the study of polynomial equations. The solvability of an equation is determined by a group of permutations of its roots; before Abel [1824] and Galois [1830] mastered this relationship, it led Lagrange [1770] and Cauchy [1812] to investigate permutations and prove forerunners of the theorems that bear their names. The term "group" was coined by Galois. Interest in groups of transformations, and in what we now call the classical groups, grew after 1850; thus, Klein's *Erlanger Programme* [1872] emphasized their role in geometry. Modern group theory began when the axiomatic method was applied to these results; Burnside's *Theory of Groups of Finite Order* [1897] marks the beginning of a new discipline, abstract algebra, in that structures are defined by axioms, and the nature of their elements is irrelevant.

Today, groups are one of the fundamental structures of algebra; they underlie most of the other objects we shall encounter (rings, fields, modules, algebras) and are widely used in other branches of mathematics. Group theory is also an active area of research with major recent achievements.

This chapter contains the definitions and basic examples and properties of semigroups, groups, subgroups, homomorphisms, free groups, and presentations. Its one unusual feature is Light's test of associativity, that helps with presentations. The last section (free products) may be skipped.

1. Semigroups

Semigroups are sets with an associative binary operation. This section contains simple properties and examples that will be useful later.

Definition. A binary operation on a set S is a mapping of the Cartesian product $S \times S$ into S.

For example, addition and multiplication of real numbers are binary operations on the set $\mathbb{R}$ of all real numbers. The set $\mathbb{N}$ of all natural numbers 1, 2, ..., n, ..., the set $\mathbb{Z}$ of all integers, the set $\mathbb{Q}$ of all rational numbers, and the set $\mathbb{C}$ of all complex numbers have similar operations. Addition and multiplication of matrices also provide binary operations on the set $M_n(\mathbb{R})$ of all

$n \times n$ matrices with coefficients in $\mathbb{R}$, for any given integer $n > 0$. Some size restriction is necessary here, since arbitrary matrices cannot always be added or multiplied, whereas a binary operation $S \times S \longrightarrow S$ must be defined at every $(x, y) \in S \times S$ (for every $x, y \in S$). (General matrix addition and multiplication are *partial* operations, not always defined.)

More generally, an *n-ary operation* on a set S is a mapping of the Cartesian product $S^n = S \times S \times \cdots \times S$ of n copies of S into S. Most operations in algebra are binary, but even in this chapter we encounter two other types. The *empty* Cartesian product S^0 is generally defined as one's favorite one-element set, perhaps $\{0\}$ or $\{\emptyset\}$; a 0-ary or *constant* operation on a set S is a mapping $f : \{0\} \longrightarrow S$ and simply selects one element $f(0)$ of S. The Cartesian product S^1 is generally defined as S itself; a 1-ary operation or *unary* operation on S is a mapping of S into S (a *transformation* of S).

For binary operations $f : S \times S \longrightarrow S$, two notations are in wide use. In the *additive* notation, $f(x, y)$ is denoted by $x + y$; then f is an *addition*. In the *multiplicative* notation, $f(x, y)$ is denoted by xy or by $x \cdot y$; then f is a *multiplication*. In this chapter we mostly use the multiplicative notation.

Definition. Let S be a set with a binary operation, written multiplicatively. An identity element *of S is an element e of S such that $ex = x = xe$ for all $x \in S$.*

Readers will easily show that an identity element, if it exists, is unique. In the multiplicative notation, we usually denote the identity element, if it exists, by 1. Almost all the examples above have identity elements.

Products. A binary multiplication provides products only of two elements. Longer products, with terms $x_1, x_2, \ldots, x_n$, must break into products of two shorter products, with terms $x_1, x_2, \ldots, x_k$ and $x_{k+1}, x_{k+2}, \ldots, x_n$ for some $1 \leq k < n$. It is convenient also to define 1-term products and empty products:

Definition. Let S be a set with a binary operation, written multiplicatively. Let $n \geq 1$ ($n \geq 0$, if an identity element exists) and let $x_1, x_2, \ldots, x_n \in S$.

If $n = 1$, then $x \in S$ is a product of $x_1, x_2, \ldots, x_n$ *(in that order)* if and only *if $x = x_1$. If S has an identity element 1 and $n = 0$, then $x \in S$ is a* product of $x_1, x_2, \ldots, x_n$ *(in that order)* if and only if $x = 1$.

If $n \geq 2$, then $x \in S$ is a product of $x_1, x_2, \ldots, x_n$ *(in that order)* if and only if, *for some $1 \leq k < n$, x is a product $x = yz$ of a product y of $x_1, \ldots, x_k$ (in that order) and a product z of $x_{k+1}, \ldots, x_n$ (in that order).*

Our definition of empty products is not an exercise in Zen Buddhism (even though its contemplation might lead to enlightenment). Empty products are defined as 1 because if we multiply, say, xy by an empty product, that adds no new term, the result should be xy.

In the definition of products with $n = 2$ terms, necessarily $k = 1$, so that $x \in S$ is a product of x_1 and x_2 (in that order) if and only if $x = x_1 x_2$.

If $n = 3$, then $k = 1$ or $k = 2$, and $x \in S$ is a product of x_1, x_2, x_3 (in that order) if and only if $x = yz$, where either $y = x_1$ and $z = x_2 x_3$ (if $k = 1$), or $y = x_1 x_2$ and $z = x_3$ (if $k = 2$); that is, either $x = x_1 (x_2 x_3)$ or $x = (x_1 x_2) x_3$. Readers will work out the cases $n = 4, 5$.

Associativity avoids unseemly proliferations of products.

Definition. A binary operation on a set S (written multiplicatively) is associative when $(xy)z = x(yz)$ for all $x, y, z \in S$.

Thus, associativity states that products with three terms do not depend on the placement of parentheses. This extends to all products: more courageous readers will write a proof of the following property:

Proposition **1.1.** *Under an associative multiplication, all products of n given elements $x_1, x_2, \ldots, x_n$ (in that order) are equal.*

Then *the* product of $x_1, x_2, \ldots, x_n$ (in that order) is denoted by $x_1 x_2 \cdots x_n$.

An even stronger result holds when terms can be permuted.

Definition. A binary operation on a set S (written multiplicatively) is commutative when $xy = yx$ for all $x, y \in S$.

Recall that a *permutation* of $1, 2, \ldots, n$ is a bijection of $\{1, 2, \ldots, n\}$ onto $\{1, 2, \ldots, n\}$. Readers who are familiar with permutations may prove the following:

Proposition **1.2.** *Under a commutative and associative multiplication, $x_{\sigma(1)} x_{\sigma(2)} \cdots x_{\sigma(n)} = x_1 x_2 \cdots x_n$ for every permutation σ of $1, 2, \ldots, n$.*

Propositions 1.1 and 1.2 are familiar properties of sums and products in $\mathbb{N}$, $\mathbb{Q}$, $\mathbb{R}$, and $\mathbb{C}$. Multiplication in $M_n(\mathbb{R})$, however, is associative but not commutative (unless $n = 1$).

Definitions. A semigroup *is an ordered pair of a set S, the* underlying set *of the semigroup, and one associative binary operation on S. A semigroup with an identity element is a* monoid. *A semigroup or monoid is* commutative *when its operation is commutative.*

It is customary to denote a semigroup and its underlying set by the same letter, when this creates no ambiguity. Thus, $\mathbb{Z}$, $\mathbb{Q}$, $\mathbb{R}$, and $\mathbb{C}$ are commutative monoids under addition and commutative monoids under multiplication; the multiplicative monoid $M_n(\mathbb{R})$ is not commutative when $n > 1$.

Powers are a particular case of products.

Definition. Let S be a semigroup (written multiplicatively). Let $a \in S$ and let $n \geq 1$ be an integer ($n \geq 0$ if an identity element exists). The nth power a^n of a is the product $x_1 x_2 \cdots x_n$ in that $x_1 = x_2 = \cdots = x_n = a$.

Propositions 1.1 and 1.2 readily yield the following properties:

Proposition **1.3.** *In a semigroup* S *(written multiplicatively) the following properties hold for all* $a \in S$ *and all integers* $m, n \geqq 1$ ($m, n \geqq 0$ *if an identity element exists):*

(1) $a^m a^n = a^{m+n}$;

(2) $(a^m)^n = a^{mn}$;

(3) *if there is an identity element* 1, *then* $a^0 = 1 = 1^n$;

(4) *if* S *is commutative, then* $(ab)^n = a^n b^n$ *(for all* $a, b \in S$*).*

Subsets are multiplied as follows.

Definition. In a set S *with a multiplication, the* product *of two subsets* A *and* B *of* S *is* $AB = \{ ab \mid a \in A, \ b \in B \}$.

In other words, $x \in AB$ if and only if $x = ab$ for some $a \in A$ and $b \in B$. Readers will easily prove the following result:

Proposition **1.4.** *If the multiplication on a set* S *is associative, or commutative, then so is the multiplication of subsets of* S.

The additive notation. In a semigroup whose operation is denoted additively, we denote the identity element, if it exists, by 0; the product of $x_1, x_2, \ldots, x_n$ (in that order) becomes their *sum* $x_1 + x_2 + \cdots + x_n$; the nth power of $a \in S$ becomes the *integer multiple* na (the sum $x_1 + x_2 + \cdots + x_n$ in that $x_1 = x_2 = \cdots = x_n = a$); the product of two subsets A and B becomes their *sum* $A + B$. Propositions 1.1, 1.2, and 1.3 become as follows:

Proposition **1.5.** *In an additive semigroup* S, *all sums of* n *given elements* $x_1, x_2, \ldots, x_n$ *(in that order) are equal; if* S *is commutative, then all sums of* n *given elements* $x_1, x_2, \ldots, x_n$ *(in any order) are equal.*

Proposition **1.6.** *In an additive semigroup* S *the following properties hold for all* $a \in S$ *and all integers* $m, n \geq 1$ ($m, n \geqq 0$ *if an identity element exists):*

(1) $ma + na = (m + n) a$;

(2) $m(na) = (mn) a$;

(3) *if there is an identity element* 0, *then* $0a = 0 = n0$;

(4) *if* S *is commutative, then* $n(a + b) = na + nb$ *(for all* $a, b \in S$*).*

Light's test. Operations on a set S with few elements (or with few kinds of elements) can be conveniently defined by a square *table*, whose rows and columns are labeled by the elements of S, in that the row of x and column of y intersect at the product xy (or sum $x + y$).

Example **1.7.**

	a	b	c	d
a	a	b	c	b
b	b	c	a	c
c	c	a	b	a
d	b	c	a	c

For example, the table of Example 1.7 above defines an operation on the set $\{a, b, c, d\}$, in that, say, $da = b$, $db = c$, etc.

Commutativity is shown in such a table by symmetry about the main diagonal. For instance, Example 1.7 is commutative. Associativity, however, is a different kettle of beans: the 4 elements of Example 1.7 beget 64 triples (x, y, z), each with two products $(xy)z$ and $x(yz)$ to compare. This chore is made much easier by *Light's associativity test* (from Clifford and Preston [1961]).

Light's test constructs, for each element y, a *Light's table* of the binary operation $(x, z) \longmapsto (xy)z$: the column of y, that contains all products xy, is used to label the rows; the row of xy is copied from the given table and contains all products $(xy)z$. The row of y, that contains all the products yz, is used to label the columns. If the column labeled by yz in Light's table coincides with the column of yz in the original table, then $(xy)z = x(yz)$ for all x.

Definition. If, for every z, the column labeled by yz in Light's table coincides with the column of yz in the original table, then the element y passes Light's test. Otherwise, y fails Light's test.

In Example 1.7, $y = d$ passes Light's test: its Light's table is

d	b	c	a	c
b	b	c	a	c
c	c	a	b	a
a	a	b	c	b
c	c	a	b	a

On the other hand, in the following example (table on left), a fails Light's test: the column of b in Light's table of a does not match the column of b in the original table. The two mismatches indicate that $a(aa) \neq (aa)a$ and $b(aa) \neq (ba)a$:

	a	b	c
a	b	c	c
b	a	c	c
c	c	c	c

a	b	c	c
b	a	c	c
a	b	c	c
c	c	c	c

Example   Light's table of a

Associativity requires that *every* element pass Light's test. But some elements can usually be skipped, due to the following result, left to readers:

Proposition **1.8.** *Let S be a set with a multiplication and let X be a subset of S. If every element of S is a product of elements of X, and every element of X passes Light's test, then every element of S passes Light's test (and the operation on S is associative).*

In Example 1.7, $d^2 = c$, $dc = a$, and $da = b$, so that a, b, c, d all are products of d's; since d passes Light's test, Example 1.7 is associative.

Free semigroups. One useful semigroup F is constructed from an arbitrary set X so that $X \subseteq F$ and every element of F can be written uniquely as a product of elements of X. The elements of F are all finite sequences $(x_1, x_2, \ldots, x_n)$ of elements of X. The multiplication on F is *concatenation*:

$$(x_1, x_2, \ldots, x_n)(y_1, y_2, \ldots, y_m) = (x_1, x_2, \ldots, x_n, y_1, y_2, \ldots, y_m).$$

It is immediate that concatenation is associative. The *empty* sequence $()$ is an identity element. Moreover, every sequence can be written uniquely as a product of one-term sequences:

$$(x_1, x_2, \ldots, x_n) = (x_1)(x_2) \cdots (x_n).$$

If every element x of X is identified with the corresponding one-term sequence (x), then $X \subseteq F$ and every element of F can be written uniquely as a product of elements of X. The usual notation makes this identification transparent by writing every sequence $(x_1, x_2, \ldots, x_n)$ as a product or *word* $x_1 x_2 \cdots x_n$ in the *alphabet* X. (This very book can now be recognized as a long dreary sequence of words in the English alphabet.)

Definition. The free semigroup *on a set* X *is the semigroup of all finite nonempty sequences of elements of* X. *The* free monoid *on a set* X *is the semigroup of all finite (possibly empty) sequences of elements of* X.

For instance, the free monoid on a one-element set $\{x\}$ consists of all words $1, x, xx, xxx, \ldots, xx \cdots x, \ldots$, that is, all powers of x, no two of that are equal. This semigroup is commutative, by Proposition 1.12. Free semigroups on larger alphabets $\{x, y, \ldots\}$ are not commutative, since the sequences xy and yx are different when x and y are different. Free monoids are a basic tool of mathematical linguistics, and of the theory of computation.

Free commutative semigroups. The free commutative semigroup C on a set X is constructed so that $X \subseteq C$, C is a commutative semigroup, and every element of C can be written uniquely, up to the order of the terms, as a product of elements of X. At this time we leave the general case to interested readers and assume that X is finite, $X = \{x_1, x_2, \ldots, x_n\}$. In the commutative semigroup C, a product of elements of X can be rewritten as a product of positive powers of distinct elements of X, or as a product $x_1^{a_1} x_2^{a_2} \cdots x_n^{a_n}$ of nonnegative powers of all the elements of X. These products look like monomials and are multiplied in the same way:

$$\left(x_1^{a_1} x_2^{a_2} \cdots x_n^{a_n}\right)\left(x_1^{b_1} x_2^{b_2} \cdots x_n^{b_n}\right) = x_1^{a_1+b_1} x_2^{a_2+b_2} \cdots x_n^{a_n+b_n}.$$

Formally, the free commutative monoid C on $X = \{x_1, x_2, \ldots, x_n\}$ is the set of all mappings $x_i \longmapsto a_i$ that assign to each $x_i \in X$ a nonnegative integer a_i; these mappings are normally written as *monomials* $x_1^{a_1} x_2^{a_2} \cdots x_n^{a_n}$, and multiplied as above. The identity element is $x_1^0 x_2^0 \cdots x_n^0$. Each $x_i \in X$ may be identified with the monomial $x_1^0 \cdots x_{i-1}^0 x_i^1 x_{i+1}^0 \cdots x_n^0$; then every monomial $x_1^{a_1} x_2^{a_2} \cdots x_n^{a_n}$ is a product of nonnegative powers $x_1^{a_1}, x_2^{a_2}, \ldots, x_n^{a_n}$ of $x_1, x_2, \ldots, x_n$, uniquely up to the order of the terms.

Definition. The free commutative monoid *on a finite set* $X = \{x_1, x_2, \ldots, x_n\}$ *is the semigroup of all monomials* $x_1^{a_1} x_2^{a_2} \cdots x_n^{a_n}$ *(with nonnegative integer exponents); the* free commutative semigroup *on* $X = \{x_1, x_2, \ldots, x_n\}$ *is the semigroup of all monomials* $x_1^{a_1} x_2^{a_2} \cdots x_n^{a_n}$ *with positive degree* $a_1 + a_2 + \cdots + a_n$.

For instance, the free commutative monoid on a one-element set $\{x\}$ consists of all (nonnegative) powers of x: $1 = x^0, x, x^2, \ldots, x^n, \ldots$, no two of that are equal; this monoid is also the free monoid on $\{x\}$.

Exercises

1. Write all products of x_1, x_2, x_3, x_4 (in that order), using parentheses as necessary.

2. Write all products of x_1, x_2, x_3, x_4, x_5 (in that order).

3. Count all products of $x_1, \ldots, x_n$ (in that order) when $n = 6$; $n = 7$; $n = 8$.

*4. Prove the following: in a semigroup, all products of $x_1, x_2, \ldots, x_n$ (in that order) are equal.

5. Show that a binary operation has at most one identity element (so that an identity element, if it exists, is unique).

*6. Prove the following: in a commutative semigroup, all products of $x_1, x_2, \ldots, x_n$ (in any order) are equal. (This exercise requires some familiarity with permutations.)

7. Show that multiplication in $M_n(\mathbb{R})$ is not commutative when $n > 1$.

8. Find two 2×2 matrices A and B (with real entries) such that $(AB)^2 \neq A^2 B^2$.

9. In a semigroup (written multiplicatively) multiplication of subsets is associative.

10. Show that the semigroup of subsets of a monoid is also a monoid.

11. Show that products of subsets distribute unions: for all subsets A, B, A_i, B_j,

$$\left(\bigcup_{i \in I} A_i\right) B = \bigcup_{i \in I} (A_i B) \quad \text{and} \quad A\left(\bigcup_{j \in J} B_j\right) = \bigcup_{j \in J} (A B_j).$$

12. Let S be a set with a binary operation (written multiplicatively) and let X be a subset of S. Prove the following: if every element of S is a product of elements of X, and every element of X passes Light's test, then every element of S passes Light's test.

13,14,15. Test for associativity:

	a b c d		a b c d		a b c d
a	a b a b	a	a b a b	a	a b c d
b	a b a b	b	b a d c	b	b a d c
c	c d c d	c	a b c d	c	c d c d
d	c d c d	d	d c d c	d	d c d c
	Exercise 13		Exercise 14		Exercise 15

16. Construct a free commutative monoid on an arbitrary (not necessarily finite) set.

2. Groups

This section gives the first examples and properties of groups.

Definition. A group *is an ordered pair of a set* G *and one binary operation on that set* G *such that*

(1) *the operation is associative;*

(2) *there is an identity element;*

(3) (*in the multiplicative notation*) *every element* x *of* G *has an* inverse (*there is an element* y *of* G *such that* $xy = yx = 1$).

In this definition, the set G is the *underlying set* of the group. It is customary to denote a group and its underlying set by the same letter. We saw in Section 1 that the identity element of a group is unique; readers will easily show that inverses are unique (an element of a group has only one inverse in that group).

In the multiplicative notation the inverse of x is denoted by x^{-1}. In the additive notation, the identity element is denoted by 0; the inverse of x becomes its *opposite* (the element y such that $x + y = y + x = 0$) and is denoted by $-x$.

Groups can be defined more compactly as monoids in that every element has an inverse (or an opposite). Older definitions started with a fourth axiom, that every two elements of a group have a unique product (or sum) in that group. We now say that a group has a binary operation. When showing that a bidule is a group, however, it is sensible to first make sure that the bidule does have a binary operation, that is, that every two elements of the bidule have a unique product (or sum) in that bidule. (*Bidule* is the author's name for unspecified mathematical objects.)

Examples. Number systems provide several examples of groups. $(\mathbb{Z}, +)$, $(\mathbb{Q}, +)$, $(\mathbb{R}, +)$, and $(\mathbb{C}, +)$ all are groups. But $(\mathbb{N}, +)$ is not a group, and $\mathbb{Z}$, $\mathbb{Q}$, $\mathbb{R}$, $\mathbb{C}$ are not groups under multiplication, since their element 0 has no inverse. However, nonzero rational numbers, nonzero real numbers, nonzero complex numbers, all constitute groups under multiplication; so do positive rational numbers, positive real numbers, and complex numbers with absolute value 1.

The set of all $n \times n$ matrices (with entries in $\mathbb{R}$, or in any given field) is a group under addition, but not under multiplication; however, invertible $n \times n$ matrices constitute a group under multiplication. So do, more generally, invertible linear transformations of a vector space into itself.

In algebraic topology, the homotopy classes of paths from x to x in a space X constitute the *fundamental group* $\pi_1(X, x)$ of X at x.

The *permutations* of a set X (the bijections of X onto itself) constitute a group under composition, the *symmetric group* S_X on X. The symmetric group S_n on $\{1, 2, \ldots, n\}$ is studied in some detail in the next chapter.

Small groups may be defined by tables. If the identity element is listed first, then the row and column labels of a table duplicate its first row and column, and are usually omitted. For example, the *Klein four-group* (Viergruppe) $V_4 = \{1, a, b, c\}$ is defined by either table below:

V_4	1	a	b	c
1	1	a	b	c
a	a	1	c	b
b	b	c	1	a
c	c	b	a	1

1	a	b	c
a	1	c	b
b	c	1	a
c	b	a	1

Readers will verify that V_4 is indeed a group.

Dihedral groups. Euclidean geometry relies for "equality" on *isometries*, that are permutations that preserve distances. In the Euclidean plane, isometries can be classified into translations (by a fixed vector), rotations about a point, and symmetries about a straight line. If an isometry sends a geometric configuration onto itself, then the inverse isometry also sends that geometric configuration onto itself, so that isometries with this property constitute a group under composition, the *group of isometries* of the configuration, also called the *group of rotations and symmetries* of the configuration if no translation is involved. These groups are used in crystallography, and in quantum mechanics.

Definition. The dihedral group D_n *of a regular polygon with* $n \geq 2$ *vertices is the group of rotations and symmetries of that polygon.*

A regular polygon P with $n \geq 2$ vertices has a center and has n axes of symmetry that intersect at the center. The isometries of P onto itself are the n symmetries about these axes and the n rotations about the center by multiples of $2\pi/n$. In what follows, we number the vertices counterclockwise $0, 1, \ldots, n-1$, and number the axes of symmetry counterclockwise, $0, 1, \ldots, n-1$, so that vertex 0 lies on axis 0; s_i denotes the symmetry about axis i and r_i denotes the rotation by $2\pi i/n$ about the center. Then $D_n = \{r_0, r_1, \ldots, r_{n-1}, s_0, s_1, \ldots, s_{n-1}\}$; the identity element is $r_0 = 1$. It is convenient to define r_i and s_i for every integer i so that $r_{i+n} = r_i$ and $s_{i+n} = s_i$ for all i. (This amounts to indexing modulo n.)

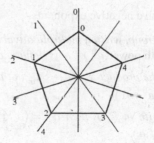

Compositions can be found as follows. First, $r_i \circ r_j = r_{i+j}$ for all i and j. Next, geometry tells us that following the symmetry about a straight line

L by the symmetry about a straight line L' that intersects L amounts to a rotation about the intersection by twice the angle from L to L'. Since the angle from axis j to axis i is $\pi\,(i-j)/n$, it follows that $s_i \circ s_j = r_{i-j}$. Finally, $s_i \circ s_i = s_j \circ s_j = 1$; hence $s_j = s_i \circ r_{i-j}$ and $s_i = r_{i-j} \circ s_j$, equivalently $s_i \circ r_k = s_{i-k}$ and $r_k \circ s_j = s_{k+j}$, for all i, j, k. This yields a (compact) composition table for D_n:

D_n	r_j	s_j
r_i	r_{i+j}	s_{i+j}
s_i	s_{i-j}	r_{i-j}

Properties. Groups inherit all the properties of semigroups and monoids in Section 1. Thus, for any $n \geq 0$ elements $x_1, \ldots, x_n$ of a group (written multiplicatively) all products of $x_1, \ldots, x_n$ (in that order) are equal (Proposition 1.1); multiplication of subsets

$$AB = \{\, ab \mid a \in A,\ b \in B \,\}$$

is associative (Proposition 1.3). But groups have additional properties.

Proposition **2.1.** *In a group, written multiplicatively, the* cancellation laws *hold:* $xy = xz$ *implies* $y = z$, *and* $yx = zx$ *implies* $y = z$. *Moreover, the equations* $ax = b$, $ya = b$ *have unique solutions* $x = a^{-1} b$, $y = b a^{-1}$.

Proof. $xy = xz$ implies $y = 1y = x^{-1} xy = x^{-1} xz = 1z = z$, and similarly for $yx = zx$. The equation $ax = b$ has at most one solution $x = a^{-1} ax = a^{-1} b$, and $x = a^{-1} b$ is a solution since $a\,a^{-1} b = 1b = b$. The equation $ya = b$ is similar. $\square$

Proposition **2.2.** *In a group, written multiplicatively,* $(x^{-1})^{-1} = x$ *and* $(x_1 x_2 \cdots x_n)^{-1} = x_n^{-1} \cdots x_2^{-1} x_1^{-1}$.

Proof. In a group, $uv = 1$ implies $v = 1v = u^{-1} uv = u^{-1}$. Hence $x^{-1} x = 1$ implies $x = (x^{-1})^{-1}$. We prove the second property when $n = 2$ and leave the general case to our readers: $x\,y\,y^{-1}\,x^{-1} = x\,1\,x^{-1} = 1$; hence $y^{-1} x^{-1} = (xy)^{-1}$. $\square$

Powers in a group can have negative exponents.

Definition. Let G be a group, written multiplicatively. Let $a \in G$ and let n be an arbitrary integer. The nth power a^n *of a is defined as follows:*

(1) *if $n \geq 0$, then a^n is the product $x_1 x_2 \cdots x_n$ in that $x_1 = x_2 = \cdots = x_n = a$ (in particular, $a^1 = a$ and $a^0 = 1$);*

(2) *if $n \leq 0$, $n = -m$ with $m \geq 0$, then $a^n = (a^m)^{-1}$ (in particular, the -1 power a^{-1} is the inverse of a).*

Propositions 1.3 and 2.2 readily yield the following properties:

Proposition **2.3.** *In a group G (written multiplicatively) the following proper-ties hold for all $a \in S$ and all integers m, n:*

(1) $a^0 = 1$, $a^1 = a$;

(2) $a^m a^n = a^{m+n}$;

(3) $(a^m)^n = a^{mn}$;

(4) $(a^n)^{-1} = a^{-n} = (a^{-1})^n$.

The proof makes an awful exercise, inflicted upon readers for their own good.

Corollary **2.4.** *In a finite group, the inverse of an element is a positive power of that element.*

Proof. Let G be a finite group and let $x \in G$. Since G is finite, the powers x^n of x, $n \in \mathbb{Z}$, cannot be all distinct; there must be an equality $x^m = x^n$ with, say, $m < n$. Then $x^{n-m} = 1$, $x \, x^{n-m-1} = 1$, and $x^{-1} = x^{n-m-1} = x^{n-m-1} x^{n-m}$ is a positive power of x. $\square$

The additive notation. Commutative groups are called *abelian*, and the addi-tive notation is normally reserved for abelian groups.

As in Section 1, in the additive notation, the identity element is denoted by 0; the product of $x_1, x_2, \ldots, x_n$ becomes their sum $x_1 + x_2 + \cdots + x_n$; the product of two subsets A and B becomes their sum

$$A + B = \{ a + b \mid a \in A, \ b \in B \}.$$

Proposition 2.1 yields the following:

Proposition **2.5.** *In an abelian group G (written additively), $-(-x) = x$ and $-(x_1 + x_2 + \cdots + x_m) = (-x_1) + (-x_2) + \cdots + (-x_m)$.*

In the additive notation, the nth power of $a \in S$ becomes the *integer multiple* na: if $n \geqq 0$, then na is the sum $x_1 + x_2 + \cdots + x_n$ in that $x_1 = x_2 = \cdots = x_n = a$; if $n = -m \leqq 0$, then na is the sum $-(x_1 + x_2 + \cdots + x_m) = (-x_1) + (-x_2) + \cdots + (-x_m)$ in which $x_1 = x_2 = \cdots = x_m = -a$. By 1.3, 2.3:

Proposition **2.6.** *In an abelian group G (written additively) the following properties hold for all $a, b \in G$ and all integers m, n:*

(1) $ma + na = (m + n) a$;

(2) $m(na) = (mn) a$;

(3) $0a = 0 = n0$;

(4) $-(na) = (-n) a = n(-a)$;

(5) $n(a + b) = na + nb$.

Exercises

1. Show that an element of a group has only one inverse in that group.

*2. Let S be a semigroup (written multiplicatively) in which there is a *left identity element* e (an element e such that $ex = x$ for all $x \in S$) relative to which every element of S has a *left inverse* (for each $x \in S$ there exists $y \in S$ such that $yx = e$). Prove that S is a group.

*3. Let S be a semigroup (written multiplicatively) in which the equations $ax = b$ and $ya = b$ have a solution for every $a, b \in S$. Prove that S is a group.

*4. Let S be a finite semigroup (written multiplicatively) in which the cancellation laws hold (for all $x, y, z \in S$, $xy = xz$ implies $y = z$, and $yx = zx$ implies $y = z$). Prove that S is a group. Give an example of an infinite semigroup in which the cancellation laws hold, but which is not a group.

5. Verify that the Klein four-group V_4 is indeed a group.

6. Draw a multiplication table of S_3.

7. Describe the group of isometries of the sine curve (the graph of $y = \sin x$): list its elements and construct a (compact) multiplication table.

8. Compare the (detailed) multiplication tables of D_2 and V_4.

9. For which values of n is D_n commutative?

10. Prove the following: in a group G, $a^m a^n = a^{m+n}$, for all $a \in G$ and $m, n \in \mathbb{Z}$.

11. Prove the following: in a group G, $(a^m)^n = a^{mn}$, for all $a \in G$ and $m, n \in \mathbb{Z}$.

12. Prove the following: a finite group with an even number of elements contains an even number of elements x such that $x^{-1} = x$. State and prove a similar statement for a finite group with an odd number of elements.

3. Subgroups

A subgroup of a group G is a subset of G that inherits a group structure from G. This section contains general properties, up to Lagrange's theorem.

Definition. A subgroup of a group G (written multiplicatively) is a subset H of G such that

(1) $1 \in H$;

(2) $x \in H$ *implies* $x^{-1} \in H$;

(3) $x, y \in H$ *implies* $xy \in H$.

By (3), the binary operation on G has a restriction to H (under which the product of two elements of H is the same as their product in G). By (1) and (2), this operation makes H a group; the identity element of H is that of G, and an element of H has the same inverse in H as in G. This group H is also called a *subgroup* of G.

Examples show that a subset that is closed under multiplication is not necessarily a subgroup. But every group has, besides its binary operation, a constant operation that picks out the identity element, and a unary operation $x \longmapsto x^{-1}$. A subgroup is a subset that is closed under all three operations.

The multiplication table of $V_4 = \{1, a, b, c\}$ shows that $\{1, a\}$ is a subgroup of V_4; so are $\{1, b\}$ and $\{1, c\}$. In D_n the rotations constitute a subgroup. Every group G has two obvious subgroups, G itself and the trivial subgroup $\{1\}$, also denoted by 1.

In the additive notation, a subgroup of an abelian group G is a subset H of G such that $0 \in H$, $x \in H$ implies $-x \in H$, and $x, y \in H$ implies $x + y \in H$. For example, $(\mathbb{Z}, +)$ is a subgroup of $(\mathbb{Q}, +)$; $(\mathbb{Q}, +)$ is a subgroup of $(\mathbb{R}, +)$; $(\mathbb{R}, +)$ is a subgroup of $(\mathbb{C}, +)$. On the other hand, $(\mathbb{N}, +)$ is not a subgroup of $(\mathbb{Z}, +)$ (even though $\mathbb{N}$ is closed under addition).

We denote the relation "H is a subgroup of G" by $H \leq G$. (The notation $H < G$ is more common; we prefer $H \leq G$, on the grounds that G is a subgroup of itself.)

Proposition 3.1. *A subset H of a group G is a subgroup if and only if $H \neq \emptyset$ and $x, y \in H$ implies $xy^{-1} \in H$.*

Proof. These conditions are necessary by (1), (2), and (3). Conversely, assume that $H \neq \emptyset$ and $x, y \in H$ implies $xy^{-1} \in H$. Then there exists $h \in H$ and $1 = hh^{-1} \in H$. Next, $x \in H$ implies $x^{-1} = 1x^{-1} \in H$. Hence $x, y \in H$ implies $y^{-1} \in H$ and $xy = x(y^{-1})^{-1} \in H$. Therefore H is a subgroup. $\square$

Proposition 3.2. *A subset H of a finite group G is a subgroup if and only if $H \neq \emptyset$ and $x, y \in H$ implies $xy \in H$.*

The case of $\mathbb{N} \subseteq \mathbb{Z}$ shows the folly of using this criterion in infinite groups.

Proof. If $H \neq \emptyset$ and $x, y \in H$ implies $xy \in H$, then $x \in H$ implies $x^n \in H$, for all $n > 0$ and $x^{-1} \in H$, by 2.4; hence $x, y \in H$ implies $y^{-1} \in H$ and $xy^{-1} \in H$, and H is a subgroup by 3.1. Conversely, if H is a subgroup, then $H \neq \emptyset$ and $x, y \in H$ implies $xy \in H$. $\square$

Generators. Our next result yields additional examples of subgroups.

Proposition 3.3. *Let G be a group and let X be a subset of G. The set of all products in G (including the empty product and one-term products) of elements of X and inverses of elements of X is a subgroup of G; in fact, it is the smallest subgroup of G that contains X.*

Proof. Let $H \subseteq G$ be the set of all products of elements of X and inverses of elements of X. Then H contains the empty product 1; $h \in H$ implies $h^{-1} \in H$, by 2.2; and $h, k \in H$ implies $hk \in H$, since the product of two products of elements of X and inverses of elements of X is another such product. Thus H is a subgroup of X. Also, H contains all the elements of X, which are one-term products of elements of X. Conversely, a subgroup of G that contains all the elements of X also contains their inverses and contains all products of elements of X and inverses of elements of X. $\square$

Definitions. The subgroup $\langle X \rangle$ of a group G generated by a subset X of G is the set of all products in G (including the empty product and one-term *products) of elements of X and inverses of elements of X. A group G is generated by a subset X when* $\langle X \rangle = G$.

Thus, $G = \langle X \rangle$ when every element of G is a product of elements of X and inverses of elements of X. For example, the dihedral group D_n of a polygon is generated (in the notation of Section 2) by $\{r_1, s_0\}$: indeed, $r_i = r_1^i$, and $s_i = r_i \circ s_0$, so that every element of D_n is a product of r_1's and perhaps one s_0.

Corollary **3.4.** *In a finite group G, the subgroup* $\langle X \rangle$ *of G generated by a subset X of G is the set of all products in G of elements of X.*

Proof. This follows from 3.3: if G is finite, then the inverses of elements of X are themselves products of elements of X, by 2.4. $\square$

Proposition **3.5.** *Let G be a group and let* $a \in G$. *The set of all powers of a is a subgroup of G; in fact, it is the subgroup generated by* $\{a\}$.

Proof. That the powers of a constitute a subgroup of G follows from the parts $a^0 = 1$, $(a^n)^{-1} = a^{-n}$, and $a^m a^n = a^{m+n}$ of 2.3. Also, nonnegative powers of a are products of a's, and negative powers of a are products of a^{-1}'s, since $a^{-n} = (a^{-1})^n$. $\square$

Definitions. The cyclic subgroup generated by *an element a of a group is the set* $\langle a \rangle$ *of all powers of a (in the additive notation, the set of all integer multiples of a). A group or subgroup is* cyclic *when it is generated by a single element.*

Proposition 3.5 provides a strategy for finding the subgroups of any given finite group. First list all cyclic subgroups. Subgroups with two generators are also generated by the union of two cyclic subgroups (which is closed under inverses). Subgroups with three generators are also generated by the union of a subgroup with two generators and a cyclic subgroup; and so forth. If the group is not too large this quickly yields all subgroups, particularly if one makes use of Lagrange's theorem (Corollary 3.14 below).

Infinite groups are quite another matter, except in some particular cases:

Proposition **3.6.** *Every subgroup of* $\mathbb{Z}$ *is cyclic, generated by a unique nonnegative integer.*

Proof. The proof uses integer division. Let H be a subgroup of (the additive group) $\mathbb{Z}$. If $H = 0$ $(= \{0\})$, then H is cyclic, generated by 0. Now assume that $H \neq 0$, so that H contains an integer $m \neq 0$. If $m < 0$, then $-m \in H$; hence H contains a positive integer. Let n be the smallest positive integer that belongs to H. Every integer multiple of n belongs to H. Conversely, let $m \in H$. Then $m = nq + r$ for some $q, r \in \mathbb{Z}$, $0 \leq r < n$. Since H is a subgroup, $qn \in H$ and $r = m - qn \in H$. Now, $0 < r < n$ would contradict the choice of n; therefore $r = 0$, and $m = qn$ is an integer multiple of n. Thus H is the set of all integer multiples of n and is cyclic, generated by $n > 0$. (In particular, $\mathbb{Z}$ itself is generated by 1.) Moreover, n is the unique positive generator of H, since larger multiples of n generate smaller subgroups. $\square$

Properties.

Proposition 3.7. *In a group G, a subgroup of a subgroup of G is a subgroup of G.*

Proposition 3.8. *Every intersection of subgroups of a group G is a subgroup of G.*

The proofs are exercises. By itself, Proposition 3.8 implies that given a subset X of a group G, there is a smallest subgroup of G that contains X. Indeed, there is at least one subgroup of G that contains X, namely, G itself. Then the intersection of all the subgroups of G that contain X is a subgroup of G by 3.8, contains X, and is contained in every subgroup of G that contains X. This argument, however, does not describe the subgroup in question.

Unions of subgroups, on the other hand, are in general not subgroups; in fact, the union of two subgroups is a subgroup if and only if one of the two subgroups is contained in the other (see the exercises). But some unions yield subgroups.

Definition. A chain of subsets of a set S is a family $(C_i)_{i \in I}$ of subsets of S such that, for every $i, j \in I$, $C_i \subseteq C_j$ or $C_j \subseteq C_i$.

Definition. A directed family of subsets of a set S is a family $(D_i)_{i \in I}$ of subsets of S such that, for every $i, j \in I$, there is some $k \in I$ such that $D_i \subseteq D_k$ and $D_j \subseteq D_k$.

For example, every chain is a directed family. Chains, and directed families, are defined similarly in any partially ordered set (not necessarily the partially ordered set of all subsets of a set S under inclusion). Readers will prove the following:

Proposition 3.9. *The union of a nonempty directed family of subgroups of a group G is a subgroup of G. In particular, the union of a nonempty chain of subgroups of a group G is a subgroup of G.*

Cosets. We now turn to individual properties of subgroups.

Proposition 3.10. *If H is a subgroup of a group, then $HH = Ha = aH = H$ for every $a \in H$.*

Here aH and Ha are products of subsets: aH is short for $\{a\}H$, and Ha is short for $H\{a\}$.

Proof. In the group H, the equation $ax = b$ has a solution for every $b \in H$. Therefore $H \subseteq aH$. But $aH \subseteq H$ since $a \in H$. Hence $aH = H$. Similarly, $Ha = H$. Finally, $H \subseteq aH \subseteq HH \subseteq H$. $\square$

Next we show that subgroups partition groups into subsets of equal size.

Definitions. Relative to a subgroup H of a group G, the left coset *of an element x of G is the subset xH of G; the* right coset *of an element x of G*

is the subset Hx *of* G. *These sets are also called left and right* cosets *of* H. □

For example, H is the left coset and the right coset of every $a \in H$, by 3.10.

Proposition **3.11.** *Let* H *be a subgroup of a group* G. *The left cosets of* H *constitute a partition of* G; *the right cosets of* H *constitute a partition of* G.

Proof. Define a binary relation $\mathcal{R}$ on G by

$$x \, \mathcal{R} \, y \text{ if and only if } xy^{-1} \in H.$$

The relation $\mathcal{R}$ is reflexive, since $xx^{-1} = 1 \in H$; symmetric, since $xy^{-1} \in H$ implies $yx^{-1} = (xy^{-1})^{-1} \in H$; and transitive, since $xy^{-1} \in H$, $yz^{-1} \in H$ implies $xz^{-1} = (xy^{-1})(yz^{-1}) \in H$. Thus $\mathcal{R}$ is an equivalence relation, and equivalence classes modulo $\mathcal{R}$ constitute a partition of G. Now, $x \, \mathcal{R} \, y$ if and only if $x \in Hy$; hence the equivalence class of y is its right coset. Therefore the right cosets of H constitute a partition of G. Left cosets of H arise similarly from the equivalence relation, $x \, \mathcal{L} \, y$ if and only if $y^{-1}x \in H$. □

In an abelian group G, $xH = Hx$ for all x, and the partition of G into left cosets of H coincides with its partition into right cosets. The exercises give an example in which the two partitions are different.

Proposition **3.12.** *The number of left cosets of a subgroup is equal to the number of its right cosets.*

Proof. Let G be a group and $H \leq G$. Let $a \in G$. If $y \in aH$, then $y = ax$ for some $x \in H$ and $y^{-1} = x^{-1}a^{-1} \in Ha^{-1}$. Conversely, if $y^{-1} \in Ha^{-1}$, then $y^{-1} = ta^{-1}$ for some $t \in H$ and $y = at^{-1} \in aH$. Thus, when $A = aH$ is a left coset of H, then

$$A' = \{ y^{-1} \mid y \in A \}$$

is a right coset of H, namely $A' = Ha^{-1}$; when $B = Hb = Ha^{-1}$ is a right coset of H, then $B' = \{ x^{-1} \mid x \in B \}$ is a left coset of H, namely aH. We now have mutually inverse bijections $A \longmapsto A'$ and $B \longmapsto B'$ between the set of all left cosets of H and the set of all right cosets of H. □

Definition. The index $[\,G : H\,]$ *of a subgroup* H *of a group* G *is the* (cardinal) *number of its left cosets, and also the number of its right cosets.*

The number of elements of a finite group is of particular importance, due to our next result. The following terminology is traditional.

Definition. The order *of a group* G *is the* (cardinal) *number* $|G|$ *of its elements.*

Proposition **3.13.** *If* H *is a subgroup of a group* G, *then* $|G| = [\,G : H\,]\,|H|$.

Corollary **3.14** (Lagrange's Theorem). *In a finite group* G, *the order and index of a subgroup divide the order of* G.

Proof. Let $H \leq G$ and let $a \in G$. By definition, $aH = \{ ax \mid x \in H \}$, and the cancellation laws show that $x \longmapsto ax$ is a bijection of H onto aH. Therefore $|aH| = |H|$: all left cosets of H have order $|H|$. Since the different left cosets of H constitute a partition, the number of elements of G is now equal to the number of different left cosets times their common number of elements: $|G| = [G:H]\,|H|$. If $|G|$ is finite, then $|H|$ and $[G:H]$ divide $|G|$. $\square$

For instance, a group of order 9 has no subgroup of order 2. A group G whose order is a prime number has only two subgroups, G itself and $1 = \{1\}$. The original version of Lagrange's theorem applied to functions $f(x_1, \ldots, x_n)$ whose arguments are permuted: when $x_1, \ldots, x_n$ are permuted in all possible ways, the number of different values of $f(x_1, \ldots, x_n)$ is a divisor of $n!$

At this point it is not clear whether, conversely, a divisor of $|G|$ is necessarily the order of a subgroup of G. Interesting partial answers to this question await us in the next chapter.

Exercises

1. Let $G = D_n$ and $H = \{ 1, s_0 \}$. Show that the partition of G into left cosets of H is different from its partition into right cosets when $n \geq 3$.

2. Prove that every intersection of subgroups of a group G is a subgroup of G.

3. Find a group with two subgroups whose union is not a subgroup.

4. Let A and B be subgroups of a group G. Prove that $A \cup B$ is a subgroup of G if and only if $A \subseteq B$ or $B \subseteq A$.

5. Show that the union of a nonempty directed family of subgroups of a group G is a subgroup of G.

6. Find all subgroups of V_4.

7. Find all subgroups of D_3.

8. Find all subgroups of D_4.

9. Can you think of subsets of $\mathbb{R}$ that are groups under the multiplication on $\mathbb{R}$? and similarly for $\mathbb{C}$?

10. Find other generating subsets of D_n.

11. Show that every group of prime order is cyclic.

12. A subgroup M of a finite group G is *maximal* when $M \neq G$ and there is no subgroup $M \subsetneqq H \subsetneqq G$. Show that every subgroup $H \neq G$ of a finite group is contained in a maximal subgroup.

13. Show that $x \in G$ lies in the intersection of all maximal subgroups of G if and only if it has the following property: if $X \subseteq G$ contains x and generates G, then $X \setminus \{ x \}$ generates G. (The intersection of all maximal subgroups of G is the *Frattini subgroup* of G.)

14. In a group G, show that the intersection of a left coset of $H \leq G$ and a left coset of $K \leq G$ is either empty or a left coset of $H \cap K$.

15. Show that the intersection of two subgroups of finite index also has finite index.

16. By the previous exercises, the left cosets of subgroups of finite index of a group G constitute a basis (of open sets) of a topology on G. Show that the multiplication on G is continuous. What can you say of G as a topological space?

4. Homomorphisms

Homomorphisms of groups are mappings that preserve products. They allow different groups to relate to each other.

Definition. A homomorphism *of a group A into a group B (written multiplicatively) is a mapping φ of A into B such that $\varphi(xy) = \varphi(x)\,\varphi(y)$ for all $x, y \in A$.* $\square$

If A is written additively, then $\varphi(xy)$ becomes $\varphi(x + y)$; if B is written additively, then $\varphi(x)\,\varphi(y)$ becomes $\varphi(x) + \varphi(y)$. For example, given an element a of a group G, the power map $n \longmapsto a^n$ is a homomorphism of $\mathbb{Z}$ into G. The natural logarithm function is a homomorphism of the multiplicative group of all positive reals into $(\mathbb{R}, +)$. If H is a subgroup of a group G, then the inclusion mapping $\iota : H \longrightarrow G$, defined by $\iota(x) = x$ for all $x \in H$, is the *inclusion homomorphism* of H into G.

In algebraic topology, continuous mappings of one space into another induce homomorphisms of their fundamental groups at corresponding points.

Properties. Homomorphisms compose:

Proposition **4.1.** *If $\varphi : A \longrightarrow B$ and $\psi : B \longrightarrow C$ are homomorphisms of groups, then so is $\psi \circ \varphi : A \longrightarrow C$. Moreover, the identity mapping 1_G on a group G is a homomorphism.*

Homomorphisms preserve identity elements, inverses, and powers, as readers will gladly verify. In particular, homomorphisms of groups preserve the constant and unary operation as well as the binary operation.

Proposition **4.2.** *If $\varphi : A \longrightarrow B$ is a homomorphism of groups (written multiplicatively), then $\varphi(1) = 1$, $\varphi(x^{-1}) = \big(\varphi(x)\big)^{-1}$, and $\varphi(x^n) = \big(\varphi(x)\big)^n$, for all $x \in A$ and $n \in \mathbb{Z}$.*

Homomorphisms also preserve subgroups:

Proposition **4.3.** *Let $\varphi : A \longrightarrow B$ be a homomorphism of groups. If H is a subgroup of A, then $\varphi(H) = \{\,\varphi(x) \mid x \in H\,\}$ is a subgroup of B. If J is a subgroup of B, then $\varphi^{-1}(J) = \{\,x \in A \mid \varphi(x) \in J\,\}$ is a subgroup of A.*

The subgroup $\varphi(H)$ is the *direct image* of $H \leq A$ under φ, and the subgroup $\varphi^{-1}(J)$ is the *inverse image* or *preimage* of $J \leq B$ under φ. The notation $\varphi^{-1}(J)$ should not be read to imply that φ is bijective, or that $\varphi^{-1}(J)$ is the direct image of J under some misbegotten map φ^{-1}.

Two subgroups of interest arise from 4.3:

Definitions. Let $\varphi : A \longrightarrow B$ *be a homomorphism of groups. The* image *or* range *of* φ *is*

$$\mathrm{Im}\, \varphi \,=\, \{\, \varphi(x) \mid x \in A \,\}.$$

The kernel *of* φ *is*

$$\mathrm{Ker}\, \varphi \,=\, \{\, x \in A \mid \varphi(x) = 1 \,\}.$$

In the additive notation, $\mathrm{Ker}\, \varphi = \{\, x \in A \mid \varphi(x) = 0 \,\}$. By 4.3, $\mathrm{Im}\, \varphi = \varphi(G)$ and $\mathrm{Ker}\, \varphi = \varphi^{-1}(1)$ are subgroups of B and A respectively.

The kernel $K = \mathrm{Ker}\, \varphi$ has additional properties. Indeed, $\varphi(x) = \varphi(y)$ implies $\varphi(y\,x^{-1}) = \varphi(y)\,\varphi(x)^{-1} = 1$, $y\,x^{-1} \in K$, and $y \in Kx$. Conversely, $y \in Kx$ implies $y = kx$ for some $k \in K$ and $\varphi(y) = \varphi(k)\,\varphi(x) = \varphi(x)$. Thus, $\varphi(x) = \varphi(y)$ if and only if $y \in Kx$. Similarly, $\varphi(x) = \varphi(y)$ if and only if $y \in xK$. In particular, $Kx = xK$ for all $x \in A$.

Definition. A subgroup N *of a group* G *is* normal *when* $xN = Nx$ *for all* $x \in G$.

This concept is implicit in Galois [1830]. The left cosets of a normal subgroup coincide with its right cosets and are simply called *cosets*.

For instance, all subgroups of an abelian group are normal. Readers will verify that D_n has a normal subgroup, which consists of its rotations, and already know, having diligently worked all exercises, that $\{\, 1,\, s_0 \,\}$ is not a normal subgroup of D_n when $n \geqq 3$. In general, we have obtained the following:

Proposition **4.4.** *Let* $\varphi : A \longrightarrow B$ *be a homomorphism of groups. The image of* φ *is a subgroup of* B. *The kernel* K *of* φ *is a normal subgroup of* A. *Moreover,* $\varphi(x) = \varphi(y)$ *if and only if* $y \in xK = Kx$.

We denote the relation "N is a normal subgroup of G" by $N \trianglelefteq G$. (The notation $N \triangleleft G$ is more common; the author prefers $N \trianglelefteq G$, on the grounds that G is a normal subgroup of itself.) The following result, gladly proved by readers, is often used as the definition of normal subgroups.

Proposition **4.5.** *A subgroup* N *of a group* G *is normal if and only if* $xNx^{-1} \subseteq N$ *for all* $x \in G$.

Special kinds of homomorphisms. It is common practice to call an injective homomorphism a *monomorphism*, and a surjective homomorphism an *epimorphism*. This terminology is legitimate in the case of groups, though not in general. The author prefers to introduce it later.

Readers will easily prove the next result:

Proposition **4.6.** *If* φ *is a bijective homomorphism of groups, then the inverse bijection* φ^{-1} *is also a homomorphism of groups.*

Definitions. An isomorphism *of groups is a bijective homomorphism of groups. Two groups* A *and* B *are* isomorphic *when there exists an isomorphism of* A *onto* B *; this relationship is denoted by* $A \cong B$.

By 4.1, 4.6, the isomorphy relation $\cong$ is reflexive, symmetric, and transitive. Isomorphy would like to be an equivalence relation; but groups are not allowed to organize themselves into a set (see Section A.3).

Philosophical considerations give isomorphism a particular importance. Abstract algebra studies groups but does not care what their elements look like. Accordingly, isomorphic groups are regarded as instances of the same "abstract" group. For example, the dihedral groups of various triangles are all isomorphic, and are regarded as instances of *the* "abstract" dihedral group D_3.

Similarly, when a topological space X is path connected, the fundamental groups of X at various points are all isomorphic to each other; topologists speak of *the* fundamental group $\pi_1(X)$ of X.

Definitions. An endomorphism *of a group* G *is a homomorphism of* G *into* G *; an* automorphism *of a group* G *is an isomorphism of* G *onto* G.

Using Propositions 4.1 and 4.6 readers will readily show that the endomorphisms of a group G constitute a monoid $\text{End}(G)$ under composition, and that the automorphisms of G constitute a group $\text{Aut}(G)$.

Quotient groups. Another special kind of homomorphism consists of projections to quotient groups and is constructed as follows from normal subgroups.

Proposition **4.7.** *Let* N *be a normal subgroup of a group* G*. The cosets of* N *constitute a group under the multiplication of subsets, and the mapping* $x \longmapsto xN = Nx$ *is a surjective homomorphism, whose kernel is* N.

Proof. Let S temporarily denote the set of all cosets of N. Multiplication of subsets of G is associative and induces a binary operation on S, since $xN\, yN = xyNN = xyN$. The identity element is N, since $NxN = xNN = xN$. The inverse of xN is $x^{-1}N$, since $xN\, x^{-1}N = x\, x^{-1}\, NN = N = x^{-1}N\, xN$. Thus S is a group. The surjection $x \longmapsto xN = Nx$ is a homomorphism, since $xN\, yN = xyN$; its kernel is N, since $xN = N$ if and only if $x \in N$. $\square$

Definitions. Let N *be a normal subgroup of a group* G*. The group of all cosets of* N *is the* quotient group G/N *of* G *by* N*. The homomorphism* $x \longmapsto xN = Nx$ *is the* canonical projection *of* G *onto* G/N.

For example, in any group G, $G \trianglelefteq G$ (with $Gx = xG = G$ for all $x \in G$), and G/G is the trivial group; $1 \trianglelefteq G$ (with $1x = x1 = \{x\}$ for all $x \in G$), and the canonical projection is an isomorphism $G \cong G/1$.

For a more interesting example, let $G = \mathbb{Z}$. Every subgroup N of $\mathbb{Z}$ is normal and is, by 3.6, generated by a unique nonnegative integer n (so that $N = \mathbb{Z}n$). If $n = 0$, then $\mathbb{Z}/N \cong \mathbb{Z}$; but $n > 0$ yields a new group:

Definition. For every positive integer n, the additive group $\mathbb{Z}_n$ of the integers modulo n is the quotient group $\mathbb{Z}/\mathbb{Z}n$.

The group $\mathbb{Z}_n$ is also denoted by $\mathbb{Z}(n)$. Its elements are the different cosets $\overline{x} = x + \mathbb{Z}n$ with $x \in \mathbb{Z}$. Note that $\overline{x} = \overline{y}$ if and only if x and y are congruent modulo n, whence the name "integers modulo n".

Proposition 4.8. $\mathbb{Z}_n$ is a cyclic group of order n, with elements $\overline{0}, \overline{1}, \ldots, \overline{n-1}$ and addition

$$\overline{i} + \overline{j} = \begin{cases} \overline{i+j} & \text{if } i+j < n, \\ \overline{i+j-n} & \text{if } i+j \geq n. \end{cases}$$

Proof. The proof uses integer division. For every $x \in \mathbb{Z}$ there exist unique q and r such that $x = qn + r$ and $0 \leq r < n$. Therefore every coset $\overline{x} = x + \mathbb{Z}n$ is the coset of a unique $0 \leq r < n$. Hence $\mathbb{Z}_n = \{\overline{0}, \overline{1}, \ldots, \overline{n-1}\}$, with the addition above. We see that $r\overline{1} = \overline{r}$, so that $\mathbb{Z}_n$ is cyclic, generated by $\overline{1}$. $\square$

In general, the order of G/N is the index of N in G: $|G/N| = [G:N]$; if G is finite, then $|G/N| = |G|/|N|$. The subgroups of G/N are quotient groups of subgroups of G:

Proposition 4.9. Let N be a normal subgroup of a group G. Every subgroup of G/N is the quotient H/N of a unique subgroup H of G that contains N.

Proof. Let $\pi : G \longrightarrow G/N$ be the canonical projection and let B be a subgroup of G/N. By 4.3,

$$A = \pi^{-1}(B) = \{a \in G \mid aN \in B\}$$

is a subgroup of G and contains $\pi^{-1}(1) = \operatorname{Ker} \pi = N$. Now, N is a subgroup of A, and is a normal subgroup of A since $aN = Na$ for all $a \in A$. The elements aN of A/N all belong to B by definition of A. Conversely, if $xN \in B$, then $x \in A$ and $xN \in A/N$. Thus $B = A/N$.

Assume that $B = H/N$, where $H \leq G$ contains N. If $h \in H$, then $hN \in H/N = B$ and $h \in A$. Conversely, if $a \in A$, then $aN \in B = H/N$, $aN = hN$ for some $h \in H$, and $a \in hN \subseteq H$. Thus $H = A$. $\square$

We prove a stronger version of 4.9; the exercises give an even stronger version.

Proposition 4.10. Let N be a normal subgroup of a group G. Direct and inverse image under the canonical projection $G \longrightarrow G/N$ induce a one-to-one correspondence, which preserves inclusion and normality, between subgroups of G that contain N and subgroups of G/N.

Proof. Let $\mathcal{A}$ be the set of all subgroups of G that contain N; let $\mathcal{B}$ be the set of all subgroups of G/N; let $\pi : G \longrightarrow G/N$ be the canonical projection. By 4.16 and its proof, $A \longmapsto A/N$ is a bijection of $\mathcal{A}$ onto $\mathcal{B}$, and the inverse bijection is $B \longmapsto \pi^{-1}(B)$, since $B = A/N$ if and only if $A = \pi^{-1}(B)$. Both bijections preserve inclusions (e.g., $A_1 \subseteq A_2$ implies $A_1/N \subseteq A_2/N$ when $N \subseteq A_1$); the exercises imply that they preserve normality. $\square$

Exercises

1. Let $\varphi : A \longrightarrow B$ be a homomorphism of groups (written multiplicatively). Show that $\varphi(1) = 1$, $\varphi(x^{-1}) = \left(\varphi(x)\right)^{-1}$, and $\varphi(x^n) = \left(\varphi(x)\right)^n$, for all $x \in A$ and $n \in \mathbb{Z}$.

2. Let $\varphi : A \longrightarrow B$ be a homomorphism of groups and let $H \leqq A$. Show that $\varphi(H) \leqq B$.

3. Let $\varphi : A \longrightarrow B$ be a homomorphism of groups and let $H \leqq B$. Show that $\varphi^{-1}(H) \leqq A$.

4. Show that the following are equivalent when $N \leqq G$: (i) $xN = Nx$ for all $x \in G$; (ii) $Nx\,Ny \subseteq Nxy$ for all $x, y \in G$; (iii) $xNx^{-1} \subseteq N$ for all $x \in G$.

5. Let $\varphi : A \longrightarrow B$ be a homomorphism of groups. Show that $N \trianglelefteq B$ implies $\varphi^{-1}(N) \trianglelefteq A$.

6. Let $\varphi : A \longrightarrow B$ be a surjective homomorphism of groups. Show that $N \trianglelefteq A$ implies $\varphi(N) \trianglelefteq B$.

7. Give an example that $N \trianglelefteq A$ does not necessarily imply $\varphi(N) \trianglelefteq B$ when $\varphi : A \longrightarrow B$ is an arbitrary homomorphism of groups.

8. Prove that every subgroup of index 2 is normal.

9. Prove that every intersection of normal subgroups of a group G is a normal subgroup of G.

10. Prove that the union of a nonempty directed family of normal subgroups of a group G is a normal subgroup of G.

11. Show that $G = D_4$ contains subgroups A and B such that $A \trianglelefteq B$ and $B \trianglelefteq G$ but not $A \trianglelefteq G$.

12. Let the group G be generated by a subset X. Prove the following: if two homomorphisms $\varphi, \psi : G \longrightarrow H$ agree on X (if $\varphi(x) = \psi(x)$ for all $x \in X$), then $\varphi = \psi$ ($\varphi(x) = \psi(x)$ for all $x \in G$).

13. Find all homomorphisms of D_2 into D_3.

14. Find all homomorphisms of D_3 into D_2.

15. Show that $D_2 \cong V_4$.

16. Show that $D_3 \cong S_3$.

17. Find all endomorphisms of V_4.

18. Find all automorphisms of V_4.

19. Find all endomorphisms of D_3.

20. Find all automorphisms of D_3.

21. Let $\varphi : A \longrightarrow B$ be a homomorphism of groups. Show that φ induces an order-preserving one-to-one correspondence between the set of all subgroups of A that contain $\mathrm{Ker}\,\varphi$ and the set of all subgroups of B that are contained in $\mathrm{Im}\,\varphi$.

5. The Isomorphism Theorems

This section contains further properties of homomorphisms and quotient groups.

Factorization. Quotient groups provide our first example of a *universal property*. This type of property becomes increasingly important in later chapters.

Theorem **5.1** (Factorization Theorem). *Let N be a normal subgroup of a group G. Every homomorphism of groups $\varphi : G \longrightarrow H$ whose kernel contains N factors uniquely through the canonical projection $\pi : G \longrightarrow G/N$ (there exists a homomorphism $\psi : G/N \longrightarrow H$ unique such that $\varphi = \psi \circ \pi$):*

$$G \xrightarrow{\ \pi\ } G/N$$
$$\varphi \searrow \quad \downarrow \psi$$
$$H$$

Proof. We use the formal definition of a mapping $\psi : A \longrightarrow B$ as a set of ordered pairs (a, b) with $a \in A$, $b \in B$, such that (i) for every $a \in A$ there exists $b \in B$ such that $(a, b) \in \psi$, and (ii) if $(a_1, b_1) \in \psi$, $(a_2, b_2) \in \psi$, and $a_1 = a_2$, then $b_1 = b_2$. Then $\psi(a)$ is the unique $b \in B$ such that $(a, b) \in \psi$.

Since $\operatorname{Ker} \varphi$ contains N, $x^{-1}y \in N$ implies $\varphi(x^{-1})\,\varphi(y) = \varphi(x^{-1}y) = 1$, so that $xN = yN$ implies $\varphi(x) = \varphi(y)$. As a set of ordered pairs,

$$\psi = \{\, (xN, \varphi(x)) \mid x \in G \,\}.$$

In the above, (i) holds by definition of G/N, and we just proved (ii); hence ψ is a mapping. (Less formally one says that ψ is *well defined* by $\psi(xN) = \varphi(x)$.) By definition, $\psi(xN) = \varphi(x)$, so $\psi \circ \pi = \varphi$. Also, ψ is a homomorphism:

$$\psi(xN\,yN) = \psi(xyN) = \varphi(xy) = \varphi(x)\,\varphi(y) = \psi(xN)\,\psi(yN).$$

To show that ψ is unique, let $\chi : G/N \longrightarrow H$ be a homomorphism such that $\chi \circ \pi = \varphi$. Then $\chi(xN) = \varphi(x) = \psi(xN)$ for all $xN \in G/N$ and $\chi = \psi$. $\square$

The homomorphism theorem is also called the *first isomorphism theorem*.

Theorem **5.2** (Homomorphism Theorem). *If $\varphi : A \longrightarrow B$ is a homomorphism of groups, then*

$$A/\operatorname{Ker} \varphi \cong \operatorname{Im} \varphi;$$

in fact, there is an isomorphism $\theta : A/\operatorname{Ker} f \longrightarrow \operatorname{Im} f$ unique such that $\varphi = \iota \circ \theta \circ \pi$, where $\iota : \operatorname{Im} f \longrightarrow B$ is the inclusion homomorphism and $\pi : A \longrightarrow A/\operatorname{Ker} f$ is the canonical projection:

$$
\begin{array}{ccc}
A & \xrightarrow{\ \varphi\ } & B \\
\pi \downarrow & & \uparrow \iota \\
A/\operatorname{Ker} \varphi & \underset{\theta}{\dashrightarrow} & \operatorname{Im} \varphi
\end{array}
$$

Proof. Let $\psi : A \longrightarrow \operatorname{Im} \varphi$ be the same mapping as φ (the same set of ordered pairs) but viewed as a homomorphism of A onto $\operatorname{Im} \varphi$. Then $\operatorname{Ker} \psi = \operatorname{Ker} \varphi$; by 5.1, ψ factors through π: $\psi = \theta \circ \pi$ for some homomorphism $\theta : A/K \longrightarrow \operatorname{Im} \varphi$, where $K = \operatorname{Ker} \varphi$. Then $\theta(xK) = \psi(x) = \varphi(x)$ for all $x \in A$ and $\varphi = \iota \circ \theta \circ \pi$. Moreover, θ, like ψ, is surjective; θ is injective since $\theta(xK) = 1$ implies $\varphi(x) = 1$, $x \in \operatorname{Ker} \varphi = K$, and $xK = 1$ in A/K. If $\zeta : A/\operatorname{Ker} f \longrightarrow \operatorname{Im} f$ is another isomorphism such that $\varphi = \iota \circ \zeta \circ \pi$, then

$$\zeta(xK) = \iota\big(\zeta\big(\pi(x)\big)\big) = \varphi(x) = \iota\big(\theta\big(\pi(x)\big)\big) = \theta(xK)$$

for all $x \in A$, and $\zeta = \theta$. (This also follows from uniqueness in 5.1.) $\square$

The homomorphism theorem implies that every homomorphism is a composition of three basic types of homomorphism: inclusion homomorphisms of subgroups; isomorphisms; and canonical projections to quotient groups.

Corollary 5.3. *Let $\varphi : A \longrightarrow B$ be a homomorphism. If φ is injective, then $A \cong \operatorname{Im} \varphi$. If φ is surjective, then $B \cong A/\operatorname{Ker} \varphi$.*

Proof. If φ is injective, then $\operatorname{Ker} \varphi = 1$ and $A \cong A/\operatorname{Ker} \varphi \cong \operatorname{Im} \varphi$. If φ is surjective, then $B = \operatorname{Im} \varphi \cong A/\operatorname{Ker} \varphi$. $\square$

We illustrate the use of Theorem 5.2 with a look at cyclic groups. We saw that the additive groups $\mathbb{Z}$ and $\mathbb{Z}_n$ are cyclic. Up to isomorphism, $\mathbb{Z}$ and $\mathbb{Z}_n$ are the only cyclic groups:

Proposition 5.4. *Let G be a group and let $a \in G$. If $a^m \neq 1$ for all $m \neq 0$, then $\langle a \rangle \cong \mathbb{Z}$; in particular, $\langle a \rangle$ is infinite. Otherwise, there is a smallest positive integer n such that $a^n = 1$, and then $a^m = 1$ if and only if n divides m, and $\langle a \rangle \cong \mathbb{Z}_n$; in particular, $\langle a \rangle$ is finite of order n.*

Proof. The power map $p : m \longmapsto a^m$ is a homomorphism of $\mathbb{Z}$ into G. By 5.1, $\langle a \rangle = \operatorname{Im} p \cong \mathbb{Z}/\operatorname{Ker} p$. By 3.6, $\operatorname{Ker} p$ is cyclic, $\operatorname{Ker} p = \mathbb{Z}n$ for some unique nonnegative integer n. If $n = 0$, then $\langle a \rangle \cong \mathbb{Z}/0 \cong \mathbb{Z}$, and $a^m = 1$ ($a \in \operatorname{Ker} p$) if and only if $m = 0$. If $n > 0$, then $\langle a \rangle \cong \mathbb{Z}/\mathbb{Z}n = \mathbb{Z}_n$, and $a^m = 1$ if and only if m is a multiple of n. $\square$

Definition. The order *of an element a of a group G is infinite if $a^m \neq 1$ for all $m \neq 0$; otherwise, it is the smallest positive integer n such that $a^n = 1$.* $\square$

Equivalently, the order of a is the order of $\langle a \rangle$. Readers will be careful that $a^n = 1$ does not imply that a has order n, only that the order of a divides n.

Corollary 5.5. *Any two cyclic groups of order n are isomorphic.*

We often denote "the" cyclic group of order n by C_n.

Corollary 5.6. *Every subgroup of a cyclic group is cyclic.*

This follows from Propositions 5.4 and 3.6; the details make a pretty exercise. More courageous readers will prove a stronger result:

Proposition 5.7. *In a cyclic group G of order n, every divisor d of n is the order of a unique cyclic subgroup of G, namely $\{ x \in G \mid x^d = 1 \}$.*

The isomorphism theorems. The isomorphisms theorems are often numbered so that Theorem 5.2 is the first isomorphism theorem. Then Theorems 5.8 and 5.9 are the second and third isomorphism theorems.

Theorem 5.8 (First Isomorphism Theorem). *Let A be a group and let B, C be normal subgroups of A. If $C \subseteq B$, then C is a normal subgroup of B, B/C is a normal subgroup of A/C, and*

$$A/B \cong (A/C)/(B/C);$$

in fact, there is a unique isomorphism $\theta : A/B \longrightarrow (A/C)/(B/C)$ such that $\theta \circ \rho = \tau \circ \pi$, where $\pi : A \longrightarrow A/C$, $\rho : A \longrightarrow A/B$, and $\tau : A/C \longrightarrow (A/C)/(B/C)$ are the canonical projections:

$$
\begin{array}{ccc}
A & \xrightarrow{\ \pi\ } & A/C \\
{\scriptstyle\rho}\downarrow & {\scriptstyle\sigma}\swarrow & \downarrow{\scriptstyle\tau} \\
A/B & \xrightarrow[\ \theta\ ]{} & (A/C)/(B/C)
\end{array}
$$

Proof. By 5.1, ρ factors through π: $\rho = \sigma \circ \pi$ for some homomorphism $\sigma : A/C \longrightarrow A/B$; namely, $\sigma : aC \longmapsto aB$. Like ρ, σ is surjective. We show that $\operatorname{Ker} \sigma \doteq B/C$. First, $C \trianglelefteq B$, since $C \trianglelefteq A$. If $bC \in B/C$, where $b \in B$, then $\sigma(bC) = bB = 1$ in A/B. Conversely, if $\sigma(aC) = 1$, then $aB = B$ and $a \in B$. Thus $\operatorname{Ker} \sigma = \{ bC \mid b \in B \} = B/C$; in particular, $B/C \trianglelefteq A/C$. By 5.2, $A/B = \operatorname{Im} \sigma \cong (A/C)/\operatorname{Ker} \sigma = (A/C)/(B/C)$. In fact, Theorem 5.2 yields an isomorphism $\theta : A/B \longrightarrow (A/C)/(B/C)$ such that $\theta \circ \sigma = \tau$, and then $\theta \circ \rho = \tau \circ \pi$; since ρ is surjective, θ is unique with this property. $\square$

Theorem 5.9 (Second Isomorphism Theorem). *Let A be a subgroup of a group G, and let N be a normal subgroup of G. Then AN is a subgroup of G, N is a normal subgroup of AN, $A \cap N$ is a normal subgroup of A, and*

$$AN/N \cong A/(A \cap N);$$

in fact, there is an isomorphism $\theta : A/(A \cap N) \longrightarrow AN/N$ unique such that $\theta \circ \rho = \pi \circ \iota$, where $\pi : AN \longrightarrow AN/N$ and $\rho : A \longrightarrow A/(A \cap N)$ are the canonical projections and $\iota : A \longrightarrow AN$ is the inclusion homomorphism:

$$
\begin{array}{ccc}
A & \xrightarrow{\ \rho\ } & A/(A \cap N) \\
{\scriptstyle\iota}\downarrow & {\scriptstyle\varphi}\searrow & \downarrow{\scriptstyle\theta} \\
AN & \xrightarrow[\ \pi\ ]{} & AN/N
\end{array}
$$

In particular, $|AN|/|N| = |A|/|A \cap N|$ when G is finite.

Proof. We show that $AN \leqq G$. First, $1 \in AN$. Since $N \trianglelefteq G$, $NA = AN$; hence $an \in AN$ (with $a \in A$, $n \in N$) implies $(an)^{-1} = n^{-1} a^{-1} \in NA = AN$. Finally, $AN\,AN = AANN = AN$.

Now, $N \trianglelefteq AN$. Let $\varphi = \pi \circ \iota$. Then $\varphi(a) = aN \in AN/N$ for all $a \in A$, and φ is surjective. Moreover, $\varphi(a) = 1$ if and only if $a \in N$, so that Ker $\varphi = A \cap N$; in particular, $A \cap N \trianglelefteq N$. By 5.2, $AN/N = \text{Im } \varphi \cong A/\text{Ker } \varphi = A/(A \cap N)$; in fact, there is a unique isomorphism $\theta : A/(A \cap N) \longrightarrow AN/N$ such that $\theta \circ \rho = \varphi = \pi \circ \iota$. $\square$

Theorem 5.9 implies that the intersection of two normal subgroups of finite index also has finite index. Consequently, the cosets of normal subgroups of finite index constitute a basis of open sets for a topology (see the exercises).

Exercises

1. Let $\varphi : A \longrightarrow B$ and $\psi : A \longrightarrow C$ be homomorphisms of groups. Prove the following: if ψ is surjective, then φ factors through ψ if and only if Ker $\psi \subseteq$ Ker φ, and then φ factors uniquely through ψ.

2. Show that the identity homomorphism $1_{2\mathbb{Z}} : 2\mathbb{Z} \longrightarrow 2\mathbb{Z}$ does not factor through the inclusion homomorphism $\iota : 2\mathbb{Z} \longrightarrow \mathbb{Z}$ (there is no homomorphism $\varphi : \mathbb{Z} \longrightarrow 2\mathbb{Z}$ such that $1_{2\mathbb{Z}} = \varphi \circ \iota$) even though Ker $\iota \subseteq$ Ker $1_{2\mathbb{Z}}$. (Of course, ι is not surjective.)

3. Let $\varphi : A \longrightarrow C$ and $\psi : B \longrightarrow C$ be homomorphisms of groups. Prove the following: if ψ is injective, then φ factors through ψ ($\varphi = \psi \circ \chi$ for some homomorphism $\chi : A \longrightarrow B$) if and only if Im $\varphi \subseteq$ Im ψ, and then φ factors uniquely through ψ.

4. Show that the additive group $\mathbb{R}/\mathbb{Z}$ is isomorphic to the multiplicative group of all complex numbers of modulus 1.

5. Show that the additive group $\mathbb{Q}/\mathbb{Z}$ is isomorphic to the multiplicative group of all complex roots of unity (all complex numbers $z \neq 0$ of finite order in $\mathbb{C}\backslash\{0\}$).

6. Prove that every subgroup of a cyclic group is cyclic.

7. Let $C_n = \langle c \rangle$ be a cyclic group of finite order n. Show that every divisor d of n is the order of a unique subgroup of C_n, namely $\langle c^{n/d} \rangle = \{ x \in C_n \mid x^d = 1 \}$.

8. Show that every divisor of $|D_n|$ is the order of a subgroup of D_n.

9. Find the order of every element of D_4.

10. List the elements of S_4 and find their orders.

11. Show that the complex nth roots of unity constitute a cyclic group. Show that $\omega_k = \cos(2\pi k/n) + i \sin(2\pi k/n)$ generates this cyclic group if and only if k and n are relatively prime (then ω_k is a *primitive* nth root of unity).

12. Let A and B be subgroups of a finite group G. Show that $|AB| = |A||B|/|A \cap B|$.

13. Find a group G with subgroups A and B such that AB is not a subgroup.

14. If G is a finite group, $H \leq G$, $N \trianglelefteq G$, and $|N|$ and $[G : N]$ are relatively prime, then show that $H \subseteq N$ if and only if $|H|$ divides $|N|$. (Hint: consider HN.)

15. Show that, in a group G, the intersection of two normal subgroups of G of finite index is a normal subgroup of G of finite index.

16. Let A and B be cosets of (possibly different) normal subgroups of finite index of a group G. Show that $A \cap B$ is either empty or a coset of a normal subgroup of G of finite index.

17. By the previous exercise, cosets of normal subgroups of finite index of a group G constitute a basis of open sets of a topology, the *profinite topology* on G. What can you say about this topology?

6. Free Groups

This section and the next construct groups that are generated by a given set. The free groups in this section are implicit in Dyck [1882]; the name seems due to Nielsen [1924].

In a group G generated by a subset X, every element of G is a product of elements of X and inverses of elements of X, by 3.3. But the elements of G are not written uniquely in this form, since, for instance, $1 = x\,x^{-1} = x^{-1}\,x$ for every $x \in X$: some *relations* between the elements of X (equalities between products of elements of X and inverses of elements of X) always hold in G.

The *free group* on a set X is generated by X with as few relations as possible between the elements of X. Products of elements of X and inverses of elements of X can be *reduced* by deleting all $x\,x^{-1}$ and $x^{-1}\,x$ subproducts until none is left. The free group on X consists of formal reduced products, multiplied by concatenation and reduction. That it has as few relations as possible is shown by a universal property. The details follow.

Reduction. Let X be an arbitrary set. Let X' be a set that is disjoint from X and comes with a bijection $x \longmapsto x'$ of X onto X'. (Once our free group is constructed, x' will be the inverse of x.) It is convenient to denote the inverse bijection $X' \longrightarrow X$ by $y \longmapsto y'$, so that $(x')' = x$ for all $x \in X$, and $(y')' = y$ for all $y \in Y = X \cup X'$. *Words* in the alphabet Y are finite, possibly empty sequences of elements of Y, and represent products of elements of X and inverses of elements of X. The free monoid on Y is the set W of all such words, multiplied by concatenation.

Definition. A word $a = (a_1, a_2, \ldots, a_n) \in W$ is reduced when $a_{i+1} \neq a_i'$ for all $1 \leq i < n$.

For example, the empty word and all one-letter words are reduced, for want of consecutive letters. If $X = \{x, y, z, \ldots\}$, then (x, y, z) and (x, x, x) are reduced, but (x, y, y', z) is not reduced.

Reduction deletes subsequences (a_i, a_i') until a reduced word is reached.

Definitions. In W, we write $a \overset{1}{\longrightarrow} b$ when $a = (a_1, a_2, \ldots, a_n)$, $a_{i+1} = a_i'$, and $b = (a_1, \ldots, a_{i-1}, a_{i+2}, \ldots, a_n)$, for some $1 \leq i < n$;

we write $a \overset{k}{\longrightarrow} b$ when $k \geq 0$ and $a \overset{1}{\longrightarrow} a' \overset{1}{\longrightarrow} a'' \overset{1}{\longrightarrow} \cdots \overset{1}{\longrightarrow} a^{(k)} = b$ for some $a', a'', \ldots, a^{(k)} \in W$ (when $a = b$, if $k = 0$);

we write $a \longrightarrow b$ when $a \overset{k}{\longrightarrow} b$ for some $k \geq 0$.

If a is reduced, then $a \xrightarrow{k} b$ implies $k = 0$, since there is no $a \xrightarrow{1} c$, and $a \longrightarrow b$ implies $a = b$.

Lemma 6.1. *For every word $a \in W$ there is a reduction $a \longrightarrow b$ to a reduced word b.*

Proof. By induction on the length of a. If a is reduced, then $b = a$ serves. Otherwise, $a \xrightarrow{1} c$ for some $c \in W$, $c \longrightarrow b$ for some reduced $b \in W$ since c is shorter than a, and then $a \longrightarrow b$. $\square$

We show that the word b in Lemma 6.3 is unique.

Lemma 6.2. *If $a \xrightarrow{1} b$ and $a \xrightarrow{1} c \neq b$, then $b \xrightarrow{1} d$, $c \xrightarrow{1} d$ for some d.*

Proof. Let $a = (a_1, a_2, \ldots, a_n)$. We have $a_{i+1} = a_i'$ and

$$b = (a_1, \ldots, a_{i-1}, a_{i+2}, \ldots, a_n),$$

for some $1 \leq i < n$; also, $a_{j+1} = a_j'$ and

$$c = (a_1, \ldots, a_{j-1}, a_{j+2}, \ldots, a_n),$$

for some $1 \leq j < n$. Since $b \neq c$ we have $i \neq j$ and may assume $i < j$. If $j = i + 1$, then $a_i = a_{i+1}' = a_{j+1} = a_{i+2}$, $(a_{i-1}, a_{i+2}, a_{i+3}) = (a_{j-2}, a_{j-1}, a_{j+2})$, and $b = c$; hence $j \geq i + 2$. Then a_i and a_{i+1} are consecutive letters of c, a_j and a_{j+1} are consecutive letters of b, and

$$d = (a_1, \ldots, a_{i-1}, a_{i+2}, \ldots, a_{j-1}, a_{j+2}, \ldots, a_n)$$

serves (or $d = (a_1, \ldots, a_{i-1}, a_{j+2}, \ldots, a_n)$, if $j = i + 2$.) $\square$

Lemma 6.3. *If $a \longrightarrow b$ and $a \longrightarrow c$, then $b \longrightarrow d$ and $c \longrightarrow d$ for some d.*

Proof. Say $a \xrightarrow{k} b$ and $a \xrightarrow{\ell} c$. The result is trivial if $k = 0$ or if $\ell = 0$.

We first prove 6.3 when $\ell = 1$, by induction on k. We have $a \xrightarrow{1} c$. If $k \leq 1$, then 6.3 holds, by 6.2. Now let $k > 1$, so that $a \xrightarrow{1} u \xrightarrow{k-1} b$ for some u. If $u = c$, then $d = b$ serves. Otherwise, $u \xrightarrow{1} v$ and $c \xrightarrow{1} v$ for some v:

$$
\begin{array}{ccccc}
a & \xrightarrow{1} & u & \longrightarrow & b \\
\downarrow & \scriptstyle 1 & \downarrow & & \downarrow \\
c & \xrightarrow{1} & v & \longrightarrow & d
\end{array}
$$

by 6.2. The induction hypothesis, applied to $u \xrightarrow{k-1} b$ and $u \xrightarrow{1} v$, then yields $b \longrightarrow d$ and $c \xrightarrow{1} v \longrightarrow d$ for some d.

Now, 6.3 holds when $\ell \leq 1$; the general case is proved by induction on ℓ:

$$
\begin{array}{ccccc}
a & \xrightarrow{1} & u & \longrightarrow & c \\
\downarrow & & \downarrow & & \downarrow \\
b & \longrightarrow & v & \longrightarrow & d
\end{array}
$$

If $\ell > 1$, then $a \longrightarrow b$ and $a \overset{1}{\longrightarrow} u \overset{\ell-1}{\longrightarrow} c$ for some u. By the case $\ell = 1$, $b \longrightarrow v$ and $u \longrightarrow v$ for some v. The induction hypothesis, applied to $u \longrightarrow v$ and $u \overset{\ell-1}{\longrightarrow} c$, then yields $b \longrightarrow v \longrightarrow d$ and $c \longrightarrow d$ for some d. $\square$

Lemma 6.4. *For every word $a \in W$ there is a unique reduced word b such that $a \longrightarrow b$.*

Proof. If $a \longrightarrow b$ and $a \longrightarrow c$, with b and c reduced, then, in Lemma 6.3, $b \longrightarrow d$ and $c \longrightarrow d$ imply $b = d = c$. $\square$

Definition. The reduction red a *of $a \in W$ is the unique reduced word b such that $a \longrightarrow b$.* $\square$

Construction. The free group on X is now within reach.

Proposition 6.5. *Under the operation $a \cdot b = $ red (ab), the set F_X of all reduced words in X is a group.*

Proof. If $a \overset{1}{\longrightarrow} b$, then $ac \overset{1}{\longrightarrow} bc$ and $ca \overset{1}{\longrightarrow} cb$ for all $c \in W$. Hence $a \longrightarrow b$ implies $ac \longrightarrow bc$ and $ca \longrightarrow cb$ for all $c \in W$. If now $a, b, c \in W$ are reduced, then $ab \longrightarrow a \cdot b$ and $bc \longrightarrow b \cdot c$ yield

$$abc \longrightarrow (a \cdot b)c \longrightarrow (a \cdot b) \cdot c \text{ and } abc \longrightarrow a(b \cdot c) \longrightarrow a \cdot (b \cdot c).$$

Hence $(a \cdot b) \cdot c = a \cdot (b \cdot c)$, by 6.4.

The empty word $1 = ()$ is reduced and is the identity element of F_X, since $1 \cdot a = $ red $(1a) = $ red $a = a$ and $a \cdot 1 = $ red $a = a$ when a is reduced.

The inverse of a reduced word $a = (a_1, a_2, \ldots, a_n)$ is, not surprisingly,

$$a^{-1} = (a_n', a_{n-1}', \ldots, a_1');$$

indeed, a^{-1} is reduced, since $a_i' \neq (a_{i-1}')'$ for all $i > 1$, and $a a^{-1} \longrightarrow 1$, $a^{-1} a \longrightarrow 1$. Thus F_X is a group. $\square$

In particular, the inverse of a one-letter word (y) is (y'). The first part of the proof implies that concatenation followed by reduction also yields products of three or more terms in F_X.

Definition. The free group *on a set X is the group F_X in Proposition 6.5, which consists all reduced words in X.*

Readers will enjoy showing that $F_X \cong \mathbb{Z}$ when X has just one element.

Properties. The free group on X should be generated by X. Strictly speaking, X is not a subset of F_X. However, there is a *canonical injection* $\eta : X \longrightarrow F_X$, $x \longmapsto (x)$, which is conveniently extended to $Y = X \cup X'$ so that $\eta : x' \longmapsto (x')$; then F_X is generated by $\eta(X)$:

Proposition 6.6. *If $a = (a_1, a_2, \ldots, a_n)$ is a reduced word in X, then*

$$a = \eta(a_1) \cdot \eta(a_2) \cdot \cdots \cdot \eta(a_n).$$

In particular, F_X is generated by $\eta(X)$.

Proof. If $a = (a_1, a_2, \ldots, a_n)$ is reduced, then concatenating the one-letter words $(a_1), (a_2), \ldots, (a_n)$ yields a reduced word; hence

$$a = (a_1) \cdot (a_2) \cdot \cdots \cdot (a_n) = \eta(a_1) \cdot \eta(a_2) \cdot \cdots \cdot \eta(a_n).$$

We saw that $\eta(x') = \eta(x)^{-1}$ for all $x \in X$; hence every $a \in F_X$ is a product of elements of $\eta(X)$ and inverses of elements of $\eta(X)$. $\square$

Theorem 6.7. *Let $\eta : X \longrightarrow F_X$ be the canonical injection. For every mapping f of X into a group G, there is a homomorphism φ of F_X into G unique such that $f = \varphi \circ \eta$, namely*

$$\varphi(a_1, a_2, \ldots, a_n) = f(a_1) f(a_2) \cdots f(a_n).$$

$$X \xrightarrow{\ \eta\ } F_X$$
$$f \searrow \quad \downarrow \varphi$$
$$G$$

Proof. We show uniqueness first. Let $\varphi : F_X \longrightarrow G$ be a homomorphism such that $f = \varphi \circ \eta$. Extend f to X' so that $f(x') = f(x)^{-1}$ for all $x \in X$. For every $x \in X$, we have $\varphi(\eta(x)) = f(x)$ and $\varphi(\eta(x')) = \varphi(\eta(x)^{-1}) = f(x)^{-1} = f(x')$. If now $a = (a_1, a_2, \ldots, a_n)$ is reduced, then necessarily

$$\varphi(a) = \varphi(\eta(a_1) \cdot \eta(a_2) \cdot \cdots \cdot \eta(a_n)) = f(a_1) f(a_2) \cdots f(a_n),$$

since φ is a homomorphism. Hence φ is unique.

It remains to show that the mapping $\varphi : F_X \longrightarrow G$ defined for every reduced word $a = (a_1, a_2, \ldots, a_n)$ by

$$\varphi(a) = f(a_1) f(a_2) \cdots f(a_n)$$

is a homomorphism of groups. First we can extend φ to all of W by using the formula above for every word, reduced or not. Then $\varphi(ab) = \varphi(a)\varphi(b)$ for all $a, b \in W$. Also $a \xrightarrow{1} b$ implies $\varphi(a) = \varphi(b)$: indeed, if $a = (a_1, a_2, \ldots, a_n)$, $a_{i+1} = a_i'$, and $b = (a_1, \ldots, a_{i-1}, a_{i+2}, \ldots, a_n)$, for some $1 \leqq i < n$, then

$$\varphi(a) = f(a_1) \cdots f(a_{i-1}) f(a_i) f(a_{i+1}) f(a_{i+2}) \cdots f(a_n)$$
$$= f(a_1) \cdots f(a_{i-1}) f(a_{i+2}) \cdots f(a_n) = \varphi(b),$$

since $f(a_{i+1}) = f(a_i)^{-1}$. Therefore $a \longrightarrow b$ implies $\varphi(a) = \varphi(b)$. If now a and b are reduced, then $\varphi(a \cdot b) = \varphi(ab) = \varphi(a)\varphi(b)$. $\square$

Corollary 6.8. *If the group G is generated by a subset X, then there is a surjective homomorphism of F_X onto G.*

Proof. By Theorem 6.7, there is a homomorphism $\varphi : F_X \longrightarrow G$ such that $\varphi \circ \eta$ is the inclusion mapping $X \longrightarrow G$; then $\operatorname{Im} \varphi = G$, since $\operatorname{Im} \varphi$ contains every generator $x = \varphi(\eta(x))$ of G. $\square$

Notation. The construction of F_X is clearer when X and F_X are kept separate, but once F_X is constructed, the usual practice is to identify $x \in X$ and $\eta(x) = (x) \in F_X$, to identify $x' \in X'$ and $x^{-1} = (x)^{-1} = \eta(x') \in F_X$, and to write the elements of F_X as words rather than sequences (for instance, $abb^{-1}c$ instead of (a, b, b', c)). This notation is used in all subsequent sections. Then $X \subseteq F_X$, $\eta : X \longrightarrow F_X$ is an inclusion mapping, and F_X is generated by X.

With these identifications, the universal property in Theorem 6.7 states that every mapping f of X into a group G can be extended uniquely to a homomorphism φ of F_X into G. If $X \subseteq G$, then φ sends the typical element of F_X, which is a product of elements of X and inverses of elements of X, onto the same product but calculated in G. Hence every relation between the elements of X (every equality between products of elements of X and inverses of elements of X) that holds in F_X also holds in every group G that contains X. Thus, F_X has as few relations as possible between the elements of X.

Exercises

1. In an alphabet with two elements, how many reduced words are there of length 4? of length n?

2. Show that, in F_X, $a \cdot b \cdot \cdots \cdot h = \text{red}\,(ab \cdots h)$.

3. Show that $F_X \cong \mathbb{Z}$ if X has just one element.

4. Prove that the universal property in Theorem 6.7 characterizes the free group on X up to isomorphism. (Let F be a group and let $j : X \longrightarrow F$ be a mapping. Assume that for every mapping f of X into a group G, there is a homomorphism φ of F into G unique such that $f = \varphi \circ j$. Show that $F \cong F_X$.)

5. Show that every mapping $f : X \longrightarrow Y$ induces a homomorphism $F_f : F_X \longrightarrow F_Y$ unique such that $F_f \circ \eta_X = \eta_Y \circ f$ (where η_X, η_Y are the canonical injections). Moreover, if f is the identity on X, then F_f is the identity on F_X; if $g \circ f$ is defined, then $F_{g \circ f} = F_g \circ F_f$.

6. Locate a statement of *Kurosh's theorem* on subgroups of free groups.

7. Define homomorphisms of semigroups and prove a universal property of the free semigroup on a set X.

8. Prove a universal property of the free commutative semigroup on a set X.

7. Presentations

Presentations, also called definitions by generators and relations, construct groups that are generated by a given set whose elements satisfy given relations. These groups are often too large to be defined by multiplication tables. Presentations were first considered by Dyck [1882].

Relations. Informally, a group relation between elements of a set X is an equality between products of elements of X and inverses of elements of X. Free groups provide formal models of all such products, and a formal definition:

Definition. A group relation *between the elements of a set* X *is an ordered pair* (u, v) *of elements of* F_X.

These are called *group* relations because there are similar but different relations for rings, modules, and other bidules.

Relations (u, v) are normally written as equalities $u = v$. This should cause no confusion: if u and v are actually equal in F_X, then the relation $u = v$ is trivial and is not likely to be considered; normally, u and v are different in F_X and it is obvious that $u = v$ is a relation and not an equality.

In order for a relation $u = v$ to hold in group G, the elements of X have to be carried, kicking and screaming, into G, by a mapping of X into G.

Definition. A group relation (u, v) *between the elements of a set* X *holds in a group* G *via a mapping* f *of* X *into* G *when* $\varphi(u) = \varphi(v)$, *where* $\varphi : F_X \longrightarrow G$ *is the unique homomorphism that extends* f.

The definition of relations makes most sense when $X \subseteq G$ and f is the inclusion mapping (in which case mention of f is usually omitted). Then φ sends products of elements of X and inverses of elements of X, as calculated in F_X, to the same products but calculated in G; and the relation $u = v$ holds in G if and only if the products u and v are equal when calculated in G.

For example, the relation $a^8 = 1$ holds in a cyclic group $G = \langle a \rangle$ of order 8, in which a has order 8. Formally, f is the inclusion mapping $X = \{ a \} \longrightarrow G$; the free group F on X is cyclic and generated by a; φ sends a^n, as calculated in F, to a^n as calculated in G; the relation $a^8 = 1$ holds in G since a^8 and 1 are equal in G.

In general, relations of type $w = 1$ suffice: indeed, $u = v$ holds if and only if $uv^{-1} = 1$ holds, since $\varphi(u) = \varphi(v)$ if and only if $\varphi(uv^{-1}) = \varphi(1)$.

Construction. Given a group G and a subset X of G, readers will show, as an exercise, that there exists a smallest normal subgroup of G that contains X. This provides a way to construct a group in which given relations must hold.

Definition. Given a set X *and a set* R *of group relations between elements of* X, *the group* $\langle X \mid R \rangle$ *is the quotient of the free group* F_X *by the smallest normal subgroup* N *of* F_X *that contains all* $u\, v^{-1}$ *with* $(u, v) \in R$.

The group $\langle X \mid R \rangle = F_X/N$ comes with a *canonical mapping* $\iota : X \longrightarrow \langle X \mid R \rangle$, the composition $\iota = \pi \circ \eta$ of the inclusion mapping $\eta : X \longrightarrow F_X$ and canonical projection $\pi : F_X \longrightarrow F_X/N$.

Proposition **7.1.** *Let* R *be a set of group relations between elements of a set* X. *Every relation* $(u, v) \in R$ *holds in* $\langle X \mid R \rangle$ *via the canonical mapping* $\iota : X \longrightarrow \langle X \mid R \rangle$; *moreover,* $\langle X \mid R \rangle$ *is generated by* $\iota(X)$.

Proof. The canonical projection $\pi : F_X \longrightarrow \langle X \mid R \rangle$ is a homomorphism that extends ι to F_X, since $\pi \circ \eta = \iota$; therefore it is *the* homomorphism that extends ι to F_X. If $(u, v) \in R$, then $u\,v^{-1} \in N = \mathrm{Ker}\,\pi$, $\pi(u) = \pi(v)$, and (u, v) holds in $\langle X \mid R \rangle$ via ι.

Every element g of $\langle X \mid R \rangle$ is the image under φ of an element a of F; a is a product of elements of X and inverses of elements of X; hence g is a product of elements of $\varphi(X) = \iota(X)$ and inverses of elements of $\iota(X)$. $\square$

Definitions. $\langle X \mid R \rangle$ *is the* (largest) *group generated by* X *subject to every relation* $(u, v) \in R$. *The elements of* X *are the* generators *of* $\langle X \mid R \rangle$, *and the relations* $(u, v) \in R$ *are its* defining relations.

This terminology is traditional but unfortunate. Indeed, $\langle X \mid R \rangle$ is generated by $\iota(X)$, not X. The canonical mapping ι is usually injective on examples, since superfluous generators have been eliminated, but has no reason to be injective in general; thus, X cannot be a priori identified with a subset of $\langle X \mid R \rangle$. Even when ι is injective and X may be identified with a subset of $\langle X \mid R \rangle$, X can generate barrels of groups in which every relation $(u, v) \in R$ holds; $\langle X \mid R \rangle$ is merely the largest (see Theorem 7.2 below). These considerations should be kept in mind when one refers to $\langle X \mid R \rangle$ as *the* group generated by X subject to every relation in R; $\langle X \mid R \rangle$ should be thought of as the *largest* group generated by X subject to every relation in R.

For example, the relation $a^8 = 1$ holds in a cyclic group $C_8 = \langle a \rangle$ of order 8; in a cyclic group $C_4 = \langle a \rangle$ of order 4; in a cyclic group $C_2 = \langle a \rangle$ of order 2; and in the trivial group $1 = \{ a \}$. But only C_8 is $\langle a \mid a^8 = 1 \rangle$.

Universal property.

Theorem 7.2 (Dyck [1882]). *Let R be a set of group relations between elements of a set X. If f is a mapping of X into a group G, and every relation $(u, v) \in R$ holds in G via f, then there exists a homomorphism $\psi : \langle X \mid R \rangle \longrightarrow G$ unique such that $f = \psi \circ \iota$ (where $\iota : X \longrightarrow \langle X \mid R \rangle$ is the canonical mapping). If G is generated by $f(X)$, then φ is surjective.*

$$X \xrightarrow{\ \iota\ } \langle X \mid R \rangle$$
$$f \searrow \quad \downarrow \psi$$
$$G$$

In particular, when a group G is generated by X, and every relation $(u, v) \in R$ holds in G, then there is a surjective homomorphism $\langle X \mid R \rangle \longrightarrow G$, and G is isomorphic to a quotient group of $\langle X \mid R \rangle$. In this sense $\langle X \mid R \rangle$ is the largest group generated by X subject to every relation in R.

Proof. Let N be the smallest normal subgroup of F_X that contains all $u\,v^{-1}$ with $(u, v) \in R$. By 6.7 there is a unique homomorphism $\varphi : F_X \longrightarrow G$ that extends f. Since every $(u, v) \in R$ holds in G via f, we have $\varphi(u) = \varphi(v)$ and $u\,v^{-1} \in \mathrm{Ker}\,\varphi$, for all $(u, v) \in R$. Therefore $\mathrm{Ker}\,\varphi \supseteq N$. By 5.1,

φ factors uniquely through the canonical projection $\pi : F_X \longrightarrow F_X/N = \langle X \mid R \rangle$: there exists a homomorphism $\psi : \langle X \mid R \rangle \longrightarrow G$ unique such that $\psi \circ \pi = \varphi$:

$$X \xrightarrow{\iota} F_X \xrightarrow{\kappa} \langle X \mid R \rangle$$

Then also $\psi \circ \iota = f$. Moreover, ψ is the only homomorphism of $\langle X \mid R \rangle$ into G such that $\psi \circ \iota = f$: if $\chi \circ \iota = f$, then ψ and χ agree on every generator $\iota(x)$ of $\langle X \mid R \rangle$ and therefore agree on all of $\langle X \mid R \rangle$.

If G is generated by $f(X)$, then Im $\psi = G$, since Im ψ contains every generator $f(x) = \psi\big(\iota(x)\big)$ of G. $\square$

Presentations. We now turn to examples.

Definition. A presentation *of a group* G *is an isomorphism of some* $\langle X \mid R \rangle$ *onto* G. $\square$

A presentation of a group G completely specifies G but provides no description, and needs to be supplemented by a list of elements and, if G is not too large, a multiplication table. The usual procedure is to play with the defining relations until no more equalities pop up between the potential elements of G. Then one must make sure that all such equalities have been found. Alas, inequalities in $\langle X \mid R \rangle$ can be obtained only from its universal property. In practice this means that the desired group must be constructed by some other method in order to prove its isomorphy to $\langle X \mid R \rangle$. Examples will illustrate several methods.

Proposition **7.3.** $D_n \cong \langle a, b \mid a^n = b^2 = 1, \ bab = a^{-1} \rangle$.

Proof. Let $G = \langle a, b \mid a^n = b^2 = 1, \ bab = a^{-1} \rangle$. The elements of G are products of a's, b's, a^{-1}'s, and b^{-1}'s. Since a and b have finite order, the elements of G are in fact products of a's and b's. Using the equality $ba = a^{n-1}b$, every product of a's and b's can be rewritten so that all a's precede all b's. Since $a^n = b^2 = 1$, every element of G is now a product of fewer than n a's followed by fewer than 2 b's; in other words, $G = \{\, a^i, \ a^i b \mid 0 \leq i < n \,\}$. In particular, G has at most $2n$ elements; we do not, however, know whether these elements are distinct in G; we might have found an equality between them if we had tried harder, or been more clever, or if, like our Lord in Heaven, we had enough time to list *all* consequences of the defining relations.

We do, however, know that G is supposed to be isomorphic to D_n, and this provides the required alternate construction of G. We know that D_n is, in the notation of Section I.2, generated by r_1 and s_0. Moreover, in D_n, the equalities $r_1^n = s_0^2 = 1$ and $s_0 r_1 s_0 = r_{-1} = r_1^{-1}$ hold, so that the defining relations of G hold in D_n via $f : a \longmapsto r_1, \ b \longmapsto s_0$. By 7.2, f induces a surjective homomorphism $\theta : G \longrightarrow D_n$. Hence G has at least $2n$ elements. Therefore

G has exactly $2n$ elements (the elements a^i, $a^i b$, $0 \leqq i < n$, are distinct in G); θ is bijective; and $G \cong D_n$. $\square$

In the same spirit our reader will verify that a cyclic group $C_n = \langle a \rangle$ of order n has the presentation $C_n \cong \langle a \mid a^n = 1 \rangle$.

Example **7.4.** List the elements and construct a multiplication table of the *quaternion group*

$$Q = \langle a, b \mid a^4 = 1, \ b^2 = a^2, \ bab^{-1} = a^{-1} \rangle.$$

Solution. As in the case of D_n, the elements of Q are products of a's and b's, which can be rewritten, using the relation $ba = a^3 b$, so that all a's precede all b's. Since $a^4 = 1$ and $b^2 = a^2$, at most three a's and at most one b suffice. Hence $Q = \{ 1, a, a^2, a^3, b, ab, a^2 b, a^3 b \}$. In particular, Q has at most eight elements.

The possible elements of Q multiply as follows: $a^i a^j = a^{i+j}$ for all $0 \leqq i, j \leqq 3$ (with $a^{i+j} = a^{i+j-4}$ if $i + j \geqq 4$); $ba = a^3 b$, $ba^2 = b^3 = a^2 b$; $ba^3 = a^2 ba = a^2 a^3 b = ab$, so that $ba^i = a^{4-i} b$ for all $0 \leqq i \leqq 3$; $a^i b a^j b = a^i a^{4-j} b^2 = a^{i+6-j}$ for all i, j. This yields a multiplication table:

1	a	a^2	a^3	b	ab	$a^2 b$	$a^3 b$
a	a^2	a^3	1	ab	$a^2 b$	$a^3 b$	b
a^2	a^3	1	a	$a^2 b$	$a^3 b$	b	ab
a^3	1	a	a^2	$a^3 b$	b	ab	$a^2 b$
b	$a^3 b$	$a^2 b$	ab	a^2	a	1	a^3
ab	b	$a^3 b$	$a^2 b$	a^3	a^2	a	1
$a^2 b$	ab	b	$a^3 b$	1	a^3	a^2	a
$a^3 b$	$a^2 b$	ab	b	a	1	a^3	a^2

The quaternion group Q.

It remains to show that the eight possible elements of Q are distinct, so that the multiplication table above is indeed that of Q. The table itself can always serve as a construction of last resort. It defines a binary operation on a set G with eight elements, bizarrely named $1, a, a^2, a^3, b, ab, a^2 b, a^3 b$; but these are actual products in G. Inspection of the table shows that 1 is an identity element, and that every element has an inverse. Associativity can be verified by Light's test in Section 1. Since all elements of G are products of a's and b's, only a and b need to be tested, by 1.8. The table below (next page) shows that a passes. Readers will verify that b also passes. Hence G is a group.

The multiplication table shows that, in G, $bab^{-1} = a^3 b \, a^2 b = a^3 = a^{-1}$, so that the defining relations $a^4 = 1$, $b^2 = a^2$, $bab^{-1} = a^{-1}$ of Q hold in G, via $f : a \longmapsto a$, $b \longmapsto b$. By 7.2, f induces a homomorphism $\theta : Q \longrightarrow G$,

a	a	a^2	a^3	1	ab	a^2b	a^3b	b
a	a	a^2	a^3	1	ab	a^2b	a^3b	b
a^2	a^2	a^3	1	a	a^2b	a^3b	b	ab
a^3	a^3	1	a	a^2	a^3b	b	ab	a^2b
1	1	a	a^2	a^3	b	ab	a^2b	a^3b
a^3b	a^3b	a^2b	ab	b	a	1	a^3	a^2
b	b	a^3b	a^2b	ab	a^2	a	1	a^3
ab	ab	b	a^3b	a^2b	a^3	a^2	a	1
a^2b	a^2b	ab	b	a^3b	1	a^3	a^2	a

Light's table of a.

which is surjective since G is generated by a and b. Hence Q has at least eight elements. Therefore Q has exactly eight elements; the elements a^i, $a^i b$, $0 \leqq i < 3$, are distinct in Q; and θ is bijective, so that $Q \cong G$ and the multiplication table above is that of Q.

An alternate method of construction for Q (from which Q actually originates) is provided by the *quaternion algebra*, which we denote by $\mathbb{H}$ after its discoverer Hamilton [1843]. $\mathbb{H}$ is a vector space over $\mathbb{R}$, with a basis $\{1, i, j, k\}$ whose elements multiply so that 1 is an identity element; $i^2 = j^2 = k^2 = -1$; $ij = k$, $jk = i$, $ki = j$; and $ji = -k$, $kj = -i$, $ik = -j$. In general,

$$(a+bi + cj + dk)(a' + b'i + c'j + d'k)$$
$$= (aa' - bb' - cc' - dd') + (ab' + ba' + cd' - dc')i$$
$$+ (ac' + ca' + db' - bd')j + (ad' + da' + bc' - cb')k.$$

In $\mathbb{H}$ there is a group $G = \{\pm 1, \pm i, \pm j, \pm k\}$ in which $i^4 = 1$, $j^2 = i^2$, and $jij^{-1} = ji(-j) = -jk = -i = i^{-1}$. Thus, the defining relations of Q hold in this group G, via $f : a \longmapsto i$, $b \longmapsto j$. Then $Q \cong G$, as above. $\square$

Other examples, which make fine exercises, are

$$T = \langle a, b \mid a^6 = 1, b^2 = a^3, bab^{-1} = a^{-1} \rangle$$

and the group

$$A_4 \cong \langle a, b \mid a^3 = 1, b^2 = 1, aba = ba^2b \rangle,$$

one of the *alternating groups*, about which more will be said in the next chapter.

Exercises

1. The *conjugates* of an element x of a group G are the elements axa^{-1} of G, where $a \in G$. Given a group G and a subset X of G, show that there exists a smallest normal subgroup of G that contains X, which consists of all products of conjugates of elements of X and inverses of elements of X.

2. Show that $\langle\, X \mid R \,\rangle$ is determined, up to isomorphism, by its universal property.

3. Show that a cyclic group C_n of order n has the presentation $C_n \cong \langle\, a \mid a^n = 1 \,\rangle$.

4. Find all groups with two generators a and b in which $a^4 = 1$, $b^2 = a^2$, and $bab^{-1} = a^{-1}$.

5. Write a proof that isomorphic groups have the same number of elements of order k, for every $k \geqq 1$.

6. Show that $Q \ncong D_4$.

7. List the elements and draw a multiplication table of $T = \langle\, a, b \mid a^6 = 1,\ b^2 = a^3,\ bab^{-1} = a^{-1} \,\rangle$; prove that you have the required group.

8. Show that a group is isomorphic to T in the previous exercise if and only if it has two generators a and b such that a has order 6, $b^2 = a^3$, and $bab^{-1} = a^{-1}$.

9. List the elements and draw a multiplication table of the group $A_4 \cong \langle\, a, b \mid a^3 = 1,\ b^2 = 1,\ aba = ba^2b \,\rangle$; prove that you have the required group.

10. Show that no two of D_6, T, and A_4 are isomorphic.

11. Show that A_4 does not have a subgroup of order 6.

12. List the elements and draw a multiplication table of the group $\langle\, a, b \mid a^2 = 1,\ b^2 = 1,\ (ab)^3 = 1 \,\rangle$; prove that you have the required group. Do you recognize this group?

13. List the elements and draw a (compact) multiplication table of the group $\langle\, a, b \mid a^2 = 1,\ b^2 = 1 \,\rangle$; prove that you have the required group.

14. Show that a group is isomorphic to D_n if and only if it has two generators a and b such that a has order n, b has order 2, and $bab^{-1} = a^{-1}$.

15. The elements of $\mathbb{H}$ can be written in the form $a + v$, where $a \in \mathbb{R}$ and v is a three-dimensional vector. What is $(a + v)(a' + v')$?

16. Prove that multiplication on $\mathbb{H}$ is associative.

17. Let $|a + bi + cj + dk| = \sqrt{a^2 + b^2 + c^2 + d^2}$. Prove that $|hh'| = |h||h'|$ for all $h,\ h' \in \mathbb{H}$.

18. Show that $\mathbb{H}\backslash\{0\}$ is a group under multiplication (this makes the ring $\mathbb{H}$ a *division ring*).

8. Free Products

This section may be skipped. The free product of two groups A and B is the largest group that is generated by $A \cup B$ in a certain sense. Its construction was devised by Artin in the 1920s. Free products occur in algebraic topology when two path-connected spaces X and Y have just one point in common; then the fundamental group $\pi_1(X \cup Y)$ of $X \cup Y$ is the free product of $\pi_1(X)$ and $\pi_1(Y)$.

In a group G that is generated by the union $A \cup B$ of two subgroups, every element is a product of elements of $A \cup B$. But the elements of G cannot be written uniquely in this form, since, for instance, a product of elements of A can always be replaced by a single element of A. Thus, in G, there are always relations of sorts between the elements of $A \cup B$ (equalities between products of elements of $A \cup B$). Even more relations exist if $A \cap B \neq 1$, a situation which is considered at the end of this section.

The *free product* of A and B is constructed so that there are as few relations as possible between the elements of $A \cup B$ (in particular, $A \cap B = 1$ in the free product). Then a product of elements of $A \cup B$ can always be *reduced* by replacing all subproducts of elements of A by single elements, and similarly for B, until no such subproduct is left. The free product of A and B consists of formal reduced products, multiplied by concatenation and reduction. That it has as few relations as possible between the elements of $A \cup B$ is shown by its universal property. A similar construction yielded free groups in Section 6. The free product of any number of groups is constructed in the same fashion, as adventurous readers will verify.

Reduction. In what follows, A and B are groups. If $A \cap B \neq 1$ we replace A and B by isomorphic groups A' and B' such that $A' \cap B' = 1$: for instance, $A' = \{1\} \cup ((A \backslash \{1\}) \times \{0\})$ and $B' = \{1\} \cup ((B \backslash \{1\}) \times \{1\})$, with operations carried from A and B by the bijections $\theta : A \longrightarrow A'$ and $\zeta : B \longrightarrow B'$: $xy = \theta(\theta^{-1}(x)\, \theta^{-1}(y))$ for all $x, y \in A'$, and similarly for B'; then $A' \cong A$, $B' \cong B$, and $A' \cap B' = \{1\}$. Hence we may assume from the start that $A \cap B = 1$.

Words in the alphabet $A \cup B$ are finite nonempty sequences of elements of $A \cup B$. Let W be the free semigroup on $A \cup B$: the set of all such nonempty words, multiplied by concatenation. For clarity's sake we write words as sequences during construction; in the usual notation, the word $(x_1, x_2, \ldots, x_n)$ is written as a product $x_1 x_2 \cdots x_n$.

Definition. A word $x = (x_1, x_2, \ldots, x_n) \in W$ in the alphabet $A \cup B$ is reduced when it does not contain consecutive letters x_i, x_{i+1} such that $x_i, x_{i+1} \in A$ or $x_i, x_{i+1} \in B$.

Thus a word is reduced when it does not contain consecutive letters from the same group. For example, the empty word, and all one-letter words, are reduced. If $a, a' \in A$ and $b \in B$, then aba' is reduced, as long as $a, a', b \neq 1$, but $aa'b$ is not reduced. In general, when $x = (x_1, x_2, \ldots, x_n)$ is reduced and $n > 1$, then $x_1, x_2, \ldots, x_n \neq 1$, and elements of A alternate with elements of B in the sequence $x_1, x_2, \ldots, x_n$.

The reduction process replaces consecutive letters from the same group by their product in that group, until a reduced word is reached.

Definitions. In W we write $x \xrightarrow{1} y$ when $x = (x_1, x_2, \ldots, x_n)$, $x_i, x_{i+1} \in A$ or $x_i, x_{i+1} \in B$, and $y = (x_1, \ldots, x_{i-1}, x_i x_{i+1}, x_{i+2}, \ldots, x_n)$, for some $1 \leq i < n$;

we write $x \xrightarrow{k} y$ *when* $k \geqq 0$ *and* $x \xrightarrow{1} x' \xrightarrow{1} x'' \xrightarrow{1} \cdots \xrightarrow{1} x^{(k)} = y$ *for some* $x', x'', \ldots, x^{(k)} \in W$ *(when* $x = y$, *if* $k = 0$*);*

we write $x \longrightarrow y$ *when* $x \xrightarrow{k} y$ *for some* $k \geqq 0$.

Lemma 8.1. *For every word* $x \in W$ *there is a reduction* $x \longrightarrow y$ *to a reduced word* y.

We show that the word y in Lemma 8.1 is unique, so that all different ways of reducing a word yield the same reduced word.

Lemma 8.2. *If* $x \xrightarrow{1} y$ *and* $x \xrightarrow{1} z \neq y$, *then* $y \xrightarrow{1} t$, $z \xrightarrow{1} t$ *for some* t.

Proof. By definition, $x = (x_1, x_2, \ldots, x_n)$, $x_i, x_{i+1} \in A$ or $x_i, x_{i+1} \in B$ for some i,

$$y = (x_1, \ldots, x_{i-1}, x_i x_{i+1}, x_{i+2}, \ldots, x_n),$$

$x_j, x_{j+1} \in A$ or $x_j, x_{j+1} \in B$ for some j, and

$$z = (x_1, \ldots, x_{j-1}, x_j x_{j+1}, x_{j+2}, \ldots, x_n).$$

Then $i \neq j$, since $y \neq z$. We may assume that $i < j$. If $i + 1 < j$, then x_i, x_{i+1} are consecutive letters of z, x_j, x_{j+1} are consecutive letters of y, and

$$t = (x_1, \ldots, x_{i-1}, x_i x_{i+1}, x_{i+2}, \ldots, x_{j-1}, x_j x_{j+1}, x_{j+2}, \ldots, x_n)$$

serves. If $i + 1 = j$, and if $x_i, x_{i+1} \in A$ and $x_{i+1}, x_{i+2} \in B$, or if $x_i, x_{i+1} \in B$ and $x_{i+1}, x_{i+2} \in A$, then $x_{i+1} = 1$ and

$$y = z = (x_1, \ldots, x_{i-1}, x_i, x_{i+2}, \ldots, x_n),$$

contradicting $y \neq z$; therefore $x_i, x_{i+1}, x_{i+2} \in A$ or $x_i, x_{i+1}, x_{i+2} \in B$, and

$$t = (x_1, \ldots, x_{i-1}, x_i x_{i+1} x_{i+2}, x_{i+3}, \ldots, x_n)$$

serves. □

As in Section 6 we now have the following:

Lemma 8.3. *If* $x \longrightarrow y$ *and* $x \longrightarrow z$, *then* $y \longrightarrow t$ *and* $z \longrightarrow t$ *for some* t.

Lemma 8.4. *For every word* $x \in W$ *there is a unique reduced word* y *such that* $x \longrightarrow y$.

Definition. The reduction red x *of* $x \in W$ *is the unique reduced word* y *such that* $x \longrightarrow y$.

Construction. The free product of A and B is now defined as follows.

Proposition 8.5. *If* $A \cap B = 1$, *then the set* $A \amalg B$ *of all reduced nonempty words in* $A \cup B$ *is a group under the operation* $x \cdot y = \text{red}\,(xy)$.

Proof. Associativity is proved as in Proposition 6.5. The one-letter word $1 = (1)$ is reduced and is the identity element of $A \amalg B$, since $1 \cdot x = \text{red}\,(1x) =$

red $x = x$ and $x \cdot 1 = $ red $x = x$ when x is reduced. The inverse of a reduced word $x = (x_1, x_2, \ldots, x_n)$ is $x^{-1} = (x_n^{-1}, x_{n-1}^{-1}, \ldots, x_1^{-1})$: indeed, x^{-1} is reduced, since, like x, it does not contain consecutive letters from the same group, and $x\,x^{-1} \longrightarrow 1$, $x^{-1}x \longrightarrow 1$. Thus $A \amalg B$ is a group. $\square$

Definition. Let A and B be groups such that $A \cap B = 1$. The free product *of A and B is the group $A \amalg B$ in Proposition 8.5, which consists of all reduced words in $A \cup B$.*

Readers will gladly show that $F_{X \cup Y} \cong F_X \amalg F_Y$ when X and Y are disjoint, and that $A \cong A'$, $B \cong B'$ implies $A \amalg B \cong A' \amalg B'$ when $A' \cap B' = 1$.

The free product $A \amalg B$ comes with canonical *injections* $\iota : A \longrightarrow A \amalg B$ and $\kappa : B \longrightarrow A \amalg B$, which send an element of A or B to the corresponding one-letter word.

Proposition **8.6.** $\operatorname{Im} \iota \cong A$, $\operatorname{Im} \kappa \cong B$, $\operatorname{Im} \iota \cap \operatorname{Im} \kappa = 1$, *and $A \amalg B$ is generated by* $\operatorname{Im} \iota \cup \operatorname{Im} \kappa$.

Proof. $\operatorname{Im} \iota \cong A$ and $\operatorname{Im} \kappa \cong B$ since ι and κ are injective; $\operatorname{Im} \iota \cap \operatorname{Im} \kappa = 1$ since $A \cap B = 1$; $A \amalg B$ is generated by $\operatorname{Im} \iota \cup \operatorname{Im} \kappa$ since every reduced word is a product of one-letter words. $\square$

Notation. The usual practice is to identify the elements of A or B and the corresponding one-letter words; then A and B are subgroups of $A \amalg B$, and the latter is generated by $A \cup B$. Also, products in $A \amalg B$ are usually written multiplicatively, e.g., xy rather than $x \cdot y$. Various other symbols are used instead of $\amalg$.

Universal property. By Proposition 8.6, the free product of A and B is also the free product of $\operatorname{Im} \iota$ and $\operatorname{Im} \kappa$ (up to isomorphism). If $A \cap B \neq 1$, then the free product of A and B is defined (up to isomorphism) as the free product of any $A' \cong A$ and $B' \cong B$ such that $A' \cap B' = 1$, with injections $A \longrightarrow A' \longrightarrow A' \amalg B'$ and $B \longrightarrow B' \longrightarrow A' \amalg B'$.

$A \amalg B$ is the "largest" group generated by $A \cup B$, in the following sense:

Proposition **8.7.** *Let A and B be groups. For every group G and homomorphisms $\varphi : A \longrightarrow G$ and $\psi : B \longrightarrow G$, there is a unique homomorphism $\chi : A \amalg B \longrightarrow G$ such that $\chi \circ \iota = \varphi$ and $\chi \circ \kappa = \psi$, where $\iota : A \longrightarrow A \amalg B$ and $\kappa : B \longrightarrow A \amalg B$ are the canonical injections:*

$$X \xrightarrow{\;\iota\;} F_X \xrightarrow{\;\kappa\;} \langle X \mid R \rangle$$
$$\varphi \searrow \quad \downarrow{\chi} \quad \swarrow \psi$$
$$G$$

In particular, there is a homomorphism of $A \amalg B$ onto any group that is generated by $A \cup B$.

Proof. We may assume that $A \cap B = 1$, as readers will verify. Then it is convenient to combine ι and κ into a single mapping $\lambda : A \cup B \longrightarrow A \amalg B$, and to combine φ and ψ into a single mapping $\omega : A \cup B \longrightarrow G$. Now, every reduced word $x = (x_1, x_2, \ldots, x_n)$ is a product of one-letter words $x = \lambda(x_1) \cdot \lambda(x_2) \cdot \cdots \cdot \lambda(x_n)$. If $\chi \circ \iota = \varphi$ and $\chi \circ \kappa = \psi$, equivalently $\chi \circ \lambda = \omega$, then $\chi(x) = \omega(x_1) \omega(x_2) \cdots \omega(x_n)$. Hence χ is unique.

Conversely, as in the proof of 6.7, define a mapping $\xi : W \longrightarrow G$ by

$$\xi(x_1, x_2, \ldots, x_n) = \omega(x_1) \omega(x_2) \cdots \omega(x_n),$$

for every $(x_1, x_2, \ldots, x_n) \in W$. Then $\xi(xy) = \xi(x) \xi(y)$ for all $x, y \in W$. Moreover, $\xi(x) = \xi(y)$ when $x \overset{1}{\longrightarrow} y$: if, say, $x = (x_1, x_2, \ldots, x_n)$ has x_i, $x_{i+1} \in A$, so that

$$y = (x_1, \ldots, x_{i-1}, x_i x_{i+1}, x_{i+2}, \ldots, x_n),$$

then $\omega(x_i x_{i+1}) = \omega(x_i) \omega(x_{i+1})$, since φ is a homomorphism, hence $\xi(x) = \xi(y)$. Therefore $x \longrightarrow y$ implies $\xi(x) = \xi(y)$. If now x and y are reduced, then $\omega(x \cdot y) = \omega(xy) = \omega(x) \omega(y)$. Hence the restriction χ of ω to $A \amalg B \subseteq W$ is a homomorphism. Moreover, $\chi \circ \lambda = \omega$. $\square$

Free products with amalgamation. If A and B are groups with a common subgroup $A \cap B = H$, then the union $A \cup B$ is a *group amalgam*, and it is a property of groups that any group amalgam $A \cup B$ can be embedded into a group G, so that A and B are subgroups of G and G is generated by $A \cup B$. The "largest" such group is the free product with amalgamation of A and B (which amalgamates H). This generalization of free products is due to Schreier. Free products with amalgamation occur in algebraic topology when two spaces X and Y have a common subspace $Z = X \cap Y$; under the proper hypotheses, the fundamental group $\pi_1(X \cup Y)$ of $X \cup Y$ is the free product with amalgamation of $\pi_1(X)$ and $\pi_1(Y)$ amalgamating $\pi_1(Z)$.

We sketch the general construction without proofs. Given A and B with $A \cap B = H$ we consider nonempty words in the alphabet $A \cup B$. A word is *reduced* when it does not contain consecutive letters from the same group. Every element of the free product with amalgamation can be written as a reduced word, but this representation is not unique (unless $H = 1$): for instance, if $a \in A \backslash H$, $b \in B \backslash H$, and $h \in H$, then (ah, b) and (a, hb) are reduced, but should represent the same element. Thus, the elements of the free product with amalgamation must be equivalence classes of reduced words.

In detail, two reduced words are *equivalent* when one can be transformed into the other in finitely many steps, where a step replaces consecutive letters ah and b (or bh and a) by a and hb (or by b and ha), or vice versa. A given word can now be reduced in several different ways, but it can be shown that all the resulting reduced words are equivalent. More generally, equivalent words reduce to equivalent reduced words. Equivalence classes of reduced words are then multiplied as follows: $\text{cls } x \cdot \text{cls } y = \text{cls } z$, where xy reduces to z.

With this multiplication, equivalence classes of reduced words constitute a group, the *free product with amalgamation* P of A and B *amalgamating* H, also denoted by $A \amalg_H B$. It comes with canonical injections $\iota : A \longrightarrow P$ and $\kappa : B \longrightarrow P$ that have the following properties:

Proposition 8.8. *Let P be the free product with amalgamation of two groups A and B amalgamating a common subgroup $H = A \cap B$. The canonical injections $\iota : A \longrightarrow P$ and $\kappa : B \longrightarrow P$ are injective homomorphisms and agree on H; moreover, $\mathrm{Im}\, \iota \cong A$, $\mathrm{Im}\, \kappa \cong B$, $\mathrm{Im}\, \iota \cap \mathrm{Im}\, \kappa = \iota(H) = \kappa(H)$, and P is generated by $\mathrm{Im}\, \iota \cup \mathrm{Im}\, \kappa$.*

Proposition 8.9. *Let P be the free product with amalgamation of two groups A and B amalgamating a common subgroup $H = A \cap B$. For every group G and homomorphisms $\varphi : A \longrightarrow G$ and $\psi : B \longrightarrow G$ that agree on H, there is a unique homomorphism $\chi : P \longrightarrow G$ such that $\chi \circ \iota = \varphi$ and $\chi \circ \kappa = \psi$, where $\iota : A \longrightarrow P$ and $\kappa : B \longrightarrow P$ are the canonical injections.*

The free product with amalgamation of groups $(A_i)_{i \in I}$ with a common subgroup $A_i \cap A_j = H$ is constructed similarly, and has a similar universal property.

Exercises

1. Show that $F_{X \cup Y} \cong F_X \amalg F_Y$ when X and Y are disjoint.

2. Show that $A \cong A'$, $B \cong B'$ implies $A \amalg B \cong A' \amalg B'$ when $A' \cap B' = 1$.

3. Given presentations of A and B, find a presentation of $A \amalg B$.

4. Suppose that $A \cong A'$, $B \cong B'$, and $A' \cap B' = 1$, so that $A \amalg B = A' \amalg B'$, with injections $\iota : A \longrightarrow A' \xrightarrow{\iota'} A' \amalg B'$ and $\kappa : B \longrightarrow B' \xrightarrow{\kappa'} A' \amalg B'$. Show that the universal property of ι' and κ' yields a similar universal property of ι and κ.

5. Show that $A \amalg B$ is uniquely determined, up to isomorphism, by its universal property.

6. Show that $(A \amalg B) \amalg C \cong A \amalg (B \amalg C)$ (use the universal property).

*7. Construct a free product of any family of groups $(A_i)_{i \in I}$; then formulate and prove its universal property.

*8. In the construction of free products with amalgamation, verify that equivalent words reduce to equivalent reduced words.

9. Prove the universal property of free products with amalgamation.

II
Structure of Groups

This chapter studies how finite groups are put together. Finite abelian groups decompose into direct products of cyclic groups. For finite groups in general, one method, based on the Sylow theorems and further sharpened in the last two sections, leads in Section 5 to the determination of all groups of order less than 16. The other method, composition series, yields interesting classes of groups.

Sections 2, 8, 10, 11, and 12 may be skipped.

1. Direct Products

Direct products are an easy way to construct larger groups from smaller ones. This construction yields all finite abelian groups.

Definition. The direct product *of two groups G_1 and G_2 is their Cartesian product $G_1 \times G_2$, also denoted by $G_1 \oplus G_2$, together with the* componentwise *operation: in the multiplicative notation,*

$$(x_1, x_2)(y_1, y_2) = (x_1 y_1, x_2 y_2).$$

Readers will verify that $G_1 \times G_2$ is indeed a group.

In algebraic topology, direct products of groups arise from direct products of spaces: when X and Y are path connected, then $\pi_1(X \times Y) \cong \pi_1(X) \times \pi_1(Y)$.

The direct product $G_1 \times G_2 \times \cdots \times G_n$ of n groups, also denoted by $G_1 \oplus G_2 \oplus \cdots \oplus G_n$, is defined similarly when $n \geq 2$ as the Cartesian product $G_1 \times G_2 \times \cdots \times G_n$ with componentwise multiplication

$$(x_1, x_2, \ldots, x_n)(y_1, y_2, \ldots, y_n) = (x_1 y_1, x_2 y_2, \ldots, x_n y_n).$$

It is convenient to let $G_1 \times G_2 \times \cdots \times G_n$ be the trivial group if $n - 0$ and be just G_1 if $n = 1$. In all cases, $|G_1 \times G_2 \times \cdots \times G_n| = |G_1| |G_2| \cdots |G_n|$. Longer direct products are associative; for instance,

$$(G_1 \times G_2 \times \cdots \times G_n) \times G_{n+1} \cong G_1 \times G_2 \times \cdots \times G_n \times G_{n+1}.$$

Direct sums. Next, we give conditions under which a group splits into a direct product.

Proposition **1.1.** *A group G is isomorphic to the direct product $G_1 \times G_2$ of two groups G_1, G_2 if and only if it contains normal subgroups $A \cong G_1$ and $B \cong G_2$ such that $A \cap B = 1$ and $AB = G$.*

Proof. The direct product $G_1 \times G_2$ comes with *projections* $\pi_1 : G_1 \times G_2 \longrightarrow G_1$, $(x_1, x_2) \longmapsto x_1$ and $\pi_2 : G_1 \times G_2 \longrightarrow G_2$, $(x_1, x_2) \longmapsto x_2$, which are homomorphisms, since the operation on $G_1 \times G_2$ is componentwise. Hence

$$\mathrm{Ker}\, \pi_1 = \{\, (x_1, x_2) \in G_1 \times G_2 \mid x_1 = 1 \,\} \text{ and}$$
$$\mathrm{Ker}\, \pi_2 = \{\, (x_1, x_2) \in G_1 \times G_2 \mid x_2 = 1 \,\}$$

are normal subgroups of $G_1 \times G_2$. We see that $(x_1, 1) \longmapsto x_1$ is an isomorphism of $\mathrm{Ker}\, \pi_2$ onto G_1 and that $(1, x_2) \longmapsto x_2$ is an isomorphism of $\mathrm{Ker}\, \pi_1$ onto G_2. Moreover, $\mathrm{Ker}\, \pi_2 \cap \mathrm{Ker}\, \pi_1 = \{(1, 1)\} = 1$, and $(\mathrm{Ker}\, \pi_2)(\mathrm{Ker}\, \pi_1) = G_1 \times G_2$, since every $(x_1, x_2) \in G_1 \times G_2$ is the product $(x_1, x_2) = (x_1, 1)(1, x_2)$ of $(x_1, 1) \in \mathrm{Ker}\, \pi_2$ and $(1, x_2) \in \mathrm{Ker}\, \pi_1$.

If now $\theta : G_1 \times G_2 \longrightarrow G$ is an isomorphism, then $A = \theta(\mathrm{Ker}\, \pi_2)$ and $B = \theta(\mathrm{Ker}\, \pi_1)$ are normal subgroups of G, $A \cong \mathrm{Ker}\, \pi_2 \cong G_1$, $B \cong \mathrm{Ker}\, \pi_1 \cong G_2$, $A \cap B = 1$, and $AB = G$.

Conversely, assume that $A \trianglelefteq G$, $B \trianglelefteq G$, $A \cap B = 1$, and $AB = G$. Then every element g of G is a product $g = ab$ of some $a \in A$ and $b \in B$. Moreover, if $ab = a'b'$, with $a, a' \in A$ and $b, b' \in B$, then $a'^{-1}a = b'b^{-1} \in A \cap B$ yields $a'^{-1}a = b'b^{-1} = 1$ and $a = a'$, $b = b'$. Hence the mapping $\theta : (a, b) \longmapsto ab$ of $A \times B$ onto G is a bijection. We show that θ is an isomorphism.

For all $a \in A$ and $b \in B$, we have $aba^{-1}b^{-1} = a(ba^{-1}b^{-1}) \in A$ and $aba^{-1}b^{-1} = (aba^{-1})b^{-1} \in B$, since $A, B \trianglelefteq G$; hence $aba^{-1}b^{-1} = 1$ and $ab = ba$. (Thus, A and B *commute elementwise*.) Therefore

$$\theta((a, b)(a', b')) = \theta(aa', bb') = aa'bb' = aba'b' = \theta(a, b)\,\theta(a', b'). \qquad \square$$

Definition. A group G is the (internal) direct sum *$G = A \oplus B$ of two subgroups A and B when $A, B \trianglelefteq G$, $A \cap B = 1$, $AB = G$.*

Then $G \cong A \times B$, by 1.1. For example, $V_4 = \{\, 1, a, b, c \,\}$ is the direct sum of $A = \{\, 1, a \,\}$ and $B = \{\, 1, b \,\}$.

The proof of 1.1 shows that direct products contain a certain amount of commutativity, and its conditions $A, B \trianglelefteq G$, $A \cap B = 1$, $AB = G$ are rather stringent. Hence comparatively few groups split into nontrivial direct products.

Abelian groups. In abelian groups, however, all subgroups are normal and the conditions in Proposition 1.1 reduce to $A \cap B = 1$, $AB = G$ ($A + B = G$, in the additive notation). Subgroups with these properties are common in finite abelian groups. In fact, all finite abelian groups are direct sums of cyclic groups:

Theorem **1.2.** *Every finite abelian group is isomorphic to the direct product of cyclic groups whose orders are positive powers of prime numbers, and these cyclic groups are unique, up to order of appearance and isomorphism.*

Early versions of this result are due to Schering [1868], Kronecker [1870], and Frobenius and Stickelberger [1878]. We postpone the proof until the more general results in Section VIII.6.

Theorem 1.2 readily yields all finite abelian groups of given order n (up to isomorphism). If G is a direct product of cyclic groups of orders $p_1^{k_1}$, $p_2^{k_2}$, ..., $p_r^{k_r}$, for some $k_1, k_2, ..., k_r > 0$ and some not necessarily distinct primes $p_1, p_2, ..., p_r$, then G has order $n = p_1^{k_1} p_2^{k_2} \cdots p_r^{k_r}$. This equality must match the unique factorization of n into positive powers of distinct primes.

First let G be a *p-group* (a group of order $p^k > 1$ for some prime p). A *partition* of a positive integer k is a sequence $k_1 \geq k_2 \geq \cdots \geq k_r > 0$ such that $k = k_1 + k_2 + \cdots + k_r$. If $n = p^k$ is a positive power of a prime p, then, in the equality $n = p_1^{k_1} p_2^{k_2} \cdots p_r^{k_r}$, all p_i are equal to p, and the positive exponents k_i, when numbered in descending order, constitute a partition of k. Hence abelian groups of order p^k correspond to partitions of k: to a partition $k = k_1 + k_2 + \cdots + k_r$ corresponds the direct product $C_{p^{k_1}} \oplus C_{p^{k_2}} \oplus \cdots \oplus C_{p^{k_r}}$ of cyclic groups of orders p^{k_1}, $p_2^{k_2}$, ..., p^{k_r}.

For example, let $n = 16 = 2^4$. We find five partitions of 4: $4 = 4$; $4 = 3 + 1$; $4 = 2 + 2$; $4 = 2 + 1 + 1$; and $4 = 1 + 1 + 1 + 1$. Hence there are, up to isomorphism, five abelian groups of order 16:

$$C_{16}; \quad C_8 \oplus C_2; \quad C_4 \oplus C_4; \quad C_4 \oplus C_2 \oplus C_2; \quad \text{and } C_2 \oplus C_2 \oplus C_2 \oplus C_2.$$

Now let the abelian group G of arbitrary order n be the direct product of cyclic groups of orders $p_1^{k_1}$, $p_2^{k_2}$, ..., $p_r^{k_r}$. Classifying the terms of this product by distinct prime divisors of n shows that G is a direct product of p-groups, one for each prime divisor p of n:

Corollary 1.3. *Let $p_1, ..., p_r$ be distinct primes. An abelian group of order $p_1^{k_1} p_2^{k_2} \cdots p_r^{k_r}$ is a direct sum of subgroups of orders $p_1^{k_1}$, $p_2^{k_2}$, ..., $p_r^{k_r}$.*

Abelian groups of order n are therefore found as follows: write n as a product of positive powers p^k of distinct primes; for each p find all abelian p-groups of order p^k, from the partitions of k; the abelian groups of order n are the direct products of these p-groups, one for each prime divisor p of n.

For example let $n = 200 = 2^3 \cdot 5^2$. There are three partitions of 3: $3 = 3$, $3 = 2 + 1$, and $3 = 1 + 1 + 1$, which yield three 2-groups of order 8:

$$C_8; \quad C_4 \oplus C_2; \quad \text{and } C_2 \oplus C_2 \oplus C_2.$$

The two partitions of 2, $2 = 2$ and $2 = 1 + 1$, yield two 5-groups of order 25:

$$C_{25} \quad \text{and} \quad C_5 \oplus C_5.$$

Hence there are, up to isomorphism, $3 \times 2 = 6$ abelian groups of order 200:

$$C_8 \oplus C_{25}; \ C_8 \oplus C_5 \oplus C_5; \ C_4 \oplus C_2 \oplus C_{25}; \ C_4 \oplus C_2 \oplus C_5 \oplus C_5;$$
$$C_2 \oplus C_2 \oplus C_2 \oplus C_{25}; \ \text{and} \ C_2 \oplus C_2 \oplus C_2 \oplus C_5 \oplus C_5.$$

For another example, "the" cyclic group C_n of order $n = p_1^{k_1} \, p_2^{k_2} \, \cdots \, p_r^{k_r}$, where $p_1, \ldots, p_r$ are distinct primes, is a direct sum of cyclic subgroups of orders $p_1^{k_1}, \ p_2^{k_2}, \ \ldots, \ p_r^{k_r}$. This also follows from the next result:

Proposition 1.4. *If m and n are relatively prime, then $C_{mn} \cong C_m \times C_n$.*

Proof. Let $C_{mn} = \langle c \rangle$ be cyclic of order mn. Then c^n has order m (since $(c^m)^k = 1$ if and only if mn divides mk, if and only if n divides k) and c^m has order n. The subgroups

$$A = \langle c^n \rangle \cong C_m \ \text{and} \ B = \langle c^m \rangle \cong C_n$$

have the following properties. First, $c^k \in A$ if and only if n divides k: if $c^k = c^{nt}$ for some t, then $k - nt$ is a multiple of mn and k is a multiple of n. Similarly, $c^k \in B$ if and only if m divides k. If now $c^k \in A \cap B$, then m and n divide k, mn divides k, and $c^k = 1$; thus $A \cap B = 1$. Also $AB = C_{mn}$: since m and n are relatively prime, there exist integers u and v such that $mu + nv = 1$; for every k,

$$c^k = c^{kmu+knv} = c^{nkv} \, c^{mku},$$

where $c^{nkv} \in A$ and $c^{mku} \in B$. Hence $C_{mn} \cong C_m \times C_n$, by 1.1. $\square$

The abelian groups of order 200 may now be listed as follows: $C_8 \oplus C_{25} \cong C_{200}$; $C_8 \oplus C_5 \oplus C_5 \cong C_{40} \oplus C_5$; $C_4 \oplus C_2 \oplus C_{25} \cong C_{100} \oplus C_2$; etc.

Euler's ϕ function. These results yield properties of Euler's function ϕ.

Definition. Euler's function $\phi(n)$ *is the number of integers $1 \leq k \leq n$ that are relatively prime to n.*

If p is prime, then $\phi(p) = p - 1$; more generally, if $n = p^m$, every p th number $1 \leq k \leq p^m$ is a multiple of p, so that $\phi(p^m) = p^m - (p^m/p) = p^m (1 - 1/p)$.

Proposition 1.5. *A cyclic group of order n has exactly $\phi(n)$ elements of order n.*

Proof. Let $G = \langle a \rangle$ be cyclic of order n. Let $1 \leq k \leq n$. The order of a^k divides n, since $(a^k)^n = (a^n)^k = 1$. We show that a^k has order n if and only if k and n are relatively prime. If $\gcd(k, n) = d > 1$, then $(a^k)^{n/d} = (a^n)^{k/d} = 1$ and a^k has order at most $n/d < n$. But if $\gcd(k, n) = 1$, then a^k has order n: if $(a^k)^m = 1$, then n divides km and n divides m. $\square$

Properties of cyclic groups, such as Proposition 1.4, now provide nifty proofs of purely number-theoretic properties of ϕ.

Proposition **1.6.** *If m and n are relatively prime, then* $\phi(mn) = \phi(m)\,\phi(n)$.

Proof. By 1.4 a cyclic group C_{mn} of order mn is, up to isomorphism, the direct product of a cyclic group C_m of order m and a cyclic group C_n of order n. In $C_m \times C_n$, $(x, y)^k = 1$ if and only if $x^k = 1$ and $y^k = 1$, so that the order of (x, y) is the least common multiple of the orders of x and y (which is the product of the orders of x and y, since the latter divide m and n and are relatively prime). It follows that (x, y) has order mn if and only if x has order m and y has order n. Hence $\phi(mn) = \phi(m)\,\phi(n)$, by 1.5. $\square$

Corollary **1.7.** $\phi(n) = n \prod_{p \text{ prime, } p|n} (1 - 1/p)$.

Proof. This follows from 1.6 and $\phi(p^m) = p^m (1 - 1/p)$, since n is a product of relatively prime powers of primes. $\square$

Proposition **1.8.** $\sum_{d|n} \phi(d) = n$.

Proof. Let $G = \langle c \rangle$ be a cyclic group of order n. By I.5.7, every divisor d of n is the order of a unique cyclic subgroup of G, namely $D = \{ x \in G \mid x^d = 1 \}$. Since D is cyclic of order d, G has exactly $\phi(d)$ elements of order d. Now, every element of G has an order that is some divisor of n; hence $n = \sum_{d|n} \phi(d)$. $\square$

Exercises

1. Verify that the direct product of two groups is a group.

2. Define the direct product of any family of groups, and verify that it is a group.

3. Prove the following universal property of the direct product $A \times B$ of two groups and its projections $\pi : A \times B \longrightarrow A$, $\rho : A \times B \longrightarrow B$: for every homomorphisms $\varphi : G \longrightarrow A$, $\psi : G \longrightarrow B$ of a group G, there is a homomorphism $\chi : G \longrightarrow A \times B$ unique such that $\pi \circ \chi = \varphi$ and $\rho \circ \chi = \psi$.

4. Show that the direct product of two groups is characterized, up to isomorphism, by the universal property in the previous exercise.

5. Find all abelian groups of order 35.

6. Find all abelian groups of order 36.

7. Find all abelian groups of order 360.

8. Prove directly that no two of the groups C_8, $C_4 \oplus C_2$, and $C_2 \oplus C_2 \oplus C_2$ are isomorphic.

A group G is *indecomposable* when $G \neq 1$, and $G = A \oplus B$ implies $A = 1$ or $B = 1$.

9. Prove that D_5 is indecomposable.

10. Prove that D_4 is indecomposable.

11. Prove directly that a cyclic group of order p^k, with p prime, is indecomposable.

2. The Krull-Schmidt Theorem.

This section may be skipped. The Krull-Schmidt theorem, also known as the Krull-Schmidt-Remak theorem, is a uniqueness theorem for decompositions into direct products, due to Remak [1911], Schmidt [1912], Krull [1925], and, in the general case, Schmidt [1928].

Direct sums. We begin with the following generalization of Proposition 1.1:

Proposition **2.1.** *A group G is isomorphic to the direct product $G_1 \times G_2 \times \cdots \times G_n$ of groups G_1, G_2, ..., G_n if and only if it contains normal subgroups $A_i \cong G_i$ such that $A_1 A_2 \cdots A_n = G$ and $(A_1 A_2 \cdots A_i) \cap A_{i+1} = 1$ for all $i < n$. Then every element g of G can be written uniquely in the form $g = a_1 a_2 \cdots a_n$ with $a_i \in A_i$; $a_i \in A_i$ and $a_j \in A_j$ commute whenever $i \neq j$; and the mapping $(a_1, a_2, ..., a_n) \longmapsto a_1 a_2 \cdots a_n$ is an isomorphism of $A_1 \times A_2 \times \cdots \times A_n$ onto G.*

Proof. This is trivial if $n = 1$, and Proposition 1.1 is the case $n = 2$.

Since the operation on $G_1 \times G_2 \times \cdots \times G_n$ is componentwise,

$$G_k' = \{ (1, ..., 1, x_k, 1, ..., 1) \in G_1 \times G_2 \times \cdots \times G_n \mid x_k \in G_k \}$$

is a normal subgroup of $G_1 \times G_2 \times \cdots \times G_n$, for every $1 \leq k \leq n$. Moreover, $\iota_k : x_k \longmapsto (1, ..., 1, x_k, 1, ..., 1)$ is an isomorphism of G_k onto G_k'. Also

$$G_1' G_2' \cdots G_k' = \{ (x_1, ..., x_n) \in G_1 \times \cdots \times G_n \mid x_i = 1 \text{ for all } i > k \}.$$

Hence $(G_1' G_2' \cdots G_k') \cap G_{k+1}' = 1$ for all $k < n$ and $G_1' G_2' \cdots G_n' = G$. (In fact, $(G_1' \cdots G_{k-1}' G_{k+1}' \cdots G_n') \cap G_k' = 1$ for all k.) Finally,

$$(x_1, ..., x_n) = \iota_1(x_1) \iota_2(x_2) \cdots \iota_n(x_n),$$

so that every element $(x_1, ..., x_n)$ of $G_1 \times G_2 \times \cdots \times G_n$ can be written uniquely in the form $x_1' x_2' \cdots x_n'$ with $x_i' \in G_i'$ for all i; $x_i' \in G_i'$ and $x_j' \in G_j'$ commute whenever $i \neq j$; and the mapping $(x_1', x_2', ..., x_n') \longmapsto x_1' x_2' \cdots x_n'$ is an isomorphism of $G_1' \times G_2' \times \cdots \times G_n'$ onto G.

If now $\theta : G_1 \times G_2 \times \cdots \times G_n \longrightarrow G$ is an isomorphism, then $A_k = \theta(G_k')$ is a normal subgroup of G, $A_k \cong G_k' \cong G_k$, $(A_1 A_2 \cdots A_k) \cap A_{k+1} = 1$ for all $k < n$, and $A_1 A_2 \cdots A_n = G$. (In fact, $(A_1 \cdots A_{k-1} A_{k+1} \cdots A_n) \cap A_k = 1$ for all k.) Moreover, every element g of G can be written uniquely in the form $g = a_1 a_2 \cdots a_n$ with $a_i \in A_i$; $a_i \in A_i$ and $a_j \in A_j$ commute whenever $i \neq j$; and the mapping $(a_1, a_2, ..., a_n) \longmapsto a_1 a_2 \cdots a_n$ is an isomorphism of $A_1 \times A_2 \times \cdots \times A_n$ onto G.

The converse is proved by induction on n. We may assume that $n > 2$. Let G contain normal subgroups $A_i \cong G_i$ such that $(A_1 A_2 \cdots A_i) \cap A_{i+1} = 1$ for all $i < n$ and $A_1 A_2 \cdots A_n = G$. Then $A = A_1 A_2 \cdots A_{n-1} \trianglelefteq G$, since A_1, A_2, ..., $A_{n-1} \trianglelefteq G$, and $A_n \trianglelefteq G$, $A \cap A_n = 1$, $A A_n = G$. Hence

$G \cong A \times A_n$ by 1.1, $A \cong G_1 \times G_2 \times \cdots \times G_{n-1}$ by the induction hypothesis, and $G \cong (G_1 \times G_2 \times \cdots \times G_{n-1}) \times G_n \cong G_1 \times G_2 \times \cdots \times G_n$. $\square$

Definition. A group G is the *(internal) direct sum* $G = A_1 \oplus A_2 \oplus \cdots \oplus A_n$ of subgroups A_1, A_2, ..., A_n when $A_i \trianglelefteq G$ for all i, $(A_1 A_2 \cdots A_i) \cap A_{i+1} = 1$ for all $i < n$, and $A_1 A_2 \cdots A_n = G$.

Then $G \cong A_1 \times A_2 \times \cdots \times A_n$ by 2.1 and, as noted in the proof of 2.1, $(A_1 \cdots A_{i-1} A_{i+1} \cdots A_n) \cap A_i = 1$ for all i; every element g of G can be written uniquely in the form $g = a_1 a_2 \cdots a_n$ with $a_i \in A_i$, and the mapping $(a_1, a_2, ..., a_n) \longmapsto a_1 a_2 \cdots a_n$ is an isomorphism of $A_1 \times A_2 \times \cdots \times A_n$ onto G. Particular cases of direct sums include the empty direct sum 1 (then $n = 0$ and $G = A_1 A_2 \cdots A_n$ is the empty product), direct sums with one term A_1 (then $G = A_1$), and the direct sums with two terms in Section 1.

Finite groups decompose into direct sums of smaller groups until the latter can be decomposed no further. In detail:

Definition. A group G is *indecomposable when $G \neq 1$, and $G = A \oplus B$ implies $A = 1$ or $B = 1$.*

Then every finite group is a direct sum of indecomposable subgroups. We prove a somewhat more general statement.

Definition. A group G has *finite length when every chain of normal subgroups of G is finite.*

Proposition **2.2.** *Every group of finite length is a direct sum of (finitely many) indecomposable subgroups.*

Proof. Assume that there is a group G of finite length that is not a direct sum of indecomposable subgroups. Call a normal subgroup B of G *bad* when $G = A \oplus B$ for some subgroup A, but B is not a direct sum of indecomposable subgroups. For instance, $G = 1 \oplus G$ is bad. Since G has finite length, there must exist a *minimal* bad subgroup (a bad subgroup M with no bad subgroup $B \subsetneqq M$): otherwise, G is not minimal and there is a bad subgroup $B_1 \subsetneqq G$; B_1 is not minimal and there is a bad subgroup $B_2 \subsetneqq B_1$; B_2 is not minimal and there is a bad subgroup $B_3 \subsetneqq B_2$; and there is an infinite chain of (bad) normal subgroups of G, which is more than any group of finite length can tolerate.

Now, M is not trivial and is not indecomposable (since M is not a direct sum of zero or one indecomposable subgroups). Therefore $M = C \oplus D$ for some $C, D \neq 1$. Then $G = A \oplus C \oplus D$ for some subgroup A, so that $C, D \trianglelefteq G$. Then $C, D \subsetneqq M$, so C and D are not bad; C and D are direct sums of indecomposable subgroups; then so is M, which is the required contradiction. $\square$

Proposition 2.2 holds more generally for groups whose normal subgroups satisfy the descending chain condition (defined in Section A.1).

Main result. The Krull-Schmidt theorem states that the direct sum decomposition in Proposition 2.2 is unique, up to isomorphism and indexing. In fact, a stronger statement holds:

Theorem **2.3** (Krull-Schmidt). *If a group G of finite length is a direct sum*

$$G = G_1 \oplus G_2 \oplus \cdots \oplus G_m = H_1 \oplus H_2 \oplus \cdots \oplus H_n$$

of indecomposable subgroups $G_1, \ldots, G_m$ *and* $H_1, \ldots, H_n$, *then* $m = n$ *and* $H_1, \ldots, H_n$ *can be indexed so that* $H_i \cong G_i$ *for all* $i \leqq n$ *and*

$$G = G_1 \oplus \cdots \oplus G_k \oplus H_{k+1} \oplus \cdots \oplus H_n$$

for every $k < n$.

The last part of the statement is the Krull-Schmidt *exchange property*. Theorem 2.3 is often stated as follows: if a group G of finite length is isomorphic to two direct products $G \cong G_1 \times G_2 \times \cdots \times G_m \cong H_1 \times H_2 \times \cdots \times H_n$ of indecomposable subgroups $G_1, G_2, \ldots, G_m$ and $H_1, H_2, \ldots, H_n$, then $m = n$ and $H_1, H_2, \ldots, H_n$ can be indexed so that $H_i \cong G_i$ for all $i \leqq n$ and $G \cong G_1 \times \cdots \times G_k \times H_{k+1} \times \cdots \times H_n$ for all $k < n$.

Normal endomorphisms. Recall that an endomorphism of a group G is a homomorphism of G into G. The proof of Theorem 2.3 requires properties of endomorphisms, and some patience.

In this proof we write endomorphisms as left operators. Endomorphisms compose: if η and ζ are endomorphisms of G, then so is $\eta\zeta : x \longmapsto \eta(\zeta x)$. Thus the set $\mathrm{End}\,(G)$ of all endomorphisms of G becomes a monoid.

An endomorphism η of a group G is *normal* when $\eta(gxg^{-1}) = g(\eta x)g^{-1}$ for all $x, g \in G$ (in other words, when η commutes with all inner automorphisms). Then both $\mathrm{Im}\,\eta$ and $\mathrm{Ker}\,\eta$ are normal subgroups.

Lemma **2.4.** *If* G *has finite length, then a normal endomorphism of* G *is injective if and only if it is surjective, if and only if it is bijective.*

Proof. Let $\eta \in \mathrm{End}\,(G)$ be normal. For every $n > 0$, η^n is normal, so that $\mathrm{Im}\,\eta^n$ and $\mathrm{Ker}\,\eta^n$ are normal subgroups of G. The descending sequence

$$\mathrm{Im}\,\eta \supseteq \mathrm{Im}\,\eta^2 \supseteq \cdots \supseteq \mathrm{Im}\,\eta^n \supseteq \mathrm{Im}\,\eta^{n+1} \supseteq \cdots$$

cannot be infinite, since G has finite length; therefore $\mathrm{Im}\,\eta^n = \mathrm{Im}\,\eta^{n+1}$ for some n. For every $x \in G$ we now have $\eta^n x = \eta^{n+1} y$ for some $y \in G$; if η is injective, this implies $x = \eta y$, and η is surjective.

Similarly, the ascending sequence

$$\mathrm{Ker}\,\eta \subseteq \mathrm{Ker}\,\eta^2 \subseteq \cdots \subseteq \mathrm{Ker}\,\eta^n \subseteq \mathrm{Ker}\,\eta^{n+1} \subseteq \cdots$$

cannot be infinite; therefore $\mathrm{Ker}\,\eta^n = \mathrm{Ker}\,\eta^{n+1}$ for some n. If η is surjective, then for every $x \in \mathrm{Ker}\,\eta$ we have $x = \eta^n y$ for some $y \in G$, so that $\eta^{n+1} y = \eta x = 1$, $y \in \mathrm{Ker}\,\eta^{n+1} = \mathrm{Ker}\,\eta^n$, and $x = \eta^n y = 1$; thus η is injective. $\square$

Lemma **2.5.** *If* G *has finite length and* η *is a normal endomorphism of* G, *then* $G = \mathrm{Im}\,\eta^n \oplus \mathrm{Ker}\,\eta^n$ *for some* $n > 0$.

Proof. As in the proof of Lemma 2.4, the sequences

$$\operatorname{Im} \eta \supseteq \cdots \supseteq \operatorname{Im} \eta^n \supseteq \cdots \quad \text{and} \quad \operatorname{Ker} \eta \subseteq \cdots \subseteq \operatorname{Ker} \eta^n \subseteq \cdots$$

cannot be infinite, so that $\operatorname{Im} \eta^k = \operatorname{Im} \eta^{k+1}$ for some k and $\operatorname{Ker} \eta^m = \operatorname{Ker} \eta^{m+1}$ for some m. Applying η to $\operatorname{Im} \eta^k = \operatorname{Im} \eta^{k+1}$ yields $\operatorname{Im} \eta^n = \operatorname{Im} \eta^{n+1}$ for all $n \geq k$; similarly, $\operatorname{Ker} \eta^n = \operatorname{Ker} \eta^{n+1}$ for all $n \geq m$. Therefore $\operatorname{Im} \eta^n = \operatorname{Im} \eta^{2n}$ and $\operatorname{Ker} \eta^n = \operatorname{Ker} \eta^{2n}$ both hold when n is large enough.

If now $x \in \operatorname{Im} \eta^n \cap \operatorname{Ker} \eta^n$, then $x = \eta^n y$ for some y, $\eta^{2n} y = \eta^n x = 1$, $y \in \operatorname{Ker} \eta^{2n} = \operatorname{Ker} \eta^n$, and $x = \eta^n y = 1$. Thus $\operatorname{Im} \eta^n \cap \operatorname{Ker} \eta^n = 1$.

For any $x \in G$ we have $\eta^n x \in \operatorname{Im} \eta^n = \operatorname{Im} \eta^{2n}$ and $\eta^n x = \eta^{2n} y$ for some y. Then $x = (\eta^n y)(\eta^n y^{-1}) x$, with $\eta^n y \in \operatorname{Im} \eta^n$ and $(\eta^n y^{-1}) x \in \operatorname{Ker} \eta^n$, since $\eta^n \big((\eta^n y^{-1}) x \big) = (\eta^{2n} y)^{-1} \eta^n x = 1$. Thus $(\operatorname{Im} \eta^n)(\operatorname{Ker} \eta^n) = G$. $\square$

If G is indecomposable, then the direct sum in Lemma 2.5 is trivial. Call an endomorphism η of a group G *nilpotent* when $\operatorname{Im} \eta^n = 1$ for some $n > 0$.

Lemma 2.6. *If G is an indecomposable group of finite length, then every normal endomorphism of G is either nilpotent or an automorphism.*

Proof. By 2.5, either $\operatorname{Im} \eta^n = 1$ and η is nilpotent, or $\operatorname{Ker} \eta^n = 1$, and then $\operatorname{Ker} \eta = 1$ and η is bijective by 2.4. $\square$

Pointwise products. The group operation on G induces a partial operation $\cdot$ on $\operatorname{End}(G)$: the pointwise product $\eta \cdot \zeta$ of η and $\zeta \in \operatorname{End}(G)$ is defined in $\operatorname{End}(G)$ if and only if the mapping $\xi : x \longmapsto (\eta x)(\zeta x)$ is an endomorphism, and then $\eta \cdot \zeta = \xi$. Longer products are defined similarly, when possible. The following properties are straightforward:

Lemma 2.7. *$\eta \cdot \zeta$ is defined in $\operatorname{End}(G)$ if and only if ηx and ζy commute for every $x, y \in G$. If η and ζ are normal and $\eta \cdot \zeta$ is defined, then $\eta \cdot \zeta$ is normal. If $\eta_1, \eta_2, \ldots, \eta_n \in \operatorname{End}(G)$, and $\eta_i x$ commutes with $\eta_j y$ for every $x, y \in G$ and every $i \neq j$, then $\eta_1 \cdot \eta_2 \cdot \cdots \cdot \eta_n$ is defined in $\operatorname{End}(G)$, and $\eta_1 \cdot \eta_2 \cdot \cdots \cdot \eta_n = \eta_{\sigma 1} \cdot \eta_{\sigma 2} \cdot \cdots \cdot \eta_{\sigma n}$ for every permutation σ.*

Some distributivity always holds in $\operatorname{End}(G)$: $\xi(\eta \cdot \zeta) = (\xi\eta) \cdot (\xi\zeta)$ and $(\eta \cdot \zeta)\xi = (\eta\xi) \cdot (\zeta\xi)$, if $\eta \cdot \zeta$ is defined. (If G is abelian, written additively, then $\eta \cdot \zeta$ is always defined and is denoted by $\eta + \zeta$, and $\operatorname{End}(G)$ is a ring.)

Lemma 2.8. *Let $\eta_1, \eta_2, \ldots, \eta_n$ be normal endomorphisms of an indecomposable group G of finite length. If $\eta_i x$ commutes with $\eta_j y$ for every $x, y \in G$ and every $i \neq j$, and every η_i is nilpotent, then $\eta_1 \cdot \eta_2 \cdot \cdots \cdot \eta_n$ is nilpotent.*

Proof. We prove this when $n = 2$; the general case follows by induction on n. Assume that $\eta, \zeta \in \operatorname{End}(G)$ are normal and nilpotent, and that $\alpha = \eta \cdot \zeta$ is defined but not nilpotent. Then α is an automorphism, by 2.6. Let $\varphi = \eta\alpha^{-1}$ and $\psi = \zeta\alpha^{-1}$. Then φ and ψ are nilpotent by 2.6, since they are not automorphisms. Also $\varphi \cdot \psi = (\eta \cdot \zeta)\alpha^{-1} = 1_G$. Hence

$$\varphi\varphi \cdot \varphi\psi = \varphi(\varphi \cdot \psi) = \varphi = (\varphi \cdot \psi)\varphi = \varphi\varphi \cdot \psi\varphi.$$

Under pointwise multiplication this implies $\varphi\psi = \psi\varphi$. Therefore $(\varphi \cdot \psi)^n$ can be calculated as in the Binomial theorem: $(\varphi \cdot \psi)^n$ is a pointwise product with $\binom{n}{i}$ terms $\varphi^i \psi^j$ for every $i + j = n$. By 2.7, this pointwise product can be calculated in any order, since $\varphi\psi = \psi\varphi$ and every $\varphi x \in \operatorname{Im} \eta$ commutes with every $\psi y \in \operatorname{Im} \zeta$.

Now, η and ζ are nilpotent: $\operatorname{Im} \eta^k = \operatorname{Im} \zeta^\ell = 1$ for some $k, \ell > 0$. If $i + j = n \geq k + \ell$, then either $i \geq k$ and $\varphi^i \psi^j x \in \operatorname{Im} \eta^i = 1$, or $j \geq \ell$ and $\varphi^i \psi^j x = \psi^j \varphi^i x \in \operatorname{Im} \zeta^j = 1$ (or both), for all $x \in G$; hence $\varphi^i \psi^j x = 1$ for all $x \in G$ and $\operatorname{Im} (\varphi \cdot \psi)^n = 1$, contradicting $\varphi \cdot \psi = 1_G$. $\square$

Direct sums. Direct sum decompositions come with normal endomorphisms. Let $G = G_1 \oplus G_2 \oplus \cdots \oplus G_m$, so that every $x \in G$ can be written uniquely in the form $x = x_1 x_2 \cdots x_n$ with $x_i \in G_i$ for all i. For every k let η_k be the mapping $\eta_k : x_1 x_2 \cdots x_n \longmapsto x_k \in G$; η_k can also be obtained by composing the isomorphism $G \cong G_1 \times G_2 \times \cdots \times G_m$, the projection $G_1 \times G_2 \times \cdots \times G_m \longrightarrow G_k$, and the inclusion homomorphism $G_k \longrightarrow G$, and is therefore an endomorphism of G, the kth *projection endomorphism* of the direct sum $G = G_1 \oplus G_2 \oplus \cdots \oplus G_m$. The following properties are immediate:

Lemma 2.9. *In any direct sum decomposition $G = G_1 \oplus G_2 \oplus \cdots \oplus G_m$, the projection endomorphisms η_1, η_2, ..., η_n are normal endomorphisms; $\operatorname{Im} \eta_k = G_k$; $\eta_k x = x$ for all $x \in G_k$; $\eta_k x = 1$ for all $x \in G_i$ if $i \neq k$; $\eta_i x$ commutes with $\eta_j y$ for every $x, y \in G$ and every $i \neq j$; $\eta_1 \cdot \eta_2 \cdot \cdots \cdot \eta_n$ is defined in $\operatorname{End}(G)$; and $\eta_1 \cdot \eta_2 \cdot \cdots \cdot \eta_n = 1_G$.*

Lemma 2.10. *Let $G = A \oplus B$. Every normal subgroup of A is a normal subgroup of G. If η is a normal endomorphism of G and $\eta A \subseteq A$, then the restriction $\eta_{|A}$ of η to A is a normal endomorphism of A.*

Proof. Let $a \in A$ and $b \in B$. The inner automorphism $x \longmapsto abxb^{-1}a^{-1}$ of G has a restriction to A, which is the inner automorphism $x \longmapsto axa^{-1}$ of A, since b commutes with every element of A. Therefore every normal subgroup of A is normal in G. Moreover, if η commutes with every inner automorphism of G, then $\eta_{|A}$ commutes with every inner automorphism of A. $\square$

Proof of 2.3. Armed with these results we assail Theorem 2.3. Let G be a group of finite length that is a direct sum

$$G = G_1 \oplus G_2 \oplus \cdots \oplus G_m = H_1 \oplus H_2 \oplus \cdots \oplus H_n$$

of indecomposable subgroups G_1, G_2, ..., G_m and H_1, H_2, ..., H_n. We prove by induction on k that the following hold for all $k \leq n$:

(1) $k \leq m$, and H_1, H_2, ..., H_n can be indexed so that

(2) $H_i \cong G_i$ for all $i \leq k$ and

(3) $G = G_1 \oplus \cdots \oplus G_k \oplus H_{k+1} \oplus \cdots \oplus H_n$.

With $k = n$, (1) yields $n \leqq m$; exchanging G's and H's then yields $n = m$. The other parts of Theorem 2.3 follow from (2) when $k = n$, and from (3).

There is nothing to prove if $k = 0$. Let $0 < k \leqq n$; assume that (1), (2), (3) hold for $k - 1$. By 2.10, all G_i and H_j have finite length. Let $\eta_1, \eta_2, \ldots, \eta_m$ be the projection endomorphisms of the direct sum $G = G_1 \oplus G_2 \oplus \cdots \oplus G_m$, and let $\zeta_1, \zeta_2, \ldots, \zeta_n$ be the projection endomorphisms of the direct sum $G = G_1 \oplus \cdots \oplus G_{k-1} \oplus H_k \oplus \cdots \oplus H_n$ in the induction hypothesis.

By 2.9, $\eta_1 \cdot \cdots \cdot \eta_m = 1_G$. Hence

$$\zeta_k = \zeta_k (\eta_1 \cdot \cdots \cdot \eta_m) = \zeta_k \eta_1 \cdot \cdots \cdot \zeta_k \eta_m.$$

If $k > m$, then $\eta_i x \in G_i$ for all $x \in G$ and Im $\zeta_k \eta_i = 1$ for all $i \leqq m < k$, by 2.9; hence $H_k = $ Im $\zeta_k = 1$, a contradiction since H_k is indecomposable. Therefore $m \leqq k$ and (1) holds for k.

Similarly, $\zeta_1 \cdot \cdots \cdot \zeta_n = 1_G$, $\eta_k \zeta_j G_k \subseteq G_k$ for all j, and Im $\eta_k \zeta_j = 1$ for all $j < k$, for then $\zeta_j x \in G_j$ for all $x \in G$. Hence

$$\eta_k = \eta_k (\zeta_1 \cdot \cdots \cdot \zeta_m) = \eta_k \zeta_1 \cdot \cdots \cdot \eta_k \zeta_m = \eta_k \zeta_k \cdot \cdots \cdot \eta_k \zeta_m.$$

Now, every $(\eta_k \zeta_j)_{|G_k}$ is a normal endomorphism of G_k, by 2.10, and $\eta_{k|G_k} = (\eta_k \zeta_k)_{|G_k} \cdot \cdots \cdot (\eta_k \zeta_m)_{|G_k}$ is the identity on G_k and is not nilpotent; by 2.8, $(\eta_k \zeta_j)_{|G_k}$ is not nilpotent for some $j \geqq k$. The groups $H_k, \ldots, H_n$ can be indexed so that $(\eta_k \zeta_k)_{|G_k}$ is not nilpotent.

We show that $G_k \cong H_k$. By 2.6, $(\eta_k \zeta_k)_{|G_k}$ is an automorphism of G_k. Hence $\eta_k \zeta_k$ is not nilpotent. Then Im $\eta_k (\zeta_k \eta_k)^n \zeta_k = $ Im $(\eta_k \zeta_k)^{n+1} \neq 1$ for all n and $\zeta_k \eta_k$ is not nilpotent. Hence $(\zeta_k \eta_k)_{|H_k}$ is not nilpotent, since $(\zeta_k \eta_k)^n x = 1$ for all $x \in H_k$ would imply $(\zeta_k \eta_k)^n \zeta_k \eta_k x = 1$ for all $x \in G$, and $(\zeta_k \eta_k)_{|H_k}$ is an automorphism of H_k by 2.6. Then $\eta_{k|H_k}$ is injective and $\zeta_k G_k = H_k$, since $\zeta_k G_k \supseteq \zeta_k \eta_k H_k = H_k$. Similarly, $\zeta_{k|G_k}$ is injective and $\eta_k H_k = G_k$. Hence $\eta_{k|H_k}$ is an isomorphism of H_k onto G_k (and $\zeta_{k|G_k}$ is an isomorphism of G_k onto H_k). Thus (2) holds for k.

Let $K = G_1 \cdots G_{k-1} H_{k+1} \cdots H_n$. Since $G = G_1 \oplus \cdots \oplus G_{k-1} \oplus H_k \oplus \cdots \oplus H_n$ by the induction hypothesis, we have $K = G_1 \oplus \cdots \oplus G_{k-1} \oplus H_{k+1} \oplus \cdots \oplus H_n$. Also $\zeta_k K = 1$, since $\zeta_k G_i = \zeta_k H_j = 1$ when $i < k < j$. Since $\zeta_{k|G_k}$ is injective this implies $K \cap G_k = 1$. Hence $K G_k = K \oplus G_k$.

Now, $\eta_{k|H_k} : H_k \longrightarrow G_k$ is an isomorphism, and $\eta_k x = 1$ when $x \in G_i$ or $x \in H_j$ and $i < k < j$. Hence $\theta = \zeta_1 \cdot \cdots \cdot \zeta_{k-1} \cdot \eta_k \cdot \zeta_{k+1} \cdot \cdots \cdot \zeta_n$ is an isomorphism

$$\theta : G \ = \ G_1 \oplus \cdots \oplus G_{k-1} \oplus H_k \oplus H_{k+1} \oplus \cdots \oplus H_n$$
$$\longrightarrow \ G_1 \oplus \cdots \oplus G_{k-1} \oplus G_k \oplus H_{k+1} \oplus \cdots \oplus H_n \ = \ KG_k.$$

Viewed as an endomorphism of G, θ is normal and injective; hence θ is surjective, by 2.4, $G = KG_k$, and (3) holds for k. $\square$

3. Group Actions

It has been said that groups make a living by acting on sets. This section contains basic properties of group actions and their first applications, including the class equation and nice results on p-groups.

Definition. A left group action *of a group G on a set X is a mapping $G \times X \longrightarrow X$, $(g, x) \longmapsto g \cdot x$, such that $1 \cdot x = x$ and $g \cdot (h \cdot x) = (gh) \cdot x$, for all $g, h \in G$ and $x \in X$. Then G acts on the left on X.*

In some cases $g \cdot x$ is denoted by gx or by $^g x$. A *right* group action $X \times G \longrightarrow X$, $(x, g) \longmapsto x \cdot g$, must satisfy $x \cdot 1 = x$ and $(x \cdot g) \cdot h = x \cdot gh$ for all x, g, h; $x \cdot g$ may be denoted by xg or by x^g.

For example, the symmetric group S_X of all permutations of a set X acts on X *by evaluation*: $\sigma \cdot x = \sigma(x)$. Every group G acts on itself *by left multiplication*: $g \cdot x = gx$. Every subgroup of G acts on G by left multiplication.

Properties.

Proposition **3.1.** *In a (left) group action of a group G on a set X, the action $\sigma_g : x \longmapsto g \cdot x$ of $g \in G$ is a permutation of X; moreover, $g \longmapsto \sigma_g$ is a homomorphism of G into the symmetric group S_X.*

Thus, a group always *acts by permutations*.

Proof. By definition, σ_1 is the identity mapping on X, and $\sigma_g \circ \sigma_h = \sigma_{gh}$ for all $g, h \in G$. In particular, $\sigma_g \circ \sigma_{g^{-1}} = 1_X = \sigma_{g^{-1}} \circ \sigma_g$, so that σ_g and $\sigma_{g^{-1}}$ are mutually inverse bijections. Thus $\sigma_g \in S_X$. The equality $\sigma_g \circ \sigma_h = \sigma_{gh}$ shows that $\sigma : g \longmapsto \sigma_g$ is a homomorphism. $\square$

Our tireless readers will show that there is in fact a one-to-one correspondence between left actions of G on X and homomorphisms $G \longrightarrow S_X$.

Corollary **3.2** (Cayley's Theorem). *Every group G is isomorphic to a subgroup of the symmetric group S_G.*

Proof. Let G act on itself by left multiplication. The homomorphism $\sigma : G \longrightarrow S_G$ in 3.1 is injective: if $\sigma_g = 1_G$, then $gx = x$ for all $x \in G$ and $g = 1$. Hence $G \cong \text{Im } \sigma \leq S_G$. $\square$

Proposition **3.3.** *Let the group* G *act* (*on the left*) *on a set* X. *The relation*

$$x \equiv y \text{ if and only if } y = g \cdot x \text{ for some } g \in G$$

is an equivalence relation on X.

Proof. The relation $\equiv$ is reflexive since $1 \cdot x = x$, symmetric since $y = g \cdot x$ implies $x = g^{-1} \cdot (g \cdot x) = g^{-1} \cdot y$, and transitive since $y = g \cdot x$, $z = h \cdot y$ implies $z = hg \cdot x$. $\square$

Definition. In a left group action of a group G *on a set* X, *the* orbit *of* $x \in X$ *is* $\{ y \in G \mid y = g \cdot x \text{ for some } g \in G \}$.

By 3.3 the different orbits of the elements of X constitute a partition of X. For instance, if a subgroup of G acts on G by left multiplication, then the orbit of an element x of G is its right coset Hx. In the action on the Euclidean plane of the group of all rotations about the origin, the orbits are circles centered at the origin and resemble the orbits of the planets about the Sun.

Next we look at the size of the orbits.

Definition. In a left group action of a group G *on a set* X, *the* stabilizer $S(x)$ *of* $x \in X$ *is the subgroup* $S(x) = \{ g \in G \mid g \cdot x = x \}$ *of* G.

The stabilizer $S(x)$ is a subgroup since $1 \cdot x = x$, $g \cdot x = x$ implies $x = g^{-1} \cdot (g \cdot x) = g^{-1} \cdot x$, and $g \cdot x = h \cdot x = x$ implies $gh \cdot x = g \cdot (h \cdot x) = x$.

Proposition **3.4.** *The order of the orbit of an element is equal to the index of its stabilizer.*

Proof. Let G act on X. Let $x \in X$. The surjection $\widehat{x} : g \longmapsto g \cdot x$ of G onto the orbit of x induces a one-to-one correspondence between the elements of the orbit of x and the classes of the equivalence relation induced on G by $\widehat{x}$. The latter are the left cosets of $S(x)$, since $g \cdot x = h \cdot x$ is equivalent to $x = g^{-1}h \cdot x$ and to $g^{-1}h \in S(x)$. Hence the order (number of elements) of the orbit of x equals the number of left cosets of $S(x)$. $\square$

For example, let a subgroup H of G act on G by left multiplication. All stabilizers are trivial ($S(x) = 1$). The order of every orbit (the order of every right coset of H) is the index in H of the trivial subgroup, that is, the order of H.

Action by inner automorphisms. For a more interesting example we turn to inner automorphisms. Recall that an automorphism of a group G is an isomorphism of G onto G. The automorphisms of G constitute a group under composition, the *automorphism group* $\text{Aut}\,(G)$ of G.

Proposition **3.5.** *For every element* g *of a group* G, *the mapping* $\alpha_g : x \longmapsto gxg^{-1}$ *is an automorphism of* G; *moreover,* $g \longmapsto \alpha_g$ *is a homomorphism of* G *into* $\text{Aut}\,(G)$.

Proof. First, $(gxg^{-1})(gyg^{-1}) = gxyg^{-1}$ for all $x, y \in G$, so that α_g is a homomorphism. Also, α_1 is the identity mapping 1_G on G, and $\alpha_g \circ \alpha_h = \alpha_{gh}$

for all $g, h \in G$, since $g\left(hxh^{-1}\right)g^{-1} = (gh)\,x\,(gh)^{-1}$ for all $x \in G$. In particular, $\alpha_g \circ \alpha_{g^{-1}} = 1_G = \alpha_{g^{-1}} \circ \alpha_g$, so that α_g and $\alpha_{g^{-1}}$ are mutually inverse bijections. Hence α_g is an automorphism of G. The equality $\alpha_g \circ \alpha_h = \alpha_{gh}$ shows that $g \longmapsto \alpha_g$ is a homomorphism. $\square$

Definition. An inner automorphism *of a group G is an automorphism* $x \longmapsto gxg^{-1}$ *for some $g \in G$.*

The proofs of Propositions 3.5 and 3.1 are suspiciously similar. This mystery can be solved if we detect a homomorphism of G into $\mathrm{Aut}\,(G) \subseteq S_G$ in 3.5, a clue to an action of G on itself, in which $g \cdot x = \alpha_g(x) = g\,x\,g^{-1}$.

Definition. The action of a group G on itself by inner automorphisms *is defined by* $g \cdot x = g\,x\,g^{-1}$ *for all $g, x \in G$.*

The product gxg^{-1} is also denoted by ${}^g x$; the notation $x^g = g^{-1}xg$ is also in use. We see that ${}^1 x = x$ and that ${}^g({}^h x) = ghxh^{-1}g^{-1} = {}^{gh}x$, so that $g \cdot x = g\,x\,g^{-1}$ is indeed a group action.

Definitions. In the action of a group G on itself by inner automorphisms, the orbits are the conjugacy classes *of G; two elements are* conjugate *when they belong to the same conjugacy class.*

Thus, x and y are conjugate in G when $y = gxg^{-1}$ for some $g \in G$. By 3.3, conjugacy is an equivalence relation. The conjugacy class of x is trivial ($gxg^{-1} = x$ for all g) if and only if x lies in the *center* of G:

Definition. The center *of a group G is*

$$Z(G) = \{\, g \in G \mid gxg^{-1} = x \text{ for all } x \in G \,\}.$$

Equivalently, $Z(G) = \{\, g \in G \mid gx = xg \text{ for all } x \in G \,\}$.

Proposition 3.6. *$Z(G)$ and all its subgroups are normal subgroups of G.*

Proof. If $z \in Z$, then $gzg^{-1} = z$ for all $g \in G$. Hence $gHg^{-1} = H$ for all $H \leqq Z$. $\square$

In general, the order of a conjugacy class is the index of a stabilizer:

Definition. The centralizer *in G of an element x of a group G is*

$$C_G(x) = \{\, g \in G \mid gxg^{-1} = x \,\}.$$

Equivalently, $C(x) = \{\, g \in G \mid gx = xg \,\}$. In our action of G on itself, $C(x)$ is the stabilizer of x and is therefore a subgroup; in fact, it is the largest subgroup of G whose center contains x (see the exercises). Proposition 3.4 yields the next result:

Proposition 3.7. *The number of conjugates of an element of a group G is the index of its centralizer in G.*

Proposition **3.8** (The Class Equation). *In a finite group* G,

$$|G| = \sum |C| = |Z(G)| + \sum_{|C|>1} |C|.$$

The first sum has one term for each conjugacy class C; *the second sum has one term for each nontrivial conjugacy class* C.

Proof. First, $|G| = \sum |C|$, since the conjugacy classes constitute a partition of G. Now, the conjugacy class of x is trivial ($|C| = 1$) if and only if $x \in Z(G)$; hence there are $|Z(G)|$ trivial conjugacy classes and

$$|G| = \sum |C| = \sum_{|C|=1} |C| + \sum_{|C|>1} |C| = |Z(G)| + \sum_{|C|>1} |C|. \;\square$$

p-**groups.** A *p-group* is a group whose order is a power of a prime p. The class equation yields properties of these groups.

Proposition **3.9.** *Every nontrivial p-group has a nontrivial center.*

Proof. By 3.7, $|C|$ divides $|G| = p^n$ for every conjugacy class C. In particular, p divides $|C|$ when $|C| > 1$. In the class equation, p divides $|G|$ and p divides $\sum_{|C|>1} |C|$; hence p divides $|Z(G)|$ and $|Z(G)| \geqq p$. $\square$

Groups of order p are cyclic. The next result yields groups of order p^2:

Proposition **3.10.** *Every group of order* p^2, *where* p *is prime, is abelian.*

By 1.2, the groups of order p^2 are, up to isomorphism, C_{p^2} and $C_p \oplus C_p$. Groups of order p^3 are not necessarily abelian, as shown by D_4 and Q.

Proof. Readers will delight in proving that $G/Z(G)$ cyclic implies G abelian. If now $|G| = p^2$, then $|Z(G)| > 1$, so that $|Z(G)| = p$ or $|Z(G)| = p^2$. If $|Z(G)| = p^2$, then $G = Z(G)$ is abelian. If $|Z(G)| = p$, then $|G/Z(G)| = p$, $G/Z(G)$ is cyclic, and again G is abelian (and $|Z(G)| \neq p$). $\square$

Exercises

1. Show that there is a one-to-one correspondence between the left actions of a group G on a set X and the homomorphisms $G \longrightarrow S_X$.

2. Explain how the original statement of Lagrange's theorem (when $x_1, \ldots, x_n$ are permuted in all possible ways, the number of different values of $f(x_1, \ldots, x_n)$ is a divisor of $n!$) relates to orbits and stabilizers.

3. Let G be a group. Prove the following: for every $g \in G$, the mapping $\alpha_g :$ $x \longmapsto gx\,g^{-1}$ is an automorphism of G; moreover, $g \longmapsto \alpha_g$ is a homomorphism of G into $\mathrm{Aut}\,(G)$.

4. Explain why the inner automorphisms of a group G constitute a group under composition, which is isomorphic to $G/Z(G)$.

5. Find the center of the quaternion group Q.

6. Find the center of D_4.

7. Find the center of D_n.

8. List the conjugacy classes of D_4.

9. List the conjugacy classes of Q.

10. Let G be a group and let $x \in G$. Prove the following: the centralizer of x in G is the largest subgroup H of G such that $x \in Z(H)$.

11. Show that, in a finite group of order n, an element of order k has at most n/k conjugates.

12. Prove the following: if $G/Z(G)$ is cyclic, then G is abelian.

A *characteristic* subgroup of a group G is a subgroup that is invariant under all automorphisms (a subgroup H such that $\alpha(H) = H$ for all $\alpha \in \text{Aut}(G)$). In particular, a characteristic subgroup is invariant under inner automorphisms and is normal.

13. Show that the center of a group G is a characteristic subgroup of G.

14. Prove that every characteristic subgroup of a normal subgroup of a group G is a normal subgroup of G, and that every characteristic subgroup of a characteristic subgroup of a group G is a characteristic subgroup of G.

15. Let N be a characteristic subgroup of a group G. Prove that, if $N \leqq K \leqq G$ and K/N is a charateristic subgroup of G/N, then K is a characteristic subgroup of G.

4. Symmetric Groups

In this section we study the symmetric group S_n on the set $\{1, 2, ..., n\}$.

We write permutations as left operators (σx instead of $\sigma(x)$), and the operation on S_n (composition) as a multiplication ($\sigma \tau$ instead of $\sigma \circ \tau$). We follow custom in specifying a permutation by its table of values

$$\sigma = \begin{pmatrix} 1 & 2 & ... & n \\ \sigma 1 & \sigma 2 & ... & \sigma n \end{pmatrix}.$$

Transpositions. Readers probably know that every permutation is a product of transpositions; we include a proof for the sake of completeness.

Definition. Let $a, b \in \{1, 2, ..., n\}$, $a \neq b$. The transposition $\tau = (a\ b)$ *is the permutation defined by $\tau a = b$, $\tau b = a$, and $\tau x = x$ for all $x \neq a, b$.*

Proposition **4.1.** *Every permutation is a product of transpositions.*

Proof. By induction on n. Proposition 4.1 is vacuous if $n = 1$. Let $n > 1$ and $\sigma \in S_n$. If $\sigma n = n$, then, by the induction hypothesis, the restriction of σ to $\{1, 2, ..., n-1\}$ is a product of transpositions; therefore σ is a product of transpositions. If $\sigma n = j \neq n$, then $(n\ j)\sigma n = n$, $(n\ j)\sigma$ is a product of transpositions $(n\ j)\sigma = \tau_1 \tau_2 \cdots \tau_r$, and so is $\sigma = (n\ j)\tau_1 \tau_2 \cdots \tau_r$. $\square$

By 4.1, S_n is generated by all transpositions; in fact, S_n is generated by the transpositions $(1\ 2)$, $(2\ 3)$, ..., $(n-1\ n)$ (see the exercises).

There is a uniqueness statement of sorts for Proposition 4.1:

Proposition **4.2.** *If* $\sigma = \tau_1 \tau_2 \cdots \tau_r = \upsilon_1 \upsilon_2 \cdots \upsilon_s$ *is a product of transpositions* $\tau_1, \tau_2, \ldots, \tau_r$ *and* $\upsilon_1, \upsilon_2, \ldots, \upsilon_s$, *then* $r \equiv s \pmod 2$.

Equivalently, a product of an even number of transpositions cannot equal a product of an odd number of transpositions.

Proof. This proof uses the ring R of all polynomials with n indeterminates $X_1, \ldots, X_n$, with integer (or real) coefficients. Let S_n act on R by

$$\sigma \cdot f(X_1, \ldots, X_n) = f(X_{\sigma 1}, X_{\sigma 2}, \ldots, X_{\sigma n}).$$

We see that $1 \cdot f = f$ and $\sigma \cdot (\tau \cdot f) = (\sigma \tau) \cdot f$, so that the action of S_n on R is a group action. Also, the action of σ preserves sums and products in R.

Let $\tau = (a\ b)$, where we may assume that $a < b$, and let

$$p(X_1, X_2, \ldots, X_n) = \prod_{1 \leq i < j \leq n} (X_i - X_j).$$

Then $\tau \cdot p$ is the product of all $\tau \cdot (X_i - X_j) = X_{\tau i} - X_{\tau j}$ with $i < j$, and

$$\tau \cdot (X_i - X_j) = \begin{cases} X_i - X_j & \text{if } i, j \neq a, b, \quad (1) \\ X_b - X_a = -(X_a - X_b) & \text{if } i = a \text{ and } j = b, (2) \\ X_b - X_j = -(X_j - X_b) & \text{if } i = a < j < b, \quad (3) \\ X_b - X_j & \text{if } i = a < b < j, \quad (4) \\ X_a - X_j & \text{if } a < i = b < j, \quad (5) \\ X_i - X_b & \text{if } i < j = a < b, \quad (6) \\ X_i - X_a & \text{if } i < a < j = b, \quad (7) \\ X_i - X_a = -(X_a - X_i) & \text{if } a < i < j = b. \quad (8) \end{cases}$$

Inspection shows that every term of $p = \prod_{1 \leq i < j \leq n} (X_i - X_j)$ appears once in $\tau \cdot p$, though perhaps with a minus sign. Hence $\tau \cdot p = \pm p$. The minus signs in $\tau \cdot p$ come from case (2), one minus sign; case (3), one minus sign for each $a < j < b$; and case (8), one minus sign for each $a < i < b$. This adds up to an odd number of minus signs; therefore $\tau \cdot p = -p$.

If now σ is a product of r transpositions, then $\sigma \cdot p = (-1)^r p$. If σ is also a product of s transpositions, then $\sigma \cdot p = (-1)^s p$ and $(-1)^r = (-1)^s$. $\square$

Proposition 4.2 gives rise to the following definitions.

Definitions. A permutation is even when it is the product of an even number of transpositions, odd when it is the product of an odd number of transpositions.

Counting transpositions in products shows that the product of two even permutations and the product of two odd permutations are even, whereas the product of an even permutation and an odd permutation, and the product of an odd permutation and an even permutation, are odd.

Definition. The sign *of a permutation* σ *is*

$$\operatorname{sgn}\sigma \;=\; \begin{cases} +1 & \text{if } \sigma \text{ is even,} \\ -1 & \text{if } \sigma \text{ is odd.} \end{cases}$$

By the above, sgn $(\sigma\tau) = (\operatorname{sgn}\sigma)(\operatorname{sgn}\tau)$, so that, when $n \geq 2$, sgn is a homomorphism of S_n onto the multiplicative group $\{+1, -1\}$; its kernel consists of all even permutations and is a normal subgroup of S_n of index 2.

Definition. The alternating group A_n *is the normal subgroup of* S_n *that consists of all even permutations.*

Cycles. Cycles are a useful type of permutation:

Definitions. Given $2 \leq k \leq n$ and distinct elements a_1, a_2, ..., a_k of $\{1, 2, ..., n\}$, the k-cycle $(a_1\, a_2\, ...\, a_k)$ is the permutation γ defined by

$$\gamma a_i = a_{i+1} \text{ for all } 1 \leq i < k,\ \gamma a_k = a_1,\ \gamma x = x \text{ for all } x \neq a_1, ..., a_k.$$

A permutation is a cycle *when it is a k-cycle for some $2 \leq k \leq n$.*

In other words, $(a_1\, a_2\, ...\, a_k)$ permutes a_1, a_2, ..., a_k circularly, and leaves the other elements of $\{1, 2, ..., n\}$ fixed. Transpositions are 2-cycles (not bicycles). The permutation $\sigma = \bigl(\begin{smallmatrix} 1 & 2 & 3 & 4 \\ 4 & 2 & 1 & 3 \end{smallmatrix}\bigr)$ is a 3-cycle, $\sigma = (1\,4\,3)$.

In general, a k-cycle $\gamma = (a_1\, a_2\, ...\, a_k)$ has order k in S_n, since $\gamma^k = 1$ but $\gamma^h a_1 = a_{h+1} \neq a_1$ if $1 \leq h < k$.

Proposition **4.3.** A_n *is generated by all 3-cycles.*

Proof. First, $(a\, b\, c) = (a\, b)(c\, b)$ for all distinct a, b, c, so that 3-cycles are even and A_n contains all 3-cycles. Now we show that every even permutation is a product of 3-cycles. It is enough to show that every product $(a\, b)(c\, d)$ of two transpositions is a product of 3-cycles.

Let $a \neq b$, $c \neq d$. If $\{a, b\} = \{c, d\}$, then $(a\, b)(c\, d) = 1$. If $\{a, b\} \cap \{c, d\}$ has just one element, then we may assume that $b = d$, $a \neq c$, and then $(a\, b)(c\, d) = (a\, b)(c\, b) = (a\, b\, c)$. If $\{a, b\} \cap \{c, d\} = \varnothing$, then $(a\, b)(c\, d) = (a\, b)(c\, b)(b\, c)(d\, c) = (a\, b\, c)(b\, c\, d)$. $\square$

The cycle structure. Next, we analyze permutations in terms of cycles.

Definitions. The support *of a permutation* σ *is the set* $\{x \mid \sigma x \neq x\}$. *Two permutations are* disjoint *when their supports are disjoint.*

Thus, x is *not* in the support of σ if and only if it is a *fixed point* of σ (if $\sigma x = x$). The support of a k-cycle $(a_1\, a_2\, ...\, a_k)$ is the set $\{a_1, a_2, ..., a_k\}$.

Lemma **4.4.** *Disjoint permutations commute.*

Proof. Let σ and τ be disjoint. If x is not in the support of σ or τ, then $\sigma\tau x = x\tau\sigma x$. If x is in the support of σ, then so is σx, since $\sigma x \neq x$ implies $\sigma\sigma x \neq \sigma x$; then $\sigma\tau x = \sigma x = \tau\sigma x$, since σ and τ are disjoint. Similarly, if x is in the support of τ, then $\sigma\tau x = \tau x = \tau\sigma x$. $\square$

Proposition **4.5.** *Every permutation is a product of pairwise disjoint cycles, and this decomposition is unique up to the order of the terms.*

Proof. Given $\sigma \in S_n$, let $\mathbb{Z}$ act on $X = \{1, 2, ..., n\}$ by $m \cdot x = \sigma^m x$. This is a group action since $\sigma^0 = 1$ and $\sigma^\ell \sigma^m = \sigma^{\ell+m}$. It partitions X into orbits. We see that $\sigma x = x$ if and only if the orbit of x is trivial; hence the support of σ is the disjoint union of the nontrivial orbits.

By 3.4 the order of the orbit A of $a \in X$ is the index of the stabilizer $S(a) = \{m \in \mathbb{Z} \mid \sigma^m a = a\}$ of a. Hence A has k elements if and only if $S(a) = \mathbb{Z}k$, if and only if k is the least positive integer such that $\sigma^k a = a$. Then $\sigma a, ..., \sigma^{k-1} a \neq a$. In fact, $a, \sigma a, ..., \sigma^{k-1} a$ are all distinct: otherwise, $\sigma^i a = \sigma^j a$ for some $0 \leq i, j < k$ with, say, $i < j$, and $\sigma^{j-i} a = a$ with $0 < j - i < k$. Therefore $A = \{a, \sigma a, ..., \sigma^{k-1} a\}$. Moreover, σ and the k-cycle $\gamma_A = (a, \sigma a, ..., \sigma^{k-1} a)$ agree on A, if A is not trivial.

The cycles γ_A, where A ranges over all nontrivial orbits, are pairwise disjoint, and their product, in any order by 4.4, is σ: if the orbit B of x is trivial, then $\sigma x = x = \gamma_A x$ for all nontrivial A; otherwise, $\sigma x = \gamma_B x$ and $\gamma_A x = x$, $\gamma_A \gamma_B x = \gamma_B x$ for all $A \neq B$.

Conversely, assume that σ is a product of pairwise disjoint cycles $\sigma = \gamma_1 \gamma_2 \cdots \gamma_r$. Let A_i be the support of γ_i. By the hypothesis, the sets A_i are pairwise disjoint and nontrivial. If $x \notin A_1 \cup \cdots \cup A_r$, then $\gamma_i x = x$ for all i, $\sigma x = x$, and the orbit of x is trivial. If $x \in A_i$, then $\gamma_j x = x$ for all $j \neq i$, and $\sigma x = \gamma_i x \in A_i$, so that $\sigma^h x = \gamma_i^h x$ for all h, A_i is the orbit of x, and $\gamma_i = \gamma_{A_i}$. Thus $A_1, ..., A_r$ are the nontrivial orbits, and the cycles $\gamma_1, ..., \gamma_r$ are the cycles γ_A above with A nontrivial. $\square$

The proof of Proposition 4.5 provides an algorithm that decomposes any permutation into a product of pairwise disjoint cycles, in finitely many steps: apply σ repeatedly to $1, 2, ..., n$ to get the orbits. For example, let

$$\sigma = \begin{pmatrix} 1 & 2 & 3 & 4 & 5 & 6 & 7 & 8 & 9 \\ 7 & 2 & 8 & 1 & 4 & 3 & 5 & 6 & 9 \end{pmatrix}.$$

We have $\sigma 1 = 7$, $\sigma 7 = 5$, $\sigma 5 = 4$, $\sigma 4 = 1$; $\sigma 2 = 2$; $\sigma 3 = 8$, $\sigma 8 = 6$, $\sigma 6 = 3$; and $\sigma 9 = 9$. Therefore $\sigma = (1\ 7\ 5\ 4)(3\ 8\ 6)$.

Definition. *The* cycle structure *of a permutation $\sigma \in S_n$ is the sequence $k_1 + k_2 + \cdots + k_r$ in which $r \geq 0$, $k_1 \geq k_2 \geq \cdots \geq k_r \geq 2$, and the decomposition of σ into a product of pairwise disjoint cycles consists of a k_1-cycle, a k_2-cycle, ..., and a k_r-cycle.* $\square$

The plus signs are symbolic; $k_1 + k_2 + \cdots + k_r$ is the author's notation. Often, enough 1's are added that $k_1 + \cdots + k_r + 1 + \cdots + 1$ becomes a partition of n.

For example, the cycle structure of $\sigma = \begin{pmatrix} 1 & 2 & 3 & 4 & 5 & 6 & 7 & 8 & 9 \\ 7 & 2 & 8 & 1 & 4 & 3 & 5 & 6 & 9 \end{pmatrix}$ is $4 + 3$, since

$\sigma = (1\ 7\ 5\ 4)(3\ 8\ 6)$. Readers will verify that σ is odd and has order 12. In general, the order of a permutation is readily ascertained from its cycle structure (see the exercises).

Conjugates. With Proposition 4.5 we can find conjugacy classes in S_n.

Lemma **4.6.** *If* $\gamma = (a_1\ a_2\ \dots\ a_k)$ *is a* k*-cycle, then so is* $\sigma \gamma \sigma^{-1} = (\sigma a_1\ \sigma a_2\ \dots\ \sigma a_k)$.

Proof. If $x \neq \sigma a_1, \sigma a_2, \dots, \sigma a_k$, then $\sigma^{-1} x \neq a_1, a_2, \dots, a_k$, $\gamma \sigma^{-1} x = \sigma^{-1} x$, and $\sigma \gamma \sigma^{-1} x = x$. But if $x = \sigma a_i$, where $i < k$, then $\sigma \gamma \sigma^{-1} x = \sigma \gamma a_i = \sigma a_{i+1}$; similarly, $\sigma \gamma \sigma^{-1}(\sigma a_k) = \sigma a_1$. $\square$

Proposition **4.7.** *Two permutations are conjugate if and only if they have the same cycle structure.*

Proof. Let σ be a product of disjoint cycles $\sigma = \gamma_1 \gamma_2 \cdots \gamma_r$. Each γ_i is a k_i-cycle for some $k_i \geq 2$; by 4.4 we may assume that $k_1 \geq k_2 \geq \cdots \geq k_r$, and then the cycle structure of σ is $k_1 + k_2 + \cdots + k_r$. By 4.6, $\alpha \gamma_i \alpha^{-1}$ is a k_i-cycle like γ_i and

$$\alpha \sigma \alpha^{-1} = (\alpha \gamma_1 \alpha^{-1})(\alpha \gamma_2 \alpha^{-1}) \cdots (\alpha \gamma_r \alpha^{-1})$$

is a product of cycles whose supports are the images under α of the supports of $\gamma_1, \gamma_2, \dots, \gamma_r$ and are therefore pairwise disjoint. Therefore the cycle structure of $\alpha \sigma \alpha^{-1}$ is $k_1 + k_2 + \cdots + k_r$, the same as that of σ.

Conversely, let σ and τ have the same cycle structure $k_1 + k_2 + \cdots + k_r$. Then σ and τ are products of r pairwise disjoint cycles

$$\sigma = \gamma_1 \gamma_2 \cdots \gamma_r \text{ and } \tau = \delta_1 \delta_2 \cdots \delta_r,$$

in which γ_i and δ_i are k_i-cycles, $\gamma_i = (a_1\ a_2\ \dots\ a_{k_i})$, $\delta_i = (b_1\ b_2\ \dots\ b_{k_i})$. Let θ_i be the bijection of $\{a_1, a_2, \dots, a_{k_i}\}$ onto $\{b_1, b_2, \dots, b_{k_i}\}$ that sends a_t to b_t. The permutations σ and τ have $n - (k_1 + k_2 + \cdots + k_r)$ fixed points; let θ_0 be any bijection of the set of fixed points of σ onto that of τ. The set of fixed points of σ and the supports of $\gamma_1, \gamma_2, \dots, \gamma_r$ constitute a partition of $\{1, 2, \dots, n\}$; the set of fixed points of τ and the supports of $\delta_1, \delta_2, \dots, \delta_r$ also constitute a partition of $\{1, 2, \dots, n\}$. Therefore the bijections $\theta_0, \theta_1, \dots, \theta_r$ can be pasted together into a bijection θ of $\{1, 2, \dots, n\}$ onto $\{1, 2, \dots, n\}$. Then $\theta \in S_n$ and 4.6 yields $\theta \gamma_i \theta^{-1} = \delta_i$ for all i, by the choice of θ_i; hence $\theta \sigma \theta^{-1} = \tau$. $\square$

Proposition 4.7 sets up a one-to-one correspondence between conjugacy classes of S_n and cycle structures (or between the former and partitions of n). As an example we list the conjugacy classes of S_4 and determine their orders. With $n = 4$ the possible cycle structures are $4, 3, 2+2, 2$, and the empty sum. Readers who like combinatorics will verify that S_4 has six 4-cycles, eight 3-cycles, six transpositions, three products of disjoint transpositions, and one empty product of disjoint cycles, for a total of $6 + 8 + 6 + 3 + 1 = 24$ elements.

The orders of centralizers are found by similar methods. For example, let $\sigma = (1\ 7\ 5\ 4)(3\ 8\ 6) \in S_9$. By the uniqueness in 4.7, $\alpha\sigma\alpha^{-1} = \sigma$ if and only if $\alpha\ (1\ 7\ 5\ 4)\ \alpha^{-1} = (1\ 7\ 5\ 4)$ and $\alpha\ (3\ 8\ 6)\ \alpha^{-1} = (3\ 8\ 6)$; by 4.6, if and only if $(\alpha1\ \alpha7\ \alpha5\ \alpha4) = (1\ 7\ 5\ 4)$ and $(\alpha3\ \alpha8\ \alpha6) = (3\ 8\ 6)$. These conditions imply that α also permutes the fixed points 2 and 9 of σ. Now, there are four permutations of 1, 7, 5, 4 such that $(\alpha1\ \alpha7\ \alpha5\ \alpha4) = (1\ 7\ 5\ 4)$; three permutations of 3, 8, 6 such that $(\alpha3\ \alpha8\ \alpha6) = (3\ 8\ 6)$; and two permutations of 2, 9. These can be combined in all possible ways to yield elements of the centralizer. Therefore the centralizer of σ has $4 \times 3 \times 2 = 24$ elements. The conjugacy class of σ then has $9!/24 = 15120$ elements, by 3.7.

Exercises

1. Show that S_n is generated by $(1\ 2)$, $(2\ 3)$, ..., $(n-1\ n)$.

2. Show that S_n is generated by $(1\ 2)$ and $(1\ 2\ \cdots\ n)$.

3. Show that $S_4 \cong \langle a, b \mid a^4 = 1,\ b^2 = 1,\ (ba)^3 = 1 \rangle$.

4. Show that $A_4 \cong \langle a, b \mid a^3 = 1,\ b^2 = 1,\ aba = ba^2b \rangle$.

5. Devise a presentation of S_n.

6. Verify that a k-cycle is even when k is odd and odd when k is even.

7. Show that A_4 has a normal subgroup of order 4.

8. How many k-cycles are there in S_n?

9. Write $\sigma = \left(\begin{smallmatrix} 1\ 2\ 3\ 4\ 5\ 6\ 7\ 8 \\ 7\ 5\ 6\ 4\ 2\ 8\ 3\ 1 \end{smallmatrix}\right)$ as a product of pairwise disjoint cycles. Is σ even or odd? What is the order of σ?

10. What is the order of the centralizer of $\sigma = \left(\begin{smallmatrix} 1\ 2\ 3\ 4\ 5\ 6\ 7\ 8 \\ 7\ 5\ 6\ 4\ 2\ 8\ 3\ 1 \end{smallmatrix}\right)$? of its conjugacy class?

11. Write $\sigma = \left(\begin{smallmatrix} 1\ 2\ 3\ 4\ 5\ 6\ 7\ 8 \\ 8\ 4\ 7\ 2\ 1\ 6\ 3\ 5 \end{smallmatrix}\right)$ as a product of pairwise disjoint cycles. Is σ even or odd? What is the order of σ?

12. What is the order of the centralizer of $\sigma = \left(\begin{smallmatrix} 1\ 2\ 3\ 4\ 5\ 6\ 7\ 8 \\ 8\ 4\ 7\ 2\ 1\ 6\ 3\ 5 \end{smallmatrix}\right)$? of its conjugacy class?

13. Prove the following: if the cycle structure of σ is $k_1 + k_2 + \cdots + k_r$, then the order of σ is the least common multiple of k_1, k_2, ..., k_r.

14. Show that $Z(S_n) = 1$ when $n \geq 3$.

15. Make sure that the author did not pull a fast one when listing the orders of the conjugacy classes of S_4.

16. List all conjugacy classes of S_5 and their orders.

17. List all conjugacy classes of A_4 and their orders. (Warning: even permutations that are conjugate in S_4 are not necessarily conjugate in A_4.)

18. List all conjugacy classes of A_5 and their orders. (Warning: even permutations that are conjugate in S_5 are not necessarily conjugate in A_5.)

19. Show that A_5 has no normal subgroup $N \neq 1$, A_5.

5. The Sylow Theorems

The Sylow theorems (named after Sylow [1872]) are a basic tool of finite group theory. They state that certain subgroups exist and give some of their properties.

First theorem. The first Sylow theorem is a partial converse of Lagrange's theorem.

Theorem **5.1** (First Sylow Theorem). *Let G be a finite group and let p be a prime number. If p^k divides the order of G, then G has a subgroup of order p^k.*

Proof. First we prove a particular case: if G is abelian and p divides $|G|$, then G has a subgroup of order p. Readers will easily derive this statement from Theorem 1.2 but may prefer a direct proof. If $|G| = p$, then G itself serves. Otherwise, $|G| > p$ and we proceed by induction on $|G|$. Let $a \in G$, $a \neq 1$. If the order of a is a multiple mp of p, then a^m has order p and G has a subgroup $\langle a^m \rangle$ of order p. Otherwise, p does not divide the order of $A = \langle a \rangle$. Hence p divides the order of G/A. By the induction hypothesis, G/A has a subgroup of order p: $bA \in G/A$ has order p in G/A for some $b \in G$. Now, the order of bA in G/A divides the order of b in G, since $b^m = 1$ implies $(bA)^m = 1$ in G/A. Therefore the order of b is a multiple of p and as above, G has a subgroup of order p.

Now let G be any finite group. Theorem 5.1 is true when $|G| = 1$; we prove by induction on $|G|$ that if any p^k divides $|G|$, then G has a subgroup of order p^k. We may assume that $p^k > 1$.

If p divides $|Z(G)|$, then by the above, $Z(G)$ has a subgroup A of order p. Then $A \trianglelefteq G$ by 3.6. If p^k divides $|G|$, then p^{k-1} divides $|G/A| < |G|$; by the induction hypothesis, G/A has a subgroup B/A of order p^{k-1}, where $A \leq B \leq G$, and then $B \leq G$ has order p^k.

If $p^k > 1$ divides $|G|$ but p does not divide $|Z(G)|$ then in the class equation, $|G| = |Z(G)| + \sum_{|C|>1} |C|$, p cannot divide every $|C| > 1$, since p divides $|G|$ but not $|Z(G)|$; hence some $|C| > 1$ is not a multiple of p. By 3.7, $|C|$ is the index of the centralizer $C(x)$ of any $x \in C$; hence p^k divides $|C(x)| = |G|/|C|$. Now, $|C(x)| < |G|$, since $|C| > 1$; by the induction hypothesis, $C(x) \leq G$ has a subgroup of order p^k. $\square$

Corollary **5.2** (Cauchy's Theorem). *A finite group whose order is divisible by a prime p contains an element of order p.*

Cauchy's theorem implies an equivalent definition of p-groups:

Corollary **5.3.** *Let p be a prime number. The order of a finite group G is a power of p if and only if the order of every element of G is a power of p.*

Normalizers. The next Sylow theorems are proved by letting G act on its subgroups by inner automorphisms. For each $g \in G$, $x \longmapsto gxg^{-1}$ is an (inner) automorphism of G, and $H \leqq G$ implies $gHg^{-1} \leqq G$. This defines a group action $g \cdot H = gHg^{-1}$ of G on the set of all its subgroups.

Definitions. In the action by inner automorphisms of a group G on its subgroups, the orbits are the conjugacy classes *of subgroups of G; two subgroups of G are* conjugate *when they belong to the same conjugacy class.*

Thus, H and K are conjugate when $K = gHg^{-1}$ for some $g \in G$.

The number of conjugates of a subgroup is the index of a stabilizer:

Definition. The normalizer *in G of a subgroup H of a group G is*

$$N_G(H) = \{ g \in G \mid gHg^{-1} = H \}.$$

Equivalently, $N(H) = \{ g \in G \mid gH = Hg \}$. In the action of G on its subgroups, $N(H)$ is the stabilizer of H and is therefore a subgroup; in fact, it is the largest subgroup of G in which H is normal (see the exercises). Hence:

Proposition **5.4.** *The number of conjugates of a subgroup of a group G is the index of its normalizer in G.*

The second and third theorems. These theorems give properties of p-subgroups of maximal order.

Definition. Let p be prime. A Sylow p-subgroup *of a finite group G is a subgroup of order p^k, where p^k divides $|G|$ and p^{k+1} does not divide $|G|$.*

The existence of Sylow p-subgroups is ensured by Theorem 5.1.

Proposition **5.5.** *If a Sylow p-subgroup of a finite group G is normal in G, then it is the largest p-subgroup of G and the only Sylow p-subgroup of G.*

Proof. Let the Sylow p-subgroup S be normal in G. If T is a p-subgroup of G, then $ST \leqq G$ and $|ST| = |S|\,|T|/|S \cap T| \geqq |S|$, by I.5.9. Hence $|ST| = |S|$, by the choice of S, so that $T \subseteq ST = S$. □

Theorem **5.6** (Second Sylow Theorem). *Let p be a prime number. The number of Sylow p-subgroups of a finite group G divides the order of G and is congruent to 1 modulo p.*

Theorem **5.7** (Third Sylow Theorem). *Let p be a prime number. All Sylow p-subgroups of a finite group are conjugate.*

Sylow [1872] proved Theorems 5.6 and 5.7 in the following form: all Sylow p-subgroups of a finite group of permutations are conjugate, and their number is congruent to 1 modulo p. By Cayley's theorem, this must also hold in every finite group. Like Sylow, we prove the two theorems together.

Proof. Let S be a Sylow p-subgroup. A conjugate of a Sylow p-subgroup is a Sylow p-subgroup; therefore S acts on the set $\mathcal{S}$ of all Sylow p-subgroups by inner automorphisms. Under this action, $\{S\}$ is an orbit, since $aSa^{-1} = S$ for all $a \in S$. Conversely, if $\{T\}$ is a trivial orbit, then $aTa^{-1} = T$ for all $a \in S$ and $S \subseteq N_G(T)$; then 5.5 applies to $T \trianglelefteq N_G(T)$ and yields $S = T$. Thus $\{S\}$ is the only trivial orbit. The orders of the other orbits are indexes in S of stabilizers and are multiples of p. Hence $|\mathcal{S}| \equiv 1 \pmod{p}$.

Suppose that $\mathcal{S}$ contains two distinct conjugacy classes $\mathcal{C}'$ and $\mathcal{C}''$ of subgroups. Any $S \in \mathcal{C}'$ acts on $\mathcal{C}'$ and $\mathcal{C}'' \subseteq \mathcal{S}$ by inner automorphisms. Then the trivial orbit $\{S\}$ is in $\mathcal{C}'$; by the above, $|\mathcal{C}'| \equiv 1$ and $|\mathcal{C}''| \equiv 0 \pmod{p}$. But any $T \in \mathcal{C}''$ also acts on $\mathcal{C}' \cup \mathcal{C}''$ by inner automorphisms; then the trivial orbit $\{T\}$ is in $\mathcal{C}''$, so that $|\mathcal{C}''| \equiv 1$ and $|\mathcal{C}'| \equiv 0 \pmod{p}$. This blatant contradiction shows that $\mathcal{S}$ cannot contain two distinct conjugacy classes of subgroups. Therefore $\mathcal{S}$ is a conjugacy class. Then $|\mathcal{S}|$ divides $|G|$, by 5.4. $\square$

Theorem 5.7 has the following corollary:

Corollary **5.8.** *A Sylow p-subgroup is normal if and only if it is the only Sylow p-subgroup.*

The use of Theorems 5.6 and 5.7 may be shown by an example. Let G be a group of order 15. The divisors of 15 are 1, 3, 5, and 15; its prime divisors are 3 and 5. Since 1 is the only divisor of 15 that is congruent to 1 $(\mathrm{mod}\ 3)$, G has only one Sylow 3-subgroup S; since 1 is the only divisor of 15 that is congruent to 1 $(\mathrm{mod}\ 5)$, G has only one Sylow 5-subgroup T. Now, $S \cong C_3$ and $T \cong C_5$ are cyclic; $S, T \trianglelefteq G$ by 5.8; $S \cap T = 1$, since $|S \cap T|$ must divide $|S|$ and $|T|$; and $|ST| = |S||T|/|S \cap T| = 15$, so that $ST = G$. By 1.1, 1.4, $G \cong C_3 \times C_5 \cong C_{15}$. Thus, every group of order 15 is cyclic.

Further results. The list of Sylow theorems sometimes includes the next three results, which are of use in later sections.

Proposition **5.9.** *In a finite group, every p-subgroup is contained in a Sylow p-subgroup.*

Proof. As above, a p-subgroup H of a finite group G acts by inner automorphisms on the set $\mathcal{S}$ of all Sylow p-subgroups. Since $|\mathcal{S}| \equiv 1 \pmod{p}$ there is at least one trivial orbit $\{S\}$. Then $hSh^{-1} = S$ for all $h \in H$ and $H \subseteq N_G(S)$. Now, S is a Sylow p-subgroup of $N_G(S)$, and $H \subseteq S$, by 5.5 applied to $S \trianglelefteq N_G(S)$. $\square$

In particular, the maximal p-subgroups are the Sylow p-subgroups.

Proposition **5.10.** *In a finite group, a subgroup that contains the normalizer of a Sylow p-subgroup is its own normalizer.*

Proof. Let S be a Sylow p-subgroup of a finite group G, and let H be a subgroup of G that contains $N_G(S)$. Let $a \in N_G(H)$. Then $aHa^{-1} = H$, so that S and aSa^{-1} are Sylow p-subgroups of H. By 5.7, S and aSa^{-1} are conjugate in H: $S = haSa^{-1}h^{-1}$ for some $h \in H$. Then $ha \in N_G(S) \subseteq H$ and $a \in H$. Hence $N_G(H) = H$. $\square$

Proposition **5.11.** *A p-subgroup of a finite group that is not a Sylow p-subgroup is not its own normalizer.*

Proof. Let H be a p-subgroup of a finite group G. If H is not a Sylow p-subgroup, then p divides $[G:H]$. Now, $H \leqq N_G(H)$, and $[G:N_G(H)]$ divides $[G:H]$. If p does not divide $[G:N_G(H)]$, then $[G:N_G(H)] < [G:H]$ and $H \subsetneqq N_G(H)$. Now assume that p divides $[G:N_G(H)]$.

The subgroup H acts by inner automorphisms on its conjugacy class $\mathcal{C}$. Then $\{H\}$ is a trivial orbit. Since p divides $|\mathcal{C}| = [G:N_G(H)]$, there must be another trivial orbit $\{K\} \neq \{H\}$. Then $hKh^{-1} = K$ for all $h \in H$ and $H \subseteq N_G(K)$; hence $K \subsetneqq N_G(K)$. Since there is an inner automorphism of G that takes K to H, this implies $H \subsetneqq N_G(H)$. $\square$

Corollary **5.12.** *In a finite p-group, every subgroup of index p is normal.*

Exercises

1. Use Theorem 1.2 to show that a finite abelian group whose order is a multiple of a prime p has a subgroup of order p.

2. Use Theorem 1.2 to prove the following: when G is a finite abelian group, every divisor of $|G|$ is the order of a subgroup of G.

3. Prove the following: when $H \leqq G$, then $N_G(H)$ is the largest subgroup of G such that $H \trianglelefteq N_G(H)$.

4. Show that A_4 does not contain a subgroup of order 6.

5. Show that, in a group of order $n \leqq 11$, every divisor of n is the order of a subgroup.

6. Find the Sylow subgroups of S_4.

7. Find the Sylow subgroups of S_5.

8. Show that every group G of order 18 has a normal subgroup $N \neq 1, G$.

9. Show that every group G of order 30 has a normal subgroup $N \neq 1, G$.

10. Show that every group G of order 56 has a normal subgroup $N \neq 1, G$.

11. Find all groups of order 33.

12. Find all groups of order 35.

13. Find all groups of order 45.

14. Prove the following: if p^{k+1} divides $|G|$, then every subgroup of G of order p^k is normal in a subgroup of order p^{k+1}.

6. Small Groups

In this section we construct all groups of order at most 15. (Finding the 14 groups of order 16 is more difficult.)

General results. For every prime p, we saw that every group of order p is cyclic, and that every group of order p^2 is abelian (3.10).

Proposition **6.1.** *Let* p *be prime. A group of order* $2p$ *is cyclic or dihedral.*

Proof. A group of order 4 is abelian and either cyclic or isomorphic to $V_4 \cong D_2$. Now let $p > 2$. By the Sylow theorems, a group G of order $2p$ has a Sylow p-subgroup A of order p and a Sylow 2-subgroup B of order 2; A and B are cyclic, $A = \langle a \rangle \cong C_p$, $B = \langle b \rangle \cong C_2$. Moreover, $A \trianglelefteq G$, since A has index 2, $A \cap B = 1$, and $G = AB$, since $|G| = |A||B| = |AB|$. Then G is generated by $\{a, b\}$, and a, b satisfy $a^p = 1$, $b^2 = 1$, and $bab^{-1} = a^k$ for some k, since $bab^{-1} \in A$. Since $b^2 = 1$, we have $a = bbab^{-1}b^{-1} = ba^k b^{-1} = (bab^{-1})^k = (a^k)^k = a^{k^2}$; hence p divides $k^2 - 1 = (k-1)(k+1)$. Since p is prime, p divides $k - 1$ or $k + 1$.

If p divides $k - 1$, then $bab^{-1} = a^k = a$ and $ba = ab$; hence G is abelian, $B \trianglelefteq G$, and $G = A \oplus B \cong C_p \oplus C_2 \cong C_{2p}$ is cyclic. If p divides $k + 1$, then $bab^{-1} = a^k = a^{-1}$, the defining relations of D_p in I.7.3 hold in G, and there is a homomorphism θ of D_p into G, which is surjective since G is generated by a and b; θ is an isomorphism, since $|D_p| = |G| = 2p$. Thus $G \cong D_p$. $\square$

Proposition **6.2.** *If* $p > q$ *are primes, and* q *does not divide* $p - 1$, *then every group of order* pq *is cyclic.*

For instance, we saw in Section 5 that every group of order 15 is cyclic. But D_3 has order $6 = 3 \times 2$, where 2 divides $3 - 1$, and is not cyclic.

Proof. By the Sylow theorems, a group G of order pq has a Sylow p-subgroup P of order p and a Sylow q-subgroup Q of order q, both of which are cyclic. Among the divisors 1, p, q, pq of pq, only 1 is congruent to 1 (mod p), since $q < p$, and only 1 is congruent to 1 (mod q), since q does not divide $p - 1$. Therefore $P, Q \trianglelefteq G$. Moreover, $P \cap Q = 1$ and $PQ = G$, since $|G| = |P||Q| = |PQ|$. Hence $G = P \oplus Q \cong C_p \oplus C_q \cong C_{pq}$. $\square$

We now know all groups of the following orders:

Order	Type
1, 2, 3, 5, 7, 11, 13	cyclic;
4, 9	abelian (3.10);
6, 10, 14	cyclic or dihedral (6.1);
15	cyclic (6.2).

Groups of order 8. Up to isomorphism, there are three abelian groups of order 8, C_8, $C_4 \oplus C_2$, $C_2 \oplus C_2 \oplus C_2$, and at least two nonabelian groups, $D_4 = \langle a, b \mid a^4 = 1, b^2 = 1, bab^{-1} = a^{-1} \rangle$ and $Q = \langle a, b \mid a^4 = 1, b^2 = a^2, bab^{-1} = a^{-1} \rangle$; the exercises have shown that $D_4 \not\cong Q$.

Proposition **6.3.** *A nonabelian group of order* 8 *is isomorphic to either* D_4 *or* Q.

Proof. Let G be a nonabelian group of order 8. No element of G has order 8, since G is not cyclic, and the elements of G cannot all have order 1 or 2: if $x = x^{-1}$ for all $x \in G$, then $xy = (xy)^{-1} = y^{-1}x^{-1} = yx$ for all $x, y \in G$. Therefore G has an element a of order 4. Then $A = \langle a \rangle$ is a subgroup of G of order 4; $A \trianglelefteq G$ since A has index 2.

The group G is generated by a and any $b \notin A$, since $A \subsetneqq \langle a, b \rangle \subseteq G$. Now, $b^2 \in A$, since Ab has order 2 in G/A. Also, $b^2 \neq a, a^3$: otherwise, b has order 8. Hence $b^2 = 1$ or $b^2 = a^2$. Moreover, $bab^{-1} \in A$ has order 4 like a; $bab^{-1} \neq a$, otherwise, G is abelian; hence $bab^{-1} = a^3 = a^{-1}$.

The coup de grace is now administered as in the proof of 6.1. If $b^2 = 1$, then the defining relations of D_4 hold in G; hence there is a homomorphism θ of D_4 onto G, which is an isomorphism since both groups have order 8; thus $G \cong D_4$. If $b^2 = a^2$, then the defining relations of Q, etc., etc., and $G \cong Q$. $\square$

Groups of order 12. Up to isomorphism, there are two abelian groups of order 12, $C_4 \oplus C_3 \cong C_{12}$ and $C_2 \oplus C_2 \oplus C_3$, and at least three nonabelian groups, $D_6 = \langle a, b \mid a^6 = 1, b^2 = 1, bab^{-1} = a^{-1} \rangle$, $T = \langle a, b \mid a^6 = 1, b^2 = a^3, bab^{-1} = a^{-1} \rangle$ (from Section I.7), and A_4; the exercises have shown that D_6, Q, and A_4 are not isomorphic to each other.

Proposition 6.4. *A nonabelian group of order* 12 *is isomorphic to either* D_4 *or* T *or* A_4.

Proof. A nonabelian group G of order 12 has a subgroup P of order 3. Then G acts by left multiplication on the set of all four left cosets of P: $g \cdot xP = gxP$. By 3.1, this group action induces a homomorphism of G into S_4, whose kernel K is a normal subgroup of G. Moreover, $K \subseteq P$, since $gxP = xP$ for all x implies $g \in P$; hence $K = 1$ or $K = P$.

If $K = 1$, then G is isomorphic to a subgroup H of S_4 of order 12. Let $\gamma \in S_4$ be a 3-cycle. Since H has index 2, two of $1, \gamma, \gamma^2$ must be in the same left coset of H. Hence $\gamma \in H$, or $\gamma^2 \in H$ and $\gamma = \gamma^4 \in H$. Thus H contains all 3-cycles. Hence $H = A_4$, by 4.3, and $G \cong A_4$.

If $K = P$, then $P \trianglelefteq G$, P is the only Sylow 3-subgroup of G, and G has only two elements of order 3. If $c \in P$, $c \neq 1$, then c has at most two conjugates and its centralizer $C_G(c)$ has order 6 or 12. By Cauchy's theorem, $C_G(c)$ contains an element d of order 2. Then $cd = dc$, since $d \in C_G(c)$, and $a = cd$ has order 6. Then $A = \langle a \rangle$ is a subgroup of G of order 6; $A \trianglelefteq G$ since A has index 2.

As in the proof of 6.3, G is generated by a and any $b \notin A$. Now, $bab^{-1} \in A$ has order 6 like a; $bab^{-1} \neq a$, otherwise, G is abelian; hence $bab^{-1} = a^5 = a^{-1}$. Also, $b^2 \in A$, since Ab has order 2 in G/A; $b^2 \neq a, a^5$, otherwise,

b has order 12 and G is cyclic; $b^2 \neq a^2$, a^4, since b^2 commutes with b but $ba^2b^{-1} = a^{-2}$ yields $ba^2 = a^4b$. Hence $b^2 = 1$ or $b^2 = a^3$. Then, as in the proof of 6.3, $G \cong D_6$ or $G \cong T$. $\square$

Summary. The groups of order 1 to 15 are, up to isomorphism:

Order	Groups:
1	1;
2	C_2;
3	C_3;
4	C_4, $C_2 \oplus C_2 \cong V_4$;
5	C_5;
6	C_6, $D_3 \cong S_3$;
7	C_7;
8	C_8, $C_4 \oplus C_2$, $C_2 \oplus C_2 \oplus C_2$, D_4, Q;
9	C_9, $C_3 \oplus C_3$;
10	C_{10}, D_5;
11	C_{11};
12	C_{12}, $C_2 \oplus C_2 \oplus C_3$, D_6, T, A_4;
13	C_{13};
14	C_{14}, D_7;
15	C_{15}.

Exercises

1. To which group of order 12 is $C_2 \oplus D_3$ isomorphic?

Nonabelian groups in the following exercises should be specified by presentations.

2. Find all groups of order 51.

3. Find all groups of order 21.

4. Find all groups of order 39.

5. Find all groups of order 55.

6. Find all groups of order 57.

7. Find all groups of order 93.

7. Composition Series

Analysis by normal series is another tool for the study of finite groups.

Definitions. A normal series of a group G is a finite ascending sequence A_0, $A_1, \ldots A_m$ of subgroups of G such that $1 = A_0 \trianglelefteq A_1 \trianglelefteq A_2 \trianglelefteq \cdots \trianglelefteq A_m = G$; then m is the length of the series.

The subgroups that appear in normal series are called *subnormal*; they need not be normal (see the exercises). Normal series are sometimes called *subnormal series*. Some infinite sequences of subgroups are also called series.

For example, every group G has a trivial normal series $1 \leq G$. We saw in Section 3 that S_n has a nontrivial normal series $1 \leq A_n \leq S_n$.

Definition. The factors *of a normal series* $1 = A_0 \leq A_1 \leq A_2 \leq \cdots \leq A_m = G$ *are the quotient groups* A_i / A_{i-1} $(1 \leq i \leq m)$.

Definition. Two normal series $\mathcal{A} : 1 = A_0 \leq A_1 \leq A_2 \leq \cdots \leq A_m = G$ *and* $\mathcal{B} : 1 = B_0 \leq B_1 \leq B_2 \leq \cdots \leq B_n = G$ *are* equivalent *when* $m = n$ *and there is a permutation* σ *such that* $A_i / A_{i-1} \cong B_{\sigma i} / B_{\sigma i - 1}$ *for all* $i > 0$. $\square$

In other words, two normal series are equivalent when they have the same length and, up to isomorphism and indexing, the same factors. For instance, a cyclic group $C = \langle c \rangle$ of order 6 has two equivalent normal series $1 \leq \{1, c^3\} \leq C$ and $1 \leq \{1, c^2, c^4\} \leq C$.

One may think of a normal series $1 = A_0 \leq A_1 \leq A_2 \leq \cdots \leq A_m = G$ as analyzing the group G as somehow assembled from simpler groups, the factors A_1 / A_0, A_2 / A_1, ..., A_m / A_{m-1}. Reconstructing G from these factors is more difficult and is discussed in Section 12. For now, our philosophy is to ignore reconstruction difficulties and make the factors as simple as possible.

Refinement adds terms to a normal series to obtain smaller, simpler factors.

Definition. A refinement *of a normal series* $\mathcal{A}$ ($1 = A_0 \leq A_1 \leq A_2 \leq \cdots \leq A_m = G$ *is a normal series* $\mathcal{B} : 1 = B_0 \leq B_1 \leq B_2 \leq \cdots \leq B_n = G$ *such that every* A_i *is one of the* B_j's.

For example, D_4 has a normal series $\mathcal{A} : 1 \leq R \leq D_4$, where $R = \{r_0, r_1, r_2, r_3\}$ in our notation. The normal series $\mathcal{B} : 1 \leq \{r_0, r_1\} \leq R \leq D_4$ is a refinement of $\mathcal{A}$.

In general, refinement replaces each interval $A_{i-1} \leq A_i$ by a sequence $A_{i-1} = B_j \leq B_{j+1} \leq \cdots \leq B_k = A_i$. By I.4.9, the new factors B_h / B_{h-1} are the factors of a normal series $1 \leq B_{j+1} / B_j \leq \cdots \leq B_k / B_j$ of A_i / A_{i-1}; this analyzes the original factors A_i / A_{i-1} into smaller and simpler factors.

Refinements exhibit a kind of convergence.

Theorem 7.1 (Schreier [1928]) *Any two normal series of a group have equivalent refinements.*

Proof. Let $\mathcal{A} : 1 = A_0 \leq A_1 \leq A_2 \leq \cdots \leq A_m = G$ and $\mathcal{B} : 1 = B_0 \leq B_1 \leq B_2 \leq \cdots \leq B_n = G$ be two normal series of a group G. Let $C_{mn} = D_{mn} = G$; for every $0 \leq i < m$ and $0 \leq j < n$, let

$$C_{ni+j} = A_i (A_{i+1} \cap B_j) \text{ and } D_{mj+i} = B_j (B_{i+1} \cap A_i).$$

This defines C_k and D_k for every $0 \leq k \leq mn$, since every $0 \leq k < mn$ can be written uniquely in the form $k = ni + j$ with $0 \leq i < m$ and $0 \leq j < n$, and can be written uniquely in the form $k = mj' + i'$ with $0 \leq i' < m$ and $0 \leq j' < n$. Thus $\sigma : ni + j \longmapsto mj + i$ is a permutation of $\{ 0, 1, \ldots, mn - 1 \}$.

We see that $A_i = C_{ni} \subseteq C_{ni+1} \subseteq \cdots \subseteq C_{ni+n} = A_{i+1}$ and $B_j = D_{mj} \subseteq D_{mj+1} \subseteq \cdots \subseteq D_{mj+m} = B_{j+1}$, for all $0 \leq i < m$ and $0 \leq j < n$; in particular, $C_0 = D_0 = 1$. Hence $C_k \subseteq C_{k+1}$ and $D_k \subseteq D_{k+1}$ for all k; if we can show that $\mathcal{C} : C_0, C_1, \ldots, C_{mn}$ and $\mathcal{D} : D_0, D_1, \ldots, D_{mn}$ are normal series, they will be refinements of $\mathcal{A}$ and $\mathcal{B}$.

That $\mathcal{C}$ and $\mathcal{D}$ are normal series follows from Zassenhaus's lemma 7.2 below, applied to $A = A_i$, $A' = A_{i+1}$, $B = B_j$, and $B' = B_{j+1}$: by this lemma, $C_{ni+j} = A(A' \cap B)$, $C_{ni+j+1} = A(A' \cap B')$, $D_{mj+i} = B(B' \cap A)$, and $D_{mj+i+1} = B(B' \cap A')$ are subgroups of G; $C_{ni+j} = A(A' \cap B) \trianglelefteq A(A' \cap B') = C_{ni+j+1}$; $D_{mj+i} = B(B' \cap A) \trianglelefteq B(B' \cap A') = D_{mj+i+1}$; and

$$C_{ni+j+1}/C_{ni+j} = A(A' \cap B')/A(A' \cap B)$$
$$\cong B(B' \cap A')/B(B' \cap A) = D_{mj+i+1}/D_{mj+i}.$$

Therefore $\mathcal{C}$ and $\mathcal{D}$ are normal series, and are refinements of $\mathcal{A}$ and $\mathcal{B}$; moreover, $C_{k+1}/C_k \cong D_{\sigma k+1}/D_{\sigma k}$ for all $0 \leq k < mn$, where σ is our earlier permutation of $\{ 0, 1, \ldots, mn - 1 \}$, so that $\mathcal{C}$ and $\mathcal{D}$ are equivalent. $\square$

Lemma **7.2** (Zassenhaus [1934]). *If $A \trianglelefteq A' \leq G$ and $B \trianglelefteq B' \leq G$, then $A(A' \cap B)$, $A(A' \cap B')$, $B(B' \cap A)$, and $B(B' \cap A')$ are subgroups of G; $A(A' \cap B) \trianglelefteq A(A' \cap B')$; $B(B' \cap A) \trianglelefteq B(B' \cap A')$; and*

$$A(A' \cap B')/A(A' \cap B) \cong B(B' \cap A')/B(B' \cap A).$$

This is often called the *Butterfly lemma*, after its subgroup inclusion pattern:

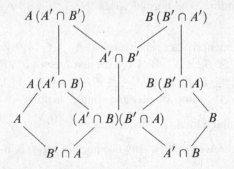

Proof. $A(A' \cap B)$ and $A(A' \cap B')$ are subgroups of A', since $A \trianglelefteq A'$. Also $A' \cap B \trianglelefteq A' \cap B'$, since $B \trianglelefteq B'$.

Let $x = ab \in A(A' \cap B')$ and $y = cd \in A(A' \cap B)$, with $a, c \in A$, $b \in A' \cap B'$, $d \in A' \cap B$. Then $xcx^{-1} \in A$, since $x \in A'$ and $A \trianglelefteq A'$;

$bdb^{-1} \in A' \cap B$, since $b \in A' \cap B'$ and $A' \cap B \trianglelefteq A' \cap B'$; $xdx^{-1} = abdb^{-1}a^{-1} \in A(A' \cap B)A = A(A' \cap B)$, since $A \trianglelefteq A'$; and $xyx^{-1} = xcx^{-1} xdx^{-1} \in A(A' \cap B)$. Thus $A(A' \cap B) \trianglelefteq A(A' \cap B')$.

Let $S = A' \cap B'$, $T = A(A' \cap B)$, and $U = A(A' \cap B')$. Then $S \leqq U$, $T \trianglelefteq U$, and

$$ST = TS = A(A' \cap B)(A' \cap B') = A(A' \cap B') = U.$$

We find $S \cap T$. First, $A \cap B' \subseteq S$, $A \cap B' \subseteq A \subseteq T$, $A' \cap B \subseteq S$, and $A' \cap B \subseteq T$, so that $(A \cap B')(A' \cap B) \subseteq S \cap T$. Conversely, if $t \in S \cap T$, then $t \in A' \cap B'$ and $t = ab$ for some $a \in A$ and $b \in A' \cap B$; then $b \in B'$, $a = tb^{-1} \in B'$, and $s = ab \in (A \cap B')(A' \cap B)$. Thus $S \cap T = (A \cap B')(A' \cap B)$.

By the second Isomorphism theorem (I.5.9), $T \trianglelefteq ST$, $S \cap T \trianglelefteq S$, and $ST/T \cong S/(S \cap T)$. Hence $(A \cap B')(A' \cap B) = S \cap T \trianglelefteq S = A' \cap B'$ and

$$A(A' \cap B')/A(A' \cap B) = ST/T$$
$$\cong S/(S \cap T) = (A' \cap B')/(A \cap B')(A' \cap B).$$

Exchanging A's and B's in the above yields that $B(B' \cap A)$ and $B(B' \cap A')$ are subgroups of G, $B(B' \cap A) \trianglelefteq B(B' \cap A')$, and

$$B(B' \cap A')/B(B' \cap A) \cong (A' \cap B')/(A \cap B')(A' \cap B).$$

Hence $A(A' \cap B')/A(A' \cap B) \cong B(B' \cap A')/B(B' \cap A)$. $\square$

Composition series. A composition series is a normal series without proper refinements, hence with the simplest possible factors:

Definition. A composition series *of a group* G *is a normal series* $\mathcal{A} : 1 = A_0 \trianglelefteq A_1 \trianglelefteq A_2 \trianglelefteq \cdots \trianglelefteq A_m = G$ *of* G *such that, for every* $1 \leq i \leq m$, $A_{i-1} \ntrianglelefteq A_i$ *and there is no subgroup* B *such that* $A_{i-1} \ntrianglelefteq B \ntrianglelefteq A_i$.

By I.4.9, subgroups B such that $A_{i-1} \ntrianglelefteq B \ntrianglelefteq A_i$ correspond to normal subgroups $N \neq 1$, A_i/A_{i-1} of A_i/A_{i-1}. Hence a composition series is a normal series in which every factor A_i/A_{i-1} is nontrivial and has no normal subgroup $N \neq 1$, A_i/A_{i-1}. We state this as follows.

Definition. A group G is simple *when* $G \neq 1$ *and* G *has no normal subgroup* $N \neq 1$, G.

Proposition 7.3. *A normal series is a composition series if and only if all its factors are simple.*

For instance, a cyclic group $C = \langle c \rangle$ of order 6 has two composition series $1 \trianglelefteq \{1, c^3\} \trianglelefteq C$ and $1 \trianglelefteq \{1, c^2, c^4\} \trianglelefteq C$.

Proposition 7.4. *Every finite group has a composition series.*

However, not every group has a composition series (see the exercises).

Proof. In a group G of order n, every strictly ascending normal series $\mathcal{A} : 1 = A_0 \subsetneqq A_1 \subsetneqq A_2 \subsetneqq \cdots \subsetneqq A_m = G$ has length $m \leqq n$. Hence G has a strictly ascending normal series of maximal length, i.e., a composition series. $\square$

The following theorem was proved by Hölder [1889]. Jordan [1869] had shown earlier that different compositions series have factors of the same order.

Theorem **7.5**. (Jordan-Hölder). *Any two composition series of a group are equivalent.*

Proof. Let $\mathcal{A} : 1 = A_0 \subsetneqq A_1 \subsetneqq A_2 \subsetneqq \cdots \subsetneqq A_m = G$ and $\mathcal{B} : 1 = B_0 \subsetneqq B_1 \subsetneqq B_2 \subsetneqq \cdots \subsetneqq B_n = G$ be two composition series of a group G. By Schreier's theorem (7.1), $\mathcal{A}$ and $\mathcal{B}$ have equivalent refinements $\mathcal{C}$ and $\mathcal{D}$. Since $\mathcal{A}$ is a composition series, $\mathcal{C}$ is obtained from $\mathcal{A}$ by adding equalities; its factors are the (nontrivial) factors of $\mathcal{A}$ and a bunch of trivial factors. Similarly, the factors of $\mathcal{D}$ are the (nontrivial) factors of $\mathcal{B}$ and another bunch of trivial factors. The permutation σ such that $C_k/C_{k-1} \cong D_{\sigma k}/D_{\sigma(k-1)}$ for all $k > 0$ sends the nontrivial factors of $\mathcal{C}$ onto the nontrivial factors of $\mathcal{D}$, and sends the factors of $\mathcal{A}$ onto the factors of $\mathcal{B}$; therefore $\mathcal{A}$ and $\mathcal{B}$ are equivalent. $\square$

By 7.5, when a group G has a composition series, all composition series of G have the same factors, up to isomorphism and order of appearance.

Definition. The simple factors *of a group G that has a composition series are the factors of any composition series of G.*

For instance, the simple factors of a cyclic group $C = \langle c \rangle$ of order 6 are one cyclic group of order 2 and one cyclic group of order 3, as shown by either of the composition series $1 \trianglelefteq \{1, c^3\} \trianglelefteq C$ and $1 \trianglelefteq \{1, c^2, c^4\} \trianglelefteq C$.

Simple groups. Analysis by composition series shows that simple groups are a basic building block of finite groups in general.

One of the great achievements of late twentieth century mathematics is the *Classification theorem*, which lists all finite simple groups; its proof set a new record for length, and is being published in installments (Gorenstein et al. [1994 up]), some 2100 pages as of this writing. With 26 exceptions, finite simple groups fall into some 18 infinite families. We can produce two such families now; a third is constructed in the next section.

Proposition **7.6**. *A finite abelian group is simple if and only if it is a cyclic group of prime order.*

This follows from, say, Theorem 1.2.

Nonabelian simple groups arise from composition series of sufficiently large groups. Dihedral groups are unsuitable for this (see the exercises) but S_n has a normal series $1 \trianglelefteq A_n \trianglelefteq S_n$, which is a composition series if $n \geqq 5$:

Proposition **7.7**. A_n *is simple for all $n \geqq 5$.*

The group $A_3 \cong C_3$ is simple, too, but A_4 has a normal subgroup of order 4 and is not simple (see the exercises in Section 3).

Proof. The simplicity of A_5 is proved by counting the elements of its conjugacy classes, which readers will verify consist of:

12 5-cycles;

12 more 5-cycles;

20 3-cycles;

15 products of two disjoint transpositions; and

1 identity element.

A normal subgroup of A_5 is the union of $\{\,1\,\}$ and other conjugacy classes. These unions have orders 1, 13, 16, 21, 25, 28, and over 30, none of which is a proper divisor of $|A_5| = 60$; therefore A_5 has no normal subgroup $N \neq 1$, A_5.

The simplicity of A_n when $n > 5$ is proved by induction on n. Let $N \neq 1$ be a normal subgroup of A_n. We want to show that $N = A_n$.

First we show that N is *transitive*: for every $i, j \in \{\,1, 2, ..., n\,\}$, there exists $\sigma \in N$ such that $\sigma i = j$. Since $N \neq 1$ we have $\sigma k \neq k$ for some $\sigma \in N$ and $k \in \{\,1, 2, ..., n\,\}$. For any $i \in \{\,1, 2, ..., n\,\}$ we can rig an even permutation α such that $\alpha k = k$ and $\alpha \sigma k = i$; then $\alpha \sigma \alpha^{-1} \in N$ and $\alpha \sigma \alpha^{-1} k = i$. If now $i, j \in \{\,1, 2, ..., n\,\}$, there exist $\mu, \nu \in N$ such that $\mu k = i$, $\nu k = j$, and then $\nu \mu^{-1} \in N$ and $\nu \mu^{-1} i = j$.

Next we show that some $\sigma \in N$, $\sigma \neq 1$, has a fixed point ($\sigma k = k$ for some k). As above, we have $\sigma k = j \neq k$ for some $\sigma \in N$. Then $\sigma j \neq j$. Let $i \neq j$, k, σj. If $\sigma i = i$, σ serves. Otherwise, i, σi, j, σj are all different; since $n \geq 6$ we can concoct an even permutation α such that $\alpha j = k$, $\alpha \sigma j = j$, $\alpha i = i$, and $\alpha \sigma i \neq i$, j, k, σi. Then $\mu = \alpha \sigma \alpha^{-1} \in N$, $\mu k = \alpha \sigma j = j = \sigma k$, and $\mu i = \alpha \sigma i \neq \sigma i$. Hence $\nu = \sigma^{-1} \mu \in N$, $\nu \neq 1$ since $\nu i \neq i$, and $\nu k = k$.

Let k be a fixed point of some $\sigma \in N$, $\sigma \neq 1$. Let $B = \{\, \alpha \in A_n \mid \alpha k = k \,\}$. Then $N \cap B \neq 1$ and $N \cap B \trianglelefteq B$. Since $B \cong A_{n-1}$ is simple, by the induction hypothesis, this implies $N \cap B = B$ and $B \subseteq N$. If $\alpha \in A$, then $\mu k = \alpha k$ for some $\mu \in N$, since N is transitive, so that $\mu^{-1} \alpha \in B \subseteq N$ and $\alpha = \mu\,(\mu^{-1}\alpha) \in N$. Thus $N = A_n$. $\square$

Exercises

1. Show that D_4 has a normal series with a term that is not a normal subgroup of D_4.

2. Show that A_4 has a normal series with a term that is not a normal subgroup of A_4.

3. Let $N \trianglelefteq G$. Show that normal series of N and G/N can be "pieced together" to yield a normal series of G.

4. Let $\mathcal{A} : 1 = A_0 \trianglelefteq A_1 \trianglelefteq A_2 \trianglelefteq \cdots \trianglelefteq A_m = G$ be a normal series. Explain how normal series of all the factors A_i/A_{i-1} give rise to a refinement of $\mathcal{A}$.

5. Show that $\mathbb{Z}$ does not have a composition series.

6. Prove the following: if $N \trianglelefteq G$ and G/N have composition series, then G has a composition series.

7. Prove the following: if $N \trianglelefteq G$ and G has a composition series, then N and G/N have composition series. (Hint: first show that N appears in a composition series of G.)

8. Prove the following: if G has a composition series, then every strictly ascending normal series of G can be refined into a composition series.

9. Find all composition series and simple factors of D_4.

10. Find all composition series and simple factors of A_4.

11. Find all composition series and simple factors of D_5.

12. Prove that an abelian group has a composition series if and only if it is finite.

13. Prove that all abelian groups of order n have the same simple factors.

14. Show that the simple factors of D_n are all abelian.

15. Show that $1 \trianglelefteq A_n \trianglelefteq S_n$ is the only composition series of S_n when $n \geq 5$.

16. Show that a group of order p^n, where p is prime, has a composition series of length n.

17. Let G be a group of order n and let m be the length of its composition series. Show that $m \leq \log_2 n$. Show that the equality $m = \log_2 n$ occurs for arbitrarily large values of n.

The following exercise is more like a small research project.

*18. Without using results from later sections, show that there is no nonabelian simple group of order less than 60.

8. The General Linear Group

This section can be skipped. The general linear group is one of the *classical groups* whose study in the nineteenth century eventually gave rise to today's group theory. Its normal series yields new simple groups.

Definition. Let V be a vector space of finite dimension $n \geq 2$ over a field K. The general linear group $GL(V)$ of V *is the group of all invertible linear transformations of V into V.*

Given a basis of V, every linear transformation of V into V has a matrix; hence $GL(V)$ is isomorphic to the multiplicative group $GL(n, K)$ of all invertible $n \times n$ matrices with coefficients in K. In particular, all vector spaces of dimension n over K have isomorphic general linear groups. The group $GL(n, K)$ is also called a *general linear group*.

The Special Linear Group. We construct a normal series of $GL(V)$. First, determinants provide a homomorphism of $GL(V)$ into the multiplicative group K^* of all nonzero elements of K. Its kernel is a normal subgroup of $GL(V)$.

Definition. Let V be a vector space of finite dimension $n \geqq 2$ over a field K. The special linear group $SL(V)$ *of V is the group of all linear transformations of V into V whose determinant is 1.*

The multiplicative group $SL(n, K)$ of all $n \times n$ matrices with coefficients in K and determinant 1 is isomorphic to $SL(V)$ and is also called a *special linear group*. Matrices with arbitrary nonzero determinants are readily constructed; hence the determinant homomorphism $GL(V) \longrightarrow K^*$ is surjective and the Homomorphism theorem yields the following result:

Proposition **8.1.** $SL(V) \trianglelefteq GL(V)$ *and* $GL(V) / SL(V) \cong K^*$.

Centers. Since K^* is abelian, any nonabelian simple factor of $GL(V)$ must come from $SL(V)$, in fact must come from $SL(V) / Z\big(SL(V)\big)$. To find the center of $SL(V)$ we use elementary transformations.

Readers may recall that an *elementary* $n \times n$ matrix E is obtained from the identity matrix by adding one nonzero entry outside the main diagonal:

$$E = \begin{pmatrix} 1 & & & & & \\ & \ddots & & & a & \\ & & 1 & & & \\ & & & \ddots & & \\ & & & & & 1 \end{pmatrix}$$

If E has $a \neq 0$ in row k and column $\ell \neq k$, then multiplying a matrix on the left by E adds a times row ℓ to row k, which is the basic step in Gauss-Jordan reduction. Multiplying on the left by E^{-1} reverses this step; thus E^{-1} is the elementary matrix with $-a$ in row k and column ℓ.

Definition. A linear transformation $T : V \longrightarrow V$ is elementary *when there exists a basis $b_1, b_2, \ldots, b_n$ of V such that $Tb_1 = b_1 + b_2$ and $Tb_i = b_i$ for all $i \geqq 2$.*

Readers will show that T is elementary if and only if its matrix in some basis of V is elementary.

Proposition **8.2.** *For a linear transformation $T \in GL(V)$ the following are equivalent:* (i) $T : V \longrightarrow V$ *is elementary;* (ii) $\det T = 1$ *and*

$$F(T) = \{ x \in V \mid Tx = x \} = \mathrm{Ker}\,(T - 1)$$

has dimension $\dim V - 1$*;* (iii) $\det T = 1$ *and* $\mathrm{Im}\,(T - 1)$ *has dimension 1. In particular, $SL(V)$ contains all elementary transformations.*

Proof. Let T be elementary, so that there is a basis $b_1, b_2, \ldots, b_n$ of V such that $Tb_1 = b_1 + b_2$ and $Tb_i = b_i$ for all $i \geqq 2$. The matrix of T in that basis is triangular, with 1's on the main diagonal; hence $\det T = 1$. Also $F(T)$ is the subspace generated by $b_2, \ldots, b_n$ and has dimension $n - 1$; $\mathrm{Im}\,(T - 1)$ is the subspace generated by b_2 and has dimension 1.

Conversely, assume that $\det T = 1$ and $F(T)$ has dimension $\dim V - 1$, equivalently, that $\det T = 1$ and $\dim \operatorname{Im}(T-1) = \dim V - \dim \operatorname{Ker}(T-1) = 1$. Let $b_1 \notin F(T)$; then $V = Kb_1 \oplus F(T)$ and $Tb_1 = cb_1 + v \neq b_1$, for some $c \in K$ and $v \in F(T)$. For any basis $b_2, \ldots, b_n$ of $F(T)$, the matrix of T in the basis $b_1, b_2, \ldots, b_n$ of V is triangular, since $Tb_i = b_i$ for all $i \geq 2$, with $c, 1, \ldots, 1$ on the main diagonal; hence $c = \det T = 1$ and $T(b_1) = b_1 + v$ with $v \in F(T)$, $v \neq 0$. There is a basis $b_2, \ldots, b_n$ of $F(T)$ in which $b_2 = v$. Then $Tb_1 = b_1 + b_2$. Thus T is elementary. $\square$

Proposition **8.3.** *For a linear transformation $T \in GL(V)$ the following are equivalent:* (i) *T is in the center of $GL(V)$;* (ii) *T commutes with every elementary transformation;* (iii) *$Tx \in Kx$ for all $x \in V$;* (iv) *$T = \lambda 1_V$ for some $\lambda \in K$, $\lambda \neq 0$. Hence $Z(GL(V)) \cong K^*$.*

A linear transformation $T \in SL(V)$ is in the center of $SL(V)$ if and only if it is in the center of $GL(V)$, if and only if $T = \lambda 1_V$ for some $\lambda \in K$ such that $\lambda^n = 1$. Hence $Z(SL(V))$ is isomorphic to the multiplicative group of all nth roots of 1 in K.

Proof. Let $T \in GL(V)$. We see that (i) implies (ii).

Assume (ii). If $b_1, b_2, \ldots, b_n$ is a basis of V, then there is an elementary transformation E such that $Eb_1 = b_1 + b_2$ and $Eb_i = b_i$ for all $i \geq 2$. Then $\operatorname{Im}(E-1) = Kb_2$ and

$$Tb_2 = T(E-1)b_1 = (E-1)Tb_2 = ab_2$$

for some $a \in K$. For every $x \in V$, $x \neq 0$, there is a basis of V in which $b_2 = x$; hence $Tx = ax$ for some $a \in K$, and (ii) implies (iii).

If (iii) holds, then in any basis $b_1, b_2, \ldots, b_n$ of V we have $Tb_i = a_i b_i$ for some $a_i \in K$. Now,

$$a_i b_i + a_j b_j = T(b_i + b_j) = a(b_i + b_j)$$

for some $a \in K$; hence $a_i = a = a_j$, for every i, j, and $T = \lambda 1_V$, where $\lambda = a_1 = a_2 = \cdots = a_n \neq 0$, since $\lambda^n = \det T \neq 0$. Thus (iii) implies (iv). Finally, (iv) implies (i), since scalar multiples $T = \lambda 1_V$ of the identity transformation on V commute with every linear transformation.

Now let $T \in SL(V)$. If $T \in Z(SL(V))$, then T commutes with every elementary transformation; hence $T \in Z(GL(V))$, and $T = \lambda 1_V$, where $\lambda^n = \det T = 1$. $\square$

The Projective Special Linear Group. We noted that any nonabelian simple factor of $GL(V)$ or $SL(V)$ must come from $SL(V)/Z(SL(V))$.

Definition. Let V be a vector space of finite dimension $n \geq 2$ over a field K. The projective special linear group *$PSL(V)$ or $PSL(n, K)$ is the quotient group $SL(V)/Z(SL(V))$.*

We digress to show that projective linear groups are groups of transformations of projective spaces, much as linear groups are groups of transformations of vector spaces. For every vector space V over a field K a *projective space* P over K is constructed a follows: the relation

$$x \sim y \text{ if and only if } x = \lambda y \text{ for some } \lambda \in K, \lambda \neq 0,$$

is an equivalence relation on V; P is the set of all equivalence classes $[x]$ of nonzero vectors $x \in V$.

In the Euclidean plane, the motivation for projective spaces lies in the projection of one straight line L to another straight line L', not parallel to L, from one point not on L or L'. This projection is almost a bijection of L onto L', except that one point of L disappears at infinity in the direction of L', and one point of L' seems to arrive from infinity in the direction of L. If every straight line could be completed by the addition of a point at infinity, then projection of one straight line onto another, whether parallel or from a point, would always be bijective:

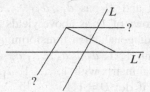

This is precisely what happens in the projective plane P (over $\mathbb{R}$), which is the projective space of $V = \mathbb{R}^3$. In P there are two kinds of points: points $[(a, b, c)]$ with $c \neq 0$, which may be written in the form $[(x, y, 1)]$ and identified with points (x, y) of $\mathbb{R}^2$, and *points at infinity* $[(x, y, 0)]$. A straight line in P is the set of all points $[(x, y, z)]$ that satisfy a linear equation $ax + by + cz = 0$ with $(a, b, c) \neq 0$. In P, the points at infinity constitute a straight line, $z = 0$; every other straight line consists of a straight line in $\mathbb{R}^2$ plus one point at infinity. Two straight lines in P always intersect; parallel lines in $\mathbb{R}^2$ intersect at infinity when completed to straight lines in P.

In general, an invertible linear transformation $T \in GL(V)$ of V induces a *projective transformation* $[T]$ of the corresponding projective space P, which is well defined by $[T][x] = [Tx]$. Readers will easily deduce from Proposition 8.3 that $[T] = [U]$ if and only if $T = \lambda U$ for some $\lambda \in K$, $\lambda \neq 0$. Hence the group of all projective transformations of P is isomorphic to $GL(V) / Z(GL(V))$. Similarly, linear transformations $T \in SL(V)$ induce a group of projective transformations of P, which is isomorphic to $SL(V) / Z(SL(V))$; this is the projective special linear group $PSL(V)$.

Elementary transformations. Our main result concerns the simplicity of $PSL(V)$. The proof requires two properties of elementary transformations.

Proposition **8.4.** *For every vector space V of finite dimension $n \geq 2$ over a field K, the group $SL(V)$ is generated by all elementary transformations.*

Proof. We use matrices. As long as rows are not permuted or multiplied by scalars, Gauss-Jordan reduction is equivalent to left multiplication by elementary matrices. Therefore, when M is an invertible matrix, there are elementary matrices $E_1, \ldots, E_n$ such that $E_1 \cdots E_n M$ is diagonal. Since the inverse of an elementary matrix is elementary, M is the product of elementary matrices and a diagonal matrix D. Moreover, $\det M = \det D$, since elementary matrices have determinant 1. We claim that, when a diagonal matrix D has determinant 1, there are elementary matrices $E_1, \ldots, E_n$ such that $E_1 \cdots E_n M$ is the identity matrix. Then D is a product of elementary matrices, and Proposition 8.4 is proved.

The claim is proved if $n = 2$ by the Gauss-Jordan reduction:

$$D = \begin{pmatrix} a & 0 \\ 0 & b \end{pmatrix} \longrightarrow \begin{pmatrix} a & 0 \\ a & b \end{pmatrix} \longrightarrow \begin{pmatrix} 0 & -b \\ a & b \end{pmatrix} \longrightarrow \begin{pmatrix} 0 & -b \\ a & 0 \end{pmatrix}$$

$$\longrightarrow \begin{pmatrix} 1 & -b \\ a & 0 \end{pmatrix} \longrightarrow \begin{pmatrix} 1 & -b \\ 0 & ab \end{pmatrix} \longrightarrow \begin{pmatrix} 1 & 0 \\ 0 & ab \end{pmatrix},$$

where $ab = 1$. If $n > 2$ and D has $d_1, d_2, \ldots, d_n$ on the diagonal, then transforming the first two rows of D as above yields a diagonal matrix with 1, $d_1 d_2, d_3, \ldots, d_n$ on the diagonal; then transforming rows 2 and 3 as above yields a diagonal matrix with $1, 1, d_1 d_2 d_3, \ldots, d_n$ on the diagonal; repeating this process yields a diagonal matrix with $1, \ldots, 1, d_1 d_2 \ldots d_n$ on the diagonal, which is the identity matrix if $\det D = 1$. $\square$

Proposition 8.5. *Elementary transformations constitute a conjugacy class of* $GL(V)$; *if* $\dim V \geqq 3$, *they constitute a conjugacy class of* $SL(V)$.

Proof. Let E be an elementary transformation and let $T E T^{-1}$ be a conjugate of E in $GL(V)$. There is a basis $b_1, b_2, \ldots, b_n$ of V such that $Eb_1 = b_1 + b_2$ and $Eb_i = b_i$ for all $i \geqq 2$. Then $Tb_1, Tb_2, \ldots, Tb_n$ is a basis of V and $T E T^{-1} Tb_1 = Tb_1 + Tb_2$, $T E T^{-1} Tb_i = Tb_i$ for all $i \geqq 2$; hence $T E T^{-1}$ is elementary. Thus, all conjugates of E are elementary.

Conversely, let E and E' be elementary transformations. There is a basis $b_1, b_2, \ldots, b_n$ of V such that $Eb_1 = b_1 + b_2$ and $Eb_i = b_i$ for all $i \geqq 2$, and a basis $b'_1, b'_2, \ldots, b'_n$ of V such that $E'b'_1 = b'_1 + b'_2$ and $E'b'_i = b'_i$ for all $i \geqq 2$. Let T be the linear transformation such that $Tb_i = b'_i$ for all i; T is invertible and $T E T^{-1} = E'$, since $T E T^{-1} b'_i = E'b'_i$ for all i. Thus E and E' are conjugate in $GL(V)$.

To prove conjugacy in $SL(V)$ we need $T \in SL(V)$. Let $d = \det T \neq 0$. If $n \geqq 3$, then $b''_1 = b'_1, \ldots, b''_{n-1} = b'_{n-1}$, $b''_n = d^{-1} b'_n$ is still a basis of V, and $E'b''_1 = b''_1 + b''_2$, $E''b''_i = b''_i$ for all $i \geqq 2$. Let T' be the linear transformation such that $T'b_i = b''_i$ for all i; T' is invertible, $\det T' = d^{-1} \det T = 1$, and $T' E T'^{-1} = E'$; hence E and E' are conjugate in $SL(V)$. $\square$

The last part of the proof breaks down if $n = 2$; in fact, elementary transformations need not be conjugate in $SL(V)$ when $\dim V = 2$ (see the exercises).

Main result:

Theorem **8.6.** $PSL(V)$ *is simple when* $\dim V \geq 3$.

Proof. A nontrivial normal subgroup of $PSL(V)$ comes from a normal subgroup $Z(SL(V)) \subsetneqq N \subseteq SL(V)$ of $SL(V)$. We show that N must contain an elementary transformation; then $N = SL(V)$, by 8.5 and 8.4.

Since N properly contains $Z(SL(V))$, some $A \in N$ does not commute with some elementary transformation D, by 8.3. Then $B = ADA^{-1}D^{-1} \neq 1$ and $B \in N$. Also $F(ADA^{-1}) \cap F(D) \subseteq F(B)$; since ADA^{-1} and D are elementary, we have $\dim F(ADA^{-1}) = \dim F(D) = n - 1$, by 8.2, and

$$\dim F(B) \geq \dim\left(F(ADA^{-1}) \cap F(D)\right)$$
$$= \dim F(ADA^{-1}) + \dim F(D) - \dim\left(F(ADA^{-1}) + F(D)\right) \geq n - 2.$$

Hence $F(B)$ has dimension $n - 1$ or $n - 2$. If $F(B)$ has dimension $n - 1$, then B is elementary by 8.2, and we are done.

Assume that $F(B) = \text{Ker}(B - 1)$ has dimension $n - 2$. Then $\text{Im}(B - 1)$ has dimension 2. Since V has dimension at least 3, $\text{Im}(B - 1)$ is contained in a subspace U of V (a hyperplane) of dimension $n - 1$. Then $BU \subseteq (B-1)U + U \subseteq U$ and $BU = U$.

For every $u \in U$, $u \neq 0$, and $v \in V \backslash U$ there is an elementary transformation E such that $F(E) = U$ and $Ev = u + v$; in particular, $\text{Im}(E - 1) = Ku$. Then $C = BEB^{-1}E^{-1} \in N$; also $U \subseteq F(C)$, since $x \in U$ implies $B^{-1}x \in U = F(E)$ and $Cx = BEB^{-1}x = BB^{-1}x = x$. Hence $F(C) = U$ or $F(C) = V$. We show that u and v can be chosen so that $C \neq 1$; then C is elementary.

If $F(B) \not\subseteq U$, choose $v \in F(B) \backslash U$ and $u \in U \backslash F(B)$. Then $Bu \neq u$, $BEB^{-1}v = BEv = Bv + Bu \neq v + u = Eu$, $BEB^{-1} \neq E$, and $C \neq 1$.

If $F(B) \subseteq U$, then B has a restriction B' to U and $F(B') = F(B) \neq 0, U$; by 8.4, $B' \notin Z(SL(U))$ and $Bu = B'u \notin Ku$ for some $u \in U$. Then BEB^{-1} is elementary by 8.5; also, $BEB^{-1}(Bv) = BEv = Bu + Bv$, so that $\text{Im}(BEB^{-1} - 1) = KBu \neq Ku = \text{Im}(E - 1)$, $BEB^{-1} \neq E$, and $C \neq 1$.

In either case $C \in N$ is elementary. $\square$

The case when $\dim V = 2$ is more complex. Then $PSL(V)$ is not simple when K has only 2 or 3 elements (see below). The following result makes a substantial exercise:

Theorem **8.7.** $PSL(V)$ *is simple when* $\dim V = 2$ *and* $|K| \geq 4$.

Orders. If K is finite, then so are $GL(n, K)$, $SL(n, K)$, and $PSL(n, K)$.

Proposition **8.8.** *If* K *has* q *elements, then* $|GL(n, K)| = \prod_{0 \leq i < n} (q^n - q^i)$; $|SL(n, K)| = |GL(n, K)|/(q - 1)$; *and* $|PSL(n, K)| = |SL(n, K)|/r$, *where* r *is the number of* n*th roots of* 1 *in* K.

Proof. An $n \times n$ matrix M is invertible if and only if its columns constitute a basis of the vector space K^n. If $|K| = q$, there are q^n possible columns and $q^n - 1$ ways to choose the first column of M, which must not be the zero column; there are $q^n - q$ ways to choose the second column, which must not be one of the q scalar multiples of the first column; there are $q^n - q^2$ ways to choose the third column, which must not be one of the q^2 linear combinations of the first two; ...; and there are $q^n - q^{n-1}$ ways to choose the last column, which must not be one of the q^{n-1} linear combinations of the first $n - 1$ columns. Hence $|GL(n, K)| = (q^n - 1)(q^n - q)(q^n - q^2) \cdots (q^n - q^{n-1})$.

Then $|GL(n, K)| / |SL(n, K)| = q - 1$, by 8.1; if r is the number of nth roots of 1 in K, then $|PSL(n, K)| = |SL(n, K)|/r$, by 8.3. $\square$

We will show in Chapter V that a finite field K is uniquely determined, up to isomorphism, by its number q of elements, which must be a power of a prime; then $GL(n, K)$, $SL(n, K)$, and $PSL(n, K)$ are usually denoted by $GL(n, q)$, $SL(n, q)$, and $PSL(n, q)$.

If $q = 2$, then $K \cong \mathbb{Z}_2$, the field of integers modulo 2, and has one square root of unity. Hence $|GL(2, 2)| = (4 - 1)(4 - 2) = 6$ and $|PSL(2, 2)| = |SL(2, 2)| = 6/(2 - 1) = 6$. If $q = 3$, then $K \cong \mathbb{Z}_3$ has 2 square roots of unity, $|GL(2, 3)| = (9 - 1)(9 - 3) = 48$, $|SL(2, 3)| = 48/(3 - 1) = 24$, and $|PSL(2, 2)| = 24/2 = 12$. Hence $PSL(2, 2)$ and $PSL(2, 3)$ are not simple.

On the other hand, $|GL(3, 2)| = (8 - 1)(8 - 2)(8 - 4) = 168$ and $|PSL(3, 2)| = |SL(3, 2)| = 168/(2 - 1) = 168$, since $K \cong \mathbb{Z}_2$. The simple group $PSL(3, 2)$ is not abelian (it is not of prime order) and is not one of the simple alternating groups (whose orders are $60, 360, 2520, \ldots$). Thus, the groups $PSL(n, q)$ constitute a new family of simple groups.

Exercises

1. Show that a linear transformation is elementary if and only if its matrix in some basis is elementary (has 1's on the main diagonal and exactly one nonzero entry off the main diagonal).

2. Show that in $SL(2, K)$, every elementary matrix is conjugate to some $\begin{pmatrix} 1 & a \\ 0 & 1 \end{pmatrix}$.

3. Show that $\begin{pmatrix} 1 & a \\ 0 & 1 \end{pmatrix}$ and $\begin{pmatrix} 1 & b \\ 0 & 1 \end{pmatrix}$, where $a, b \neq 0$, are conjugate in $SL(2, K)$ if and only if $ab^{-1} = c^2$ for some $c \in K$.

4. Show that $PSL(2, 2) \cong D_3$.

5. To which known group is $PSL(2, 3)$ isomorphic?

6. Draw a table showing $|PSL(n, q)|$ for all $n, q = 2, 3, 4, 5$.

7. Let $|K| = q = p^k$ and let S be the set of all upper triangular $n \times n$ matrices with 1's on the main diagonal. Show that S is a Sylow p-subgroup of $SL(n, K)$.

8. Show that $Z\big(GL(V)\big) \subsetneqq N \trianglelefteq GL(V)$ implies $SL(V) \subseteq N$.

*9. Prove that $PSL(V)$ is simple when $\dim V = 2$ and $|K| \geqq 4$.

9. Solvable Groups

Solvable groups are a large class of groups with remarkable properties. Their connection with polynomial equations is explained in Chapter V. The Hall theorems at the end of this section may be skipped.

Definition. A solvable group is a group with a normal series whose factors are abelian.

Solvable groups are sometimes called *metabelian*. Abelian groups are solvable; readers will easily show that D_n is solvable, and that all groups of order less than 60 are solvable. On the other hand, nonabelian simple groups are not solvable, since the single factor in their one normal series is not abelian; thus, A_n (when $n \geq 5$) and the simple groups in Section 8 are not solvable.

The first major step of the Classification theorem, the *Feit and Thompson theorem* [1963], states that all nonabelian finite simple groups have even orders; equivalently, every group of odd order is solvable.

The commutator series. The commutator series is the smallest descending sequence of subgroups with abelian factors. It provides an alternate definition of solvable groups, and is constructed as follows.

Definitions. The commutator of two elements x, y is $xyx^{-1}y^{-1}$; the commutator subgroup or derived group of a group G is the subgroup G' of G generated by all commutators.

The commutator $xyx^{-1}y^{-1}$ is traditionally denoted by $[x, y]$. In algebraic topology, the derived group of $\pi_1(X)$ is the first homology group $H_1(X)$ of X.

Proposition 9.1. G' is a normal subgroup of G; in fact, G' is the smallest normal subgroup N of G such that G/N is abelian.

Proof. The inverse of a commutator $xyx^{-1}y^{-1}$ is a commutator, and a conjugate of a commutator is again a commutator:

$$a\, xyx^{-1}y^{-1}\, a^{-1} = axa^{-1}\, aya^{-1}\, (axa^{-1})^{-1}\, (aya^{-1})^{-1}.$$

Hence every $x \in G'$ is a product of commutators $x = c_1 c_2 \cdots c_n$, and then $axa^{-1} = ac_1 a^{-1}\, ac_2 a^{-1} \cdots ac_n a^{-1} \in G'$ for all $a \in G$. Thus $G' \trianglelefteq G$.

Next, $xyx^{-1}y^{-1} \in G'$ for all $x, y \in G$; hence $G'xy = G'yx$ and G/G'

is abelian. Conversely, if $N \trianglelefteq G$ and G/N is abelian, then $Nxy = Nyx$ and $xyx^{-1}y^{-1} \in N$ for all $x, y \in G$, and $G' \subseteq N$. $\square$

Readers will prove the following universal property: every homomorphism of G into an abelian group factors uniquely through the projection $G \longrightarrow G/G'$.

Definition. The commutator series *of a group G is the sequence*

$$G \trianglerighteq G' \trianglerighteq G'' \trianglerighteq \cdots \trianglerighteq G^{(k)} \trianglerighteq G^{(k+1)} \trianglerighteq \cdots$$

in which $G^{(0)} = G$ and $G^{(k+1)} = (G^{(k)})'$ for all $k \geqq 0$.

The group $G^{(k)}$ is the kth *derived group* of G; it is normal in $G^{(k-1)}$ by 9.1. The commutator series is not a normal series, but it becomes one if some $G^{(r)} = 1$ and the tail $G^{(r+1)} \trianglerighteq \cdots$ is chopped off (preferably under anesthesia).

Proposition 9.2. *A group G is solvable if and only if $G^{(r)} = 1$ for some $r \geqq 0$.*

Proof. If $G^{(r)} = 1$, then $1 = G^{(r)} \trianglelefteq G^{(r-1)} \trianglelefteq \cdots \trianglelefteq G' \trianglelefteq G$ is a normal series whose factors are abelian, by 9.1. Conversely, assume that G has a normal series $1 = A_0 \trianglelefteq A_1 \trianglelefteq \cdots \trianglelefteq A_m = G$ whose factors A_i/A_{i-1} are all abelian. Then G/A_{m-1} is abelian; by 9.1, $G' \subseteq A_{m-1}$. In general, A_{m-k}/A_{m-k-1} is abelian, so $G^{(k)} \subseteq A_{m-k}$ implies $G^{(k+1)} \subseteq A'_{m-k} \subseteq A_{m-k-1}$, by 9.1. Induction then yields $G^{(k)} \subseteq A_{m-k}$ for all $k \leqq m$, in particular $G^{(m)} = 1$. $\square$

Proposition 9.2 is often used as a definition of solvable groups.

Properties. The class of all solvable groups has three basic properties that can be proved either from the definition or from 9.2, and make fine exercises.

Proposition 9.3. *Every subgroup of a solvable group is solvable.*

Proposition 9.4. *Every quotient group of a solvable group is solvable.*

Proposition 9.5. *If $N \trianglelefteq G$ and G/N are solvable, then G is solvable.*

These properties yield further examples.

Proposition 9.6. *Every finite p-group is solvable.*

Proof. That a group G of order p^n is solvable is proved by induction on n. If $n \leqq 2$, then G is abelian, hence solvable, by 3.10. In general, G has a subgroup N of order p^{n-1}, which is normal by 5.12; then N and G/N are solvable, by the induction hypothesis, and G is solvable, by 9.5. $\square$

Proposition 9.7. *Every group of order $p^n q$ (where p and q are primes) is solvable.*

In Chapter IX we prove a stronger result, *Burnside's $p^m q^n$ theorem*: every group of order $p^m q^n$, where p, q are primes, is solvable. Proposition 9.7 and its proof may be skipped.

Proof. We may assume that $p \neq q$. The proof is by induction on n.

Let S be a Sylow p-subgroup of G. If $S \trianglelefteq G$, then G/S is cyclic, since $|G/S| = q$ is prime, S is solvable by 9.6, and G is solvable by 9.5.

Now assume that S is not normal. Then $S \subseteq N_G(S) \subsetneqq G$; since $[G:S] = q$ is prime, this implies $N_G(S) = S$, and S has $[G:N_G(S)] = q$ conjugates. Thus there are q Sylow p-subgroups.

If the q Sylow p-subgroups of G are pairwise disjoint ($S \cap T = 1$ when $S \neq T$), then G has $q(p^n - 1)$ elements whose order is a positive power of p, leaving at most q elements whose order is a power of q. Therefore G has only one Sylow q-subgroup Q, and $Q \trianglelefteq G$. Then Q is cyclic, G/Q is solvable by 9.6, and G is solvable by 9.5. In particular, 9.7 holds when $n = 1$.

Now assume that the q Sylow p-subgroups of G are not pairwise disjoint. Then there are Sylow p-subgroups S and T such that $S \cap T \neq 1$, and one can choose S and T so that $M = S \cap T$ has the greatest possible number of elements. By Lemma 9.8 below, $H = N_G(M)$ has more than one Sylow p-subgroup; M is the intersection of all the Sylow p-subgroups of H; and every Sylow p-subgroup of H is contained in a unique Sylow p-subgroup of G. Now, the number of Sylow p-subgroups of H divides $p^n q$ but is not divisible by p; hence H has q Sylow p-subgroups. Since G also has q Sylow p-subgroups, M is contained in every Sylow p-subgroup of G. Therefore M is the intersection of all the Sylow p-subgroups of G. Since the latter are all conjugate, this implies that $M \trianglelefteq G$. Now, $|M| = p^k$, where $1 \leq k < n$. Hence G/M is solvable, by the induction hypothesis; M is solvable by 9.6; and G is solvable, by 9.5. $\square$

Lemma 9.8. *Let M be the intersection of two distinct Sylow p-subgroups of a group G. If M has the greatest possible number of elements, then $H = N_G(M)$ has more than one Sylow p-subgroup; M is the intersection of all the Sylow p-subgroups of H; and every Sylow p-subgroup of H is contained in a unique Sylow p-subgroup of G.*

Proof. We have $M \subsetneqq S$ for some Sylow p-subgroup S of G. By 5.11, $M \subsetneqq N_S(M) = H \cap S$. Now, $N_S(M) \subseteq S$ is a p-subgroup of H and is by 5.9 contained in a Sylow p-subgroup P of H, which is in turn contained in a Sylow p-subgroup T of G. Then $M \subsetneqq N_S(M) \subseteq S \cap T$ and $S = T$ by the choice of M. Hence $P \subseteq H \cap S = N_S(M)$ and $N_S(M) = P$ is a Sylow p-subgroup of H. Since $M \trianglelefteq N_G(M) = H$, M is contained in every conjugate of P and is contained in every Sylow p-subgroup of H.

We also have $M = S \cap T$ for some Sylow p-subgroups $S \neq T$ of G. Then $M \subseteq N_S(M) \cap N_T(M) \subseteq S \cap T$ and $M = N_S(M) \cap N_T(M)$. Then $N_S(M) \neq N_T(M)$, since $M \subsetneqq N_S(M), N_T(M)$. By the above, applied to S and to T, M is the intersection of two distinct Sylow p-subgroups of H. Therefore H has more than one Sylow p-subgroup, and M is the intersection of all the Sylow p-subgroups of H.

Finally, let P be any Sylow p-subgroup of H. By 5.9, P is contained in a Sylow p-subgroup S of G, but P is not contained in two distinct Sylow p-subgroups S and T of G: otherwise, $M \subsetneqq P \subseteq S \cap T$ contradicts the choice of M. $\square$

The Hall Theorems. This part may be skipped. The three theorems below, due to Hall [1928], are stronger versions of the Sylow theorems that hold in solvable groups. First we prove a lemma.

Lemma **9.9.** *Every nontrivial finite solvable group contains a nontrivial abelian normal p-subgroup for some prime p.*

Proof. Let G be be a finite solvable group. There is a smallest integer $r > 0$ such that $G^{(r)} = 1$. Then $A = G^{(r-1)}$ is a nontrivial abelian normal subgroup of G. Some prime p divides $|A| > 1$; let N be the set of all elements of A whose order is a power of p. Then $N \neq 1$, $N \leqq A$, and N is a p-group. If $x \in N$ and $g \in G$, then $gxg^{-1} \in A$ and the order of gxg^{-1} is a power of p, so that $gxg^{-1} \in N$; thus $N \trianglelefteq G$. $\square$

The proof of the first theorem also uses Schur's theorem, proved in Section 12 by other methods: if m and n are relatively prime, then a group of order mn that contains an abelian normal subgroup of order n also contains a subgroup of order m.

Theorem **9.10.** *Let m and n be relatively prime. Every solvable group of order mn contains a subgroup of order m.*

Proof. Let G be solvable of order mn. If m is a power of a prime, then 9.10 follows from the first Sylow theorem. Otherwise, we proceed by induction on $|G|$. By 9.9, G contains contains a nontrivial abelian normal subgroup N of order $p^k > 1$ for some prime p. Now, p^k divides $|G| = mn$; since m and n are relatively prime, either p^k divides m, or p^k divides n.

If p^k divides m, then $|G/N| = (m/p^k) n$, where m/p^k and n are relatively prime and $|G/N| < |G|$. By the induction hypothesis, G/N has a subgroup H/N of order m/p^k, where $N \subseteq H \leqq G$; then $|H| = m$.

If p^k divides n, then $|G/N| = (n/p^k) m$, where n/p^k and m are relatively prime and $|G/N| < |G|$. By the induction hypothesis, G/N has a subgroup H/N of order m, where $N \subseteq H \leqq G$. Then $|H| = mp^k$. Now, $N \trianglelefteq H$, N is abelian, and N has order p^k, which is relatively prime to m; by Schur's theorem, H has a subgroup of order m, and then so does G. $\square$

The subgroups of G of order m are the *Hall subgroups* of G.

Lemma **9.11.** *Let m and n be relatively prime and let G be a group of order mn with an abelian normal subgroup of order n. All subgroups of G of order m are conjugate.*

Proof. Let $|G| = mn$ and let $N \trianglelefteq G$, with $|N| = n$ and N abelian. Let

A and B be subgroups of G of order m. Since m and n are relatively prime we have $A \cap N = B \cap N = 1$; hence $AN = BN = G$. Therefore every coset of N intersects A in exactly one element, and similarly for B. The element of $Nx \cap B$ can then be written as $u_x x$ for some unique $u_x \in N$. Then $u_a (au_b a^{-1}) ab = (u_a a)(u_b b) \in B$ for all $a, b \in A$ and

$$u_{ab} = u_a \, au_b a^{-1}.$$

Let $v = \prod_{b \in A} u_b \in N$. Since N is abelian,

$$v = \prod_{b \in A} u_{ab} = \prod_{b \in A} (u_a \, au_b a^{-1}) = u_a^m \, ava^{-1}$$

for all $a \in A$. We also have $u_a^n = 1$, since $|N| = n$. Now, $qm + rn = 1$ for some $q, r \in \mathbb{Z}$, since m and n are relatively prime; hence $u_a = u_a^{qm+rn} = u_a^{qm}$, $w = v^q = u_a^{mq} (ava^{-1})^q = u_a \, awa^{-1}$, and $u_a a = waw^{-1}$ for all $a \in A$. Therefore $B = wAw^{-1}$ is a conjugate of A. $\square$

Theorem 9.12. *In a solvable group of order mn, where m and n are relatively prime, all subgroups of order m are conjugate.*

Proof. Let G be solvable of order mn. If m is a power of a prime, then 9.12 follows from the third Sylow theorem. Otherwise, we proceed by induction on $|G|$. By 9.9, G contains an abelian normal subgroup N of order $p^k > 1$ for some prime p, and p^k divides m or n. Let $A, B \leq G$ have order m.

Assume that p^k divides m. Then $|NA| = |A| (|N|/|A \cap N|) = mp^h$ for some $h \leq k$. Now, $mp^h = |NA|$ divides $mn = |G|$; since p^h and n are relatively prime this implies $p^h = 1$. Hence $|NA| = |A|$ and $N \subseteq A$. Similarly, $N \subseteq B$. By the induction hypothesis, A/N and B/N are conjugate in G/N: $B/N = (Nx)(A/N)(Nx)^{-1}$ for some $x \in G$. Then

$$B = \bigcup_{b \in B} Nb = \bigcup_{a \in A} (Nx)(Na)(Nx)^{-1}$$
$$= \bigcup_{a \in A} Nxax^{-1} = N(xAx^{-1}) = xAx^{-1},$$

since $N = xNx^{-1} \subseteq xAx^{-1}$. Thus A and B are conjugate in G.

Now assume that p^k divides n. Then $A \cap N = B \cap N = 1$; hence $|NA| = |NB| = p^k m$, and the subgroups $NA/N \cong A/(A \cap N)$ and $NB/N \cong B/(B \cap N)$ of G/N have order m. By the induction hypothesis, NA/N and NB/N are conjugate in G/N. As above, it follows that NA and NB are conjugate in G: $NB = xNAx^{-1}$ for some $x \in G$. Then B and xAx^{-1} are subgroups of NB of order m. Hence B and xAx^{-1} are conjugate in NB: this follows from the induction hypothesis if $p^k < n$, from Lemma 9.11 if $p^k = n$. Therefore A and B are conjugate in G. $\square$

Theorem 9.13. *In a solvable group of order mn, where m and n are relatively prime, every subgroup whose order divides m is contained in a subgroup of order m.*

Proof. Let G be solvable of order mn. If m is a power of a prime, then 9.13 follows from 5.9. Otherwise, we proceed by induction on $|G|$. By 9.9, G contains an abelian normal subgroup N of order $p^k > 1$ for some prime p, and p^k divides m or n. Let H be a subgroup of G whose order ℓ divides m.

Assume that p^k divides m. Then $|NH/N| = |H|/|H \cap N|$ divides m, is relatively prime to n, and divides $|G/N| = (m/p^k)n$. By the induction hypothesis, H/N is contained in a subgroup K/N of G/N of order m/p^k, where $N \subseteq K \leqq G$; then H is contained in the subgroup K of G of order m.

Assume that p^k divides n. Then $H \cap N = 1$ and $|NH| = p^k \ell$. Hence $|NH/N| = \ell$ divides m, is relatively prime to n, and divides $|G/N| = (n/p^k)m$. By the induction hypothesis, NH/N is contained in a subgroup K/N of G/N of order m, where $N \subseteq K \leqq G$; then $|K| = p^k m$ and $H \subseteq NH \subseteq K$. If $p^k < n$, then $|K| < |G|$ and H is contained in a subgroup of K of order m, by the induction hypothesis.

Now assume that $p^k = n$. Let A be a subgroup of G of order m. Then $A \cap N = 1$, $|NA| = |N||A| = |G|$, and $NA = G$. Hence $|A \cap NH| = |A||NH|/|ANH| = mp^k\ell/mn = \ell$. Thus H and $K = A \cap NH$ are subgroups of NH of order ℓ. By 9.12, H and K are conjugate in NH: $H = xKx^{-1}$ for some $x \in NH$. Then H is contained in the subgroup xAx^{-1} of G, which has order m. $\square$

Exercises

1. Find the commutator series of S_4.

2. Find the commutator series of A_4.

3. Show that D_n is solvable.

4. Show that every group of order less than 60 is solvable.

5. Show that G' is a *fully invariant* subgroup of G ($\eta G' \subseteq G'$ for every endomorphism η of G).

6. Show that $G^{(k)}$ is a fully invariant subgroup of G, for every $k \geqq 0$.

7. Show that G/G' has the following universal property: every homomorphism of G into an abelian group factors uniquely through the projection $G \longrightarrow G/G'$.

8. Show that a group that has a composition series is solvable if and only if all its simple factors are abelian.

9. Prove that every subgroup of a solvable group is solvable.

10. Prove that every quotient group of a solvable group is solvable.

11. Prove the following: if $N \trianglelefteq G$ and G/N are solvable, then G is solvable.

12. Show that S_n is solvable if and only if $n \leq 4$.

13. Find the commutator series of S_n.

10. Nilpotent Groups

Nilpotent groups are a class of solvable groups with even more striking properties.

Definition. A normal series $1 = C_0 \trianglelefteq C_1 \trianglelefteq \cdots \trianglelefteq C_m = G$ *is* central *when* $C_i \trianglelefteq G$ *and* $C_{i+1}/C_i \subseteq Z(G/C_i)$ *for all* $0 \leq i < m$.

Central normal series are also called just *central series*. A central normal series has abelian factors, but a normal series with abelian factors need not be central; the exercises give a counterexample.

Definition. A group is nilpotent *when it has a central normal series.*

In particular, abelian groups are nilpotent, and nilpotent groups are solvable. The converses are not true; we shall see that D_4 is nilpotent but not abelian, and that D_3 and D_5 are solvable but not nilpotent.

Two central series. Nilpotent groups have two explicit central normal series.

Definition. The descending central series *of a group* G *is the sequence*

$$G \supseteq G^1 \supseteq \cdots \supseteq G^k \supseteq G^{k+1} \supseteq \cdots$$

in which $G^0 = G$, *and* G^{k+1} *is the subgroup generated by all commutators* $xyx^{-1}y^{-1}$ *with* $x \in G$ *and* $y \in G^k$.

In particular, $G^1 = G'$. The descending central series yields a central normal series if some $G^r = 1$ and subsequent terms are removed (or fall off):

Proposition **10.1.** $G^k \trianglelefteq G$ *and* $G^k/G^{k+1} \subseteq Z(G/G^{k+1})$, *for all* k.

Proof. The proof is by induction on k. First, $G^0 = G \trianglelefteq G$, and $G^0/G^1 = G/G' \subseteq Z(G/G^1)$ since G/G' is abelian by 9.1.

Now assume that $G^k \trianglelefteq G$. As in the proof of 9.1, the inverse of the commutator $xyx^{-1}y^{-1}$ of x and y is the commutator of y and x; a conjugate

$$a\,xyx^{-1}y^{-1}a^{-1} = axa^{-1}\,aya^{-1}\,(axa^{-1})^{-1}\,(aya^{-1})^{-1}$$

of $xyx^{-1}y^{-1}$ is the commutator of a conjugate of x and a conjugate of y. Hence every $g \in G^{k+1}$ is a product $g = c_1, \ldots, c_n$ of commutators $xyx^{-1}y^{-1}$ of $x \in G$ and $y \in G^k$, and commutators $xyx^{-1}y^{-1}$ of $x \in G_k$ and $y \in G$; then $aga^{-1} = ac_1a^{-1} \cdots ac_na^{-1}$ is a product of similar commutators. Thus $G^{k+1} \trianglelefteq G$. For all $x \in G$ and $y \in G^k$, $xyx^{-1}y^{-1} \in G^{k+1}$; hence $G^{k+1}xy = G^{k+1}yx$ and $G^{k+1}y \in Z(G/G^{k+1})$. Thus $G^k/G^{k+1} \subseteq Z(G/G^{k+1})$. $\square$

The other series ascends by way of centers and is constructed as follows.

Proposition **10.2.** *Every group G has unique normal subgroups $Z_k(G)$ such that $Z_0(G) = 1$ and $Z_{k+1}(G) / Z_k(G) = Z(G/Z_k(G))$ for all $k \geqq 0$.*

Proof. First, $Z_0(G) = 1$ is normal in G. If $Z_k \trianglelefteq G$, then $Z(G/Z_k(G))$ is a normal subgroup of $G/Z_k(G)$; by I.4.9, there is a unique normal subgroup $Z_{k+1}(G) \supseteq Z_k(G)$ of G such that $Z(G/Z_k(G)) = Z_{k+1}(G) / Z_k(G)$. $\square$

In particular, $Z_1(G) = Z(G)$ is the center of G.

Definition. The ascending central series *of a group G is the sequence*

$$1 = Z_0(G) \trianglelefteq Z_1(G) \trianglelefteq \cdots \trianglelefteq Z_k(G) \trianglelefteq Z_{k+1}(G) \trianglelefteq \cdots$$

constructed in Proposition 10.2.

The ascending central series yields a central normal series if some $Z_r(G) = G$ and subsequent terms are removed (or just abandoned by the wayside).

Proposition **10.3.** *A group G is nilpotent if and only if $G^r = 1$ for some $r \geqq 0$, if and only if $Z_r(G) = G$ for some $r \geqq 0$.*

Proof. If $G^r = 1$ for some $r \geqq 0$, or if $Z_r(G) = G$ for some $r \geqq 0$, then truncating the descending central series, or the ascending central series, yields a central normal series, and G is nilpotent.

Conversely, assume that G has a central normal series $1 = C_0 \trianglelefteq C_1 \trianglelefteq \cdots \trianglelefteq C_m = G$. We prove by induction on k than $G^k \subseteq C_{m-k}$ and $C_k \subseteq Z_k(G)$ for all $0 \leqq k \leqq m$; hence $G^m = 1$ and $Z_m(G) = G$. (Thus, the ascending and descending central series are in this sense the "fastest" central series.)

We have $G^{m-m} = G = C_m$. Assume that $G^{m-j} \subseteq C_j$, where $j > 0$. Let $x \in G$ and $y \in G^{m-j} \subseteq C_j$. Since $C_{j-1} y \in C_j/C_{j-1} \subseteq Z(G/C_{j-1})$, we have $C_{j-1} xy = C_{j-1} yx$ and $xyx^{-1}y^{-1} \in C_{j-1}$. Thus C_{j-1} contains every generator of G^{m-j+1}; hence $G^{m-j+1} \subseteq C_{j-1}$.

We also have $Z_0(G) = 1 = C_0$. Assume that $C_k \subseteq Z_k = Z_k(G)$, where $k < m$. Then $G/Z_k \cong (G/C_k)/(Z_k/C_k)$ and there is a surjective homomorphism $\pi : G/C_k \longrightarrow G/Z_k$ with kernel Z_k/C_k, namely $\pi : C_k x \longmapsto Z_k x$. Since π is surjective, π sends the center of G/C_k into the center of G/Z_k:

$$\pi (C_{k+1}/C_k) \subseteq \pi Z(G/C_k) \subseteq Z(G/Z_k) = Z_{k+1}/Z_k;$$

hence $Z_k x \in Z_{k+1}/Z_k$ for all $x \in C_{k+1}$, and $C_{k+1} \subseteq Z_{k+1}$. $\square$

In fact, we have shown that $G^r = 1$ if and only if $Z_r(G) = G$; the least such r is the *nilpotency index* of G.

Properties. Nilpotent groups, as a class, have basic properties that can be proved either from the definition or from 10.3, and make wonderful exercises.

Proposition **10.4.** *Every subgroup of a nilpotent group is nilpotent.*

Proposition **10.5.** *Every quotient group of a nilpotent group is nilpotent.*

Proposition **10.6.** *If $N \subseteq Z(G)$ and G/N is nilpotent, then G is nilpotent.*

Proposition **10.7.** *If A and B are nilpotent, then $A \oplus B$ is nilpotent.*

Armed with these properties we now determine all nilpotent finite groups.

Proposition **10.8.** *Every finite p-group is nilpotent.*

Proof. That a group G of order p^n is nilpotent is proved by induction on n. If $n \leq 2$, then G is abelian, hence nilpotent, by 3.10. If $n > 2$, then G has a nontrivial center, by 3.9; then $G/Z(G)$ is nilpotent, by the induction hypothesis, and G is nilpotent, by 10.6. $\square$

Proposition **10.9.** *A finite group is nilpotent if and only if all its Sylow subgroups are normal, if and only if it is isomorphic to a direct product of p-groups (for various primes p).*

Proof. The ascending central series of any group G has the following property: if $Z_k \subseteq H \leq G$, then $Z_{k+1} \subseteq N_G(H)$. Indeed, let $x \in Z_{k+1}$ and $y \in H$. Since $Z_k x \in Z_{k+1}/Z_k \subseteq C(G/Z_k)$ we have $Z_k xy = Z_k yx$, so that $xyx^{-1}y^{-1} \in Z_k$ and $xyx^{-1} = (xyx^{-1}y^{-1})y \in H$. Thus $x \in N_G(H)$.

Now let G be a finite group. Let S be a Sylow p-subgroup of G. By 5.10, $N_G(S)$ is its own normalizer. Hence $Z_0 = 1 \subseteq N_G(S)$, and $Z_k \subseteq N_G(S)$ implies $Z_{k+1} \subseteq N_G(N_G(S)) = N_G(S)$ by the above, so that $Z_k \subseteq N_G(S)$ for all k. If G is nilpotent, then $N_G(S) = G$, by 10.3, and $S \trianglelefteq G$.

Next, assume that every Sylow subgroup of G is normal. Let $p_1, p_2, \ldots, p_m$ be the prime divisors of $|G|$. Then G has one Sylow p_i-subgroup S_i for every p_i. We have $|G| = |S_1||S_2| \cdots |S_m|$; hence $G = S_1 S_2 \cdots S_m$. Moreover, $(S_1 \cdots S_i) \cap S_{i+1} = 1$ for all $i < m$, since $|S_{i+1}|$ and $|S_1 \cdots S_i|$ are relatively prime. Hence $G \cong S_1 \times S_2 \times \cdots \times S_m$, by 2.1.

Finally, if G is isomorphic to a direct product of p-groups, then G is nilpotent, by 10.8 and 10.7. $\square$

In particular, D_4 and Q are nilpotent, by 10.8, but the solvable groups D_3 and D_5 are not nilpotent, by 10.9. If G is a nilpotent finite group, readers will easily deduce from 10.9 that every divisor of $|G|$ is the order of a subgroup of G. This property does not extend to solvable groups; for instance, the solvable group A_4 of order 12 does not have a subgroup of order 6.

Exercises

1. Give an example of a normal series that is not central but whose factors are abelian.

2. Show that $Z_k(G)$ is a characteristic subgroup of G ($\alpha \, Z_k(G) = Z_k(G)$ for every automorphism α of G).

3. Show that $Z_k(G)$ is a *fully invariant* subgroup of G ($\eta \, Z_k(G) \subseteq Z_k(G)$ for every endomorphism η of G).

4. Find the ascending central series of S_n.

5. Find the descending central series of S_n.

6. Prove that $G^r = 1$ if and only if $Z_r(G) = G$.

7. Prove that every subgroup of a nilpotent group is nilpotent.

8. Prove that every quotient group of a nilpotent group is nilpotent.

9. Prove the following: if $N \subseteq Z(G)$ and G/N is nilpotent, then G is nilpotent.

10. Prove the following: if A and B are nilpotent, then $A \oplus B$ is nilpotent.

11. Find a group G with a normal subgroup N such that N and G/N are nilpotent but G is not nilpotent.

12. Prove the following: when G is a nilpotent finite group, every divisor of $|G|$ is the order of a subgroup of G.

13. A *maximal* subgroup of a group G is a subgroup $M \subsetneqq G$ such that there exists no subgroup $M \subsetneqq H \subsetneqq G$. Prove that a finite group G is nilpotent if and only if every maximal subgroup of G is normal, if and only if every maximal subgroup of G contains G'.

11. Semidirect Products

Semidirect products are direct products in which the componentwise operation is twisted by a group-on-group action. The exercises give some applications.

Definition. A group B acts on a group A by automorphisms when there is a group action of B on the set A such that the action of every $b \in B$ is an automorphism of the group A.

In what follows, A and B are written multiplicatively, and we use the left exponential notation $(b, a) \longmapsto {}^b a$ for actions of B on A. Then B acts on A by automorphisms if and only if

$$ {}^1 a = a, \quad {}^b({}^{b'} a) = {}^{bb'} a, \quad {}^b(aa') = {}^b a \, {}^b a' $$

for all $a, a' \in A$ and $b, b' \in B$; the first two laws ensure a group action, and the last law ensures that the action of $b \in B$ on A (the permutation $a \longmapsto {}^b a$) is a homomorphism, hence an automorphism, of A. Then

$$ {}^b 1 = 1, \quad {}^b(a^n) = ({}^b a)^n $$

for all $a \in A$, $b \in B$, and $n \in \mathbb{Z}$.

For example, the action ${}^g x = gxg^{-1}$ of a group G on itself by inner automorphisms is, felicitously, an action by automorphisms.

Proposition 11.1. Let the group B act on a group A by automorphisms. The mapping $\varphi : b \longmapsto \varphi(b)$ defined by $\varphi(b): a \longmapsto {}^b a$ is a homomorphism of B into the group $\mathrm{Aut}\,(A)$ of all automorphisms of A.

Proof. By definition, every $\varphi(b)$ is an automorphism of A; moreover, $\varphi(1)$ is the identity mapping on A, and $\varphi(b) \circ \varphi(b') = \varphi(bb')$ for all $b, b' \in B$. □

In fact, there is a one-to-one correspondence between actions of B on A by automorphisms and homomorphisms $B \longrightarrow \mathrm{Aut}\,(A)$. It is convenient to denote an action of B on A and the corresponding homomorphism of B into $\mathrm{Aut}\,(A)$ by the same letter.

Definition. Given two groups A and B and an action φ of B on A by automorphisms, the semidirect product $A \rtimes_{\varphi} B$ *is the Cartesian product $A \times B$ with the multiplication defined for all $a, a' \in A$ and $b, b' \in B$ by*

$$(a, b)\,(a', b') = (a^{\,b}a', bb').$$

When φ is known without ambiguity, $A \rtimes_{\varphi} B$ is denoted by $A \rtimes B$.

Readers will verify that $A \rtimes_{\varphi} B$ is indeed a group. If B acts trivially on A ($^{b}a = a$ for all a and b), then $A \rtimes_{\varphi} B$ is the Cartesian product $A \times B$ with componentwise multiplication, as in Section 1. The exercises give examples of semidirect products that are not direct products.

Internal characterization. Proposition 1.1 on internal direct sums extends to semidirect products: in fact, 1.1 is the case where $B \trianglelefteq G$.

Proposition 11.2. *A group G is isomorphic to a semidirect product $G_1 \rtimes G_2$ of two groups G_1, G_2 if and only if it contains subgroups $A \cong G_1$ and $B \cong G_2$ such that $A \trianglelefteq G$, $A \cap B = 1$, and $AB = G$.*

Proof. $G_1 \rtimes G_2$ comes with a *projection* $\pi : G_1 \rtimes G_2 \longrightarrow G_2$, $(x_1, x_2) \longmapsto x_2$, which is a homomorphism, since the operation on $G_1 \rtimes G_2$ is componentwise in the second component. Hence

$$\mathrm{Ker}\,\pi = \{\,(x_1, x_2) \in G_1 \times G_2 \mid x_2 = 1\,\}$$

is a normal subgroup of $G_1 \times G_2$; and $(x_1, 1) \longmapsto x_1$ is an isomorphism of $\mathrm{Ker}\,\pi$ onto G_1. There is also an *injection* $\iota : G_2 \longrightarrow G_1 \rtimes G_2$, $x_2 \longmapsto (1, x_2)$, which is a homomorphism since $^{x}1 = 1$ for all $x \in G_2$; hence

$$\mathrm{Im}\,\iota = \{\,(x_1, x_2) \in G_1 \times G_2 \mid x_1 = 1\,\}$$

is a subgroup of $G_1 \rtimes G_2$ and is isomorphic to G_2. Moreover, $\mathrm{Ker}\,\pi \cap \mathrm{Im}\,\iota = \{\,(1, 1)\,\} = 1$, and $(\mathrm{Ker}\,\pi)\,(\mathrm{Im}\,\iota) = G_1 \times G_2$, since every $(x_1, x_2) \in G_1 \times G_2$ is the product $(x_1, x_2) = (x_1, 1)(1, x_2)$ of $(x_1, 1) \in \mathrm{Ker}\,\pi$ and $(1, x_2) \in \mathrm{Im}\,\iota$.

If now $\theta : G_1 \rtimes G_2 \longrightarrow G$ is an isomorphism, then $A = \theta\,(\mathrm{Ker}\,\pi)$ and $B = \theta\,(\mathrm{Im}\,\iota)$ are subgroups of G, $A \cong \mathrm{Ker}\,\pi \cong G_1$, $B \cong \mathrm{Im}\,\iota \cong G_2$, $A \trianglelefteq G$, $A \cap B = 1$, and $AB = G$.

Conversely, assume that $A \trianglelefteq G$, $B \leqq G$, $A \cap B = 1$, and $AB = G$. Every element g of G can then be written uniquely as a product $g = ab$ for some $a \in A$ and $b \in B$: if $ab = a'b'$, with $a, a' \in A$ and $b, b' \in B$, then

$a'^{-1}a = b'b^{-1} \in A \cap B$ yields $a'^{-1}a = b'b^{-1} = 1$ and $a = a'$, $b = b'$. Hence the mapping $\theta : (a, b) \longmapsto ab$ of $A \times B$ onto G is a bijection.

Like G, B acts on A by inner automorphisms: $^b a = bab^{-1}$. Products ab then multiply as follows: $(ab)(a'b') = a\, ba'b^{-1}\, bb' = a\, {}^b a'\, bb'$, and $\theta : (a, b) \longmapsto ab$ is an isomorphism of $A \rtimes B$ onto G. $\square$

Exercises

1. Verify that $A \rtimes_\varphi B$ is a group.

2. Show that D_n is a semidirect product of a cyclic group of order n by a cyclic group of order 2, which is not a direct product if $n > 2$.

In the following exercises C_n is cyclic of order n.

3. Find $\mathrm{Aut}\,(C_3)$.

4. Find $\mathrm{Aut}\,(C_4)$.

5. Find $\mathrm{Aut}\,(C_5)$.

6. Show that $\mathrm{Aut}\,(C_p)$ is cyclic of order $p - 1$ when p is prime.

7. Find all semidirect products of C_4 by C_2.

8. Find all semidirect products of C_3 by C_4.

9. Given presentations of A and B, set up a presentation of $A \rtimes_\varphi B$.

10. Prove the following: for any group G there exists a group H such that $G \trianglelefteq H$ and every automorphism of G is induced by an inner automorphism of H.

Nonabelian groups in the following exercises should be specified by presentations.

11. Find all groups of order 21.

12. Find all groups of order 39.

13. Find all groups of order 55.

14. Find all groups of order 57.

15. Find all groups of order 20.

16. Find all groups of order 28.

17. Find all groups of order 44.

18. Find all groups of order 52.

In the remaining exercises, $p > 2$ is a prime number.

19. Show that $C_p \oplus C_p$ has $(p^2 - 1)(p^2 - p)$ automorphisms. (Hint: $C_p \oplus C_p$ is also a vector space over $\mathbb{Z}_p$.)

20. Construct all automorphisms α of $C_p \oplus C_p$ such that $\alpha^2 = 1$.

21. Construct all groups of order $2p^2$. (Give presentations.)

12. Group Extensions

Group extensions are more complex than semidirect products but also more general, since they require only one normal subgroup. They lead to the beautiful results at the end of this section, and to the Hall theorems in Section 9.

Definition. Informally, a group extension of a group G by a group Q is a group E with a normal subgroup N such that $G \cong N$ and $E/N \cong Q$. Composing these isomorphisms with the inclusion homomorphism $N \longrightarrow E$ and canonical projection $E \longrightarrow E/N$ yields an *injection* $G \longrightarrow E$ whose image is N and a *projection* $E \longrightarrow Q$ whose kernel is N. Our formal definition of group extensions includes these homomorphisms:

Definition. A group extension $G \xrightarrow{\kappa} E \xrightarrow{\rho} Q$ of a group G by a group Q consists of a group E, an injective homomorphism $\kappa : G \longrightarrow E$, and a surjective homomorphism $\rho : E \longrightarrow Q$, such that $\mathrm{Im}\,\kappa = \mathrm{Ker}\,\rho$. □

Then $N = \mathrm{Im}\,\kappa = \mathrm{Ker}\,\rho$ is a normal subgroup of E, $G \cong N$, and $E/N \cong Q$. For example, every group E with a normal subgroup N is an extension of N by E/N; every semidirect product $A \rtimes B$ of groups is an extension of A by B, with injection $a \longmapsto (a, 1)$ and projection $(a, b) \longmapsto b$.

Group extensions need be considered only up to isomorphism, more precisely, up to isomorphisms that respect injection and projection, or *equivalences*:

Definitions. An equivalence of group extensions $G \xrightarrow{\kappa} E \xrightarrow{\rho} Q$ and $G \xrightarrow{\lambda} F \xrightarrow{\sigma} Q$ of G by Q is an isomorphism $\theta : E \longrightarrow F$ such that $\theta \circ \kappa = \lambda$ and $\rho = \sigma \circ \theta$.

$$
\begin{array}{ccc}
G & \xrightarrow{\kappa} E & \xrightarrow{\rho} Q \\
\| & \theta\downarrow & \| \\
G & \xrightarrow{\lambda} F & \xrightarrow{\sigma} Q
\end{array}
$$

Two group extensions E and F are equivalent *when there is an equivalence $E \longrightarrow F$ of group extensions.*

Readers will show that equivalence of group extensions is reflexive, symmetric, and transitive (and would be an equivalence relation if group extensions were allowed to constitute a set).

Schreier's Theorem. Given two groups G and Q, Schreier's theorem constructs all extensions of G by Q. This construction is of theoretical interest, even though it does not lend itself to computing examples by hand. In case $N \trianglelefteq E$, Schreier's construction is based on the arbitrary selection of one element in each coset of N; this creates a bijection of $N \times E/N$ onto E.

Definition. A cross-section of a group extension $G \xrightarrow{\kappa} E \xrightarrow{\rho} Q$ is a family $p = (p_a)_{a \in Q}$ such that $\rho(p_a) = a$ for all $a \in Q$.

"Cross-section" is the author's terminology; various other names are in use.

In the above, the inverse image $\rho^{-1}(a)$ of any $a \in Q$ is a coset of $N = \operatorname{Im} \kappa = \operatorname{Ker} \rho$; thus, a cross-section of E selects one element in every coset of N, and is actually a cross-section of the partition of E into cosets of N. Every element of E belongs to only one coset and can now be written uniquely in the form np_a with $n \in N$ and $a \in Q$ (then $a = \rho(np_a)$):

Lemma 12.1. *Let* $G \xrightarrow{\kappa} E \xrightarrow{\rho} Q$ *be a group extension and let* p *be a cross-section of* E. *Every element of* E *can be written in the form* $\kappa(x)\, p_a$ *for some unique* $x \in G$ *and* $a \in Q$ (*then* $a = \rho(\kappa(x)\, p_a)$).

Lemma 12.1 provides a bijection $(x, a) \longmapsto \kappa(x)\, p_a$ of $G \times Q$ onto E. Now we put every product $\kappa(x)\, p_a\, \kappa(y)\, p_b$ in the form $\kappa(z)\, p_c$. We start with the simpler products $p_a\, \kappa(y)$ and $p_a\, p_b$.

Definitions. Let $G \xrightarrow{\kappa} E \xrightarrow{\rho} Q$ *be a group extension, and let* p *be a cross-section of* E. *The set action* $(a, x) \longmapsto {}^a x$ *of the set* Q *on the group* G *relative to* p *is defined by*

$$p_a\, \kappa(x) = \kappa({}^a x)\, p_a.$$

The factor set $s = (s_{a,b})_{a,b \in Q}$ *of* E *relative to* p *is defined by*

$$p_a\, p_b = \kappa(s_{a,b})\, p_{ab}.$$

Since $\rho(p_a\, \kappa(x)) = a$ and $\rho(p_a\, p_b) = ab$, it follows from Lemma 12.1 that $p_a\, \kappa(x) = \kappa({}^a x)\, p_a$ and $p_a\, p_b = \kappa(s_{a,b})\, p_{ab}$ for some unique ${}^a x,\, s_{a,b} \in G$. Thus the definitions above make sense. "Set action" is the author's terminology for the action of Q on A, which, sadly, is usually not a group action.

With the set action and factor set we can compute

$$(\kappa(x)\, p_a)\, (\kappa(y)\, p_b) = \kappa(x)\, \kappa({}^a y)\, p_a\, p_b = \kappa(x\, {}^a y\, s_{a,b})\, p_{ab}.$$

This suggests the operation

$$(x, a)\, (y, b) = (x\, {}^a y\, s_{a,b},\, ab) \tag{M}$$

on the set $G \times Q$; then $E \cong G \times Q$; in particular, $G \times Q$ is a group.

Now we determine when (M) makes $G \times Q$ a group extension of G by Q.

Lemma 12.2. *Relative to any cross-section,* Q *acts on* G *by automorphisms; in particular,* ${}^a(xy) = {}^a x\, {}^a y$ *and* ${}^a 1 = 1$ *for all* $x, y \in G$ *and* $a \in Q$. *Moreover, the cross-section* p *can be chosen so that* $p_1 = 1$, *and then*

$${}^1 x = x \quad \text{and} \quad s_{a,1} = 1 = s_{1,a} \tag{N}$$

for all $x \in G$ *and* $a \in Q$.

This is straightforward. The automorphism $x \longmapsto {}^a x$ is induced on G, via κ, by the inner automorphism $x \longmapsto p_a\, x\, p_a^{-1}$, which has a restriction to $\operatorname{Im} \kappa$; (N) is the *normalization* condition.

Lemma 12.3. *If* (N) *holds, then* (M) *is associative if and only if*

$$^{a}(^{b}x)\, s_{a,b} = s_{a,b}\,^{ab}x \quad \text{and} \quad s_{a,b}\, s_{ab,c} = {}^{a}s_{b,c}\, s_{a,bc}, \tag{A}$$

for all $x \in G$ *and* $a, b, c \in Q$.

Condition (A) is the *associativity* condition; the first part of (A) shows that the set action of Q on G is a group action only up to inner automorphisms.

Proof. By (M) and 12.2,

$$
\begin{aligned}
((x,a)(y,b))(z,c) &= (x\,^{a}y\, s_{a,b}, ab)(z,c) \\
&= (x\,^{a}y\, s_{a,b}\,^{ab}z\, s_{ab,c}, (ab)c), \\
(x,a)((y,b)(z,c)) &= (x,a)(y\,^{b}z\, s_{b,c}, bc) \\
&= (x\,^{a}(y\,^{b}z\, s_{b,c})\, s_{a,bc}, a(bc)) \\
&= (x\,^{a}y\,^{a}(^{b}z)\,^{a}s_{b,c}\, s_{a,bc}, a(bc)).
\end{aligned}
$$

Hence (M) is associative if and only if

$$x\,^{a}y\, s_{a,b}\,^{ab}z\, s_{ab,c} = x\,^{a}y\,^{a}(^{b}z)\,^{a}s_{b,c}\, s_{a,bc} \tag{A*}$$

holds for all $x, y, z \in G$ and $a, b, c \in Q$. With $x = y = z = 1$, (A^{*}) yields $s_{a,b}\, s_{ab,c} = {}^{a}s_{b,c}\, s_{a,bc}$. With $x = y = 1$ and $c = 1$, (A^{*}) yields $s_{a,b}\,^{ab}z = {}^{a}(^{b}z)\, s_{a,b}$, since $s_{ab,1} = {}^{a}s_{b,1} = 1$ by 12.2. Thus (A^{*}) implies (A). Conversely, (A) implies the following equalities and implies (A^{*}):

$$x\,^{a}y\, s_{a,b}\,^{ab}z\, s_{ab,c} = x\,^{a}y\,^{a}(^{b}z)\, s_{a,b}\, s_{ab,c} = x\,^{a}y\,^{a}(^{b}z)\,^{a}s_{b,c}\, s_{a,bc}. \qquad \square$$

In what follows we denote by φ the mapping $a \longmapsto \varphi(a)$, $\varphi(a): x \longmapsto {}^{a}x$ of Q into $\mathrm{Aut}\,(G)$, which encapsulates the set action of Q on G; then ${}^{a}x$ can be denoted by ${}^{a}_{\varphi}x$ to avoid ambiguity.

Theorem 12.4 (Schreier [1926]). *Let* G *and* Q *be groups, and let* $s : Q \times Q \longrightarrow G$ *and* $\varphi : Q \longrightarrow \mathrm{Aut}\,(G)$ *be mappings such that* (N) *and* (A) *hold. Then* $E(s, \varphi) = G \times Q$ *with multiplication* (M), *injection* $x \longmapsto (x, 1)$, *and projection* $(x, a) \longmapsto a$, *is a group extension of* G *by* Q. *Conversely, if* E *is a group extension of* G *by* Q, *and* s, φ *are the factor set and set action of* E *relative to a cross-section of* E, *then* E *is equivalent to* $E(s, \varphi)$.

Proof. If s and φ satisfy (N) and (A), then (M) is associative by 12.3, and $(1,1)$ is an identity element of $E(s, \varphi)$. Moreover, every element (y, b) of $E(s, \varphi)$ has a left inverse: if $q = b^{-1}$ and $x = ({}^{a}y\, s_{a,b})^{-1}$, then (M) yields $(x, a)(y, b) = (1, 1)$. Therefore $E(s, \varphi)$ is a group. By (M) and (N), $\lambda : x \longmapsto (x, 1)$ and $\sigma : (x, a) \longmapsto a$ are homomorphisms, and we see that $\mathrm{Im}\,\lambda = \mathrm{Ker}\,\sigma$. Thus $E(s, \varphi)$ is a group extension of G by Q.

Conversely, let $G \xrightarrow{\kappa} E \xrightarrow{\rho} Q$ be a group extension of G by Q. Choose a cross-section p of E such that $p_1 = 1$ and let φ and s be the corresponding set action and factor set. We saw that $\theta : (x, a) \longmapsto \kappa(x)\, p_a$ is an isomorphism of $E(s, \varphi)$ onto E. Moreover, (N) and (A) hold, by 12.2 and 12.3. Finally, $\theta\,(\lambda(x)) = \kappa(x)\, p_1 = \kappa(x)$ and $\rho\,(\theta(x, a)) = a = \sigma\,(x, a)$, for all $x \in G$ and $a \in Q$. Thus E is equivalent to $E(s, \varphi)$: $\square$

$$
\begin{array}{ccccc}
G & \xrightarrow{\ \lambda\ } & E(s, \varphi) & \xrightarrow{\ \sigma\ } & Q \\[2pt]
\| & & \theta\downarrow & & \| \\[2pt]
G & \xrightarrow{\ \kappa\ } & E & \xrightarrow{\ \rho\ } & Q
\end{array}
$$

Equivalence. We complete Theorem 12.4 with a criterion for equivalence.

Proposition **12.5.** $E(s, \varphi)$ *and* $E(t, \psi)$ *are equivalent if and only if there exists a mapping* $u : a \longmapsto u_a$ *of* Q *into* G *such that*

$$
u_1 = 1, \quad {}^a_{\varphi}x = u_a\, {}^a_{\psi}x\, u_a^{-1}, \quad \text{and} \quad s_{a,b} = u_a\, {}^a_{\psi}u_b\, t_{a,b}\, u_{ab}^{-1}, \tag{E}
$$

for all $x \in G$ *and* $a, b \in Q$.

Proof. Let $\theta : E(s, \varphi) \longrightarrow E(t, \psi)$ be an equivalence of group extensions:

$$
\begin{array}{ccccc}
G & \longrightarrow & E(s, \varphi) & \longrightarrow & Q \\[2pt]
\| & & \theta\downarrow & & \| \\[2pt]
G & \longrightarrow & E(t, \psi) & \longrightarrow & Q
\end{array}
$$

We have $\theta\,(x, 1) = (x, 1)$ and $\theta\,(1, a) = (u_a,\, a)$ for some $u_a \in G$, since θ respects injections and projections. Then $u_1 = 1$, since $\theta\,(1, 1) = (1, 1)$, and

$$
\theta\,(x, a) = \theta\big((x, 1)(1, a)\big) = (x, 1)(u_a,\, a) = (xu_a,\, a)
$$

by (N). Since θ is a homomorphism,

$$
(x\, {}^a_{\varphi}y\, s_{a,b}\, u_{ab},\, ab) = \theta\big((x, a)(y, b)\big)
$$
$$
= \theta\,(x, a)\,\theta\,(y, b) = (xu_a\, {}^a_{\psi}y\, {}^a_{\psi}u_b\, t_{a,b},\, ab)
$$

by (M), and

$$
x\, {}^a_{\varphi}y\, s_{a,b}\, u_{ab} = xu_a\, {}^a_{\psi}y\, {}^a_{\psi}u_b\, t_{a,b} \tag{E*}
$$

for all x, y, a, b. With $x = y = 1$, (E^*) yields $s_{a,b} = u_a\, {}^a_{\psi}u_b\, t_{a,b}\, u_{ab}^{-1}$. Hence

$$
x\, {}^a_{\varphi}y\, u_a\, {}^a_{\psi}u_b\, t_{a,b} = x\, {}^a_{\varphi}y\, s_{a,b}\, u_{ab} = xu_a\, {}^a_{\psi}y\, {}^a_{\psi}u_b\, t_{a,b}
$$

and ${}^a_{\varphi}y = u_a\, {}^a_{\psi}y\, u_a^{-1}$. Thus (E) holds.

Conversely, (E) implies (E^*), by the same calculation. Then $\theta : (x, a) \longmapsto (xu_a,\, a)$ is a homomorphism, and, clearly, an equivalence of group extensions. $\square$

Split extensions.

Proposition **12.6.** *For a group extension* $G \xrightarrow{\kappa} E \xrightarrow{\rho} Q$ *the following conditions are equivalent:*

(1) *There exists a homomorphism* $\mu : Q \longrightarrow E$ *such that* $\rho \circ \mu = 1_Q$.

(2) *There is a cross-section of* E *relative to which* $s_{a,b} = 1$ *for all* $a, b \in Q$.

(3) E *is equivalent to a semidirect product of* G *by* Q.

(4) *Relative to any cross-section of* E *there exists a mapping* $u : a \longmapsto u_a$ *of* Q *into* E *such that* $u_1 = 1$ *and* $s_{a,b} = {}^a u_b \, u_a \, u_{ab}^{-1}$ *for all* $a, b \in Q$.

A group extension *splits* when it satisfies these conditions.

Proof. (1) implies (2). If (1) holds, then $p_a = \mu(a)$ is a cross-section of E, relative to which $s_{a,b} = 1$ for all a, b, since $\mu(a) \, \mu(b) = \mu(ab)$.

(2) implies (3). If $s_{a,b} = 1$ for all a, b, then $\varphi : Q \longrightarrow \mathrm{Aut}\,(G)$ is a homomorphism, by (A), and (M) shows that $E(s, \varphi) = G \rtimes_\varphi Q$. Then E is equivalent to $E(s, \varphi)$, by Schreier's theorem.

(3) implies (4). A semidirect product $G \rtimes_\psi Q$ of G by Q is a group extension $E(t, \psi)$ in which $t_{a,b} = 1$ for all a, b. If E is equivalent to $G \rtimes_\psi Q$, then, relative to any cross-section of E, $E(s, \varphi)$ and $E(t, \psi)$ are equivalent, and (E) yields $s_{a,b} = u_a \, {}^a_\psi u_b \, t_{a,b} \, u_{ab}^{-1} = {}^a_\varphi u_b \, u_a \, u_{ab}^{-1}$ for all $a, b \in Q$.

(4) implies (1). If $s_{a,b} = {}^a u_b \, u_a \, u_{ab}^{-1}$ for all $a, b \in Q$, then $u_a^{-1} \, {}^a(u_b^{-1}) \, s_{a,b} = u_{ab}^{-1}$ and $\mu : a \longmapsto \kappa(u_a^{-1}) \, p_a$ is a homomorphism, since

$$\mu(a) \, \mu(b) = \kappa\big(u_a^{-1} \, {}^a(u_b^{-1}) \, s_{a,b}\big) \, p_{ab} = \kappa(u_{ab}^{-1}) \, p_{ab} = \mu(ab). \ \square$$

Extensions of abelian groups. Schreier's theorem becomes much nicer if G is abelian. Then (A) implies ${}^a({}^b x) = {}^{ab} x$ for all a, b, x, so that the set action of Q on G is a group action. Equivalently, $\varphi : Q \longrightarrow \mathrm{Aut}\,(G)$ is a homomorphism. Theorem 12.4 then simplifies as follows.

Corollary **12.7.** *Let* G *be an abelian group, let* Q *be a group, let* $s : Q \times Q \longrightarrow G$ *be a mapping, and let* $\varphi : Q \longrightarrow \mathrm{Aut}\,(G)$ *be a homomorphism, such that*

$$s_{a,1} = 1 = s_{1,a} \quad and \quad s_{a,b} \, s_{ab,c} = {}^a s_{b,c} \, s_{a,bc}$$

for all $a, b, c \in Q$. *Then* $E(s, \varphi) = G \times Q$ *with multiplication* (M), *injection* $x \longmapsto (x, 1)$, *and projection* $(x, a) \longmapsto a$ *is a group extension of* G *by* Q. *Conversely, every group extension* E *of* G *by* Q *is equivalent to some* $E(s, \varphi)$.

If G is abelian, then condition (F) implies ${}^a_\varphi x = {}^a_\psi x$ for all a and x, so that $\varphi = \psi$. Thus, equivalent extensions share the same action, and Proposition 12.5 simplifies as follows.

Corollary **12.8.** *If* G *is abelian, then* $E(s, \varphi)$ *and* $E(t, \psi)$ *are equivalent if and only if* $\varphi = \psi$ *and there exists a mapping* $u : a \longmapsto u_a$ *of* Q *into* G *such* *that*

$$u_1 = 1 \quad and \quad s_{a,b} = u_a \, {}^a u_b \, u_{ab}^{-1} \, t_{a,b} \quad for \ all \ a, b \in Q.$$

Corollaries 12.7 and 12.8 yield an abelian group whose elements are essentially the equivalence classes of group extensions of G by Q with a given action φ. Two factor sets s and t are *equivalent* when condition (E) holds. If G is abelian, then factor sets can be multiplied pointwise: $(s \cdot t)_{a,b} = s_{a,b} \, t_{a,b}$, and the result is again a factor set, by 12.7. Under pointwise multiplication, factor sets $s : Q \times Q \longrightarrow G$ then constitute an abelian group $Z_\varphi^2(Q, G)$. *Split* factor sets (factor sets $s_{a,b} = u_a \, {}^a u_b \, u_{ab}^{-1}$ with $u_1 = 1$) constitute a subgroup $B_\varphi^2(Q, G)$ of $Z_\varphi^2(Q, G)$. By 12.8, two factor sets are equivalent if and only if they lie in the same coset of $B_\varphi^2(Q, G)$; hence equivalence classes of factor sets constitute an abelian group $H_\varphi^2(Q, G) = Z_\varphi^2(Q, G) / B_\varphi^2(Q, G)$, the *second cohomology group* of Q with coefficients in G. (The cohomology of groups is defined in full generality in Section XII.7; it has become a major tool of group theory.)

The abelian group $H_\varphi^2(Q, G)$ *classifies* extensions of G by Q, meaning that there is a one-to-one correspondence between elements of $H_\varphi^2(Q, G)$ and equivalence classes of extensions of G by Q with the action φ. (These equivalence classes would constitute an abelian group if they were sets and could be allowed to belong to sets.)

Hölder's Theorem. As a first application of Schreier's theorem we find all extensions of one cyclic group by another.

Theorem **12.9** (Hölder). *A group G is an extension of a cyclic group of order m by a cyclic group of order n if and only if G is generated by two elements a and b such that a has order m, $b^n = a^t$, $b^i \notin \langle a \rangle$ when $0 < i < n$, and $bab^{-1} = a^r$, where $r^n \equiv 1$ and $rt \equiv t \pmod{m}$. Such a group exists for every choice of integers r, t with these properties.*

Proof. First let $G = \langle a, b \rangle$, where a has order m, $b^n = a^t$, $b^i \notin \langle a \rangle$ when $0 < i < n$, and $bab^{-1} = a^r$, where $r^n \equiv 1$ and $rt \equiv 1 \pmod{m}$. Then $A = \langle a \rangle$ is cyclic of order m. Since b has finite order, every element of G is a product of a's and b's, and it follows from $bab^{-1} = a^r$ that $A \trianglelefteq G$. Then G/A is generated by Ab; since $b^n \in A$ but $b^i \notin A$ when $0 < i < n$, Ab has order n in G/A, and G/A is cyclic of order n. Thus G is an extension of a cyclic group of order m by a cyclic group of order n.

Conversely, assume that G is an extension of a cyclic group of order m by a cyclic group of order n. Then G has a normal subgroup A that is cyclic of order m, such that G/A is cyclic of order n. Let $A = \langle a \rangle$ and $G/A = \langle Ab \rangle$, where $a, b \in G$. The elements of G/A are A, Ab, ..., Ab^{n-1}; therefore G is generated by a and b. Moreover, a has order m, $b^n = a^t$ for some t, $b^i \notin \langle a \rangle$ when $0 < i < n$, and $bab^{-1} = a^r$ for some r, since $A \trianglelefteq G$. Then

$a^{rt} = ba^t b^{-1} = bb^n b^{-1} = a^t$ and $rt \equiv t \pmod{m}$. Also $b^2 ab^{-2} = ba^r b^{-1} = (a^r)^r = a^{r^2}$ and, by induction, $b^k ab^{-k} = a^{r^k}$; hence $a = b^n ab^{-n} = a^{r^n}$ and $r^n \equiv 1 \pmod{m}$.

In the above, $1, b, \ldots, b^{n-1}$ is a cross-section of G. The corresponding action is $^{Ab^j} a^i = b^j a^i b^{-j} = a^{ir^j}$. If $0 \leq i, j < n$, then $b^j b^k = b^{j+k}$ if $j + k < n$, $b^j b^k = a^t b^{j+k-n}$ if $j + k \geq n$; this yields the corresponding factor set. This suggests a construction of G for any suitable m, n, r, t.

Assume that $m, n > 0$, $r^n \equiv 1$, and $rt \equiv t \pmod{m}$. Let $A = \langle a \rangle$ be cyclic of order m and let $C = \langle c \rangle$ be cyclic of order n. Since $r^n \equiv 1 \pmod{m}$, r and m are relatively prime and $\alpha : a^i \longmapsto a^{ir}$ is an automorphism of A. Also, $\alpha^j(a^i) = a^{ir^j}$ for all j; in particular, $\alpha^n(a^i) = a^{ir^n} = a^i$. Hence $\alpha^n = 1_A$ and there is a homomorphism $\varphi : C \longrightarrow \mathrm{Aut}(A)$ such that $\varphi(c) = \alpha$. The action of C on A, written $^j a^i = {}^{c^j}_\varphi a^i$, is $^j a^i = \alpha^j(a^i) = a^{ir^j}$.

Define $s : C \times C \longrightarrow A$ as follows: for all $0 \leq j, k < n$,

$$s_{j,k} = s_{c^j, c^k} = \begin{cases} 1 & \text{if } j + k < n, \\ a^t & \text{if } j + k \geq n. \end{cases}$$

Then $s_{1, c^k} = 1 = s_{c^j, 1}$. We show that $s_{c^j, c^k} s_{c^j c^k, c^\ell} = {}^{c^j} s_{c^k, c^\ell} s_{c^j, c^k c^\ell}$ (equivalently, $s_{j,k} s_{c^j c^k, \ell} = {}^j s_{k, \ell} s_{j, c^k c^\ell}$) for all $0 \leq j, k, \ell < n$.

If $j + k + \ell < n$, then $s_{j,k} s_{j+k, \ell} = 1 = {}^j s_{k, \ell} s_{j, k+\ell}$.

If $j + k < n$, $k + \ell < n$, and $j + k + \ell \geq n$, then $s_{j,k} s_{j+k, \ell} = a^t = {}^j s_{k, \ell} s_{j, k+\ell}$.

Since $rt \equiv t \pmod{m}$, we have $^j a^t = a^{tr^j} = a^t$. If now $j + k < n$ and $k + \ell \geq n$, then $c^k c^\ell = c^{k+\ell-n}$, $j + k + \ell - n < n$, since $(j + k) + \ell < 2n$, and $s_{j,k} s_{j+k, \ell} = a^t = {}^j a^t = {}^j s_{k, \ell} s_{j, k+\ell-n}$.

If $j + k \geq n$ and $k + \ell < n$, then similarly $c^j c^k = c^{j+k-n}$, $j + k + \ell - n < n$, and $s_{j,k} s_{j+k-n, \ell} = a^t = {}^j s_{k, \ell} s_{j, k+\ell}$.

If $j + k \geq n$, $k + \ell \geq n$, and $j + k + \ell < 2n$, then $j + k + \ell - n < n$ and $s_{j,k} s_{j+k-n, \ell} = a^t = {}^j a^t = {}^j s_{k, \ell} s_{j, k+\ell-n}$.

Finally, if $j + k \geq n$, $k + \ell \geq n$, and $j + k + \ell \geq 2n$, then $j + k + \ell - n \geq n$ and $s_{j,k} s_{j+k-n, \ell} = a^t a^t = {}^j a^t a^t = {}^j s_{k, \ell} s_{j, k+\ell-n}$.

It now follows from 12.7 that $E(s, \varphi)$ is an extension of A by C. $\square$

Readers will verify that $\langle a, b \mid a^m = 1, \ b^n = a^t, \ bab^{-1} = a^r \rangle$ is a presentation of the group G in Theorem 12.9.

The Schur-Zassenhaus Theorem. We begin with the Schur part:

Theorem **12.10** (Schur). *If m and n are relatively prime, then a group of order mn that contains an abelian normal subgroup of order n also contains a subgroup of order m.*

Schur's theorem is often stated as follows: if m and n are relatively prime, then every group extension of an abelian group of order n by a group of order m splits. The two statements are equivalent: if E has a normal subgroup G of order n and a subgroup H of order m, then $G \cap H = 1$, so that $GH = E$, E is a semidirect product of G and H, and E splits, as a group extension of G; conversely, a split extension of a group G by a group Q of order m is isomorphic to a semidirect product of G and Q and contains a subgroup that is isomorphic to Q and has order m.

Proof. Let G be a group of order mn with an abelian normal subgroup N of order n. Then G is an extension of N by a group $Q = G/N$ of order m. Let s be any factor set of this extension. For every $a \in Q$ let $t_a = \prod_{c \in Q} s_{a,c}$. Then $t_1 = 1$, by (N). Since Q is a group, $\prod_{c \in Q} s_{a,bc} = \prod_{d \in Q} s_{a,d} = t_a$. Since N is abelian, applying $\prod_{c \in Q}$ to $s_{a,b} \, s_{ab,c} = {}^a s_{b,c} \, s_{a,bc}$ yields

$$s_{a,b}^m \, t_{ab} = {}^a t_b \, t_a.$$

We also have $s_{a,b}^n = t_a^n = 1$, since $|N| = n$. Now, $qm + rn = 1$ for some $q, r \in \mathbb{Z}$, since m and n are relatively prime; hence

$$s_{a,b} = s_{a,b}^{qm+rn} = {}^a t_b^q \, t_a^q \, (t_{ab}^q)^{-1}$$

and 12.8 (with $u_c = t_c^q$) shows that the extension splits. $\square$

If G is abelian and $|G| = n$, $|Q| = m$ are relatively prime, Schur's theorem implies that every extension of G by Q has a subgroup of order m. Any two such subgroups are conjugate, by 9.11.

Zassenhaus [1937] extended Schur's theorem as follows:

Theorem **12.11** (Schur-Zassenhaus). *If m and n are relatively prime, then a group of order mn that contains a normal subgroup of order n also contains a subgroup of order m.*

If $|G| = mn$ and G has a normal subgroup N of order n and a subgroup H of order m (relatively prime to n), then, as with Schur's theorem, G is a semidirect product of N and H, and the extension G of N splits.

Proof. Let $N \trianglelefteq G$ with $|N| = n$, $|G| = mn$. If N is abelian, then 12.11 follows from Schur's theorem. The general case is proved by induction on n. If $n = 1$, then 12.11 holds, trivially. Now let $n > 1$.

Let p be a prime divisor of n. Then p divides $|G|$. If S is a Sylow p-subgroup of G, then the order of $SN/N \cong S/(S \cap N)$ is a power of p and divides $m = |G/N|$; since m and n are relatively prime, this implies $|SN/N| = 1$ and $S \subseteq N$. Thus N contains every Sylow p-subgroup of G. Hence G and N have the same Sylow p-subgroups.

Since all Sylow p-subgroups are conjugates, G has $[G : N_G(S)]$ Sylow p-subgroups, and N has $[N : N_N(S)]$ Sylow p-subgroups. Hence

$$|G|/|N_G(S)| = [G : N_G(S)] = [N : N_N(S)] = |N|/|N_N(S)|$$

and

$$[N_G(S) : N_N(S)] = |N_G(S)|/|N_N(S)| = |G|/|N| = m.$$

Now, $N_N(S) = N \cap N_G(S) \trianglelefteq N_G(S)$, since $N \trianglelefteq G$, and $S \trianglelefteq N_G(S)$. Hence $N_N(S)/S \trianglelefteq N_G(S)/S$,

$$[N_G(S)/S : N_N(S)/S] = [N_G(S) : N_N(S)] = m,$$

and $|N_N(S)/S|$ is relatively prime to m, since $N_N(S) \leqq N$. Moreover, $|N_N(S)/S| < |N| = n$, since $|S| > 1$. By the induction hypothesis, $N_G(S)/S$ has a subgroup K/S of order m, where $S \subseteq K \leqq N_G(S)$.

By 3.9, the center Z of S is not trivial: $|Z| > 1$. Diligent readers know that Z is a characteristic subgroup of S, so that $S \trianglelefteq K$ implies $Z \trianglelefteq K$. Then $S/Z \trianglelefteq K/Z$, $[K/Z : S/Z] = [K : S] = m$, and $|S/Z|$ is relatively prime to m since it divides $|N| = n$. Moreover, $|S/Z| < n$. By the induction hypothesis, K/Z has a subgroup L/Z of order m, where $Z \subseteq L \leqq K$.

Now, $Z \trianglelefteq L$, $[L : Z] = m$, and $|Z|$ is relatively prime to m since it divides $|S|$ and $|N| = n$. By the abelian case, L contains a subgroup of order m; hence so does G. $\square$

It is known that in Theorem 12.11 all subgroups of order m are conjugate. We proved this in two particular cases: when the normal subgroup is abelian (Lemma 9.11) and when the group is solvable (Theorem 9.12).

Exercises

1. Show that equivalence of group extensions is reflexive, symmetric, and transitive.

2. Find a cross-section of $E(s, \varphi)$ relative to which s is the factor set and φ is the set action.

3. Show that $E(s, \varphi)$ and $E(t, \psi)$ are equivalent if and only if there exist a group extension E and two cross-sections of E relative to which s, φ and t, ψ are the factor set and set action of E.

4. Find all extensions of C_3 by C_2 (up to equivalence).

5. Find all extensions of C_4 by C_2 (up to equivalence).

6. Find all extensions of C_p by C_q (up to equivalence) when p and q are distinct primes.

7. Find all extensions of C_3 by C_4 (up to equivalence).

8. Let the group G be generated by two elements a and b such that a has order m, $b^n = a^t$, $b^i \notin \langle a \rangle$ when $0 < i < n$, and $bab^{-1} = a^r$, where $r^n \equiv 1$ and $rt \equiv t$ (mod m), as in Hölder's theorem. Show that $\langle a, b \mid a^m = 1, b^n = a^t, bab^{-1} = a^r \rangle$ is a presentation of G.

*9. Devise presentations for all groups of order 16.

Nonabelian groups in the following exercises should be specified by presentations.

10. Find all groups of order 30.

11. Find all groups of order 42.

12. Find all groups of order 70.

13. Find all groups of order 105.

III
Rings

Rings are our second major algebraic structure; they marry the complexity of semigroups and the good algebraic properties of abelian groups.

Gauss [1801] studied the arithmetic properties of complex numbers $a + bi$ with $a, b \in \mathbb{Z}$, and of polynomials with integer coefficients. From this start ring theory expanded in three directions. Sustained interest in more general numbers and their properties finally led Dedekind [1871] to state the first formal definition of rings, fields, ideals, and prime ideals, though only for rings and fields of algebraic integers. The quaternions, discovered by Hamilton [1843], were generalized by Pierce [1864] and others into another type of rings: vector spaces with bilinear multiplications (see Chapter XIII). Growing interest in curves and surfaces defined by polynomial equations led Hilbert [1890], [1893] and others to study rings of polynomials. Modern ring theory began in the 1920s with the work of Noether, Artin, and Krull (see Chapters VII and IX).

This chapter contains general properties of rings and polynomials, with some emphasis on arithmetic properties. It requires basic properties of groups and homomorphisms (Sections I.1 through I.5), and makes occasional use of Zorn's lemma and of the ascending chain condition in the appendix. Sections 7, 9, and 12 may be skipped; Section 11 may be covered later (but before Chapter VII).

1. Rings

This section contains the definition and first examples and properties of rings.

Definition. A ring is an ordered triple $(R, +, \cdot)$ *of a set* R *and two binary operations on* R, *an addition and a multiplication, such that*

(1) $(R, +)$ *is an abelian group;*

(2) $(R, \cdot)$ *is a semigroup (the multiplication is associative);*

(3) *the multiplication is distributive:* $x(y + z) = xy + xz$ *and* $(y + z)x = yx + zx$ *for all* $x, y, z \in R$.

Definition. A ring with identity is a ring whose multiplicative semigroup $(R, \cdot)$ *has an identity element.*

The identity element of a ring R with identity $(R, \cdot))$ is generally denoted by 1, whereas the identity element of the underlying abelian group $(R, +)$ is the *zero element* of the ring R and is denoted by 0.

Rings with identity are also called *rings with unity*; many definitions also require $1 \neq 0$. Rings as defined above are often called *associative rings*; *nonassociative rings* have only properties (1) and (3).

Examples. $\mathbb{Z}$ (short for: $(\mathbb{Z}, +, \cdot)$, with the usual addition and multiplication) is a ring with identity; so are $\mathbb{Q}$, $\mathbb{R}$, $\mathbb{C}$, the quaternion algebra $\mathbb{H}$, and the ring $\mathbb{Z}_n$ of integers modulo n (also constructed in the next section).

Polynomials provide major examples, which we study in Sections 5 and 6.

In a vector space V over a field K, the *endomorphisms* of V (the linear transformations $V \longrightarrow V$) constitute a ring with identity $\text{End}_K(V)$, whose addition is pointwise and multiplication is composition. A related example is the ring $M_n(K)$ of $n \times n$ matrices over K, with the usual matrix addition and multiplication.

Proposition **1.1.** *The set* $\text{End}(A)$ *of all endomorphisms of an abelian group A is a ring with identity.*

In the additive notation for A, addition on $\text{End}(A)$ is pointwise $((\eta + \zeta)x = \eta x + \zeta x)$; multiplication on $\text{End}(A)$ is composition $((\eta \zeta)x = \eta(\zeta x))$. Readers will cheerfully verify properties (1), (2), and (3).

Properties. Calculations in rings follow the familiar rules for addition, subtraction, and multiplication of numbers, except that multiplication in a ring might not be commutative, and there is in general no division.

Sections I.1 and I.2 provide basic properties of sums, opposites, products, integer multiples, and powers. Properties that are specific to rings come from distributivity. Readers will happily supply proofs (sometimes, not so happily).

Proposition **1.2.** *In a ring R,* $\left(\sum_i x_i\right)\left(\sum_j y_j\right) = \sum_{i,j} x_i y_j$, *for all* $x_1, \ldots, x_m,\, y_1, \ldots, y_n \in R$.

Proposition **1.3.** *In a ring R,* $(mx)(ny) = (mn)(xy)$, *in particular,* $(mx)y = x(my) = m(xy)$, *for all $m, n \in \mathbb{Z}$ and $x, y \in R$. If R is a ring with identity, then $nx = (n1)x$ for all $n \in \mathbb{Z}$ and $x \in R$.*

Subtraction may be defined in any abelian group by $x - y = x + (-y)$, and it satisfies $x - x = 0$ and $x - (y - z) = (x - y) + z$ for all x, y, z.

Proposition **1.4.** *In a ring R,* $x(y - z) = xy - xz$ *and* $(y - z)x = yx - zx$, *for all $x, y, z \in R$. In particular,* $x0 = 0 = 0x$ *for all $x \in R$.*

Definition. A ring is commutative *when its multiplication is commutative.*

The familiar rules for addition, subtraction, and multiplication of numbers hold in commutative rings. So does the following result:

Proposition **1.5** (Binomial Theorem). *In a commutative ring* R,

$$(x + y)^n = \sum_{0 \leq i \leq n} \binom{n}{i} x^i \, y^{n-i}, \text{ where } \binom{n}{i} = \frac{n!}{i! \, (n - i)!}.$$

In fact, 1.5 works in every ring, as long as $xy = yx$.

Infinite sums are defined in any abelian group as follows. Without a topology we don't have limits of finite sums, so our "infinite" sums are not really infinite.

Definition. A property P holds for almost all *elements i of a set I when* $\{ i \in I \mid P \text{ does not hold} \}$ *is finite.*

Definition. The sum $\sum_{i \in I} x_i$ *of elements* $(x_i)_{i \in I}$ *of an abelian group A is defined in A when* $x_i = 0$ *for almost all $i \in I$, and then* $\sum_{i \in I} x_i = \sum_{i \in I, \, x_i \neq 0} x_i$.

Thus the "arbitrary" sum $\sum_{i \in I} x_i$ is a finite sum to which any number of zeros have been added, a fine example of window dressing.

Proposition **1.6.** *In a ring,* $\left(\sum_i x_i \right) \left(\sum_j y_j \right) = \sum_{i,j} x_i \, y_j$, *whenever* $x_i = 0$ *for almost all $i \in I$ and $y_j = 0$ for almost all $j \in J$.*

Homomorphisms. Homomorphisms of rings are mappings that preserve sums and products:

Definitions. A homomorphism *of a ring R into a ring S is a mapping φ of R into S that preserves sums and products:* $\varphi(x + y) = \varphi(x) + \varphi(y)$ *and* $\varphi(xy) = \varphi(x) \, \varphi(y)$ *for all $x, y \in R$.*

If R and S are rings with identity, a homomorphism *of rings with identity of R into S also preserves the identity element:* $\varphi(1) = 1$.

For example, in any ring R with identity, the mapping $n \longmapsto n1$ is a homomorphism of rings with identity of $\mathbb{Z}$ into R; this follows from I.2.6 and 1.3.

A homomorphism of rings also preserves the zero element ($\varphi(0) = 0$), integer multiples ($\varphi(nx) = n \, \varphi(x)$), all sums and products (including infinite sums), differences, and powers ($\varphi(x^n) = \varphi(x)^n$).

Homomorphisms compose: when $\varphi : R \longrightarrow S$ and $\psi : S \longrightarrow T$ are homomorphisms of rings, then so is $\psi \circ \varphi : R \longrightarrow T$. Moreover, the identity mapping 1_R on a ring R is a homomorphism. Homomorphisms of rings with identity have similar properties.

It is common practice to call an injective homomorphism a *monomorphism*, and a surjective homomorphism an *epimorphism*. In the case of epimorphisms of rings the author finds this terminology illegitimate and prefers to avoid it.

Definition. An isomorphism *of rings is a bijective homomorphism of rings.*

If φ is a bijective homomorphism of rings, then the inverse bijection φ^{-1} is also a homomorphism of rings. Two rings R and S are *isomorphic,* $R \cong S$, when there exists an isomorphism of R onto S. As in Section I.2, we regard isomorphic rings as instances of the same "abstract" ring.

Adjoining an identity. Homomorphisms of rings are studied in more detail in Section 3. We consider them here to show that every ring R can be embedded into a ring with identity. The new ring must contain an identity element 1, all its integer multiples $n1$, and all sums $x + n1$ with $x \in R$. The next result basically says that these sums suffice.

Proposition 1.7. *For every ring R, the set $R^1 = R \times \mathbb{Z}$, with operations*

$$(x, m) + (y, n) = (x + y, \, m + n), \quad (x, m)(y, n) = (xy + nx + my, \, mn),$$

is a ring with identity. Moreover, $\iota : x \longmapsto (x, 0)$ is an injective homomorphism of R into R^1.

The proof is straightforward but no fun, and left to our poor, abused readers.

The ring R^1 has a universal property, which will be useful in Chapter VIII.

Proposition 1.8. *Every homomorphism φ of R into a ring S with identity factors uniquely through $\iota : R \longrightarrow R^1$ (there is a homomorphism $\psi : R^1 \longrightarrow S$ of rings with identity, unique such that $\varphi = \psi \circ \iota$).*

$$
\begin{array}{ccc}
R & \overset{\iota}{\longrightarrow} & R^1 \\
& {\scriptstyle \varphi} \searrow & \downarrow {\scriptstyle \psi} \\
& & S
\end{array}
$$

Proof. In R^1, the identity element is $(0, 1)$ and $(x, n) = (x, 0) + n\,(0, 1)$. If now $\psi(0, 1) = 1$ and $\psi \circ \iota = \varphi$, then necessarily $\psi\,(x, n) = \varphi(x) + n1 \in S$; hence ψ is unique. Conversely, it is straightforward that the mapping $\psi : (x, n) \longmapsto \varphi(x) + n1$ is a homomorphism with all required properties. $\square$

If now φ is a homomorphism of R into an arbitrary ring S, then applying Proposition 1.8 to $R \overset{\varphi}{\longrightarrow} S \longrightarrow S^1$ yields a homomorphism $\psi : R^1 \longrightarrow S^1$ of rings with identity; in this sense every homomorphism of rings is induced by a homomorphism of rings with identity.

Some properties are lost in the embedding of R into R^1 (see the exercises), but in most situations an identity element may be assumed, for instance when one studies rings and their homomorphisms *in general*. We make this assumption in all later sections. The important examples of rings at the beginning of this section all have identity elements.

Exercises

1. In the definition of a ring with identity, show that one may omit the requirement that the addition be commutative. [Assume that $(R, +, \cdot)$ satisfies (2), (3), that $(R, \cdot)$ has an identity element, and that $(R, +)$ is a group. Show that $(R, +)$ is abelian.]

2. Verify that $\mathrm{End}\,(A)$ is a ring when A is an abelian group.

3. A *unit* of a ring R with identity is an element u of R such that $uv = vu = 1$ for some $v \in R$. Show that v is unique (given u). Show that the set of all units of R is a group under multiplication.

4. Let R be a ring with identity. Show that u is a unit of R if and only if $xu = uy = 1$ for some $x, y \in R$.

5. Show that $\overline{x} \in \mathbb{Z}_n$ is a unit of $\mathbb{Z}_n$ if and only if x and n are relatively prime.

6. Prove that $x^{\phi(n)} \equiv 1 \pmod{n}$ whenever x and n are relatively prime. (ϕ is Euler's ϕ function.)

7. A *Gauss integer* is a complex number $a + ib$ in which a and b are integers. Show that Gauss integers constitute a ring. Find the units.

8. Show that complex numbers $a + ib\sqrt{2}$ in which a and b are integers constitute a ring. Find the units.

9. Show that $\left(\sum_{i \in I} x_i\right) + \left(\sum_{i \in I} y_i\right) = \sum_{i \in I} (x_i + y_i)$ holds in every ring, when $x_i = 0$ for almost all $i \in I$ and $y_i = 0$ for almost all $i \in I$.

10. Show that $\left(\sum_{i \in I} x_i\right)\left(\sum_{j \in J} y_j\right) = \sum_{(i,j) \in I \times J} x_i y_j$ holds in every ring, when $x_i = 0$ for almost all $i \in I$ and $y_j = 0$ for almost all $j \in J$.

11. Let R be a ring. Show that $R^1 = R \times \mathbb{Z}$, with operations

$$(x, m) + (y, n) = (x + y, \; m + n), \quad (x, m)(y, n) = (xy + nx + my, \; mn),$$

is a ring with identity.

12. A ring R is *regular* (also called *von Neumann regular*) when there is for every $a \in R$ some $x \in R$ such that $axa = a$. Prove that R^1 can never be regular.

13. Let R be a ring with identity. Show that R can be embedded into $\operatorname{End}(R, +)$. (Hence every ring can be embedded into the endomorphism ring of an abelian group.)

2. Subrings and Ideals

From this point on, all rings are rings with identity, and all homomorphisms of rings are homomorphisms of rings with identity.

Subrings of a ring R are subsets of R that inherit a ring structure from R.

Definition. A subring of a ring R [with identity] is a subset S of R such that S is a subgroup of $(R, +)$, is closed under multiplication ($x, y \in S$ implies $xy \in S$), and contains the identity element.

For example, every ring is a subring of itself. In any ring R, the integer multiples of 1 constitute a subring, by Proposition 1.3; on the other hand, the trivial subgroup $0 = \{0\}$ is not a subring of R, unless $R = 0$.

Let S be a subring of R. The operations on R have restrictions to S that make S a ring in which the sum and product of two elements of S are the same as their sum and product in R. This ring S is also called a *subring* of R.

Readers will show that every intersection of subrings of a ring R is a subring of R. Consequently, there is for every subset X of R a smallest subring S of R that contains X; the exercises give a description of S.

Ideals of a ring are subgroups that *admit multiplication*:

Definitions. An ideal *of a ring R is a subgroup I of $(R, +)$ such that $x \in I$ implies $xy \in I$ and $yx \in I$ for all $y \in R$. A* proper *ideal also satisfies $I \neq R$.*

The definition of an ideal often includes the condition $I \neq R$.

For example, every subgroup $n\mathbb{Z}$ of $\mathbb{Z}$ is also an ideal of $\mathbb{Z}$ (proper if $n \neq \pm 1$). Every ring R is an (improper) ideal of itself and has a trivial ideal $0 = \{0\}$.

Properties. Our first property makes an easy exercise:

Proposition **2.1.** *Every intersection of ideals of a ring R is an ideal of R.*

By 2.1 there is for every subset S of R a smallest ideal of R that contains S, namely the intersection of all the ideals of R that contain S.

Definitions. The ideal (S) *of a ring R* generated *by a subset S of R is the smallest ideal of R that contains S. A* principal *ideal is an ideal generated by a single element.*

Proposition **2.2.** *In a ring R [with identity], the ideal (S) generated by a subset S is the set of all finite sums of elements of the form xsy, with $s \in S$ and $x, y \in R$. If R is commutative, then (S) is the set of all finite linear combinations of elements of S with coefficients in R.*

Proof. An ideal that contains S must also contain all elements of the form xsy with $s \in S$ and $x, y \in R$, and all finite sums of such elements. We show that the set I of all such sums is an ideal of R. First, I contains the empty sum 0; I is closed under sums by definition, and is closed under opposites since $-(xsy) = (-x)sy$. Hence I is a subgroup of $(R, +)$. Moreover, $(xsy)r = xs(yr)$, for all $r \in R$; hence $i \in I$ implies $ir \in I$. Similarly, $i \in I$ implies $ri \in I$, for all $r \in R$. Thus I is an ideal of R; then $I = (S)$.

If R is commutative, then $xsy = (xy)s$ and (S) is the set of all finite sums $x_1 s_1 + \cdots + x_n s_n$ with $n \geq 0$, $x_1, \ldots, x_n \in R$, and $s_1, \ldots, s_n \in S$. $\square$

Proposition **2.3.** *In a commutative ring R [with identity], the principal ideal generated by $a \in R$ is the set $(a) = Ra$ of all multiples of a.*

This follows from Proposition 2.2: by distributivity, a linear combination $x_1 a + \cdots + x_n a$ of copies of a is a multiple $(x_1 + \cdots + x_n)a$ of a. Propositions 2.3 and I.3.6 yield a property of $\mathbb{Z}$:

Proposition **2.4.** *Every ideal of $\mathbb{Z}$ is principal, and is generated by a unique nonnegative integer.*

A union of ideals is not generally an ideal, but there are exceptions:

Proposition **2.5.** *The union of a nonempty directed family of ideals of a ring* R *is an ideal of* R. *In particular, the union of a nonempty chain of ideals of a ring* R *is an ideal of* R.

Proposition 2.5 implies that *Zorn's lemma* in Section A.2 can be applied to ideals. Zorn's lemma states that a nonempty partially ordered set in which every nonempty chain has an upper bound must contain a *maximal* element (an element m such that $m < x$ holds for no other element x). In a ring, "maximal ideal" is short for "maximal proper ideal":

Definition. A maximal *ideal of a ring* R *is an ideal* $M \neq R$ *of* R *such that there is no ideal* I *of* R *such that* $M \subsetneq I \subsetneq R$.

Proposition **2.6.** *In a ring* R [*with identity*], *every proper ideal is contained in a maximal ideal.*

Proof. An ideal that contains the identity element must contain all its multiples and is not proper. Hence an ideal is proper if and only if it does not contain the identity element. Therefore the union of a nonempty chain of proper ideals, which is an ideal by 2.5, is a proper ideal.

Given an ideal $I \neq R$ we now apply Zorn's lemma to the set $\mathcal{S}$ of all proper ideals of R that contain I, partially ordered by inclusion. Every nonempty chain in $\mathcal{S}$ has an upper bound in $\mathcal{S}$, namely its union. Also, $\mathcal{S} \neq \varnothing$, since $I \in \mathcal{S}$. By Zorn's lemma, $\mathcal{S}$ has a maximal element M. Then M is a maximal (proper) ideal that contains I. $\square$

Finally, we note that the union $I \cup J$ of two ideals always admits multiplication. By 2.2, the ideal generated by $I \cup J$ is the set of all finite sums of elements of $I \cup J$, that is, the sum $I + J$ of I and J as subsets.

Definition. *The* sum *of two ideals* I *and* J *of a ring* R *is their sum as subsets:* $I + J = \{ x + y \mid x \in I, \ y \in J \}$.

Equivalently, $I + J$ is the smallest ideal of R that contains both I and J. More generally, every union $\bigcup_{i \in I} J_i$ of ideals J_i admits multiplication; hence the ideal it generates is the set of all finite sums of elements of $\bigcup_{i \in I} J_i$, which can be simplified so that all terms come from different ideals.

Definition. *The* sum *of ideals* $(J_i)_{i \in I}$ *of a ring* R *is*

$$\textstyle\sum_{i \in I} J_i = \{ \sum_{i \in I} x_i \mid x_i \in J_i \text{ and } x_i = 0 \text{ for almost all } i \in I \}.$$

Equivalently, $\sum_{i \in I} J_i$ is the smallest ideal of R that contains every J_i.

Proposition **2.7.** *Every sum of ideals of a ring* R *is an ideal of* R.

Exercises

All rings in the following exercises have an identity element.

1. Show that every intersection of subrings of a ring R is a subring of R.

2. Show that the union of a nonempty directed family of subrings of a ring R is a subring of R.

3. Show that the smallest subring of a ring R that contains a subset X of R is the set of all sums of products of elements of X and opposites of such products.

4. Show that every intersection of ideals of a ring R is an ideal of R.

5. Show that the union of a nonempty directed family of ideals of a ring R is an ideal of R.

6. Let I and J be ideals of a ring R. Show that $I \cup J$ is an ideal of R if and only if $I \subseteq J$ or $J \subseteq I$.

7. An element x of a ring is *nilpotent* when $x^n = 0$ for some $n > 0$. Show that the nilpotent elements of a commutative ring R constitute an ideal of R.

8. Let $n > 0$. Show that the ideal $n\mathbb{Z}$ of $\mathbb{Z}$ is maximal if and only if n is prime.

9. Polynomials in two variables (with real coefficients) constitute a ring R under the usual operations. Let I be the set of all polynomials $f \in R$ whose constant coefficient is 0. Show that I is a maximal ideal of R. Show that I is not a principal ideal.

The *product* AB of two ideals A and B of a ring is the ideal generated by their product as subsets. (Both products are denoted by AB, but, in a ring, the product of two ideals is their product as ideals, not their product as subsets.)

10. Show that the product AB of two ideals A and B of a ring R is the set of all finite sums $a_1 b_1 + \cdots + a_n b_n$ in which $n \geq 0$, $a_1, \ldots, a_n \in A$, and $b_1, \ldots, b_n \in B$.

11. Show that the product of ideals is associative.

12. Show that the product of ideals distributes sums: $A(B + C) = AB + AC$ and $(B + C)A = BA + CA$; for extra credit, $A\left(\sum_{i \in I} B_i\right) = \sum_{i \in I} (AB_i)$ and $\left(\sum_{i \in I} A_i\right)B = \sum_{i \in I} (A_i B)$.

3. Homomorphisms

This section extends to rings the wonderful properties of group homomorphisms in Sections I.4 and I.5.

Subrings and ideals. Homomorphisms of rings (defined in Section 1) are mappings that preserve sums, products, and identity elements. Homomorphisms also preserve subrings, and, to some extent, ideals.

Proposition 3.1. *Let* $\varphi : R \longrightarrow S$ *be a homomorphism of rings. If A is a subring of R, then*

$$\varphi(A) = \{\, \varphi(x) \mid x \in A \,\}$$

is a subring of S. If B is a subring of S, then

$$\varphi^{-1}(B) = \{\, x \in R \mid \varphi(x) \in B \,\}$$

is a subring of R.

If A is an ideal of R and φ is surjective, then $\varphi(A)$ is an ideal of S. If B is an ideal of S, then $\varphi^{-1}(B)$ is an ideal of R.

Readers will happily concoct proofs for these statements, and show that non-surjective homomorphisms do not necessarily send ideals to ideals. In Proposition 3.1, $\varphi(A)$ is the *direct image* of $A \subseteq R$ under φ and $\varphi^{-1}(B)$ is the *inverse image* of $B \subseteq S$ under φ. Two subsets of interest arise as particular cases:

Definitions. *Let $\varphi : R \longrightarrow S$ be a homomorphism of rings. The* image *or* range *of φ is*

$$\mathrm{Im}\, \varphi = \{ \varphi(x) \mid x \in R \}.$$

The kernel *of φ is*

$$\mathrm{Ker}\, \varphi = \{ x \in R \mid \varphi(x) = 0 \}.$$

Propositions 3.1 and I.4.4 yield the following result:

Proposition 3.2. *Let $\varphi : R \longrightarrow S$ be a homomorphism of rings. The image of φ is a subring of S. The kernel K of φ is an ideal of R. Moreover, $\varphi(x) = \varphi(y)$ if and only if $x - y \in K$.*

Conversely, every subring S of a ring R is the image of the *inclusion homomorphism* $x \longmapsto x$ of S into R.

Quotient rings. Ideals yield quotient rings and projection homomorphisms.

Proposition 3.3. *Let I be an ideal of a ring R. The cosets of I in the abelian group $(R, +)$ constitute a ring R/I. In R/I, the sum of two cosets is their sum as subsets, so that $(x + I) + (y + I) = (x + y) + I$; the product of two cosets is the coset that contains their product as subsets, so that $(x + I)(y + I) = xy + I$. The mapping $x \longmapsto x + I$ is a surjective homomorphism of rings, whose kernel is I.*

Proof. R/I is already an abelian group, by I.4.7. If $x + I$, $y + I \in R/I$, then the product $(x + I)(y + I)$ of subsets is contained in the single coset $xy + I$, since $(x + i)(y + j) = xy + xj + iy + ij \in xy + I$ for all $i, j \in I$. Hence multiplication in R/I can be defined as above. It is immediate that R/I is now a ring; the identity element of R/I is $1 + I$. $\square$

Definitions. *Let I be an ideal of a ring R. The ring of all cosets of I is the* quotient ring R/I *of R by I. The homomorphism $x \longmapsto x + I$ is the* canonical projection *of R onto R/I.*

For example, every ring R is an ideal of itself and R/R is the trivial ring $\{0\}$; 0 is an ideal of R and the canonical projection is an isomorphism $R \cong R/0$.

For a more interesting example, let $R = \mathbb{Z}$. By 2.4, every ideal I of $\mathbb{Z}$ is principal, and is generated by a unique nonnegative integer n. If $n = 0$, then $I = 0$ and $\mathbb{Z}/I \cong \mathbb{Z}$; if $n > 0$, then the additive group $\mathbb{Z}_n$ becomes a ring (which readers probably know already):

Definition. For every positive integer n, the ring $\mathbb{Z}_n$ of the integers modulo n is the quotient ring $\mathbb{Z}/\mathbb{Z}n$.

In general, the subrings of R/I are quotients of subrings of R, and similarly for ideals (in the sense that $A/I = \{ a + I \mid a \in A \}$ when $A \subseteq R$):

Proposition 3.4. Let I be an ideal of a ring R. Every subring of R/I is the quotient S/I of a unique subring S of R that contains I. Every ideal of R/I is the quotient J/I of a unique ideal J of R that contains I.

This follows from I.4.9. Theorem I.5.1 also extends to quotient rings:

Theorem 3.5 (Factorization Theorem). *Let I be an ideal of a ring R. Every homomorphism of rings $\varphi : R \longrightarrow S$ whose kernel contains I factors uniquely through the canonical projection $\pi : R \longrightarrow R/I$ (there exists a homomorphism $\psi : R/I \longrightarrow S$ unique such that $\varphi = \psi \circ \pi$).*

$$
\begin{array}{ccc}
R & \xrightarrow{\ \pi\ } & R/I \\
& {\scriptstyle \varphi}\searrow & \downarrow{\scriptstyle \psi} \\
& & S
\end{array}
$$

Proof. By I.5.1 there is a homomorphism of abelian groups ψ of $(R/I, +)$ into $(S, +)$ unique such that $\varphi = \psi \circ \pi$; equivalently, $\psi(x + I) = \varphi(x)$ for all $x \in R$. Now, ψ is a homomorphism of rings. Indeed,

$$\psi\big((x + I)(y + I)\big) = \psi(xy + I) = \varphi(xy) = \varphi(x)\,\varphi(y) = \psi(x + I)\,\psi(y + I)$$

for all $x + I$, $y + I \in R/I$, and $\psi(1) = \psi(1 + I) = \varphi(1) = 1$. $\square$

The homomorphism theorem. Theorem I.5.2 also extends to rings; so do the isomorphism theorems in Section I.5 (see the exercises).

Theorem 3.6 (Homomorphism Theorem). *If $\varphi : R \longrightarrow S$ is a homomorphism of rings, then*

$$R/\mathrm{Ker}\,\varphi \;\cong\; \mathrm{Im}\,\varphi;$$

in fact, there is an isomorphism $\theta : R/\mathrm{Ker}\, f \longrightarrow \mathrm{Im}\, f$ unique such that $\varphi = \iota \circ \theta \circ \pi$, where $\iota : \mathrm{Im}\, f \longrightarrow S$ is the inclusion homomorphism and $\pi : R \longrightarrow R/\mathrm{Ker}\, f$ is the canonical projection.

$$
\begin{array}{ccc}
R & \xrightarrow{\ \varphi\ } & S \\
{\scriptstyle \pi}\downarrow & & \uparrow{\scriptstyle \iota} \\
R/\mathrm{Ker}\,\varphi & \dashrightarrow[\theta] & \mathrm{Im}\,\varphi
\end{array}
$$

Proof. By I.5.2 there is an isomorphism of abelian groups $\theta : (R/\mathrm{Ker}\, f, +) \longrightarrow (\mathrm{Im}\, f, +)$ unique such that $\varphi = \iota \circ \theta \circ \pi$; equivalently, $\theta(x + \mathrm{Ker}\,\varphi) = \varphi(x)$ for all $x \in R$. As in the proof of 3.5, this implies that θ is a homomorphism of rings, hence is an isomorphism. $\square$

Our first application of the homomorphism theorem is the following result.

Proposition **3.7.** *Let R be a ring [with identity]. There is a unique homomorphism of rings of $\mathbb{Z}$ into R. Its image is the smallest subring of R; it consists of all integer multiples of the identity element of R, and is isomorphic either to $\mathbb{Z}$ or to $\mathbb{Z}_n$ for some unique $n > 0$.*

Proof. If $\varphi : \mathbb{Z} \longrightarrow R$ is a homomorphism of rings [with identity], then $\varphi(1) = 1$ and $\varphi(n) = \varphi(n1) = n1 \in R$ for all $n \in \mathbb{Z}$. Hence φ is unique. Conversely, we saw that the mapping $\varphi : n \longmapsto n1$ is a homomorphism of rings of $\mathbb{Z}$ into R. Then $\operatorname{Im} \varphi$, which is the set of all integer multiples of the identity element of R, is a subring of R; it is the smallest such subring, since a subring of R must contain the identity element and all its integer multiples.

By the homomorphism theorem, $\operatorname{Im} \varphi \cong \mathbb{Z}/I$ for some ideal I of $\mathbb{Z}$. By 2.4, I is principal, $I = n\mathbb{Z}$ for some $n \geq 0$. If $n = 0$, then $\operatorname{Im} \varphi \cong \mathbb{Z}/0 \cong \mathbb{Z}$. If $n > 0$, then $\operatorname{Im} \varphi \cong \mathbb{Z}_n$; then $n > 0$ is unique with this property, since, say, the rings $\mathbb{Z}_n$ all have different numbers of elements. $\square$

The unique integer $n > 0$ in Proposition 3.7 is also the smallest $m > 0$ such that $m1 = 0$ and the smallest $m > 0$ such that $mx = 0$ for all $x \in R$.

Definition. The characteristic *of a ring R [with identity] is 0 if $n1 \neq 0$ in R for all $n > 0$; otherwise, it is the smallest integer $n > 0$ such that $n1 = 0$.*

For example, $\mathbb{Z}_n$ has characteristic n.

Exercises

1. Let $\varphi : R \longrightarrow S$ be a homomorphism of rings and let A be a subring of R. Show that $\varphi(A)$ is a subring of B.

2. Let $\varphi : R \longrightarrow S$ be a homomorphism of rings and let B be a subring of S. Show that $\varphi^{-1}(B)$ is a subring of R.

3. Let $\varphi : R \longrightarrow S$ be a surjective homomorphism of rings and let I be an ideal of R. Show that $\varphi(I)$ is an ideal of S.

4. Find a homomorphism $\varphi : R \longrightarrow S$ of commutative rings and an ideal I of R such that $\varphi(I)$ is not an ideal of S.

5. Let $\varphi : R \longrightarrow S$ be a homomorphism of rings and let J be an ideal of S. Show that $\varphi^{-1}(J)$ is an ideal of R.

6. Let R be a ring and let I be an ideal of R. Show that every ideal of R/I is the quotient J/I of a unique ideal J of R that contains I.

7. Let R be a ring and let I be an ideal of R. Show that quotient by I is a one-to-one correspondence, which preserves inclusions, between ideals of R that contain I and ideals of R/I.

8. Let $I \subseteq J$ be ideals of a ring R. Show that $(R/I)/(J/I) \cong R/J$.

9. Let S be a subring of a ring R and let I be an ideal of R. Show that $S + I$ is a subring of R, I is an ideal of $S + I$, $S \cap I$ is an ideal of S, and $(S + I)/I \cong S/(S \cap I)$.

4. Domains and Fields

Domains and fields are major types of rings.

Definition. A domain *is a commutative ring* $R \neq 0$ *[with identity] in which* $x, y \neq 0$ *implies* $xy \neq 0$.

Equivalently, a ring R is a domain when $R \backslash \{0\}$ is a commutative monoid under multiplication. For example, $\mathbb{Z}, \mathbb{Q}, \mathbb{R}$, and $\mathbb{C}$ are domains. In fact, $\mathbb{Q}, \mathbb{R}$, and $\mathbb{C}$ have a stronger property:

Definition. A field *is a commutative ring* $F \neq 0$ *such that* $F \backslash \{0\}$ *is a group under multiplication.*

Domains and fields may also be defined as follows. A *zero divisor* of a commutative ring R is an element $x \neq 0$ of R such that $xy = 0$ for some $y \neq 0$, $y \in R$. A commutative ring $R \neq 0$ is a domain if and only if R has no zero divisor. A *unit* of a ring R [with identity] is an element u of R such that $uv = vu = 1$ for some $v \in R$; then v is a unit, the *inverse* u^{-1} of u. Units cannot be zero divisors. A commutative ring $R \neq 0$ is a field if and only if every nonzero element of R is a unit.

Proposition 4.1. *Let* $n > 0$. *The ring* $\mathbb{Z}_n$ *is a domain if and only if* n *is prime, and then* $\mathbb{Z}_n$ *is a field.*

Proof. If $n > 0$ is not prime, then either $n = 1$, in which case $\mathbb{Z}_n = 0$, or $n = xy$ for some $1 < x, y < n$, in which case $\overline{x}\,\overline{y} = \overline{0}$ in $\mathbb{Z}_n$ and $\mathbb{Z}_n$ has a zero divisor. In either case $\mathbb{Z}_n$ is not a domain.

Now let n be prime. If $1 \leqq x < n$, then n and x are relatively prime and $ux + vn = 1$ for some $u, v \in \mathbb{Z}$. Hence $\overline{x}\,\overline{u} = \overline{1}$ in $\mathbb{Z}_n$ and $\overline{x}$ is a unit. Thus $\mathbb{Z}_n$ is a field. $\square$

Domains are also called *integral domains*, and the term "domain" is sometimes applied to noncommutative rings without zero divisors. A noncommutative ring $R \neq 0$ such that $R \backslash \{0\}$ is a group under multiplication (equivalently, in which every nonzero element is a unit) is a *division ring*.

Properties. The cancellation law holds as follows in every domain:

Proposition 4.2. *In a domain,* $xy = xz$ *implies* $y = z$, *when* $x \neq 0$.

Proof. If $x(y - z) = 0$ and $x \neq 0$, then $y - z = 0$: otherwise, x would be a zero divisor. $\square$

Proposition 4.3. *The characteristic of a domain is either* 0 *or a prime number.*

Proof. The smallest subring of a domain has no zero divisors; by 3.7, 4.1, it is isomorphic either to $\mathbb{Z}$, or to $\mathbb{Z}_p$ for some prime p. $\square$

Domains of characteristic $p \neq 0$ have an amusing property:

Proposition **4.4.** *In a commutative ring R of prime characteristic p, $(x + y)^p = x^p + y^p$ and $(x - y)^p = x^p - y^p$, for all $x, y \in R$.*

Proof. By the binomial theorem, $(x + y)^p = \sum_{0 \leq i \leq p} \binom{p}{i} x^i y^{p-i}$, where $\binom{p}{i} = \frac{p!}{i!(p-i)!}$. If $0 < i < p$, then p divides $p!$ but does not divide $i!$ or $(p - i)!$; hence p divides $\binom{p}{i}$, $\binom{p}{i} r = 0$ for all $r \in R$, and

$$(x + y)^p = \sum_{0 \leq i \leq p} \binom{p}{i} x^i y^{p-i} = \sum_{i=0, p} \binom{p}{i} x^i y^{p-i} = x^p + y^p.$$

Then $(x - y)^p = x^p + (-1)^p y^p$; if p is odd, then $x^p + (-1)^p y^p = x^p - y^p$, whereas, if $p = 2$, then $x^p + (-1)^p y^p = x^p + y^p = x^p - y^p$. $\square$

Prime and maximal ideals have quotient rings that are domains and fields.

Definition. A prime *ideal of a commutative ring R is an ideal $\mathfrak{p} \neq R$ such that $xy \in \mathfrak{p}$ implies $x \in \mathfrak{p}$ or $y \in \mathfrak{p}$.*

Proposition **4.5.** *If $\mathfrak{a}$ is an ideal of a commutative ring R [with identity], then $R/\mathfrak{a}$ is a domain if and only if $\mathfrak{a}$ is a prime ideal.*

The proof is an exercise.

Proposition **4.6.** *If $\mathfrak{a}$ is an ideal of a commutative ring R [with identity], then $R/\mathfrak{a}$ is a field if and only if $\mathfrak{a}$ is a maximal ideal.*

Proof. A field F has no proper ideal $\mathfrak{c} \neq 0$: indeed, if $x \in \mathfrak{c}$, $x \neq 0$, then $1 \in \mathfrak{c}$ since x is a unit, and $\mathfrak{c} = F$. Conversely, let $R \neq 0$ be a commutative ring with no proper ideal $\mathfrak{c} \neq 0$. For every $x \in R$, $x \neq 0$, we have $1 \in Rx = R$, so that x is a unit. Hence R is a field.

If now $\mathfrak{a}$ is an ideal of a commutative ring R, then $R/\mathfrak{a}$ is a field if and only if $R/\mathfrak{a} \neq 0$ and $R/\mathfrak{a}$ has no ideal $0 \subsetneqq \mathfrak{c} \subsetneqq R/\mathfrak{a}$, if and only if $\mathfrak{a} \neq R$ and R has no ideal $\mathfrak{a} \subsetneqq \mathfrak{b} \subsetneqq R$, by 3.4. $\square$

Corollary **4.7.** *In a commutative ring [with identity], every maximal ideal is prime.*

Corollary **4.8.** *An ideal $\mathbb{Z}n$ of $\mathbb{Z}$ is prime if and only if n is prime, and then $\mathbb{Z}n$ is maximal.*

As in the proof of Proposition 4.6, a field has no proper ideal $\mathfrak{a} \neq 0$. If now φ is a homomorphism of fields, then $\operatorname{Ker} \varphi$ is a proper ideal, since $\varphi(1) = 1 \neq 0$, so that $\operatorname{Ker} \varphi = 0$ and φ is injective:

Proposition **4.9.** *Every homomorphism of fields is injective.*

Fields of fractions. A subring of a domain is a domain. Conversely, we show that every domain is (up to isomorphism) a subring of a field.

A field that contains a domain R must also contain the inverses of nonzero elements of R and all products xy^{-1} with $x, y \in R$, $y \neq 0$. The latter add and multiply like fractions x/y: $(xy^{-1})(zt^{-1}) = (xz)(yt)^{-1}$ and $(xy^{-1}) + (zt^{-1}) = (xt + yz)(yt)^{-1}$; moreover, $xy^{-1} = zt^{-1}$ if and only if $xt = yz$.

This suggests the following construction. Let R be a domain. Define a binary relation $\sim$ on $R \times (R \backslash \{0\})$ by

$$(x, y) \sim (z, t) \text{ if and only if } xt = yz.$$

It is immediate that $\sim$ is an equivalence relation. The equivalence class of $(x, y) \in R \backslash \{0\}$ is a *fraction*, x/y or $\frac{x}{y}$. Readers will verify that operations on the quotient set $Q(R) = (R \times (R \backslash \{0\}))/\sim$ are well defined by

$$(x/y) + (z/t) = (xt + yz)/yt \text{ and } (x/y)(z/t) = xz/yt$$

and that the following holds:

Proposition **4.10.** *For every domain R, $Q(R)$ is a field and $\iota : x \longmapsto x/1$ is an injective homomorphism.*

Then R is isomorphic to the subring $\operatorname{Im} \iota$ of the field $Q(R)$. It is common practice to identify $x \in R$ and $\iota(x) = x/1 \in Q(R)$; then ι is an inclusion homomorphism and R is a subring of $Q(R)$.

Definition. If R is a domain, then $Q(R)$ is the field of fractions, *or* field of quotients, *or* quotient field, *of R.* $\square$

For instance, if $R = \mathbb{Z}$, then $Q(R) \cong \mathbb{Q}$. Thus, Proposition 4.10 generalizes the construction of rational numbers from integers.

The field of fractions of a domain has a universal property:

Proposition **4.11.** *Let R be a domain. Every injective homomorphism φ of R into a field F factors uniquely through $\iota : R \longmapsto Q(R)$: $\varphi = \psi \circ \iota$ for some unique homomorphism $\psi : Q(R) \longrightarrow F$, namely $\psi(x/y) = \varphi(x) \varphi(y)^{-1}$.*

Proof. Every homomorphism ψ of fields preserves inverses: if $x \neq 0$, then $\psi(x) \psi(x^{-1}) = \psi(1) = 1$, so that $\psi(x) \neq 0$ and $\psi(x^{-1}) = \psi(x)^{-1}$.

In $Q(R)$, $(x/y)^{-1} = y/x$, when $x, y \neq 0$; hence $x/y = (x/1)(1/y) = \iota(x) \iota(y)^{-1}$ when $y \neq 0$. If now $\varphi : R \longrightarrow F$ is injective, then $y \neq 0$ implies $\varphi(y) \neq 0$ and $\psi \circ \iota = \varphi$ implies $\psi(x/y) = \psi(\iota(x)) \psi(\iota(y)^{-1}) = \varphi(x) \varphi(y)^{-1}$.

Taking this hint we observe that $x/y = z/t$ implies $xt = yz$, $\varphi(x) \varphi(t) = \varphi(y) \varphi(z)$, and, in the field F, $\varphi(x) \varphi(y)^{-1} = \varphi(z) \varphi(t)^{-1}$ (since $\varphi(y)$, $\varphi(t) \neq 0$). Therefore a mapping ψ of $Q(R)$ into F is well defined by

$$\psi(x/y) = \varphi(x) \varphi(y)^{-1}.$$

Moreover, $\psi(1) = \varphi(1)\,\varphi(1)^{-1} = 1$,

$$\psi\big((x/y)+(z,t)\big) = \varphi(xt + yz)\,\varphi(yt)^{-1}$$
$$= \varphi(x)\,\varphi(t)\,\varphi(y)^{-1}\,\varphi(t)^{-1} + \varphi(y)\,\varphi(z)\,\varphi(y)^{-1}\,\varphi(t)^{-1}$$
$$= \varphi(x)\,\varphi(y)^{-1} + \varphi(z)\,\varphi(t)^{-1} = \psi(x/y) + \psi(z/t),$$

since $\varphi(yt)^{-1} = \varphi(y)^{-1}\,\varphi(t)^{-1}$, and

$$\psi\big((x/y)\,(z/t)\big) = \varphi(x)\,\varphi(y)\,\varphi(z)^{-1}\,\varphi(t)^{-1} = \psi(x/y)\,\psi(z/t),$$

whenever $y, t \neq 0$. Thus ψ is a homomorphism. By the beginning of the proof, ψ is the only homomorphism such that $\psi \circ \iota = \varphi$. $\square$

If R is identified with a subring of $Q(R)$, then every injective homomorphism of R into a field F extends uniquely to a [necessarily injective] homomorphism of $Q(R)$ into F; hence $Q(R)$ is, up to isomorphism, the smallest field that contains R as a subring.

From 4.11 we deduce an "internal" characterization of $Q(R)$:

Proposition **4.12.** *Let R be a subring of a field K. The identity on R extends to an isomorphism $K \cong Q(R)$ if and only if every element of K can be written in the form ab^{-1} for some $a, b \in R,\ b \neq 0$.*

Proof. This condition is necessary since every element of $Q(R)$ can be written in the form $a/b = ab^{-1}$ for some $a, b \in R,\ b \neq 0$. Conversely, by 4.11, the inclusion homomorphism $R \longrightarrow K$ extends to a homomorphism $\theta : Q(R) \longrightarrow K$, which is injective by 4.6, and surjective if every element of K can be written in the form $ab^{-1} = \theta\,(a/b)$. $\square$

Exercises

1. Let $\mathfrak{a}$ be an ideal of a commutative ring R. Prove that $R/\mathfrak{a}$ is a domain if and only if $\mathfrak{a}$ is a prime ideal.

2. Let $n > 0$. Give a direct proof that the ideal $\mathbb{Z}n$ of $\mathbb{Z}$ is prime if and only if n is prime, and then $\mathbb{Z}n$ is maximal.

3. Give a direct proof that every maximal ideal of a commutative ring is a prime ideal.

4. Show that the field of fractions of a domain is completely determined, up to isomorphism, by its universal property.

5. Let S be a monoid that is commutative and cancellative. Construct a group of fractions of S. State and prove its universal property.

5. Polynomials in One Variable

A polynomial in one indeterminate X should be a finite linear combination $a_0 + a_1 X + \cdots + a_n X^n$ of powers of X. Unfortunately, this natural concept of polynomial s leads to a circular definition: one needs a set of polynomials in order to make linear combinations in it. Our formal definition of polynomials must therefore seem somewhat unnatural. It specifies a polynomial by its coefficients:

Definition. Let $M = \{\, 1, X, \ldots, X^n, \ldots \,\}$ *be the free monoid on* $\{X\}$. *A polynomial over a ring* R *[with identity] in the* indeterminate X *is a mapping* $A : X^n \longmapsto a_n$ *of* M *into* R *such that* $a_n = 0$ *for almost all* $n \geqq 0$. *The set of all polynomials in* X *over* R *is denoted by* $R[X]$.

In this definition, M can be replaced by any monoid. The resulting ring $R[M]$ is a *semigroup ring* (a *group ring* if M is a group). The exercises give details of this construction.

We quickly define operations on $R[X]$, so that we can return to the usual notation $A = a_0 + a_1 X + \cdots + a_n X^n$. Polynomials are added pointwise,

$$A + B = C \text{ when } c_n = a_n + b_n \text{ for all } n \geqq 0,$$

and multiplied by the usual rule,

$$AB = C \text{ when } c_n = \textstyle\sum_{i+j=n} a_i b_j \text{ for all } n \geqq 0.$$

Proposition **5.1.** *For every ring* R *[with identity]*, $R[X]$, *with the operations above, is a ring.*

Proof. For each $A \in R[X]$ there exists some $m \geqq 0$ such that $a_k = 0$ for all $k > m$: otherwise, $\{\, k \geqq 0 \mid a_k \neq 0 \,\}$ is not finite. If $a_k = 0$ for all $k > m$, and $b_k = 0$ for all $k > n$, then $a_k + b_k = 0$ for all $k > \max(m, n)$; hence $\{\, k \geqq 0 \mid a_k + b_k \neq 0 \,\}$ is finite and $A + B \in R[X]$. If $c_k = \sum_{i+j=k} a_i b_j$ for all $k \geqq 0$, then $c_k = 0$ for all $k > m + n$, since $a_i = 0$ if $i > m$ and $b_j = 0$ if $i + j = k$ and $i \leqq m$ (then $j > n$); hence $\{\, k \geqq 0 \mid c_k \neq 0 \,\}$ is finite and $AB \in R[X]$. Thus the operations on $R[X]$ are well defined.

It is immediate that $(R[X], +)$ is an abelian group. The identity element of $R[X]$ is the polynomial 1 with coefficients $1_n = 0$ for all $n > 0$ and $1_0 = 1$. Multiplication on $R[X]$ inherits distributivity from the multiplication on R; associativity is an exercise. $\square$

By custom, $r \in R$ is identified with the *constant* polynomial with coefficients $a_n = 0$ for all $n > 0$ and $a_0 = r$; X is identified with the polynomial with coefficients $a_n = 0$ for all $n \neq 1$ and $a_1 = 1$. This allows a more natural notation:

Proposition **5.2.** $A = \sum_{n \geqq 0} a_n X^n$, *for every* $A \in R[X]$; *if* $a_i = 0$ *for all* $i > n$, *then* $A = a_0 + a_1 X + \cdots + a_n X^n$.

Proof. The infinite sum $\sum_{n \geqq 0} a_n X^n$ exists since $a_n X^n = 0$ for almost all n. Its coefficients are found as follows. By induction on k, X^k has coefficients $a_n = 0$ if $n \neq k$, $a_k = 1$. Then $r X^k$ has coefficients $a_n = 0$ if $n \neq k$, $a_k = r$, for every $r \in R$. Hence $\sum_{n \geq 0} a_n X^n$ has the same coefficients as A. If $a_i = 0$ for all $i > n$, then $A = \sum_{i \geq 0} a_i X^i = \sum_{0 \leq i \leq n} a_i X^i$. $\square$

Operations on polynomials can now be carried out in the usual way.

Definitions. The degree deg A of a polynomial $A \neq 0$ is the largest integer $n \geq 0$ such that $a_n \neq 0$; then a_n is the leading coefficient *of A.*

The degree of the zero polynomial 0 is sometimes left undefined or is variously defined as $-1 \in \mathbb{Z}$ or as $-\infty$, as long as $\deg 0 < \deg A$ for all $A \neq 0$.

A polynomial A has degree at most n if and only if $a_i = 0$ for all $i > n$, if and only if A can be written in the form $A = a_0 + a_1 X + \cdots + a_n X^n$. The following properties are straightforward:

Proposition **5.3.** *For all $A, B \neq 0$ in $R[X]$:*

(1) $\deg (A + B) \leq \max (\deg A, \deg B)$;

(2) *if $\deg A \neq \deg B$, then $\deg (A + B) = \max (\deg A, \deg B)$;*

(3) $\deg (AB) \leq \deg A + \deg B$;

(4) *if R has no zero divisors, then $\deg (AB) = \deg A + \deg B$.*

In particular, if R is a domain, then $R[X]$ is a domain.

Corollary **5.4.** *If R has no zero divisors, then the units of $R[X]$ are the units of R.*

Polynomial division. In $R[X]$, polynomial or long division of A by $B \neq 0$ requires repeated division by the leading coefficient b_n of B. For good results b_n should be a unit of R, for then division by b_n is just multiplication by b_n^{-1} and has a unique result. In particular, polynomial division of A by B works if B is monic (its leading coefficient b_n is 1), and for all $B \neq 0$ if R is a field.

Proposition **5.5.** *Let $B \in R[X]$ be a nonzero polynomial whose leading coefficient is a unit of R. For every polynomial $A \in R[X]$ there exist polynomials $Q, S \in R[X]$ such that $A = BQ + S$ and $\deg S < \deg B$; moreover, Q and S are unique.*

Proof. First we assume that B is monic and prove existence by induction on $\deg A$. Let $\deg B = n$. If $\deg A < n$, then $Q = 0$ and $S = A$ serve. Now let $\deg A = m \geq n$. Then $B a_m X^{m-n}$ has degree m and leading coefficient a_m. Hence $A - B a_m X^{m-n}$ has degree less than m. By the induction hypothesis, $A - B a_m X^{m-n} = BQ_1 + S$ for some $Q_1, S \in R[X]$ such that $\deg S < \deg B$. Then $A = B(a_m X^{m-n} + Q_1) + S$.

In general, the leading coefficient b_n of B is a unit of R; then $B b_n^{-1}$ is monic, and $A = B b_n^{-1} Q + S$ with $\deg S < \deg B$, for some Q and S.

Uniqueness follows from the equality $\deg (BC) = \deg B + \deg C$, which holds for all $C \neq 0$ since the leading coefficient of B is not a zero divisor. Let

$A = BQ_1 + S_1 = BQ_2 + S_2$, with $\deg S_1$, $\deg S_2 < \deg B$. If $Q_1 \neq Q_2$, then $S_1 - S_2 = B(Q_2 - Q_1)$ has degree $\deg B + \deg(Q_2 - Q_1) \geqq \deg B$, contradicting $\deg S_1$, $\deg S_2 < \deg B$; hence $Q_1 = Q_2$, and then $S_1 = S_2$. $\square$

Evaluation. Polynomials $A \in R[X]$ can be *evaluated* at elements of R:

Definition. If $A = a_0 + a_1 X + \cdots + a_n X^n \in R[X]$ and $r \in R$, then $A(r) = a_0 + a_1 r + \cdots + a_n r^n \in R$.

The polynomial A itself is often denoted by $A(X)$. A polynomial $A(X) = a_0 + a_1 X + \cdots + a_n X^n \in R[X]$ can also be evaluated at any element of a larger ring S; for example, at another polynomial $B \in R[X]$, the result being $A(B) = a_0 + a_1 B + \cdots + a_n B^n \in R[X]$. This operation, *substitution*, is discussed in the exercises.

In general, $(A + B)(r) = A(r) + B(r)$, but readers should keep in mind that $(AB)(r) = A(r)\,B(r)$ requires some commutativity.

Proposition 5.6. *If R is commutative, then evaluation at $r \in R$ is a homomorphism of $R[X]$ into R. More generally, if R is a subring of S and $s \in S$ commutes with every element of R, then evaluation at s is a homomorphism of $R[X] \subseteq S[X]$ into S.*

The commutativity condition in this result is necessary (see the exercises).

Proof. For all $A, B \in R[X]$, $(A + B)(s) = A(s) + B(s)$ and

$$A(s)\,B(s) = \left(\sum_i a_i s^i\right)\left(\sum_j b_j s^j\right) = \sum_{i,j}\left(a_i s^i\, b_j s^j\right)$$
$$= \sum_{i,j}\left(a_i\, b_j\, s^i\, s^j\right) = \sum_k \left(\sum_{i+j=k} a_i\, b_j\right) s^k = (AB)(s),$$

since every s^i commutes with every b_j. Also $1(s) = 1$. $\square$

Roots. A *root* of a polynomial $A \in R[X]$ is an element r (of R, or of a larger ring) such that $A(r) = 0$.

Proposition 5.7. *Let $r \in R$ and $A \in R[X]$. If R is commutative, then A is a multiple of $X - r$ if and only if $A(r) = 0$.*

Proof. By polynomial division, $A = (X - r)\,B + S$, where B and S are unique with $\deg S < 1$. Then S is constant. Evaluating at r yields $S = A(r)$, by 5.6. Hence $X - r$ divides A if and only if $A(r) = 0$. $\square$

Definitions. Let $r \in R$ be a root of $A \in R[X]$. The multiplicity of r is the largest integer $m > 0$ such that $(X - r)^m$ divides A; r is a simple *root when it has multiplicity 1, a* multiple *root otherwise.*

For example, i is a simple root of $X^2 + 1 = (X - i)(X + i) \in \mathbb{C}[X]$ and a multiple root (with multiplicity 2) of $X^4 + 2X^2 + 1 = (X - i)^2 (X + i)^2 \in \mathbb{C}[X]$.

To detect multiple roots we use a derivative:

Definition. The formal derivative *of* $A(X) = \sum_{n \geq 0} a_n X^n \in K[X]$ *is* $A'(X) = \sum_{n \geq 1} n a_n X^{n-1} \in K[X]$.

Without a topology on R, this is only a formal derivative, without an interpretation as a limit. Yet readers will prove some familiar properties:

Proposition **5.8.** *For all* $A, B \in K[X]$ *and* $n > 0$, $(A + B)' = A' + B'$, $(AB)' = A'B + AB'$, *and* $(A^n)' = nA^{n-1}A'$.

Derivatives detect multiple roots as follows:

Proposition **5.9.** *If* R *is commutative, then a root* $r \in R$ *of a polynomial* $A \in R[X]$ *is simple if and only if* $A'(r) \neq 0$.

Proof. If r has multiplicity m, then $A = (X - r)^m B$, where $B(r) \neq 0$: otherwise $(X - r)^{m+1}$ divides A by 5.7. If $m = 1$, then $A' = B + (X - r) B'$ and $A'(r) = B(r) \neq 0$. If $m > 1$, then $A' = m (X - r)^{m-1} B + (X - \alpha)^m B'$ and $A'(r) = 0$. $\square$

Homomorphisms.

Proposition **5.10.** *Every homomorphism of rings* $\varphi : R \longrightarrow S$ *induces a homomorphism of rings* $A \longmapsto {}^\varphi A$ *of* $R[X]$ *into* $S[X]$, *namely,*

$${}^\varphi\!\left(a_0 + a_1 X + \cdots + a_n X^n\right) = \varphi(a_0) + \varphi(a_1) X + \cdots + \varphi(a_n) X^n.$$

The next result, of fundamental importance in the next chapter, is a universal property that constructs every ring homomorphism $\psi : R[X] \longrightarrow S$. Necessarily the restriction φ of ψ to R is a ring homomorphism; and $\psi(X)$ commutes with every $\varphi(r)$, since, in $R[X]$, X commutes with all constants.

Theorem **5.11.** *Let* R *and* S *be rings and let* $\varphi : R \longrightarrow S$ *be a homomorphism of rings. Let* s *be an element of* S *that commutes with* $\varphi(r)$ *for every* $r \in R$ *(an arbitrary element if* S *is commutative). There is a unique homomorphism of rings* $\psi : R[X] \longrightarrow S$ *that extends* φ *and sends* X *to* s, *namely* $\psi(A) = {}^\varphi A (s)$.

$$R \overset{\subseteq}{\longrightarrow} R[X]$$
$$\varphi \searrow \quad \downarrow \psi$$
$$S$$

Proof. The mapping ψ is a homomorphism since it is the composition of the homomorphisms $A \longmapsto {}^\varphi A$ in 5.10 and $B \longmapsto B(s)$, $(\operatorname{Im} \varphi)[X] \longrightarrow S$ in 5.6. We see that ψ extends φ and sends X to s. By 5.2, a homomorphism with these properties must send $A = a_0 + a_1 X + \cdots + a_n X^n \in R[X]$ to $\varphi(a_0) + \varphi(a_1) s + \cdots + \varphi(a_n) s^n = {}^\varphi A (s)$. $\square$

Propositions 5.6 and 5.10 are particular cases of Theorem 5.11.

The field case. The ring $K[X]$ has additional properties when K is a field.

Proposition **5.12.** *For every field* K : $K[X]$ *is a domain; every ideal of* $K[X]$ *is principal; in fact, every nonzero ideal of* $K[X]$ *is generated by a unique monic polynomial.*

Proof. The trivial ideal $0 = \{0\}$ is generated by the zero polynomial 0. Now, let $\mathfrak{A} \neq 0$ be a nonzero ideal of $K[X]$. There is a polynomial $B \in \mathfrak{A}$ such that $B \neq 0$ and B has the least possible degree. Dividing B by its leading coefficient does not affect its degree, so we may assume that B is monic. We have $(B) \subseteq \mathfrak{A}$. Conversely, if $A \in \mathfrak{A}$, then $A = BQ + R$ for some $Q, R \in K[X]$ with $\deg R < \deg B$. Since $R = A - BQ \in \mathfrak{A}$, $R \neq 0$ would contradict the choice of B; therefore $R = 0$ and $A = BQ \in (B)$. Hence $\mathfrak{A} = (B)$.

If $\mathfrak{A} = (B) = (C) \neq 0$, then $C = BQ_1$ and $B = CQ_2$ for some $Q_1, Q_2 \in K[X]$; hence $\deg B = \deg C$ and Q_1, Q_2 are constants. If C is monic like B, leading coefficients show that $Q_1 = Q_2 = 1$, so that $C = B$. $\square$

Since $K[X]$ is a domain, it has a field of fractions.

Definitions. Let K is a field. The field of fractions of $K[X]$ is the field of rational fractions $K(X)$. *The elements of $K(X)$ are* rational fractions *in one indeterminate X with coefficients in K.*

In $K(X)$, rational fractions are written as quotients, A/B or $\frac{A}{B}$, with $A, B \in K[X]$, $B \neq 0$. By definition, $A/B = C/D$ if and only if $AD = BC$, and

$$\frac{A}{B} + \frac{C}{D} = \frac{AD + BC}{BD}, \quad \frac{A}{B}\frac{C}{D} = \frac{AC}{BD}.$$

Rational fractions can be evaluated: when $F = A/B \in K(X)$ and $x \in K$, then $F(x) = A(x)\, B(x)^{-1} \in K$ is defined if $B(x) \neq 0$ and depends only on the fraction A/B and not on the polynomials A and B themselves (as long as $F(x)$ is defined). The evaluation mapping $x \longmapsto F(x)$ has good properties, but stops short of being a homomorphism, as pesky denominators keep having roots.

Section 9 brings additional properties of rational fractions in one variable.

Exercises

1. Verify that the multiplication on $R[X]$ is associative.

2. Let R be a commutative ring and let $b \in R$. Prove that the equation $bx = c$ has a unique solution in R for every $c \in R$ if and only if b is a unit.

3. Let $A \in R[X]$ have degree $n \geq 0$ and let $B \in R[X]$ have degree at least 1. Prove the following: if the leading coefficient of B is a unit of R, then there exist unique polynomials $Q_0, Q_1, \ldots, Q_n \in R[X]$ such that $\deg Q_i < \deg B$ for all i and $A = Q_0 + Q_1 B + \cdots + Q_n B^n$.

4. Let R be a subring of S. Show that evaluation at $s \in S$, $A \longmapsto A(s)$, is a ring homomorphism if and only if s commutes with every element of R.

5. Find an example of a ring R, an element $r \in R$, and polynomials $A, B \in R[X]$ such that $(AB)(r) \neq A(r)\, B(r)$.

6. Let M be a maximal ideal of R. Show that $M + (X)$ is a maximal ideal of $R[X]$.

7. Verify that $(AB)' = A'B + AB'$ for every $A, B \in R[X]$.

8. Verify that $(A^n)' = nA^{n-1}A'$ for every $n > 0$ and $A \in R[X]$.

9. Show that every polynomial $A \in R[X]$ has a kth derivative $A^{(k)}$ for every $k > 0$. If $A = \sum_{n \geq 0} a_n X^n$ has degree k, then show that $A^{(k)}(0) = k!\, a_k$.

10. Let R be commutative, with characteristic either 0 or greater than m. Show that a root r of $A \in R[X]$ has multiplicity m if and only if $A^{(k)}(r) = 0$ for all $k < m$ and $A^{(m)}(r) \neq 0$. Show that the hypothesis about the characteristic of R cannot be omitted from this result.

11. Let R be a domain and let Q be its field of fractions. Show that the field of fractions of $R[X]$ is isomorphic to $Q(X)$.

Substitution in $R[X]$ is defined as follows: if $A(X) = a_0 + a_1 X + \cdots + a_n X^n \in R[X]$ and $B \in R[X]$, then $A(B) = a_0 + a_1 B + \cdots + a_n B^n \in R$. The notation $A \circ B$ is also used for $A(B)$, since $A(B)(r) = A(B(r))$ for all $r \in R$ when R is commutative.

12. Show that substitution is an associative operation on $R[X]$.

13. Show that $A \longmapsto A(B)$ is a homomorphism of rings, when R is commutative.

14. Prove the following: if R has no zero divisors, then $A \longmapsto A(B)$ is a homomorphism of rings for every $B \in R[X]$ if and only if R is commutative.

Let R be a ring and let M be a monoid. The *semigroup ring* $R[M]$ is the ring of all mappings $a : m \longmapsto a_m$ of M into R such that $a_m = 0$ for almost all $m \in M$, added pointwise, $(a + b)_m = a_m + b_m$, and multiplied by $(ab)_m = \sum_{x,y \in M,\ xy=m} a_x b_y$.

15. Verify that $R[M]$ is a ring.

16. Explain how every $a \in R[M]$ can be written uniquely as a finite linear combination of elements of M with coefficients in R.

17. State and prove a universal property for $R[M]$.

6. Polynomials in Several Variables

A polynomial in n indeterminates X_1, X_2, ..., X_n should be a finite linear combination of monomials $X_1^{k_1} X_2^{k_2} \cdots X_n^{k_n}$. But, as before, this natural concept makes a poor definition. The formal definition specifies polynomials by their coefficients; this readily accommodates infinitely many indeterminates.

Definition. A monomial in the family $(X_i)_{i \in I}$ (of indeterminates) is a possibly infinite product $\prod_{i \in I} X_i^{k_i}$ with integer exponents $k_i \geq 0$ such that $k_i = 0$ for almost all i; then $\prod_{i \in I} X_i^{k_i}$ is the finite product $\prod_{i \in I,\, k_i \neq 0} X_i^{k_i}$.

It is convenient to denote $\prod_{i \in I} X_i^{k_i}$ by X^k, where $k = (k_i)_{i \in I}$. Monomials are multiplied by adding exponents componentwise: $X^k X^\ell = X^{k+\ell}$, where

$(k + \ell)_i = k_i + \ell_i$ for all $i \in I$; the result is a monomial, since $k_i + \ell_i \neq 0$ implies $k_i \neq 0$ or $\ell_i \neq 0$, so that $\{ i \in I \mid k_i + \ell_i \neq 0 \}$ is finite.

Definition. The free commutative monoid *on a set, written as a family* $(X_i)_{i \in I}$, *is the set* M *of all monomials* $X^k = \prod_{i \in I} X_i^{k_i}$, *where* $k_i \geqq 0$ *and* $k_i = 0$ *for almost all* i, *with multiplication* $X^k X^\ell = X^{k+\ell}$.

The identity element of M is the empty product, $1 = \prod_{i \in I} X_i^{k_i}$, in which $k_i = 0$ for all i. If $I = \{ 1, 2, \ldots, n \}$, then M is the free commutative monoid on $X_1, X_2, \ldots, X_n$ in Section I.1.

Definition. Let M *be the free commutative monoid on a family* $(X_i)_{i \in I}$. *A* polynomial in the *indeterminates* $(X_i)_{i \in I}$ *over a ring* R *[with identity] is a mapping* $A : X^k \longmapsto a_k$ *of* M *into* R *such that* $a_k = 0$ *for almost all* $k \in M$. *The set of all such polynomials is denoted by* $R[(X_i)_{i \in I}]$.

If $I = \{ 1, 2, \ldots, n \}$, where $n \geqq 1$, then $R[(X_i)_{i \in I}]$ is denoted by $R[X_1, \ldots, X_n]$. If $n = 1$, then $R[X_1, \ldots, X_n]$ is just $R[X]$; the notations $R[X, Y]$, $R[X, Y, Z]$ are commonly used when $n = 2$ or $n = 3$.

$R[(X_i)_{i \in I}]$ and $R[X_1, \ldots, X_n]$ are semigroup rings, as in the Section 5 exercises. They are often denoted by $R[X]$ and $R[x]$ when the indeterminates are well understood; we'll stick with $R[(X_i)_{i \in I}]$ and $R[X_1, \ldots, X_n]$.

Polynomials are added pointwise,

$$A + B = C \text{ when } c_k = a_k + b_k \text{ for all } k,$$

and multiplied by the usual rule,

$$AB = C \text{ when } c_m = \sum_{k+\ell=m} a_k b_\ell \text{ for all } m.$$

Proposition **6.1.** *For every ring* R *[with identity],* $R[(X_i)_{i \in I}]$, *with the operations above, is a ring.*

Proof. Let $A, B \in R[(X_i)_{i \in I}]$. Since $a_k + b_k \neq 0$ implies $a_k \neq 0$ or $b_k \neq 0$, the set $\{ k \in M \mid a_k + b_k \neq 0 \}$ is finite and $A + B \in R[(X_i)_{i \in I}]$. If $c_m = \sum_{k+\ell=m} a_k b_\ell$ for all m, then, similarly, $c_m \neq 0$ implies $a_k, b_\ell \neq 0$ for some k, ℓ; therefore $\{ m \mid c_m \neq 0 \}$ is finite and $AB \in R[(X_i)_{i \in I}]$. Thus the operations on $R[(X_i)_{i \in I}]$ are well defined.

It is immediate that $R[(X_i)_{i \in I}]$ is an abelian group under addition. The identity element of $R[(X_i)_{i \in I}]$ is the polynomial 1 with coefficients $a_k = 1$ if $k_i = 0$ for all i, $a_k = 0$ otherwise. Multiplication on $R[(X_i)_{i \in I}]$ inherits distributivity from the multiplication on R; associativity is an exercise. $\square$

Each element r of R is identified with the *constant* polynomial r with coefficients $a_k = r$ if $k_i = 0$ for all i, $a_k = 0$ otherwise; and each indeterminate X_i is identified with the polynomial X_i with coefficients $a_k = 1$ if $k_i = 1$ and $k_j = 0$ for all $j \neq i$, $a_k = 0$ otherwise. This allows a more natural notation:

Proposition 6.2. $A = \sum_k a_k X^k$, *for every* $A \in R[(X_i)_{i \in I}]$.

Proof. The infinite sum $\sum_k a_k X^k$ exists since $a_k X^k = 0$ for almost all k. We find its coefficients. By induction on m_i, the coefficient of X^k in $X_i^{m_i}$ is 1 if $k_i = m_i$ and $k_j = 0$ for all $j \neq i$, otherwise 0. Hence the coefficient of X^k in X^m is 1 if $k = m$, 0 if $k \neq m$. Then the coefficient of X^k in rX^m is r if $k = m$, 0 if $k \neq m$. Hence $\sum_m a_m X^m$ has the same coefficients as A. $\square$

Operations on polynomials can now be carried out as usual.

The ring $R[X_1, ..., X_n]$ is often defined by induction. This is useful in proving properties of $R[X_1, ..., X_n]$.

Proposition 6.3. $R[X_1, ..., X_n] \cong (R[X_1, ..., X_{n-1}])[X_n]$ *when* $n \geq 2$.

Proof. Every polynomial in $R[X_1, ..., X_n]$ can be rearranged by increasing powers of X_n, and thereby written uniquely in the form $A_0 + A_1 X_n + \cdots + A_q X_n^q$, with $A_1, ..., A_q \in R[X_1, ..., X_{n-1}]$. This bijection of $R[X_1, ..., X_n]$ onto $(R[X_1, ..., X_{n-1}])[X_n]$ preserves sums and products, since $R[X_1, ..., X_n]$ is a ring and X_n commutes with every $B \in R[X_1, ..., X_{n-1}]$. $\square$

Degrees. The degree of a monomial is its total degree. Monomials also have a degree in each indeterminate.

Definitions. The degree *of a monomial* $X^k = \prod_{i \in I} X_i^{k_i}$ *is* $\deg X^k = \sum_{i \in I} k_i$. *The* degree $\deg A$ *of a nonzero polynomial* $A = \sum_k a_k X^k$ *is the largest* $\deg X^k$ *such that* $a_k \neq 0$.

The degree in X_j *of a monomial* $X^k = \prod_{i \in I} X_i^{k_i}$ *is* $\deg_{X_j} X^k = k_j$. *The* degree in X_j, $\deg_{X_j} A$, *of a nonzero polynomial* $A = \sum_k a_k X^k$ *is the largest* $\deg_{X_j} X^k$ *such that* $a_k \neq 0$. $\square$

Readers will verify the following properties:

Proposition 6.4. *For all* $A, B \neq 0$ *in* $R[(X_i)_{i \in I}]$:

(1) $\deg (A + B) \leq \max (\deg A, \deg B)$;

(2) *if* $\deg A \neq \deg B$, *then* $\deg (A + B) = \max (\deg A, \deg B)$;

(3) $\deg (AB) \leq \deg A + \deg B$;

(4) *if* R *has no zero divisors, then* $\deg (AB) = \deg A + \deg B$.

In particular, if R *is a domain, then* $R[(X_i)_{i \in I}]$ *is a domain.*

Degrees in one indeterminate have similar properties.

Corollary 6.5. If R *has no zero divisors, then the units of* $R[(X_i)_{i \in I}]$ *are the units of* R.

Polynomial division in $R[(X_i)_{i \in I}]$ is considered in Section 12.

Homomorphisms. Polynomials in several indeterminates can be *evaluated*:

Definition. If $A = \sum_k \left(a_k \prod_{i \in I} X_i^{k_i}\right) \in R[(X_i)_{i \in I}]$ and $(r_i)_{i \in I}$ is a family of elements of R, then $A\left((r_i)_{i \in I}\right) = \sum_k \left(a_k \prod_{i \in I} r_i^{k_i}\right) \in R$.

In this formula, the possibly infinite product $\prod_{i \in I} s_i^{k_i}$ denotes the finite product $\prod_{i \in I,\, k_i \neq 0} s_i^{k_i}$. More generally, a polynomial $A \in R[(X_i)_{i \in I}]$ can be evaluated at elements of a larger ring $S \supseteq R$, for instance, at a family of polynomials in some $R[(Y_j)_{j \in J}]$; the details of this operation, *substitution*, are left to interested readers, or to those who rightly fear idleness.

If $I = \{\, 1,\, 2,\, \ldots,\, n \,\}$, then $A\left((r_i)_{i \in I}\right)$ is denoted by $A(r_1, \ldots, r_n)$. The polynomial A itself is often denoted by $A\left((X_i)_{i \in I}\right)$, or by $A(X_1, \ldots, X_n)$. As in Section 5, $(A + B)\left((r_i)_{i \in I}\right) = A\left((r_i)_{i \in I}\right) + B\left((r_i)_{i \in I}\right)$, but $(AB)\left((r_i)_{i \in I}\right) = A\left((r_i)_{i \in I}\right) B\left((r_i)_{i \in I}\right)$ requires commutativity:

Proposition **6.6.** *If R is commutative, then evaluation at $(r_i)_{i \in I} \in R$ is a homomorphism of $R[X]$ into R. More generally, if R is a subring of S and $(s_i)_{i \in I}$ are elements of S that commute with each other and with every element of R, then evaluation at $(s_i)_{i \in I}$ is a homomorphism of $R[(X_i)_{i \in I}] \subseteq S[(X_i)_{i \in I}]$ into S.*

This is proved like 5.6; we encourage our tireless readers to provide the details. Homomorphisms of rings also extend to their polynomial rings, as in Section 5.

Proposition **6.7.** *Every homomorphism of rings $\varphi : R \longrightarrow S$ extends uniquely to a homomorphism of rings $A \longmapsto {}^\varphi A$ of $R[(X_i)_{i \in I}]$ into $S[(X_i)_{i \in I}]$ that sends every X_i to itself, namely*

$$^\varphi\left(\textstyle\sum_k a_k X^k\right) = \sum_k \varphi(a_k)\, X^k.$$

The universal property of $R[(X_i)_{i \in I}]$ constructs every ring homomorphism $\psi : R[(X_i)_{i \in I}] \longrightarrow S$. Necessarily, the restriction $\varphi : R \longrightarrow S$ of ψ to R is a ring homomorphism, and the elements $\psi(X_i)$ of S commute with each other and with every $\varphi(r)$, since, in $R[(X_i)_{i \in I}]$, the monomials X_i commute with each other and with all constants.

Theorem **6.8.** *Let R and S be rings and let $\varphi : R \longrightarrow S$ be a homomorphism of rings. Let $(s_i)_{i \in I}$ be elements of S that commute with each other and with $\varphi(r)$ for every $r \in R$ (arbitrary elements of S if S is commutative). There is a unique homomorphism of rings $\psi : R[(X_i)_{i \in I}] \longrightarrow S$ that extends φ and sends X_i to s_i for every i, namely $\psi\left(A((X_i)_{i \in I})\right) = {}^\varphi A\left((s_i)_{i \in I}\right).$*

$$R \overset{\subseteq}{\longrightarrow} R[(X_i)_{i \in I}]$$
$$\varphi \searrow \quad \downarrow \psi$$
$$S$$

This is proved like Theorem 5.11. Propositions 6.6 and 6.7 are particular cases of Theorem 6.8.

Rational fractions. We now let R be a field K. Then $K[(X_i)_{i \in I}]$ is a domain, by 6.4, and has a field of fractions:

Definitions. Let K be a field. The field of fractions of $K[(X_i)_{i \in I}]$ is the field of rational fractions $K((X_i)_{i \in I})$. The elements of $K((X_i)_{i \in I})$ are rational fractions in the indeterminates $(X_i)_{i \in I}$ over the field K. $\square$

If $I = \{1\,2, \ldots, n\}$, where $n \geq 1$, then $K((X_i)_{i \in I})$ is denoted by $K(X_1, \ldots, X_n)$. If $n = 1$, $K(X_1, \ldots, X_n)$ is just $K(X)$; $K(X_1, X_2)$ and $K(X_1, X_2, X_3)$ are more commonly denoted by $K(X, Y)$ and $K(X, Y, Z)$.

In $K((X_i)_{i \in I})$, rational fractions are written as quotients, A/B or $\frac{A}{B}$, with $A, B \in K[(X_i)_{i \in I}]$, $B \neq 0$. By definition, $A/B = C/D$ if and only if $AD = BC$, and

$$\frac{A}{B} + \frac{C}{D} = \frac{AD + BC}{BD}, \quad \frac{A}{B}\frac{C}{D} = \frac{AC}{BD}.$$

The field $K(X_1, \ldots, X_n)$ can also be defined by induction:

Proposition **6.9.** $K(X_1, \ldots, X_n) \cong \big(K(X_1, \ldots, X_{n-1})\big)(X_n)$ *when* $n \geq 2$.

As in the one-variable case, rational fractions can be evaluated: when $F = A/B \in K((X_i)_{i \in I})$ and $x_i \in K$ for all $i \in I$, then

$$F\big((x_i)_{i \in I}\big) = A\big((x_i)_{i \in I}\big) \, B\big((x_i)_{i \in I}\big)^{-1} \in K$$

is defined if $B\big((x_i)_{i \in I}\big) \neq 0$, and, when defined, depends only on the fraction A/B and not on the polynomials A and B themselves. The mapping $(x_i)_{i \in I} \longmapsto F\big((x_i)_{i \in I}\big)$, defined wherever possible, is a *rational function*.

Exercises

1. Give a direct proof that multiplication in $R[(X_i)_{i \in I}]$ is associative.

2. Let M be a maximal ideal of R. Show that $M + \big((X_i)_{i \in I}\big)$ is a maximal ideal of $R[(X_i)_{i \in I}]$.

3. Let K be a field. Show that $K[X_1, X_2]$ has ideals that are not principal.

4. Flesh out a detailed proof of the statement that the bijection of $R[X_1, \ldots, X_n]$ onto $\big(R[X_1, \ldots, X_{n-1}]\big)[X_n]$, obtained by rearranging polynomials in $R[X_1, \ldots, X_n]$ by increasing powers of X_n, "preserves sums and products since $R[X_1, \ldots, X_n]$ is a ring".

5. A polynomial $A \in K[(X_i)_{i \in I}]$ is *homogeneous* when all its monomials have the same degree ($\deg X^k = \deg X^\ell$ whenever a_k, $a_\ell \neq 0$). Show that every polynomial in $K[(X_i)_{i \in I}]$ can be written uniquely as a sum of homogeneous polynomials.

6. Prove the universal property of $R[(X_i)_{i \in I}]$.

7. Use induction on n to prove the universal property of $R[X_1, \ldots, X_n]$.

8. Show that $\mathbb{Z}[(X_i)_{i \in I}]$ is the *free commutative ring* [with identity] on $(X_i)_{i \in I}$ in the sense that every mapping of $(X_i)_{i \in I}$ into a commutative ring R extends uniquely to a homomorphism of $\mathbb{Z}[(X_i)_{i \in I}]$ into R.

9. Let R be a domain and let Q be its field of fractions. Show that the field of fractions of $R[(X_i)_{i \in I}]$ is isomorphic to $Q((X_i)_{i \in I})$.

10. Show that $K(X_1, ..., X_n) \cong \big(K(X_1, ..., X_{n-1}) \big)(X_n)$ when $n \geq 2$.

*11. Define substitution in $R[[(X_i)_{i \in I}]]$ and establish its main properties.

*12. Define a polynomial ring in which the indeterminates $(X_i)_{i \in I}$ commute with constants but not with each other. State and prove its universal property. Does this yield "free" rings?

7. Formal Power Series

This section can be skipped. Power series lose some of their charm when transplanted to algebra: they can still be added and multiplied, but, without a topology, there don't have sums; they become *formal* power series.

Definition. Let $M = \{ 1, X, ..., X^n, ... \}$ *be the free monoid on* $\{X\}$. *A formal power series* $A = \sum_{n \geq 0} a_n X^n$ *in the* indeterminate X over *a ring* R [*with identity*] *is a mapping* $A : X^n \longmapsto a_n$ *of* M *into* R.

Power series are added pointwise,

$$A + B = C \text{ when } c_n = a_n + b_n \text{ for all } n \geq 0,$$

and multiplied by the usual rule,

$$AB = C \text{ when } c_n = \sum_{i+j=n} a_i\, b_j \text{ for all } n \geq 0.$$

The following result is straightforward:

Proposition **7.1.** *If* R *is a ring, then formal power series over* R *in the indeterminate* X *constitute a ring* $R[[X]]$.

At this point, $A = \sum_{n \geq 0} a_n X^n$ is not an actual sum in $R[[X]]$ (unless A is a polynomial). But we shall soon find a way to add series in $R[[X]]$.

Order. Power series do not have degrees, but they have something similar.

Definition. The order ord A *of a formal power series* $A = \sum_{n \geq 0} a_n X^n \neq 0$ *is the smallest integer* $n \geq 0$ *such that* $a_n \neq 0$.

The order of the zero series 0 is sometimes left undefined; we define it as ∞, so that ord $0 >$ ord A for all $A \neq 0$. Thus $A = \sum_{n \geq 0} a_n X^n$ has order at least n if and only if $a_k = 0$ for all $k < n$, if and only if it is a multiple of X^n. The following properties are straightforward:

Proposition **7.2.** *For all* $A, B \neq 0$ *in* $R[[X]]$:

(1) ord $(A + B) \geq$ min (ord A, ord B);

(2) *if* ord $A \neq$ ord B, *then* ord $(A + B) =$ min (ord A, ord B);

(3) ord $(AB) \geq$ ord $A +$ ord B;

(4) *if R has no zero divisors, then* ord $(AB) =$ ord $A +$ ord B.

In particular, if R is a domain, then $R[[X]]$ is a domain.

Sums. Certain series can now be added in $R[[X]]$ in a purely algebraic fashion (but for which the exercises give a topological interpretation).

Definition. A sequence T_0, T_1, ..., T_k, ... of formal power series $T_k = \sum_{n \geq 0} t_{k,n} X^n \in R[[X]]$ is addible, or summable, in $R[[X]]$ when, for every $n \geq 0$, T_k has order at least n for almost all k. Then the sum $S = \sum_{k \geq 0} T_k$ is the power series with coefficients $s_n = \sum_{k \geq 0} t_{k,n}$.

If ord $T_k \geq n$ for almost all k, then $t_{k,n} = 0$ for almost all k, and the infinite sum $s_n = \sum_{k \geq 0} t_{k,n}$ is defined in R.

In particular, T_0, T_1, ..., T_k, ... is addible whenever ord $T_k \geq k$ for all k. For example, for any $A = \sum_{n \geq 0} a_n X^n \in R[[X]]$, the sequence a_0, $a_1 X$, ..., $a_n X^n$, ... is addible, since ord $a_n X^n \geq n$. Its sum is A. Thus $A = \sum_{n \geq 0} a_n X^n$ is now an actual sum in $R[[X]]$.

Proposition **7.3.** *If R is commutative, then $A = \sum_{n \geq 0} a_n X^n$ is a unit of $R[[X]]$ if and only if a_0 is a unit of R.*

Proof. If A is a unit of $R[[X]]$, then $AB = 1$ for some $B \in R[[X]]$, $a_0 b_0 = 1$, and a_0 is a unit of R.

We first prove the converse when $a_0 = 1$. Let $A = 1 - T$. Then ord $T \geq 1$, and ord $T^n \geq n$, by 7.2. Hence the sequence 1, T, ..., T^n, ... is addible. We show that $B = \sum_{k \geq 0} T^k$ satisfies $AB = 1$.

Let $B_n = 1 + T + \cdots + T^n$. Then $B - B_n = \sum_{k > n} T^k$ and ord $(B - B_n) > n$, since ord $T^k > n$ when $k > n$. By 7.2, ord $(AB - AB_n) > n$. Now,

$$AB_n = (1 - T)(1 + T + \cdots + T^n) = 1 - T^{n+1}.$$

Hence ord $(AB_n - 1) > n$. By 7.2, ord $(AB - 1) > n$. This holds for all $n \geq 0$; therefore $AB = 1$, and A is a unit of $R[[X]]$.

If now a_0 is any unit, then $A a_0^{-1}$, which has constant coefficient 1, is a unit of $R[[X]]$, $A a_0^{-1} B = 1$ for some $B \in R[[X]]$, and A is a unit of $R[[X]]$. $\square$

Formal Laurent series. A Laurent series is a power series with a few additional negative terms.

Definition. Let $G = \{ X^n \mid n \in \mathbb{Z} \}$ be the free group on $\{X\}$. A formal Laurent series $A = \sum_n a_n X^n$ in the indeterminate X over a ring R is a mapping $A : X^n \longmapsto a_n$ of G into R such that $a_n = 0$ for almost all $n < 0$.

Equivalently, a Laurent series $A = \sum_n a_n X^n$ looks like (and will soon be) the sum of a polynomial $\sum_{n < 0} a_n X^n$ in X^{-1} and a power series $\sum_{n \geq 0} a_n X^n$.

Laurent series are added pointwise,

$$A + B = C \text{ when } c_n = a_n + b_n \text{ for all } n \in \mathbb{Z},$$

and multiplied by the usual rule,

$$AB = C \text{ when } c_n = \sum_{i+j=n} a_i\, b_j \text{ for all } n \in \mathbb{Z}.$$

The following result is straightforward:

Proposition **7.4.** *For every ring* R *[with identity], the Laurent series over* R *with one indeterminate* X *constitute a ring* $R((X))$.

The *order* ord A of a Laurent series $A = \sum_n a_n X^n \neq 0$ is the smallest integer n such that $a_n \neq 0$; as before, we let ord $0 = \infty$. Thus, a Laurent series $A = \sum_{n \geq 0} a_n X^n$ has order at least $n \in \mathbb{Z}$ if and only if $a_k = 0$ for all $k < n$, if and only if it is the product of X^n and a power series.

Readers will easily extend Proposition 7.2 to Laurent series, but may be more interested in the following result.

Proposition **7.5.** *For every field* K, $K((X))$ *is a field; in fact,* $K((X))$ *is isomorphic to the field of fractions of* $K[[X]]$.

Proof. Let $A = \sum_n a_n X^n \in K((X))$, $A \neq 0$. Then A has order $m \in \mathbb{Z}$ and $A = X^m B$ for some $B \in K[[X]]$ whose constant term is $b_0 = a_m \neq 0$. By 7.3, B is a unit of $K[[X]]$: $BC = 1$ for some $C \in K[[X]]$. Then $A X^{-m} C = 1$ and A is a unit of $K((X))$. Thus $K((X))$ is a field. Moreover, $K[[X]]$ is a subring of $K((X))$, and a Laurent series $A \in K((X))$ either has order $m \geq 0$ and belongs to $K[[X]]$, or has order $m < 0$ and can be written as $A = X^m B = B (X^{-m})^{-1}$ with $B, X^{-m} \in K[[X]]$; by 4.12, $K((X))$ is isomorphic to the field of fractions of $K[[X]]$. $\square$

Exercises

1. Verify that multiplication on $R[[X]]$ is associative.

2. Let R be commutative and let $\mathfrak{m}$ be a maximal ideal of R. Show that $\mathfrak{m} + (X)$ is a maximal ideal of $R[[X]]$.

Substitution in $R[[X]]$ substitutes a power series C of order at least 1 into any power series $A = \sum_{n \geq 0} a_n X^n$ to yield a power series $A \circ C$ or $A(C) = \sum_{n \geq 0} a_n C^n$.

3. Show that substitution is a well-defined operation on $R[[X]]$, which is associative when R is commutative.

4. Show that, in $R[[X]]$, $(A + B) \circ C = (A \circ C) + (B \circ C)$, and, if R is commutative, $(AB) \circ C = (A \circ C)(B \circ C)$, whenever C has order at least 1.

5. Show that $R[[X]]$ is a metric space, in which $d(A, B) = 2^{-\text{ord}\,(A-B)}$ if $A \neq B$, $d(A, B) = 0$ if $A = B$. Show that the operations on $R[[X]]$ are continuous.

6. Show that the metric space $R[[X]]$ is the completion of $R[X]$.

7. Let $T_0, T_1, \ldots, T_k, \ldots$ be an addible sequence. Show that $\sum_k T_k$ is the sum of a series in the metric space $R[[X]]$.

8. Let K be the field of fractions of a domain R. Show that $K((X))$ is the field of fractions of $R[[X]]$.

9. Let R be commutative. Show that a Laurent series of order n is a unit of $R((X))$ if and only if its coefficient a_n is a unit of R.

10. Let K be a field. Describe the homomorphism $K(X) \longrightarrow K((X))$, whose existence is guaranteed by 7.5, that expands rational fractions into Laurent series.

*11. Set up a theory of formal power series in several variables.

8. Principal Ideal Domains

This section extends the main arithmetic properties of $\mathbb{Z}$ to all principal ideal domains, including polynomial rings $K[X]$ where K is a field.

Definition. A principal ideal domain or PID *is a domain (a commutative ring with identity and no zero divisors) in which every ideal is principal.*

We already have some examples: by 2.4, 5.12, $\mathbb{Z}$ is a PID, and so is $K[X]$ for every field K. On the other hand, polynomial rings with more than one indeterminate are not PIDs (see the exercises).

Representatives. By 2.3, every ideal $\mathfrak{a}$ of a PID R is the set $\mathfrak{a} = (a) = Ra$ of all multiples of some $a \in R$; thus $x \in \mathfrak{a}$ if and only if $a \mid x$ (a divides x). Here, a is unique up to multiplication by a unit:

Lemma **8.1.** *In a domain R, $Ra = Rb$ if and only if $a = ub$ for some unit u.*

Proof. If u is a unit, then $Ru = R$ and $Rub = Rb$. Conversely, if $Ra = Rb$, then $a = ub$, $b = va$ for some $u, v \in R$; if $a = 0$, then $b = 0$ and $a = 1b$; otherwise, $uva = a \neq 0$ implies $uv = 1$, so that u is a unit. $\square$

In Lemma 8.1, the equivalence relation $Ra = Rb$ partitions R into equivalence classes; equivalent elements are often called *associates*, and we call the equivalence classes *associate classes*. Uniqueness in various results can be achieved by selecting one *representative* element in each associate class.

Proposition **8.2.** *In a domain R, every principal ideal is generated by a unique representative element.*

In $\mathbb{Z}$, the units are ± 1, and nonnegative integers serve as representative elements; Proposition 2.4 already states that every ideal of $\mathbb{Z}$ is generated by a unique nonnegative integer. By 5.4, the units of $K[X]$ are the nonzero elements of K; monic polynomials, together with 0, serve as representative elements; in Proposition 5.12, every nonzero ideal of $K[X]$ is already generated by a unique monic polynomial. Fortunately, these manic representatives do not assemble to pass laws.

Properties. We now extend to PIDs the basic arithmetic properties of integers. The main property has to do with elements that are sometimes called prime (as in $\mathbb{Z}$), sometimes called irreducible (as in $K[X]$).

Definitions. An element p of a domain R is prime *when p is not zero or a unit, and $p \mid ab$ implies $p \mid a$ or $p \mid b$ (equivalently, $ab \in Rp$ implies $a \in Rp$ or $b \in Rp$). An element q of a domain R is* irreducible *when q is not zero or a unit, and $q = ab$ implies that a is a unit or b is a unit.*

Proposition **8.3.** *In a principal ideal domain R, the following conditions on an element $p \in R$ are equivalent:* (i) *p is irreducible;* (ii) *p is prime;* (iii) *Rp is a nonzero prime ideal;* (iv) *Rp is a nonzero maximal ideal.*

In case $R = \mathbb{Z}$, this is Corollary 4.8. In general, Proposition 8.3 implies the following: when u is a unit of R, then p is irreducible if and only if up is irreducible.

Proof. (iv) implies (iii), by 4.7; (iii) implies (ii) trivially; and (ii) implies (i): if p is prime and $p = ab$, then p divides, say, a; since a already divides p, b is a unit, by 8.1.

We show that (i) implies (iv). Assume that Rp is contained in an ideal $\mathfrak{a} = Ra$ of R. Then $p = ab$ for some $b \in R$. By (i), either a is a unit, and then $\mathfrak{a} = R$, or b is a unit, and then $\mathfrak{a} = Rp$. $\square$

The main property of PIDs can now be stated in two equivalent forms.

Theorem **8.4A.** *In a principal ideal domain R, every element, other than 0 and units, is a nonempty product of irreducible elements. If furthermore two nonempty products $p_1 p_2 \cdots p_m = q_1 q_2 \cdots q_n$ of irreducible elements are equal, then $m = n$ and the terms can be indexed so that $Rp_i = Rq_i$ for all i.*

Theorem **8.4B.** *Every nonzero element of R can be written as the product $u \, p_1^{k_1} \, p_2^{k_2} \cdots p_n^{k_n}$ of a unit and of positive powers of distinct representative irreducible elements, which are unique up to the order of the terms.*

Proof. We prove the first statement, which implies the second. Assume that R has *bad* elements: elements, other than 0 and units, that are not products of irreducible elements. The *bad* principal ideals generated by bad elements then constitute a nonempty set $\mathcal{B}$ of ideals of R. We show that $\mathcal{B}$ has a maximal element Rb. Otherwise, let $Rb_1 \in \mathcal{B}$. Since Rb_1 is not maximal there exists $Rb_1 \subsetneqq Rb_2 \in \mathcal{B}$. Since Rb_2 is not maximal there exists $Rb_2 \subsetneqq Rb_3 \in \mathcal{B}$. This constructs a chain of ideals $Rb_1 \subsetneqq \cdots \subsetneqq Rb_n \subsetneqq Rb_{n+1} \subsetneqq \cdots$. Then $\mathfrak{b} = \bigcup_{n>0} Rb_n$ is an ideal of R. Since R is a PID, $\mathfrak{b}$ is generated by some $b \in R$. Then $b \in Ra_n$ for some n, and $(b) \subseteq Rb_n \subsetneqq Rb_{n+1} \subsetneqq \mathfrak{b} = (b)$. This contradiction shows that $\mathcal{B}$ has a maximal element Rm, where m is bad. (Readers who are already familiar with Noetherian rings will easily recognize this part of the proof.)

Now, m, which is bad, is not 0, not a unit, and not irreducible. Hence

$m = ab$ for some $a, b \in R$, neither of which is 0 or a unit. Then $Rm \subsetneqq Ra$ and $Rm \subsetneqq Rb$. Hence a and b cannot be bad and are products of irreducible elements. But then so is $m = ab$. This contradiction shows that every element of R, other than 0 and units, is a product of irreducible elements.

Next, assume that $p_1 p_2 \cdots p_m = q_1 q_2 \cdots q_n$, where $m, n > 0$ and all p_i, q_j are irreducible. We prove by induction on $m + n \geq 2$ that $m = n$ and the elements p_i, q_j can be reindexed so that $Rp_i = Rq_i$ for all i. This is clear if $m = n = 1$. Now assume, say, $m > 1$. Then p_m divides $q_1 q_2 \cdots q_n$; since p_m is prime by 10.3, p_m divides some q_k: $q_k = up_m$ for some $u \in R$. Since q_k is irreducible, u is a unit and $Rq_k = Rp_m$. The elements q_j can be reindexed so that $k = n$; then $Rq_n = Rp_m$ and $q_n = up_m$.

The equality $p_1 p_2 \cdots p_m = q_1 q_2 \cdots q_n$ now yields $p_1 p_2 \cdots p_{m-1} = u q_1 q_2 \cdots q_{n-1}$. Hence $n > 1$: otherwise, $p_1 p_2 \cdots p_{m-1} = u$ and p_1, ..., p_{n-1} are units, a contradiction. Now, uq_1 is irreducible; by the induction hypothesis, $m - 1 = n - 1$, and the remaining terms can be reindexed so that $Rp_1 = Ruq_1 = Rq_1$ and $Rp_i = Rq_i$ for all $1 < i < m$. $\square$

Least common multiples and greatest common divisors can be defined in any domain, but do not necessarily exist.

Definitions. In a domain, an element m is a least common multiple *or* l.c.m. *of two elements a and b when m is a multiple of a and of b, and every multiple of both a and b is also a multiple of m; an element d is a* greatest common divisor *or* g.c.d. *of two elements a and b when d divides a and b, and every element that divides a and b also divides d.*

Any two l.c.m.s of a and b must be multiples of each other, and similarly for g.c.d.s; by 8.1, the l.c.m. and g.c.d. of a and b, when they exist, are unique up to multiplication by a unit. They are often denoted by $[a, b]$ and (a, b); the author prefers $\text{lcm}(a, b)$ and $\text{gcd}(a, b)$.

In a PID, l.c.m.s and g.c.d.s arise either from ideals or from 8.4.

Proposition **8.5.** *In a principal ideal domain R, every $a, b \in R$ have a least common multiple and a greatest common divisor. Moreover, $m = \text{lcm}(a, b)$ if and only if $Rm = Ra \cap Rb$, and $d = \text{gcd}(a, b)$ if and only if $Rd = Ra + Rb$. In particular, $d = \text{gcd}(a, b)$ implies $d = xa + yb$ for some $x, y \in R$.*

Proof. By definition, $m = \text{lcm}(a, b)$ (m is an l.c.m. of a and b) if and only if $m \in Ra \cap Rb$, and $c \in Ra \cap Rb$ implies $c \in Rm$; if and only if $Rm = Ra \cap Rb$. An l.c.m. exists since the ideal $Ra \cap Rb$ must be principal.

Similarly, $d = \text{gcd}(a, b)$ if and only if $a, b \in Rd$, and $a, b \in Rc$ implies $c \in Rd$, if and only if Rd is the smallest principal ideal of R that contains both Ra and Rb. The latter is $Ra + Rb$, since every ideal of R is principal. Hence $d = \text{gcd}(a, b)$ if and only if $Rd = Ra + Rb$, and then $d = xa + yb$ for some $x, y \in R$. A g.c.d. exists since the ideal $Ra + Rb$ must be principal. $\square$

Readers may now define l.c.m.s and g.c.d.s of arbitrary families $(a_i)_{i \in I}$ and use similar arguments to prove their existence in PIDs.

In a PID, the l.c.m. and g.c.d. of a and $b \in R$ can also be obtained from 8.4. Write a and b as products $a = u \, p_1^{k_1} \, p_2^{k_2} \, \cdots \, p_m^{k_m}$ and $b = v \, q_1^{\ell_1} \, q_2^{\ell_2} \, \cdots \, q_n^{\ell_n}$ of a unit and positive powers of distinct representative irreducible elements. Merge the sequences $p_1, \ldots, p_m$ and $q_1, \ldots, q_n$, so that a and b are products $a = u \, p_1^{a_1} \, p_2^{a_2} \, \cdots \, p_n^{a_n}$ and $b = v \, p_1^{b_1} \, p_2^{b_2} \, \cdots \, p_n^{b_n}$ of a unit and nonnegative powers of the same distinct representative irreducible elements. Readers may establish the following properties:

Proposition **8.6.** *In a principal ideal domain, let* $a = u \, p_1^{a_1} \, p_2^{a_2} \, \cdots \, p_n^{a_n}$ *and* $b = v \, p_1^{b_1} \, p_2^{b_2} \, \cdots \, p_n^{b_n}$ *be products of a unit and nonnegative powers of the same distinct representative irreducible elements. Then:*

(1) *a divides b if and only if* $a_i \leqq b_i$ *for all* i.

(2) $c = p_1^{c_1} \, p_2^{c_2} \, \cdots \, p_n^{c_n}$ *is an l.c.m. of a and b if and only if* $c_i = \max(a_i, b_i)$ *for all* i.

(3) $d = p_1^{d_1} \, p_2^{d_2} \, \cdots \, p_n^{d_n}$ *is a g.c.d. of a and b if and only if* $d_i = \min(a_i, b_i)$ *for all* i.

(4) $\mathrm{lcm}\,(a, b) \gcd(a, b) = wab$ *for some unit* w.

For instance, if $R = \mathbb{Z}$ and $a = 24 = 2^3 \cdot 3$, $b = 30 = 2 \cdot 3 \cdot 5$, then $\mathrm{lcm}\,(a, b) = 2^3 \cdot 3 \cdot 5 = 120$ and $\gcd(a, b) = 2 \cdot 3 = 6$.

The following properties make fine exercises:

Proposition **8.7.** *In a PID, if* $\gcd(a, b) = \gcd(a, c) = 1$, *then* $\gcd(a, bc) = 1$; *if a divides bc and* $\gcd(a, b) = 1$, *then a divides c.*

Irreducible polynomials. Now, let K be a field. Theorem 8.4 yields the following property of $K[X]$:

Corollary **8.8.** *Let* K *be a field. In* $K[X]$, *every nonzero polynomial is the product of a constant and positive powers of distinct monic irreducible polynomials, which are unique up to the order of the terms.*

What are these irreducible polynomials? The answer reveals profound differences between various fields. We begin with a general result, left to readers.

Proposition **8.9.** *Let* K *be a field. In* $K[X]$:

(1) *every polynomial of degree 1 is irreducible;*

(2) *an irreducible polynomial of degree at least 2 has no root in* K;

(3) *a polynomial of degree 2 or 3 with no root in* K *is irreducible.*

On the other hand, $(X^2 + 1)^2 \in \mathbb{R}[X]$ has no root in $\mathbb{R}$ but is not irreducible.

Equipped with Proposition 8.9 we clean up the cases $K = \mathbb{C}$ and $K = \mathbb{R}$.

Proposition **8.10.** *A polynomial over* $\mathbb{C}$ *is irreducible if and only if it has degree* 1.

Proposition 8.10 is often stated as follows:

Theorem **8.11** (Fundamental Theorem of Algebra). *Every nonconstant polynomial over* $\mathbb{C}$ *has a root in* $\mathbb{C}$.

This result is due to Gauss [1799]. In 1799, algebra was primarily concerned with polynomial equations, and Theorem 8.11 was indeed of fundamental importance.

Complex analysis provides the best proof of Theorem 8.11 (a much more algebraic proof is given in Section VI.2). Assume that $f \in \mathbb{C}[X]$ has no root in $\mathbb{C}$. Then the function $g(z) = 1/f(z)$ is holomorphic on all of $\mathbb{C}$. If f has degree 1 or more, then $|g(z)| \longrightarrow 0$ when $z \longrightarrow \infty$, so that the larger values of $|g(z)|$ all occur inside some closed disk D; since $|g(z)|$ is continuous it has a maximum value on the compact set D, which is also its maximum value on all of $\mathbb{C}$. This also holds if f is constant. The Maximum principle now implies that g is constant, and then so is f.

Proposition **8.12.** *A polynomial over* $\mathbb{R}$ *is irreducible if and only if it has either degree* 1, *or degree* 2 *and no root in* $\mathbb{R}$.

Proof. Polynomials with these properties are irreducible, by 8.9. Conversely, let $f \in \mathbb{R}[X]$, $f \neq 0$. As a polynomial over $\mathbb{C}$, f is, by 8.8 and 8.10, the product of a constant and monic polynomials of degree 1:

$$f(X) = a_n (X - r_1)(X - r_2) \cdots (X - r_n).$$

Then $n = \deg f$, a_n is the leading coefficient of f, and $r_1, \ldots, r_n$ are the (not necessarily distinct) roots of f in $\mathbb{C}$. Since f has real coefficients, complex conjugation yields

$$f(X) = \overline{f}(X) = a_n (X - \overline{r}_1)(X - \overline{r}_2) \cdots (X - \overline{r}_n),$$

Then $\{ r_1, \ldots, r_n \} = \{ \overline{r}_1, \ldots, \overline{r}_n \}$, for f has only one such factorization. Therefore the roots of f consist of real roots and pairs of nonreal complex conjugate roots. Hence f is the product of a_n, polynomials $X - r \in \mathbb{R}[X]$ with $r \in \mathbb{R}$, and polynomials

$$(X - z)(X - \overline{z}) = X^2 - (z + \overline{z}) X + z\overline{z} \in \mathbb{R}[X]$$

with $z \in \mathbb{C} \backslash \mathbb{R}$ and no root in R. If f is irreducible in $\mathbb{R}[X]$, then f has either degree 1, or degree 2 and no root in $\mathbb{R}$. $\square$

The case $K = \mathbb{Q}$ is more complicated and is left to Section 10. We now turn to the finite fields $K = \mathbb{Z}_p$.

Proposition **8.13.** *For every field* K, $K[X]$ *contains infinitely many monic irreducible polynomials.*

Proof. This proof is due to Euclid, who used a similar argument to show that $\mathbb{Z}$ contains infinitely many primes. We show that no finite sequence $q_1, q_2, \ldots, q_n$ can contain every monic irreducible polynomial of $K[X]$. Indeed, $f = 1 + q_1 q_2 \cdots q_n$ is not constant and is by 8.8 a multiple of a monic irreducible polynomial q. Then $q \neq q_1, q_2, \ldots, q_n$: otherwise, q divides $1 = f - q_1 q_2 \cdots q_n$. $\square$

If K is finite, then $K[X]$ has irreducible polynomials of arbitrarily high degree, since there are only finitely many polynomials of degree at most n.

Irreducible polynomials of low degree are readily computed when $K = \mathbb{Z}_p$ and p is small. For example, let $K = \mathbb{Z}_2$. Let $f \in \mathbb{Z}_2[X]$, $f \neq 0$. The coefficients of f are either 0 or 1; hence f has no root in $\mathbb{Z}_2$ if and only if its constant coefficient is 1 and it has an odd number of nonzero terms. Then

$$X, \ X + 1, \ X^2 + X + 1, \ X^3 + X + 1, \text{ and } X^3 + X^2 + 1$$

are irreducible, by 8.9, and all other polynomials of degree 2 or 3 have roots in $\mathbb{Z}_2$. Next there are four polynomials of degree 4 with no roots: $X^4 + X + 1$, $X^4 + X^2 + 1$, $X^4 + X^3 + 1$, and $X^4 + X^3 + X^2 + X + 1$. If one of these is not irreducible, then it is a product of irreducible polynomials of degree 2 (degree 1 is out, for lack of roots) and must be $(X^2 + X + 1)(X^2 + X + 1) = X^4 + X^2 + 1$ (by 4.4). This leaves three irreducible polynomials of degree 4:

$$X^4 + X + 1, \ X^4 + X^3 + 1, \text{ and } X^4 + X^3 + X^2 + X + 1.$$

Exercises

1. Show that no polynomial ring with more than one indeterminate is a PID.

2. A *Gauss integer* is a complex number $x + iy$ in which x and y are integers. Show that the ring R of all Gauss integer is a PID. (You may wish to first prove the following: for every $a, b \in R$, $b \neq 0$, there exist $q, r \in R$ such that $a = bq + r$ and $|r| < |b|$.)

3. A ring R is *Euclidean* when there exists a mapping $\varphi : R\backslash\{0\} \longrightarrow \mathbb{N}$ with the following division property: for every $a, b \in R$, $b \neq 0$, there exist $q, r \in R$ such that $a = bq + r$ and either $r = 0$ or $\varphi(r) < \varphi(b)$. Prove that every Euclidean domain is a PID.

4. Show that every family of elements of a PID has an l.c.m. (which may be 0).

5. Show that every family $(a_i)_{i \in I}$ of elements of a PID has a g.c.d. d, which can be written in the form $d = \sum_{i \in I} x_i a_i$ for some $x_i \in R$ (with $x_i = 0$ for almost all $i \in I$).

6. Prove Proposition 8.6.

7. In a PID, show that $\gcd(a, b) = \gcd(a, c) = 1$ implies $\gcd(a, bc) = 1$.

8. Prove the following: in a PID, if a divides bc and $\gcd(a, b) = 1$, then a divides c.

9. Let K be a field. Prove that, in $K[X]$, a polynomial of degree 2 or 3 is irreducible if and only if it has no root in K.

10. Write $X^5 + X^3 - X^2 - 1 \in \mathbb{R}[X]$ as a product of irreducible polynomials.

11. Write $X^4 + 1 \in \mathbb{R}[X]$ as a product of irreducible polynomials.

12. Find all irreducible polynomials of degree 5 in $\mathbb{Z}_2[X]$.

13. Find all monic irreducible polynomials of degree up to 3 in $\mathbb{Z}_3[X]$. (Readers who are blessed with long winter evenings can try degree 4.)

9. Rational Fractions

A first application of principal ideal domains is the decomposition of rational fractions into a sum of partial fractions, a perennial favorite of calculus students.

Let K be a field. A *partial fraction* is a rational fraction $f/q^r \in K(X)$ in which q is monic and irreducible, $r \geq 1$, and $\deg f < \deg q$. Then f, q, and r are unique (see the exercises). The main result of this section is the following:

Theorem **9.1.** *Every rational fraction over a field can be written uniquely as the sum of a polynomial and partial fractions with distinct denominators.*

The proof of Theorem 9.1 has three parts. The first part reduces rational fractions and ejects the polynomial part. A rational fraction $f/g \in K(X)$ is in *reduced form* when g is monic and $\gcd(f, g) = 1$.

Lemma **9.2.** *Every rational fraction can be written uniquely in reduced form.*

Proof. Given f/g, divide f and g by the leading coefficient of g and then by a monic g.c.d. of f and g; the result is in reduced form.

Let $f/g = p/q$, $fq = gp$, with g, q monic and $\gcd(f, g) = \gcd(p, q) = 1$. Then q divides gp; since $\gcd(p, q) = 1$, q divides g, by 8.7. Similarly, g divides q. Since q and g are monic, $q = g$. Then $p = f$. $\square$

We call a rational fraction f/g *polynomial-free* when $\deg f < \deg g$.

Lemma **9.3.** *Every rational fraction can be written uniquely as the sum of a polynomial and a polynomial-free fraction in reduced form.*

Proof. By 9.2 we may start with a rational fraction f/g in reduced form. Polynomial division yields $f = gq + r$ with $q, r \in K[X]$ and $\deg r < \deg g$. Then $f/g = q + r/g$; r/g is polynomial-free and is in reduced form, since g is monic and $\gcd(r, g) = \gcd(f, g) = 1$. Conversely let $f/g = p + s/h$, with $p \in K[X]$, $\deg s < \deg h$, h monic, and $\gcd(s, h) = 1$. Then $f/g = (ph + s)/h$. Both fractions are in reduced form; hence $g = h$ and $f = ph + s = pg + s$, by 9.2. Uniqueness in polynomial division then yields $p = q$ and $s = r$. $\square$

The second part of the proof breaks a reduced polynomial-free fraction f/g into a sum of reduced polynomial-free fractions a/q^k, in which q is irreducible. (These are not quite partial fractions, since $\deg a < \deg q^k$, rather than $\deg a < \deg q$.)

Lemma **9.4.** *If $\deg f < \deg gh$ and $\gcd(g, h) = 1$, then there exist unique polynomials a, b such that $\deg a < \deg g$, $\deg b < \deg h$, and $f/(gh) = (a/g) + (b/h)$. If $\gcd(f, gh) = 1$, then $\gcd(a, g) = \gcd(b, h) = 1$.*

Proof. Since $\gcd(g, h) = 1$, there exist polynomials s, t such that $gs + ht = f$. Polynomial division yields $t = gp + a$, $s = hq + b$, where $\deg a < \deg g$ and $\deg b < \deg h$. Then $f = gh(p + q) + ah + bg$, with $\deg(ah + bg) < \deg gh$, and $p + q = 0$: otherwise, $\deg f \geq \deg gh$, contradicting the hypothesis. Hence $f = ah + bg$, and $f/(gh) = (a/g) + (b/h)$. If $\gcd(f, gh) = 1$, then a polynomial that divides a and g, or divides b and h, also divides $f = ah + bg$ and gh; hence $\gcd(a, g) = \gcd(b, h) = 1$.

Now assume that $f/(gh) = (c/g) + (d/h)$, with $\deg c < \deg g$, $\deg d < \deg h$. Then $ch + dg = f = ah + bg$ and $(c - a)h = (b - d)g$. Hence g divides $c - a$ and h divides $b - d$, by 8.7, since $\gcd(g, h) = 1$. But $\deg(c - a) < \deg g$, $\deg(b - d) < \deg h$; therefore $c - a = b - d = 0$. $\square$

Lemma **9.5.** *If* $\deg f < \deg g$ *and* $\gcd(f, g) = 1$, *then there exist unique integers* $n \geq 0$, $k_1, \ldots, k_n > 0$ *and unique polynomials* $a_1, \ldots, a_n$, $q_1, \ldots, q_n$ *such that* $q_1, \ldots, q_n$ *are distinct monic irreducible polynomials,* $\deg a_i < \deg q_i^{k_i}$ *for all* i, $\gcd(a_i, q_i) = 1$ *for all* i, *and*

$$\frac{f}{g} = \frac{a_1}{q_1^{k_1}} + \cdots + \frac{a_n}{q_n^{k_n}}.$$

If g is monic in Lemma 9.5, readers will see that $g = q_1^{k_1} q_2^{k_2} \cdots q_n^{k_n}$ is the unique factorization of g into a product of positive powers of distinct monic irreducible polynomials; then 9.5 follows from 9.4 by induction on n.

The last part of the proof breaks reduced polynomial-free fractions a/q^k, in which q is monic and irreducible, into sums of partial fractions.

Lemma **9.6.** *If* $\deg q > 0$, $k > 0$, *and* $\deg a < \deg q^k$, *then there exist unique polynomials* $a_1, \ldots, a_k$ *such that* $\deg a_i < \deg q$ *for all* i *and*

$$\frac{a}{q^k} = \frac{a_1}{q} + \frac{a_2}{q^2} + \cdots + \frac{a_k}{q^k}.$$

Readers will easily prove Lemma 9.6 by induction on k, using polynomial division.

Theorem 9.1 now follows from Lemmas 9.3, 9.5, and 9.6. The proof provides a general procedure, which can be used on examples: given f/g, first divide f by g to obtain an equality $f/g = p + r/g$, where p is a polynomial and r/g is polynomial free; use the factorization of g as a product of positive powers of irreducible polynomials to set up a decomposition of r/g as a sum of partial fractions; expansion, substitution, and lucky guesses yield the numerators.

For instance, consider $\dfrac{X^4 + 1}{X^3 + X^2 + X} \in \mathbb{Z}_2(X)$. Polynomial division yields $X^4 + 1 = (X^3 + X^2 + X)(X + 1) + (X + 1)$; hence

$$\frac{X^4 + 1}{X^3 + X^2 + X} = X + 1 + \frac{X + 1}{X^3 + X^2 + X}.$$

Now, $X^3 + X^2 + X = X(X^2 + X + 1)$, and we have seen that X and $X^2 + X + 1$ are irreducible in $\mathbb{Z}_2[X]$. Hence

$$\frac{X+1}{X^3 + X^2 + X} = \frac{a}{X} + \frac{bX + c}{X^2 + X + 1}$$

for some unique $a, b, c \in \mathbb{Z}_2$. Expansion yields

$$X + 1 = a(X^2 + X + 1) + (bX + c)X = (a + b)X^2 + (a + c)X + a,$$

whence $a = 1$, $a + c = 1$, $c = 0$, $a + b = 0$, and $b = 1$; we might also have seen that $X + 1 = (X^2 + X + 1) + (X)(X)$. Hence

$$\frac{X^4 + 1}{X^3 + X^2 + X} = X + 1 + \frac{1}{X} + \frac{X}{X^2 + X + 1}.$$

Exercises

1. Prove the following: if $f/p^r = g/q^s$, with p, q monic irreducible, $r, s \geqq 1$, and $\deg f < \deg p$, $\deg g < \deg q$, then $f = g$, $p = q$, and $r = s$.

2. Write a proof of Lemma 9.5.

3. Let $\deg q > 0$, $k > 0$, and $\deg a < \deg q^k$. Show that there exist unique polynomials $a_1, \ldots, a_k$ such that $\deg a_i < \deg q$ for all i and $\dfrac{a}{q^k} = \dfrac{a_1}{q} + \dfrac{a_2}{q^2} + \cdots + \dfrac{a_k}{q^k}$.

4. Write $\dfrac{X^5 + 1}{X^4 + X^2} \in \mathbb{Z}_2(X)$ as the sum of a polynomial and partial fractions.

5. Write $\dfrac{X^5 + 1}{X^4 + X^2} \in \mathbb{Z}_3(X)$ as the sum of a polynomial and partial fractions.

6. Write $\dfrac{1}{X^5 + X^3 + X} \in \mathbb{Z}_2(X)$ as a sum of partial fractions.

10. Unique Factorization Domains

These domains share the main arithmetic properties of PIDs and include polynomial rings $K[X_1, \ldots, X_n]$ over a field K and polynomial rings over a PID.

Definition. A unique factorization domain or UFD is a domain R (a commutative ring with identity and no zero divisors) in which (1) every element, other than 0 and units, is a nonempty product of irreducible elements of R; and (2) if two nonempty products $p_1 p_2 \cdots p_m = q_1 q_2 \cdots q_n$ of irreducible elements of R are equal, then $m = n$ and the terms can be indexed so that $Rp_i = Rq_i$ for all i.

Equivalently, a UFD is a domain in which every nonzero element can be written uniquely, up to the order of the terms, as the product $u\, p_1^{k_1}\, p_2^{k_2} \cdots p_n^{k_n}$ of a unit and of positive powers of distinct representative irreducible elements.

By Theorem 8.4, every PID is a UFD; in particular, $\mathbb{Z}$ and $K[X]$ are UFDs for every field K. UFDs that are not PIDs will arrive in five minutes.

In a UFD, any two elements a and b have an l.c.m. and a g.c.d., which can be found as in Section 8 from their factorizations, once a and b are rewritten as products $a = u\, p_1^{a_1}\, p_2^{a_2}\, \cdots\, p_n^{a_n}$ and $b = v\, p_1^{b_1}\, p_2^{b_2}\, \cdots\, p_n^{b_n}$ of a unit and nonnegative powers of the same distinct representative irreducible elements:

Proposition 10.1. *In a unique factorization domain, let* $a = u\, p_1^{a_1}\, p_2^{a_2}\, \cdots\, p_n^{a_n}$ *and* $b = v\, p_1^{b_1}\, p_2^{b_2}\, \cdots\, p_n^{b_n}$ *be products of a unit and nonnegative powers of the same distinct representative irreducible elements. Then:*

(1) *a divides b if and only if* $a_i \leqq b_i$ *for all* i.

(2) $c = p_1^{c_1}\, p_2^{c_2}\, \cdots\, p_n^{c_n}$ *is a least common multiple of a and b if and only if* $c_i = \max\,(a_i,\, b_i)$ *for all* i.

(3) $d = p_1^{d_1}\, p_2^{d_2}\, \cdots\, p_n^{d_n}$ *is a greatest common divisor of a and b if and only if* $d_i = \min\,(a_i,\, b_i)$ *for all* i.

(4) $\mathrm{lcm}\,(a, b)\, \gcd\,(a, b) = wab$ *for some unit* w.

On the other hand, in a UFD, the g.c.d. of a and b is not necessarily in the form $xa + yb$. Proposition 10.1 is proved like its particular case Proposition 8.6. More generally, every family of elements has a g.c.d., and every finite family of elements has an l.c.m.; the proofs of these statements make nifty exercises. The same methods yield two more results:

Proposition 10.2. *In a UFD, an element is prime if and only if it is irreducible.*

Proposition 10.3. *In a UFD, if* $\gcd\,(a, b) = \gcd\,(a, c) = 1$, *then* $\gcd\,(a, bc) = 1$; *if a divides bc and* $\gcd\,(a, b) = 1$, *then a divides c.*

This result is proved like its particular case Proposition 8.7.

Polynomials. Our main result was first proved by Gauss [1801] for $\mathbb{Z}[X]$.

Theorem 10.4. *If R is a unique factorization domain, then* $R[X]$ *is a unique factorization domain.*

Hence (by induction on n) $\mathbb{Z}[X_1, ..., X_n]$ and $K[X_1, ..., X_n]$ are UFDs (for any field K). This provides examples of UFDs that are not PIDs. Actually, Theorem 10.4 holds for any number of indeterminates, so that $\mathbb{Z}[(X_i)_{i \in I}]$ and $K[(X_i)_{i \in I}]$ are UFDs (see the exercises).

The proof of Theorem 10.4 uses the quotient field Q of R, and studies irreducible polynomials to show how $R[X]$ inherits unique factorization from $Q[X]$.

Definition. A polynomial p over a unique factorization domain R is primitive *when no irreducible element of R divides all the coefficients of p.* □

Equivalently, $p_0 + \cdots + p_n X^n$ is primitive when $\gcd\,(p_0, ..., p_n) = 1$, or when no irreducible element divides all p_i.

Lemma 10.5. *Every nonzero polynomial* $f(X) \in Q(X)$ *can be written in the* form $f(X) = t f^*(X)$, where $t \in Q$, $t \neq 0$, and $f^*(X) \in R[X]$ is primitive; moreover, t and f^* are unique up to multiplication by units of R.

Proof. We have $f(X) = (a_0/b_0) + (a_1/b_1) X + \cdots + (a_n/b_n) X^n$, where a_i, $b_i \in R$ and $b_i \neq 0$. Let b be a common denominator (for instance, $b = b_0 b_1 \cdots b_n$). Then $f(X) = (1/b)(c_0 + c_1 X + \cdots + c_n X^n)$ for some $c_i \in R$. Factoring out $a = \gcd(c_0, c_1, \ldots, c_n)$ yields $f(X) = (a/b) f^*(X)$, where f^* is primitive.

Assume that $(a/b) g(X) = (c/d) h(X)$, where g, h are primitive. Since g and h are primitive, ad is a g.c.d. of the coefficients of $ad\,g(X)$, and bc is a g.c.d. of the coefficients of $bc\,h(X)$; hence $bc = adu$ for some unit u of R, so that $g(X) = u\,h(X)$ and $(a/b) u = c/d$ in Q. $\square$

Lemma **10.6** (Gauss). *If f and $g \in R[X]$ are primitive, then fg is primitive.*

Proof. Let $f(X) = a_0 + a_1 X + \cdots + a_m X^m$ and $g(X) = b_0 + b_1 X + \cdots + b_n X^n$, so that $(fg)(X) = c_0 + c_1 X + \cdots + c_{m+n} X^{m+n}$, where $c_k = \sum_{i+j=k} a_i b_j$. We show that no irreducible element divides all c_k.

Let $p \in R$ be irreducible. Since f and g are primitive, p divides neither all a_i nor all b_j. Let k and ℓ be smallest such that p does not divide a_k or b_ℓ. Then p divides a_i for all $i < k$, and divides b_j for all $j < \ell$. By 10.2, p does not divide $a_k b_\ell$; but p divides $a_i b_j$ whenever $i < k$ and whenever $i + j = k + \ell$ and $i > k$, for then $j < \ell$. Therefore p does not divide $c_{k+\ell}$. $\square$

Corollary **10.7.** *In Lemma 10.5, f is irreducible in $Q[X]$ if and only if f^* is irreducible in $R[X]$.*

Proof. We may assume that $\deg f \geq 1$. If f is not irreducible, then f has a factorization $f = gh$ in $Q[X]$ where $\deg g$, $\deg h \geq 1$. Let $g(X) = v g^*(X)$, $h(X) = w h^*(X)$, with $g^*, h^* \in R[X]$ primitive, as in 10.5. Then $t f^*(X) = f(X) = vw g^*(X) h^*(X)$. By 10.6, $g^* h^*$ is primitive; hence $f^*(X) = u g^*(X) h^*(X)$ for some unit u of R, by 10.5, and f^* is not irreducible. Conversely, if f^* is not irreducible, then neither is $f(X) = t f^*(X)$. $\square$

Lemma **10.8.** *In $R[X]$, every polynomial, other than 0 and units of R, is a nonempty product of irreducible elements of R and irreducible primitive polynomials. Hence the irreducible elements of $R[X]$ are the irreducible elements of R and the irreducible primitive polynomials.*

Proof. Assume that $f \in R[X]$ is not zero and not a unit of R. Let d be a g.c.d. of the coefficients of f. Then $f(X) = d f^*(X)$, where $f^* \in R[X]$ is primitive; moreover, d and f^* are not 0 and are not both units of R. Now, d is either a unit or a product of irreducible elements of R, and f^* is either a unit of R or not constant. If f^* is not constant, then f^* is, in $Q[X]$, a product $f = q_1 q_2 \cdots q_n$ of irreducible polynomials $q_i \in Q[X]$. By 10.5, $q_i(X) = t_i q_i^*(X)$ for some $0 \neq t_i \in Q$ and primitive $q_i^* \in R[X]$.

Then $f^*(X) = t_1 \cdots t_n q_1^*(X) \cdots q_n^*(X)$. By 10.6, $q_1^* \cdots q_n^*$ is primitive. Hence 10.5 yields $f^*(X) = u q_1^*(X) \cdots q_n^*(X)$ for some unit u of R (namely, $u = t_1 \cdots t_n$), with $u q_1^*, q_2^*, \ldots, q_n^*$ primitive and irreducible by 10.7. $\square$

To prove Theorem 10.4 we still need to show the following: if two nonempty products

$$p_1 \cdots p_k p_{k+1} \cdots p_m = q_1 \cdots q_\ell q_{\ell+1} \cdots q_n$$

of irreducible elements of $R[X]$ are equal, then $m = n$ and the terms can be indexed so that $(p_i) = (q_i)$ for all i. By 10.8 we may arrange that $p_1, \ldots, p_k$, $q_1, \ldots, q_\ell$ are irreducible elements of R and $p_{k+1}, \ldots, p_m, q_{\ell+1}, \ldots, q_n \in R[X]$ are irreducible primitive polynomials.

Let $a = p_1 \cdots p_k$ and $b = q_1 \cdots q_\ell \in R$, $f = p_{k+1} \cdots p_m$ and $g = q_{\ell+1} \cdots q_n \in R[X]$, so that $af = bg$. By 10.7, f and g are primitive. Hence $f = ug$, $au = b$ for some unit u of R, by 10.5. Since R is a UFD and $au = b$, we have $k = \ell$ and $p_1, \ldots, p_k, q_1, \ldots, q_\ell$ can be reindexed so that $p_i = u_i q_i$ for all $i \leqq k$, where u_i is a unit of R.

Since $Q[X]$ is a UFD and $f = ug$, we also have $m - k = n - \ell$ and p_{k+1}, $\ldots, p_m, q_{\ell+1} = q_{k+1}, \ldots, q_n$ can be reindexed so that $p_j(X) = u_j q_j(X)$ for all $j > k$, where u_j is a unit of Q. In fact, u_j is a unit of R. Indeed, let $u_j = c/d$, where $c, d \in R$. Then $c p_j(X) = d q_j(X)$. Since p_j and q_j are both primitive, taking the g.c.d. of the coefficients on both sides yields $c = ud$ for some unit u of R. Hence $u_j = c/d = u$ is a unit of R. We now have $m = n$ and $(p_i) = (q_i)$ in $R[X]$ for all i. $\square$

Irreducibility tests. Let R be a UFD and let Q be its quotient field. By Corollary 10.7, the irreducible polynomials of $Q[X]$ are determined by those of $R[X]$. For instance, the irreducible polynomials of $\mathbb{Q}[X]$ are determined by those of $\mathbb{Z}[X]$. We now give two sufficient conditions for irreducibility. The first is essentially due to Eisenstein [1850]; the exercises give a generalization.

Proposition **10.9** (Eisenstein's Criterion). *Let R be a UFD and let $f(X) = a_0 + a_1 X + \cdots + a_n X^n \in R[X]$. If f is primitive and there exists an irreducible element p of R such that p divides a_i for all $i < n$, p does not divide a_n, and p^2 does not divide a_0, then f is irreducible.*

Proof. Suppose that $f = gh$; let $g(X) = b_0 + b_1 X + \cdots + b_r X^r$ and $h(X) = c_0 + c_1 X + \cdots + c_s X^s \in R[X]$, where $r = \deg g$ and $s = \deg h$. Then $a_k = \sum_{i+j=k} b_i c_j$ for all k; in particular, $a_0 = b_0 c_0$. Since p^2 does not divide a_0, p does not divide both b_0 and c_0. But p divides a_0, so p divides, say, b_0, but not c_0. Also, p does not divide b_r, since p does not divide $a_n = b_r c_s$. Hence there is a least $k \leqq r$ such that p does not divide b_k, and then p divides b_i for all $i < k$. Now p divides every term of $\sum_{i+j=k} b_i c_j$ except for $b_k c_0$. Hence p does not divide a_k. Therefore $k = n$; since $k \leqq r \leqq r + s = n$ this implies $r = n$, and h is constant. $\square$

For example, $f = 3X^3 + 4X - 6 \in \mathbb{Z}[X]$ is irreducible in $\mathbb{Z}[X]$: indeed, f is primitive, 2 divides all the coefficients of f except the leading coefficient, and 4 does not divide the constant coefficient. By 10.7, f is also irreducible in $\mathbb{Q}[X]$, and so is $\frac{5}{6}f = \frac{5}{2}X^3 + \frac{5}{3}X - 5$.

Proposition **10.10.** *Let R be a domain, let $\mathfrak{a}$ be an ideal of R, and let $\pi : R \longmapsto R/\mathfrak{a}$ be the projection. If $f \in R[X]$ is monic and ${}^{\pi}f$ is irreducible in $(R/\mathfrak{a})[X]$, then f is irreducible in $R[X]$.*

Readers will delight in proving this. For instance, $f = X^3 + 2X + 4$ is irreducible in $\mathbb{Z}[X]$: if $\pi : \mathbb{Z} \longrightarrow \mathbb{Z}_3$ is the projection, then ${}^{\pi}f = X^3 - X + 1$ is irreducible in $\mathbb{Z}_3[X]$, since it has degree 3 and no root in $\mathbb{Z}_3$.

Exercises

1. Show that every family $(a_i)_{i \in I}$ of elements of a UFD has a g.c.d.

2. Show that every finite family of elements of a UFD has an l.c.m.

3. Does *every* family of elements of a UFD have an l.c.m.?

4. Find a UFD with two elements a and b whose g.c.d. cannot be written in the form $xa + yb$.

5. In a UFD, show that $\gcd(a, b) = \gcd(a, c) = 1$ implies $\gcd(a, bc) = 1$.

6. Prove the following: in a UFD, if a divides bc and $\gcd(a, b) = 1$, then a divides c.

7. Prove the following: in a UFD, an element is prime if and only if it is irreducible.

8. Prove the following stronger version of Lemma 10.5: when R is a UFD and Q its field of fractions, every nonzero polynomial $f(X) \in Q(X)$ can be written in the form $f(X) = (a/b) f^*(X)$, where $a, b \in R$, $\gcd(a, b) = 1$, and $f^*(X) \in R[X]$ is primitive; moreover, a, b, and f^* are unique up to multiplication by units of R.

9. Prove Proposition 10.10: Let R be a domain, let $\mathfrak{a}$ be an ideal of R, and let $\pi : R \longmapsto R/\mathfrak{a}$ be the projection. If $f \in R[X]$ is monic and ${}^{\pi}f$ is irreducible in $(R/\mathfrak{a})[X]$, then f is irreducible in $R[X]$.

10. Show that $X^3 - 10$ is irreducible in $\mathbb{Q}[X]$.

11. Show that $X^3 + 3X^2 - 6X + 3$ is irreducible in $\mathbb{Q}[X]$.

12. Show that $X^3 + 3X^2 - 6X + 9$ is irreducible in $\mathbb{Q}[X]$.

13. Show that $X^3 - 3X + 4$ is irreducible in $\mathbb{Q}[X]$.

14. Prove the following generalization of Eisenstein's criterion. Let R be a domain and let $f(X) = a_0 + a_1 X + \cdots + a_n X^n \in R[X]$. If f is primitive (if the only common divisors of $a_0, \ldots, a_n$ are units) and there exists a prime ideal $\mathfrak{p}$ of R such that $a_i \in \mathfrak{p}$ for all $i < n$, $a_n \notin \mathfrak{p}$, and a_0 is not the product of two elements of $\mathfrak{p}$, then f is irreducible.

*15. Prove the following: when R is a UFD, then $R[(X_i)_{i \in I}]$ is a UFD.

11. Noetherian Rings

Noetherian rings are named after Emmy Noether, who initiated the study of these rings in [1921]. In this section we define Noetherian rings and prove that $K[X_1, ..., X_n]$ is Noetherian for every field K.

Definition. Applied to the ideals of a commutative ring R, the *ascending chain condition*, or *a.c.c.*, has three equivalent forms:

(a) every infinite ascending sequence $\mathfrak{a}_1 \subseteq \mathfrak{a}_2 \subseteq \cdots \subseteq \mathfrak{a}_n \subseteq \mathfrak{a}_{n+1} \subseteq \cdots$ of ideals of R *terminates*: there exists $N > 0$ such that $\mathfrak{a}_n = \mathfrak{a}_N$ for all $n \geq N$;

(b) there is no infinite strictly ascending sequence $\mathfrak{a}_1 \subsetneqq \mathfrak{a}_2 \subsetneqq \cdots \subsetneqq \mathfrak{a}_n \subsetneqq \mathfrak{a}_{n+1} \subsetneqq \cdots$ of ideals of R;

(c) every nonempty set $\mathcal{S}$ of ideals of R has a maximal element (an element $\mathfrak{s}$ of $\mathcal{S}$, not necessarily a maximal ideal of R, such that there is no $\mathfrak{s} \subsetneqq \mathfrak{a} \in \mathcal{S}$).

Indeed, (a) implies (b), since a strictly ascending infinite sequence cannot terminate. If the nonempty set $\mathcal{S}$ in (c) has no maximal element, then there exists some $\mathfrak{a}_1 \in \mathcal{S}$; since $\mathfrak{a}_1$ is not maximal in $\mathcal{S}$ there exists some $\mathfrak{a}_1 \subsetneqq \mathfrak{a}_2 \in \mathcal{S}$; this continues indefinitely and begets a strictly ascending infinite sequence. Hence (b) implies (c). Finally, (c) implies (a), since some $\mathfrak{a}_N$ must be maximal, and then $\mathfrak{a}_N \subsetneqq \mathfrak{a}_n$ is impossible when $n \geq N$. Section A.1 has a more general but entirely similar proof of the equivalence of (a), (b), and (c).

Definition. A commutative ring is Noetherian *when its ideals satisfy the ascending chain condition.*

For example, $\mathbb{Z}$ is Noetherian, by 11.1 below; $K[X]$ and $K[X_1, ..., X_n]$ are Noetherian for every field K, by 11.3 below.

In a ring, the a.c.c. has a fourth equivalent form. Recall that the ideal $\mathfrak{a}$ of R generated by a subset S of R consists of all linear combinations of elements of S with coefficients in R. Hence $\mathfrak{a}$ is finitely generated (as an ideal) if and only if there exist $a_1, ..., a_n \in \mathfrak{a}$ such that $\mathfrak{a} = \{ r_1 a_1 + \cdots + r_n a_n \mid r_1, ..., r_n \in R \}$. The set $\{ a_1, ..., a_n \}$ is traditionally called a *basis* of $\mathfrak{a}$, even though the elements of $\mathfrak{a}$ need not be writable uniquely in the form $r_1 a_1 + \cdots + r_n a_n$.

Proposition 11.1. A commutative ring R is Noetherian if and only if every ideal of R is finitely generated (as an ideal).

Proof. Let $\mathfrak{a}$ be an ideal of R. Let $\mathcal{S}$ be the set of all finitely generated ideals of R contained in $\mathfrak{a}$. Then $\mathcal{S}$ contains principal ideals and is not empty. If R is Noetherian, then $\mathcal{S}$ has a maximal element $\mathfrak{s}$ by (c). Then $\mathfrak{s} \subseteq \mathfrak{s} + (a) \in \mathcal{S}$ for every $a \in \mathfrak{a}$, since $\mathfrak{s} + (a) \subseteq \mathfrak{a}$ and $\mathfrak{s} + (a)$ is finitely generated, by a and the generators of $\mathfrak{s}$. Since $\mathfrak{s}$ is maximal in $\mathcal{S}$ it follows that $\mathfrak{s} = \mathfrak{s} + (a)$ and $a \in \mathfrak{s}$. Hence $\mathfrak{a} = \mathfrak{s}$ and $\mathfrak{a}$ is finitely generated.

Conversely, assume that every ideal of R is finitely generated. Let $\mathfrak{a}_1 \subseteq \mathfrak{a}_2 \subseteq \cdots \subseteq \mathfrak{a}_n \subseteq \mathfrak{a}_{n+1} \subseteq \cdots$ be ideals of R. Then $\mathfrak{a} = \bigcup_{n>0} \mathfrak{a}_n$ is an ideal of R and is finitely generated, by, say, $a_1, \ldots, a_k$. Then $a_i \in \mathfrak{a}_{n_i}$ for some $n_i > 0$. If $N \geqq n_1, \ldots, n_k$, then $\mathfrak{a}_N$ contains $a_1, \ldots, a_k$; hence $\mathfrak{a} \subseteq \mathfrak{a}_N$, and $\mathfrak{a} \subseteq \mathfrak{a}_N \subseteq \mathfrak{a}_n \subseteq \mathfrak{a}$ shows that $\mathfrak{a}_n = \mathfrak{a}_N$ for all $n \geqq N$. $\square$

The main result in this section is basically due to Hilbert [1890].

Theorem **11.2** (Hilbert Basis Theorem). *Let R be a commutative ring with identity. If R is Noetherian, then $R[X]$ is Noetherian.*

Proof. Let $\mathfrak{A}$ be an ideal of $R[X]$. We construct a finite set of generators of $\mathfrak{A}$. For every $n \geq 0$ let

$$\mathfrak{a}_n = \{\, r \in R \mid rX^n + a_{n-1}X^{n-1} + \cdots + a_0 \in \mathfrak{A} \text{ for some } a_{n-1}, \ldots, a_0 \in R \,\}.$$

Then $\mathfrak{a}_n$ is an ideal of R, since $\mathfrak{A}$ is an ideal of $R[X]$, and $\mathfrak{a}_n \subseteq \mathfrak{a}_{n+1}$, since $f(X) \in \mathfrak{A}$ implies $Xf(X) \in \mathfrak{A}$. Since R is Noetherian, the ascending sequence $\mathfrak{a}_0 \subseteq \mathfrak{a}_1 \subseteq \cdots \subseteq \mathfrak{a}_n \subseteq \mathfrak{a}_{n+1} \subseteq \cdots$ terminates at some $\mathfrak{a}_m$ ($\mathfrak{a}_n = \mathfrak{a}_m$ for all $n \geqq m$). Also, each ideal $\mathfrak{a}_k$ has a finite generating set S_k, by 11.1.

For each $s \in S_k$ there exists $g_s = sX^k + a_{k-1}X^{k-1} + \cdots + a_0 \in \mathfrak{A}$. We show that $\mathfrak{A}$ coincides with the ideal $\mathfrak{B}$ generated by all g_s with $s \in S_0 \cup S_1 \cup \cdots \cup S_m$; hence $\mathfrak{A}$ is finitely generated, and $R[X]$ is Noetherian. Already $\mathfrak{B} \subseteq \mathfrak{A}$, since every $g_s \in \mathfrak{A}$. The converse implication, $f \in \mathfrak{A}$ implies $f \in \mathfrak{B}$, is proved by induction on $\deg f$. First, $0 \in \mathfrak{B}$. Now let $f = a_nX^n + \cdots + a_0 \in \mathfrak{A}$ have degree $n \geqq 0$. Then $a_n \in \mathfrak{a}_n$.

If $n \leqq m$, then $a_n = r_1 s_1 + \cdots + r_k s_k$ for some $r_1, \ldots, r_n \in R$ and $s_1, \ldots, s_k \in S_n$; then $g = r_1 g_{s_1} + \cdots + r_k g_{s_k} \in \mathfrak{B}$ has degree at most n, and the coefficient of X^n in g is $r_1 s_1 + \cdots + r_k s_k = a_n$. Hence $f - g \in \mathfrak{A}$ has degree less than n. Then $f - g \in \mathfrak{B}$, by the induction hypothesis, and $f \in \mathfrak{B}$.

If $n > m$ then $a_n \in \mathfrak{a}_n = \mathfrak{a}_m$ and $a_n = r_1 s_1 + \cdots + r_k s_k$ for some $r_1, \ldots, r_n \in R$ and $s_1, \ldots, s_k \in S_m$; then $g = r_1 g_{s_1} + \cdots + r_k g_{s_k} \in \mathfrak{B}$ has degree at most m, and the coefficient of X^m in g is $r_1 s_1 + \cdots + r_k s_k = a_n$. Hence $X^{n-m}g \in \mathfrak{B}$ has degree at most n, and the coefficient of X^n in g is a_n. As above, $f - X^{n-m}g \in \mathfrak{A}$ has degree less than n, $f - X^{n-m}g \in \mathfrak{B}$ by the induction hypothesis, and $f \in \mathfrak{B}$. $\square$

Corollary **11.3.** $K[X_1, \ldots, X_n]$ *is Noetherian, for every field K and $n > 0$.*

This corollary is also known as the Hilbert basis theorem; the case $K = \mathbb{C}$ was Hilbert's original statement, "every ideal of $\mathbb{C}[X_1, \ldots, X_n]$ has a finite basis".

Corollary **11.4.** *Let $R \subseteq S$ be commutative rings. If R is Noetherian, and S is generated by R and finitely many elements of S, then S is Noetherian.*

We leave this to our readers, who deserve some fun.

Exercises

1. Let R be a Noetherian ring. Prove that every quotient ring of R is Noetherian.

2. Find a commutative ring that is not Noetherian.

3. Let $R \subseteq S$ be commutative rings. Suppose that R is Noetherian and that S is generated by R and finitely many elements of S. Prove that S is Noetherian.

*4. Let M be the free commutative monoid on a finite set $\{ X_1, ..., X_n \}$ (which consists of all monomials $X^m = X_1^{m_1} \cdots X_n^{m_n}$ with nonnegative integer exponents). A *congruence* on M is an equivalence relation $\mathcal{C}$ on M such that $X^a \mathcal{C} X^b$ implies $X^a X^c \mathcal{C} X^b X^c$ for all $X^c \in M$. Prove *Rédei's theorem* [1956]: the congruences on M satisfy the ascending chain condition. (Hint: relate congruences on M to ideals of $\mathbb{Z}[X_1, ..., X_n]$.)

*5. Prove the following: if R is Noetherian, then the power series ring $R[[X]]$ is Noetherian. You may want to adjust the proof of Theorem 11.2, using

$$\mathfrak{a}_n = \{ r \in R \mid rX^n + a_{n+1}X^{n+1} + \cdots \in \mathfrak{A} \text{ for some } a_{n+1}, ... \text{ in } R \}.$$

12. Gröbner Bases

This section may be skipped or covered later with Section VIII.9. Gröbner bases are carefully chosen generating sets of ideals of $K[X_1, ..., X_n]$. The basic properties in this section are due to Gröbner [1939] and Buchberger [1965].

Monomial orders. The definition of Gröbner bases requires polynomial division in n indeterminates. When K is a field, polynomial division in $K[X]$ is possible because monomials in one indeterminate are naturally ordered, $1 < X < \cdots < X^m < \cdots$ Polynomial division in $K[X_1, ..., X_n]$ is made possible by suitable total orders on the monomials $X^m = X_1^{m_1} \cdots X_n^{m_n}$ of $K[X_1, ..., X_n]$.

Definition. A monomial order *on* $K[X_1, ..., X_n]$ *is a total order on its monomials such that* $X^a \geqq 1$ *for all* X^a, *and* $X^a < X^b$ *implies* $X^a X^c < X^b X^c$.

Monomial orders are often called *term orders*. The author prefers "monomials" for products $X_1^{m_1} \cdots X_n^{m_n}$ and "terms" for their scalar multiples $aX_1^{m_1} \cdots X_n^{m_n}$. There is only one monomial order $1 < X < \cdots < X^m < \cdots$ on $K[X]$, but in general monomial orders can be constructed in several ways. This gives Gröbner bases great flexibility.

Definitions. In the lexicographic order *on* $K[X_1, ..., X_n]$ *with* $X_1 > X_2 > \cdots > X_n$, $X^a < X^b$ *if and only if there exists* $1 \leqq k \leqq n$ *such that* $a_i = b_i$ *for all* $i < k$ *and* $a_k < b_k$.

In the degree lexicographic order *on* $K[X_1, ..., X_n]$ *with* $X_1 > X_2 > \cdots > X_n$, $X^a < X^b$ *if and only if either* $\deg X^a < \deg X^b$, *or* $\deg X^a = \deg X^b$ *and there exists* $1 \leqq k \leqq n$ *such that* $a_i = b_i$ *for all* $i < k$ *and* $a_k < b_k$ ($\deg X^m = m_1 + \cdots + m_n$ *is the total degree of* X^m).

In the degree reverse lexicographic order *on* $K[X_1, ..., X_n]$ *with* $X_1 > X_2 > \cdots > X_n$, $X^a < X^b$ *if and only if either* $\deg X^a < \deg X^b$, *or* $\deg X^a = \deg X^b$ *and there exists* $1 \leqq k \leqq n$ *such that* $a_i = b_i$ *for all* $i > k$ *and* $a_k > b_k$.

Readers will show that the above are monomial orders:

Proposition **12.1.** *The lexicographic order, degree lexicographic order, and degree reverse lexicographic order are monomial orders on $K[X_1, ..., X_n]$.*

In any monomial order, $X^a X^b > X^a$ whenever $X^b \neq 1$; hence $X^c \geqq X^a$ whenever X^c is a multiple of X^a. We also note the following property:

Proposition **12.2.** *In any monomial order, there is no infinite strictly decreasing sequence $X^{m_1} > X^{m_2} > \cdots > X^{m_k} > X^{m_{k+1}} > \cdots$.*

By 12.2, every nonempty set S of monomials has a least element (otherwise S would contain an infinite strictly decreasing sequence).

Proof. Suppose that $X^{m_1} > X^{m_2} > \cdots > X^{m_k} > X^{m_{k+1}} > \cdots$ By 12.3 below, the ideal of $K[X_1, ..., X_n]$ generated by all X^{m_i} is generated by finitely many X^{m_i}'s. Let X^t be the least of these. Every X^{m_k} is a linear combination of monomials $X^m \geqq X^t$ and is a multiple of some $X^m \geqq X^t$; hence $X^{m_k} \geqq X^t$ for all k. On the other hand X^t is a linear combination of monomials X^{m_k} and is a multiple of some X^{m_ℓ}; hence $X^t \geqq X^{m_\ell}$ for some ℓ. Then $X^t = X^{m_\ell}$, and $X^{m_\ell} > X^{m_{\ell+1}}$ is not possible. $\square$

Lemma **12.3.** *An ideal of $K[X_1, ..., X_n]$ that is generated by a set S of monomials is generated by a finite subset of S.*

Proof. By the Hilbert basis theorem, the ideal (S) generated by S is generated by finitely many polynomials $f_1, ..., f_r$. Every nonzero term of f_j is a multiple of some $X^s \in S$. Let T be the set of all $X^s \in S$ that divide a nonzero term of some f_j; then T is finite, (T) contains every f_j, and $(T) = (S)$. $\square$

Polynomial division. With a monomial order on $K[X_1, ..., X_n]$, the monomials that appear in a nonzero polynomial $f = \sum_m a_m X^m \in K[X_1, ..., X_n]$ can be arranged in decreasing order, and f acquires a leading term:

Definitions. Relative to a monomial order, the leading monomial *of a nonzero polynomial $f = \sum_m a_m X^m \in K[X_1, ..., X_n]$ is the greatest monomial* ldm $f = X^m$ *such that $a_m \neq 0$, and then the* leading coefficient *of f is* ldc $f = a_m$ *and the* leading term *of f is* ldt $f = a_m X^m$.

Other notations are in use for ldt f, for instance, in (f). Polynomial division in $K[X_1, ..., X_n]$ can now be carried out as usual, except that one can divide a polynomial by several others, and the results are not unique.

Proposition **12.4.** *Let K be a field. Let $f, g_1, ..., g_k \in K[X_1, ..., X_n]$, $g_1, ..., g_k \neq 0$. Relative to any monomial order on $K[X_1, ..., X_n]$, there exist $q_1, ..., q_k, r \in K[X_1, ..., X_n]$ such that*

$$f = g_1 q_1 + \cdots + g_k q_k + r,$$

ldm $(g_i q_i) \leqq$ ldm f *for all i,* ldm $r \leqq$ ldm f, *and none of* ldm $g_1, ...,$ ldm g_k *divides a nonzero term of the remainder r.*

Proof. Let $f_0 = f$. If none of $\operatorname{ldm} g_1, \ldots, \operatorname{ldm} g_k$ divides a nonzero term of f, then $q_1 = \cdots = q_k = 0$ and $r = f$ serve. Otherwise, there is a greatest monomial X^m that appears in a term $a_m X^m \neq 0$ of f and is a multiple of some $\operatorname{ldm} g_j$; in particular, $X^m \leq \operatorname{ldm} f$. Then X^m no longer appears in $f_1 = f - (a_m X^m / \operatorname{ldt} g_j) g_j$: it has been replaced by lesser terms. Repeating this step yields a sequence $f_0, f_1, \ldots$ in which X^m decreases at each step. By 12.2, this is a finite sequence. The last f_s serves as r and has the form $r = f - g_1 q_1 - \cdots - g_k q_k$. Every q_i is a sum of terms $a_m X^m / \operatorname{ldt} g_i$, where $X^m \leq \operatorname{ldm} f$; hence $\operatorname{ldm}(g_i q_i) \leq \operatorname{ldm} f$ and $\operatorname{ldm} r \leq \operatorname{ldm} f$. □

The proof of Proposition 12.4 provides a practical procedure for polynomial division. For an example, let us divide $f = X^2 Y - Y$ by $g_1 = XY - X$ and $g_2 = X^2 - Y \in \mathbb{C}[X]$, using the degree lexicographic order with $X > Y$. Then $\operatorname{ldm} g_1 = XY$, $\operatorname{ldm} g_2 = X^2$; X^2 divides $X^2 Y$, so that

$$f_1 = (X^2 Y - Y) - (X^2 Y / X^2)(X^2 - Y) = Y^2 - Y.$$

Since XY and Y^2 do not divide Y^2 or Y, division stops here, with $f = Y g_2 + (Y^2 - Y)$. We see that the remainder $r = Y^2 - Y$ is not 0, even though $f = X g_1 + g_2$ lies in the ideal (g_1, g_2) generated by g_1 and g_2.

Gröbner bases. The *membership problem* for ideals of $K[X_1, \ldots, X_n]$ is, does a given polynomial f belong to the ideal $(g_1, \ldots, g_k)$ generated by given polynomials $g_1, \ldots, g_k$? We just saw that unfettered polynomial division does not provide a reliable solution. This is where Gröbner bases come in.

Definition. Let K be a field, let $\mathfrak{A}$ be an ideal of $K[X_1, \ldots, X_n]$, and let $<$ be a monomial order on $K[X_1, \ldots, X_n]$. Let $\operatorname{ldm} \mathfrak{A}$ be the ideal of $K[X_1, \ldots, X_n]$ generated by all $\operatorname{ldm} f$ with $f \in \mathfrak{A}$. Nonzero polynomials $g_1, \ldots, g_k \in K[X_1, \ldots, X_n]$ constitute a Gröbner basis of $\mathfrak{A}$, relative to $<$, when $g_1, \ldots, g_k$ generate $\mathfrak{A}$ and $\operatorname{ldm} g_1, \ldots, \operatorname{ldm} g_k$ generate $\operatorname{ldm} \mathfrak{A}$.

Proposition **12.5.** *Let K be a field, let $\mathfrak{A}$ be an ideal of $K[X_1, \ldots, X_n]$, let $g_1, \ldots, g_k$ be a Gröbner basis of $\mathfrak{A}$ relative to a monomial order $<$, and let $f \in K[X_1, \ldots, X_n]$. All divisions of f by $g_1, \ldots, g_k$ (using $<$) yield the same remainder r, and $f \in \mathfrak{A}$ if and only if $r = 0$.*

Proof. Let $f \in \mathfrak{A}$. Let r be the remainder in a division of f by $g_1, \ldots, g_k$. Then $r \in \mathfrak{A}$. If $r \neq 0$, then $\operatorname{ldt} r \in \operatorname{ldm} \mathfrak{A}$ is a linear combination of $\operatorname{ldm} g_1, \ldots, \operatorname{ldm} g_k$ and is a multiple of some $\operatorname{ldm} g_j$, contradicting 12.4. Therefore $r = 0$. Conversely, if $r = 0$, then $f \in (g_1, \ldots, g_k) = \mathfrak{A}$. If now r_1 and r_2 are remainders in divisions of f by $g_1, \ldots, g_k$, then $r_1 - r_2 \in \mathfrak{A}$, and no $\operatorname{ldm} g_j$ divides a nonzero term of $r_1 - r_2$; as above, this implies $r_1 - r_2 = 0$. □

Buchberger's algorithm. We now assume that $K[X_1, \ldots, X_n]$ has a monomial order and find an effective way to construct Gröbner bases. Together with Proposition 12.5, this will solve the membership problem. First we prove that Gröbner bases exist:

Proposition **12.6.** *Every ideal of $K[X_1, ..., X_n]$ has a Gröbner basis.*

Proof. Let $\mathfrak{A}$ be an ideal of $K[X_1, ..., X_n]$. By 12.3, ldm $\mathfrak{A}$ is generated by ldm g_1, ..., ldm g_k for some nonzero g_1, ..., $g_k \in \mathfrak{A}$. Let $f \in \mathfrak{A}$. As in the proof of 12.5, let r be the remainder in a division of f by g_1, ..., g_k. Then $r \in \mathfrak{A}$. If $r \neq 0$, then ldt $r \in$ ldm $\mathfrak{A}$ is a linear combination of ldm g_1, ..., ldm g_k and is a multiple of some ldm g_j, contradicting 12.4. Therefore $r = 0$ and $f \in (g_1, ..., g_k)$. Hence $\mathfrak{A} = (g_1, ..., g_k)$. $\square$

Proposition **12.7** (Buchberger's Criterion). *Let K be a field and let $g_1, ..., g_k \in K[X_1, ..., X_n]$ be nonzero polynomials. Let $\ell_{ij} = \operatorname{lcm}(\operatorname{ldm} g_i, \operatorname{ldm} g_j)$, let $d_{i,j} = (\ell_{ij}/\operatorname{ldt} g_i)\, g_i - (\ell_{ij}/\operatorname{ldt} g_j)\, g_j$, and let $r_{i,j}$ be the remainder in a polynomial division of $d_{i,j}$ by g_1, ..., g_k. Then g_1, ..., g_k is a Gröbner basis of $(g_1, ..., g_k)$ if and only if $r_{i,j} = 0$ for all $i < j$, and then $r_{i,j} = 0$ for all i, j.*

Proof. The ideals $\mathfrak{A} = (g_1, ..., g_k)$ and $(\operatorname{ldm} g_1, ..., \operatorname{ldm} g_k)$ and polynomials $d_{i,j}$ do not change when g_1, ..., g_k are divided by their leading coefficients; hence we may assume that g_1, ..., g_k are monic.

If g_1, ..., g_k is a Gröbner basis of $\mathfrak{A}$, then $r_{i,j} = 0$ by 12.5, since $d_{i,j} \in \mathfrak{A}$.

The converse follows from two properties of the polynomials $d_{i,j}$. Let ldt $g_i = X^{m_i}$, so that $d_{i,j} = (\ell_{ij}/X^{m_i})\, g_i - (\ell_{ij}/X^{m_j})\, g_j$.

(1) If $g_i' = X^{t_i} g_i$, $g_j' = X^{t_j} g_j$, and $\ell_{ij}' = \operatorname{lcm}(X^{t_i+m_i}, X^{t_j+m_j})$, then

$$
\begin{aligned}
d_{i,j}' &= (\ell_{ij}'/\operatorname{ldt} g_i')\, g_i' - (\ell_{ij}'/\operatorname{ldt} g_j')\, g_j' \\
&= (\ell_{ij}'/X^{t_i} X^{m_i})\, X^{t_i} g_i - (\ell_{ij}'/X^{t_j} X^{m_j})\, X^{t_j} g_j = (\ell_{ij}'/\ell_{ij})\, d_{i,j}.
\end{aligned}
$$

(2) If ldm $g_i = X^m$ for all i and $\operatorname{ldm}(a_1 g_1 + \cdots + a_k g_k) < X^m$, where $a_1, ..., a_k \in K$, then $a_1 + \cdots + a_k = 0$, $d_{i,j} = g_i - g_j$, and

$$
\begin{aligned}
a_1 g_1 + \cdots + a_k g_k &= a_1 (g_1 - g_2) + (a_1 + a_2)(g_2 - g_3) \\
&\quad + \cdots + (a_1 + \cdots + a_{k-1})(g_{k-1} - g_k) + (a_1 + \cdots + a_k)\, g_k \\
&= a_1 d_{1,2} + (a_1 + a_2)\, d_{2,3} + \cdots + (a_1 + \cdots + a_{k-1})\, d_{k-1,k}.
\end{aligned}
$$

Now assume that $r_{ij} = 0$ for all $i < j$. Then $r_{i,j} = 0$ for all i and j, since $d_{i,i} = 0$ and $d_{j,i} = -d_{i,j}$. Every nonzero $f \in \mathfrak{A}$ is a linear combination $f = p_1 g_1 + \cdots + p_k g_k$, where $p_1, ..., p_k \in K[X_1, ..., X_n]$. Let X^m be the greatest of all ldm $(p_j g_j)$. Choose p_1, ..., p_k so that X^m is minimal.

Assume that X^m does not appear in f. We may number g_1, ..., g_k so that X^m is the leading monomial of the first products $p_1 g_1$, ..., $p_h g_h$. Then $h \geq 2$: otherwise, X^m cannot be canceled in the sum $p_1 g_1 + \cdots + p_k g_k$ and appears in f. Also $\operatorname{ldm}(p_1 g_1 + \cdots + p_h g_h) < X^m$, since X^m does not appear in f or in $p_{h+1} g_{h+1} + \cdots + p_k g_k$.

Let ldt $p_j = a_j X^{t_j}$ and $g'_j = X^{t_j} g_j$. Then $\mathrm{ldm}\,(g'_j) = X^m$ for all $j \leqq k$ and $\mathrm{ldm}\,(a_1 g'_1 + \cdots + a_h g'_h) < X^m$. By (2) and (1),

$$a_1 g'_1 + \cdots + a_h g'_h = c_1 d'_{1,2} + \cdots + c_{h-1} d'_{h-1,h}$$

for some $c_1, \ldots, c_{h-1} \in K$, where $d'_{i,j} = g'_i - g'_j = (X^m / \ell_{ij})\, d_{i,j}$. Now, $\mathrm{ldm}\, d'_{i,j} < X^m$ when $i < j \leqq h$, since $\mathrm{ldm}\, g'_i = \mathrm{ldm}\, g'_j = X^m$. By the hypothesis, every $d_{i,j}$ can be divided by $g_1, \ldots, g_k$ with zero remainder when $i < j$; so can $d'_{i,j}$ and $c_1 d'_{1,2} + \cdots + c_{k-1} d'_{h-1,h}$, and 12.4 yields

$$a_1 X^{t_1} g_1 + \cdots + a_h X^{t_h} g_h = c_1 d'_{1,2} + \cdots + c_{h-1} d'_{h-1,h} = q'_1 g_1 + \cdots + q'_k g_k \,,$$

where $q'_1, \ldots, q'_k \in K[X_1, \ldots, X_n]$ and $\mathrm{ldm}\,(q'_i g_i) \leqq X^m$ for all i, since $\mathrm{ldm}\,(c_1 d'_{1,2} + \cdots + c_{h-1} d'_{h-1,h}) < X^m$. Since $a_j X^{t_j} = \mathrm{ldt}\, p_j$ this implies

$$p_1 g_1 + \cdots + p_h g_h = q_1 g_1 + \cdots + q_k g_k \,,$$

where $q_1, \ldots, q_k \in K[X_1, \ldots, X_n]$ and $\mathrm{ldm}\,(q_i g_i) < X^m$ for all i. Then

$$f = p_1 g_1 + \cdots + p_h g_h + \cdots + p_k g_k = p'_1 g_1 + \cdots + p'_k g_k \,,$$

where $p'_1, \ldots, p'_k \in K[X_1, \ldots, X_n]$ and $\mathrm{ldm}\,(p'_i g_i) < X^m$ for all i, a gross contradiction of the minimality of X^m.

Therefore X^m appears in f. Hence every nonzero $f \in \mathfrak{A}$ is a linear combination $f = p_1 g_1 + \cdots + p_k g_k$ in which $\mathrm{ldm}\,(p_j g_j) \leqq \mathrm{ldm}\, f$ for all j. Then $\mathrm{ldt}\, f$ is a linear combination of those $\mathrm{ldt}\,(p_j g_j)$ such that $\mathrm{ldm}\,(p_j g_j) = \mathrm{ldm}\, f$; hence $\mathrm{ldm}\, f \in (\mathrm{ldm}\, g_1, \ldots, \mathrm{ldm}\, g_k)$. Thus $\mathrm{ldm}\, g_1, \ldots, \mathrm{ldm}\, g_k$ generate $\mathrm{ldm}\, \mathfrak{A}$. $\square$

Proposition 12.7 yields an effective procedure for finding Gröbner bases, which together with Proposition 12.5 solves the ideal membership problem (without raising membership fees).

Proposition 12.8 (Buchberger's Algorithm). *Let K be a field and let $g_1, \ldots, g_k \in K[X_1, \ldots, X_n]$ be nonzero polynomials. Compute a sequence B of polynomials as follows. Start with $B = g_1, \ldots, g_k$. Compute all polynomials $r_{i,j}$ with $i < j$ of B as in Proposition 12.7 and add one $r_{i,j} \neq 0$ to B in case one is found. Repeat until no $r_{i,j} \neq 0$ is found. Then B is a Gröbner basis of the ideal $(g_1, \ldots, g_k)$.*

Proof. Let $\mathfrak{A} = (g_1, \ldots, g_k)$. Since $r_{i,j}$ is the remainder of some $d_{i,j} \in \mathfrak{A}$ in a division by $g_1, \ldots, g_k$, we have $r_{i,j} \in \mathfrak{A}$, but, if $r_{i,j} \neq 0$, no $\mathrm{ldm}\, g_t$ divides $\mathrm{ldm}\, r_{i,j}$ and $\mathrm{ldm}\, r_{i,j} \notin (\mathrm{ldm}\, g_1, \ldots, \mathrm{ldm}\, g_k)$. Hence $(\mathrm{ldm}\, g_1, \ldots, \mathrm{ldm}\, g_k)$ increases with each addition to B. Since $K[X_1, \ldots, X_n]$ is Noetherian, the procedure terminates after finitely many additions; then B is a Gröbner basis of $\mathfrak{A}$, by 12.7. $\square$

Example **12.9.** Let $g_1 = XY - X$, $g_2 = Y - X^2 \in \mathbb{C}[X, Y]$. Use the lexicographic order with $Y > X$. Start with $B = g_1, g_2$.

We have $\operatorname{ldm} g_1 = XY$, $\operatorname{ldm} g_2 = Y$, $\ell_{12} = XY$, and

$$d_{1,2} = (\ell_{12}/\operatorname{ldt} g_1)\, g_1 - (\ell_{12}/\operatorname{ldt} g_2)\, g_2 = g_1 - Xg_2 = X^3 - X = r_{1,2}\,,$$

since XY and Y divide no term of $X^3 - X$.

Now let $B = g_1$, g_2, $g_3 = X^3 - X$. We have $\operatorname{ldm} g_1 = XY$, $\operatorname{ldm} g_2 = Y$, and $\operatorname{ldm} g_3 = X^3$. As before, $d_{1,2} = X^3 - X = g_3$, but now division yields $r_{1,2} = 0$. Also $\ell_{13} = X^3 Y$ and

$$d_{1,3} = (\ell_{13}/\operatorname{ldt} g_1)\, g_1 - (\ell_{13}/\operatorname{ldt} g_3)\, g_3 = X^2 g_1 - Y g_3 = XY - X^3\,;$$

XY divides $\operatorname{ldt}(XY - X^3) = XY$, so $d_{1,3} = g_1 - X^3 + X$; then X^3 divides $\operatorname{ldt}(-X^3 + X)$, so $d_{1,3} = g_1 - g_3$ and $r_{1,3} = 0$. Finally, $\ell_{23} = X^3 Y$ and

$$d_{2,3} = (\ell_{23}/\operatorname{ldt} g_2)\, g_2 - (\ell_{23}/\operatorname{ldt} g_3)\, g_3 = X^3 g_2 - Y g_3 = XY - X^5\,;$$

XY divides $\operatorname{ldt}(XY - X^5) = XY$, so

$$d_{2,3} = g_1 + (X - X^5) = g_1 - X^2 g_3 + (X - X^3) = g_1 - (X^2 + 1)\, g_3$$

and $r_{2,3} = 0$. The procedure ends; and $B = g_1$, g_2, g_3 is a Gröbner basis of (g_1, g_2).

The polynomial $f = X^3 + Y$ does not belong to (g_1, g_2): using the same lexicographic order, division by g_1, g_2, g_3 yields $f = g_2 + (X^3 + X^2) = g_2 + g_3 + (X^2 + X)$, with remainder $X^2 + X \neq 0$. On the other hand,

$$X^2 Y - Y = (X^2 - 1)\, g_2 + X^4 - X^2 = (X^2 - 1)\, g_2 + X g_3 \in (g_1, g_2).\ \square$$

Exercises

1. Show that the lexicographic order on $K[X_1, \dots, X_n]$ is a monomial order.

2. Show that the degree lexicographic order on $K[X_1, \dots, X_n]$ is a monomial order.

3. Show that the degree reverse lexicographic order on $K[X_1, \dots, X_n]$ is a monomial order.

4. Using the lexicographic order with $X > Y$, find all quotients and remainders when $f = 2X^3 Y^3 + 4Y^2$ is divided by $g_1 = 2XY^2 + 3X + 4Y^2$ and $g_2 = Y^2 - 2Y - 2$ in $\mathbb{C}[X, Y]$.

5. Using the lexicographic order with $X > Y$, find all quotients and remainders when $f = 2X^3 Y^3 + 4Y^2$ is divided by $g_1 = 2XY^2 + 3X + 4Y^2$, $g_2 = Y^2 - 2Y - 2$, and $g_3 = XY$ in $\mathbb{C}[X, Y]$.

6. Let $\mathfrak{A}$ be the ideal of $K[X_1, \dots, X_n]$ generated by nonzero polynomials $g_1, \dots, g_s$. Let a monomial order be given. Suppose that $f \in \mathfrak{A}$ if and only if, in every division of f by $g_1, \dots, g_s$, the remainder is 0. Show that $g_1, \dots, g_s$ is a Gröbner basis of $\mathfrak{A}$.

7. Using the lexicographic order with $X > Y$, find a Gröbner basis of the ideal $(2XY^2 + 3X + 4Y^2,\ Y^2 - 2Y - 2)$ of $\mathbb{C}[X, Y]$. Does $f = 2X^3 Y^3 + 4Y^2$ belong to this ideal?

8. Using the lexicographic order with $X > Y$, find a Gröbner basis of the ideal $(2XY^2 + 3X + 4Y^2, \ Y^2 - 2Y - 2, \ XY)$ of $\mathbb{C}[X, Y]$. Does $f = 2X^3Y^3 + 4Y^2$ belong to this ideal?

9. Using the lexicographic order with $X > Y$, find a Gröbner basis of the ideal $(X^2 + Y^2 + 1, \ X^2Y + 2XY + X)$ of $\mathbb{Z}_5[X, Y]$.

IV
Field Extensions

Fields are our third major algebraic structure. Their history may be said to begin with Dedekind [1871], who formulated the first clear definition of a field, albeit limited to fields of algebraic numbers. Steinitz [1910] wrote the first systematic abstract treatment. Today's approach is basically due to Artin, on whose lectures van der Waerden's *Moderne Algebra* [1930] is partly based.

Up to isomorphism, fields relate to each other by inclusion; hence the study of fields is largely that of field extensions. This chapter gives general properties of fields, field extensions, and algebraic extensions, plus some properties of transcendental extensions. The emphasis is on general structure results, that tell how extensions can be constructed from simpler extensions. Deeper properties of algebraic extensions will be found in the next chapter.

All this requires a couple of calls on Zorn's lemma, and makes heavy use of Chapter III. Sections 6, 7, and 9 may be skipped at first reading.

The few rings that have trespassed into this chapter all have identity elements.

1. Fields

A *field* is a commutative ring (necessarily a domain) whose nonzero elements constitute a group under multiplication. Chapter III established a few properties of fields. This section brings additional elementary properties, pertaining to homomorphisms, the characteristic, roots of unity, subrings, and subfields.

A *subring* of a field F is a subset S of F such that S is an additive subgroup of F, is closed under multiplication ($x, y \in S$ implies $xy \in S$), and contains the identity element; so that S inherits a ring structure from F. Subfields are similar:

Definition. A subfield of a field F is a subset K of F such that K is an additive subgroup of F and $K \backslash \{0\}$ is a multiplicative subgroup of $F \backslash \{0\}$.

Equivalently, K is a subfield of F if and only if (i) $0, 1 \in K$; (ii) $x, y \in K$ implies $x - y \in K$; and (iii) $x, y \in K$, $y \neq 0$ implies $xy^{-1} \in K$. Then $x, y \in K$ implies $x + y \in K$ and $xy \in K$, so that K inherits an addition and a multiplication from F, and K is a field under these inherited operations; this field K is also called a *subfield* of F.

For example, $\mathbb{Q}$ is a subfield of $\mathbb{R}$, and $\mathbb{R}$ is a subfield of $\mathbb{C}$.

Homomorphisms of fields are homomorphisms of rings with identity:

Definition. A homomorphism *of a field* K *into a field* L *is a mapping* $\varphi :$ $K \longrightarrow L$ *such that* $\varphi(1) = 1$, $\varphi(x + y) = \varphi(x) + \varphi(y)$, *and* $\varphi(xy) = \varphi(x)\,\varphi(y)$, *for all* $x, y \in K$.

For instance, when K is a subfield of F, the inclusion mapping $K \longrightarrow F$ is a homomorphism of fields, the *inclusion homomorphism* of K into F.

An *isomorphism* of fields is a bijective homomorphism of fields; then the inverse bijection is also an isomorphism.

Proposition **1.1.** *Every homomorphism of fields is injective.*

This is Proposition III.4.9. Consequently, a homomorphism of a field K into a field L induces a homomorphism of multiplicative groups of $K \backslash \{0\}$ into $L \backslash \{0\}$, and preserves powers and inverses. Proposition 1.1 has a another consequence:

Proposition **1.2** (Homomorphism Theorem). *If* $\varphi : K \longrightarrow L$ *is a field homomorphism, then* $\operatorname{Im} \varphi$ *is a subfield of* L *and* $K \cong \operatorname{Im} \varphi$.

Thus, up to isomorphism, the basic relationship between fields is inclusion. Inclusions between fields are studied in later sections.

For future use we note the following particular case of Proposition III.5.7:

Proposition **1.3.** *Every field homomorphism* $\varphi : K \longrightarrow L$ *induces a ring homomorphism* $f \longmapsto {}^{\varphi}f$ *of* $K[X]$ *into* $L[X]$; *if* $f(X) = a_0 + a_1 X + \cdots + a_n X^n$, *then* ${}^{\varphi}f(X) = \varphi(a_0) + \varphi(a_1)X + \cdots + \varphi(a_n)X^n$.

The characteristic. By Proposition III.3.7 there is for any field K a unique homomorphism of rings of $\mathbb{Z}$ into R. Its image is the smallest subring of K; it consists of all integer multiples of the identity element of K, and is isomorphic either to $\mathbb{Z}$ or to $\mathbb{Z}_n$ for some unique $n > 0$, the *characteristic* of K.

Proposition **1.4.** *The characteristic of a field is either* 0 *or a prime number.*

Proposition **1.5.** *Every field* K *has a smallest subfield, which is isomorphic to* $\mathbb{Q}$ *if* K *has characteristic* 0, *to* $\mathbb{Z}_p$ *if* K *has characteristic* $p \neq 0$.

Proofs. If K has characteristic $p \neq 0$, then p is prime, by III.4.3; hence the smallest subring of K is a field, by III.4.1, and is the smallest subfield of K. If K has characteristic 0, then, by III.4.11, the injection $m \longmapsto m1$ of $\mathbb{Z}$ into K extends to a homomorphism φ of the quotient field $\mathbb{Q} = Q(\mathbb{Z})$ into K, namely $\varphi(m/n) = m1\,(n1)^{-1}$. By 1.2, $\operatorname{Im} \varphi \cong \mathbb{Q}$ is a subfield of K; it is the smallest subfield of K since every subfield of K must contain 1 and every element $m1\,(n1)^{-1}$ of $\operatorname{Im} \varphi$. $\square$

Roots of unity.

Definition. An element r of a field K is an nth root of unity *when $r^n = 1$.*

For example, the nth roots of unity in $\mathbb{C}$ are all $e^{2ik\pi/n}$ with $k = 0, 1, \ldots,$ $n - 1$; they constitute a cyclic group under multiplication, generated by $e^{2i\pi/n}$, or by any $e^{2ik\pi/n}$ in which k is relatively prime to n. All fields share this property:

Proposition **1.6.** *Every finite multiplicative subgroup of a field is cyclic.*

Such a subgroup consists of roots of unity, since its elements have finite order.

Proof. Let K be a field and let G be a finite subgroup of the multiplicative group $K \backslash \{0\}$. Write $|G|$ as a product $|G| = p_1^{k_1} p_2^{k_2} \cdots p_r^{k_r}$ of positive powers of distinct primes. By II.1.3, G is a direct sum of subgroups $H_1, \ldots, H_r$ of orders $p_1^{k_1}, p_2^{k_2}, \ldots, p_r^{k_r}$.

Let $p = p_i$ be a prime divisor of $|G|$; then $H = H_i = \{ x \in G \mid x^{p^j} = 1$ for some $j \geqq 0 \}$. In H there is an element c of maximal order p^k. Then $x^{p^k} = 1$ for all $x \in H$. In the field K, the equation $X^{p^k} = 1$ has at most p^k solutions; hence $|H| \leqq p^k$. On the other hand, $\langle c \rangle \subseteq H$ already has p^k elements. Therefore $H = \langle c \rangle$. Thus $H_1, \ldots, H_r$ are cyclic. Since their orders are relatively prime, $G = H_1 \oplus H_2 \oplus \cdots \oplus H_r$ is cyclic, by II.1.4. $\square$

By 1.6, the nth roots of unity of any field constitute a cyclic group under multiplication; its generators are *primitive* nth roots of unity:

Definition. A primitive nth root of unity *in a field K is a generator of the cyclic multiplicative group of all nth roots of unity.*

Subfields. Subfields have a number of general properties.

Proposition **1.7.** *Every intersection of subfields of a field F is a subfield of F.*

The proof is an exercise. On the other hand, a union of subfields is not in general a subfield, a notable exception being the union of a nonempty chain, or of a nonempty directed family. Readers will prove a more general property:

Proposition **1.8.** *The union of a nonempty directed family of fields is a field. In particular, the union of a nonempty directed family of subfields of a field F is a subfield of F.*

By 1.7 there is for every subset S of a field F a smallest subfield of F that contains S, the subfield of F *generated* by S. The next result describes the subfield generated by the union of S and a subfield K of F; this yields the subfield generated by just S, if K is the smallest subfield of F.

Proposition **1.9.** *Let K be a subfield of a field F and let S be a subset of F.*

The subring $K[S]$ of F generated by $K \cup S$ is the set of all finite linear combinations with coefficients in K of finite products of powers of elements of S.

The subfield $K(S)$ of F generated by $K \cup S$ is the set of all $ab^{-1} \in F$ with $a, b \in K[S]$, $b \neq 0$, and is isomorphic to the field of fractions of $K[S]$.

Proof. Let $(X_s)_{s \in S}$ be a family of indeterminates, one for each $s \in S$. By III.6.6 there is an evaluation homomorphism $\varphi : K[(X_s)_{s \in S}] \longrightarrow F$:

$$\varphi\left(\sum_k \left(a_k \prod_{s \in S} X_s^{k_s}\right)\right) = \sum_k \left(a_k \prod_{s \in S} s^{k_s}\right),$$

where $\prod_{s \in S} s^{k_s}$ denotes the finite product $\prod_{s \in S,\, k_s \neq 0} s^{k_s}$. Then $\operatorname{Im} \varphi$ is a subring of F, which contains K and S and consists of all finite linear combinations with coefficients in K of finite products of powers of elements of S. All these linear combinations must belong to any subring of F that contains K and S, so $\operatorname{Im} \varphi$ is the smallest such subring.

By III.4.11 the inclusion homomorphism of $K[S]$ into F extends to a homomorphism ψ of the quotient field $Q(K[S])$ into F, which sends a/b to $ab^{-1} \in F$ for all $a, b \in K[S]$, $b \neq 0$. Hence

$$\operatorname{Im} \psi = \{ ab^{-1} \in F \mid a, b \in K[S],\ b \neq 0 \} \cong Q(K[S])$$

is a subfield of F, which contains $K[S]$ and $K \cup S$. Moreover, any subfield that contains K and S must contain $K[S]$ and all $ab^{-1} \in F$ with $a, b \in K[S]$, $b \neq 0$; hence $\operatorname{Im} \psi$ is the smallest such subfield. $\square$

The notation $K[S]$, $K(S)$ is traditional, but readers should keep in mind that $K[S]$ is not a polynomial ring, even though its elements look like polynomials, and that $K(S)$ is not a field of rational fractions, even though its elements look like rational fractions. Moreover, $K[S]$ and $K(S)$ depend on F, not just on K and S. If $S = \{ s_1, \dots, s_n \}$ is finite, then $K[S]$ and $K(S)$ are denoted by $K[s_1, \dots, s_n]$ and $K(s_1, \dots, s_n)$.

Proposition 1.9 implies some useful properties.

Corollary **1.10.** *Let F be a field, let K be a subfield of F, let S be a subset of F, and let $x, \alpha_1, \dots, \alpha_n \in F$.*

(1) $x \in K[\alpha_1, \dots, \alpha_n]$ *if and only if* $x = f(\alpha_1, \dots, \alpha_n)$ *for some polynomial* $f \in K[X_1, \dots, X_n]$.

(2) $x \in K(\alpha_1, \dots, \alpha_n)$ *if and only if* $x = r(\alpha_1, \dots, \alpha_n)$ *for some rational fraction* $r \in K(X_1, \dots, X_n)$.

(3) $x \in K[S]$ *if and only if* $x \in K[\alpha_1, \dots, \alpha_n]$ *for some* $\alpha_1, \dots, \alpha_n \in S$.

(4) $x \in K(S)$ *if and only if* $x \in K(\alpha_1, \dots, \alpha_n)$ *for some* $\alpha_1, \dots, \alpha_n \in S$.

Composites. The compositum or composite is another operation on subfields, a worthy alternative to unions.

Definition. The composite $\prod_{i \in I} K_i$ *of a nonempty family $(K_i)_{i \in I}$ of subfields of a field F is the subfield of F generated by $\bigcup_{i \in I} K_i$.* $\square$

If $I = \{1, 2, \ldots, n\}$ is finite, then $\prod_{i \in I} K_i$ is denoted by $K_1 K_2 \cdots K_n$. Regarding this traditional notation, readers should keep in mind that a composite is not a product of subsets, and that it depends on the larger field F. The author pledges to avoid confusion by never multiplying subfields as subsets.

Proposition 1.9 yields the following description of composites:

Proposition **1.11.** *Let* $(K_i)_{i \in I}$ *be a nonempty family of subfields of a field* F. *Then* $x \in \prod_{i \in I} K_i$ *if and only if* $x = ab^{-1} \in F$ *for some* $a, b \in R$, $b \neq 0$, *where* R *is the set of all finite sums of finite products of elements of* $\bigcup_{i \in I} K_i$.

In particular, $x \in F$ *is in the composite* KL *of two subfields* K *and* L *of* F *if and only if* $x = ab^{-1} \in F$ *for some* $a, b \in R$, $b \neq 0$, *where* R *is the set of all finite sums of products of an element of* K *and an element of* L.

Proof. We have $\prod_{i \in I} K_i = K_0 (\bigcup_{i \in I} K_i)$, where K_0 is the smallest subfield of F. Multiplying an element of K_0 by a finite product of powers of elements of $\bigcup_{i \in I} K_i$ yields a finite product of elements of $\bigcup_{i \in I} K_i$; hence, in 1.9, linear combinations with coefficients in K_0 of finite products of powers of elements of $\bigcup_{i \in I} K_i$ are just finite sums of finite products of elements of $\bigcup_{i \in I} K_i$. In the case of two subfields K and L, a finite product of elements of $K \cup L$ is the product of an element of K and an element of L. $\square$

In the case of two subfields K and L, the composite KL is generated by $K \cup L$, so that $KL = K(L) = L(K)$ and Proposition 1.11 follows directly from Proposition 1.9.

Exercises

1. Prove that every intersection of subfields of a field K is a subfield of K.

2. Prove that the union of a nonempty directed family of fields is a field.

3. Let K, L, M be subfields of a field F. Show that $(KL)M = K(LM)$.

4. Let L be a subfield of a field F and let $(K_i)_{i \in I}$ be a nonempty directed family of subfields of F. Show that $\left(\bigcup_{i \in I} K_i \right) L = \bigcup_{i \in I} (K_i L)$.

2. Extensions

This section contains basic properties of field extensions.

Definition. A field extension of a field K *is a field* F *of which* K *is a subfield.*

We write this relationship as an inclusion $K \subseteq E$ when it is understood that K and E are fields.

A field extension of a field K can also be defined as a field F together with a homomorphism of K into F. The two definitions are fully equivalent up to isomorphisms. If K is a subfield of E and $E \cong F$, then there is a homomorphism of K into F. Conversely, if $\varphi : K \longrightarrow F$ is a homomorphism, then K is isomorphic to the subfield $\text{Im } \varphi$ of F.

Moreover, when $\varphi : K \longrightarrow F$ is a homomorphism of fields, there is a field $E \cong F$ that contains K as a subfield. To see this, cut $\text{Im } \varphi$ from F and attach K in its place to make a disjoint union $E = K \cup (F \setminus \text{Im } \varphi)$. Electrify this monster to life as a field through the bijection $\theta : E \longrightarrow F$ that takes $x \in K$ to $\varphi(x) \in F$ and is the identity on $F \setminus \text{Im } \varphi$: define sums and products in E by $x + y = \theta^{-1}\big(\theta(x) + \theta(y)\big)$, $xy = \theta^{-1}\big(\theta(x)\,\theta(y)\big)$; then E is a field like F, θ is an isomorphism, and K is a subfield of E, since φ is a homomorphism. This construction can be used with most bidules; the author calls it *surgery*.

K**-homomorphisms** let extensions of a field K relate to each other.

Definition. Let $K \subseteq E$ and $K \subseteq F$ be field extensions of K. A K-homomorphism of E into F is a field homomorphism $\varphi : E \longrightarrow F$ that is the identity on K ($\varphi(x) = x$ for all $x \in K$).

The inclusion homomorphism $E \longrightarrow F$ in a *tower* $K \subseteq E \subseteq F$ of extensions is a K-homomorphism. Conversely, if $K \subseteq E, F$ and $\varphi : E \longrightarrow F$ is a K-homomorphism, then there is a K-isomorphism $E \cong \text{Im } \varphi \subseteq F$.

Definitions. A K-isomorphism is a bijective K-homomorphism. A K-automorphism of a field extension $K \subseteq E$ is a K-isomorphism of E onto E.

We view K-isomorphic extensions as avatars of the same "abstract" extension.

Degree. The first property of any field extension $K \subseteq E$ is that it is a vector space over K, in which scalar multiplication is just multiplication in E. In this light, K-homomorphisms are (in particular) linear transformations.

In Chapter VIII we show that any two bases of a vector space V have the same number of elements, the *dimension* of V (which is an infinite cardinal number if V does not have a finite basis).

Definitions. The degree $[E : K]$ *of a field extension $K \subseteq E$ is its dimension as a vector space over K. A field extension $K \subseteq E$ is* finite *when it has finite degree and is* infinite *otherwise.* $\square$

For example, $\mathbb{C}$ is a finite extension of $\mathbb{R}$, with $[\mathbb{C} : \mathbb{R}] = 2$, but $\mathbb{R}$ is an infinite extension of $\mathbb{Q}$ (in fact, $[\mathbb{R} : \mathbb{Q}] = |\mathbb{R}|$). Readers will remember that finite extensions are not usually finite in their number of elements, only in their dimension. The traditional terminology "degree" originated in a number of cases in which the degree of an extension is the degree of a related polynomial.

Proposition 2.1. If $K \subseteq E \subseteq F$, then $[F : K] = [F : E][E : K]$.

Proof. Let $(\alpha_i)_{i \in I}$ be a basis of E over K and let $(\beta_j)_{j \in J}$ be a basis of F over E. Every element of F is a linear combination of β_j's with coefficients in E, which are themselves linear combinations of α_i's with coefficients in K.

Hence every element of F is a linear combination of $\alpha_i \beta_j$'s with coefficients in K. Moreover, $(\alpha_i \beta_j)_{(i,j) \in I \times J}$ is a linearly independent family in F, viewed as a vector space over K: if $\sum_{(i,j) \in I \times J} x_{i,j} \alpha_i \beta_j = 0$ (with $x_{i,j} = 0$ for almost all (i,j)), then $\sum_{j \in J} \left(\sum_{i \in I} x_{i,j} \alpha_i \right) \beta_j = 0$, $\sum_{i \in I} x_{i,j} \alpha_i = 0$ for all j, and $x_{i,j} = 0$ for all i, j. Thus $(\alpha_i \beta_j)_{(i,j) \in I \times J}$ is a basis of F over K and $[F:K] = |I \times J| = |I| \, |J| = [F:E] \, [E:K]$. $\square$

Simple extensions are easily constructed and serve as basic building blocks for field extensions in general.

Definitions. A field extension $K \subseteq E$ is finitely generated *when* $E = K(\alpha_1, \ldots, \alpha_n)$ for some $\alpha_1, \ldots, \alpha_n \in E$. A field extension $K \subseteq E$ is simple *when* $E = K(\alpha)$ for some $\alpha \in E$; then α is a primitive element of E.

For example, the field of rational fractions $K(X)$ is a simple extension of K; the indeterminate X is a primitive element. Unlike simple groups, simple extensions may have proper subfields (see the exercises).

Let $K \subseteq E$ be a field extension. For each $\alpha \in E$, Proposition III.5.6 provides an evaluation homomorphism $f \longmapsto f(\alpha)$ of $K[X]$ into E. Its kernel is an ideal of $K[X]$ and is either 0 or generated by a unique monic polynomial.

Proposition 2.2. Let $K \subseteq E$ be a field extension and let $\alpha \in E$.

Either $f(\alpha) \neq 0$ for every nonzero polynomial $f(X) \in K[X]$, in which case there is a K-isomorphism $K(\alpha) \cong K(X)$;

or $f(\alpha) = 0$ for some nonzero polynomial $f(X) \in K[X]$, in which case there is a unique monic irreducible polynomial q such that $q(\alpha) = 0$; then $f(\alpha) = 0$ if and only if q divides f, $K[\alpha] = K(\alpha) \cong K[X]/(q)$, $[K(\alpha):K] = \deg q$, and $1, \alpha, \ldots, \alpha^{n-1}$ is a basis of $K(\alpha)$ over K, where $n = \deg q$.

Proof. Let $\psi : K[X] \longrightarrow E$, $f(X) \longmapsto f(\alpha)$ be the evaluation homomorphism. By 1.10, $\operatorname{Im} \psi = K[\alpha] \subseteq E$.

If $\operatorname{Ker} \psi = 0$, then $K[\alpha] \cong K[X]$; by 1.9 and III.4.11, $K(\alpha)$ is K-isomorphic to the quotient field of $K[\alpha]$, and $K(\alpha) \cong K(X)$.

Otherwise, by III.5.12, the nonzero ideal $\operatorname{Ker} \psi$ of $K[X]$ is generated by a unique monic polynomial q. Then $f(\alpha) = 0$ if and only if q divides f, and $K[\alpha] \cong K[X]/\operatorname{Ker} \psi = K[X]/(q)$. Now, $K[\alpha] \subseteq E$ is a domain; hence (q) is a prime ideal. In the PID $K[X]$ this implies that q is irreducible and that (q) is a maximal ideal. Hence $K[\alpha] \cong K[X]/(q)$ is a field; therefore $K(\alpha) = K[\alpha]$. If $p \in K[X]$ is monic irreducible and $p(\alpha) = 0$, then q divides p and $q = p$.

Let $n = \deg q > 0$. For every $f \in K[X]$, we have $f = qg + r$, where $\deg r < n = \deg q$. Then $f(\alpha) = r(\alpha)$, and every element $f(\alpha)$ of $K[\alpha]$ is a linear combination of $1, \alpha, \ldots, \alpha^{n-1}$ with coefficients in K. Moreover, $1, \alpha, \ldots, \alpha^{n-1}$ are linearly independent over K: if $r(\alpha) = 0$, where $r \in K[X]$

and $\deg r < n$, then q divides r and $r = 0$: otherwise, $\deg r \geqq \deg q = n$. Thus $1, \alpha, \ldots, \alpha^{n-1}$ is a basis of $K[\alpha]$ (as a vector space) over K. $\square$

Algebraic and transcendental elements. Proposition 2.2 leads to the following classification.

Definitions. Let $K \subseteq E$ be a field extension. An element α of E is algebraic over K *when $f(\alpha) = 0$ for some nonzero polynomial $f(X) \in K[X]$. Otherwise, α is* transcendental over K.

Equivalently, α is algebraic over K if and only if $[K(\alpha) : K]$ is finite. For example, every element of K is algebraic over K in any extension of K. Every complex number is algebraic over $\mathbb{R}$; $1 + \sqrt{3}$ and $\sqrt[3]{2} \in \mathbb{R}$ are algebraic over $\mathbb{Q}$. It has been shown by other methods that e and $\pi \in \mathbb{R}$ are transcendental over $\mathbb{Q}$; in fact, most real numbers are transcendental over $\mathbb{Q}$ (see Section A.5).

Definitions. Let α be algebraic over K. The unique monic irreducible polynomial $q = \mathrm{Irr}\,(\alpha : K) \in K[X]$ such that $q(\alpha) = 0$ is the irreducible polynomial of α over K; *the* degree *of α over K is the degree of $\mathrm{Irr}\,(\alpha : K)$.

For example, $\mathrm{Irr}\,(i : \mathbb{R}) = X^2 + 1$ and i has degree 2 over $\mathbb{R}$. Also, $\sqrt[3]{2} \in \mathbb{R}$ is algebraic over $\mathbb{Q}$; $\mathrm{Irr}\,(\sqrt[3]{2} : \mathbb{Q}) = X^3 - 2$ (irreducible in $\mathbb{Q}[X]$ by Eisenstein's criterion), and $\sqrt[3]{2}$ has degree 3 over $\mathbb{Q}$.

Finite simple extensions. We complete 2.2 with two more results.

Proposition **2.3.** *Let K be a field and let $q \in K[X]$ be irreducible. Up to isomorphism, $E = K[X]/(q)$ is a simple field extension of K: $E = K(\alpha)$, where $\alpha = X + (q)$. Moreover, $[E : K] = \deg q$ and $q = \mathrm{Irr}\,(\alpha : K)$.*

Kronecker [1887] had a very similar construction.

Proof. By III.8.3, (q) is a maximal ideal of $K[X]$; hence $E = K[X]/(q)$ is a field. Then $x \longmapsto x + (q)$ is a homomorphism of K into E; we may identify $x \in K$ and $x + (q) \in E$, and then E is an extension of K.

Let $\alpha = X + (q) \in E$. By the universal property of $K[X]$ there is a unique homomorphism of $K[X]$ into E that sends X to α and every $x \in K$ to $x + (q) = x$. Since the evaluation homomorphism $f(X) \longmapsto f(\alpha)$ and the canonical projection $K[X] \longrightarrow E$ have these properties, they coincide, and $f(X) + (q) = f(\alpha)$ for all $f \in K[X]$. Hence $E = K[\alpha]$, by 1.10, $K[\alpha]$ is a field, and $E = K(\alpha)$. Also $q(\alpha) = q + (q) = 0$ in E, so that α is algebraic over K and $\mathrm{Irr}\,(\alpha : K) = q$. Then $[E : K] = \deg q$, by 2.2. $\square$

Thus every irreducible polynomial $q \in K[X]$ has a root in some extension of K. For example, $\mathbb{R}[X]/(X^2 + 1)$ is a simple extension $\mathbb{R}(\alpha)$ of $\mathbb{R}$, with a basis $1, \alpha$ over $\mathbb{R}$ by 2.2 in which $\alpha^2 + 1 = 0$. Hence $\mathbb{R}[X]/(X^2 + 1) \cong \mathbb{C}$. This provides a construction of $\mathbb{C}$ that does not require any overt adjunction.

Finite simple extensions inherit from polynomial rings a very useful universal property, which constructs field homomorphisms $K(\alpha) \longrightarrow L$.

Proposition 2.4. *Let α be algebraic over K and let $q = \mathrm{Irr}\,(\alpha : K)$. If $\psi : K(\alpha) \longrightarrow L$ is a field homomorphism and φ is the restriction of ψ to K, then $\psi(\alpha)$ is a root of ${}^{\varphi}q$ in L. Conversely, for every field homomorphism $\varphi : K \longrightarrow L$ and every root β of ${}^{\varphi}q$ in L, there exists a unique field homomorphism $\psi : K(\alpha) \longrightarrow L$ that extends φ and sends α to β.*

$$
\begin{array}{ccc}
K & \overset{\subseteq}{\longrightarrow} & K(\alpha) \\
& {}_{\varphi}\searrow & \downarrow {}_{\psi} \\
& & L
\end{array}
$$

Proof. Let $\psi : K(\alpha) \longrightarrow L$ be a field homomorphism. Its restriction φ to K is a field homomorphism. For each $f(X) = a_0 + a_1 X + \cdots + a_m X^m \in K[X]$, we have

$$
\begin{aligned}
\psi\big(f(\alpha)\big) &= \psi\big(a_0 + a_1 \alpha + \cdots + a_m \alpha^m\big) \\
&= \varphi(a_0) + \varphi(a_1)\,\psi(\alpha) + \cdots + \varphi(a_m)\,\psi(\alpha)^m = {}^{\varphi}f\big(\psi(\alpha)\big).
\end{aligned}
$$

Hence $q(\alpha) = 0$ yields ${}^{\varphi}q\big(\psi(\alpha)\big) = 0$. Thus $\psi(\alpha)$ is a root of ${}^{\varphi}q$ in L.

Conversely, let $\beta \in L$ be a root of ${}^{\varphi}q$. Since $K(\alpha) \cong K[X]/(q)$ by 2.3, we may assume that $K(\alpha) = K[X]/(q)$ and $\alpha = X + (q)$. By the universal property of $K[X]$, φ extends to a unique homomorphism $\chi : K[X] \longrightarrow L$ that sends X to β, namely $\chi : f \longmapsto {}^{\varphi}f(\beta)$. Then $\chi(q) = {}^{\varphi}q(\beta) = 0$; hence $(q) \subseteq \mathrm{Ker}\,\chi$.

$$
\begin{array}{ccccc}
K & \overset{\subseteq}{\longrightarrow} & K[X] & \overset{\pi}{\longrightarrow} & K[X]/(q) = K(\alpha) \\
& {}_{\varphi}\searrow & \downarrow {}_{\chi} & \swarrow {}_{\psi} & \\
& & L & &
\end{array}
$$

By the Factorization theorem (III.3.5) χ factors uniquely through the projection $\pi : K[X] \longrightarrow K[X]/(q)$: $\chi = \psi \circ \pi$ for some unique homomorphism $\psi : K(\alpha) \longrightarrow L$. Then ψ extends φ and sends α to β; ψ is the only homomorphism with these properties, since $1, \alpha, \ldots, \alpha^{n-1}$ is a basis of $K(\alpha)$. $\square$

Infinite simple extensions have a similar property (see the exercises).

Exercises

1. Show that every field extension is a directed union of finitely generated extensions.

2. Show that $\alpha = 1 + \sqrt{5} \in \mathbb{R}$ is algebraic over $\mathbb{Q}$; find $\mathrm{Irr}\,(\alpha : \mathbb{Q})$.

3. Show that $\alpha = \sqrt{2} + \sqrt{3} \in \mathbb{R}$ is algebraic over $\mathbb{Q}$; find $\mathrm{Irr}\,(\alpha : \mathbb{Q})$.

4. Show that $\alpha = \sqrt{2} + i\sqrt{3} \in \mathbb{C}$ is algebraic over $\mathbb{Q}$; find $\mathrm{Irr}\,(\alpha : \mathbb{Q})$.

5. Show that the simple extension $E = \mathbb{Q}(\sqrt[6]{2}) \subseteq \mathbb{R}$ of $\mathbb{Q}$ has intermediate fields $\mathbb{Q} \subsetneqq F \subsetneqq E$.

6. Show that the simple extension $K(X)$ of K has intermediate fields $K \subsetneqq F \subsetneqq K(X)$.

7. Construct a field with four elements; draw its addition and multiplication tables.

8. Construct a field with eight elements; draw its addition and multiplication tables.

9. Construct a field with nine elements; draw its addition and multiplication tables.

10. Prove the following. Let α be transcendental over K. If $\psi : K(\alpha) \longrightarrow L$ is a field homomorphism, then $\psi(\alpha)$ is transcendental over $\psi(K)$. Conversely, if $\varphi : K \longrightarrow L$ is a field homomorphism and $\beta \in L$ is transcendental over $\varphi(K)$, then there exists a unique field homomorphism $\psi : K(\alpha) \longrightarrow L$ that extends φ and sends α to β.

3. Algebraic Extensions

This section contains basic properties of the class of algebraic extensions. Transcendental extensions are considered in Sections 8 and 9.

Definitions. A field extension $K \subseteq E$ is algebraic, *and E is* algebraic over K, *when every element of E is algebraic over K. A field extension $K \subseteq E$ is* transcendental, *and E is* transcendental over K, *when some element of E is trancendental over K.*

For example, $\mathbb{C}$ is an algebraic extension of $\mathbb{R}$ and $\mathbb{R}$ is a transcendental extension of $\mathbb{Q}$.

Algebraic extensions have a number of basic properties that make wonderful and highly recommended exercises.

Proposition 3.1. Every finite field extension is algebraic.

Proposition 3.2. If $E = K(\alpha_1, \ldots, \alpha_n)$ and every α_i is algebraic over K, then E is finite (hence algebraic) over K.

Proof. We give this proof as an example. Let $E = K(\alpha_1, \ldots, \alpha_n)$, where all α_i are algebraic over K. We prove by induction on n that E is finite over K. If $n = 0$, then $E = K$ is finite over K. If $n > 0$, then $F = K(\alpha_1, \ldots, \alpha_{n-1})$ is finite over K by the induction hypothesis; α_n is algebraic over F, since $f(\alpha_n) = 0$ for some nonzero $f \in K[X] \subseteq F[X]$; hence $E = F(\alpha_n)$ is finite over F by 2.2, and E is finite over K, by 2.1. $\square$

Proposition 3.3. If every $\alpha \in S$ is algebraic over K, then $K(S)$ is algebraic over K.

Proposition 3.4. Let $K \subseteq E \subseteq F$ be fields. If F is algebraic over K, then E is algebraic over K and F is algebraic over E.

Proposition 3.5 (Tower Property). Let $K \subseteq E \subseteq F$ be fields. If E is algebraic over K, and F is algebraic over E, then F is algebraic over K.

Proposition 3.6. If E is algebraic over K and the composite EF exists, then EF is algebraic over KF.

Proposition 3.7. Every composite of algebraic extensions of a field K is an algebraic extension of K.

Here are some applications of these results.

Proposition **3.8.** *If* E *is finite over* K *and the composite* EF *exists, then* EF *is finite over* KF. *Hence the composite of finitely many finite extensions of* K *is a finite extension of* K.

Proof. We prove the first statement and leave the second as an exercise. Let $\alpha_1, \ldots, \alpha_n$ be a basis of E over K. Then $E = K(\alpha_1, \ldots, \alpha_n)$ and every α_i is algebraic over K. Hence $EF = KF(\alpha_1, \ldots, \alpha_n)$, every α_i is algebraic over KF by 3.6, and EF is finite over KF by 3.2. $\square$

Proposition **3.9.** *In any field extension* $K \subseteq E$, *the elements that are algebraic over* K *constitute a field.*

Proof. First, 0 and $1 \in K$ are algebraic over K. Now let $\alpha, \beta \in E$ be algebraic over K. By 3.3, $K(\alpha, \beta) \subseteq E$ is algebraic over K. Hence $\alpha - \beta \in K(\alpha, \beta)$ and $\alpha \beta^{-1} \in K(\alpha, \beta)$ are algebraic over K. $\square$

For example, the set of all algebraic real numbers (over $\mathbb{Q}$) is a field.

Exercises

Prove the following:

1. If every $\alpha \in S$ is algebraic over K, then $K(S)$ is algebraic over K.

2. If $K \subseteq E \subseteq F$ are fields and F is algebraic over K, then E is algebraic over K and F is algebraic over E.

3. If $K \subseteq E \subseteq F$ are fields, E is algebraic over K, and F is algebraic over E, then F is algebraic over K. (Hint: every $\alpha \in F$ is algebraic over $K(\alpha_0, \alpha_1, \ldots, \alpha_n)$, where $\alpha_0, \alpha_1, \ldots, \alpha_n$ are the coefficients of $\mathrm{Irr}\,(\alpha : E)$.)

4. If E is algebraic over K, then the composite EF, if it exists, is algebraic over KF.

5. Every composite of algebraic extensions of K is an algebraic extension of K.

6. The composite of finitely many finite extensions of K is a finite extension of K.

4. The Algebraic Closure

In this section we show that every field has a greatest algebraic extension, its algebraic closure, which is unique up to isomorphism.

Algebraically closed fields have no proper algebraic extensions:

Proposition **4.1.** *For a field* K *the following properties are equivalent:*

(1) *the only algebraic extension of* K *is* K *itself;*

(2) *in* $K[X]$, *every irreducible polynomial has degree 1;*

(3) *every nonconstant polynomial in* $K[X]$ *has a root in* K.

Proof. (1) implies (2): when $q \in K[X]$ is irreducible, then $E = K[X]/(q)$ has degree $[E : K] = \deg q$, by 2.3; hence (1) implies $\deg q = 1$.

(2) implies (3) since every nonconstant polynomial $f \in K[X]$ is a nonempty product of irreducible polynomials.

(3) implies (1): when α is algebraic over K, then $q = \mathrm{Irr}\,(\alpha : K)$ has a root r in K; hence $q = X - r$, and $q(\alpha) = 0$ yields $\alpha = r \in K$. $\square$

Definition. A field is algebraically closed *when it satisfies the equivalent conditions in Proposition* 4.1. $\square$

For instance, the fundamental theorem of algebra (Theorem III.8.11) states that $\mathbb{C}$ is algebraically closed. The fields $\mathbb{R}$, $\mathbb{Q}$, $\mathbb{Z}_p$ are not algebraically closed, but $\mathbb{R}$ and $\mathbb{Q}$ can be embedded into the algebraically closed field $\mathbb{C}$.

Algebraically closed fields have an interesting homomorphism property.

Theorem 4.2. *Every homomorphism of a field K into an algebraically closed field can be extended to every algebraic extension of K.*

Proof. Let E be an algebraic extension of K and let φ be a homomorphism of K into an algebraically closed field L. If $E = K(\alpha)$ is a simple extension of K, and $q = \mathrm{Irr}\,(\alpha : K)$, then ${}^{\varphi}q \in L[X]$ has a root in L, since L is algebraically closed, and φ can be extended to E by 2.4.

The general case uses Zorn's lemma. Let $\mathcal{S}$ be the set of all ordered pairs (F, ψ) in which F is a subfield of E, $K \subseteq F \subseteq E$, and $\psi : F \longrightarrow L$ is a homomorphism that extends φ ($\psi(x) = \varphi(x)$ for all $x \in K$). For instance, $(K, \varphi) \in \mathcal{S}$. Partially order $\mathcal{S}$ by $(F, \psi) \leq (G, \chi)$ if and only if F is a subfield of G and χ extends ψ. Let $\mathcal{C} = (F_i, \psi_i)_{i \in I}$ be a nonempty chain of $\mathcal{S}$. Then $F = \bigcup_{i \in I} F_i$ is a subfield of E, by 1.8. A mapping $\psi : F \longrightarrow L$ is well defined by $\psi(x) = \psi_i(x)$ whenever $x \in F$ is in F_i: if $x \in F_i \cap F_j$, then, say, $(F_i, \psi_i) \leq (F_j, \psi_j)$, ψ_j extends ψ_i, and $\psi_j(x) = \psi_i(x)$. Then ψ extends every ψ_i, and is a homomorphism since any $x, y \in F$ belong to some F_i and ψ_i is a homomorphism. Hence $(F, \psi) \in \mathcal{S}$, $(F_i, \psi_i) \leq (F, \psi)$ for all $i \in I$, and $\mathcal{C}$ has an upper bound in $\mathcal{S}$.

By Zorn's lemma, $\mathcal{S}$ has a maximal element (M, μ). If $M \neq E$, then any $\alpha \in E \backslash M$ is algebraic over M, since E is an algebraic extension of $K \subseteq M$, and μ can be extended to the simple algebraic extension $M(\alpha)$ of M, contradicting the maximality of M. So $M = E$, and μ extends φ to E. $\square$

The proof of Theorem 4.2 is a standard argument that generally provides maximal extensions of bidule homomorphisms (see the exercises). The homomorphism in Theorem 4.2 can generally be extended in several ways; already, in Proposition 2.4, φ can usually be extended in several ways, since ${}^{\varphi}q$ usually has several roots. This phenomenon is studied in more detail in the next section.

Embeddings. The main result of this section is that every field K can be embedded into an algebraically closed field $\overline{K}$ that is algebraic over K; and then every algebraic extension of K can be embedded in $\overline{K}$, by Theorem 4.2.

Lemma **4.3.** *Every field* K *has an algebraic extension that contains a root of every nonconstant polynomial with coefficients in* K.

Proof. (Kempf) For any finitely many nonconstant polynomials $f_1, \ldots, f_n \in K[X]$, we note that K has an algebraic extension in which every f_i has a root: repeated applications of Propositions 2.3 to irreducible factors of $f_1, \ldots, f_n$ yield an extension of K in which every f_i has a root, which is algebraic over K by 3.3.

Now write the set of all nonconstant polynomials $f \in K[X]$ as a family $(f_i)_{i \in I}$. Form the polynomial ring $K[(X_i)_{i \in I}]$, using the same index set I, and let $\mathfrak{A}$ be the ideal of $K[(X_i)_{i \in I}]$ generated by all $f_i(X_i)$.

We show that $\mathfrak{A} \neq K[(X_i)_{i \in I}]$. Otherwise, $1 \in \mathfrak{A}$ and $1 = \sum_{j \in J} u_j\, f_j(X_j)$ for some finite subset J of I and polynomials $u_j \in K[(X_i)_{i \in I}]$. Since J is finite, K has an algebraic extension E in which every f_j has a root α_j. The universal property of $K[(X_i)_{i \in I}]$ yields a homomorphism φ of $K[(X_i)_{i \in I}]$ into E such that $\varphi(x) = x$ for all $x \in K$, $\varphi(X_i) = 0$ for all $i \in I \backslash J$, and $\varphi(X_j) = \alpha_j$ for all $j \in J$. Then $\varphi\big(f_j(X_j)\big) = f_j(\alpha_j)$ and $1 = \varphi(1) = \sum_{j \in J} \varphi(u_j)\, \varphi\big(f_j(X_j)\big) = 0$. This is the required contradiction.

Now, $\mathfrak{A} \neq K[(X_i)_{i \in I}]$ is contained in a maximal ideal $\mathfrak{M}$ of $K[(X_i)_{i \in I}]$. Then $F = K[(X_i)_{i \in I}]/\mathfrak{M}$ is a field. We now follow the proof of Proposition 2.3.

There is a homomorphism $x \longmapsto x + \mathfrak{M}$ of K into F. We may identify $x \in K$ and $x + \mathfrak{M} \in F$; then F is an extension of K. Let $\alpha_i = X_i + \mathfrak{M} \in F$. By uniqueness in the universal property of $K[(X_i)_{i \in I}]$, the canonical projection $K[(X_i)_{i \in I}] \longrightarrow F$ coincides with the evaluation homomorphism $f\big((X_i)_{i \in I}\big) \longmapsto f\big((\alpha_i)_{i \in I}\big)$, since both send X_i to α_i for all i and send every $x \in K$ to $x + \mathfrak{M} = x$. Thus, $f\big((X_i)_{i \in I}\big) + \mathfrak{M} = f\big((\alpha_i)_{i \in I}\big)$ for all $f \in K[(X_i)_{i \in I}]$. Hence $F = K[(\alpha_i)_{i \in I}]$, by 1.10, $K[(\alpha_i)_{i \in I}]$ is a field, and $F = K((\alpha_i)_{i \in I})$. Also $f_i(\alpha_i) = f_i(X_i) + \mathfrak{M} = 0$ in F, so that every α_i is algebraic over K; hence F is algebraic over K, by 3.3. $\square$

Another proof of Lemma 4.3 is given in Section A.4 (see the exercises for that section).

Theorem **4.4.** *Every field* K *has an algebraic extension* $\overline{K}$ *that is algebraically closed. Moreover,* $\overline{K}$ *is unique up to* K-*isomorphism.*

Proof. There is a very tall tower of fields

$$K = E_0 \subseteq E_1 \subseteq \cdots \subseteq E_n \subseteq E_{n+1} \subseteq \cdots$$

in which E_{n+1} is the algebraic extension of E_n in Lemma 4.3, which contains a root of every nonconstant polynomial with coefficients in E_n. Then every E_n is algebraic over K, by 3.5, and $\overline{K} = \bigcup_{n \geq 0} E_n$, which is a field by 1.8, is an algebraic extension of K. Then $\overline{K}$ is algebraically closed: when $f \in \overline{K}[X]$ is not constant, the finitely many coefficients of f all lie in some E_n and f has a root in $E_{n+1} \subseteq \overline{K}$.

Let L be an algebraically closed, algebraic extension of K. By 4.2, there is a K-homomorphism $\varphi : \overline{K} \longrightarrow L$. Then $\mathrm{Im}\, \varphi \cong \overline{K}$ is algebraically closed, and L is algebraic over $\mathrm{Im}\, \varphi$ by 3.4; therefore $L = \mathrm{Im}\, \varphi$ and φ is a K-isomorphism. $\square$

Definition. An algebraic closure *of a field K is an algebraic extension $\overline{K}$ of K that is algebraically closed.*

The field $\overline{K}$ in this definition is also called *the* algebraic closure of K, since Theorem 4.4 ensures that all algebraic closures of K are K-isomorphic. For example, $\mathbb{C}$ is 'the' algebraic closure of $\mathbb{R}$, since it is algebraically closed and algebraic over $\mathbb{R}$.

The algebraic closure $\overline{K}$ of K can be characterized in several ways:

(1) $\overline{K}$ is an algebraically closed, algebraic extension of K (by definition);

(2) $\overline{K}$ is a maximal algebraic extension of K (if $\overline{K} \subseteq E$ and E is algebraic over K, then $\overline{K} = E$);

(3) $\overline{K}$ is, up to K-isomorphism, the largest algebraic extension of K (if E is algebraic over K, then E is K-isomorphic to a subfield of $\overline{K}$, by 4.2);

(4) $\overline{K}$ is a minimal algebraically closed extension of K (if $K \subseteq L \subseteq \overline{K}$ and L is algebraically closed, then $L = \overline{K}$);

(5) $\overline{K}$ is, up to K-isomorphism, the smallest algebraically closed extension of K (if $K \subseteq L$ and L is algebraically closed, then $\overline{K}$ is K-isomorphic to a subfield of L, by 4.2).

By (3) we may limit the study of algebraic extensions to the intermediate fields $K \subseteq E \subseteq \overline{K}$ of any algebraic closure of K:

Corollary **4.5.** *For every algebraic extension E of K, $\overline{E}$ is an algebraic closure of K; hence E is K-isomorphic to an intermediate field $K \subseteq F \subseteq \overline{K}$ of any algebraic closure of K.*

Finally, we note the following properties.

Proposition **4.6.** *Every K-endomorphism of $\overline{K}$ is a K-automorphism.*

Proof. Let $\varphi : \overline{K} \longrightarrow \overline{K}$ is a K-homomorphism. As in the proof of 4.4, $\mathrm{Im}\, \varphi \cong \overline{K}$ is algebraically closed, $\overline{K}$ is algebraic over $\mathrm{Im}\, \varphi$, by 3.4, hence $\overline{K} = \mathrm{Im}\, \varphi$ and φ is a K-isomorphism. $\square$

Proposition **4.7.** *If $K \subseteq E \subseteq \overline{K}$ is an algebraic extension of K, then every K-homomorphism of E into $\overline{K}$ extends to a K-automorphism of $\overline{K}$.*

Proof. By 4.2, every K-homomorphism of E into $\overline{K}$ extends to a K-endomorphism of $\overline{K}$, which is a K-automorphism of $\overline{K}$ by 4.6. $\square$

Exercises

1. Let G be a group, let H be a subgroup of G, and let $\varphi : H \longrightarrow J$ be a homomorphism of groups. Show that there is a pair (M, μ) such that $H \leq M \leq G$, $\mu : M \longrightarrow J$ is a homomorphism that extends φ, and (M, μ) is maximal with these properties.

2. Show that every algebraically closed field is infinite.

3. Let A be the field of all complex numbers that are algebraic over $\mathbb{Q}$. Show that A is an algebraic closure of $\mathbb{Q}$.

5. Separable Extensions

An algebraic extension $K \subseteq E$ is separable when the irreducible polynomials of its elements are separable (have no multiple roots). This section relates polynomial separability to the number of K-homomorphisms of E into $\overline{K}$.

Separable polynomials. Let $f \in K[X]$ be a nonconstant polynomial with coefficients in a field K. Viewed as a polynomial with coefficients in any algebraic closure $\overline{K}$ of K, f factors uniquely (up to the order of the terms) into a product of positive powers of irreducible polynomials of degree 1:

$$f(X) = a (X - \alpha_1)^{m_1} (X - \alpha_2)^{m_2} \cdots (X - \alpha_r)^{m_r};$$

then $a \in K$ is the leading coefficient of f, $r > 0$, $m_1, \ldots, m_r > 0$, $\alpha_1, \ldots, \alpha_r \in \overline{K}$ are the distinct roots of f in $\overline{K}$, and m_i is the *multiplicity* of α_i. Recall that a root α_i of f is *multiple* when it has multiplicity $m_i > 1$.

Definition. A polynomial $f \in K[X]$ is separable *when it has no multiple root in $\overline{K}$.*

For example, $f(X) = X^4 + 2X^2 + 1 \in \mathbb{R}[X]$ factors as $f(X) = (X^2 + 1)^2 = (X - i)^2 (X + i)^2$ in $\mathbb{C}[X]$ and has two multiple roots in $\overline{\mathbb{R}} = \mathbb{C}$; it is not separable. But $X^2 + 1 \in \mathbb{R}[X]$ is separable. Readers will show, however, that an irreducible polynomial is not necessarily separable.

Proposition 5.1. Let $q \in K[X]$ be irreducible.

(1) *If K has characteristic 0, then q is separable.*

(2) *If K has characteristic $p \neq 0$, then all roots of q in $\overline{K}$ have the same multiplicity, which is a power p^m of p, and there exists a separable irreducible polynomial $s \in K[X]$ such that $q(X) = s(X^{p^m})$.*

Proof. We may assume that q is monic. If q has a multiple root α in $\overline{K}$, then $q'(\alpha) = 0$ by III.5.9. Now, α is algebraic over K, with $q = \mathrm{Irr}\,(\alpha : K)$ since $q(\alpha) = 0$; hence q divides q', and $q' = 0$, since $\deg q' < \deg q$. But $q' \neq 0$ when K has characteristic 0, since q is not constant; hence q is separable.

Now let K have characteristic $p \neq 0$. If $q(X) = \sum_{n \geq 0} a_n X^n$ has a multiple root, then, as above, $q'(X) = \sum_{n \geq 1} n a_n X^{n-1} = 0$; hence $a_n = 0$ whenever n

is not a multiple of p and q contains only powers of X^p. Thus $q(X) = r(X^p)$ for some $r \in K[X]$; r is, like q, monic and irreducible in $K[X]$ (if r had a nontrivial factorization, then so would q), and $\deg r < \deg q$. If r is not separable, then $r(X) = t(X^p)$, and $q(X) = t(X^{p^2})$, where t is monic and irreducible in $K[X]$ and $\deg t < \deg r < \deg q$. This process must stop; then $q = s(X^{p^m})$, where $s \in K[X]$ is monic, irreducible, and separable.

Write $s(X) = (X - \beta_1)(X - \beta_2) \cdots (X - \beta_n)$, where $\beta_1, \ldots, \beta_n$ are the distinct roots of s in $\overline{K}$. Since $\overline{K}$ is algebraically closed there exist $\alpha_1, \ldots, \alpha_n \in \overline{K}$ such that $\beta_i = \alpha_i^{p^m}$ for all i; in particular, $\alpha_1, \ldots, \alpha_n$ are distinct. In K and $\overline{K}$, $(x - y)^p = x^p - y^p$ for all x, y, by III.4.4, so that

$$q(X) = s(X^{p^m}) = \prod_i (X^{p^m} - \alpha_i^{p^m}) = \prod_i (X - \alpha_i)^{p^m} ;$$

hence the roots of q in $\overline{K}$ are $\alpha_1, \ldots, \alpha_n$, and all have multiplicity p^m. $\square$

The separability degree. We now relate polynomial separability to the number of K-homomorphisms into $\overline{K}$.

Definition. The separability degree $[E:K]_s$ *of an algebraic extension $K \subseteq E$ is the number of K-homomorphisms of E into an algebraic closure $\overline{K}$ of K.*

By 3.11, $[E:K]_s$ does not depend on the choice of $\overline{K}$. If E is a simple extension of K, then Propositions 2.12 and 5.1 yield the following properties:

Proposition 5.2. *If α is algebraic over K, then $[K(\alpha):K]_s$ is the number of distinct roots of $\mathrm{Irr}\,(\alpha:K)$ in $\overline{K}$. Hence $[K(\alpha):K]_s \leq [K(\alpha):K]$; if K has characteristic $p \neq 0$, then $[K(\alpha):K] = p^m [K(\alpha):K]_s$ for some $m \geq 0$; and $[K(\alpha):K]_s = [K(\alpha):K]$ if and only if $\mathrm{Irr}\,(\alpha:K)$ is separable.*

We now look at algebraic extensions in general.

Proposition 5.3 (Tower Property). *If F is algebraic over K and $K \subseteq E \subseteq F$, then $[F:K]_s = [F:E]_s \, [E:K]_s$.*

Proof. By 3.13 we may assume that $E \subseteq \overline{K}$. Let $\varphi : E \longrightarrow \overline{K}$ be a K-homomorphism. By 3.14, there is a K-automorphism σ of $\overline{K}$ that extends φ. If now $\psi : F \longrightarrow \overline{K}$ is an E-homomorphism, then $\sigma \circ \psi$ is a K-homomorphism that extends φ. Conversely, if $\chi : F \longrightarrow \overline{K}$ is a K-homomorphism that extends φ, then $\psi = \sigma^{-1} \circ \chi$ is an E-homomorphism:

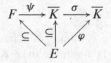

Hence there are $[F:E]_s$ K-homomorphisms of F into $\overline{K}$ that extend φ. The K-homomorphisms of F into $\overline{K}$ can now be partitioned into $[E:K]_s$ equivalence classes according to their restrictions to E. Each class has $[F:E]_s$ elements; hence there are $[E:K]_s \, [F:E]_s$ K-homomorphisms of F into $\overline{K}$. $\square$

Proposition **5.4.** *For every finite extension* E *of* K, $[E:K]_s \leqq [E:K]$; *if* K *has characteristic* 0, *then* $[E:K]_s = [E:K]$; *if* K *has characteristic* $p \neq 0$, *then* $[E:K] = p^m [E:K]_s$ *for some* $m \geqq 0$.

Proof. $E = K(\alpha_1, ..., \alpha_n)$ for some $\alpha_1, ..., \alpha_n \in E$, which yields a tower $K = E_0 \subseteq E_1 \subseteq \cdots \subseteq E_n = E$ of simple extensions $E_i = K(\alpha_1, ..., \alpha_i) = E_{i-1}(\alpha_i)$. For every i, $[E_i : E_{i-1}]_s \leqq [E_i : E_{i-1}]$, by 5.2; hence $[E:K]_s \leqq [E:K]$, by 5.3 and 2.4. The other two parts are proved similarly. □

Separable extensions.

Definitions. An element α *is* separable over K *when* α *is algebraic over* K *and* $\mathrm{Irr}\,(\alpha : K)$ *is separable. An algebraic extension* E *of* K *is* separable, *and* E *is* separable over K, *when every element of* E *is separable over* K.

Proposition 5.1 yields examples:

Proposition **5.5.** *If* K *has characteristic* 0, *then every algebraic extension of* K *is separable.*

The main property of separable algebraic extensions $K \subseteq E$ is that the number of K-homomorphisms of E into $\overline{K}$ is readily determined.

Proposition **5.6.** *For a finite extension* $K \subseteq E$ *the following conditions are equivalent:*

(1) E *is separable over* K *(every element of* E *is separable over* K);

(2) E *is generated by finitely many separable elements;*

(3) $[E:K]_s = [E:K]$.

Proof. (1) implies (2), since $E = K(\alpha_1, ..., \alpha_n)$ for some $\alpha_1, ..., \alpha_n \in E$.

(2) implies (3). Let $E = K(\alpha_1, ..., \alpha_n)$, where $\alpha_1, ..., \alpha_n$ are separable over K. Then $K = E_0 \subseteq E_1 \subseteq \cdots \subseteq E_n = E$, where $E_i = K(\alpha_1, ..., \alpha_i) = E_{i-1}(\alpha_i)$ when $i > 0$. Let $q = \mathrm{Irr}\,(\alpha_i : K)$ and $q_i = \mathrm{Irr}\,(\alpha_i : E_{i-1})$. Then $q \in K[X] \subseteq E_{i-1}[X]$ and $q(\alpha_i) = 0$; hence q_i divides q and is separable. Then $[E_i : E_{i-1}]_s = [E_i : E_{i-1}]$, by 5.2, and $[E:K]_s = [E:K]$, by 5.3 and 2.4.

(3) implies (1). Assume $[E:K]_s = [E:K]$ and let $\alpha \in E$. By 5.3 and 2.4,

$$[E:K(\alpha)]_s [K(\alpha):K]_s = [E:K]_s = [E:K] = [E:K(\alpha)][K(\alpha):K].$$

Since $[E:K(\alpha)]_s \leqq [E:K(\alpha)]$ and $[K(\alpha):K]_s \leqq [K(\alpha):K]$, this implies $[E:K(\alpha)]_s = [E:K(\alpha)]$ and $[K(\alpha):K]_s = [K(\alpha):K]$. Hence α is separable over K, by 5.2. (This argument requires $[E:K]$ finite.) □

Properties. The following properties make nifty exercises.

Proposition **5.7.** *If every* $\alpha \in S$ *is separable over* K, *then* $K(S)$ *is separable over* K.

Proposition **5.8.** *Let* $K \subseteq E \subseteq F$ *be algebraic extensions. If* F *is separable over* K, *then* E *is separable over* K *and* F *is separable over* E.

Proposition **5.9** (Tower Property). *Let* $K \subseteq E \subseteq F$ *be algebraic extensions. If* E *is separable over* K *and* F *is separable over* E, *then* F *is separable over* K.

Proposition **5.10.** *If* E *is algebraic and separable over* K *and the composite* EF *exists, then* EF *is separable over* KF.

Proposition **5.11.** *Every composite of algebraic separable extensions of a field* K *is a separable extension of* K.

We give some applications of these results.

Proposition **5.12** (Primitive Element Theorem). *Every finite separable extension is simple.*

Proof. Let E be a finite separable extension of a field K. If K is finite, then E is finite, the multiplicative group $E \backslash \{0\}$ is cyclic by 1.6, and E is singly generated as an extension.

Now let K be infinite. We show that every finite separable extension $E = K(\alpha, \beta)$ of K with two generators is simple; then so is every finite separable extension $K(\alpha_1, \ldots, \alpha_k)$ of K, by induction on k. Let $n = [E : K] = [E : K]_s$ and $\varphi_1, \ldots, \varphi_n$ be the K-homomorphisms of E into $\overline{K}$. Let

$$f(X) = \prod_{i < j} \left(\varphi_i \alpha + (\varphi_i \beta) X - \varphi_j \alpha - (\varphi_j \beta) X \right) \in \overline{K}[X].$$

Since K is infinite we cannot have $f(t) = 0$ for all $t \in K$; hence $f(t) \neq 0$ for some $t \in K$. Then $\varphi_1 (\alpha + \beta t), \ldots, \varphi_n (\alpha + \beta t)$ are all distinct. Hence there are at least n K-homomorphisms of $K(\alpha + \beta t)$ into $\overline{K}$ and $[K(\alpha + \beta t) : K] \geqq n$. Therefore $[K(\alpha + \beta t) : K] = [E : K]$ and $E = K(\alpha + \beta t)$. $\square$

Proposition **5.13.** *If* E *is separable over* K *and* $\mathrm{Irr}\,(\alpha : K)$ *has degree at most* n *for every* $\alpha \in E$, *then* E *is finite over* K *and* $[E : K] \leqq n$.

Proof. Choose $\alpha \in E$ so that $m = \deg \mathrm{Irr}\,(\alpha : K)$ is maximal. For every $\beta \in E$ we have $K(\alpha, \beta) = K(\gamma)$ for some $\gamma \in E$, by 5.12. Then $\deg \mathrm{Irr}\,(\gamma : K) \leqq m$ and $[K(\gamma) : K] \leqq m$. Since $K(\gamma)$ contains $K(\alpha)$ and $[K(\alpha) : K] = m$, it follows that $K(\gamma) = K(\alpha)$. Hence $K(\alpha)$ contains every $\beta \in E$; that is, $E = K(\alpha)$, and then $[E : K] = m \leqq n$. $\square$

Exercises

1. Find an irreducible polynomial that is not separable. (Hint: coefficients need to be in an infinite field of nonzero characteristic.)

2. Prove the following: if $E = K(S)$ is algebraic over K and every $\alpha \in S$ is separable over K, then $K(S)$ is separable over K.

3. Let $K \subseteq E \subseteq F$ be algebraic extensions. Prove the following: if F is separable over K, then E is separable over K and F is separable over E.

4. Let $K \subseteq E \subseteq F$ be algebraic extensions. Prove the following: if E is separable over K and F is separable over E, then F is separable over K.

5. Let $K \subseteq E$ be an algebraic extension. Prove the following: if E is separable over K and the composite EF exists, then EF is separable over KF.

6. Prove the following: every composite of algebraic separable extensions of a field K is a separable extension of K.

6. Purely Inseparable Extensions

In a purely inseparable extension of a field K, only the elements of K are separable over K. This section contains basic properties and examples, with applications to perfect fields, and may be skipped at first reading.

Definition. An algebraic extension $K \subseteq E$ is purely inseparable, and E is purely inseparable over K, when no element of $E \backslash K$ is separable over K.

One reason for our interest in purely inseparable extensions is that every algebraic extension is a purely inseparable extension of a separable extension (moreover, some extensions are separable extensions of purely inseparable extensions; see Proposition V.2.10).

Proposition 6.1. For every algebraic extension E of K, $S = \{ \alpha \in E \mid \alpha$ is separable over $K \}$ is a subfield of E, S is separable over K, and E is purely inseparable over S.

Proof. First, 0 and $1 \in K$ are separable over K. If $\alpha, \beta \in E$ are separable over K, then $K(\alpha, \beta)$ is separable over K by 5.7 and $\alpha - \beta$, $\alpha \beta^{-1} \in K(\alpha, \beta)$ are separable over K. Thus S is a subfield of E. Clearly S is separable over K. If $\alpha \in E$ is separable over S, then $S(\alpha)$ is separable over K by 5.7, 5.9, and $\alpha \in S$. $\square$

By 5.5, purely inseparable extensions are trivial unless K has characteristic $p \neq 0$. Then $(\alpha - \beta)^{p^m} = \alpha^{p^m} - \beta^{p^m}$ for all $m > 0$ and $\alpha, \beta \in \overline{K}$, by III.4.4, so that every $a \in K$ has a unique p^mth root in $\overline{K}$ and a polynomial in the form $X^{p^m} - a \in K[X]$ has only one root in $\overline{K}$. This provides the following example.

Proposition 6.2. If K has characteristic $p \neq 0$, then $K^{1/p^\infty} = \{ \alpha \in \overline{K} \mid \alpha^{p^m} \in K$ for some $m \geq 0 \}$ is a purely inseparable field extension of K.

Proof. We have $K \subseteq K^{1/p^\infty}$, in particular $0, 1 \in K^{1/p^\infty}$. If $\alpha, \beta \in K^{1/p^\infty}$, then α^{p^m}, $\beta^{p^m} \in K$ when m is large enough, and then $(\alpha - \beta)^{p^m} = \alpha^{p^m} - \beta^{p^m} \in K$ and $(\alpha \beta^{-1})^{p^m} = (\alpha^{p^m})(\beta^{p^m})^{-1} \in K$. Thus K^{1/p^∞} is a subfield of $\overline{K}$. If $\alpha \in K^{1/p^\infty} \backslash K$, then α is algebraic over K and $\mathrm{Irr}\,(\alpha : K)$ divides some $X^{p^m} - a \in K[X]$; hence $\mathrm{Irr}\,(\alpha : K)$ has only one root in $\overline{K}$ and

α is not separable. $\square$

This example leads to equivalent definitions of purely inseparable extensions.

Lemma **6.3.** *If K has characteristic $p \neq 0$, and α is algebraic over K, then $\alpha^{p^n} \in K$ for some $n \geq 0$ if and only if $\mathrm{Irr}(\alpha : K) = X^{p^m} - a$ for some $m \geq 0$, and $a \in K$.*

Proof. Let $q = \mathrm{Irr}(\alpha : K)$. By 5.1, $q(X) = s(X^{p^m})$ for some $m \geq 0$ and separable monic irreducible polynomial $s \in K[X]$. If $\alpha^{p^n} = b \in K$ for some $n \geq 0$, then q divides $X^{p^n} - b$, q has only one root in $\overline{K}$, and s has only one root in $\overline{K}$; since s is separable this implies $s(X) = X - a$ for some $a \in K$ and $q(X) = s(X^{p^m}) = X^{p^m} - a$. The converse holds since $q(\alpha) = 0$. $\square$

Definition. If K has characteristic $p \neq 0$, then α is purely inseparable *over K when $\alpha^{p^n} \in K$ for some $n \geq 0$, equivalently when α is algebraic over K and $\mathrm{Irr}(\alpha : K) = X^{p^m} - a$ for some $m \geq 0$ and $a \in K$.*

Proposition **6.4.** *Let K have characteristic $p \neq 0$ and let E be an algebraic extension of K. The following conditions are equivalent:*

(1) *E is purely inseparable over K (no $\alpha \in E \backslash K$ is separable over K);*

(2) *every element of E is purely inseparable over K;*

(3) *there exists a K-homomorphism of E into $K^{1/p^{\infty}}$;*

(4) *$[E : K]_s = 1$.*

Proof. (1) implies (2). Assume that E is purely inseparable over K. Let $\alpha \in E$ and $q = \mathrm{Irr}(\alpha : K)$. By 5.1, $q(X) = s(X^{p^m})$ for some $m \geq 0$ and separable monic irreducible polynomial $s \in K[X]$. Then $s(\alpha^{p^m}) = 0$, $s = \mathrm{Irr}(\alpha^{p^m} : K)$, and α^{p^m} is separable over K. If E is purely inseparable over K, then $\alpha^{p^m} \in K$.

(2) implies (3). By 3.9 there is a K-homomorphism $\varphi : E \longrightarrow \overline{K}$. If $\alpha \in E$, then $\alpha^{p^m} \in K$ by (2), $\varphi(\alpha^{p^m}) \in K$, and $\varphi(\alpha) \in K^{1/p^{\infty}}$. Thus φ is a K-homomorphism of E into $K^{1/p^{\infty}}$.

(3) implies (4). Let $\varphi : E \longrightarrow K^{1/p^{\infty}}$ and $\psi : E \longrightarrow \overline{K}$ be K-homomorphisms. Since K has characteristic $p \neq 0$, every element of K has a unique p^mth root in $\overline{K}$. If $\alpha \in E$, then $\varphi(\alpha^{p^m}) \in K$ for some $m \geq 0$, equivalently, $\alpha^{p^m} \in K$, since φ is injective; then $\psi(\alpha^{p^m}) = \alpha^{p^m}$, $\psi(\alpha)$ is, like $\varphi(\alpha)$, the unique p^mth root of α^{p^m} in $\overline{K}$, and $\psi(\alpha) = \varphi(\alpha)$. Hence there is only one K-homomorphism of E into $\overline{K}$.

(4) implies (1). If $\alpha \in E$ is separable over K, then there are $n = [K(\alpha) : K]$ distinct K-homomorphisms of $K(\alpha)$ into $\overline{K}$, which by 3.9 extend to at least n

distinct K-homomorphisms of E into $\overline{K}$; hence (4) implies $n = 1$ and $\alpha \in K$. Thus E is purely inseparable over K. $\square$

Properties. By Proposition 6.4, K^{1/p^∞} is, up to K-isomorphism, the largest purely inseparable extension of K. This costs the following results some of their charm. The proofs are exercises.

Proposition **6.5.** *If every* $\alpha \in S$ *is purely inseparable over* K, *then* $K(S)$ *is purely inseparable over* K.

Proposition **6.6.** *Let* $K \subseteq E \subseteq F$ *be algebraic extensions. If* F *is purely inseparable over* K, *then* E *is purely inseparable over* K *and* F *is purely inseparable over* E.

Proposition **6.7** (Tower Property). *Let* $K \subseteq E \subseteq F$ *be algebraic extensions. If* E *is purely inseparable over* K *and* F *is purely inseparable over* E, *then* F *is purely inseparable over* K.

Proposition **6.8.** *If* E *is algebraic and purely inseparable over* K *and the composite* EF *exists, then* EF *is purely inseparable over* KF.

Proposition **6.9.** *Every composite of algebraic purely inseparable extensions of a field* K *is a purely inseparable extension of* K.

Exercises

1. Let $\alpha \in \overline{K}$. Show that $\alpha \in K^{1/p^\infty}$ if and only if $\sigma\alpha = \alpha$ for every K-automorphism σ of $\overline{K}$.

2. Find properties of the *inseparability degree* $[E:K]_i = [E:K]/[E:K]_s$.

3. Prove the following: if every $\alpha \in S$ is purely inseparable over K, then $K(S)$ is purely inseparable over K.

4. Let $K \subseteq E \subseteq F$ be algebraic extensions. Prove the following: if F is purely inseparable over K, then E is purely inseparable over K and F is purely inseparable over E.

5. Let $K \subseteq E \subseteq F$ be algebraic extensions. Prove the following: if E is purely inseparable over K and F is purely inseparable over E, then F is purely inseparable over K.

6. Let $K \subseteq E$ be an algebraic extension. Prove the following: if E is purely inseparable over K and the composite EF exists, then EF is purely inseparable over KF.

7. Prove that every composite of purely inseparable extensions of a field K is purely inseparable over K.

7. Resultants and Discriminants

The resultant of two polynomials detects from their coefficients when they have a common root. Similarly, the discriminant of a polynomial f detects from the coefficients of f whether f is separable. This section can be skipped, though the formulas for discriminants are quoted in Section V.5.

The resultant. K denotes a field in what follows.

Definition. Let $f(X) = a_m (X - \alpha_1) \cdots (X - \alpha_m)$ and $g(X) = b_n (X - \beta_1)$ $\cdots (X - \beta_n)$ *be polynomials of degrees* m *and* n *with coefficients in a field* K *and roots* $\alpha_1, \ldots, \alpha_m, \beta_1, \ldots, \beta_n$ *in* $\overline{K}$. *The* resultant *of* f *and* g *is*

$$
\begin{aligned}
\mathrm{Res}\,(f, g) &= a_m^n b_n^m \prod_{i,j} (\alpha_i - \beta_j) \\
&= a_m^n \prod_i g(\alpha_i) = (-1)^{mn} b_n^m \prod_j f(\beta_j).
\end{aligned}
$$

The terms a_m^n and b_m^n will ensure that $\mathrm{Res}\,(f, g) \in K$. Our interest in the resultant stems from the next two results.

Proposition 7.1. *If* K *is a field and* $f, g \in K[X]$, *then* $\mathrm{Res}\,(f, g) = 0$ *if and only if* f *and* g *have a common root in* $\overline{K}$.

Next, we calculate $\mathrm{Res}\,(f, g)$ from the coefficients of f and g. Hence the resultant of f and g tells from their coefficients whether f and g have a common root in $\overline{K}$.

Proposition 7.2. *Let* K *be a field and* $f(X) = a_m X^m + \cdots + a_0$, $g(X) = b_n X^n + \cdots + b_0 \in K[X]$. *If* a_m, $b_n \neq 0$, *then*

$$
\mathrm{Res}\,(f, g) = \begin{vmatrix}
a_m & \cdots & \cdots & a_0 & & & \\
 & \ddots & & & \ddots & & \\
 & & a_m & \cdots & \cdots & a_0 \\
b_n & \cdots & \cdots & b_0 & & & \\
 & \ddots & & & \ddots & & \\
 & & b_n & \cdots & \cdots & b_0
\end{vmatrix} \in K.
$$

In this determinant, each of the first n rows is the row of coefficients of f, padded with zeros to length $m + n$; the last m rows are constructed similarly from the coefficients of g. For example, if $f = aX^2 + bX + c$ and $g = dX + e$, then

$$
\mathrm{Res}\,(f, g) = \begin{vmatrix}
a & b & c \\
d & e & 0 \\
0 & d & e
\end{vmatrix} = ae^2 - bde + cd^2.
$$

Proof. If a polynomial $p \in \mathbb{Z}[X_1, \ldots, X_n]$ becomes 0 when $X_j \neq X_i$ is substituted for X_i, then p is divisible by $X_i - X_j$: if, say, $i = 1$, then polynomial division in $\mathbb{Z}[X_2, \ldots, X_n][X_1]$ yields $p = (X_1 - X_j) q + r$, where r has degree less than 1 in X_1, that is, X_1 does not appear in r; if p becomes 0 when X_j is substituted for X_1, then so does r; but this substitution does not change r, so $r = 0$.

For the rest of the proof we replace the coefficients and roots of f and g by indeterminates. Let $F = A_m X^m + \cdots + A_0 \in \mathbb{Z}[A_m, ..., A_0, X]$, $G = B_n X^n + \cdots + B_0 \in \mathbb{Z}[B_n, ..., B_0, X]$, and

$$
P = \begin{vmatrix}
A_m & \cdots & & \cdots & A_0 & & \\
& \ddots & & & & \ddots & \\
& & A_m & \cdots & & \cdots & A_0 \\
B_n & \cdots & & \cdots & B_0 & & \\
& \ddots & & & & \ddots & \\
& & B_n & \cdots & & \cdots & B_0
\end{vmatrix} \in \mathbb{Z}[A_m, ..., A_0, B_n, ..., B_0].
$$

Then $f(X) = F(a_m, ..., a_0, X)$, $g(X) = G(b_n, ..., b_0, X)$, and the determinant D in the statement is $D = P(a_m, ..., a_0, b_n, ..., b_0)$.

Expansion shows that the coefficient of X^{m-k} in

$$
A_m (X - R_1) \cdots (X - R_m) \in \mathbb{Z}[A_m, R_1, ..., R_m][X]
$$

is $(-1)^k A_m \, \mathfrak{s}_k(R_1, ..., R_m)$, where $\mathfrak{s}_k \in \mathbb{Z}[R_1, ..., R_m]$ is homogeneous of degree k (all its monomials have degree k); $\mathfrak{s}_1, ..., \mathfrak{s}_m$ are the *elementary symmetric polynomials* in m variables, studied in greater detail in Section V.8. Similarly, the coefficient of X^{n-k} in

$$
B_n (X - S_1) \cdots (X - S_n) \in \mathbb{Z}[B_n, S_1, ..., S_n][X]
$$

is $(-1)^k B_n \, \mathfrak{t}_k(S_1, ..., S_m)$, where $\mathfrak{t}_k \in \mathbb{Z}[S_1, ..., S_n]$ is homogeneous of degree k. In particular,

$$
a_{m-k} = (-1)^k a_m \, \mathfrak{s}_k(\alpha_1, ..., \alpha_m) \text{ and } b_{n-k} = (-1)^k b_n \, \mathfrak{t}_k(\beta_1, ..., \beta_n).
$$

Let Φ be the ring homomorphism

$$
\Phi : \mathbb{Z}[A_m, ..., A_0, B_n, ..., B_0, X] \longrightarrow \mathbb{Z}[A_m, B_n, R_1, ..., R_m, S_1, ..., S_n, X]
$$

such that

$$
\Phi(A_{m-k}) = (-1)^k A_m \, \mathfrak{s}_k(R_1, ..., R_m)
$$

and

$$
\Phi(B_{n-k}) = (-1)^k B_n \, \mathfrak{t}_k(S_1, ..., S_n)
$$

for every $k > 0$; Φ substitutes $(-1)^k A_m \, \mathfrak{s}_k(R_1, ..., R_m)$ for A_{m-k} and $(-1)^k B_n \, \mathfrak{t}_k(S_1, ..., S_n)$ for B_{n-k}. By the above,

$$
\Phi(F) = A_m (X - R_1) \cdots (X - R_m)
$$

and

$$\Phi(G) = B_n (X - S_1) \cdots (X - S_n).$$

We show that $\Phi(P) = A_m^n B_n^m \prod_{i,j} (R_i - S_j)$.

Let $A_t = 0$ if $t < 0$ or $t > m$, $B_t = 0$ if $t < 0$ or $t > n$. The entry $C_{r,c}$ of P in row r and column c is A_{m+r-c} if $r \leqq n$, B_{r-c} if $r > n$. Hence

$$P = \sum_\sigma \left(\operatorname{sgn} \sigma \, C_{1,\sigma 1} \cdots C_{m+n, \, \sigma(m+n)} \right)$$
$$= \sum_\sigma \left(\operatorname{sgn} \sigma \prod_{1 \leq r \leq n} A_{m+r-\sigma r} \prod_{n < r \leq m+n} B_{r-\sigma r} \right).$$

Upon substituting

$$A_{m-k} = (-1)^k A_m \, \mathfrak{s}_k(R_1, ..., R_m), \quad B_{n-k} = (-1)^k B_n \, \mathfrak{t}_k(S_1, ..., S_m),$$

which are homogeneous of degree $k + 1$, the typical term

$$\prod_{1 \leq r \leq n} A_{m+r-\sigma r} \prod_{n < r \leq m+n} B_{n-(n+\sigma r-r)}$$

of P becomes a polynomial that is a multiple of $A_m^n B_n^m$ and is homogeneous of degree $\sum_{1 \leq r \leq n} (\sigma r - r + 1) + \sum_{n < r \leq m+n} (n + \sigma r - r + 1) = n + mn + m$. Hence $\Phi(P)$ is divisible by $A_m^n B_n^m$ and is homogeneous of degree $mn + m + n$.

Now, consider the homogeneous system of linear equations

$$X^{n-1} F = A_m X^{m+n-1} + \cdots + A_0 X^{n-1} = 0,$$
$$X^{n-2} F = A_m X^{m+n-2} + \cdots + A_0 X^{n-2} = 0,$$
$$\vdots$$
$$F = A_m X^m + \cdots + A_0 X^0 = 0,$$
$$X^{m-1} G = B_n X^{m+n-1} + \cdots + B_0 X^{m-1} = 0,$$
$$\vdots$$
$$G = B_m X^n + \cdots + B_0 X^0 = 0,$$

in which X^{m+n-1}, X^{m+n-2}, ..., X^0 are squatting in the unknowns' locations. The determinant of this system is P. Substituting

$$A_{m-k} = (-1)^k A_m \, \mathfrak{s}_k(R_1, ..., R_m), \quad B_{n-k} = (-1)^k B_n \, \mathfrak{t}_k(S_1, ..., S_n)$$

for every $k > 0$ yields a system of linear equations whose determinant is $\Phi(P)$. Since $\Phi(F)(R_i) = \Phi(G)(S_j) = 0$, further substituting $R_i = S_j$ yields a system whose determinant is 0, since it has a nontrivial solution S_j^{m+n-1}, S_j^{m+n-2}, ..., S_j^0. Thus $\Phi(P)$ becomes 0 when S_j is substituted for R_i. Therefore $\Phi(P)$ is divisible by $R_i - S_j$, for every i and j.

Poor $\Phi(P)$ is now divisible by

$$R = A_m^n B_n^m \prod_{1 \leq i \leq m, \, 1 \leq j \leq n} (R_i - S_j).$$

Since both $\Phi(P)$ and R are homogeneous of degree $mn + m + n$, then $\Phi(P) = tR$ for some $t \in \mathbb{Z}$. Now, $A_m^n B_0^m$ is a term of P; hence

$$A_m^n (-1)^{mn} B_n^m \, \mathfrak{t}_n(S_1, \ldots, S_n)^m = (-1)^{mn} A_m^n B_n^m \, (S_1 \cdots S_n)^m$$

is a term of $\Phi(P)$. Since this is also a term of R, it follows that $\Phi(P) = R$. In other words, substituting

$$A_{m-k} = (-1)^k A_m \, \mathfrak{s}_k(R_1, \ldots, R_m), \quad B_{n-k} = (-1)^k B_n \, \mathfrak{t}_k(S_1, \ldots, S_n)$$

in P for every $k > 0$ yields $R = A_m^n B_n^m \prod_{i,j} (R_i - S_j)$. Hence, substituting

$$a_{m-k} = (-1)^k a_m \, \mathfrak{s}_k(\alpha_1, \ldots, \alpha_m), \quad b_{n-k} = (-1)^k b_n \, \mathfrak{t}_k(\beta_1, \ldots, \beta_n)$$

for every $k > 0$ in $D = P(a_m, \ldots, a_0, b_n, \ldots, b_0)$ yields

$$D = a_m^n b_n^m \prod_{i,j} (\alpha_i - \beta_j) = \mathrm{Res}\,(f, g). \quad \square$$

Discriminants. We still let K be a field. The discriminant of $f \in K[X]$ detects from the coefficients of f whether f is separable. Discriminants also turn up in the solution of polynomial equations of low degree.

Definition. Let $f(X) = a_n (X - \alpha_1) \cdots (X - \alpha_n)$ be a polynomial of degree $n \geq 1$ with coefficients in a field K and not necessarily distinct roots $\alpha_1, \ldots, \alpha_n$ in $\overline{K}$. The discriminant of f is

$$\mathrm{Dis}\,(f) = a_n^{2n-2} \prod_{1 \leq i < j \leq n} (\alpha_i - \alpha_j)^2.$$

If $n = 1$, then $\mathrm{Dis}\,(f) = a_1^0 = 1$. In general, the term a_n^{2n-2} will ensure that $\mathrm{Dis}\,(f) \in K$. Permutations of $\alpha_1, \ldots, \alpha_n$ may change the signs of individual differences $\alpha_i - \alpha_j$ but do not affect the product $\prod_{i<j} (\alpha_i - \alpha_j)^2$; hence $\mathrm{Dis}\,(f)$ depends only on f and not on the numbering of its roots.

Proposition 7.3. Let K be a field. A nonconstant polynomial $f \in K[X]$ is separable over K if and only if $\mathrm{Dis}\,(f) \neq 0$.

The next result relates discriminants to resultants.

Proposition 7.4. Let K be a field. If $f \in K[X]$ has degree $n \geq 2$ and leading coefficient a_n, then $\mathrm{Res}\,(f, f') = (-1)^{n(n-1)/2} a_n \mathrm{Dis}\,(f)$.

Proof. In $\overline{K}[X]$, $f(X) = a_n \prod_i (X - \alpha_i)$. By III.5.11,

$$f'(X) = a_n \sum_i \left(\prod_{j \neq i} (X - \alpha_j) \right).$$

Hence $f'(\alpha_i) = a_n \prod_{j \neq i} (\alpha_i - \alpha_j)$ and

$$\mathrm{Res}\,(f, f') = a_n^{n-1} \prod_{1 \leq i \leq n} f'(\alpha_i) = a_n^{n-1} a_n^n \prod_{1 \leq i, j \leq n, \, j \neq i} (\alpha_i - \alpha_j)$$

$$= a_n^{2n-1} (-1)^{n(n-1)/2} \prod_{1 \leq i < j \leq n} (\alpha_i - \alpha_j)^2. \quad \square$$

Combining Propositions 7.2 and 7.4 yields a determinant formula for discriminants:

Proposition **7.5.** *Let K be a field. If $f = a_n X^n + \cdots + a_0 \in K[X]$ has degree $n \geqq 2$, then*

$$
\mathrm{Dis}\,(f) \;=\; (-1)^{n(n-1)/2}\,\frac{1}{a_n}\,
\begin{vmatrix}
a_n & \cdots & \cdots & a_0 & & & \\
 & \ddots & & & \ddots & & \\
 & & a_n & \cdots & \cdots & a_0 \\
n a_n & \cdots & \cdots & a_1 & & & \\
 & \ddots & & & \ddots & & \\
 & & n a_n & \cdots & \cdots & a_1
\end{vmatrix}
\;\in K.
$$

In this determinant, each of the first n rows is the row of coefficients of f, padded with zeros to length $2n - 1$; the last $n - 1$ rows are constructed similarly from the coefficients of f'.

For example, if $f = aX^2 + bX + c$, then

$$
\mathrm{Res}\,(f, f') \;=\;
\begin{vmatrix}
a & b & c \\
2a & b & 0 \\
0 & 2a & b
\end{vmatrix}
= 4a^2 c - ab^2;
$$

hence $\mathrm{Dis}\,(f) = b^2 - 4ac$. If K does not have characteristic 2, readers may derive this formula directly from the roots of f.

For $f = X^3 + pX + q$, readers will verify that

$$
\mathrm{Res}\,(f, f') \;=\;
\begin{vmatrix}
1 & 0 & p & q & 0 \\
0 & 1 & 0 & p & q \\
3 & 0 & p & 0 & 0 \\
0 & 3 & 0 & p & 0 \\
0 & 0 & 3 & 0 & p
\end{vmatrix}
= 4p^3 + 27q^2;
$$

hence $\mathrm{Dis}\,(f) = -4p^3 - 27q^2$.

Exercises

In the following exercises, K denotes a field.

1. When do $X^2 + aX + b$ and $X^2 + pX + q \in K[X]$ have a common root in $\overline{K}$?

2. Find the roots of $f = aX^2 + bX + c \in K[X]$ in $\overline{K}$ in case K does not have characteristic 2, and deduce that $\mathrm{Dis}\,(f) = b^2 - 4ac$. What happens when K has characteristic 2?

3. Verify that $X^3 + pX + q$ has discriminant $-4p^3 - 27q^2$.

4. Find the discriminant of $X^4 + pX^2 + qX + r$.

8. Transcendental Extensions

We now turn to transcendental extensions, find their general structure, and prove a dimension property. This section may be covered immediately after Section 3.

Totally transcendental extensions have as few algebraic elements as possible.

Definition. A field extension $K \subseteq E$ is totally transcendental, *and E is* totally transcendental over K, *when every element of $E \backslash K$ is trancendental over K.*

Proposition 8.1. *For every field K, $K((X_i)_{i \in I})$ is totally transcendental over K.*

Proof. First we show that $K(X)$ is totally transcendental over K. For clarity's sake we prove the equivalent result that $K(\chi) \cong K(X)$ is totally transcendental over K when χ is transcendental over K. Let $\alpha \in K(\chi)$, so that $\alpha = f(\chi)/g(\chi)$ for some $f, g \in K[X]$, $g \neq 0$. If $\alpha \notin K$, then $\alpha \, g(X) \notin K[X]$, $\alpha \, g(X) \neq f(X)$, and $\alpha \, g(X) - f(X) \neq 0$ in $K(\alpha)[X]$. But $\alpha \, g(\chi) - f(\chi) = 0$, so χ is algebraic over $K(\alpha)$. Hence $K(\chi) = K(\alpha)(\chi)$ is finite over $K(\alpha)$. Therefore $[\,K(\alpha) : K\,]$ is infinite: otherwise, $[\,K(\chi) : K\,]$ would be finite. Hence α is transcendental over K.

That $K[X_1, ..., X_n]$ is totally transcendental over K now follows by induction on n. Let $\alpha \in K(X_1, ..., X_n)$ be algebraic over K. Then $\alpha \in K(X_1, ..., X_{n-1})(X_n)$ is algebraic over $K(X_1, ..., X_{n-1})$. By the case $n = 1$, $\alpha \in K(X_1, ..., X_{n-1})$, and the induction hypothesis yields $\alpha \in K$.

Finally, let $\alpha = f/g \in K((X_i)_{i \in I})$ be algebraic over K. The polynomials f and g have only finitely many nonzero terms. Hence $\alpha \in K((X_i)_{i \in J})$ for some finite subset J of I. Therefore $\alpha \in K$. $\square$

A field extension is *purely transcendental* when it is K-isomorphic to some $K((X_i)_{i \in I})$. By 8.1, purely transcendental extensions are totally transcendental.

Proposition 8.2. *Every field extension is a totally transcendental extension of an algebraic extension.*

Proof. In any field extension $K \subseteq E$, the set $A = \{\, \alpha \in E \mid \alpha$ is algebraic over $K\,\}$ is a subfield of E by 3.6, and contains K. Hence A is an algebraic extension of K. If now $\alpha \in E$ is algebraic over A, then $A(\alpha)$ is algebraic over A by 3.3, $A(\alpha)$ is algebraic over K by 3.5, α is algebraic over K, and $\alpha \in A$; thus E is a totally transcendental extension of A. $\square$

For example, $\mathbb{R}$ is a totally transcendental extension of its field of algebraic numbers. We now show that every field extension is also an algebraic extension of a totally transcendental extension.

Algebraic independence. Elements are algebraically independent when they do not satisfy polynomial relations:

Definitions. A family $(\alpha_i)_{i \in I}$ *of elements of a field extension* $K \subseteq E$ *is* algebraically independent over K *when* $f\left((\alpha_i)_{i \in I}\right) \neq 0$ *for every nonzero polynomial* $f \in K[(X_i)_{i \in I}]$. *A subset* S *of a field extension* $K \subseteq E$ *is* algebraically independent over K *when it is algebraically independent over* K *as a family* $(s)_{s \in S}$.

For instance, $\{\alpha\}$ is algebraically independent over K if and only if α is transcendental over K; in $K((X_i)_{i \in I})$, $(X_i)_{i \in I}$ is algebraically independent over K. In general, an *algebraically dependent* family $(\alpha_i)_{i \in I}$ satisfies a nontrivial polynomial relation $f\left((\alpha_i)_{i \in I}\right) = 0$ (where $f \in K[(X_i)_{i \in I}]$, $f \neq 0$).

By III.6.6 there is an evaluation homomorphism $\varphi : K[(X_i)_{i \in I}] \longrightarrow E$, $f \longmapsto f\left((\alpha_i)_{i \in I}\right)$. Then $\text{Im } \varphi = K[(\alpha_i)_{i \in I}]$, by 1.13. We see that $(\alpha_i)_{i \in I}$ is algebraically independent over K if and only if φ is injective. Then $K[(\alpha_i)_{i \in I}]$ $\cong K[(X_i)_{i \in I}]$, whence $K((\alpha_i)_{i \in I}) \cong K((X_i)_{i \in I})$; in particular, $K((\alpha_i)_{i \in I})$ is totally transcendental over K, by 8.1.

The next lemmas show how algebraically independent subsets can be constructed by successive adjunction of elements. Their proofs make fine exercises.

Lemma **8.3.** *If* S *is algebraically independent over* K *and* β *is transcendental over* $K(S)$, *then* $S \cup \{\beta\}$ *is algebraically independent over* K.

Lemma **8.4.** S *is algebraically independent over* K *if and only if* β *is transcendental over* $K(S \backslash \{\beta\})$ *for every* $\beta \in S$.

Transcendence bases. Algebraic independence resembles linear independence and yields bases in much the same way.

Lemma **8.5.** *For a subset* S *of a field extension* $K \subseteq E$ *the following conditions are equivalent:*

(1) S *is a maximal algebraically independent subset;*

(2) S *is algebraically independent over* K *and* E *is algebraic over* $K(S)$;

(3) S *is minimal such that* E *is algebraic over* $K(S)$.

Proof. (1) and (2) are equivalent: by 8.3, if no $S \cup \{\beta\}$ with $\beta \notin S$ is algebraically independent, then every $\beta \in E \backslash S$ is algebraic over $K(S)$; conversely, if every $\beta \in E$ is algebraic over $K(S)$, then no $S \cup \{\beta\}$ with $\beta \notin S$ is algebraically independent.

(2) implies (3). Let S be algebraically independent over K and let E be algebraic over $K(S)$. If $T \subseteq S$ and E is algebraic over $K(T)$, then T is algebraically independent over K, T is a maximal algebraically independent subset since (2) implies (1), and $T = S$.

(3) implies (2). Assume that E is algebraic over $K(S)$ and that S is not algebraically independent over K. By 8.4, 3.3, 3.5, some $\beta \in S$ is algebraic over $K(S \backslash \{\beta\})$; then $K(S)$ is algebraic over $K(S \backslash \{\beta\})$ and E is algebraic over $K(S \backslash \{\beta\})$. Hence S is not minimal such that E is algebraic over $K(S)$. $\square$

Definition. A transcendence base *of a field extension* $K \subseteq E$ *is a subset of* E *that satisfies the equivalent conditions in Lemma 8.5.* □

For example, $(X_i)_{i \in I}$ is a transcendence base of $K((X_i)_{i \in I})$.

Theorem **8.6.** *Every field extension* $K \subseteq E$ *has a transcendence base; in fact, when* $S \subseteq T \subseteq E$, S *is algebraically independent over* K, *and* E *is algebraic over* $K(T)$, *then* E *has a transcendence base* $S \subseteq B \subseteq T$ *over* K.

Proof. Readers will verify that the union of a chain of algebraically independent subsets is algebraically independent. The existence of a maximal algebraically independent subset then follows from Zorn's lemma.

More generally, let $S \subseteq T \subseteq E$, where S is algebraically independent over K and E is algebraic over $K(T)$. Let $\mathcal{A}$ be the set of all algebraically independent subsets A such that $S \subseteq A \subseteq T$. Then $\mathcal{A} \neq \varnothing$, and, by the above, every nonempty chain in $\mathcal{A}$ has an upper bound in $\mathcal{A}$. By Zorn's lemma, $\mathcal{A}$ has a maximal element B. If $\beta \in T \backslash B$, then β is algebraic over $K(B)$: otherwise, $B \cup \{\beta\}$ is algebraically independent by 8.3 and B is not maximal in $\mathcal{A}$. By 3.3, 3.5, $K(T)$ is algebraic over $K(B)$ and E is algebraic over $K(B)$. Hence B is a transcendence base of E. □

If B is a transcendence base of a field extension $K \subseteq E$, then E is algebraic over $K(B)$, and $K(B)$ is totally transcendental over K; thus, every field extension is an algebraic extension of a totally transcendental extension.

Theorem **8.7.** *In a field extension, all transcendence bases have the same number of elements.*

Theorem 8.7 is similar to the statement that all bases of a vector space have the same number of elements, and is proved in much the same way. First we establish an *exchange property*.

Lemma **8.8.** *Let* B *and* C *be transcendence bases of a field extension* E *of* K. *For every* $\beta \in B$ *there exists* $\gamma \in C$ *such that* $(B \backslash \{\beta\}) \cup \{\gamma\}$ *is a transcendence base of* E *over* K, *and either* $\gamma = \beta$ *or* $\gamma \notin B$.

Proof. If $\beta \in C$, then $\gamma = \beta$ serves. Now let $\beta \notin C$. If every $\gamma \in C$ is algebraic over $K(B \backslash \{\beta\})$, then, by 3.3, 3.5, $K(C)$ is algebraic over $K(B \backslash \{\beta\})$, and E, which is algebraic over $K(C)$, is algebraic over $K(B \backslash \{\beta\})$, contradicting 8.4. Therefore some $\gamma \in C$ is transcendental over $K(B \backslash \{\beta\})$. Then $\gamma \notin B \backslash \{\beta\}$; in fact, $\gamma \notin B$ since $\gamma \neq \beta$. By 8.3, $B' = (B \backslash \{\beta\}) \cup \{\gamma\}$ is algebraically independent over K

Since B is a maximal algebraically independent subset, $B' \cup \{\beta\} = B \cup \{\gamma\}$ is not algebraically independent over K, and β is algebraic over $K(B')$ by 8.3. By 3.3, 3.5, $K(B)$ is algebraic over $K(B')$, and E, which is algebraic over $K(B)$, is algebraic over $K(B')$. □

We now prove 8.7. Let B and C be transcendence bases of $K \subseteq E$.

Assume that C is finite, with $n = |C|$ elements. If $B = \{\beta_1, \ldots, \beta_n, \beta_{n+1}, \ldots\}$ has more than n elements, then repeated applications of 8.8 yield transcendence bases $\{\gamma_1, \beta_2, \ldots, \beta_n, \beta_{n+1}, \ldots\}$, $\{\gamma_1, \gamma_2, \beta_3, \ldots, \beta_n, \beta_{n+1}, \ldots\}$, $\ldots$, $\{\gamma_1, \ldots, \gamma_n, \beta_{n+1}, \ldots\}$. But C is a maximal algebraically independent subset. Hence B has at most n elements. Exchanging B and C then yields $|B| = |C|$.

Now assume that C is infinite. Then B is infinite. In this case we use a cardinality argument. Every $\beta \in B$ is algebraic over $K(C)$. Hence β is algebraic over $K(C_\beta)$ for some finite subset C_β of C: indeed, $f(\beta) = 0$ for some polynomial $f \in K(C)[X]$, and C_β need only include all the elements of C that appear in the coefficients of f. Then every $\beta \in B$ is algebraic over $K(C')$, where $C' = \bigcup_{\beta \in B} C_\beta \subseteq C$. By 3.3, 3.5, $K(B)$ is algebraic over $K(C')$, and E is algebraic over $K(C')$. Since C is minimal with this property, it follows that $C = C' = \bigcup_{\beta \in B} C_\beta$. Thus C is the union of $|B|$ finite sets and $|C| \leqq |B| \aleph_0 = |B|$, by A.5.9. Exchanging B and C yields $|B| = |C|$. $\square$

Definition. The transcendence degree $\mathrm{tr.d.}\,(E : K)$ *of an extension* $K \subseteq E$ *is the number of elements of its transcendence bases.* $\square$

For instance, E is algebraic over K if and only if $\mathrm{tr.d.}\,(E : K) = 0$. The example of $K((X_i)_{i \in I})$ shows that $\mathrm{tr.d.}\,(E : K)$ can be any cardinal number.

Exercises

1. Show that the union of a chain of algebraically independent subsets is algebraically independent.

2. Prove the following: if S is algebraically independent over K and β is transcendental over $K(S)$, then $S \cup \{\beta\}$ is algebraically independent over K.

3. Prove that S is algebraically independent over K if and only if β is transcendental over $K(S \backslash \{\beta\})$ for every $\beta \in S$.

4. Let $K \subseteq E \subseteq F$ be field extensions. Show that
$$\mathrm{tr.d.}\,(F : K) = \mathrm{tr.d.}\,(F : E) + \mathrm{tr.d.}\,(E : K).$$

9. Separability

The definition of separability in Section 5 works for algebraic extensions only. This section brings a definition that is suitable for all extensions, devised by MacLane [1939]. We begin with a new relationship between field extensions, called linear disjointness, used in MacLane's definition.

Linearly disjoint extensions. Readers will prove our first result.

Proposition **9.1.** *Let* $K \subseteq E \subseteq L$ *and* $K \subseteq F \subseteq L$ *be fields. The following conditions are equivalent:*

(1) $(\alpha_i)_{i \in I} \in E$ *linearly independent over* K *implies* $(\alpha_i)_{i \in I}$ *linearly independent over* F;

(2) $(\beta_j)_{j \in J} \in F$ *linearly independent over* K *implies* $(\beta_j)_{j \in J}$ *linearly independent over* E;

(3) $(\alpha_i)_{i \in I} \in E$ *and* $(\beta_j)_{j \in J} \in F$ *linearly independent over* K *implies* $(\alpha_i \beta_j)_{(i,j) \in I \times J} \in L$ *linearly independent over* K.

Definition. *Two field extensions* $K \subseteq E \subseteq L$, $K \subseteq F \subseteq L$ *are* linearly disjoint *over* K *when they satisfy the equivalent conditions in Proposition 9.1.*

Linear disjointness can be established in several other ways.

Proposition 9.2. *Let* $K \subseteq E \subseteq L$ *and* $K \subseteq F \subseteq L$ *be fields. Let* E *be the quotient field of a ring* $K \subseteq R \subseteq E$ *(for instance, let* $R = E$ *). If*

(1) $\alpha_1, \ldots, \alpha_n \in R$ *linearly independent over* K *implies* $\alpha_1, \ldots, \alpha_n$ *linearly independent over* F, *or if*

(2) *there is a basis of* R *over* K *that is linearly independent over* F,

then E *and* F *are linearly disjoint over* K.

Proof. Assume (1) and let $(\alpha_i)_{i \in I} \in E$ be linearly independent over K. Then $(\alpha_j)_{j \in J}$ is linearly independent over K for every finite subset J of I. If J is finite, then there exists $r \in R$, $r \neq 0$, such that $r\alpha_j \in R$ for all $j \in J$. Since $R \subseteq E$ has no zero divisors, $(r\alpha_j)_{j \in J}$ is linearly independent over K. By (1), $(r\alpha_j)_{j \in J}$ is linearly independent over F. Hence $(\alpha_j)_{j \in J}$ is linearly independent over F, for every finite subset J of I, and $(\alpha_i)_{i \in I}$ is linearly independent over F. Thus E and F are linearly disjoint over K.

Now assume that there is a basis B of R over K that is linearly independent over F. Let $(\alpha_i)_{i \in I} \in R$ be a finite family that is linearly independent over K. All α_i lie in the subspace V of R generated by a finite subfamily $(\beta_j)_{j \in J}$ of B. Hence $(\alpha_i)_{i \in I}$ is contained in a finite basis $(\alpha_h)_{h \in H}$ of V. We show that $(\alpha_h)_{h \in H}$ is linearly independent over F: since $(\alpha_h)_{h \in H}$ and $(\beta_j)_{j \in J}$ are bases of V there is an invertible matrix $C = (c_{hj})_{h \in H, \, j \in J}$ with entries in K such that $\alpha_h = \sum_{i \in I} c_{hj} \beta_j$ for all h; if now $\sum_h x_h \alpha_h = 0$ for some $x_h \in F$, then $\sum_{h,j} x_h c_{hj} \beta_j = 0$, $\sum_h x_h c_{hj} = 0$ for all j since $(\beta_j)_{j \in J}$ is linearly independent over F, and $x_h = 0$ for all h since C is invertible. In particular, $(\alpha_i)_{i \in I}$ is linearly independent over F. Thus (1) holds. Hence E and F are linearly disjoint over K. $\square$

Corollary 9.3. *If* $K \subseteq E \subseteq L$ *and* $\alpha_1, \ldots, \alpha_n \in L$ *are algebraically independent over* E, *then* E *and* $K(\alpha_1, \ldots, \alpha_n)$ *are linearly disjoint over* K.

Proof. $K(\alpha_1, \ldots, \alpha_n) \cong K(X_1, \ldots, X_n)$ is the quotient field of $K[\alpha_1, \ldots, \alpha_n] \cong K[X_1, \ldots, X_n]$, and the monomials $\alpha_1^{m_1} \alpha_2^{m_2} \cdots \alpha_n^{m_n}$ constitute a basis of $K[\alpha_1, \ldots, \alpha_n]$ over K. The monomials $\alpha_1^{m_1} \alpha_2^{m_2} \cdots \alpha_n^{m_n}$ are linearly independent over E, since $\alpha_1, \ldots, \alpha_n$ are algebraically independent over E. By part (2) of 9.2, $K(\alpha_1, \ldots, \alpha_n)$ and E are linearly disjoint over K. $\square$

Proposition 9.4. *Let* $K \subseteq E \subseteq L$ *and* $K \subseteq F \subseteq F' \subseteq L$ *be fields. If* E

and F are linearly disjoint over K, and EF and F' are linearly disjoint over F, then E and F' are linearly disjoint over K.

Proof. Take bases $(\alpha_i)_{i \in I}$ of E over K, $(\beta_j)_{j \in J}$ of F over K, and $(\gamma_h)_{h \in H}$ of F' over F. Then $(\beta_j \gamma_h)_{j \in J, h \in H}$ is a basis of F' over K. If E and F are linearly disjoint over K, then $(\alpha_i)_{i \in I}$ is linearly independent over F. If also EF and F' are linearly disjoint over F, then $(\alpha_i \gamma_h)_{i \in I, h \in H}$ is linearly independent over F. Therefore $(\alpha_i \beta_j \gamma_h)_{i \in I, j \in J, h \in H}$ is linearly independent over K: if $\sum_{i,j,h} a_{ijh} \alpha_i \beta_j \gamma_h = 0$, where $a_{ijh} \in K$, then $\sum_{i,h} \left(\sum_j a_{ijh} \beta_j \right) \alpha_i \gamma_h = 0$, $\sum_j a_{ijh} \beta_j = 0$ for all i, h, and $a_{ijh} = 0$ for all i, j, h. Hence $(\alpha_i)_{i \in I}$ is linearly independent over F'. $\square$

Finally, we note two cases of linear disjointness. Let K have characteristic $p \neq 0$. Let $K^{1/p^\infty} = \{ \alpha \in \overline{K} \mid \alpha^{p^r} \in K$ for some $r \geqq 0 \}$. Up to K-isomorphism, K^{1/p^∞} is the largest purely inseparable extension of K, by 6.2, 6.4.

Proposition 9.5. *If K has characteristic $p \neq 0$ and E is purely transcendental over K, then E and K^{1/p^∞} are linearly disjoint over K.*

Proof. Let $E = K((\chi_i)_{i \in I}) \cong K((X_i)_{i \in I})$, where $(\chi_i)_{i \in I}$ are algebraically independent over K. Both E and K^{1/p^∞} are contained in $\overline{K}((\chi_i)_{i \in I}) \cong \overline{K}((X_i)_{i \in I})$, and E is the field of quotients of $R = K[(\chi_i)_{i \in I}] \cong K[(X_i)_{i \in I}]$. The monomials $m = \prod_{i \in I} \chi_i^{m_i}$ constitute a basis of R over K. Suppose that $\alpha_1 m_1 + \cdots + \alpha_k m_k = 0$ for some $\alpha_1, \ldots, \alpha_k \in K^{1/p^\infty}$ and some distinct monomials $m_1, \ldots, m_k$. Then $\alpha_1^{p^r}, \ldots, \alpha_k^{p^r} \in K$ for some $r \geqq 0$. Since $x \longmapsto x^{p^r}$ is an injective homomorphism, $m_1^{p^r}, \ldots, m_k^{p^r}$ are distinct monomials and $\alpha_1^{p^r} m_1^{p^r} + \cdots + \alpha_k^{p^r} m_k^{p^r} = 0$; hence $\alpha_1^{p^r} = \cdots = \alpha_k^{p^r} = 0$ and $\alpha_1 = \cdots = \alpha_k = 0$. Thus the monomials $m = \prod_{i \in I} \chi_i^{m_i}$ are linearly independent over K^{1/p^∞}; by 9.2, E and K^{1/p^∞} are linearly disjoint over K. $\square$

Proposition 9.6. *If K has characteristic $p \neq 0$ and E is algebraic over K, then E is separable over K if and only if E and K^{1/p^∞} are linearly disjoint over K.*

Proof. First we prove this when E is a simple extension.

Let $\alpha \in \overline{K}$ be separable over K. Then $q = \mathrm{Irr}\,(\alpha : K(\alpha^p))$ divides $X^p - \alpha^p = (X - \alpha)^p$ in $\overline{K}[X]$, since $(X^p - \alpha^p)(\alpha) = 0$, and $q = (X - \alpha)^k$ for some $k \leqq p$. But q is separable, so $k = 1$ and $\alpha \in K(\alpha^p)$. Thus $K(\alpha) = K(\alpha^p)$. Hence $K(\alpha) = K(\alpha^p) = K(\alpha^{p^2}) = \cdots = K(\alpha^{p^r})$ for all $r \geqq 0$.

Now, $K(\alpha)$ has a basis $1, \alpha, \ldots, \alpha^{n-1}$ over K. Since $K(\alpha^{p^r}) = K(\alpha)$ has the same degree, $1, \alpha^{p^r}, \ldots, \alpha^{(n-1)p^r}$ is a basis of $K(\alpha^{p^r})$ over K. Hence

1, α, ..., α^{n-1} are linearly independent over K^{1/p^∞} : if γ_0, ..., $\gamma_{n-1} \in$ K^{1/p^∞} and $\gamma_0 + \gamma_1 \alpha + \cdots + \gamma_{n-1} \alpha^{n-1} = 0$, then $\gamma_0^{p^r}$, ..., $\gamma_{n-1}^{p^r} \in K$ for some $r \geqq 0$, $\gamma_0^{p^r} + \gamma_1^{p^r} \alpha^{p^r} + \cdots + \gamma_{n-1}^{p^r} \alpha^{(n-1)p^r} = (\gamma_0 + \gamma_1 \alpha + \cdots + \gamma_{n-1} \alpha^{n-1})^{p^r} = 0$, $\gamma_0^{p^r} = \gamma_1^{p^r} = \cdots = \gamma_{n-1}^{p^r} = 0$, and $\gamma_0 = \gamma_1 = \cdots = \gamma_{n-1} = 0$. Therefore $K(\alpha)$ and K^{1/p^∞} are linearly disjoint over K, by part (2) of 9.2.

Conversely, assume that $K(\alpha)$ and K^{1/p^∞} are linearly disjoint over K (where $\alpha \in \overline{K}$). Let $\alpha \in E$ and $\mathrm{Irr}\,(\alpha : K) = q(X) = a_0 + a_1 X + \cdots + a_n X^n$, with $a_n \neq 0$. Then 1, α, ..., α^{n-1} are linearly independent over K, and over K^{1/p^∞}. As above, $a_i = \gamma_i^p$ for some $\gamma_i \in K^{1/p^\infty}$. If $q' = 0$, then $a_i = 0$ whenever i is not a multiple of p; $q(X) = a_0 + a_p X^p + \cdots + a_{kp} X^{kp}$;

$$(\gamma_0 + \gamma_p \alpha + \cdots + \gamma_{kp} \alpha^k)^p = \gamma_0^p + \gamma_p^p \alpha^p + \cdots + \gamma_{kp}^p \alpha^{kp}$$
$$= a_0 + a_p \alpha^p + \cdots + a_{kp} \alpha^{kp} = q(\alpha) = 0;$$

$\gamma_0 + \gamma_p \alpha + \cdots + \gamma_{kp} \alpha^k = 0$; $\gamma_0 = \gamma_p = \cdots = \gamma_{kp} = 0$, since 1, α_1, ..., α^{n-1} are linearly independent over K^{1/p^∞}; and $q(X) = 0$. Therefore $q' \neq 0$. Hence the irreducible polynomial q is separable, and $\alpha \in E$ is separable over K.

Now let E be algebraic over K. We may assume that $E \subseteq \overline{K}$. If E and K^{1/p^∞} are linearly disjoint over K, then every $\alpha \in E$ is separable over K, since $K(\alpha) \subseteq E$ and K^{1/p^∞} are linearly disjoint over K. Conversely, if E is separable over K and α_1, ..., $\alpha_n \in E$ are linearly independent over K, then $K(\alpha_1, ..., \alpha_n) = K(\alpha)$ for some $\alpha \in E$ by 5.12, $K(\alpha)$ and K^{1/p^∞} are linearly disjoint over K, and α_1, ..., α_n are linearly independent over K^{1/p^∞}; hence E and K^{1/p^∞} are linearly disjoint over K, by part (1) of 9.2. $\square$

Separability. We now turn to the general definition of separable extensions.

Definition. A transcendence base B of a field extension $K \subseteq E$ is separating when E is separable (algebraic) over $K(B)$.

Separable algebraic extensions, and purely transcendental extensions, ought to be separable. Hence an extension with a separating transcendence base, which is an algebraic separable extension of a purely transcendental extension, also ought to be separable. Since directed unions of separable extensions ought to be separable, an extension in which every finitely generated intermediate field has a separating transcendence base ought to be separable as well. On the other hand, 9.5 and 9.6 suggest that separability over K could be defined by linear disjointness from K^{1/p^∞}, when K has characteristic $p \neq 0$. MacLane's theorem states that this yields the same class of extensions.

Theorem **9.7** (MacLane [1939]). *Let K be a field of characteristic $p \neq 0$. For a field extension $K \subseteq E$ the following conditions are equivalent:*

(1) *every finitely generated intermediate field $F = K(\alpha_1, \ldots, \alpha_n) \subseteq E$ has a separating transcendence base;*

(2) *E and K^{1/p^∞} are linearly disjoint over K;*

(3) *E and $K^{1/p} = \{\alpha \in \overline{K} \mid \alpha^p \in K\}$ are linearly disjoint over K.*

Moreover, in (1), there is a separating transcendence base $B \subseteq \{\alpha_1, \ldots, \alpha_n\}$.

In 9.7, the inclusion homomorphism $K \longrightarrow \overline{E}$ extends to a field homomorphism $\overline{K} \longrightarrow \overline{E}$; hence we may assume that $\overline{K} \subseteq \overline{E}$, so that E, $K^{1/p^\infty} \subseteq \overline{E}$.

Proof. (1) implies (2). By (1) of 9.2 we need only show that every finitely generated subfield $K \subseteq F \subseteq E$ is linearly disjoint from K^{1/p^∞} over K. By (1), F has a separating transcendence base B. By 1.11, $K^{1/p^\infty} K(B) \subseteq K(B)^{1/p^\infty}$. Now, $K(B) \subseteq F$ and K^{1/p^∞} are linearly disjoint over K, by 9.5; F and $K^{1/p^\infty} K(B) \subseteq K(B)^{1/p^\infty}$ are linearly disjoint over $K(B)$, by 9.6; hence F and K^{1/p^∞} are linearly disjoint over K, by 9.4.

(2) implies (3) since $K^{1/p} \subseteq K^{1/p^\infty}$.

(3) implies (1). We prove by induction on n that every finitely generated subfield $F = K(\alpha_1, \ldots, \alpha_n) \subseteq E$ has a separating transcendence base $B \subseteq \{\alpha_1, \ldots, \alpha_n\}$. There is nothing to prove if $n = 0$. Assume that $n > 0$. By 8.6, $\alpha_1, \ldots, \alpha_n$ contains a transcendence base, which we may assume is $\{\alpha_1, \ldots, \alpha_r\}$, where $r = \mathrm{tr.d.}\,(F:K) \leqq n$. If $r = n$, then $\{\alpha_1, \ldots, \alpha_n\}$ is a separating transcendence base of F. Hence we may further assume that $r < n$.

Since $\alpha_1, \ldots, \alpha_{r+1}$ are algebraically dependent over K, there is a nonzero polynomial $f \in K[X_1, \ldots, X_{r+1}]$ such that $f(\alpha_1, \ldots, \alpha_{r+1}) = 0$. Choose f so that its degree is as small as possible and, with this degree, its number of terms is as small as possible. Then f is irreducible. Let $f = c_1 m_1 + \cdots + c_k m_k$, where $c_1, \ldots, c_k \in K$ and $m_1, \ldots, m_k \in K[X_1, \ldots, X_n]$ are monomials. Then $c_1, \ldots, c_k \neq 0$, by the choice of f.

Suppose that every exponent that appears in f, and in $m_1, \ldots, m_k$, is a multiple of p. Then $f(X_1, \ldots, X_{r+1}) = g(X_1^p, \ldots, X_{r+1}^p)$ for some $g \in K[X_1, \ldots, X_{r+1}]$; similarly, $m_i(X_1, \ldots, X_{r+1}) = \ell_i(X_1^p, \ldots, X_{r+1}^p)$ for some monomial $\ell_i \in K[X_1, \ldots, X_{r+1}]$; and every c_i has a pth root $\gamma_i \in \overline{K}$. Hence

$$f(X_1, \ldots, X_{r+1}) = \sum_i \gamma_i^p \ell_i(X_1^p, \ldots, X_{r+1}^p)$$
$$= \left(\sum_i \gamma_i \ell_i(X_1, \ldots, X_{r+1})\right)^p,$$

with $\gamma_i \in K^{1/p}$, and $\sum_i \gamma_i \ell_i(\alpha_1, \ldots, \alpha_{r+1}) = 0$, so that $\ell_1(\alpha_1, \ldots, \alpha_{r+1})$, $\ldots, \ell_k(\alpha_1, \ldots, \alpha_{r+1})$ are linearly dependent over $K^{1/p}$. However, $\ell_1(\alpha_1, \ldots,$

$\alpha_{r+1}), \ldots, \ell_k(\alpha_1, \ldots, \alpha_{r+1})$ are linearly independent over K: otherwise, one of the $\ell_i(X_1^p, \ldots, X_{r+1}^p)$ could be replaced in f by a linear combination of the others, yielding a polynomial $g \in K[X_1, \ldots, X_{r+1}]$ such that $g(\alpha_1, \ldots, \alpha_{r+1}) = 0$, with lower degree than f or with the same degree but fewer terms. Our supposition thus contradicts either (3) or the choice of f. Therefore one of $X_1, \ldots, X_{r+1}$ appears in f with an exponent that is not a multiple of p.

Suppose that, say, X_1 appears in f with an exponent that is not a multiple of p. Let $g(X) = f(X, \alpha_2, \ldots, \alpha_{r+1}) \in K(\alpha_2, \ldots, \alpha_{r+1})[X]$. Then $g(\alpha_1) = 0$, and F is algebraic over $K(\alpha_2, \ldots, \alpha_{r+1})$. By 8.6, 8.7, $\{\alpha_2, \ldots, \alpha_{r+1}\}$ is a transcendence base of F; hence $g \in K(\alpha_2, \ldots, \alpha_{r+1})[X] \cong K[X_1, \ldots, X_{r+1}]$ is irreducible, since $f \in K[X_1, \ldots, X_{r+1}]$ is irreducible. Moreover, $g' \neq 0$, since X appears in g with an exponent that is not a multiple of p; therefore g, which is irreducible, is separable. The equality $g(\alpha_1) = 0$ then shows that α_1 is algebraic and separable over $K(\alpha_2, \ldots, \alpha_{r+1})$. Hence $F = K(\alpha_1, \ldots, \alpha_n)$ is algebraic and separable over $K(\alpha_2, \ldots, \alpha_n)$. By the induction hypothesis, $K(\alpha_2, \ldots, \alpha_n)$ has a separating transcendence base $B \subseteq \{\alpha_2, \ldots, \alpha_n\}$, and then B is a separating transcendence base of F.

The other case, in which X_{r+1} appears in f with an exponent that is not a multiple of p, is similar but simpler. Then $\{\alpha_1, \ldots, \alpha_r\}$ is already a transcendence base of F. Then $g(\alpha_{r+1}) = 0$, where $g(X) = f(\alpha_1, \ldots, \alpha_r, X)$. As above, g is irreducible and separable. Hence $F = K(\alpha_1, \ldots, \alpha_n)$ is algebraic and separable over $K(\alpha_1, \ldots, \alpha_r, \alpha_{r+2}, \ldots, \alpha_n)$. By the induction hypothesis, the latter has a separating transcendence base $B \subseteq \{\alpha_1, \ldots, \alpha_r, \alpha_{r+2} \ldots, \alpha_n\}$, and B is a separating transcendence base of F. $\square$

Definition. A field extension E of K is separable, *and E is* separable over K, *when every finitely generated subfield $K \subseteq F$ of E has a separating transcendence base.*

By 9.7, E is separable over K if and only if either K has characteristic 0, or K has characteristic $p \neq 0$ and E is linearly disjoint from K^{1/p^∞}.

The class of separable extensions has several desirable properties. Separable algebraic extensions are separable in the previous sense, by 9.5. If E is purely transcendental over K, then E is separable over K, by 9.6. If E is separable over K, then every intermediate field $K \subseteq F \subseteq E$ is separable over K.

Proposition **9.8** (Tower Property). *If F is separable over K, and E is separable over F, then E is separable over K.*

The proof is an easy exercise, using 9.4. One might hope for one more tower property: if E is separable over K and $K \subseteq F \subseteq E$, then E is separable over F. Alas, this is false in general; readers will find a counterexample.

Exercises

1. Let $K \subseteq E \subseteq L$, $K \subseteq F \subseteq L$ be fields. Prove the following: if E is algebraic over K, and F is purely transcendental over K, then E and F are linearly disjoint over K.

2. Prove the following: if K is perfect, then every field extension of K is separable.

3. Prove the following: if F is separable over K, and E is separable over F, then E is separable over K.

4. Show that a directed union of separable extensions of K is a separable extension of K.

5. Let $K \subseteq E \subseteq L$, $K \subseteq F \subseteq L$ be fields. Prove the following: if E are F are linearly disjoint over K, and $\alpha_1, \ldots, \alpha_n \in E$ are algebraically independent over K, then $\alpha_1, \ldots, \alpha_n$ are algebraically independent over F.

6. Let $K \subseteq E \subseteq L$, $K \subseteq F \subseteq L$ be fields. Prove the following: if E is separable over K, and $\alpha_1, \ldots, \alpha_n \in E$ algebraically independent over K implies $\alpha_1, \ldots, \alpha_n$ algebraically independent over F, then EF is separable over K.

7. Find a separable extension $K \subseteq E$ with an intermediate field $K \subseteq F \subseteq E$ such that E is not separable over F.

8. Find a separable extension $K \subseteq E$ that does not have a separating transcendence base. (You may want to try $K(X, X^{1/p}, \ldots, X^{1/p^r}, \ldots)$, where K has characteristic $p \neq 0$ and X is transcendental over K.)

V
Galois Theory

Algebra began when quadratic equations were solved by al-Khowarizmi. Its next step was the solution of third and fourth degree equations, published by Cardano in [1545]. Equations of degree 5, however, resisted all efforts at similar solutions, until Abel [1824] and Galois [1830] proved that no such solution exists. Abel's solution did not hold the germs of future progress, but Galois's ideas initiated the theory that now bears his name, even though Galois himself lacked a clear definition of fields. The modern version has remained virtually unchanged since Artin's lectures in the 1920s.

Galois theory provides a one-to-one correspondence between intermediate fields $K \subseteq F \subseteq E$ of suitable extensions and subgroups of their groups of K-automorphisms. This allows group theory to apply to fields. For instance, a polynomial equation is solvable by radicals if and only if the corresponding group is solvable (as defined in Section II.9).

Sections II.7, II.9, and IV.1 through IV.5 are a necessary foundation. Sections 4 and 9 may be skipped.

1. Splitting Fields

The splitting field of a set of polynomials is the field generated by their roots in some algebraic closure. This section contains basic properties of splitting fields, and the determination of all finite fields.

Splitting fields. We saw in Section IV.2 that every polynomial with coefficients in a field K has a root in some field extension of K. A polynomial splits in an extension when it has all its roots in that extension:

Definition. A polynomial $f \in K[X]$ splits in a field extension E of K when it has a factorization $f(X) = a(X - \alpha_1)(X - \alpha_2) \cdots (X - \alpha_n)$ in $E[X]$.

In the above, $a \in K$ is the leading coefficient of f, n is the degree of f, and $\alpha_1, \ldots, \alpha_n \in E$ are the (not necessarily distinct) roots of f in E. For example, every polynomial $f \in K[X]$ splits in the algebraic closure $\overline{K}$ of K.

Definition. Let K be a field. A splitting field over K of a polynomial

$f \in K[X]$ is a field extension E of K such that f splits in E and E is generated over K by the roots of f. A splitting field over K of a set $\mathcal{S} \subseteq K[X]$ of polynomials is a field extension E of K such that every $f \in \mathcal{S}$ splits in E and E is generated over K by the roots of all $f \in \mathcal{S}$.

In particular, splitting fields are algebraic extensions, by 3.3. Every set $\mathcal{S} \subseteq K[X]$ of polynomials has a splitting field, which is generated over K by the roots of all $f \in \mathcal{S}$ in $\overline{K}$, and which we show is unique up to K-isomorphism.

Lemma 1.1. *If E and F are splitting fields of $\mathcal{S} \subseteq K[X]$ over K, and $F \subseteq \overline{K}$, then $\varphi E = F$ for every K-homomorphism $\varphi : E \longrightarrow \overline{K}$.*

Proof. Every $f \in \mathcal{S}$ has unique factorizations $f(X) = a(X - \alpha_1)(X - \alpha_2) \cdots (X - \alpha_n)$ in $E[X]$ and $f(X) = a(X - \beta_1)(X - \beta_2) \cdots (X - \beta_n)$ in $F[X] \subseteq \overline{K}[X]$. Since φ is the identity on K, $f = {}^{\varphi}f = a(X - \varphi\alpha_1)(X - \varphi\alpha_2) \cdots (X - \varphi\alpha_n)$ in $\overline{K}[X]$; therefore $\varphi\{\alpha_1, ..., \alpha_n\} = \{\beta_1, ..., \beta_n\}$. Thus φ sends the set R of all roots of all $f \in \mathcal{S}$ in E onto the set S of all roots of all $f \in \mathcal{S}$ in F. By IV.1.9, φ sends $E = K(R)$ onto $K(S) = F$. $\square$

With $\mathcal{S} = \{f\}$, the proof of Lemma 1.1 shows that every K-homomorphism $F \longrightarrow \overline{K}$ permutes the roots of f. This phenomenon is explored in later sections. By IV.4.2, every splitting field has a K-homomorphism into $\overline{K}$; hence Lemma 1.1 yields a uniqueness result:

Proposition 1.2. *Every set $\mathcal{S} \subseteq K[X]$ of polynomials has a splitting field $E \subseteq \overline{K}$ over K; moreover, all splitting fields of $\mathcal{S}$ over K are K-isomorphic.*

Accordingly, we speak of *the* splitting field of $\mathcal{S}$ over K.

Finite fields. A finite field F has prime characteristic $p \neq 0$ and is a finite extension of $\mathbb{Z}_p$; hence F has order $|F| = p^n$ for some $n = [F : \mathbb{Z}_p] > 0$.

Theorem 1.3. *For every prime p and every $n > 0$ there is, up to isomorphism, exactly one field F of order p^n; F is a splitting field of $X^{p^n} - X$ over $\mathbb{Z}_p$, and all its elements are roots of $X^{p^n} - X$.*

Proof. Let F be a field of order p^n. By IV.1.6, the multiplicative group $F^* = F \backslash \{0\}$ is cyclic; since $|F^*| = p^n - 1$ we have $x^{p^n - 1} = 1$ for all $x \in F^*$ and $x^{p^n} = x$ for all $x \in F$. Thus the elements of F are roots of $f(X) = X^{p^n} - X$; since f has at most p^n roots, F consists of all the roots of f. Hence F is a splitting field of f over $\mathbb{Z}_p$, and is unique up to isomorphism.

Conversely, let F be a splitting field of $f(X) = X^{p^n} - X$ over $\mathbb{Z}_p$. Then F has characteristic p. The roots of f in F constitute a subfield of F: 0 and 1 are roots of f, and when α, β are roots of f, then so are $\alpha - \beta$ and $\alpha \beta^{-1}$, since $(\alpha - \beta)^{p^n} = \alpha^{p^n} - \beta^{p^n} = \alpha - \beta$ by III.4.4 and $(\alpha \beta^{-1})^{p^n} = \alpha^{p^n} \beta^{-p^n} = \alpha \beta^{-1}$. Since F is generated by roots of f it follows that F consists of roots of f. Now, all roots of f are simple by III.5.12, since $f'(X) = -1$; therefore

f has p^n roots in F, and F has p^n elements. $\square$

The field of order $q = p^n$ is the *Galois field* $GF(q)$, after Galois [1830], who showed that "imaginary roots modulo p" of the equation $X^{p^n} - X = 0$ can be added and multiplied. Properties of Galois fields make entertaining exercises.

Exercises

1. What is the splitting field of $X^3 - 2$ over $\mathbb{Q}$?

2. What is the splitting field of $X^4 + 5X^2 + 6$ over $\mathbb{Q}$?

3. Set up addition and multiplication tables for $GF(4)$.

4. Set up addition and multiplication tables for $GF(8)$.

5. Let K be a field of characteristic $p \neq 0$. Show that K contains a subfield of order p^n if and only if $X^{p^n} - X$ splits in K, and then K contains only one subfield of order p^n.

6. Show that a field of order p^n contains a subfield of order p^m if and only if m divides n.

7. Let L and M be subfields of a field K of orders p^ℓ and p^m, respectively. Show that $L \cap M$ has order p^d, where $d = \gcd(\ell, m)$.

2. Normal Extensions

A normal extension is the splitting field of a set of polynomials. This section contains basic properties, with applications to perfect fields.

Definition. By IV.4.4, IV.4.5, every algebraic extension of K is contained in an algebraic closure $\overline{K}$ of K, which is unique up to K-isomorphism. Normal extensions are defined by the following equivalent properties.

Proposition **2.1.** *For an algebraic extension* $K \subseteq E \subseteq \overline{K}$ *the following conditions are equivalent:*

(1) *E is the splitting field over K of a set of polynomials;*

(2) $\varphi E = E$ *for every K-homomorphism* $\varphi : E \longrightarrow \overline{K}$;

(3) $\varphi E \subseteq E$ *for every K-homomorphism* $\varphi : E \longrightarrow \overline{K}$;

(4) $\sigma E = E$ *for every K-automorphism* σ *of* $\overline{K}$;

(5) $\sigma E \subseteq E$ *for every K-automorphism* σ *of* $\overline{K}$;

(6) *every irreducible polynomial* $q \in K[X]$ *with a root in E splits in E.*

Proof. (1) implies (2) by 1.1; (2) implies (3) and (4) implies (5); (2) implies (4), and (3) implies (5), since every K-automorphism of $\overline{K}$ induces a K-homomorphism of E into $\overline{K}$.

(5) implies (6). Let $q \in K[X]$ be irreducible, with a root α in E. We may assume that q is monic; then $q = \mathrm{Irr}\,(\alpha : K)$. For every root β of q in $\overline{K}$,

IV.2.4 yields a K-homomorphism φ of $K(\alpha) \subseteq E$ into $\overline{K}$ that sends α to β. By IV.4.5, φ extends to a K-automorphism σ of $\overline{K}$. Then $\beta = \sigma\alpha \in E$ by (5). Thus E contains every root of q in $\overline{K}$; hence q splits in E.

(6) implies (1). E is a splitting field of $\mathcal{S} = \{ \mathrm{Irr}\,(\alpha : K) \in K[X] \mid \alpha \in E \}$: every $q = \mathrm{Irr}\,(\alpha : K) \in \mathcal{S}$ has a root α in E and splits in E, by (6); moreover, E consists of all the roots of all $q \in \mathcal{S}$. $\square$

Definition. A normal extension of a field K is an algebraic extension of K that satisfies the equivalent conditions in Proposition 2.1 for some algebraic closure of K.

Conjugates. Normal extensions can also be defined as follows.

Definitions. Let K be a field. A conjugate of $\alpha \in \overline{K}$ over K is the image of α under a K-automorphism of $\overline{K}$. A conjugate of an algebraic extension $E \subseteq \overline{K}$ of K is the image of E under a K-automorphism of $\overline{K}$.

For example, an $\mathbb{R}$-automorphism σ of $\mathbb{C}$ must satisfy $(\sigma i)^2 + 1 = \sigma\,(i^2 + 1) = 0$; therefore, either $\sigma i = i$ and σ is the identity on $\mathbb{C}$, or $\sigma i = -i$ and σ is ordinary complex conjugation. Hence a complex number z has two conjugates over $\mathbb{R}$, itself and its ordinary conjugate $\overline{z}$.

Proposition 2.2. Over a field K, the conjugates of $\alpha \in \overline{K}$ are the roots of $\mathrm{Irr}\,(\alpha : K)$ in $\overline{K}$.

Proof. If σ is a K-automorphism of $\overline{K}$, then $\sigma\alpha$ is a root of $q = \mathrm{Irr}\,(\alpha : K)$, since $q(\sigma\alpha) = {}^{\sigma}q(\sigma\alpha) = \sigma\,q(\alpha) = 0$. Conversely, if β is a root of q in $\overline{K}$, then there is by IV.2.4 a K-homomorphism φ of $K(\alpha) \subseteq E$ into $\overline{K}$ that sends α to β, which IV.4.7 extends to a K-automorphism σ of $\overline{K}$. $\square$

Proposition 2.3. For an algebraic extension $K \subseteq E \subseteq \overline{K}$ the following conditions are equivalent:

(1) *E is a normal extension of K;*

(2) *E contains all conjugates over K of all elements of E;*

(3) *E has only one conjugate.*

Proof. (3) is part (4) of Proposition 2.1, and (2) is, by 2.2, equivalent to part (6) of 2.1. $\square$

Properties. The class of normal extensions has some basic properties, for which readers will easily cook up proofs.

Proposition 2.4. If F is normal over K and $K \subseteq E \subseteq F$, then F is normal over E.

Proposition 2.5. If E is normal over K and the composite EF exists, then EF is normal over KF.

Proposition 2.6. Every composite of normal extensions of K is a normal extension of K.

Proposition **2.7.** *Every intersection of normal extensions $E \subseteq \overline{K}$ of K is a normal extension of K.*

One might expect two additional tower statements: if F is normal over K and $K \subseteq E \subseteq F$, then E is normal over K; if $K \subseteq E \subseteq F$, E is normal over K, and F is normal over E, then F is normal over K. Both statements are false; the next sections will explain why.

By 2.7, there is for every algebraic extension $K \subseteq E \subseteq \overline{K}$ of K a smallest normal extension $N \subseteq \overline{K}$ of K that contains E, namely, the intersection of all normal extensions $N \subseteq \overline{K}$ of K that contain E.

Proposition **2.8.** *The smallest normal extension $N \subseteq \overline{K}$ of K that contains an algebraic extension $E \subseteq \overline{K}$ of K is the composite of all conjugates of E.*

Proof. A normal extension of K that contains E contains all conjugates of E by 2.1 and contains their composite. Conversely, the composite of all conjugates of E is normal over K, since a K-automorphism of $\overline{K}$ permutes the conjugates of E and therefore leaves their composite unchanged. $\square$

Proposition **2.9.** *Every finite (respectively separable, finite separable) extension $E \subseteq \overline{K}$ of a field K is contained in a finite (separable, finite separable) normal extension of K.*

Proof. If $E \subseteq \overline{K}$ is finite, then, by IV.5.4, there are only finitely many K-homomorphisms of E into $\overline{K}$. Since the restriction to E of a K-automorphism of $\overline{K}$ is a K-homomorphism, E has only finitely many conjugates; their composite F is a finite extension of K by IV.3.5, and is normal over K by 2.8. If in general E is separable over K, then so are the conjugates $\sigma E \cong E$ of E, and so is their composite, by IV.5.11. $\square$

The remaining results of this section require purely inseparable extensions (see Section IV.6) and may be skipped at first reading.

Proposition **2.10.** *If $E \subseteq \overline{K}$ is a normal extension of K, then*

$$F = \{\, \alpha \in E \mid \sigma\alpha = \alpha \text{ for every } K\text{-automorphism } \sigma \text{ of } \overline{K} \,\}$$

is a purely inseparable extension of K, and E is a separable extension of F.

Proof. First, F is a subfield of E and $K \subseteq F$. If $\alpha \in F$, then every K-homomorphism φ of $K(\alpha)$ into $\overline{K}$ extends to a K-automorphism of $\overline{K}$, by IV.4.7; hence $\varphi(\alpha) = \alpha$ and φ is the identity on $K(\alpha)$. Thus $[K(\alpha) : K]_s = 1$. Hence $\alpha \in K(\alpha)$ is purely inseparable over K, and F is purely inseparable over K.

Now let $\alpha \in E$. Let $\varphi_1, \ldots, \varphi_n$ be the distinct K-homomorphisms of $K(\alpha)$ into $\overline{K}$; one of these, say φ_1, is the inclusion homomorphism $K(\alpha) \longrightarrow \overline{K}$. Since every φ_i extends to a K-automorphism of $\overline{K}$, we have $\varphi_i E \subseteq E$ and $\varphi_i \alpha \in E$ for all i; moreover, $\varphi_1 \alpha, \ldots, \varphi_n \alpha$ are distinct: if $\varphi_i \alpha = \varphi_j \alpha$, then $\varphi_i = \varphi_j$, since $K(\alpha)$ is generated by α. Let

$$f(X) = (X - \varphi_1\alpha)(X - \varphi_2\alpha) \cdots (X - \varphi_n\alpha) \in E[X].$$

Then $f(\alpha) = 0$, since $\varphi_1\alpha = \alpha$, and f is separable, since $\varphi_1\alpha, \ldots, \varphi_n\alpha$ are distinct. If σ is a K-automorphism of $\overline{K}$, then $\sigma\varphi_1, \ldots, \sigma\varphi_n$ are distinct K-homomorphisms of $K(\alpha)$ into $\overline{K}$, $\{\sigma\varphi_1, \ldots, \sigma\varphi_n\} = \{\varphi_1, \ldots, \varphi_n\}$, σ permutes $\varphi_1\alpha, \ldots, \varphi_n\alpha$, and $^\sigma f = f$. Therefore all coefficients of f are in F and $f \in F[X]$. Since $f(\alpha) = 0$, Irr $(\alpha : F)$ divides f and is separable; hence α is separable over F. Thus E is separable over F. $\square$

Perfect fields constitute a really nice class of fields, to which our new knowledge of normal and purely inseparable extensions can now be applied.

Definition. A field K is perfect *when either K has characteristic 0, or K has characteristic $p \neq 0$ and every element of K has a pth root in K.* $\square$

Proposition **2.11.** *Finite fields and algebraically closed fields are perfect.*

Proof. Algebraically closed fields are supremely perfect. If K is a finite field, then the characteristic of K is some prime $p \neq 0$, $\pi : x \longmapsto x^p$ is injective by III.4.4; therefore π is surjective and K is perfect. $\square$

Lemma **2.12.** *A perfect field has no proper purely inseparable extension.*

Proof. By IV.5.5 we may assume that K has characteristic $p \neq 0$. If K is perfect, then K contains the pth root of every $a \in K$ in $\overline{K}$; by induction, K contains the p^mth root of every $a \in K$ in $\overline{K}$. Therefore, only the elements of K are purely inseparable over K. $\square$

Proposition **2.13.** *Every algebraic extension of a perfect field is separable.*

Proof. Let K be perfect and let $E \subseteq \overline{K}$ be an algebraic extension of K. By 2.8, E is contained in a normal extension N of K, which by 2.10 is a separable extension of a purely inseparable extension F of K. By 2.12, $F = K$; hence $E \subseteq N$ is separable over K. $\square$

Proposition **2.14.** *Every algebraic extension of a perfect field is perfect.*

The proof is an exercise. (Readers may *not* groan.)

Exercises

1. Find the conjugates of $\sqrt[3]{2}$ over $\mathbb{Q}$.

2. Find the conjugates of $\sqrt{2} + \sqrt{3}$ over $\mathbb{Q}$.

Prove the following:

3. If F is normal over K and $K \subseteq E \subseteq F$, then F is normal over E.

4. If E is normal over K and the composite EF exists, then EF is normal over KF.

5. If E and F are normal over K, then $E \cap F$ is normal over K.

6. Every intersection of normal extensions $E \subseteq \overline{K}$ of K is a normal extension of K.

7. If E and F are normal over K, then EF (if it exists) is normal over K.

8. Every composite of normal extensions of a field K is a normal extension of K.

9. A field K is perfect if and only if $K^{1/p^{\infty}} = K$.

10. $K(X)$ is not perfect when K has characteristic $p \neq 0$.

11. A field K is perfect if and only if every algebraic extension of K is separable.

12. Every algebraic extension of a perfect field is perfect.

3. Galois Extensions

A Galois extension is a normal and separable extension. The main result of this section is a one-to-one correspondence between the intermediate fields of a Galois extension and the subgroups of its group of K-automorphisms.

Definition. A Galois extension of a field K is a normal and separable extension E of K; then E is Galois over K.

If K has characteristic 0, then every normal extension of K is a Galois extension of K; for instance, $\overline{K}$ is Galois over K. A finite field of characteristic p is a Galois extension of $\mathbb{Z}_p$.

The basic properties of Galois extensions follow from those of normal and separable extensions:

Proposition 3.1. If F is Galois over K and $K \subseteq E \subseteq F$, then F is Galois over E.

Proposition 3.2. If F is Galois over K and $E \subseteq F$ is normal over K, then E is Galois over K.

Proposition 3.3. If E is Galois over K and the composite EF exists, then EF is Galois over KF.

Proposition 3.4. Every composite of Galois extensions of K is a Galois extension of K.

Proposition 3.5. Every intersection of Galois extensions $E \subseteq \overline{K}$ of K is a Galois extension of K.

The fundamental theorem. This main result relates two constructions.

Definition. The Galois group $\mathrm{Gal}\,(E:K)$ of a Galois extension E of a field K, also called the Galois group of E over K, is the group of all K-automorphisms of E.

For example, the Galois group of $\mathbb{C} = \overline{\mathbb{R}}$ over $\mathbb{R}$ has two elements, the identity on $\mathbb{C}$ and complex conjugation.

Proposition 3.6. If E is Galois over K, then $\left| \mathrm{Gal}\,(E:K) \right| = [E:K]$.

Proof. If $E \subseteq \overline{K}$ is normal over K, then every K-homomorphism of E into $\overline{K}$ sends E onto E and is (as a set of ordered pairs) a K-automorphism of E. Hence $\left| \mathrm{Gal}\,(E:K) \right| = [E:K]_s = [E:K]$ when E is separable over K. $\square$

Definition. Let E be a field and let G be a group of automorphisms of E. The fixed field *of G is* $\mathrm{Fix}_E(G) = \{\,\alpha \in E \mid \sigma\alpha = \alpha$ for all $\sigma \in G\,\}$.

We see that $\mathrm{Fix}_E(G)$ is a subfield of E. For example, if $G = \mathrm{Gal}(\mathbb{C}:\mathbb{R})$, then $\mathrm{Fix}_\mathbb{C}(G) = \mathbb{R}$. A similar result holds whenever G is finite:

Proposition **3.7** (Artin). *If G is a finite group of automorphisms of a field E, then E is a finite Galois extension of* $F = \mathrm{Fix}_E(G)$ *and* $\mathrm{Gal}(E:F) = G$.

Proof. Let $\alpha \in E$. Since G is finite, $G\alpha$ is a finite set, $G\alpha = \{\,\alpha_1, ..., \alpha_n\,\}$, where $n \leqq |G|$, $\alpha_1, ..., \alpha_n \in E$ are distinct, and, say, $\alpha_1 = \alpha$. Let $f_\alpha(X) = (X - \alpha_1)(X - \alpha_2)\cdots(X - \alpha_n) \in E[X]$. Then $f_\alpha(\alpha) = 0$ and f_α is separable. Moreover, every $\sigma \in G$ permutes $\alpha_1, ..., \alpha_n$, so that $^\sigma f_\alpha = f_\alpha$; therefore $f_\alpha \in F[X]$. Hence α is algebraic over F, $\mathrm{Irr}(\alpha:F)$ divides f_α, and α is separable over F. Thus E is algebraic and separable over F. (This also follows from 2.10.) In fact, E is finite over F, with $[E:F] \leqq |G|$ by IV.5.13, since $\deg \mathrm{Irr}(\alpha:F) \leqq \deg f_\alpha \leqq |G|$ for every $\alpha \in E$. We see that E is a splitting field of the polynomials $f_\alpha \in F[X]$; hence E is normal over F.

By 3.6, $\big|\mathrm{Gal}(E:F)\big| = [E:F] \leqq |G|$. But every $\sigma \in G$ is an F-automorphism of E, so that $G \subseteq \mathrm{Gal}(E:F)$. Therefore $\mathrm{Gal}(E:F) = G$. $\square$

Proposition **3.8.** *If E is a Galois extension of K, then the fixed field of* $\mathrm{Gal}(E:K)$ *is K.*

Proof. Let $G = \mathrm{Gal}(E:K)$. Then $K \subseteq \mathrm{Fix}_E(G)$. Conversely, let $\alpha \in \mathrm{Fix}_E(G)$. By IV.4.5, there is an algebraic closure $\overline{K} \supseteq E$. By IV.4.7, every K-homomorphism φ of $K(\alpha)$ into $\overline{K}$ extends to a K-automorphism σ of $\overline{K}$; since E is normal over K, ψ has a restriction τ to E, which is a K-automorphism of E. Hence $\varphi\alpha = \tau\alpha = \alpha$, and φ is the inclusion homomorphism of $K(\alpha)$ into $\overline{K}$. Thus $[K(\alpha):K]_s = 1$. Since $K(\alpha) \subseteq E$ is separable over K, this implies $K(\alpha) = K$ and $\alpha \in K$. (Alternately, $\mathrm{Fix}_E(G)$ is purely inseparable over K, by 2.10; hence $\mathrm{Fix}_E(G) = K$.) $\square$

Propositions 3.1, 3.7, and 3.8 yield the fundamental theorem:

Theorem **3.9** (Fundamental Theorem of Galois Theory). *Let E be a finite Galois extension of a field K.*

If F is a subfield of E that contains K, then E is a finite Galois extension of F and F is the fixed field of $\mathrm{Gal}(E:F)$.

If H is a subgroup of $\mathrm{Gal}(E:K)$, *then $F = \mathrm{Fix}_E(H)$ is a subfield of E that contains K, and* $\mathrm{Gal}(E:F) = H$.

This defines a one-to-one correspondence between intermediate fields $K \subseteq F \subseteq E$ and subgroups of $\mathrm{Gal}(E:K)$.

The hypothesis that E is finite over K cannot be omitted in Theorem 3.9. What happens when E is infinite over K is considered in the next section.

Properties. We complete Theorem 3.9 with the following properties.

Proposition **3.10.** *Let* F_1, F_2, F_3 *be intermediate fields of a finite Galois extension* E *of* K, *with Galois groups* H_1, H_2, H_3.

(1) $F_1 \subseteq F_2$ *if and only if* $H_1 \supseteq H_2$;

(2) $F_1 = F_2 F_3$ *if and only if* $H_1 = H_2 \cap H_3$;

(3) $F_1 = F_2 \cap F_3$ *if and only if* H_1 *is the subgroup generated by* $H_2 \cup H_3$;

(4) *when* $E \subseteq \overline{K}$, *then* F_1 *and* F_2 *are conjugate if and only if* H_1 *and* H_2 *are conjugate in* $\mathrm{Gal}\,(E : K)$.

Proof. We prove (4) and leave (1), (2), (3) as exercises. First, F_1 and F_2 are conjugate if and only if $\tau F_2 = F_1$ for some $\tau \in \mathrm{Gal}\,(E : K)$: indeed, τ can be extended to a K-automorphism σ of $\overline{K}$; conversely, if $\sigma F_2 = F_1$ for some K-automorphism σ of $\overline{K}$, then σ has a restriction τ to the normal extension E, τ is a K-automorphism of E, and $\tau F_2 = F_1$.

If now $K \subseteq F \subseteq E$ and $\sigma, \tau \in \mathrm{Gal}\,(E : K)$, then $\sigma \in \mathrm{Gal}\,(E : \tau F)$ if and only if $\sigma \tau \alpha = \tau \alpha$ for all $\alpha \in F$; equivalently, $\tau^{-1} \sigma \tau$ is an F-automorphism, or $\sigma \in \tau \, \mathrm{Gal}\,(E : F)\, \tau^{-1}$. Thus $\mathrm{Gal}\,(E : \tau F) = \tau \, \mathrm{Gal}\,(E : F)\, \tau^{-1}$. If, conversely, $\mathrm{Gal}\,(E : F_3) = \tau \, \mathrm{Gal}\,(E : F)\, \tau^{-1}$ for some $\tau \in \mathrm{Gal}\,(E : K)$, then $F_3 = \tau F$. $\square$

The next two properties resemble the Isomorphism theorems for groups.

Proposition **3.11.** *If* E *is a finite Galois extension of* K, *then an intermediate field* $K \subseteq F \subseteq E$ *is normal over* K *if and only if* $\mathrm{Gal}\,(E : F)$ *is normal in* $\mathrm{Gal}\,(E : K)$, *and then* $\mathrm{Gal}\,(F : K) \cong \mathrm{Gal}\,(E : K) / \mathrm{Gal}\,(E : F)$.

Proof. By part (4) of 3.10, F is normal over K (F has only one conjugate) if and only if $\mathrm{Gal}\,(E : F)$ is normal in $\mathrm{Gal}\,(E : K)$. Now let F be normal over K. By 3.2, F is Galois over K. Hence every $\sigma \in \mathrm{Gal}\,(E : K)$ has a restriction $\sigma_{|F}$ to F, which is a K-automorphism of F. Then $\Phi : \sigma \longmapsto \sigma_{|F}$ is a homomorphism of $\mathrm{Gal}\,(E : K)$ into $\mathrm{Gal}\,(F : K)$, which is surjective, since every K-automorphism of F extends to a K-automorphism of $\overline{K}$ whose restriction to the normal extension E is a K-automorphism of E; and $\mathrm{Ker}\,\Phi = \mathrm{Gal}\,(E : F)$. $\square$

Proposition **3.12.** *If* E *is a finite Galois extension of* K, F *is a field extension of* K, *and the composite* EF *is defined, then* EF *is a finite Galois extension of* F, E *is a finite Galois extension of* $E \cap F$, *and* $\mathrm{Gal}\,(EF : F) \cong \mathrm{Gal}\,(E : E \cap F)$.

Proof. By 3.3, 3.1, EF is a Galois extension of F and E is a Galois extension of $E \cap F \subseteq E$; E is finite over $E \cap F$ since E is finite over $K \subseteq E \cap F$, and EF is finite over F by IV.3.8.

Since E is normal over $E \cap F$, every F-automorphism σ of EF has a restriction to E, which is an $E \cap F$-automorphism since σ is the identity on F. This yields a homomorphism $\Theta : \sigma \longmapsto \sigma_{|E}$ of $\mathrm{Gal}\,(EF : F)$ into $\mathrm{Gal}\,(E : E \cap F)$. Since EF is generated by $E \cup F$, a K-homomorphism of EF is uniquely determined by its restrictions to E and F; therefore Θ is injective.

If $\alpha \in E$, then $\sigma_{|E}\,\alpha = \alpha$ for all $\sigma \in \mathrm{Gal}\,(EF : F)$ if and only if $\sigma\alpha = \alpha$ for all $\sigma \in \mathrm{Gal}\,(EF : F)$, if and only if $\alpha \in F$, by 3.8. Thus $E \cap F$ is the fixed field of $\mathrm{Im}\,\Theta \subseteq \mathrm{Gal}\,(E : E \cap F)$; by 3.9, $\mathrm{Im}\,\Theta = \mathrm{Gal}\,(E : E \cap F)$. $\square$

Exercises

1. Let F_1, F_2 be intermediate fields of a finite Galois extension E of K, with Galois groups H_1, H_2. Show that $F_1 \subseteq F_2$ if and only if $H_1 \supseteq H_2$.

2. Let F_1, F_2, F_3 be intermediate fields of a finite Galois extension E of K, with Galois groups H_1, H_2, H_3. Show that $F_1 = F_2\,F_3$ if and only if $H_1 = H_2 \cap H_3$.

3. Let F_1, F_2, F_3 be intermediate fields of a finite Galois extension E of K, with Galois groups H_1, H_2, H_3. Show that $F_1 = F_2 \cap F_3$ if and only if H_1 is the subgroup generated by $H_2 \cup H_3$.

4. A *Galois connection* between two partially ordered sets X and Y is a pair of order reversing mappings $F : X \longrightarrow Y$, $G : Y \longrightarrow X$ ($x' \leqq x''$ implies $Fx' \geqq Fx''$, $y' \leqq y''$ implies $Gy' \geqq Gy''$) such that $FGy \geqq y$ and $GFx \geqq x$ for all x, y. Show that F and G induce mutually inverse, order reversing bijections between $\{\, x \in X \mid GFx = x \,\}$ and $\{\, y \in Y \mid FGy = y \,\}$.

5. Let F be a finite field of order p^n. Show that $\mathrm{Gal}\,(F : \mathbb{Z}_p)$ is cyclic of order p^{n-1}.

6. Let K be a field of characteristic 0 and let $\varepsilon \in \overline{K}$ be a root of unity ($\varepsilon^n = 1$ for some $n > 0$). Show that $K(\varepsilon)$ is Galois over K and that $\mathrm{Gal}\,(K(\varepsilon) : K)$ is abelian.

4. Infinite Galois Extensions

This section may be skipped. It contains Krull's theorem that extends the fundamental theorem of Galois theory to infinite Galois extensions.

Galois groups. Krull's theorem places a topology on Galois groups, whose construction is based on certain properties of these groups.

Proposition **4.1.** *Let* E *be Galois over* K *and let* $K \subseteq F \subseteq E$. *Then* $[\,\mathrm{Gal}\,(E : K) : \mathrm{Gal}\,(E : F)\,] = [\,F : K\,]$. *Moreover,* $\mathrm{Gal}\,(E : F)$ *is normal in* $\mathrm{Gal}\,(E : K)$ *if and only if* F *is normal over* K.

Proof. By 3.1, E is Galois over F. Every K-homomorphism of F into $\overline{K} \supseteq E$ is the restriction to F of a K-automorphism of E. Now, σ and $\tau \in \mathrm{Gal}\,(E : K)$ have the same restriction to F if and only if $\sigma^{-1}\tau$ is the identity on F, if and only if $\sigma^{-1}\tau \in \mathrm{Gal}\,(E : F)$. Hence there is a one-to-one correspondence between left cosets of $\mathrm{Gal}\,(E : F)$ and K-homomorphisms of F into $\overline{K}$, and $[\,\mathrm{Gal}\,(E : K) : \mathrm{Gal}\,(E : F)\,] = [\,F : K\,]_s = [\,F : K\,]$. The rest of the statement is left to readers. $\square$

Proposition **4.2.** *Let* E *be a Galois extension of* K *and let* $\mathcal{F}$ *be the set of all Galois groups* $\mathrm{Gal}\,(E : F) \subseteq \mathrm{Gal}\,(E : K)$ *of finite extensions* $F \subseteq E$ *of* K.

(1) *Every $H \in \mathcal{F}$ has finite index in* $\mathrm{Gal}\,(E:K)$.

(2) $\bigcap_{H \in \mathcal{F}} H = 1$.

(3) *$\mathcal{F}$ is closed under finite intersections.*

(4) *Every $H \in \mathcal{F}$ contains a normal subgroup $N \in \mathcal{F}$ of* $\mathrm{Gal}\,(E:K)$.

Proof. (1) follows from 4.1.

(2). Let $\sigma \in \bigcap_{H \in \mathcal{F}} H$. If $\alpha \in E$, then $K(\alpha) \subseteq E$ is finite over K, $\mathrm{Gal}\,(E:K(\alpha)) \in \mathcal{F}$, $\sigma \in \mathrm{Gal}\,(E:K(\alpha))$, and $\sigma\alpha = \alpha$. Hence $\sigma = 1_E$.

(3). Let $H_1 = \mathrm{Gal}\,(E:F_1)$ and $H_2 = \mathrm{Gal}\,(E:F_2)$, where F_1, $F_2 \subseteq E$ are finite over K. Then $F_1 F_2$ is finite over K, by IV.3.8, and $\mathrm{Gal}\,(E:F_1 F_2) = H_1 \cap H_2$, since $\sigma \in \mathrm{Gal}\,(E:K)$ is the identity on $F_1 F_2$ if and only if σ is the identity on F_1 and the identity on F_2. Hence $H_1 \cap H_2 \in \mathcal{F}$.

(4). Every finite extension $F \subseteq E \subseteq \overline{K}$ of K is contained in a finite normal extension N of K, namely the composite of all conjugates of F, and $N \subseteq E$ since every conjugate of F is contained in [a conjugate of] E. Then $\mathrm{Gal}\,(E:N) \trianglelefteq \mathrm{Gal}\,(E:K)$ by 4.1 and $\mathrm{Gal}\,(E:N) \subseteq \mathrm{Gal}\,(E:F)$. $\square$

By 4.2, the trivial subgroup of a Galois group is the intersection of normal subgroups of finite index. Hence not every group is a Galois group (see the exercises). But we shall see in Section 7 that every finite group is a Galois group.

The Krull topology. Let X and Y be sets and let M be a set of mappings of X into Y. For every $f \in M$ and finite subset S of X let

$$V(f, S) = \{\, g \in M \mid g(s) = f(s) \text{ for all } s \in S \,\}.$$

If $h \in V(f, S) \cap V(g, T)$, then $V(f, S) \cap V(g, T) = V(h,\ S \cup T)$. Hence the sets $V(f, S)$ constitute a basis for a topology, the *finite topology* on M.

Proposition **4.3.** *Let E be a Galois extension of K. Let*

$\mathcal{N}$ *be the set of all cosets of normal subgroups $N \in \mathcal{F}$, let*

$\mathcal{L}$ *be the set of all left cosets of subgroups $H \in \mathcal{F}$, and let*

$\mathcal{R}$ *be the set of all right cosets of subgroups $H \in \mathcal{F}$.*

Then $\mathcal{N}$ is a basis for the finite topology on $\mathrm{Gal}\,(E:K)$, and so are $\mathcal{L}$ and $\mathcal{R}$.

Proof. Let $H = \mathrm{Gal}\,(E:F) \in \mathcal{F}$, where $F \subseteq E$ is finite over K. Then $F = K(S)$ for some finite subset S of E. If $\sigma, \tau \in \mathrm{Gal}\,(E:K)$, then $\tau \in V(\sigma, S)$ if and only if $\sigma\alpha = \tau\alpha$ for all $\alpha \in K(S)$, if and only if $\sigma^{-1}\tau \in H$. Thus $V(\sigma, S) = \sigma H$. Hence $\mathcal{L}$ is a basis of the finite topology on $\mathrm{Gal}\,(E:K)$.

If $A, B \in \mathcal{N}$ and $\sigma \in A \cap B$, then $A = \sigma M$, $B = \sigma N$ for some normal subgroups $M, N \in \mathcal{F}$, and $A \cap B = \sigma\,(M \cap N) \in \mathcal{N}$, since $M \cap N \in \mathcal{F}$ by 4.2. Hence $\mathcal{N}$ is the basis of a topology. Now, $\mathcal{L}$ and $\mathcal{N}$ are bases of the same topology: $\mathcal{N} \subseteq \mathcal{L}$; conversely, every subgroup $H \in \mathcal{F}$ contains a normal subgroup $N \in \mathcal{F}$, by 4.2. Similarly, $\mathcal{R}$ and $\mathcal{N}$ are bases of the same topology. $\square$

The finite topology on $\mathrm{Gal}\,(E:K)$ is also known as the *Krull topology*. Its open sets are unions of members of $\mathcal{N}$, equivalently, unions of members of $\mathcal{L}$ (or $\mathcal{R}$). Unlike the author's office door, every subgroup $H \in \mathcal{F}$ is both open and closed (since its complement is a union of left cosets of H).

If $\mathrm{Gal}\,(E:K)$ is finite, then $\{1\}$ is open and the finite topology is the discrete topology. The general case is as follows:

Proposition **4.4.** *In the finite topology,* $\mathrm{Gal}\,(E:K)$ *is compact Hausdorff and totally disconnected.*

Proof. Let $G = \mathrm{Gal}\,(E:K)$. Let $\sigma, \tau \in G$, $\sigma \neq \tau$. Then $\sigma^{-1}\tau \notin H$ for some $H \in \mathcal{F}$, since $\bigcap_{H \in \mathcal{F}} H = 1$ by 4.2, and then σH, $\tau H \in \mathcal{L}$ are disjoint. Hence G is Hausdorff. Also, σH and $G \backslash \sigma H$ are both open ($G \backslash \sigma H$ is a union of left cosets of H) and $\sigma \in \sigma H$, $\tau \in G \backslash \sigma H$. Hence G is totally disconnected.

That G is compact follows from Tychonoff's theorem. We give a direct proof: we show that every ultrafilter $\mathcal{U}$ on G converges to some $\sigma \in G$.

Every $\alpha \in E$ belongs to a finite extension $F \subseteq E$ of K (e.g., to $K(\alpha)$). Then $H = \mathrm{Gal}\,(E:F) \in \mathcal{F}$ has finite index, G is the union of finitely many left cosets of H, and $\tau H \in \mathcal{U}$ for some $\tau \in G$, since $\mathcal{U}$ is an ultrafilter. Assume that $\alpha \in F, F'$, where $F, F' \subseteq E$ are finite over K, and τH, $\tau' H' \in \mathcal{U}$, where $H = \mathrm{Gal}\,(E:F)$ and $H' = \mathrm{Gal}\,(E:F')$. Then $\tau H \cap \tau' H' \in \mathcal{U}$ contains some $\upsilon \in G$, $\upsilon^{-1}\tau \in H = \mathrm{Gal}\,(E:F)$, $\upsilon^{-1}\tau' \in H' = \mathrm{Gal}\,(E:F')$, and $\upsilon^{-1}\tau\alpha = \alpha = \upsilon^{-1}\tau'\alpha$, since $\alpha \in F \cap F'$. Hence $\tau\alpha = \tau'\alpha$. Therefore a mapping $\sigma : E \longrightarrow E$ is well defined by $\sigma\alpha = \tau\alpha$ whenever $\alpha \in F$ and $\tau H \in \mathcal{U}$, where $F \subseteq E$ is finite over K, $\tau \in G$, and $H = \mathrm{Gal}\,(E:F)$.

If $\alpha, \beta \in E$, then $F = K(\alpha, \beta)$ is finite over K, $H = \mathrm{Gal}\,(E:F) \in \mathcal{F}$, and $\tau H \in \mathcal{U}$ for some $\tau \in G$. Hence $\sigma\alpha = \tau\alpha$, $\sigma\beta = \tau\beta$, and $\sigma\,(\alpha + \beta) = \sigma\alpha + \sigma\beta$, $\sigma\,(\alpha\beta) = (\sigma\alpha)(\sigma\beta)$. Also $\sigma x = \tau x = x$ for all $x \in K$. Thus σ is a K-endomorphism of E. Since E is normal over K, $\sigma E = E$, and $\sigma \in G$.

Let $H = \mathrm{Gal}\,(E:F) \in \mathcal{F}$, where $F \subseteq E$ is finite over K. As above, $\tau H \in \mathcal{U}$ for some $\tau \in G$, and then $\sigma\alpha = \tau\alpha$ for all $\alpha \in F$, $\tau^{-1}\sigma \in \mathrm{Gal}\,(E:F) = H$, and $\sigma H = \tau H \in \mathcal{U}$. Thus $\mathcal{U}$ contains every neighborhood of σ. $\square$

Krull's theorem. The one-to-one correspondence in Krull's theorem is between the intermediate fields of a Galois extension and the closed subgroups of its Galois group, under the finite topology. The next result explains why.

Proposition **4.5.** *If* E *is a Galois extension of* K *and* H *is subgroup of* $\mathrm{Gal}\,(E:K)$, *then* E *is a Galois extension of* $F = \mathrm{Fix}_E\,(H)$ *and* $\mathrm{Gal}\,(E:F)$ *is the closure of* H *in* $\mathrm{Gal}\,(E:K)$.

Proof. By 3.1, E is Galois over F. Let $\sigma \in \overline{H}$ and $\alpha \in F$. Then $K(\alpha) \subseteq F$ is finite over K, $U = \mathrm{Gal}\,(E:K(\alpha)) \in \mathcal{F}$, σU is open, and there exists $\tau \in H \cap \sigma U$. Then $\tau^{-1}\sigma \in U$, $\tau^{-1}\sigma\alpha = \alpha$, and $\sigma\alpha = \tau\alpha = \alpha$. Thus $\sigma \in \mathrm{Gal}\,(E:F)$.

Conversely, let $\sigma \in \mathrm{Gal}\,(E:F)$. Let $U = \mathrm{Gal}\,(E:L) \in \mathcal{F}$, where $L \subseteq E$ is finite over K. Then LF is finite over F; by 2.9, LF is contained in a finite normal extension $N \subseteq E$ of F (the composite of all conjugates of LF, all of which are contained in E). Then N is a finite Galois extension of F. Restriction to N is a homomorphism $\Phi : \tau \longmapsto \tau_{|N}$ of $\mathrm{Gal}\,(E:F)$ into $\mathrm{Gal}\,(N:F)$. Now, $F = \mathrm{Fix}_E\,(H)$; therefore $F = \mathrm{Fix}_N\,(\Phi(H))$. In the finite Galois extension N of F this implies $\Phi(H) = \mathrm{Gal}\,(N:F)$. Then $\Phi(\sigma) = \Phi(\tau)$ for some $\tau \in H$, whence $\sigma_{|N} = \tau_{|N}$, $\sigma_{|L} = \tau_{|L}$, $\sigma^{-1}\tau \in \mathrm{Gal}\,(E:L) = U$, and $\tau \in \sigma U \cap H$. Then every neighborhood of σ intersects H, and $\sigma \in \overline{H}$. $\square$

Krull's theorem follows from Propositions 3.1, 4.5, and 3.8:

Theorem 4.6 (Krull). *Let E be a Galois extension of a field K.*

If F is a subfield of E that contains K, then E is a Galois extension of F and F is the fixed field of $\mathrm{Gal}\,(E:F)$.

If H is a closed subgroup of $\mathrm{Gal}\,(E:K)$ in the finite topology, then $F = \mathrm{Fix}_E\,(H)$ is a subfield of E that contains K, and $\mathrm{Gal}\,(E:F) = H$.

This defines a one-to-one correspondence between intermediate fields $K \subseteq F \subseteq E$ and closed subgroups of $\mathrm{Gal}\,(E:K)$.

If E is finite over K, then $\mathrm{Gal}\,(E:K)$ has the discrete topology, every subgroup is closed, and Krull's theorem reduces to Theorem 3.9. Readers will easily extend Propositions 3.10, 3.11, and 3.12 to arbitrary Galois extensions.

An example. This example, from McCarthy [1966], has uncountably many subgroups of finite index, only countably many of which are closed. Thus, a Galois group may have comparatively few closed subgroups; subgroups of finite index need not be closed in the finite topology; and the finite topology has fewer open sets than the profinite topology mentioned at the end of Section I.5.

Let $E \subseteq \mathbb{C}$ be generated over $\mathbb{Q}$ by the square roots of all primes $p \geq 3$; E is the splitting field of the set of all polynomials $X^2 - p$, and is Galois over $\mathbb{Q}$.

Let $G = \mathrm{Gal}\,(E:\mathbb{Q})$. Since $\mathrm{Irr}\,(\sqrt{p}:\mathbb{Q}) = X^2 - p$, $\sqrt{p}$ has only two conjugates over $\mathbb{Q}$, $\sqrt{p}$ and $-\sqrt{p}$. Hence $\sigma\sqrt{p} = \sqrt{p}$ or $\sigma\sqrt{p} = -\sqrt{p}$, for every $\mathbb{Q}$-automorphism σ of E. Conversely, for every subset S of P, there is a $\mathbb{Q}$-automorphism σ of E such that $\sigma\sqrt{p} = -\sqrt{p}$ for all $p \in S$ and $\sigma\sqrt{p} = \sqrt{p}$ for all $p \notin S$. Therefore $|G| = 2^{\aleph_0}$ and G is uncountable.

We also have $\sigma^2 = 1$ for every $\mathbb{Q}$-automorphism σ of E. Therefore G is abelian, and is a vector space over $\mathbb{Z}_2$. Let B be a basis of G over $\mathbb{Z}_2$. Then B is uncountable, since G is. For every $\beta \in B$, $B \setminus \{\beta\}$ generates a subgroup of G of index 2. Therefore G has uncountably many subgroups of finite index.

On the other hand, E is, like all algebraic extensions of $\mathbb{Q}$, countable. If $F \subseteq E$ is finite over $\mathbb{Q}$, then $F = \mathbb{Q}(\alpha)$ for some $\alpha \in E$, by IV.5.12. Therefore there are only countably many finite extensions $F \subseteq E$ of Q. By 4.6, G has only countably many closed subgroups of finite index.

Exercises

1. Given a Galois extension E of K and $K \subseteq F \subseteq E$, show that $\mathrm{Gal}\,(E:F)$ is normal in $\mathrm{Gal}\,(E:K)$ if and only if F is normal over K, and then $\mathrm{Gal}\,(F:K) \cong \mathrm{Gal}\,(E:K)/\mathrm{Gal}\,(E:F)$.

2. In any group, show that the intersection of two subgroups of finite index is a subgroup of finite index.

3. In any group, show that every subgroup of finite index contains a normal subgroup of finite index.

4. In a group G, show that the identity is the intersection of normal subgroups of finite index if and only if G can be embedded into (is isomorphic to a subgroup of) a direct product of finite groups. (These groups are called *profinite*).

5. Show that the additive group $\mathbb{Q}$ is not profinite.

6. Use Tychonoff's theorem to prove that $\mathrm{Gal}\,(E:K)$ is compact in the finite topology.

7. In a Galois group, show that the multiplication $(\sigma, \tau) \longmapsto \sigma\tau$ and inversion $\sigma \longmapsto \sigma^{-1}$ are continuous in the finite topology.

8. Let F_1, F_2, F_3 be intermediate fields of a Galois extension E of K, with Galois groups H_1, H_2, H_3. Show that $F_1 = F_2 F_3$ if and only if $H_1 = H_2 \cap H_3$.

9. Let F_1, F_2, F_3 be intermediate fields of a Galois extension E of K, with Galois groups H_1, H_2, H_3. Show that $F_1 = F_2 \cap F_3$ if and only if H_1 is the closure of the subgroup generated by $H_2 \cup H_3$.

10. Let E be a Galois extension of K and let F be a field extension of K such that the composite EF is defined. Show that EF is a Galois extension of F, E is a Galois extension of $E \cap F$, and $\mathrm{Gal}\,(EF:F) \cong \mathrm{Gal}\,(E:E \cap F)$. Is this isomorphism continuous? a homeomorphism?

5. Polynomials

In this section we look at the splitting fields of polynomials of degree at most 4. This provides concrete examples of Galois groups. The material on polynomial equations may be skipped, but it shows a nice interplay between ancient results and modern Galois theory.

General results. We begin with abitrary polynomials.

Definition. The Galois group $\mathrm{Gal}\,(f:K)$ *of a polynomial* $f \in K[X]$ *over a field* K *is the group of* K-*automorphisms of its splitting field over* K. $\square$

If $E \subseteq \overline{K}$ is the splitting field of $f \in K[X]$ over K, then E is finite over K and its group G of K-automorphisms is finite; by 3.7, E is a finite

Galois extension of $F = \text{Fix}_E(G)$ and $\text{Gal}(f : K) = G = \text{Gal}(E : F)$. Then $E = F(\alpha)$ for some $\alpha \in E$, by IV.5.12 and E is the splitting field over F of the separable irreducible polynomial $\text{Irr}(\alpha : F)$. Thus the Galois group of any polynomial is the Galois group (perhaps over a larger field) of a finite Galois extension, and of a separable irreducible polynomial.

If f is separable over K, then its roots in $\overline{K}$ are separable over K, its splitting field E is separable over K, E is a Galois extension of K, and $\text{Gal}(f : K) = \text{Gal}(E : K)$.

Proposition **5.1.** *Let* $\alpha_1, \ldots, \alpha_n$ *be the distinct roots of* $f \in K[X]$ *in* $\overline{K}$. *Every* $\tau \in \text{Gal}(f : K)$ *permutes the roots of* f *in* $\overline{K}$; *hence* $\text{Gal}(f : K)$ *is isomorphic to a subgroup* G *of the symmetric group* S_n. *If* f *is separable and irreducible, then* n *divides* $|G|$ *and* G *is a transitive subgroup of* S_n.

Proof. Let E be the splitting field of f. If τ is a K-automorphism of E and $f(\alpha) = 0$, then $f(\tau\alpha) = {}^\tau f(\tau\alpha) = \tau f(\alpha) = 0$; hence τ permutes the roots of f and induces a permutation $\sigma \in S_n$ such that $\tau\alpha_i = \alpha_{\sigma i}$ for all i. Since E is generated by $\alpha_1, \ldots, \alpha_n$, τ is uniquely determined by σ, and the mapping $\varphi : \tau \longmapsto \sigma$ is injective; φ is a homomorphism since $\tau\tau'\alpha_i = \tau\alpha_{\sigma'i} = \alpha_{\sigma\sigma'i}$. Hence $\text{Gal}(f : K)$ is isomorphic to the subgroup $G = \text{Im}\,\varphi$ of S_n.

If f is separable and irreducible, then f has degree n, $f = \text{Irr}(\alpha_i : K)$ for every i, $K(\alpha_i) \subseteq E$ has degree n over K, and $\text{Gal}(f : K) = \text{Gal}(E : K)$ has a subgroup $\text{Gal}(E : K(\alpha_i))$ of index n. Hence n divides $|\text{Gal}(f : K)| = |G|$. For every i, j there is a K-automorphism τ of E such that $\tau\alpha_i = \alpha_j$; hence G is transitive (for every i, j there is some $\sigma \in G$ such that $\sigma i = j$). $\square$

For a separable and irreducible polynomial $f \in K[X]$, 5.1 implies the following. If f has degree 2, then $\text{Gal}(f : K) \cong S_2$ is cyclic of order 2. If f has degree 3, then either $\text{Gal}(f : K) \cong S_3$, or $\text{Gal}(f : K) \cong A_3$ is cyclic of order 3.

Example. The Galois group G and splitting field $E \subseteq \mathbb{C}$ of $f(X) = X^3 - 2$ over $\mathbb{Q}$ can be analyzed in some detail. First, f is irreducible, by Eisenstein's criterion. The complex roots of f are $\rho = \sqrt[3]{2} \in \mathbb{R}$, $j\rho$, and $j^2\rho$, where $j = -1/2 + i\sqrt{3}/2$ is a primitive cube root of unity. Hence $E = \mathbb{Q}(\rho, j\rho, j^2\rho) = \mathbb{Q}(\rho, j)$, and E has an intermediate field $\mathbb{Q}(\rho) \subseteq \mathbb{R}$. We see that $[\mathbb{Q}(\rho) : \mathbb{Q}] = 3$ and $[E : \mathbb{Q}(\rho)] = 2$. Hence $[E : \mathbb{Q}] = 6$ and $G = \text{Gal}(E : \mathbb{Q}) \cong S_3$, by 5.1. Next, S_3 is generated by the 3-cycle $(1\,2\,3)$ and the transposition $(2\,3)$; hence G is generated by γ and τ, where

$$\gamma\rho = j\rho,\ \gamma(j\rho) = j^2\rho,\ \gamma(j^2\rho) = \rho,\ \gamma j = j,$$
$$\tau\rho = \rho,\ \tau(j\rho) = j^2\rho,\ \tau(j^2\rho) = j\rho,\ \tau j = j^2,$$

and $G = \{1, \gamma, \gamma^2, \tau, \gamma\tau, \gamma^2\tau\}$. The subgroups of G are 1, G, and

$$\{1, \tau\},\ \{1, \gamma\tau\},\ \{1, \gamma^2\tau\},\ \{1, \gamma, \gamma^2\}.$$

Hence E has four intermediate fields $\mathbb{Q} \subsetneqq F \subsetneqq E$. The fixed field F of $\{1, \tau\}$ contains ρ and has degree 3 over $\mathbb{Q}$, since $[E:F] = |\text{Gal}(E:F)| = 2$; hence it is $\mathbb{Q}(\rho)$. Similarly, the fixed field of $\{1, \gamma\tau\}$ is $\mathbb{Q}(j^2\rho)$, which has degree 3 over $\mathbb{Q}$; the fixed field of $\{1, \gamma^2\tau\}$ is $\mathbb{Q}(j\rho)$, which has degree 3 over $\mathbb{Q}$; and the fixed field of $\{1, \gamma, \gamma^2\}$ is $\mathbb{Q}(j)$, which has degree 2 over $\mathbb{Q}$ and is normal over $\mathbb{Q}$ since $\{1, \gamma, \gamma^2\} \trianglelefteq G$. $\square$

Polynomials of degree 3. Let $f(X) = a_n(X - \alpha_1) \cdots (X - \alpha_n)$ be a polynomial of degree $n \geq 1$ with coefficients in a field K and not necessarily distinct roots $\alpha_1, \ldots, \alpha_n$ in $\overline{K}$. The *discriminant* of f is

$$\text{Dis}(f) = a_n^{2n-2} \prod_{1 \leq i < j \leq n} (\alpha_i - \alpha_j)^2.$$

Some properties of discriminants, for instance, $\text{Dis}(f) \in K$, are proved in Section IV.7.

Proposition 5.2. If $f \in K[X]$ and the field K does not have characteristic 2, then $\text{Gal}(f:K)$ induces an odd permutation if and only if $\text{Dis}(f)$ does not have a square root in K.

Proof. The splitting field E of f contains all α_i and contains $\text{Dis}(f)$. We see that $\text{Dis}(f) = d^2$, where $d = a_n^{n-1} \prod_{1 \leq i < j \leq n} (\alpha_i - \alpha_j)$. If $\tau \in \text{Gal}(f:K)$ transposes two roots, then $\tau d = -d$; hence $\tau d = d$ whenever τ induces an even permutation, and $\tau d = -d \neq d$ whenever τ induces an odd permutation. If $d \in K$, equivalently if $\text{Dis}(f)$ has a square root in K (which must be d or $-d$), then no $\tau \in \text{Gal}(f:K)$ induces an odd permutation. If $d \notin K$, then $\tau d \neq d$ for some $\tau \in \text{Gal}(f:K)$, since K is the fixed field of $\text{Gal}(f:K)$, and some $\tau \in \text{Gal}(f:K)$ induces an odd permutation. $\square$

Corollary 5.3. Let $f \in K[X]$ be a separable irreducible polynomial of degree 3. If $\text{Dis}(f)$ has a square root in K, then $\text{Gal}(f:K) \cong A_3$; otherwise, $\text{Gal}(f:K) \cong S_3$.

Proof. We saw that $\text{Gal}(f:K)$ is isomorphic to either A_3 or S_3. $\square$

The discriminant of $X^3 + pX + q$ is known to be $-4p^3 - 27q^2$. For example, $f(X) = X^3 - 2 \in \mathbb{Q}[X]$ is irreducible by Eisenstein's criterion; $\text{Dis}(f) = -27 \times 2^2 = -108$ does not have a square root in $\mathbb{Q}$; therefore $\text{Gal}(f:\mathbb{Q}) \cong S_3$. In this example, the roots of f in $\overline{\mathbb{Q}} \subseteq \mathbb{C}$ are reached by first adjoining to $\mathbb{Q}$ a square root of -108 (equivalently, a square root of -3), then a cube root of 2; this corresponds to the structure $\mathbb{Q} \subsetneqq \mathbb{Q}(j) \subsetneqq E$ of the splitting field and to the structure $S_3 \trianglerighteq A_3 \trianglerighteq 1$ of the Galois group.

Cardano's formula. Cardano's sixteenth century method [1545] yields formulas for the roots of polynomials of degree 3, and an explicit way to reach them by successive adjunctions of square roots and cube roots.

Let K be a field that does not have characteristic 2 or 3, and let $f(X) = aX^3 + bX^2 + cX + d \in K[X]$, where $a \neq 0$. The general equation $f(x) = 0$

is first simplified by the substitution $x = y - b/3a$, which puts it in the form $g(y) = a(y^3 + py + q) = 0$, where $p, q \in K$, since p and q are rational functions of a, b, c, d. Note that $\text{Dis}(f) = \text{Dis}(g)$, since f and g have the same leading coefficient and differences between roots.

To solve the equation $g(x) = x^3 + px + q = 0$, let $x = u + v$ to obtain

$$(u + v)^3 + p(u + v) + q = u^3 + v^3 + (3uv + p)(u + v) + q = 0.$$

If $3uv + p = 0$, then u^3 and v^3 satisfy $u^3 v^3 = -p^3/27$ and $u^3 + v^3 = -q$, and are the roots of the *resolvent* polynomial $(X - u^3)(X - v^3) = X^2 + qX - p^3/27 \in K[X]$:

$$u^3 = \frac{-q + \sqrt{q^2 + 4p^3/27}}{2}, \quad v^3 = \frac{-q - \sqrt{q^2 + 4p^3/27}}{2},$$

and we obtain *Cardano's formula*:

Proposition 5.4. *If K does not have characteristic 2 or 3 and $p, q \in K$, then the roots of $X^3 + pX + q$ in $\overline{K}$ are*

$$u + v = \sqrt[3]{\frac{-q + \sqrt{q^2 + 4p^3/27}}{2}} + \sqrt[3]{\frac{-q - \sqrt{q^2 + 4p^3/27}}{2}},$$

where the cube roots are chosen so that $uv = -p/3$.

Equations of degree 4. The following method solves equations of degree 4 and yields explicit formulas that construct the roots by successive adjunctions of square roots and cube roots. Cardano had a simpler solution, but it does not relate as well to Galois groups.

Let $f(X) = aX^4 + bX^3 + cX^2 + dX + e \in K[X]$, where $a \neq 0$ and K does not have characteristic 2. Simplify the equation $f(x) = 0$ by the substitution $x = y - b/4a$, which puts it in the form $g(y) = a(y^4 + py^2 + qy + r) = 0$, where $p, q, r \in K$ are rational functions of a, b, c, d, e. The roots $\alpha_1, \ldots, \alpha_4$ of f and $\beta_1, \ldots, \beta_4$ of g in $\overline{K}$ are related by $\alpha_i = -b/4a + \beta_i$ for all i. In particular, $\text{Dis}(f) = \text{Dis}(g)$.

In $\overline{K}[X]$, $g(X) = a(X^4 + pX^2 + qX + r) = a(X - \beta_1)(X - \beta_2)(X - \beta_3)(X - \beta_4)$, whence $\sum_i \beta_i = 0$, $\sum_{i<j} \beta_i \beta_j = p$, $\sum_{i<j<k} \beta_i \beta_j \beta_k = -q$, $\beta_1 \beta_2 \beta_3 \beta_4 = r$. Let

$$u = -(\beta_1 + \beta_2)(\beta_3 + \beta_4) = (\beta_1 + \beta_2)^2,$$
$$v = -(\beta_1 + \beta_3)(\beta_2 + \beta_4) = (\beta_1 + \beta_3)^2,$$
$$w = -(\beta_1 + \beta_4)(\beta_2 + \beta_3) = (\beta_1 + \beta_4)^2;$$

equivalently,

$$u = -\left(\alpha_1 + \alpha_2 + \frac{b}{2a}\right)\left(\alpha_3 + \alpha_4 + \frac{b}{2a}\right) = \left(\alpha_1 + \alpha_2 + \frac{b}{2a}\right)^2,$$

$$v = -\left(\alpha_1 + \alpha_3 + \frac{b}{2a}\right)\left(\alpha_2 + \alpha_4 + \frac{b}{2a}\right) = \left(\alpha_1 + \alpha_3 + \frac{b}{2a}\right)^2,$$

$$w = -\left(\alpha_1 + \alpha_4 + \frac{b}{2a}\right)\left(\alpha_2 + \alpha_3 + \frac{b}{2a}\right) = \left(\alpha_1 + \alpha_4 + \frac{b}{2a}\right)^2.$$

Tedious computations, which our reader will probably not forgive, yield

$$u + v + w = -2p, \ uv + uw + vw = p^2 - 4r, \ uvw = q^2.$$

Hence u, v, and w are the roots of the *resolvent* polynomial of f and g,

$$s(X) = (X - u)(X - v)(X - w) = X^3 + 2pX^2 + (p^2 - 4r)X - q^2 \in K[X].$$

Note that $u - v = (\beta_1 - \beta_4)(\beta_2 - \beta_3)$, $u - w = (\beta_1 - \beta_3)(\beta_2 - \beta_4)$, $v - w = (\beta_1 - \beta_2)(\beta_3 - \beta_4)$, so that $\mathrm{Dis}(s) = \prod_{i<j}(\beta_i - \beta_j)^2$ and $\mathrm{Dis}(f) = \mathrm{Dis}(g) = a^6 \mathrm{Dis}(s)$.

Now, $\beta_1 + \beta_2 = u'$ is a square root of u, $\beta_1 + \beta_3 = v'$ is a square root of v, $\beta_1 + \beta_4 = w'$ is a square root of w, and

$$u'v'w' = (\beta_1 + \beta_2)(\beta_1 + \beta_3)(\beta_1 + \beta_4) = \beta_1^2 \sum_i \beta_i + \sum_{i<j<k} \beta_i \beta_j \beta_k = -q.$$

Finally, $u' + v' + w' = 3\beta_1 + \beta_2 + \beta_3 + \beta_4 = 2\beta_1$; similarly, $u' - v' - w' = 2\beta_2$, $-u' + v' - w' = 2\beta_3$, $-u' - v' + w' = 2\beta_4$ and we obtain formulas for $\beta_1, \beta_2, \beta_3, \beta_4$:

Proposition 5.5. *If K does not have characteristic 2 and $p, q, r \in K$, then the roots of $X^4 + pX^2 + qX + r$ in $\overline{K}$ are*

$$\beta_1 = \frac{1}{2}(u' + v' + w'), \quad \beta_2 = \frac{1}{2}(u' - v' - w'),$$

$$\beta_3 = \frac{1}{2}(-u' + v' - w'), \quad \beta_4 = \frac{1}{2}(-u' - v' + w'),$$

where u', v', w' are square roots of the roots u, v, w of the resolvent $s(X) = X^3 + 2pX^2 + (p^2 - 4r)X - q^2$, chosen so that $u'v'w' = -q$.

If K does not have characteristic 2 or 3, then Cardano's formula for u, v, and w yields explicit formulas for $\beta_1, \ldots, \beta_4$ and explicit formulas for $\alpha_1, \ldots, \alpha_4$, showing that they can be reached from K by successive adjunctions of square roots and cube roots.

Polynomials of degree 4. The Galois groups of polynomials of degree 4 reflect the construction of their roots in Proposition 5.5.

Proposition 5.6. *Let $f(X) = aX^4 + bX^3 + cX^2 + dX + e \in K[X]$ be a separable irreducible polynomial of degree 4, where the field K does not have* *characteristic* 2. *Let* $F \subseteq \overline{K}$ *be the splitting field of its resolvent. Then* $[F:K]$ *divides 6 and:*

(1) *If* $[F:K] = 6$, *then* $\mathrm{Gal}(f:K) \cong S_4$.

(2) *If* $[F:K] = 3$, *then* $\mathrm{Gal}(f:K) \cong A_4$.

(3) *If* $[F:K] = 2$, *then* $\mathrm{Gal}(f:K) \cong D_4$ *if* f *is irreducible over* F, *otherwise* $\mathrm{Gal}(f:K)$ *is cyclic of order 4.*

(4) *If* $[F:K] = 1$, *then* $\mathrm{Gal}(f:K) \cong V_4$.

Proof. The resolvent s of f is separable, since $a^6 \mathrm{Dis}(s) = \mathrm{Dis}(f) \neq 0$. Hence its roots u, v, w are all distinct. Let $E \subseteq \overline{K}$ and $F = K(u, v, w) \subseteq \overline{K}$ be the splitting fields of f and s. By 5.5, $E \subseteq K(u', v', w')$, where u', v', w' are square roots of u, v, w such that $u'v'w' \in K$. Hence $E \subseteq F(u', v')$ and $[E:F] \leqq 4$. Also F is Galois over K; $\mathrm{Gal}(s:K)$ is isomorphic to a subgroup of S_3, and $[F:K] = |\mathrm{Gal}(s:K)|$ divides 6.

Before tackling parts (1) through (4), we look at S_4. The normal subgroup $V = \{1, (1\ 2)(3\ 4), (1\ 3)(2\ 4), (1\ 4)(2\ 3)\}$ of S_4 is isomorphic to V_4. The centralizer C of $(1\ 2)(3\ 4) \in V$ consists of all permutations σ such that either $\sigma\{1, 2\} = \{1, 2\}$, $\sigma\{3, 4\} = \{3, 4\}$, or $\sigma\{1, 2\} = \{3, 4\}$, $\sigma\{3, 4\} = \{1, 2\}$. Since $(1\ 2)(3\ 4)$ has three conjugates in S_4, C has eight elements and consists of $(1\ 2)$, $(3\ 4)$, $(1\ 3\ 2\ 4)$, $(1\ 4\ 2\ 3)$, and the elements of V, all of which commute with $(1\ 2)(3\ 4)$. Thus C is a Sylow 2-subgroup of S_4. We see that $C \cong D_4$. The centralizers of $(1\ 3)(2\ 4)$ and $(1\ 4)(2\ 3)$ are the other Sylow 2-subgroups of S_4 and consist of similar permutations. Hence $\sigma \in S_4$ commutes with every element of V if and only if $\sigma \in V$.

By 5.1, $\mathrm{Gal}(f:K)$ is isomorphic to a subgroup G of S_4: every $\tau \in \mathrm{Gal}(f:K)$ permutes the roots $\alpha_1, \ldots, \alpha_4$ of f and induces a permutation $\sigma \in S_4$ such that $\tau\alpha_i = \alpha_{\sigma i}$. The equalities

$$u = -\left(\alpha_1 + \alpha_2 + \frac{b}{2a}\right)\left(\alpha_3 + \alpha_4 + \frac{b}{2a}\right) = \left(\alpha_1 + \alpha_2 + \frac{b}{2a}\right)^2,$$

$$v = -\left(\alpha_1 + \alpha_3 + \frac{b}{2a}\right)\left(\alpha_2 + \alpha_4 + \frac{b}{2a}\right) = \left(\alpha_1 + \alpha_3 + \frac{b}{2a}\right)^2,$$

$$w = -\left(\alpha_1 + \alpha_4 + \frac{b}{2a}\right)\left(\alpha_2 + \alpha_3 + \frac{b}{2a}\right) = \left(\alpha_1 + \alpha_4 + \frac{b}{2a}\right)^2$$

show that τ also permutes u, v, and w. If $\sigma \in V$, then the same equalities show that $\tau u = u$, $\tau v = v$, and $\tau w = w$. Conversely, if $\tau u = u$, $\tau v = v$, and $\tau w = w$, then $\tau u = u \neq v, w$, whence $\sigma\{1, 2\} = \{1, 2\}$, $\sigma\{3, 4\} = \{3, 4\}$ and σ commutes with $(1\ 2)(3\ 4)$; similarly, σ commutes with $(1\ 3)(2\ 4)$ and $(1\ 4)(2\ 3)$; hence $\sigma \in V$. Therefore $\tau \in \mathrm{Gal}(E:F)$ if and only if $\sigma \in V$, and $\mathrm{Gal}(E:F) \cong G \cap V$. By 3.11, $\mathrm{Gal}(F:K) \cong \mathrm{Gal}(E:K)/\mathrm{Gal}(E:F) \cong G/(G \cap V)$.

By 5.1, $G \cong \mathrm{Gal}(f:K)$ is a transitive subgroup G of S_4, whose order is divisible by 4. Hence $|G| = 4, 8, 12$, or 24. If $|G| = 24$, then $\mathrm{Gal}(f:K) \cong S_4$

and $[F:K] = 6$, since $[E:K] = 24$, $[E:F] \leqq 4$, and $[F:K] \leqq 6$.

If $|G| = 12$, then $G = A_4$, since A_4 is the only subgroup of S_4 of order 12; hence $\mathrm{Gal}\,(f:K) \cong A_4$, $V \subseteq G$, $\mathrm{Gal}\,(F:K) \cong G/V$ has order 3, and $[F:K] = 3$.

If $|G| = 8$, then G is one of the three Sylow 2-subgroups of S_4; hence $\mathrm{Gal}\,(f:K) \cong D_4$, $V \subseteq G$, $\mathrm{Gal}\,(F:K) \cong G/V$ has order 2, and $[F:K] = 2$. Since V is transitive and $V \subseteq G$, there exists, for every i, some $\tau \in \mathrm{Gal}\,(f:K)$ such that $\tau\alpha_1 = \alpha_i$ and τ induces some $\sigma \in V$; then $\tau \in \mathrm{Gal}\,(E:F)$; hence α_i is a root of $\mathrm{Irr}\,(\alpha_1:F)$. Thus $\mathrm{Irr}\,(\alpha_1:F)$ has four distinct roots in E and degree at least 4. Since $f(\alpha_1) = 0$, it follows that f is proportional to $\mathrm{Irr}\,(\alpha_1:F)$, and is irreducible in $F[X]$.

Finally, let $|G| = 4$. If $G = V$, then $\mathrm{Gal}\,(f:K) \cong V_4$, $\mathrm{Gal}\,(F:K) = 1$, and $[F:K] = 1$. Otherwise, G is cyclic, generated by a 4-cycle, $\mathrm{Gal}\,(f:K)$ is cyclic of order 4, $\mathrm{Gal}\,(F:K) \cong G/(G \cap V)$ has order 2, and $[F:K] = 2$; then $G \cap V \cong \mathrm{Gal}\,(E:F)$ is not transitive, so f is not irreducible over F, by 5.1. $\square$

Exercises

1. Find fields $K \subseteq E \subseteq F$ such that F is normal over K but E is not normal over K.

2. Let $f \in \mathbb{R}[X]$ have degree 3. How does the sign of $\mathrm{Dis}\,(f)$ relate to the number of real roots of f?

3. Let $f \in \mathbb{R}[X]$ have degree 4. How does the sign of $\mathrm{Dis}\,(f)$ relate to the number of real roots of f?

In the following exercises, find the Galois group of the given polynomial over the given field, and all intermediate fields of its splitting field.

4. $X^3 - X - 1$, over $\mathbb{Q}$.

5. $X^3 - 10$, over $\mathbb{Q}$.

6. $(X^2 - 2)(X^2 - 3)$, over $\mathbb{Q}$.

7. $X^4 - 3$, over $\mathbb{Q}$.

In the following exercises, find the Galois group of the given polynomial over the given field.

8. $X^3 - X - 1$, over $\mathbb{Q}(\sqrt{-23})$.

9. $X^3 - 10$, over $\mathbb{Q}(\sqrt{2})$.

10. $X^3 - 10$, over $\mathbb{Q}(\sqrt{-3})$.

11. $X^4 - 3$, over $\mathbb{Q}(\sqrt{3})$.

12. $X^4 - 3$, over $\mathbb{Q}(\sqrt{-3})$.

13. $X^4 + X + 3$, over $\mathbb{Q}$.

The following exercises are for the last part of this section, which may have been skipped.

14. Use Cardano's formula to find the roots of $X^3 - 3X + 1$ in $\mathbb{C}$.

15. Verify that $u + v + w = -2p$, $uv + uw + vw = p^2 - 4r$, and $uvw = q^2$, in the proof of 5.5.

16. Cardano solved the equation $x^4 + px^2 + qx + r = 0$ by rewriting it as $(x^2 + y)^2 = \cdots$ and choosing y so that the right hand side is a perfect square. Fill in the details.

17. Find the roots of $X^4 + 3X + 3$ in $\mathbb{C}$.

18. Find the Galois group of $X^4 + 3X + 3$ over $\mathbb{Q}$.

6. Cyclotomy

The Greek roots of the word "cyclotomy" mean "circle" and "cut", as in cutting the unit circle into n equal arcs. This section uses Galois theory rather than scissors to study complex roots of unity, their irreducible polynomials, and the fields they generate over $\mathbb{Q}$. This yields more examples of Galois groups. Applications include Wedderburn's theorem on finite division rings; a particular case of Dirichlet's Theorem on primes in arithmetic progressions; and a proof that every finite abelian group is the Galois group of a finite extension of $\mathbb{Q}$.

Except for Proposition 6.5, which is quoted in Section 9, this material will not be used later.

Cyclotomic polynomials. Recall that the nth roots of unity in $\mathbb{C}$ are the complex numbers $\varepsilon_k = \cos(2\pi k/n) + i \sin(2\pi k/n)$, $0 \leq k < n$. The nth root of unity ε_k is primitive if and only if k and n are relatively prime, so that there are $\phi(n)$ primitive nth roots of unity, where ϕ is Euler's function.

Definition. The nth cyclotomic polynomial is the product $\Phi_n(X) \in \mathbb{C}[X]$ of all $X - \varepsilon$ in which ε is a primitive nth root of unity.

For example, $\Phi_1(X) = X - 1$; $\Phi_2(X) = X + 1$; $\Phi_3(X) = (X - j)(X - \bar{j}) = X^2 + X + 1$; $\Phi_4(X) = (X - i)(X + i) = X^2 + 1$.

Cyclotomic polynomials have some basic properties.

Proposition 6.1. *For all integers $n, q \geq 2$, $\Phi_n(q) \in \mathbb{R}$ and $\Phi_n(q) > q - 1$.*

Proof. The number $\Phi_n(q)$ is the product of $\phi(n)$ complex numbers $q - \varepsilon$, where $|\varepsilon| = 1$, $\varepsilon \neq 1$. Hence $|q - \varepsilon| > q - 1 \geq 1$ and $|\Phi_n(q)| > (q - 1)^{\phi(n)} \geq q - 1$. Moreover, $\Phi_n(q)$ is positive real, since the numbers $q - \varepsilon$ are conjugate in pairs or real. $\square$

Proposition 6.2. $X^n - 1 = \prod_{d|n} \Phi_d(X)$.

Proof. If ε has order d (if $\varepsilon^d = 1$ and $\varepsilon^k \neq 1$ for all $0 < k < d$), then d divides n and ε is a primitive dth root of unity. Classifying by order yields

$$X^n - 1 = \prod_{\varepsilon^n = 1}(X - \varepsilon) = \prod_{d|n}\left(\prod_{\varepsilon \text{ has order } d}(X - \varepsilon)\right) = \prod_{d|n} \Phi_d(X). \square$$

Since Φ_n has degree $\phi(n)$, Proposition 6.2 implies $n = \sum_{d|n} \phi(d)$, which is II.1.8. With 6.2, Φ_n can be computed recursively. For instance,

$$\Phi_6(X) = (X^6 - 1)/(X - 1)(X + 1)(X^2 + X + 1) = X^2 - X + 1.$$

Proposition 6.3. Φ_n *is monic and has integer coefficients.*

Proof. By induction. First, $\Phi_1(X) = X - 1$ is monic and has integer coefficients. If $n > 1$, then polynomial division in $\mathbb{Z}[X]$ of $X^n - 1 \in \mathbb{Z}[X]$ by the monic polynomial $\prod_{d \mid n,\, d < n} \Phi_d \in \mathbb{Z}[X]$ yields $\Phi_n(X)$, by 6.2. $\square$

Proposition 6.4. *For all $n > 0$, Φ_n is irreducible in $\mathbb{Q}[X]$.*

Proof. Assume that Φ_n is not irreducible in $\mathbb{Q}[X]$. Then $\Phi_n \in \mathbb{Z}[X]$ is not irreducible in $\mathbb{Z}[X]$ and $\Phi_n(X) = q(X)\,r(X)$ for some nonconstant $q, r \in \mathbb{Z}[X]$. We may assume that q is irreducible. Since Φ_n is monic, the leading coefficients of q and r are ± 1, and we may also assume that q and r are monic. The nonconstant polynomials q and r have complex roots ε and ζ, respectively, that are primitive nth roots of unity since they are also roots of Φ_n. Hence $\zeta = \varepsilon^k$ for some $k > 0$ (since ε is primitive) and k is relatively prime to n (since ζ is primitive). Choose ε and ζ so that k is as small as possible. Then $k > 1$: otherwise, $\zeta = \varepsilon$ is a multiple root of Φ_n.

Let p be a prime divisor of k. Then p does not divide n, ε^p is primitive, and $\Phi_n(\varepsilon^p) = 0$. If $q(\varepsilon^p) = 0$, then $\zeta = (\varepsilon^p)^{k/p}$ contradicts the choice of ε and ζ. Therefore $r(\varepsilon^p) = 0$. But $k \geq p$ is as small as possible, so $k = p$. Moreover, $q(X)$ divides $r(X^p)$ in $\mathbb{Q}[X]$, since $q = \mathrm{Irr}\,(\varepsilon : \mathbb{Q})$ and $r(\varepsilon^p) = 0$, so that $r(X^p) = q(X)\,s(X)$ for some $s \in \mathbb{Q}[X]$. Since q is monic, polynomial division in $\mathbb{Z}[X]$ yields $s \in \mathbb{Z}[X]$, so that q divides $r(X^p)$ in $\mathbb{Z}[X]$.

The projection $a \longmapsto \overline{a}$ of $\mathbb{Z}$ onto $\mathbb{Z}_p$ induces a homomorphism $f \longmapsto \overline{f}$ of $\mathbb{Z}[X]$ into $\mathbb{Z}_p[X]$: if $r(X) = r_m X^m + \cdots + r_0$, then $\overline{r}(X) = X^m + \overline{r}_{m-1} X^{m-1} + \cdots + \overline{r}_0$. By 1.3, $\overline{a}^p = \overline{a}$ for all $\overline{a} \in \mathbb{Z}_p$, so that

$$\overline{r}(X)^p = X^{mp} + \overline{r}_{m-1}^p X^{(m-1)p} + \cdots + \overline{r}_0^p = \overline{r}(X^p).$$

Hence $\overline{q}$ divides $\overline{r}^p$, and $\overline{q}$, $\overline{r}$ have a common irreducible divisor $\overline{t} \in \mathbb{Z}_p[X]$. Then $\overline{t}^2$ divides $\overline{q}\,\overline{r}$, which divides $\overline{f}(X) = X^n - \overline{1} \in \mathbb{Z}_p[X]$ since $qr = \Phi_n$ divides $X^n - 1$ by 6.2; hence $\overline{f}$ has a multiple root in $\overline{\mathbb{Z}}_p$. But $\overline{f}'(X) = nX^{n-1} \neq 0$, since p does not divide n, so that $\overline{f}$ and $\overline{f}'$ have no common root in $\overline{\mathbb{Z}}_p$, and $\overline{f}$ is separable. This is the required contradiction. $\square$

Definition. The nth cyclotomic field *is $\mathbb{Q}(\varepsilon_n) \subseteq \mathbb{C}$, where $\varepsilon_n \in \mathbb{C}$ is a primitive nth root of unity.*

Proposition 6.5. *The field $\mathbb{Q}(\varepsilon_n)$ is a Galois extension of $\mathbb{Q}$; $[\mathbb{Q}(\varepsilon_n) : \mathbb{Q}] = \phi(n)$; and $\mathrm{Gal}\,(\mathbb{Q}(\varepsilon_n) : \mathbb{Q})$ is isomorphic to the group of units U_n of $\mathbb{Z}_n$.*

Proof. First, $\mathbb{Q}(\varepsilon_n)$, which contains all complex nth roots of unity, is a splitting field of $X^n - 1$ and is Galois over $\mathbb{Q}$. Next, $\Phi_n(\varepsilon_n) = 0$, whence

$\Phi_n = \mathrm{Irr}\,(\varepsilon_n : \mathbb{Q})$, by 6.4; hence $[\,\mathbb{Q}(\varepsilon_n) : \mathbb{Q}\,] = \phi(n)$.

The group U_n consists of all $\overline{k} \in \mathbb{Z}_n$ such that k and n are relatively prime. Let $C = \langle\, \varepsilon_n \,\rangle$ be the multiplicative group of all complex nth roots of unity, which is cyclic of order n. An endomorphism of C sends ε_n to some ε_n^k, then sends ε_n^i to ε_n^{ki}, and is an automorphism if and only if k and n are relatively prime. Hence the group $\mathrm{Aut}\,(C)$ of automorphisms of C is isomorphic to U_n. Now, every $\sigma \in \mathrm{Gal}\,(\mathbb{Q}(\varepsilon_n) : \mathbb{Q})$ permutes the roots of $X^n - 1$ and induces an automorphism $\sigma_{|C}$ of C. This yields a homomorphism $\sigma \longmapsto \sigma_{|C}$ of $\mathrm{Gal}\,(\mathbb{Q}(\varepsilon_n) : \mathbb{Q})$ into $\mathrm{Aut}\,(C) \cong U_n$, which is injective, since $\mathbb{Q}(\varepsilon_n)$ is generated by ε, hence bijective, since $|\mathrm{Gal}\,(\mathbb{Q}(\varepsilon_n) : \mathbb{Q})| = \phi(n) = |U_n|$. $\square$

The exercises give additional properties of $\mathbb{Q}(\varepsilon_n)$.

Division rings. A *division ring* is a ring in which every nonzero element is a unit. Commutative division rings are fields. The quaternion algebra $\mathbb{H}$ in Section I.7 is a division ring but not a field. In Section VIII.5 we show that a finitely generated vector space over a division ring has a finite basis, and all its bases have the same number of elements; impatient readers may prove this now.

Theorem 6.6 (Wedderburn [1905]). *Every finite division ring is a field.*

Proof. Let D be a finite division ring. The center $K = \{\, x \in D \mid xy = yx$ for all $y \in D \,\}$ of D is a subfield of D. Let n be the dimension of D as a vector space over K. We prove that $n = 1$.

Let $|K| = q$, so that $|D| = q^n$. The center of $D \backslash \{0\}$ has $q - 1$ elements. The centralizer of $a \in D \backslash \{0\}$ is $L \backslash \{0\}$, where $L = \{\, x \in D \mid xa = ax \,\}$. Now, L is a subring of D, and a division ring, and L contains K. Hence D is a vector space over L and L is a vector space over K. Readers will verify that $\dim_K D = (\dim_L D)(\dim_K L)$, so that $d = \dim_K L$ divides n. Then $|L| = q^d$, the centralizer of a has $q^d - 1$ elements, and the conjugacy class of a has $(q^n - 1)/(q^d - 1)$ elements. Moreover, $q^d < q^n$ when $a \notin K$, for then $L \subsetneqq D$. Hence the class equation of the multiplicative group $D \backslash \{0\}$ reads

$$q^n - 1 = (q - 1) + \sum \frac{q^n - 1}{q^d - 1},$$

where the sum has one term for each nontrivial conjugacy class, in which $d < n$ and $d \mid n$. Now, $q^n - 1 = \prod_{d \mid n} \Phi_d(q)$, by 6.2. If $d < n$ and $d \mid n$, then

$$
\begin{aligned}
q^n - 1 &= \Phi_n(q) \prod_{c \mid n,\, c < n} \Phi_c(q) \\
&= \Phi_n(q) \prod_{c \mid d} \Phi_c(q) \prod_{c \mid n,\, c < n,\, c \nmid d} \Phi_c(q) \\
&= \Phi_n(q) \,(q^d - 1) \prod_{c \mid n,\, c < n,\, c \nmid d} \Phi_c(q),
\end{aligned}
$$

and $\Phi_n(q)$ divides $q^n - 1$ and $(q^n - 1)/(q^d - 1)$. Therefore $\Phi_n(q)$ divides $q - 1$. But $\Phi_n(q) > q - 1$ when $n > 1$, by 6.1. Hence $n = 1$. $\square$

Dirichlet's theorem. The general form of this theorem, due to Dirichlet [1837], states that every arithmetic progression contains infinitely many primes. We use cyclotomic polynomials to prove a particular case:

*Theorem **6.7*** (Dirichlet). *For every positive integer n there are infinitely many prime numbers $p \equiv 1 \pmod n$.*

Proof. We start with a lemma.

*Lemma **6.8.** Let p be prime and $m, n > 0$. If p divides $\Phi_n(m)$, then p does not divide m, and either p divides n or $p \equiv 1 \pmod n$.*

Proof. By 6.2, $\Phi_n(m)$ divides $m^n - 1$; hence p divides $m^n - 1$, and does not divide m. Let k be the order of $\overline{m}$ in the multiplicative group $\mathbb{Z}_p \backslash \{0\}$; k divides $|\mathbb{Z}_p \backslash \{0\}| = p - 1$, and divides n, since $\overline{m}^n = \overline{1}$ in $\mathbb{Z}_p$. Let $\ell = n/k$.

If $\ell = 1$, then $n = k$ divides $p - 1$ and $p \equiv 1 \pmod n$.

Let $\ell > 1$. Since every divisor of k is a divisor of n, $\prod_{d|n,\, d<n} \Phi_d(X) = f(X) \prod_{d|k} \Phi_d(X)$, where f is a product of cyclotomic polynomials. Hence

$$m^{k\ell} - 1 = \Phi_n(m) \prod_{d|n,\, d<n} \Phi_d(m) = \Phi_n(m)\,(m^k - 1)\, f(m)$$

by 6.2. Therefore p, which divides $\Phi_n(m)$, divides

$$(m^{k\ell} - 1)/(m^k - 1) = (m^k)^{\ell-1} + (m^k)^{\ell-2} + \cdots + 1.$$

Now, $m^k \equiv 1 \pmod p$; hence p divides ℓ and $n = k\ell$. $\square$

We now prove 6.7. We may assume $n > 1$. For every $k \geq 2$ we have $\Phi_{kn}(kn) > kn - 1 > 1$ by 6.1, and $\Phi_{kn}(kn)$ has a prime divisor p. By 6.8, p does not divide kn; hence $p \equiv 1 \pmod{kn}$ and $p > kn$. Thus there are arbitrarily large primes $p \equiv 1 \pmod n$. $\square$

The proof of 6.7 shows algebra coming to the aid of number theory. Number theory now comes to the aid of algebra.

*Proposition **6.9.*** *Every finite abelian group is the Galois group of a finite extension of $\mathbb{Q}$.*

Proof. A finite abelian group G is a direct sum $G = C_{n_1} \oplus C_{n_2} \oplus \cdots \oplus C_{n_r}$ of cyclic groups of orders $n_1, \ldots, n_r$. By 6.7 there exist distinct primes $p_1, \ldots, p_r$ such that $p_i \equiv 1 \pmod{n_1 n_2 \cdots n_r}$ for all i. Let $n = p_1 p_2 \cdots p_r$.

By 6.5, $\mathrm{Gal}\,(\mathbb{Q}(\varepsilon_n) : \mathbb{Q}) \cong U_n$. If k and ℓ are relatively prime, then $\mathbb{Z}_{k\ell} \cong \mathbb{Z}_k \oplus \mathbb{Z}_\ell$; since $(u, v) \in \mathbb{Z}_k \times \mathbb{Z}_\ell$ is a unit if and only if u and v are units, then $U_{k\ell} \cong U_k \oplus U_\ell$. Therefore $\mathrm{Gal}\,(\mathbb{Q}(\varepsilon_n) : \mathbb{Q}) \cong U_n \cong U_{p_1} \oplus U_{p_2} \oplus \cdots \oplus U_{p_r}$. Now, U_{p_i} is cyclic of order $p_i - 1$, since $\mathbb{Z}_{p_i}$ is a field, and n_i divides $p_i - 1$; hence U_{p_i} has a subgroup H_i of index n_i. Then $U_{p_i}/H_i \cong C_{n_i}$ and $U_{p_1} \oplus U_{p_2} \oplus \cdots \oplus U_{p_r}$ has a subgroup $H_1 \oplus H_2 \oplus \cdots \oplus H_r$ such that $(U_{p_1} \oplus U_{p_2} \oplus \cdots \oplus U_{p_r})/(H_1 \oplus H_2 \oplus \cdots \oplus H_r) \cong G$. Therefore

$\mathrm{Gal}\,(\mathbb{Q}(\varepsilon_n):\mathbb{Q})$ has a subgroup H such that $\mathrm{Gal}\,(\mathbb{Q}(\varepsilon_n):\mathbb{Q})/H \cong G$; then the fixed field F of H is a finite Galois extension of $\mathbb{Q}$ and $G \cong \mathrm{Gal}\,(F:\mathbb{Q})$. $\square$

Exercises

1. Find Φ_n for all $n \leqq 10$.

2. Find Φ_{12} and Φ_{18}.

3. Show that $\Phi_n(0) = \pm 1$, and that $\Phi_n(0) = 1$ if $n > 1$ is odd.

4. Show that $\Phi_{2n}(X) = \Phi_n(-X)$ when $n > 1$ is odd.

Readers who have polynomial division software and long winter evenings can now formulate and disprove a conjecture that all coefficients of Φ_n are 0 or ± 1.

5. Let p be prime. Show that $\Phi_{np}(X) = \Phi_n(X^p)$ if p divides n, $\Phi_{np}(X) = \Phi_n(X^p)/\Phi_n(X)$ if p does not divide n.

6. Let n be divisible by p^2 for some prime p. Show that the sum of all complex primitive nth roots of unity is 0.

7. Show that $\mathbb{Q}(\varepsilon_m)\,\mathbb{Q}(\varepsilon_n) = \mathbb{Q}(\varepsilon_{\mathrm{lcm}\,(m,n)})$.

8. Show that $\mathbb{Q}(\varepsilon_m) \cap \mathbb{Q}(\varepsilon_n) = \mathbb{Q}(\varepsilon_{\gcd\,(m,n)})$. (You may want to use 3.11.)

9. Find the least $n > 0$ such that $\mathrm{Gal}\,(\mathbb{Q}(\varepsilon_n):\mathbb{Q})$ is not cyclic.

10. Let $D \subseteq E \subseteq F$ be division rings, each a subring of the next. Show that $\dim_K D = (\dim_L D)(\dim_K L)$.

*11. Prove that a finitely generated vector space over a division ring has a finite basis, and that all its bases have the same number of elements.

7. Norm and Trace

The norm and trace are functions defined on every finite field extension. In this section we establish their basic properties and use the results to construct all Galois extensions with cyclic Galois groups.

Definition. Recall that a linear transformation T of a finite-dimensional vector space V has a determinant and a trace, which are the determinant and trace (sum of all diagonal entries) of the matrix of T in any basis of V. If

$$c(X) = \det\,(T - XI) = (-1)^n X^n + (-1)^{n-1} c_{n-1} X^{n-1} + \cdots + c_0$$

is the characteristic polynomial of T, then the determinant of T is c_0 and its trace is c_{n-1}. In particular, the determinant and trace of the matrix of T in a basis of V do not depend on the choice of a basis.

A finite extension E of a field K is a finite-dimensional vector space over K, and multiplication by $\alpha \in E$ is a linear transformation $\gamma \longmapsto \alpha\gamma$ of E.

Definitions. Let E be a finite extension of a field K. The norm $\mathrm{N}_K^E(\alpha)$ and trace $\mathrm{Tr}_K^E(\alpha)$ of $\alpha \in E$ over K are the determinant and trace of the linear transformation $T_\alpha : \gamma \longmapsto \alpha\gamma$ of E.

Both $\mathrm{N}_K^E(\alpha)$ and $\mathrm{Tr}_K^E(\alpha)$ are elements of K. When $K \subseteq E$ is the only extension in sight we denote $\mathrm{N}_K^E(\alpha)$ and $\mathrm{Tr}_K^E(\alpha)$ by $\mathrm{N}(\alpha)$ and $\mathrm{Tr}\,(\alpha)$.

For example, in the finite extension $\mathbb{C}$ of $\mathbb{R}$, multiplication by $z = a + bi$ is a linear transformation $x + iy \longmapsto (ax - by) + i(bx + ay)$ with matrix $\begin{pmatrix} a & -b \\ b & a \end{pmatrix}$ in the basis $\{\,1,\, i\,\}$; hence z has norm $a^2 + b^2 = z\bar{z}$ and trace $2a = z + \bar{z}$.

In general, the norm and trace of $\alpha \in E$ can also be computed from the K-homomorphisms of E into $\overline{K}$, and from the conjugates of α. First we show:

Lemma 7.1. *If E is finite over K and $\alpha \in E$, then* $\det\,(T_\alpha - XI) = (-1)^n\, q(X)^\ell$, *where* $n = [\,E : K\,]$, $q = \mathrm{Irr}\,(\alpha : K)$, *and* $\ell = [\,E : K(\alpha)\,]$.

Proof. We have $T_{a\beta} = aT_\beta$, $T_{\beta+\gamma} = T_\beta + T_\gamma$, and $T_{\beta\gamma} = T_\beta T_\gamma$, for all $a \in K$ and $\beta, \gamma \in E$. Hence $f(T_\alpha) = T_{f(\alpha)}$ for every $f \in K[X]$. In particular, $q(T_\alpha) = T_{q(\alpha)} = 0$. (Thus, q is the minimal polynomial of T_α.)

Choose a basis of E over K. The matrix M of T_α in the chosen basis can be viewed as a matrix with coefficients in $\overline{K}$, and the characteristic polynomial $c(X) = \det\,(T_\alpha - XI)$ of T_α is also the characteristic polynomial of M. In $K[X]$, c is the product of its leading coefficient $(-1)^n$ and monic irreducible polynomials $r_1, \ldots, r_\ell \in K[X]$, for some ℓ. If $\lambda \in \overline{K}$ is a root of r_j, then $c(\lambda) = 0$, λ is an eigenvalue of M, $Mv = \lambda v$ for some $v \neq 0$, $M^i v = \lambda^i v$, $f(M)v = f(\lambda)v$ for every $f \in K[X]$, $q(\lambda)v = q(M)v = 0$, and $q(\lambda) = 0$. Therefore $r_j = \mathrm{Irr}\,(\lambda : K) = q$. Hence $c = (-1)^n\, q^\ell$. Then $\ell = \deg c / \deg q = [\,E : K\,]/[\,K(\alpha) : K\,] = [\,E : K(\alpha)\,]$. $\square$

Proposition 7.2. *Let E be a finite extension of K of degree n. Let $\alpha_1, \ldots, \alpha_r \in \overline{K}$ be the distinct conjugates of $\alpha \in E$, and let $\varphi_1, \ldots, \varphi_t$ be the distinct K-homomorphisms of E into $\overline{K}$. Then r and t divide n and*

$$\mathrm{N}_K^E(\alpha) = (\alpha_1 \cdots \alpha_r)^{n/r} = ((\varphi_1\alpha) \cdots (\varphi_t\alpha))^{n/t} \in K,$$
$$\mathrm{Tr}_K^E(\alpha) = \frac{n}{r}\,(\alpha_1 + \cdots + \alpha_r) = \frac{n}{t}\,(\varphi_1\alpha + \cdots + \varphi_t\alpha) \in K.$$

The norm and trace are often defined by these formulas.

Proof. The conjugates of α are the roots of $q = \mathrm{Irr}\,(\alpha : K)$, which by IV.5.1 all have the same multiplicity m. Hence

$$q(X) = (X - \alpha_1)^m \cdots (X - \alpha_r)^m$$
$$= X^{rm} - m\,(\alpha_1 + \cdots + \alpha_r)\,X^{rm-1} + \cdots + (-1)^{rm}\,(\alpha_1 \cdots \alpha_r)^m.$$

Then $[\,K(\alpha) : K\,] = rm$ divides n and $\ell = [\,E : K(\alpha)\,] = n/rm$. By 7.1, $c(X) = \det\,(T_\alpha - XI) = (-1)^n\, q(X)^\ell$. The constant coefficient of c is

$$\mathrm{N}(\alpha) = (-1)^n\,(-1)^{rm\ell}\,(\alpha_1 \cdots \alpha_r)^{m\ell} = (\alpha_1 \cdots \alpha_r)^{n/r},$$

since $rm\ell = n$. The trace of α is $(-1)^{n-1}$ times the coefficient of X^{n-1} in c:

$$\mathrm{Tr}\,(\alpha) = (-1)^{n-1}\,(-1)^n\,(-\ell)\,m\,(\alpha_1 + \cdots + \alpha_r) = \frac{n}{r}\,(\alpha_1 + \cdots + \alpha_r).$$

Next, $t = [\,E : K\,]_s$, which divides n by IV.5.2. Since q has r distinct roots in $\overline{K}$, there are r K-homomorphisms of $K(\alpha)$ into $\overline{K}$, that send α to $\alpha_1, \ldots, \alpha_r$. Each can be extended to E in $k = [\,E : K(\alpha)\,]_s$ ways. Hence $t = kr$; $(\varphi_1 \alpha)\cdots(\varphi_t \alpha) = (\alpha_1 \cdots \alpha_r)^k$; and $\varphi_1 \alpha + \cdots + \varphi_t \alpha = k\,(\alpha_1 + \cdots + \alpha_r)$. This completes the proof since $(n/r)/k = n/t$. $\square$

Proposition 7.2 becomes simpler in some cases, as readers will easily verify.

Corollary **7.3.** *Let E be finite over K and let $\alpha \in E$.*

(1) *If $\alpha \in K$, then $\mathrm{N}_K^E(\alpha) = \alpha^n$ and $\mathrm{Tr}_K^E(\alpha) = n\alpha$, where $n = [\,E : K\,]$;*

(2) *if $E = K(\alpha)$ is separable over K, then $\mathrm{N}_K^E(\alpha)$ is the product of the conjugates of α, and $\mathrm{Tr}_K^E(\alpha)$ is their sum;*

(3) *if E is not separable over K, then $\mathrm{Tr}_K^E(\alpha) = 0$;*

(4) *if E is Galois over K, with Galois group G, then $\mathrm{N}_K^E(\alpha) = \prod_{\sigma \in G} \sigma\alpha$ and $\mathrm{Tr}_K^E(\alpha) = \sum_{\sigma \in G} \sigma\alpha$.*

Properties.

Proposition **7.4.** *If E is finite over K, then $\mathrm{N}_K^E(\alpha\beta) = \mathrm{N}_K^E(\alpha)\,\mathrm{N}_K^E(\beta)$ and $\mathrm{Tr}_K^E(\alpha + \beta) = \mathrm{Tr}_K^E(\alpha) + \mathrm{Tr}_K^E(\beta)$, for all $\alpha, \beta \in E$.*

Proof. In Proposition 7.2, $\varphi_1, \ldots, \varphi_t$ arc homomorphisms. $\square$

Proposition **7.5** (Tower Property). *If $K \subseteq E \subseteq F$ are finite over K, then $\mathrm{N}_K^F(\alpha) = \mathrm{N}_K^E(\mathrm{N}_E^F(\alpha))$ and $\mathrm{Tr}_K^F(\alpha) = \mathrm{Tr}_K^E(\mathrm{Tr}_E^F(\alpha))$, for all $\alpha \in E$.*

Proof. We may assume that $F \subseteq \overline{K}$ and choose $\overline{E} = \overline{K}$. Let $m = [\,E : K\,]$ and $n = [\,F : E\,]$, let $\varphi_1, \ldots, \varphi_t$ be the distinct K-homomorphisms of E into $\overline{K}$, and let $\psi_1, \ldots, \psi_u$ be the distinct E-homomorphisms of F into $\overline{E} = \overline{K}$. As in the proof that $[\,F : K\,]_s = [\,E : K\,]_s\,[\,F : E\,]_s$, let $\sigma_1, \ldots, \sigma_t$ be K-automorphisms of $\overline{K}$ that extend $\varphi_1, \ldots, \varphi_t$. If $\chi : F \longrightarrow \overline{K}$ is a K-homomorphism, then $\chi_{|E} = \varphi_i$ for some i, $\sigma_i^{-1} \chi : F \longrightarrow \overline{K}$ is an E-homomorphism, $\sigma_i^{-1} \chi = \psi_j$ for some j, and $\chi = \sigma_i \psi_j$. Thus the K-homomorphisms of F into $\overline{K}$ are the tu distinct maps $\sigma_i \psi_j$. We now use 7.2: since $\mathrm{N}_E^F(\alpha) \in E$,

$$\mathrm{N}_K^F(\alpha) = \Big(\prod_{i,j} \sigma_i \psi_j \alpha\Big)^{mn/tu} = \Big(\prod_i \sigma_i \big(\prod_j \psi_j \alpha\big)^{n/u}\Big)^{m/t}$$

$$= \Big(\prod_i \sigma_i\, \mathrm{N}_E^F(\alpha)\Big)^{m/t} = \Big(\prod_i \varphi_i\, \mathrm{N}_E^F(\alpha)\Big)^{m/t} = \mathrm{N}_K^E(\mathrm{N}_E^F(\alpha)),$$

and similarly for $\mathrm{Tr}_K^F(\alpha)$. $\square$

Hilbert's Theorem 90 is the key that opens the door behind which lie cyclic extensions. Like many good theorems, it requires a lemma.

Lemma **7.6.** *Let E and F be field extensions of K. Distinct K-homomorphisms of E into F are linearly independent over F.*

Proof. Assume that there is an equality $\gamma_1\varphi_1 + \cdots + \gamma_n\varphi_n = 0$, in which $n > 0$, $\gamma_1, \ldots, \gamma_n \in F$ are not all 0, and $\varphi_1, \ldots, \varphi_n$ are distinct K-homomorphisms of E into F. Among all such equalities there is one in which n is as small as possible. Then $\gamma_i \neq 0$ for all i and $n \geq 2$. Since $\varphi_n \neq \varphi_1$ we have $\varphi_n\alpha \neq \varphi_1\alpha$ for some $\alpha \in E$. Then

$$\gamma_1\,(\varphi_1\alpha)(\varphi_1\beta) + \cdots + \gamma_n\,(\varphi_n\alpha)(\varphi_n\beta) \;=\; \gamma_1\,\varphi_1(\alpha\beta) + \cdots + \gamma_n\varphi_n(\alpha\beta) = 0 \;\text{ and }$$

$$\gamma_1\,(\varphi_n\alpha)(\varphi_1\beta) + \cdots + \gamma_n\,(\varphi_n\alpha)(\varphi_n\beta) \;=\; 0$$

for all $\beta \in E$. Subtracting the second sum from the first yields

$$\gamma_1\,(\varphi_1\alpha - \varphi_n\alpha)(\varphi_1\beta) + \cdots + \gamma_{n-1}\,(\varphi_{n-1}\alpha - \varphi_n\alpha)(\varphi_{n-1}\beta) \;=\; 0$$

for all $\beta \in E$ and a shorter equality

$$\gamma_1\,(\varphi_1\alpha - \varphi_n\alpha)\,\varphi_1 + \cdots + \gamma_{n-1}\,(\varphi_{n-1}\alpha - \varphi_n\alpha)\,\varphi_{n-1} \;=\; 0$$

with a nonzero coefficient $\gamma_1\,(\varphi_1\alpha - \varphi_n\alpha)$. This is the required contradiction. $\square$

Lemma **7.7** (Hilbert's Theorem 90 [1897]). *Let E be a finite Galois extension of K. If $\mathrm{Gal}\,(E:K)$ is cyclic, $\mathrm{Gal}\,(E:K) = \langle\,\tau\,\rangle$, then, for any $\alpha \in E$:*

(1) $\mathrm{N}_K^E(\alpha) = 1$ *if and only if* $\alpha = \tau\gamma/\gamma$ *for some* $\gamma \in E$, $\gamma \neq 0$.

(2) $\mathrm{Tr}_K^E(\alpha) = 0$ *if and only if* $\alpha = \tau\gamma - \gamma$ *for some* $\gamma \in E$.

Proof. If $\gamma \in E$, $\gamma \neq 0$, then

$$\mathrm{N}(\tau\gamma) \;=\; \prod_{\sigma \in G}\sigma\tau\gamma \;=\; \prod_{\sigma \in G}\sigma\gamma \;=\; \mathrm{N}(\gamma)$$

by 7.3, where $G = \mathrm{Gal}\,(E:K)$; hence $\mathrm{N}(\tau\gamma/\gamma) = 1$, by 7.4.

Conversely, assume that $\mathrm{N}(\alpha) = 1$. Let $[\,E:K\,] = n$. Then $\mathrm{Gal}\,(E:K) = \{\,1,\ \tau,\ \ldots,\ \tau^{n-1}\,\}$. By 7.6, $1, \tau, \ldots, \tau^{n-1}$ are linearly independent over K; therefore $1 + \alpha\tau + \alpha\,(\tau\alpha)\,\tau^2 + \cdots + \alpha\,(\tau\alpha) \cdots (\tau^{n-2}\alpha)\,\tau^{n-1} \neq 0$ and

$$\delta \;=\; \beta + \alpha\tau\beta + \alpha\,(\tau\alpha)\,(\tau^2\beta) + \cdots + \alpha\,(\tau\alpha)\cdots(\tau^{n-2}\alpha)\,(\tau^{n-1}\beta) \;\neq\; 0$$

for some $\beta \in E$. If $\mathrm{N}(\alpha) = \alpha\,(\tau\alpha)\cdots(\tau^{n-2}\alpha)\,(\tau^{n-1}\alpha) = 1$, then

$$\alpha\,(\tau\delta) \;=\; \alpha\tau\beta + \alpha\,(\tau\alpha)\,(\tau^2\beta) + \cdots + \alpha\,(\tau\alpha)\cdots(\tau^{n-1}\alpha)\,(\tau^n\beta) \;=\; \delta,$$

since $\tau^n = 1$; hence $\alpha = \tau\gamma/\gamma$, where $\gamma = \delta^{-1}$.

Similarly, if $\gamma \in E$, then

$$\mathrm{Tr}\,(\tau\gamma) \;=\; \sum_{\sigma \in G}\sigma\tau\gamma \;=\; \sum_{\sigma \in G}\sigma\gamma \;=\; \mathrm{Tr}\,(\gamma)$$

by 7.3, where $G = \mathrm{Gal}\,(E:K)$; hence $\mathrm{Tr}\,(\tau\gamma - \gamma) = 0$, by 7.4.

Conversely, assume that $\mathrm{Tr}\,(\alpha) = 0$. Since $1, \tau, \ldots, \tau^{n-1}$ are linearly independent over K, we have $1 + \tau + \cdots + \tau^{n-1} \neq 0$ and $\mathrm{Tr}\,(\beta) = \beta + \tau\beta + \cdots +$

$\tau^{n-1}\beta \neq 0$ for some $\beta \in E$. Let

$$\delta = \alpha\tau\beta + (\alpha + \tau\alpha)(\tau^2\beta) + \cdots + (\alpha + \tau\alpha + \cdots + \tau^{n-2}\alpha)(\tau^{n-1}\beta).$$

If $\text{Tr}(\alpha) = \alpha + \tau\alpha + \cdots + \tau^{n-1}\alpha = 0$, then

$$\begin{aligned}
\tau\delta = {} & (\tau\alpha)(\tau^2\beta) + (\tau\alpha + \tau^2\alpha)(\tau^3\beta) \\
& + \cdots + (\tau\alpha + \tau^2\alpha + \cdots + \tau^{n-2}\alpha)(\tau^{n-1}\beta) - \alpha\beta.
\end{aligned}$$

Hence $\delta - \tau\delta = \alpha\tau\beta + \alpha\tau^2\beta + \cdots + \alpha\tau^{n-1}\beta + \alpha\beta = \alpha\,\text{Tr}(\beta)$ and $\alpha = \tau\gamma - \gamma$, where $\gamma = -\delta/\text{Tr}(\beta)$. $\square$

Cyclic extensions are extensions with finite cyclic Galois groups. These extensions generally arise through the adjunction of an nth root.

Definition. A cyclic extension is a finite Galois extension whose Galois group is cyclic.

Proposition 7.8. *Let $n > 0$. Let K be a field whose characteristic is either 0 or not a divisor of n, and that contains a primitive nth root of unity.*

If E is a cyclic extension of K of degree n, then $E = K(\alpha)$, where $\alpha^n \in K$.

If $E = K(\alpha)$, where $\alpha^n \in K$, then E is a cyclic extension of K, $m = [E:K]$ divides n, and $\alpha^m \in K$.

Proof. By the hypothesis, K contains a primitive nth root of unity $\varepsilon \in \overline{K}$.

Let E be cyclic over K of degree n and $\text{Gal}(E:K) = \langle \tau \rangle$. Since $\text{N}(\varepsilon) = \varepsilon^n = 1$ we have $\tau\alpha = \varepsilon\alpha$ for some $\alpha \in E$, $\alpha \neq 0$, by 7.7. Then $\tau(\alpha^n) = (\tau\alpha)^n = \alpha^n$; hence $\sigma(\alpha^n) = \alpha^n$ for all $\sigma \in \text{Gal}(E:K)$ and $\alpha^n \in K$. Since α has n conjugates α, $\tau\alpha = \varepsilon\alpha$, ..., $\tau^{n-1}\alpha = \varepsilon^{n-1}\alpha$, there are n K-homomorphisms of $K(\alpha)$ into $\overline{K}$, $[K(\alpha):K] = [E:K]$, and $K(\alpha) = E$.

Now let $E = K(\alpha)$, where $\alpha^n = c \in K$. We may assume that $E \subseteq \overline{K}$ and that $\alpha \notin K$. In $\overline{K}$, the roots of $X^n - c \in K[X]$ are α, $\varepsilon\alpha$, ..., $\varepsilon^{n-1}\alpha$. Hence $X^n - c$ is separable, its splitting field is E, and E is Galois over K.

If $\sigma \in \text{Gal}(E:K)$, then $\sigma\alpha$ is a root of $X^n - c$ and $\sigma\alpha = \varepsilon^i\alpha$ for some i. This provides a homomorphism of $\text{Gal}(E:K)$ into the multiplicative group of all nth roots of unity, which is injective since α generates E. The latter group is cyclic of order n; hence $\text{Gal}(E:K)$ is cyclic and its order m divides n. Let $\text{Gal}(E:K) = \langle \tau \rangle$ and $\tau\alpha = \varepsilon^j\alpha$; then ε^j has order m, $\tau(\alpha^m) = (\tau\alpha)^m = \alpha^m$, $\sigma\alpha^m = \alpha^m$ for all $\sigma \in \text{Gal}(E:K)$, and $\alpha^m \in K$. $\square$

Primitive nth roots of unity are readily adjoined if needed in 7.8:

Proposition 7.9. *A root of unity $\varepsilon \in \overline{K}$ is a primitive nth root of unity for some $n > 0$; if K has characteristic $p \neq 0$, then p does not divide n; $K(\varepsilon)$ is a Galois extension of K of degree at most n; and $\text{Gal}(K(\varepsilon):K)$ is abelian.*

The proof is an enjoyable exercise. In 7.9, it may happen that $[K(\varepsilon):K] < n$, and that $\text{Gal}(K(\varepsilon):K)$ is not cyclic; this makes more fine exercises.

If K has characteristic $p \neq 0$, then the identity $(\alpha - \beta)^p = \alpha^p - \beta^p$ shows that pth roots are unique, but not separable, and are quite incapable of generating cyclic extensions of K of degree p. Hence 7.8 fails in this case. To obtain cyclic extensions of degree p we must replace $X^p - c$ by another polynomial:

Proposition **7.10** (Artin-Schreier). *Let K be a field of characteristic $p \neq 0$.*

If E is a cyclic extension of K of degree p, then $E = K(\alpha)$, where $\alpha^p - \alpha \in K$.

If $E = K(\alpha)$, where $\alpha^p - \alpha \in K$, $\alpha \notin K$, then E is a cyclic extension of K of degree p.

Proof. Let E be cyclic over K of degree p and $\mathrm{Gal}(E:K) = \langle \tau \rangle$. Since $\mathrm{Tr}(1) = p1 = 0$ we have $\tau\alpha - \alpha = 1$ for some $\alpha \in E$, by 7.7. Then $\tau^i \alpha = \alpha + i$ for all i; hence α has p conjugates $\tau^i \alpha = \alpha + i$, $0 \leq i < p$; there are p K-homomorphisms of $K(\alpha)$ into $\overline{K}$; $[K(\alpha):K] = [E:K]$; and $K(\alpha) = E$.

Now let $E = K(\alpha)$, where $c = \alpha^p - \alpha \in K$ and $\alpha \notin K$. We may assume that $E \subseteq \overline{K}$. Since K has characteristic p, $(\alpha + 1)^p - (\alpha + 1) = \alpha^p - \alpha = c$; therefore the roots of $X^p - X - c \in K[X]$ are α, $\alpha + 1$, ..., $\alpha + p - 1$. Hence $X^p - X - c$ is separable, its splitting field is E, and E is Galois over K. Moreover, $\mathrm{Irr}(\alpha:K)$ divides $X^p - X - c$; hence $[E:K] \leq p$. We have $\tau\alpha = \alpha + 1$ for some $\tau \in \mathrm{Gal}(E:K)$; then $\tau^i\alpha = \alpha + i \neq \alpha$ for all $i = 1, 2, ..., p - 1$, $\tau^p\alpha = \alpha$, and τ has order p in $\mathrm{Gal}(E:K)$. Therefore $\mathrm{Gal}(E:K) = \langle \tau \rangle$, $\mathrm{Gal}(E:K)$ has order p, and $[E:K] = p$. $\square$

Exercises

1. Show that $\mathrm{Tr}_K^E(\alpha) = 0$ for all $\alpha \in E$ when E is not separable over K.

2. Show that $\mathrm{Tr}_K^E(\alpha) \neq 0$ for some $\alpha \in E$ when E is separable over K.

3. Find $\mathrm{N}_{\mathbb{Q}}^E(\alpha)$ and $\mathrm{Tr}_{\mathbb{Q}}^E(\alpha)$ when $\alpha \in E = \mathbb{Q}(\sqrt{n}) \subseteq \mathbb{R}$, where $n > 0$.

4. Find $\mathrm{N}_{\mathbb{Q}}^E(\alpha)$ and $\mathrm{Tr}_{\mathbb{Q}}^E(\alpha)$ when $\alpha \in E = \mathbb{Q}(i\sqrt{n}) \subseteq \mathbb{C}$, where $n > 0$.

5. Find the units of $\mathbb{Q}[i\sqrt{n}] \subseteq \mathbb{C}$, where n is a positive integer.

6. Let $\alpha = \sqrt{2} + \sqrt{3}$ and $E = \mathbb{Q}(\alpha) \subseteq \mathbb{R}$. Find $\mathrm{N}_{\mathbb{Q}}^E(\alpha)$ and $\mathrm{Tr}_{\mathbb{Q}}^E(\alpha)$.

7. Let $\alpha = \sqrt{2} + i\sqrt{3}$ and $E = \mathbb{Q}(\alpha) \subseteq \mathbb{C}$. Find $\mathrm{N}_{\mathbb{Q}}^E(\alpha)$ and $\mathrm{Tr}_{\mathbb{Q}}^E(\alpha)$.

8. Show that a root of unity is a primitive nth root of unity for some $n > 0$, where p does not divide n if the characteristic is $p \neq 0$.

9. Show that $K(\varepsilon)$ is a Galois extension of K of degree at most n when $\varepsilon \in \overline{K}$ is a primitive nth root of unity.

10. Show that $\mathrm{Gal}(K(\varepsilon):K)$ is abelian when $\varepsilon \in \overline{K}$ is a root of unity.

11. Show that $[K(\varepsilon):K] < n$ may happen when $\varepsilon \in \overline{K}$ is a primitive nth root of unity. (The author buried an example somewhere in this book but lost the map.)

12. Show that $\mathrm{Gal}(K(\varepsilon):K)$ need not be cyclic when $\varepsilon \in \overline{K}$ is a primitive nth root of unity.

8. Solvability by Radicals

"Radical" is a generic name for nth roots. A polynomial equation is solvable by radicals when its solutions can be reached by successive adjunctions of nth roots (for various n). We saw in Section 5 that equations of degree at most 4 generally have this property. This section gives a more refined definition of solvability by radicals, and relates it to the solvability of Galois groups. (In fact, solvable groups are named for this relationship.) The main application is Abel's theorem that general equations of degree 5 or more are not solvable by radicals.

Solvability. By 7.8, extensions generated by an nth root coincide with cyclic extensions, except in characteristic $p \neq 0$, where Proposition 7.10 shows that roots of polynomials $X^p - X - c$ are a better choice than pth roots (roots of $X^p - c$). Accordingly, we formally define radicals as follows.

Definitions. An element α of a field extension of K is radical *over K when either $\alpha^n \in K$ for some $n > 0$ and the characteristic of K does not divide n, or $\alpha^p - \alpha \in K$ where $p \neq 0$ is the characteristic of K. A* radical extension *of K is a simple extension $E = K(\alpha)$, where α is radical over K.*

Definitions. A field extension $K \subseteq E$ is solvable by radicals, *and E is* solvable by radicals *over K, when there exists a tower of radical extensions $K = F_0 \subseteq F_1 \subseteq \cdots \subseteq F_r$ such that $E \subseteq F_r$. A polynomial is* solvable by radicals *over K when its splitting field is solvable by radicals over K.*

Thus, E is solvable by radicals over K when every element of E "can be reached from K by successive adjunctions of radicals"; a polynomial is solvable by radicals over K when its roots have this property. More precisely, in a tower $K = F_0 \subseteq F_1 \subseteq \cdots \subseteq F_r$ of radical extensions, the elements of F_1 are polynomial functions with coefficients in K of some $\alpha \in F_1$ that is radical over K; the elements of F_2 are polynomial functions with coefficients in F_1 of some $\beta \in F_2$ that is radical over F_1; and so forth. We saw in Section 5 that the roots of a polynomial of degree at most 4 can be written in this form, except perhaps when K has characteristic 2 or 3; then polynomials of degree at most 4 are solvable by radicals.

Readers will gain familiarity with these radical new concepts by proving their basic properties:

Proposition 8.1. *If F is solvable by radicals over K and $K \subseteq E \subseteq F$, then E is solvable by radicals over K and F is solvable by radicals over E.*

If $K \subseteq E \subseteq F$, E is solvable by radicals over K, and F is solvable by radicals over E, then F is solvable by radicals over K.

If E is radical over K and the composite EF exists, then EF is radical over KF.

If E is solvable by radicals over K and the composite EF exists, then EF is solvable by radicals over KF.

The main result of this section is the following:

Theorem **8.2.** *An extension of a field K is solvable by radicals if and only if it is contained in a finite Galois extension of K whose Galois group is solvable.*

Proof. Let E be solvable by radicals over K, so that $E \subseteq F_r$ for some tower $K = F_0 \subseteq F_1 \subseteq \cdots \subseteq F_r$ of radical extensions $F_i = F_{i-1}(\alpha_i)$, where α_i is radical over F_{i-1}: either $\alpha_i^{n_i} \in F_{i-1}$ for some $n_i > 0$ and the characteristic of K does not divide n_i, or $\alpha_i^p - \alpha_i \in F_{i-1}$, where $p \neq 0$ is the characteristic of K and we let $n_i = p$. We may assume that $F_r \subseteq \overline{K}$. We construct a better tower. First we adjoin to K a carefully chosen root of unity, so that we can use 7.8; then we adjoin conjugates of $\alpha_1, \ldots, \alpha_n$ to obtain a normal extension.

Let $m = n_1 n_2 \cdots n_r$ if K has characteristic 0; if K has characteristic $p \neq 0$, let $n_1 n_2 \cdots n_r = p^t m$, where p does not divide m. In either case, if the characteristic of K does not divide n_i, then n_i divides m. Let $\varepsilon \in \overline{K}$ be a primitive mth root of unity. Then $K(\varepsilon)$ contains a primitive ℓth root of unity for every divisor ℓ of m, namely $\varepsilon^{m/\ell}$. The composite $K(\varepsilon) F_r$ is a finite extension of K and is contained in a finite normal extension N of K; N is the composite of all conjugates of $K(\varepsilon) F_r = K(\varepsilon, \alpha_1, \ldots, \alpha_r)$ and is generated over K by all conjugates of $\varepsilon, \alpha_1, \ldots, \alpha_r$. Let $\varphi_0, \varphi_1, \ldots, \varphi_{n-1}$ be the K-homomorphisms of F_r into $\overline{K}$. The conjugates of α_i are all $\varphi_j \alpha_i$. Let

$$K \subseteq K(\varepsilon) = L_0 \subseteq L_1 \subseteq \cdots \subseteq L_s,$$

where $s = nr$ and $L_{jr+i} = L_{jr+i-1}(\varphi_j \alpha_i)$ for all $1 \leq i \leq r$ and $0 \leq j < n$. Then L_s is generated over K by all the conjugates of $\varepsilon, \alpha_1, \ldots, \alpha_r$, since $K(\varepsilon)$ already contains all the conjugates of ε, and $L_s = N$. Moreover, $\varphi_j F_i \subseteq L_{jr+i}$: indeed, $F_0 = K \subseteq L_{jr}$, and $\varphi_j F_{i-1} \subseteq L_{jr+i-1}$ implies

$$\varphi_j F_i = \varphi_j F_{i-1}(\alpha_i) = (\varphi_j F_{i-1})(\varphi_j \alpha_i) \subseteq L_{jr+i-1}(\varphi_j \alpha_i) = L_{jr+i} .$$

Since α_i is radical over F_{i-1}, $\varphi_j \alpha_i$ is radical over $\varphi_j F_{i-1}$ and is radical over L_{jr+i-1}. Hence every $L_{k-1} \subseteq L_k$ is a radical extension; so is $K \subseteq K(\varepsilon)$. Finally, $E \subseteq F_r = K(\alpha_1, \ldots, \alpha_r) \subseteq N$. We now have a tower of radical extensions that ends with a normal extension $L_s = N \supseteq E$.

Let $\beta_k = \varphi_j \alpha_i$, where $k = jr + i$. If $\alpha_i^{n_i} \in F_{i-1}$, where the characteristic of K does not divide n_i, then $\beta_k^{n_i} \in L_{k-1}$ and $K(\varepsilon) \subseteq L_{k-1}$ contains a primitive n_ith root of unity, since n_i divides m; by 7.8, L_k is Galois over L_{k-1} and $\mathrm{Gal}(L_k : L_{k-1})$ is cyclic. If $\alpha_i^p - \alpha_i \in F_{i-1}$, where $p \neq 0$ is the characteristic of K, then $\beta_k^p - \beta_k \in \varphi_j F_{i-1} \subseteq L_{k-1}$; again L_k is Galois over L_{k-1} and $\mathrm{Gal}(L_k : L_{k-1})$ is cyclic, by 7.10. Finally, $K(\varepsilon)$ is Galois over K and $\mathrm{Gal}(K(\varepsilon) : K)$ is abelian, by 7.9. Therefore N is separable over K, and is a Galois extension of K. By 3.11, the intermediate fields $K \subseteq L_0 \subseteq \cdots \subseteq N$ yield a normal series

$$1 = \mathrm{Gal}(N : L_s) \trianglelefteq \mathrm{Gal}(N : L_{s-1}) \trianglelefteq \cdots \trianglelefteq \mathrm{Gal}(N : L_0) \trianglelefteq \mathrm{Gal}(N : K)$$

whose factors are abelian since they are isomorphic to $\text{Gal}\,(L_k : L_{k-1})$ and $\text{Gal}\,(L_0 : K)$. Therefore $\text{Gal}\,(N : K)$ is solvable.

For the converse we show that a finite Galois extension $E \subseteq \overline{K}$ of K with a solvable Galois group is solvable by radicals over K; then every extension $K \subseteq F \subseteq E$ is solvable by radicals, by 8.1. Let $n = [\,E : K\,]$.

Again we first adjoin a primitive mth root of unity ε to K, where $m = n$ if K has characteristic 0 and $p^t m = n$ if K has characteristic $p \neq 0$ and p does not divide m. As above, $F = K(\varepsilon)$ contains a primitive ℓth root of unity for every divisor ℓ of m. By 3.11, EF is a finite Galois extension of F, and $\text{Gal}\,(EF : F) \cong \text{Gal}\,(E : E \cap F) \leq \text{Gal}\,(E : K)$; hence $[\,EF : F\,]$ divides n, $\text{Gal}\,(EF : F)$ is solvable, and $\text{Gal}\,(EF : F)$ has a composition series

$$1 = H_0 \trianglelefteq H_1 \trianglelefteq \cdots \trianglelefteq H_{r-1} \trianglelefteq H_r = \text{Gal}\,(EF : F)$$

whose factors H_i / H_{i-1} are cyclic of prime orders. This yields a tower

$$F = F_r \subseteq F_{r-1} \subseteq \cdots \subseteq F_1 \subseteq F_0 = EF$$

of fixed fields $F_i = \text{Fix}_{EF}\,(H_i)$; by 3.11, F_{i-1} is a Galois extension of F_i and $\text{Gal}\,(F_{i-1} : F_i) \cong H_i / H_{i-1}$ is cyclic of prime order p_i. If p_i is not the characteristic of K, then p_i divides n, p_i divides m, F contains a primitive p_ith root of unity, and F_{i-1} is a radical extension of F_i, by 7.8. If p_i is the characteristic of K, then again F_{i-1} is a radical extension of F_i, by 7.10. Hence EF is solvable by radicals over F, and so is $E \subseteq EF$, by 8.1. $\square$

Abel's theorem states, roughly, that there is no formula that computes the roots of a polynomial of degree 5 or more from its coefficients, using only sums, products, quotients, and nth roots. For a more precise statement, define:

Definition. The general polynomial of degree n *over a field* K *is*

$$g(X) = A_n X^n + A_{n-1} X^{n-1} + \cdots + A_0 \in K(A_0, A_1, \ldots, A_n)[X],$$

where A_0, A_1, ..., A_n *are indeterminates.* $\square$

The *general equation of degree* n is $A_n X^n + A_{n-1} X^{n-1} + \cdots + A_0 = 0$. The general equation of degree 2 is solvable by radicals when K does not have characteristic 2: the formula

$$X_1, X_2 = \frac{-B \pm \sqrt{B^2 - 4AC}}{2A}$$

for the roots of $AX^2 + BX + C$ shows that they lie in a radical extension of $K(A, B, C)$. Longer but similar formulas in Section 5 show that the general polynomials of degree 3 and 4 are solvable by radicals when K does not have characteristic 2 or 3. For the general polynomial, solvability by radicals expresses the idea that there is a formula that computes the roots of a polynomial from its coefficients, using only field operations and radicals (nth roots, and the roots of polynomials $X^p - X - c$ in case K has characteristic $p \neq 0$).

Theorem **8.3** (Abel [1824]). *The general polynomial of degree n is solvable by radicals if and only if $n \leqq 4$.*

Abel's theorem follows from the relationship between coefficients and roots, and from properties of the elementary symmetric polynomials.

Definitions. A polynomial $f \in K[X_1, ..., X_n]$ or rational fraction $f \in K(X_1, ..., X_n)$ is symmetric when $f(X_{\sigma 1}, X_{\sigma 2}, ..., X_{\sigma n}) = f(X_1, ..., X_n)$ for every permutation $\sigma \in S_n$. The elementary symmetric polynomials $\mathfrak{s}_0$, $\mathfrak{s}_1$, ..., $\mathfrak{s}_n$ in $X_1, ..., X_n$ are $\mathfrak{s}_0 = 1$ and

$$\mathfrak{s}_k(X_1, ..., X_n) = \sum_{1 \leqq i_1 < i_2 < \cdots < i_k \leqq n} X_{i_1} X_{i_2} \cdots X_{i_k}, \quad k = 1, 2, ..., n.$$

That $\mathfrak{s}_1$, ..., $\mathfrak{s}_n$ are symmetric can be proved directly but also follows from:

Proposition **8.4.** *In $K(X_1, ..., X_n)[X]$,*

$$(X - X_1)(X - X_2) \cdots (X - X_n) = \sum_{0 \leq k \leq n} (-1)^k \mathfrak{s}_k(X_1, ..., X_n) X^{n-k}.$$

Proof. Expanding $(X - X_1)(X - X_2) \cdots (X - X_n)$ yields a sum whose terms are all products $t_1 t_2 \cdots t_n$ in which, for every $1 \leqq i \leqq n$, either $t_i = X$ or $t_i = -X_i$. A product $t_1 t_2 \cdots t_n$ in which $t_i = -X_i$ happens k times equals $(-1)^k X_{i_1} X_{i_2} \cdots X_{i_k} X^{n-k}$ for some $1 \leqq i_1 < i_2 < \cdots < i_k \leqq n$. The sum of all such products is $(-1)^k \mathfrak{s}_k(X_1, ..., X_n) X^{n-k}$. $\square$

Proposition **8.5.** *For any field K, $K(X_1, ..., X_n)$ is a Galois extension of $K(\mathfrak{s}_1, ..., \mathfrak{s}_n)$, whose Galois group is isomorphic to the symmetric group S_n.*

Proof. For every $\sigma \in S_n$, $\overline{\sigma} : f(X_1, ..., X_n) \longmapsto f(X_{\sigma 1}, X_{\sigma 2}, ..., X_{\sigma n})$ is an automorphism of $K(X_1, ..., X_n)$. Then $G = \{ \overline{\sigma} \mid \sigma \in S_n \}$ is a finite group of automorphisms of $E = K(X_1, ..., X_n)$. By 3.6, E is a finite Galois extension of $S = \text{Fix}_E(G)$ and $\text{Gal}(E : S) = G \cong S_n$.

Now, S consists of all symmetric rational fractions. Hence $\mathfrak{s}_1, ..., \mathfrak{s}_n \in S$ and $L = K(\mathfrak{s}_1, ..., \mathfrak{s}_n) \subseteq S$. By 8.4, $f(X) = (X - X_1)(X - X_2) \cdots (X - X_n) \in L[X]$. Hence E is a splitting field of the separable polynomial f over L, and is Galois over L. An L-automorphism of E must permute the roots of f and is uniquely determined by its values at $X_1, ..., X_n$; therefore there are at most $n!$ L-automorphisms of E, and $[E : L] \leqq n!$. But $L \subseteq S$ and $[E : S] = n!$. Therefore $[E : L] = [E : S]$, $L = S$, and $\text{Gal}(E : L) \cong S_n$. $\square$

Corollary **8.6.** *Every symmetric rational fraction of X_1, ..., X_n is a rational function of the elementary symmetric polynomials in X_1, ..., X_n.*

Proof. This follows from the equality $S = L$ in the proof of 8.5. $\square$

Corollary **8.7.** *The elementary symmetric polynomials $\mathfrak{s}_1$, ..., $\mathfrak{s}_n$ are algebraically independent over K in $K(X_1, ..., X_n)$.*

Proof. By IV.8.6, $K(X_1, ..., X_n)$ has transcendence degree n and a transcendence base $B \subseteq \{ \mathfrak{s}_1, ..., \mathfrak{s}_n \}$; hence $B = \{ \mathfrak{s}_1, ..., \mathfrak{s}_n \}$. $\square$

Corollary **8.8.** *Every finite group is isomorphic to a Galois group.*

Proof. This follows from Proposition 8.5 since, by Cayley's theorem II.3.2, every finite group is isomorphic to a subgroup of some S_n . $\square$

We now return to the general polynomial of degree n,

$$g(X) = A_n X^n + A_{n-1} X^{n-1} + \cdots + A_0 \in K(A_0, A_1, ..., A_n)[X].$$

Proposition **8.9.** *The Galois group of the general polynomial of degree n is isomorphic to the symmetric group S_n .*

Proof. We show that the general polynomial of degree n can also be defined by its roots. Let $S = K(A)(\mathfrak{s}_1, ..., \mathfrak{s}_n) \subseteq K(A)(X_1, ..., X_n)$ and

$$f(X) = A (X - X_1)(X - X_2) \cdots (X - X_n) \in K(A, X_1, ..., X_n)[X].$$

By 8.4, the coefficients $a_0, a_1, ..., a_{n-1}$ of f are

$$a_{n-k} = (-1)^k A \mathfrak{s}_k(X_1, ..., X_n) \in S.$$

Hence $f \in S[X]$, and $K(A)(X_1, ..., X_n)$ is a splitting field of f over S. By 8.5, $K(A)(X_1, ..., X_n)$ is Galois over S, and its Galois group is isomorphic to S_n. Thus the Galois group of f is isomorphic to S_n. Now, $a_0, a_1, ..., a_{n-1}$ are algebraically independent over $K(A)$, by 8.7. Hence there is an isomorphism $S = K(a_0, ..., a_{n-1}, A) \cong K(A_0, A_1, ..., A_n)$ that sends f to g. Therefore the Galois groups of f and g are isomorphic. $\square$

Abel's theorem now follows from Proposition 8.9 and Theorem 8.2, since we saw in Section II.9 that S_n is solvable if and only if $n \leqq 4$.

Exercises

1. Show that every extension of degree 2 is a radical extension, except perhaps in characteristic 2.

2. Let K have characteristic $p \neq 0$ and let $c \in K$. Show that $X^p - X - c \in K[X]$ either splits in K or is irreducible in $K[X]$.

3. Prove the following: if F is solvable by radicals over K and $K \subseteq E \subseteq F$, then E is solvable by radicals over K and F is solvable by radicals over E.

4. Prove the following: if $K \subseteq E \subseteq F$, E is solvable by radicals over K, and F is solvable by radicals over E, then F is solvable by radicals over K.

5. Prove the following: if E is radical over K and the composite EF exists, then EF is radical over KF.

6. Prove the following: if E is solvable by radicals over K and the composite EF exists, then EF is solvable by radicals over KF.

7. Find fields $K \subseteq E \subseteq F$ such that F is normal over K but E is not normal over K.

8. Find fields $K \subseteq E \subseteq F$ such that E is normal over K and F is normal over E but F is not normal over K.

9. Geometric Constructions

The geometric constructions in question are procedures in plane Euclidean geometry that construct points and other figures using only a straightedge and a compass; with these antique instruments one can draw a straight line through any two points and a circle with any center and radius. Ancient geometers devised constructions by straightedge and compass for many specific tasks, but, in certain cases which became, in their time, famous problems, no construction could be found, despite repeated efforts. This section explains why.

Constructibility. Constructions by straightedge and compass need at least two points P and Q to start, from which points, straight lines, and circles can be constructed. In our definition, "constructible" is short for "constructible from P and Q by straightedge and compass"; a *line* is a straight line or a circle.

Definition. *Let P and Q be two points in the Euclidean plane.*

(a) *P and Q are constructible;*

(b) *if $A \neq B$ and C are constructible points, then the straight line AB and the circle with radius AB and center C are constructible;*

(c) *intersections of constructible lines are constructible;*

(d) *a point or line is constructible when it can be obtained from P and Q by finitely many applications of (b) and (c).*

For example, when $B, C \neq A$ are constructible points, the fourth point of the parallelogram $ABCD$ is constructible: the circle with radius AB and center C is constructible; the circle with radius AC and center B is constructible; their intersections, which include D, are constructible. In particular, the straight line through C that is parallel to AB is constructible. Felicitously, this construction works even when A, B, and C lie on the same straight line.

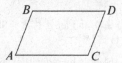

Constructibility becomes clearer with a Cartesian system of coordinates that represents each point in the Euclidean plane by a pair of real numbers, or by a single complex number. We put the origin O at P and choose axes and units of length so Q is represented by the complex number 1.

Definition. *A complex number is constructible (from 0 and 1) when the corresponding point in the Euclidean plane is constructible (from P and Q).*

Proposition 9.1. *Constructible complex numbers constitute a subfield of $\mathbb{C}$.*

Proof. The numbers 0 and 1 are constructible. Let $a, b \in \mathbb{C}$ be constructible and let A, B be the corresponding points. Then $a + b$ corresponds to the fourth point C of the parallelogram $OABC$ and is constructible; $a - b$ corresponds to the fourth point D of the parallelogram $BOAD$ and is constructible.

The product ab corresponds to a point C such that the triangles OQA and OBC are similar. The point R on OB such that $OR = OQ = 1$ is constructible. Then the triangle ORD, which is equal to OQA, is constructible. The straight line through B that is parallel to RD is constructible. It intersects OD at C, so C is constructible:

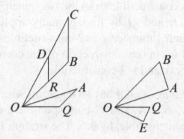

The point E that corresponds to a/b when $b \neq 0$ is likewise constructible, since the triangles OQE and OBA are similar. $\square$

In particular, rational numbers are constructible. Readers will enjoy proving the following properties:

Lemma **9.2.** *If z^2 is constructible, then z is constructible. If $z = x + iy$, where $x, y \in \mathbb{R}$, then z is constructible if and only if x and y are constructible.*

Main result. In analytic geometry, intersections of straight lines and circles are found by solving linear and quadratic equations. Hence constructible complex numbers are algebraic over $\mathbb{Q}$. Our main result tells the complete story:

Theorem **9.3.** *A complex number is constructible (from 0 and 1) if and only if it is algebraic over $\mathbb{Q}$ and its degree is a power of 2.*

Proof. Call a complex number 2-*constructible* when it is algebraic over $\mathbb{Q}$ and its degree over $\mathbb{Q}$ is a power of 2.

Lemma **9.4.** *A complex number z is 2-constructible if and only if it belongs to a finite extension of $\mathbb{Q}$ whose degree is a power of 2, if and only if it belongs to a finite normal extension of $\mathbb{Q}$ whose degree is a power of 2.*

The proof is an exercise. In particular, 2-constructible complex numbers are those that can be reached from $\mathbb{Q}$ by successive adjunctions of square roots. We want to show that a complex number is constructible if and only if it is 2-constructible. Not surprisingly, these two kinds of numbers have the same basic properties:

Lemma **9.5.** *The 2-constructible complex numbers constitute a subfield of $\mathbb{C}$.*

Proof. Let $a, b \in \mathbb{C}$ be 2-constructible. By 9.4, a and b belong to finite normal extensions E and F of $\mathbb{Q}$ whose degrees are powers of 2. By 3.2, 3.11, the composite EF is a Galois extension of $\mathbb{Q}$, whose degree is a power of 2 since the orders of $\mathrm{Gal}\,(F:\mathbb{Q})$ and $\mathrm{Gal}\,(EF:F) \cong \mathrm{Gal}\,(E:E \cap F) \leqq \mathrm{Gal}\,(E:\mathbb{Q})$ are powers of 2. Hence $a - b \in EF$ and $a/b \in EF$ are 2-constructible. $\square$

Lemma **9.6.** *If z^2 is 2-constructible, then z is 2-constructible. If $x, y \in \mathbb{R}$, then $z = x + iy$ is 2-constructible if and only if x and y are 2-constructible.*

We leave this as another exercise and prove 9.3. By definition, a point or line (straight line or circle) is constructible from the two given points P and Q when it can be obtained from P and Q by finitely many applications of (b) and (c). That a constructible complex number z is 2-constructible is shown by induction on the number of uses of (c) in the construction of the corresponding point. If (c) is not used, then z is 0 or 1 and is 2-constructible.

Call a line 2-*constructible* when it has an equation with 2-constructible coefficients. Let the points A, $B \neq A$, and C correspond to 2-constructible numbers. Their coordinates are 2-constructible, by 9.6. The straight line AB, and the circle with radius AB and center C, are 2-constructible, since their equations

$$(x_B - x_A)(y - y_A) = (y_B - y_A)(x - x_A),$$
$$(x - x_C)^2 + (y - y_C)^2 = (x_A - x_B)^2 + (y_A - y_B)^2$$

have 2-constructible coefficients, by 9.5. Now, the intersection of two 2-constructible straight lines has 2-constructible coordinates. The intersections of a 2-constructible straight line $y = ax + b$ and a 2-constructible circle $x^2 + y^2 + cx + dy + e = 0$ are found from the quadratic equation

$$x^2 + (ax + b)^2 + cx + d(ax + b) + e = 0,$$

whose coefficients and discriminant δ are 2-constructible; if $\delta > 0$, then $\sqrt{\delta}$ is 2-constructible by 9.6, and the intersections have 2-constructible coordinates. The intersections of two 2-constructible circles $x^2 + y^2 + ax + by + c = 0$ and $x^2 + y^2 + dx + ey + f = 0$ are also the intersections of $x^2 + y^2 + ax + by + c = 0$ and the 2-constructible straight line $ax + by + c = dx + ey + f$, and have 2-constructible coordinates. In each case, intersections correspond to 2-constructible complex numbers, by 9.6. This completes the induction.

Conversely, let $z \in \mathbb{C}$ be 2-constructible. By 9.4, z belongs to a finite normal extension E of $\mathbb{Q}$ whose degree is a power 2^r of 2. That z is constructible is proved by induction on r. If $r = 0$, then $z \in \mathbb{Q}$ is constructible. In general, the Galois group of E over $\mathbb{Q}$ is a finite 2-group and has a normal series whose factors are cyclic of order 2. Hence E has a tower $\mathbb{Q} = F_0 \subseteq F_1 \subseteq \cdots \subseteq F_r = E$ of extensions of degree 2. Now, $\mathbb{Q}$ does not have enough nth roots of unity to use 7.8, but we can argue as follows. Since $[E : F_{r-1}] = 2$, we have $E = F_{r-1}(\alpha)$ for some α, and $q = \mathrm{Irr}\,(\alpha:\mathbb{Q})$ has degree 2, $q(X) = X^2 + bX + c$ for some $b, c \in F_{r-1}$. Then $\alpha = \frac{1}{2}\,(-b \pm \sqrt{b^2 - 4c})$ and $E = F_{r-1}(\beta)$, where $\beta^2 = b^2 - 4c \in F_{r-1}$. Now, β^2 is constructible by the induction hypothesis, β is constructible by 9.2, and $z \in F_r$ is constructible by 9.1, since $z = x + \beta y$ for some $x, y \in F_{r-1}$ and x, y are constructible by the induction hypothesis. $\square$

Applications. With Theorem 9.3 in hand we return to Euclidean geometry.

Corollary 9.7. There is no construction by straightedge and compass that can trisect angles (split any angle into three equal parts).

Proof. If a $\pi/3$ angle could be trisected, then the complex number $\varepsilon = e^{i\pi/9}$ would be constructible. But ε is a primitive 18th root of unity; by 6.5, ε has degree $\phi(18) = 6$ over $\mathbb{Q}$, and is not constructible. $\square$

In the same spirit, readers will tackle two more problems.

Corollary 9.8. There is no construction by straightedge and compass that can duplicate cubes (construct the side of a cube whose volume is twice the volume of any given cube).

Corollary 9.9. There is no construction by straightedge and compass that can square circles (construct a square whose area is that of any given circle).

The last problem is splitting a circle into n arcs of equal lengths; equivalently, constructing a regular polygon with n sides. Its solution requires a definition.

Definition. A Fermat prime *is a prime number of the form* $2^{2^n} + 1$.

For example, $2^{2^0} + 1 = 3$, $2^{2^1} + 1 = 5$, $2^{2^2} + 1 = 17$, $2^{2^3} + 1 = 257$, and $2^{2^4} + 1 = 65537$ are Fermat primes. To the author's knowledge, no other Fermat primes have been discovered as of this writing.

Lemma 9.10. If $2^k + 1$ is prime, then k is a power of 2.

Proof. If k is not a power of 2, then $k = 2^i j$, where j is odd. Then every $m^j + 1$ is divisible by $m + 1$, and $2^k + 1 = (2^{2^i})^j + 1$ is divisible by $2^{2^i} + 1$. $\square$

Corollary 9.11 (Gauss [1801]). *A regular polygon with n sides can be constructed from its radius by straightedge and compass if and only if n is the product of a power of 2 and distinct Fermat primes.*

Thus, regular polygons with 2, 3, 4, 5, or 6 sides are constructible, but not those with 7 sides or 9 sides.

Proof. A regular polygon with n sides is constructible from its radius if and only if the primitive nth root of unity $\varepsilon_n = e^{2i\pi/n}$ is constructible. By 6.5, ε_n has degree $\phi(n)$ over $\mathbb{Q}$. Write n as the product $n = 2^m p_1^{m_1} \cdots p_r^{m_r}$ of a power of 2 and positive powers of distinct odd primes. Then

$$\phi(n) = 2^{m-1} p_1^{m_1-1} (p_1 - 1) \cdots p_r^{m_r-1} (p_r - 1).$$

Hence $\phi(n)$ is a power of 2 if and only if $m_1 = \cdots = m_r = 1$ and $p_1 - 1, \ldots, p_r - 1$ are powers of 2; equivalently, n is the product of a power of 2 and distinct odd primes, which are Fermat primes by 9.10. $\square$

Exercises

For the following exercises, give geometric solutions.

1. Show that the y-axis is constructible.

2. Prove the following: if $z = x + iy$, where $x, y \in \mathbb{R}$, then z is constructible if and only if x and y are constructible.

3. Prove the following. Let A be a constructible point. Show that there is a constructible point B such that $OB = \sqrt{OA}$. Hint: when a point R on a circle projects to a point S on a diameter CD, then $CS.DS = RS^2$.

4. Prove the following: if $z \in \mathbb{C}$ and z^2 is constructible, then z is constructible.

5. Devise a construction by straightedge and compass of a regular pentagon, given its radius.

For the following exercises, give algebraic solutions.

6. Show that a complex number z is 2-constructible if and only if it belongs to a finite extension of $\mathbb{Q}$ whose degree is a power of 2, if and only if it belongs to a finite normal extension of $\mathbb{Q}$ whose degree is a power of 2.

7. Prove the following: if $z \in \mathbb{C}$ and z^2 is 2-constructible, then z is 2-constructible.

8. Prove the following: if $z = x + iy$, where $x, y \in \mathbb{R}$, then z is 2-constructible if and only if x and y are 2-constructible.

9. Prove the following: if K does not have characteristic 2, then every extension $K \subseteq E$ of degree 2 is a radical extension.

10. Show that there is no construction by straightedge and compass that can duplicate arbitrary cubes.

11. Show that there is no construction by straightedge and compass that can square arbitrary circles. (You may take it for granted that π is transcendental.)

VI
Fields with Orders and Valuations

The results of Artin and Schreier [1926] on ordered fields, presented in Sections 1 and 2, extend known properties of $\mathbb{R}$ and $\mathbb{C}$ and give new insights into the relationship of a field to its algebraic closure. The remaining sections study valuations and completions, which have become valuable tools of algebraic geometry.

1. Ordered Fields

An ordered field is a field with a compatible total order relation. This section contains basic properties, and a universal property of $\mathbb{R}$.

Definition. An ordered field (short for "totally ordered field") is a field F together with a total order relation $\leqq$ on F such that, for all $x, y, z \in F$:

(1) $x < y$ *implies* $x + z < y + z$;

(2) *if* $z > 0$, *then* $x < y$ *implies* $xz < yz$.

For instance, $\mathbb{Q}$ and $\mathbb{R}$ are ordered fields (with their usual order relations). The exercises give other examples.

Some familiar properties of $\mathbb{Q}$ and $\mathbb{R}$ extend easily to all ordered fields:

$x > 0$ if and only if $-x < 0$ (otherwise, say, $x > 0$ and $-x > 0$, and then $0 = x + (-x) > x > 0$);

$x > y$ if and only if $x - y > 0$;

$x < y$ if and only if $-x > -y$;

$x > y > 0$ implies $y^{-1} > x^{-1} > 0$;

if $z < 0$, then $-z > 0$ and $x < y$ implies $xz = (-x)(-z) > (-y)(-z) = yz$;

$x^2 > 0$ for all $x \neq 0$, since $x < 0$ implies $x^2 = (-x)(-x) > 0$; in particular, $1 = 1^2 > 0$;

ordered fields have characteristic 0 (since $0 < 1 < 1 + 1 < 1 + 1 + 1 < \cdots$).

Thus, not every field can be an ordered field; for example, $\mathbb{C}$ cannot be ordered, since an ordered field cannot have $-1 = i^2 > 0$. In general:

Proposition **1.1.** *A field F can be ordered if and only if -1 is not a sum of squares of elements of F, if and only if 0 is not a nonempty sum of nonzero squares of elements of F.*

Proof. If $-1 = \sum_i x_i^2$, then $0 = 1 + \sum_i x_i^2$ is a sum of squares. Conversely, if $0 = \sum_i x_i^2$ with, say, $x_k \neq 0$, then $-x_k^2 = \sum_{i \neq k} x_i^2$ and $-1 = \sum_{i \neq k} (x_i/x_k)^2$ is a sum of squares. In an ordered field, squares are nonnegative, and $-1 < 0$ is not a sum of squares.

Conversely, assume that -1 is not a sum of squares. Let S be the set of all nonempty sums of squares of nonzero elements of F. Then $0 \notin S$, $-1 \notin S$, and S is closed under addition. Morover, S is closed under multiplication, since $\left(\sum_i x_i^2 \right) \left(\sum_j y_j^2 \right) = \sum_{i,j} (x_i y_j)^2$, and is a multiplicative subgroup of $F \backslash \{0\}$, since $1 \in S$, and $x = \sum_i x_i^2 \in S$ implies $x^{-1} = x/x^2 = \sum_i (x_i/x)^2$.

By Zorn's lemma there is a subset M of F that contains S, is closed under addition, is a multiplicative subgroup of $F \backslash \{0\}$ (in particular, $0 \notin M$), and is maximal with these properties. Then M, $\{0\}$, and $-M = \{ x \in F \mid -x \in M \}$ are pairwise disjoint (if $x \in M \cap (-M)$, then $0 = x + (-x) \in M$). We show that $F = M \cup \{0\} \cup (-M)$; readers will easily deduce that F becomes an ordered field, when ordered by $x < y$ if and only if $y - x \in M$.

Suppose that $a \in F$, $a \neq 0$, and $-a \notin M$. Let
$$M' = \{ x + ay \mid x, y \in M \cup \{0\}, \text{ with } x \neq 0 \text{ or } y \neq 0 \}.$$

Then $S \subseteq M \subseteq M'$; M' is closed under addition, like M, and closed under multiplication, since $x, y, z, t \in M \cup \{0\}$ implies $(x + ay)(z + at) = (xz + a^2 yt) + a(yz + xt)$ with $xz + a^2 yt$, $yz + xt \in M \cup \{0\}$ since $a^2 \in S \subseteq M$. Also, $0 \notin M'$: $x + ay \neq 0$ when $x = 0 \neq y$ or $x \neq 0 = y$, or when $x, y \in M$ (otherwise, $a = -x/y \in -M$). Moreover, $1 \in S \subseteq M'$, and $t = x + ay \in M'$ implies $t^{-1} = t/t^2 = (x/t^2) + a(y/t^2) \in M'$ (since $t^2 \in M$). Thus, M' is a multiplicative subgroup of $F \backslash \{0\}$. Therefore $M' = M$, and $a \in M$. $\square$

Archimedean fields. Since an ordered field F has characteristic 0, it contains a subfield $Q = \{ m1/n1 \mid m, n \in \mathbb{Z}, n \neq 0 \}$, which is isomorphic to $\mathbb{Q}$ as a field. In fact, Q is isomorphic to $\mathbb{Q}$ as an ordered field: we saw that $n1 > 0$ in F when $n > 0$ in $\mathbb{Z}$; hence $m1/n1 > 0$ in F when $m/n > 0$ in $\mathbb{Q}$, and $a1/b1 > c1/d1$ in F if and only if $a/b > c/d$ in $\mathbb{Q}$. We identify $\mathbb{Q}$ with Q, so that $\mathbb{Q}$ is an ordered subfield of F.

Definition. *An ordered field F is* archimedean *when every positive element of F is less than a positive integer.*

For example, $\mathbb{Q}$ and $\mathbb{R}$ are archimedean, but not every ordered field is archimedean (see the exercises). The next result finds all archimedean ordered fields.

Theorem **1.2.** *An ordered field is archimedean if and only if it is isomorphic as an ordered field to a subfield $\mathbb{Q} \subseteq F \subseteq \mathbb{R}$ of $\mathbb{R}$.*

Proof. Ordered subfields of $\mathbb{R}$ are archimedean, since $\mathbb{R}$ is archimedean.

Conversely, let F be an archimedean ordered field. We show that $\mathbb{Q} \subseteq F$ is *dense* in F, that is, between any two $x < y$ in F lies some $r \in \mathbb{Q}$. Since F is archimedean, there exist integers $\ell, m, n > 0$ such that $-\ell < x < y < m$ and $1/(y - x) < n$; then $0 < 1/n < y - x$. Now $(i/n) - \ell > x$ when $i \geq n(\ell + m)$. Hence there is a least $j > 0$ such that $(j/n) - \ell > x$. Then $(j/n) - \ell < y$, since $(i/n) - \ell \geq y$ implies $((i - 1)/n) - \ell \geq y - (1/n) > x$.

To embed F into $\mathbb{R}$ we define limits of sequences in F: $L = \lim_F x_n$ if and only if $L \in F$ and, for every positive $\varepsilon \in \mathbb{Q}$, $L - \varepsilon < x_n < L + \varepsilon$ holds for all sufficiently large n. If $\lim_F x_n$ exists, it is unique, since $\mathbb{Q}$ is dense in F. Moreover, sequences with limits in F are Cauchy sequences: if $\lim_F a_n$ exists, then, for every positive $\varepsilon \in \mathbb{Q}$, $-\varepsilon < a_m - a_n < \varepsilon$ holds for all sufficiently large m, n. Readers with a yen for analysis will easily prove the following limit laws: $\lim_F (a_n + b_n) = \left(\lim_F a_n \right) + \left(\lim_F b_n \right)$ and $\lim_F (a_n b_n) = \left(\lim_F a_n \right) \left(\lim_F b_n \right)$, whenever $\lim_F a_n$ and $\lim_F b_n$ exist; the usual arguments work since F is archimedean.

Every element x of F is the limit in F of a Cauchy sequence of rationals: since $\mathbb{Q}$ is dense in F, there exists for every $n > 0$ some $a_n \in \mathbb{Q}$ such that $x - (1/n) < a_n < x + (1/n)$; then $\lim_F a_n = x$. If $(b_n)_{n>0}$ is another sequence of rational numbers such that $\lim_F b_n = x$, then, for every positive $\varepsilon \in \mathbb{Q}$, $|a_n - b_n| < \varepsilon$ holds for all sufficiently large n, so that a_n and b_n have the same limit in $\mathbb{R}$. Hence a mapping $\lambda : F \longrightarrow \mathbb{R}$ is well defined by

$$\lambda(x) = \lim_{n \to \infty} a_n \text{ whenever } x = \lim_F a_n \text{ and } a_n \in \mathbb{Q}.$$

If $x \in \mathbb{Q}$, then $\lambda(x) = x$, since we can let $a_n = x$ for all n. Our two limits laws show that λ a homomorphism. If $x > 0$ in F, then $x^{-1} < m$ for some integer $m > 0$ and $1/m < x$; we can arrange that $x - (1/n) < a_n < x + (1/n)$ and $1/m < a_n$ for all n; then $\lambda(x) = \lim_F a_n \geq 1/m > 0$. Hence $x < y$ implies $\lambda(x) < \lambda(y)$; the converse holds, since $x \geq y$ would imply $\lambda(x) \geq \lambda(y)$. Thus F is isomorphic to $\lambda(F)$, as an ordered field, and $\mathbb{Q} \subseteq \lambda(F) \subseteq \mathbb{R}$. $\square$

Exercises

Prove the following:

1. A field F is an ordered field, and $P \subseteq F$ is its set of positive elements, if and only if $0 \notin P$, P is closed under addition and multiplication, $F = P \cup \{0\} \cup (-P)$ (where $-P = \{ x \in F \mid -x \in P \}$), and F is ordered by $x < y$ if and only if $y - x \in P$.

2. $\mathbb{Q}$ can be made into an ordered field in only one way.

3. $\mathbb{R}$ can be made into an ordered field in only one way.

4. $\mathbb{Q}(\sqrt{2}) \subseteq \mathbb{R}$ can be made into an ordered field in exactly two ways.

5. If F is an ordered field, then so is $F(X)$, when $f/g > 0$ if and only if $a/b > 0$, where a and b are the leading coefficients of f and g; but $F(X)$ is not archimedean.

6. If F is an ordered field, then so is $F(X)$, when $f/g > 0$ if and only if $f/g = X^n f_0/g_0$, where $n \in \mathbb{Z}$ and $f_0(0) > 0$, $g_0(0) > 0$; but $F(X)$ is not archimedean.

7. Let F be an archimedean ordered field. Without using Theorem 1.2, show that $\lim_F (a_n + b_n) = \left(\lim_F a_n\right) + \left(\lim_F b_n\right)$ whenever $\lim_F a_n$ and $\lim_F b_n$ exist.

8. Let F be an archimedean ordered field. Without using Theorem 1.2, show that $\lim_F (a_n b_n) = \left(\lim_F a_n\right) \left(\lim_F b_n\right)$ whenever $\lim_F a_n$ and $\lim_F b_n$ exist.

2. Real Fields

This section studies fields that can be ordered. A number of properties of $\mathbb{R}$ extend to these fields. The Artin-Schreier theorem, which concludes the section, throws some light on the relationship between a field and its algebraic closure.

Formally real fields are fields that can be ordered:

Definition. A field F is formally real when there is a total order relation on F that makes F an ordered field.

By 1.2, a field F is formally real if and only if -1 is not a sum of squares of elements of F. For example, every subfield of $\mathbb{R}$ is formally real. If F is formally real, then so is $F(X)$ (see the exercises for Section 1).

Proposition 2.1. If F is a formally real field and $\alpha^2 \in F$, $\alpha^2 > 0$, then $F(\alpha)$ is formally real.

Proof. We may assume that $\alpha \notin F$. Then every element of $F(\alpha)$ can be written in the form $x + \alpha y$ for some unique $x, y \in F$. If $\alpha^2 > 0$ in F, then $F(\alpha)$ is formally real, since

$$-1 = \sum_i (x_i + \alpha y_i)^2 = \sum_i (x_i^2 + \alpha^2 y_i^2) + \alpha \sum_i (2 x_i y_i)$$

for some $x_i, y_i \in F$ would imply $-1 = \sum_i (x_i^2 + \alpha^2 y_i^2) \geqq 0$. $\square$

On the other hand, $\mathbb{C} = \mathbb{R}(i)$ is not formally real.

Proposition 2.2. If F is a formally real field, then every finite extension of F of odd degree is formally real.

Proof. This is proved by induction on n, simultaneously for all F and all $E \supseteq F$ of odd degree $n = [E : F]$. There is nothing to prove if $n = 1$. Let $n > 1$. If $\alpha \in E \backslash F$ and $F \subsetneqq F(\alpha) \subsetneqq E$, then $[F(\alpha) : F]$ and $[E : F(\alpha)]$ are odd, since they divide n; hence $F(\alpha)$ and E are formally real, by the induction hypothesis. Now let $E = F(\alpha)$. Then $q = \mathrm{Irr}\,(\alpha : F)$ has odd degree n; the elements of E can be written in the form $f(\alpha)$ with $f \in F[X]$ and $\deg f < n$.

If E is not formally real, then $-1 = \sum_i f_i(\alpha)^2$, where $f_i \in F[X]$ and $\deg f_i < n$. Hence q divides $1 + \sum_i f_i^2$ and $1 + \sum_i f_i^2 = qg$ for some

$g \in F[X]$. Since the leading coefficients of all f_i^2 are all positive in F, $\deg\left(1 + \sum_i f_i^2\right) = \max_i \deg\left(f_i^2\right)$ is even and less than $2n$. Since $\deg q = n$ is odd, $\deg g$ is odd and less than n, and g has an irreducible factor r whose degree is odd and less than n. Now, r has a root $\beta \in \overline{F}$ and $1 + \sum_i f_i(\beta)^2 = q(\beta)g(\beta) = 0$ in $F(\beta)$, so that $F(\beta)$ is not formally real. This contradicts the induction hypothesis since $[F(\beta):F] = \deg r$ is odd and less than n. $\square$

Real closed fields. We saw that every field has a maximal algebraic extension (its algebraic closure). Formally real field have a similar property.

Definition. A field R is real closed *when it is formally real and there is no formally real algebraic extension $E \supsetneq R$.*

For example, $\mathbb{R}$ is real closed: up to $\mathbb{R}$-isomorphism, the only algebraic extension $E \supsetneq \mathbb{R}$ of $\mathbb{R}$ is $\mathbb{C}$, which is not formally real.

Theorem 2.3. A formally real field R is real closed if and only if (i) *every positive element of R is a square in R, and* (ii) *every polynomial of odd degree in $R[X]$ has a root in R; and then $\overline{R} = R(i)$, where $i^2 = -1$.*

Proof. Readers will enjoy proving that real closed fields have properties (i) and (ii). Now let R be a formally real field in which (i) and (ii) hold. Then R has no finite extension $E \supsetneq R$ of odd degree: if $[E:R]$ is odd and $\alpha \in E$, $q = \mathrm{Irr}\,(\alpha:R)$, then $\deg q = [R(\alpha):R]$ divides $[F:R]$ and is odd, q has a root in R by (ii), $\deg q = 1$, and $\alpha \in R$, so that $E = R$.

Let $C = R(i) \subseteq \overline{R}$, where $i^2 = -1$. We show that every element $a + bi$ of C is a square in C. First, every $a \in R$ is a square in C, by (i) (if $a < 0$, then $-a = y^2$ and $a = (iy)^2$ for some $y \in R$). Now let $b \neq 0$. We have $a + bi = (x + yi)^2$ if and only if $x^2 - y^2 = a$ and $2xy = b$, and $2 > 0$ in R since R is formally real. With $y = b/2x$ the first equation reads $x^4 - ax^2 - \frac{b^2}{4} = 0$. This quadratic equation in x^2 has two solutions s_1, $s_2 \in R$, since its discriminant is $a^2 + b^2 > 0$. Moreover, $s_1 s_2 = -\frac{b^2}{4} < 0$, so that s_1, say, is positive. Then $s_1 = x^2$ for some $x \in R$, and then $a + bi = (x + ib/2x)^2$.

Then every quadratic polynomial $f \in C[X]$ has a root in C, since its discriminant has a square root in C. Hence $C[X]$ contains no irreducible polynomial of degree 2. Then C has no extension $C \subseteq E$ of degree 2: otherwise, $E = C(\alpha)$ for any $\alpha \in E \backslash C$ and $\mathrm{Irr}\,(\alpha:C)$ would be irreducible of degree 2.

We show that $\overline{C} = C$; this implies $\overline{R} = C$. If $\alpha \subset \overline{C}$, then α and its conjugates over R generate a finite Galois extension E of C, which is also a finite Galois extension of R. Then $G = \mathrm{Gal}\,(E:R)$ has even order $|G| = [E:R] = 2[E:C]$. If S is a Sylow 2-subgroup of G and $F = \mathrm{Fix}_E\,(S)$ is its fixed field, then $[F:R] = [G:S]$ is odd, which we saw implies $F = R$; hence $G = S$ is a 2-group. Then $\mathrm{Gal}\,(E:C)$ is a 2-group. If $C \subsetneq E$, then $\mathrm{Gal}\,(E:C)$ has a subgroup of index 2, whose fixed field F' has degree 2 over C, which we saw cannot happen; therefore $E = C$, and $\alpha \in C$.

If now $R \subsetneqq E \subseteq \overline{R}$ is a proper algebraic extension of R, then $E = \overline{R}$ and E is not formally real, since -1 is a square in C. Hence R is real closed. $\square$

Wher $R = \mathbb{R}$, this argument gives a more algebraic proof that $\mathbb{C}$ is algebraically closed, based on properties (i) and (ii) of $\mathbb{R}$.

Corollary **2.4.** *A real closed field R can be made into an ordered field in only one way.*

Proof. Namely, $x \leqq y$ if and only if $y - x = a^2$ for some $a \in R$, by 2.3. $\square$

Corollary **2.5.** *If R is real closed, then $f \in R[X]$ is irreducible if and only if either f has degree 1, or f has degree 2 and no root in R.*

This is proved like the similar property of $\mathbb{R}$, as readers will happily verify.

Corollary **2.6.** *The field of all algebraic real numbers is real closed.*

Proof. "Algebraic" real numbers are algebraic over $\mathbb{Q}$; they constitute a field A. This field A has properties (i) and (ii) in Theorem 2.3, since real numbers that are algebraic over A are algebraic over $\mathbb{Q}$. $\square$

The exercises give other properties of $\mathbb{R}$ that extend to all real closed fields.

Real closure. We now find maximal real closed algebraic extensions.

Definition. A real closure *of an ordered field F is a real closed field that is algebraic over F, and whose order relation induces the order relation on F.*

Proposition **2.7.** *Every ordered field has a real closure.*

Proof. Let F be an ordered field. The subfield E of $\overline{F}$ generated by all square roots of positive elements of F is formally real: if $-1 = \sum_i \beta_i^2$ in E, then $-1 = \sum_i \beta_i^2$ in $F(\alpha_1, \ldots, \alpha_n)$ for some square roots $\alpha_1, \ldots, \alpha_n$ of positive elements of F, and $F(\alpha_1, \ldots, \alpha_n)$ is not formally real, contradicting 2.1.

By Zorn's lemma there is a subfield $E \subseteq R$ of $\overline{F}$ that is formally real and is maximal with this property. Then R is real closed, since a proper algebraic extension of R is, up to R-isomorphism, contained in $\overline{R}$ and cannot be formally real by the maximality of R; R is algebraic over F; and the order relation on R induces the order relation on F: positive elements of F are squares in $E \subseteq R$ and are positive in R, whence negative elements of F are negative in R. $\square$

It is known that every order preserving homomorphism of F into a real closed field R extends to an order preserving homomorphism of any real closure of F into R; hence any two real closures of F are isomorphic as ordered fields.

The Artin-Schreier theorem is another characterization of real closed fields.

Theorem **2.8** (Artin-Schreier [1926]). *For a field $K \neq \overline{K}$ the following conditions are equivalent:* (1) *K is real closed;* (2) *$[\overline{K} : K]$ is finite;* (3) *there is an upper bound for the degrees of irreducible polynomials in $K[X]$.*

Thus, if $[\overline{K} : K]$ is finite, then $[\overline{K} : K] = 2$; either the irreducible polynomials in $K[X]$ have arbitrarily high degrees, or all have degree at most 2.

Proof. We start with three lemmas.

Lemma 2.9. *If there is an upper bound for the degrees of irreducible polynomials in* $K[X]$, *then* K *is perfect.*

Proof. If K is not perfect, then K has characteristic $p \neq 0$, and some $c \in K$ is not a pth power in K. We show that $f(X) = X^{p^r} - c \in K[X]$ is irreducible for every $r \geqq 0$. In $K[X]$, f is a product $f = q_1 q_2 \cdots q_k$ of monic irreducible polynomials $q_1, \ldots, q_k$. Let α be a root of f in $\overline{K}$. Then $\alpha^{p^r} = c$ and $f(X) = X^{p^r} - \alpha^{p^r} = (X - \alpha)^{p^r}$. Hence $q_i = (X - \alpha)^{t_i}$ for some $t_i > 0$. If $t = \min(t_1, \ldots, t_r)$, then $q = (X - \alpha)^t$ is irreducible and divides $q_1, \ldots, q_k$; therefore $q_1 = \cdots = q_k = q$ and $f = q^k$. In particular, $\alpha^{kt} = c$, $p^r = kt$, and k is a power of p. But k is not a multiple of p, since $\alpha^t \in K$ and $c = (\alpha^t)^k$ is not a pth power in K. Therefore $k = 1$, and $f = q$ is irreducible. $\square$

Lemma 2.10. *If* F *is a field in which* -1 *is a square, then* $\overline{F}$ *is not a Galois extension of* F *of prime degree.*

Proof. We show that a Galois extension $E \subseteq \overline{F}$ of F of prime degree p (necessarily a cyclic extension) cannot be algebraically closed.

If F has characteristic p, then, by V.7.10, $E = F(\alpha)$, where $c = \alpha^p - \alpha \in F$; $\mathrm{Irr}(\alpha : F) = X^p - X - c$; and $1, \alpha, \ldots, \alpha^{p-1}$ is a basis of E over F. Let $\beta = b_0 + b_1 \alpha + \cdots + b_{p-1} \alpha^{p-1} \in E$, where $b_0, \ldots, b_{p-1} \in F$. Then

$$\begin{aligned} \beta^p &= b_0^p + b_1^p \alpha^p + \cdots + b_{p-1}^p \alpha^{(p-1)p} \\ &= b_0^p + b_1^p (\alpha + c) + \cdots + b_{p-1}^p (\alpha + c)^{p-1} \end{aligned}$$

and $\beta^p - \beta - c\alpha^{p-1} = a_0 + a_1 \alpha + \cdots + a_{p-1} \alpha^{p-1}$, where $a_{p-1} = b_{p-1}^p - b_{p-1} - c$. Hence $\beta^p - \beta - c\alpha^{p-1} \neq 0$: otherwise, $a_{p-1} = 0$ and $X^p - X - c$ disgraces irreducibility by having a root b_{p-1} in F. Thus $X^p - X - c\alpha^{p-1} \in E[X]$ has no root in E.

Now assume that F does not have characteristic p. We may also assume that E contains a primitive pth root of unity ε: otherwise, E is not algebraically closed. Then ε is a root of $(X^p - 1)/(X - 1) \in F[X]$ and $[F(\varepsilon) : F] < p$. Therefore $[F(\varepsilon) : F] = 1$ and $\varepsilon \in F$. Then V.7.8 yields $E = F(\alpha)$, where $\alpha^p \in F$ and $\alpha \notin F$.

Assume that α has a pth root β in E. Let $\sigma \in \mathrm{Gal}(E : F)$, $\zeta = (\sigma\beta)/\beta \in E$, and $\eta = (\sigma\zeta)/\zeta \in E$. Then $\beta^{p^2} = \alpha^p \in F$, $(\sigma\beta)^{p^2} = \beta^{p^2}$, $(\zeta^p)^p = 1$, $\zeta^p \in F$, $(\sigma\zeta)^p = \zeta^p$, $\eta^p = 1$, and $\eta \in F$. Now $\sigma\beta = \zeta\beta$ and $\sigma\zeta = \eta\zeta$; by induction, $\sigma^k \beta = \eta^{k(k-1)/2} \zeta^k \beta$ for all k, since this equality implies $\sigma^{k+1} \beta = \eta^{k(k-1)/2} (\eta^k \zeta^k)(\zeta\beta) = \eta^{k(k+1)/2} \zeta^{k+1} \beta$. Then $\eta^{p(p-1)/2} \zeta^p = 1$, since $\sigma^p = 1$. If p is odd, then p divides $p(p-1)/2$ and $\eta^{p(p-1)/2} = 1$. If $p = 2$, then $\zeta^4 = 1$ and $\zeta^2 = \pm 1$; if $\zeta^2 = 1$, then $\eta^{p(p-1)/2} = 1$; if $\zeta^2 = -1$, then $\zeta \in F$, since F contains a square root of -1, and again $\eta^{p(p-1)/2} = \eta = (\sigma\zeta)/\zeta = 1$. In every case, $\zeta^p = 1$. Hence $\zeta = \varepsilon^t \in F$, $\sigma\beta = \beta$, and $\sigma\alpha = \sigma\beta^p = \beta^p = \alpha$. But $\alpha \notin F$, so $\sigma\alpha \neq \alpha$ for some $\sigma \in \mathrm{Gal}\,(E:F)$. Therefore $X^p - \alpha \in E[X]$ has no root in E. $\square$

Lemma **2.11.** *If* $[\overline{K}:K] = n$ *is finite, then every irreducible polynomial in* $K[X]$ *has degree at most* n, K *is perfect, and* $\overline{K} = K(i)$, *where* $i^2 = -1$.

Proof. Every irreducible polynomial $q \in K[X]$ has a root α in $\overline{K}$; then $q = \mathrm{Irr}\,(\alpha:K)$ has degree $[K(\alpha):K] \leq n$. Then K is perfect, by 2.9, and $\overline{K}$ is Galois over K. Let $i \in \overline{K}$ be a root of $X^2 + 1 \in K[X]$. If $K(i) \subsetneqq \overline{K}$, then $\overline{K}$ is Galois over $K(i)$ and $\mathrm{Gal}\,(\overline{K}:K(i))$ has a subgroup H of prime order; then $\overline{K}$ is Galois over the fixed field F of H, of prime degree $[\overline{K}:F] = |H|$, contradicting 2.10. Therefore $K(i) = \overline{K}$. $\square$

We now prove Theorem 2.8. By 2.3, (1) implies (2); (2) implies (3), by 2.11.

(3) implies (2). If every irreducible polynomial in $K[X]$ has degree at most n, then K is perfect by 2.9 and $\overline{K}$ is separable over K. Moreover, every element of $\overline{K}$ has degree at most n over K; hence $[\overline{K}:K] \leq n$, by IV.6.13.

(2) implies (1). Assume that $[\overline{K}:K]$ is finite. By 2.11, K is perfect and $\overline{K} = K(i)$, where $i^2 = -1$. Then $i \notin K$, since $K \neq \overline{K}$. Every $z = x + iy \in \overline{K}$ has two conjugates, z and $\overline{z} = x - iy$, and $z\overline{z} = x^2 + y^2 \in K$. For every $x, y \in K$, $x + iy = u^2$ for some $u \in \overline{K}$, and then $x^2 + y^2 = u^2 \overline{u}^2 = (u\overline{u})^2$ is a square in K. Hence, in K, every sum of squares is a square. Now, -1 is not a square in K, since $i \notin K$. Hence K is formally real; K is real closed, since the only algebraic extension $K \subsetneqq E = \overline{K}$ of K is not formally real. $\square$

Exercises

1. Let R be a real closed field, with an order relation that makes it an ordered field. Show that every positive element of R is a square in R.

2. Let R be a real closed field. Show that every polynomial $f \in R[X]$ of odd degree has a root in R.

3. Prove the following: if R is real closed, then $f \in R[X]$ is irreducible if and only if either f has degree 1, or f has degree 2 and no root in R.

4. Prove the following: if R is real closed, $f \in R[X]$, and $f(a)f(b) < 0$ for some $a < b$ in R, then $f(r) = 0$ for some $a < r < b$. (Hint: use Corollary 2.5.)

5. Prove the following: if K and L are real closed fields, then every homomorphism of K into L is order preserving.

6. In an ordered field F, the *absolute value* of $x \in F$ is $|x| = \max(x, -x) \in F$. (This is not an absolute value as defined in Section 3.) Show that $|xy| = |x||y|$ and that $|x + y| \leq |x| + |y|$.

7. Prove the following: when R is real closed and $f(X) = a_0 + a_1 X + \cdots + a_n X^n \in R[X]$, then every root of f in R lies in the interval $-M \leq x \leq M$, where $M = \max(1, |a_0| + |a_1| + \cdots + |a_n|)$.

8. Show that $X^p - a \in F[X]$ is irreducible when p is prime, the field F does not have characteristic p and does not contain a pth root of a, but F contains a primitive pth root of unity and a square root of -1.

9. Let C be an algebraically closed field. Prove that $\text{Aut}(C)$ has no finite subgroup of order greater than 2.

3. Absolute Values

Absolute values on fields are generalizations of the familiar absolute values on $\mathbb{Q}$, $\mathbb{R}$, and $\mathbb{C}$. They yield further insight into these fields as well as new constructions and examples. The general definition is due to Kürschak [1913]. This section contains general definitions and properties.

Definition. An absolute value v on a field F is a mapping $v : F \longrightarrow \mathbb{R}$, $x \longmapsto |x|_v$, such that:

(a) $|x|_v \geq 0$ for all $x \in F$, and $|x|_v = 0$ if and only if $x = 0$;

(b) $|xy|_v = |x|_v |y|_v$ for all $x, y \in F$;

(c) $|x + y|_v \leq |x|_v + |y|_v$ for all $x, y \in F$.

Absolute values are also called *real valuations* or *real-valued valuations*, especially in the nonarchimedean cases discussed below. We denote $|x|_v$ by $|x|$ when v is known.

Examples include the familiar absolute values on $\mathbb{Q}$ and $\mathbb{R}$, and the absolute value or modulus on $\mathbb{C}$. Every field F also has a *trivial* absolute value t, $|x|_t = 1$ for all $x \neq 0$. For less trivial examples let K be any field. Readers will verify that

$$|f/g|_\infty = \begin{cases} 2^{\deg f - \deg g} & \text{if } f \neq 0, \\ 0 & \text{if } f = 0 \end{cases}$$

is well defined and is an absolute value v_∞ on $K(X)$. Similarly, an absolute value v_0 on $K(X)$ is well defined by

$$|f/g|_0 = \begin{cases} 2^{\text{ord } g - \text{ord } f} & \text{if } f \neq 0, \\ 0 & \text{if } f = 0, \end{cases}$$

where the *order* ord f of $f(X) = a_0 + a_1 X + \cdots + a_n X^n \neq 0$ is the smallest $n \geq 0$ such that $a_n \neq 0$. In these definitions, 2 can be replaced by any positive constant (by Proposition 3.1 below). For every $a \in K$,

$$|f(X)/g(X)|_a = |f(X-a)/g(X-a)|_0$$

is another absolute value v_a on $K(X)$. These absolute values are trivial on K.

By (a) and (c), an absolute value v on any field F induces a distance function

$$d(x, y) = |x - y|_v$$

on F, which makes F a metric space. Readers will verify that the operations on F, and v itself, are continuous in the resulting topology. The completion of F as a metric space is considered in the next section.

Equivalence of absolute values is defined by the following result.

Proposition **3.1.** *Let v and w be absolute values on a field F. The following conditons are equivalent:*

(1) v *and* w *induce the same topology on* F;

(2) $|x|_v < 1$ *if and only if* $|x|_w < 1$;

(3) *there exists* $c > 0$ *such that* $|x|_w = \left(|x|_v\right)^c$ *for all* $x \in F$.

Proof. (1) implies (2). Assume $|x|_v < 1$. Then $\lim_{n \to \infty} |x^n|_v = 0$ and $\lim_{n \to \infty} x^n = 0$ in the topology induced by v and w. Hence the open set $\{ x \in F ; |x|_w < 1 \}$ contains some x^n; $|x^n|_w < 1$ for some n; and $|x|_w < 1$. Exchanging v and w yields the converse implication.

(2) implies (3). If (2) holds, then $|x|_v > 1$ if and only if $|x|_w > 1$, and $|x|_v = 1$ if and only if $|x|_w = 1$. In particular, v is trivial if and only if w is trivial, in which case (3) holds. Now assume that v and w are not trivial. Then $|a|_v > 1$ for some $a \in F$, $|a|_w > 1$, and $|a|_w = \left(|a|_v\right)^c$ for some $c > 0$.

We show that $|x|_w = \left(|x|_v\right)^c$ for all $x \in F$. We may assume that $|x|_v \neq 0, 1$. Then $|x|_v = \left(|a|_v\right)^t$ for some $t \in \mathbb{R}$. If $m/n \in \mathbb{Q}$ and $m/n < t$, then $\left(|a|_v\right)^{m/n} < |x|_v$, $|a^m|_v < |x^n|_v$, $|a^m/x^n|_v < 1$, $|a^m/x^n|_w < 1$, and $\left(|a|_w\right)^{m/n} < |x|_w$. Similarly, $m/n > t$ implies $\left(|a|_w\right)^{m/n} > |x|_w$. Therefore $\left(|a|_w\right)^t = |x|_w$, and $|x|_w = \left(|a|_w\right)^t = \left(|a|_v\right)^{ct} = \left(|x|_v\right)^c$.

(3) implies (1). If (3) holds, then the metric spaces on F defined by v and w have the same open disks, and therefore the same topology. $\square$

Definition. Two absolute values on a field are equivalent *when they satisfy the equivalent conditions in Proposition* 3.1.

Archimedean absolute values.

Definition. An absolute value v on a field F is archimedean *when there is no $x \in F$ such that $|n|_v \leq |x|_v$ for every positive integer n.*

Here $|n|_v$ is short for $|n1|_v$, where $n1 \in F$. The usual absolute values on $\mathbb{Q}$, $\mathbb{R}$, and $\mathbb{C}$ are archimedean, but not the absolute values v_0 and v_∞ on $K(X)$.

Proposition **3.2.** *For an absolute value v the following are equivalent:*

(1) v *is nonarchimedean;*

(2) $|n|_v \leqq 1$ *for all $n \in \mathbb{Z}$, $n > 1$;*

(3) $|n|_v \leqq 1$ *for all $n \in \mathbb{Z}$;*

(d) $|x + y|_v \leqq \max (|x|_v, |y|_v)$ *for all $x, y \in F$.*

Proof. (1) implies (2). If $|m| > 1$ for some $m > 1$, then $\lim_{k \to \infty} |m^k| = \infty$ and there is no $x \in F$ such that $|n| \leqq |x|$ for all $n \in \mathbb{Z}$.

(2) implies (3) since $|-1|^2 = |1| = 1$ and $|-1| = 1$, whence $|-m| = |m|$.

(3) implies (d). If (3) holds, then, for all $n > 0$,

$$|x + y|^n = \left| \sum_{0 \leqq k \leqq n} \binom{n}{k} x^k y^{n-k} \right|$$
$$\leqq \sum_{0 \leqq k \leqq n} |x|^k |y|^{n-k} \leqq (n + 1) \max (|x|^n, |y|^n).$$

Hence $|x + y| \leqq (n + 1)^{1/n} \max (|x|, |y|)$ for all n. This yields (d), since $\lim_{n \to \infty} (n + 1)^{1/n} = 1$.

(d) implies (1): (d) implies $|n| = |1 + \cdots + 1| \leqq |1| = 1$ for all $n > 0$. $\square$

The next result is an exercise:

Corollary **3.3.** *A field of characteristic $p \neq 0$ has no archimedean absolute value.*

Absolute values on $\mathbb{Q}$. Besides the usual absolute value, $\mathbb{Q}$ has an absolute value v_p for every prime p, which readers will verify is well defined by

$$|m/n|_p = \begin{cases} p^{-k} & \text{if } m/n = p^k t/u \neq 0, \text{ where } p \text{ does not divide } t \text{ or } u, \\ 0 & \text{if } m/n = 0. \end{cases}$$

It turns out that these are essentially all absolute values on $\mathbb{Q}$.

Proposition **3.4** (Ostrowski [1918]). *Every nontrivial absolute value on $\mathbb{Q}$ is equivalent either to the usual absolute value or to v_p for some unique prime p.*

Proof. Let v be a nontrivial nonarchimedean absolute value on $\mathbb{Q}$. By 3.2, $|n|_v \leqq 1$ for all $n \in \mathbb{Z}$. If $|n|_v = 1$ for all $0 \neq n \in \mathbb{Z}$, then $|m/n|_v |n|_v = |m|_v$ implies $|x|_v = 1$ for all $0 \neq x \in \mathbb{Q}$ and v is trivial. Therefore

$$P = \{ n \in \mathbb{Z}; |n|_v < 1 \} \neq 0.$$

In fact, P is a prime ideal of $\mathbb{Z}$: indeed, P is an ideal by (b), (d), $1 \notin P$, and $m, n \notin P$ implies $|m|_v = |n|_v = 1$ and $mn \notin P$. Hence P is generated by a prime p. Let $c = |p|_v < 1$. If $m/n = p^k t/u \neq 0$, where p does not divide t

or u, then $|t|_v = |u|_v = 1$ and $|m/n|_v = c^k$. Hence v is equivalent to v_p; p is unique since v_p and v_q are not equivalent when $p \neq q$ (e.g., $|p|_q = 1$).

Now let v be an archimedean absolute value on $\mathbb{Q}$. We show that $|n|_v > 1$ for all $n > 1$. Let $m, n \in \mathbb{Z}$, $m, n > 1$. Then $n^k \leq m < n^{k+1}$ for some $k \geq 0$, $k \leq \log_n m$. Repeated division by n yields $0 \leq r_1, \ldots, r_k < n$ such that $m = r_0 + r_1 n + \cdots + r_k n^k$. Let $t = \max\left(1, |n|_v\right)$. Since $|r_i|_v \leq r_i < n$,

$$|m|_v \leq (k+1)\, n\, t^k \leq (1 + \log_n m)\, n\, t^{\log_n m}.$$

This inequality holds for all $m > 1$ and also holds for any $m^r > 1$; hence $\left(|m|_v\right)^r = |m^r|_v \leq (1 + r\log_n m)\, n\, t^{r\log_n m}$ and

$$|m|_v \leq \left((1 + r\log_n m)\, n\right)^{1/r} t^{\log_n m},$$

for all $r > 0$. Since $\lim_{r \to \infty} \left((1 + r\log_n m)\, n\right)^{1/r} = 1$, we obtain

$$|m|_v \leq t^{\log_n m} = \left(\max\left(1, |n|_v\right)\right)^{\log_n m}$$

for all $m, n > 1$. By 3.2, $|m|_v > 1$ for some $m > 1$; therefore $|n|_v > 1$ for all $n > 1$. Then $|m|_v \leq \left(|n|_v\right)^{\log_n m} = \left(|n|_v\right)^{\ln m/\ln n}$, $\left(|m|_v\right)^{1/\ln m} = \left(|n|_v\right)^{1/\ln n}$, and $\left(\ln|m|_v\right)/\ln m = \left(\ln|n|_v\right)/\ln n$, for all $m, n > 1$. Hence $c = \left(\ln|n|_v\right)/\ln n$ does not depend on n (as long as $n > 1$). Then $|n|_v = n^c$ for all $n \geq 1$, and $|x|_v = \left(|x|\right)^c$ for all $x \in \mathbb{Q}$. $\square$

Exercises

1. Prove that $\left||x|_v - |y|_v\right| \leq |x - y|_v$, for every absolute value v.

2. Define absolute values on a domain; show that every absolute value on a domain extends uniquely to an absolute value on its quotient field.

3. Verify that v_∞ is a nonarchimedean absolute value on $K(X)$.

4. Verify that v_0 is a nonarchimedean absolute value on $K(X)$.

5. Verify that v_p is a nonarchimedean absolute value on $\mathbb{Q}$ for every prime p.

6. Verify that the operations on F (including $x \longmapsto x^{-1}$, where $x \neq 0$) are continuous in the topology induced by an absolute value.

7. Show that every absolute value is continuous in its own topology.

8. Prove that a field of characteristic $p \neq 0$ has no archimedean absolute value.

9. Let v be an absolute value on a field F. Show that $\left(|x|_v\right)^c$ is an absolute value on F for every constant $0 < c < 1$ (for every $c > 0$ if v is nonarchimedean).

10. Let $0 \neq x \in \mathbb{Q}$. Show that $|x| \prod_{p \text{ prime}} |x|_p = 1$. (First show that $|x|_p = 1$ for almost all primes p.)

11. Let K be any field. Prove that every nontrivial absolute value on $K(X)$ that is trivial on K is equivalent either to v_∞, or to a suitably defined v_q for some unique monic irreducible polynomial $q \in K[X]$.

4. Completions

As metric spaces have completions, so do fields with absolute values (Kürschak [1913]). The construction of $\mathbb{R}$ from $\mathbb{Q}$ by Cauchy sequences is an example. Completions also yield a new field, the field of p-adic numbers, and its ring of p-adic integers, first constructed by Hensel [1897].

In what follows, F is a field with an absolute value. A *Cauchy sequence* in F is a sequence $a = (a_n)_{n>0}$ of elements of F such that, for every positive real number ε, $\left| a_m - a_n \right| < \varepsilon$ holds for all sufficiently large m and n. In F, every sequence that has a limit is a Cauchy sequence.

Definition. A field F is complete *with respect to an absolute value when it is complete as a metric space (when every Cauchy sequence of elements of F has a limit in F).*

For example, $\mathbb{R}$ and $\mathbb{C}$ are complete, but not $\mathbb{Q}$.

Construction. The main result of this section is the following:

*Theorem **4.1.** Let F be a field with an absolute value v. There exists a field extension $\widehat{F} = \widehat{F}_v$ of F and an absolute value $\widehat{v}$ on $\widehat{F}$ that extends v, such that $\widehat{F}$ is complete with respect to $\widehat{v}$ and F is dense in $\widehat{F}$.*

Proof. As a set, $\widehat{F}$ is the completion of F as a metric space. Let C be the set of all Cauchy sequences of elements of F. Termwise sums and products

$$(a_n)_{n>0} + (b_n)_{n>0} = (a_n + b_n)_{n>0}, \quad (a_n)_{n>0} \, (b_n)_{n>0} = (a_n b_n)_{n>0}$$

of Cauchy sequences are Cauchy sequences (see the exercises). Hence C is a commutative ring; its identity element is the constant sequence, $a_n = 1$ for all n.

Let $\mathfrak{z}$ be the set of all $a = (a_n)_{n>0} \in C$ such that $\lim a_n = \lim_{n \to \infty} a_n = 0$, equivalently $\lim \left| a_n \right| = 0$. Readers will verify that $\mathfrak{z}$ is an ideal of C. We show that $\mathfrak{z}$ is a maximal ideal of C. Let $\mathfrak{a} \supsetneqq \mathfrak{z}$ be an ideal of C. Let $a \in \mathfrak{a} \backslash \mathfrak{z}$. Then $\lim \left| a_n \right| > 0$ and there exists $\delta > 0$ such that $\left| a_n \right| \geqq \delta$ for all sufficiently large n. Let $b_n = 1/a_n$ if $a_n \neq 0$, $b_n = 1$ if $a_n = 0$. If m and n are sufficiently large, then $\left| a_m \right| \geqq \delta$, $\left| a_n \right| \geqq \delta$, and

$$\left| b_m - b_n \right| = \left| \frac{1}{a_m} - \frac{1}{a_m} \right| = \left| \frac{a_n - a_m}{a_m \, a_n} \right| \leqq \frac{\left| a_n - a_m \right|}{\delta^2};$$

Hence $b = (b_n)_{n>0}$ is a Cauchy sequence. We see that $\lim (a_n b_n) = 1$, so that $c = ab - 1 \in \mathfrak{z}$. Hence $1 = ab + c \in \mathfrak{a}$ and $\mathfrak{a} = C$.

We show that the field $\widehat{F} = C/\mathfrak{z}$ has the required properties. For every $x \in F$ there is a constant sequence $\overline{x} = (x_n)_{n>0}$ in which $x_n = x$ for all n. This yields a homomorphism $x \longmapsto \overline{x} + \mathfrak{z}$ of F into $\widehat{F}$. Hence F becomes a subfield of $\widehat{F}$ when we identify $x \in F$ and $\overline{x} + \mathfrak{z} \in \widehat{F}$.

If $a = (a_n)_{n>0}$ is a Cauchy sequence in F, then $\left(\left| a_n \right| \right)_{n \geq 0}$ is a Cauchy sequence in $\mathbb{R}$, since $||a_n| - |b_n|| \leqq |a_n - b_n|$. Hence $\lim |a_n|$ exists in $\mathbb{R}$. If $a - b \in \mathfrak{z}$, then $\lim |a_n - b_n| = 0$ and the same inequality $||a_n| - |b_n|| \leqq |a_n - b_n|$ implies $\lim |a_n| = \lim |b_n|$. Therefore a mapping $\widehat{v} : \widehat{F} \longrightarrow \mathbb{R}$ is well defined by $\widehat{v}(a + \mathfrak{z}) = \lim |a_n|$, whenever $a = (a_n)_{n>0}$ is a Cauchy sequence in F. It is immediate that $\widehat{v}$ is an absolute value on $\widehat{F}$. If $x \in F$, then $\widehat{v}(x) = \lim |x| = |x|$; hence $\widehat{v}$ extends v.

If $a = (a_n)_{n>0}$ is a Cauchy sequence in F, then $a + \mathfrak{z} = \lim a_n$ in $\widehat{F}$, since $\widehat{v}(a + \mathfrak{z} - a_m) = \lim |a_n - a_m|$. Hence F is dense in $\widehat{F}$. Finally, let $A = (A_n)_{n>0}$ be a Cauchy sequence in $\widehat{F}$. Since F is dense in $\widehat{F}$ there exists for every $n > 0$ some $a_n \in F$ such that $\widehat{v}(A_n - a_n) < 1/n$. Then $a = (a_n)_{n>0}$ is a Cauchy sequence in $\widehat{F}$, a is a Cauchy sequence in F, and $a + \mathfrak{z} = \lim a_n = \lim A_n$ in $\widehat{F}$. Thus $\widehat{F}$ is complete. □

Definition. A completion *of a field* F *with respect to an absolute value* v *is a field extension* $\widehat{F}_v$ *of* F *with an absolute value* $\widehat{v}$ *that extends* v, *such that* $\widehat{F}$ *is complete with respect to* $\widehat{v}$ *and* F *is dense in* $\widehat{F}$.

For example, $\mathbb{R}$ is a completion of $\mathbb{Q}$ with respect to its usual absolute value; readers will show that $K((X))$ is a completion of $K(X)$ with respect to v_0. Another example, the field of p-adic numbers, is given below.

Properties. Completions have a universal property:

Proposition **4.2.** *Let* F *and* K *be fields with absolute values. If* K *is complete, then every homomorphism of* F *into* K *that preserves absolute values extends uniquely to a homomorphism of any completion of* F *into* K *that preserves absolute values.*

Proof. Let $\widehat{F}$ be a completion of F and let $\varphi : F \longrightarrow K$ be a homomorphism that preserves absolute values. Since F is dense in $\widehat{F}$, every element α of the metric space $\widehat{F}$ is the limit of a sequence $a = (a_n)_{n>0}$ of elements of F, which is a Cauchy sequence in $\widehat{F}$ and therefore a Cauchy sequence in F. Since φ preserves absolute values, $(\varphi a_n)_{n>0}$ is a Cauchy sequence and has a limit in K. This limit depends only on α: if $\alpha = \lim a_n = \lim b_n$, where $a_n, b_n \in F$, then $\lim (a_n - b_n) = 0$, $\lim (\varphi a_n - \varphi b_n) = 0$ since φ preserves absolute values, and $\lim \varphi a_n = \lim \varphi b_n$. Hence a mapping $\psi : \widehat{F} \longrightarrow K$ is well defined by $\psi \alpha = \lim \varphi a_n$ whenever $\alpha = \lim a_n$ and $a_n \in F$.

It is immediate that ψ extends φ and, from the limit laws, that ψ is a field homomorphism. Moreover, ψ preserves absolute values: $\alpha = \lim a_n$ implies $|\alpha| = \lim |a_n| = \lim |\varphi a_n| = |\psi \alpha|$, since absolute values are continuous and preserved by φ. Conversely, if $\chi : \widehat{F} \longrightarrow K$ is a field homomorphism that extends φ and preserves absolute values, then χ is continuous and $\alpha = \lim a_n$ implies $\chi \alpha = \lim \chi a_n = \lim \varphi a_n = \psi \alpha$; hence $\chi = \psi$. □

A standard universal property argument then yields uniqueness:

Proposition **4.3.** *The completion of a field with respect to an absolute value is unique up to isomorphisms that preserve absolute values.*

p-adic numbers.

Definitions. For every prime number p, $\widehat{\mathbb{Q}}_p$ *is the completion of* $\mathbb{Q}$ *with respect to* v_p; $\widehat{\mathbb{Z}}_p = \{ x \in \widehat{\mathbb{Q}}_p \, ; \, |x|_p \leqq 1 \}$; *a* p-adic number *is an element of* $\widehat{\mathbb{Q}}_p$; *a* p-adic integer *is an element of* $\widehat{\mathbb{Z}}_p$.

Other notations are in use for $\widehat{\mathbb{Q}}_p$ and $\widehat{\mathbb{Z}}_p$.

In $\widehat{\mathbb{Q}}_p$, for every $x_n \in \mathbb{Z}$, the series $\sum x_n p^n$ converges: the partial sums constitute a Cauchy sequence, since $\left| x_n p^n \right|_p \leqq p^{-n}$ and $\left| x_n p^n + x_{n+1} p^{n+1} + \cdots + x_m p^m \right|_p \leqq p^{-n}$. This yields a more concrete description of $\widehat{\mathbb{Q}}_p$:

Proposition **4.4.** *Every Laurent series* $\sum_{n \geqq m} x_n p^n$ *with coefficients* $x_n \in \mathbb{Z}$ *converges in* $\widehat{\mathbb{Q}}_p$; *every* p-adic integer x *is the sum*

$$x = x_0 + x_1 p + \cdots + x_n p^n + \cdots$$

of a unique power series with coefficients $x_n \in \mathbb{Z}$ *such that* $0 \leqq x_n < p$; *every* p-adic number $x \neq 0$ *is the sum of a unique Laurent series*

$$x = x_m p^m + x_{m+1} p^{m+1} + \cdots + x_n p^n + \cdots$$

with coefficients $x_n \in \mathbb{Z}$ *such that* $0 \leqq x_n < p$ *for all* $n \geqq m$, *and* $x_m \neq 0$; *and then* $|x|_p = p^{-m}$, *and* x *is a* p-adic integer if and only if $m \geqq 0$.

Proof. First we prove a lemma.

Lemma **4.5.** *If* $x \in \widehat{\mathbb{Q}}_p$ *and* $|x|_p \leqq p^{-m}$, *then* $\left| x - t p^m \right|_p < p^{-m}$ *for some unique integer* $0 \leqq t < p$; *if* $|x|_p = p^{-m}$, *then* $t \neq 0$.

Proof. If $|x|_p < p^{-m}$, then $t = 0$ serves. Now assume $|x|_p = p^{-m}$. Since $\mathbb{Q}$ is dense in $\widehat{\mathbb{Q}}_p$ we have $|x - y|_p < p^{-m}$ for some $y \in \mathbb{Q}$. Then $|y|_p = p^{-m}$ and $y = p^m k / \ell$, where $m \in \mathbb{Z}$ and p does not divide k or ℓ. Since $\mathbb{Z}_p$ is a field we have $k \equiv t\ell \pmod{p}$ for some $t \in \mathbb{Z}$, and can arrange that $0 \leqq t < p$. Then $0 < t < p$, since p does not divide t, and $\left| y - t p^m \right|_p = \left| p^m (k - t\ell)/\ell \right|_p < p^{-m}$, since p divides $k - t\ell$ but not ℓ. Hence $\left| x - t p^m \right|_p < p^{-m}$. If also $\left| x - u p^m \right|_p < p^{-m}$, where $0 \leqq u < p$, then $\left| t p^m - u p^m \right|_p < p^{-m}$, p divides $t - u$, and $t = u$. $\square$

We now prove Proposition 4.4. Let $|x|_p = p^{-m}$. By 4.5, $\left| x - x_m p^m \right|_p \leqq p^{-(m+1)}$ for some unique integer $0 < x_m < p$; hence

$$\left| x - x_m p^m - x_{m+1} p^{m+1} \right|_p \leqq p^{-(m+2)}$$

for some unique integer $0 \leq x_{m+1} < p$; repetition yields unique $0 \leq x_n < p$ such that $\left| x - \sum_{m \leq n \leq r} x_n p^n \right|_p < p^{-r}$, for every $r > m$. Then the series $\sum_{n \geq m} x_n p^n$ converges to x in $\widehat{\mathbb{Q}}_p$. If $x \in \widehat{\mathbb{Z}}_p$, then $m \geq 0$ and $\sum_{n \geq m} x_n p^n$ is a power series, $\sum_{n \geq 0} x_n p^n$, with $x_n = 0$ for all $n < m$.

Assume that $x = \sum_{n \geq \ell} y_n p^n$, where $0 \leq y_n < p$ for all n and $y_\ell \neq 0$. Then $\left| \sum_{n > \ell} y_n p^n \right|_p = \lim_{r \to \infty} \left| \sum_{\ell < n \leq r} y_n p^n \right|_p \leq p^{-(\ell+1)}$ and $\left| x \right|_p = \left| y_\ell p^\ell \right|_p = p^{-\ell}$. Hence $\ell = m$. Uniqueness of x_n is proved by induction: if $x_n = y_n$ for all $n < r$ (for instance, if $r = m$, or, in case $x \in \widehat{\mathbb{Z}}_p$, $r = 0$), then $\left| x - \sum_{m \leq n \leq r} y_n p^n \right|_p < p^{-r}$ and uniqueness in 4.5 yields $x_r = y_r$. $\square$

Exercises

1. In any field with an absolute value, show that the termwise sum and product of two Cauchy sequences are Cauchy sequences.

2. Prove the following: in any field with an absolute value, if $a = (a_n)_{n>0}$ converges to 0 and $b = (b_n)_{n>0}$ is a Cauchy sequence, then $ab = (a_n b_n)_{n \geq 0}$ converges to 0.

3. Let K be a field. Show that $K((X))$ is a completion of $K(X)$ with respect to v_0.

4. Show that a completion is uniquely determined by its universal property, up to isomorphisms that preserve absolute values.

5. Let F be complete with respect to a nonarchimedean absolute value. Show that a series $\sum a_n$ converges in F if and only if $\lim a_n = 0$.

6. Prove directly that every integer $x \in \mathbb{Z}$ is a sum $x = x_0 + x_1 p + \cdots + x_m p^m$ for some unique $m \geq 0$ and $x_0, \ldots, x_m \in \mathbb{Z}$ such that $0 \leq x_0, \ldots, x_m < p$ and $x_m \neq 0$.

7. Let $x = \sum_{n \geq 0} x_n p^n \in \widehat{\mathbb{Z}}_p$, where $x_n \in \mathbb{Z}$ and $0 \leq x_n < p$ for all n. Show that x is a unit in $\widehat{\mathbb{Z}}_p$ is and only if $x_0 \neq 0$.

8. Write -1 as the sum of a Laurent series in $\widehat{\mathbb{Q}}_2$.

9. Write $\frac{1}{3}$ as the sum of a Laurent series in $\widehat{\mathbb{Q}}_2$.

10. Write $\frac{1}{2}$ as the sum of a Laurent series in $\widehat{\mathbb{Q}}_3$.

11. Show that $\widehat{\mathbb{Q}}_p$ is the field of fractions of $\widehat{\mathbb{Z}}_p$.

12. Show that $p\widehat{\mathbb{Z}}_p$ is a maximal ideal of $\widehat{\mathbb{Z}}_p$. Find $\widehat{\mathbb{Z}}_p / p\widehat{\mathbb{Z}}_p$.

13. Show that $\widehat{\mathbb{Z}}_p$ is a PID with only one representative prime, and that the ideals of $\widehat{\mathbb{Z}}_p$ constitute a chain.

14. Show that every domain with an absolute value has a completion with a suitable universal property.

15. Let K be a field. Show that $K[[X]]$ is a completion of $K[X]$ with respect to v_0.

5. Extensions

Can absolute values on a field K be extended to absolute values on algebraic extensions of K? This is the extension problem for absolute values, which was solved by Ostrowski [1918], [1934]. In this section we solve the extension problem in the archimedean case. We also prove Ostrowski's theorem [1918], which determines all complete archimedean fields.

Completeness. Let E be a finite field extension of K. An absolute value $E \longrightarrow \mathbb{R}$ on E induces an absolute value $K \longrightarrow \mathbb{R}$ on K. First we show that, if K is complete, then so is E. We prove this in a more general setting.

Definition. Let K be a field with an absolute value and let V be a vector space over K. A norm *on V is a mapping $x \longmapsto \|x\|$ of V into $\mathbb{R}$ such that*

(a) $\|x\| \geqq 0$ *for all $x \in V$, and $\|x\| = 0$ if and only if $x = 0$;*

(b) $\|ax\| = |a| \, \|x\|$ *for all $a \in K$ and $x \in V$;*

(c) $\|x + y\| \leqq \|x\| + \|y\|$ *for all $x, y \in V$.*

Then V is a normed vector space *over K.*

For instance, when E is a field extension of K, viewed as a vector space over K, then an absolute value on E induces an absolute value on K and is, in particular, a norm on E. In general, a norm on V induces a distance function $d(x, y) = \|x - y\|$ on V, which makes V a metric space.

Proposition **5.1.** *Let V be a normed vector space of finite dimension over a field K with an absolute value. If K is complete, then V is complete and, in any basis $e_1, \ldots, e_n$ of V over K, (1) the ith coordinate function $\sum x_i e_i \longmapsto x_i$ is continuous; (2) a sequence $(x_k)_{k \geq 0}$, $x_k = \sum_i x_{k,i} e_i$, converges in V if and only if all its coordinate sequences $(x_{k,i})_{k \geq 0}$ converge in K; (3) a sequence is a Cauchy sequence in V if and only if all its coordinate sequences are Cauchy sequences in K.*

Proof. We start with (3). For every $x = \sum_i x_i e_i$ we have

$$\|x\| = \|\textstyle\sum_i x_i e_i\| \leqq \textstyle\sum_i \left(|x_i| \|e_i\| \right).$$

Let $(x_k)_{k \geq 0}$ be a sequence of elements of V, $x_k = \sum_i x_{k,i} e_i$. If $(x_{k,i})_{k \geq 0}$ is Cauchy in K for all i, then $(x_k)_{k \geq 0}$ is Cauchy in V, by the inequality above.

The converse is proved by induction on n. There is nothing to prove if $n = 1$. Now let $n > 1$. Assume that $(x_k)_{k \geq 0}$ is Cauchy in V, but that, say, $(x_{k,n})_{k \geq 0}$ is not Cauchy in K. Then there exists $\varepsilon > 0$ such that $|x_{i,n} - x_{j,n}|_v \geqq \varepsilon$ for arbitrarily large i and j. In particular, for every $k \geqq 0$ we have $|x_{i_k,n} - x_{j_k,n}| \geqq \varepsilon$ for some $i_k, j_k > k$; then $x_{i_k,n} - x_{j_k,n} \neq 0$ in K. Let

$$y_k = (x_{i_k,n} - x_{j_k,n})^{-1} (x_{i_k} - x_{j_k}) \in V.$$

Then $\lim_{k \to \infty} y_k = 0$, since $\left| (x_{i_k,n} - x_{j_k,n})^{-1} \right| \leqq 1/\varepsilon$, and $(y_k)_{k \geq 0}$ is Cauchy in V. Hence $(y_k - e_n)_{k \geq 0}$ is Cauchy in V. Since y_k has nth coordinate $y_{k,n} = 1$, $(y_k - e_n)_{k \geq 0}$ is a Cauchy sequence in the subspace of V spanned by $e_1, \ldots, e_{n-1}$. By the induction hypothesis, $(y_{k,i})_{k \geq 0}$ is Cauchy in K when $i < n$. Since K is complete, $(y_{k,i})_{k \geq 0}$ has a limit z_i in K when $i < n$; the sequence $(y_{k,n})_{k \geq 0}$ also has a limit $z_n = 1$. Let $z = \sum_i z_i e_i$. Then $\|y_k - z\| \leqq \sum_i (|y_{k,i} - z_i| \|e_i\|)$ and $\lim_{k \to \infty} y_k = z$. But this not possible, since $\lim_{k \to \infty} y_k = 0$ and $z \neq 0$. This contradiction proves (3).

If now every $(x_{k,i})_{k \geq 0}$ has a limit y_i in K, then $y = \sum_i y_i e_i \in V$ and $\|x_k - y\| \leqq \sum_i (|x_{k,i} - y_i| \|e_i\|)$, so that $(x_k)_{k \geq 0}$ has a limit y in V. Conversely, if $(x_k)_{k \geq 0}$ converges in V, then $(x_k)_{k \geq 0}$ is Cauchy in V and, for every i, $(x_{k,i})_{k \geq 0}$ is Cauchy in K by (3) and converges in K since K is complete, which proves (2); in fact, if $\lim_{k \to \infty} x_{k,i} = y_i$, then $\lim_{k \to \infty} x_k = \sum_i y_i e_i$ by the direct part, so that $\lim_{k \to \infty} x_{k,i}$ is the ith coordinate of $\lim_{k \to \infty} x_k$. If $(x_k)_{k \geq 0}$ is a Cauchy sequence in V, then every $(x_{k,i})_{k \geq 0}$ is Cauchy in K, by (3); every $(x_{k,i})_{k \geq 0}$ converges in K, since K is complete; and $(x_k)_{k \geq 0}$ converges in V by (2); hence V is complete.

Finally, if the ith coordinate function is not continuous at $t = \sum_i t_i e_i \in V$, then there exist $\varepsilon > 0$ and, for every $k > 0$, some $x_k \in V$ such that $\|x_k - t\| < 1/k$ and $|x_{k,i} - t_i| \geqq \varepsilon$; then $\lim_{k \to \infty} x_k = t$, whence $\lim_{k \to \infty} x_{k,i} = t_i$; this contradiction proves (1). $\square$

Uniqueness. We now return to the extension problem for absolute values.

Theorem **5.2.** *Let E be a finite extension of degree n of a field K that is complete with respect to an absolute value v. If there exists an absolute value w on E that extends v, then w is unique and*

$$|\alpha|_w = \left(\left| N_K^E(\alpha) \right|_v \right)^{1/n}$$

for all $\alpha \in E$; moreover, E is complete with respect to w.

Proof. The definition of $N(\alpha)$ shows that $N(\alpha)$ is a polynomial function of the coordinates of α in any basis of E over K, hence continuous, by 5.1. Let $\alpha \in E$, $\alpha \neq 0$, and $\beta = \alpha^n N(\alpha)^{-1}$. Then $N(\beta) = 1$ by V.7.3, since $N(\alpha) \in K$. Hence $N(\beta^k) = 1$ for all k, $N(\lim_{k \to \infty} \beta^k) = \lim_{k \to \infty} N(\beta^k) = 1$, $\lim_{k \to \infty} \beta^k \neq 0$, $|\beta|_w \geqq 1$, and $(|\alpha|_w)^n \geqq |N(\alpha)|_v$. Similarly, $(|\alpha|_w^{-1})^n \geqq |N(\alpha^{-1})|_v$; hence $(|\alpha|_w)^n = |N(\alpha)|_v$. Finally, E is complete, by 5.1. $\square$

Existence. Theorem 5.2 yields absolute values, but only in some cases.

Proposition **5.3.** *If K is a field that is complete with respect to an absolute value v, and does not have characteristic 2, then v can be extended to every finite extension of K of degree 2.*

Proof. Inspired by 5.2 we try $|\alpha|_w = \left(\left|N(\alpha)\right|_v\right)^{1/2}$. By V.6.1, a finite extension E of K of degree 2 is a Galois extension and has a nontrivial automorphism $\alpha \longmapsto \overline{\alpha}$. Then $N(\alpha) = \alpha\overline{\alpha}$. Hence w has properties (a) and (b) in Section 3. This leaves (c), $|\alpha + \beta|_w \leqq |\alpha|_w + |\beta|_w$.

Let $b = \alpha + \overline{\alpha}$ and $c = \alpha\overline{\alpha}$. Then $b, c \in K$. If $\left(|b|_v\right)^2 > 4|c|_v$, then, by 5.4 below, $(X - \alpha)(X - \overline{\alpha}) = X^2 - bX + c$ has a root in K; hence $\alpha \in K$ and $\left(|b|_v\right)^2 = \left(\left|2\alpha\right|_v\right)^2 \leqq 4\left|\alpha^2\right|_v = 4|c|_v$. Therefore $\left(|b|_v\right)^2 \leqq 4|c|_v$.

If $\beta = 1$, then (c) reads $\left|N(\alpha + 1)\right|_v = \left(\left|\alpha + 1\right|_w\right)^2 \leqq \left(|\alpha|_w + 1\right)^2 = \left|N(\alpha)\right|_v + 2\left(\left|N(\alpha)\right|_v\right)^2 + 1$; since $\left(|b|_v\right)^2 \leqq 4|c|_v = 4\left|N(\alpha)\right|_v$,

$$
\begin{aligned}
\left|N(\alpha + 1)\right|_v &= \left|(\alpha + 1)(\overline{\alpha} + 1)\right|_v = |c + b + 1|_v \\
&\leqq |c|_v + |b|_v + 1 \leqq \left|N(\alpha)\right|_v + 2\left(\left|N(\alpha)\right|_v\right)^{1/2} + 1,
\end{aligned}
$$

and (c) holds. Then (c) holds for all $\beta \neq 0$: $|\alpha + \beta|_w = |\beta|_w \left|\alpha\beta^{-1} + 1\right|_w \leqq |\beta|_w \left(\left|\alpha\beta^{-1}\right|_w + 1\right) = |\alpha|_w + |\beta|_w$. $\square$

Lemma 5.4. *Let K be a field that is complete with respect to an absolute value v, and does not have characteristic 2. If $\left(|b|_v\right)^2 > 4|c|_v$, then $X^2 - bX + c \in K[X]$ has a root in K.*

Proof. We may assume that $c \neq 0$; then $b \neq 0$. We use successive approximations $x_{n+1} = f(x_n)$ to find a root, noting that $x^2 - bx + c = 0$ if and only if $x = b - (c/x)$. Let $x_1 = \frac{1}{2}b$ and $x_{n+1} = b - (c/x_n)$. If $|x| \geqq \frac{1}{2}|b| > 0$, then

$$
|x - (c/x)| \geqq |b| - (|c|/|x|) \geqq |b| - 2(|c|/|b|) \geqq |b| - \tfrac{1}{2}|b| = \tfrac{1}{2}|b| > 0,
$$

since $|c| < \frac{1}{4}|b|$; hence $|x_n| \geqq \frac{1}{2}|b|$ for all n, $x_n \neq 0$ for all n, and x_n is well defined for all n. We show that $(x_n)_{n>0}$ is a Cauchy sequence. Let $r = 4|c|/|b|^2 < 1$. Since $|x_n| \geqq \frac{1}{2}|b|$ for all n,

$$
\begin{aligned}
|x_{n+2} - x_{n+1}| &= \left|\frac{c}{x_n} - \frac{c}{x_{n+1}}\right| = \frac{|c|\,|x_{n+1} - x_n|}{|x_n|\,|x_{n+1}|} \\
&\leqq \frac{4|c|\,|x_{n+1} - x_n|}{|b|^2} = r\,|x_{n+1} - x_n|;
\end{aligned}
$$

therefore $|x_{n+2} - x_{n+1}| \leqq r^n |x_2 - x_1|$ for all n and $(x_n)_{n>0}$ is Cauchy. Hence $(x_n)_{n>0}$ has a limit x in K; then $|x| \geqq \frac{1}{2}|b| > 0$, $x = b - (c/x)$, and $x^2 + bx + c = 0$. $\square$

Ostrowski's theorem now follows from the previous results.

Theorem 5.5 (Ostrowski [1918]). *Up to isomorphisms that preserve absolute values, $\mathbb{R}$ and $\mathbb{C}$ are the only fields that are complete with respect to an archimedean absolute value.*

Proof. Let F be complete with respect to an archimedean absolute value v. By 3.3, F has characteristic 0. Hence $\mathbb{Q} \subseteq F$, up to isomorphism. By 3.4, the valuation induced by F on $\mathbb{Q}$ is equivalent to the usual absolute value. We may replace v by an equivalent absolute value that induces the usual absolute value on $\mathbb{Q}$. Then 4.2 yields an isomorphism, that preserves absolute values, of $\mathbb{R} = \widehat{\mathbb{Q}}$ onto a subfield of F. Therefore we may assume from the start that $\mathbb{R} \subseteq F$ and that F induces the usual absolute value on $\mathbb{R}$.

If F contains an element i such that $i^2 = -1$, then $\mathbb{C} = \mathbb{R}(i) \subseteq F$; by the uniqueness in 5.2, v induces the usual absolute value on $\mathbb{C}$, since both induce the usual absolute value on $\mathbb{R}$. If F contains no element i such that $i^2 = -1$, then v extends to an absolute value w on $E = F(i)$ by 5.3, and E is complete by 5.2. Then $\mathbb{C} = \mathbb{R}(i) \subseteq E$; by 5.2, w induces the usual absolute value on $\mathbb{C}$, since both induce the usual absolute value on $\mathbb{R}$. In this case, $E = \mathbb{C}$ implies $F = \mathbb{R}$. Therefore we may assume that $\mathbb{C} \subseteq F$ and that v induces the usual absolute value on $\mathbb{C}$; we need to prove that $F = \mathbb{C}$.

Assume that $\mathbb{C} \subsetneqq F$. Let $\alpha \in F \backslash \mathbb{C}$. Let $r = \text{g.l.b.} \{ |z - \alpha| ; z \in \mathbb{C} \}$. Since the function $f(x) = |x - \alpha|$ is continuous on $\mathbb{C}$, the "disk"

$$D = \{ z \in \mathbb{C} ; |z - \alpha| \leqq r + 1 \}$$

is a closed nonempty subset of $\mathbb{C}$. Hence $r = \text{g.l.b.} \{ |z - \alpha| ; z \in D \}$. Also, D is bounded, since $x, y \in D$ implies $|x - y| = |(x - \alpha) - (y - \alpha)| \leqq |x - \alpha| + |y - \alpha| \leqq 2r + 2$. Therefore the continuous function $f(x) = |x - \alpha|$ has a minimum value on D and $|z - \alpha| = r$ for some $z \in \mathbb{C}$. Then the "circle"

$$C = \{ z \in \mathbb{C} ; |z - \alpha| = r \}$$

is nonempty, closed since f is continuous, and bounded since $C \subseteq D$. We show that C is open; since $\mathbb{C}$ is connected this provides the required contradiction.

We show that $x \in C$, $y \in \mathbb{C}$, and $|x - y| < r$ implies $y \in C$ (hence C is open). Let $\beta = \alpha - x$ and $z = y - x$, so that $|\beta| = r$ and $|z| < r$. Let $n > 0$ and ε be a primitive nth root of unity. Then

$$\beta^n - z^n = (\beta - z)(\beta - \varepsilon z) \cdots (\beta - \varepsilon^{n-1} z)$$

and $|\beta - \varepsilon^i z| = |\alpha - x - \varepsilon^i z| \geqq r$ by the choice of r. Hence

$$|\beta - z| \, r^{n-1} \leqq |\beta^n - z^n| \leqq |\beta|^n + |z|^n = r^n + |z|^n$$

and $|\beta - z| \leqq r + (|z|^n / r^{n-1})$. Since $|z| < r$, letting $n \to \infty$ yields $|\beta - z| \leqq r$. But $|\beta - z| = |\alpha - y| \geqq r$. Hence $|\alpha - y| = r$ and $y \in C$. $\square$

In addition to a neat characterization of $\mathbb{R}$ and $\mathbb{C}$, Ostrowski's theorem tells the complete story on fields that are complete with respect to an archimedean absolute value: up to isomorphism, they are subfields of $\mathbb{C}$, and their absolute values are induced by the usual absolute value on $\mathbb{C}$.

With Ostrowski's theorem, the extension problem for archimedean absolute values becomes trivial: when v is an archimedean absolute value on a field F, then, up to isomorphism, v can be extended to $\mathbb{C}$ and to every algebraic extension of F. The nonarchimedean case is considered in Section 7.

Exercises

1. Verify that addition and scalar multiplication on a normed vector space are continuous.

2. Let V be a finite-dimensional vector space over a field K that is complete with respect to an absolute value. Show that all norms on V induce the same topology on V.

3. Find all archimedean absolute values on $\mathbb{Q}(\sqrt{-5})$.

6. Valuations

Valuations were first defined in full generality by Krull [1932]. They are more general than nonarchimedean absolute values. They are also more flexible and extend more readily to field extensions. Their values are not restricted to real numbers, but are taken from the following more general objects.

Definition. A totally ordered abelian group is an ordered pair of an abelian group G together with a total order relation $\leqq$ on G such that $x < y$ implies $xz < yz$ for all $z \in G$.

For example, the multiplicative group $\mathbb{P}$ of all positive real numbers, and its subgroups, are totally ordered abelian groups with the usual order relation. When $n > 1$, readers will verify that $\mathbb{P}^n = \mathbb{P} \times \cdots \times \mathbb{P}$ is a totally ordered abelian group that is not isomorphic (as a totally ordered abelian group) to a subgroup of $\mathbb{P}$ when ordered lexicographically: $(x_1, \ldots, x_n) < (y_1, \ldots, y_n)$ if and only if there exists $k \leqq n$ such that $x_i = y_i$ for all $i < k$ and $x_k < y_k$.

Totally ordered abelian groups are also called just *ordered abelian groups*. They are often written additively (but here we prefer the multiplicative notation). In a totally ordered abelian group, $x < y$ implies $y^{-1} < x^{-1}$, since $y^{-1} > x^{-1}$ would imply $1 = xx^{-1} < xy^{-1} < yy^{-1} = 1$. Totally ordered abelian groups are torsion free, since $x > 1$ implies $1 < x < x^2 < \cdots < x^n < \cdots$.

An *isomorphism* of totally ordered abelian groups is an order preserving isomorphism ($x < y$ implies $\theta(x) < \theta(y)$); since these groups are totally ordered, the inverse bijection is also an order preserving isomorphism. For example, the natural logarithm function is an isomorphism of totally ordered abelian groups of $\mathbb{P}$ onto the additive group $(\mathbb{R}, +)$.

Definition. Let G be a totally ordered abelian group. Adjoin an element 0 to G such that $0 < g$ and $g0 = 0 = 0g$ for all $g \in G$. A valuation on a field F with values in G is a mapping $v : F \longrightarrow G \cup \{0\}$ such that

(a) $v(x) = 0$ *if and only if $x = 0$;*

(b) $v(xy) = v(x) v(y)$ for all $x, y \in F$;

(c) $v(x + y) \leqq \max (v(x), v(y))$ for all $x, y \in F$.

In (c) we have $v(x) \leqq v(y)$ or $v(y) \leqq v(x)$, since $G \cup \{0\}$ is totally ordered, and $\max (v(x), v(y))$ exists. For example, nonarchimedean absolute values are valuations with values in $\mathbb{P}$; thus, v_p is a valuation on $\mathbb{Q}$; v_∞ and v_0 are valuations on $K(X)$ for any field K.

Readers will verify that a valuation v_0 can be defined on any $K(X_1, ..., X_n)$ as follows. Let $v_0(0) = 0$. Every nonzero $f/g \in K(X_1, ..., X_n)$ can be written uniquely in the form $f/g = X_1^{m_1} X_2^{m_2} \cdots X_n^{m_n} (h/k)$, with $m_1, ..., m_n \geqq 0$ and $h(0, ..., 0), k(0, ..., 0) \neq 0$; let $v_0(f/g) = (2^{-m_1}, ..., 2^{-m_n}) \in \mathbb{P}^n$.

In general, $G_v = \{ v(x) \mid x \in F \backslash \{0\} \}$ is a subgroup of G, by (b).

Definitions. The value group *of a valuation* $v : F \longrightarrow G \cup \{0\}$ *is* $G_v = \{ v(x) \mid x \in F \backslash \{0\} \}$. *Two valuations* $v, w : F \longrightarrow G \cup \{0\}$ *are* equivalent *when there exists an order preserving isomorphism* θ *of* G_v *onto* G_w *such that* $w(x) = \theta(v(x))$ *for all* $x \neq 0$.

For every $c > 0$, $x \longmapsto x^c$ is an order preserving automorphism of $\mathbb{P}$; hence nonarchimedean absolute values that are equivalent as absolute values are equivalent as valuations. On the other hand, readers will be delighted to find that the valuation v_0 on $K(X_1, ..., X_n)$ is not equivalent to an absolute value; thus valuations are more general.

Valuation rings. Up to equivalence, valuations on a field F are determined by certain subrings of F.

Definition. The valuation ring *of a valuation* v *on a field* F *is* $\mathfrak{o}_v = \{ x \in F \mid v(x) \leqq 1 \}$.

Readers will prove the following properties:

Proposition **6.1.** *For every valuation* v *on a field* F:

(1) $\mathfrak{o}_v$ *is a subring of* F; *when* $x \in F \backslash \{0\}$, *then* $x \in \mathfrak{o}_v$ *or* $x^{-1} \in \mathfrak{o}_v$; *in particular,* F *is the quotient field of* $\mathfrak{o}_v$;

(2) *the group of units of* $\mathfrak{o}_v$ *is* $\mathfrak{u}_v = \{ x \in F \mid v(x) = 1 \}$;

(3) $\mathfrak{o}_v$ *has exactly one maximal ideal* $\mathfrak{m}_v = \{ x \in F \mid v(x) < 1 \} = \mathfrak{o}_v \backslash \mathfrak{u}_v$;

(4) *the ideals of* $\mathfrak{o}_v$ *form a chain.*

We prove a converse:

Proposition **6.2.** *Let* R *be a subring of a field* F *and let* $\mathfrak{u}$ *be the group of units of* R. *The following properties are equivalent:*

(1) R *is the valuation ring of a valuation on* F;

(2) $F = Q(R)$ *and the ideals of* R *form a chain;*

(3) *when* $x \in F \backslash \{0\}$, *then* $x \in R$ *or* $x^{-1} \in R$.

Then $G = (F\backslash\{0\})/\mathfrak{u}$ *is a totally ordered abelian group,* $v_R : x \longmapsto x\mathfrak{u}$ *is a valuation on* F, *and* R *is the valuation ring of* v_R.

Proof. (1) implies (2) by 6.1.

(2) implies (3). Let $x = a/b \in F$, where $a, b \in R$, $b \neq 0$. If $Ra \subseteq Rb$, then $a = br$ for some $r \in R$ and $x = r \in R$. If $Rb \subseteq Ra$, then $b = ar$ for some $r \in R$ and $x^{-1} = r \in R$.

(3) implies (1). The group of units $\mathfrak{u}$ of R is a subgroup of the multiplicative group $F^* = F\backslash\{0\}$. Let $G = F^*/\mathfrak{u}$. Order G by

$$x\mathfrak{u} \leqq y\mathfrak{u} \text{ if and only if } xy^{-1} \in R, \text{ if and only if } Rx \subseteq Ry.$$

Then $\leqq$ is well defined, since $x\mathfrak{u} = z\mathfrak{u}$, $y\mathfrak{u} = t\mathfrak{u}$ implies $Rx = Rz$, $Ry = Rt$; $\leqq$ is reflexive, transitive, and antisymmetric, since $xy^{-1} \in R$ and $\left(xy^{-1}\right)^{-1} = yx^{-1} \in R$ implies $xy^{-1} \in \mathfrak{u}$ and $x\mathfrak{u} = y\mathfrak{u}$; $\leqq$ is a total order on G, by (3); and $x\mathfrak{u} \leqq y\mathfrak{u}$ implies $(x\mathfrak{u})(z\mathfrak{u}) \leqq (y\mathfrak{u})(z\mathfrak{u})$. Now G has become a proud totally ordered abelian group. Let

$$v_R(x) = x\mathfrak{u} \in G$$

for all $x \in F^*$, with $v_R(0) = 0 \in G \cup \{0\}$. Then (a) and (b) hold. Property (c), $v(x + y) \leqq \max\left(v(x), v(y)\right)$, holds whenever $x = 0$ or $y = 0$; if $x, y \neq 0$ and, say, $v_R(x) \leqq v_R(y)$, then $Rx \subseteq Ry$, $R(x + y) \subseteq Ry$, and $v_R(x + y) \leqq v_R(y)$. Thus v_R is a valuation on F (with value group G); R is the valuation ring of v_R, since $v_R(x) \leqq 1 = 1\mathfrak{u}$ if and only if $x = x1^{-1} \in R$. $\square$

Definitions. A valuation ring *or* valuation domain *is a domain that satisfies the equivalent conditions in Proposition* 6.2; *then* v_R *is the valuation* induced by R. A valuation ring *of a field* F *is a subring of* F *that satisfies the equivalent conditions in Proposition* 6.2.

Proposition 6.3. *Every valuation is equivalent to the valuation induced by its valuation ring. In particular, two valuations on the same field are equivalent if and only if they have the same valuation ring.*

Proof. Let v be a valuation on a field F and let $\mathfrak{o}$ be its valuation ring. The valuations v and $v_{\mathfrak{o}}$ induce surjective homomorphisms of multiplicative groups:

where $F^* = F\backslash\{0\}$ and $\mathfrak{u}$ is the group of units of $\mathfrak{o}$. Since $\operatorname{Ker} v = \mathfrak{u} = \operatorname{Ker} v_{\mathfrak{o}}$ there is a multiplicative isomorphism $\theta : G_v \longrightarrow F^*/\mathfrak{u}$ such that $\theta \circ v = v_{\mathfrak{o}}$. If $x, y \in F^*$, then $v(x) \leqq v(y)$ is equivalent to $v(xy^{-1}) \leqq 1$, to $xy^{-1} \in \mathfrak{o}$, and to $v_{\mathfrak{o}}(x) \leqq v_{\mathfrak{o}}(y)$; therefore θ is order preserving. $\square$

Discrete valuations. Since $\mathbb{P}$ contains cyclic subgroups, every valuation whose value group is cyclic is equivalent to a nonarchimedean absolute value.

Definition. A valuation is discrete *when its value group is cyclic.*

A discrete valuation v on a field F induces a topology on F, which is induced by any equivalent discrete absolute value. The infinite cyclic group G_v has just two generators and has a unique generator $v(p) < 1$. Then every $x \in F \backslash \{0\}$ can be written uniquely in the form $x = up^k$ with $v(u) = 1$ and $k \in \mathbb{Z}$, since $v(x) = v(p^k)$ for some unique $k \in \mathbb{Z}$.

Proposition 6.4. *Let R be a domain and let $F = Q(R)$ be its quotient field. Then R is the valuation ring of a discrete valuation on F if and only if R is a principal ideal domain with a unique nonzero prime ideal.*

The proof is an exercise.

Definition. A discrete *valuation ring is a principal ideal domain with a unique nonzero prime ideal; equivalently, the valuation ring of a discrete valuation.*

For instance, the ring $\widehat{\mathbb{Z}}_p$ of p-adic integers is a discrete valuation ring. In fact, Proposition 4.4 extends wholeheartedly to all discrete valuations. In the next result, v is a discrete valuation, $v(p) < 1$ is a generator of its value group, $\mathfrak{o}$ is the valuation ring of v, and $\mathfrak{m}$ is its maximal ideal.

Proposition 6.5. *Let F be a field with a discrete valuation v. Let $\mathfrak{r}$ be a subset of $\mathfrak{o}$ with $0 \in \mathfrak{r}$ and one element in every coset of $\mathfrak{m}$ in $\mathfrak{o}$. Every element of $\mathfrak{o}$ is the sum of a unique power series $\sum_{n \geq 0} r_k p^k$ with coefficients $r_k \in \mathfrak{r}$. Every nonzero element of F is the sum of a unique Laurent series $\sum_{k \geq m} r_k p^k$ with coefficients $r_k \in \mathfrak{r}$ for all $k \geq m$, $r_m \neq 0$.*

Proof. By 6.4, $\mathfrak{m}$ is the ideal of $\mathfrak{o}$ generated by p. Let $x \in F$, $x \neq 0$. Then $x = up^m$ for some unique $u \in \mathfrak{u}$ and $m \in \mathbb{Z}$. By the choice of $\mathfrak{r}$, $u \in r_m + \mathfrak{m}$ for some unique $r_m \in \mathfrak{r}$, and $r_m \neq 0$ since $u \notin \mathfrak{m}$. Hence $u = r_m + py$ for some unique $y \in \mathfrak{o}$. Then $y = r_{m+1} + pz$ for some unique $r_{m+1} \in \mathfrak{r}$ and $z \in \mathfrak{o}$. Continuing thus yields expansions $x = r_m p^m + \cdots + r_k p^k + p^{k+1} t$ and a series $\sum_{k \geq m} r_k p^k$ that converges to x, since $v(r_k p^k + r_{k+1} p^{k+1} + \cdots) \leqq v(p)^k$ for all k. If $x \in \mathfrak{o}$, then $m \geqq 0$ and $\sum_{k \geq m} r_k p^k$ is a power series $\sum_{k \geq 0} r_k p^k$, with $r_k = 0$ for all $k < m$. Uniqueness makes a nifty exercise. $\square$

Every discrete valuation v on F is equivalent to a nonarchimedean absolute value on F and yields a completion $\widehat{F}_v$. We may assume that G_v is a subgroup of $\mathbb{P}$. Then 6.5 extends to $\widehat{F}_v$. In the next result, v is a discrete valuation, $v(p) < 1$ is a generator of its value group, $\mathfrak{o}$ is the valuation ring of v, and $\mathfrak{m}$ is its maximal ideal; $\widehat{F} = \widehat{F}_v$ and $\widehat{\mathfrak{o}}$ is its valuation ring.

Proposition 6.6. *Let F be a field with a discrete valuation v. Every Laurent series $\sum_{k \geq m} r_k p^k$ with coefficients $r_k \in \mathfrak{o}$ converges in $\widehat{F}$. Conversely, let $\mathfrak{r}$ be a subset of $\mathfrak{o}$ with $0 \in \mathfrak{r}$ and one element in every coset of $\mathfrak{m}$ in $\mathfrak{o}$. Every element of $\widehat{\mathfrak{o}}$ is the sum of a unique power series $\sum_{n \geq 0} r_k p^k$ with coefficients*

$r_k \in \mathfrak{r}$; *every nonzero element of* $\widehat{F}$ *is the sum of a unique Laurent series* $\sum_{k \geq m} r_k p^k$ *with coefficients* $r_k \in \mathfrak{r}$ *for all* $k \geq m$, $r_m \neq 0$.

Proof. First we show that v and its extension $\widehat{v}$ to $\widehat{F}$ have the same value group G_v. We may assume that G_v is a subgroup of $\mathbb{P}$. Every nonzero $x \in \widehat{F}$ is the limit of a Cauchy sequence $(x_n)_{n>0}$ of elements of F; then $\widehat{v}(x) = \lim_{n \to \infty} v(x_n) \in G_v$, since G_v is closed in $\mathbb{P}$.

We now follow the proof of Proposition 6.5. Let $x \in \widehat{F}$, $x \neq 0$. Then $\widehat{v}(x) = p^{-m}$ for some $m \in \mathbb{Z}$. Since F is dense in $\widehat{F}$ we have $\widehat{v}(x - y) < p^{-m}$ for some $y \in F$. Then $v(y) = p^{-m}$; as above, $y = r_m p^m + p^{m+1} z$ for some $r_m \in \mathfrak{r}$, $r_m \neq 0$, and some $z \in \mathfrak{o}$. Then $\widehat{v}(x - r_m p^m) \leq p^{-(m+1)}$. Hence $\widehat{v}(x - r_m p^m - r_{m+1} p^{m+1}) \leq p^{-(m+2)}$ for some $r_{m+1} \in \mathfrak{r}$; repetition yields $r_n \in \mathfrak{r}$ such that $\widehat{v}(x - \sum_{m \leq n \leq r} r_n p^n) < p^{-r}$, for every $r > m$. Then the series $\sum_{n \geq m} r_n p^n$ converges to x in $\widehat{F}$. If $x \in \widehat{\mathfrak{o}}$, then $m \geq 0$ and $\sum_{n \geq m} r_n p^n$ is a power series. Uniqueness is again an exercise. $\square$

Exercises

1. Show that $\mathbb{P}^n$ is a totally ordered abelian group when ordered lexicographically.

2. A totally ordered abelian group G is *archimedean* when for every $a, b > 1$ in G the inequality $a^n > b$ holds for some $n > 0$. Show that every subgroup of $\mathbb{P}$ is archimedean. Show that $\mathbb{P}^n$ is not archimedean when $n \geq 2$, and therefore is not isomorphic (as a totally ordered abelian group) to a subgroup of $\mathbb{P}$.

3. Show that v_0 is a valuation on $K(X_1, ..., X_n)$. Show that v_0 is not equivalent to an absolute value. (Find its value group and show that it is not isomorphic, as a totally ordered abelian group, to a subgroup of $\mathbb{P}$.)

4. Find all automorphisms of the totally ordered abelian group $\mathbb{P}$.

5. Prove that every multiplicative subgroup of $\mathbb{P}$ is either cyclic or dense in $\mathbb{P}$.

6. Let v be a valuation. Show that $v(x + y) = \max(v(x), v(y))$ when $v(x) \neq v(y)$.

7. Let v be a valuation on a field F. Show that $\mathfrak{o}_v$ has exactly one maximal ideal $\mathfrak{m}_v = \{ x \in F \mid v(x) < 1 \} = \mathfrak{o}_v \backslash \mathfrak{u}_v$, and that the ideals of $\mathfrak{o}_v$ form a chain.

8. Show that a ring is the valuation ring of a discrete valuation if and only if it is a PID with a unique nonzero prime ideal.

9. Prove that the series expansions in Propositions 6.5, 6.6 are unique.

10. Prove that a valuation ring is discrete if and only if it is Noetherian.

11. Prove that every totally ordered abelian group is the value group of a valuation.

In the following exercises, a *place* on a field F with values in a field Q is a mapping $\pi : F \longrightarrow Q \cup \{\infty\}$ such that (i) $\mathfrak{o}_\pi = \{ x \in F \mid \pi(x) \neq \infty \}$ is a subring of F; (ii) the restriction of π to $\mathfrak{o}_\pi$ is a ring homomorphism; (iii) if $\pi(x) = \infty$, then $x^{-1} \in \mathfrak{o}_\pi$ and $\pi(x^{-1}) = 0$.

12. Let π be a place on F with values in Q. Show that $\{\pi(x) \mid x \in F,\ x \neq \infty\}$ is a subfield of Q.

13. Let π be a place on F with values in Q and let ρ be a place on Q with values in L. Show that $\rho \circ \pi$ is a place on F with values on L.

14. Show that every valuation v on a field F induces a place π with values in $\mathfrak{o}_v / \mathfrak{m}_v$.

15. Define equivalence of places and show that, up to equivalence, every place is induced by a valuation.

7. Extending Valuations

In this section we consider the extension problem for valuations, including nonarchimedean absolute values: when E is a finite extension of a field K with a valuation v, can v be extended to a valuation on E? We show that v has extensions to E and prove some of their properties. This yields new properties of finite field extensions. The results in this section are due to Ostrowski [1934].

Existence. Before extending valuations we extend homomorphisms.

Theorem **7.1.** *Let R be a subring of a field K. Every homomorphism of R into an algebraically closed field L can be extended to a valuation ring of K.*

Proof. Let $\varphi : R \longrightarrow L$ be a homomorphism. By Zorn's lemma there exists a homomorphism $\psi : S \longrightarrow L$ that extends φ to a subring $S \supseteq R$ of K and is maximal in the sense that ψ cannot be extended to a subring $T \supsetneqq S$ of K. We show that S is a valuation ring of K.

Claim 1: if $\psi(a) \neq 0$, then a is a unit of S; hence $\mathfrak{m} = \operatorname{Ker} \psi$ is a maximal ideal of S; $F = \operatorname{Im} \psi \cong S/\mathfrak{m}$ is a field; and every $a \in S \backslash \mathfrak{m}$ is a unit of S.

Given $a \in S$, $\psi(a) \neq 0$, let $T = \{xa^{-k} \in K \mid x \in S,\ k \geq 0\}$; T is a subring of K, which contains S since $a^0 = 1$. If $\psi(a) \neq 0$, then $xa^{-k} = ya^{-\ell}$ implies $xa^{\ell} = ya^k$, $\psi(x)\psi(a)^{\ell} = \psi(y)\psi(a)^k$, and $\psi(x)\psi(a)^{-k} = \psi(y)\psi(a)^{-\ell}$; therefore a mapping $\chi : T \longrightarrow L$ is well defined by $\chi(xa^{-k}) = \psi(x)\psi(a)^{-k}$. Then χ is a homomorphism that extends ψ. By maximality, $T = S$. Thus $\psi(a) \neq 0$ implies $a^{-1} \in S$ (in K), so that a is a unit of S.

Claim 2: if $c \in K \backslash S$, then $m_0 + m_1 c + \cdots + m_k c^k = 1$ for some $k > 0$ and $m_0, m_1, \ldots, m_k \in \mathfrak{m}$. The subring $S[c] \subseteq K$ is the image of the evaluation homomorphism $\widehat{c} : f \longmapsto f(c)$ of $S[X]$ into K, and $\mathfrak{A} = \operatorname{Ker} \widehat{c} = \{f \in S[X] \mid f(c) = 0\}$ is an ideal of $S[X]$. Now, $\psi : S \longrightarrow F$ induces a surjective homomorphism $\overline{\psi} : S[X] \longrightarrow F[X]$, $f \longmapsto {}^{\psi}f$; then $\mathfrak{B} = \overline{\psi}(\mathfrak{A})$ is an ideal of $F[X]$ and consists of all the multiples of some $b \in F[X]$.

We show that $\mathfrak{B} = F[X]$. Assume that b is not constant. Then $b \in F[X] \subseteq L[X]$ has a root γ in the algebraically closed field L. Let $\widehat{\gamma} : F[X] \longrightarrow L$, $g \longmapsto g(\gamma)$ be the evaluation homomorphism. Then $\widehat{\gamma}(\overline{\psi}(f)) = 0$ for all

$f \in \mathfrak{A}$, since $b(\gamma) = 0$ and $\overline{\psi}(f) \in \mathfrak{B}$ is a multiple of b. Hence $\operatorname{Ker} \widehat{c} = \mathfrak{A} \subseteq$ $\operatorname{Ker}(\widehat{\gamma} \circ \overline{\psi})$ and $\widehat{\gamma} \circ \overline{\psi}$ factors through $\widehat{c} : S[X] \longrightarrow S[c]$:

$$
\begin{array}{ccc}
S[X] & \xrightarrow{\ \overline{\psi}\ } & F[X] \\
\widehat{c} \downarrow & & \downarrow \widehat{\gamma} \\
S[c] & \dashrightarrow[\chi]{} & L
\end{array}
$$

$\widehat{\gamma} \circ \overline{\psi} = \chi \circ \widehat{c}$ for some ring homomorphism $\chi : S[c] \longrightarrow L$. If $x \in S \subseteq S[X]$, then $\overline{\psi}(x) = \psi(x) \in F \subseteq F[X]$ and $\chi(x) = \chi(\widehat{c}(x)) = \widehat{\gamma}(\overline{\psi}(x)) = \psi(x)$. Thus χ extends ψ, contradicting the maximality of S. Therefore $\mathfrak{B} = F[X]$.

Since $\mathfrak{B} = F[X]$ there exists $f = a_0 + a_1 X + \cdots + a_n X^n \in \mathfrak{A}$ such that $1 = \overline{\psi}(f) = \psi(a_0) + \psi(a_1)X + \cdots + \psi(a_n)X^n$, equivalently, $1 - a_0$ and $a_1, \ldots,$ a_n are all in $\mathfrak{m} = \operatorname{Ker} \psi$. Then $f(c) = 0$ yields $1 = (1 - a_0) - a_1 c - \cdots - a_n c^n$, where $1 - a_0, -a_1, \ldots, -a_n \in \mathfrak{m}$. This proves Claim 2.

We show that S is a valuation ring. Let $c \in K$. If $c, c^{-1} \notin S$, then

$$
m_0 + m_1 c + \cdots + m_k c^k = 1 = n_0 + n_1 c^{-1} + \cdots + n_\ell c^{-\ell}
$$

for some $k, \ell \geq 0$ and $m_0, \ldots, m_k, n_0, \ldots, n_\ell \in \mathfrak{m}$, by Claim 2. We may choose these equalities so that $k \geq \ell$ and $k + \ell$ is as small as possible. Then $m_k, n_\ell \neq 0$ and $k, \ell \geq 1$ (since $1 \notin \mathfrak{m}$). Now, $1 - n_0 \notin \mathfrak{m}$ is a unit of S, by Claim 1; hence $1 - n_0 = n_1 c^{-1} + \cdots + n_\ell c^{-\ell}$, $c^k = (1 - n_0)^{-1}(n_1 c^{k-1} + \cdots + n_\ell c^{k-\ell})$, and substituting for c^k in the left hand side lowers k by 1, contradicting the minimality of $k + \ell$. Therefore $c \in S$ or $c^{-1} \in S$. $\square$

Covered by Theorem 7.1 we now approach valuations.

Theorem 7.2. *Let K be a subfield of E. Every valuation on K extends to a valuation on E.*

Proof. Let v be a valuation on K; let $\mathfrak{o}$ be the valuation ring of v, let $\mathfrak{m}$ be the maximal ideal of $\mathfrak{o}$, and let $\mathfrak{u}$ be its group of units. Let L be the algebraic closure of the field $\mathfrak{o}/\mathfrak{m}$. By 7.1, the projection $\pi : \mathfrak{o} \longrightarrow \mathfrak{o}/\mathfrak{m} \subseteq L$ extends to a homomorphism $\varphi : \mathfrak{O} \longrightarrow L$ of a valuation ring $\mathfrak{O} \supseteq \mathfrak{o}$ of E. Let $\mathfrak{M}$ be the maximal ideal of $\mathfrak{O}$ and let $\mathfrak{U}$ be its group of units.

We show that $\mathfrak{O} \cap K = \mathfrak{o}$, $\mathfrak{M} \cap K = \mathfrak{m}$, and $\mathfrak{U} \cap K = \mathfrak{u}$. If $x \in \mathfrak{m}$, then $\varphi(x) = \pi(x) = 0$, $x \notin \mathfrak{U}$, and $x \in \mathfrak{M}$. If now $x \in K \backslash \mathfrak{o}$, then $x \notin \mathfrak{u}$, $x^{-1} \in \mathfrak{m}$, $x^{-1} \in \mathfrak{M}$, and $x \notin \mathfrak{O}$; hence $\mathfrak{O} \cap K = \mathfrak{o}$. If $x \in \mathfrak{o} \backslash \mathfrak{m}$, then $x \in \mathfrak{u}$, $x \in \mathfrak{U}$, and $x \notin \mathfrak{M}$; hence $\mathfrak{M} \cap K = \mathfrak{M} \cap \mathfrak{o} = \mathfrak{m}$. Then $\mathfrak{U} \cap K = \mathfrak{u}$.

$$
\begin{array}{ccc}
K^* & \xrightarrow{\ \subseteq\ } & E^* \\
{}^{v}\swarrow \ \downarrow v_0 & & \downarrow v_{\mathfrak{O}} \\
G_v \underset{\cong}{\rightrightarrows} K^*/\mathfrak{u} & \dashrightarrow[\psi]{} & E^*/\mathfrak{U}
\end{array}
$$

The inclusion homomorphism $K^* = K \backslash \{0\} \longrightarrow E^* = E \backslash \{0\}$ now induces a homomorphism $\psi : K^*/\mathfrak{u} \longrightarrow E^*/\mathfrak{U}$, $x\mathfrak{u} \longmapsto x\mathfrak{U}$ (see the diagram above) which is injective since $\mathfrak{U} \cap K^* = \mathfrak{u}$. Moreover, $\psi(x\mathfrak{u}) \leqq \psi(y\mathfrak{u})$ in $E^*/\mathfrak{U}$ is equivalent to $x\mathfrak{u} \leqq y\mathfrak{u}$ in $K^*/\mathfrak{u}$, since both are equivalent to $xy^{-1} \in \mathfrak{O} \cap K = \mathfrak{o}$; hence ψ induces an order preserving isomorphism $K^*/\mathfrak{u} \cong \operatorname{Im} \psi$. Up to this isomorphism, the valuation $v_{\mathfrak{O}}$ on E extends the valuation $v_{\mathfrak{o}}$ on K. The given valuation v is equivalent to $v_{\mathfrak{o}}$ by 6.3 and can also be extended to E. $\square$

Properties. In what follows, v is a valuation on K; w extends v; $\mathfrak{o}$, $\mathfrak{m}$, and $\mathfrak{u}$ denote the valuation ring of v, its maximal ideal, and its group of units; $\mathfrak{O}$, $\mathfrak{M}$, and $\mathfrak{U}$ denote the valuation ring of w, its maximal ideal, and its group of units. Since w extends v we have $\mathfrak{O} \cap K = \mathfrak{o}$, $\mathfrak{M} \cap K = \mathfrak{m}$, and $\mathfrak{U} \cap K = \mathfrak{u}$.

Definition. The residue class field *of a valuation v is the quotient* $F_v = \mathfrak{o}_v/\mathfrak{m}_v$, *where $\mathfrak{m}_v$ is the maximal ideal of $\mathfrak{o}_v$.*

If w extends v, then $F_w = \mathfrak{O}/\mathfrak{M}$ is a field extension of $F_v = \mathfrak{o}/\mathfrak{m}$: since $\mathfrak{M} \cap \mathfrak{o} = \mathfrak{m}$, there is a commutative square

$$
\begin{array}{ccc}
\mathfrak{o} & \xrightarrow{\ \subseteq\ } & \mathfrak{O} \\
\downarrow & & \downarrow \\
F_v = \mathfrak{o}/\mathfrak{m} & \dashrightarrow & F_w = \mathfrak{O}/\mathfrak{M}
\end{array}
$$

where the vertical maps are projections. It is convenient to identify the *residue classes* $\bar{x} = x + \mathfrak{m} \in F_v$ and $\bar{x} = x + \mathfrak{M} \in F_w$ of every $x \in K$, so that F_v becomes a subfield of F_w.

Definitions. If E is a field extension of K and w is a valuation on E that extends v, then $\mathrm{e}(w:v) = [G_w : G_v]$ *is the* ramification index *of w over v, and* $\mathrm{f}(w:v) = [F_w : F_v]$ *is the* residue class degree *of w over v.*

These numbers $\mathrm{e}(w:v)$ and $\mathrm{f}(w:v)$ are also denoted by $\mathrm{e}(E:K)$ and $\mathrm{f}(E:K)$ (preferably when they do not depend on the choice of w).

Proposition **7.3.** *If E is a field extension of K, v is a valuation on K, and w is a valuation on E that extends v, then* $\mathrm{e}(w:v)\,\mathrm{f}(w:v) \leqq [E:K]$.

Proof. Let $(\alpha_i)_{i \in I}$ be elements of E. Let $(\beta_j)_{j \in J}$ be elements of $\mathfrak{O}$ whose residue classes are linearly independent over F_v.

Let $\gamma = \sum_{j \in J} x_j \beta_j \in E$, where $x_j \in K$, $x_j \neq 0$ for some $j \in J$. Then $m = \max_{j \in J} v(x_j) \neq 0$, $m = v(x_t)$ for some $t \in J$, and then $x_j/x_t \in \mathfrak{o}$ for all $j \in J$, with $x_t/x_t = 1 \notin \mathfrak{m}$. Since the residue classes $(\bar{\beta}_j)_{j \in J}$ are linearly independent over F_v we have $\sum_{j \in J} \overline{x_j/x_t}\, \bar{\beta}_j \neq 0$ in F_w, $\sum_{j \in J} x_j/x_t\, \beta_j \notin \mathfrak{M}$, $w(\sum_{j \in J} x_j/x_t\, \beta_j) = 1$, and $w(\gamma) = w(\sum_{j \in J} x_j \beta_j) = w(x_t) \in G_v$.

Now assume that $\sum_{i \in I,\, j \in J} x_{ij}\, \alpha_i\, \beta_j = 0$ for some $x_{ij} \in K$ such that $x_{ij} = 0$ for almost all i, j but $x_{ij} \neq 0$ for some i and j. Let $\gamma_i = \sum_{j \in J} x_{ij} \beta_j$. By the above, either $w(\gamma_i) \in G_v$ (when $x_{ij} \neq 0$ for some j), or $w(\gamma_i) = 0$. Then

$m = \max_{i \in I} w(\alpha_i \gamma_i) \neq 0$ and $m = w(\alpha_k \gamma_k)$ for some $k \in I$. If $w(\alpha_i \gamma_i) < m$ for all $i \neq k$, then $w(\sum_{i \in I} \alpha_i \gamma_i) = m$, contradicting $\sum_{i \in I} \alpha_i \gamma_i = 0$. Hence $w(\alpha_i \gamma_i) = w(\alpha_k \gamma_k) \neq 0$ for some $i \neq k$, and then $w(\alpha_i) G_v = w(\alpha_j) G_v$. Therefore, if all $w(\alpha_i)$ lie in different cosets of G_v, and the residue classes $(\overline{\beta}_j)_{j \in J}$ are linearly independent over F_v, then the elements $(\alpha_i \beta_j)_{i \in I, \, j \in J}$ are linearly independent over K. $\square$

A nonarchimedean absolute value is a valuation v such that $G_v \subseteq \mathbb{P}$. This property is not necessarily preserved under extension, except in the finite case:

Theorem 7.4. *If E is a finite extension of K, then every nonarchimedean absolute value on K extends to a nonarchimedean absolute value on E.*

Proof. By 7.2, a nonarchimedean absolute value v on K extends to a valuation w on E. By 7.3, $e = e(w : v)$ is finite. Then $g^e \in G_v$ for every $g \in G_w$, and $g \longmapsto g^e$ is a homomorphism of G_w into $G_v \subseteq \mathbb{P}$, which is injective since the totally ordered abelian group G_w is torsion free. Now, every $r \in \mathbb{P}$ has a unique eth root $\sqrt[e]{r}$ in $\mathbb{P}$, and $r \longmapsto \sqrt[e]{r}$ is an automorphism of $\mathbb{P}$. Then $x \longmapsto \sqrt[e]{w(x)^e}$ is a nonarchimedean absolute value on E that extends v. $\square$

Using the same homomorphism $g \longmapsto g^e$, readers may prove the following:

Proposition 7.5. *Let E be a finite extension of K, let v be a valuation on K, and let w be a valuation on E that extends v. If v is discrete, then w is discrete.*

Proposition 7.6. *Let E be a finite extension of K, let v be a discrete nonarchimedean absolute value on K, and let w be a discrete nonarchimedean absolute value on E that extends v. If K is complete with respect to v, then $e(w : v) \, f(w : v) = [E : K]$.*

Proof. First, w exists, by 7.4 and 7.5. Let $v(p) < 1$ generate G_v and $w(\rho) < 1$ generate G_w. Then $v(p) = w(\rho)^e$ for some $e > 0$; $\rho^e = up$ for some $u \in \mathfrak{U}$; the cosets of G_v are $G_v, w(\rho) G_v, \ldots, w(\rho^{e-1}) G_v$; and $e = e(w : v)$. Let $f = f(w : v)$ and $\beta_1, \ldots, \beta_f$ be elements of $\mathfrak{O}$ whose residue classes $\overline{\beta}_1, \ldots, \overline{\beta}_f$ constitute a basis of F_w over F_v. As in the proof of 7.3, the products $(\rho^i \beta_j)_{0 \leq i < e, \, 1 \leq j \leq f}$ are linearly independent over K.

Let $\mathfrak{r} \subseteq \mathfrak{o}$ be a set with one element in every residue class of $\mathfrak{o}$. Since $\overline{\beta}_1, \ldots, \overline{\beta}_f$ is a basis of F_w over F_v, the set $\{ r_1 \beta_1 + \cdots + r_f \beta_f \mid r_1, \ldots, r_f \in \mathfrak{r} \}$ has one element in every residue class of $\mathfrak{O}$. We now expand every nonzero element α of E, much as in the proof of 6.5. We have $w(\alpha) = w(\rho)^m$ for some $m \in \mathbb{Z}$ and $m = e\ell + i$ for some $\ell, i \in \mathbb{Z}$, $0 \leq i < e$; then $w(\alpha) = w(p^\ell \rho^i)$ and $\alpha = u p^\ell \rho^i$ for some $u \in \mathfrak{U}$, and $u = r_{\ell i 1} \beta_1 + \cdots + r_{\ell i f} \beta_f + \mu$ for some $r_{\ell i j} \in \mathfrak{r}$ and $\mu \in \mathfrak{M}$; hence

$$\alpha = \left(\sum_{1 \leq j \leq f} r_{\ell i j} p^\ell \rho^i \beta_j \right) + \alpha',$$

where $\alpha' = \mu p^\ell \rho^i$, so that $w(\alpha') < w(p^\ell \rho^i) = w(\rho^m)$ and $w(\alpha') \leqq w(\rho)^{m+1}$. Continuing in this fashion yields a series expansion

$$\alpha = \sum_{k \geqq \ell, \, 0 \leqq i < e, \, 1 \leqq j \leqq f} r_{kij} \, p^k \rho^i \beta_j .$$

Since K is complete, $\sum_{k \geqq \ell} r_{kij} p^k$ has a sum x_{ij} in K for every i, j. Then $\alpha = \sum_{i,j} x_{ij} \rho^i \beta_j$. Hence $(\rho^i \beta_j)_{0 \leqq i < e, \, 1 \leqq j \leqq f}$ is a basis of E over K. $\square$

Extensions of nonarchimedean absolute values are now counted as follows.

Theorem 7.7. Let v be a nonarchimedean absolute value on a field K and let $E = K(\alpha)$ be a finite extension of K. There are finitely many absolute values on E that extend v, one for each monic irreducible factor q_i of $\mathrm{Irr}\,(\alpha : K)$ in $\widehat{K}[X]$. Moreover, $[\,\widehat{E}_{w_i} : \widehat{K}\,] = \deg q_i$ and $\sum_i [\,\widehat{E}_{w_i} : \widehat{K}\,] \leqq [\,E : K\,]$, with $\sum_i [\,\widehat{E}_{w_i} : \widehat{K}\,] = [\,E : K\,]$ if E is separable over K.

Proof. First, $F_i = \widehat{K}[X]/(q_i)$ is a finite extension of $\widehat{K}$, $F_i = \widehat{K}(\alpha_i)$, where $\mathrm{Irr}\,(\alpha_i : K) = q_i$, and $[\,F_i : K\,] = \deg q_i$. Since $\prod_i q_i$ divides $q = \mathrm{Irr}\,(\alpha : K)$, we have $\sum_i [\,F_i : \widehat{K}\,] = \sum_i \deg q_i \leqq \deg q = [\,E : K\,]$; if E is separable over K, then $q = \prod_i q_i$ and $\sum_i [\,F_i : \widehat{K}\,] = [\,E : K\,]$.

Since $(q) \subseteq (q_i)$ in $\widehat{K}[X]$, there is a homomorphism $E \longrightarrow F_i$ such that the following square commutes:

$$
\begin{array}{ccc}
K[X] & \xrightarrow{\ \subseteq\ } & \widehat{K}[X] \\
\big\downarrow & & \big\downarrow \\
E \cong K[X]/(q) & \dashrightarrow & \widehat{K}[X]/(q_i) \cong F_i
\end{array}
$$

where the vertical maps are projections. Hence F_i is a field extension of E. By 7.4 there is an absolute value on F_i that extends the absolute value $\widehat{v}$ on $\widehat{K}$; this induces an absolute value w_i on E that extends v.

Conversely, let w be an absolute value on E that extends v. The completion $\widehat{E}_w$ contains α and $\widehat{K}$, so there is an evaluation homomorphism $\widehat{\alpha} : f \longmapsto f(\alpha)$ of $\widehat{K}[X]$ into $\widehat{E}_w$. Since α is algebraic over $\widehat{K} \supseteq K$, $\mathrm{Im}\,\widehat{\alpha} = \widehat{K}[\alpha]$ is a finite field extension of $\widehat{K}$ and is complete by 5.2. Since E is dense in $\mathrm{Im}\,\widehat{\alpha} \subseteq \widehat{E}_w$ this implies $\widehat{E}_w = \mathrm{Im}\,\widehat{\alpha} = \widehat{K}[\alpha] \cong \widehat{K}[X]/(r)$, where $r = \mathrm{Irr}\,(\alpha : \widehat{K})$. Now, r divides q, since $r(\alpha) = 0$; hence r is one of the monic irreducible factors of q in $\widehat{K}[X]$, $r = q_i$ for some i. Then $\widehat{E}_w \cong \widehat{K}[X]/(q_i) = F_i$, by a $\widehat{K}$-isomorphism that takes α to α_i. Up to this isomorphism, $\widehat{w} = w_i'$ by 5.2, since both induce the same absolute value $\widehat{v}$ on $\widehat{K}$; hence $w = w_i$. Moreover, i is unique: if $w_i = w_j$, then $\widehat{E}_{w_i} = \widehat{E}_{w_j}$ and there is a $\widehat{K}$-isomorphism $F_i \cong F_j$ that takes α_i to α_j; hence $q_i = \mathrm{Irr}\,(\alpha_i : \widehat{K}) = \mathrm{Irr}\,(\alpha_j : \widehat{K}) = q_j$ and $i = j$. $\square$

If v is discrete, then all w_i are discrete, by 7.5, and 7.6 yields $[\widehat{E}_{w_i} : \widehat{K}]$
$= e(\widehat{w}_i : \widehat{v}) \, f(\widehat{w}_i : \widehat{v})$. Readers will verify that a discrete absolute value and its
completion have the same value group and have isomorphic residue class fields;
hence $e(\widehat{w}_i : \widehat{v}) = e(w_i : v)$, $f(\widehat{w}_i : \widehat{v}) = f(w_i : v)$, and 7.7 yields the following:

Theorem **7.8.** *Let v be a discrete absolute value on a field K, let $E = K(\alpha)$ be a
finite extension of K, and let $w_1, \ldots, w_r$ be the distinct absolute values on E that
extend v. Then $\sum_i e(w_i : v) \, f(w_i : v) \leqq [E : K]$, with $\sum_i e(w_i : v) \, f(w_i : v) =
[E : K]$ if E is separable over K.*

Further improvements occur when v is discrete and E is a finite Galois extension of K (hence simple, by the primitive element theorem). Then it can be shown
that $w_1, \ldots, w_r$ are conjugate ($w_j = w_i \circ \sigma$ for some $\sigma \in \mathrm{Gal}\,(E : K)$). Hence
all w_i have the same ramification index e and the same residue class degree f
over v, and 7.8 yields $[E : K] = efr$.

In the next section we complete 7.7 and 7.8 with criteria for irreducibility in
$\widehat{K}[X]$.

Exercises

1. Let R be a subring of K and let $\varphi : R \longrightarrow L$ be a homomorphism of rings. Fill in the
details for the statement that, by Zorn's lemma, there exists a homomorphism $\psi : S \longrightarrow L$
that extends φ to a subring $S \supseteq R$ of K and is maximal in the sense that ψ cannot be
extended to a subring $T \supsetneqq S$ of K.

2. Find the valuation ring and residue class field of the valuation v_p on $\mathbb{Q}$.

3. Let K be a field. Find the valuation ring and residue class field of the valuation v_0
on $K(X)$.

4. Let w extend v and v extend u. Show that $e(w : v) \, e(v : u) = e(w : u)$ and that
$f(w : v) \, f(v : u) = f(w : u)$. Do not expect extravagant praise when you find a proof.

5. Let E be a finite extension of K, let v be a valuation on K, and let w be a valuation
on E that extends v. Show that, if v is discrete, then w is discrete.

6. Let v be a discrete nonarchimedean absolute value on a field F and let $\widehat{v}$ be the
corresponding absolute value on $\widehat{F}$. Show that v and $\widehat{v}$ have the same value group.

7. Let v be an nonarchimedean absolute value on a field F and let $\widehat{v}$ be the corresponding
absolute value on $\widehat{F}$. Show that v and $\widehat{v}$ have isomorphic residue class fields.

8. Hensel's Lemma

This result, first proved by Hensel [1904] for p-adic numbers, is now an irreducibility criterion for polynomials with coefficients in a complete field.

Primitive polynomials. First we show that the irreducible polynomials over a
field K are determined by the irreducible polynomials over any of its valuation
rings. By 6.4, a discrete valuation ring $\mathfrak{o}$ of a field K is a PID, hence a UFD. As

we saw in Section III.10, every nonzero polynomial $f \in K[X]$ can be written in the form $f(X) = a \, f^*(X)$ with $a \in K$, $a \neq 0$, and $f^* \in \mathfrak{o}[X]$ primitive, and f^* is unique up to multiplication by a unit of $\mathfrak{o}$; a product of primitive polynomials is primitive (Gauss's lemma); hence f is irreducible in $K[X]$ if and only if f^* is irreducible in $\mathfrak{o}[X]$.

These properties extend to all valuation rings.

Definition. Let $\mathfrak{o}$ be a valuation ring and let $\mathfrak{m}$ be its maximal ideal. A polynomial $f \in \mathfrak{o}[X]$ is primitive when its coefficients are not all in $\mathfrak{m}$.

Proposition 8.1. *Let $\mathfrak{o}$ be a valuation ring of a field K and let $\mathfrak{m}$ be its maximal ideal. Every nonzero polynomial $f \in K[X]$ can be written in the form $f(X) = t \, f^*(X)$, where $t \in K$, $t \neq 0$, and $f^*(X) \in \mathfrak{o}[X]$ is primitive; moreover, t and f^* are unique up to multiplication by units of $\mathfrak{o}$.*

Proof. The ring $\mathfrak{o}$ is the valuation ring of a valuation v on K. For every $f(X) = a_0 + a_1 X + \cdots + a_n X^n \in K[X]$, let $V(f) = \max\left(v(a_0), \ldots, v(a_n) \right)$. If $f \neq 0$, then $V(f) = v(a_t)$ for some t, and then $a_t \in K$, $a_t \neq 0$, and $f(X) = a_t \, f^*(X)$, where $f^* \in \mathfrak{o}[X]$ is primitive since $V(f) = v(a_t) \, V(f^*)$.

If $a \, g(X) = b \, h(X)$, where $b, c \neq 0$ and g, h are primitive, then $v(a) = V(a \, g(X)) = V(b \, h(X)) = v(b) \neq 0$, $u = a/b$ is a unit of $\mathfrak{o}$, and $h = ug$. $\square$

Lemma 8.2 (Gauss). *Let $\mathfrak{o}$ be a valuation ring of a field K. If f and $g \in \mathfrak{o}[X]$ are primitive, then fg is primitive.*

Proof. Let $f(X) = a_0 + a_1 X + \cdots + a_m X^m$ and $g(X) = b_0 + b_1 X + \cdots + b_n X^n$, so that $(fg)(X) = c_0 + c_1 X + \cdots + c_{m+n} X^{m+n}$, where $c_k = \sum_{i+j=k} a_i b_j$. Since f and g are primitive, the maximal ideal $\mathfrak{m}$ of $\mathfrak{o}$ does not contain all a_i and does not contain all b_j. Let k and ℓ be smallest such that $a_k \notin \mathfrak{m}$ and $b_\ell \notin \mathfrak{m}$. Then $a_i \in \mathfrak{m}$ for all $i < k$ and $b_j \in \mathfrak{m}$ for all $j < \ell$. Since $\mathfrak{m}$ is a prime ideal, $a_k b_\ell \notin \mathfrak{m}$; but $a_i b_j \in \mathfrak{m}$ when $i < k$, and when $i + j = k + \ell$ and $i > k$, for then $j < \ell$. Therefore $c_{k+\ell} \notin \mathfrak{m}$. $\square$

Corollary 8.3. *In Proposition 8.1, f is irreducible in $K[X]$ if and only if f^* is irreducible in $\mathfrak{o}[X]$.*

Proof. We may assume that $\deg f \geqq 2$. If f is not irreducible, then f has a factorization $f = gh$ in which $\deg g$, $\deg h \geqq 1$. Let $f(X) = a \, f^*(X)$, $g(X) = b \, g^*(X)$, $h(X) = c \, h^*(X)$, with $f^*, g^*, h^* \in \mathfrak{o}[X]$ primitive, as in 8.1. Then $a \, f^*(X) = bc \, g^*(X) \, h^*(X)$. By 8.2, $g^* h^*$ is primitive; hence $f^*(X) = u \, g^*(X) h^*(X)$ for some unit u of $\mathfrak{o}$, by 8.1, and f^* is not irreducible. Conversely, if f^* is not irreducible, then neither is $f(X) = a \, f^*(X)$. $\square$

Readers will prove the following generalization of Eisenstein's criterion:

Proposition 8.4. *If $f(X) = a_0 + a_1 X + \cdots + a_n X^n \in \mathfrak{o}[X]$, $a_0, \ldots, a_{n-1} \in \mathfrak{m}$, $a_n \notin \mathfrak{m}$, and a_0 is not the product of two elements of $\mathfrak{m}$, then f is irreducible.*

If v is discrete and $v(p) < 1$ generates its value group, then the conditions in 8.4 hold when p divides $a_0, \ldots, a_{n-1}$ but not a_n, and p^2 does not divide a_0.

Hensel's lemma. Let $F = \mathfrak{o}/\mathfrak{m}$ be the residue class field. Every polynomial $f(X) = a_0 + a_1 X + \cdots + a_n X^n$ in $\mathfrak{o}[X]$ has an image $\overline{f}(X) = \overline{a}_0 + \overline{a}_1 X + \cdots + \overline{a}_n X^n$ in $F[X]$ (where $\overline{x} = x + \mathfrak{m}$). Hensel's lemma uses factorizations of $\overline{f}$ to obtain factorizations of f when f is primitive. Equivalently, it refines "approximate" factorizations modulo $\mathfrak{m}$ into "exact" factorizations.

Theorem **8.5** (Hensel's Lemma). *Let K be complete for a valuation v and let $f \in \mathfrak{o}_v[X]$ be primitive. If*

$$\overline{f} = \overline{g}_0 \overline{h}_0$$

for some $g_0, h_0 \in \mathfrak{o}_v[X]$, where $\overline{g}_0$ is monic and $\overline{g}_0$, $\overline{h}_0$ are relatively prime, then there exist $g, h \in \mathfrak{o}_v[X]$ such that

$$f = gh, \ \overline{g} = \overline{g}_0, \ \overline{h} = \overline{h}_0, \ g \text{ is monic, and } \deg g = \deg \overline{g}.$$

Proof. Without changing $\overline{g}_0$ and $\overline{h}_0$ we may assume that g_0 and h_0 have no nonzero coefficients in $\mathfrak{m}$. Then the leading coefficients of g_0 and h_0 are units of $\mathfrak{o}$; $\deg \overline{g}_0 = \deg g_0$; and $\deg \overline{h}_0 = \deg h_0$, so that $\deg h_0 \leq \deg f - \deg g_0$. Since the leading coefficient of g_0 is a unit, we may further assume that g_0 is monic.

Since $\overline{g}_0$ and $\overline{h}_0$ are relatively prime, there exist $s, t \in \mathfrak{o}[X]$ such that $\overline{s} \, \overline{g}_0 + \overline{t} \, \overline{h}_0 = 1$. Then $\mathfrak{m}$ contains every coefficient a_i of $r_0 = f - g_0 h_0$ and b_j of $sg_0 + th_0 - 1$. Choose $c \in \mathfrak{m}$ so that $v(c) = \max_{i,j} \left(v(a_i), v(b_j) \right)$. Then $v(c) < 1$. Write $p \equiv q \pmod{a}$ when every coefficient of $p - q$ is a multiple of a in $\mathfrak{o}$. Since $a_i/c, b_j/c \in \mathfrak{o}$ for all i, j, we have

$$f \equiv g_0 h_0, \ sg_0 + th_0 \equiv 1 \pmod{c}.$$

To prove 8.5, we construct polynomials $g_n, h_n \in \mathfrak{o}[X]$ as follows, so that $f \equiv g_n h_n \pmod{c^{n+1}}$. Polynomial division by g_0 is possible in $\mathfrak{o}[X]$ since g_0 is monic; hence, given g_n and h_n, there exist $r_n, q, k \in \mathfrak{o}[X]$ such that

$$f - g_n h_n = c^{n+1} r_n, \ r_n t = g_0 q + k, \text{ and } \deg k < \deg g_0.$$

Let $\ell \in \mathfrak{o}[X]$ be obtained from $h_0 q + r_n s$ by replacing all coefficients that are multiples of c by 0, so that $\ell \equiv h_0 q + r_n s \pmod{c}$. Let

$$g_{n+1} = g_n + c^{n+1} k \text{ and } h_{n+1} = h_n + c^{n+1} \ell.$$

We prove

$$(1) \quad \begin{cases} \overline{g}_n = \overline{g}_0, \overline{h}_n = \overline{h}_0, \deg g_n = \deg g_0, \deg h_n \leq \deg f - \deg g_0, \\ g_n \text{ is monic, and } f \equiv g_n h_n \pmod{c^{n+1}} \end{cases}$$

by induction on n. First, (1) holds if $n = 0$. Assume that (1) holds for n. Then $\overline{g}_{n+1} = \overline{g}_n = \overline{g}_0$, $\overline{h}_{n+1} = \overline{h}_n = \overline{h}_0$, $\deg g_{n+1} = \deg g_n = \deg g_0$, since $\deg k <$

$\deg g_0$, and g_{n+1} is monic like g_n. Next, $1 \equiv sg_0 + th_0 \pmod{c}$ implies

$$r_n \equiv r_n sg_0 + r_n th_0 = g_0 (r_n s + h_0 q) + h_0 k \equiv g_0 \ell + h_0 k \pmod{c}, \text{ and}$$

$$\begin{aligned} f - g_{n+1} h_{n+1} &\equiv f - (g_n h_n + c^{n+1}(g_n \ell + h_n k)) \\ &= c^{n+1}(r_n - (g_n \ell + h_n k)) \\ &\equiv c^{n+1}(r_n - (g_0 \ell + h_0 k)) \equiv 0 \pmod{c^{n+2}}. \end{aligned}$$

Finally, if $\deg h_{n+1} > \deg f - \deg g_0$, then $\deg \ell > \deg f - \deg g_0$, and $\deg (g_0 \ell + h_0 k) = \deg g_0 \ell > \deg f$, since $\deg h_0 k < \deg h_0 + \deg g_0 \leqq \deg f$. But $r_n \equiv g_0 \ell + h_0 k \pmod{c}$ and $\deg r_n \leqq \deg f$; hence the leading coefficient of $g_0 l$ is a multiple of c, which contradicts the construction of ℓ. Therefore $\deg h_{n+1} \leqq \deg f - \deg g_0$. Thus (1) holds for all n.

Let $g_n(X) = a_{0n} + a_{1n}X + \cdots + a_{r-1,n}X^{r-1} + X^r$, where $r = \deg g_n = \deg g_0$. If $m, n \geqq N$, then $g_m \equiv g_n \pmod{c^N}$, $a_{im} - a_{in}$ is a multiple of c^N in $\mathfrak{o}$, and $v(a_{im} - a_{in}) \leqq v(c)^N$. Hence $(a_{in})_{n \geqq 0}$ is a Cauchy sequence and has a limit a_i in $\mathfrak{o}$. Then g_n has a limit $g(X) = a_0 + a_1 X + \cdots + a_{r-1}X^{r-1} + X^r \in \mathfrak{o}[X]$ such that $g_n \equiv g \pmod{c^n}$ for all n; moreover, $\overline{g} = \overline{g}_0$, g is monic, and $\deg g = \deg g_0$. Similarly, h_n has a limit $h \in \mathfrak{o}[X]$ such that $h_n \equiv h \pmod{c^n}$ for all n, $\overline{h} = \overline{h}_0$, and $\deg h \leqq \deg f - \deg g_0$. Then $f \equiv gh \pmod{c^n}$ for all n; therefore $f = gh$. $\square$

Corollary **8.6.** *Let K be complete for a valuation v and $f \in \mathfrak{o}_v[X]$. If $f(a) \in \mathfrak{m}_v$, $f'(a) \notin \mathfrak{m}_v$ for some $a \in \mathfrak{o}_v$, then $f(b) = 0$ for some $b \in a + \mathfrak{m}_v$.*

Proof. The polynomial f is primitive, since $f'(a) \notin \mathfrak{m}$. We have $\overline{f}(\overline{a}) = 0$, $\overline{f}'(\overline{a}) \neq 0$. Hence $\overline{f}(X) = (X - \overline{a})\overline{h}(X)$ for some $h \in \mathfrak{o}[X]$, where $X - \overline{a}$ and $\overline{h}$ are relatively prime. By 8.5, $f = gh$ for some $g, h \in \mathfrak{o}[X]$, where g is monic, $\deg g = 1$, and $\overline{g} = X - \overline{a}$. Then $g(X) = X - b$ for some $b \in \mathfrak{o}$, $\overline{b} = \overline{a}$, and $f(b) = 0$. $\square$

For example, let $f(X) = X^2 + 1 \in \widehat{\mathbb{Z}}_5[X]$. We have $|f(2)|_5 = |5|_5 < 1$, $|f'(2)|_5 = |4|_5 = 1$. By 8.6, there is a 5-adic integer x such that $x^2 = -1$.

Adapting Newton's method yields a sharper version of 8.6 for discrete valuations, also known as Hensel's lemma.

Proposition **8.7** (Hensel's Lemma). *Let K be complete for a discrete valuation v and $f \in \mathfrak{o}_v[X]$. If $v(f(a)) < v(f'(a))^2$ for some $a \in \mathfrak{o}_v$, then $f(b) = 0$ for some unique $b \in \mathfrak{o}_v$ such that $v(b - a) \leqq v(f(a))/v(f'(a))$.*

Proof. We may regard v as a nonarchimedean absolute value. Assume that $|f(a)| < |f'(a)|^2$. Let $\alpha = |f(a)|$, $\beta = |f'(a)|$, and $\gamma = |f(a)|/|f'(a)|^2 < 1$. Define $b_1 = a$ and

$$b_{n+1} = b_n - \frac{f(b_n)}{f'(b_n)}.$$

We prove by induction on n that

(2) $\qquad\qquad b_n \in \mathfrak{o}, \ |b_n - a| \leqq \alpha/\beta, \text{ and } |f(b_n)| \leqq \beta^2 \gamma^n.$

We see that (2) holds when $n = 1$. If $n > 1$, then binomial expansion yields

$$f(x + t) = f(x) + tf'(x) + t^2 g(x, t) \text{ and}$$
$$f'(x + t) = f'(x) + tf''(x) + t^2 h(x, t)$$

for some $g, h \in \mathfrak{o}[X, T]$. Hence $|b_n - a| \leqq \alpha/\beta$ implies

$$f'(b_n) - f'(a) = (b_n - a)\big(f''(a) + (b_n - a) h(a, b_n - a)\big),$$
$$|f'(b_n) - f'(a)| \leqq |b_n - a| \leqq \alpha/\beta < \beta,$$

and $|f'(b_n)| = |f'(a)| = \beta$. In particular, $f'(b_n) \neq 0$, so that b_{n+1} is defined. Let $t = -f(b_n)/f'(b_n)$. We have $f(b_{n+1}) = f(b_n + t) = t^2 g(b_n, t)$, and $|f(b_n)| \leqq \beta^2 \gamma^n$ implies

$$|f(b_{n+1})| \leqq |t|^2 = |f(b_n)|^2/|f'(b_n)|^2 \leqq \beta^2 \gamma^{2n} \leqq \beta^2 \gamma^{n+1}.$$

In particular, $|f(b_n)/f'(b_n)| < 1$, so that $b_n \in \mathfrak{o}$ implies $b_{n+1} \in \mathfrak{o}$. Also,

(3) $\qquad\qquad |b_{n+1} - b_n| = |f(b_n)|/|f'(b_n)| \leqq \beta \gamma^n,$

so that $|b_{n+1} - b_n| \leqq \beta\gamma = \alpha/\beta$ and $|b_n - a| \leqq \alpha/\beta$ implies $|b_{n+1} - a| \leqq \alpha/\beta$. Thus (2) holds for all n.

Since $\gamma < 1$, (3) shows that $(b_n)_{n>0}$ is a Cauchy sequence. Since K is complete, $(b_n)_{n>0}$ has a limit b. Then (2) yields $b \in \mathfrak{o}$, $|b - a| \leqq \alpha/\beta$, and $|f(b)| = 0$, whence $f(b) = 0$.

The equality $|f'(b_n)| = \beta$ also implies $|f'(b)| = \beta$. Let $c \in \mathfrak{o}$, $|c - a| \leqq \alpha/\beta$, and $f(c) = 0$. Then $|c - b| \leqq \alpha/\beta$ and

$$0 = f(c) - f(b) = (c - b)f'(b) + (c - b)^2 g(b, c - b).$$

If $c \neq b$, then $f'(b) = -(c - b) g(b, c - b)$, and $|f'(a)| = |f'(b)| \leqq |c - b| \leqq |f(a)|/|f'(a)|$ contradicts the hypothesis. $\square$

For example, let v be the 2-adic valuation, $K = \widehat{\mathbb{Q}}_2$, and $f(X) = X^2 + 7 \in \widehat{\mathbb{Z}}_2[X]$. We have $|f(1)| = |8| = 1/8$, $|f'(1)| = |2| = 1/2$. By 8.7, there is a unique 2-adic integer x such that $|x - 1| \leqq 1/4$ and $x^2 = -7$.

Exercises

1. Let K be complete with respect to a discrete valuation, and let $f \in \mathfrak{o}[X]$ be primitive and irreducible. Show that $\overline{f}$ is, up to a constant, a power of a single irreducible polynomial.

2. Let R be a commutative ring and let $\mathfrak{p}$ be a prime ideal of R. Prove the following: if $f(X) = a_0 + a_1 X + \cdots + a_n X^n \in R[X]$, $a_0, \ldots, a_{n-1} \in \mathfrak{p}$, $a_n \notin \mathfrak{p}$, and a_0 is not the product of two elements of $\mathfrak{p}$, then f is irreducible.

3. Let K be complete for a valuation v and $f = a_0 + a_1 X + \cdots + a_n X^n \in K[X]$ be primitive and irreducible of degree n. Show that $\max\big(v(a_0), v(a_1), \ldots, v(a_n)\big) = \max\big(v(a_0), v(a_n)\big)$.

4. Which of the polynomials $X^2 + 1$, $X^2 + 3$, $X^2 + 5$ are irreducible in $\widehat{\mathbb{Q}}_3$?

5. Find all extensions to $\mathbb{Q}(\sqrt{5})$ of the 3-adic valuation on $\mathbb{Q}$.

6. Find all extensions to $\mathbb{Q}(\sqrt{7})$ of the 3-adic valuation on $\mathbb{Q}$.

7. Find all extensions to $\mathbb{Q}(\sqrt{3})$ of the 3-adic valuation on $\mathbb{Q}$.

8. Determine all valuations on $\mathbb{Q}(i)$.

9. Filtrations and Completions

Filtrations by ideals, or by powers of an ideal, provide very general completions of commutative rings. This leads to a more general statement of Hensel's lemma, and to a universal property of power series rings.

Construction. Ring *filtrations* are infinite descending sequences of ideals:

Definition. A filtration *on a commutative ring R is an infinite descending sequence $\mathfrak{a}_1 \supseteq \mathfrak{a}_2 \supseteq \cdots \supseteq \mathfrak{a}_i \supseteq \mathfrak{a}_{i+1} \supseteq \cdots$ of ideals of R.*

For instance, if $\mathfrak{a}$ is an ideal of R, then

$$\mathfrak{a} \supseteq \mathfrak{a}^2 \supseteq \cdots \supseteq \mathfrak{a}^i \supseteq \mathfrak{a}^{i+1} \supseteq \cdots$$

is a filtration on R, the $\mathfrak{a}$-*adic filtration* on R. ($\mathfrak{a}^n$ is the ideal of R generated by all $a_1 \cdots a_n$ with $a_1, \ldots, a_n \in \mathfrak{a}$.)

Definition. The completion *of a commutative ring R relative to a filtration $\mathcal{A} : \mathfrak{a}_1 \supseteq \mathfrak{a}_2 \supseteq \cdots$ on R is the ring $\widehat{R}_\mathcal{A}$ of all infinite sequences*

$$(x_1 + \mathfrak{a}_1, \ldots, x_i + \mathfrak{a}_i, \ldots)$$

such that $x_i \in R$ and $x_i + \mathfrak{a}_i = x_j + \mathfrak{a}_i$ whenever $j \geqq i$; equivalently, $x_{i+1} \in x_i + \mathfrak{a}_i$ for all $i \geqq 1$. If $\mathfrak{a}$ is an ideal of R, then the $\mathfrak{a}$-adic completion $\widehat{R}_\mathfrak{a}$ of R is its completion relative to the $\mathfrak{a}$-adic filtration.

We see that $\widehat{R}_\mathcal{A}$ is a subring of $R/\mathfrak{a}_1 \times R/\mathfrak{a}_2 \times \cdots$. In Section XI.4 we will recognize $\widehat{R}_\mathcal{A}$ as the inverse limit of the rings $R/\mathfrak{a}_i$. One may view the rings $R/\mathfrak{a}_i$ as increasingly accurate approximations of $\widehat{R}_\mathcal{A}$.

By definition, the $\mathfrak{a}$-adic completion $\widehat{R}_\mathfrak{a}$ of R consists of all sequences

$$(x_1 + \mathfrak{a}, \ldots, x_i + \mathfrak{a}^i, \ldots)$$

such that $x_i \in R$ and $x_i + \mathfrak{a}^i = x_j + \mathfrak{a}^i$ whenever $j \geqq i$; equivalently, $x_{i+1} \in x_i + \mathfrak{a}^i$ for all $i \geqq 1$. For example, when p is a prime and $\mathfrak{p} = \mathbb{Z}p$, then $\widehat{\mathbb{Z}}_{\mathfrak{p}}$ is isomorphic to the ring $\widehat{\mathbb{Z}}_p$ of p-adic integers; when $\mathfrak{o}$ is the valuation ring of a discrete valuation v on a field F and $\mathfrak{m}$ is its maximal ideal, then $\widehat{\mathfrak{o}}_{\mathfrak{m}}$ is isomorphic to the valuation ring of $\widehat{F}_v$ (see the exercises).

Proposition 9.1. *If $R = S[X_1, ..., X_n]$, where S is a commutative ring, and $\mathfrak{a}$ is the ideal generated by $X_1, \ldots, X_n$, then $\widehat{R}_{\mathfrak{a}} \cong S[[X_1, ..., X_n]]$.*

Proof. We see that $\mathfrak{a}^i$ consists of all polynomials of order at least i. For every $f \in S[[X_1, ..., X_n]]$, let $\varphi_i(f) \in R/\mathfrak{a}^i$ be the coset of any polynomial with the same terms of degree less than i as f. Then φ_i is a homomorphism, and $\varphi_i(f) + \mathfrak{a}^i = \varphi_j(f) + \mathfrak{a}^i$ whenever $j \geqq i$; hence $\varphi : f \longmapsto (\varphi_1(f), \ldots, \varphi_i(f), \ldots)$ is a homomorphism of $S[[X_1, ..., X_n]]$ into $\widehat{R}_{\mathfrak{a}}$.

Conversely, let $g = (g_1 + \mathfrak{a}, \ldots, g_i + \mathfrak{a}^i, \ldots) \in \widehat{R}_{\mathfrak{a}}$, so that $g_i \in S[X_1, ..., X_n]$ and $g_{i+1} - g_i \in \mathfrak{a}^{i+1}$ has order at least $i + 1$ for all $i \geqq 1$. Then $(g_{i+1} - g_i)_{i>0}$ is addible in $S[[X_1, ..., X_n]]$. Let $\overline{g} = g_1 + \sum_{i>0} (g_{i+1} - g_i)$. The partial sum $g_1 + (g_2 - g_1) + \cdots + (g_i - g_{i-1})$ is g_i; hence $\overline{g}$ and g_i have the same terms of degree less than i. In particular, $g = \varphi(\overline{g})$, and $\overline{g}$ depends only on the cosets $g_i + \mathfrak{a}^i$ and not on the choice of g_i in $g_i + \mathfrak{a}^i$. Hence $g \longmapsto \overline{g}$ is a well defined homomorphism of $\widehat{R}_{\mathfrak{a}}$ into $S[[X_1, ..., X_n]]$.

We see that $\overline{\varphi(f)} = f$ for every $f \in S[[X_1, ..., X_n]]$, since f and $\overline{\varphi(f)}$ have the same terms of degree less than i as $\varphi_i(f)$, for all i. Thus φ and $g \longmapsto \overline{g}$ are mutually inverse isomorphisms. $\square$

Properties. In general, the completion $\widehat{R}_{\mathcal{A}}$ comes with a canonical homomorphism $R \longrightarrow \widehat{R}_{\mathcal{A}}$ defined as follows.

Definition. *If $\mathcal{A} : \mathfrak{a}_1 \supseteq \mathfrak{a}_2 \supseteq \cdots$ is a filtration on R, then $\iota : x \longmapsto (x + \mathfrak{a}_1, \ldots, x + \mathfrak{a}_i, \ldots)$ is the* canonical homomorphism *of R into $\widehat{R}_{\mathcal{A}}$.*

Readers will verify that ι is injective if and only if $\bigcap_{i>0} \mathfrak{a}_i = 0$; this is true in the examples above, but not in general.

Definition. *A ring R is* complete *relative to a filtration $\mathcal{A}$ when the canonical homomorphism $\iota : R \longrightarrow \widehat{R}_{\mathcal{A}}$ is an isomorphism. A ring R is* complete *relative to an ideal $\mathfrak{a}$ when the canonical homomorphism $\iota : R \longrightarrow \widehat{R}_{\mathfrak{a}}$ is an isomorphism.*

We show that $\widehat{R}_{\mathcal{A}}$ is always complete.

Proposition 9.2. *If $\mathcal{A} : \mathfrak{a}_1 \supseteq \mathfrak{a}_2 \supseteq \cdots$ is a filtration on R, then*

$$\widehat{\mathfrak{a}}_j = \{ (x_1 + \mathfrak{a}_1, \ldots, x_i + \mathfrak{a}_i \ldots) \in \widehat{R}_{\mathcal{A}} \mid x_j \in \mathfrak{a}_j \}$$

is an ideal of $\widehat{R}_{\mathcal{A}}$, $\widehat{\mathcal{A}} : \widehat{\mathfrak{a}}_1 \supseteq \widehat{\mathfrak{a}}_2 \supseteq \cdots$ is a filtration on $\widehat{R}_{\mathcal{A}}$, and $\widehat{R}_{\mathcal{A}}$ is complete relative to $\widehat{\mathcal{A}}$.

Note that $\widehat{\mathfrak{a}}_j = \{ (x_1 + \mathfrak{a}_1, \ldots, x_i + \mathfrak{a}_i, \ldots) \in \widehat{R}_A \mid x_i \in \mathfrak{a}_i \text{ for all } i \leqq j \}$; thus, $(\overline{x}_1, \ldots, \overline{x}_i, \ldots) \in \widehat{\mathfrak{a}}_j$ if and only if $\overline{x}_i = 0$ in $R/\mathfrak{a}_i$ for all $i \leqq j$.

Proof. First, $\widehat{\mathfrak{a}}_j$ is an ideal of $\widehat{R}_A$, since it is the kernel of the homomorphism $(x_1 + \mathfrak{a}_1, x_2 + \mathfrak{a}_2, \ldots) \longmapsto x_j + \mathfrak{a}_j$ of $\widehat{R}_A$ into $R/\mathfrak{a}_j$. In particular, $\widehat{R}_A/\widehat{\mathfrak{a}}_j \cong R/\mathfrak{a}_j$; the isomorphism sends $(x_1 + \mathfrak{a}_1, x_2 + \mathfrak{a}_2, \ldots) + \widehat{\mathfrak{a}}_j$ to $x_j + \mathfrak{a}_j$. The alternate description of $\widehat{\mathfrak{a}}_j$ shows that $\widehat{\mathfrak{a}}_1 \supseteq \widehat{\mathfrak{a}}_2 \supseteq \cdots$.

Let $S = \widehat{R}_A$ and $\widehat{S} = \widehat{S}_{\widehat{A}}$. If $x = (x_1 + \mathfrak{a}_1, x_2 + \mathfrak{a}_2, \ldots) \in S$ lies in $\bigcap_{i>0} \widehat{\mathfrak{a}}_j$, then $x_i \in \mathfrak{a}_i$ for all $i \leqq j$ and all j, and $x = (\mathfrak{a}_1, \mathfrak{a}_2, \ldots) = 0$ in S; hence $\iota : S \longrightarrow \widehat{S}$ is injective. Let $x = (x_1 + \widehat{\mathfrak{a}}_1, x_2 + \widehat{\mathfrak{a}}_2, \ldots) \in \widehat{S}$, so that $x_j \in \widehat{R}_A$ for all j and $x_j + \widehat{\mathfrak{a}}_j = x_k + \widehat{\mathfrak{a}}_j$ whenever $k \geqq j$. Then $x_k - x_j \in \widehat{\mathfrak{a}}_j$ when $k \geqq j$, so that $x_k = (x_{k1} + \mathfrak{a}_1, x_{k2} + \mathfrak{a}_2, \ldots)$ and $x_j = (x_{j1} + \mathfrak{a}_1, x_{j2} + \mathfrak{a}_2, \ldots)$ have the same component $x_{ki} + \mathfrak{a}_i = x_{ji} + \mathfrak{a}_i \in R/\mathfrak{a}_i$ for all $i \leqq j$. Let $y_i = x_{ii} \in R$. If $k \geqq j$, then $x_{kj} - x_{jj} \in \mathfrak{a}_j$ and $x_{kk} - x_{kj} \in \mathfrak{a}_j$, since $x_k \in \widehat{R}_A$; hence $y_k - y_j \in \mathfrak{a}_j$. Thus $y = (y_1 + \mathfrak{a}_1, y_2 + \mathfrak{a}_2, \ldots) \in \widehat{R}_A$. Moreover, $y - x_j \in \widehat{\mathfrak{a}}_j$ for all j, since $y_j - x_{jj} \in \mathfrak{a}_j$. Hence $x = (x_1 + \widehat{\mathfrak{a}}_1, x_2 + \widehat{\mathfrak{a}}_2, \ldots) = (y + \widehat{\mathfrak{a}}_1, y + \widehat{\mathfrak{a}}_2, \ldots) = \iota(y)$. Thus $\iota : S \longrightarrow \widehat{S}$ is an isomorphism. $\square$

Limits of sequences are defined in R as follows when R has a filtration.

Definitions. Relative to a filtration $\mathfrak{a}_1 \supseteq \mathfrak{a}_2 \supseteq \cdots$, $x \in R$ *is a* limit *of a sequence* $(x_n)_{n>0}$ *when for every* $i > 0$ *there exists* $N > 0$ *such that* $x - x_n \in \mathfrak{a}_i$ *for all* $n \geqq N$; *a sequence* $(x_n)_{n>0}$ *is* Cauchy *when for every* $i > 0$ *there exists* $N > 0$ *such that* $x_m - x_n \in \mathfrak{a}_i$ *for all* $m, n \geqq N$.

Readers will verify basic properties of limits, such as the limit laws.

Proposition **9.3.** *If* R *is complete (relative to a filtration), then every Cauchy sequence of elements of* R *has a unique limit in* R.

Proof. Let $(x_n)_{n>0}$ be a Cauchy sequence relative to $A : \mathfrak{a}_1 \supseteq \mathfrak{a}_2 \supseteq \cdots$. Choose $n(i)$ by induction so that $n(i + 1) \geqq n(i)$ and $x_m - x_n \in \mathfrak{a}_i$ for all $m, n \geqq n(i)$. Then $(x_{n(i)})_{i>0}$ is Cauchy, with $x_{n(j)} - x_{n(k)} \in \mathfrak{a}_i$ for all $j, k \geqq i$. Hence $\widehat{x} = (x_{n(1)} + \mathfrak{a}_1, x_{n(2)} + \mathfrak{a}_2, \ldots) \in \widehat{R}_A$. Since R is complete, $\widehat{x} = \iota(x)$ for some $x \in R$. Then $x - x_{n(i)} \in \mathfrak{a}_i$ for all i and x is a limit of $(x_{n(i)})_{i>0}$; by the choice of $n(i)$, x is also a limit of $(x_n)_{n>0}$. Readers will easily prove that the latter is unique. $\square$

The exercises give an alternate construction of $\widehat{R}_A$ by Cauchy sequences. Proposition 9.3 also yields some truly infinite sums:

Corollary **9.4.** (1) *If* R *is complete relative to a filtration* $\mathfrak{a}_1 \supseteq \mathfrak{a}_2 \supseteq \cdots$, *then every family* $(x_t)_{t \in T}$ *of elements of* R *such that*

$$\text{for every } i > 0, \ x_t \in \mathfrak{a}_i \text{ for almost all } t \in T$$

has a sum in R, *namely* $\lim_{i \to \infty} s_i$, *where* s_i *is the sum of all* $x_t \notin \mathfrak{a}_i$.

(2) *If* R *is complete relative to an ideal* $\mathfrak{a}$ *and* $a_1, \ldots, a_n \in \mathfrak{a}$, *then every power series* $\sum r_m a_1^{m1} \cdots a_n^{mn}$ *with coefficients in* R *has a sum in* R, *namely* $\lim_{i \to \infty} s_i$, *where* s_i *is the sum of all terms of degree less than* i.

Proof. (1). We see that s_i is a finite sum, and $(s_i)_{i > 0}$ is a Cauchy sequence. (2). For every i, $r_m a_1^{m1} \cdots a_n^{mn} \in \mathfrak{a}^i$ holds for almost all m, since $r_m a_1^{m1} \cdots a_n^{mn} \in \mathfrak{a}^{m_1 + \cdots + m_n} \subseteq \mathfrak{a}^i$ when $m_1 + \cdots + m_n \geqq i$. By (1), $\sum_m r_m a_1^{m1} \cdots a_n^{mn}$ exists, and $\sum_m r_m a_1^{m1} \cdots a_n^{mn} = \lim_{i \to \infty} s_i$, where s_i is the sum of all terms of degree less than i: indeed, s_i has the same limit as the sum t_i of all $r_m a_1^{m1} \cdots a_n^{mn} \notin \mathfrak{a}^i$, since $s_i - t_i \in \mathfrak{a}^i$. $\square$

Power series rings. If R is complete relative to an ideal $\mathfrak{a}$, then 9.4 yields for every $a_1, \ldots, a_n \in \mathfrak{a}$ an *evaluation* mapping of $R[[X_1, \ldots, X_n]] \longrightarrow R$, which sends $f(X_1, \ldots, X_n) = \sum_m r_m X_1^{m1} \cdots X_n^{mn}$ to

$$f(a_1, \ldots, a_n) = \sum_m r_m a_1^{m1} \cdots a_n^{mn}.$$

Readers will verify that this mapping is a homomorphism. Evaluation in turn yields a universal property:

Proposition **9.5.** *Let* R *and* S *be commutative rings. If* S *is complete relative to an ideal* $\mathfrak{b}$, *then for every ring homomorphism* $\varphi : R \longrightarrow S$ *and elements* $b_1, \ldots, b_n$ *of* $\mathfrak{b}$ *there exists a unique homomorphism* $\psi : R[[X_1, \ldots, X_n]] \longrightarrow S$ *that extends* φ *and sends* $X_1, \ldots, X_n$ *to* $b_1, \ldots, b_n$, *namely*

$$\sum r_m X_1^{m1} \cdots X_n^{mn} \longmapsto \sum \varphi(r_m) b_1^{m1} \cdots b_n^{mn}. \tag{1}$$

Compare to the universal property of polynomial rings, Theorem III.6.8.

Proof. Let $\mathfrak{a}$ be the ideal of $R[[X_1, \ldots, X_n]]$ generated by $X_1, \ldots, X_n$. For every $f = \sum_m r_m X_1^{m1} \cdots X_n^{mn} \in S[[X_1, \ldots, X_n]]$ let f_i be the sum of all terms of f of degree less than i. Then $f - f_i \in \mathfrak{a}^i$ for all i. By 9.4, the sum $\overline{f} = \sum_m \varphi(r_m) b_1^{m1} \cdots b_n^{mn}$ exists and has a similar property: $\overline{f} - \overline{f}_i \in \mathfrak{b}^i$ for all i, where $\overline{f}_i$ is the sum of all $\varphi(r_m) b_1^{m1} \cdots b_n^{mn}$ of degree $m_1 + \cdots + m_n < i$.

Let $\psi : R[[X_1, \ldots, X_n]] \longrightarrow S$ be a homomorphism that extends φ and sends $X_1, \ldots, X_n$ to $b_1, \ldots, b_n$. Then $\psi(\mathfrak{a}) \subseteq \mathfrak{b}$, $\psi(f_i) = \overline{f}_i$, and $\psi(f) - \overline{f}_i = \psi(f - f_i) \in \mathfrak{b}^i$ for all i. Since limits in S are unique, this implies $\psi(f) = \lim_{i \to \infty} \overline{f}_i = \overline{f}$. Hence ψ is unique, and given by (1).

Conversely, (1) defines a mapping $\psi : R[[X_1, \ldots, X_n]] \longrightarrow S$ that extends φ and sends $X_1, \ldots, X_n$ to $b_1, \ldots, b_n$. Then $\psi(f_i) = \overline{f}_i$ for all $f \in R[[X_1, \ldots, X_n]]$ and $i > 0$, and $\psi(f) = \lim_{i \to \infty} \overline{f}_i$. It is now straightforward that ψ is a homomorphism. $\square$

Substitution in power series follows from Proposition 9.5: for every $f_1, \ldots, f_n \in S[[X_1, \ldots, X_n]]$ of order at least 1, there is a substitution endomorphism of $S[[X_1, \ldots, X_n]]$ that sends $g(X_1, \ldots, X_n) = \sum_m s_m X_1^{m_1} \cdots X_n^{m_n}$ to

$$g(f_1, \ldots, f_n) = \sum_m s_m f_1^{m_1} \cdots f_n^{m_n}.$$

Hensel's lemma. Finally, we extend Hensel's lemma to every ring that is complete relative to an ideal. First we prove a lemma.

Lemma **9.6.** *If* $f = r_1 X + \cdots + r_i X^i + \cdots \in R[[X]]$, *then* $\eta : g(X) \longmapsto g(f)$ *is an automorphism of* $R[[X]]$ *if and only if* r_1 *is a unit of* R, *and then* $\eta^{-1} :$ $h(X) \longmapsto h(k)$ *for some* $k \in R[[X]]$ *of order* 1.

Proof. If η is an automorphism, then $X = \eta(g)$ for some g; $g = hX$ for some h, since g and $g(f)$ have the same constant term; X is a multiple of $\eta(X) = f$; and r_1 is a unit. Moreover, $k(X) = \eta^{-1}(X) \in R[[X]]$ has order at least 1, since $k(X)$ and $k(f) = X$ have the same constant term. By uniqueness in 9.5, η^{-1} and $h(X) \longmapsto h(k)$ coincide, since both are homomorphisms that send X to k. Hence $h(X) \longmapsto h(k)$ is an automorphism and k has order 1.

Conversely, assume that r_1 is a unit. If $g \neq 0$ in $R[[X]]$, then $g(X) = s_j X^j + s_{j+1} X^{j+1} + \cdots$ for some $j \geqq 0$ and $s_j \neq 0$; $g(f) = s_j f^j + s_{j+1} f^{j+1} + \cdots$ has a term of degree j with coefficient $s_j r_1^j \neq 0$, since r_1^j is a unit; and $\eta(g) \neq 0$. Thus η is injective.

Let $h = t_j X^j + t_{j+1} X^{j+1} + \cdots \in R[[X]]$ have order at least j. Since r_1^j is a unit we have $t_j = s r_1^j$ for some $s \in R$. Then $h - sf^j$ has order at least $j + 1$. If now $h \in R[[X]]$ is arbitrary, then there exist $s_0, s_1, \ldots, s_j, \ldots \in R$ such that $h - s_0$ has order at least 1, $h - (s_0 + s_1 f)$ has order at least 2, $\ldots$, and, for every j, $h - (s_0 + s_1 f + \cdots + s_j f^j)$ has order at least $j + 1$. Then $h = \sum_{j \geqq 0} s_j f^j = g(f)$, where $g = \sum_{j \geqq 0} s_j X^j$. Thus η is surjective. $\square$

We can now prove:

Theorem **9.7** (Hensel's Lemma). *Let* $f \in R[X]$, *where* R *is complete relative to an ideal* $\mathfrak{a}$. *If* $f(a) \in f'(a)^2 \, \mathfrak{a}$, *then* $f(b) = 0$ *for some* $b \in a + f'(a) \, \mathfrak{a}$; *moreover,* b *is unique if* $f'(a)$ *is not a zero divisor in* R.

Proof. Let $r = f'(a)$. We have

$$f(a + rX) = f(a) + (rX)f'(a) + (rX)^2 h(X) = f(a) + r^2 (X + X^2 h(X))$$

for some $h \in R[X]$. By 9.6, $\eta : g(X) \longmapsto g(X + X^2 h(X))$ is an automorphism of $R[[X]]$, and $\eta^{-1} : g(X) \longmapsto g(k)$, where $k \in R[[X]]$ has order 1. Hence

$$f(a + rk(X)) = \eta^{-1}\big(f(a + rX)\big) = f(a) + r^2 X.$$

Now, $f(a) = r^2 s$ for some $s \in \mathfrak{a}$. Evaluating at $-s \in \mathfrak{a}$ yields $f(a + rk(-s)) = f(a) - r^2 s = 0$. Thus $f(b) = 0$, where $b = a + rk(-s)$, and $b - a = rk(-s) \in$

$f'(a)\,\mathfrak{a}$: since $k(X)$ has order 1, $-s \in \mathfrak{a}$ implies $k(-s) \in \mathfrak{a}$.

Assume that $f(c) = 0$ for some $c \in a + f'(a)\,\mathfrak{a}$. Let $b - a = rs$, $c - a = rt$, where $s, t \in \mathfrak{a}$. Since $f(a + rX) = f(a) + r^2\,(X + X^2 h(X))$,

$$f(a) + r^2\,(s + s^2 h(s)) = f(a + rs) = 0 = f(a + rt) = f(a) + r^2\,(t + t^2 h(t)).$$

If r is not a zero divisor in R, this implies $s + s^2 h(s) = t + t^2 h(t)$. Then $\eta(k)(s) = k\big(s + s^2 h(s)\big) = k\big(t + t^2 h(t)\big) = \eta(k)(t)$. But $\eta(k) = X$, since $k = \eta^{-1}(X)$; hence $s = t$. $\square$

As a consequence of Hensel's lemma, some polynomial equations $f(X, Y) = 0$ have power series solutions $Y = g(X)$. We show this in $\mathbb{C}[[X]]$, which is complete relative to the ideal $\mathfrak{a}$ generated by X.

For example, let $R = \mathbb{C}[[X]]$ and $f(Y) = Y^3 - XY + X^3 \in R[Y]$. Then $f'(0) = -X$ is not a zero divisor in R, and $f(0) = X^3 \in (-X)^2\,\mathfrak{a}$. By Hensel's lemma, there is a unique $g(X) \in R$ such that $f\big(g(X)\big) = 0$ and $g \in 0 + f'(0)\,\mathfrak{a}$, that is, g has order at least 2.

The first terms of g are readily computed. Let $g(X) = aX^2 + bX^3 + cX^4 + dX^5 + \cdots$. Then $g^3 = a^3 X^6 + \cdots$ and

$$0 = g^3 - Xg + X^3 = (1 - a)X^3 - bX^4 - cX^5 + (a^3 - d)X^6 + \cdots;$$

hence $a = 1$, $b = c = 0$, and $d = a^3 = 1$. Thus $g(X) = X^2 + X^5 + \cdots$.

The geometric significance of this solution is that the algebraic curve $y^3 - xy + x^3 = 0$ has a branch $y = x^2 + x^5 + \cdots$ at the origin. By symmetry there is another branch $x = y^2 + y^5 + \cdots$. In effect, this separates the double point at the origin into two separate locations on the curve. The expansions also indicate the shape of the two branches.

Exercises

1. Prove that $\widehat{\mathbb{Z}}_{\mathfrak{p}}$ is isomorphic to the ring $\widehat{\mathbb{Z}}_p$ of p-adic integers when $\mathfrak{p} = \mathbb{Z}p$.

2. Let F be a field and let $\mathfrak{o}$ be the valuation ring of a discrete valuation v on F, with maximal ideal $\mathfrak{m}$. Prove that $\widehat{\mathfrak{o}}_{\mathfrak{m}}$ is isomorphic to the valuation ring of $\widehat{F}_v$. (You may want to use Proposition 6.6.)

R and S are commutative rings in all the following exercises.

3. Let $\mathcal{A} : \mathfrak{a}_1 \supseteq \mathfrak{a}_2 \supseteq \cdots$ and $\mathcal{B} : \mathfrak{b}_1 \supseteq \mathfrak{b}_2 \supseteq \cdots$ be filtrations on R. Suppose that every $\mathfrak{a}_i$ contains some $\mathfrak{b}_j$ and that every $\mathfrak{b}_j$ contains some $\mathfrak{a}_i$. Show that $\widehat{R}_{\mathcal{A}} \cong \widehat{R}_{\mathcal{B}}$.

4. Let $\mathcal{A} : \mathfrak{a}_1 \supseteq \mathfrak{a}_2 \supseteq \cdots$ be a filtration on R and let $\mathcal{B} : \mathfrak{b}_1 \supseteq \mathfrak{b}_2 \supseteq \cdots$ be a filtration on S. Let $\varphi : R \longrightarrow S$ be a ring homomorphism such that $\varphi(\mathfrak{a}_i) \subseteq \mathfrak{b}_i$ for all i. Show that φ induces a homomorphism $\widehat{\varphi} : \widehat{R}_{\mathcal{A}} \longrightarrow \widehat{S}_{\mathcal{B}}$.

5. State and prove a uniqueness property in the previous exercise. Then state and prove a universal property of $\widehat{R}_{\mathcal{A}}$.

6. Relative to a filtration $\mathfrak{a}_1 \supseteq \mathfrak{a}_2 \supseteq \cdots$ on R, show that limits of sequences are unique in R if and only if $\bigcap_{i>0} \mathfrak{a}_i = 0$.

7. Relative to a filtration $\mathfrak{a}_1 \supseteq \mathfrak{a}_2 \supseteq \cdots$ on R, show that every sequence of elements of R that has a limit in R is a Cauchy sequence.

8. Prove the following limit laws: relative to a filtration $\mathfrak{a}_1 \supseteq \mathfrak{a}_2 \supseteq \cdots$ on R, if x is a limit of $(x_n)_{n>0}$, and y is a limit of $(y_n)_{n>0}$, then $x + y$ is a limit of $(x_n + y_n)_{n>0}$ and xy is a limit of $(x_n y_n)_{n>0}$.

9. Show that R is complete (relative to a filtration on R) if and only if every Cauchy sequence of elements of R has a unique limit in R.

10. Relative to a filtration $\mathcal{A}$ on R, show that Cauchy sequences constitute a ring; show that sequences with limit zero constitute an ideal of that ring; show that the quotient ring is isomorphic to $\widehat{R}_{\mathcal{A}}$.

11. Prove the following: if R is complete relative to an ideal $\mathfrak{a}$, then $1 - a$ is a unit of R for every $a \in \mathfrak{a}$.

12. In the ring $R[[X_1, ..., X_n]]$, show that a power series is a unit if and only if its constant term is a unit of R. (You may want to use the previous exercise.)

13. Let R be complete relative to an ideal $\mathfrak{a}$. Show that evaluation at $a_1, ..., a_n \in \mathfrak{a}$ is a homomorphism of $R[[X_1, ..., X_n]]$ into R.

14. Use Hensel's lemma to show that the equation $Y^3 - X - 1 = 0$ has a solution $Y = g(X) \in \mathbb{C}[[X]]$. Then find an explicit solution.

15. Use Hensel's lemma to show that the equation $Y^2 - X^3 - X^2 = 0$ has a solution $Y = g(X) \in \mathbb{C}[[X]]$. Then find an explicit solution.

16. Use Hensel's lemma to show that the equation $Y^3 - 2XY + X^3 = 0$ has a solution $Y = g(X - 1)$, where $g \in \mathbb{C}[[X]]$ and $g(0) = 1$. Calculate the first three terms of g. What does this say about the curve $y^3 - 2xy + x^3 = 0$ near the point $(1, 1)$?

17. Let K be a field and $f \in K[X, Y]$. Suppose that $f(a, b) = 0$, $\frac{\partial f}{\partial y}(a, b) \neq 0$ for some $a, b \in K$. Prove that $f(X, g(X - a)) = 0$ for some $g \in K[[X]]$ such that $g(0) = b$.

18. Let $\mathfrak{a}_1 \supseteq \mathfrak{a}_2 \supseteq \cdots$ be a filtration on R. Show that the cosets of $\mathfrak{a}_1, \mathfrak{a}_2, \ldots$ constitute a basis for a topology on R. (This defines the *Krull topology* on R, named after the similar Krull topology on Galois groups.)

19. Relative to any filtration on R, show that the operations on R are continuous for its Krull topology.

20. Show that the Krull topology on a complete ring R is Hausdorff and totally disconnected.

*21. How much of Propositions 9.1 through 9.5 extends to not necessarily commutative rings?

VII
Commutative Rings

Commutative algebra, the study of commutative rings and related concepts, originated with Kummer's and Dedekind's study of the arithmetic properties of algebraic integers, and grew very quickly with the development of algebraic geometry, which consumes vast amounts of it. This chapter contains general properties of ring extensions, Noetherian rings, and prime ideals; takes a look at algebraic integers; and ends with a very minimal introduction to algebraic geometry.

A first reading might include only Sections 1 and 5, which have little prerequisites. Other sections use modules at a few critical places; definitions and proofs have been provided, but this chapter could be covered after Chapter VIII.

All rings in what follows are commutative rings with an identity element.

1. Primary Decomposition

This section contains basic properties of ideals and the Noether-Lasker theorem, proved in special cases by Lasker [1905] and in general by Noether [1921] in the seminal paper which introduced the ascending chain condition.

Ideals. We saw in Section III.2 that the sum and intersection of a family of ideals of a ring R are ideals of R. We define two additional operations.

Proposition **1.1.** *Let* $\mathfrak{a}$ *and* $\mathfrak{b}$ *be ideals of a commutative ring* R *and let* S *be a subset of* R.

(1) *The set* $\mathfrak{a}\mathfrak{b}$ *of all finite sums* $a_1 b_1 + \cdots a_n b_n$, *where* $n \geqq 0$, $a_1, \ldots, a_n \in \mathfrak{a}$, *and* $b_1, \ldots, b_n \in \mathfrak{b}$, *is an ideal of* R.

(2) *The set* $\mathfrak{a} : S = \{\, r \in R \mid rs \in \mathfrak{a} \text{ for all } s \in S \,\}$ *is an ideal of* R; *in particular,* $\mathfrak{a} : \mathfrak{b} = \{\, r \in R \mid r\mathfrak{a} \subseteq \mathfrak{b} \,\}$ *is an ideal of* R.

The quotient $\mathfrak{a} : S$ is also called the *transporter* of S into $\mathfrak{a}$. The notation $\mathfrak{a}\mathfrak{b}$ is traditional; one must remember that the product of $\mathfrak{a}$ and $\mathfrak{b}$ as ideals is larger than their product $\{\, ab \mid a \in \mathfrak{a},\ b \in \mathfrak{b} \,\}$ as subsets; in fact, the former is the ideal generated by the latter. Readers will verify the following properties.

Proposition **1.2.** *In a commutative ring* R, *the product of ideals is commutative and associative, and distributes sums and unions of chains. Moreover,*

$R\mathfrak{a} = \mathfrak{a}$ *and* $\mathfrak{a}\mathfrak{b} \subseteq \mathfrak{a} \cap \mathfrak{b}$, *for all ideals* $\mathfrak{a}$ *and* $\mathfrak{b}$ *of* R.

By 1.2, all products of ideals $\mathfrak{a}_1, \ldots, \mathfrak{a}_n$ (in any order) are equal. The resulting product $\mathfrak{a}_1 \cdots \mathfrak{a}_n$ is written without parentheses; readers will verify that it is the ideal generated by all products $a_1 \cdots a_n$ in which $a_i \in \mathfrak{a}_i$ for all i.

Proposition 1.3. *The following properties hold in a commutative ring* R, *for every subset* S *and ideals* $\mathfrak{a}, \mathfrak{b}, \mathfrak{c},$ *and* $\mathfrak{a}_i, \mathfrak{b}_i$ *of* R:

(1) $\mathfrak{a} : S$ *is an ideal of* R;

(2) $\mathfrak{a} \subseteq \mathfrak{a} : S$, *with* $\mathfrak{a} : S = R$ *if and only if* $S \subseteq \mathfrak{a}$;

(3) $\mathfrak{c} \subseteq \mathfrak{a} : \mathfrak{b}$ *if and only if* $\mathfrak{b}\mathfrak{c} \subseteq \mathfrak{a}$;

(4) $(\mathfrak{a} : \mathfrak{b}) : \mathfrak{c} = \mathfrak{a} : \mathfrak{b}\mathfrak{c}$;

(5) $\left(\bigcap_{i \in I} \mathfrak{a}_i \right) : S = \bigcap_{i \in I} (\mathfrak{a}_i : S)$;

(6) $\mathfrak{a} : \sum_{i \in I} \mathfrak{b}_i = \bigcap_{i \in I} (\mathfrak{a} : \mathfrak{b}_i)$.

Radicals. Recall that an ideal $\mathfrak{p}$ of a commutative ring R is *prime* when $\mathfrak{p} \neq R$ and $xy \in \mathfrak{p}$ implies $x \in \mathfrak{p}$ or $y \in \mathfrak{p}$ (Section III.4). Readers will verify that $\mathfrak{p}$ is prime if and only if, for all ideals $\mathfrak{a}$, $\mathfrak{b}$ of R, $\mathfrak{a}\mathfrak{b} \subseteq \mathfrak{p}$ implies $\mathfrak{a} \subseteq \mathfrak{p}$ or $\mathfrak{b} \subseteq \mathfrak{p}$. Moreover, $R/\mathfrak{a}$ is a domain if and only if $\mathfrak{a}$ is prime (III.4.5).

Definition. In a commutative ring R, *the* radical Rad $\mathfrak{a}$ *of an ideal* $\mathfrak{a}$ *is the intersection of all prime ideals of* R *that contain* $\mathfrak{a}$.

If $\mathfrak{a} = R$, then no prime ideal of R contains $\mathfrak{a}$ and we let the empty intersection Rad $\mathfrak{a}$ be R itself. In general, Rad $\mathfrak{a}$ is sometimes denoted by $\sqrt{\mathfrak{a}}$, and is often defined as follows:

Proposition 1.4. Rad $\mathfrak{a} = \{ x \in R \mid x^n \in \mathfrak{a} \text{ for some } n > 0 \}$.

Proof. Let $x \in R$ and let $\mathfrak{r}$ be the intersection of all prime ideals that contain $\mathfrak{a}$. If $x \in R \backslash \mathfrak{r}$, then $x \notin \mathfrak{p}$ for some prime ideal $\mathfrak{p} \subseteq \mathfrak{a}$, $x^n \notin \mathfrak{p}$ for all $n > 0$ since $\mathfrak{p}$ is prime, and $x^n \notin \mathfrak{a}$ for all $n > 0$.

Conversely, assume that $x^n \notin \mathfrak{a}$ for all $n > 0$. By Zorn's lemma there is an ideal $\mathfrak{p}$ that contains $\mathfrak{a}$, contains no x^n, and is maximal with these properties. Let $a, b \in R \backslash \mathfrak{p}$. By the choice of $\mathfrak{p}$, $\mathfrak{p} + (a)$ contains some x^m, and $\mathfrak{p} + (b)$ contains some x^n. Then $x^m = p + ra$, $x^n = q + sb$ for some $p, q \in \mathfrak{p}$ and $r, s \in R$, $x^{m+n} = pq + psb + qra + rsab \in \mathfrak{p} + (ab)$, $\mathfrak{p} + (ab) \supsetneqq \mathfrak{p}$, and $ab \notin \mathfrak{p}$. Thus $\mathfrak{p}$ is a prime ideal; since $x \notin \mathfrak{p}$, it follows that $x \notin \mathfrak{r}$. $\square$

By 1.4, the set Rad $0 = \{ x \in R \mid x^n = 0 \text{ for some } n > 0 \}$ of all nilpotent elements of R is an ideal of R, and is the intersection of all prime ideals of R; Rad 0 is the *nilradical* of R. Readers will prove the next two results:

Proposition 1.5. Rad $(\mathfrak{a}_1 \cap \cdots \cap \mathfrak{a}_n) = $ Rad $\mathfrak{a}_1 \cap \cdots \cap$ Rad $\mathfrak{a}_n$, *for all ideals* $\mathfrak{a}_1, \ldots, \mathfrak{a}_n$ *of a commutative ring* R.

Proposition 1.6. *If the ideal* Rad $\mathfrak{a}$ *is finitely generated, then* $(\text{Rad } \mathfrak{a})^n \subseteq \mathfrak{a}$ *for some* $n > 0$.

Definition. An ideal $\mathfrak{a}$ *of a commutative ring* R *is* semiprime *when it is an intersection of prime ideals.*

Thus, $\mathfrak{a}$ is semiprime when $\operatorname{Rad}\mathfrak{a} = \mathfrak{a}$; equivalently, when $x^n \in \mathfrak{a}$ implies $x \in \mathfrak{a}$. Readers will show that an ideal $\mathfrak{s}$ of R is semiprime if and only if $\mathfrak{a}^n \subseteq \mathfrak{s}$ implies $\mathfrak{a} \subseteq \mathfrak{s}$, for every $n > 0$ and ideal $\mathfrak{a}$ of R.

Definition. An ideal $\mathfrak{q}$ *of a commutative ring* R *is* primary *when* $\mathfrak{q} \neq R$ *and, for all* $x, y \in R$, $xy \in \mathfrak{q}$ *implies* $x \in \mathfrak{q}$ *or* $y^n \in \mathfrak{q}$ *for some* $n > 0$. *An ideal* $\mathfrak{q}$ *of* R *is* $\mathfrak{p}$-primary *when* $\mathfrak{q}$ *is primary and* $\operatorname{Rad}\mathfrak{q} = \mathfrak{p}$.

Readers will show that an ideal $\mathfrak{q}$ of R with radical $\mathfrak{p}$ is $\mathfrak{p}$-primary if and only if $\mathfrak{a}\mathfrak{b} \subseteq \mathfrak{q}$ implies $\mathfrak{a} \subseteq \mathfrak{q}$ or $\mathfrak{b} \subseteq \mathfrak{p}$, for every ideals $\mathfrak{a}$, $\mathfrak{b}$ of R; and that:

Proposition 1.7. (1) *The radical of a primary ideal is a prime ideal.*
(2) *The intersection of finitely many* $\mathfrak{p}$-primary *ideals is* $\mathfrak{p}$-primary.
(3) *An ideal whose radical is a maximal ideal is primary.*

Primary decomposition. We now let R be Noetherian. An ideal $\mathfrak{i}$ of R is *irreducible* (short for "intersection irreducible") when $\mathfrak{i} \neq R$ and $\mathfrak{i}$ is not the intersection of ideals $\mathfrak{a}, \mathfrak{b} \supsetneq \mathfrak{i}$.

Lemma 1.8. *Every ideal of a Noetherian ring* R *is the intersection of finitely many irreducible ideals of* R.

Proof. "Intersections" include one-term intersections and the empty intersection R. Call an ideal $\mathfrak{b}$ of R *nasty* when it is not the intersection of finitely many irreducible ideals of R. If the result is false, then the set of all nasty ideals of R is not empty; since R is Noetherian, there is a maximal nasty ideal $\mathfrak{n}$. This bad boy $\mathfrak{n}$ is not R and is not irreducible. Therefore $\mathfrak{n} = \mathfrak{a} \cap \mathfrak{b}$ for some ideals $\mathfrak{a}, \mathfrak{b} \supsetneq \mathfrak{n}$. By the maximality of $\mathfrak{n}$, $\mathfrak{a}$ and $\mathfrak{b}$ are intersections of finitely many irreducible ideals; but then so is $\mathfrak{n}$, a contradiction. $\square$

Theorem 1.9. *In a Noetherian ring, every ideal is the intersection of finitely many primary ideals.*

Proof. By 1.8 we need only show that every irreducible ideal $\mathfrak{i}$ of a Noetherian ring R is primary. Assume that $ab \in \mathfrak{i}$ and $b \notin \operatorname{Rad}\mathfrak{i}$. Let $\mathfrak{a}_n = \mathfrak{i} : b^n$. Then $a \in \mathfrak{a}_1$, $\mathfrak{i} \subseteq \mathfrak{a}_n$, $\mathfrak{a}_n$ is an ideal, and $\mathfrak{a}_n \subseteq \mathfrak{a}_{n+1}$, since $xb^n \in \mathfrak{i}$ implies $xb^{n+1} \in \mathfrak{i}$. Since R is Noetherian, the ascending sequence $(\mathfrak{a}_n)_{n>0}$ terminates; hence $\mathfrak{a}_{2n} = \mathfrak{a}_n$ if n is large enough. Let $\mathfrak{b} = \mathfrak{i} + Rb^n$. If $x \in \mathfrak{a}_n \cap \mathfrak{b}$, then $xb^n \in \mathfrak{i}$ and $x = t + yb^n$ for some $t \in \mathfrak{i}$ and $y \in R$, whence $tb^n + yb^{2n} \in \mathfrak{i}$, $yb^{2n} \in \mathfrak{i}$, $y \in \mathfrak{a}_{2n} = \mathfrak{a}_n$, $yb^n \in \mathfrak{i}$, and $x = t + yb^n \in \mathfrak{i}$. Hence $\mathfrak{a}_n \cap \mathfrak{b} = \mathfrak{i}$. Now, $\mathfrak{b} \supsetneq \mathfrak{i}$, since $b^n \notin \mathfrak{i}$. Therefore $\mathfrak{a}_n = \mathfrak{i}$; hence $\mathfrak{a}_1 = \mathfrak{i}$ and $a \in \mathfrak{i}$. $\square$

An *algebraic set* $A \subseteq K^n$ over a field K is the set of solutions of a system of polynomial equations $f(x_1, \ldots, x_n) = 0$, where f ranges through a subset S of $K[X_1, \ldots, X_n]$. Then A is also the solution set of the ideal $\mathfrak{a}$ generated by S. Hence algebraic geometry, the general study of algebraic sets and related concepts, begins with ideals of $K[X_1, ..., X_n]$. By the Hilbert basis theorem, every algebraic set can be defined by finitely many equations.

In the above, $\mathfrak{a}$ may as well be semiprime, since A is also the set of all $x \in K^n$ such that $f(x) = 0$ for all $f \in \operatorname{Rad} \mathfrak{a}$. By 1.9, $\mathfrak{a}$ is the intersection of finitely many primary ideals, and $\mathfrak{a} = \operatorname{Rad} \mathfrak{a}$ is the intersection of their radicals, that is, the intersection of finitely many prime ideals. It follows that A is the union of finitely many algebraic sets defined by prime ideals (see Section 10). Algebraic geometry can now concentrate on prime ideals of $K[X_1, ..., X_n]$. Prime ideals of Noetherian rings are studied in more detail in Sections 7 and 8.

Uniqueness. Intersections $\mathfrak{q}_1 \cap \cdots \cap \mathfrak{q}_r$ of primary ideals can be simplified in two ways: by deleting *superfluous* terms $\mathfrak{q}_i$ such that $\mathfrak{q}_1 \cap \cdots \cap \mathfrak{q}_r = \mathfrak{q}_1 \cap \cdots \cap \mathfrak{q}_{i-1} \cap \mathfrak{q}_{i+1} \cap \cdots \cap \mathfrak{q}_r$; or by replacing several terms with the same radical $\mathfrak{p}$ by their intersection, which by 1.7 is a $\mathfrak{p}$-primary ideal. An intersection $\mathfrak{q}_1 \cap \cdots \cap \mathfrak{q}_n$ of primary ideals is *reduced* when it has no superfluous term and the radicals $\operatorname{Rad} \mathfrak{q}_1, ..., \operatorname{Rad} \mathfrak{q}_r$ are distinct.

Theorem **1.10** (Noether-Lasker). *In a Noetherian ring, every ideal is a reduced intersection of finitely many primary ideals, whose radicals are unique.*

Proof. The *associated prime ideals* of an ideal $\mathfrak{a}$ are the prime ideals of the form $\mathfrak{a} : c$, where $c \notin \mathfrak{a}$. We show that in every reduced primary decomposition $\mathfrak{a} = \mathfrak{q}_1 \cap \cdots \cap \mathfrak{q}_r$ of an ideal $\mathfrak{a}$, the distinct prime ideals $\mathfrak{p}_i = \operatorname{Rad} \mathfrak{q}_i$ coincide with the associated prime ideals of $\mathfrak{a}$.

Let $1 \leq j \leq r$ and let $\mathfrak{b} = \bigcap_{i \neq j} \mathfrak{q}_i$. Then $\mathfrak{a} = \mathfrak{b} \cap \mathfrak{q}_j \subsetneqq \mathfrak{b}$. By 1.6, $\mathfrak{p}_j^n \subseteq \mathfrak{q}_j$ for some n; then $\mathfrak{b}\mathfrak{p}_j^n \subseteq \mathfrak{b} \cap \mathfrak{q}_j = \mathfrak{a}$. Let n be minimal such that $\mathfrak{b}\mathfrak{p}_j^n \subseteq \mathfrak{a}$. Then $n > 0$, $\mathfrak{b}\mathfrak{p}_j^{n-1} \not\subseteq \mathfrak{a}$, and there exists $c \in \mathfrak{b}\mathfrak{p}_j^{n-1} \backslash \mathfrak{a}$. We show that $\mathfrak{p}_j = \mathfrak{a} : c$. We have $c \in \mathfrak{b}$ and $c \notin \mathfrak{q}_j$ (otherwise, $c \in \mathfrak{a}$). Since $\mathfrak{q}_j$ is $\mathfrak{p}_j$-primary, $cx \in \mathfrak{a} \subseteq \mathfrak{q}_j$ implies $x \in \mathfrak{p}_j$, and $\mathfrak{a} : c \subseteq \mathfrak{p}_j$. Conversely, $c\mathfrak{p}_j \subseteq \mathfrak{b}\mathfrak{p}_j^n \subseteq \mathfrak{a}$, so that $\mathfrak{p}_j \subseteq \mathfrak{a} : c$. Thus $\mathfrak{p}_j = \mathfrak{a} : c$.

Conversely, let $\mathfrak{p} = \mathfrak{a} : c$ be an associated prime ideal of $\mathfrak{a}$, where $c \notin \mathfrak{a}$. Then $c \notin \mathfrak{q}_j$ for some j. Let $\mathfrak{b} = \prod_{c \notin \mathfrak{q}_i} \mathfrak{q}_i$. Then $c\mathfrak{b} \subseteq \mathfrak{q}_i$ for all i, $c\mathfrak{b} \subseteq \mathfrak{a}$, and $\mathfrak{b} \subseteq \mathfrak{p}$. Hence $\mathfrak{q}_i \subseteq \mathfrak{p}$ for some i such that $c \notin \mathfrak{q}_i$. Then $\mathfrak{p}_i = \operatorname{Rad} \mathfrak{q}_i \subseteq \mathfrak{p}$. Conversely, $cx \in \mathfrak{a} \subseteq \mathfrak{q}_i$ implies $x \in \mathfrak{p}_i$, since $\mathfrak{q}_i$ is $\mathfrak{p}_i$-primary, so that $\mathfrak{p} = \mathfrak{a} : c \subseteq \mathfrak{p}_i$. Thus $\mathfrak{p} = \mathfrak{p}_i$. $\square$

In the above, $\mathfrak{q}_1, ..., \mathfrak{q}_r$ are not in general unique. It can be proved, however, that if $\mathfrak{p}_i = \operatorname{Rad} \mathfrak{q}_i$ is minimal among $\mathfrak{p}_1, ..., \mathfrak{p}_r$, then $\mathfrak{q}_i$ is unique.

Exercises

1. Let $m, n \in \mathbb{Z}$. In the ring $\mathbb{Z}$, what is $(m)(n)$? what is $(m):(n)$?

2. Let $n \in \mathbb{Z}$. When is (n) a semiprime ideal of $\mathbb{Z}$?

3. Let $n \in \mathbb{Z}$. When is (n) a primary ideal of $\mathbb{Z}$?

R is a commutative ring in what follows.

4. Show that $\mathfrak{a}_1 \cdots \mathfrak{a}_n$ is the ideal generated by all products $a_1 \cdots a_n$ in which $a_i \in \mathfrak{a}_i$ for all i.

5. Show that the product of ideals of R is associative.

6. Show that $\mathfrak{a}\left(\sum_{i \in I} \mathfrak{b}_i\right) = \sum_{i \in I} (\mathfrak{a}\mathfrak{b}_i)$, for all ideals $\mathfrak{a}$ and $\mathfrak{b}_i$ of R.

7. Show that $\mathfrak{a}\left(\bigcup_{i \in I} \mathfrak{b}_i\right) = \bigcup_{i \in I} (\mathfrak{a}\mathfrak{b}_i)$ when $\mathfrak{a}$ and $\mathfrak{b}_i$ are ideals of R and $(\mathfrak{b}_i)_{i \in I}$ is a nonempty directed family.

8. Show that $\mathfrak{a} : S$ is an ideal of R, that $\mathfrak{a} \subseteq \mathfrak{a} : S$, and that $\mathfrak{a} : S = R$ if and only if $S \subseteq \mathfrak{a}$, for every ideal $\mathfrak{a}$ and subset S of R.

9. Show that $\mathfrak{c} \subseteq \mathfrak{a} : \mathfrak{b}$ if and only if $\mathfrak{b}\mathfrak{c} \subseteq \mathfrak{a}$, for all ideals $\mathfrak{a}, \mathfrak{b}, \mathfrak{c}$ of R.

10. Show that $(\mathfrak{a} : \mathfrak{b}) : \mathfrak{c} = \mathfrak{a} : \mathfrak{b}\mathfrak{c}$, for all ideals $\mathfrak{a}, \mathfrak{b}, \mathfrak{c}$ of R.

11. Show that $\left(\bigcap_{i \in I} \mathfrak{a}_i\right) : S = \bigcap_{i \in I} (\mathfrak{a}_i : S)$ and $\mathfrak{a} : \sum_{i \in I} \mathfrak{b}_i = \bigcap_{i \in I} (\mathfrak{a} : \mathfrak{b}_i)$, for all ideals $\mathfrak{a}, \mathfrak{b}, \mathfrak{a}_i, \mathfrak{b}_i$ of R.

12. Show that $\left(\bigcup_{i \in I} \mathfrak{a}_i\right) : S = \bigcup_{i \in I} (\mathfrak{a}_i : S)$, if $(\mathfrak{a}_i)_{i \in I}$ is a nonempty directed family and S is finite. Show that $\left(\bigcup_{i \in I} \mathfrak{a}_i\right) : \mathfrak{b} = \bigcup_{i \in I} (\mathfrak{a}_i : \mathfrak{b})$, if $(\mathfrak{a}_i)_{i \in I}$ is a nonempty directed family and $\mathfrak{b}$ is a finitely generated ideal.

13. Show that an ideal $\mathfrak{a} \neq R$ of R is prime if and only if $\mathfrak{b}\mathfrak{c} \subseteq \mathfrak{a}$ implies $\mathfrak{b} \subseteq \mathfrak{a}$ or $\mathfrak{c} \subseteq \mathfrak{a}$, for every ideals $\mathfrak{b}, \mathfrak{c}$ of R.

14. Show that an ideal $\mathfrak{a}$ of R is semiprime if and only if $\mathfrak{c}^n \subseteq \mathfrak{a}$ implies $\mathfrak{c} \subseteq \mathfrak{a}$, for every $n > 0$ and ideal $\mathfrak{c}$ of R.

15. Show that $\mathrm{Rad}\left(\bigcap_{i \in I} \mathfrak{a}_i\right) = \bigcap_{i \in I} \mathrm{Rad}\, \mathfrak{a}_i$ when I is finite. Give an example in which I is infinite and $\mathrm{Rad}\left(\bigcap_{i \in I} \mathfrak{a}_i\right) \neq \bigcap_{i \in I} \mathrm{Rad}\, \mathfrak{a}_i$.

16. Show that $(\mathrm{Rad}\, \mathfrak{a})^n \subseteq \mathfrak{a}$ for some $n > 0$ when $\mathrm{Rad}\, \mathfrak{a}$ is finitely generated.

17. Show that an ideal $\mathfrak{q}$ of R with radical $\mathfrak{p}$ is $\mathfrak{p}$-primary if and only if $\mathfrak{a}\mathfrak{b} \subseteq \mathfrak{q}$ implies $\mathfrak{a} \subseteq \mathfrak{q}$ or $\mathfrak{b} \subseteq \mathfrak{p}$, for all ideals $\mathfrak{a}, \mathfrak{b}$ of R.

18. Show that the radical of a primary ideal is a prime ideal.

19. Show that the intersection of finitely many $\mathfrak{p}$-primary ideals is $\mathfrak{p}$-primary.

20. Show that an ideal whose radical is a maximal ideal is primary.

21. Show that an ideal $\mathfrak{a}$ of a Noetherian ring has only finitely many associated prime ideals, whose intersection is $\mathrm{Rad}\, \mathfrak{a}$.

2. Ring Extensions

In this section we extend some properties of field extensions to ring extensions.

Definition. A ring extension of a commutative ring R is a commutative ring E of which R is a subring.

In particular, the identity element of R is also the identity element of all its ring extensions. A ring extension of R may also be defined, as was the case with field extensions, as a ring E with an injective homomorphism of R into E; surgery shows that this is an equivalent definition, up to isomorphisms.

Proposition **2.1.** *Let E be a ring extension of R and let S be a subset of E. The subring $R[S]$ of E generated by $R \cup S$ is the set of all linear combinations with coefficients in R of products of powers of elements of S.*

Proof. This is proved like IV.1.9. Let $(X_s)_{s \in S}$ be a family of indeterminates, one for each $s \in S$. Let $\psi : R[(X_s)_{s \in S}] \longrightarrow E$ be the evaluation homomorphism that sends X_s to s for all $s \in S$:

$$\psi\left(\sum_k \left(a_k \prod_{s \in S} X_s^{k_s}\right)\right) = \sum_k \left(a_k \prod_{s \in S} s^{k_s}\right).$$

Then $\operatorname{Im} \psi$ is a subring of E, which contains R and S and consists of all finite linear combinations with coefficients in R of finite products of powers of elements of S. Conversely, all such linear combinations belong to every subring of E that contains R and S. $\square$

Corollary **2.2.** *In a ring extension, $\alpha \in R[s_1, \ldots, s_n]$ if and only if $\alpha = f(s_1, \ldots, s_n)$ for some $f \in R[X_1, \ldots, X_n]$; $\alpha \in R[S]$ if and only if $\alpha \in R[s_1, \ldots, s_n]$ for some $s_1, \ldots, s_n \in S$.*

Definition. A ring extension E of R is finitely generated *over R when $E = R[\alpha_1, \ldots, \alpha_n]$ for some $n \geqq 0$ and $\alpha_1, \ldots, \alpha_n \in E$.*

Modules. Every field extension of a field K is a vector space over K. The corresponding concept for ring extensions, introduced here, is studied in more detail in the next chapter.

Definitions. Let R be a ring. An R-module is an abelian group M together with an action $(r, x) \longmapsto rx$ of R on M such that $r(x + y) = rx + ry$, $(r + s)x = rx + sx$, $r(sx) = (rs)x$, and $1x = x$, for all $r, s \in R$ and $x, y \in M$. A submodule of an R-module M is an additive subgroup N of M such that $x \in N$ implies $rx \in N$ for every $r \in R$.

If K is a field, then a K-module is the same as a vector space over K. Every ring extension E of R is an R-module, on which multiplication in E provides the action of R. Every intermediate ring $R \subseteq S \subseteq E$ is a submodule of E; so is every ideal of E, and every ideal of R.

Modules are the most general structure in which one can make sensible linear combinations with coefficients in a given ring. For instance, if X is a subset of an R-module M, readers will show that linear combinations of elements of X with coefficients in R constitute a submodule of M, which is the smallest submodule of M that contains X.

Definitions. Let M be an R-module. The submodule of M generated by a subset X of M is the set of all linear combinations of elements of X with coefficients in R. A submodule of M is finitely generated when it is generated (as a submodule) by a finite subset of M.

For example, in a ring extension, the subring $R[S]$ is also the submodule generated by all products of powers of elements of S, by 2.1.

Modules inherit from abelian groups the agreeable property that the quotient of an R-module by a submodule is an R-module. Readers will show that the action of R on M/N in the next result is well defined and is a module action:

Proposition 2.3. *Let N be a submodule of an R-module M. The quotient group M/N is an R-module, in which $r(x+N) = rx + N$ for all $x \in M$.*

Integral elements. Recall that, in a field extension of a field K, an element α is algebraic over K when $f(\alpha) = 0$ for some nonzero polynomial $f \in K[X]$, equivalently when $K[\alpha]$ is finite over K.

Proposition 2.4. *For an element α of a ring extension E of a commutative ring R the following conditions are equivalent:*

(1) $f(\alpha) = 0$ *for some monic polynomial $f \in R[X]$;*

(2) $R[\alpha]$ *is a finitely generated submodule of E;*

(3) α *belongs to a subring of E that is a finitely generated R-module.*

Proof. (1) implies (2). Let $f(\alpha) = 0$, where f is monic; let $n = \deg f$. We show that $1, \alpha, \ldots, \alpha^{n-1}$ generate $R[\alpha]$ as a submodule of E. Indeed, let $\beta \in R[\alpha]$. By 2.2, $\beta = g(\alpha)$ for some $g \in R[X]$. Since f is monic, g can be divided by f, and $g = fq + r$, where $\deg r < n$. Then $\beta = g(\alpha) = r(\alpha)$ is a linear combination of $1, \alpha, \ldots, \alpha^{n-1}$ with coefficients in R.

(2) implies (3). $R[\alpha]$ serves.

(3) implies (1). Let α belong to a subring F of E that is generated, as a sub-R-module of E, by $\beta_1, \ldots, \beta_n$. Since $\alpha\beta_i \in F$ there is an equality $\alpha\beta_i = x_{i1}\beta_1 + \cdots + x_{in}\beta_n$, where $x_{i1}, \ldots, s_{in} \in R$. Hence

$$-x_{i1}\beta_1 - \cdots - x_{i,i-1}\beta_{i-1} + (\alpha - x_{ii})\beta_i - x_{i,i+1}\beta_{i+1} - \cdots - x_{in}\beta_n = 0$$

for all i. By Lemma 2.5 below, applied to the R-module F, the determinant

$$D = \begin{vmatrix} \alpha - x_{11} & -x_{12} & \cdots & -x_{1n} \\ -x_{21} & \alpha - x_{22} & \cdots & -x_{2n} \\ \vdots & \vdots & \ddots & \vdots \\ -x_{n1} & x_{n2} & \cdots & \alpha - x_{nn} \end{vmatrix}$$

satisfies $D\beta_j = 0$ for all j. Since $\beta_1, \ldots, \beta_n$ generate F this implies $D\beta = 0$ for all $\beta \in F$ and $D = D1 = 0$. Expanding D shows that $D = f(\alpha)$ for some monic polynomial $f \in K[X]$. $\square$

Lemma 2.5. *Let M be an R-module and let $m_1, \ldots, m_n \in M$. If $x_{ij} \in R$ for all $i, j = 1, \ldots, n$ and $\sum_{1 \leq j \leq n} x_{ij}m_j = 0$ for all i, then $D = \det(x_{ij})$ satisfies $Dm_i = 0$ for all i.*

Proof. If R is a field this is standard linear algebra. In general, we expand D by columns, which yields cofactors c_{ik} such that $\sum_k c_{ik} x_{kj} = D$ if $i = j$, $\sum_k c_{ik} x_{kj} = 0$ if $i \neq j$; hence

$$Dm_i = \sum_k c_{ik} x_{ki} m_i = \sum_{j,k} c_{ik} x_{kj} m_j = 0 \text{ for all } i.$$

Definition. An element α of a ring extension E of R is integral over R *when it satisfies the equivalent conditions in 2.4.*

For instance, every element of R is integral over R. In $\mathbb{R}$, $\sqrt{2}$ is integral over $\mathbb{Z}$. On the other hand, $1/2 \in \mathbb{Q}$ is not integral over $\mathbb{Z}$: as a $\mathbb{Z}$-module, $\mathbb{Z}[1/2]$ is generated by $1/2$, $1/4$, ..., $1/2^n$, ...; a finitely generated submodule of $\mathbb{Z}[1/2]$ is contained in some $\mathbb{Z}[1/2^k]$, and cannot contain all $1/2^n$.

The following property makes a nifty exercise:

Proposition 2.6. If R is a domain and Q is its quotient field, then α is algebraic over Q if and only if $r\alpha$ is integral over R for some $r \in R$, $r \neq 0$.

Exercises

1. Prove the following: when X is a subset of an R-module M, the set of all linear combinations of elements of X with coefficients in R is the smallest submodule of M that contains X.

2. Prove the following: when N is a submodule of an R-module M, the quotient group M/N is an R-module, in which $r(x + N) = rx + N$ for all $x \in M$.

3. Show that $\sqrt{2} + \sqrt{3}$ is integral over $\mathbb{Z}$.

4. Show that $\sqrt{1/2}$ is not integral over $\mathbb{Z}$.

5. Prove the following: when R is a domain and Q is its quotient field, then α is algebraic over Q if and only if $r\alpha$ is integral over R for some $r \in R$, $r \neq 0$.

6. Prove the following: when R is a domain and Q is its quotient field, then $a \in R$ is a unit of R if and only if $a \neq 0$ and $1/a \in Q$ is integral over R.

3. Integral Extensions

In this section we extend some properties of algebraic extensions to integral extensions. We also establish some properties of their prime ideals and take our first look at algebraic integers.

Definition. A ring extension $R \subseteq E$ is integral, *and E is* integral over R, *when every element of E is integral over R.*

Proposition 3.1. Let E be a ring extension of a commutative ring R.

(1) *If E is a finitely generated R-module, then E is integral over R.*

(2) *If $E = R[\alpha_1, ..., \alpha_n]$ and $\alpha_1, ..., \alpha_n$ are integral over R, then E is a finitely generated R-module, hence is integral over R.*

(3) *If $E = R[S]$ and every $\alpha \in S$ is integral over R, then E is integral over R.*

Proof. (1). Every $\alpha \in E$ satisfies condition (3) in Proposition 2.4.

(2). By induction on n. If $n = 0$, then $E = R$ is integral over R. Now, let E be a finitely generated R-module; let $F = E[\alpha]$, where α is integral over R. Then α is integral over E and F is a finitely generated E-module: every element of F is a linear combination of some $\beta_1, \ldots, \beta_\ell \in F$, with coefficients in E that are themselves linear combinations with coefficients in R of some $\alpha_1, \ldots, \alpha_k$. Hence every element of F is a linear combination with coefficients in R of the $k\ell$ elements $\alpha_i \beta_j$; and F is a finitely generated R-module.

(3) follows from (2) since $\alpha \in R[S]$ implies $\alpha \in R[\alpha_1, \ldots, \alpha_n]$ for some $\alpha_1, \ldots, \alpha_n \in S$, by 1.2. $\square$

The next properties follow from Proposition 3.1 and make bonny exercises.

Proposition 3.2. *In a ring extension E of R, the elements of E that are integral over R constitute a subring of E.*

Proposition 3.3. *Let $R \subseteq E \subseteq F$ be commutative rings.*

(1) *If F is integral over R, then F is integral over E and E is integral over R.*

(2) (Tower Property) *If F is integral over E and E is integral over R, then F is integral over R.*

(3) *If F is integral over E and $R[F]$ is defined in some larger ring, then $R[F]$ is integral over $R[E]$.*

(4) *If E is integral over R and $\varphi : E \longrightarrow S$ is a ring homomorphism, then $\varphi(E)$ is integral over $\varphi(R)$.*

(5) *If E is integral over R over R and R is a field, then E is a field and is algebraic over R.*

Ideals. We show that the prime ideals of an integral extension of R are closely related to the prime ideals of R.

Definition. In a ring extension E of R, an ideal $\mathfrak{A}$ of E lies over an ideal $\mathfrak{a}$ of R when $\mathfrak{A} \cap R = \mathfrak{a}$.

Proposition 3.4. *If E is a ring extension of R and $\mathfrak{A} \subseteq E$ lies over $\mathfrak{a} \subseteq R$, then $R/\mathfrak{a}$ may be identified with a subring of $E/\mathfrak{A}$; if E is integral over R, then $E/\mathfrak{A}$ is integral over $R/\mathfrak{a}$.*

Proof. The inclusion homomorphism $R \longrightarrow E$ induces a homomorphism $R \longrightarrow E/\mathfrak{A}$ whose kernel is $\mathfrak{A} \cap R = \mathfrak{a}$, and an injective homomorphism $R/\mathfrak{a} \longrightarrow E/\mathfrak{A}$, $r + \mathfrak{a} \longmapsto r + \mathfrak{A}$. Hence $R/\mathfrak{a}$ may be identified with a subring of $E/\mathfrak{A}$. If $\alpha \in E$ is integral over R, then $\alpha + \mathfrak{A} \in E/\mathfrak{A}$ is integral over $R/\mathfrak{a}$, by part (4) of 3.3. $\square$

Proposition **3.5** (Lying Over). *Let E be an integral extension of R. For every prime ideal $\mathfrak{p}$ of R there exists a prime ideal $\mathfrak{P}$ of E that lies over $\mathfrak{p}$. In fact, for every ideal $\mathfrak{A}$ of E such that $\mathfrak{p}$ contains $\mathfrak{A} \cap R$, there exists a prime ideal $\mathfrak{P}$ of E that contains $\mathfrak{A}$ and lies over $\mathfrak{p}$.*

Proof. Let $\mathfrak{A}$ be an ideal of E such that $\mathfrak{A} \cap R \subseteq \mathfrak{p}$ (for instance, 0). In the set of all ideals $\mathfrak{B}$ of E such that $\mathfrak{A} \subseteq \mathfrak{B}$ and $\mathfrak{B} \cap R \subseteq \mathfrak{p}$, there is a maximal element $\mathfrak{P}$, by Zorn's lemma. We have $1 \notin \mathfrak{P}$, since $1 \notin \mathfrak{P} \cap R \subseteq \mathfrak{p}$. If $\alpha, \beta \in E\backslash\mathfrak{P}$, then $\mathfrak{P} + E\alpha$ contains some $s \in R\backslash\mathfrak{p}$ (otherwise, $(\mathfrak{P} + E\alpha) \cap R \subseteq \mathfrak{p}$ and $\mathfrak{P}$ is not maximal); hence $\pi + \gamma\alpha \in R\backslash\mathfrak{p}$, and $\rho + \delta\beta \in R\backslash\mathfrak{p}$, for some $\pi, \rho \in \mathfrak{P}$ and $\gamma, \delta \in E$; hence $(\pi + \gamma\alpha)(\rho + \delta\beta) \in R\backslash\mathfrak{p}$ since $\mathfrak{p}$ is prime, $\mathfrak{P} + E\alpha\beta \neq \mathfrak{P}$, and $\alpha\beta \notin \mathfrak{P}$. Thus $\mathfrak{P}$ is a prime ideal of E.

Assume that $p \in \mathfrak{p}$ and $p \notin \mathfrak{P}$. As above, $s = \pi + \gamma p \in R\backslash\mathfrak{p}$ for some $\pi \in \mathfrak{P}$ and $\gamma \in E$. Since E is integral over R, $\gamma^n + r_{n-1}\gamma^{n-1} + \cdots + r_0 = 0$ for some $n > 0$ and $r_{n-1}, \ldots, r_0 \in R$. Multiplying by p^n yields

$$(s - \pi)^n + pr_{n-1}(s - \pi)^{n-1} + \cdots + p^n r_0$$
$$= p^n \gamma^n + pr_{n-1} p^{n-1} \gamma^{n-1} + \cdots + p^n r_0 = 0.$$

Hence $s^n = pr + \delta\pi$ for some $r \in R$ and $\delta \in E$, $s^n - pr \in \mathfrak{P} \cap R \subseteq \mathfrak{p}$, and $s^n \in \mathfrak{p}$, an unbearable contradiction. Therefore $\mathfrak{p} \subseteq \mathfrak{P}$ and $\mathfrak{P} \cap R = \mathfrak{p}$. $\square$

The proof of Proposition 3.5 shows that an ideal that is maximal among the ideals lying over $\mathfrak{p}$ is necessarily a prime ideal. Conversely, a prime ideal that lies over $\mathfrak{p}$ is maximal among the ideals that lie over $\mathfrak{p}$ (see the exercises). We prove a particular case:

Proposition **3.6.** *Let E be an integral extension of R and let $\mathfrak{P}, \mathfrak{Q} \subseteq E$ be prime ideals of E that lie over $\mathfrak{p} \subseteq R$. If $\mathfrak{P} \subseteq \mathfrak{Q}$, then $\mathfrak{P} = \mathfrak{Q}$.*

Proof. Let $\alpha \in \mathfrak{Q}$. We have $f(\alpha) = 0 \in \mathfrak{P}$ for some monic polynomial $f \in R[X]$. Let $f(X) = X^n + r_{n-1}X^{n-1} + \cdots + r_0 \in R[X]$ be a monic polynomial of minimal degree $n > 0$ such that $f(\alpha) \in \mathfrak{P}$. Then $r_0 \in \mathfrak{Q} \cap R = \mathfrak{p}$, since $\alpha \in \mathfrak{Q}$, and $\alpha(\alpha^{n-1} + r_{n-1}\alpha^{n-2} + \cdots + r_1) = f(\alpha) - r_0 \in \mathfrak{P}$. Now, $\alpha^{n-1} + r_{n-1}\alpha^{n-2} + \cdots + r_1 \notin \mathfrak{P}$, by the choice of f. Therefore $\alpha \in \mathfrak{P}$. $\square$

Proposition **3.7.** *If E is an integral extension of R and the prime ideal $\mathfrak{P} \subseteq E$ lies over $\mathfrak{p} \subseteq R$, then $\mathfrak{P}$ is a maximal ideal of E if and only if $\mathfrak{p}$ is a maximal ideal of R.*

Proof. By 3.4 we may identify $R/\mathfrak{p}$ with a subring of $E/\mathfrak{P}$, and then $E/\mathfrak{P}$ is integral over $R/\mathfrak{p}$. If $\mathfrak{p}$ is maximal, then $R/\mathfrak{p}$ is a field, $E/\mathfrak{P}$ is a field by 3.3, and $\mathfrak{P}$ is maximal. But if $\mathfrak{p}$ is not maximal, then $\mathfrak{p}$ is contained in a maximal ideal $\mathfrak{m} \supsetneq \mathfrak{p}$ of R; by 3.5, a prime ideal $\mathfrak{M} \supseteq \mathfrak{P}$ of E lies over $\mathfrak{m}$; then $\mathfrak{P} \subsetneq \mathfrak{M} \neq E$ and $\mathfrak{P}$ is not maximal. $\square$

Here comes another bonny exercise:

Proposition **3.8** (Going Up). *Let E be an integral extension of R and let*

$\mathfrak{p} \subsetneq \mathfrak{q}$ *be prime ideals of* R. *For every prime ideal* $\mathfrak{P}$ *of* E *that lies over* $\mathfrak{p}$, *there exists a prime ideal* $\mathfrak{Q} \supsetneq \mathfrak{P}$ *of* E *that lies over* $\mathfrak{q}$.

Integrally closed domains. A ring has, in general, no "greatest" integral extension. A domain, however, has a largest integral extension inside its quotient field, by 3.2, which is somewhat similar to an algebraic closure.

Definitions. The integral closure *of a ring* R *in a ring extension* E *of* R *is the subring* $\overline{R}$ *of* E *of all elements of* E *that are integral over* R. *The elements of* $\overline{R} \subseteq E$ *are the* algebraic integers *of* E *(over* R*).* ,

Definition. A domain R *is* integrally closed *when its integral closure in its quotient field* $Q(R)$ *is* R *itself (when no* $\alpha \in Q(R) \backslash R$ *is integral over* R*).*

Since $R \subseteq \overline{R} \subseteq Q(R)$, we have $Q(\overline{R}) = Q(R)$. Moreover, if $\alpha \in Q(R)$ is integral over $\overline{R}$, then α is integral over R, by 3.3, so that $\overline{R}$ is integrally closed. Thus, every domain R has an integral extension $\overline{R} \subseteq Q(R)$ that is integrally closed. Integrally closed domains are also called *normal* domains; then $\overline{R} \subseteq Q(R)$ is the *normalization* of R.

Proposition 3.9. *Every unique factorization domain is integrally closed.*

Proof. Let R be a UFD and let $a/b \in Q(R)$. We may assume that a and b are relatively prime (no irreducible element of R divides both a and b). If a/b is integral over R, then $f(a/b) = 0$ for some monic polynomial $f(X) = X^n + r_{n-1}X^{n-1} + \cdots + r_0 \in R[X]$ and

$$a^n + r_{n-1}a^{n-1}b + \cdots + r_0 b^n = b^n f(a/b) = 0.$$

No irreducible element p of R divides b: otherwise, p divides a^n and p divides a, a contradiction; therefore b is a unit of R and $a/b \in R$. $\square$

By Proposition 3.9, $\mathbb{Z}$ is integrally closed, and so is $K[X_1, ..., X_n]$ for every field K. The next result yields integrally closed domains that are not UFDs.

Proposition 3.10. *Let* R *be a domain and let* E *be an algebraic extension of its quotient field. The integral closure* $\overline{R}$ *of* R *in* E *is an integrally closed domain whose quotient field is* E.

Proof. Every $\alpha \in E$ is algebraic over $Q(R)$; by 2.6, $r\alpha$ is integral over R for some $r \in R$; hence $E = Q(\overline{R})$. If $\alpha \in E$ is integral over $\overline{R}$, then α is integral over R by 3.1 and $\alpha \in \overline{R}$, so $\overline{R}$ is integrally closed. $\square$

Ready examples of integrally closed domains come from *quadratic extensions* of $\mathbb{Q}$, which are fields $\mathbb{Q}(\sqrt{m}) \subseteq \mathbb{C}$, where $m \in \mathbb{Q}$. One may assume that $m \in \mathbb{Z}$ and that m is *square free* (if n^2 divides m, then $n = 1$): indeed, $\mathbb{Q}(\sqrt{a/b}) = \mathbb{Q}(\sqrt{ab})$, since $\sqrt{ab} = b\sqrt{a/b}$; and $\mathbb{Q}(\sqrt{m}) = \mathbb{Q}(\sqrt{m/n^2})$ when n^2 divides m.

Proposition 3.11. *If* $m \in \mathbb{Z}$ *is square free and not congruent to* 1 (mod 4), *then* $\mathbb{Z}[\sqrt{m}]$ *is integrally closed; and then, for all* $x, y \in \mathbb{Q}$, $x + y\sqrt{m}$ *is an algebraic integer in* $\mathbb{Q}(\sqrt{m})$ *(over* $\mathbb{Z}$*) if and only if* $x, y \in \mathbb{Z}$.

Thus, $\mathbb{Z}[\sqrt{-5}]$ is integrally closed; readers will show that it is not a UFD. On the other hand, $\mathbb{Z}[\sqrt{5}]$ is not integrally closed (see the exercises).

Proof. We show that $\mathbb{Z}[\sqrt{m}]$ is the integral closure of $\mathbb{Z}$ in $\mathbb{Q}[\sqrt{m}]$; hence $\mathbb{Z}[\sqrt{m}]$ is integrally closed, by 3.10. First, $\mathbb{Z}[\sqrt{m}] = \{\, x + y\sqrt{m} \mid x, y \in \mathbb{Z} \,\}$, by 2.2; hence $\mathbb{Z}[\sqrt{m}]$ is integral over $\mathbb{Z}$, by 3.2.

Conversely, let $\alpha = x + y\sqrt{m} \in \mathbb{Q}[\sqrt{m}]$ be integral over $\mathbb{Z}$ (where $x, y \in \mathbb{Q}$). Since $\mathbb{Q}(\sqrt{m})$ has a $\mathbb{Q}$-automorphism that sends $\sqrt{m}$ onto $-\sqrt{m}$, then $\beta = x - y\sqrt{m}$ is integral over $\mathbb{Z}$. By 3.2, $2x = \alpha + \beta$ and $x^2 - my^2 = \alpha\beta$ are integral over $\mathbb{Z}$. Since $\mathbb{Z}$ is integrally closed this implies $u = 2x \in \mathbb{Z}$ and $x^2 - my^2 \in \mathbb{Z}$; hence $4my^2 \in \mathbb{Z}$ and $v = 2y \in \mathbb{Z}$, since m is square free.

If $x \notin \mathbb{Z}$, then u is odd, mv^2 is odd since $u^2 - mv^2 = 4(x^2 - my^2)$ is even, and m, v are odd; hence, modulo 4, $u^2 \equiv v^2 \equiv 1$ and $m \equiv mv^2 \equiv u^2 \equiv 1$, contradicting the hypothesis. Therefore $x \in \mathbb{Z}$. Hence $mv^2/4 = my^2 \in \mathbb{Z}$, 4 divides mv^2, v^2 is even since m is square free, v is even, and $y \in \mathbb{Z}$. Thus $\alpha \in \mathbb{Z}[\sqrt{m}]$. $\square$

Exercises

Prove the following:

1. If F is integral over E and E is integral over R, then F is integral over R.

2. If F is integral over E and $R[F]$ is defined in some larger ring, then $R[F]$ is integral over $R[E]$.

3. If E is integral over R and $\varphi : E \longrightarrow S$ is a ring homomorphism, then $\varphi(E)$ is integral over $\varphi(R)$.

4. In a ring extension E of R, the elements of E that are integral over R constitute a subring of E.

5. If E is integral over R, then $E[X_1, ..., X_n]$ is integral over $R[X_1, ..., X_n]$.

6. If E is integral over R and R is a field, then E is a field.

7. Let E be an integral extension of R and let $\mathfrak{p}$ be a prime ideal of R. Show that a prime ideal of E that lies over $\mathfrak{p}$ is maximal among the ideals of E that lie over $\mathfrak{p}$.

8. Prove the going up theorem: Let E be an integral extension of R and let $\mathfrak{p} \subsetneqq \mathfrak{q}$ be prime ideals of R. For every prime ideal $\mathfrak{P}$ of E that lies over $\mathfrak{p}$, there exists a prime ideal $\mathfrak{Q} \supsetneqq \mathfrak{P}$ of E that lies over $\mathfrak{q}$.

9. Find all prime ideals of $\mathbb{Z}[\sqrt{-5}]$ that lie over the prime ideal (5) of $\mathbb{Z}$.

10. Find all prime ideals of $\mathbb{Z}[\sqrt{-5}]$ that lie over the prime ideal (2) of $\mathbb{Z}$.

11. Find all prime ideals of $\mathbb{Z}[\sqrt{-5}]$ that lie over the prime ideal (3) of $\mathbb{Z}$.

12. Show that $\mathbb{Z}[\sqrt{-5}]$ is not a UFD.

13. Show that $\mathbb{Q}[\sqrt{5}]$ contains an algebraic integer $x + y\sqrt{5}$ such that $x, y \notin \mathbb{Z}$, so that $\mathbb{Z}[\sqrt{5}]$ is not integrally closed.

14. Find the algebraic integers of $\mathbb{Q}(\sqrt{m})$ when $m \in \mathbb{Z}$ is square free and $m \equiv 1$ (mod 4).

15. Let R be an integrally closed domain and let $Q = Q(R)$. Prove the following: if $f, g \in Q[X]$ are monic and $fg \in R[X]$, then $f, g \in R[X]$.

16. Let R be an integrally closed domain and let $Q = Q(R)$. Prove the following: if α is integral over R, then $\mathrm{Irr}\,(\alpha : Q) \in R[X]$.

4. Localization

A local ring is a commutative ring with only one maximal ideal; the name comes from algebraic geometry. Localization expands a ring into a local ring by adjoining inverses of some of its elements. The usefulness of this construction was recognized rather late; it was defined in domains by Grell [1927], but not in general until Uzkov [1948]. This section constructs rings of fractions, studies their ideals, and proves some useful homomorphism properties.

Rings of fractions. R still is any commutative ring [with identity].

Definitions. A multiplicative subset of a commutative ring R is a subset S of R that contains the identity element of R and is closed under multiplication. A multiplicative subset S is proper when $0 \notin S$.

Readers may write a proof of the following result.

Lemma **4.1.** *Let S be a proper multiplicative subset of a commutative ring R. The relation*

$$(a, s) \equiv (b, t) \text{ if and only if } atu = bsu \text{ for some } u \in S$$

is an equivalence relation on $R \times S$, and $S^{-1}R = (R \times S)/\!\equiv$ is a ring, with the operations

$$(a/s) + (b/t) = (at + bs)/(st), \quad (a/s)(b/t) = (ab)/(st),$$

where a/s denotes the equivalence class of (a, s).

By definition, $a/s = b/t$ if and only if $atu = bsu$ for some $u \in S$. In particular, $a/s = at/st$ for all $t \in S$; $s/s = 1 \, (= 1/1)$ for all $s \in S$; and $a/s = 0$ $(= 0/1)$ if and only if $at = 0$ for some $t \in S$.

Definition. If S is a proper multiplicative subset of a commutative ring R, then $S^{-1}R$ is the ring of fractions of R with denominators in S.

For instance, if R is a domain, then $S = R \backslash \{0\}$ is a proper multiplicative subset and $S^{-1}R$ is the field of fractions or quotient field $Q(R)$ of R.

The universal property of quotient fields extends to rings of fractions. Every ring of fraction comes with a canonical homomorphism $\iota : R \longrightarrow S^{-1}R$, $a \longmapsto a/1$, which is injective if R is a domain. Moreover, $\iota(s) = s/1$ is a unit of $S^{-1}R$ for every $s \in S$, with inverse $1/s$.

Proposition **4.2.** *Let S be a proper multiplicative subset of a commutative ring R. Every homomorphism φ of R into a ring R' in which $\varphi(s)$ is a unit for every $s \in S$ factors uniquely through $\iota : R \longrightarrow S^{-1}R$: there is a homomorphism $\psi : S^{-1}R \longrightarrow R'$ unique such that $\psi \circ \iota = \varphi$, given by $\psi(a/s) = \varphi(a)\,\varphi(s)^{-1}$ for all $a \in R$ and $s \in S$.*

$$R \overset{\iota}{\longrightarrow} S^{-1}R$$

Proof. If $a/s = b/t$, then $atu = bsu$ for some $u \in S$, $\varphi(a)\,\varphi(t)\,\varphi(u) = \varphi(b)\,\varphi(b)\,\varphi(u)$, $\varphi(a)\,\varphi(t) = \varphi(b)\,\varphi(s)$, and $\varphi(a)\,\varphi(s)^{-1} = \varphi(b)\,\varphi(t)^{-1}$. Hence a mapping $\psi : S^{-1}R \longrightarrow R'$ is well defined by: $\psi(a/s) = \varphi(a)\,\varphi(s)^{-1}$. It is immediate that ψ is a homomorphism and that $\psi \circ \iota = \varphi$.

Conversely, let $\chi : S^{-1}R \longrightarrow R'$ be a homomorphism such that $\chi \circ \iota = \varphi$. If $s \in S$, then $1/s$ is the inverse of $\iota(s)$ in $S^{-1}R$ for every $s \in S$, hence $\chi(1/s)$ is the inverse of $\chi(\iota(s)) = \varphi(s)$ in R'. Then $a/s = (a/1)(1/s)$ yields $\chi(a/s) = \chi(\iota(a))\,\chi(1/s) = \varphi(a)\,\varphi(s)^{-1} = \psi(a/s)$, and $\chi = \psi$. $\square$

The rings of fractions of a domain can be retrieved from its quotient field:

Corollary **4.3.** *If R is a domain, then $S^{-1}R$ is isomorphic to the subring $\{\, as^{-1} \mid a \in R,\ s \in S \,\}$ of $Q(R)$.*

Proof. Up to isomorphism, R is a subring of $Q(R)$, and 4.2 provides a homomorphism $\psi : S^{-1}R \longrightarrow Q(R)$ that sends $a/s \in S^{-1}R$ to as^{-1} $(= a/s$ as calculated in $Q(R))$. We see that ψ is injective. $\square$

If R is a domain, then the ring $S^{-1}R$ is usually identified with the subring $\{\, as^{-1} \mid a \in R,\ s \in S \,\}$ of $Q(R)$ in Corollary 4.3.

Ideals. We now shuttle ideals between a ring and its rings of fractions.

Definitions. Let S be a proper multiplicative subset of R.
The contraction *of an ideal $\mathfrak{A}$ of $S^{-1}R$ is $\mathfrak{A}^C = \{\, a \in R \mid a/1 \in \mathfrak{A} \,\}$.*
The expansion *of an ideal $\mathfrak{a}$ of R is $\mathfrak{a}^E = \{\, a/s \in S^{-1}R \mid a \in \mathfrak{a},\ s \in S \,\}$.*

It is immediate that $\mathfrak{A}^C = \iota^{-1}(\mathfrak{A})$ is an ideal of R and that $\mathfrak{a}^E$ is an ideal of $R_{\mathfrak{p}}$; in fact, $\mathfrak{a}^E$ is the ideal generated by $\iota(\mathfrak{a})$.

Proposition **4.4.** *For all ideals $\mathfrak{a}$, $\mathfrak{b}$ of R and $\mathfrak{A}$ of $S^{-1}R$:*

(1) $\mathfrak{a}^E = S^{-1}R$ *if and only if $\mathfrak{a} \cap S \neq \varnothing$;*

(2) *if $\mathfrak{a} = \mathfrak{A}^C$, then $\mathfrak{A} = \mathfrak{a}^E$;*

(3) $(\mathfrak{a} + \mathfrak{b})^E = \mathfrak{a}^E + \mathfrak{b}^E$, $(\mathfrak{a} \cap \mathfrak{b})^E = \mathfrak{a}^E \cap \mathfrak{b}^E$, *and* $(\mathfrak{a}\,\mathfrak{b})^E = \mathfrak{a}^E\,\mathfrak{b}^E$.

The proofs make good exercises.

Proposition **4.5.** *Let S be a proper multiplicative subset of R. Contraction and expansion induce a one-to-one correspondence between prime ideals of $S^{-1}R$ and prime ideals of R disjoint from S.*

Proof. If $\mathfrak{p} \subseteq R\backslash S$ is a prime ideal of R, then $a/s \in \mathfrak{p}^E$ implies $a/s = b/t$ for some $b \in \mathfrak{p}$, $t \in S$, $atu = bsu \in \mathfrak{p}$ for some $u \in S$, and $a \in \mathfrak{p}$, since $\mathfrak{p}$ is prime and $tu \notin \mathfrak{p}$; thus, $a/s \in \mathfrak{p}^E$ if and only if $a \in \mathfrak{p}$. Hence $\mathfrak{p}^E$ is a prime ideal of $S^{-1}R$: $1/1 \notin \mathfrak{p}^E$, and if $(a/s)(b/t) \in \mathfrak{p}^E$ and $a/s \notin \mathfrak{p}^E$, then $ab \in \mathfrak{p}$, $a \notin \mathfrak{p}$, $b \in \mathfrak{p}$, and $b/t \in \mathfrak{p}^E$. Also $\mathfrak{p} = (\mathfrak{p}^E)^C$.

Conversely, if $\mathfrak{P}$ is a prime ideal of $S^{-1}R$, then $\mathfrak{P}^C$ is a prime ideal of R: $1 \notin \mathfrak{P}^C$, since $1/1 \notin \mathfrak{P}$, and if $ab \in \mathfrak{P}^C$ and $a \notin \mathfrak{P}^C$, then $(a/1)(b/1) \in \mathfrak{P}$, $a/1 \notin \mathfrak{P}$, $b/1 \in \mathfrak{P}$, and $b \in \mathfrak{P}^C$. Moreover, $(\mathfrak{P}^C)^E = \mathfrak{P}$, by 4.4. $\square$

Proposition **4.6.** *Let S be a proper multiplicative subset of R. Contraction and expansion induce a one-to-one correspondence, which preserves radicals, between primary ideals of $S^{-1}R$ and primary ideals of R disjoint from S.*

This is proved like 4.5. The following properties also make good exercises.

Proposition **4.7.** *Let S be a proper multiplicative subset of R.*

(1) *If R is Noetherian, then $S^{-1}R$ is Noetherian.*

(2) *If E is integral over R, then $S^{-1}E$ is integral over $S^{-1}R$.*

(3) *If R is an integrally closed domain, then so is $S^{-1}R$.*

Localization. If $\mathfrak{p}$ is a prime ideal of R, then $R\backslash\mathfrak{p}$ is a proper multiplicative subset of R.

Definition. The localization *of a commutative ring R at a prime ideal $\mathfrak{p}$ is the ring of fractions $R_\mathfrak{p} = (R\backslash\mathfrak{p})^{-1}R$.*

Every commutative ring is isomorphic to a ring of fractions (see the exercises); but not every ring is isomorphic to a localization.

Proposition **4.8.** *If $\mathfrak{p}$ is a prime ideal of R, then $R_\mathfrak{p}$ has only one maximal ideal, $\mathfrak{M} = \mathfrak{p}^E = \{ a/s \in R_\mathfrak{p} \mid a \in \mathfrak{p} \}$; moreover, $x \in R_\mathfrak{p}$ is a unit if and only if $x \notin \mathfrak{M}$.*

Proof. If $a/s \in \mathfrak{M}$, then $a/s = b/t$ for some $b \in \mathfrak{p}$, $t \notin \mathfrak{p}$, $atu = bsu \in \mathfrak{p}$ for some $u \notin \mathfrak{p}$, and $a \in \mathfrak{p}$ since $\mathfrak{p}$ is a prime ideal and $tu \notin \mathfrak{p}$. Thus $a/s \in \mathfrak{M}$ if and only if $a \in \mathfrak{p}$. Now, $x - a/s \subset R_\mathfrak{p}$ is a unit if and only if $x \notin \mathfrak{M}$: if $a \notin \mathfrak{p}$, then x is a unit, and $x^{-1} = s/a$; conversely, if x is a unit, then $ab/st = 1$ for some $b, t \in R$, $t \notin \mathfrak{p}$, $abu = stu \notin \mathfrak{p}$ for some $u \notin \mathfrak{p}$, and $a \notin \mathfrak{p}$. Hence the ideal $\mathfrak{M}$ of $R_\mathfrak{p}$ is a maximal ideal. $\square$

Definition. A commutative ring is local *when it has only one maximal ideal.*

For instance, valuation rings are local, by VI.6.1; $R_\mathfrak{p}$ is local, by 4.8. In a local ring R with maximal ideal $\mathfrak{m}$, every $x \in R \backslash \mathfrak{m}$ is a unit (see the exercises).

Homomorphisms. Localization transfers properties from local rings to more general rings. We illustrate this with some nifty homomorphism properties.

Theorem **4.9.** *Every homomorphism of a ring R into an algebraically closed field L can be extended to every integral extension E of R.*

Proof. If R is a field, then E is a field, by 3.3, and E is an algebraic extension of R; we saw that 4.9 holds in that case.

$$
\begin{array}{ccc}
E & \longrightarrow & E/\mathfrak{M} \\
\subseteq \uparrow & & \subseteq \uparrow \quad \searrow \\
R & \longrightarrow & R/\mathfrak{m} \xrightarrow{\ \psi\ } L
\end{array}
$$

Now, let R be local and let $\varphi : R \longrightarrow L$ be a homomorphism whose kernel is the maximal ideal $\mathfrak{m}$ of R. Then φ factors through the projection $R \longrightarrow R/\mathfrak{m}$ and induces a homomorphism $\psi : R/\mathfrak{m} \longrightarrow L$. By 3.5, 3.7, there is a maximal ideal $\mathfrak{M}$ of E that lies over $\mathfrak{m}$. By 3.4, the field $R/\mathfrak{m}$ may be identified with a subfield of $E/\mathfrak{M}$. Then $E/\mathfrak{M}$ is algebraic over $R/\mathfrak{m}$ and $\psi : R/\mathfrak{m} \longrightarrow L$ can be extended to $E/\mathfrak{M}$. Hence φ can be extended to E.

$$
\begin{array}{ccc}
E & \xrightarrow{\ \iota\ } & S^{-1}E \\
\subseteq \uparrow & & \subseteq \uparrow \quad \searrow \\
R & \xrightarrow{\ \iota\ } & S^{-1}R \xrightarrow{\ \psi\ } L
\end{array}
$$

Finally, let $\varphi : R \longrightarrow L$ be any homomorphism. Then $\mathfrak{p} = \mathrm{Ker}\, \varphi$ is a prime ideal of R and $S = R \backslash \mathfrak{p}$ is a proper multiplicative subset of R and of E. By 4.2, $\varphi = \psi \circ \iota$ for some homomorphism $\psi : S^{-1}R = R_{\mathfrak{p}} \longrightarrow L$, namely, $\psi(a/s) = \varphi(a)\varphi(s)^{-1}$. Then $\mathrm{Ker}\, \psi = \{\, a/s \in R_{\mathfrak{p}} \mid a \in \mathrm{Ker}\, \varphi = \mathfrak{p} \,\}$ is the maximal ideal of $R_{\mathfrak{p}}$. Therefore ψ extends to $S^{-1}E$, which is integral over $S^{-1}R$ by 4.7; hence φ extends to E. $\square$

Theorem **4.10.** *Every homomorphism of a field K into an algebraically closed field L can be extended to every finitely generated ring extension of K.*

Proof. Let $\varphi : K \longrightarrow L$ be a homomorphism and let $R = K[\alpha_1, \ldots, \alpha_m]$ be a finitely generated ring extension of K.

First, assume that R is a field. We may assume that R is not algebraic over K. Let $\beta_1, \ldots, \beta_n$ be a transcendence base of R over K. Every $\alpha \in R$ is algebraic over $K(\beta_1, \ldots, \beta_n)$, so that $\gamma_k \alpha^k + \cdots + \gamma_0 = 0$ for some $k > 0$ and $\gamma_0, \ldots, \gamma_k \in K(\beta_1, \ldots, \beta_n)$, $a_k \neq 0$. Since we may multiply $\gamma_0, \ldots, \gamma_k$ by a common denominator in $K(\beta_1, \ldots, \beta_n) \cong K(X_1, \ldots, X_n)$, we may assume that $\gamma_0, \ldots, \gamma_k \in D = K[\beta_1, \ldots, \beta_n]$. Dividing by $\gamma_k \neq 0$ then shows that α

is integral over $D[1/\gamma_k]$. Applying this to $\alpha_1, \ldots, \alpha_m$ shows that $\alpha_1, \ldots, \alpha_m$ are integral over $D[1/\delta_1, \ldots, 1/\delta_m]$ for some nonzero $\delta_1, \ldots, \delta_m \in D$. Hence $\alpha_1, \ldots, \alpha_m$ are integral over $D[1/\delta]$, where $\delta = \delta_1 \cdots \delta_m \in D$, $\delta \neq 0$. Then R is integral over $D[1/\delta]$.

Now, φ extends to a homomorphism

$$\psi : D = K[\beta_1, \ldots, \beta_n] \cong K[X_1, \ldots, X_n] \longrightarrow L[X_1, \ldots, X_n].$$

Let $g = \psi(\delta)$. Since the algebraically closed field L is infinite, we have $g(x_1, \ldots, x_n) \neq 0$ for some $x_1, \ldots, x_n \in L$, as readers will show. Let $\chi = \hat{x} \circ \psi$, where $\hat{x} : L[X_1, \ldots, X_n] \longrightarrow L$ is the evaluation homomorphism $f \longmapsto f(x_1, \ldots, x_n)$. Then $\mathfrak{p} = \mathrm{Ker}\,\chi$ is a prime ideal of D and 4.2 extends χ to the local ring $D_{\mathfrak{p}}$. By 4.3 we may assume that $D_{\mathfrak{p}} = K[\beta_1, \ldots, \beta_n]_{\mathfrak{p}} \subseteq K(\beta_1, \ldots, \beta_n)$. Now, $\delta \notin \mathfrak{p}$, since $g(x_1, \ldots, x_n) \neq 0$; hence δ has an inverse in $D_{\mathfrak{p}}$ and $D[1/\delta] \subseteq D_{\mathfrak{p}}$. Hence χ extends to $D[1/\delta]$. Then χ extends to R by 4.9, since R is integral over $D[1/\delta]$.

Finally, let $R = K[\alpha_1, \ldots, \alpha_m]$ be any finitely generated ring extension of K. Let $\mathfrak{m}$ be a maximal ideal of R and let $\pi : R \longrightarrow R/\mathfrak{m}$ be the projection.

$$
\begin{array}{ccc}
R & \xrightarrow{\;\pi\;} & R/\mathfrak{m} \\
\subseteq \uparrow & & \subseteq \uparrow \qquad \searrow \\
K & \xrightarrow{\;\cong\;} & \pi(K) \longrightarrow L
\end{array}
$$

Then $R/\mathfrak{m}$ is a field, $\pi(K) \cong K$, and $R/\mathfrak{m} = \pi(K)[\pi(\alpha_1), \ldots, \pi(\alpha_m)]$ is a finitely generated ring extension of $\pi(K)$. Every homomorphism of $\pi(K)$ into L extends to $R/\mathfrak{m}$; hence every homomorphism of K into L extends to R. $\square$

Exercises

1. Let S be a proper multiplicative subset of a commutative ring R. Show that

$$(a, s) \equiv (b, t) \text{ if and only if } atu = bsu \text{ for some } u \in S$$

is an equivalence relation on $R \times S$, and that $S^{-1}R = (R \times S)/\equiv$ is a ring when

$$(a/s) + (b/t) = (at + bs)/(st), \quad (a/s)(b/t) = (ab)/(st).$$

2. Show that $S^{-1}R \cong R$ when S is contained in the group of units of R.

3. Show that the canonical homomorphism $R \longmapsto S^{-1}R$ is injective if and only if no element of S is a zero divisor.

4. Describe $\mathbb{Z}_{(p)}$.

5. Let R be a local ring and let $\mathfrak{m}$ be its maximal ideal. Show that $R \backslash \mathfrak{m}$ is the group of units of R.

6. Let R be a local ring. Prove that there is a commutative ring R' and a prime ideal $\mathfrak{p}$ of R' such that $R'_{\mathfrak{p}} \cong R$.

7. Show that $R_{\mathfrak{p}}/\mathfrak{p}^E \cong Q(R/\mathfrak{p})$.

8. Show that $\mathfrak{a}^E = R_{\mathfrak{p}}$ if and only if $\mathfrak{a} \not\subseteq \mathfrak{p}$.

9. Show that $(\mathfrak{A}^C)^E = \mathfrak{A}$ for every ideal $\mathfrak{A}$ of $S^{-1}R$.

10. Show that $(\mathfrak{a} + \mathfrak{b})^E = \mathfrak{a}^E + \mathfrak{b}^E$.

11. Show that $(\mathfrak{a} \cap \mathfrak{b})^E = \mathfrak{a}^E \cap \mathfrak{b}^E$.

12. Show that $(\mathfrak{a}\,\mathfrak{b})^E = \mathfrak{a}^E\,\mathfrak{b}^E$.

13. Prove that contraction and expansion induce a one-to-one correspondence, which preserves radicals, between primary ideals of $S^{-1}R$ and primary ideals of R disjoint from S.

14. Prove the following: if R is Noetherian, then $S^{-1}R$ is Noetherian.

15. Prove the following: if E is integral over R, then $S^{-1}E$ is integral over $S^{-1}R$, for every proper multiplicative subset of R.

16. Prove the following: if R is an integrally closed domain, then so is $S^{-1}R$.

17. Let L be an infinite field and let $f \in L[X_1, ..., X_n]$, $f \neq 0$. Show that $f(x_1, ..., x_n) \neq 0$ for some $x_1, ..., x_n \in L$. (You may want to proceed by induction on n.)

5. Dedekind Domains

Kummer and Dedekind studied rings of algebraic integers and discovered, sometime before 1871, that their ideals have better arithmetic properties than their elements. Domains with these properties are now called Dedekind domains. This section gives a few basic properties; the next section has deeper results.

Fractional ideals. First we generalize ideals as follows.

Definition. A fractional ideal of a domain R is a subset of its quotient field Q of the form $\mathfrak{a}/c = \{ a/c \in Q \mid a \in \mathfrak{a} \}$, where $\mathfrak{a}$ is an ideal of R and $c \in R$, $c \neq 0$.

Fractional ideals of R are submodules of Q. Every ideal $\mathfrak{a}$ of R is a fractional ideal, $\mathfrak{a} = \mathfrak{a}/1$. Conversely, a fractional ideal $\mathfrak{a}/c$ contained in R is an ideal of R, since it is a submodule of R; then $\mathfrak{a}/c = \mathfrak{a}:c = \{ x \in R \mid cx \in \mathfrak{a} \}$. Not all fractional ideals of R are contained in R; readers will easily find examples.

Proposition 5.1. Let R be a domain and let Q be its quotient field. Every finitely generated submodule of Q is a fractional ideal of R. If R is Noetherian, then every fractional ideal of R is finitely generated as a submodule.

Proof. If $n > 0$ and $q_1 = a_1/c_1, ..., q_n = a_n/c_n \in Q$, then

$$Rq_1 + \cdots + Rq_n = Rb_1/c + \cdots + Rb_n/c = (Rb_1 + \cdots + Rb_n)/c,$$

where $c = c_1 \cdots c_n$; hence $Rq_1 + \cdots + Rq_n$ is a fractional ideal of R. Conversely, if every ideal $\mathfrak{a}$ of R is finitely generated, $\mathfrak{a} = Rb_1 + \cdots + Rb_n$ for some $b_1, ..., b_n \in R$, then every fractional ideal $\mathfrak{a}/c = Rb_1/c + \cdots + Rb_n/c$ is a finitely generated submodule of Q. $\square$

A fractional ideal is *finitely generated* when it is finitely generated as a submodule. Readers will verify the following properties.

Proposition **5.2.** *Let* $\mathfrak{A}$ *and* $\mathfrak{B}$ *be fractional ideals of* R.

(1) $\mathfrak{A} \cap \mathfrak{B}$ *is a fractional ideal of* R.

(2) $\mathfrak{A} + \mathfrak{B} = \{\, a + b \mid a \in \mathfrak{A},\ b \in \mathfrak{B} \,\}$ *is a fractional ideal of* R.

(3) *The set* $\mathfrak{A}\mathfrak{B}$ *of all finite sums* $a_1 b_1 + \cdots + a_n b_n$, *where* $a_1, \ldots, a_n \in \mathfrak{A}$, $b_1, \ldots, b_n \in \mathfrak{B}$, *and* $n \geqq 0$, *is a fractional ideal of* R.

(4) *The multiplication of fractional ideals in* (3) *is commutative and associative.*

(5) *If* $\mathfrak{A} \neq 0$ *is finitely generated, then* $\mathfrak{B} : \mathfrak{A} = \{\, q \in Q \mid q\mathfrak{A} \subseteq \mathfrak{B} \,\}$ *is a fractional ideal of* R; *in particular,* $\mathfrak{A}' = R : \mathfrak{A} = \{\, q \in Q \mid q\mathfrak{A} \subseteq R \,\}$ *is a fractional ideal of* R.

(6) *If* $\mathfrak{A} = \mathfrak{a}/c$ *is a fractional ideal, then* $\mathfrak{A} = \mathfrak{a}c'$, *where* $c = Rc$.

Similar constructions were seen in Section 1. The notation $\mathfrak{A}\mathfrak{B}$ is traditional; readers will surely remember that the product of $\mathfrak{A}$ and $\mathfrak{B}$ as fractional ideals is larger than their product $\{\, ab \mid a \in \mathfrak{A},\ b \in \mathfrak{B} \,\}$ as subsets.

Definition. *A fractional ideal* $\mathfrak{A}$ *of* R *is* invertible *when* $\mathfrak{A}\mathfrak{B} = R$ *for some fractional ideal* $\mathfrak{B}$ *of* R.

Proposition **5.3.** (1) *Every invertible fractional ideal is finitely generated.*

(2) *A fractional ideal* $\mathfrak{A}$ *is invertible if and only if* $\mathfrak{A}$ *is finitely generated and* $\mathfrak{A}\mathfrak{A}' = R$.

(3) *Every nonzero principal ideal is invertible.*

Proof. (1). If $\mathfrak{A}\mathfrak{B} = R$, then $1 = a_1 b_1 + \cdots + a_n b_n$ for some $a_1, \ldots, a_n \in \mathfrak{A}$ and $b_1, \ldots, b_n \in \mathfrak{B}$. Then $Ra_1 + \cdots + Ra_n \subseteq \mathfrak{A}$; conversely, $a \in \mathfrak{A}$ implies $a = a_1 b_1 a + \cdots + a_n b_n a \in Ra_1 + \cdots + Ra_n$.

(2). If $\mathfrak{A}\mathfrak{B} = R$, then $\mathfrak{B} \subseteq \mathfrak{A}'$ and $\mathfrak{A}' = \mathfrak{A}'R = \mathfrak{A}'\mathfrak{A}\mathfrak{B} \subseteq R\mathfrak{B} = \mathfrak{B}$.

(3). If $\mathfrak{A} = Ra$, where $a \in R$, $a \neq 0$, then $\mathfrak{A}' = R/a$ and $\mathfrak{A}\mathfrak{A}' = R$. $\square$

Definition. Dedekind domains are defined by the following equivalent conditions. Theorem 6.2 gives additional chracterizations.

Theorem **5.4.** *For a domain* R *the following conditions are equivalent:*

(1) *every nonzero ideal of* R *is invertible (as a fractional ideal);*

(2) *every nonzero fractional ideal of* R *is invertible;*

(3) *every nonzero ideal of* R *is a product of prime ideals of* R;

(4) *every nonzero ideal of* R *can be written uniquely as a product of positive powers of distinct prime ideals of* R.

Then R *is Noetherian, and every prime ideal of* R *is maximal.*

In (3) and (4), products are finite products and include the empty product R and one-term products. The proof starts with a lemma.

Lemma **5.5.** *If* $\mathfrak{a} = \mathfrak{p}_1 \mathfrak{p}_2 \cdots \mathfrak{p}_r = \mathfrak{q}_1 \mathfrak{q}_2 \cdots \mathfrak{q}_s$ *is a product of invertible prime ideals* $\mathfrak{p}_1, \ldots, \mathfrak{p}_r$ *and* $\mathfrak{q}_1, \ldots, \mathfrak{q}_s$ *of* R, *then* $r = s$ *and* $\mathfrak{q}_1, \ldots, \mathfrak{q}_s$ *can be renumbered so that* $\mathfrak{p}_i = \mathfrak{q}_i$ *for all* i.

Proof. By induction on r. If $r = 0$, then $\mathfrak{a} = R$ and $s = 0$: otherwise, $\mathfrak{a} \subseteq \mathfrak{q}_1 \subsetneqq R$. Let $r > 0$. Then $\mathfrak{p}_r$, say, is minimal in $\{ \mathfrak{p}_1, \ldots, \mathfrak{p}_r \}$. Since $\mathfrak{p}_r$ is prime, $\mathfrak{q}_1 \cdots \mathfrak{q}_s \subseteq \mathfrak{p}_r$ implies $\mathfrak{q}_j \subseteq \mathfrak{p}_r$ for some j. Then $\mathfrak{p}_1 \cdots \mathfrak{p}_r \subseteq \mathfrak{q}_j$ implies $\mathfrak{p}_i \subseteq \mathfrak{q}_j$ for some i and $\mathfrak{p}_i = \mathfrak{q}_j = \mathfrak{p}_r$, since $\mathfrak{p}_r$ is minimal. We may renumber $\mathfrak{q}_1, \ldots, \mathfrak{q}_s$ so that $\mathfrak{q}_s = \mathfrak{p}_r$. Then multiplication by $\mathfrak{p}_r' = \mathfrak{q}_s'$ yields $\mathfrak{p}_1 \mathfrak{p}_2 \cdots \mathfrak{p}_{r-1} = \mathfrak{q}_1 \mathfrak{q}_2 \cdots \mathfrak{q}_{s-1}$; by the induction hypothesis, $r = s$ and $\mathfrak{q}_1, \ldots, \mathfrak{q}_{s-1}$ can be renumbered so that $\mathfrak{p}_i = \mathfrak{q}_i$ for all i. $\square$

Proof of 5.4. (1) implies (2). Let $\mathfrak{A} = \mathfrak{a}/c$ be a fractional ideal and let $\mathfrak{c} = Rc$. If $\mathfrak{a}$ is invertible, then $\mathfrak{a}\mathfrak{a}' = R$ and $\mathfrak{A}\mathfrak{a}'\mathfrak{c} = \mathfrak{a}\mathfrak{c}'\mathfrak{a}'\mathfrak{c} = \mathfrak{a}\mathfrak{a}'\mathfrak{c}\mathfrak{c}' = R$, by 5.2, 5.3.

(2) implies R Noetherian: by (2), every ideal of R is finitely generated as a fractional ideal, hence is finitely generated as an ideal of R.

(2) implies (3). If (3) does not hold, then R has a *bad* ideal, which is not a product of prime ideals. Since R is Noetherian by (2), R has a maximal bad (really bad) ideal $\mathfrak{b}$. Now, $\mathfrak{b}$ is not a prime ideal and $\mathfrak{b} \neq R$, since $\mathfrak{b}$ is not a one-term or empty product of prime ideals. Hence $\mathfrak{b} \subseteq \mathfrak{p}$ for some prime (actually, maximal) ideal of R, and $\mathfrak{b} \subsetneqq \mathfrak{p}$. By (2), $\mathfrak{b} = \mathfrak{b}\mathfrak{p}'\mathfrak{p}$, and $\mathfrak{b}\mathfrak{p}' \subseteq \mathfrak{p}\mathfrak{p}' \subseteq R$, so that $\mathfrak{b}\mathfrak{p}'$ is an ideal of R. Also $\mathfrak{b} = \mathfrak{b}\mathfrak{p}\mathfrak{p}' \subseteq \mathfrak{b}\mathfrak{p}'$, and $\mathfrak{b} \subsetneqq \mathfrak{b}\mathfrak{p}'$, since $\mathfrak{b}'\mathfrak{b}\mathfrak{p} = \mathfrak{p} \subsetneqq R = \mathfrak{b}'\mathfrak{b}\mathfrak{p}'\mathfrak{p} = R$. Hence $\mathfrak{b}\mathfrak{p}'$ is not bad and $\mathfrak{b} = \mathfrak{b}\mathfrak{p}'\mathfrak{p}$ is a product of prime ideals, an intolerable contradiction.

(3) implies (4). A product of prime ideals is a product of positive powers of distinct prime ideals; uniqueness follows from 5.5.

(4) implies (3). The proof of this is too short to fit in the margin.

(3) implies (1). First, (3) implies that every invertible prime ideal $\mathfrak{p}$ of R is maximal. Indeed, let $a \in R \backslash \mathfrak{p}$. By (3), $\mathfrak{p} + Ra = \mathfrak{p}_1 \mathfrak{p}_2 \cdots \mathfrak{p}_r$ and $\mathfrak{p} + Ra^2 = \mathfrak{q}_1 \mathfrak{q}_2 \cdots \mathfrak{q}_s$ are products of prime ideals. Apply the projection $x \longmapsto \overline{x}$ of R onto the domain $\overline{R} = R/\mathfrak{p}$:

$$\overline{\mathfrak{q}}_1 \cdots \overline{\mathfrak{q}}_s = \overline{\mathfrak{p} + Ra^2} = \overline{R}\overline{a}^2 = \left(\overline{R}\overline{a} \right)^2 = \left(\overline{\mathfrak{p} + Ra} \right)^2 = \overline{\mathfrak{p}}_1^2 \cdots \overline{\mathfrak{p}}_r^2.$$

Since $\overline{R}\overline{a}$ and $\overline{R}\overline{a}^2$ are invertible by 5.3 it follows that $\overline{\mathfrak{q}}_1, \ldots, \overline{\mathfrak{q}}_s$ and $\overline{\mathfrak{p}}_1, \ldots, \overline{\mathfrak{p}}_r$ are invertible. By 5.5, $s = 2r$ and $\mathfrak{q}_1, \ldots, \mathfrak{q}_s$ can be reindexed so that $\overline{\mathfrak{q}}_{2i-1} = \overline{\mathfrak{q}}_{2i} = \overline{\mathfrak{p}}_i$ for all i. Now, the projection $R \longrightarrow \overline{R} = R/\mathfrak{p}$ induces a one-to-one correspondence between the prime ideals of R that contain $\mathfrak{p}$ and the prime ideals of $\overline{R}$; hence $\mathfrak{q}_{2i-1} = \mathfrak{q}_{2i} = \mathfrak{p}_i$ for all i, and $\mathfrak{p} + Ra^2 = (\mathfrak{p} + Ra)^2$.

Now $\mathfrak{p} \subseteq (\mathfrak{p} + Ra)^2 \subseteq \mathfrak{p}^2 + Ra$. In fact, $\mathfrak{p} \subseteq \mathfrak{p}^2 + \mathfrak{p}a$, since $x \in \mathfrak{p}^2$,

$x + ya \in \mathfrak{p}$ implies $ya \in \mathfrak{p}$ and $y \in \mathfrak{p}$. Hence $\mathfrak{p} \subseteq \mathfrak{p}(\mathfrak{p} + Ra) \subseteq \mathfrak{p}$, $\mathfrak{p}(\mathfrak{p} + Ra) = \mathfrak{p}$, and $\mathfrak{p} + Ra = \mathfrak{p}'\mathfrak{p}(\mathfrak{p} + Ra) = \mathfrak{p}'\mathfrak{p} = R$. Thus $\mathfrak{p}$ is maximal.

For the coup de grace, let $\mathfrak{p} \neq 0$ be a prime ideal of R. Let $a \in \mathfrak{p}$, $a \neq 0$. By (3), $Ra = \mathfrak{p}_1 \cdots \mathfrak{p}_r$ is a product of prime ideals of R, which are invertible since Ra is invertible by 5.3. Then $\mathfrak{p}_1 \cdots \mathfrak{p}_r \subseteq \mathfrak{p}$ and $\mathfrak{p}_i \subseteq \mathfrak{p}$ for some i. But $\mathfrak{p}_i$ is maximal by the above. Hence $\mathfrak{p} = \mathfrak{p}_i$ is invertible. Thus every nonzero prime ideal of R is invertible; then (3) implies (1). $\square$

Definition. A Dedekind domain *is a domain that satisfies the equivalent conditions in Theorem 5.4.*

Principal ideals. By 5.4, every PID is a Dedekind domain. Examples of Dedekind domains that are not PIDs will be seen in the next section. First we show that Dedekind domains in general are not very far from PIDs.

Let $\mathfrak{a} \neq 0$ be an ideal. We denote by $e_{\mathfrak{a}}(\mathfrak{p})$ the exponent of $\mathfrak{p}$ in the unique expansion of $\mathfrak{a}$ as a product of positive powers of distinct prime ideals. Thus $e_{\mathfrak{a}}(\mathfrak{p}) = 0$ for almost all $\mathfrak{p}$ and $\mathfrak{a} = \prod_{\mathfrak{p} \text{ prime}} \mathfrak{p}^{e_{\mathfrak{a}}(\mathfrak{p})}$. Equivalently, $e_{\mathfrak{a}}(\mathfrak{p})$ is the largest integer $k \geq 0$ such that $\mathfrak{a} \subseteq \mathfrak{p}^k$: indeed, $\mathfrak{a} = \prod_{\mathfrak{p} \text{ prime}} \mathfrak{p}^{e_{\mathfrak{a}}(\mathfrak{p})} \subseteq \mathfrak{p}^{e_{\mathfrak{a}}(\mathfrak{p})}$; conversely, if $\mathfrak{a} \subseteq \mathfrak{p}^k$, then $\mathfrak{b} = \mathfrak{a}(\mathfrak{p}^k)' \subseteq R$, $\mathfrak{a} = \mathfrak{b}\mathfrak{p}^k$, and $k \leq e_{\mathfrak{a}}(\mathfrak{p})$.

Proposition 5.6. *Let $\mathfrak{a} \neq 0$ be an ideal of a Dedekind domain R and let $\mathfrak{p}_1, \ldots, \mathfrak{p}_n$ be distinct nonzero prime ideals of R. There exists a principal ideal $\mathfrak{b}$ such that $e_{\mathfrak{b}}(\mathfrak{p}_i) = e_{\mathfrak{a}}(\mathfrak{p}_i)$ for all i.*

Proof. Let $\mathfrak{a}_i = \mathfrak{p}_i^{e_{\mathfrak{a}}(\mathfrak{p}_i)}$ and let $\mathfrak{c}_i = \mathfrak{p}_i^{e_{\mathfrak{a}}(\mathfrak{p}_i) + 1}$. Then $\mathfrak{a} \subseteq \mathfrak{a}_i$ and $\mathfrak{a} \nsubseteq \mathfrak{c}_i$. Let $a_i \in \mathfrak{a}_i \backslash \mathfrak{c}_i$. If $\mathfrak{c}_i + \mathfrak{c}_j \neq R$, then $\mathfrak{c}_i + \mathfrak{c}_j \subseteq \mathfrak{m}$ for some maximal ideal $\mathfrak{m}$ of R, $\mathfrak{p}_i, \mathfrak{p}_j \subseteq \mathfrak{m}$, $\mathfrak{p}_i = \mathfrak{m} = \mathfrak{p}_j$ since $\mathfrak{p}_i, \mathfrak{p}_j$ are themselves maximal by 5.4, and $i = j$. Hence $\mathfrak{c}_i + \mathfrak{c}_j = R$ when $i \neq j$. By 5.8 below, there exists $b \in R$ such that $b + \mathfrak{c}_i = a_i + \mathfrak{c}_i$ for all i. Then $b \in \mathfrak{a}_i \backslash \mathfrak{c}_i$. Hence $e_{\mathfrak{a}}(\mathfrak{p}_i)$ is the largest integer $k \geq 0$ such that $Rb \subseteq \mathfrak{p}_i^k$ and $e_{Rb}(\mathfrak{p}_i) = e_{\mathfrak{a}}(\mathfrak{p}_i)$ for all i. $\square$

Proposition 5.7. *Every ideal of a Dedekind domain is generated by at most two elements.* (This is often called "generated by $1\frac{1}{2}$ elements".)

Proof. Let R be a Dedekind domain and let $\mathfrak{a} \neq 0$ be an ideal of R. Let $c \in \mathfrak{a}$, $c \neq 0$, and $\mathfrak{c} = Rc$. By 5.6 there exists a principal ideal $\mathfrak{b} = Rb$ such that $e_{\mathfrak{b}}(\mathfrak{p}) = e_{\mathfrak{a}}(\mathfrak{p})$ whenever $e_{\mathfrak{c}}(\mathfrak{p}) \neq 0$. We show that $\mathfrak{a} = \mathfrak{b} + \mathfrak{c}$.

If $e_{\mathfrak{c}}(\mathfrak{p}) = 0$, then $\mathfrak{c} \nsubseteq \mathfrak{p}$, $\mathfrak{a} \nsubseteq \mathfrak{p}$, $\mathfrak{b} + \mathfrak{c} \nsubseteq \mathfrak{p}$, and $e_{\mathfrak{a}}(\mathfrak{p}) = e_{\mathfrak{b}+\mathfrak{c}}(\mathfrak{p}) = 0$. Now, let $e_{\mathfrak{c}}(\mathfrak{p}) > 0$ and $k = e_{\mathfrak{a}}(\mathfrak{p}) = e_{\mathfrak{b}}(\mathfrak{p})$. Then $\mathfrak{a} \subseteq \mathfrak{p}^k$ and $\mathfrak{b} + \mathfrak{c} \subseteq \mathfrak{b} + \mathfrak{a} \subseteq \mathfrak{p}^k$, but $\mathfrak{b} + \mathfrak{c} \nsubseteq \mathfrak{p}^{k+1}$, since $\mathfrak{b} \nsubseteq \mathfrak{p}^{k+1}$. Hence $e_{\mathfrak{b}+\mathfrak{c}}(\mathfrak{p}) = k$. Thus $e_{\mathfrak{b}+\mathfrak{c}}(\mathfrak{p}) = e_{\mathfrak{a}}(\mathfrak{p})$ for all $\mathfrak{p}$. Therefore $\mathfrak{a} = \mathfrak{b} + \mathfrak{c} = Rb + Rc$. $\square$

Proposition 5.8 (Chinese Remainder Theorem). *Let $\mathfrak{a}_1, \ldots, \mathfrak{a}_n$ be ideals of a commutative ring R such that $\mathfrak{a}_i + \mathfrak{a}_j = R$ whenever $i \neq j$. For every $x_1, \ldots, x_n \in R$ there exists $x \in R$ such that $x + \mathfrak{a}_i = x_i + \mathfrak{a}_i$ for all i.*

By 5.8, if any two of $m_1, \ldots, m_n \in \mathbb{Z}$ are relatively prime, then for every $x_1, \ldots, x_n \in \mathbb{Z}$ there exists $x \in \mathbb{Z}$ such that $x \equiv x_i \pmod{m_i}$ for all i.

Proof. Let $\mathfrak{b}_j = \prod_{i \neq j} \mathfrak{a}_i$. If $\mathfrak{a}_j + \mathfrak{b}_j \neq R$, then $\mathfrak{a}_j + \mathfrak{b}_j$ is contained in a maximal ideal $\mathfrak{m}$ of R, $\mathfrak{a}_k \subseteq \mathfrak{m}$ for some $k \neq j$ since $\mathfrak{b}_j \subseteq \mathfrak{m}$ and $\mathfrak{m}$ is prime, and $\mathfrak{a}_j + \mathfrak{a}_k \subseteq \mathfrak{m}$, contradicting the hypothesis. Therefore $\mathfrak{a}_j + \mathfrak{b}_j = R$. Hence $x_j + \mathfrak{a}_j = y_j + \mathfrak{a}_j$ for some $y_j \in \mathfrak{b}_j$. Let $x = y_1 + \cdots + y_n$. Then $x_j + \mathfrak{a}_j = x + \mathfrak{a}_j$, since $y_i \in \mathfrak{b}_i \subseteq \mathfrak{a}_j$. $\square$

Exercises

1. Find all fractional ideals of $\mathbb{Z}$.

2. Find all fractional ideals of a PID.

In the following exercises, R is a domain and Q is its quotient field. Prove the following:

3. $\mathfrak{A} \subseteq Q$ is a fractional ideal of R if and only if $\mathfrak{A}$ is a submodule of Q and $\mathfrak{A}c \subseteq R$ for some $c \in R$, $c \neq 0$.

4. If $\mathfrak{A}$ and $\mathfrak{B}$ are fractional ideals of R, then $\mathfrak{A} \cap \mathfrak{B}$ is a fractional ideal of R.

5. Intersections of (too many) fractional ideals of R need not be fractional ideals of R.

6. If $\mathfrak{A}$ and $\mathfrak{B}$ are fractional ideals of R, then $\mathfrak{A} + \mathfrak{B} = \{\, a + b \mid a \in \mathfrak{A},\ b \in \mathfrak{B} \,\}$ is a fractional ideal of R.

7. If $\mathfrak{A}$ and $\mathfrak{B}$ are fractional ideals of R, then so is the set $\mathfrak{A}\mathfrak{B}$ of all finite sums $a_1 b_1 + \cdots a_n b_n$, where $n \geq 0$, $a_1, \ldots, a_n \in \mathfrak{A}$, and $b_1, \ldots, b_n \in \mathfrak{B}$.

8. The multiplication of fractional ideals is associative.

9. If $\mathfrak{A}$ and $\mathfrak{B}$ are fractional ideals of R and $\mathfrak{A} \neq 0$ is finitely generated, then $\mathfrak{B} : \mathfrak{A} = \{\, q \in Q \mid q\mathfrak{A} \subseteq \mathfrak{B} \,\}$ is a fractional ideal of R.

10. If $p_1, \ldots, p_r$ are irreducible elements of a PID and $(p_i) \neq (p_j)$ whenever $i \neq j$, then $(p_1^{m_1}) \cdots (p_r^{m_r}) = (p_1^{m_1} \cdots p_r^{m_r}) = (p_1^{m_1}) \cap \cdots \cap (p_r^{m_r})$.

11. If R is a Dedekind domain and $\mathfrak{a} \neq 0$, then every ideal of $R/\mathfrak{a}$ is principal.

6. Algebraic Integers

This section brings additional characterizations of Dedekind domains. Following in Dedekind's footsteps we then show that the algebraic integers of any finite extension of $\mathbb{Q}$ constitute a Dedekind domain.

First we prove the following:

Proposition **6.1.** *Every Noetherian, integrally closed domain with only one nonzero prime ideal is a PID.*

Equivalently, every Noetherian, integrally closed domain with only one nonzero prime ideal is a discrete valuation ring, as defined in Section VI.6. Conversely, PIDs are Noetherian, and are integrally closed, by Proposition 3.9.

Proof. Let R be a Noetherian, integrally closed domain with quotient field Q and only one nonzero prime ideal $\mathfrak{p}$. Then $\mathfrak{p}$ is maximal. We show:

(a): $\mathfrak{A} : \mathfrak{A} = R$ for every fractional ideal $\mathfrak{A} \neq 0$ of R. Indeed, $\mathfrak{A} : \mathfrak{A} = \{ x \in Q \mid x\mathfrak{A} \subseteq \mathfrak{A} \}$ is a subring of Q that contains R and is, by 5.1, a finitely generated R-module. By 2.4, $\mathfrak{A} : \mathfrak{A}$ is integral over R. Hence $\mathfrak{A} : \mathfrak{A} = R$.

(b): $R \subsetneqq \mathfrak{p}'$. First, $R \subseteq \mathfrak{p}'$. If $x \in \mathfrak{p}$, $x \neq 0$, then x is not a unit, $x^{-1} \in (Rx)' \backslash R$, and $(Rx)' \supsetneqq R$. Let $\mathcal{S}$ be the set of all nonzero ideals $\mathfrak{a}$ of R such that $\mathfrak{a}' \supsetneqq R$. Since R is Noetherian, $\mathcal{S}$ has a maximal element $\mathfrak{b}$. Let $a, b \in R$, $ab \in \mathfrak{b}$, $a \notin \mathfrak{b}$. Then $(\mathfrak{b} + Ra)' = R$: otherwise, $\mathfrak{b}$ is not maximal in $\mathcal{S}$. For any $t \in \mathfrak{b}' \backslash R$ we have $bt(\mathfrak{b} + Ra) \subseteq t\mathfrak{b} \subseteq R$, $bt \in (\mathfrak{b} + Ra)' = R$, $t(\mathfrak{b} + Rb) \subseteq R$, and $t \in (\mathfrak{b} + Rb)' \backslash R$; hence $(\mathfrak{b} + Rb)' \supsetneqq R$ and $b \in \mathfrak{b}$, since $\mathfrak{b}$ is maximal in $\mathcal{S}$. Thus $\mathfrak{b}$ is a prime ideal. Hence $\mathfrak{b} = \mathfrak{p}$, and $R \subsetneqq \mathfrak{p}'$.

(c): $\mathfrak{p}$ is invertible. Indeed, $\mathfrak{p} \subseteq \mathfrak{pp}' \subseteq R$, and $\mathfrak{pp}' \neq \mathfrak{p}$: otherwise, $\mathfrak{p}' \subseteq \mathfrak{p} : \mathfrak{p} = R$, contradicting (b). Therefore $\mathfrak{pp}' = R$, since $\mathfrak{p}$ is maximal.

(d): $\mathfrak{i} = \bigcap_{n>0} \mathfrak{p}^n = 0$. By (c), $\mathfrak{ip}' \subseteq \mathfrak{p}^{n+1}\mathfrak{p}' = \mathfrak{p}^n$ for all n, $\mathfrak{ip}' \subseteq \mathfrak{i}$, and $R \subsetneqq \mathfrak{p}' \subseteq \mathfrak{i} : \mathfrak{i}$ by (b); hence $\mathfrak{i} = 0$, by (a). (This also follows from Theorem 8.3.)

(e): $\mathfrak{p}$ is principal. If $\mathfrak{p}^2 = \mathfrak{p}$, then $\mathfrak{p}^n = \mathfrak{p}$ for all n, contradicting (d); therefore $\mathfrak{p}^2 \subsetneqq \mathfrak{p}$. Let $p \in \mathfrak{p} \backslash \mathfrak{p}^2$. Then $p\mathfrak{p}' \subseteq \mathfrak{pp}' = R$ and $p\mathfrak{p}' \not\subseteq \mathfrak{p}$: otherwise, $p \in p\mathfrak{p}'\mathfrak{p} \subseteq \mathfrak{p}^2$. Since $\mathfrak{p}$ is maximal, every ideal $\mathfrak{a} \neq R$ of R is contained in $\mathfrak{p}$; therefore $p\mathfrak{p}' = R$, and $\mathfrak{p} = p\mathfrak{p}'\mathfrak{p} = Rp$.

Now, let $\mathfrak{a} \neq R, 0$ be an ideal of R. Then $\mathfrak{a} \subseteq \mathfrak{p}$, since $\mathfrak{a}$ is contained in a maximal ideal, but $\mathfrak{a}$ is not contained in every $\mathfrak{p}^n$, by (d). Hence $\mathfrak{a} \subseteq \mathfrak{p}^n$ and $\mathfrak{a} \not\subseteq \mathfrak{p}^{n+1}$ for some $n > 0$. Let $a \in \mathfrak{a} \backslash \mathfrak{p}^{n+1}$. By (e), $\mathfrak{p} = Rp$ for some $p \in R$, so that $\mathfrak{p}^n = Rp^n$, $a = rp^n$ for some $r \in R$, $r \notin \mathfrak{p}$. In the local ring R this implies that r is a unit. Therefore $Rp^n = Ra \subseteq \mathfrak{a}$, and $\mathfrak{a} = \mathfrak{p}^n = Rp^n$. $\square$

We prove a more general result, essentially due to Noether [1926]:

Theorem 6.2. *For a domain R the following conditions are equivalent:*

(1) *R is a Dedekind domain;*

(2) *R is Noetherian and integrally closed, and every nonzero prime ideal of R is maximal;*

(3) *R is Noetherian and $R_{\mathfrak{p}}$ is a PID for every prime ideal $\mathfrak{p} \neq 0$ of R.*

Proof. (1) implies (2). Let R be Dedekind. If $x \in Q(R)$ is integral over R, then $R[x] \subseteq Q(R)$ is a finitely generated R-module, $R[x]$ is a fractional ideal by 5.1, $R[x]$ is invertible, and $(R[x])(R[x]) = R[x]$ yields $R[x] = R$ and $x \in R$. Thus R is integrally closed. The other parts of (2) follow from 5.4.

(2) implies (3). Let $\mathfrak{p} \neq 0$ be a prime ideal of R. Then $R_{\mathfrak{p}}$ is Noetherian and integrally closed, by 4.7, and has only one nonzero prime ideal by 4.5, since $\mathfrak{p}$ is maximal by (2). By 6.1, $R_{\mathfrak{p}}$ is a PID.

(3) implies (1). Since R is Noetherian, every ideal $\mathfrak{a} \neq 0$ of R is finitely generated, $\mathfrak{a} = Ra_1 + \cdots + Ra_n$, and $\mathfrak{a}'$ is a fractional ideal. Now, $\mathfrak{a}\mathfrak{a}' \subseteq R$. If $\mathfrak{a}\mathfrak{a}' \subsetneqq R$, then $\mathfrak{a}\mathfrak{a}'$ is contained in a maximal ideal $\mathfrak{m}$ of R. In $R_{\mathfrak{m}}$, $\mathfrak{a}^E$ is principal, by (3): $\mathfrak{a}^E = R_{\mathfrak{m}}(a/s)$ for some $a \in \mathfrak{a}$, $s \in R\backslash\mathfrak{m}$. Hence $a_i/1 = (x_i/s_i)(a/s)$ for some $x_i \in R$ and $s_i \in R\backslash\mathfrak{m}$. Then $t = s_1 \cdots s_n s \in R\backslash\mathfrak{m}$, $(t/a)\,a_i = tx_i/s_i s \in R$ for all i, $(t/a)\,\mathfrak{a} \subseteq R$, $t/a \in \mathfrak{a}'$, and $t \in a\mathfrak{a}' \subseteq \mathfrak{a}\mathfrak{a}' \subseteq \mathfrak{m}$. This disagreeable contradiction shows that $\mathfrak{a}\mathfrak{a}' = R$; thus $\mathfrak{a}$ is invertible. $\square$

Extensions. We now turn to algebraic integers. First we prove two results.

Proposition **6.3.** *Let R be an integrally closed domain and let E be a finite separable field extension of its quotient field Q. The integral closure of R in E is contained in a finitely generated submodule of E.*

Proof. By the primitive element theorem, $E = Q(\alpha)$ for some $\alpha \in E$. We may assume that α is integral over R: by 2.6, $r\alpha$ is integral over R for some $0 \neq r \in R$, and $E = Q(r\alpha)$.

Since E is separable over Q, α has $n = [E:Q]$ distinct conjugates $\alpha_1, \ldots, \alpha_n$ in the algebraic closure $\overline{Q}$ of Q, which are, like α, integral over R; and $F = Q(\alpha_1, \ldots, \alpha_n)$ is a Galois extension of Q. Let $\delta \in E$ be the Vandermonde determinant

$$\delta = \begin{vmatrix} 1 & 1 & \cdots & 1 \\ \alpha_1 & \alpha_2 & \cdots & \alpha_n \\ \vdots & \vdots & \ddots & \vdots \\ \alpha_1^{n-1} & \alpha_2^{n-1} & \cdots & \alpha_n^{n-1} \end{vmatrix} = \prod_{i>j}(\alpha_i - \alpha_j) \neq 0.$$

Expanding δ by rows yields cofactors γ_{jk} such that $\sum_j \alpha_j^i \gamma_{jk} = \delta$ if $i = k$, $\sum_j \alpha_j^i \gamma_{jk} = 0$ if $i \neq k$. Also, every Q-automorphism of F permutes $\alpha_1, \ldots, \alpha_n$, sends δ to $\pm\delta$, and leaves δ^2 fixed; hence $\delta^2 \in Q$.

Let $\beta \in E$ be integral over R. Then $\beta = f(\alpha)$ for some polynomial $f(X) = b_0 + b_1 X + \cdots + b_{n-1} X^{n-1} \in Q[X]$. The conjugates of β are all $\beta_j = f(\alpha_j)$ and are, like β, integral over R. Then $\sum_j \beta_j \gamma_{jk} = \sum_{i,j} b_i \alpha_j^i \gamma_{jk} = b_k \delta$ and $b_k \delta^2 = \sum_j \beta_j \gamma_{jk} \delta$. Now, δ and all γ_{jk} are integral over R, since $\alpha_1, \ldots, \alpha_n$ are integral over R, and so are $\beta_1, \ldots, \beta_n$; hence $b_k \delta^2$ is integral over R. Since $b_k \delta^2 \in Q$, it follows that $b_k \delta^2 \in R$. Hence $\beta = \sum_i b_i \alpha^i$ belongs to the submodule of E generated by $1/\delta^2$, α/δ^2, $\ldots$, α^{n-1}/δ^2, and the latter contains every $\beta \in E$ that is integral over R. $\square$

A module is *Noetherian* when its submodules satisfy the ascending chain condition. Readers will verify, as in Proposition III.11.1, that an R-module M is Noetherian if and only if every submodule of M is finitely generated.

Proposition **6.4** (Noether [1926]). *If R is Noetherian, then every finitely* *generated R-module is Noetherian.*

We omit the proof, since a more general result, Proposition VIII.8.3, is proved in the next chapter. A direct proof is outlined in the exercises. We can now prove our second main result:

Theorem **6.5.** *Let R be a Dedekind domain with quotient field Q. The integral closure of R in any finite field extension of Q is a Dedekind domain.*

Proof. By 3.10, 6.3, the integral closure $\overline{R}$ of R in a finite extension of Q is integrally closed and is contained in a finitely generated R-module M. Hence $\overline{R}$ is Noetherian: its ideals are submodules of M and satisfy the ascending chain condition, by 6.4. If $\mathfrak{P}$ is a nonzero prime ideal of $\overline{R}$, then $\mathfrak{p} = \mathfrak{P} \cap R$ is a prime ideal of R, and $\mathfrak{p} \neq 0$: otherwise, $\mathfrak{P} = 0$ by 3.6. Hence $\mathfrak{p}$ is maximal, and $\mathfrak{P}$ is maximal by 3.7. Therefore $\overline{R}$ is Dedekind, by 6.2. $\square$

Corollary **6.6.** *In every finite field extension of $\mathbb{Q}$, the algebraic integers constitute a Dedekind domain.*

This follows from Theorem 6.5, since $\mathbb{Z}$ is a Dedekind domain. Thus, the ideals of any ring of algebraic integers (over $\mathbb{Z}$) can be factored uniquely into products of positive powers of prime ideals, even though the algebraic integers themselves may lack a similar property.

Exercises

1. Show that an R-module M is Noetherian if and only if every submodule of M is finitely generated.

2. Give an example of a Dedekind domain that is not a UFD.

3. Show that a Dedekind domain with finitely many prime ideals is a PID.

4. Show that the direct product of two Noetherian R-modules is Noetherian.

5. If R is Noetherian, show that every finitely generated R-module M is Noetherian. (Hint: let M have n generators; construct a module homomorphism $R^n \longrightarrow M$; then use the previous exercise.)

7. Galois Groups

This section uses properties of algebraic integers to obtain elements of Galois groups (over $\mathbb{Q}$) with known cycle structures.

Proposition **7.1.** *Let R be a PID and let J be the ring of algebraic integers of a finite field extension E of $Q = Q(R)$. There exists a basis of E over Q that also generates J as an R-module. Hence $J \cong R^n$ (as an R-module), where $n = [E : Q]$.*

Here R^n is the R-module of all n-tuples $(r_1, \ldots, r_n)$ of elements of R, with componentwise addition and action of R, $(r_1, \ldots, r_n) + (s_1, \ldots, s_n) = (r_1 + s_1, \ldots, r_n + s_n)$ and $r(r_1, \ldots, r_n) = (rr_1, \ldots, rr_n)$.

Proof. By 6.3, J is contained in a finitely generated submodule M of E. Then M is a torsion free, finitely generated R-module. In Section VIII.6 we prove by other methods that $J \subseteq M$ must have a finite basis $\beta_1, \ldots, \beta_n$, which means that every element of J can be written uniquely as a linear combination $r_1\beta_1 + \cdots + r_n\beta_n$ of $\beta_1, \ldots, \beta_n$ with coefficients in R. Then $(r_1, \ldots, r_n) \longmapsto r_1\beta_1 + \cdots + r_n\beta_n$ is an isomorphism $R^n \longrightarrow J$ of R-modules.

We show that $\beta_1, \ldots, \beta_n$ is also a basis of E over Q. Indeed, if $q_1\beta_1 + \cdots + q_n\beta_n = 0$ for some $q_1, \ldots, q_n \in Q$, then $q_1, \ldots, q_n$ have a common denominator $r \in R$, $r \neq 0$, and then $rq_1\beta_1 + \cdots + rq_n\beta_n = 0$ with $rq_1, \ldots, rq_n \in R$, $rq_i = 0$ for all i, and $q_i = 0$ for all i. Moreover, $\beta_1, \ldots, \beta_n$ span E: if $\alpha \in E$, then $r\alpha \in J$ for some $0 \neq r \in R$ by 2.6, $r\alpha = r_1\beta_1 + \cdots + r_n\beta_n$ for some $r_1, \ldots, r_n \in R$, and $\alpha = (r_1/r)\beta_1 + \cdots + (r_n/r)\beta_n$. $\square$

Proposition 7.2. *Let R be an integrally closed domain, let J be the ring of algebraic integers of a finite Galois extension E of $Q(R)$, and let $\mathfrak{p}$ be a prime ideal of R. There are only finitely many prime ideals of J that lie over $\mathfrak{p}$, and they are all conjugate in E.*

Proof. Let $G = \mathrm{Gal}\,(E : Q(R))$. If α is integral over R, then $\sigma\alpha$ is integral over R for every $\sigma \in G$; hence the norm $\mathrm{N}(\alpha) = \prod_{\sigma \in G} \sigma\alpha$ is integral over R. Since $\mathrm{N}(\alpha) \in Q(R)$ this implies $\mathrm{N}(\alpha) \in R$.

Let $\mathfrak{P}$ and $\mathfrak{Q}$ be prime ideals of J that lie over $\mathfrak{p}$. We have $\mathfrak{Q} \subseteq \bigcup_{\sigma \in G} \sigma\mathfrak{P}$, since $\alpha \in \mathfrak{Q}$ implies $\mathrm{N}(\alpha) \in \mathfrak{Q} \cap R \subseteq \mathfrak{P}$ and $\sigma\alpha \in \mathfrak{P}$ for some $\sigma \in G$. By 7.3 below, $\mathfrak{Q}$ is contained in a single $\sigma\mathfrak{P}$. Then $\mathfrak{Q} = \sigma\mathfrak{P}$, by 3.6, since both lie over $\mathfrak{p}$. Since G is finite, there are only finitely many prime ideals of J that lie over $\mathfrak{p}$. $\square$

Readers will establish the following property:

Lemma 7.3. *Let $\mathfrak{p}_1, \ldots, \mathfrak{p}_n$ be prime ideals of a commutative ring R. An ideal of R that is contained in $\mathfrak{p}_1 \cup \cdots \cup \mathfrak{p}_n$ is contained in some $\mathfrak{p}_i$.*

We now let $R = \mathbb{Z}$.

Proposition 7.4. *Let J be the ring of algebraic integers of a finite Galois extension E of $\mathbb{Q}$ and let $\mathfrak{P}_1, \ldots, \mathfrak{P}_r$ be the prime ideals of J that lie over $p\mathbb{Z}$, where p is prime. All $J/\mathfrak{P}_i$ are isomorphic; $\overline{E} \cong J/\mathfrak{P}_i$ is a finite Galois extension of $\mathbb{Z}_p$; $\mathrm{Gal}\,(\overline{E} : \mathbb{Z}_p)$ is cyclic; and $|\overline{E}| = p^k$, where $kr \leq [E : \mathbb{Q}]$.*

Proof. If $\sigma \in \mathrm{Gal}\,(E : \mathbb{Q})$, then $\sigma J = J$; hence all $J/\mathfrak{P}_i$ are isomorphic, by 7.2. Moreover, $\mathfrak{P}_i$ is maximal by 3.7; hence $\overline{E} \cong J/\mathfrak{P}_i$ is a field.

The projections $J \longrightarrow J/\mathfrak{P}_i$ induce a homomorphism of rings $\varphi : J \longrightarrow J/\mathfrak{P}_1 \times \cdots \times J/\mathfrak{P}_r$, $\alpha \longmapsto (\alpha + \mathfrak{P}_1, \ldots, \alpha + \mathfrak{P}_r)$; φ is surjective, by 5.8, and $\mathrm{Ker}\,\varphi = \mathfrak{A} = \mathfrak{P}_1 \cap \cdots \cap \mathfrak{P}_r$. Hence $J/\mathfrak{A} \cong \overline{E}^r$. Now, $pJ \subseteq \mathfrak{A}$, since $p \in p\mathbb{Z} \subseteq \mathfrak{A}$. As an abelian group ($\mathbb{Z}$-module), $J \cong \mathbb{Z}^n$ by 7.1, where $n = [E : \mathbb{Q}]$. Hence $J/pJ \cong \mathbb{Z}^n/p\mathbb{Z}^n \cong (\mathbb{Z}/p\mathbb{Z})^n$ is finite, with p^n elements, $J/\mathfrak{A} \cong (J/pJ)/(\mathfrak{A}/pJ)$ is finite, and $\overline{E}$ is finite, with p^k elements for some

$k \geq 0$. Then $kr \leq n$, since $\overline{E}^r \cong J/\mathfrak{A}$ has at most p^n elements. (This inequality can also be proved by extending valuations as in Section VI.7.)

By V.1.3, the finite field $\overline{E}$ of order p^k is the splitting field of the separable polynomial $X^{p^k} - X \in \mathbb{Z}_p[X]$. Hence $\overline{E}$ is a finite Galois extension of $\mathbb{Z}_p$. Eager readers will delight in proving that $\mathrm{Gal}\,(\overline{E} : \mathbb{Z}_p)$ is cyclic. $\square$

We can now prove our main result.

Theorem 7.5. *Let $q \in \mathbb{Z}[X]$ be a monic irreducible polynomial and let p be a prime number. Let the image $\overline{q}$ of q in $\mathbb{Z}_p[X]$ be the product of irreducible polynomials $q_1, \ldots, q_s \in \mathbb{Z}_p[X]$ of degrees $d_1, \ldots, d_s$. For almost every prime p, the polynomials $q_1, \ldots, q_s$ are distinct, and then the Galois group of q over $\mathbb{Q}$ contains a product of disjoint cycles of orders $d_1, \ldots, d_s$.*

Proof. The roots $\alpha_1, \ldots, \alpha_n$ of q in $\mathbb{C}$ are integral over $\mathbb{Z}$, since q is monic. Let $E = \mathbb{Q}(\alpha_1, \ldots, \alpha_n)$ be the splitting field of q. Let J, $\mathfrak{P}_1, \ldots, \mathfrak{P}_r$, $\overline{E}$, k, and r be as in 7.4; let $\mathfrak{P} = \mathfrak{P}_1$, and let $\alpha \longmapsto \overline{\alpha} = \alpha + \mathfrak{P}$ be the projection $J \longrightarrow \overline{E} = J/\mathfrak{P}$. Then $\mathbb{Z}_p \subseteq \overline{E}$, $\overline{q}$, $q_1, \ldots, q_s \in \overline{E}[X]$, and $q(X) = (X - \alpha_1) \cdots (X - \alpha_n)$ in $J[X]$ yields

$$q_1 \cdots q_s = \overline{q} = (X - \overline{\alpha}_1) \cdots (X - \overline{\alpha}_n)$$

in $\overline{E}[X]$. Hence every q_j is the irreducible polynomial of some $\overline{\alpha}_i \in \overline{E}$.

If $q_1, \ldots, q_s$ are not distinct, then the discriminant $\prod_{i<j} (\overline{\alpha}_i - \overline{\alpha}_j)^2$ of $\overline{q}$ is zero, and the discriminant $D = \prod_{i<j} (\alpha_i - \alpha_j)^2$ of q lies in $\mathfrak{P}$. Then $D \in \mathfrak{P} \cap \mathbb{Z} = p\mathbb{Z}$ and D is an integer multiple of p. Therefore there are only finitely many primes p such that $q_1, \ldots, q_s$ are not distinct.

Now, assume that $q_1, \ldots, q_s$ are distinct. Since $\overline{E}$ is Galois over $\mathbb{Z}_p$, $q_1, \ldots, q_s$ are separable and have no multiple roots in $\overline{E}$. Moreover, $q_1, \ldots, q_s$ have no common roots in $\overline{E}$, since they are the distinct irreducible polynomials of elements of E. Therefore $\overline{\alpha}_1, \ldots, \overline{\alpha}_n$ are all distinct.

Let $G = \mathrm{Gal}\,(E : \mathbb{Q})$, $\overline{G} = \mathrm{Gal}\,(\overline{E} : \mathbb{Z}_p)$, and $H = \{\sigma \in G \mid \sigma \mathfrak{P} = \mathfrak{P}\}$ be the stabilizer of $\mathfrak{P}$. If $\sigma \in H$, then $\sigma J = J$, $\sigma \mathfrak{P} = \mathfrak{P}$, and σ induces an automorphism $\overline{\sigma} : \alpha \longmapsto \overline{\alpha} = \alpha + \mathfrak{P}$ of $J/\mathfrak{P} = \overline{E}$. Then $\varphi : \sigma \longrightarrow \overline{\sigma}$ is a homomorphism of H into $\overline{G}$. If $\overline{\sigma} = 1$, then, for all i, $\overline{\alpha}_i = \overline{\sigma}\,\overline{\alpha}_i = \overline{\sigma \alpha_i}$, $\alpha_i = \sigma \alpha_i$, since $\overline{\alpha}_1, \ldots, \overline{\alpha}_n$ are distinct, and $\sigma = 1$. Thus φ is injective and $|H| \leq |\overline{G}| = k$. But the orbit of $\mathfrak{P}$ under the action of G is $\{\mathfrak{P}_1, \ldots, \mathfrak{P}_r\}$, by 7.2; hence $[G : H] = r$ and $|H| = n/r \geq k$. Therefore $|H| = k$ and φ is an isomorphism. Thus every $\tau \in \overline{G}$ is induced (as $\tau = \overline{\sigma}$) by a unique $\sigma \in H$.

Identify every $\sigma \in H$ with the permutation of $\alpha_1, \ldots, \alpha_n$ that it induces, and every $\overline{\sigma} \in \overline{G}$ with the similar permutation of $\overline{\alpha}_1, \ldots, \overline{\alpha}_n$. Then σ and $\overline{\sigma}$ have the same cycle structure. Let $\overline{\tau}$ generate the group $\overline{G}$, which is cyclic by 7.4. Then $\overline{\tau}$ permutes the roots $\beta_1, \ldots, \beta_{d_j}$ of q_j, since $^{\overline{\tau}}q_j = q_j$. But $\overline{\tau}$ cannot permute any proper subset, say $\{\beta_1, \ldots, \beta_t\}$, of $\{\beta_1, \ldots, \beta_{d_j}\}$: otherwise, $f(X) = (X - \beta_1) \cdots (X - \beta_t)$ and $g(X) = (X - \beta_{t+1}) \cdots (X - \beta_{d_j})$ are fixed under $\overline{\tau}$, f and g are fixed under $\overline{G}$, $f, g \in \mathbb{Z}_p[X]$, and $fg = q_j$, an insult to q_j's irreducibility. Therefore $\overline{\tau}$ has a restriction to $\{\beta_1, \ldots, \beta_{d_j}\}$ that is a d_j-cycle. Thus $\overline{\tau}$ is a product of disjoint cycles of orders $d_1, \ldots, d_s$. Then $\overline{\tau}$ is induced by some $\tau \in H$ with the same cycle structure. $\square$

For example, let $q(X) = X^5 - X + 1 \in \mathbb{Z}[X]$. Readers will verify that $\overline{q}$ is irreducible in $\mathbb{Z}_3[X]$. Hence q, which is monic, is irreducible in $\mathbb{Z}[X]$. Also

$$X^5 - X + 1 = (X^2 + X + 1)(X^3 + X^2 + 1) \text{ in } \mathbb{Z}_2[X].$$

By 7.5, the Galois group G of q over $\mathbb{Q}$, viewed as a subgroup of S_5, contains a 5-cycle, and contains the product of a 2-cycle and a disjoint 3-cycle. Therefore G contains a 5-cycle and a transposition, and $G = S_5$.

Exercises

1. Let $\mathfrak{p}_1, \ldots, \mathfrak{p}_n$ be prime ideals of a commutative ring R. Prove that an ideal of R that is contained in $\mathfrak{p}_1 \cup \cdots \cup \mathfrak{p}_n$ is contained in some $\mathfrak{p}_i$.

2. Let F be a finite field of order p^n. Show that $\mathrm{Gal}\,(F : \mathbb{Z}_p)$ is cyclic of order p^{n-1}.

3. Find the Galois group of $X^4 + X + 1$ over $\mathbb{Q}$.

4. Find the Galois group of $X^4 + 2X^2 + X + 1$ over $\mathbb{Q}$.

5. Find the Galois group of $X^4 + 2X^2 + 3X + 1$ over $\mathbb{Q}$.

8. Minimal Prime Ideals

In this section we establish several finiteness properties for the prime ideals of Noetherian rings, due to Krull [1928], for use in the next section.

Artinian modules. We begin with a peek at modules that satisfy the descending chain condition.

A module M is *Artinian* when every infinite descending sequence $S_1 \supseteq S_2 \supseteq \cdots \supseteq S_n \supseteq S_{n+1} \supseteq \cdots$ of submodules of M terminates (there exists $m > 0$ such that $S_n = S_m$ for all $n \geq m$). For example, a finite-dimensional vector space over a field K is Artinian as a K-module.

In an Artinian module M, every nonempty set $\mathcal{S}$ of submodules of M has a *minimal* element (an element S of $\mathcal{S}$ such that there is no $S \supsetneq T \in \mathcal{S}$): otherwise, there exists some $S_1 \in \mathcal{S}$; since S_1 is not minimal in $\mathcal{S}$ there exists some $S_1 \supsetneq S_2 \in \mathcal{S}$; since S_2 is not minimal in $\mathcal{S}$ there exists some $S_2 \supsetneq S_3 \in \mathcal{S}$; this continues indefinitely, ruining the neighborhood with an infinite descending sequence that won't terminate.

Proposition 8.1. If N is a submodule of M, then M is Artinian if and only if N and M/N are Artinian.

Proof. Assume that N and M/N are Artinian and let $S_1 \supseteq \cdots S_n \supseteq S_{n+1} \supseteq \cdots$ be an infinite descending sequence of submodules of M. Then $S_1 \cap N \supseteq \cdots S_n \cap N \supseteq S_{n+1} \cap N \supseteq \cdots$ is an infinite descending sequence of submodules of N, and $(S_1 + N)/N \supseteq \cdots (S_n + N)/N \supseteq (S_{n+1} + N)/N \supseteq \cdots$ is an infinite descending sequence of submodules of M/N. Both sequences terminate: there exists $m > 0$ such that $S_n \cap N = S_m \cap N$ and $(S_n + N)/N = (S_m + N)/N$ for all $n \geqq m$. Then $S_n + N = S_m + N$ for all $n \geqq m$. Hence $S_n = S_m$ for all $n \geqq m$: if $x \in S_n \subseteq S_m + N$, then $x = y + t$ for some $y \in S_m$ and $t \in N$, and then $t = x - y \in S_n \cap N = S_m \cap N$ and $x = y + t \in S_m$. Thus M is Artinian. We leave the converse to enterprising readers. $\square$

Lemma 8.2. *If* $\mathfrak{m}$ *is a maximal ideal of a Noetherian ring* R, *then* $R/\mathfrak{m}^n$ *is an Artinian R-module, for every* $n > 0$.

Proof. Let $M = \mathfrak{m}^{n-1}/\mathfrak{m}^n$ ($M = R/\mathfrak{m}$, if $n = 1$). We show that M is an Artinian R-module; since $(R/\mathfrak{m}^n)/(\mathfrak{m}^{n-1}/\mathfrak{m}^n) \cong R/\mathfrak{m}^{n-1}$, it then follows from 8.1, by induction on n, that $R/\mathfrak{m}^n$ is Artinian. Since $\mathfrak{m}M = 0$, the action of R on M induces a module action of $R/\mathfrak{m}$ on M, which is well defined by $(r + \mathfrak{m})x = rx$. Then M has the same submodules as an R-module and as an $R/\mathfrak{m}$-module. Since R is Noetherian, $\mathfrak{m}^{n-1}$ is a finitely generated R-module and M is a finitely generated as an R-module, and as an $R/\mathfrak{m}$-module. But $R/\mathfrak{m}$ is a field; hence M is a finite-dimensional vector space over $R/\mathfrak{m}$ and M is Artinian as an $R/\mathfrak{m}$-module and as an R-module. $\square$

The intersection theorem.

Theorem 8.3 (Krull Intersection Theorem [1928]). *Let* $\mathfrak{a} \neq R$ *be an ideal of a Noetherian ring* R *and let* $\mathfrak{i} = \bigcap_{n>0} \mathfrak{a}^n$. *Then* $\mathfrak{a}\mathfrak{i} = \mathfrak{i}$ *and* $(1 - a)\mathfrak{i} = 0$ *for some* $a \in \mathfrak{a}$. *If* R *is a domain, or if* R *is local, then* $\mathfrak{i} = 0$.

If $\mathfrak{i} = 0$, then R embeds into its $\mathfrak{a}$-adic completion in Section VI.9.

Proof. Let $\mathfrak{q}$ be a primary ideal that contains $\mathfrak{a}\mathfrak{i}$ and let $\mathfrak{p}$ be its radical. Then $\mathfrak{p}^n \subseteq \mathfrak{q}$ for some n, by 1.6, and $\mathfrak{i} \subseteq \mathfrak{q}$: otherwise, $\mathfrak{a} \subseteq \mathfrak{p}$, since $\mathfrak{q}$ is primary, and $\mathfrak{i} \subseteq \mathfrak{a}^n \subseteq \mathfrak{q}$ anyway. Since $\mathfrak{a}\mathfrak{i}$ is an intersection of primary ideals, by 1.9, this implies $\mathfrak{i} \subseteq \mathfrak{a}\mathfrak{i}$ and $\mathfrak{i} = \mathfrak{a}\mathfrak{i}$. Lemma 8.4 below then yields $(1 - a)\mathfrak{i} = 0$ for some $a \in \mathfrak{a}$. Then $1 - a \neq 0$. If R is a domain, then $(1 - a)\mathfrak{i} = 0$ implies $\mathfrak{i} = 0$. If R is local, then $1 - a$ is a unit and again $(1 - a)\mathfrak{i} = 0$ implies $\mathfrak{i} = 0$. $\square$

Lemma 8.4. *Let* $\mathfrak{a}$ *be an ideal of a commutative ring* R *and let* M *be a finitely generated R-module. If* $\mathfrak{a}M = M$, *then* $(1 - a)M = 0$ *for some* $a \in \mathfrak{a}$.

Proof. $\mathfrak{a}M$ is the set of all sums $a_1 x_1 + \cdots + a_n x_n$ in which $a_1, \ldots, a_n \in \mathfrak{a}$ and $x_1, \ldots, x_n \in M$. If M is generated by $e_1, \ldots, e_m$, then every element x of $\mathfrak{a}M$ is a sum $a_1 e_1 + \cdots + a_m e_m$, where $a_1, \ldots, a_m \in \mathfrak{a}$.

Since $M = \mathfrak{a}M$, there are equalities $e_i = \sum_j a_{ij} e_j$ in which $a_{ij} \in \mathfrak{a}$ for all i, j. Then $\sum_j b_{ij} e_j = 0$ for all i, where $b_{ij} = 1 - a_{ij}$ if $i = j$, $b_{ij} = -a_{ij}$

otherwise. By 2.5, the determinant

$$
D = \begin{vmatrix} b_{11} & b_{12} & \cdots & b_{1n} \\ b_{21} & b_{22} & \cdots & b_{2n} \\ \vdots & \vdots & \ddots & \vdots \\ b_{n1} & b_{n2} & \cdots & b_{nn} \end{vmatrix} = \begin{vmatrix} 1-a_{11} & -a_{12} & \cdots & -a_{1n} \\ -a_{21} & 1-a_{22} & \cdots & -a_{2n} \\ \vdots & \vdots & \ddots & \vdots \\ -a_{n1} & -a_{n2} & \cdots & 1-a_{nn} \end{vmatrix}
$$

satisfies $De_i = 0$ for all i. Now $D = 1 - a$ for some $a \in \mathfrak{a}$. Then $(1-a)\,e_i = 0$ for all i, and $(1-a)\,x = 0$ for all $x \in M$. $\square$

Corollary **8.5** (Nakayama's Lemma). *Let* $\mathfrak{a}$ *be an ideal of a commutative ring* R *and let* M *be a finitely generated R-module. If* $\mathfrak{a}$ *is contained in every maximal ideal of* R *and* $\mathfrak{a}M = M$, *then* $M = 0$.

This makes a fine exercise. A more general version is proved in Section IX.5.

Prime ideals. We now turn to prime ideals. Let $\mathfrak{a}$ be an ideal of R. A prime ideal $\mathfrak{p}$ is *minimal over* $\mathfrak{a}$, or an *isolated prime ideal* of $\mathfrak{a}$, when it is minimal among all prime ideals of R that contain $\mathfrak{a}$.

Proposition **8.6.** *Let* $\mathfrak{a} \neq R$ *be an ideal of a Noetherian ring* R. *There exists a prime ideal of* R *that is minimal over* $\mathfrak{a}$; *in fact, every prime ideal that contains* $\mathfrak{a}$ *contains a prime ideal that is minimal over* $\mathfrak{a}$. *Moreover, there are only finitely many prime ideals of* R *that are minimal over* $\mathfrak{a}$.

Proof. By 1.9, $\mathfrak{a}$ is the intersection $\mathfrak{a} = \mathfrak{q}_1 \cap \cdots \cap \mathfrak{q}_r$ of finitely many primary ideals with radicals $\mathfrak{p}_1, \ldots, \mathfrak{p}_r$. A prime ideal that contains $\mathfrak{a}$ contains $\mathfrak{q}_1 \cdots \mathfrak{q}_r \subseteq \mathfrak{a}$, contains some $\mathfrak{q}_i$, and contains its radical $\mathfrak{p}_i$. Hence $\mathfrak{p}$ is minimal over $\mathfrak{a}$ if and only if $\mathfrak{p}$ is minimal among $\mathfrak{p}_1, \ldots, \mathfrak{p}_r$. $\square$

Lemma **8.7.** *Let* R *be a Noetherian local ring with maximal ideal* $\mathfrak{m}$. *If* $a \in \mathfrak{m}$ *and* R/Ra *is Artinian as an R-module, then there is at most one prime ideal* $\mathfrak{p}$ *of* R *that does not contain* a, *namely the nilradical* $\mathfrak{p} = \mathrm{Rad}\, 0$ *of* R.

Proof. Let $\mathfrak{p}$ be a prime ideal of R that does not contain a. For every $n > 0$ let $\mathfrak{p}^{(n)} = ((\mathfrak{p}^E)^n)^C$, as calculated in $R_{\mathfrak{p}}$; $\mathfrak{p}^{(n)}$ is the nth *symbolic power* of $\mathfrak{p}$. Since $x \in \mathfrak{p}^{(n)}$ is equivalent to $x/1 = (\mathfrak{p}^E)^n = (\mathfrak{p}^n)^E$, $x/1 = y/s$ for some $y \in \mathfrak{p}^n$ and $s \in R \backslash \mathfrak{p}$, and $stx \in \mathfrak{p}^n$ for some $s, t \in R \backslash \mathfrak{p}$, we have

$$
\mathfrak{p}^{(n)} = \{\, x \in R \mid sx \in \mathfrak{p}^n \text{ for some } s \in R \backslash \mathfrak{p}\,\}.
$$

Hence $\mathfrak{p}^{(1)} = \mathfrak{p}$, since $\mathfrak{p}$ is prime, and $\mathfrak{p}^{(n)} \supseteq \mathfrak{p}^{(n+1)}$ for all n. Since $\mathrm{Rad}\,(\mathfrak{p}^E)^n = \mathfrak{p}^E$ is the maximal ideal of $R_{\mathfrak{p}}$, $(\mathfrak{p}^n)^E = (\mathfrak{p}^E)^n$ is $\mathfrak{p}^E$-primary by 1.10 and $\mathfrak{p}^{(n)}$ is $\mathfrak{p}$-primary by 4.6 (this can also be proved directly).

We show that the descending sequence $\mathfrak{p} = \mathfrak{p}^{(1)} \supseteq \cdots \supseteq \mathfrak{p}^{(n)} \supseteq \mathfrak{p}^{(n+1)} \supseteq \cdots$ terminates. Let $\mathfrak{b} = \mathfrak{p} \cap Ra$. Then $\mathfrak{p}/\mathfrak{b} = \mathfrak{p}/(\mathfrak{p} \cap Ra) \cong (\mathfrak{p} + Ra)/Ra \subseteq R/Ra$ is an Artinian R-module, and the descending sequence

$$
\mathfrak{p}/\mathfrak{b} = \mathfrak{p}^{(1)}/\mathfrak{b} \supseteq \cdots \supseteq (\mathfrak{p}^{(n)} + \mathfrak{b})/\mathfrak{b} \supseteq (\mathfrak{p}^{(n+1)} + \mathfrak{b})/\mathfrak{b} \supseteq \cdots
$$

terminates: there is some $m > 0$ such that $\mathfrak{p}^{(n)} + \mathfrak{b} = \mathfrak{p}^{(m)} + \mathfrak{b}$ for all $n \geqq m$. If $n \geqq m$, then $M = \mathfrak{p}^{(m)}/\mathfrak{p}^{(n)}$ is a finitely generated R-module, and $aM = M$, since $x \in \mathfrak{p}^{(m)} \subseteq \mathfrak{p}^{(n)} + Ra$ implies $x = y + ra$ for some $y \in \mathfrak{p}^{(n)}$ and $r \in R$, $ra = x - y \in \mathfrak{p}^{(m)}$, $r \in \mathfrak{p}^{(m)}$ since $a \notin \mathfrak{p}$, $x = y + ra \in \mathfrak{p}^{(n)} + \mathfrak{p}^{(m)}a$, and $x + \mathfrak{p}^{(n)} = a\,(r + \mathfrak{p}^{(n)}) \in aM$. Hence $RaM = M$, $\mathfrak{m}M = 0$ since $Ra \subseteq \mathfrak{m}$, and $M = 0$ by 8.5. Thus $\mathfrak{p}^{(n)} = \mathfrak{p}^{(m)}$ for all $n \geqq m$.

On the other hand, $\bigcap_{n>0} (\mathfrak{p}^E)^n = 0$, by 8.3, applied to the local ring $R_{\mathfrak{p}}$. Hence $\mathfrak{p}^{(m)} = \bigcap_{n>0} \mathfrak{p}^{(m)} = 0$ and $\mathfrak{p} = \mathrm{Rad}\,\mathfrak{p}^{(m)} = \mathrm{Rad}\,0$. $\square$

The main result in this section is Krull's Hauptidealsatz (*principal ideal theorem*). Every commutative ring has a least prime ideal, its nilradical $\mathrm{Rad}\,0$. A prime ideal $\mathfrak{p}$ *has height at most* 1 when there is at most one prime ideal $\mathfrak{q} \subsetneqq \mathfrak{p}$: when either $\mathfrak{p} = \mathrm{Rad}\,0$ or $\mathrm{Rad}\,0$ is the only prime ideal $\mathfrak{q} \subsetneqq \mathfrak{p}$. In a domain, $\mathrm{Rad}\,0 = 0$ and a prime ideal of height at most 1 is either 0 or a minimal nonzero prime ideal. (Heights are defined in general in the next section.)

Theorem **8.8** (Krull's Hauptidealsatz [1928]). *In a Noetherian ring, a prime ideal that is minimal over a principal ideal has height at most 1.*

Proof. First let $\mathfrak{p} = \mathfrak{m}$ be the maximal ideal of a Noetherian local ring R. Assume that $\mathfrak{m}$ is minimal over a principal ideal Ra of R. Then $\mathrm{Rad}\,Ra \subseteq \mathfrak{m}$, $\mathfrak{m} = \mathrm{Rad}\,Ra$, and $\mathfrak{m}^n \subseteq Ra$ for some $n > 0$ by 1.6. Now, $R/\mathfrak{m}^n$ is an Artinian R-module, by 8.2. Hence $R/Ra \cong (R/\mathfrak{m}^n)/(Ra/\mathfrak{m}^n)$ is Artinian, by 8.1. Since $\mathfrak{p}$ is minimal over Ra, a prime ideal $\mathfrak{q} \subsetneqq \mathfrak{p}$ cannot contain a; by 8.7, there is at most one prime ideal $\mathfrak{q} \subsetneqq \mathfrak{p}$.

Now, let R be Noetherian and $\mathfrak{p}$ be a prime ideal of R that is minimal over a principal ideal Ra. Then $R_{\mathfrak{p}}$ is a Noetherian local ring, $\mathfrak{p}^E$ is the maximal ideal of $R_{\mathfrak{p}}$, and $\mathfrak{p}^E$ is minimal over $(Ra)^E$ by 4.5. Hence $\mathfrak{p}^E$ has height at most 1 in $R_{\mathfrak{p}}$. Then $\mathfrak{p}$ has height at most 1 in R, by 4.5 again. $\square$

Readers will easily prove that, conversely, a prime ideal of height at most one is minimal over a principal ideal.

Exercises

1. Prove the following: let N be a submodule of an R-module M; if M is Artinian, then N and M/N are Artinian.

2. Show that every ideal $\mathfrak{a} \neq R$ of a commutative ring R has an isolated prime ideal (even if R is not Noetherian). (First show that the intersection of a nonempty chain of prime ideals is a prime ideal.)

3. Prove *Nakayama's lemma*: Let $\mathfrak{a}$ be an ideal of a commutative ring R and let M be a finitely generated R-module. If $\mathfrak{a}$ is contained in every maximal ideal of R and $\mathfrak{a}M = M$, then $M = 0$.

4. Give a direct proof that $\mathfrak{p}^{(n)} = \{\, x \in R \mid sx \in \mathfrak{p}^n \text{ for some } s \in R \backslash \mathfrak{p} \,\}$ is a $\mathfrak{p}$-primary ideal, whenever $\mathfrak{p}$ is a prime ideal of a commutative ring R.

5. Use Nakayama's lemma to prove that $\mathfrak{p}^{(n)} = \mathfrak{p}^{(n+1)}$ implies $\mathfrak{p}^{(n)} = 0$, when $\mathfrak{p}$ is a prime ideal of a local ring.

6. Let R be a Noetherian ring and let $\mathfrak{p}$ be a prime ideal of R of height at most 1. Show that $\mathfrak{p}$ is minimal over a principal ideal of R.

7. Let R be a Noetherian domain. Show that R is a UFD if and only if every prime ideal of R of height at most 1 is principal. (You may want to show that irreducible elements of R are prime, then follow the proof of Theorem III.8.4.)

9. Krull Dimension

In this section we prove that prime ideals of Noetherian rings have finite height (Krull [1928]). This leads to a dimension concept for Noetherian rings.

Definition. In a commutative ring R, the height hgt $\mathfrak{p}$ *of a prime ideal $\mathfrak{p}$ is the least upper bound of the lengths of strictly descending sequences $\mathfrak{p} = \mathfrak{p}_0 \supsetneqq \mathfrak{p}_1 \supsetneqq \cdots \supsetneqq \mathfrak{p}_m$ of prime ideals of R.*

Thus $\mathfrak{p}$ has height at most n if and only if every strictly descending sequence $\mathfrak{p} = \mathfrak{p}_0 \supsetneqq \mathfrak{p}_1 \supsetneqq \cdots \supsetneqq \mathfrak{p}_m$ of prime ideals has length at most n. For instance, the least prime ideal $\mathrm{Rad}\, 0$ of R has height 0; $\mathfrak{p}$ has height at most 1 if and only if $\mathfrak{p} \supsetneqq \mathrm{Rad}\, 0$ and there is no prime ideal $\mathfrak{p} \supsetneqq \mathfrak{q} \supsetneqq \mathrm{Rad}\, 0$.

The height of $\mathfrak{p}$ is also called its *codimension*. We saw in Section 1 that an algebraic set $A \subseteq \mathbb{C}^n$ is the union of finitely many algebraic sets defined by prime ideals, known as *algebraic varieties*. We shall prove that a prime ideal $\mathfrak{P}$ of $\mathbb{C}[X, Y, Z]$ has height 0 ($\mathfrak{P} = 0$), 1 (for instance, if $\mathfrak{P} = (q)$, where q is irreducible), 2, or 3 (if $\mathfrak{P}$ is maximal); the corresponding algebraic varieties are $\mathbb{C}^3$, algebraic surfaces, algebraic curves (intersections of two surfaces, defined by two equations), and single points.

To prove Krull's theorem we start with an inclusion avoidance lemma:

Lemma 9.1. Let $\mathfrak{p}_0, \mathfrak{p}_1, \ldots, \mathfrak{p}_m, \mathfrak{q}_1, \ldots, \mathfrak{q}_n$ be prime ideals of a Noetherian ring R. If $\mathfrak{p}_0 \supsetneqq \mathfrak{p}_1 \supsetneqq \cdots \supsetneqq \mathfrak{p}_m$ and $\mathfrak{p}_0 \nsubseteq \mathfrak{q}_1 \cup \cdots \cup \mathfrak{q}_n$, then $\mathfrak{p}_0 \supsetneqq \mathfrak{p}'_1 \supsetneqq \cdots \supsetneqq \mathfrak{p}'_{m-1} \supsetneqq \mathfrak{p}_m$ for some prime ideals $\mathfrak{p}'_1, \ldots, \mathfrak{p}'_{m-1} \nsubseteq \mathfrak{q}_1 \cup \cdots \cup \mathfrak{q}_n$.

Proof. By induction on m. There is nothing to prove if $m \leq 1$. If $m \geq 2$, then the induction hypothesis yields $\mathfrak{p}_0 \supsetneqq \mathfrak{p}'_1 \supsetneqq \cdots \supsetneqq \mathfrak{p}'_{m-2} \supsetneqq \mathfrak{p}_{m-1}$ for some prime ideals $\mathfrak{p}'_1, \ldots, \mathfrak{p}'_{m-2} \nsubseteq \mathfrak{q}_1 \cup \cdots \cup \mathfrak{q}_n$. Then $\mathfrak{p}'_{m-2} \nsubseteq \mathfrak{p}_m \cup \mathfrak{q}_1 \cup \cdots \cup \mathfrak{q}_n$, by 7.3. Let $a \in \mathfrak{p}'_{m-2} \setminus (\mathfrak{p}_m \cup \mathfrak{q}_1 \cup \cdots \cup \mathfrak{q}_n)$. By 8.6, $\mathfrak{p}'_{m-2}$ contains a prime ideal $\mathfrak{p}'_{m-1}$ that is minimal over $Ra + \mathfrak{p}_m$. Then $\mathfrak{p}'_{m-1} \nsubseteq \mathfrak{q}_1 \cup \cdots \cup \mathfrak{q}_n$ and $\mathfrak{p}'_{m-1} \supsetneqq \mathfrak{p}_m$. Finally, $\mathfrak{p}'_{m-2} \supsetneqq \mathfrak{p}'_{m-1}$: in the Noetherian ring $R/\mathfrak{p}_m$, $\mathfrak{p}'_{m-2}/\mathfrak{p}_m$ has height at least 2, since $\mathfrak{p}'_{m-2} \supsetneqq \mathfrak{p}_{m-1} \supsetneqq \mathfrak{p}_m$, whereas $\mathfrak{p}'_{m-1}/\mathfrak{p}_m$ is minimal over $(Ra + \mathfrak{p}_m)/\mathfrak{p}_m$ and has height at most 1 by 8.8. $\square$

Krull's theorem, also called *principal ideal theorem* like Theorem 8.8, is the following result.

Theorem **9.2** (Krull [1928]). *In a Noetherian ring, every prime ideal* $\mathfrak{p}$ *has finite height; in fact, if* $\mathfrak{p}$ *is minimal over an ideal with* r *generators, then* $\mathfrak{p}$ *has height at most* r.

Proof. Let R be Noetherian; let $\mathfrak{a} = Rx_1 + \cdots + Rx_r$ be an ideal of R with r generators $x_1, \ldots, x_r$, and let $\mathfrak{p}$ be a prime ideal of R that is minimal over $\mathfrak{a}$. We prove by induction on r that hgt $\mathfrak{p} \leqq r$. Hence hgt $\mathfrak{p} \leqq r$ when $\mathfrak{p}$ has r generators, since $\mathfrak{p}$ is minimal over itself. (Conversely, every prime ideal of height r is minimal over some ideal with r generators; see the exercises.)

Theorem 8.8 is the case $r = 1$. Assume $r \geq 2$. Let $\mathfrak{b} = Rx_1 + \cdots + Rx_{r-1}$. If $\mathfrak{p}$ is minimal over $\mathfrak{b}$, then hgt $\mathfrak{p} \leqq r - 1$, by the induction hypothesis. Otherwise, there are only finitely many prime ideals $\mathfrak{q}_1, \ldots, \mathfrak{q}_n$ of R that are minimal over $\mathfrak{b}$, by 8.6, and $\mathfrak{p} \not\subseteq \mathfrak{q}_1 \cup \cdots \cup \mathfrak{q}_n$, by 7.3. Let $\mathfrak{p} = \mathfrak{p}_0 \supsetneqq \mathfrak{p}_1 \supsetneqq \cdots \supsetneqq \mathfrak{p}_m$ be a strictly decreasing sequence of prime ideals.

By 9.1 we may assume that $\mathfrak{p}_1, \ldots, \mathfrak{p}_{m-1} \not\subseteq \mathfrak{q}_1 \cup \cdots \cup \mathfrak{q}_n$. In the Noetherian ring $R/\mathfrak{b}$, the prime ideals that are minimal over $0 = \mathfrak{b}/\mathfrak{b}$ are $\mathfrak{q}_1/\mathfrak{b}, \ldots, \mathfrak{q}_n/\mathfrak{b}$; and $\mathfrak{p}/\mathfrak{b}$, which is minimal over the principal ideal $(Rx_r + \mathfrak{b})/\mathfrak{b}$ but not over $\mathfrak{b}/\mathfrak{b} = 0$, has height 1, by 8.8. Hence $\mathfrak{p}$ is minimal over $\mathfrak{p}_{m-1} + \mathfrak{b}$: if $\mathfrak{q}$ is a prime ideal and $\mathfrak{p} \supseteq \mathfrak{q} \supsetneqq \mathfrak{p}_{m-1} + \mathfrak{b}$, then $\mathfrak{q}/\mathfrak{b} \not\subseteq \mathfrak{q}_1/\mathfrak{b} \cup \cdots \cup \mathfrak{q}_n/\mathfrak{b}$, since $\mathfrak{p}_{m-1} \not\subseteq \mathfrak{q}_1 \cup \cdots \cup \mathfrak{q}_n$, $\mathfrak{q}/\mathfrak{b}$ has height at least 1, $\mathfrak{p}/\mathfrak{b} = \mathfrak{q}/\mathfrak{b}$ since $\mathfrak{p}/\mathfrak{b}$ has height 1, and $\mathfrak{p} = \mathfrak{q}$. Then $\mathfrak{p}/\mathfrak{p}_{m-1}$ is minimal over $(\mathfrak{p}_{m-1} + \mathfrak{b})/\mathfrak{p}_{m-1}$. Since $(\mathfrak{p}_{m-1} + \mathfrak{b})/\mathfrak{p}_{m-1}$ has $r - 1$ generators, hgt $\mathfrak{p}/\mathfrak{p}_{m-1} \leqq r - 1$, by the induction hypothesis, and

$$\mathfrak{p}/\mathfrak{p}_{m-1} = \mathfrak{p}_0/\mathfrak{p}_{m-1} \supsetneqq \mathfrak{p}_1/\mathfrak{p}_{m-1} \supsetneqq \cdots \supsetneqq \mathfrak{p}_{m-1}/\mathfrak{p}_{m-1}$$

implies $m - 1 \leqq r - 1$. Hence $m \leqq r$. $\square$

Definitions. The spectrum *of a commutative ring is the set of its prime ideals, partially ordered by inclusion. The* Krull dimension *or* dimension dim R *of* R *is the least upper bound of the heights of the prime ideals of* R.

Thus a ring R has dimension at most n if every prime ideal of R has height at most n. Readers will verify that the height of a prime ideal $\mathfrak{p}$ of R is also the dimension of $R_{\mathfrak{p}}$, and that dim $R \geq 1 + \dim R/\mathfrak{p}$ when $\mathfrak{p} \neq \text{Rad}\, 0$.

We now turn to polynomial rings.

Lemma **9.3.** *Let* R *be a domain and let* $\mathfrak{P}$ *be a prime ideal of* $R[X]$. *If* $\mathfrak{P} \cap R = 0$, *then* $\mathfrak{P}$ *has height at most* 1.

Proof. Let Q be the quotient field of R and let $S = R \backslash \{0\}$, so that $S^{-1}R = Q$. Since every $r \in R \backslash 0$ is a unit in Q and in $Q[X]$, 4.2 yields an injective homomorphism $\theta : S^{-1}(R[X]) \longrightarrow Q[X]$, which sends $(a_0 + \cdots + a_n X^n)/r \in S^{-1}(R[X])$ to $(a_0/r) + \cdots + (a_n/r)X^n \in Q[X]$. If $g(X) = q_0 + q_1 X + \cdots + q_n X^n \in Q[X]$, then rewriting $q_0, q_1, \ldots, q_n$ with a common denominator puts g in the form $g = f/r$ for some $f \in R[X]$ and $r \in R$; hence θ is an isomorphism. Thus $S^{-1}(R[X]) \cong Q[X]$ is a PID.

Now, let $\mathfrak{P} \cap R = 0$ and let $0 \neq \mathfrak{Q} \subseteq \mathfrak{P}$ be a prime ideal of $R[X]$. Then $\mathfrak{Q} \cap R = 0$, and $\mathfrak{Q}^E \subseteq \mathfrak{P}^E$ are nonzero prime ideals of the PID $S^{-1}(R[X])$, by 4.5. Hence $\mathfrak{Q}^E$ is a maximal ideal, $\mathfrak{Q}^E = \mathfrak{P}^E$, and $\mathfrak{Q} = \mathfrak{P}$ by 4.5. $\square$

Theorem 9.4. *If R is a Noetherian domain, then* $\dim R[X] = 1 + \dim R$.

Proof. First, (X) is a prime ideal of $R[X]$ and $R[X]/(X) \cong R$; hence $\dim R[X] \geq 1 + \dim R[X]/(X) = \dim R + 1$. In particular, $\dim R[X]$ is infinite when $\dim R$ is infinite, and we may assume that $n = \dim R$ is finite. We prove by induction on n that $\dim R[X] \leq n + 1$. If $n = 0$, then 0 is a maximal ideal of R, R is a field, $R[X]$ is a PID, and $\dim R[X] = 1$. Now, let $n > 0$ and

$$\mathfrak{P}_0 \supsetneqq \mathfrak{P}_1 \supsetneqq \cdots \supsetneqq \mathfrak{P}_m$$

be prime ideals of $R[X]$. We want to show that $m \leq n + 1$.

Since $n \geq 1$ we may assume that $m \geq 2$. We may also assume that $\mathfrak{P}_{m-1} \cap R \neq 0$. Indeed, suppose that $\mathfrak{P}_{m-1} \cap R = 0$. Then $\mathfrak{P}_{m-2} \cap R \neq 0$ by 9.3 and there exists $0 \neq a \in \mathfrak{P}_{m-2} \cap R$. Now, $\mathfrak{P}_{m-2}$ has height at least 2 and is not minimal over (a), by 8.8. Hence $\mathfrak{P}_{m-2}$ properly contains a prime ideal $\mathfrak{Q}$ that is minimal over (a), by 8.6. Then $\mathfrak{P}_{m-2} \supsetneqq \mathfrak{Q} \supsetneqq 0$, with $\mathfrak{Q} \cap R \neq 0$.

Now, $\mathfrak{p} = \mathfrak{P}_{m-1} \cap R$ is a nonzero prime ideal of R. Then $\dim R/\mathfrak{p} \leq \dim R - 1 = n - 1$. By the induction hypothesis, $\dim (R/\mathfrak{p})[X] \leq n$. The projection $R \longrightarrow R/\mathfrak{p}$ induces a surjective homomorphism $R[X] \longrightarrow (R/\mathfrak{p})[X]$ whose kernel is a nonzero prime ideal $\mathfrak{P}$ of $R[X]$, which consists of all $f \in R[X]$ with coefficients in $\mathfrak{p}$. Then $\mathfrak{P} \subseteq \mathfrak{P}_{m-1}$, since $\mathfrak{p} \subseteq \mathfrak{P}_{m-1}$; $\dim R[X]/\mathfrak{P} = \dim (R/\mathfrak{p})[X] \leq n$; and the sequence

$$\mathfrak{P}_0/\mathfrak{P} \supsetneqq \mathfrak{P}_1/\mathfrak{P} \supsetneqq \cdots \supsetneqq \mathfrak{P}_{m-1}/\mathfrak{P}$$

has length $m - 1 \leq n$, so that $m \leq n + 1$. $\square$

Theorem 9.4 implies $\dim R[X_1, ..., X_n] = \dim R + n$ and the following result:

Corollary 9.5. *If K is a field, then* $K[X_1, ..., X_n]$ *has dimension n.*

Exercises

1. Let $\mathfrak{p}$ be a prime ideal of height r in a Noetherian ring R. Show that $\mathfrak{p}$ is minimal over an ideal $\mathfrak{a}$ of R that has r generators. (You may construct $a_1, ..., a_r \in \mathfrak{p}$ so that a prime ideal that is minimal over $Ra_1 + \cdots + Ra_k$ has height k for every $k \leq r$.)

2. Show that $\dim R = \dim R/\mathfrak{a}$ whenever $\mathfrak{a} \subseteq \operatorname{Rad} 0$.

3. Give examples of Noetherian rings of dimension 1.

4. Show that the height of a prime ideal $\mathfrak{p}$ of R is also the dimension of $R_\mathfrak{p}$.

5. Show that $\dim R \geq 1 + \dim R/\mathfrak{p}$ when $\mathfrak{p}$ is a prime ideal and $\mathfrak{p} \neq \operatorname{Rad} 0$.

6. Show that $\dim R = \dim S$ when S is integral over R.

10. Algebraic Sets

This section contains a few initial properties of algebraic sets, with emphasis on overall structure and relationship to ideals.

Definitions. Let K be a field and let $\overline{K}$ be its algebraic closure. The zero set of a set $S \subseteq K[X_1, ..., X_n]$ of polynomials is

$$\mathcal{Z}(S) = \{\, x = (x_1, \ldots, x_n) \in \overline{K}^n \mid f(x) = 0 \text{ for all } f \in S \,\}.$$

An algebraic set *in $\overline{K}^n$ with coefficients in K is the zero set of a set of polynomials $S \subseteq K[X_1, ..., X_n]$.*

The algebraic sets defined above are also called *affine* algebraic sets because they are subsets of the *affine space* $\overline{K}^n$. *Projective* algebraic sets are subsets of the projective space $\overline{P}^n$ on $\overline{K}$ that are defined by sets of homogeneous equations in $n + 1$ variables. They differ from affine sets by their points at infinity.

The straight line $x + y - 4 = 0$ and circle $x^2 + y^2 - 10 = 0$ are algebraic sets in $\mathbb{C}^2$ with coefficients in $\mathbb{R}$. Algebraic sets in $\mathbb{C}^2$ with a single nontrivial equation (the zero sets of single nonconstant polynomials) are *algebraic curves*; they have been studied at some depth, in part because of their relationship to Riemann surfaces. But $\mathbb{C}^2$ contains other kinds of algebraic sets: $\mathcal{Z}(\{\,(X^2 + Y^2 - 10)(X + Y - 4)\,\})$ is the union of two algebraic curves; $\mathcal{Z}(\{\,X^2 + Y^2 - 10,\ X + Y - 4\,\})$ consists of two points, $(1, 3)$ and $(3, 1)$; the empty set $\emptyset = \mathcal{Z}(\{1\})$ and $\mathbb{C}^2 = \mathcal{Z}(\{0\})$ are algebraic sets.

Proposition **10.1.** *Every algebraic set is the zero set of an ideal.*

Proof. Let $S \subseteq K[X_1, ..., X_n]$ and let $\mathfrak{a}$ be the ideal generated by S. If $f(x) = 0$ for all $f \in S$, then $(u_1 f_1 + \cdots + u_m f_m)(x) = 0$ for all $u_1 f_1 + \cdots + u_m f_m \in \mathfrak{a}$; hence $\mathcal{Z}(S) = \mathcal{Z}(\mathfrak{a})$. $\square$

Proposition **10.2.** *Every intersection of algebraic sets is an algebraic set. The union of finitely many algebraic sets is an algebraic set.*

Proof. First, $\bigcap_{i \in I} \mathcal{Z}(S_i) = \mathcal{Z}(\bigcup_{i \in I} S_i)$. Hence $\bigcap_{i \in I} \mathcal{Z}(\mathfrak{a}_i) = \mathcal{Z}(\bigcup_{i \in I} \mathfrak{a}_i)$ $= \mathcal{Z}(\sum_{i \in I} \mathfrak{a}_i)$ for all ideals $(\mathfrak{a}_i)_{i \in I}$ of $K[X_1, ..., X_n]$). Next, let $A = \mathcal{Z}(\mathfrak{a})$ and $B = \mathcal{Z}(\mathfrak{b})$, where $\mathfrak{a}$ and $\mathfrak{b}$ are ideals of $K[X_1, ..., X_n]$. Then $A \cup B \subseteq \mathcal{Z}(\mathfrak{a} \cap \mathfrak{b})$. Conversely, if $x \in \mathcal{Z}(\mathfrak{a} \cap \mathfrak{b})$ and $x \notin A$, then $f(x) \neq 0$ for some $f \in \mathfrak{a}$; but $f(x) g(x) = 0$ for every $g \in \mathfrak{b}$, since $fg \in \mathfrak{a} \cap \mathfrak{b}$; hence $g(x) = 0$ for all $g \in \mathfrak{b}$, and $x \in B$. Thus $A \cup B = \mathcal{Z}(\mathfrak{a} \cap \mathfrak{b})$. $\square$

The proof of Proposition 10.2 shows that sums of ideals yield intersections of algebraic sets, and that finite intersections of ideals yield finite unions of algebraic sets.

The Nullstellensatz. "Nullstellensatz" means "Theorem of Zero Points". The last hundred years have brought stronger and deeper versions. The original version below was proved by Hilbert [1893] in the case $K = \mathbb{C}$.

Theorem **10.3** (Hilbert's Nullstellensatz). *Let* K *be a field, let* $\mathfrak{a}$ *be an ideal of* $K[X_1, ..., X_n]$, *and let* $f \in K[X_1, ..., X_n]$. *If every zero of* f *in* $\overline{K}^n$ *is a zero of* $\mathfrak{a}$, *then* $\mathfrak{a}$ *contains a power of* f.

Proof. Assume that $\mathfrak{a}$ contains no power of f. By 1.4 there is a prime ideal $\mathfrak{p}$ of $K[X_1, ..., X_n]$ that contains $\mathfrak{a}$ but not f. Then $R = K[X_1, ..., X_n]/\mathfrak{p}$ is a domain. The projection $\pi : K[X_1, ..., X_n] \longrightarrow R$ induces an isomorphism $K \cong \pi K \subseteq R$ and a homomorphism $g \longmapsto {}^{\pi}g$ of $K[X_1, ..., X_n]$ into $(\pi K)[X_1, ..., X_n]$. Let $\alpha_1 = \pi X_1, ..., \alpha_n = \pi X_n \in R$. Then $y = {}^{\pi}f(\alpha_1, ..., \alpha_n) = \pi f \neq 0$ in R, since $f \notin \mathfrak{p}$. By 4.10, the homomorphism $\pi K \cong K \subseteq \overline{K}$ extends to a homomorphism ψ of $(\pi K)[\alpha_1, ..., \alpha_n, 1/y] \subseteq Q(R)$ into $\overline{K}$. Then $\psi \pi$ is the identity on K, $(\psi y)(\psi(1/y)) = 1$,

$$f(\psi \alpha_1, ..., \psi \alpha_n) = {}^{\psi \pi}f(\psi \alpha_1, ..., \psi \alpha_n) = \psi y \neq 0, \text{ but}$$

$$g(\psi \alpha_1, ..., \psi \alpha_n) = {}^{\psi \pi}g(\psi \pi X_1, ..., \psi \pi X_n) = \psi \pi g = 0$$

for all $g \in \mathfrak{a} \subseteq \mathfrak{p}$. Thus $(\psi \alpha_1, ..., \psi \alpha_n) \in \overline{K}^n$ is a zero of $\mathfrak{a}$ but not of f. $\square$

The following consequences of Theorem 10.3 make fun exercises.

Corollary **10.4.** *Every proper ideal of* $K[X_1, ..., X_n]$ *has a zero in* $\overline{K}^n$.

Corollary **10.5.** *An ideal* $\mathfrak{m}$ *is a maximal ideal of* $K[X_1, ..., X_n]$ *if and only if* $\mathfrak{m} = \{ f \in K[X_1, ..., X_n] \mid f(x) = 0 \}$ *for some* $x \in \overline{K}^n$.

The main consequence of the Nullstellensatz is that every algebraic set A is the zero set of a unique semiprime ideal, namely

$$\mathfrak{I}(A) = \{ f \in K[X_1, ..., X_n] \mid f(x) = 0 \text{ for all } x \in A \}.$$

Corollary **10.6.** *The mappings* $\mathfrak{I}$ *and* $\mathfrak{Z}$ *induce an order reversing one-to-one correspondence between algebraic sets in* $\overline{K}^n$ *and semiprime ideals of* $K[X_1, ..., X_n]$.

Proof. For every $A \subseteq \overline{K}^n$, $\mathfrak{I}(A)$ is a semiprime ideal, since $f^n(x) = 0$ implies $f(x) = 0$. By definition, $A \subseteq \mathfrak{Z}(\mathfrak{I}(A))$ and $\mathfrak{a} \subseteq \mathfrak{I}(\mathfrak{Z}(\mathfrak{a}))$. If $A = \mathfrak{Z}(\mathfrak{a})$ is an algebraic set, then $\mathfrak{a} \subseteq \mathfrak{I}(A)$, $\mathfrak{Z}(\mathfrak{I}(A)) \subseteq \mathfrak{Z}(\mathfrak{a}) = A$, and $\mathfrak{Z}(\mathfrak{I}(A)) = A$. If $\mathfrak{a}$ is semiprime, then $\mathfrak{I}(\mathfrak{Z}(\mathfrak{a})) \subseteq \operatorname{Rad} \mathfrak{a} = \mathfrak{a}$, by 10.3, and $\mathfrak{I}(\mathfrak{Z}(\mathfrak{a})) = \mathfrak{a}$. $\square$

Algebraic sets do not correspond as nicely to ideals of $K[X_1, ..., X_n]$ when the zero set is defined as a subset of K^n rather than $\overline{K}^n$. For instance, if $K = \mathbb{R}$, then, in $\mathbb{R}^2$, the curve $x^2 + y^2 - 1 = 0$ is a circle; the curve $x^2 + y^2 + 1 = 0$ is empty, and is the zero set of widely different ideals. But both are circles in $\mathbb{C}^2$.

Algebraic varieties. In the Noetherian ring $K[X_1, ..., X_n]$, every ideal $\mathfrak{a}$ is a reduced intersection $\mathfrak{a} = \mathfrak{q}_1 \cap \cdots \cap \mathfrak{q}_r$ of primary ideals $\mathfrak{q}_1, ..., \mathfrak{q}_r$ with unique radicals $\mathfrak{p}_1, ..., \mathfrak{p}_r$ (Theorem 1.10). If $\mathfrak{a}$ is semiprime, then taking radicals yields $\mathfrak{a} = \operatorname{Rad} \mathfrak{a} = \mathfrak{p}_1 \cap \cdots \cap \mathfrak{p}_r$; thus every semiprime ideal is an irredundant finite intersection of unique prime ideals. By Corollary 10.6, every algebraic sets is an irredundant finite union of unique algebraic sets defined by prime ideals.

Definition. An algebraic set $A \subseteq \overline{K}^n$ is irreducible, or is an algebraic variety, when $A \neq \emptyset$ and A is the zero set of a prime ideal. $\square$

Equivalently, A is an algebraic variety when $A \neq \emptyset$ and A is not the union of two nonempty algebraic sets $B, C \subsetneqq A$. For instance, $\emptyset$ and $\overline{K}^n$ are algebraic varieties; so is any single point of $\overline{K}^n$, by Corollary 10.5. Algebraic varieties $A \subseteq \overline{K}^n$ are also called *affine* algebraic varietys.

Corollary 10.7. The mappings $\mathfrak{I}$ and $\mathfrak{Z}$ induce an order reversing one-to-one correspondence between algebraic varieties in $\overline{K}^n$ and prime ideals of $K[X_1, ..., X_n]$.

Corollary 10.8. Every algebraic set in $\overline{K}^n$ is uniquely an irredundant finite union of algebraic varieties.

Algebraic geometry can now focus on algebraic varieties, equivalently, on prime ideals of $K[X_1, ..., X_n]$.

The height of prime ideals in $K[X_1, ..., X_n]$ (Theorem 9.4) yields a dimension for algebraic varieties:

Definition. The dimension $\dim A$ of an algebraic variety $A \subseteq \overline{K}^n$ is the length of the longest strictly decreasing sequence $A = A_0 \supsetneqq A_1 \supsetneqq \cdots \supsetneqq A_r$ of nonempty algebraic varieties contained in A; equivalently, $n - \mathrm{hgt}\, \mathfrak{I}(A)$.

For instance, single points of $\overline{K}^n$ have dimension 0; straight lines and (non-degenerate) circles in $\mathbb{C}^2$ have dimension 1; irreducible algebraic surfaces in $\mathbb{C}^3$ have dimension 2; and $\mathbb{C}^3$ itself has dimension 3.

The Zariski topology. It turns out that dimension and Corollary 10.8 are purely topological phenomena. Since $\emptyset$ and $\overline{K}^n$ are algebraic sets, Proposition 10.2 implies that the algebraic sets of $\overline{K}^n$ are the closed sets of a topology on $\overline{K}^n$. Zariski [1944] originated this type of topology; Weil [1952] applied it to algebraic sets.

Definition. The Zariski topology on an algebraic set $A \subseteq \overline{K}^n$ is the topology whose closed sets are the algebraic sets $B \subseteq A$.

Equivalently, the *Zariski topology* on an algebraic set $A \subseteq \overline{K}^n$ is the topology induced on A by the Zariski topology on the algebraic set $\overline{K}^n$. With this topology, an algebraic set $A \subseteq \overline{K}^n$ is *Noetherian* in the sense that its open sets satisfy the ascending chain condition. Readers will verify that the following properties hold in every Noetherian topological space: every closed subset is uniquely an irredundant union of irreducible closed subsets; every closed subset C has a dimension, which is the least upper bound of the lengths of strictly decreasing sequences $A = C_0 \supsetneqq C_1 \supsetneqq \cdots \supsetneqq C_m$ of closed subsets.

By Corollary 10.5 the points of $\overline{K}^n$ correspond to the maximal ideals of $K[X_1, ..., X_n]$. The exercises extend the Zariski topology to the entire spectrum of $K[X_1, ..., X_n]$, in fact, to the spectrum of any commutative ring.

Exercises

1. Let K be a field. Show that every proper ideal of $K[X_1, ..., X_n]$ has a zero in $\overline{K}^n$.

2. Let K be a field. Show that $\mathfrak{m}$ is a maximal ideal of $K[X_1, ..., X_n]$ if and only if $\mathfrak{m} = \{ f \in K[X_1, ..., X_n] \mid f(x) = 0 \}$ for some $x \in \overline{K}^n$.

3. Let K be a field. Show that the prime ideals of height 1 in $K[X_1, ..., X_n]$ are the principal ideals generated by irreducible polynomials.

4. Show that $\overline{K}^n$ is compact (though not Hausdorff) in the Zariski topology.

5. Prove the following: in a Noetherian topological space, every closed subset is uniquely an irredundant union of irreducible closed subsets. (A closed subset C is *irreducible* when $C \neq \emptyset$ and C is not the union of nonempty closed subsets $A, B \subsetneq C$.)

In the following exercises, R is any commutative ring.

6. Show that the sets $\{ \mathfrak{p} \mid \mathfrak{p}$ is a prime ideal of R and $\mathfrak{p} \supseteq \mathfrak{a} \}$, where $\mathfrak{a}$ is an ideal of R, are the closed sets of a topology (the *Zariski topology*) on the spectrum of R.

7. Verify that the Zariski topology on the spectrum of $K[X_1, ..., X_n]$ induces the Zariski topology on $\overline{K}^n$ when the elements of $\overline{K}^n$ are identified with the maximal ideals of $K[X_1, ..., X_n]$.

8. Prove the following: when $R = K[X]$, where K is an algebraically closed field, a proper subset of the spectrum is closed in the Zariski topology if and only if it is finite.

9. Show that the sets $\{ \mathfrak{p} \mid \mathfrak{p}$ is a prime ideal of R and $a \notin \mathfrak{p} \}$, where $a \in R$, constitute a basis of open sets of the Zariski topology on the spectrum of R.

10. Show that the spectrum of R is compact in the Zariski topology.

11. Regular Mappings

In this section we define isomorphisms of algebraic varieties, and construct, for every algebraic variety A, a ring $C(A)$ that determines A up to isomorphism. This recasts algebraic geometry as the study of suitable rings.

The coordinate ring. We begin with $C(A)$ and define isomorphisms later. Let $A \subseteq \overline{K}^n$ be an algebraic set. Every polynomial $f \in K[X_1, ..., X_n]$ induces a polynomial mapping $x \longmapsto f(x)$, also denoted by f, and a mapping $f_{|A}$: $x \longmapsto f(x)$ of A into $\overline{K}$. For instance, every $a \in K$ induces a constant mapping $x \longmapsto a$ on A, which may be identified with a; $X_i \in K[X_1, ..., X_n]$ induces the *coordinate function* $(x_1, ..., x_n) \longmapsto x_i$.

Definitions. Let $A \subseteq \overline{K}^n$ be an algebraic set. A polynomial function of A is a mapping of A into $\overline{K}$ that is induced by a polynomial $f \in K[X_1, ..., X_n]$. The coordinate ring of A is the ring $C(A)$ of all such mappings.

The operations on $C(A)$ are pointwise addition and multiplication. The elements of K may be identified with constant functions, and then $C(A)$ becomes a ring extension of K. Then $C(A)$ is generated over K by the coordinate functions, whence its name.

Proposition 11.1. *If K is a field and $A \subseteq \overline{K}^n$ is an affine algebraic set, then $C(A) \cong K[X_1, ..., X_n]/\mathfrak{I}(A)$. Hence $C(A)$ is a commutative, finitely generated ring extension of K; its nilradical is 0; if A is an algebraic variety, then $C(A)$ is a domain. Conversely, every commutative, finitely generated ring extension R of K with trivial nilradical is isomorphic to the coordinate ring of an affine algebraic set; if R is a domain, then R is isomorphic to the coordinate ring of an algebraic variety.*

Ring extensions of a field K are also known as *K-algebras* (see Chapter XIII). Commutative, finitely generated ring extensions of K with zero nilradical are also called *affine rings* over K.

Proof. Two polynomials $f, g \in K[X_1, ..., X_n]$ induce the same polynomial function on A if and only if $f(x) - g(x) = 0$ for all $x \in A$, if and only if $f - g \in \mathfrak{I}(A)$. Therefore $C(A) \cong K[X_1, ..., X_n]/\mathfrak{I}(A)$. In particular, $C(A)$ is generated, as a ring extension of K, by the coordinate functions of A; the nilradical $\mathrm{Rad}\, 0$ of $C(A)$ is trivial, since $\mathfrak{I}(A)$ is a semiprime ideal; if A is an algebraic variety, then $\mathfrak{I}(A)$ is a prime ideal and $C(A)$ is a domain.

Conversely, let R be a commutative ring extension of K with trivial nilradical, which is finitely generated, as a ring extension of K, by some $r_1, ..., r_n \in R$. By the universal property of $K[X_1, ..., X_n]$ there is a homomorphism $\varphi : K[X_1, ..., X_n] \longrightarrow R$ that sends $X_1, \ldots, X_n$ to $r_1, \ldots, r_n$. Since R is generated by $r_1, \ldots, r_n$, φ is surjective, and $R \cong K[X_1, ..., X_n]/\mathrm{Ker}\, \varphi$. Moreover, $\mathrm{Ker}\, \varphi$ is a semiprime ideal, since $\mathrm{Rad}\, 0 = 0$ in R. Hence $\mathrm{Ker}\, \varphi = \mathfrak{I}(A)$ for some algebraic set $A = \mathcal{Z}(\mathrm{Ker}\, \varphi) \subseteq \overline{K}^n$, and then $R \cong C(A)$. If R is a domain, then $\mathrm{Ker}\, \varphi$ is a prime ideal and A is an algebraic variety. $\square$

The points of A correspond to the maximal ideals of $C(A)$:

Proposition 11.2. *Let A be an algebraic set. For every $x \in A$ let $\mathfrak{m}_x = \{ f \in C(A) \mid f(x) = 0 \}$. The mapping $x \longmapsto \mathfrak{m}_x$ is a bijection of A onto the set of all maximal ideals of $C(A)$.*

Proof. By 10.5 the mapping $x \longmapsto \mathfrak{I}(\{x\}) = \{ f \in K[X_1, ..., X_n] \mid f(x) = 0 \}$ is a bijection of $\overline{K}^n$ onto the set of all maximal ideals of $K[X_1, ..., X_n]$. By 10.6, $x \in A = \mathcal{Z}(\mathfrak{I}(A))$ if and only if $\mathfrak{I}(A) \subseteq \mathfrak{I}(\{x\})$. Hence $x \longmapsto \mathfrak{I}(\{x\})$ is a bijection of A onto the set of all maximal ideals $\mathfrak{m} \supseteq \mathfrak{I}(A)$ of $K[X_1, ..., X_n]$. The latter correspond to the maximal ideals of $K[X_1, ..., X_n]/\mathfrak{I}(A)$ and to the maximal ideals of $C(A)$. The composite bijection is $x \longmapsto \mathfrak{m}_x$. $\square$

If A is a variety, then the domain $C(A)$ has a quotient field Q, which is a finitely generated field extension of K; the elements of Q are *rational functions* of A into $\overline{K}$. Valuations on Q are a flexible way to define "locations" on A;

the exercises give some details when A is an algebraic curve.

Proposition **11.3.** *For every algebraic variety* A, $\dim A = \dim C(A)$.

It is known that the dimension of $C(A)$ also equals the transcendence degree (over K) of its quotient field. Section XII.9 gives a third approach to dimension.

Proof. First, $C(A)$ is Noetherian, by III.11.4. By 10.7, $\dim A$ is the length of the longest strictly descending sequence $\mathfrak{P}_0 \gneqq \mathfrak{P}_1 \gneqq \cdots \gneqq \mathfrak{P}_m = \mathfrak{I}(A)$ of prime ideals of $K[X_1, ..., X_n]$. As in the proof of 11.2, these correspond to strictly descending sequences of prime ideals of $C(A)$. $\square$

Propositions 11.2 and 11.3 describe the points and dimension of A in a way that depends only on the ring $C(A)$ and not on the embedding of A in $\overline{K}^n$.

Regular mappings. By 11.2, isomorphisms $C(A) \cong C(B)$ induce bijections $A \longrightarrow B$. We investigate these bijections.

Definition. Let K be a field and let $A \subseteq \overline{K}^m$, $B \subseteq \overline{K}^n$ be algebraic varieties. A mapping $F = (f_1, ..., f_n) \colon A \longrightarrow B \subseteq \overline{K}^n$ is regular when its components $f_1, ..., f_n \colon A \longrightarrow \overline{K}$ are polynomial functions.

Regular mappings are also called *polynomial mappings, morphisms* of algebraic varieties, and, in older texts, *rational transformations*. For example, let $A \subseteq \mathbb{C}^2$ be the parabola $y = x^2$ and let $B \subseteq \mathbb{C}^2$ be the x-axis; the usual projection $A \longrightarrow B$ is a regular mapping, since it may be described by polynomials as $(x, y) \longmapsto (x, y - x^2)$. Regular mappings into $\overline{K}$ are the same as polynomial functions.

Readers will verify that regular mappings compose: if $F : A \longrightarrow B$ and $G : B \longrightarrow C$ are regular, then $G \circ F : A \longrightarrow C$ is regular. In particular, when $F : A \longrightarrow B$ is regular and $g : B \longrightarrow \overline{K}$ is a polynomial mapping of B, then $g \circ F : A \longrightarrow \overline{K}$ is a polynomial mapping of A. This defines a mapping $C(F) \colon C(B) \longrightarrow C(A)$, which preserves pointwise addition, pointwise multiplication, and constants. Thus $C(F)$ is a homomorphism of K-algebras (in this case, a homomorphism of rings that is the identity on K).

Proposition **11.4.** *Let K be a field and let $A \subseteq \overline{K}^m$, $B \subseteq \overline{K}^n$ be algebraic varieties. Every regular mapping F of A into B induces a homomorphism $C(F)$ of $C(B)$ into $C(A)$. Conversely, every homomorphism of $C(B)$ into $C(A)$ is induced by a unique regular mapping of A into B.*

Proof. Let $\varphi : C(B) \longrightarrow C(A)$ be a homomorphism of K-algebras. Let $q_j :$ $B \longrightarrow K$, $(y_1, ..., y_n) \longmapsto y_j$ be the jth coordinate function of B. The polynomial function $B \longrightarrow \overline{K}$ induced by $g = \sum_k c_k Y_1^{k_1} \cdots Y_n^{k_n} \in K[Y_1, ..., Y_n]$ sends $y = (y_1, ..., y_n) \in B$ to

$$g(y_1, ..., y_n) = g(q_1(y), ..., q_n(y))$$
$$= \sum_k c_k \, q_1(y)^{k_1} \cdots q_n(y)^{k_n} = \left(\sum_k c_k \, q_1^{k_1} \cdots q_n^{k_n} \right)(y)$$

and coincides with $g(q_1, \ldots, q_n)$ as calculated in $C(B)$.

Now, $\overline{\varphi} = (\varphi(q_1), \ldots, \varphi(q_n))$ is a regular mapping of A into $\overline{K}^n$, since $\varphi(q_j) \in C(A)$ for all j. If $x \in A$ and $g = \sum_k c_k Y_1^{k_1} \cdots Y_n^{k_n} \in \mathfrak{I}(B)$, then $g(y) = 0$ for all $y \in B$, $\sum_k c_k q_1^{k_1} \cdots q_n^{k_n} = 0$ in $C(B)$ by the above, and

$$
\begin{aligned}
g\big(\overline{\varphi}(x)\big) &= g\big(\varphi(q_1)(x), \ldots, \varphi(q_n)(x)\big) \\
&= \sum_k c_k \, \varphi(q_1)(x)^{k_1} \cdots \varphi(q_n)(x)^{k_n} \\
&= \big(\sum_k c_k \, \varphi(q_1)^{k_1} \cdots \varphi(q_n)^{k_n}\big)(x) \\
&= \varphi\big(\sum_k c_k \, q_1^{k_1} \cdots q_n^{k_n}\big)(x) = 0,
\end{aligned}
$$

since φ is a homomorphism. Hence $\overline{\varphi}(x) \in \mathcal{Z}(\mathfrak{I}(B)) = B$. Thus $\overline{\varphi}$ is a regular mapping of A into B. Similarly, $g\big(\overline{\varphi}(x)\big) = \varphi\big(g(q_1, \ldots, q_n)\big)(x)$ for every $g \in K[Y_1, \ldots Y_n]$ and $x \in A$; thus φ is the homomorphism induced by $\overline{\varphi}$.

Finally, let $F : A \longrightarrow B$ be a regular mapping. If F induces φ, then $q_j \circ F = \varphi(q_j) = q_j \circ \overline{\varphi}$ for all j, and $F = \overline{\varphi}$, since $F(x)$ and $\overline{\varphi}(x)$ have the same coordinates for every $x \in A$. $\square$

Definition. *Two algebraic varieties A and B are* isomorphic *when there exist mutually inverse regular bijections $A \longrightarrow B$ and $B \longrightarrow A$.*

For example, in $\mathbb{C}^2$, the parabola $A : y = x^2$ and x-axis $B : y = 0$ are isomorphic, since the regular mappings $(x, y) \longmapsto (x, y - x^2)$ of A onto B and $(x, y) \longmapsto (x, x^2)$ of B onto A are mutually inverse bijections. Similarly, A and $\mathbb{C}$ are isomorphic. In general, Proposition 11.4 yields the following result:

*Proposition **11.5**. Over a given algebraically closed field, two algebraic varieties are isomorphic if and only if their coordinate rings are isomorphic.*

Isomorphisms of algebraic varieties preserve their structure as algebraic sets: for instance, isomorphic varieties are homeomorphic (with their Zariski topologies) and their algebraic subsets are organized in the same fashion (see the exercises). Much as abstract algebra studies groups and rings only up to isomorphism, disregarding the nature of their elements, so does algebraic geometry study algebraic varieties only up to isomorphism, without much regard for their embeddings into affine spaces. This redefines algebraic geometry as the study of affine rings and domains.

Exercises

1. Let $F : A \longrightarrow B$ and $G : B \longrightarrow C$ be regular mappings. Verify that $G \circ F : A \longrightarrow C$ is a regular mapping.

2. Show that every regular mapping $A \longrightarrow B$ is continuous (in the Zariski topologies).

3. Let A be an algebraic set over an algebraically closed field K. The Zariski topology on the spectrum of $C(A)$ induces a topology on the set $\mathcal{M}$ of all maximal ideals of $C(A)$. Show that A and $\mathcal{M}$ are homeomorphic.

4. The algebraic subsets of an algebraic variety A constitute a partially ordered set when ordered by inclusion. Show that isomorphic algebraic varieties have isomorphic partially ordered sets of algebraic subsets.

5. Show that every straight line in $\overline{K}^2$ is isomorphic to $\overline{K}$.

6. Let A be an algebraic curve through the origin, with equation $F(x, y) = 0$, where $F \in \mathbb{C}[X, Y]$ is irreducible and $F(0, 0) = 0$. Let $y = s(x)$ be a power series solution of $F(x, y) = 0$ as in Section VI.9, where $s \in \mathbb{C}[[X]]$ and $s(0) = 0$. Let $f, g \in \mathbb{C}[X, Y]$ induce the same polynomial function on A. Show that the series $f(x, s(x))$ and $g(x, s(x))$ have the same order; hence there is a discrete valuation v on the quotient field of $C(A)$ such that $v(f/g) = 2^{\operatorname{ord} g - \operatorname{ord} f}$ when $f, g \neq 0$ in $C(A)$.

7. Let R be a domain with quotient field Q. Prove the following: if $x \in Q$ belongs to $R_{\mathfrak{m}} \subseteq Q$ for every maximal ideal $\mathfrak{m}$ of R, then $x \in R$.

8. Let $A \subseteq K^n$ be an algebraic variety, where K is algebraically closed. Call a function $f : A \longrightarrow K$ *rational* when every $x \in A$ has an open neighborhood $V \subseteq A$ on which f is induced by a rational fraction $g/h \in K(X)$ (for all $y \in V$, $h(y) \neq 0$ and $f(y) = g(y)/h(y)$). Show that the rational functions of A constitute a domain R. Then show that $R = C(A)$. (You may want to show that $R \subseteq C(A)_{\mathfrak{m}_x}$ for every $x \in A$, then use the previous exercise.)

VIII
Modules

"Module" is a nineteenth century name for abelian groups; an abelian group on which a ring R acts thus became an R-module. The usefulness of this concept was quickly appreciated; the major results of this chapter already appear in van der Waerden's *Moderne Algebra* [1930]. Modules have gained further importance with the development of homological algebra.

This chapter contains basic properties of modules, submodules, homomorphisms, direct products and sums, and free modules; the structure theorem for finitely generated modules over PIDs; and its applications to abelian groups and linear algebra. The last section may be skipped. Mild emphasis on free modules and bases should help readers compare modules to vector spaces.

1. Definition

We saw that groups act on sets. Rings, not inclined to be left behind, act on abelian groups; the resulting structures are modules.

Definitions. Let R be a ring (not necessarily with an identity element). A left R-module is an abelian group M together with a left action $(r, x) \longmapsto rx$ of R on M, the left R-module structure *on M, such that*

(1) $r(sx) = (rs)x$, *and*

(2) $(r + s)x = rx + sx$, $r(x + y) = rx + ry$

for all $r \in R$ and $x, y \in M$. If R has an identity element, then a left R-module M is unital *when*

(3) $1x = x$ *for all $x \in M$.*

The notation $_R M$ generally indicates that M is a left R-module.

Readers who encounter modules for the first time should keep in mind the following three basic examples.

A vector space over a field K is (exactly) a unital left K-module.

Every abelian group A is a unital $\mathbb{Z}$-module, in which nx is the usual integer multiple, $nx = x + x + \cdots + x$ when $n > 0$.

Every ring R acts on itself by left multiplication. This makes R a left R-module, denoted by ${}_R R$ to distinguish it from the ring R. If R has an identity element, then ${}_R R$ is unital.

The last two examples show that modules are more general and complex than vector spaces, even though their definitions are similar.

Some elementary properties of vector spaces hold in every left R-module M: $r0 = 0$, $0x = 0$, $(r - s)x = rx - sx$, and $r(x - y) = rx - ry$, for all $r \in R$ and $x, y \in M$. Left R-modules are also tailored for the formation of *linear combinations* $r_1 x_1 + \cdots + r_n x_n$ with coefficients in R. But, in a module, $r_1 x_1 + \cdots + r_n x_n = 0$ does not make x_1 a linear combination of $x_2, \ldots, x_n$ when $r_1 \neq 0$ (see the exercises).

More generally, (apparently) infinite sums are defined in any left R-module M, as in any abelian group: $\sum_{i \in I} x_i$ is defined in M when $x_i = 0$ for almost all $i \in I$ (when $\{ i \in I \mid x_i \neq 0 \}$ is finite), and then $\sum_{i \in I} x_i = \sum_{i \in I, \, x_i \neq 0} x_i$ (as in Section III.1). *Infinite linear combinations* are defined similarly: $\sum_{i \in I} r_i x_i$ is defined in a left R-module M when $r_i x_i = 0$ for almost all $i \in I$ (for instance, when $r_i = 0$ for almost all i, or when $x_i = 0$ for almost all i) and then $\sum_{i \in I} r_i x_i = \sum_{i \in I, \, r_i x_i \neq 0} r_i x_i$.

Equivalent definition. Module structures on an abelian group A can also be defined as ring homomorphisms. Recall (Proposition III.1.1) that, when A is an abelian group, the endomorphisms of A (written on the left) constitute a ring $\mathrm{End}_{\mathbb{Z}}(A)$ with an identity element, under pointwise addition and composition. The notation $\mathrm{End}_{\mathbb{Z}}(A)$ specifies that we regard A as a mere abelian group (a $\mathbb{Z}$-module), not as a module over some other ring.

Proposition 1.1. Let A be an abelian group and let R be a ring. There is a one-to-one correspondence between left R-module structures $R \times A \longrightarrow A$ on A and ring homomorphisms $R \longrightarrow \mathrm{End}_{\mathbb{Z}}(A)$; and unital left R-module structures correspond to homomorphisms of rings with identity.

Proof. Let A is an R-module. The action α_r of $r \in R$ on A, $\alpha_r(x) = rx$, is an endomorphism of A, since $r(x + y) = rx + ry$ for all $x, y \in A$. Then $r \longmapsto \alpha_r$ is a ring homomorphism of R into $\mathrm{End}_{\mathbb{Z}}(A)$, since $r(sx) = (rs)x$ and $(r + s)x = rx + sx$ ($\alpha_r \circ \alpha_s = \alpha_{rs}$ and $\alpha_r + \alpha_s = \alpha_{r+s}$) for all $r, s \in R$ and $x \in A$.

Conversely, if $\alpha : R \longrightarrow \mathrm{End}_{\mathbb{Z}}(A)$ is a ring homomorphism, then the action $rx = \big(\alpha(r)\big)(x)$ of R on A is an R-module structure on A: for all $r, s \in R$ and $x, y \in A$, $r(x + y) = rx + ry$ holds since $\alpha(r)$ is an endomorphism, and $r(sx) = (rs)x$, $(r + s)x = rx + sx$ hold since $\alpha(rs) = \alpha(r) \circ \alpha(s)$ and $\alpha(r + s) = \alpha(r) + \alpha(s)$; and then $\alpha_r = \alpha(r)$ for all $r \in R$. Then A is a unital R-module if and only if $1x = x$ for all $x \in A$, if and only if $\alpha(1)$ is the identity on A, which is the identity element of $\mathrm{End}_{\mathbb{Z}}(A)$. $\square$

We note two consequences of Proposition 1.1; the exercises give others. For every R-module M, the kernel of $R \longrightarrow \text{End}_{\mathbb{Z}}(M)$ is an ideal of R:

Definitions. The annihilator *of a left R-module M is the ideal* $\text{Ann}(M) = \{\, r \in R \mid rx = 0 \text{ for all } x \in M \,\}$ *of R. A left R-module is* faithful *when its annihilator is* 0.

By Propositions III.1.7, III.1.8, every ring R can be embedded into a ring with identity R^1 such that every ring homomorphism of R into a ring S with identity extends uniquely to a homomorphism of rings with identity of R^1 into S (that sends the identity element of R^1 to the identity element of S). By 1.1:

Corollary **1.2.** *Let A be an abelian group and let R be a ring. There is a one-to-one correspondence between left R-module structures on A and unital left* R^1*-module structures on A.* □

Thus every left R-module can be made uniquely into a unital left R^1-module. Consequently, the *general* study of modules can be limited to unital modules (over rings with identity). We do so in later sections.

Right modules. Conservative minded rings prefer to act on the right:

Definitions. Let R be a ring. A right R-module *is an abelian group M together with a right action* $(r, x) \longmapsto rx$ *of R on M such that*

(1*) $(xr)s = x(rs)$, *and*

(2*) $x(r + s) = xr + xs$, $(x + y)r = xr + yr$

for all $r \in R$ *and* $x, y \in M$. *If R has an identity element, a right R-module M is* unital *when*

(3*) $x1 = x$ *for all* $x \in M$.

For example, every ring R acts on itself on the right by right multiplication; this makes R a right R-module R_R. The notation M_R generally indicates that M is a right R-module. The relationship between right module structures and ring homomorphisms is explained in the exercises.

The following construction reduces right modules to left modules. The multiplication on a ring R has an *opposite* multiplication $*$, $r * s = sr$, that, together with the addition on R, satisfies all ring axioms.

Definition. The opposite ring R^{op} *of a ring R has the same underlying set and addition as R, and the opposite multiplication.*

Proposition **1.3.** *Every right R-module is a left* R^{op}*-module, and conversely. Every unital right R-module is a unital left* R^{op}*-module, and conversely.*

Proof. Let M be a right R-module. Define a left action of R^{op} on M by $rx = xr$, for all $r \in R$ and $x \in M$. Then $r(sx) = (xs)r = x(sr) = (sr)x = (r * s)x$, and (1) follows from (1*) (once the multiplication on R has been reversed). Axioms (2) and, in the unital case, (3) follow from (2*) and (3*). Thus M is a left R^{op}-module. The converse is similar. □

If R is commutative, then $R^{op} = R$ and 1.3 becomes 1.4 below; then we usually refer to left or right R-modules as just R-modules.

Corollary **1.4.** *If R is commutative, then every left R-module is a right R-module, and conversely.*

Submodules are subsets that inherit a module structure.

Definition. A submodule *of a left R-module M is an additive subgroup A of M such that $x \in A$ implies $rx \in A$ for all $r \in R$.*

A submodule of a right R-module M is an additive subgroup A of M such that $x \in A$ implies $xr \in A$ for all $r \in R$. The relation "A is a submodule of M" is often denoted by $A \leqq M$. Then the addition and action of R on M induce an addition and action of R on A, under which A is a (left or right) R-module; this module is also called a *submodule* of M.

For example, $0 = \{0\}$ and M itself are submodules of M. The submodules of a vector space are its subspaces. The submodules of an abelian group ($\mathbb{Z}$-module) are its subgroups (since they are closed under integer multiplication).

Definition. A left ideal *of a ring R is a submodule of $_R R$. A right ideal *of a ring R is a submodule of R_R.*

An ideal of R is a subset that is both a left ideal and a right ideal; hence the ideals of R are often called *two-sided* ideals. If R is commutative, then ideals, left ideals, and right ideals all coincide.

Submodules have a number of unsurprising properties, whose boring proofs we happily dump on our readers.

Proposition **1.5.** *Let M be a module. Every intersection of submodules of M is a submodule of M. The union of a nonempty directed family of submodules of M is a submodule of M.* $\square$

By 1.5 there is for every subset S of a module M a smallest submodule of M that contains S.

Definition. In a module M, the smallest submodule of M that contains a subset S of M is the submodule of M generated by S. $\square$

This submodule can be described as follows.

Proposition **1.6.** *Let M be a unital left R-module. The submodule of M generated by a subset S of M is the set of all linear combinations of elements of S with coefficients in R.*

In particular, when M is a unital left R-module, the submodule of M generated by a finite subset $\{a_1, \ldots, a_n\}$ of elements of M is the set of all linear combinations $r_1 a_1 + \cdots + r_n a_n$ with $r_1, \ldots, r_n \in R$; such a submodule is *finitely generated*. The *cyclic* submodule of M generated by a single element x of M is the set $Rx = \{rx \mid r \in R\}$. These descriptions become more complicated when M is not unital (see the exercises).

The *sum* of a family $(A_i)_{i\in I}$ of submodules is their sum as subgroups:

$$\sum_{i\in I} A_i = \{\sum_{i\in I} a_i \mid a_i \in A_i \text{ for all } i, \text{ and } a_i = 0 \text{ for almost all } i\};$$

equivalently, $\sum_{i\in I} A_i$ is the submodule generated by the union $\bigcup_{i\in I} A_i$.

Proposition 1.7. *A sum of submodules of a module M is a submodule of M.*

The last operation on submodules is multiplication by left ideals.

Definition. Let A be a submodule of a left R-module M and let L be a left ideal of R. The product LA is the set of all linear combinations of elements of A with coefficients in L. □

The notation LA is traditional. Readers will remember that LA is *not* the set product $\{\ell a \mid \ell \in L, a \in A\}$ of L and A; rather, it is the submodule generated by the set product of L and A.

Proposition 1.8. *Let M be a left R-module, let A, $(A_i)_{i\in I}$ be submodules of M, and let L, L', $(L_i)_{i\in I}$ be left ideals of R. The product LA is a submodule of M. Moreover:*

(1) $LA \subseteq A$;

(2) $L(L'A) = (LL')A$;

(3) $L(\sum_{i\in I} A_i) = \sum_{i\in I}(LA_i)$ and $(\sum_{i\in I} L_i)A = \sum_{i\in I}(L_iA)$.

In particular, the product of two left ideals of R is a left ideal of R.

Exercises

1. Verify that the equalities $r0 = 0$ and $(r-s)x = rx - sx$ hold in every left R-module.

2. Show that $rx = 0$ may happen in a module even when $r \neq 0$ and $x \neq 0$.

3. Show that the endomorphisms of an abelian group A constitute a ring with an identity element, under pointwise addition and composition.

4. Show that every abelian group has a unique unital left $\mathbb{Z}$-module structure.

5. Let $n > 0$ and let A be an abelian group. When does there exist a unital left $\mathbb{Z}_n$-module structure on A? and, if so, is it unique?

6. Let A be an abelian group. When does there exist a unital left $\mathbb{Q}$-module structure on A? and, if so, is it unique?

7. Let $\varphi : R \longrightarrow S$ be a homomorphism of rings with identity and let A be a unital left S-module. Make A a unital left R-module.

8. Let M be a unital left R-module and let I be a two-sided ideal of R such that $I \subseteq \text{Ann}(M)$. Make M a unital R/I-module. Formulate and prove a converse.

9. Let V be a vector space and let $T : V \longrightarrow V$ be a linear transformation. Make V a unital $K[X]$-module in which $Xv = T(v)$ for all $v \in V$. Formulate and prove a converse.

10. Fill in the details in the following. Let A be an abelian group. The ring of endomorphisms of A, written on the right, is the opposite ring $\text{End}_{\mathbb{Z}}(A)^{\text{op}}$ of its ring of endomorphisms $\text{End}_{\mathbb{Z}}(A)$ written on the left. Right R-module structures on an abelian

group A correspond to ring homomorphisms of R into this ring $\text{End}_{\mathbb{Z}}(A)^{\text{op}}$. This provides another proof of Proposition 1.3.

11. Let M be a left R-module. Show that M has the same submodules as an R-module and as an R^1-module.

12. Show that every intersection of submodules of a module M is a submodule of M.

13. Show that the union of a nonempty directed family of submodules of a module M is a submodule of M.

14. Let S be a subset of a unital left R-module M. Show that the submodule of M generated by S is the set of all linear combinations of elements of S with coefficients in R.

15. Let S be a subset of a not necessarily unital left R-module M. Describe the submodule of M generated by S.

16. Show that every sum of submodules of a module M is a submodule of M.

17. Show that $L(L'A) = (LL')A$, whenever A is a submodule of a left R-module M and L, L' are left ideals of R.

18. Show that $L\sum_{i \in I} A_i = \sum_{i \in I} (LA_i)$, whenever $(A_i)_{i \in I}$ are submodules of a left R-module M and L is a left ideal of R.

19. Show that $\left(\sum_{i \in I} L_i\right)A = \sum_{i \in I} (L_i A)$, whenever A is a submodule of a left R-module M and $(L_i)_{i \in I}$ are left ideals of R.

20. Let A, B, C be submodules of some module. Prove the following: if $A \subseteq C$, then $(A + B) \cap C = A + (B \cap C)$.

2. Homomorphisms

Module homomorphisms are homomorphisms of additive groups, that also preserve the ring action.

Definition. Let A and B be left R-modules. A homomorphism $\varphi : A \longrightarrow B$ of left R-modules is a mapping $\varphi : A \longrightarrow B$ such that $\varphi(x + y) = \varphi(x) + \varphi(y)$ and $\varphi(rx) = r\,\varphi(x)$, for all $x, y \in A$ and $r \in R$.

A homomorphism of right R-modules is a mapping φ such that $\varphi(x + y) = \varphi(x) + \varphi(y)$ and $\varphi(xr) = \varphi(x)\,r$, for all x, y, r. Module homomorphisms preserve all sums and linear combinations: $\varphi\left(\sum_i r_i x_i\right) = \sum_i (r_i\,\varphi(x_i))$, whenever $r_i x_i = 0$ for almost all i. Homomorphisms of vector spaces are also called *linear transformations*.

An *endomorphism* of a module M is a module homomorphism of M into M. Injective module homomorphisms are also called *monomorphisms*; surjective module homomorphisms are also called *epimorphisms*. An *isomorphism* of modules is a homomorphism of modules that is bijective; then the inverse bijection is also an isomorphism.

Properties. Homomorphisms of modules inherit all the felicitous properties of homomorphisms of abelian groups. We show this for left R-modules; homomorphisms of right R-modules have the same properties, since a homomorphism of right R-modules is also a homomorphism of left R^{op}-modules.

The identity mapping 1_M on any left R-module M is a module homomorphism. Module homomorphisms compose: if $\varphi : A \longrightarrow B$ and $\psi : B \longrightarrow C$ are homomorphisms of left R-modules, then $\psi \circ \varphi : A \longrightarrow C$ is a homomorphism of left R-modules.

Module homomorphisms can be added pointwise: when $\varphi, \psi : A \longrightarrow B$ are homomorphisms of left R-modules, then $\varphi + \psi : A \longrightarrow B$, defined by $(\varphi + \psi)(x) = \varphi(x) + \psi(x)$ for all $x \in A$, is a homomorphism of left R-modules. This property is used extensively in Chapters XI and XII.

Module homomorphisms preserve submodules:

Proposition 2.1. *Let $\varphi : A \longrightarrow B$ be a module homomorphism. If C is a submodule of A, then $\varphi(C) = \{\, \varphi(x) \mid x \in C \,\}$ is a submodule of B. If D is a submodule of B, then $\varphi^{-1}(D) = \{\, x \in A \mid \varphi(x) \in D \,\}$ is a submodule of A.*

Here $\varphi(C)$ is the *direct image* of $C \leqq A$ under φ, and $\varphi^{-1}(D)$ is the *inverse image* or *preimage* of D under φ. The notation $\varphi^{-1}(D)$ does not imply that φ is bijective, or that $\varphi^{-1}(D)$ is the direct image of D under a spurious map φ^{-1}.

Two submodules of interest arise from Proposition 2.1:

Definitions. *Let $\varphi : A \longrightarrow B$ be a module homomorphism. The* image *or* range *of φ is* $\mathrm{Im}\,\varphi = \{\, \varphi(x) \mid x \in A \,\} = \varphi(A)$. *The* kernel *of φ is* $\mathrm{Ker}\,\varphi = \{\, x \in A \mid \varphi(x) = 0 \,\} = \varphi^{-1}(0)$.

Quotient modules. Conversely, submodules give rise to homomorphisms. If A is a submodule of M, then the inclusion mapping $A \longrightarrow M$ is a module homomorphism. Moreover, there is a quotient module M/A, that comes with a projection $M \longrightarrow M/A$.

Proposition 2.2. *Let M be a left R-module and let A be a submodule of M. The quotient group M/A is a left R-module, in which $r\,(x + A) = rx + A$ for all $r \in R$ and $x \in M$. If M is unital, then M/A is unital. The* projection *$x \longmapsto x + A$ is a homomorphism of left R-modules, whose kernel is A.*

Proof. Since every subgroup of the abelian group M is normal, there is a quotient group M/A, in which cosets of A are added as subsets, in particular $(x + A) + (y + A) = (x + y) + A$ for all $x, y \in M$. Since A is a submodule, the action of $r \in R$ sends a coset $x + A = \{\, x + a \mid a \in A \,\}$ into a single coset, namely the coset of rx: if $y \in x + A$, then $y = x + a$ for some $a \in A$ and $ry = rx + ra \in rx + A$. Hence an action of R on M/A is well defined by $r\,(x + A) = rx + A$, the coset that contains all ry with $y \in x + A$. The projection $M \longrightarrow M/A$ preserves this action of R; hence the module axioms, (1), (2), and, in the unital case, (3), hold in M/A, since they hold in M. We see that the projection $M \longrightarrow M/A$ is a module homomorphism. $\square$

Definition. Let A be a submodule of a module M. The module of all cosets of A is the quotient module M/A *of M by A.*

Submodules of a quotient module M/A are quotients of submodules of M:

Proposition 2.3. *If A is a submodule of a module M, then $C \longmapsto C/A$ is an inclusion preserving, one-to-one correspondence between submodules of M that contain A and submodules of M/A.*

Proof. We saw (Proposition I.4.10) that direct and inverse image under the projection induce a one-to-one correspondence, which preserves inclusion, between subgroups of M that contain A, and subgroups of M/A. Direct and inverse image preserve submodules, by 2.1. □

Like quotient groups, quotient modules have a most useful universal property.

Theorem 2.4 (Factorization Theorem). *Let A be a left R-module and let B be a submodule of A. Every homomorphism of left R-modules $\varphi : A \longrightarrow C$ whose kernel contains B factors uniquely through the canonical projection $\pi : A \longrightarrow A/B$ ($\varphi = \psi \circ \pi$ for some unique module homomorphism $\psi : A/B \longrightarrow C$):*

$$
\begin{array}{ccc}
A & \xrightarrow{\ \pi\ } & A/B \\
 & \varphi \searrow & \downarrow \psi \\
 & & C
\end{array}
$$

Proof. By the corresponding property of abelian groups (Theorem I.5.1), $\varphi = \psi \circ \pi$ for some unique homomorphism $\psi : A/B \longrightarrow C$ of abelian groups. Then $\psi(x + B) = \varphi(x)$ for all $x \in A$. Hence ψ is a module homomorphism. □

Readers will prove a useful stronger version of Theorem 2.4:

Theorem 2.5 (Factorization Theorem). *If $\varphi : A \longrightarrow B$ and $\rho : A \longrightarrow C$ are module homomorphisms, ρ is surjective, and $\mathrm{Ker}\,\rho \subseteq \mathrm{Ker}\,\varphi$, then φ factors uniquely through ρ.*

The homomorphism and isomorphism theorems now hold for modules.

Theorem 2.6 (Homomorphism Theorem). *If $\varphi : A \longrightarrow B$ is a homomorphism of left R-modules, then*

$$
A/\mathrm{Ker}\,\varphi \cong \mathrm{Im}\,\varphi;
$$

in fact, there is an isomorphism $\theta : A/\mathrm{Ker}\,f \longrightarrow \mathrm{Im}\,f$ unique such that $\varphi = \iota \circ \theta \circ \pi$, where $\iota : \mathrm{Im}\,f \longrightarrow B$ is the inclusion homomorphism and $\pi : A \longrightarrow A/\mathrm{Ker}\,f$ is the canonical projection.

Thus every module homomorphism is the composition of an inclusion homomorphism, an isomorphism, and a canonical projection to a quotient module:

$$
\begin{array}{ccc}
A & \xrightarrow{\;\varphi\;} & B \\
{\scriptstyle\pi}\downarrow & & \uparrow{\scriptstyle\iota} \\
A/\mathrm{Ker}\,\varphi & \xdashrightarrow[\;\theta\;]{} & \mathrm{Im}\,\varphi
\end{array}
$$

Theorem 2.7 (First Isomorphism Theorem). *If A is a left R-module and $B \supseteq C$ are submodules of A, then*

$$
A/B \cong (A/C)/(B/C);
$$

in fact, there is a unique isomorphism $\theta : A/B \longrightarrow (A/C)/(B/C)$ such that $\theta \circ \rho = \tau \circ \pi$, where $\pi : A \longrightarrow A/C$, $\rho : A \longrightarrow A/B$, and $\tau : A/C \longrightarrow (A/C)/(B/C)$ are the canonical projections:

$$
\begin{array}{ccc}
A & \xrightarrow{\;\pi\;} & A/C \\
{\scriptstyle\rho}\downarrow & & \downarrow{\scriptstyle\tau} \\
A/B & \xrightarrow[\;\theta\;]{} & (A/C)/(B/C)
\end{array}
$$

Theorem 2.8 (Second Isomorphism Theorem). *If A and B are submodules of a left R-module, then*

$$
(A + B)/B \cong A/(A \cap B);
$$

in fact, there is an isomorphism $\theta : A/(A \cap B) \longrightarrow (A + B)/B$ unique such that $\theta \circ \rho = \pi \circ \iota$, where $\pi : A + B \longrightarrow (A + B)/B$ and $\rho : A \longrightarrow A/(A \cap B)$ are the canonical projections and $\iota : A \longrightarrow A + B$ is the inclusion homomorphism:

$$
\begin{array}{ccc}
A & \xrightarrow{\;\rho\;} & A/(A \cap B) \\
{\scriptstyle\iota}\downarrow & & \downarrow{\scriptstyle\theta} \\
A + B & \xrightarrow[\;\pi\;]{} & (A + B)/B
\end{array}
$$

Proofs. In Theorem 2.6, there is by Theorem I.5.2 a unique isomorphism θ of abelian groups such that $\varphi = \iota \circ \theta \circ \pi$. Then $\theta(a + \mathrm{Ker}\,\varphi) = \varphi(a)$ for all $a \in A$. Therefore θ is a module homomorphism. Theorems 2.7 and 2.8 are proved similarly. $\square$

As in Section I.5, the isomorphisms theorems are often numbered so that 2.6 is the first isomorphism theorem. Then our first and second isomorphism theorems, 2.7 and 2.8, are the second and third isomorphism theorems.

As another application of Theorem 2.6 we construct all cyclic modules.

Proposition 2.9. *A unital left R-module is cyclic if and only if it is isomorphic to R/L $(= {}_R R/L)$ for some left ideal L of R. If $M = Rm$ is cyclic, then $M \cong R/\mathrm{Ann}\,(m)$, where*

$$
\mathrm{Ann}\,(m) = \{\, r \in R \mid rm = 0 \,\}
$$

is a left ideal of R. If R is commutative, then $\mathrm{Ann}\,(Rm) = \mathrm{Ann}\,(m)$.

Proof. Let $M = Rm$ be cyclic. Then $\varphi : r \longmapsto rm$ is a module homomorphism of ${}_R R$ onto M. By 2.6, $M \cong R/\text{Ker } \varphi$, and we see that $\text{Ker } \varphi = \text{Ann}\,(m)$. In particular, $\text{Ann}\,(m)$ is a left ideal of R. Moreover, $\text{Ann}\,(M) \subseteq \text{Ann}\,(m)$; if R is commutative, then, conversely, $sm = 0$ implies $s(rm) = 0$ for all $r \in R$, and $\text{Ann}\,(m) = \text{Ann}\,(M)$.

Conversely, if L is a left ideal of R, then R/L is cyclic, generated by $1 + L$, since $r + L = r\,(1 + L)$ for every $r \in R$. Hence any $M \cong R/L$ is cyclic. $\square$

The left ideal $\text{Ann}\,(m)$ is the *annihilator* of m. In any left R-module M, $\text{Ann}\,(m)$ is a left ideal of R; moreover, $\text{Ann}\,(M) = \bigcap_{m \in M} \text{Ann}\,(m)$.

Exercises

1. Let $\varphi : A \longrightarrow B$ be a homomorphism of left R-modules. Show that $\varphi\big(\varphi^{-1}(C)\big) = C \cap \text{Im } \varphi$, for every submodule C of B.

2. Let $\varphi : A \longrightarrow B$ be a homomorphism of left R-modules. Show that $\varphi^{-1}\big(\varphi(C)\big) = C + \text{Ker } \varphi$, for every submodule C of A.

3. Let $\varphi : A \longrightarrow B$ be a homomorphism of left R-modules. Show that direct and inverse image under φ induce a one-to-one correspondence, which preserves inclusion, between submodules of A that contain $\text{Ker } \varphi$, and submodules of $\text{Im } \varphi$.

4. Let M be a [unital] left R-module and let I be a two-sided ideal of R. Make M/IM an R/I-module.

5. Let R be a (commutative) domain. Show that all nonzero principal ideals of R are isomorphic (as R-modules).

6. Let A and B be submodules of M. Show by an example that $A \cong B$ does not imply $M/A \cong M/B$.

7. Let R be a ring with an identity element. If $x, y \in R$ and $xR = yR$, then show that $Rx \cong Ry$ (as left R-modules); in fact, there is an isomorphism $Rx \longrightarrow Ry$ that sends x to y.

8. Let $\varphi : A \longrightarrow B$ and $\psi : B \longrightarrow C$ be module homomorphisms. Show that $\psi \circ \varphi = 0$ if and only if φ factors through the inclusion homomorphism $\text{Ker } \psi \longrightarrow B$.

9. Let $\varphi : A \longrightarrow B$ and $\psi : B \longrightarrow C$ be module homomorphisms. Show that $\psi \circ \varphi = 0$ if and only if ψ factors through the projection $B \longrightarrow B/\text{Im } \varphi$.

10. If $\varphi : A \longrightarrow B$ and $\rho : A \longrightarrow C$ are module homomorphisms, ρ is surjective, and $\text{Ker } \rho \subseteq \text{Ker } \varphi$, then show that φ factors uniquely through ρ.

3. Direct Sums and Products

Direct sums and products construct modules from simpler modules, and their universal properties help build diagrams. The definitions and basic properties in this section are stated for left modules but apply to right modules as well.

Direct products. The direct product of a family of modules is their Cartesian product, with componentwise operations:

Definition. The direct product *of a family* $(A_i)_{i \in I}$ *of left R-modules is their Cartesian product* $\prod_{i \in I} A_i$ *(the set of all families* $(x_i)_{i \in I}$ *such that* $x_i \in A_i$ *for all i) with componentwise addition and action of R:*

$$(x_i)_{i \in I} + (y_i)_{i \in I} = (x_i + y_i)_{i \in I}, \quad r(x_i)_{i \in I} = (rx_i)_{i \in I}.$$

It is immediate that these operations make $\prod_{i \in I} A_i$ a left R-module. If $I = \emptyset$, then $\prod_{i \in I} A_i = \{0\}$. If $I = \{1\}$, then $\prod_{i \in I} A_i \cong A_1$. If $I = \{1, 2, ..., n\}$, then $\prod_{i \in I} A_i$ is also denoted by $A_1 \times A_2 \times \cdots \times A_n$.

The direct product $\prod_{i \in I} A_i$ comes with a *projection* $\pi_j : \prod_{i \in I} A_i \longrightarrow A_j$ for every $j \in I$, which sends $(x_i)_{i \in I}$ to its j component x_j, and is a homomorphism; in fact, the left R-module structure on $\prod_{i \in I} A_i$ is the only module structure such that every projection is a module homomorphism.

The direct product and its projections have a universal property:

Proposition 3.1. *Let* M *and* $(A_i)_{i \in I}$ *be left R-modules. For every family* $(\varphi_i)_{i \in I}$ *of module homomorphisms* $\varphi_i : M \longrightarrow A_i$ *there exists a unique module homomorphism* $\varphi : M \longrightarrow \prod_{i \in I} A_i$ *such that* $\pi_i \circ \varphi = \varphi_i$ *for all* $i \in I$:

$$\begin{array}{ccc} & & \prod_{i \in I} A_i \\ & \varphi \nearrow & \downarrow \pi_i \\ M & \xrightarrow{\varphi_i} & A_i \end{array}$$

The proof is an exercise. By 3.1, every family of homomorphisms $\varphi_i : A_i \longrightarrow B_i$ induces a homomorphism $\varphi = \prod_{i \in I} \varphi_i$ unique such that every square

$$\begin{array}{ccc} \prod_{i \in I} A_i & \xrightarrow{\varphi} & \prod_{i \in I} B_i \\ \pi_i \downarrow & & \downarrow \rho_i \\ A_i & \xrightarrow{\varphi_i} & B_i \end{array}$$

commutes ($\rho_i \circ \varphi = \varphi_i \circ \pi_i$ for all i), where π_i and ρ_i are the projections; namely, $\varphi((x_i)_{i \in I}) = (\varphi_i(x_i))_{i \in I}$. This can also be shown directly. If $I = \{1, 2, ..., n\}$, then $\prod_{i \in I} \varphi_i$ is also denoted by $\varphi_1 \times \varphi_2 \times \cdots \times \varphi_n$.

External direct sums.

Definition. The direct sum, *or* external direct sum, *of a family* $(A_i)_{i \in I}$ *of left R-modules is the following submodule of* $\prod_{i \in I} A_i$:

$$\bigoplus_{i \in I} A_i = \{ (x_i)_{i \in I} \in \prod_{i \in I} A_i \mid x_i = 0 \text{ for almost all } i \in I \}$$

It is immediate that this defines a submodule. If $I = \emptyset$, then $\bigoplus_{i \in I} A_i = 0$. If $I = \{1\}$, then $\bigoplus_{i \in I} A_i \cong A_1$. If $I = \{1, 2, ..., n\}$, then $\bigoplus_{i \in I} A_i$ is also denoted by $A_1 \oplus A_2 \oplus \cdots \oplus A_n$, and coincides with $A_1 \times A_2 \times \cdots \times A_n$.

The direct sum $\bigoplus_{i \in I} A_i$ comes with an *injection* $\iota_j : A_j \longrightarrow \bigoplus_{i \in I} A_i$ for every $j \in I$, defined by its components: for all $x \in A_j$,

$$\iota_j(x)_j = x \in A_j, \quad \iota_j(x)_i = 0 \in A_i \text{ for all } i \neq j.$$

Happily, every ι_j is an injective homomorphism.

The direct sum and its injections have a universal property:

Proposition **3.2.** *Let M and $(A_i)_{i \in I}$ be left R-modules. For every family $(\varphi_i)_{i \in I}$ of module homomorphisms $\varphi_i : A_i \longrightarrow M$ there exists a unique module homomorphism $\varphi : \bigoplus_{i \in I} A_i \longrightarrow M$ such that $\varphi \circ \iota_i = \varphi_i$ for all $i \in I$, namely $\varphi((x_i)_{i \in I}) = \sum_{i \in I} \varphi_i(x_i)$:*

$$\begin{array}{ccc} & \bigoplus_{i \in I} A_i & \\ \iota_i \uparrow & & \searrow{\varphi} \\ A_i & \xrightarrow{\varphi_i} & M \end{array}$$

Proof. First we prove a quick lemma.

Lemma **3.3.** *If $x = (x_i)_{i \in I} \in \bigoplus_{i \in I} A_i$, then $x = \sum_{i \in I} \iota_i(x_i)$; moreover, x can be written uniquely in the form $x = \sum_{i \in I} \iota_i(y_i)$, where $y_i \in A_i$ for all i and $y_i = 0$ for almost all i.*

Proof. Let $x = (x_i)_{i \in I} \in \bigoplus_{i \in I} A_i$, so that $x_i \in A_i$ for all $i \in I$ and $J = \{ i \in I \mid x_i \neq 0 \}$ is finite. Then $y = \sum_{i \in I} \iota_i(x_i)$ is defined and $y = \sum_{i \in J} \iota_i(x_i)$; hence $y_j = \sum_{i \in J} \iota_i(x_i)_j = x_j$ if $j \in J$, $y_j = \sum_{i \in J} \iota_i(x_i)_j = 0$ otherwise. Thus $y = x$, and $x = \sum_{i \in I} \iota_i(x_i)$. If $x = \sum_{i \in I} \iota_i(y_i)$, with $y_i \in A_i$ for all i and $y_i = 0$ for almost all i, then $y = (y_i)_{i \in I} \in \bigoplus_{i \in I} A_i$, $y = \sum_{i \in I} \iota_i(y_i)$ by the above, $y = x$, and $y_i = x_i$ for all i. $\square$

Now, the homomorphism φ in Proposition 3.2 must satisfy

$$\varphi((x_i)_{i \in I}) = \varphi(\textstyle\sum_{i \in I} \iota_i(x_i)) = \sum_{i \in I} \varphi(\iota_i(x_i)) = \sum_{i \in I} \varphi_i(x_i)$$

for all $x = (x_i)_{i \in I} \in \bigoplus_{i \in I} A_i$, by 3.3; therefore φ is unique. On the other hand, $\sum_{i \in I} \varphi_i(x_i)$ is defined for every $x = (x_i)_{i \in I} \in \bigoplus_{i \in I} A_i$; hence the equality $\varphi((x_i)_{i \in I}) = \sum_{i \in I} \varphi_i(x_i)$ defines a mapping $\varphi : \bigoplus_{i \in I} A_i \longrightarrow M$. Then φ is a module homomorphism and $\varphi \circ \iota_j = \varphi_j$ for every $j \in I$. $\square$

If $I = \{ 1, 2, \ldots, n \}$ is finite, then $A_1 \oplus A_2 \oplus \cdots \oplus A_n = A_1 \times A_2 \times \cdots \times A_n$ is blessed with two universal properties.

In general, Proposition 3.2 implies that every family of homomorphisms $\varphi_i : A_i \longrightarrow B_i$ induces a homomorphism $\varphi = \bigoplus_{i \in I} \varphi_i$ unique such that every square

$$\begin{array}{ccc} \bigoplus_{i \in I} A_i & \xrightarrow{\varphi} & \bigoplus_{i \in I} B_i \\ \iota_i \uparrow & & \uparrow \kappa_i \\ A_i & \xrightarrow{\varphi_i} & B_i \end{array}$$

commutes ($\varphi \circ \iota_i = \kappa_i \circ \varphi_i$ for all i), where ι_i and κ_i are the injections; namely, $\varphi((x_i)_{i \in I}) = (\varphi_i(x_i))_{i \in I}$. If $I = \{ 1, 2, \ldots, n \}$, then $\bigoplus_{i \in I} \varphi_i$ is also denoted by $\varphi_1 \oplus \varphi_2 \oplus \cdots \oplus \varphi_n$, and coincides with $\varphi_1 \times \varphi_2 \times \cdots \times \varphi_n$.

Direct sums have another characterization in terms of homomorphisms:

Proposition 3.4. *Let* $(M_i)_{i \in I}$ *be left R-modules. A left R-module M is isomorphic to* $\bigoplus_{i \in I} M_i$ *if and only if there exist module homomorphisms* $\mu_i :$ $M_i \longrightarrow M$ *and* $\rho_i : M \longrightarrow M_i$ *for every* $i \in I$ *such that* (1) $\rho_i \circ \mu_i = 1_{M_i}$ *for all i ;* (2) $\rho_i \circ \mu_j = 0$ *whenever* $i \neq j$ *;* (3) *for every* $x \in M$, $\rho_i(x) = 0$ *for almost all i ; and* (4) $\sum_{i \in I} \mu_i \circ \rho_i = 1_M$.

The sum in part (4) is a pointwise sum; we leave the details to our readers.

Internal direct sums. Direct sums of modules can also be characterized in terms of submodules rather than homomorphisms.

Proposition 3.5. *Let* $(M_i)_{i \in I}$ *be left R-modules. For a left R-module M the following conditions are equivalent:*

(1) $M \cong \bigoplus_{i \in I} M_i$;

(2) *M contains submodules* $(A_i)_{i \in I}$ *such that* $A_i \cong M_i$ *for all i and every element of M can be written uniquely as a sum* $\sum_{i \in I} a_i$, *where* $a_i \in A_i$ *for all i and* $a_i = 0$ *for almost all i ;*

(3) *M contains submodules* $(A_i)_{i \in I}$ *such that* $A_i \cong M_i$ *for all i, M =* $\sum_{i \in I} A_i$, *and* $A_j \cap (\sum_{i \neq j} A_i) = 0$ *for all j .*

Proof. (1) implies (2). By 3.3, $\bigoplus_{i \in I} M_i$ contains submodules $M_i' = \iota_i(M_i) \cong M_i$ such that every element of $\bigoplus_{i \in I} M_i$ can be written uniquely as a sum $\sum_{i \in I} a_i$, where $a_i \in M_i'$ for all i and $a_i = 0$ for almost all i. If $\theta : \bigoplus_{i \in I} M_i \longrightarrow M$ is an isomorphism, then the submodules $A_i = \theta(M_i')$ of M have similar properties.

(2) implies (3). By (2), $M = \sum_{i \in I} A_i$; moreover, if $x \in A_j \cap (\sum_{i \neq j} A_i)$, then x is a sum $x = \sum_{i \in I} a_i'$ in which $a_j' = x$, $a_i' = 0$ for all $i \neq j$, and a sum $x = \sum_{i \in I} a_i''$ in which $a_j'' = 0 \in A_j$, $a_i'' \in A_i$ for all i, $a_i'' = 0$ for almost all i ; by (2), $x = a_j' = a_j'' = 0$.

(3) implies (2). By (3), $M = \sum_{i \in I} A_i$, so that every element of M is a sum $\sum_{i \in I} a_i$, where $a_i \in A_i$ for all i and $a_i = 0$ for almost all i. If $\sum_{i \in I} a_i' = \sum_{i \in I} a_i''$ (where a_i', $a_i'' \in A_i$, etc.), then, for every $j \in I$, $a_j'' - a_j' = \sum_{i \neq j} (a_i' - a_i'') \in A_j \cap (\sum_{i \neq j} A_i)$ and $a_j'' = a_j'$ by (3).

(2) implies (1). By 3.2, the inclusion homomorphisms $A_i \longrightarrow M$ induce a module homomorphism $\theta : \bigoplus_{i \in I} A_i \longrightarrow M$, namely $\theta((a_i)_{i \in I}) = \sum_{i \in I} a_i$. Then θ is bijective, by (2). The isomorphisms $M_i \cong A_i$ then induce an isomorphism $\bigoplus_{i \in I} M_i \cong \bigoplus_{i \in I} A_i \cong M$. $\square$

Definition. A left R-module M is the internal direct sum $M = \bigoplus_{i \in I} A_i$ *of submodules* $(A_i)_{i \in I}$ *when every element of M can be written uniquely as a sum* $\sum_{i \in I} a_i$, *where* $a_i \in A_i$ *for all i and* $a_i = 0$ *for almost all i .*

Equivalently, $M = \bigoplus_{i \in I} A_i$ when $M = \sum_{i \in I} A_i$ and $A_j \cap \left(\sum_{i \neq j} A_i \right) = 0$ for all j. One also says that the sum $\sum_{i \in I} A_i$ is *direct*.

By Proposition 3.5, internal and external direct sums differ only by isomorphisms: if M is an external direct sum of modules $(M_i)_{i \in I}$, then M is an internal direct sum of submodules $A_i \cong M_i$; if M is an internal direct sum of submodules $(A_i)_{i \in I}$, then M is isomorphic to the external direct sum $\bigoplus_{i \in I} A_i$. The same notation is used for both; it should be clear from context which kind is meant.

If I is a totally ordered set, then the condition $A_j \cap \left(\sum_{i \neq j} A_i \right) = 0$ for all j in Proposition 3.5 can be replaced by the apparently weaker condition $A_j \cap \left(\sum_{i < j} A_i \right) = 0$ for all j, as readers will easily show. We state two particular cases of interest:

Corollary 3.6. An R-module M is a direct sum $M = A_1 \oplus \cdots \oplus A_n$ if and only if $M = A_1 + \cdots + A_n$ and $A_j \cap (A_1 + \cdots + A_{j-1}) = 0$ for all $1 < j \leqq n$.

Corollary 3.7. An R-module M is a direct sum $M = A \oplus B$ if and only if $A + B = M$ and $A \cap B = 0$; and then $M/A \cong B$, $M/B \cong A$.

The exercises also give some associativity properties of internal direct sums.

Definition. A direct summand *of a module M is a submodule A of M such that $A \oplus B = M$ for some submodule B of M.*

Corollary 3.7 characterizes direct summands. Readers may prove another characterization:

Proposition 3.8. A submodule is a direct summand of a module M if and only if there exists an endomorphism η of M such that $\eta \circ \eta = \eta$ and $\operatorname{Im} \eta = A$.

Exercises

1. Show that the direct product of a family of modules, and its projections, are characterized up to isomorphism by their universal property.

2. Prove the following: if A_i is a submodule of M_i for every $i \in I$, then $\prod_{i \in I} A_i$ is a submodule of $\prod_{i \in I} M_i$, and $\left(\prod_{i \in I} M_i \right) / \left(\prod_{i \in I} A_i \right) \cong \prod_{i \in I} (M_i / A_i)$.

3. Prove the following associativity property of direct products: if $I = \bigcup_{j \in J} I_j$ is a partition of I, then $\prod_{i \in I} A_i \cong \prod_{j \in J} \left(\prod_{i \in I_j} A_i \right)$, for every family $(A_i)_{i \in I}$ of left R-modules.

4. Show that the direct sum of a family of modules, and its injections, are characterized up to isomorphism by their universal property.

5. Prove the following: if A_i is a submodule of M_i for every $i \in I$, then $\bigoplus_{i \in I} A_i$ is a submodule of $\bigoplus_{i \in I} M_i$, and $\left(\bigoplus_{i \in I} M_i \right) / \left(\bigoplus_{i \in I} A_i \right) \cong \bigoplus_{i \in I} (M_i / A_i)$.

6. Prove the following: if $A_i \cong B_i$ for all i, then $\bigoplus_{i \in I} A_i \cong \bigoplus_{i \in I} B_i$.

7. Show by an example that $A \oplus B \cong A' \oplus B'$ does not imply $B \cong B'$ even when $A \cong A'$.

8. Prove the following associativity property of direct sums: if $I = \bigcup_{j \in J} I_j$ is a partition of I, then $\bigoplus_{i \in I} A_i \cong \bigoplus_{j \in J} \left(\bigoplus_{i \in I_j} A_i \right)$, for every family $(A_i)_{i \in I}$ of left R-modules.

9. Show that a family $(\varphi_i)_{i \in I}$ of module homomorphisms $\varphi_i : A \longrightarrow B$ can be added pointwise if, for every $a \in A$, $\varphi_i(a) = 0$ for almost all i, and then the pointwise sum is a module homomorphism $\varphi = \sum_{i \in I} \varphi_i : A \longrightarrow B$.

10. Let M and $(M_i)_{i \in I}$ be left R-modules. Show that $M \cong \bigoplus_{i \in I} M_i$ if and only if there exist module homomorphisms $\mu_i : M_i \longrightarrow M$ and $\rho_i : M \longrightarrow M_i$ for every $i \in I$, such that (1) $\rho_i \circ \mu_i = 1_{M_i}$ for all i; (2) $\rho_i \circ \mu_j = 0$ whenever $i \neq j$; (3) for every $x \in M$, $\rho_i(x) = 0$ for almost all i; and (4) $\sum_{i \in I} \mu_i \circ \rho_i = 1_M$ (pointwise, as in the previous exercise).

11. Let $\mu : A \longrightarrow B$ and $\sigma : B \longrightarrow A$ be module homomorphisms such that $\sigma \circ \mu = 1_A$. Show that $B = \operatorname{Im} \mu \oplus \operatorname{Ker} \sigma$.

12. Let I be totally ordered. Show that a module M is the internal direct sum of submodules $(A_i)_{i \in I}$ if and only if $M = \sum_{i \in I} A_i$ and $A_j \cap \left(\sum_{i < j} A_i \right) = 0$ for all $j \in I$.

13. Show that a sum $\sum_{i \in I} A_i$ of submodules is direct if and only if every finite subsum $\sum_{i \in J} A_i$ is direct.

14. Let A, B, C be submodules of some module. Prove the following: if $A \subseteq C$ and $A \cap B = 0$, then $(A \oplus B) \cap C = A \oplus (B \cap C)$.

15. Let A, B, C be submodules of some module. Prove the following: if $A \cap B = 0$ and $(A + B) \cap C = 0$, then $B \cap C = 0$ and $A \cap (B + C) = 0$ (in other words, if the sum $(A + B) + C$ is direct, then so is the sum $A + (B + C)$).

16. Prove the following associativity property of internal direct sums: if $(A_i)_{i \in I}$ are submodules of some module and $I = \bigcup_{j \in J} I_j$ is a partition of I, then $\sum_{i \in I} A_i$ is direct if and only if $\sum_{i \in I_j} A_i$ is direct for every $j \in J$ and $\sum_{j \in J} \left(\sum_{i \in I_j} A_i \right)$ is direct.

17. Give an example of a submodule that is not a direct summand.

18. Prove that a submodule A of a module M is a direct summand of M if and only if there exists an endomorphism η of M such that $\eta \circ \eta = \eta$ and $\operatorname{Im} \eta = A$.

4. Free Modules

A module is free when it has a basis. Free modules share a number of properties with vector spaces.

Definition. In what follows, all rings have an identity element, and all modules are unital. Most definitions and results are stated for left modules but apply equally to right modules.

Bases of modules are defined like bases of vector spaces. They can be regarded as subsets or as families. Indeed, every set S can be written as a family $(x_i)_{i \in I}$ in which $x_i \neq x_j$ when $i \neq j$; for instance, let $I = S$ and $x_s = s$ for all $s \in S$. Conversely, every family $(x_i)_{i \in I}$ gives rise to a set $\{ x_i \mid i \in I \}$, and the mapping $i \longmapsto x_i$ is bijective if $x_i \neq x_j$ whenever $i \neq j$; the bases and linearly independent families defined below have this property.

Definitions. Let M be a left R-module. A subset S of M is linearly independent *(over R) when $\sum_{s \in S} r_s s = 0$, with $r_s \in R$ for all s and $r_s = 0$ for almost all s, implies $r_s = 0$ for all s. A family $(e_i)_{i \in I}$ of elements of M is* linearly independent *(over R) when $\sum_{i \in I} r_i e_i = 0$, with $r_i \in R$ for all i and $r_i = 0$ for almost all i, implies $r_i = 0$ for all i.*

Definitions. A basis *of a left R-module M is a linearly independent subset or family of M that generates (spans) M. A module is* free *when it has a basis, and is then* free on *that basis.*

Readers will verify that a basis of a module M is, in particular, a maximal linearly independent subset of M, and a minimal generating subset of M; moreover, Zorn's lemma ensures that every module has maximal generating subsets. But a maximal generating subset of a module is not necessarily a basis; in fact, some modules have no basis at all (not like the theorems in this book).

*Proposition **4.1**. A family $(e_i)_{i \in I}$ of elements of a [unital] left R-module M is a basis of M if and only if every element of M can be written uniquely as a linear combination $x = \sum_{i \in I} x_i e_i$ (with $x_i \in R$ for all i and $x_i = 0$ for almost all i).*

Proof. By 1.6, $(e_i)_{i \in I}$ generates M if and only if every element of M is a linear combination $x = \sum_{i \in I} x_i e_i$. If this expression is unique for all x, then $\sum_{i \in I} r_i e_i = 0$ implies $r_i = 0$ for all i, and $(e_i)_{i \in I}$ is linearly independent. Conversely, if $(e_i)_{i \in I}$ is a basis, then $x = \sum_{i \in I} x_i e_i = \sum_{i \in I} y_i e_i$ implies $\sum_{i \in I} (x_i - y_i) e_i = 0$ and $x_i - y_i = 0$ for all i. $\square$

If $(e_i)_{i \in I}$ is a basis of M and $x = \sum_{i \in I} x_i e_i$, then the scalars $x_i \in R$ are the *coordinates* of $x \in M$ in the basis $(e_i)_{i \in I}$. The uniqueness in Proposition 4.1 implies that the i coordinate of rx is rx_i, and that the i coordinate of $x + y$ is $x_i + y_i$, so that $x \longmapsto x_i$ is a module homomorphism $M \longrightarrow {}_R R$.

Proposition 4.1 suggests that bases are related to direct sums. The details are as follows. Let $\bigoplus_{i \in I} {}_R R$ denote the direct sum of $|I|$ copies of ${}_R R$ (the direct sum $\bigoplus_{i \in I} M_i$ in which $M_i = {}_R R$ for all i).

*Proposition **4.2**. Let M be a [unital] left R-module.*

(1) *If $(e_i)_{i \in I}$ is a basis of M, then there is an isomorphism $M \cong \bigoplus_{i \in I} {}_R R$ that assigns to every element of M its coordinates in the basis $(e_i)_{i \in I}$.*

(2) *Every $\bigoplus_{i \in I} {}_R R$ has a* canonical basis *$(e_i)_{i \in I}$, in which the components of e_j are $(e_j)_i = 1$ if $i = j$, $(e_j)_i = 0$ if $i \neq j$.*

(3) *M is free if and only if $M \cong \bigoplus_{i \in I} {}_R R$ for some set I.*

In particular, an abelian group is free (as a $\mathbb{Z}$-module) if and only if it is a direct sum of copies of $\mathbb{Z}$.

Proof. (1): If $x = \sum_{i \in I} x_i\, e_i$ in M, then $(x_i)_{i \in I} \in \bigoplus_{i \in I}\ {}_R R$ by 4.1. The mapping $(x_i)_{i \in I} \longmapsto x$ is bijective, by 4.1; the inverse bijection $x \longmapsto (x_i)_{i \in I}$ is a module homomorphism, by the remarks following 4.1.

(2). We see that $e_i = \iota_i(1)$, where $\iota_i :\ {}_R R \longrightarrow \bigoplus_{i \in I}\ {}_R R$ is the i injection. By 3.3, every $x \in \bigoplus_{i \in I}\ {}_R R$ can be written uniquely in the form $\sum_{i \in I} \iota_i(r_i) = \sum_{i \in I} r_i\, e_i$, with $r_i \in R$ and $r_i = 0$ for almost all i. By 4.1, $(e_i)_{i \in I}$ is a basis. (3) follows from (1) and (2). $\square$

Proposition 4.2 shows that some modules are free. In fact, when supplemented by surgery, Proposition 4.2 implies that every set is a basis of some free module:

Corollary 4.3. *Given any set X, there exists a left R-module that is free on X, and it is unique up to isomorphism.*

Readers will prove this when they feel like, well, cutting up. We invoked Corollary 4.3 in Chapter III when we constructed polynomial rings.

Corollary 4.4. *Every free left R-module M has a right R-module structure, which depends on the choice of a basis $(e_i)_{i \in I}$ of M, in which $\left(\sum_{i \in I} x_i\, e_i\right) r = \sum_{i \in I} x_i\, r\, e_i$.*

Proof. First, $\bigoplus_{i \in I}\ {}_R R$ is a right R-module, on which R acts component-wise, $\left((x_i)_{i \in I}\right) r = (x_i\, r)_{i \in I}$. If $(e_i)_{i \in I}$ is a basis of M, then the isomorphism $\bigoplus_{i \in I}\ {}_R R \cong M$ that sends $(x_i)_{i \in I}$ to $\sum_{i \in I} x_i\, e_i$ transfers this right R-module structure from $\bigoplus_{i \in I}\ {}_R R$ to M, so that $\left(\sum_{i \in I} x_i\, e_i\right) r = \sum_{i \in I} x_i\, r\, e_i$ in M. $\square$

In Corollary 4.4, M is free as a right R-module, with the same basis. The right action of R on $\bigoplus R$ is canonical, but the right R-module structure on M depends on the choice of a basis, since the isomorphism $\bigoplus R \cong M$ does. If R is commutative, however, then the right action of R on M coincides with its left action and does not depend on the choice of a basis.

Universal property. Free modules have a universal property:

Proposition 4.5. *Let $X = (e_i)_{i \in I}$ be a basis of a left R-module M. Every mapping f of X into a left R-module N extends uniquely to a module homomorphism φ of M into N, namely $\varphi\left(\sum_{i \in I} x_i\, e_i\right) = \sum_{i \in I} x_i\, f(e_i)$:*

If N is generated by $f(X)$, then φ is surjective.

Proof. If φ extends f to M, then $\varphi\left(\sum_{i \in I} x_i\, e_i\right) = \sum_{i \in I} x_i\, \varphi(e_i) = \sum_{i \in I} x_i\, f(e_i)$ for all $x = \sum_{i \in I} x_i\, e_i \in M$; hence φ is unique. Conversely, it is immediate that the mapping $\varphi : M \longrightarrow N$ defined by $\varphi\left(\sum_{i \in I} x_i\, e_i\right) = \sum_{i \in I} x_i\, f(e_i)$ is a module homomorphism and extends f. $\square$

Corollary **4.6.** *For every left R-module M there exists a surjective homomorphism* $F \longrightarrow M$ *where F is free; if M is generated by a subset X one may choose F free on X.*

Homomorphisms. The relationship between linear transformations of vector spaces and matrices extends to free modules. We do this for modules with finite bases, a restriction that is easily removed (see the exercises). For reasons which will soon become clear we begin with right modules.

Proposition **4.7.** *Let A and B be free right R-modules with bases* $e_1, \ldots, e_n$ *and* $f_1, \ldots, f_m$ *respectively. There is a one-to-one correspondence between module homomorphisms of A into B and* $m \times n$ *matrices with entries in R.*

Proof. Let $\varphi : A \longrightarrow B$ be a module homomorphism. Then $\varphi(e_j) = \sum_i f_i r_{ij}$ for some unique $r_{ij} \in R$, $i = 1, \ldots, m$, $j = 1, \ldots, n$. This defines an $m \times n$ matrix $M(\varphi) = (r_{ij})$ with entries in R, the *matrix of* φ *in* the given bases, in which the jth column holds the coordinates of $\varphi(e_j)$. This matrix determines φ, since $\varphi(\sum_j e_j x_j) = \sum_j (\sum_i f_i r_{ij}) x_j = \sum_i f_i (\sum_j r_{ij} x_j)$.

Conversely, if $M = (r_{ij})$ is an $m \times n$ matrix with entries in R, then the mapping $\varphi : A \longrightarrow B$ defined by $\varphi(\sum_j e_j x_j) = \sum_i f_i (\sum_j r_{ij} x_j)$ is a module homomorphism with matrix M in the given bases. $\square$

In the above, the matrix of φ is constructed in the usual way, with the coordinates of $\varphi(e_j)$ in the jth column; the coordinates of $\varphi(\sum_j e_j x_j)$ are computed by the usual matrix multiplication of $M(\varphi)$ by the column matrix (x_j).

Proposition **4.8.** *If* $\varphi, \psi : A \longrightarrow B$ *and* $\chi : B \longrightarrow C$ *are homomorphisms of free right R-modules with finite bases, then, in any given bases of A, B, and C,* $M(\varphi + \psi) = M(\varphi) + M(\psi)$ *and* $M(\chi \circ \varphi) = M(\chi) M(\varphi)$.

Proof. We prove the last equality. Let $e_1, \ldots, e_n, f_1, \ldots, f_m, g_1, \ldots, g_\ell$ be bases of A, B, and C, respectively. Let $\varphi(e_j) = \sum_i f_i r_{ij}$ and $\chi(f_i) = \sum_g h s_{hi}$. Then $M(\varphi) = (r_{ij})$, $M(\chi) = (s_{hi})$, $\chi(\varphi(e_j)) = \sum_i (\sum_h g_h s_{hi}) r_{ij}$ $= \sum_h g_h (\sum_i s_{hi} r_{ij})$, and $M(\chi \circ \varphi) = (\sum_i s_{hi} r_{ij}) = M(\chi) M(\varphi)$. $\square$

Module homomorphisms can be added and composed. Hence the endomorphisms of any right R-module M constitute a ring $\mathrm{End}_R(M)$, under pointwise addition and composition. Proposition 4.8 describes this ring when M is free:

Corollary **4.9.** *If M is a free right R-module and M has a basis with n elements, then* $\mathrm{End}_R(M)$ *is isomorphic to the ring* $M_n(R)$ *of all* $n \times n$ *matrices with entries in R.*

We now consider left R-modules. Left R-modules are right R^{op}-modules, and a homomorphism of left R-modules is also a homomorphism of right R^{op}-modules. Moreover, a module that is free as a left R-module is also free as a right R^{op}-module, with the same basis. If now A and B are free left R-modules with bases $e_1, \ldots, e_n$ and $f_1, \ldots, f_m$ respectively, then, by 4.7, there is a one-to-one correspondence between module homomorphisms of A into B and $m \times n$ matrices with entries in R^{op}. The latter look just like matrices with entries in R but are multiplied differently: the hj entry in $(s_{hi})(r_{ij})$ is now $\sum_i r_{ij} s_{hi}$ when calculated in R. This rule is likely to increase sales of headache medicines. The author prefers to stick with R^{op}. By 4.9:

Corollary **4.10.** *If M is a free left R-module and M has a basis with n elements, then $\operatorname{End}_R(M) \cong M_n(R^{op})$.*

Rank. All bases of a vector space have the same number of elements (the next section has a proof). Modules in general do not have this property; the exercises give a counterexample.

Definition. If all bases of a free module M have the same number of elements, then that number is the rank *of M,* rank M.

Proposition 4.11. *Let M be a free left R-module with an infinite basis. All bases of M have the same number of elements.*

Proof. The proof is a cardinality argument. Let X and Y be bases of M. Every $x \in X$ is a linear combination of elements of a finite subset Y_x of Y. Hence the submodule of M generated by $\bigcup_{x \in X} Y_x$ contains X and is all of M. Therefore $Y = \bigcup_{x \in X} Y_x$: otherwise, some element of $Y \setminus \bigcup_{x \in X} Y_x$ is a linear combination of elements of $\bigcup_{x \in X} Y_x$ and Y is not linearly independent. Hence $|Y| \leq \aleph_0 |X|$; if X is infinite, then $|Y| \leq \aleph_0 |X| = |X|$, by A.5.9.

Similarly, X is the union of $|Y|$ finite sets. If X is infinite, then Y is infinite, and $|X| \leq \aleph_0 |Y| = |Y|$; therefore $|X| = |Y|$. $\square$

Proposition 4.12. *Let M be a free left R-module. If R is commutative, then all bases of M have the same number of elements.*

Proof. The proof uses quotient rings of R. Let $(e_i)_{i \in I}$ be a basis of M and let $\mathfrak{a}$ be an ideal of R. Then $\mathfrak{a}M$ is generated by products rx with $r \in \mathfrak{a}$, $x \in M$, whose coordinates are all in $\mathfrak{a}$. Conversely, if $x_i \in \mathfrak{a}$ for all i, then $\sum_i x_i e_i \in \mathfrak{a}M$. Thus $x = \sum_i x_i e_i \in \mathfrak{a}M$ if and only if $x_i \in \mathfrak{a}$ for all i.

Since $\mathfrak{a}$ is an ideal of R, readers will verify that $M/\mathfrak{a}M$ is an $R/\mathfrak{a}$-module, in which $(r + \mathfrak{a})(x + \mathfrak{a}M) = rx + \mathfrak{a}M$. Then $(e_i + \mathfrak{a}M)_{i \in I}$ is a basis of $M/\mathfrak{a}M$: every element of $M/\mathfrak{a}M$ is a linear combination of $e_i + \mathfrak{a}M$'s, and if $\sum_{i \in I} (r_i + \mathfrak{a})(e_i + \mathfrak{a}M) = 0$ in $M/\mathfrak{a}M$, then $\sum_{i \in I} r_i e_i \in \mathfrak{a}M$, $r_i \in \mathfrak{a}$ for all i by the above, and $r_i + \mathfrak{a} = 0$ in $R/\mathfrak{a}$ for all i.

Now, let $\mathfrak{u}$ be a maximal ideal of R. Then $R/\mathfrak{a}$ is a field, $M/\mathfrak{a}M$ is a vector space over $R/\mathfrak{a}$, and all bases of $M/\mathfrak{a}M$ (over $R/\mathfrak{a}$) have the same number of elements. Therefore all bases of M (over R) have that same number of elements. (By 4.11 we need only consider finite bases in this argument.) $\square$

Exercises

1. Show that the $\mathbb{Z}$-module $\mathbb{Q}$ has no basis, even though all its maximal linearly independent subsets have the same number of elements.

2. Show that every left R-module M has a maximal linearly independent subset, and that a maximal linearly independent subset of M generates a submodule S that is *essential* in M ($S \cap A \neq 0$ for every submodule $A \neq 0$ of M).

3. Fill in the details of the surgical procedure that, given any set X, constructs a left R-module in which X is a basis.

4. Prove that a direct sum of free left R-modules is free.

5. Show that a module that is free on a given basis is characterized, up to isomorphism, by its universal property.

In the following three exercises, I and J are sets; an $I \times J$ matrix is a rectangular array $(r_{ij})_{i \in I, \, j \in J}$; an $I \times J$ matrix is *column finitary* when every column has only finitely many nonzero entries (for every $j \in J$, $r_{ij} = 0$ for almost all $i \in I$).

6. Prove the following: when A and B are free right R-modules with bases $(e_j)_{j \in J}$ and $(f_i)_{i \in I}$ respectively, there is a one-to-one correspondence between module homomorphisms of A into B and column finitary $I \times J$ matrices with entries in R.

7. Explain how column finitary matrices can be multiplied. Prove directly that this multiplication is associative.

8. Prove the following: when $\varphi : A \longrightarrow B$ and $\psi : B \longrightarrow C$ are homomorphisms of free right R-modules, then, in any given bases of A, B, and C, $M(\chi \circ \varphi) = M(\chi) \, M(\varphi)$. (In particular, when M is a free right R-module with a basis $(e_i)_{i \in I}$, then $\mathrm{End}_R(M)$ is isomorphic to the ring $M_I(R)$ of all column finitary $I \times I$ matrices with entries in R.)

*9. Cook up a definition of "the left R-module generated by a set X subject to set $\mathcal{R}$ of defining relations", and prove its universal property.

10. Let $R = \mathrm{End}_K(V)$, where V is a vector space over a field K with an infinite basis e_0, e_1, ..., e_n, ... Let α and $\beta \in R$ be the linear transformations (K-endomorphisms) of V such that $\alpha(e_{2n}) = e_n$, $\alpha(e_{2n+1}) = 0$, and $\beta(e_{2n}) = 0$, $\beta(e_{2n+1}) = e_n$ for all $n \geqq 0$. Show that $\{1\}$ and $\{\alpha, \beta\}$ are bases of $_R R$.

11. Prove that the module $_R R$ in the previous exercise has a basis with m elements, for every integer $m > 0$.

5. Vector Spaces

In this section we give a more general definition of vector spaces and prove their dimension property.

Definitions. A division ring *is a ring with identity in which every nonzero element is a unit.* A vector space *is a unital module over a division ring.*

A commutative division ring is a field; the quaternion algebra $\mathbb{H}$ is a division ring, but is not commutative. Division rings are still sometimes called *skew fields*. If D is a division ring, then so is D^{op}; hence we need only consider left D-modules in what follows.

In a module over a division ring, if $r_1 x_1 + \cdots + r_n x_n = 0$ and $r_1 \neq 0$, then $x_1 = r_1^{-1} r_2 x_2 + \cdots + r_1^{-1} r_n x_n$ is a linear combination of x_2, ..., x_n. As a result, the main properties of bases and dimension in vector spaces over a field extend, with little change in the proofs, to vector spaces over division rings.

Lemma 5.1. *Let X be a linearly independent subset of a vector space V. If $y \in V \backslash X$, then $X \cup \{y\}$ is linearly independent if and only if y is not a linear combination of elements of X.*

Proof. If $X \cup \{y\}$ is not linearly independent, then $r_y\, y + \sum_{x \in X} r_x\, x = 0$, where r_x, r_y are not all zero. Then $r_y \neq 0$: otherwise, X is not linearly independent. Therefore y is a linear combination of elements of X. $\square$

Theorem 5.2. *Every vector space has a basis.*

Proof. Let V be a left D-module, where D is a division ring. The union $X = \bigcup_{i \in I} X_i$ of a nonempty chain $(X_i)_{i \in I}$ of linearly independent subsets of V is linearly independent: if $\sum_{x \in X} r_x\, x = 0$, with $r_x \in D$, $r_x = 0$ for almost all $x \in X$, then the finite set $\{\, x \in X \mid r_x \neq 0 \,\}$ is contained in some X_i, whence $r_x = 0$ for all x since X_i is linearly independent. Also the empty set is linearly independent. By Zorn's lemma there exists a maximal linearly independent subset M of V. Then M generates V, by 5.1. $\square$

Intrepid readers will show that Theorem 5.2 characterizes division rings. Tweaking the proof of 5.2 readily yields a slightly more general result:

Proposition 5.3. *Let V be a vector space. If $Y \subseteq V$ generates V and $X \subseteq Y$ is linearly independent, then V has a basis $X \subseteq B \subseteq Y$.*

There is an *exchange property* for bases:

Lemma 5.4. *Let V be a vector space and let X, Y be bases of V. For every $x \in X$ there exists $y \in Y$ such that $(X \backslash \{x\}) \cup \{y\}$ is a basis of V.*

Proof. If $x \in Y$, then $y = x$ serves. Assume that $x \notin Y$, and let S be the subspace (submodule) of V generated by $X \backslash \{x\}$. If $Y \subseteq S$, then $S = V$, $x \in S$, and x is a linear combination of elements of $X \backslash \{x\}$, causing X to lose its linear independence. Therefore $Y \not\subseteq S$ and some $y \in Y$ is not a linear combination of elements of $X \backslash \{x\}$. Then $y \notin X \backslash \{x\}$; in fact, $y \notin X$, since $x \notin Y$. By 5.1, $X' = (X \backslash \{x\}) \cup \{y\}$ is linearly independent.

Now, x is a linear combination of elements of X': otherwise, $X' \cup \{x\} = X \cup \{y\}$ is linearly independent by 5.1, even though $y \notin X$ is a linear combination of elements of X. Therefore every element of X is a linear combination of elements of X', and X' generates V. $\square$

The exchange property implies our second main result:

Theorem 5.5. *All bases of a vector space have the same number of elements.*

Proof. Let X and Y be bases of a vector space V. If X is infinite, or if Y is infinite, then $|X| = |Y|$, by 4.11. Now, assume that X and Y are finite. Repeated applications of Lemma 5.4 construct a basis of V in which every element of X has been replaced by an element of Y. This is possible only if $|X| \leqq |Y|$. Exchanging the roles of X and Y then yields $|Y| \leqq |X|$, whence $|X| = |Y|$. $\square$

Definition. The dimension $\dim V$ *of a vector space* V *is the number of elements of its bases.*

Familiar properties of subspaces extend to vector spaces over division rings. The proofs will delight readers who miss the simpler pleasures of linear algebra.

Proposition **5.6.** *Every subspace* S *of a vector space* V *is a direct summand; moreover,* $\dim S + \dim V/S = \dim V$.

Proposition **5.7.** *Let* S *be a subspace of a vector space* V. *If* $\dim V$ *is finite and* $\dim S = \dim V$, *then* $S = V$.

Exercises

1. Prove the following: let M be a module over a division ring; when $Y \subseteq M$ generates M and $X \subseteq Y$ is linearly independent, then M has a basis $X \subseteq B \subseteq Y$.

2. Prove the following: every subspace S of a vector space V is a direct summand; moreover, $\dim S + \dim V/S = \dim V$.

3. Prove the following: let S be a subspace of a vector space V; if $\dim V$ is finite and $\dim S = \dim V$, then $S = V$.

4. Prove the following: when S and T are subspaces of a vector space, then $\dim(S \cap T) + \dim(S + T) = \dim S + \dim T$.

5. Prove the following: let V and W be vector spaces over the same division ring; when $T : V \longrightarrow W$ is a linear transformation, then $\dim \operatorname{Im} T + \dim \operatorname{Ker} T = \dim V$.

6. Let $D \subseteq E \subseteq F$ be division rings, each a subring of the next. Show that $[F : D] = [F : E][E : D]$. (As in the case of fields, $[E : D]$ denotes the dimension of E as a vector space over D.)

In the following exercises R is a ring with an identity element.

7. Show that R is a division ring if and only if it has no left ideal $L \neq 0, R$.

8. Suppose that R has a two-sided ideal that is also a maximal left ideal. Prove that all bases of a free left R-module have the same number of elements.

*9. Prove that R is a division ring if and only if every left R-module is free.

6. Modules over Principal Ideal Domains

A number of properties of abelian groups extend to modules over a principal ideal domain. In this section we construct finitely generated modules. The next section has applications to linear algebra.

Free modules. First we look at submodules of free modules.

Theorem **6.1.** *Let* R *be a principal ideal domain and let* F *be a free* R-module. *Every submodule of* F *is free, with rank at most* $\operatorname{rank} F$.

Proof. Let X be a basis of F and let M be a submodule of F. Every subset Y of X generates a submodule F_Y of F.

Assume that $Y \neq X$ and that $M_Y = F_Y \cap M$ has a basis B. Let $Z = Y \cup \{x\}$, where $x \in X \backslash Y$, so that $M_Z = \{ rx + t \in M \mid r \in R,\ t \in F_Y \}$. The basic step of the proof enlarges B to a basis of M_Z. Let

$$\mathfrak{a} = \{ r \in R \mid rx + t \in M_Z \text{ for some } t \in F_Y \}.$$

If $\mathfrak{a} = 0$, then $M_Z = M_Y$ and B is a basis of M_Z. Now, let $\mathfrak{a} \neq 0$. Since $\mathfrak{a}$ is an ideal of R we have $\mathfrak{a} = Ra$ for some $a \in R$, $a \neq 0$, and $ax + p = c \in M_Z$ for some $p \in F_Y$. We show that $C = B \cup \{c\}$ is a basis of M_Z. If

$$r_c c + \sum_{b \in B} r_b b = 0,$$

where $r_b,\ r_c \in R$ and $r_b = 0$ for almost all $b \in B$, then $r_c c \in M_Y$ and $r_c ax = r_c c - r_c p \in F_Y$. Since X is linearly independent, this implies $r_c a = 0$ and $r_c = 0$. Then $\sum_{b \in B} r_b b = 0$ and $r_b = 0$ for all $b \in B$. Thus C is linearly independent (and $c \notin B$). Next let $rx + t \in M_Z$, where $r \in R$ and $t \in F_Y$. Then $r = sa$ for some $s \in R$, $(rx + t) - sc = t - sp \in F_Y \cap M = M_Y$, and $(rx + t) - sc$ is a linear combination of elements of B. Thus $C \subseteq M_Z$ generates M_Z, and is a basis of M_Z. Note that $|B| \leq |Y|$ implies $|C| = |B| + 1 \leq |Y| + 1 = |Z|$.

We now apply Zorn's lemma to the set $\mathcal{S}$ of all pairs (Y, B) such that $Y \subseteq X$, B is a basis of M_Y, and $|B| \leq |Y|$. We have $\mathcal{S} \neq \varnothing$, since $(\varnothing, \varnothing) \in \mathcal{S}$. Partially order $\mathcal{S}$ by $(Y, B) \leq (Z, C)$ if and only if $Y \subseteq Z$ and $B \subseteq C$. If $(Y_i,\ B_i)_{i \in I}$ is a chain of elements of $\mathcal{S}$, then $(Y, B) = \left(\bigcup_{i \in I} Y_i,\ \bigcup_{i \in I} B_i \right) \in \mathcal{S}$: indeed, $F_Y = \bigcup_{i \in I} F_{Y_i}$, since a linear combination of elements of Y is a linear combination of finitely many elements of Y and is a linear combination of some Y_i; hence $M_Y = \bigcup_{i \in I} M_{Y_i}$; M_Y is generated by B, since a linear combination of elements of B is a linear combination of elements of some B_i; and B is linearly independent, for the same reason. Therefore Zorn's lemma applies to $\mathcal{S}$ and begets a maximal element (Y, B) of $\mathcal{S}$. The beginning of the proof shows that (Y, B) is not maximal when $Y \neq X$. Therefore $Y = X$, and then B is a basis of $M_X = M$ and $|B| \leq |X|$. $\square$

Theorem 6.1 can be sharpened when F is finitely generated.

Theorem 6.2. *Let R be a principal ideal domain; let F be a free R-module of finite rank n and let M be a submodule of F. There exist a basis $e_1, \ldots, e_n$ of F, an integer $0 \leq r \leq n$, and nonzero elements $a_1, \ldots, a_r$ of R, such that $a_{i+1} \in Ra_i$ for all $i < r$ and $a_1 e_1, \ldots, a_r e_r$ is a basis of M.*

Moreover, $r = \operatorname{rank} M$ is unique; it can be arranged that $a_1, \ldots, a_r$ are representative elements, and then $a_1, \ldots, a_r$ are unique, by 6.3.

Proof. The proof is by induction on the rank n of F. The result holds if $n \leq 1$, since R is a PID, or if $M = 0$; hence we may assume that $n \geq 2$ and $M \neq 0$.

If Theorem 6.2 holds, then so does the following: if $\varphi : F \longrightarrow {}_RR$ is a module homomorphism, then $\varphi(s) \in Ra_1$ for all $s \in M$, and $\varphi(M) \subseteq Ra_1$. Moreover, $\varphi(M) = Ra_1$ when $\varphi : \sum_{i \in I} x_i e_i \longmapsto x_1$. Thus Ra_1 is the largest ideal of R of the form $\varphi(M)$. With this hint we begin by finding a_1.

Since R is Noetherian, the set of all ideals of R of the form $\varphi(M)$, where $\varphi : F \longrightarrow {}_RR$ is a homomorphism, has a maximal element $\mu(M)$. Let $\mu(M) = Ra$, where $a \in R$, and let $a = \mu(m)$, where $m \in M$.

We show that $\varphi(m) \in Ra$ for every homomorphism $\varphi : F' \longrightarrow {}_RR$. Indeed, $Ra + R\varphi(m) = Rd$ for some $d = ua + v\varphi(m) \in R$. Then $\psi : u\mu + v\varphi : F \longrightarrow {}_RR$ is a homomorphism and $\psi(m) = ua + v\varphi(m) = d$; hence $\mu(M) = Ra \subseteq Rd \subseteq \psi(M)$, $Ra = Rd$ by the choice of μ, and $\varphi(m) \in Ra$.

In any basis $e_1, \ldots, e_n$ of F, $\varphi_j : \sum_i x_i e_i \longmapsto x_j$ is a homomorphism of F into ${}_RR$. Since $M \neq 0$, we have $\varphi_j(M) \neq 0$ for some j, whence $\mu(M) \neq 0$ and $a \neq 0$. By the above, $\varphi_j(m) \in Ra$ for all j; hence $m = ae$ for some $e \in F$. Then $a = \mu(m) = a\,\mu(e)$ and $\mu(e) = 1$. Hence $Re \cap \text{Ker } \mu = 0$, and $F = Re + \text{Ker } \mu$, since $x - \mu(x)e \in \text{Ker } \mu$ for all $x \in F$. Thus $F = Re \oplus \text{Ker } \mu$. If $x \in M$, then $\mu(x) \in Ra$, $\mu(x)e \in Rae = Rm \subseteq M$, and $x - \mu(x)e \in \text{Ker } \mu \cap M$; hence $M = Rm \oplus (\text{Ker } \mu \cap M)$.

Now, $\text{Ker } \mu$ has rank at most n, by 6.1. In fact, $\text{Ker } \mu$ has rank $n - 1$: if $e_2, \ldots, e_k$ is a basis of $\text{Ker } \mu$, then $e, e_2, \ldots, e_k$ is a basis of F, since $F = Re \oplus \text{Ker } \mu$ and $re = 0$ implies $r = \mu(re) = 0$; hence $k = n$. By the induction hypothesis, there exist a basis $e_2, \ldots, e_n$ of $\text{Ker } \mu$, an integer $1 \leq r \leq n$, and nonzero elements $a_2, \ldots, a_r$ of R, such that $a_{i+1} \in Ra_i$ for all $i < r$ and $a_2 e_2, \ldots, a_r e_r$ is a basis of $\text{Ker } \mu \cap M$. Then $e, e_2, \ldots, e_n$ is a basis of F, and $ae, a_2 e_2, \ldots, e_n$ is a basis of M, since $M = Rae \oplus (\text{Ker } \mu \cap M)$ and $rae = 0$ implies $r = 0$. It remains to show that $a_2 \in Ra$.

As above, $Ra + Ra_2 = Rd$ for some $d = ua + va_2 \in R$. By 4.5 there is a homomorphism $\varphi : F \longrightarrow {}_RR$ such that $\varphi(e) = \varphi(e_2) = 1$ and $\varphi(e_i) = 0$ for all $i > 2$. Then $d = ua + va_2 = \varphi(uae + va_2 e_2) \in \varphi(M)$ and $\mu(M) = Ra \subseteq Rd \subseteq \varphi(M)$. Therefore $Ra = Rd$ and $a_2 \in Ra$. $\square$

Finitely generated modules. Theorem 6.2 implies that every finitely generated module over a PID is a direct sum of cyclic modules. We state this in two essentially equivalent forms, that also include uniqueness statements.

Theorem **6.3A.** *Let R be a principal ideal domain. Every finitely generated R-module M is the direct sum*

$$M \cong F \oplus R/Ra_1 \oplus \cdots \oplus R/Ra_s$$

of a finitely generated free R-module F and cyclic R-modules $R/Ra_1, \ldots, R/Ra_s$ with annihilators $R \supsetneqq Ra_1 \supseteq \cdots \supseteq Ra_s \supsetneqq 0$. Moreover, the rank of F, the number s, and the ideals $Ra_1, \ldots, Ra_s$ are unique.

Theorem **6.3B.** *Let R be a principal ideal domain. Every finitely generated*

R-module M is the direct sum

$$M \cong F \oplus R/Rp_1^{k_1} \oplus \cdots \oplus R/Rp_t^{k_t}$$

of a finitely generated free R-module F and cyclic R-modules $R/Rp_1^{k_1}$, ..., $R/Rp_t^{k_t}$ whose annihilators $Rp_1^{k_1}$, ..., $Rp_t^{k_t}$ are generated by positive powers of prime elements of R. Moreover, the rank of F, the number t, and the ideals $Rp_1^{k_1}$, ..., $Rp_t^{k_t}$ are unique, up to their order of appearance.

In Theorem 6.3A, we can arrange that a_1, ..., a_s are representative elements; then a_1, ..., a_s are unique. Similarly, in Theorem 6.3B we can arrange that p_1, ..., p_t are representative primes; then p_1, ..., p_t and k_1, ..., k_t are unique. If $R = \mathbb{Z}$, Theorem 6.3 becomes the *fundamental theorem of finitely generated abelian groups*, a particular case of which, Theorem II.1.2, was seen in Chapter II.

Proof. Existence follows from 6.2. By 4.6, there is a surjective homomorphism $\varphi : F' \longrightarrow M$, where F' is a finitely generated free R-module. By 6.2, there exist a basis e_1, ..., e_n of F', an integer $0 \leq r \leq n$, and nonzero elements a_1, ..., a_r of R, such that $Ra_1 \supseteq \cdots \supseteq Ra_r$ and $a_1 e_1$, ..., $a_r e_r$ is a basis of Ker φ. Then

$$F' = Re_1 \oplus \cdots \oplus Re_r \oplus Re_{r+1} \oplus \cdots \oplus Re_n,$$

$$\text{Ker } \varphi = Ra_1 e_1 \oplus \cdots \oplus Ra_r e_r \oplus 0 \oplus \cdots \oplus 0, \text{ and}$$

$$M \cong F'/\text{Ker } \varphi \cong Re_1/Ra_1 e_1 \oplus \cdots \oplus Re_r/Ra_r e_r \oplus Re_{r+1} \oplus \cdots \oplus Re_n.$$

Deleting zero terms from this decomposition yields

$$M \cong Re_k/Ra_k e_k \oplus \cdots \oplus Re_r/Ra_r e_r \oplus F,$$

where $R \supsetneq Ra_k \supseteq \cdots \supseteq Ra_r$ and F is free of rank $n - r$. Finally, $Re_i/Ra_i e_i \cong R/Ra_i$, since $r \longmapsto re_i$ is an isomorphism of $_RR$ onto Re_i and sends Ra_i to $Ra_i e_i$. By 2.12, R/Ra_i is a cyclic R-module with annihilator Ra_i. ∎

Torsion. Completing the proof of Theorem 6.3 requires additional definitions.

Definitions. An element x of an R-module is torsion *when* Ann $(x) \neq 0$ *(when $rx = 0$ for some $r \in R$, $r \neq 0$),* torsion-free *when* Ann $(x) = 0$ *(when $rx = 0$ implies $r = 0$). A module is* torsion *when all its elements are torsion,* torsion-free *when all its nonzero elements are torsion-free.*

For example, finite abelian groups are torsion (as $\mathbb{Z}$-modules); when R has no zero divisors, free R-modules $F \cong \bigoplus {}_RR$ are torsion-free.

Proposition 6.4. If R is a domain, then the torsion elements of an R-module M constitute a submodule $T(M)$ of M, and $M/T(M)$ is torsion-free.

The proof is an exercise. The submodule $T(M)$ is the *torsion part* of M. Torsion modules are analyzed as follows.

Proposition **6.5.** *Let* R *be a principal ideal domain and let* P *be a set of representative prime elements of* R. *Every torsion* R-*module* M *is a direct sum* $M = \bigoplus_{p \in P} M(p)$, *where*

$$M(p) = \{x \in M \mid p^k x = 0 \text{ for some } k > 0\}.$$

Proof. Let $x \in M$. If $ax = 0$ and $a = bc \neq 0$, where $b, c \in R$ are relatively prime, then $1 = ub + vc$ for some $u, v \in R$ and $x = ubx + vcx$, where $c(ubx) = 0$ and $b(vcx) = 0$. If now $a = u\, p_1^{k_1} \cdots p_r^{k_r}$ is the product of a unit and positive powers of distinct representative primes, then $u\, p_1^{k_1} \cdots p_{r-1}^{k_{r-1}}$ and $p_r^{k_r}$ are relatively prime, and it follows, by induction on r, that $x = x_1 + \cdots + x_r$, where $p_i^{k_i} x_i = 0$, so that $x_i \in M(p_i)$. Hence $M = \sum_{p \in P} M(p)$.

Let $p \in P$ and $x \in M(p) \cap \left(\sum_{q \in P, \, q \neq p} M(q) \right)$. Then $p^k x = 0$ and $x = \sum_{q \in P, \, q \neq p} x_q$, where $x_q \in M(q)$ for all $q \neq p$, $q^{k_q} x_q = 0$ for some $k_q > 0$, and $x_q = 0$ for almost all q. Let $a = \prod_{q \in P, \, q \neq p, \, x_q \neq 0} q^{k_q}$. Then $ax_q = 0$ for all $q \neq p$ and $ax = 0$. As above, $1 = ua + vp^k$ for some $u, v \in R$, so that $x = uax + vp^k x = 0$. Thus $M(p) \cap \left(\sum_{q \in P, \, q \neq p} M(q) \right) = 0$ for all $p \in P$, and $M = \bigoplus_{p \in P} M(p)$. $\square$

Proposition 6.5 yields decompositions of cyclic modules:

Proposition **6.6.** *Let* R *be a principal ideal domain and let* $M \cong R/Ra$ *be a cyclic* R-*module, where* $a = u\, p_1^{k_1} \cdots p_r^{k_r} \in R$ *is the product of a unit and positive powers of distinct representative prime elements of* R. *Then* $M(p_i) \cong R/Rp_i^{k_i}$ *and* $M \cong R/Rp_1^{k_1} \oplus \cdots \oplus R/Rp_r^{k_r}$.

Proof. We have $ax = 0$ in $M = R/Ra$, for all $x \in M$. Let $x \in M(p)$, with $p^k x = 0$. If $p \neq p_1, \ldots, p_r$, then a and p^k are relatively prime, $1 = ua + vp^k$ for some $u, v \in R$, and $x = uax + vp^k x = 0$. Thus $M(p) = 0$.

Now, let $1 \leqq i \leqq r$ and $b = \prod_{j \neq i} p_j^{k_j}$. If $x = r + Ra \in M(p_i)$, then $p_i^k x = 0$ in $M = R/Ra$ for some $k > 0$, $p_i^k r \in Ra$, b divides $p_i^k r$, and b divides r. Conversely, if b divides r, then $r \in Rb$, $p_i^{k_i} r \in Ra$, $p_i^{k_i} x = 0$ in $M = R/Ra$, and $x \in M(p_i)$. Thus $M(p_i) = Rb/Ra$. Since R is a domain, $r \longmapsto br$ is a module isomorphism of $_R R$ onto Rb, that sends $Rp_i^{k_i}$ onto $Rp_i^{k_i} b = Ra$. Hence $M(p_i) = Rb/Ra \cong R/Rp_i^{k_i}$. $\square$

The existence part of Theorem 6.3B follows from Proposition 6.6 and the existence part of Theorem 6.3A.

Uniqueness. We first prove uniqueness when $M = M(p)$ in Theorem 6.3B.

Lemma **6.7.** *Let* R *be a PID and let* $p \in R$ *be a prime element. If* $M \cong R/Rp^{k_1} \oplus \cdots \oplus R/Rp^{k_t}$ *and* $0 < k_1 \leqq \cdots \leqq k_t$, *then the numbers* t

and $k_1, \ldots, k_t$ are uniquely determined by M.

Proof. If A is an R-module, then A/pA is an R/Rp-module, since $(Rp)(A/pA) = 0$, and A/pA is a vector space over the field R/Rp. Moreover, $p(A \oplus B) = pA \oplus pB$, so that $(A \oplus B)/p(A \oplus B) \cong A/pA \oplus B/pB$. Since $(R/Rp^k)/p(R/Rp^k) = (R/Rp^k)/(Rp/Rp^k) \cong R/Rp$, it follows from $M \cong R/Rp^{k_1} \oplus \cdots \oplus R/Rp^{k_t}$ that M/pM is a direct sum of t copies of R/Rp. Hence $t = \dim M/pM$ is uniquely determined by M.

We use induction on k_t to show that $k_1, \ldots, k_t$ are uniquely determined by M. First, $M \cong R/Rp^{k_1} \oplus \cdots \oplus R/Rp^{k_t}$, with $k_1 \leq \cdots \leq k_t$, implies $\mathrm{Ann}\,(M) = Rp^{k_t}$; hence k_t is uniquely determined by M. Since $p\,(R/Rp^k) = Rp/Rp^k \cong R/Rp^{k-1}$, we have

$$pM \cong p\bigl(R/Rp^{k_1} \oplus \cdots \oplus R/Rp^{k_t}\bigr)$$
$$\cong pR/Rp^{k_1} \oplus \cdots \oplus pR/Rp^{k_t} \cong R/Rp^{k_1-1} \oplus \cdots \oplus R/Rp^{k_t-1}.$$

Deleting zero terms yields a direct sum

$$pM \cong R/Rp^{k_{s+1}-1} \oplus \cdots \oplus R/Rp^{k_t-1},$$

where $s \geq 0$, $k_1 = \cdots = k_s = 1$, and $0 < k_{s+1} - 1 \leq \cdots \leq k_t - 1$. By the induction hypothesis, $t - s$ and $k_{s+1} - 1, \ldots, k_t - 1$ are uniquely determined by M. Therefore s and $k_1, \ldots, k_t$ are uniquely determined by M. $\square$

If M is torsion, uniqueness in Theorem 6.3B now follows from Lemma 6.7. Let

$$M \cong R/Rp_1^{k_1} \oplus \cdots \oplus R/Rp_t^{k_t},$$

where $p_1, \ldots, p_t$ are representative primes and $k_1, \ldots, k_t > 0$. For every prime p, $(A \oplus B)(p) = A(p) \oplus B(p)$, as readers will verify. Hence

$$M(p) \cong \bigoplus_i (R/Rp_i^{k_i})(p) = \bigoplus_{p_i = p} R/Rp^{k_i},$$

by 6.6. By 6.7, the number of primes $p_i = p$, and the corresponding exponents k_i, are uniquely determined by M, up to their order of appearance. Since this holds for every prime p, the number t, the representative primes $p_1, \ldots, p_t$, and their exponents $k_1, \ldots, k_t$, are uniquely determined by M, up to their order of appearance. If in the direct sum $M \cong R/Rp_1^{k_1} \oplus \cdots \oplus R/Rp_t^{k_t}$ the primes $p_1, \ldots, p_t$ are arbitrary, they can be replaced by representative primes without changing the ideals $Rp_i^{k_i}$, which are therefore uniquely determined by M, up to their order of appearance.

Uniqueness in Theorem 6.3A is proved as follows when M is torsion. Let

$$M \cong R/Ra_1 \oplus \cdots \oplus R/Ra_s,$$

where $R \supsetneq Ra_1 \supseteq \cdots \supseteq Ra_s \supsetneq 0$. Then a_s is a product of units and positive powers $p_i^{k_{is}}$ of representative prime elements $p_1, \ldots, p_r$ of R. Since

$a_1, \ldots, a_{s-1}$ divide a_s, every a_j is a product of units and positive powers $p_i^{k_{ij}}$ of $p_1, \ldots, p_r$; moreover, $0 \leqq k_{i1} \leqq \cdots \leqq k_{is}$, since $Ra_j \supseteq Ra_{j+1}$ for all $j < s$. By 6.5, M is isomorphic to the direct sum of all the cyclic modules $R/Rp_i^{k_{ij}}$. Therefore the primes $p_1, \ldots, p_r$ and their exponents k_{ij} are uniquely determined by M, up to their order of appearance. Hence $a_1, \ldots, a_s$ are uniquely determined by M up to multiplication by units, and the ideals $Ra_1, \ldots, Ra_s$ are uniquely determined by M.

Finally, let $M \cong F \oplus R/Ra_1 \oplus \cdots \oplus R/Ra_s = M'$ be any finitely generated R-module, where F is free and $a_1, \ldots, a_s \neq 0$, as in Theorem 6.3A. Every nonzero element of F is torsion-free, and every element of $T = R/Ra_1 \oplus \cdots \oplus R/Ra_s$ is torsion. If $f \in F$ and $t \in T$, then $x = f + t$ is torsion if and only if $f = 0$; thus $T(M') = T$ and $M' = F \oplus T(M')$. Hence $T \cong T(M)$, $F \cong M'/T(M') \cong M/T(M)$, and the rank of F are uniquely determined by M, up to isomorphism (in fact, rank $F = $ rank M). The uniqueness of the ideals $Ra_1, \ldots, Ra_s$ then follows from the torsion case. Uniqueness in Theorem 6.3B follows similarly from the torsion case. $\square$

Exercises

1. Let R be a PID and let M be an R-module that is generated by r elements. Show that every submodule of R can be generated by at most r elements.

2. Let R be a PID. Show that every submodule of a cyclic R-module is cyclic.

3. Let R be a domain. Show that the torsion elements of an R-module M constitute a submodule $T(M)$ of M, and that $M/T(M)$ is torsion-free.

4. Let R be a domain. Show that $Ra/Rab \cong R/Rb$ whenever $a, b \in R$, $a, b \neq 0$.

5. Let A and B be modules over a PID R. Show that $(A \oplus B)(p) = A(p) \oplus B(p)$ for every prime p of R.

6. Let R be a PID and let M be a finitely generated R-module. Show that $M(p) = 0$ for almost all representative primes p.

7. What is $\text{Ann}(M)$ when R is a PID and $M \cong R/Rp_1^{k_1} \oplus \cdots \oplus R/Rp_t^{k_t}$ is a finitely generated torsion R-module?

8. Let R be a commutative ring such that every submodule of a free R-module is free. Prove that R is a PID.

7. Jordan Form of Matrices

In this section, V is a finite-dimensional vector space over a field K, and $T : V \longrightarrow V$ is a linear transformation. We use results in the previous section to show that T has a matrix in Jordan form when K is algebraically closed. Jordan [1870] proved this when $K = \mathbb{C}$.

K[X]-modules. First we show that the linear transformation $T : V \longrightarrow V$ makes V a $K[X]$-module. This gives access to the properties in Section 6.

Proposition 7.1. *Let V be a vector space over a field K. Every linear transformation $T : V \longrightarrow V$ induces a $K[X]$-module structure on V, in which $(a_0 + a_1 X + \cdots + a_n X^n)\, v = a_0 v + a_1 T v + \cdots + a_n T^n v$.*

Proof. The ring $\operatorname{End}_K(V)$ of all linear transformations of V comes with a homomorphism $\sigma : K \longrightarrow \operatorname{End}_K(V)$ that assigns to $a \in K$ the *scalar* transformation $\sigma(a)\colon v \longmapsto av$. Now, T commutes with every scalar transformation. By the universal property of $K[X]$, there is a homomorphism $\varphi : K[X] \longrightarrow \operatorname{End}_K(V) \subseteq \operatorname{End}_{\mathbb{Z}}(V)$ that extends σ and sends X to T, namely $\varphi(a_0 + a_1 X + \cdots + a_n X^n) = \sigma(a_0) + \sigma(a_1)T + \cdots + \sigma(a_n)T^n$. This makes V a $K[X]$-module, in which $(a_0 + a_1 X + \cdots + a_n X^n)\, v = \varphi(a_0 + a_1 X + \cdots + a_n X^n)(v) = a_0 v + a_1 T v + \cdots + a_n T^n v$, as in the statement. $\square$

If $f(X) = a_0 + a_1 X + \cdots + a_n X^n \in K[X]$, then $a_0 + a_1 T + \cdots + a_n T^n = f(T)$ in $\operatorname{End}_K(V)$; with this notation, $f(X)v = f(T)v$.

Proposition 7.2. *In the $K[X]$-module structure on V induced by T, a submodule of V is a subspace S of V such that $TS \subseteq S$, and then the $K[X]$-module structure on S induced by V coincides with the $K[X]$-module structure on S induced by the restriction $T_{|S}$ of T to S.*

Proof. Either way, $Xs = Ts = T_{|S}s$ for all $s \in S$. (Then $f(T)_{|S} = f(T_{|S})$ for all $f \in K[X]$.) $\square$

Proposition 7.3. *If V is finite-dimensional, then the ideal $\operatorname{Ann}(V)$ of $K[X]$ is generated by a unique monic polynomial $m_T \in K[X]$; then $m_T(T) = 0$, and $f(T) = 0$ if and only if m_T divides f, for all $f \in K[X]$.*

Proof. $\operatorname{Ann}(V) = \{\, f \in K[X] \mid f(T) = 0 \,\}$ is an ideal of $K[X]$. Moreover, $\operatorname{Ann}(V) \neq 0$: $\operatorname{Ann}(V)$ is the kernel of the homomorphism $\varphi : f \longmapsto f(T)$ of $K[X]$ into $\operatorname{End}_K(V)$, which cannot be injective since $\dim_K \operatorname{End}_K(V)$ is finite but $\dim_K K[X]$ is infinite. Since $K[X]$ is a PID, the nonzero ideal $\operatorname{Ann}(V)$ is generated by a unique monic polynomial m_T. $\square$

Definition. In Proposition 7.3, m_T is the minimal polynomial *of T.*

We now bring in the heavy artillery. If V is finite-dimensional, then V is finitely generated as a $K[X]$-module (since V is already finitely generated as a K-module). Moreover, V is torsion, since $m_T(X)\, v = 0$ for all $v \in V$. By Theorem 6.3B, V is a direct sum of cyclic submodules $S_1, \ldots, S_t$ whose annihilators are generated by positive powers $q_1^{k_1}, \ldots, q_t^{k_t}$ of prime elements (irreducible polynomials) $q_1, \ldots, q_t \in K[X]$, moreover, we can arrange that $q_1, \ldots, q_t$ are monic, and then the number t, the polynomials $q_1, \ldots, q_t$, and the positive exponents $k_1, \ldots, k_t$ are unique, up to their order of appearance. By 7.2, 7.3, $q_i^{k_i}$ is the minimal polynomial of $T_{|S_i}$. Moreover, $f(T) = 0$ if and only if $f(T_{|S_i}) = 0$ for all i, since $V = \bigoplus_i S_i$. Thus we obtain the following result:

Theorem 7.4. *Let V be a finite-dimensional vector space over a field K,* *and let* $T : V \longrightarrow V$ *be a linear transformation. As a* $K[X]$*-module,* V *is the direct sum* $V = S_1 \oplus \cdots \oplus S_t$ *of cyclic submodules* $S_1, \ldots, S_t$, *such that the minimal polynomial of* $T_{|S_i}$ *is a positive power* $q_i^{k_i}$ *of a monic irreducible polynomial* $q_i \in K[X]$. *The number* t, *the polynomials* $q_1, \ldots, q_t$, *and the positive exponents* $k_1, \ldots, k_t$ *are unique, up to their order of appearance, and the minimal polynomial of* T *is the least common multiple of* $q_1^{k_1}, \ldots, q_t^{k_t}$.

Cyclic modules. We now take a closer look at cyclic $K[X]$-modules.

Lemma **7.5.** *If* $\dim_K V = n$ *is finite and* $V = K[X]e$ *is a cyclic* $K[X]$*-module, then* $\deg m_T = n$ *and* $e, Te, \ldots, T^{n-1}e$ *is a basis of* V *over* K.

Proof. Let $m = \deg m_T$. Every element of V has the form $f(X)e$ for some $f \in K[X]$. Now, $f = m_T q + r$, where $\deg r < m$, and $f(X)e = r(X)e$. Hence every element of V is a linear combination of $e, Te, \ldots, T^{m-1}e$ with coefficients in K. If $a_0 e + a_1 Te + \cdots + a_{m-1}T^{m-1}e = 0$, where $a_0, a_1, \ldots, a_{m-1} \in K$, and $g(X) = a_0 + a_1 X + \cdots + a_{m-1}X^{m-1} \in K[X]$, then $g(X)e = 0$, $g(X)v = 0$ for all $v = f(X)e \in V$, $g(T) = 0$, g is a multiple of m_T, and $g = 0$, since $\deg g < \deg m_T$; hence $a_0 = a_1 = \cdots = a_{m-1} = 0$. Therefore $e, Te, \ldots, T^{m-1}e$ is a basis of V. In particular, $m = \dim_K V$. $\square$

If K is algebraically closed, then the monic irreducible polynomials q_i in Theorem 7.4 have degree 1 and the minimal polynomials $q_i^{k_i}$ have the form $(X - \lambda_i)^{k_i}$, where $\lambda_i \in K$. In this case, Lemma 7.5 simplifies and yields a triangular matrix.

Lemma **7.6.** *If* $\dim_K V = n$ *is finite,* V *is a cyclic* $K[X]$*-module, and* $m_T = (X - \lambda)^m$ *for some* $\lambda \in K$, *then* $m = n$ *and* V *has a basis* $e_1, \ldots, e_n$ *over* K *such that* $Te_1 = \lambda e_1$ *and* $Te_i = \lambda e_i + e_{i-1}$ *for all* $i > 1$.

Proof. First, $m = n$, by 7.5. Let $V = K[X]e$, where $e \in V$. Let $e_i = (T - \lambda)^{n-i}e$ for all $i = 1, 2, \ldots, n$. By the binomial theorem, $e_{n-i} = (T - \lambda)^i e$ is a linear combination of $T^i e, T^{i-1}e, \ldots, e$, in which the coeficient of $T^i e$ is 1. Hence the matrix of $e_n, \ldots, e_1$ in the basis $e, Te, \ldots, T^{n-1}e$ is upper triangular, with 1's on the diagonal, and is invertible. Therefore $e_1, \ldots, e_n$ is a basis of V. If $i > 1$, then $(T - \lambda)e_i = e_{i-1}$. If $i = 1$, then $(T - \lambda)e_1 = (T - \lambda)^n e = 0$. $\square$

In Lemma 7.6, the matrix of T in the basis $e_1, \ldots, e_n$ is

$$
\begin{pmatrix}
\lambda & 1 & 0 & \ldots & 0 & 0 \\
0 & \lambda & 1 & \ddots & 0 & 0 \\
0 & 0 & \lambda & \ddots & 0 & 0 \\
\vdots & \vdots & \ddots & \ddots & 1 & 0 \\
0 & 0 & \ldots & 0 & \lambda & 1 \\
0 & 0 & \ldots & 0 & 0 & \lambda
\end{pmatrix}.
$$

A square matrix in this form is a *Jordan block* with λ on the diagonal. Jordan blocks are also defined with 1's below the diagonal rather than above.

Jordan form. A square matrix is in *Jordan form* when it consists of Jordan blocks arranged along the diagonal, with zeros elsewhere:

$$\begin{pmatrix} J_1 & 0 & \cdots & 0 \\ 0 & J_2 & \cdots & 0 \\ \vdots & & \ddots & \vdots \\ 0 & 0 & \cdots & J_t \end{pmatrix}.$$

If K is algebraically closed and $\dim_K V$ is finite, then, by 7.4, V is a direct sum $V = S_1 \oplus \cdots \oplus S_t$ of cyclic submodules S_1, ..., S_t, such that the minimal polynomial of $T_{|S_i}$ is a positive power $(X - \lambda_i)^{k_i}$ of a monic irreducible polynomial $q_i = X - \lambda_i \in K[X]$; moreover, the number t, the polynomials q_1, ..., q_t, and the positive exponents k_1, ..., k_t are unique, up to their order of appearance. By 7.6, S_i has a basis B_i over K in which the matrix of $T_{|S_i}$ is a Jordan block with eigenvalue λ_i. Then $B_1 \cup \cdots \cup B_t$ is a basis of V in which the matrix of T is in Jordan form, and we have proved our main result:

Theorem 7.7. Let V be a finite-dimensional vector space over an algebraically closed field K, and let $T : V \longrightarrow V$ be a linear transformation. There exists a basis of V in which the matrix of T is in Jordan form. Moreover, all such matrices of T contain the same Jordan blocks.

Readers may prove the following properties:

Corollary 7.8. If the matrix of T is in Jordan form, then:

(1) *the diagonal entries are the eigenvalues of T;*

(2) *the minimal polynomial of T is $(X - \lambda_1)^{\ell_1} \cdots (X - \lambda_r)^{\ell_r}$, where λ_1, ..., λ_r are the distinct eigenvalues of T and ℓ_i is the size of the largest Jordan block with λ_i on the diagonal;*

(3) *when λ is an eigenvalue of T, the dimension of the corresponding eigenspace equals the number of Jordan blocks with λ on the diagonal.*

Corollary 7.9. Let V be a finite-dimensional vector space over an algebraically closed field K. A linear transformation $T : V \longrightarrow V$ is diagonalizable (there is a basis of V in which the matrix of T is diagonal) if and only if its minimal polynomial is separable.

Theorem 7.7 has another consequence that concerns the *characteristic polynomial* c_T of T, $c_T(X) = \det(T - XI)$ (where I is the identity on V).

Theorem 7.10 (Cayley-Hamilton). *If V is a finite-dimensional vector space over a field K, then $c_T(T) = 0$ for every linear transformation T of V.*

Proof. If K is algebraically closed, then V has a basis in which the matrix of T is in Jordan form. Then $c_T(X) = (X - \lambda_1)^{n_1} \cdots (X - \lambda_r)^{n_r}$, where

$\lambda_1, \ldots, \lambda_r$ are the distinct eigenvalues of T and n_i is the number of appearances of λ_i on the diagonal. By 7.8, m_T divides c_T, and $c_T(T) = 0$.

In general, the characteristic polynomial c_T of T is also the characteristic polynomial c_M of its matrix M in any basis of V. Now, c_M does not change when M is viewed as a matrix with entries in the algebraic closure $\overline{K}$ of K, rather than as a matrix with entries in K. Hence $c_M(M) = 0$, and $c_T(T) = 0$. $\square$

Exercises

1. State the theorem obtained by using Theorem 6.3A rather than Theorem 6.3B in the proof of Theorem 7.4.

2. Define the minimal polynomial m_A of an $n \times n$ matrix A with coefficients in a field K. Prove that m_A does not change when K is replaced by one of its field extensions. (You may want to use the previous exercise.)

3. Let $\dim_K V = n$ be finite and let $T : V \longrightarrow V$ be a linear transformation whose matrix in some basis of V is a Jordan block, with λ on the diagonal. Show that V is a cyclic $K[X]$-module and that $m_T(X) = (X - \lambda)^n$.

4. Having studied this section, what can you say about nilpotent linear transformations? (T is *nilpotent* when $T^m = 0$ for some $m > 0$.)

5. Let the matrix of T be in Jordan form. Show that the minimal polynomial of T is $(X - \lambda_1)^{\ell_1} \cdots (X - \lambda_r)^{\ell_r}$, where $\lambda_1, \ldots, \lambda_r$ are the distinct eigenvalues of T and ℓ_i is the size of the largest Jordan block with λ_i on the diagonal.

6. Let the matrix of T be in Jordan form, and let λ be an eigenvalue of T. Show that the dimension of the corresponding eigenspace equals the number of Jordan blocks with λ on the diagonal.

7. Let $\dim_K V$ be finite. Show that a linear transformation of V is diagonalizable if and only if its minimal polynomial is separable.

8. Chain Conditions

This section contains basic properties of Noetherian modules, Artinian modules, and modules of finite length, for use in the next chapter. As before, all rings have an identity element and all modules are unital.

Noetherian modules. Applied to the submodules of a left R-module M, the *ascending chain condition* or *a.c.c.* has three equivalent forms:

(a) every infinite ascending sequence $A_1 \subseteq \cdots \subseteq A_n \subseteq A_{n+1} \subseteq \cdots$ of submodules of M *terminates*: there exists $N > 0$ such that $A_n = A_N$ for all $n \geq N$;

(b) there is no infinite strictly ascending sequence $A_1 \subsetneqq \cdots \subsetneqq A_n \subsetneqq A_{n+1} \subsetneqq \cdots$ of submodules of M;

(c) every nonempty set $\mathcal{S}$ of submodules of M has a maximal element (an element S of $\mathcal{S}$ such that there is no $S \subsetneqq A \in \mathcal{S}$).

The equivalence of (a), (b), and (c) is proved as in Section III.11 (see also Section A.1).

Definition. A module is Noetherian *when its submodules satisfy the ascending chain condition.*

For example, a commutative ring R is Noetherian if and only if the module ${}_R R$ is Noetherian.

Definition. A ring R is left Noetherian *when the module ${}_R R$ is Noetherian. A ring R is* right Noetherian *when the module ${}_R R$ is Noetherian.*

Equivalently, R is left (right) Noetherian when its left (right) ideals satisfy the ascending chain condition.

In a module, the a.c.c. has a fourth equivalent form, proved like III.11.1:

Proposition 8.1. A module M is Noetherian if and only if every submodule of M is finitely generated.

Proposition 8.2. If N is a submodule of a module M, then M is Noetherian if and only if N and M/N are Noetherian.

Proof. Let N and M/N be Noetherian and let $A_1 \subseteq A_2 \subseteq \cdots \subseteq A_n \subseteq A_{n+1} \subseteq \cdots$ be an infinite ascending sequence of submodules of M. Then

$$A_1 \cap N \subseteq A_2 \cap N \subseteq \cdots \subseteq A_n \cap N \subseteq A_{n+1} \cap N \subseteq \cdots$$

is an infinite ascending sequence of submodules of N, and

$$(A_1 + N)/N \subseteq (A_2 + N)/N \subseteq \cdots \subseteq (A_n + N)/N \subseteq (A_{n+1} + N)/N \subseteq \cdots$$

is an infinite ascending sequence of submodules of M/N. Since N and M/N are Noetherian, both sequences terminate: there exists $m > 0$ such that $A_n \cap N = A_m \cap N$ and $(A_n + N)/N = (A_m + N)/N$ for all $n \geq m$. Then $A_n + N = A_m + N$ for all $n \geq m$. It follows that $A_n = A_m$ for all $n \geq m$: indeed, $x \in A_n$ implies $x \in A_m + N$, $x = y + z$ for some $y \in A_m$ and $z \in N$, $z = x - y \in A_n \cap N = A + m \cap N$, and $x = y + z \in A_m$. Therefore M is Noetherian. The converse is an exercise. $\square$

Proposition 8.3. If R is left Noetherian, then every finitely generated left R-module is Noetherian.

Proof. A finitely generated free left R-module F is the direct sum $F = R^{(n)}$ of n copies of ${}_R R$ and is Noetherian, by induction on n: $R^{(1)} = {}_R R$ is Noetherian, and if $R^{(n)}$ is Noetherian, then so is $R^{(n+1)}$ by 8.2, since $R^{(n)} \subseteq R^{(n+1)}$ and $R^{(n+1)}/R^{(n)} \cong {}_R R$ are Noetherian. If now M is a finitely generated left R-module, then $M \cong F/N$ for some finitely generated free R-module F and submodule N of F, by 4.6, and M is Noetherian, by 8.2. $\square$

Artinian modules. Applied to the submodules of a left R-module M, the *descending chain condition* or *d.c.c.* has three equivalent forms:

(a) every infinite descending sequence $A_1 \supseteq \cdots \supseteq A_n \supseteq A_{n+1} \supseteq \cdots$ of submodules of M *terminates*: there exists $N > 0$ such that $A_n = A_N$ for all $n \geq N$;

(b) there is no infinite strictly descending sequence $A_1 \supsetneq \cdots \supsetneq A_n \supsetneq A_{n+1} \supsetneq \cdots$ of submodules of M;

(c) every nonempty set $\mathcal{S}$ of submodules of M has a minimal element (an element S of $\mathcal{S}$ such that there is no $S \supsetneq A \in \mathcal{S}$).

Readers will prove the equivalence of (a), (b), and (c), as in Section III.11, by reversing inclusions; again, a more general proof is given in Section A.1.

Definitions. A module is Artinian *when its submodules satisfy the descending chain condition. A ring R is* left Artinian *when the module $_RR$ is Artinian. A ring R is* right Artinian *when the module R_R is Artinian.*

Equivalently, R is left (right) Artinian when its left (right) ideals satisfy the descending chain condition. Artinian rings and modules are named after Emil Artin, who pioneered the use the d.c.c. in rings. Finite abelian groups are Artinian $\mathbb{Z}$-modules; the next chapter has examples of Artinian rings. The following properties make fine exercises:

Proposition **8.4.** *If N is a submodule of a module M, then M is Artinian if and only if N and M/N are Artinian.*

Proposition **8.5.** *If R is left Artinian, then every finitely generated left R-module is Artinian.*

Modules of finite length. We saw in Section 2 that the isomorphism theorems for groups extend to every module. So does the Jordan-Hölder theorem. Rather than repeating proofs, we only sketch the main results, and leave the details to our more intrepid readers.

In a module, normal series are replaced by finite chains of submodules, also called *series*. The *length* of a series $A_0 \subsetneq A_1 \subsetneq \cdots \subsetneq A_n$ is its number n of intervals; its *factors* are the quotient modules A_i/A_{i-1}. A *refinement* of a series is a series that contains it (as a set of submodules). Two series are *equivalent* when they have the same length and, up to isomorphism and order of appearance, the same factors. Schreier's theorem has a module form:

Theorem **8.6.** *Any two series of a module have equivalent refinements.*

A *composition series* of a module M is a finite chain of submodules that is also a maximal chain of submodules (has no proper refinement). Not every module is blessed with a composition series. But modules that are so blessed, are also blessed with a Jordan-Hölder theorem:

Theorem **8.7.** *Any two composition series of a module are equivalent.*

Definition. A module M is of finite length *when it has a composition series; then the* length *of M is the common length of all its composition series.*

For example, finite abelian groups are of finite length as $\mathbb{Z}$-modules. Modules with compositions series also turn up in the next chapter.

Proposition **8.8.** *The following conditions on a module M are equivalent:*

(1) M *is both Noetherian and Artinian;*

(2) *every chain of submodules of M is finite;*

(3) M *is of finite length (M has a composition series).*

Then (4) *all chains of submodules of M have length at most n, where n is the length of M.*

Proof. (1) implies (2). Suppose that M is Artinian and contains an infinite chain C of submodules. Since M is Artinian, C has a minimal element A_1, which is in fact the least element of C since C is a chain. Then $C \backslash \{A_1\}$ is an infinite chain and has a least element $A_2 \supsetneqq A_1$. Continuing thus builds up an infinite strictly ascending sequence of submodules of M. Therefore, if M is Artinian and Noetherian, then every chain of submodules of M is finite.

(2) implies (3). The union of a chain $(C_i)_{i \in I}$ of chains C_i of submodules of M is a chain of submodules: if $A, B \in \bigcup_{i \in I} C_i$, then $A, B \in C_i$ for some i, and $A \subseteq B$ or $B \subseteq A$. By Zorn's lemma, M has a maximal chain of submodules, which is a composition series if (2) holds.

(3) implies (4) and (2). Let n be the length of a composition series. By Schreier's theorem, any series of M and the composition series have equivalent refinements; hence every series of M can be refined to a composition series. Therefore every finite chain of submodules of M has length at most n, and there cannot be an infinite chain of submodules.

(2) implies (1). An infinite strictly ascending or descending sequence of submodules would contradict (2). □

Modules of finite length have another property. A module M is *indecomposable* when $M \neq 0$ and M has no proper direct summand; equivalently, when $M \neq 0$, and $M = A \oplus B$ implies $A = 0$ or $B = 0$.

Proposition **8.9.** *Every module of finite length is a direct sum of finitely many indecomposable submodules.*

This is proved like II.2.2, a fun exercise for our readers. More courageous readers will wade through the proof of the Krull-Schmidt theorem (Theorem II.2.3) and adapt it to prove the module version, which is a uniqueness property for Proposition 8.9:

Theorem **8.10.** *If a module M of finite length is a direct sum*

$$M = A_1 \oplus A_2 \oplus \cdots \oplus A_m = B_1 \oplus B_2 \oplus \cdots \oplus B_n$$

of indecomposable submodules $A_1, \ldots, A_m$ and $B_1, \ldots, B_n$, then $m = n$ and $B_1, \ldots, B_n$ can be indexed so that $B_i \cong A_i$ for all $i \leq n$ and, for all $k < n$,

$$M = A_1 \oplus \cdots \oplus A_k \oplus B_{k+1} \oplus \cdots \oplus B_n.$$

Exercises

In the following exercises, all rings have an identity element and all modules are unital. Prove the following:

1. A module M is Noetherian if and only if every submodule of M is finitely generated.

2. Every submodule of a Noetherian module is Noetherian.

3. Every quotient module of a Noetherian module is Noetherian.

4. Every submodule of an Artinian module is Artinian.

5. Every quotient module of an Artinian module is Artinian.

6. If N is a submodule of M, and N, M/N are Artinian, then M is Artinian.

7. If R is a left Artinian ring, every finitely generated left R-module is Artinian.

8. Let E be an infinite extension of a field K. Let R be the set of all matrices $\begin{pmatrix} \alpha & \beta \\ 0 & c \end{pmatrix}$ with $\alpha, \beta \in E$ and $c \in K$. Show that R is a subring of $M_2(E)$. Show that R is left Noetherian and left Artinian, but is neither right Noetherian nor right Artinian.

9. A direct sum of finitely many left R-modules of finite length is a left R-module of finite length.

10. Every finitely generated torsion module over a PID is of finite length.

*11. Any two series of a module have equivalent refinements.

12. Any two composition series of a module are equivalent (use the previous exercise).

13. Every module of finite length is a direct sum of finitely many indecomposable submodules.

14. If R is a PID, then $_R R$ is indecomposable, and R/Rp^k is indecomposable when $p \in R$ is prime and $k > 0$.

*15. Prove the module version of the Krull-Schmidt theorem.

9. Gröbner Bases

Gröbner bases of ideals of $K[X_1, ..., X_n]$ were seen in Section III.12. Similar bases are readily found for submodules of finitely generated free modules over $K[X_1, ..., X_n]$. Applications include a test for membership in a submodule, given some generators; the determination of all linear relations between these generators; and, in Chapter XII, effective computation of free resolutions. We have omitted some proofs that nearly duplicate similar proofs in Section III.12.

Monomial orders. In what follows, K is a field, $R = K[X_1, ..., X_n]$, and F is a free R-module with a finite basis $E = \{\varepsilon_1, ..., \varepsilon_r\}$. A *monomial* of F is an element $X^m \varepsilon_j$ of F, where X^m is a monomial of R and $\varepsilon_j \in E$. For instance, $K[X_1, ..., X_n]$ is a free R-module, with basis $\{1\}$; its monomials are the usual monomials. In general, the elements of F resemble polynomials in that they are linear combinations, with coefficients in K, of monomials of F. We denote monomials of F by Greek letters α, β, ...

Definition. A monomial order *on* F over *a monomial order* $<$ *on* $K[X_1,$ $..., X_n]$ *is a total order on the monomials of* F *such that* $X^a < X^b$ *implies* $X^a \alpha < X^b \alpha$, *and* $\alpha < \beta$ *implies* $X^m \alpha < X^m \beta$.

Monomial orders on F are readily constructed from monomial orders on R:

Proposition 9.1. *Let* $<$ *be a monomial order on* $K[X_1, ..., X_n]$. *A monomial order on* F *is defined by* $X^a \varepsilon_i < X^b \varepsilon_j$ *if and only if either* $X^a < X^b$ *in* $K[X_1, ..., X_n]$, *or* $X^a = X^b$ *and* $i < j$.

The proof is an exercise. Note that, in any monomial order, $\alpha \geqq \beta$ whenever $\alpha = X^m \beta$ is a multiple of β.

Proposition 9.2. *Every monomial order on* F *satisfies the descending chain condition.*

This is proved like Proposition III.12.2, using the following lemma.

Lemma 9.3. *An submodule of* F *that is generated by a set* S *of monomials is generated by a finite subset of* S.

Proof. First, F is a Noetherian module, by 8.3. Hence the submodule M generated by S is generated by finitely many $f_1, ..., f_t \in F$. Every nonzero term of f_i is a multiple of some $\sigma \in S$. Let T be the set of all $\sigma \in S$ that divide a nonzero term of some f_i. Then $T \subseteq S$ is finite and the submodule generated by S is all of M, since it contains every f_i. $\square$

Gröbner bases. In all that follows we assume that F has a monomial order, which is used in all subsequent operations.

Definitions. Let $f = \sum_\alpha a_\alpha \alpha \in F$, $f \neq 0$. *The* leading monomial *of* f *is the greatest monomial* $\operatorname{ldm} f = \lambda$ *such that* $a_\lambda \neq 0$; *then the* leading coefficient *of* f *is* $\operatorname{ldc} f = a_\lambda$ *and the* leading term *of* f *is* $\operatorname{ldt} f = a_\lambda \lambda$.

Other notations, for instance, in (f), are in use for $\operatorname{ldt} f$. An element of F can now be divided by several others; as before, the results are not unique.

Proposition 9.4. *Let* $f, g_1, ..., g_k \in F$, $g_1, ..., g_k \neq 0$. *There exist* $q_1, ..., q_k \in K[X_1, ..., X_n]$ *and* $r \in F$ *such that*

$$f = q_1 g_1 + \cdots + q_k g_k + r,$$

$\operatorname{ldm}(q_i g_i) \leqq \operatorname{ldm} f$ *for all* i, $\operatorname{ldm} r \leqq \operatorname{ldm} f$, *and none of* $\operatorname{ldm} g_1, ..., \operatorname{ldm} g_k$ *divides a nonzero term of the remainder* r.

This is proved like Proposition III.12.4. Division in $K[X_1, ..., X_n]$ is a particular case.

The *membership problem* for submodules of F is, does $f \in F$ belong to the submodule generated by $g_1, ..., g_k \in F$? We saw in Section III.12 that unbridled division does not provide a reliable test for membership. This is where

Gröbner bases come in: as in Chapter III, division by a Gröbner basis is a good membership test (see Proposition 9.6 below).

Definition. Let M be a submodule of F. Let $\operatorname{ldm} M$ be the submodule of F generated by all $\operatorname{ldm} f$ with $f \in M$. Nonzero elements $g_1, \ldots, g_k$ of F constitute a Gröbner basis of M (relative to the given monomial order) when $g_1, \ldots, g_k$ generate M and $\operatorname{ldm} g_1, \ldots, \operatorname{ldm} g_k$ generate $\operatorname{ldm} M$.

Proposition 9.5. *Let M be a submodule of $\dot{F}$. If $g_1, \ldots, g_k \in M$ and $\operatorname{ldm} g_1, \ldots, \operatorname{ldm} g_k$ generate $\operatorname{ldm} M$, then $g_1, \ldots, g_k$ is a Gröbner basis of M.*

Proof. We need only show that $g_1, \ldots, g_k$ generate M. Let $f \in M$. By 9.4, $f = \sum_i q_i g_i + r$, where none of $\operatorname{ldm} g_1, \ldots, \operatorname{ldm} g_k$ divides a nonzero term of r. Then $r = 0$: otherwise, $r \in M$, $\operatorname{ldm} r \in \operatorname{ldm} M$, $\operatorname{ldm} r = \sum_i p_i \operatorname{ldm} g_i$ for some $p_1, \ldots, p_k \in K[X_1, \ldots, X_n]$, $\operatorname{ldm} r$ must appear in some $p_i \operatorname{ldm} g_i$, and $\operatorname{ldm} r$ is a multiple of $\operatorname{ldm} g_i$. $\square$

Proposition 9.6. *Let $g_1, \ldots, g_k$ be a Gröbner basis of a submodule M of F. All divisions of $f \in F$ by $g_1, \ldots, g_k$ yield the same remainder r, and $f \in M$ if and only if $r = 0$.*

Proof. This is proved like Proposition III.12.5. Let r be the remainder in a division of f by $g_1, \ldots, g_k$. If $r = 0$, then $f \in M$. Conversely, if $f \in M$, then $r = 0$: otherwise, $r \in M$, $\operatorname{ldm} r \in \operatorname{ldm} M$ is a linear combination of $\operatorname{ldm} g_1, \ldots, \operatorname{ldm} g_k$, and $\operatorname{ldm} r$ is a multiple of some $\operatorname{ldm} g_j$. If now r_1 and r_2 are remainders in divisions of any $f \in F$ by $g_1, \ldots, g_k$, then $r_1 - r_2 \in M$, and no $\operatorname{ldm} g_j$ divides a nonzero term of $r_1 - r_2$. As above, this implies $r_1 - r_2 = 0$: otherwise, $r_1 - r_2 \in M$, $\operatorname{ldm}(r_1 - r_2) \in \operatorname{ldm} M$ is a linear combination of $\operatorname{ldm} g_1, \ldots, \operatorname{ldm} g_k$, and $\operatorname{ldm}(r_1 - r_2)$ is a multiple of some $\operatorname{ldm} g_j$. $\square$

Buchberger's algorithm. We now give an effective procedure, which constructs a Gröbner basis of a submodule from any finite set of generators. First we extend Buchberger's criterion to F.

In F, two monomials $\alpha = X^a \varepsilon_i$ and $\beta = X^b \varepsilon_j$ have a common monomial multiple if and only if $\varepsilon_i = \varepsilon_j$. Thus the *least common multiple* of $\alpha = X^a \varepsilon_i$ and $\beta = X^b \varepsilon_j$ is defined if and only if $\varepsilon_i = \varepsilon_j$ (if and only if α and β are multiples of the same ε_i), and then $\operatorname{lcm}(\alpha, \beta) = \operatorname{lcm}(X^a, X^b) \varepsilon_i$.

Proposition 9.7 (Buchberger's Criterion). *Let $g_1, \ldots, g_k \neq 0$ generate a submodule M of F. For all i, j such that $\lambda_{ij} = \operatorname{lcm}(\operatorname{ldm} g_i, \operatorname{ldm} g_j)$ is defined, let $d_{i,j} = (\lambda_{ij}/\operatorname{ldt} g_i) g_i - (\lambda_{ij}/\operatorname{ldt} g_j) g_j$ and let $r_{i,j}$ be the remainder in a division of $d_{i,j}$ by $g_1, \ldots, g_k$. Then $g_1, \ldots, g_k$ is a Gröbner basis of M if and only if $r_{i,j} = 0$ for all possible $i < j$, and then $r_{i,j} = 0$ for all possible i, j.*

Proof. We may assume that $\operatorname{ldc} g_1 = \cdots = \operatorname{ldc} g_k = 1$, since the submodules

M and $\operatorname{ldm} M$ generated by $g_1, \ldots, g_k$ and $\operatorname{ldm} g_1, \ldots, \operatorname{ldm} g_k$ do not change when $g_1, \ldots, g_k$ are divided by their leading coefficients.

If $g_1, \ldots, g_k$ is a Gröbner basis of M, then $r_{i,j} = 0$ by 9.6, since $d_{i,j} \in M$.

The converse follows from two properties of $d_{i,j}$. Let $\operatorname{ldt} g_i = \lambda_i$, so that $d_{i,j} = (\lambda_{ij}/\lambda_i)\, g_i - (\lambda_{ij}/\lambda_j)\, g_j$, when λ_{ij} is defined.

(1) If $g_i' = X^{t_i} g_i$, $g_j' = X^{t_j} g_j$, and $\lambda_{ij}' = \operatorname{lcm}(X^{t_i}\lambda_i,\, X^{t_j}\lambda_j)$, then

$$
\begin{aligned}
d_{i,j}' &= (\lambda_{ij}'/\operatorname{ldt} g_i')\, g_i' - (\lambda_{ij}'/\operatorname{ldt} g_j')\, g_j' \\
&= (\lambda_{ij}'/X^{t_i}\lambda_i)\, X^{t_i} g_i - (\lambda_{ij}'/X^{t_j}\lambda_j)\, X^{t_j} g_j = (\lambda_{ij}'/\lambda_{ij})\, d_{i,j}.
\end{aligned}
$$

(2) If $\operatorname{ldm} g_i = \lambda$ for all i and $\operatorname{ldm}(a_1 g_1 + \cdots + a_k g_k) < \lambda$, where $a_1, \ldots, a_k \in K$ and $k \geqq 2$, then $d_{i,j} = g_i - g_j$ for all $i, j \leqq k$ and

$$
\begin{aligned}
& a_1 g_1 + \cdots + a_k g_k \\
&= a_1 (g_1 - g_2) + (a_1 + a_2)(g_2 - g_3) + \cdots + (a_1 + \cdots + a_{k-1})(g_{k-1} - g_k) \\
&\quad + (a_1 + \cdots + a_k)\, g_k \\
&= a_1 d_{1,2} + (a_1 + a_2) d_{2,3} + \cdots + (a_1 + \cdots + a_{k-1}) d_{k-1,k},
\end{aligned}
$$

since $\operatorname{ldm}(a_1 g_1 + \cdots + a_k g_k) < \lambda$ implies $a_1 + \cdots + a_k = 0$.

Now, assume that $r_{ij} = 0$ for all possible $i < j$. Then $r_{i,j} = 0$ for all possible i and j, since $d_{i,i} = 0$ and $d_{j,i} = -d_{i,j}$. Every nonzero $f \in M$ is a linear combination $f = p_1 g_1 + \cdots + p_k g_k$ with coefficients $p_1, \ldots, p_k \in R$. Let μ be the greatest $\operatorname{ldm}(p_j g_j)$. By 9.3, every nonempty set of monomials of F has a minimal element. Choose $p_1, \ldots, p_k$ so that μ is minimal.

Suppose that μ does not appear in f. Number $g_1, \ldots, g_k$ so that μ is the leading monomial of the first products $p_1 g_1, \ldots, p_h g_h$. Then $k \geqq 2$: otherwise, μ is not canceled in the sum $p_1 g_1 + \cdots + p_k g_k$. Also $\operatorname{ldm}(p_1 g_1 + \cdots + p_k g_k) < \mu$, since μ does not appear in f or in $p_{k+1} g_{k+1} + \cdots + p_k g_k$.

Let $\operatorname{ldt} p_j = a_j X^{t_j}$ (in the underlying monomial order on R) and $g_j' = X^{t_j} g_j$. Then $\operatorname{ldm} g_j' = \operatorname{ldm}(p_j g_j)$; hence $\operatorname{ldm} g_j' = \mu$ for all $j \leqq h$ and $\operatorname{ldm}(a_1 g_1' + \cdots + a_h g_h') < \mu$. By (2) and (1),

$$
a_1 g_1' + \cdots + a_h g_h' = c_1 d_{1,2}' + \cdots + c_{h-1} d_{h-1,h}'
$$

for some $c_1, \ldots, c_{h-1} \in K$, where

$$
d_{i,j}' = g_i' - g_j' = (\mu/\lambda_{ij})\, d_{i,j}.
$$

Note that $\operatorname{ldm} d_{i,j}' < \mu$ when $i < j \leqq h$, since $\operatorname{ldm} g_i' = \operatorname{ldm} g_j' = \mu$. By the hypothesis, every $d_{i,j}$ can be divided by $g_1, \ldots, g_k$ with zero remainder when $i < j$; so can $d_{i,j}'$ and $c_1 d_{1,2}' + \cdots + c_{k-1} d_{h-1,h}'$, and 9.4 yields

$$
a_1 X^{t_1} g_1 + \cdots + a_h X^{t_h} g_h = c_1 d_{1,2}' + \cdots + c_{h-1} d_{h-1,h}' = q_1' g_1 + \cdots + q_k' g_k,
$$

where $q'_1, \ldots, q'_k \in R$ and $\operatorname{ldm}(q'_i g_i) \leqq \mu$ for all i, since $\operatorname{ldm}(c_1 d'_{1,2} + \cdots + c_{h-1} d'_{h-1,h}) < \mu$. Since $a_j X^{t_j} = \operatorname{ldt} p_j$ this implies

$$p_1 g_1 + \cdots + p_h g_h = q_1 g_1 + \cdots + q_k g_k,$$

where $q_1, \ldots, q_k \in R$ and $\operatorname{ldm}(q_i g_i) < \mu$ for all i. Then

$$f = p_1 g_1 + \cdots + p_h g_h + \cdots + p_k g_k = p'_1 g_1 + \cdots + p'_k g_k,$$

where $p'_1, \ldots, p'_k \in R$ and $\operatorname{ldm}(p'_i g_i) < \mu$ for all i, contradicting the minimality of μ. Therefore μ appears in f.

Every nonzero $f \in M$ is now a linear combination $f = p_1 g_1 + \cdots + p_k g_k$ in which $\operatorname{ldm}(p_j g_j) \leqq \operatorname{ldm} f$ for all j. Then $\operatorname{ldt} f$ is a linear combination of those $\operatorname{ldt}(p_j g_j)$ such that $\operatorname{ldm}(p_j g_j) = \operatorname{ldm} f$, and $\operatorname{ldm} f$ is a linear combination of $\operatorname{ldm} g_1, \ldots, \operatorname{ldm} g_k$. Thus $\operatorname{ldm} g_1, \ldots, \operatorname{ldm} g_k$ generate $\operatorname{ldm} M$. $\square$

Proposition **9.8** (Buchberger's Algorithm). *Let* $g_1, \ldots, g_k \neq 0$ *generate a submodule* M *of* F. *Compute a sequence* B *of elements of* F *as follows. Start with* $B = g_1, \ldots, g_k$. *Compute all polynomials* $r_{i,j}$ *with* $i < j$ *of* B *as in 9.7 and add one* $r_{i,j} \neq 0$ *to* B *in case one is found. Repeat until none is found. Then* B *is a Gröbner basis of* M.

Proof. Let L be the submodule generated by $\operatorname{ldm} g_1, \ldots, \operatorname{ldm} g_k$. Since $r_{i,j}$ is the remainder of some $d_{i,j} \in M$ in a division by $g_1, \ldots, g_k$ we have $r_{i,j} \in M$, but, if $r_{i,j} \neq 0$, then no $\operatorname{ldm} g_t$ divides $\operatorname{ldm} r_{i,j}$ and $\operatorname{ldm} r_{i,j} \notin L$. Hence L increases with each addition to B. Therefore the procedure terminates after finitely many additions; then B is a Gröbner basis of M, by 9.7. $\square$

Syzygies. A syzygy of $g_1, \ldots, g_k \in F$ is a formal linear relation between $g_1, \ldots, g_k$, with coefficients in R. In detail, let G be the free R-module with basis $\zeta_1, \ldots, \zeta_k$. Let $\psi : G \longrightarrow F$ be the module homomorphism such that $\psi \zeta_i = g_i$ for all i, $\psi\left(\sum_i p_i \zeta_i\right) = \sum_i p_i g_i$. A *syzygy* of $g_1, \ldots, g_k$ is an element of $\operatorname{Ker} \psi$, and $\operatorname{Ker} \psi$ is the *syzygy submodule* of $g_1, \ldots, g_k$. Thus $\sum_i p_i \zeta_i$ is a syzygy of $g_1, \ldots, g_k$ if and only if $\sum_i p_i g_i = 0$.

Syzygies of monomials are readily found:

Lemma **9.9.** *The syzygy submodule* S *of monomials* $\alpha_1, \ldots, \alpha_k \in F$ *is generated by all* $s_{i,j} = (\lambda_{ij}/\alpha_i)\,\zeta_i - (\lambda_{ij}/\alpha_j)\,\zeta_j$, *where* $\lambda_{ij} = \operatorname{lcm}(\alpha_i, \alpha_j)$.

Proof. First, $s_{i,j} \in S$, since $(\lambda_{ij}/\alpha_i)\,\alpha_i - (\lambda_{ij}/\alpha_j)\,\alpha_j = 0$.

For every monomial γ of F let

$$S_\gamma = \left\{ \sum_h a_h X^{m_h} \zeta_h \in S \mid \gamma = X^{m_h} \alpha_h,\ a_h \in K \right\}.$$

We show that $S = \sum_\gamma S_\gamma$. Let $s = \sum_h p_h \zeta_h \in S$, so that $\sum_h p_h \alpha_h = 0$. In $\sum_h p_h \alpha_h$, the coefficients of every γ add to 0: if $p_h = \sum_m a_{h,m} X^m$, then

$$\sum_h \left(\sum_m (a_{h,m} X^m \alpha_h \mid X^m \alpha_h = \gamma) \right) = 0$$

and $s_\gamma = \sum_h \left(\sum_m (a_{h,m} X^m \mid X^m \alpha_h = \gamma) \right) \zeta_h \in S_\gamma$. Then $s = \sum_\gamma s_\gamma$. Thus $S = \sum_\gamma S_\gamma$. (In fact, this is a direct sum.)

To conclude the proof we show that every $s = \sum_h a_h X^{m_h} \zeta_h \in S_\gamma$ is a linear combination of $s_{i,j}$'s. The proof is by induction on the number of nonzero terms of s. Let $s \neq 0$. Since $\sum_h a_h X^{m_h} \alpha_h = 0$ there exists $i \neq j$ such that $a_i, a_j \neq 0$ and $X^{m_i} \alpha_i = X^{m_j} \alpha_j$. Then λ_{ij} exists and $\gamma = X^{m_i} \alpha_i = X^{m_j} \alpha_j$ is a multiple of λ_{ij}. Hence X^{m_i} is a multiple of λ_{ij}/α_i, $X^{m_i} = \beta \lambda_{ij}/\alpha_i$, and $s - a_i \beta s_{i,j}$ has fewer nonzero terms than s. $\square$

In general, the computations in Buchberger's algorithm yield syzygies of $g_1, \ldots, g_k \in F$: when λ_{ij} exists, then $r_{i,j}$ is the remainder in a division

$$(\lambda_{ij}/\mathrm{ldt}\, g_i)\, g_i - (\lambda_{ij}/\mathrm{ldt}\, g_j)\, g_j = d_{i,j} = \sum_h q_{i,j,h}\, g_h + r_{i,j},$$

where $\mathrm{ldm}\,(q_{i,j,h}\, g_h) \leqq \mathrm{ldm}\, d_{i,j}$ for all h, by 9.4. If $r_{i,j} = 0$, then

$$s_{i,j} = (\lambda_{ij}/\mathrm{ldt}\, g_i)\, \zeta_i - (\lambda_{ij}/\mathrm{ldt}\, g_j)\, \zeta_j - \sum_h q_{i,j,h}\, \zeta_h$$

is a syzygy.

Theorem **9.10** (Schreyer [1980]). *Let $g_1, \ldots, g_k$ be a Gröbner basis of a submodule M of F. Let G be the free R-module with basis $\zeta_1, \ldots, \zeta_k$. The syzygy submodule S of $g_1, \ldots, g_k$ is generated by all*

$$s_{i,j} = (\lambda_{ij}/\mathrm{ldt}\, g_i)\, \zeta_i - (\lambda_{ij}/\mathrm{ldt}\, g_j)\, \zeta_j - \sum_h q_{i,j,h}\, \zeta_h$$

such that $i < j$ and λ_{ij} exists; in fact, these elements constitute a Gröbner basis of S relative to the monomial order on G in which $X^a \zeta_i < X^b \zeta_j$ if and only if either $\mathrm{ldm}\,(X^a g_i) < \mathrm{ldm}\,(X^b g_j)$, or $\mathrm{ldm}\,(X^a g_i) = \mathrm{ldm}\,(X^b g_j)$ and $i > j$.

Proof. Readers will verify that $<$ is a monomial order on G.

For the rest of the proof we may assume that $g_1, \ldots, g_k$ are monic. Indeed, let a_h be the leading coefficient of g_h. Then $g'_h = a_h^{-1} g_h$ is monic, and $\sum_h p_h g_h = 0$ if and only if $\sum_h a_h p_h g'_h = 0$: the syzygy submodule S' of $g'_1, \ldots, g'_k$ is the image of S under the automorphism $\theta : \sum_h f_h \zeta_h \longmapsto \sum_h a_h f_h \zeta_h$ of G. We see that $\theta s_{i,j} = s'_{i,j}$, and that $g_1, \ldots, g_k$ and $g'_1, \ldots, g'_k$ induce the same monomial order on G. Hence the $s_{i,j}$ with $i < j$ constitute a Gröbner basis of S if and only if the $s'_{i,j}$ with $i < j$ constitute a Gröbner basis of S'.

Therefore we may assume that $g_1, \ldots, g_k$ are monic. Then

$$d_{i,j} = (\lambda_{ij}/\mathrm{ldm}\, g_i)\, g_i - (\lambda_{ij}/\mathrm{ldm}\, g_j)\, g_j$$

and

$$s_{i,j} = (\lambda_{ij}/\mathrm{ldm}\, g_i)\, \zeta_i - (\lambda_{ij}/\mathrm{ldm}\, g_j)\, \zeta_j - \sum_h q_{i,j,h}\, \zeta_h.$$

Let λ_{ij} exist, where $i < j$. In F, $(\lambda_{ij}/\mathrm{ldm}\, g_i)\, g_i$ and $(\lambda_{ij}/\mathrm{ldm}\, g_j)\, g_j$ have the same leading monomial λ_{ij}, which is greater than the remaining monomials of g_i and g_j. Hence $\mathrm{ldm}\, d_{i,j} < \lambda_{ij}$ and $\mathrm{ldm}\,(q_{i,j,h}\, g_h) \leqq \mathrm{ldm}\, d_{i,j} < \lambda_{ij}$ for all h. Since $i < j$ it follows that $\mathrm{ldm}\, s_{i,j} = (\lambda_{ij}/\mathrm{ldm}\, g_i)\, \zeta_i$.

Let $s = \sum_h p_h \zeta_h \neq 0$ be a syzygy, so that $p_1, \ldots, p_k \in R$ and $\sum_h p_h g_h = 0$. Then $\mathrm{ldm}\, s = \mathrm{ldm}\, p_i \zeta_i$ for some i. Let H be the set of all $h \leqq k$ such that $\mathrm{ldm}\, p_h\, g_h = \mathrm{ldm}\, p_i\, g_i$ in F. If $h \in H$, then $h \geqq i$: otherwise, $\mathrm{ldm}\, p_h\, \zeta_h > \mathrm{ldm}\, p_i\, \zeta_i$, whereas $\mathrm{ldm}\, p_i\, \zeta_i = \mathrm{ldm}\, s$; thus i is the least element of H. If $h \notin H$, then similarly $\mathrm{ldm}\, p_h\, g_h \neq \mathrm{ldm}\, p_i\, g_i$, and $\mathrm{ldm}\, p_h\, g_h < \mathrm{ldm}\, p_i\, g_i$. Hence $\sum_{h \in H} \mathrm{ldm}\, p_h\, \mathrm{ldm}\, g_h = 0$: otherwise, $\sum_h p_h\, g_h \neq 0$.

This yields a syzygy $t = \sum_{h \in H} \mathrm{ldm}\, p_h\, \zeta_h$ of the monomials $\mathrm{ldm}\, g_h$ with $h \in H$. By 9.9, t belongs to the submodule of G generated by all $(\lambda_{jh}/\mathrm{ldm}\, g_j)\, \zeta_j - (\lambda_{jh}/\mathrm{ldm}\, g_h)\, \zeta_h$ with $j > h \in H$. We have $\mathrm{ldm}\, t = \mathrm{ldm}\, p_i\, \zeta_i = \mathrm{ldm}\, s$, since $\mathrm{ldm}\, p_h\, g_h = \mathrm{ldm}\, p_i\, g_i$ and $h \geqq i$ for all $h \in H$, and $\mathrm{ldm}\, t$ is a multiple of

$$\mathrm{ldm}\, \big((\lambda_{jh}/\mathrm{ldm}\, g_j)\, \zeta_j - (\lambda_{jh}/\mathrm{ldm}\, g_h)\, \zeta_h\big) = (\lambda_{jh}/\mathrm{ldm}\, g_h)\, \zeta_h$$

for some $j > h$; hence $h = i$ and $\mathrm{ldm}\, s = \mathrm{ldm}\, t$ is a multiple of $\mathrm{ldm}\, s_{i,j} = (\lambda_{ij}/\mathrm{ldm}\, g_j)\, \zeta_j$ for some $j > i$. Thus the monomials $\mathrm{ldm}\, s_{i,j}$ with $i < j$ generate $\mathrm{ldm}\, S$; since $s_{i,j} \in S$, the $s_{i,j}$ with $i < j$ constitute a Gröbner basis of S, by 9.7. $\square$

Example. In Example III.12.9 we saw that $g_1 = XY - X$, $g_2 = Y - X^2$, $g_3 = X^3 - X$ constitute a Gröbner basis of the ideal of $\mathbb{C}[X, Y]$ generated by g_1 and g_2, relative to the lexicographic order with $Y > X$ on $\mathbb{C}[X, Y]$. We saw that

$$\mathrm{ldt}\, g_1 = XY, \ \mathrm{ldt}\, g_2 = Y, \ \mathrm{ldt}\, g_3 = X^3,$$

and

$$d_{1,2} = g_1 - Xg_2 = X^3 - X = g_3,$$
$$d_{1,3} = X^2 g_1 - Y g_3 = XY - X^3 = g_1 - g_3,$$
$$d_{2,3} = X^3 g_2 - Y g_3 = XY - X^5 = g_1 - (X^2 + 1)\, g_3.$$

By 9.10,

$$s_{1,2} = \zeta_1 - X\zeta_2 - \zeta_3,$$
$$s_{1,3} = (X^2\zeta_1 - Y\zeta_3) - (\zeta_1 - \zeta_3) = (X^2 - 1)\,\zeta_1 - (Y - 1)\,\zeta_3,$$
$$s_{2,3} = (X^3\zeta_2 - Y\zeta_3) - (\zeta_1 - (X^2 + 1)\,\zeta_3) = -\zeta_1 + X^3\zeta_2 - (Y - X^2 - 1)\,\zeta_3$$

generate all syzygies of g_1, g_2, g_3. In other words, every relation $p_1 g_1 + p_2 g_2 + p_3 g_3 = 0$, where $p_1, p_2, p_3 \in \mathbb{C}[X, Y]$, is a consequence of the relations

$$g_1 - Xg_2 - g_3 = 0,$$
$$(X^2 - 1)\,g_1 - (Y - 1)\,g_3 = 0, \quad \text{and}$$
$$-g_1 + X^3 g_2 - (Y - X^2 - 1)\,g_3 = 0.$$

In fact, $s_{1,2}$, $s_{1,3}$, $s_{2,3}$ is a Gröbner basis of the syzygy submodule of g_1, g_2, g_3 for a suitable monomial order.

The syzygies of g_1 and g_2 are found by eliminating g_3. We saw that $g_3 = g_1 - Xg_2$. Hence every relation $p_1\,g_1 + p_2\,g_2 = 0$, where p_1, $p_2 \in \mathbb{C}[X, Y]$, is a consequence of the relations $g_1 - Xg_2 - (g_1 - Xg_2) = 0$, $(X^2 - 1)\,g_1 - (Y - 1)(g_1 - Xg_2) = 0$, and $-g_1 + X^3 g_2 - (Y - X^2 - 1)(g_1 - Xg_2) = 0$; equivalently, of $(X^2 - Y)\,g_1 - (XY - X)\,g_2 = 0$. (This is not surprising, since g_1 and g_2 are relatively prime in $\mathbb{C}[X, Y]$.) Equivalently, the syzygies of g_1 and g_2 are generated by

$$\zeta_1 - X\zeta_2 - (\zeta_1 - X\zeta_2) = 0,$$
$$(X^2 - 1)\,\zeta_1 - (Y - 1),(\zeta_1 - X\zeta_2) = (X^2 - Y)\,\zeta_1 - (XY - X)\,\zeta_2, \quad \text{and}$$
$$-\zeta_1 + X^3\zeta_2 - (Y - X^2 - 1)\,(\zeta_1 - X\zeta_2) = (Y - X^2)\,\zeta_1 + (XY - X)\,\zeta_2.$$

Exercises

In the following exercises, K is a field, $R = K[X_1, ..., X_n]$, and F is a free R-module with a finite basis ε_1, ..., ε_r.

1. Let M and N be submodules of F generated by monomials α_1, ..., α_k and β_1, ..., β_ℓ. Show that $M \cap N$ is generated by all $\operatorname{lcm}(\alpha_i, \beta_j)$.

2. Let M be a submodule of F generated by monomials α_1, ..., α_k. Let $X^m \in R$. Find generators of $\{\, f \in F \mid X^m f \in M \,\}$.

3. Let $<$ be a monomial order on R. Show that the relation $X^a \varepsilon_i < X^b \varepsilon_j$ if and only if either $X^a < X^b$ in R, or $X^a = X^b$ and $i < j$, is a monomial order on F over $<$.

4. Let $<$ be a monomial order on R. Show that the relation $X^a \varepsilon_i < X^b \varepsilon_j$ if and only if either $i < j$, or $i = j$ and $X^a < X^b$ in R, is a monomial order on F over $<$.

5. Show that every monomial order on F satisfies the descending chain condition.

6. Let $<$ be a monomial order on F and let g_1, ..., $g_k \in F$, g_1, ..., $g_k \neq 0$. Let G be the free R-module with basis ζ_1, ..., ζ_s. Show that a monomial order on G is defined by $X^a \zeta_i < X^b \zeta_j$ if and only if either $\operatorname{ldm}(X^a g_i) < \operatorname{ldm}(X^b g_j)$, or $\operatorname{ldm}(X^a g_i) = \operatorname{ldm}(X^b g_j)$ and $i > j$.

7. Given a monomial order on F, prove the following: for every f, g_1, ..., $g_k \in F$, g_1, ..., $g_k \neq 0$, there exist q_1, ..., $q_k \in R$ and $r \in F$ such that $f = q_1 g_1 + \cdots + q_k g_k + r$, $\operatorname{ldm}(q_i g_i) \leqq \operatorname{ldm} f$ for all i, $\operatorname{ldm} r \leqq \operatorname{ldm} f$, and none of $\operatorname{ldm} g_1$, ..., $\operatorname{ldm} g_k$ divides a nonzero term of r.

8. Without using Buchberger's algorithm, show that, relative to any monomial order on F, every submodule of F has a Gröbner basis. (You may be inspired by the proof of Proposition III.12.6.)

9. Let $<$ be a monomial order on F and let M be the submodule of F generated by $g_1, \ldots, g_k \in F$, $g_1, \ldots, g_k \neq 0$. Suppose that $f \in M$ if and only if, in any division of f by $g_1, \ldots, g_k$ (using the given monomial order) the remainder is 0. Show that $g_1, \ldots, g_k$ is a Gröbner basis of M.

In the following exercises, $K = \mathbb{C}$; use the lexicographic order with $Y > X$.

10. Find all syzygies of the polynomials $2XY^2 + 3X + 4Y^2$, $Y^2 - 2Y - 2$, XY. (First find a Gröbner basis.)

11. Find all syzygies of the polynomials $2XY^2 + 3X + 4Y^2$, $Y^2 - 2Y - 2$, X^2Y. (First find a Gröbner basis.)

IX
Semisimple Rings and Modules

The main result of this chapter is the Artin-Wedderburn theorem, which constructs the rings traditionally called semisimple Artinian. Wedderburn called a ring semisimple when it has no nonzero nilpotent ideal and considered in [1907] the particular case of finite-dimensional ring extensions of $\mathbb{C}$. Artin [1927] showed that Wedderburn's result depends only on the descending chain condition; this gave birth to noncommutative ring theory.

We follow current terminology, which increasingly calls semisimple Artinian rings just "semisimple". Sections 1,2,3 give the module proof of the Artin-Wedderburn theorem, and Sections 5,6 connect it with the nil and Jacobson radicals. Sections 7,8,9 are a brief introduction to representations of finite groups, and, like Section 4 on primitive rings, may be skipped.

In this chapter, all rings have an identity element and all modules are unital. Exercises indicate how this restriction can be removed from the Artin-Wedderburn theorem.

1. Simple Rings and Modules

Simple R-modules occur as factors of composition series and are readily constructed from R. This section also studies a class of simple rings: matrix rings over a division ring.

Definition. A module S is simple *when $S \neq 0$ and S has no submodule $M \neq 0, S$.* □

For example, an abelian group is simple as a $\mathbb{Z}$-module if and only if it is simple as a group, if and only if it is cyclic of prime order. A vector space is simple if and only if it has dimension 1.

Simple R-modules are readily constructed from R:

Proposition 1.1. A left R-module is simple if and only if it is isomorphic to $_RR/L$ for some maximal left ideal L of R.

Proof. A simple module is necessarily cyclic, generated by any nonzero element. If M is cyclic, $M \cong {}_RR/L$ for some left ideal L of R, the submodules of M correspond to the submodules $L \subseteq L' \subseteq R$ of $_R R$; hence M is simple if and only if L is maximal. $\square$

When M is a left R-module, we denote by $\mathrm{End}_R(M)$ the ring of all module endomorphisms of M, written on the left.

Proposition **1.2** (Schur's lemma). *If S and T are a simple left R-modules, then every homomorphism of S into T is either 0 or an isomorphism. In particular, $\mathrm{End}_R(S)$ is a division ring.*

Proof. If $\varphi : S \longrightarrow T$ is not 0, then $\mathrm{Ker}\, \varphi \subsetneqq S$ and $0 \subsetneqq \mathrm{Im}\, \varphi \subseteq T$, whence $\mathrm{Ker}\, \varphi = 0$, $\mathrm{Im}\, \varphi = T$, and φ is an isomorphism. In particular, every nonzero endomorphism of S is a unit in the endomorphism ring $\mathrm{End}_R(S)$. $\square$

Simple modules can also be constructed from minimal left ideals, when there are enough of the latter.

Definition. A left ideal L of a ring R is minimal *when $L \neq 0$ and there is no left ideal $0 \subsetneqq L' \subsetneqq L$; equivalently, when L is simple as a left R-module.*

Proposition **1.3.** *If S is a simple left R-module, L is a minimal left ideal of R, and $LS \neq 0$, then $S \cong L$. If R is a sum of minimal left ideals $(L_i)_{i \in I}$, then every simple left R-module is isomorphic to some L_i.*

Proof. If $LS \neq 0$, then $Ls \neq 0$ for some $s \in S$, $\ell \longmapsto \ell s$ is a nonzero homomorphism of L into S, and $L \cong S$ by 1.2. Now, let $R = \sum_{i \in I} L_i$ be a sum of minimal left ideals L_i. If $L_i S = 0$ for all i, then $S = RS = \sum_{i \in I} L_i S = 0$; but $S \neq 0$, since S is simple; therefore $S \cong L_i$ for some i. $\square$

Matrix rings provide the first nontrivial examples of simple rings.

Definition. A ring R [with identity] *is* simple *when $R \neq 0$ and R has no two-sided ideal $I \neq 0, R$.*

Fields and division rings are simple (as rings, not in real life). Rings $M_n(R)$ of $n \times n$ matrices provide other examples, due to the following property.

Proposition **1.4.** *Every two-sided ideal of $M_n(R)$ has the form $M_n(I)$ for some unique two-sided ideal I of R.*

Proof. If I is an ideal of R, then $M_n(I)$, which consists of all matrices with entries in I, is an ideal of $M_n(R)$. Conversely, let J be an ideal of $M_n(R)$. Let I be the set of all $(1, 1)$ entries of matrices in J. Then I is an ideal of R. We show that $J = M_n(I)$.

Let E_{ij} be the matrix whose (i, j) entry is 1 and all other entries are 0. Then $E_{ij} A E_{k\ell} = a_{jk} E_{i\ell}$ for every $A = (a_{ij}) \in M_n(R)$. Hence $A \in J$ implies $a_{ij} E_{11} = E_{1i} A E_{j1} \in J$ and $a_{ij} \in I$ for all i, j; thus $J \subseteq M_n(I)$. Conversely, if $r \in I$, then $r = c_{11}$ for some $C = (c_{ij}) \in J$, and $r E_{ij} = E_{i1} C E_{1j} \in J$ for all i, j; hence $A = (a_{ij}) \in M_n(I)$ implies $A = \sum_{i, j} a_{ij} E_{ij} \in J$. $\square$

Corollary **1.5.** *If D is a division ring, then $M_n(D)$ is simple.*

The exercises give other examples of simple rings.

Proposition **1.6.** *For every ring R, $M_n(R)^{\mathrm{op}} \cong M_n(R^{\mathrm{op}})$.*

Proof. Matrices A, B over a field can be transposed, and then $(AB)^t = B^t A^t$. Matrices A, B over an arbitrary ring R can also be transposed, and if A^t, B^t are regarded as matrices over R^{op}, then $(AB)^t = B^t A^t$ still holds. In particular, $A \longmapsto A^t$ is an isomorphism of $M_n(R)^{\mathrm{op}}$ onto $M_n(R^{\mathrm{op}})$. $\square$

Proposition **1.7.** *If D is a division ring, then $M_n(D)$ is a direct sum of n minimal left ideals; hence $M_n(D)$ is left Noetherian and left Artinian.*

By Proposition 1.6, $M_n(D)$ is also right Noetherian and right Artinian.

Proof. Let L_i be the set of all $n \times n$ matrices $M \in M_n(D)$ whose entries are all 0 outside the ith column. We see that L_i is a left ideal of $M_n(D)$ and that $M_n(D) = \bigoplus_i L_i$. Readers will verify that L_i is a minimal left ideal. Hence $M_n(D)$ (as a left module over itself) has a composition series

$$0 \subsetneqq L_1 \subsetneqq L_1 \oplus L_2 \subsetneqq \cdots \subsetneqq L_1 \oplus \cdots \oplus L_n = M_n(D). \square$$

Proposition **1.8.** *If $R = M_n(D)$, where D is a division ring, then all simple left R-modules are isomorphic; every simple left R-module S is faithful and has dimension n over D; moreover, $\mathrm{End}_R(S) \cong D^{\mathrm{op}}$.*

Proof. Identify D with the subring of $R = M_n(D)$ that consists of all scalar matrices (scalar multiples of the identity matrix). Then every left R-module becomes a D-module. In particular, every left ideal of R is a D-module, on which D acts by scalar multiplication. Let L_i be the set of all $n \times n$ matrices whose entries are all 0 outside the ith column. Let E_{ij} denote the matrix whose (i, j) entry is 1 and all other entries are 0. Then $E_{1i}, \ldots, E_{ni}$ is a basis of L_i over D. We saw that L_i is a minimal left ideal of R, and that $R = \bigoplus_i L_i$. By 1.3, every simple left R-module is isomorphic to some L_i.

For every matrix $A \in M_n(D)$ we have $AE_{ij} \in L_j$, and the j-column of AE_{ij} is the ith column of A. Hence $A \longmapsto AE_{ij}$ is an isomorphism $L_i \cong L_j$ (of left R-modules). Moreover, if $AB = 0$ for every $B \in L_j$, then $AE_{ij} = 0$ for all i, every column of A is 0, and $A = 0$; thus L_j is a faithful R-module.

If $d \in D$, then right multiplication by [the scalar matrix] d is an R-endomorphism $\eta_d : A \longmapsto Ad$ of L_1, since $(AB)d = A(Bd)$ for all A and B. Conversely, let $\eta \in \mathrm{End}_R(L_1)$. Let d be the $(1, 1)$ entry of ηE_{11}. For all $A \in L_1$, $\eta A = \eta(AE_{11}) = A \eta E_{11} = Ad$. Hence $d \longmapsto \eta_d$ is a bijection of D onto $\mathrm{End}_R(L_1)$. We see that $\eta_{dd'} = \eta_{d'} \circ \eta_d$. Thus $\mathrm{End}_R(L_1) \cong D^{\mathrm{op}}$. $\square$

Proposition **1.9.** *Let D and D' be division rings. If $M_n(D) \cong M_{n'}(D')$, then $n = n'$ and $D \cong D'$.*

Proof. Let $R = M_n(D)$ and $R' = M_{n'}(D')$. By 1.7, $_R R$ is of length n; therefore $_{R'} R'$ is of length n and $n = n'$. If $\theta : R' \longrightarrow R$ is an isomorphism, then a simple left R-module S is also a simple left R'-module, in which $r'x = \theta(r')x$ for all x and r'; then the R-endomorphisms of S coincide with its R'-endomorphisms. By 1.8, $D \cong \operatorname{End}_R(S)^{\mathrm{op}} = \operatorname{End}_{R'}(S)^{\mathrm{op}} \cong D'$. $\square$

Exercises

1. Show that the following properties are equivalent for a ring R [with identity]: (i) $_RR$ is simple; (ii) R_R is simple; (iii) R is a division ring.

2. Show that $M_m\big(M_n(R)\big) \cong M_{mn}(R)$.

3. Let D be a division ring. Show that the set L_i of all $n \times n$ matrices $M \in M_n(D)$ whose entries are all 0 outside the ith column is a minimal left ideal of $M_n(D)$.

4. Let D be a division ring and let $R_n = M_{2^n}(D)$. Identify a $2^n \times 2^n$ matrix $M \in R_n$ with the $2^{n+1} \times 2^{n+1}$ matrix $\begin{pmatrix} M & 0 \\ 0 & M \end{pmatrix} \in R_{n+1}$, so that R_n becomes a subring of R_{n+1}. Show that $R = \bigcup_{n>0} R_n$ is simple. Show that R is not left Artinian.

5. Let V be an infinite-dimensional vector space over a division ring D. Let $R = \operatorname{End}_D(V)$ and let F be the two-sided ideal of all linear transformations of V of finite rank. Show that R/F is simple. Show that R/F is not left Artinian.

2. Semisimple Modules

A semisimple R-module is a direct sum of simple modules. These modules are readily constructed from R and have interesting properties.

Definition. Semisimple modules are defined by the following equivalent conditions.

Proposition **2.1.** *For a module M the following properties are equivalent:*

(1) *M is a direct sum of simple submodules;*

(2) *M is a sum of simple submodules;*

(3) *every submodule of M is a direct summand.*

Proof. (1) implies (2).

(2) implies (1) and (3). Let M be a sum $M = \sum_{i \in I} S_i$ of simple submodules S_i. Let N be a submodule of M. Let $\mathcal{S}$ be the set of all subsets J of I such that the sum $N + \sum_{i \in J} S_i$ is direct; equivalently, such that $N \cap \sum_{i \in J} S_i = 0$ and $S_i \cap \big(N + \sum_{j \in J,\, j \neq i} S_j\big) = 0$ for all $i \in J$. Then $\varnothing \in \mathcal{S}$. Moreover, the union of a chain of elements of $\mathcal{S}$ is an element of $\mathcal{S}$, since $x \in \sum_{i \in J} S_i$ if and only if $x \in \sum_{i \in K} S_i$ for some finite subset K of J, and similarly for $x \in \sum_{j \in J,\, j \neq i} S_j$. By Zorn's lemma, $\mathcal{S}$ has a maximal element.

We show that J cannot be maximal if $N + \sum_{i \in J} S_i \subsetneqq M$. Indeed, $N + \sum_{i \in J} S_i \subsetneqq M$ implies $S_k \nsubseteq N + \sum_{i \in J} S_i$ for some $k \in I$. Then $\big(N + \sum_{i \in J} S_i\big) \cap S_k \subsetneqq S_k$ and $\big(N + \sum_{i \in J} S_i\big) \cap S_k = 0$, since S_k is simple. Hence $k \notin J$, the sum $\big(N + \sum_{i \in J} S_i\big) + S_k$ is direct, the sum

$N + \sum_{i \in J \cup \{k\}} S_i$ is direct, and J is not maximal. If therefore J is a maximal element of $\mathcal{S}$, then $N + \sum_{i \in J} S_i = M$; since this is a direct sum, N is a direct summand of M. The case $N = 0$ yields $M = \sum_{i \in J} S_i = \bigoplus_{i \in J} S_i$.

(3) implies (2). First we show that a cyclic submodule $Ra \neq 0$ of M contains a simple submodule. The mapping $\varphi : r \longmapsto ra$ is a module homomorphism of $_R R$ onto Ra, whose kernel is a left ideal of R and is contained in a maximal left ideal L of R. Then $La = \varphi(L)$ is a maximal submodule of Ra, and Ra/La is simple. By (3), $M = La \oplus N$ for some submodule N of M. Then $Ra = La \oplus (Ra \cap N)$: indeed, $La \cap (Ra \cap N) = 0$, and $Ra = La + (Ra \cap N)$, since every $x \in Ra$ is · the sum $x = y + n$ of some $y \in La$ and $n \in N$, with $n = x - y \in Ra \cap N$. Hence $Ra \cap N \cong Ra/La$ is a simple submodule of Ra.

Now, let N be the sum of all the simple submodules of M. Then $M = N \oplus N'$ for some submodule N' of M, and $N' = 0$: otherwise, N' contains a cyclic submodule $Ra \neq 0$, N' has a simple submodule S, and $N \cap N' \supseteq S \neq 0$, contradicting $M = N \oplus N'$. $\square$

Definition. A semisimple *module is a direct sum of simple submodules.*

Semisimple modules are also called *completely reducible*. Vector spaces are semisimple. But $\mathbb{Z}$ is not semisimple as a $\mathbb{Z}$-module.

Properties. Readers will enjoy proving the following properties.

Proposition 2.2. (1) *A direct sum of semisimple left R-modules is semisimple.*

(2) *Every submodule of a semisimple module is semisimple.*

(3) *Every quotient module of a semisimple module is semisimple.*

Next we prove some properties of endomorphism rings, for use in the next section. First we look at endomorphisms of finite direct sums. Readers may work out similar results for infinite direct sums.

Proposition 2.3. *Let $A_1, \ldots, A_m$, $B_1, \ldots, B_n$, $C_1, \ldots, C_p$ be left R-modules. There is a one-to-one correspondence between module homomorphisms $\varphi : \bigoplus_j B_j \longrightarrow \bigoplus_i A_i$ and $m \times n$ matrices (φ_{ij}) of module homomorphisms $\varphi_{ij} : B_j \longrightarrow A_i$. If $\psi : \bigoplus_k C_k \longrightarrow \bigoplus_j B_j$, then the matrix that corresponds to $\varphi \circ \psi$ is the product of the matrices that correspond to φ and ψ.*

$$
\begin{array}{ccc}
\bigoplus_j B_j & \xrightarrow{\ \varphi\ } & \prod_i A_i \\[4pt]
{\scriptstyle \iota_j}\big\uparrow & & \big\downarrow{\scriptstyle \pi_i} \\[4pt]
B_j & \xrightarrow[\ \varphi_{ij}\ ]{} & A_i
\end{array}
$$

Proof. Let $\iota_j : B_j \longrightarrow \bigoplus_j B_j$ be the jth injection and let $\pi_i : \bigoplus_i A_i = \prod_i A_i \longrightarrow A_i$ be the ith projection. By the universal properties of $\bigoplus_j B_j$ and $\prod_i A_i$ there is for every matrix (φ_{ij}) of homomorphisms $\varphi_{ij} : B_j \longrightarrow A_i$ a unique homomorphism $\varphi : \bigoplus_j B_j \longrightarrow \bigoplus_i A_i$ such that $\varphi_{ij} = \pi_i \circ \varphi \circ \iota_j$ for all i, j. This provides the required one-to-one correspondence.

$$
\begin{array}{ccccc}
\bigoplus_k C_k & \xrightarrow{\psi} & \bigoplus_j B_j & \xrightarrow{\varphi} & \bigoplus_i A_i \\[4pt]
{\scriptstyle \kappa_k} \uparrow & & {\scriptstyle \rho_j} \downarrow\uparrow {\scriptstyle \iota_j} & & \downarrow {\scriptstyle \pi_i} \\[4pt]
C_k & \xrightarrow{\psi_{jk}} & B_j & \xrightarrow{\varphi_{ij}} & A_i
\end{array}
$$

Let $\psi : \bigoplus_k C_k \longrightarrow \bigoplus_j B_j$, let $\kappa_k : C_k \longrightarrow \bigoplus_k C_k$ be the kth injection and let $\rho_j : \bigoplus_j B_j = \prod_j B_j \longrightarrow B_j$ be the jth projection. Since $\sum_j \iota_j \circ \rho_j$ is the identity on $\bigoplus_j B_j$, we have

$$
\pi_i \circ \varphi \circ \psi \circ \kappa_k = \pi_i \circ \varphi \circ \left(\sum_j \iota_j \circ \rho j\right) \circ \psi \circ \kappa_k = \sum_j \varphi_{ij} \circ \psi_{jk}
$$

for all i, k; thus the matrix of $\varphi \circ \psi$ is obtained by a matrix multiplication, in which entries are added and multiplied by pointwise addition and composition. $\square$

In particular, homomorphisms of one finitely generated free left R-module into another correspond to matrices of endomorphisms of $_R R$. In this case, Proposition 2.3 reduces to Proposition VIII.4.7, since $\mathrm{End}_R({_R R}) \cong R^{\mathrm{op}}$ by Corollary VIII.4.10.

Corollary **2.4.** *If S is a simple left R-module and $D = \mathrm{End}_R S$, then $\mathrm{End}_R(S^n) \cong M_n(D)$ for all $n > 0$. (S^n is the direct sum of n copies of S.)*

Proposition **2.5.** *Let M be a left R-module. If M is a direct sum of finitely many simple submodules, then $\mathrm{End}_R(M)$ is isomorphic to the direct product of finitely many rings of matrices $M_{n_i}(D_i)$ over division rings D_i.*

Proof. Let $M = \bigoplus_k S_k$ be the direct sum of finitely many simple submodules S_k. Grouping together the modules S_j that are isomorphic to each other rewrites M as a direct sum $M \cong S_1^{n_1} \oplus \cdots \oplus S_r^{n_r}$, where $n_i > 0$ and no two S_i are isomorphic. By 2.3, $\mathrm{End}_R(M)$ is isomorphic to a ring of $r \times r$ matrices (η_{ij}), whose entries are module homomorphisms $\eta_{ij} : S_j^{n_j} \longrightarrow S_i^{n_i}$, added and multiplied by pointwise addition and composition. If $i \neq j$, then η_{ij} corresponds, by 2.3 again, to an $n_i \times n_j$ matrix of module homomorphisms $S_j \longrightarrow S_i$, which are all 0 by 1.2, since S_i and S_j are not isomorphic; hence $\eta_{ij} = 0$. Thus the matrix (η_{ij}) is diagonal, with diagonal entries $\eta_{ii} \in \mathrm{End}_R(S_i^{n_i}) \cong M_{n_i}(D_i)$, by 2.4. Therefore $\mathrm{End}_R(M) \cong M_{n_1}(D_1) \times \cdots \times M_{n_r}(D_r)$. $\square$

Products. Finally, we look at direct products of rings. The direct products in Proposition 2.5 are "external" but rings also have "internal" direct products.

Proposition **2.6.** *Let R be a ring [with identity]. If R is isomorphic to a direct product $R_1 \times \cdots \times R_n$ of finitely many rings, then R has two-sided ideals $A_1, \ldots, A_n \neq 0$ such that $A_i \cong R_i$ (as a ring) for all i, $R = \sum_i A_i$, and $A_i A_j = 0$ whenever $i \neq j$. Conversely, if $(R_i)_{i \in I}$ are nonzero two-sided* *ideals of R such that $R = \sum_i R_i$, and $R_i R_j = 0$ whenever $i \neq j$, then I is finite; every R_i is a ring; the identity element of R is the sum of the identity elements of the rings R_i; and $R \cong \prod_{i \in I} R_i$.*

Proof. If $R = R_1 \times \cdots \times R_n$, then the sets

$$A_i = \{ (x_1, \ldots, x_n) \in R_1 \times \cdots \times R_n \mid x_j = 0 \text{ for all } j \neq i \}$$

have all the properties in the statement.

Conversely, let $(R_i)_{i \in I}$ be nonzero two-sided ideals of R such that $R = \sum_i R_i$, and $R_i R_j = 0$ whenever $i \neq j$. We have $1 = \sum_i e_i$, where $e_i \in R_i$ and $e_i = 0$ for almost all i. Then $x \in R_i$ implies $x = \sum_j x e_j = x e_i$, since $R_i R_j = 0$ when $j \neq i$. Similarly, $x \in R_i$ implies $x = \sum_j e_j x = e_i x$. In particular, $e_i \neq 0$ for all i, since $e_i = 0$ would imply $R_i = 0$. Therefore I is finite. Also R_i, which is an additive subgroup of R and is closed under multiplication, has an identity element e_i, and is a ring (though not a subring of R, unless $|I| = 1$). The sum $R = \sum_i R_i$ is direct, since $x \in R_i \cap \sum_{j \neq i} R_j$ implies $x = e_i x = 0$. Hence every element of R can be written uniquely as a sum $x = \sum_i x_i$, where $x_i \in R_i$. These sums add and multiply componentwise, since $R_i R_j = 0$ when $j \neq i$, so that $R \cong \prod_{i \in I} R_i$. $\square$

It is common practice to write $R = R_1 \times \cdots \times R_n$ when $R_1, \ldots, R_n$ are ideals of R such that $R = \sum_{i \in I} R_i$ and $R_i R_j = 0$ whenever $i \neq j$. This notation does not distinguish between "internal" and "external" direct products; as with direct sums, the distinction should be clear from context.

Proposition 2.7. *In a direct product $R = R_1 \times \cdots \times R_n$ of rings [with identity], every left (right, two-sided) ideal of R_i is a left (right, two-sided) ideal of R; every minimal left (right, two-sided) ideal of R_i is a minimal left (right, two-sided) ideal of R; every minimal left (right, two-sided) ideal of R is a minimal left (right, two-sided) ideal of some R_i.*

Proof. If L is a left ideal of R_j, then $RL = (\sum_i R_i) L = \sum_i R_i L = R_j L \subseteq L$ and L is a left ideal of R. Conversely, a left ideal of R that is contained in R_j is a left ideal of R_j. If now L is a minimal [nonzero] left ideal of R_j, then L is a minimal left ideal of R, since any left ideal $0 \subsetneqq L' \subsetneqq L$ of R would be a left ideal of R_j; if $L \subseteq R_j$ is a minimal left ideal of R, then L is a minimal left ideal of R_j, since any left ideal $0 \subsetneqq L' \subsetneqq L$ of R_j would be a left ideal of R. Conversely, if L is a minimal left ideal of R, then $R_i L \neq 0$ for some i, since $\sum_i R_i L = RL \neq 0$, $0 \neq R_i L \subseteq R_i \cap L \subseteq L$, and $L = R_i \cap L \subseteq R_i$ for some R_i. Right and two-sided ideals are handled similarly. $\square$

Proposition 2.8. *If $R = R_1 \times \cdots \times R_n$ is a direct product of rings, then the simple left R-modules are the simple left R_i-modules of the rings R_i.*

Proof. Let S be a simple R-module. Then $\sum_i R_i S = RS = S \neq 0$, $R_i S \neq 0$ for some i, $R_i S = R R_i S$ is a submodule of S, and $R_i S = S$. Let e_i be the identity element of R_i. If $j \neq i$, then $R_j S = R_j R_i S = 0$. Hence $e_i x = 1x = x$ for all $x \in S$, since $1 = \sum_{i \in I} e_i$ by 2.6. Thus S is a [unital] R_i-module. Since $R_j S = 0$ for all $j \neq i$, S has the same submodules as an R-module and as an R_i-module, and S is a simple R_i-module.

Conversely, let S be a simple R_i-module. Let R act on S so that $R_j S = 0$ for all $j \neq i$. Then S is an R-module. As above, S has the same submodules as an R-module and as an R_i-module, and S is a simple R-module. $\square$

Exercises

1. When is an abelian group semisimple (as a $\mathbb{Z}$-module)?

2. Prove that every quotient module of a semisimple module is semisimple.

3. Prove that every submodule of a semisimple module is semisimple.

4. Show that a semisimple module is of finite length if and only if it is finitely generated.

5. Show that a module M is semisimple if and only if every cyclic submodule of M is semisimple.

6. Find a module M with a submodule N such that N and M/N are semisimple but M is not semisimple.

7. Let $R = R_1 \times \cdots \times R_n$ be a direct product of rings. Show that every two-sided ideal of R_i is a two-sided ideal of R; every minimal two-sided ideal of R_i is a minimal two-sided ideal of R; and every minimal two-sided ideal of R is a minimal two-sided ideal of some R_i.

8. How would you extend Propositions 2.3 and 2.5 to arbitrary direct sums?

3. The Artin-Wedderburn Theorem

The Artin-Wedderburn theorem constructs all rings R, called semisimple, such that every left R-module is semisimple. It has remained a fundamental result of ring theory.

As in the rest of this chapter, all rings have an identity element, and all modules are unital. (The main result holds without this restriction; see the exercises.)

Definition. A ring R is semisimple *when every left R-module is semisimple.*

By rights these rings should be called *left* semisimple; but we shall show that R is semisimple if and only if every right R-module is semisimple.

Division rings are semisimple. More elaborate examples arise from the next result.

Proposition 3.1. A ring R is semisimple if and only if the module $_R R$ is semisimple, if and only if R is a direct sum of minimal left ideals, and then R is a direct sum of finitely many minimal left ideals.

Proof. If every left R-module is semisimple, then so is $_RR$. Conversely, if $_RR$ is semisimple, then, by 2.2, every free left R-module $F \cong \bigoplus {}_RR$ is semisimple, and every left R-module is semisimple, since it is isomorphic to a quotient module of a free module.

By definition, $_RR$ is semisimple if and only if it is a direct sum of simple submodules, and a simple submodule of $_RR$ is a minimal [nonzero] left ideal. Now, a direct sum $_RR = \bigoplus_{i \in I} L_i$ of nonzero left ideals is necessarily finite. Indeed, the identity element of R is a sum $1 = \sum_{i \in I} e_i$, where $e_i \in L_i$ for all i and $e_i = 0$ for almost all i. If $x \in L_j$, then $\sum_{i \in I} x e_i = x \in L_j$, with $x e_i \in L_i$, which in the direct sum implies $x e_j = x$. Hence $e_j = 0$ implies $L_j = 0$, and $L_i = 0$ for almost all i. If $L_i \neq 0$ for all $i \in I$, then I is finite. $\square$

Proposition 3.1 yields additional semisimple rings. Matrix rings $M_n(D)$ over a division ring are semisimple, by 1.7. More generally, all direct products of such rings are semisimple:

Proposition 3.2. A direct product of finitely many semisimple rings is a semisimple ring.

Proof. If $R_1, \ldots, R_n$ are semisimple, then every R_i is a sum of minimal left ideals of R_i, which by 2.7 are minimal left ideals of $R = R_1 \times \cdots \times R_n$; hence R is a sum of minimal left ideals. $\square$

The Artin-Wedderburn theorem can now be stated and proved.

Theorem 3.3 (Artin-Wedderburn). *A ring R is semisimple if and only if it is isomorphic to a direct product $M_{n_1}(D_1) \times \cdots \times M_{n_s}(D_s)$ of finitely many matrix rings over division rings $D_1, \ldots, D_s$.*

In particular, semisimple rings are direct products of simple rings.

Proof. If R is semisimple, then $R^{\mathrm{op}} \cong \mathrm{End}_R({}_RR) \cong M_{n_1}(D_1) \times \cdots \times M_{n_s}(D_s)$ for some division rings $D_1, \ldots, D_s$, by VIII.4.10 and 2.5. Hence

$$R \cong M_{n_1}(D_1)^{\mathrm{op}} \times \cdots \times M_{n_s}(D_s)^{\mathrm{op}} \cong M_{n_1}(D_1^{\mathrm{op}}) \times \cdots \times M_{n_s}(D_s^{\mathrm{op}}),$$

by 1.6, where $D_1^{\mathrm{op}}, \ldots, D_s^{\mathrm{op}}$ are division rings. $\square$

Corollary 3.4. A ring R is semisimple if and only if R^{op} is semisimple.

Proof. If $R \cong M_{n_1}(D_1) \times \cdots \times M_{n_r}(D_s)$ is semisimple, where $D_1, \ldots, D_s$ are division rings, then $R^{\mathrm{op}} \cong M_{n_1}(D_1^{\mathrm{op}}) \times \cdots \times M_{n_s}(D_s^{\mathrm{op}})$ is semisimple. $\square$

Corollary 3.5. Every semisimple ring is left Noetherian, left Artinian, right Noetherian, and right Artinian.

Proof. By 3.1, $_RR$ is a finite direct sum $_RR = L_1 \oplus \cdots \oplus L_n$ of finitely many simple submodules. Hence $_RR$ has a composition series

$$0 \subsetneqq L_1 \subsetneqq L_1 \oplus L_2 \subsetneqq \cdots \subsetneqq L_1 \oplus \cdots \oplus L_n = R$$

and R is left Noetherian and left Artinian. So is R^{op}, by 3.4, and R is right Noetherian and right Artinian. $\square$

Simple modules. We complete Theorem 3.3 by a look at modules and by some uniqueness properties. If R is semisimple, then every simple left R-module is isomorphic to a minimal left ideal of R, by Proposition 1.3; hence every left R-module is a direct sum of copies of minimal left ideals of R. This construction didn't even hurt, and it yields all R-modules.

Proposition **3.6.** *If* $R \cong M_{n_1}(D_1) \times \cdots \times M_{n_s}(D_s)$ *is a semisimple ring, where* $D_1, \ldots, D_s$ *are division rings, then every simple left R-module is isomorphic to a minimal left ideal of some* $M_{n_i}(D_i)$; *hence there are exactly s isomorphy classes of simple left R-modules.*

Proof. By 2.6, 3.3, $R = R_1 \times \cdots \times R_s$, where $R_i \cong M_{n_i}(D_i)$. By 2.7, L is a minimal left ideal of R if and only if L is a minimal left ideal of some R_i. By 1.8, all minimal left ideals of R_i are isomorphic as R_i-modules, hence also as R-modules. But minimal left ideals L_i of R_i and L_j of R_j are not isomorphic as R-modules when $i \neq j$: by 1.8, $\mathrm{Ann}_{R_i}(L_i) = 0$, so that $\mathrm{Ann}_R(L_i) = \bigoplus_{j \neq i} R_j$; hence $\mathrm{Ann}_R(L_i) \neq \mathrm{Ann}_R(L_j)$ when $i \neq j$. $\square$

Corollary **3.7.** *Let R be semisimple and let* $S_1, \ldots, S_s$ *are, up to isomorphism, all the distinct simple left R-modules (so that every simple left R-module is isomorphic to exactly one of* $S_1, \ldots, S_s$). *Every left R-module is isomorphic to a direct sum* $S_1^{m_1} \oplus \cdots \oplus S_s^{m_s}$, *for some unique cardinal numbers* $m_1, \ldots, m_s$.

Proof. Up to isomorphism, $R = R_1 \times \cdots \times R_s$, where $R_i \cong M_{n_i}(D_i)$, and $S_1, \ldots, S_s$ can be numbered so that S_i is isomorphic to a minimal left ideal of R_i. Then $R_j S_i = 0$ whenever $i \neq j$. Every R-module is a direct sum of simple modules, m_i of which are isomorphic to S_i. In the direct sum $M \cong S_1^{m_1} \oplus \cdots \oplus S_s^{m_s}$, $S_i^{m_i} \cong R_i M$ is unique up to isomorphism. Then m_i is unique, since $S_i^{m_i}$ has dimension $m_i n_i$ over D_i and n_i is finite. $\square$

The uniqueness of $m_1, \ldots, m_s$ in Corollary 3.7 also follows from the Jordan-Hölder theorem, or from the Krull-Schmidt theorem, if $m_1, \ldots, m_s$ are finite.

Theorem **3.8.** *For a ring R the following properties are equivalent:*

(1) *R is simple and semisimple;*

(2) *R is semisimple and all simple left R-modules are isomorphic;*

(3) *$R \cong M_n(D)$ for some $n > 0$ and some division ring $D \cong \mathrm{End}_R^{\mathrm{op}}(S)$, where S is a simple left R-module;*

(4) *R is left Artinian and there exists a faithful, simple left R-module;*

(5) *R is simple and left Artinian.*

Proof. (1) implies (3), and (2) implies (3). Let $R \cong M_{n_1}(D_1) \times \cdots \times M_{n_s}(D_s)$ be semisimple, where $D_1, \ldots, D_s$ are division rings. If R is simple, then $s = 1$. If all simple left R-modules are isomorphic, then $s = 1$, by 3.6.

(3) implies (1), (2), and (5), by 1.5, 1.7, and 1.8.

(5) implies (4). Since R is left Artinian, R has a minimal left ideal L, which is a simple left R-module. Then $\operatorname{Ann}(L)$ is an ideal of R, $\operatorname{Ann}(L) \neq R$ since $1 \notin \operatorname{Ann}(L)$, $\operatorname{Ann}(L) = 0$ since R is simple, and L is faithful.

(4) implies (2). Let S be a faithful simple left R-module. Since R is left Artinian, there is a module homomorphism $\varphi : {}_R R \longrightarrow S^n$ whose kernel is minimal among all kernels of module homomorphisms ${}_R R \longrightarrow S^m$, where $m > 0$ is finite. If $\operatorname{Ker} \varphi \neq 0$, then $\varphi(r) = 0$ for some $0 \neq r \in R$, $rs \neq 0$ for some $s \in S$ since S is faithful, $\psi : x \longmapsto \big(\varphi(x), xs\big)$ is a homomorphism of ${}_R R$ into $S^n \oplus S$, and $\operatorname{Ker} \psi \subsetneqq \operatorname{Ker} \varphi$. This sneaky contradiction shows that $\operatorname{Ker} \varphi = 0$. Hence ${}_R R$ is isomorphic to a submodule of S^n, and is semisimple by 2.2. If L is a minimal left ideal of R, then $LS \neq 0$ since S is faithful, and $L \cong S$; hence every simple left R-module is isomorphic to S, by 1.3. $\square$

In Theorem 3.3, R is a product of simple Artinian rings $R_i \cong M_{n_i}(D_i)$. These rings can now be constructed from R:

Proposition 3.9. *Let R be semisimple and let $S_1, \ldots, S_s$ be, up to isomorphism, all the distinct simple left R-modules. Let R_i be the sum of all the minimal left ideals $L \cong S_i$ of R. Then R_i is a two-sided ideal of R, R_i is a simple left Artinian ring, and $R = R_1 \times \cdots \times R_s$.*

Proof. Let L be a minimal left ideal of R. If $a \in R$ and $La \neq 0$, then the left ideal La is a minimal left ideal: if $A \subseteq La$ is a nonzero left ideal, then so is $L' = \{ x \in L \mid xa \in A \}$, whence $L' = L$ and $A = L'a = La$. Since $x \longmapsto xa$ is a nonzero module homomorphism of L onto La, 1.2 yields $La \cong L$. Therefore R_i is a two-sided ideal. Then $R_i R_j = 0$ when $i \neq j$, since $LL' = 0$ when L and L' are minimal left ideals and $L \not\cong L'$, by 1.3; and $R = \sum_i R_i$, since R is the sum of its minimal left ideals. Hence $R = R_1 \times \cdots \times R_s$, by 2.6.

By 2.7, R_i is a sum of minimal left ideals. Hence R_i is semisimple. Moreover, all minimal left ideals of R_i are isomorphic as left R-modules, hence also as left R_i-modules. Therefore R_i is simple and left Artinian, by 3.8. $\square$

The rings $R_1, \ldots, R_s$ in Proposition 3.9 are the *simple components* of R. Readers will enjoy proving their uniqueness:

Proposition 3.10. *Let R be a direct product $R = R_1 \times \cdots \times R_s$ of simple left Artinian rings $R_1, \ldots, R_s$. The ideals $R_1, \ldots, R_s$ of R are unique, up to their order of appearance.*

The uniqueness statement for Theorem 3.3 now follows from Propositions 3.10 and 1.9:

Corollary 3.11. *If $M_{n_1}(D_1) \times \cdots \times M_{n_s}(D_s) \cong M_{n_1'}(D_1') \times \cdots \times M_{n_t'}(D_t')$,* *where* $D_1, \ldots, D_s, D'_1, \ldots, D'_t$ *are division rings, then* $s = t$ *and* $D'_1, \ldots, D'_t$ *can be reindexed so that* $n_i = n'_i$ *and* $D_i \cong D'_i$ *for all* i.

Exercises

In the following exercises, all rings have an identity element, and all modules are unital. An *idempotent* in a ring R is an element e of R such that $e^2 = e$.

1. Prove that a left ideal of a ring R is a direct summand of $_R R$ if and only if it is generated by an idempotent.

2. Prove that a ring R is semisimple if and only if every left ideal of R is generated by an idempotent.

3. A ring R is *von Neumann regular* when for every $a \in R$ there exists $x \in R$ such that $axa = a$. Prove that every semisimple ring is von Neumann regular.

4. Let R be a direct product $R = R_1 \times \cdots \times R_n$ of simple rings $R_1, \ldots, R_n$. Show that every two-sided ideal of R is a direct sum of some of the R_i.

5. Let R be a direct product $R = R_1 \times \cdots \times R_n$ of simple rings $R_1, \ldots, R_n$. Show that the ideals $R_1, \ldots, R_n$ of R are unique, up to their order of appearance.

6. Show that a commutative ring is semisimple if and only if it is isomorphic to a direct product of finitely many fields.

7. Prove the following: if R is semisimple, then $M_n(R)$ is semisimple.

8. Prove that a semisimple ring without zero divisors is a division ring.

9. Prove the following: in a semisimple ring R, $xy = 1$ implies $yx = 1$.

The ring R in the following exercises does not necessarily have an identity element.

10. Show that R is semisimple if and only if R^1 is semisimple, and then R and R^1 have the same simple left modules.

11. Show that R is semisimple if and only if R is isomorphic to a direct product $M_{n_1}(D_1) \times \cdots \times M_{n_s}(D_s)$ of finitely many matrix rings over division rings $D_1, \ldots, D_s$. (In particular, R has an identity element anyway. Thus, the Artin-Wedderburn theorem holds even if R is not assumed to have an identity element. You may want to show that R is a two-sided ideal of R^1, and use one of the previous exercises.)

4. Primitive Rings

This section may be skipped at first reading, but is quoted in Section 7. A ring is primitive when it has a faithful simple module. The Jacobson density theorems [1945b] in this section extend properties of simple Artinian rings to all primitive rings. They also provide an alternate proof of the Artin-Wedderburn theorem.

In this section, all rings have an identity element, and all modules are unital.

Endomorphisms. We begin with further properties of endomorphism rings.

Let R be a ring and let M be a left R-module. Section 2 made use of the ring $\mathrm{End}_R(M)$ of R-endomorphisms of M, written on the left. If the endomorphisms of M are written on the right, then $\operatorname{End}_R(M)$ becomes the opposite ring $\operatorname{End}_R^{op}(M)$. The operations $x(\eta + \zeta) = x\eta + x\zeta$, $x(\eta\zeta) = (x\eta)\zeta$, $x1 = x$ on $E = \operatorname{End}_R^{op}(M)$ show that M is a right E-module. As a right E-module, M has an ring of endomorphisms $\operatorname{End}_E(M)$ (written on the left).

Proposition **4.1.** *For every left R-module M there is a canonical homomorphism $\Phi : R \longrightarrow \operatorname{End}_E(M)$, defined by $\Phi(r): x \longmapsto rx$.*

Proof. Since $(rx)\eta = r(x\eta)$ for all $r \in R$, $x \in M$, $\eta \in E$, the action $\Phi(r): x \longmapsto rx$ of r on M is an E-endomorphism. It is immediate that Φ is a ring homomorphism. $\square$

In general, Φ is neither injective nor surjective. We see that Φ is injective if and only if M is faithful. Readers will show that Φ is an isomorphism when $M = {}_R R$. If M is semisimple, then the Jacobson density theorem for semisimple modules states that Φ is not far from surjective:

Theorem **4.2** (Jacobson Density Theorem). *Let M be a semisimple left R-module and let $E = \operatorname{End}_R^{op}(M)$. For every $\xi \in \operatorname{End}_E(M)$ and $x_1, \ldots, x_n \in M$ there exists $r \in R$ such that $\xi x_i = rx_i$ for all i.*

Proof. We first prove 4.2 when $n = 1$. Since M is semisimple we have $M = Rx \oplus N$ for some submodule N of M. Then the projection $\pi : M \longrightarrow Rx$ may be viewed as an R-endomorphism of M, such that $(rx)\pi = rx$ for all $rx \in Rx$. Hence $\pi \in E$ and $\xi x = \xi(x\pi) = (\xi x)\pi = rx$ for some $r \in R$.

For the general case, let M^n be the direct sum of n copies of M and $F = \operatorname{End}_R^{op}(M^n)$. Then M^n is semisimple and $\xi^n : (x_1, \ldots, x_n) \longmapsto (\xi x_1, \ldots, \xi x_n)$ belongs to $\operatorname{End}_F(M^n)$, by 4.3 below. By the case $n = 1$, applied to M^n, there is for every $(x_1, \ldots, x_n) \in M^n$ some $r \in R$ such that $\xi^n(x_1, \ldots, x_n) = r(x_1, \ldots, x_n)$. $\square$

Lemma **4.3.** *Let M be a left R-module, let $E = \operatorname{End}_R^{op}(M)$, let $n > 0$, and let $F = \operatorname{End}_R^{op}(M^n)$. If $\xi : M \longrightarrow M$ is an E-endomorphism, then $\xi^n : M^n \longrightarrow M^n$, $(x_1, \ldots, x_n) \longmapsto (\xi x_1, \ldots, \xi x_n)$, is an F-endomorphism; moreover, $\xi \longmapsto \xi^n$ is an isomorphism $\operatorname{End}_E(M) \cong \operatorname{End}_F(M^n)$.*

Proof. Let $\iota_i : M \longrightarrow M^n$ and $\pi_j : M^n \longrightarrow M$ be the injections and projections. Let $\eta \in F$. As in the proof of 1.4, every $\eta : M^n \longrightarrow M^n$ is determined by a matrix (η_{ij}), where $\eta_{ij} = \pi_j \circ \eta \circ \iota_i$, namely,

$$(x_1, \ldots, x_n)\eta = \left(\sum_i x_i \eta_{i1}, \ldots, \sum_i x_i \eta_{in} \right),$$

since $(x_1, \ldots, x_n) = \sum_{i \in I} \iota_i x_i$ and $x\eta_{ij}$ is the j component of $(\iota_i x)\eta$. If $\xi \in \operatorname{End}_E(M)$, then $(\xi x)\eta_{ij} = \xi(x\eta_{ij})$ for all $x \in M$; hence $(\xi^n y)\eta = \xi^n(y\eta)$ for all $y \in M^n$ and $\eta \in F$, and $\xi^n \in \operatorname{End}_F(M^n)$. The second part of the statement is not needed for Theorem 4.2; we leave it to our readers. $\square$

Theorem 4.2 suggests the following definition.

Definition. Let M be an E-module. A subset S of $\mathrm{End}_E(M)$ *is* dense *in* $\mathrm{End}_E(M)$ *when, for every* $\xi \in \mathrm{End}_E(M)$ *and* $x_1, \ldots, x_n \in M$, *there exists* $s \in S$ *such that* $\xi x_i = s x_i$ *for all* i.

With this definition, Theorem 4.2 reads, *when M is a semisimple left R-module and* $E = \mathrm{End}_R^{\mathrm{op}}(M)$, *and* $\Phi : R \longrightarrow \mathrm{End}_E(M)$ *is the canonical homomorphism, then* $\Phi(R)$ *is dense in* $\mathrm{End}_E(M)$.

When M is viewed as discrete, the compact-open topology on the set of all transformations of M induces a topology on $\mathrm{End}_E(M)$, with basic open sets

$$U(a_1, \ldots, a_n, b_1, \ldots, b_n) = \{ \xi \in \mathrm{End}_E(M) \mid \xi a_i = b_i \text{ for all } i \},$$

where $n > 0$ and $a_1, \ldots, a_n, b_1, \ldots, b_n \in M$. Readers may verify that a subset of $\mathrm{End}_E(M)$ is dense as above if and only if it is dense in this topology.

We note two of the many consequences of Theorem 4.2; a third is given below.

Corollary **4.4.** *If D is a division ring, then the center of* $M_n(D)$ *consists of all scalar matrices whose diagonal entry is in the center of D.*

Proof. We prove this for D^{op}. Let I denote the identity matrix. Let V be a left D-module with a basis $e_1, \ldots, e_n$, so that $\mathrm{End}_D(V) \cong M_n(D^{\mathrm{op}})$. Let $E = \mathrm{End}_D^{\mathrm{op}}(V)$. The isomorphism $\mathrm{End}_D(V) \cong M_n(D^{\mathrm{op}})$ assigns to every $\eta \in \mathrm{End}_D(V)$ its matrix in the basis $e_1, \ldots, e_n$. If C is in the center of $M_n(D^{\mathrm{op}})$, then the corresponding endomorphism $\xi \in \mathrm{End}_D(V)$ commutes with every $\eta \in \mathrm{End}_D(V)$, and $\xi \in \mathrm{End}_E(V)$. By 4.2 there exists $r \in D$ such that $\xi e_i = r e_i$ for all i. Hence the matrix C of ξ is the scalar matrix rI, and r is in the center of D^{op}, since rI commutes with all scalar matrices. Conversely, if r is in the center of D^{op}, then rI is in the center of $M_n(D^{\mathrm{op}})$. $\square$

If D is a field, then Corollary 4.4 also follows from Proposition II.8.3.

Corollary **4.5** (Burnside [1905]). *Let K be an algebraically closed field, let V be a finite-dimensional vector space over K, and let R be a subring of* $\mathrm{End}_K(V)$ *that contains all scalar transformations* $v \longmapsto av$ *with* $a \in K$. *If V is a simple R-module, then* $R = \mathrm{End}_K(V)$.

Proof. Identify $a \in K$ with the scalar transformation $aI : v \longmapsto av$, so that K becomes a subfield of $\mathrm{End}_K(V)$ that is contained in the center of $\mathrm{End}_K(V)$ (in fact, K is the center of $\mathrm{End}_K(V)$, by 4.4) and $K \subseteq R$. Let

$$D = \mathrm{End}_R(V) = \{ \delta \in \mathrm{End}_{\mathbb{Z}}(V) \mid \delta \text{ commutes with every } \eta \in R \}.$$

Since V is simple, $D = \mathrm{End}_R(V)$ is a division ring; D consists of linear transformations, since $\delta \in D$ must commute with every $aI \in R$; $K \subseteq D$, since every aI commutes with all $\eta \in \mathrm{End}_K(V)$; K is contained in the center of D; and D has finite dimension over K, since $\mathrm{End}_K(V)$ has finite dimension over K. Then $D = K$: if $\alpha \in D$, then K and α generate a commutative subring $K[\alpha]$ of D, α is algebraic over K since $K[\alpha]$ has finite dimension over K, and $\alpha \in K$ since K is algebraically closed. Thus $\mathrm{End}_R(V) = K$.

Now, $K = \mathrm{End}_R^{\mathrm{op}}(V)$. Let $e_1, \ldots, e_n$ be a basis of V over K. For every $\eta \in \mathrm{End}_K(V)$, there exists, by 4.2, some $\rho \in R$ such that $\eta e_i = \rho e_i$ for all i. Then $\eta = \rho$. Thus R is all of $\mathrm{End}_K(V)$. $\square$

Primitive rings. Fortified with Theorem 4.2 we turn to primitive rings.

Definition. A ring R is left primitive (right primitive) *when there exists a faifthful simple left (right) R-module.*

This definition is due to Jacobson [1945a]. Readers will verify that simple rings are left primitive, but that not all left primitive rings are simple. The other Jacobson density theorem applies to primitive rings.

Theorem **4.6** (Jacobson Density Theorem). *A ring R is left primitive if and only if it is isomorphic to a dense subring of $\mathrm{End}_D(V)$ for some division ring D and right D-module V.*

Proof. If S is a faithful simple left R-module, then $D = \mathrm{End}_R^{\mathrm{op}}(S)$ is a division ring by 1.2, the canonical homomorphism $\Phi : R \longrightarrow \mathrm{End}_D(S)$ is injective since S is faithful, and $\Phi(R) \cong R$ is dense in $\mathrm{End}_D(S)$, by 4.2.

Conversely, let D be a division ring, let V be a right D-module, and let R be a dense subring of $\mathrm{End}_D(V)$. Then V is a left R-module; V is faithful since $\eta v = 0$ for all $v \in V$ implies $\eta = 0$, when $\eta \in \mathrm{End}_D(V)$. If $x, y \in V$, $x \neq 0$, then x is part of a basis of V over D, some linear transformation $\eta \in \mathrm{End}_D(V)$ sends x to y, and $y = \eta x = \rho x$ for some $\rho \in R$ since R is dense. Thus $Rx = V$ for every $0 \neq x \in V$, and V is a simple R-module. $\square$

Theorem 4.6 yields another proof that simple left Artinian rings are isomorphic to matrix rings over division rings. We prove a more general result.

Theorem **4.7.** *Let R be a left primitive ring and let S be a faithful simple left R-module, so that $D = \mathrm{End}_R^{\mathrm{op}}(S)$ is a division ring.*

(1) *If R is left Artinian, then $n = \dim_D S$ is finite and $R \cong M_n(D)$.*

(2) *If R is not left Artinian, then $\dim_D S$ is infinite and for every $n > 0$ there exists a subring R_n of R with a surjective homomorphism $R_n \longrightarrow M_n(D)$.*

Proof. As in the proof of Theorem 4.6, the canonical homomorphism $\Phi : R \longrightarrow \mathrm{End}_D(S)$ is injective, and $\Phi(R)$ is dense in $\mathrm{End}_D(S)$.

(1). If a basis of S contains an infinite sequence $e_1, \ldots, e_n, \ldots$, then

$$L_n = \{ r \in R \mid re_i = 0 \text{ for all } i \leq n \}$$

is a left ideal of R, $L_n \supseteq L_{n+1}$. Also, there exists $\eta \in \mathrm{End}_D(S)$ such that $\eta e_i = 0$ for all $i \leq n$ and $\eta e_{n+1} = e_{n+1}$; hence there exists $r \in R$ such that $re_i = \eta e_i = 0$ and $re_{n+1} = \eta e_{n+1} \neq 0$, and $L_n \supsetneq L_{n+1}$. This cannot be allowed if R is left Artinian. Therefore S has a finite basis $e_1, \ldots, e_n$ over D. For every $\eta \in \mathrm{End}_D(S)$ there exists $r \in R$ such that $\eta e_i = re_i$ for all i, and then $\eta x = rx$ for all $x \in S$. Thus $\Phi : R \longrightarrow \mathrm{End}_D(S)$ is surjective and $R \cong \mathrm{End}_D(S) \cong M_n(D)$.

(2). Now, assume that R is not left Artinian. Then $\dim_D S$ is infinite: otherwise, $R \cong M_n(D)$ as above and R is left Artinian. Hence any basis of S contains an infinite sequence $e_1, \ldots, e_n, \ldots$. Let S_n be the submodule of S_D generated by $e_1, \ldots, e_n$. Then $R_n = \{ r \in R \mid r S_n \subseteq S_n \}$ is a subring of R,

$$I_n = \{ r \in R \mid r e_i = 0 \text{ for all } i \leq n \} = \{ r \in R \mid r S_n = 0 \}$$

is a two-sided ideal of R_n, and S_n is a left R_n/I_n-module, in which $(r + I_n) x = rx$ for all $r \in R_n$ and $x \in S_n$. Then S_n has the same endomorphism ring D as an R-module and as an R_n/I_n-module. Moreover, $\Phi : R_n/I_n \longrightarrow \operatorname{End}_D(S_n)$ is surjective: every $\eta \in \operatorname{End}_D(S_n)$ extends to a D-endomorphism of S; hence there exists $r \in R$ such that $\eta e_i = r e_i = (r + I_n) e_i$ for all $i \leq n$. This provides a surjection $R_n \longrightarrow R_n/I_n \longrightarrow \operatorname{End}_D(S_n) \cong M_n(D)$. $\square$

Exercises

1. Let $M = {}_R R$ and $E = \operatorname{End}_R^{\operatorname{op}}(M)$. Show that the canonical homomorphism $R \longrightarrow \operatorname{End}_E(M)$ is an isomorphism.

2. Let M be a left R-module, $E = \operatorname{End}_R^{\operatorname{op}}(M)$, $n > 0$, and $F = \operatorname{End}_R^{\operatorname{op}}(M^n)$. Show that $\xi \longmapsto \xi^n$ is an isomorphism $\operatorname{End}_E(M) \longrightarrow \operatorname{End}_F(M^n)$ (where $\xi^n : (x_1, \ldots, x_n) \longmapsto (\xi x_1, \ldots, \xi x_n)$). (By Lemma 4.3 you only need to show that this construction yields every $\omega \in \operatorname{End}_F(M^n)$. You may note that ω must commute with every $\iota_i \circ \pi_k$.)

3. Let $R = \mathbb{Z}$, $M = \mathbb{Q}$, $E = \operatorname{End}_R^{\operatorname{op}}(M)$, and $\Phi : R \longrightarrow \operatorname{End}_E(M)$. Show that $\Phi(R)$ is not dense in $\operatorname{End}_E(M)$.

4. Verify that a dense subset of $\operatorname{End}_E(M)$ is dense in the topology induced by the compact-open topology on the set of all transformations of M, where M is discrete.

5. Show that every simple ring is left primitive.

6. Let V be an infinite-dimensional vector space over a division ring D. Show that $\operatorname{End}_D(V)$ is left primitive. Show that $\operatorname{End}_D(V)$ is not simple or left Artinian.

7. Prove that a commutative ring is left primitive if and only if it is a field.

8. If R is left primitive, show that $M_n(R)$ is left primitive.

9. If R is left primitive and $e \in R$ is idempotent, show that eRe is left primitive.

10. Let M be a D-module. A subring S of $\operatorname{End}_D(M)$ is *transitive* when there exists for every $x, y \in M$, $x \neq 0$ some $\eta \in S$ such that $\eta x = y$. Prove the following: if D is a division ring, then every transitive subring of $\operatorname{End}_D(M)$ is left primitive.

11. Show that a left Artinian ring is left primitive if and only if it is simple.

5. The Jacobson Radical

Jacobson [1945a] discovered this radical, which provides a, well, radically different approach to semisimplicity. This section contains general properties. As before, all rings have an identity element, and all modules are unital.

Definition. The Jacobson radical $J(R)$ *of a ring R is the intersection of all its maximal left ideals.*

By rights, $J(R)$ should be called the *left* Jacobson radical of R, but we shall prove that $J(R)$ is also the intersection of all the maximal right ideals of R. We begin with simpler properties.

Proposition **5.1.** *In a ring R, $J(R)$ is the intersection of all the annihilators of simple left R-modules; hence $J(R)$ is a two-sided ideal of R.*

Proof. If L is a maximal left ideal of R, then $S = {}_R R/L$ is a simple left R-module and $\operatorname{Ann}(S) \subseteq L$. Hence the intersection of all $\operatorname{Ann}(S)$ is contained in $J(R)$. Conversely, let $r \in J(R)$ and let S be a simple left R-module. If $x \in S$, $x \neq 0$, then ${}_R R/\operatorname{Ann}(x) \cong Rx = S$ is simple, $\operatorname{Ann}(x)$ is a maximal left ideal of R, $r \in \operatorname{Ann}(x)$, and $rx = 0$. Hence $rx = 0$ for all $x \in S$ and $r \in \operatorname{Ann}(S)$. $\square$

By Proposition 5.1, the elements of $J(R)$ are "close to 0" in that they have the same effect on simple modules. These elements are also "close to 0" in the following senses.

Lemma **5.2.** *If $x \in R$, then $x \in J(R)$ if and only if $1 + tx$ has a left inverse for every $t \in R$.*

Proof. If $x \notin J(R)$, then $x \notin L$ for some maximal left ideal L, $L + Rx = R$, $1 = \ell + rx$ for some $\ell \in L$ and $r \in R$, and $1 - rx \in L$ has no left inverse, since all its left multiples are in L. Conversely, if some $1 + tx$ has no left inverse, then $R(1 + tx) \neq R$, $R(1 + tx)$ is contained in a maximal left ideal L of R, and $x \notin L$, since $1 + tx \in L$ and $1 \notin L$; hence $x \notin J(R)$. $\square$

Proposition **5.3.** *In a ring R, $J(R)$ is the largest two-sided ideal I of R such that $1 + x$ is a unit of R for all $x \in I$; hence $J(R) = J(R^{\mathrm{op}})$.*

Proof. If $x \in J(R)$, then, by 5.2, $1 + x$ has a left inverse y, whence $y = 1 - yx$ and y has a left inverse z. Since y already has a right inverse $1 + x$, it follows that $1 + x = z$, and y is a two-sided inverse of $1 + x$. Thus $J(R)$ is one of the two-sided ideals I of R such that $1 + x$ is a unit of R for all $x \in I$. Moreover, $J(R)$ contains every such ideal I: when $x \in I$, then, for all $t \in R$, $tx \in I$ and $1 + tx$ has a left inverse, whence $x \in J(R)$, by 5.2. $\square$

By Proposition 5.3, $J(R)$ is also the intersection of all maximal right ideals of R, and the intersection of all the annihilators of simple right R-modules.

The following properties are easy exercises:

Proposition **5.4.** $J(R_1 \times \cdots \times R_n) = J(R_1) \times \cdots \times J(R_n)$, *for all rings $R_1, \ldots, R_n$.*

Proposition **5.5.** $J\big(R/J(R)\big) = 0$.

A *radical* in ring theory assigns to a ring R a two-sided ideal $\operatorname{Rad} R$ with nice properties, one of which must be that $\operatorname{Rad}(R/\operatorname{Rad} R) = 0$. The Jacobson radical has this property; so does the nilradical $\{\, x \in R \mid x^n = 0 \text{ for some } n > 0 \,\}$ of a commutative ring R.

Definitions. An element r of a ring R is nilpotent *when $r^n = 0$ for some $n > 0$. A left (right, two-sided) ideal N of a ring R is* nilpotent *when $N^n = 0$ for some $n > 0$.*

Proposition **5.6.** *In a ring R, $J(R)$ contains all nilpotent left or right ideals of R. If R is commutative, then $J(R)$ contains all nilpotent elements of R.*

Proof. Let N be a nilpotent left ideal and let S be a simple left R-module. If $NS \neq 0$, then $NS = S$ and $S = NS = N^2 S = \cdots = N^n S = 0$, a contradiction; therefore $NS = 0$ and $N \subseteq \mathrm{Ann}\,(S)$. Hence $N \subseteq J(R)$, by 5.1. Then $J(R)$ contains every nilpotent right ideal, by 5.3. If R is commutative, then $r \in R$ nilpotent implies Rr nilpotent, and $J(R)$ contains every nilpotent element. $\square$

Thus $J(R)$ contains the nilradical of R when R is commutative. In general, there may be plenty of nilpotent elements outside of $J(R)$ (see the exercises).

Last, but not least, are two forms of Nakayama's lemma (proved by Nakayama as a student, according to Nagata [1962]):

Proposition **5.7** (Nakayama's Lemma). *Let M be a finitely generated left R-module. If $J(R)M = M$, then $M = 0$.*

Proof. Assume $M \neq 0$. Since M is finitely generated, the union of a chain $(N_i)_{i \in I}$ of proper submodules of M is a proper submodule of M: otherwise, the finitely many generators of M all belong to some N_i, and then $N_i = M$. By Zorn's lemma, M has a maximal (proper) submodule N. Then M/N is simple and $J(R)(M/N) = 0$ by 5.1. Hence $J(R)M \subseteq N$ and $J(R)M \neq M$. $\square$

Proposition **5.8** (Nakayama's Lemma). *Let N be a submodule of a finitely generated left R-module M. If $N + J(R)M = M$, then $N = M$.*

Proof. First, M/N is finitely generated. If $N + J(R)M = M$, then $J(R)(M/N) = M/N$ and $M/N = 0$, by 5.7. $\square$

The exercises give some neat applications of Nakayama's lemma.

Exercises

1. Show that $x \in J(R)$ if and only if $1 + rxs$ is a unit for every $r, s \in R$.

2. Find $J(\mathbb{Z})$.

3. Find $J(\mathbb{Z}_n)$. When is $J(\mathbb{Z}_n) = 0$?

4. Let D be a division ring. Show that $J\big(M_n(D)\big) = 0$. Show that $J(R)$ does not necessarily contain every nilpotent element of R, and may in fact contain no nonzero nilpotent element of R.

5. Find a commutative ring in which the Jacobson radical strictly contains the nilradical.

6. Show that $J(R)$ contains no idempotent $e^2 = e \neq 0$.

7. Show that $J(R_1 \times \cdots \times R_n) = J(R_1) \times \cdots \times J(R_n)$, for all rings $R_1, \ldots, R_n$.

8. Show that $J\big(R/J(R)\big) = 0$, for every ring R.

9. Let $\varphi : R \longrightarrow S$ be a ring homomorphism. Show that $\varphi\big(J(R)\big) \subseteq J(S)$. Give an example in which $\varphi\big(J(R)\big) \subsetneqq J(S)$.

10. A left or right ideal is *nil* when all its elements are nilpotent. If R is left Artinian, show that $J(R)$ contains every nil left or right ideal of R.

In the next two exercises, an element r of a ring R is *quasiregular* when $1 - r$ is a unit of R, and *left quasiregular* when $1 - r$ has a left inverse.

11. Prove that every nilpotent element is quasiregular.

12. Prove that $J(R)$ is the largest left ideal L of R such that every element of L is left quasiregular.

13. Prove the following: if a ring R has only one maximal left ideal L, then L is a two-sided ideal and a maximal right ideal.

14. Let $e^2 = e$ be an idempotent of a ring R. Show that $J(eRe) = J(R) \cap eRe$. (Hint: when S is a simple left R-module, either $eS = 0$ or S is a simple eRe-module.)

15. Let R be a ring. Show that a matrix $A \in M_n(R)$ is in $J\big(M_n(R)\big)$ if and only if every entry of A is in $J(R)$. (Hint: when S is a simple left R-module, matrix multiplication makes S^n a simple left $M_n(R)$-module.)

16. Explain how every simple left R-module is also a simple left $R/J(R)$-module, and vice versa.

17. Let L be a left ideal of R. Prove that $L \subseteq J(R)$ if and only if, for every finitely generated left R-module M, $LM = 0$ implies $M = 0$.

In the following exercises, M is a finitely generated left R-module and $\overline{M} = M/J(R)M$.

18. Show that every module homomorphism $\varphi : M \longrightarrow M'$ induces a module homomorphism $\overline{\varphi} : \overline{M} \longrightarrow \overline{M}'$ that makes a commutative square with the projections

$$
\begin{array}{ccc}
M & \xrightarrow{\ \varphi\ } & M' \\
\downarrow & & \downarrow \\
\overline{M} & \dashrightarrow{\ \overline{\varphi}\ } & \overline{M}'
\end{array}
$$

19. Prove the following: if $\overline{\varphi}$ is surjective, then φ is surjective.

20. Prove the following: if $\overline{x}_1, \ldots, \overline{x}_n$ generate $\overline{M}$, then $x_1, \ldots, x_n$ generate M.

21. Let $\mathfrak{m}$ be a maximal ideal of a commutative ring R. Prove the following: if A is a finitely generated R-module, and $x_1, \ldots, x_n$ is a minimal generating subset of A, then $x_1 + \mathfrak{m}A, \ldots, x_n + \mathfrak{m}A$ is a basis of $A/\mathfrak{m}A$ over $R/\mathfrak{m}$.

6. Artinian Rings

Following Jacobson [1945a,b], this section introduces the classical definition of semisimplicity and relates it to the module definition and to the Jacobson radical, with further applications to ring theory.

Definition. A ring R is Jacobson semisimple *when $J(R) = 0$.*

Jacobson [1945a] called these rings *semiprimitive*. Current terminology tends to name semisimplicity of the $\mathrm{Rad}\, R = 0$ kind after the name or initial of the radical, keeping unadorned semisimplicity for the semisimple rings and modules in our Sections 2 and 3.

Semisimple rings are Jacobson semisimple, by Theorem 6.1 below; left primitive rings are Jacobson semisimple, by Proposition 5.1. Jacobson semisimple rings abound, since $R/J(R)$ is Jacobson semisimple for every ring R.

Wedderburn called a ring semisimple when it has no nonzero nilpotent ideal. The main result in this section is the following:

*Theorem **6.1.** A ring R is semisimple if and only if R is left Artinian and $J(R) = 0$, if and only if R is left Artinian and has no nonzero nilpotent ideal.*

Proof. The last two conditions are equivalent, by 6.2 below.

By 3.5, 3.3, a semisimple ring R is left Artinian, and is isomorphic to a direct product $M_{n_1}(D_1) \times \cdots \times M_{n_s}(D_s)$ of finitely many matrix rings over division rings $D_1, \ldots, D_s$. Readers will have verified that $J(M_n(D)) = 0$ when D is a division ring; hence $J(R) \cong J(M_{n_1}(D_1)) \times \cdots \times J(M_{n_s}(D_s)) = 0$, by 5.4.

Conversely, let R be left Artinian. Then $J(R)$ is the intersection of finitely many maximal left ideals of R. Indeed, let $\mathcal{S}$ be the set of all intersections of finitely many maximal left ideals of R. Since R is left Artinian, $\mathcal{S}$ has a minimal element J. For every maximal left ideal L of R we now have $J \supseteq J \cap L \in \mathcal{S}$, whence $J = J \cap L \subseteq L$. Therefore $J = J(R)$. If $J(R) = 0$, then 0 is the intersection of finitely many maximal left ideals $L_1, \ldots, L_n$ of R. The projections ${}_R R \longrightarrow R/L_i$ induce a module homomorphism $\varphi : {}_R R \longrightarrow R/L_1 \times \cdots \times R/L_n$, $r \longmapsto (r + L_1, \ldots, r + L_n)$, which is injective since $\ker \varphi = L_1 \cap \cdots \cap L_n = 0$. Hence ${}_R R$ is isomorphic to a submodule of the semisimple module $R/L_1 \oplus \cdots \oplus R/L_n$, and ${}_R R$ is semisimple. $\square$

*Lemma **6.2.** If R is left Artinian, then $J(R)$ is nilpotent, and is the greatest nilpotent left ideal of R and the greatest nilpotent right ideal of R.*

Proof. Let $J = J(R)$. Since R is left Artinian, the descending sequence $J \supseteq J^2 \supseteq \cdots \supseteq J^n \supseteq J^{n+1} \supseteq \cdots$ terminates at some J^m ($J^n = J^m$ for all $n \geqq m$). Suppose that $J^m \neq 0$. Then the nonempty set $\{ L \mid L$ is a left ideal of R and $J^m L \neq 0 \}$ has a minimal element L. We have $J^m a \neq 0$ for some $a \in L$, $a \neq 0$. Then $J^m a \subseteq L$, $J^m(J^m a) = J^m a \neq 0$, and $J^m a = L$ by the choice of L. Hence $a = xa$ for some $x \in J^m$. But then $1 - x$ has a left inverse, by 5.2, and $(1 - x) a = 0$ implies $a = 0$, a red contradiction. Therefore J is nilpotent. By 5.6, J contains every nilpotent left or right ideal of R. $\square$

Applications. If R is left Artinian, then so is $R/J(R)$, and then $R/J(R)$ is semisimple, by Theorem 6.1. Thus R has a nilpotent ideal $J(R)$ such that

$R/J(R)$ has a known structure. This yields properties of left Artinian rings. We give two examples.

Proposition 6.3. *If R is left Artinian, then a left R-module M is semisimple if and only if $J(R)M = 0$.*

Proof. Let $J = J(R)$. We have $JS = 0$ for every simple R-module S, since $J \subseteq \text{Ann}(S)$; hence $JM = 0$ whenever M is semisimple. Conversely, assume that $JM = 0$. Then M is a left R/J-module, in which $(r + J)x = rx$ for all $x \in M$, and with the same submodules as $_R M$. Since R/J is semisimple, every submodule of $_{R/J} M$ is a direct summand, every submodule of $_R M$ is a direct summand, and M is semisimple. $\square$

Theorem 6.4 (Hopkins-Levitzki). *If R is left Artinian, then for a left R-module M the following properties are equivalent:* (i) *M is Noetherian;* (ii) *M is Artinian;* (iii) *M is of finite length.*

Proof. If M is semisimple, then (i), (ii), and (iii) are equivalent, since a Noetherian or Artinian module cannot be the direct sum of infinitely many simple submodules.

In general, let $J = J(R)$. Let M be Noetherian (or Artinian). By 6.2, $J^n = 0$ for some $n > 0$, which yields a descending sequence

$$M \supseteq JM \supseteq J^2 M \supseteq \cdots \supseteq J^n M = 0.$$

For every $i < n$, $J^i M \subseteq M$ is Noetherian (or Artinian) and $J^i M / J^{i+1} M$ is Noetherian (or Artinian). But $J^i M / J^{i+1} M$ is semisimple, by 6.3. Hence every $J^i M / J^{i+1} M$ has a composition series. Then M has a composition series. $\square$

Corollary 6.5 (Hopkins [1939]). *Every left Artinian ring is left Noetherian.*

Exercises

1. A left or right ideal is *nil* when all its elements are nilpotent. If R is left Artinian, show that every nil left or right ideal of R is nilpotent.

2. Let $L_1, \ldots, L_n$ be left ideals of a ring R. Prove the following: if $L_1, \ldots, L_n$ are nilpotent, then $L_1 + \cdots + L_n$ is nilpotent.

3. Let R be a ring and let U be its group of units. Suppose that $U \cup \{0\}$ is a subring of R. Show that R is Jacobson semisimple.

4. Show that $D[X]$ is Jacobson semisimple when D is a division ring.

5. Show that every von Neumann regular ring is Jacobson semisimple. (A ring R is *von Neumann regular* when for every $a \in R$ there exists $x \in R$ such that $axa = a$.)

6. Show that a ring R is Jacobson semisimple if and only if it there exists a faithful semisimple left R-module.

7. Prove the following: if $R/J(R)$ is semisimple, then a left R-module M is semisimple if and only if $J(R)M = 0$.

8. Prove the following: if $J(R)$ is nilpotent and $R/J(R)$ is semisimple, then for a left R-module M the following properties are equivalent: (i) M is Noetherian; (ii) M is

Artinian; (iii) M is of finite length. [If R is left Artinian, then $J(R)$ is nilpotent and $R/J(R)$ is semisimple, but the converse implication is known to be false.]

9. Show that a left Artinian ring contains no nilpotent element if and only if it is isomorphic to the direct product of finitely many division rings.

10. Prove the following: if R is left Artinian and $a \in R$ is not a right zero divisor, then a is a unit of R. (Hint: right multiplication by a is a module homomorphism $_RR \longrightarrow {_RR}$.)

11. Prove the following: in a left Artinian ring, $xy = 1$ implies $yx = 1$.

7. Representations of Groups

Representations of groups were first considered in the late nineteenth century; the first systematic studies are due to Schur [1904] and Burnside [1905]. This section defines group representations, gives their basic properties, and explains their relationship with semisimplicity.

Matrix representations. As defined originally, a representation of a group G is a representation of G by matrices or by linear transformations:

Definitions. A representation of a group G over a field K is a homomorphism of G into the group $GL(n, K)$ of invertible $n \times n$ matrices with entries in K, or into the group $GL(V)$ of invertible linear transformations of a vector space V over K. The dimension of V is the degree *or* dimension *of the representation.*

For instance, in any dimension a group G has a *trivial* representation $g \longmapsto 1$. In general, let V be a vector space with basis G; left multiplication by $g \in G$ permutes the basis and extends to an invertible linear transformation of V; this yields the *regular* representation of G, which has dimension $|G|$.

A representation ρ of $\mathbb{Z}$ is determined by the single linear transformation or matrix $\rho(1)$. In general, however, group representations of a group G are typically more difficult to classify than single matrices. As with single linear transformations, we look for bases in which the matrices of all $\rho(g)$ consist of simpler diagonal blocks.

Definition. Two representations $\rho_1 : G \longrightarrow GL(V_1)$, $\rho_2 : G \longrightarrow GL(V_2)$ are equivalent *when there is an invertible linear transformation $T : V_1 \longrightarrow V_2$ such that $\rho_2(g) = T \circ \rho_1(g) \circ T^{-1}$ for all $g \in G$.*

Thus, two representations of $\mathbb{Z}$ are equivalent if and only if their matrices are similar. In general, representations need only be classified up to equivalence.

Definition. The direct sum *of representations $\rho_i : G \longrightarrow GL(V_i)$ is the representation $\rho = \bigoplus_{i \in I} \rho_i : G \longrightarrow GL\left(\bigoplus_{i \in I} V_i\right)$ that assigns to $g \in G$ the linear transformation $(v_i)_{i \in I} \longmapsto (\rho_i(g) v_i)_{i \in I}$ of $\bigoplus_{i \in I} V_i$.*

Given a basis B_i of every V_i, the disjoint union $B = \bigcup_{i \in I} B_i$ is a basis of

$\bigoplus_{i \in I} V_i$, in which the matrix of $\rho(g)$ consists of diagonal blocks

$$\begin{pmatrix} M_i(g) & 0 & \cdots \\ 0 & M_j(g) & \cdots \\ \vdots & \vdots & \ddots \end{pmatrix},$$

where $M_i(g)$ is the matrix of $\rho_i(g)$ in the basis B_i.

For example, the trivial representation of G is a direct sum of trivial representations of dimension 1. Less trivially, when ρ is a representation of $\mathbb{Z}$, putting the matrix $\rho(1)$ in Jordan form finds an equivalent representation that is a direct sum of simpler representations; for instance, $\rho(1)$ is diagonalizable if and only if ρ is a direct sum of representations of dimension 1.

Definition. A representation $\rho : G \longrightarrow GL(V)$ is irreducible when $V \neq 0$ and $\rho = \rho_1 \oplus \rho_2$ implies $V_1 = 0$ or $V_2 = 0$ (where $\rho_1 : G \longrightarrow GL(V_1)$ and $\rho_2 : G \longrightarrow GL(V_2)$).

Proposition **7.1.** *Every finite-dimensional representation is a direct sum of irreducible representations.*

Proof. This is shown by induction on the dimension of V: if ρ is not irreducible, then $\rho = \rho_1 \oplus \rho_2$, where $V = V_1 \oplus V_2$ and V_1, V_2 have lower dimension than V. $\square$

Every representation of dimension 1 is irreducible, but not every irreducible representation has dimension 1: for instance, when a matrix is not diagonalizable, the corresponding representation of $\mathbb{Z}$ is a direct sum of irreducible representations, not all of which can have dimension 1. Our goal is now to classify irreducible representations, up to equivalence.

The group algebra. Modules over algebras give a different view of representations. An *algebra* over K, or *K-algebra* (with an identity element) is a vector space over K with a bilinear associative multiplication for which there is an identity element. For example, the matrix ring $M_n(K)$ is a K-algebra; more generally, when V is a vector space over K, the linear transformations of V into itself constitute a K-algebra $\mathrm{End}_K(V)$, in which the multiplication is composition.

Now, let G be a group. Let $K[G]$ be a vector space in which G is a basis (constructed perhaps as in Section VIII.4), so that every element of $K[G]$ is a linear combination $r = \sum_{g \in G} x_g\, g$ for some unique $x_g \in K$ (with $x_g = 0$ for almost all g, in case G is infinite). Define a multiplication on $K[G]$ by

$$\left(\sum_{g \in G} x_g\, g\right)\left(\sum_{h \in G} y_h\, h\right) = \sum_{k \in G} z_k\, k,$$

where $z_k = \sum_{g,h \in G,\ gh=k} x_g\, y_h$. Skeptical readers will verify that this multiplication is well defined, associative, and bilinear, and induces the given multiplication on G. The identity element 1 of G is also the identity element of $K[G]$. Thus $K[G]$ is a K-algebra.

Definition. If G is a group and K is a field, then $K[G]$ is the group ring or group algebra of G over K.

Similar constructions were used in Chapter III to define polynomial rings. We note that $K[G]$ contains a subfield $\{\, a1 \mid a \in K \,\}$ that consists of scalar multiples of its identity element and is isomorphic to K. Multiplication by $a1$ on either side is just scalar multiplication by a. In particular, $a1$ is *central* in $K[G]$: $a1$ commutes with every $x \in K[G]$, since the multiplication on $K[G]$ is bilinear. We identify $a1$ with a, so that K becomes a central subfield of $K[G]$.

A *homomorphism* of K-algebras is a linear transformation that preserves products and identity elements. The group algebra $K[G]$ has a universal property, which produces this very kind of homomorphism:

Proposition 7.2. *Every multiplicative homomorphism of a group G into a K-algebra A extends uniquely to an algebra homomorphism of $K[G]$ into A.*

Proof. Let $\varphi : G \longrightarrow A$ be a multiplicative homomorphism (meaning $\varphi(gh) = \varphi(g)\,\varphi(h)$ for all $g, h \in G$, and $\varphi(1) = 1$). Since G is a basis of $K[G]$, φ extends uniquely to a linear transformation $\overline{\varphi} : K[G] \longrightarrow A$, namely $\overline{\varphi}(\sum_{g \in G} a_g\, g) = \sum_{g \in G} a_g\, \varphi(g)$. Readers will easily verify that $\overline{\varphi}$ preserves products; already $\overline{\varphi}(1) = \varphi(1) = 1$. $\square$

In what follows we denote $\overline{\varphi}$ by just φ, so that φ extends to $K[G]$ by $\varphi(\sum_{g \in G} a_g\, g) = \sum_{g \in G} a_g\, \varphi(g)$.

Our astute readers have probably guessed what comes next:

Proposition 7.3. *There is a one-to-one correspondence between representations of a group G over a field K and $K[G]$-modules. Two representations are equivalent if and only if the corresponding modules are isomorphic.*

Proof. By 7.2, a representation $\rho : G \longrightarrow GL(V) \subseteq \mathrm{End}_K(V)$ extends uniquely to an algebra homomorphism $\rho : K[G] \longrightarrow \mathrm{End}_K(V)$ that makes V a $K[G]$-module. Conversely, let V be a $K[G]$-module. Since K is a subfield of $K[G]$, the $K[G]$-module structure $\rho : K[G] \longrightarrow \mathrm{End}_{\mathbb{Z}}(V)$ on V induces a K-module structure on V, in which $av = \rho(a)(v)$ for all $a \in K$ and $v \in V$. Then every $\rho(x)$ is a linear transformation: $\rho(x)(av) = \rho(xa)(v) = \rho(ax)(v) = a\,\rho(x)(v)$, since every $a \in K$ is central in $K[G]$. Similarly, $\rho(ax) = a\,\rho(x)$. Hence ρ is an algebra homomorphism $K[G] \longrightarrow \mathrm{End}_K(V)$. If $g \in G$, then $\rho(g)$ is invertible, since $\rho(g) \circ \rho(g^{-1}) = 1 = \rho(g^{-1}) \circ \rho(g)$. Thus $\rho_{|G} : G \longrightarrow GL(V)$ is a representation of G; by 7.2, the corresponding $K[G]$-module structure on V is ρ itself. We now have our one-to-one correspondence.

Let $\rho_1 : G \longrightarrow GL(V_1)$ and $\rho_2 : G \longrightarrow GL(V_2)$ be equivalent, so that there is an invertible linear transformation $T : V_1 \longrightarrow V_2$ such that

$\rho_2(g) = T \circ \rho_1(g) \circ T^{-1}$ for all $g \in G$, equivalently $T(gv) = T(\rho_1(g)(v)) = \rho_2(g)(T(v)) = gT(v)$ for all $g \in G$, $v \in V_1$. Then T is a $K[G]$-module isomorphism:

$$T\big((\textstyle\sum_{g \in G} x_g\, g)v\big) = T\big(\textstyle\sum_{g \in G} x_g\, (gv)\big) = \textstyle\sum_{g \in G} x_g T(gv)$$
$$= \textstyle\sum_{g \in G} x_g\, gT(v) = \big(\textstyle\sum_{g \in G} x_g\, g\big)T(v)$$

for all $v \in V_1$ and $\sum_{g \in G} x_g\, g \in K[G]$. Conversely, if $T : V_1 \longrightarrow V_2$ is a $K[G]$-module isomorphism, then T is a linear transformation and $T(gv) = gT(v)$ for all g and v, so that ρ_1 and ρ_2 are equivalent. $\square$

The calculation $(\sum_{g \in G} x_g\, g)v = \sum_{g \in G} x_g\, (gv)$ shows that the $K[G]$-module structure on V is determined by its vector space structure and the action of G. Accordingly, left $K[G]$-modules are usually called just G-*modules* (it being understood that they already are vector spaces over K). A submodule of a G-module V is a subspace W that is closed under the action of G ($v \in W$ implies $gv \in W$ for all $g \in G$).

Maschke's theorem. If $\rho : G \longrightarrow GL(V)$ is a direct sum $\rho = \bigoplus_{i \in I} \rho_i$ of representations $\rho_i : G \longrightarrow GL(V_i)$, then $V = \bigoplus_{i \in I} V_i$ as a vector space, and

$$g\,(v_i)_{i \in I} = \rho(g)(v_i)_{i \in I} = \big(\rho_i(g)\,(v_i)\big)_{i \in I} = (gv_i)_{i \in I}$$

for all $g \in G$, so that $V = \bigoplus_{i \in I} V_i$ as a G-module. Conversely, direct sums of G-modules yield direct sums of representations. Hence a group representation is irreducible if and only if the corresponding G-module V is *indecomposable* ($V \neq 0$, and $V = V_1 \oplus V_2$ implies $V_1 = 0$ or $V_2 = 0$). Our goal is now the classification of indecomposable G-modules, up to isomorphism. This does not seem very tractable. Fortunately, there is Maschke's theorem:

Theorem **7.4** (Maschke [1898]). *Let G be a finite group and let K be a field. If K has characteristic 0, or if K has characteristic $p \neq 0$ and p does not divide the order of G, then $K[G]$ is semisimple.*

Proof. We show that every submodule W of a G-module V is a direct summand of V. We already have $V = W \oplus W'$ (as a vector space) for some subspace W'. The projection $\pi : V \longrightarrow W$ is a linear transformation and is the identity on W. Define $\varphi : V \longrightarrow W$ by

$$\varphi(v) = \frac{1}{n} \textstyle\sum_{g \in G} g^{-1} \pi(gv),$$

where $n = |G| \neq 0$ in K by the choice of K. Then φ is a linear transformation; φ is the identity on W: for all $w \in W$,

$$\varphi(w) = \frac{1}{n} \textstyle\sum_{g \in G} g^{-1} \pi(gw) = \frac{1}{n} \textstyle\sum_{g \in G} g^{-1} gw = w; \text{ and}$$

$$\varphi(hv) = \frac{1}{n} \textstyle\sum_{g \in G} g^{-1} \pi(ghv) = \frac{1}{n} \textstyle\sum_{g \in G} h\,(gh)^{-1} \pi(ghv)$$

$$= \frac{1}{n} \textstyle\sum_{k \in G} h\,k^{-1} \pi(kv) = h\,\varphi(v).$$

for all $h \in G$ and $v \in V$, so that φ is a module homomorphism. Then W is a direct summand of V (as a G-module) by VIII.3.8. $\square$

Corollary 7.5. *Let G be a finite group and let K be a field whose characteristic does not divide the order of G.*

(1) *Up to isomorphism, there are only finitely many simple G-modules $S_1, \ldots, S_s$, and they all have finite dimension over K.*

(2) *Every G-module is isomorphic to a direct sum $S_1^{m_1} \oplus \cdots \oplus S_s^{m_s}$ for some unique cardinal numbers $m_1, \ldots, m_s$.*

(3) *Up to equivalence, there are only finitely many irreducible representations of G, and they all have finite dimension over K.*

Proof. That there are only finitely many simple G-modules follows from 3.6. By 1.3, every simple G-module is isomorphic to a minimal left ideal of $K[G]$, and has finite dimension over K, like $K[G]$. Then (2) follows from 3.7, and (3) follows from (1) and 7.3. $\square$

One can say more when K is algebraically closed:

Proposition 7.6. *Let G be a finite group and let K be an algebraically closed field whose characteristic does not divide the order of G. Let the nonisomorphic simple G-modules have dimensions $d_1, \ldots, d_s$ over K. The simple components of $K[G]$ are isomorphic to $M_{d_1}(K), \ldots, M_{d_s}(K)$, and $\sum_i d_i^2 = |G|$.*

Proof. By 7.4, $K[G]$ is semisimple. The simple components $R_1, \ldots, R_r$ of $K[G]$ are simple left Artinian rings $R_i \cong M_{n_i}(D_i)$, where $D_1, \ldots, D_r$ are division rings, as well as two-sided ideals of R. Hence every R_i is a vector space over K, and its multiplication is bilinear; then R_i, like $K[G]$, has a central subfield that consists of all scalar multiples of its identity element and is isomorphic to K. Then $M_{n_i}(D_i)$ has a central subfield that etc. etc. By 4.4, this central subfield consists of scalar matrices, with entries in the center of D_i. Therefore D_i too has a central subfield K_i that consists of all scalar multiples of its identity element and is isomorphic to K. Up to this isomorphism, K_i and K induce the same vector space structure on D_i. Hence D_i, like R_i, has finite dimension over K. By 7.7 below, $D_i = K_i$; hence $R_i \cong M_{n_i}(K)$.

By 3.6, every simple left R-module is isomorphic to a minimal left ideal of some R_i, which has dimension n_i over K by 1.8. Hence $d_i = n_i$ for all i. Then $K[G] \cong M_{d_1}(K) \times \cdots \times M_{d_s}(K)$ has dimension $|G| = \sum_i d_i^2$. $\square$

Lemma 7.7. *Let D be a division ring that has finite dimension over a central subfield K. If K is algebraically closed, then $D = K$.*

Proof. If $\alpha \in D$, then K and α generate a commutative subring $K[\alpha]$ of D, α is algebraic over K since $K[\alpha]$ has finite dimension over K, and $\alpha \in K$ since K is algebraically closed. $\square$

Corollary 7.8. *Let G be a finite group and let K be an algebraically closed* *field whose characteristic does not divide the order of G. If G is abelian, then every irreducible representation of G over K has dimension 1.*

The proof is an exercise. More generally, when G is finite, every conjugacy class C of G has a sum $\sum_{g \in C} g$ in $K[G]$. These sums are linearly independent over K, since the conjugacy classes of G constitute a partition of G. Readers will enjoy showing that they constitute a basis (over K) of the center of $K[G]$.

Theorem 7.9. Let G be a finite group with s conjugacy classes and let K be an algebraically closed field whose characteristic does not divide the order of G. Up to equivalence, G has s distinct irreducible representations over K.

Proof. We look at the center $Z(K[G])$ of $K[G]$. By 7.6, $K[G] \cong R_1 \times \cdots \times R_s$, where s is now the number of distinct irreducible representations of G and $R_i \cong M_{n_i}(K)$. By 4.4, $Z(M_{n_i}(K)) \cong K$; hence $Z(K[G]) \cong K^s$. Therefore G has s conjugacy classes. □

Proposition 7.10. Let G be a finite group and let K be an algebraically closed field whose characteristic does not divide the order of G. Let $\rho : G \longrightarrow \mathrm{End}_K(S)$ be an irreducible representation and let c be the sum of a conjugacy class. Then $\rho : K[G] \longrightarrow \mathrm{End}_K(S)$ is surjective, and $\rho(c)$ is a scalar linear transformation.

Proof. First, S is a simple $K[G]$-module. Let $E = \mathrm{End}^{\mathrm{op}}_{K[G]}(S)$. Since K is central in $K[G]$, the scalar linear transformations $\lambda 1 : x \longmapsto \lambda x$ are $K[G]$-endomorphisms of S and constitute a subfield $K' \cong K$ of E. By Schur's lemma, E is a division ring; then $E = K'$ by 7.7, so that E contains only scalar linear transformations. Hence $\mathrm{End}_E(S) = \mathrm{End}_K(S)$.

Let $e_1, \ldots, e_d$ be a basis of S over K. If $T \in \mathrm{End}_K(S) = \mathrm{End}_E(S)$, then Theorem 4.2 yields $x \in K[G]$ such that $T(e_i) = xe_i$ for all i. Then $T(v) = xv$ for all $x \in S$, that is, $T = \rho(x)$. Since c commutes with every $x \in K[G]$, $\rho(c)$ commutes with every $\rho(x)$, and commutes with every $T \in \mathrm{End}_K(S)$. Hence $\rho(c)$ is a scalar linear transformation, by 4.4. □

Exercises

1. Find all irreducible representations of $\mathbb{Z}$ (up to equivalence), over an algebraically closed field.

2. True or false: there exists a group G with an irreducible representation over $\mathbb{C}$ of dimension n for every $n > 0$.

3. Find all irreducible representations of $\mathbb{Z}_n$ (up to equivalence) over an algebraically closed field.

4. Verify that the multiplication on $K[G]$ is well defined, associative, and bilinear, and that the identity element of G is also the identity element of $K[G]$.

5. Let G be a totally ordered group (a group with a total order $\leqq$ such that $x \leqq y$ implies $xz \leqq yz$ and $zx \leqq zy$ for all z). Show that $K[G]$ has no zero divisors.

6. Show that every K-algebra (with an identity element) contains a central subfield that is isomorphic to K.

7. Find a division ring D that has finite dimension over an algebraically closed subfield K but is not equal to K.

8. Let G be a finite group and let K be a field. Show that the sums of the conjugacy classes of G constitute a basis of the center of $K[G]$.

9. Let G be a finite abelian group and let K be an algebraically closed field whose characteristic does not divide the order of G. Without using Theorem 7.9, show that every irreducible representation of G over K has dimension 1.

In the next problems, C_n is a cyclic group of order n.

10. Show that $K[C_n] \cong K[X]/(X^n - 1)$, for every field K.

11. Write $\mathbb{C}[C_n]$ as a direct sum of minimal left ideals. (You may want to use the previous exercise and the Chinese remainder theorem.)

12. Write $\mathbb{Q}[C_n]$ as a direct sum of minimal left ideals.

13. Let $C_2 = \{e, a\}$ be cyclic of order 2. Show that $\mathbb{Z}_2[C_2]$ is not semisimple. (You may want to show that the ideal of $\mathbb{Z}_2[C_2]$ generated by $e + a$ is not a direct summand.)

14. Let H be a normal subgroup of a group G and let K be any field. Prove *Clifford's theorem*: if V is a simple G-module, then V is a semisimple H-module. (You may want to look at $\sum_{g \in G} gW$, where W is a simple sub-H-module of V.)

15. Let G be a finite group and let K be a field of characteristic $p \neq 0$. Show that every normal p-subgroup N of G acts trivially on every simple G-module. Show that the simple G-modules are the same as the simple G/N-modules. (You may want to use induction on $|N|$, the previous exercise, and the fact that $|Z(N)| > 1$ when $|N| > 1$.)

16. Let G be a finite group and let K be a field of characteristic $p \neq 0$. Let N be the intersection of all the Sylow p-subgroups of G. Show that $g \in N$ if and only if $g - 1 \in J(K[G])$. Show that $J(K[G])$ is the ideal of $K[G]$ generated by all $g - 1$ with $g \in N$. (Hint: $K[G/N]$ is semisimple.)

8. Characters

Characters are mappings that arise from representations. They have become a major tool of finite group theory. This section contains a few basics.

Definition. Every linear transformation T of a finite-dimensional vector space V has a trace $\mathrm{Tr}\, T$, which is the trace of its matrix in any basis of V.

Definitions. Let ρ be a finite-dimensional representation of a group G over a field K. The character of ρ is the mapping $\chi_\rho : G \longrightarrow K$ defined by

$$\chi_\rho(g) = \mathrm{Tr}\, \rho(g).$$

A character of G over K is the character of a finite-dimensional representation of G; an irreducible character of G over K is the character of an irreducible finite-dimensional representation of G.

For example, the trivial representation $\tau(g) = 1$ of dimension d has a constant *trivial character* $\chi_\tau(g) = d$ for all $g \in G$. The *regular character* of a finite group G is the character χ_ρ of its regular representation ρ; readers will verify that $\chi_\rho(1) = |G|$ and $\chi_\rho(g) = 0$ for all $g \neq 1$. A homomorphism of G into the multiplicative group $K \backslash \{0\}$ is a *one-dimensional character*.

In general, $\chi_\rho(1) = \mathrm{Tr}\ 1$ is the dimension of ρ. Moreover, χ_ρ is a *class function* (constant on every conjugacy class): $\chi_\rho(hgh^{-1}) = \chi_\rho(g)$ for all $g, h \in G$, since $\rho(g)$ and $\rho(hgh^{-1}) = \rho(h) \circ \rho(g) \circ \rho(h)^{-1}$ have the same trace.

Proposition 8.1. *If* $\rho = \bigoplus_{j \in J} \rho_j$, *then* $\chi_\rho = \sum_{j \in J} \chi_{\rho_j}$. *Hence every character is a pointwise sum of irreducible characters.*

From now on, G is finite and the characteristic of K does not divide the order of G. We use the following notation: $R_1, \ldots, R_s$ are the simple components of $K[G]$; e_i is the identity element of R_i; S_i is a minimal left ideal of R_i, so that $K[G]$ acts on S_i by left multiplication in $K[G]$; $d_i = \dim_K S_i$; ρ_i is the corresponding irreducible representation $\rho_i : G \longrightarrow \mathrm{End}_K(S_i)$ ($\rho_i(x)s = xs$ is the product in $K[G]$, for all $x \in K[G]$ and all $s \in S_i$); and χ_i is the character of ρ_i. Up to isomorphism, $S_1, \ldots, S_s$ are all the simple G-modules; hence $\rho_1, \ldots, \rho_s$ are, up to equivalence, all the irreducible representations and $\chi_1, \ldots, \chi_s$ are all the irreducible characters.

Let χ be the character of a finite-dimensional representation ρ. By Corollary 7.5, the corresponding G-module is isomorphic to a direct sum $S_1^{m_1} \oplus \cdots \oplus S_s^{m_s}$ for some nonnegative integers $m_1, \ldots, m_s$. Hence ρ is equivalent to a direct sum of m_1 copies of $\rho_1, \ldots, m_s$ copies of ρ_s, and $\chi = m_1 \chi_1 + \cdots + m_s \chi_s$. For example, the regular character has the following expansion:

Proposition 8.2. *If* K *is an algebraically closed field whose characteristic does not divide* $|G|$, *then the regular character is* $\chi_r = \sum_i d_i \chi_i$.

Proof. By 7.6, $R_i \cong M_{d_i}(K) \cong S_i^{d_i}$; hence $K[G] \cong S_1^{d_1} \oplus \cdots \oplus S_s^{d_s}$, as a G-module, and $\chi_r = \sum_i d_i \chi_i$. $\square$

In addition, every character χ extends to a linear mapping $\sum_{g \in G} x_g g \longmapsto \sum_{g \in G} x_g \chi(g)$ of $K[G]$ into K, also denoted by χ. Then

$$\chi_\rho(x) = \sum_{g \in G} x_g \, \mathrm{Tr}\, \rho(g) = \mathrm{Tr} \left(\sum_{g \in G} x_g \, \rho(g) \right) = \mathrm{Tr}\, \rho(x)$$

for all $x = \sum_{g \in G} x_g g \in K[G]$. We note the following properties.

Lemma 8.3. *If the characteristic of* K *does not divide* $|G|$, *then:*

(1) $\chi_i(x) = 0$ *when* $x \in R_j$ *and* $j \neq i$;

(2) $\chi_i(e_j) = 0$ *if* $j \neq i$ *and* $\chi_i(e_i) = \chi_i(1) = d_i$;

(3) $\chi_i(e_i x) = \chi_i(x)$ *for all* $x \in K[G]$.

Proof. If $x \in R_j$ and $j \neq i$, then $\rho_i(x)(s) = xs = 0$ for all $s \in S_i$,

$\rho_i(x) = 0$, and $\chi_i(x) = \text{Tr}\,\rho_i(x) = 0$. In particular, $\chi_i(e_j) = 0$ when $j \neq i$. Since $1 = \sum_j e_j$ by 2.4, this implies $\chi_i(e_i) = \chi_i(1) = d_i$. Finally, $\rho_i(e_i)$ is the identity on S_i, since e_i is the identity element of R_i; hence, for all $x \in K[G]$,

$$\chi_i(e_i x) = \text{Tr}\,\rho_i(e_i x) = \text{Tr}\,\rho_i(e_i)\,\rho_i(x) = \text{Tr}\,\rho_i(x) = \chi_i(x).\ \square$$

Main properties. Equivalent representations have the same character, since $T \circ \rho(g) \circ T^{-1}$ and $\rho(g)$ have the same trace when T is invertible. The converse holds when K has characteristic 0, so that $\chi_1, \ldots, \chi_s$ determine all finite-dimensional representations.

Theorem **8.4.** *If G is finite and K has characteristic 0:*

(1) *the irreducible characters of G are linearly independent over K;*

(2) *every character of G can be written uniquely as a linear combination of irreducible characters with nonnegative integer coefficients;*

(3) *two finite-dimensional representations of G are equivalent if and only if they have the same character.*

Proof. If $a_i \in K$ and $\sum_i a_i \chi_i = 0$, then $d_i a_i = \sum_j a_j \chi_j(e_i) = 0$ for all i; since K has characteristic 0 this implies $a_i = 0$ for all i.

Let χ be the character of a representation ρ. By 7.5, the corresponding G-module V is isomorphic to a direct sum $S_1^{m_1} \oplus \cdots \oplus S_s^{m_s}$ for some nonnegative integers $m_1, \ldots, m_s$. Then $\chi = m_1 \chi_1 + \cdots + m_s \chi_s$, and $\chi(e_i) = m_i \chi_i(e_i) = m_i d_i$ by 8.3. Therefore $V \cong S_1^{m_1} \oplus \cdots \oplus S_s^{m_s}$ is uniquely determined, up to isomorphism, by χ, and then ρ is uniquely determined by χ, up to equivalence. $\square$

Readers will show that (2) and (3) need not hold in characteristic $p \neq 0$.

The next result is an *orthogonality* property of the irreducible characters. The exercises give another orthogonality property.

Theorem **8.5.** *If G is a finite group and K is an algebraically closed field whose characteristic does not divide $|G|$, then $d_i \neq 0$ in K and*

$$\sum_{g \in G} \chi_i(g)\,\chi_j(g^{-1}) = \begin{cases} |G| & \text{if } i = j, \\ 0 & \text{if } i \neq j. \end{cases}$$

In case $K = \mathbb{C}$ we shall see in the next section that $\chi(g^{-1}) = \overline{\chi(g)}$, so that $\chi_1, \ldots, \chi_s$ are indeed pairwise orthogonal for a suitable complex inner product.

Proof. Let e_{ig} be the g coordinate of e_i in $K[G]$, so that $e_i = \sum_{g \in G} e_{ig}\,g$. Since $\chi_r(1) = |G|$ and $\chi_r(g) = 0$ for all $g \neq 1$ we have

$$\chi_r(e_i\,g^{-1}) = \chi_r\big(\sum_{h \in G} e_{ih}\,hg^{-1}\big) = \sum_{h \in G} e_{ih}\,\chi_r(hg^{-1}) = e_{ig}\,|G|.$$

By 8.2, 8.3,

$$e_{ig}\,|G| = \chi_r(e_i\,g^{-1}) = \sum_j d_j \chi_j(e_i\,g^{-1}) = d_i \chi_i(e_i\,g^{-1}) = d_i \chi_i(g^{-1}).$$

Since $|G| \neq 0$ in K, this implies

$$e_i = \sum_{g \in G} e_{ig} \, g = \frac{d_i}{|G|} \sum_{g \in G} \chi_i(g^{-1}) \, g.$$

Hence $d_i \neq 0$ in K, since $e_i \neq 0$. Then

$$\chi_j(e_i) = \sum_{g \in G} \frac{d_i}{|G|} \chi_i(g^{-1}) \, \chi_j(g).$$

If $j \neq i$, then $\chi_j(e_i) = 0$ by 8.3 and $\sum_{g \in G} \chi_i(g^{-1}) \chi_j(g) = 0$. If $j = i$, then $\chi_j(e_i) = d_i$ by 8.3 and $\sum_{g \in G} \chi_i(g^{-1}) \chi_j(g) = |G|$. $\square$

Exercises

In the following exercises, G is a finite group and K is a field whose characteristic does not divide the order of G. Characters are characters of G over K.

1. Show that $\chi_r(1) = |G|$ and $\chi_r(g) = 0$ for all $g \neq 1$ (χ_r is the regular character).

2. Show that the irreducible characters are linearly independent over K even when K has characteristic $p \neq 0$.

3. If K has characteristic $p \neq 0$, show that two representations that have the same character need not be equivalent.

4. Prove the following: if g and h are not conjugate, then $\sum_i \chi_i(g) \chi_i(h) = 0$; otherwise, $\sum_i \chi_i(g) \chi_i(h)$ is the order of the centralizer of g (and of the centralizer of h).

5. Find all irreducible characters of V_4 over $\mathbb{C}$.

6. Find all irreducible characters of S_3 over $\mathbb{C}$.

7. Find all irreducible characters of the quaternion group Q over $\mathbb{C}$. (It may be useful to note that $Q/Z(Q) \cong V_4$.)

8. Show that every normal subgroup of G is the kernel of some representation of G over K.

9. If K is algebraically closed, show that the values of any character are sums of roots of unity in K. (You may use restrictions to cyclic subgroups.)

9. Complex Characters

This section brings additional properties of complex characters, and a proof of Burnside's theorem that groups of order $p^m q^n$ are solvable for all primes p, q.

Character values. Representations and characters over $\mathbb{C}$ are *complex* representations and characters. Complex characters have all properties in Section 8. They also have a few properties of their own.

Proposition 9.1. Let G be a finite group and let χ be the character of a complex representation $\rho : G \longrightarrow GL(V)$ of dimension d. For every $g \in G$:

(1) $\rho(g)$ is diagonalizable;

(2) $\chi(g)$ is a sum of d roots of unity, and is integral over $\mathbb{Z}$;

(3) $\chi(g^{-1}) = \overline{\chi(g)}$;

(4) $|\chi(g)| \leqq d$; $\chi(g) = d$ if and only if $\rho(g) = 1$; $|\chi(g)| = d$ if and only if $\rho(g) = \lambda 1$ for some $\lambda \in \mathbb{C}$, and then λ is a root of unity.

Proof. (1). Let $H = \langle g \rangle$. By 7.8 the representation $\rho_{|H} : H \longrightarrow GL(V)$ is a direct sum of representations of dimension 1. Hence V, as an H-module, is a direct sum of submodules of dimension 1 over $\mathbb{C}$, and has a basis $e_1, \ldots, e_d$ over $\mathbb{C}$ that consists of eigenvectors of every $\rho_{|H}(h)$. The matrix of $\rho(g) = \rho_{|H}(g)$ in that basis is a diagonal matrix

$$
\begin{pmatrix}
\zeta_1 & 0 & \cdots & 0 \\
0 & \zeta_2 & \cdots & 0 \\
\vdots & \vdots & \ddots & \ddots \\
0 & 0 & \cdots & \zeta_d
\end{pmatrix}.
$$

(2), (3). Since g has finite order k in G we have $\rho(g)^k = \rho(g^k) = 1$ and $\zeta_1, \ldots, \zeta_d$ are kth roots of unity. Hence $\chi(g) = \mathrm{Tr}\, \rho(g) = \zeta_1 + \cdots + \zeta_d$ is a sum of d kth roots of unity. Since kth roots of unity are integral over $\mathbb{Z}$, so is $\chi(g)$. Moreover, the matrix of $\rho(g^{-1}) = \rho(g)^{-1}$ in the same basis is

$$
\begin{pmatrix}
\zeta_1^{-1} & 0 & \cdots & 0 \\
0 & \zeta_2^{-1} & \cdots & 0 \\
\vdots & \vdots & \ddots & \ddots \\
0 & 0 & \cdots & \zeta_d^{-1}
\end{pmatrix}.
$$

Now, $\zeta_i^{-1} = \overline{\zeta_i}$, since ζ_i is a kth root of unity; hence $\chi(g^{-1}) = \overline{\chi(g)}$.

(4) follows from (2): $|\zeta_i| = 1$ for all i, hence $|\chi(g)| = |\zeta_1 + \cdots + \zeta_d| \leqq d$, with equality if and only if $\zeta_1 = \cdots = \zeta_d$, equivalently, $\rho(g) = \zeta 1$ where $\zeta = \zeta_1 = \cdots = \zeta_d$ is a root of unity. Then $\chi(g) = \zeta d$; in particular, $\chi(g) = d$ if and only if $\zeta_1 = \cdots = \zeta_d = 1$, if and only if $\rho(g) = 1$. $\square$

Conjugacy classes. We keep the notation in Section 8. By Theorem 7.9, G has s conjugacy classes $C_1, \ldots, C_s$, on which $\chi_1, \ldots, \chi_s$ are constant. Let $c_j = \sum_{g \in C_j} g \in K[G]$. By Proposition 7.10, $\rho_i(c_j)$ is a scalar linear transformation $v \longmapsto c_{ij} v$. Then $\chi_i(c_j) = \mathrm{Tr}\, \rho_i(c_j) = d_i c_{ij}$. Since χ_i is constant on C_j we have $\chi_i(g) = d_i c_{ij} / |C_j|$ for all $g \in C_j$.

Lemma 9.2. c_{ij} is integral over $\mathbb{Z}$.

Proof. First, $c_j c_k = \left(\sum_{h' \in C_j} h' \right) \left(\sum_{h'' \in C_k} h'' \right) = \sum_{g \in G} n_g\, g$, where n_g is the number of ordered pairs $(h', h'') \in C_j \times C_k$ such that $h' h'' = g$. If g and g' are conjugate, then $n_g = n_{g'}$; hence $c_j c_k$ is a linear combination of $c_1, \ldots, c_s$

with (nonnegative) integer coefficients. Then $\rho_i(c_j)\,\rho_i(c_k) = \rho_i(c_j c_k)$ is a linear combination of $\rho_i(c_1)$, ..., $\rho_i(c_s)$ with integer coefficients, and $c_{ij}\,c_{ik}$ is a linear combination of c_{i1}, ..., c_{is} with integer coefficients.

Let A be the additive subgroup of $\mathbb{C}$ generated by 1 and c_{i1}, ..., c_{is}. By the above, A is closed under multiplication, and is a ring; A is also a finitely generated $\mathbb{Z}$-module; hence A is integral over $\mathbb{Z}$, by VII.3.1. $\square$

Proposition **9.3.** *If d_i and $|C_j|$ are relatively prime, then either $|\chi_i(g)| = d_i$ for all $g \in C_j$, or $\chi_i(g) = 0$ for all $g \in C_j$.*

Proof. Assume $|\chi_i(g)| < d_i$, where $g \in C_j$. Let $\alpha = \chi_i(g)/d_i = c_{ij}/|C_j|$. Then $|\alpha| < 1$. Also, $ud_i + v|C_j| = 1$ for some $u, v \in \mathbb{Z}$; hence $\alpha = ud_i\alpha + v|C_j|\alpha = u\chi_i(g) + vc_{ij}$ is integral over $\mathbb{Z}$, by 9.1, 9.2.

There is a finite Galois extension E of $\mathbb{Q}$ that contains α. If $\sigma \in \mathrm{Gal}\,(E:\mathbb{Q})$, then $\sigma\alpha$ is integral over $\mathbb{Z}$; moreover, $\sigma\chi_i(g)$ is, like $\chi_i(g)$, a sum of d_i roots of unity; hence $|\sigma\chi_i(g)| \leq d_i$ and $|\sigma\alpha| \leq 1$. Then $N(\alpha) = \prod_{\sigma\in\mathrm{Gal}\,(E:\mathbb{Q})} \sigma\alpha$ is integral over $\mathbb{Z}$ and $|N(\alpha)| < 1$. But $N(\alpha) \in \mathbb{Q}$; since $\mathbb{Z}$ is integrally closed, $N(\alpha) \in \mathbb{Z}$, $N(\alpha) = 0$, $\alpha = 0$, and $\chi_i(g) = 0$. $\square$

Burnside's $p^m q^n$ theorem now follows from the properties above.

Theorem **9.4** (Burnside). *Let p and q be prime numbers. Every group of order $p^m q^n$ is solvable.*

Proof. It is enough to show that simple groups of order $p^m q^n$ are abelian.

Assume that G is a simple nonabelian group of order $p^m q^n$. Since p-groups are solvable we may assume that $p \neq q$ and that $m, n > 0$. Number χ_1, ..., χ_s so that χ_1 is the trivial character $\chi_1(g) = 1$ for all g.

Let $Z_i = \{\, g \in G \mid |\chi_i(g)| = d_i \,\}$. Since the center of $GL(V)$ consists of all scalar linear transformations, $Z_i = \{\, g \in G \mid \rho_i(g) \in Z(GL(S_i)) \,\}$ by 9.1, and Z_i is a normal subgroup of G. If $Z_i = G$, then $|\chi_i(g)| = d_i$ for all $g \in G$,

$$|G|\,d_i^2 = \sum_{g\in G} \chi_i(g)\,\overline{\chi_i(g)} = \sum_{g\in G} \chi_i(g)\,\chi_i(g)^{-1} = |G|$$

by 8.5, $d_i = 1$, and $\rho_i : G \longrightarrow \mathbb{C}\backslash\{0\}$. Now, $\mathrm{Ker}\,\rho_i \neq 1$, since G is not abelian; therefore $\mathrm{Ker}\,\rho_i = G$, and $\chi_i = \chi_1$. Thus $Z_i = 1$ for all $i > 1$.

Let S be a Sylow q-subgroup of G. There exists $h \in Z(S)$, $h \neq 1$, and then $h \notin Z_i$ when $i > 1$. The centralizer of h contains S; its index is a power $p^k > 1$ of p, and the conjugacy class of h has p^k elements. If $i > 1$ and p does not divide d_i, then $\chi_i(h) = 0$ by 9.3, since $h \notin Z_i$. By 8.2,

$$0 = \chi_r(h) = \sum_i d_i\chi_i(h) = 1 + \sum_{i>1,\ p|d_i} d_i\chi_i(h) = 1 + p\alpha,$$

where α is integral over $\mathbb{Z}$ since every $\chi_i(h)$ is integral over $\mathbb{Z}$. But $\mathbb{Z}$ is integrally closed, so $\alpha = -1/p \in \mathbb{Q}\backslash\mathbb{Z}$ cannot be integral over $\mathbb{Z}$; this is the long awaited contradiction. $\square$

Exercises

In the following exercises, G is a finite group; all representations and characters are over $\mathbb{C}$.

1. Let F be the vector space of all class functions of G into $\mathbb{C}$. Show that $\langle \alpha, \beta \rangle = \frac{1}{|G|} \sum_{g \in G} \alpha(g) \overline{\beta(g)}$ is a complex inner product on F. Show that $\chi_1, \ldots, \chi_s$ is an orthonormal basis of F.

2. Show that $\langle x, y \rangle = \frac{1}{|G|} \sum_i \chi_i(x) \overline{\chi_j(y)}$ is a complex inner product on $Z(\mathbb{C}[G])$. Show that $c_1, \ldots, c_s$ is an orthonormal basis of $Z(\mathbb{C}[G])$.

3. In the previous two exercises, show that F and $Z(\mathbb{C}[G])$ are dual spaces, and that $\frac{1}{d_1}\chi_1, \ldots, \frac{1}{d_s}\chi_s$ and $c_1, \ldots, c_s$ are dual bases, under the pairing $\langle \alpha, x \rangle = \alpha(x)$.

4. Show that a character χ is irreducible if and only if $\langle \chi, \chi \rangle = 1$.

5. Let χ be a character such that $\chi(g) = 0$ for all $g \neq 1$. Show that χ is an integer multiple of the regular character.

6. Show that G is abelian if and only if every irreducible character of G has dimension 1.

7. Find all irreducible representations of S_4. (You may want to consider a vector space with basis $e_1, \ldots, e_4$, on which S_4 permutes $e_1, \ldots, e_4$.)

X
Projectives and Injectives

This chapter contains some basic tools and definitions of module theory: exact sequences, pullbacks and pushouts, projective and injective modules, and injective hulls, with applications to rings. Sections 5 and 6 may be skipped.

As before, all rings have an identity element; all modules are unital. Results are generally stated for left R-modules but apply equally to right R-modules.

1. Exact Sequences

This section introduces exact sequences and proofs by "diagram chasing". It can be covered much earlier, immediately after Section VIII.2.

Definition. A finite or infinite sequence $\cdots M_i \xrightarrow{\varphi_i} M_{i+1} \xrightarrow{\varphi_{i+1}} M_{i+2} \cdots$ *of module homomorphisms is* null *when* $\varphi_{i+1} \circ \varphi_i = 0$ *for all* i.

Thus the sequence $A \xrightarrow{\varphi} B \xrightarrow{\psi} C$ is null if and only if $\operatorname{Im} \varphi \subseteq \operatorname{Ker} \psi$, if and only if φ factors through the inclusion homomorphism $\operatorname{Ker} \psi \longrightarrow B$, if and only if ψ factors through the projection $B \longrightarrow \operatorname{Coker} \varphi$ defined as follows:

Definition. The cokernel *of a module homomorphism* $\varphi : A \longrightarrow B$ *is the quotient* $\operatorname{Coker} \varphi = B/\operatorname{Im} \varphi$.

Definition. A finite or infinite sequence $\cdots M_i \xrightarrow{\varphi_i} M_{i+1} \xrightarrow{\varphi_{i+1}} M_{i+2} \cdots$ *of module homomorphisms is* exact *when* $\operatorname{Im} \varphi_i = \operatorname{Ker} \varphi_{i+1}$ *for all* i.

Exact sequences first appear in Hurewicz [1941] (see MacLane [1963]). If $A \xrightarrow{\varphi} B \xrightarrow{\psi} C$ is exact, then B contains a submodule $\operatorname{Im} \varphi \cong A/\operatorname{Ker} \varphi$ such that $B/\operatorname{Im} \varphi = B/\operatorname{Ker} \psi \cong \operatorname{Im} \psi$; this provides information about the size and structure of B.

We note some particular kinds of exact sequences:

$0 \longrightarrow A \xrightarrow{\varphi} B$ is exact if and only if φ is injective;

$A \xrightarrow{\varphi} B \longrightarrow 0$ is exact if and only if φ is surjective;

$0 \longrightarrow A \xrightarrow{\varphi} B \longrightarrow 0$ is exact if and only if φ is an isomorphism.

Exact sequences $0 \longrightarrow A \longrightarrow B \longrightarrow C$ are sometimes called *left exact*; exact sequences $A \longrightarrow B \longrightarrow C \longrightarrow 0$ are sometimes called *right exact*. These sequences have useful factorization properties.

Lemma **1.1.** *If* $0 \longrightarrow A \xrightarrow{\mu} B \xrightarrow{\varphi} C$ *is exact, then every homomorphism* ψ *such that* $\varphi \circ \psi = 0$ *factors uniquely through* μ:

$$
\begin{array}{ccc}
 & M & \\
\chi \swarrow & \downarrow \psi & \\
0 \longrightarrow A \xrightarrow{\mu} B & \xrightarrow{\varphi} & C
\end{array}
$$

Proof. We have $\operatorname{Im} \psi \subseteq \operatorname{Ker} \varphi = \operatorname{Im} \mu$. Since μ is injective there is a unique mapping $\chi : M \longrightarrow A$ such that $\psi(x) = \mu\big(\chi(x)\big)$ for all $x \in M$. Then χ is a module homomorphism, like ψ and μ. $\square$

Lemma **1.2.** *If* $A \xrightarrow{\varphi} B \xrightarrow{\sigma} C \longrightarrow 0$ *is exact, then every homomorphism* ψ *such that* $\psi \circ \varphi = 0$ *factors uniquely through* σ:

$$
\begin{array}{ccc}
A \xrightarrow{\varphi} B & \xrightarrow{\sigma} C \longrightarrow 0 \\
\psi \downarrow \quad \swarrow \chi & \\
M &
\end{array}
$$

Proof. This follows from Theorem VIII.2.5, since $\operatorname{Ker} \sigma = \operatorname{Im} \varphi \subseteq \operatorname{Ker} \psi$. $\square$

A *short exact sequence* is an exact sequence $0 \longrightarrow A \longrightarrow B \longrightarrow C \longrightarrow 0$. Then B contains a submodule $B' \cong A$ such that $B/B' \cong C$. All exact sequences are "compositions" of short exact sequences (see the exercises).

Diagram chasing. Exact sequences lend themselves to a method of proof, *diagram chasing*, in which elements are "chased" around a commutative diagram.

Lemma **1.3** (Short Five Lemma). *In a commutative diagram with exact rows,*

$$
\begin{array}{ccccccccc}
0 & \longrightarrow & A & \xrightarrow{\mu} & B & \xrightarrow{\rho} & C & \longrightarrow & 0 \\
 & & \alpha \downarrow & & \beta \downarrow & & \gamma \downarrow & & \\
0 & \longrightarrow & A' & \xrightarrow{\mu'} & B' & \xrightarrow{\rho'} & C' & \longrightarrow & 0
\end{array}
$$

if α *and* γ *are isomorphisms, then so is* β.

Proof. Assume that $\beta(b) = 0$. Then $\gamma\big(\rho(b)\big) = \rho'\big(\beta(b)\big) = 0$ and $\rho(b) = 0$. By exactness, $b = \mu(a)$ for some $a \in A$. Then $\mu'\big(\alpha(a)\big) = \beta\big(\mu(a)\big) = \beta(b) = 0$. Hence $\alpha(a) = 0$, $a = 0$, and $b = \mu(a) = 0$. Thus β is injective.

Let $b' \in B'$. Then $\rho'(b') = \gamma(c)$ for some $c \in C$, and $c = \rho(b)$ for some $b \in B$. Hence $\rho'\big(\beta(b)\big) = \gamma\big(\rho(b)\big) = \gamma(c) = \rho'(b')$. Thus $b' - \beta(b) \in \operatorname{Ker} \rho'$; by exactness, $b' - \beta(b) = \mu'(a')$ for some $a' \in A'$. Then $a' = \alpha(a)$ for some $a \in A$; hence $b' = \beta(b) + \mu'\big(\alpha(a)\big) = \beta\big(b + \mu(a)\big)$. Thus β is surjective. $\square$

The proof of Lemma 1.3 does not use all the hypotheses. The exercises give sharper versions, as well as other diagram lemmas, including the nine lemma:

Lemma **1.4** (Nine Lemma). *In a commutative diagram with exact columns:*

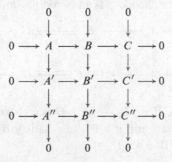

(1) *if the first two rows are exact, then the last row is exact.*

(2) *if the last two rows are exact, then the first row is exact.*

Split exact sequences. For every pair of left R-modules A and C there is a short exact sequence $0 \longrightarrow A \overset{\iota}{\longrightarrow} A \oplus C \overset{\pi}{\longrightarrow} C \longrightarrow 0$, where ι is the injection and π is the projection. The resulting short exact sequences are characterized as follows, up to isomorphism.

Proposition **1.5.** *For a short exact sequence* $0 \longrightarrow A \overset{\mu}{\longrightarrow} B \overset{\rho}{\longrightarrow} C \longrightarrow 0$ *the following conditions are equivalent:*

(1) μ *splits* ($\sigma \circ \mu = 1_A$ *for some homomorphism* $\sigma : B \longrightarrow A$);

(2) ρ *splits* ($\rho \circ \nu = 1_C$ *for some homomorphism* $\nu : C \longrightarrow B$);

(3) *there is an isomorphism* $B \cong A \oplus C$ *such that the diagram*

$$
\begin{array}{ccccccccc}
0 & \longrightarrow & A & \overset{\mu}{\longrightarrow} & B & \overset{\rho}{\longrightarrow} & C & \longrightarrow & 0 \\
 & & \| & & \downarrow{\scriptstyle\cong} & & \| & & \\
0 & \longrightarrow & A & \underset{\iota}{\longrightarrow} & A \oplus C & \underset{\pi}{\longrightarrow} & C & \longrightarrow & 0
\end{array}
$$

commutes, where ι *is the injection and* π *is the projection.*

Proof. (1) implies (3). Assume that $\sigma \circ \mu = 1_A$. There is a homomorphism $\theta : B \longrightarrow A \oplus C$ such that $\pi' \circ \theta = \sigma$ and $\pi \circ \theta = \rho$ (where $\pi' : A \oplus C \longrightarrow A$ is the projection), namely $\theta h = \big(\sigma(h),\, \rho(h)\big)$. Then $\theta\big(\mu(a)\big) = \big(\sigma(\mu(a)),\, \rho(\mu(a))\big) = (a, 0)$ and $\theta \circ \mu = \iota$. Hence θ is an isomorphism, by 1.3.

$$
\begin{array}{ccccccccc}
0 & \longrightarrow & A & \overset{\mu}{\underset{\sigma}{\rightleftarrows}} & B & \overset{\rho}{\longrightarrow} & C & \longrightarrow & 0 \\
 & & \| & & \downarrow{\scriptstyle\theta} & & \| & & \\
0 & \longrightarrow & A & \overset{\pi'}{\underset{\iota}{\rightleftarrows}} & A \oplus C & \underset{\pi}{\longrightarrow} & C & \longrightarrow & 0
\end{array}
$$

(2) implies (3). Assume that $\rho \circ \nu = 1_C$. There is a homomorphism ζ : $A \oplus C \longrightarrow B$ such that $\mu = \zeta \circ \iota$ and $\nu = \zeta \circ \iota'$ (where $\iota' : C \longrightarrow A \oplus C$ is the injection), namely $\zeta(a, c) = \mu(a) + \nu(c)$. Then $\rho(\zeta(a, c)) = \rho(\mu(a)) + \rho(\nu(c)) = c$ and $\rho \circ \zeta = \pi$; hence ζ is an isomorphism, by 1.3.

$$
\begin{array}{ccccccccc}
0 & \longrightarrow & A & \overset{\mu}{\longrightarrow} & B & \underset{\nu}{\overset{\rho}{\rightleftarrows}} & C & \longrightarrow & 0 \\
& & \| & & \big\uparrow{\scriptstyle\zeta} & & \| & & \\
0 & \longrightarrow & A & \underset{\iota}{\longrightarrow} & A \oplus C & \underset{\pi}{\overset{\iota'}{\rightleftarrows}} & C & \longrightarrow & 0
\end{array}
$$

(3) implies (1) and (2). If $\theta : B \longrightarrow A \oplus C$ is an isomorphism and $\theta \circ \mu = \iota$, $\pi \circ \theta = \rho$, then $\sigma = \pi' \circ \theta$ and $\nu = \theta^{-1} \circ \iota'$ satisfy $\sigma \circ \mu = \pi' \circ \iota = 1_A$ and $\rho \circ \nu = \pi \circ \iota' = 1_C$, by VIII.3.4. $\square$

Exercises

Given a commutative square $\beta \circ \varphi = \psi \circ \alpha$:

$$
\begin{array}{ccc}
A & \overset{\varphi}{\longrightarrow} & B \\
{\scriptstyle\alpha}\big\downarrow & & \big\downarrow{\scriptstyle\beta} \\
C & \underset{\psi}{\longrightarrow} & D
\end{array}
$$

1. Show that φ and ψ induce a homomorphism $\operatorname{Ker} \alpha \longrightarrow \operatorname{Ker} \beta$.

2. Show that φ and ψ induce a homomorphism $\operatorname{Coker} \alpha \longrightarrow \operatorname{Coker} \beta$.

3. Explain how any exact sequence $A \overset{\varphi}{\longrightarrow} B \overset{\psi}{\longrightarrow} C$ can be recoved by "composing" the short exact sequences $0 \longrightarrow \operatorname{Ker} \varphi \longrightarrow A \longrightarrow \operatorname{Im} \varphi \longrightarrow 0$, $0 \longrightarrow \operatorname{Im} \varphi \longrightarrow B \longrightarrow \operatorname{Im} \psi \longrightarrow 0$, and $0 \longrightarrow \operatorname{Im} \psi \longrightarrow C \longrightarrow C/\operatorname{Im} \psi \longrightarrow 0$.

4. Show that a module M is semisimple if and only if every short exact sequence $0 \longrightarrow A \longrightarrow M \longrightarrow C \longrightarrow 0$ splits.

Given a commutative diagram with exact rows:

$$
\begin{array}{ccccccccc}
0 & \longrightarrow & A & \longrightarrow & B & \longrightarrow & C & \longrightarrow & 0 \\
& & {\scriptstyle\alpha}\big\downarrow & & {\scriptstyle\beta}\big\downarrow & & {\scriptstyle\gamma}\big\downarrow & & \\
0 & \longrightarrow & A' & \longrightarrow & B' & \longrightarrow & C' & \longrightarrow & 0
\end{array}
$$

5. Show that $0 \longrightarrow \operatorname{Ker} \alpha \longrightarrow \operatorname{Ker} \beta \longrightarrow \operatorname{Ker} \gamma$ is exact.

6. Show that $\operatorname{Coker} \alpha \longrightarrow \operatorname{Coker} \beta \longrightarrow \operatorname{Coker} \gamma \longrightarrow 0$ is exact.

(Five Lemma). Given a commutative diagram with exact rows:

$$
\begin{array}{ccccccccc}
A & \longrightarrow & B & \longrightarrow & C & \longrightarrow & D & \longrightarrow & E \\
{\scriptstyle\alpha}\big\downarrow & & {\scriptstyle\beta}\big\downarrow & & {\scriptstyle\gamma}\big\downarrow & & {\scriptstyle\delta}\big\downarrow & & {\scriptstyle\varepsilon}\big\downarrow \\
A' & \longrightarrow & B' & \longrightarrow & C' & \longrightarrow & D' & \longrightarrow & E'
\end{array}
$$

7. If α is surjective and β, δ are injective, show that γ is injective.

8. If ε is injective and β, δ are surjective, show that γ is surjective.

(Nine Lemma). Given a commutative diagram with exact columns:

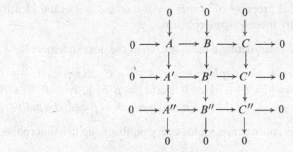

9. If the first two rows are exact, show that the last row is exact.

10. If the last two rows are exact, show that the first row is exact.

2. Pullbacks and Pushouts

Pullbacks and pushouts are commutative squares with universal properties.

Definition. A pullback of left R-modules is a commutative square $\alpha \circ \beta' = \beta \circ \alpha'$ with the following universal property: for every commutative square $\alpha \circ \varphi = \beta \circ \psi$ (with the same α and β) there exists a unique homomorphism χ such that $\varphi = \beta' \circ \chi$ and $\psi = \alpha' \circ \chi$:

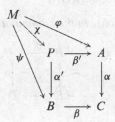

Readers will verify that the following squares are pullbacks:

$$A \oplus B \longrightarrow A \qquad A \cap B \overset{\subseteq}{\longrightarrow} A \qquad \varphi^{-1}A \longrightarrow A$$
$$\downarrow \qquad\qquad \downarrow \qquad \subseteq\downarrow \qquad\quad \downarrow\subseteq \qquad \subseteq\downarrow \qquad\quad \downarrow\subseteq$$
$$B \longrightarrow 0 \qquad B \underset{\subseteq}{\longrightarrow} C \qquad B \underset{\varphi}{\longrightarrow} C$$

Proposition 2.1. *For every pair of homomorphisms $\alpha : A \longrightarrow C$ and $\beta : B \longrightarrow C$ of left R-modules, there exists a pullback $\alpha \circ \beta' = \beta \circ \alpha'$, and it is unique up to isomorphism.*

Proof. Uniqueness follows from the universal property. Let $\alpha \circ \beta' = \beta \circ \alpha'$ and $\alpha \circ \beta'' = \beta \circ \alpha''$ be pullbacks. There exist homomorphisms θ and ζ such that $\beta'' = \beta' \circ \theta$, $\alpha'' = \alpha' \circ \theta$ and $\beta' = \beta'' \circ \zeta$, $\alpha' = \alpha'' \circ \zeta$. Then $\alpha' = \alpha' \circ \theta \circ \zeta$ and $\beta' = \beta' \circ \theta \circ \zeta$; by uniqueness in the universal property of $\alpha \circ \beta' = \beta \circ \alpha'$,

$\theta \circ \zeta$ is the identity. Similarly, $\alpha'' = \alpha'' \circ \zeta \circ \theta$ and $\beta'' = \beta'' \circ \zeta \circ \theta$; by uniqueness in the universal property of $\alpha \circ \beta'' = \beta \circ \alpha''$, $\zeta \circ \theta$ is the identity. Thus θ and ζ are mutually inverse isomorphisms.

Existence is proved by constructing a pullback, as in the next statement. $\square$

Proposition **2.2.** *Given homomorphisms* $\alpha : A \longrightarrow C$ *and* $\beta : B \longrightarrow C$ *of left R-modules, let* $P = \{ (a, b) \in A \oplus B \mid \alpha(a) = \beta(b) \}$, $\alpha' : P \longrightarrow B$, $(a, b) \longmapsto b$, *and* $\beta' : P \longrightarrow A$, $(a, b) \longmapsto a$; *then* $\alpha \circ \beta' = \beta \circ \alpha'$ *is a pullback.*

By Proposition 2.1, this construction yields every pullback, up to isomorphism.

Proof. First, $\alpha \circ \beta' = \beta \circ \alpha'$, by the choice of P. Assume $\alpha \circ \varphi = \beta \circ \psi$, where $\varphi : M \longrightarrow A$ and $\psi : M \longrightarrow B$ are module homomorphisms. Then $(\varphi(m), \psi(m)) \in P$ for all $m \in M$. Hence $\chi : m \longmapsto (\varphi(m), \psi(m))$ is a homomorphism of M into P, and $\varphi = \beta' \circ \chi$, $\psi = \alpha' \circ \chi$. If $\chi' : M \longrightarrow P$ is another homomorphism such that $\varphi = \beta' \circ \chi'$ and $\psi = \alpha' \circ \chi'$, then the components of every $\chi'(m)$ are $\varphi(m)$ and $\psi(m)$; hence $\chi' = \chi$. $\square$

Properties. The following properties can be proved either from the definition of pullbacks or from Proposition 2.2, and make nifty exercises.

Proposition **2.3** (Transfer). *In a pullback* $\alpha \circ \beta' = \beta \circ \alpha'$:

(1) *if* α *is injective, then* α' *is injective;*

(2) *if* α *is surjective, then* α' *is surjective.*

Proposition **2.4** (Juxtaposition). *In the commutative diagram*

$$
\begin{array}{ccccc}
Q & \xrightarrow{\gamma'} & P & \xrightarrow{\beta'} & A \\
{\scriptstyle \alpha''}\downarrow & & {\scriptstyle \alpha'}\downarrow & & \downarrow{\scriptstyle \alpha} \\
C & \xrightarrow{\gamma} & B & \xrightarrow{\beta} & D
\end{array}
$$

(1) *if* $\alpha \circ \beta' = \beta \circ \alpha'$ *and* $\alpha' \circ \gamma' = \gamma \circ \alpha''$ *are pullbacks, then* $\alpha \circ (\beta' \circ \gamma') = (\beta \circ \gamma) \circ \alpha''$ *is a pullback;*

(2) *if* $\alpha \circ \beta' = \beta \circ \alpha'$ *and* $\alpha \circ (\beta' \circ \gamma') = (\beta \circ \gamma) \circ \alpha''$ *are pullbacks, then* $\alpha' \circ \gamma' = \gamma \circ \alpha''$ *is a pullback.*

Pushouts. It is a peculiarity of modules that reversing all arrows in a definition or construction usually yields an equally interesting definition or construction. Pushouts are obtained from pullbacks in just this fashion:

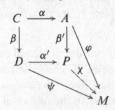

Definition. A pushout *of left R-modules is a commutative square* $\beta' \circ \alpha = \alpha' \circ \beta$ *with the following universal property: for every commutative square* $\varphi \circ \alpha = \psi \circ \beta$ *(with the same* α *and* β*) there exists a unique homomorphism* χ *such that* $\varphi = \chi \circ \beta'$ *and* $\psi = \chi \circ \alpha'$.

Readers will verify that the following squares are pushouts:

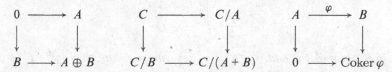

Proposition **2.5.** *For every pair of homomorphisms* $\alpha : C \longrightarrow A$ *and* $\beta : C \longrightarrow B$ *of left R-modules, there exists a pushout* $\beta' \circ \alpha = \alpha' \circ \beta$, *and it is unique up to isomorphism.*

Uniqueness follows from the universal property, and existence from a construction:

Proposition **2.6.** *Given homomorphisms* $\alpha : C \longrightarrow A$ *and* $\beta : C \longrightarrow B$ *of left R-modules, let* $K = \{ (\alpha(c), -\beta(c)) \in A \oplus B \mid c \in C \}$, $P = (A \oplus B)/K$, $\alpha' = \pi \circ \kappa$, *and* $\beta' = \pi \circ \iota$, *where* $\pi : A \oplus B \longrightarrow P$ *is the projection and* $\iota : A \longrightarrow A \oplus B$, $\kappa : B \longrightarrow A \oplus B$ *are the injections. Then* $\beta' \circ \alpha = \alpha' \circ \beta$ *is a pushout.*

$$
\begin{array}{ccc}
C & \xrightarrow{\alpha} & A \\
{\scriptstyle\beta}\downarrow & & \downarrow{\scriptstyle\iota} \\
B & \xrightarrow{\kappa} A \oplus B \xrightarrow{\pi} & P \\
& {\scriptstyle\omega}\searrow & \downarrow{\scriptstyle\chi} \\
& & M
\end{array}
$$

Proof. First, $\mathrm{Ker}\,\pi = K = \mathrm{Im}\,(\iota \circ \alpha - \kappa \circ \beta)$; hence $\beta' \circ \alpha = \pi \circ \iota \circ \alpha = \pi \circ \kappa \circ \beta = \alpha' \circ \beta$. Assume that $\varphi \circ \alpha = \chi \circ \beta$, where $\varphi : A \longrightarrow M$ and $\psi : B \longrightarrow M$ are module homomorphisms. Let $\omega : A \oplus B \longrightarrow M$ be the unique homomorphism such that $\varphi = \omega \circ \iota$ and $\psi = \omega \circ \kappa$. Then $\omega(\iota(\alpha(c))) = \varphi(\alpha(c)) = \psi(\beta(c)) = \omega(\kappa(\beta(c)))$ for all $c \in C$. Hence $\mathrm{Ker}\,\pi = K \subseteq \mathrm{Ker}\,\omega$ and ω factors through π: $\omega = \chi \circ \pi$ for some unique homomorphism $\chi : P \longrightarrow M$. Then $\chi \circ \beta' = \chi \circ \pi \circ \iota = \omega \circ \iota = \varphi$ and $\chi \circ \alpha' = \chi \circ \pi \circ \kappa = \omega \circ \kappa = \psi$. Moreover, χ is unique with these properties: if $\chi' \circ \beta' = \varphi$ and $\chi' \circ \alpha' = \psi$, then $\chi' \circ \pi \circ \iota = \varphi = \chi \circ \pi \circ \iota$ and $\chi' \circ \pi \circ \kappa = \psi = \chi \circ \pi \circ \kappa$; hence $\chi' \circ \pi = \chi \circ \pi$ and $\chi' = \chi$. $\square$

By Proposition 2.5, this construction yields every pushout, up to isomorphism.

Properties. The following properties can be proved either from the definition of pushouts or from Proposition 2.6, and make cool exercises.

Proposition **2.7** (Transfer). *In a pushout* $\beta' \circ \alpha = \alpha' \circ \beta'$:

(1) *if α is injective, then α' is injective;*

(2) *if α is surjective, then α' is surjective.*

Proposition 2.8 (Juxtaposition). *In the commutative diagram*

$$D \xrightarrow{\beta} B \xrightarrow{\gamma} C$$
$$\alpha \downarrow \qquad \downarrow \alpha' \qquad \downarrow \alpha''$$
$$A \xrightarrow{\beta'} P \xrightarrow{\gamma'} Q$$

(1) *if $\beta' \circ \alpha = \alpha' \circ \beta$ and $\gamma' \circ \alpha' = \alpha'' \circ \gamma$ are pushouts, then $(\gamma' \circ \beta') \circ \alpha = \alpha'' \circ (\gamma \circ \beta)$ is a pushout;*

(2) *if $\beta' \circ \alpha = \alpha' \circ \beta$ and $(\gamma' \circ \beta') \circ \alpha = \alpha'' \circ (\gamma \circ \beta)$ are pushouts, then $\gamma' \circ \alpha' = \alpha'' \circ \gamma$ is a pushout.*

Exercises

$$
\begin{array}{ccc}
A \cap B \xrightarrow{\subseteq} A & \mathrm{Ker}\,\varphi \longrightarrow 0 & \varphi^{-1}(A) \longrightarrow A \\
\subseteq \downarrow \qquad \downarrow \subseteq & \subseteq \downarrow \qquad \downarrow & \subseteq \downarrow \qquad \downarrow \subseteq \\
B \xrightarrow{\subseteq} C & A \xrightarrow{\varphi} B & B \xrightarrow{\varphi} C \\
\text{Square 1} & \text{Square 2} & \text{Square 3}
\end{array}
$$

1. Let A and B be submodules of C. Show that Square 1 is a pullback.

2. Let $\varphi : A \longrightarrow B$ be a module homomorphism. Show that Square 2 is a pullback.

3. Let $\varphi : B \longrightarrow C$ be a module homomorphism and let A be a submodule of C. Show that Square 3 is a pullback.

4. Let $\alpha \circ \beta' = \beta \circ \alpha'$ be a pullback. Prove the following: if α is injective, then α' is injective.

5. Let $\alpha \circ \beta' = \beta \circ \alpha'$ be a pullback. Prove the following: if α is surjective, then α' is surjective.

Given a commutative diagram:

$$Q \xrightarrow{\gamma'} P \xrightarrow{\beta'} A$$
$$\alpha'' \downarrow \qquad \downarrow \alpha' \qquad \downarrow \alpha$$
$$C \xrightarrow{\gamma} B \xrightarrow{\beta} D$$

6. Let $\alpha \circ \beta' = \beta \circ \alpha'$ and $\alpha' \circ \gamma' = \gamma \circ \alpha''$ be pullbacks. Show that $\alpha \circ (\beta' \circ \gamma') = (\beta \circ \gamma) \circ \alpha''$ is a pullback.

7. Let $\alpha \circ \beta' = \beta \circ \alpha'$ and $\alpha \circ (\beta' \circ \gamma') = (\beta \circ \gamma) \circ \alpha''$ be pullbacks. Show that $\alpha' \circ \gamma' = \gamma \circ \alpha''$ is a pullback.

$$
\begin{array}{ccc}
C \longrightarrow C/A & A \xrightarrow{\varphi} B \\
\downarrow \qquad \vdots & \downarrow \qquad \downarrow \\
C/B \dashrightarrow C/(A+B) & 0 \longrightarrow \mathrm{Coker}\,\varphi \\
\text{Square 4} & \text{Square 5}
\end{array}
$$

8. Let A and B be submodules of C. Construct Square 4 and show that it is a pushout.

9. Let $\varphi : A \longrightarrow B$ be a module homomorphism. Show that Square 5 is a pushout.

10. Let $\beta' \circ \alpha = \alpha' \circ \beta'$ be a pushout. Prove the following: if α is surjective, then α' is surjective.

11. Let $\beta' \circ \alpha = \alpha' \circ \beta'$ be a pushout. Prove the following: if α is injective, then α' is injective.

Given a commutative diagram:

$$
\begin{array}{ccccc}
D & \xrightarrow{\beta} & B & \xrightarrow{\gamma} & C \\
\downarrow{\scriptstyle\alpha} & & \downarrow{\scriptstyle\alpha'} & & \downarrow{\scriptstyle\alpha''} \\
A & \xrightarrow{\beta'} & P & \xrightarrow{\gamma'} & Q
\end{array}
$$

12. Let $\beta' \circ \alpha = \alpha' \circ \beta$ and $\gamma' \circ \alpha' = \alpha'' \circ \gamma$ be pushouts. Show that $(\gamma' \circ \beta') \circ \alpha = \alpha'' \circ (\gamma \circ \beta)$ is a pushout.

13. Let $\beta' \circ \alpha = \alpha' \circ \beta$ and $(\gamma' \circ \beta') \circ \alpha = \alpha'' \circ (\gamma \circ \beta)$ be pushouts. Show that $\gamma' \circ \alpha' = \alpha'' \circ \gamma$ is a pushout.

14. Given a short exact sequence $0 \longrightarrow A \longrightarrow B \longrightarrow C \longrightarrow 0$ and a homomorphism $C' \longrightarrow C$, construct a commutative diagram with exact rows:

$$
\begin{array}{ccccccccc}
0 & \longrightarrow & A & \dashrightarrow & B' & \dashrightarrow & C' & \longrightarrow & 0 \\
& & \| & & \downarrow & & \downarrow & & \\
0 & \longrightarrow & A & \longrightarrow & B & \longrightarrow & C & \longrightarrow & 0
\end{array}
$$

15. Given a short exact sequence $0 \longrightarrow A \longrightarrow B \longrightarrow C \longrightarrow 0$ and a homomorphism $A \longrightarrow A'$, construct a commutative diagram with exact rows:

$$
\begin{array}{ccccccccc}
0 & \longrightarrow & A & \longrightarrow & B & \longrightarrow & C & \longrightarrow & 0 \\
& & \downarrow & & \downarrow & & \| & & \\
0 & \longrightarrow & A' & \dashrightarrow & B' & \dashrightarrow & C & \longrightarrow & 0
\end{array}
$$

16. Define pullbacks of sets and prove their existence and uniqueness, as in Propositions 2.1 and 2.2.

17. Define pushouts of not necessarily abelian groups. Show that examples includes free products of two groups amalgamating a subgroup.

3. Projective Modules

Projective modules are an important class of modules. Their effective use began with Cartan and Eilenberg [1956]. This section contains basic properties.

Definition. A left R-module P is projective *when every homomorphism of P can be factored or* lifted *through every epimorphism: if $\varphi : P \longrightarrow N$ and $\rho : M \longrightarrow N$ are homomorphisms, and ρ is surjective, then $\varphi = \rho \circ \psi$ for some homomorphism $\psi : P \longrightarrow M$:*

$$
\begin{array}{ccc}
 & P & \\
 {}^{\psi}\swarrow & \big\downarrow{}^{\varphi} & \\
M \xrightarrow{\ \rho\ } N & \longrightarrow & 0
\end{array}
$$

The proof of 3.1 below shows that this factorization is in general far from unique (ψ need not be unique in the above).

Vector spaces are projective. More generally, so are free modules:

Proposition 3.1. *Every free module is projective.*

Proof. Let $\varphi : F \longrightarrow N$ and $\rho : M \longrightarrow N$ be homomorphisms and let $(e_i)_{i \in I}$ be a basis of F. If ρ is surjective, there is for every $i \in I$ some $m_i \in M$ such that $\varphi(e_i) = \rho(m_i)$. There is a homomorphism $\psi : F \longrightarrow M$ such that $\psi(e_i) = m_i$ for all i. Then $\rho\big(\psi(e_i)\big) = \varphi(e_i)$ for all i and $\rho \circ \psi = \varphi$. $\square$

If P is projective, then every epimorphism $\rho : M \longrightarrow P$ splits ($\rho \circ \mu = 1_P$ for some $\mu : P \longrightarrow M$), since 1_P can be lifted through ρ.

Proposition 3.2. *A left R-module P is projective if and only if every epimorphism $M \longrightarrow P$ splits, if and only if every short exact sequence $0 \longrightarrow A \longrightarrow B \longrightarrow P \longrightarrow 0$ splits.*

Proof. By 1.5, $0 \longrightarrow A \longrightarrow B \longrightarrow P \longrightarrow 0$ splits if and only if $B \longrightarrow P$ splits. Assume that every epimorphism $M \longrightarrow P$ splits. Let $\varphi : P \longrightarrow N$ and $\rho : M \longrightarrow N$ be homomorphisms, with ρ surjective. In the pullback $\varphi \circ \rho' = \rho \circ \varphi'$, ρ' is surjective by 2.3, hence splits: $\rho' \circ \nu = 1_P$ for some $\nu : P \longrightarrow Q$; then $\rho \circ \varphi' \circ \nu = \varphi \circ \rho' \circ \nu = \varphi$ and φ can be lifted through ρ. $\square$

$$
\begin{array}{ccc}
Q & \underset{\nu}{\overset{\rho'}{\rightleftarrows}} & P \\
{}^{\varphi'}\big\downarrow & & \big\downarrow{}^{\varphi} \\
M & \xrightarrow{\ \rho\ } & N
\end{array}
$$

Corollary 3.3. *A ring R is semisimple if and only if every short exact sequence of left R-modules splits, if and only if every left R-module is projective.*

Proof. A left R-module B is semisimple if and only if every submodule of B is a direct summand, if and only if every short exact sequence $0 \longrightarrow A \longrightarrow B \longrightarrow C \longrightarrow 0$ splits. $\square$

Corollary 3.3 readily yields projective modules that are not free. Readers will establish two more basic properties:

Proposition 3.4. *Every direct summand of a projective module is projective.*

Proposition 3.5. *Every direct sum of projective left R-modules is projective.*

In particular, every finite product of projective modules is projective. This property does not extend to infinite products; for instance, the direct product of countably many copies of $\mathbb{Z}$ is not a projective $\mathbb{Z}$-module (see the exercises).

Corollary **3.6.** *A module is projective if and only if it is isomorphic to a direct summand of a free module.*

Proof. For every module P there is by VIII.4.6 an epimorphism $\rho : F \longrightarrow P$ where F is free; if P is projective, then the exact sequence $0 \longrightarrow \operatorname{Ker} \varphi \longrightarrow F \longrightarrow P \longrightarrow 0$ splits, and P is isomorphic to a direct summand of F. Conversely, every direct summand of a free module is projective, by 3.1 and 3.4. $\square$

Corollary **3.7.** *If R is a PID, then an R-module is projective if and only if it is free.*

Proof. Every submodule of a free R-module is free, by Theorem VIII.6.1. $\square$

Local rings share this property; the exercises prove a particular case.

Exercises

1. Let F be free with basis $(e_i)_{i \in I}$; let $\varphi : F \longrightarrow N$ and $\rho : M \longrightarrow N$ be homomorphisms, with ρ surjective. In how many ways can φ be lifted through ρ?

2. Show that every direct summand of a projective module is projective.

3. Show that every direct sum of projective left R-modules is projective.

4. Show that a ring R is semisimple if and only if every cyclic left R-module is projective.

5. Let $m > 1$ and $n > 1$ be relatively prime. Show that $\mathbb{Z}_m$ is a projective, but not free, $\mathbb{Z}_{mn}$-module.

6. Give another example of a projective module that is not free.

7. Let R be a local ring with maximal ideal $\mathfrak{m}$. Prove the following: if A is a finitely generated R-module, and $x_1, \ldots, x_n$ is a minimal generating subset of A, then $x_1 + \mathfrak{m}A$, $\ldots, x_n + \mathfrak{m}A$ is a basis of $A/\mathfrak{m}A$ over $R/\mathfrak{m}$. (Invoke Nakayama.)

8. Let R be a local ring. Prove that every finitely generated projective R-module is free. (You may wish to use the previous exercise.)

9. Show that the direct product $A = \mathbb{Z} \times \mathbb{Z} \times \cdots \times \mathbb{Z} \times \cdots$ of countably many copies of $\mathbb{Z}$ is not a free abelian group (hence not projective). (Let B be the subgroup of all sequences $(x_1, x_2, \ldots, x_n, \ldots) \in A$ such that, for every $k > 0$, x_n is divisible by 2^k for almost all n. Show that B is not countable and that $B/2B$ is countable, whence B is not free.)

4. Injective Modules

Injective modules are another important class of modules, first considered by Baer [1940]. Their systematic use began with Cartan and Eilenberg [1956]. This section contains basic properties, and applications to abelian groups.

Reversing arrows in the definition of projective modules yields the following:

Definition. A left R-module J is injective *when every homomorphism into J can be factored or extended through every monomorphism: if $\varphi : M \longrightarrow J$ and $\mu : M \longrightarrow N$ are module homomorphisms, and μ is injective, then $\varphi = \psi \circ \mu$ for some homomorphism $\psi : N \longrightarrow J$:*

$$
\begin{array}{ccc}
 & J & \\
\varphi \uparrow & & \searrow \psi \\
0 \longrightarrow M & \underset{\mu}{\longrightarrow} & N
\end{array}
$$

This factorization need not be unique (ψ need not be unique in the above).

If J is injective, then every monomorphism $\mu : J \longrightarrow M$ splits ($\rho \circ \mu = 1_J$ for some $\rho : M \longrightarrow J$), since 1_J can be extended through μ .

Proposition **4.1.** *For a left R-module* J *the following conditions are equivalent:*

(1) J *is injective;*

(2) *every monomorphism* $J \longrightarrow M$ *splits;*

(3) *every short exact sequence* $0 \longrightarrow J \longrightarrow B \longrightarrow C \longrightarrow 0$ *splits;*

(4) J *is a direct summand of every left R-module* $M \supseteq J$.

Proof. We prove that (2) implies (1); the other implications are clear. Assume (2). Let $\varphi : M \longrightarrow J$ and $\mu : M \longrightarrow N$ be homomorphisms, with μ injective. In the pushout $\varphi' \circ \mu = \mu' \circ \varphi$, μ' is injective, by 2.7. Hence μ' splits: $\rho \circ \mu' = 1_J$ for some $\rho : P \longrightarrow J$. Then $\rho \circ \varphi' \circ \mu = \rho \circ \mu' \circ \varphi = \varphi$ and φ can be extended through μ . $\square$

$$
\begin{array}{ccc}
J & \underset{\rho}{\overset{\mu'}{\rightleftarrows}} & P \\
\varphi \uparrow & & \uparrow \varphi' \\
M & \underset{\mu}{\longrightarrow} & N
\end{array}
$$

Vector spaces are injective and, as in Corollary 3.3, we have:

Corollary **4.2.** *A ring* R *is semisimple if and only if every short exact sequence of left R-modules splits, if and only if every left R-module is injective.*

Readers will easily establish two basic properties:

Proposition **4.3.** *Every direct summand of an injective module is injective.*

Proposition **4.4.** *Every direct product of injective left R-modules is injective.*

In particular, every finite direct sum of injective modules is injective. This property does not extend to infinite direct sums; see Theorem 4.12 below.

Baer's criterion. The next result provides more interesting examples.

Proposition **4.5** (Baer's Criterion). *For a left R-module* J *the following conditions are equivalent:*

(1) J *is injective;*

(2) *every module homomorphism of a left ideal of* R *into* J *can be extended to* $_R R$;

(3) *for every module homomorphism φ of a left ideal L of R into J, there exists $m \in J$ such that $\varphi(r) = rm$ for all $r \in L$.*

Proof. (1) implies (2), and readers will happily show that (2) and (3) are equivalent. We show that (2) implies (1): every module homomorphism $\varphi : M \longrightarrow J$ extends through every monomorphism $\mu : M \longrightarrow N$. Then M is isomorphic to a submodule of N; we may assume that $M \subseteq N$ and that μ is the inclusion homomorphism.

We show that φ has a maximal extension to a submodule of N. Let $\mathcal{S}$ be the set of all ordered pairs (A, α) in which A is a submodule of N that contains M, and $\alpha : A \longrightarrow J$ is a module homomorphism that extends φ. Then $(M, \varphi) \in \mathcal{S}$ and $\mathcal{S} \neq \emptyset$. Order $\mathcal{S}$ by $(A, \alpha) \leq (B, \beta)$ if and only if $A \subseteq B$ and β extends α. In $\mathcal{S}$ every nonempty chain $(A_i, \alpha_i)_{i \in I}$ has an upper bound (A, α), where $A = \bigcup_{i \in I} A_i$ and $\alpha(x) = \alpha_i(x)$ whenever $x \in A_i$. By Zorn's lemma, $\mathcal{S}$ has a maximal element (C, γ). We show that $(A, \alpha) \in \mathcal{S}$ is not maximal if $A \neq N$; hence $C = N$, and γ extends φ to all of N.

Given $(A, \alpha) \in \mathcal{S}$ with $A \neq N$, let $b \in N \backslash A$ and $B = A + Rb$. Then

$$L = \{ r \in R \mid rb \in A \}$$

is a left ideal of R, and $r \longmapsto \alpha(rb)$ is a module homomorphism of L into J. By (2) there is a homomorphism $\chi : {}_R R \longrightarrow J$ such that $\chi(r) = \alpha(rb)$ for all $r \in L$. A homomorphism $\beta : B \longrightarrow J$ is then well defined by

$$\beta(rb + a) = \chi(r) + \alpha(a)$$

for all $r \in R$ and $a \in A$: if $rb + a = r'b + a'$, then $(r - r')b = a' - a$, $r - r' \in L$, $\chi(r - r') = \alpha(rb - r'b) = \alpha(a' - a)$, and $\chi(r) + \alpha(a) = \chi(r') + \alpha(a')$. Then β extends α, $(A, \alpha) < (B, \beta)$, and (A, α) is not maximal. $\square$

Proposition 4.5 gives a simple criterion for injectivity in case R is a PID.

Definition. A left R-module M is divisible *when the equation $rx = m$ has a solution $x \in M$ for every $0 \neq r \in R$ and $m \in M$.*

Proposition 4.6. *If R is a domain, then every injective R-module is divisible.*

Proof. Let J be injective. Let $a \in J$ and $0 \neq r \in R$. Since R is a domain, every element of Rr can be written in the form tr for some unique $t \in R$. Hence $\varphi : tr \longmapsto ta$ is a module homomorphism of Rr into J. By 4.5 there exists $m \in J$ such that $\varphi(s) = sm$ for all $s \subset Rr$. Then $a = \psi(r) = rm$. $\square$

Proposition 4.7. *If R is a PID, then an R-module is injective if and only if it is divisible.*

Proof. Let M be a divisible R-module. Let Rr be any (left) ideal of R and let $\varphi : Rr \longrightarrow M$ be a module homomorphism. If $r = 0$, then $\varphi(s) = s0$ for all $s \in Rr$. Otherwise $\varphi(r) = rm$ for some $m \in M$, since M is divisible. Then $\varphi(tr) = trm$ for all $tr \in Rr$. Hence M is injective, by 4.5. $\square$

Abelian groups. By 4.7, an abelian group is injective (as a $\mathbb{Z}$-module) if and only if it is divisible. For example, the additive group $\mathbb{Q}$ of all rationals is divisible. For another example, define for every prime number p an additive group $\mathbb{Z}_{p^\infty} = \langle a_1, \ldots, a_n, \ldots \mid pa_1 = 0, \ pa_n = a_{n-1} \text{ for all } n > 1 \rangle$.

Proposition **4.8.** *The group $\mathbb{Z}_{p^\infty}$ is the union of cyclic subgroups $C_1 \subseteq C_2 \subseteq \cdots \subseteq C_n \subseteq \cdots$ of orders p, p^2, $\ldots$, p^n, $\ldots$ and is a divisible abelian group.*

Proof. First we find a model of $\mathbb{Z}_{p^\infty}$. Let U be the multiplicative group of all complex numbers of modulus 1. Let $\alpha_n = e^{2i\pi/p^n} \in U$. Then $\alpha_1^p = 1$ and $\alpha_n^p = \alpha_{n-1}$ for all $n > 1$. Hence there is a homomorphism $\varphi : \mathbb{Z}_{p^\infty} \longrightarrow U$ such that $\varphi(a_n) = \alpha_n$ for all n. (Readers will show that φ is injective.)

By induction, $p^n a_n = pa_1 = 0$ for all $n \geq 1$, so a_n has order at most p^n. On the other hand, $\varphi(p^{n-1} a_n) = \alpha_n^{p^{n-1}} = \alpha_1 \neq 1$. Therefore a_n has order p^n and $C_n = \langle a_n \rangle \subseteq \mathbb{Z}_{p^\infty}$ is cyclic of order p^n. Also $C_n \subseteq C_{n+1}$, since $a_n = pa_{n+1}$. Since $a_i, a_j \in C_j$ if $i \leq j$, the generators a_n commute with each other, and $\mathbb{Z}_{p^\infty}$ is abelian. An element of $\mathbb{Z}_{p^\infty}$ is a linear combination $x = \sum_{n>0} x_n a_n$ with coefficients $x_n \in \mathbb{Z}$, $x_n = 0$ for almost all n; then $x \in C_m$ when $x_n = 0$ for all $n > m$. Hence $\mathbb{Z}_{p^\infty} = \bigcup_{n>0} C_n$.

Let $0 \neq m \in \mathbb{Z}$; write $m = p^k \ell$, where p does not divide ℓ. Let $x \in \mathbb{Z}_{p^\infty}$. Then $x \in C_n$ for some n and $x = ta_n = p^k ta_{n+k} = p^k y$ is divisible by p^k. Next, $p^{n+k} y = tp^{n+k} a_{n+k} = 0$ and $up^{n+k} + v\ell = 1$ for some $u, v \in \mathbb{Z}$; hence $y = up^{n+k} y + v\ell y = \ell v y$ is divisible by ℓ and $x = p^k \ell v y$ is divisible by m. $\square$

Theorem **4.9.** *An abelian group is divisible if and only if it is a direct sum of copies of $\mathbb{Q}$ and $\mathbb{Z}_{p^\infty}$ (for various primes p).*

Proof. Direct sums of copies of $\mathbb{Q}$ and $\mathbb{Z}_{p^\infty}$ are divisible. Conversely, let A be a divisible abelian group. The torsion part

$$T = \{ x \in A \mid nx = 0 \text{ for some } n \neq 0 \}$$

of A is divisible, since $n \neq 0$ and $nx = t \in T$ implies $x \in T$. By 4.7, T is injective. Hence $A = T \oplus D$, where $D \cong A/T$ is torsion-free, and divisible like every quotient group of A. In D every equation $nx = b$ (where $n \neq 0$) has a unique solution. Hence D is a $\mathbb{Q}$-module, in which $(m/n)b$ is the unique solution of $nx = mb$. Therefore D is a direct sum of copies of $\mathbb{Q}$.

By VIII.6.5, T is a direct sum of p-groups: $T = \bigoplus_{p \text{ prime}} T(p)$, where

$$T(p) = \{ x \in T \mid p^k x = 0 \text{ for some } k > 0 \}.$$

Every $T(p)$ is divisible, like every quotient group of T. To complete the proof we show that a divisible abelian p-group P is a direct sum of copies of $\mathbb{Z}_{p^\infty}$.

First we show that every element $b \neq 0$ of P belongs to a subgroup $B \cong \mathbb{Z}_{p^\infty}$ of P. Let $p^m > 1$ be the order of b. Define $b_1, \ldots, b_n, \ldots$ as follows. If

$n \leqq m$, let $b_n = p^{m-n}b$; in particular, $b_m = b$. Since P is divisible, we may choose $b_{m+1}, b_{m+2}, \ldots$ such that $b_n = pb_{n+1}$ for all $n \geqq m$. Then $b_n = pb_{n+1}$ for all n. Since b_m has order p^m it follows that b_n has order p^n for every n. Let B be the subgroup of P generated by $b_1, \ldots, b_n, \ldots$. Since $pb_1 = 0$ and $pb_n = b_{n-1}$ for all $n > 1$, there is a homomorphism φ of $\mathbb{Z}_{p^\infty}$ onto B such that $\varphi(a_n) = b_n$ for all n. Then φ is injective on every $C_n = \langle a_n \rangle \subseteq \mathbb{Z}_{p^\infty}$, since b_n has order p^n. Hence φ is an isomorphism and $B \cong \mathbb{Z}_{p^\infty}$.

Let $\mathcal{S}$ be the set of all sets $\mathcal{D}$ of subgroups $B \cong \mathbb{Z}_{p^\infty}$ of P such that the sum $\sum_{B \in \mathcal{D}} B$ is direct ($B \cap (\sum_{C \in \mathcal{D}, C \neq B} C) = 0$ for all $B \in \mathcal{D}$). By Zorn's lemma, $\mathcal{S}$ has a maximal element $\mathcal{M}$. Then $M = \sum_{B \in \mathcal{M}} B = \bigoplus_{B \in \mathcal{M}} B$ is divisible, hence injective, and $P = M \oplus D$ for some subgroup D of P. Now, $D \cong P/M$ is a divisible p-group. If $D \neq 0$, then D contains a subgroup $C \cong \mathbb{Z}_{p^\infty}$, and then $\sum_{B \in \mathcal{M} \cup \{C\}} B$ is direct, contradicting the maximality of $\mathcal{M}$. Therefore $D = 0$, and $P = \bigoplus_{B \in \mathcal{M}} B$ is a direct sum of copies of $\mathbb{Z}_{p^\infty}$. $\square$

Theorem 4.9 affects all abelian groups, due to the following property:

Proposition 4.10. *Every abelian group can be embedded into a divisible abelian group.*

Proof. For every abelian group A there is an epimorphism $F \longrightarrow A$ where F is a free abelian group. Now, F is a direct sum of copies of $\mathbb{Z}$ and can be embedded into a direct sum D of copies of $\mathbb{Q}$. Then D is divisible, like $\mathbb{Q}$. By 2.7, in the pushout below, $D \longrightarrow B$ is surjective, so that B is divisible, and $A \longrightarrow B$ is injective. $\square$

$$
\begin{array}{ccc}
D & \longrightarrow & B \\
\uparrow & & \uparrow \\
F & \longrightarrow & A
\end{array}
$$

We shall further refine this result. In Section 5 we embed A into a divisible abelian group B so that every $0 \neq a \in A$ has a nonzero integer multiple in B. In Section XI.2 we use other methods to extend Theorem 4.10 to all modules:

Theorem 4.11. *Every module can be embedded into an injective module.*

Noetherian rings. Our last result is due to Bass (cf. Chase [1961]).

Theorem 4.12. *A ring R is left Noetherian if and only if every direct sum of injective left R-modules is injective.*

Proof. Assume that every direct sum of injective left R-modules is injective, and let $L_1 \subseteq L_2 \subseteq \cdots \subseteq L_n \subseteq \cdots$ be an ascending sequence of left ideals of R. Then $L = \bigcup_{n>0} L_n$ is a left ideal of R. By 4.11 there is a monomorphism of R/L_n into an injective R-module J_n. By the hypothesis, $J = \bigoplus_{n>0} J_n$ is injective. Construct a module homomorphism $\varphi : L \longrightarrow J$ as follows. Let φ_n be the homomorphism $R \longrightarrow R/L_n \longrightarrow J_n$. If $x \in L$, then $x \in L_n$ for some n, and then $\varphi_k(x) = 0$ for all $k \geqq n$. Let $\varphi(x) = (\varphi_k(x))_{k>0} \in J$.

Since J is injective, φ extends to a module homomorphism $\psi : {}_R R \longrightarrow J$. Then $\psi(1) = (t_k)_{k>0} \in J$ for some $t_k \in J_k$, $t_k = 0$ for almost all k. Hence $\psi(1) \in \bigoplus_{k<n} J_k$ for some n. Then $\varphi(x) = \psi(x1) = x\,\psi(1) \in \bigoplus_{k<n} J_k$ for all $x \in L$; in particular, $\varphi_n(x) = 0$ and $x \in L_n$, since $R/L_n \longrightarrow J_n$ is injective. Thus $L = L_n$, and then $L_k = L_n$ for all $k \geqq n$.

Conversely, assume that R is left Noetherian, and let $J = \bigoplus_{i \in I} J_i$ be a direct sum of injective R-modules J_i. Let L be a left ideal and let $\varphi : L \longrightarrow J$ be a module homomorphism. Since R is left Noetherian, L is finitely generated: $L = Rr_1 + \cdots + Rr_n$ for some $r_1, \ldots, r_n \in L$. Each $\varphi(r_k)$ has finitely many nonzero components in $\bigoplus_{i \in I} J_i$ and belongs to a finite direct sum $\bigoplus_{i \in S_k} J_i$. Hence there is a finite direct sum $\bigoplus_{i \in S} J_i$, $S = S_1 \cup \cdots \cup S_n$, that contains every $\varphi(r_k)$ and contains $\varphi(L)$. Since S is finite, $\bigoplus_{i \in S} J_i$ is injective, by 4.4; hence $\varphi : L \longrightarrow \bigoplus_{i \in S} J_i$ extends to a module homomorphism ${}_R R \longrightarrow \bigoplus_{i \in S} J_i \subseteq \bigoplus_{i \in I} J_i$. Hence $\bigoplus_{i \in I} J_i$ is injective, by 4.5. $\square$

Exercises

1. Show that every direct summand of an injective module is injective.

2. Show that every direct sum of injective left R-modules is injective.

3. Let L be a left ideal of R and let $\varphi : L \longrightarrow M$ be a module homomorphism. Show that φ can be extended to ${}_R R$ if and only if there exists $m \in M$ such that $\varphi(r) = rm$ for all $r \in L$.

4. Let J be an injective left R-module. Let $a \in J$ and $r \in R$ satisfy $\mathrm{Ann}\,(r) \subseteq \mathrm{Ann}\,(a)$ (if $t \in R$ and $tr = 0$, then $ta = 0$). Prove that $a = rx$ for some $x \in J$.

5. Show that the quotient field of a Noetherian domain R is an injective R-module.

6. Find all subgroups of $\mathbb{Z}_{p^\infty}$.

7. Show that $\mathbb{Z}_{p^\infty}$ is indecomposable.

8. Let U be the multiplicative group of all complex numbers of modulus 1. Show that $U(p) \cong \mathbb{Z}_{p^\infty}$.

9. Show that the additive group $\mathbb{Q}/\mathbb{Z}$ is isomorphic to the direct sum $\bigoplus_p \mathbb{Z}_{p^\infty}$ with one term for every prime p.

*10. Can you extend Theorem 4.9 to modules over any PID?

5. The Injective Hull

In this section we show that every module has, up to isomorphism, a smallest injective extension. We show this by comparing injective extensions to another kind of extensions, *essential* extensions.

Definition. A submodule S of a left R-module M is essential when $S \cap T \neq 0$ for every submodule $T \neq 0$ of M.

Essential submodules are also called *large*. Readers may prove the following:

Proposition **5.1.** *If* $A \subseteq B$ *are submodules of* C, *then* A *is essential in* C *if and only if* A *is essential in* B *and* B *is essential in* C.

A monomorphism $\varphi : A \longrightarrow B$ is *essential* when $\operatorname{Im} \varphi$ is an essential submodule of B.

Proposition **5.2.** *If* μ *is an essential monomorphism, and* $\varphi \circ \mu$ *is injective, then* φ *is injective.*

Proof. If $\varphi \circ \mu$ is injective, then $\operatorname{Ker} \varphi \cap \operatorname{Im} \mu = 0$; hence $\operatorname{Ker} \varphi = 0$. $\square$

Definition. An *essential extension of a left R-module* A *is a left R-module* B *such that* A *is an essential submodule of* B; *more generally, a left R-module* B *with an essential monomorphism* $A \longrightarrow B$.

These two definitions are equivalent up to isomorphisms: if A is an essential submodule of B, then the inclusion homomorphism $A \longrightarrow B$ is an essential monomorphism; if $\mu : A \longrightarrow B$ is an essential monomorphism, then A is isomorphic to the essential submodule $\operatorname{Im} \mu$ of B; moreover, using surgery, one can construct a module $B' \cong B$ in which A is an essential submodule.

Proposition **5.3.** *If* $\mu : A \longrightarrow B$ *and* $v : B \longrightarrow C$ *are monomorphisms, then* $v \circ \mu$ *is essential if and only if* μ *and* v *are essential.*

This follows from Proposition 5.1; the details make nifty exercises.

Proposition **5.4.** *A left R-module* J *is injective if and only if* J *has no proper essential extension* $J \subsetneqq M$, *if and only if every essential monomorphism* $J \longrightarrow M$ *is an isomorphism.*

Proof. Let J be injective. If $J \subseteq M$, then J is a direct summand of M, $M = J \oplus N$; then $N \cap J = 0$; if J is essential in M, then $N = 0$ and $J = M$. If in turn J has no proper essential extension, and $\mu : J \longrightarrow M$ is an essential monomorphism, then $\operatorname{Im} \mu \cong J$ has no proper essential extension, hence $M = \operatorname{Im} \mu$ and μ is an isomorphism.

Finally, assume that every essential monomorphism $J \longrightarrow A$ is an isomorphism. We show that J is a direct summand of every module $M \supseteq J$. By Zorn's lemma there is a submodule K of M maximal such that $J \cap K = 0$. Readers will verify that the projection $M \longrightarrow M/K$ induces an essential monomorphism $\mu : J \longrightarrow M/K$. By the hypothesis, μ is an isomorphism; hence $J + K = M$ and $M = J \oplus K$. $\square$

Proposition **5.5.** *Let* $\mu : M \longrightarrow N$ *and* $v : M \longrightarrow J$ *be monomorphisms. If* μ *is essential and* J *is injective, then* $v = \kappa \circ \mu$ *for some monomorphism* $\kappa : N \longrightarrow J$.

Proof. Since J is injective, there exists a homomorphism $\kappa : N \longrightarrow J$ such that $v = \kappa \circ \mu$, which is injective by 5.2. $\square$

By 5.5, every essential extension of M is, up to isomorphism, contained in every injective extension of M. This leads to the main result in this section.

Theorem **5.6.** *Every left R-module M is an essential submodule of an injective R-module, which is unique up to isomorphism.*

Proof. By Theorem 4.11, M is a submodule of an injective module K. Let $\mathcal{S}$ be the set of all submodules $M \subseteq S \subseteq K$ of K in which M is essential (for instance, M itself). If $(S_i)_{i \in I}$ is a chain in $\mathcal{S}$, then $S = \bigcup_{i \in I} S_i \in \mathcal{S}$: if $N \neq 0$ is a submodule of S, then $S_i \cap N \neq 0$ for some i, and then $M \cap N = M \cap S_i \cap N \neq 0$ since M is essential in S_i; thus M is essential in S. By Zorn's lemma, $\mathcal{S}$ has a maximal element J. If J had a proper essential extension, then by 5.5 J would have a proper essential extension $J \subsetneqq J' \subseteq K$ and would not be maximal; therefore J is injective, by 5.4.

Now, assume that M is essential in two injective modules J and J'. The inclusion monomorphisms $\mu : M \longrightarrow J$ and $\nu : M \longrightarrow J'$ are essential. By 5.5 there is a monomorphism $\theta : J \longrightarrow J'$ such that $\nu = \theta \circ \mu$. Then θ is essential, by 5.3, and is an isomorphism by 5.4. $\square$

Definition. *The injective hull of a left R-module M is the injective module, unique up to isomorphism, in which M is an essential submodule.*

The injective hull or *injective envelope* $E(M)$ of M can be characterized in several ways: $E(M)$ is injective and an essential extension of M, by definition; $E(M)$ is a maximal essential extension of M, by 5.4; in fact, $E(M)$ is, up to isomorphism, the largest essential extension of M, by 5.5; $E(M)$ is a minimal injective extension of M, by 5.4, 5.1; and $E(M)$ is, up to isomorphism, the smallest injective extension of M, by 5.5. The exercises give some examples.

Exercises

1. Let $\mu : A \longrightarrow B$ and $\nu : B \longrightarrow C$ be essential monomorphisms. Show that $\nu \circ \mu$ is essential.

2. Let $\mu : A \longrightarrow B$ and $\nu : B \longrightarrow C$ be monomorphisms. If $\nu \circ \mu$ is essential, then show that μ and ν are essential.

3. Show that every nonzero ideal of a [commutative] domain R is essential in R.

4. Show that every domain R is essential (as an R-module) in its quotient field.

5. Let N be a submodule of M. Show that there is a submodule C of M maximal such that $C \cap N = 0$. Show that $N + C$ is essential in M. Show that N is, up to isomorphism, essential in M/C.

6. Prove the following: if A_i is an essential submodule of B_i for all $i = 1, 2, \ldots, n$, then $A_1 \oplus A_2 \oplus \cdots \oplus A_n$ is essential in $B_1 \oplus B_2 \oplus \cdots \oplus B_n$.

7. Show that $\bigoplus_{i \in I} M_i$ need not be essential in $\prod_{i \in I} M_i$.

8. Show that $\mathbb{Q}$ is the injective hull of $\mathbb{Z}$ (as a $\mathbb{Z}$-module).

9. Show that $\mathbb{Z}_{p^\infty}$ is the injective hull of $\mathbb{Z}_{p^n}$, when $n > 0$.

10. Show that $E(A \oplus B) \cong E(A) \oplus E(B)$.

11. If R is a Noetherian domain, show that the quotient field of R serves as $E(_R R)$.

6. Hereditary Rings

Hereditary rings generalize PIDs. This section contains basic properties, with applications to Dedekind domains.

Definition. A ring R is left hereditary *when every submodule of a projective left R-module is projective.*

For example, PIDs are left hereditary: if R is a PID, then every projective R-module is free, by 3.7, and every submodule of a free R-module is free. Semisimple rings are left hereditary. Crouching in the exercises lies a ring that is left hereditary but not right hereditary.

Proposition **6.1.** *A ring* R *is left hereditary if and only if every left ideal of* R *is projective, and then every submodule of a free left R-module is isomorphic to a direct sum of left ideals of* R.

Proof. If R is left hereditary, then every left ideal of R is projective (as an R-module), since $_R R$ is projective. Conversely, assume that every left ideal of R is projective. We show that every submodule of a free module is isomorphic to a direct sum of left ideals of R; then submodules of free modules are projective, and R is left hereditary.

Let S be a submodule of a free R-module F with basis $(e_i)_{i \in I}$. We may assume that the set I is well ordered (Theorem A.2.4). For every $i \in I$, let $F_i = \bigoplus_{t < i} R e_t$ be the free submodule of F with basis $(e_t)_{t < i}$; let $S_i = S \cap F_i$ and $S_i' = S \cap (F_i + R e_i)$. Every $x \in S_i' \subseteq F_i + R e_i$ is a sum $x = y + r e_i$ for some unique $y \in F_i$ and $r \in R$. Then $\xi : x \longmapsto r$ is a module homomorphism of S_i' into $_R R$. Since Ker $\xi = S_i$, there is an exact sequence

$$0 \longrightarrow S_i \longrightarrow S_i' \longrightarrow L \longrightarrow 0,$$

in which $L = \operatorname{Im} \xi$ is a left ideal of R. This sequence splits, since L is projective, and $S_i' = S_i \oplus T_i$, where $T_i \subseteq S_i' \subseteq F$ and $T_i \cong L$ is isomorphic to a left ideal of R. We show that $S = \bigoplus_{i \in I} T_i$.

First we prove by induction on $k \in I$ that $S_k = \sum_{i < k} T_i$ for all k. If $x \in S_k$, $x \neq 0$, then $x = \sum_{i < k} r_i e_i$, where $r_i \in R$ for all $i < k$, $r_i = 0$ for almost all $i < k$, and $r_i \neq 0$ for some $i < k$. There is a greatest $j < k$ such that $r_j \neq 0$. Then $x \in S_j'$, $x = y + m$ for some $y \in S_j$ and $m \in T_j$, $y \in \sum_{i < j} T_i$ by the induction hypothesis, and $x = y + m \in \sum_{i < k} T_i$.

If now $x \in T_j \cap \left(\sum_{i < j} T_i \right)$, then $x \in T_j \cap S_j = 0$. If $x \in S$, then $x \in S_k$ for some $k \in I$ and $x \in \sum_{i < k} T_i \subseteq \sum_{i \in I} T_i$. Hence $S = \bigoplus_{i \in I} T_i$. $\square$

We give another characterization, based on the following lemma.

Lemma **6.2.** (1) *A module* P *is projective if and only if every module homomorphism* $P \longrightarrow B$ *factors through every epimorphism* $J \longrightarrow B$ *in which* J *is injective.*

(2) *A module J is injective if and only if every module homomorphism $A \longrightarrow J$ factors through every monomorphism $A \longrightarrow P$ in which P is projective.*

Proof. We prove (1) and let readers reverse arrows to prove (2). Assume that every module homomorphism $P \longrightarrow B$ factors through every epimorphism $J \longrightarrow B$ in which J is injective. Let $\varphi : P \longrightarrow B$ be a homomorphism and let $\sigma : A \longrightarrow B$ be any epimorphism. Let $K = \operatorname{Ker} \sigma$. Let $\mu : A \longrightarrow J$ be a monomorphism into an injective module J and let $\pi : J \longrightarrow J/\mu(K)$ be the projection. Since μ is injective, $\pi\big(\mu(a)\big) = 0$ if and only if $\mu(a) \in \mu(K)$, if and only if $a \in K$; thus $\operatorname{Ker}(\pi \circ \mu) = K = \operatorname{Ker} \sigma$ and $\pi \circ \mu = \nu \circ \sigma$ for some homomorphism $\nu : B \longrightarrow J/\mu(K)$, by 1.2, which completes the diagram:

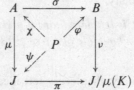

Then ν is injective: if $\nu\big(\sigma(a)\big) = 0$, then $\pi\big(\mu(a)\big) = 0$, $\mu(a) \in \mu(K)$, $a \in K$, and $\sigma(a) = 0$; since σ is surjective this implies $\operatorname{Ker} \nu = 0$.

By the hypothesis, $\nu \circ \varphi$ factors through π: $\nu \circ \varphi = \pi \circ \psi$ for some homomorphism $\psi : P \longrightarrow J$. Now, $\operatorname{Im} \psi \subseteq \operatorname{Im} \mu$: for every $p \in P$, $\varphi(p) = \sigma(a)$ for some $a \in A$, whence $\pi\big(\psi(p)\big) = \nu\big(\varphi(p)\big) = \nu\big(\sigma(a)\big) = \pi\big(\mu(a)\big)$, $\psi(p) - \mu(a) \in \operatorname{Ker} \pi = \mu(K)$, and $\psi(p) \in \operatorname{Im} \mu$. Hence ψ factors through μ: $\psi = \mu \circ \chi$ for some homomorphism $\chi : P \longrightarrow A$. Then $\nu \circ \varphi = \pi \circ \mu \circ \chi = \nu \circ \sigma \circ \chi$ and $\varphi = \sigma \circ \chi$, since ν is injective. $\square$

Proposition 6.3. *A ring R is left hereditary if and only if every quotient of an injective left R-module is injective.*

Proof. Assume that R is left hereditary. Let J be an injective module and let $\sigma : J \longrightarrow B$ be an epimorphism. To prove that B is injective we use 6.2 and show that every homomorphism $\varphi : A \longrightarrow B$ factors through every monomorphism $\mu : A \longrightarrow P$ in which P is projective:

$$P \xleftarrow{\ \mu\ } A$$
$$\chi \downarrow \quad \swarrow_{\psi} \quad \downarrow \varphi$$
$$J \xrightarrow{\ \sigma\ } B$$

Now, A is projective, since R is hereditary; hence $\varphi = \sigma \circ \psi$ for some homomorphism $\psi : A \longrightarrow J$. Since J is injective, $\psi = \chi \circ \mu$ for some homomorphism $\chi : P \longrightarrow J$. Hence $\varphi = \sigma \circ \chi \circ \mu$ factors through μ.

The converse is similar. Assume that every quotient of an injective module is injective. Let A be a submodule of a projective module P and let $\mu : A \longrightarrow P$ be the inclusion homomorphism. To prove that A is projective we use 6.2 and show that every homomorphism $\varphi : A \longrightarrow B$ factors through every epimorphism

$\sigma : J \longrightarrow B$ in which J is injective. Now, $B \cong J/\mathrm{Ker}\,\sigma$ is injective by the hypothesis; hence $\varphi = \psi \circ \mu$ for some homomorphism $\psi : P \longrightarrow B$. Since P is projective, $\psi = \sigma \circ \chi$ for some homomorphism $\chi : P \longrightarrow J$ (see the diagram below). Hence $\varphi = \sigma \circ \chi \circ \mu$ factors through σ. $\square$

$$P \xleftarrow{\ \mu\ } A$$
$$\chi \Big\downarrow \ \searrow \psi \ \ \Big\downarrow \varphi$$
$$J \xrightarrow{\ \sigma\ } B$$

Dedekind domains were defined in Section VII.5. Here we give other characterizations. Recall that a fractional ideal of a domain R is a set $\mathfrak{A} = \mathfrak{a}/c = \{ x/c \in Q \mid x \in \mathfrak{a} \}$ where $\mathfrak{a}$ is an ideal of R, $c \in R$, $c \neq 0$, and Q is the quotient field of R. Then $x \longmapsto xc$ is a module isomorphism of $\mathfrak{A}$ onto $\mathfrak{a}$.

Proposition **6.4.** *A nonzero fractional ideal of a domain R is invertible if and only if it is projective as an R-module.*

Proof. Let $\mathfrak{A} \neq 0$ be a fractional ideal of R. Let $\mathfrak{A}\mathfrak{B} = R$ for some fractional ideal $\mathfrak{B}$. As in the proof of VII.5.2, $1 = a_1 b_1 + \cdots + a_n b_n$ for some $a_1, \ldots, a_n \in \mathfrak{A}$ and $b_1, \ldots, b_n \in \mathfrak{B}$. Then $\mathfrak{A} = Ra_1 + \cdots + Ra_n$, since $a = ab_1 a_1 + \cdots + ab_n a_n$ for all $a \in \mathfrak{A}$. Let F be a free R-module with basis $e_1, \ldots, e_n$. Then $\rho : F \longrightarrow \mathfrak{A}$, $r_1 e_1 + \cdots + r_n e_n \longmapsto r_1 a_1 + \cdots + r_n a_n$ and $\mu : \mathfrak{A} \longrightarrow F$, $a \longmapsto ab_1 e_1 + \cdots + ab_n e_n$ are module homomorphisms. Since $\rho \circ \mu = 1_{\mathfrak{A}}$, the sequence $0 \longrightarrow \mathrm{Ker}\,\rho \longrightarrow F \longrightarrow \mathfrak{A} \longrightarrow 0$ splits, $\mathfrak{A}$ is isomorphic to a direct summand of F, and $\mathfrak{A}$ is projective.

Conversely, assume that $\mathfrak{A}$ is projective. There exist a free R-module F, with a basis $(e_i)_{i \in I}$, and homomorphisms $\rho : F \longrightarrow \mathfrak{A}$, $\mu : \mathfrak{A} \longrightarrow F$ such that $\rho \circ \mu = 1_{\mathfrak{A}}$. For every $a \in \mathfrak{A}$ let $\mu(a) = \sum_{i \in I} a_i e_i$, where $a_i \in R$. Then $a \longmapsto a_i$ is a module homomorphism for every i.

Let $a, b \in \mathfrak{A}$, $b \neq 0$. We have $\sum_{i \in I} (b_i/b)\,\rho(e_i) = 1$, since

$$b = \rho\big(\mu(b)\big) = \rho\big(\textstyle\sum_{i \in I} b_i e_i\big) = \textstyle\sum_{i \in I} b_i \rho(e_i) = b\big(\textstyle\sum_{i \in I} (b_i/b)\,\rho(e_i)\big).$$

Also, $ra, sc \in R$ for some $r, s \in R$, $r, s \neq 0$. Since $a \longmapsto a_i$ is a module homomorphism, $r\,(sab)_i = (rsab)_i = s\,(rab)_i$ and

$$ab_i = ra\,b_i/r = (rab)_i/r = (sab)_i/s = sb\,a_i/s = b\,a_i.$$

Hence $ab_i/b = a_i \in R$ for all $a \in \mathfrak{A}$ and $b_i/b \in \mathfrak{A}' = \{ x \in Q \mid x\mathfrak{A} \subseteq R \}$ for all i. Therefore $1 = \sum_{i \in I} (b_i/b)\,\rho(e_i) \in \mathfrak{A}'\mathfrak{A}$ and $\mathfrak{A}'\mathfrak{A} = R$. $\square$

Since every fractional ideal of R is isomorphic to an ideal of R, Proposition 6.4 begets the following result:

Proposition **6.5.** *For a domain R the following conditions are equivalent:* (i) *R is a Dedekind domain;* (ii) *every fractional ideal of R is projective;* (iii) *every ideal of R is projective;* (iv) *R is [left] hereditary.*

Proposition **6.6.** *A domain R is a Dedekind domain if and only if every divisible R-module is injective.*

Proof. Injective modules are divisible, by 4.6, and quotients of divisible modules are divisible. If divisible R-modules are injective, then quotients of injective R-module are injective, R is hereditary by 6.3, and R is Dedekind by 6.5.

Conversely, let R be a Dedekind domain and let M be a divisible R-module. Let $\mathfrak{a}$ be an ideal of R and let $\varphi : \mathfrak{a} \longrightarrow M$ be a module homomorphism. Since $\mathfrak{a}$ is invertible, $1 = a_1 b_1 + \cdots + a_n b_n$ for some $a_1, \ldots, a_n \in \mathfrak{a}$ and $b_1, \ldots, b_n \in \mathfrak{a}'$. We may assume that $a_i \neq 0$ for all i. Then $\varphi(a_i) = a_i m_i$ for some $m_i \in M$. If $a \in \mathfrak{a}$, then $ab_i \in \mathfrak{a}\mathfrak{a}' = R$ and

$$\varphi(a) = \varphi\left(\sum_i a_i b_i a\right) = \sum_i (ab_i)\,\varphi(a_i) = \sum_i ab_i a_i m_i = am,$$

where $m = \sum_i a_i b_i m_i \in M$ does not depend on a. By 4.5, M is injective. $\square$

Exercises

1. Let R be the ring of all matrices $\begin{pmatrix} a & b \\ 0 & c \end{pmatrix}$ in which $a, b \in \mathbb{Q}$ and $c \in \mathbb{Z}$. Show that R is left hereditary but not right hereditary.

2. Prove that a module J is injective if and only if every module homomorphism $A \longrightarrow J$ factors through every monomorphism $A \longrightarrow P$ in which P is projective.

3. Prove that an ideal of a UFD is projective if and only if it is principal; hence a Dedekind domain is a UFD if and only if it is a PID.

A ring R is *left semihereditary* when every finitely generated left ideal of R is projective.

4. Show that every valuation domain is left semihereditary.

5. Show that every von Neumann regular ring is left semihereditary.

6. Prove the following: if R is left semihereditary, then every finitely generated submodule of a free left R-module is isomorphic to a direct sum of finitely many finitely generated left ideals of R.

7. Prove that a ring R is left semihereditary if and only if every finitely generated submodule of a projective left R-module is projective.

XI
Constructions

This chapter introduces basic module constructions: groups of homomorphisms, direct and inverse limits, tensor products, and completions.

Sections 3, 4, 7, 8, and 9 may be skipped at first reading. As before, all rings have an identity element; all modules are unital; results that are generally stated for left R-modules apply equally to right R-modules.

1. Groups of Homomorphisms

This section studies groups and modules of homomorphisms, with emphasis on their functorial properties.

Definition. Let M and A be left R-modules. Recall that module homomorphisms φ, $\psi : M \longrightarrow A$ can be added pointwise: $(\varphi + \psi)(x) = \varphi(x) + \psi(x)$, the result being another homomorphism. With this operation, the module homomorphisms of M into A become the elements of an abelian group.

Definition. If M and A are left R-modules, or if M and A are right R-modules, then $\mathrm{Hom}_R(M, A)$ is the abelian group of all module homomorphisms of M into A, under pointwise addition.

For example, $\mathrm{Hom}_{\mathbb{Z}}(\mathbb{Z}, A) \cong A$ for every abelian group A. If R is a field, then $\mathrm{Hom}_R(M, A)$ is the abelian group of all linear transformations of M into A, under pointwise addition.

Let M, A, and B be left R-modules. Every module homomorphism $\varphi : A \longrightarrow B$ induces a mapping

$$\varphi_* = \mathrm{Hom}_R(M, \varphi). \ \mathrm{Hom}_R(M, A) \longrightarrow \mathrm{Hom}_R(M, B), \ \alpha \longmapsto \varphi \circ \alpha :$$

$$A \xrightarrow{\ \varphi\ } B$$
$$\alpha \uparrow \quad \nearrow \varphi \circ \alpha$$
$$M$$

The following properties are clear:

Proposition **1.1.** *If* $\varphi : A \longrightarrow B$ *is a homomorphism of left R-modules, then* $\varphi_* = \mathrm{Hom}_R(M, \varphi)$: $\mathrm{Hom}_R(M, A) \longrightarrow \mathrm{Hom}_R(M, B)$ *is a homomorphism of abelian groups. Moreover:*

(1) *if* φ *is the identity on* A, *then* φ_* *is the identity on* $\mathrm{Hom}_R(M, A)$;

(2) $(\psi \circ \varphi)_* = \psi_* \circ \varphi_*$ *whenever* $\varphi : A \longrightarrow B$ *and* $\psi : B \longrightarrow C$;

(3) $(\varphi + \psi)_* = \varphi_* + \psi_*$ *whenever* $\varphi, \psi : A \longrightarrow B$.

Similarly, when M, N, and A are left R-modules, every module homomorphism $\varphi : M \longrightarrow N$ induces a mapping

$$\varphi^* = \mathrm{Hom}_R(M, \varphi): \mathrm{Hom}_R(N, A) \longrightarrow \mathrm{Hom}_R(M, A), \; \alpha \longmapsto \alpha \circ \varphi :$$

$$
\begin{array}{ccc}
 & A & \\
\alpha \circ \varphi \nearrow & & \nwarrow \alpha \\
M & \underrightarrow{\varphi} & N
\end{array}
$$

Proposition **1.2.** *If* $\varphi : M \longrightarrow N$ *is a homomorphism of left R-modules, then* $\varphi^* = \mathrm{Hom}_R(\varphi, A)$: $\mathrm{Hom}_R(N, A) \longrightarrow \mathrm{Hom}_R(M, A)$ *is a homomorphism of abelian groups. Moreover:*

(1) *if* φ *is the identity on* M, *then* φ^* *is the identity on* $\mathrm{Hom}_R(M, A)$;

(2) $(\psi \circ \varphi)^* = \varphi^* \circ \psi^*$ *whenever* $\varphi : M \longrightarrow N$ *and* $\psi : N \longrightarrow P$;

(3) $(\varphi + \psi)^* = \varphi^* + \psi^*$ *whenever* $\varphi, \psi : M \longrightarrow N$.

Proposition **1.3.** *For every* $\varphi : A \longrightarrow B$ *and* $\psi : M \longrightarrow N$ *the following square commutes:*

$$
\begin{array}{ccc}
\mathrm{Hom}_R(M, A) & \xrightarrow{\mathrm{Hom}_R(M,\varphi)} & \mathrm{Hom}_R(M, B) \\
\mathrm{Hom}_R(\psi,A) \uparrow & & \uparrow \mathrm{Hom}_R(\psi,B) \\
\mathrm{Hom}_R(N, A) & \xrightarrow[\mathrm{Hom}_R(N,\varphi)]{} & \mathrm{Hom}_R(N, B)
\end{array}
$$

Proof. For every $\alpha : N \longrightarrow A$, $\varphi_*\big(\psi^*(\alpha)\big) = \varphi \circ \alpha \circ \psi = \psi^*\big(\varphi_*(\alpha)\big)$. $\square$

Functors. Propositions 1.1, 1.2, and 1.3 can be expressed more compactly using a language that will be defined less informally in Section XVI.2.

A *functor* from bidules to doohickeys is a construction that assigns to every bidule B a doohickey $F(B)$, and to every homomorphism $\varphi : B \longrightarrow C$ of bidules a homomorphism $F(\varphi): F(B) \longrightarrow F(C)$ of doohickeys, so that $F(1_B) = 1_{F(B)}$, and $F(\psi \circ \varphi) = F(\psi) \circ F(\varphi)$ whenever $\psi \circ \varphi$ is defined.

Functors are also called *covariant functors*. A *contravariant functor* from bidules to doohickeys is a construction that assigns to every bidule B a doohickey $F(B)$, and to every homomorphism $\varphi : B \longrightarrow C$ of bidules a homomorphism $F(\varphi): F(C) \longrightarrow F(B)$ of doohickeys, so that $F(1_B) = 1_{F(B)}$, and $F(\psi \circ \varphi) = F(\varphi) \circ F(\psi)$ whenever $\psi \circ \varphi$ is defined.

If homomorphisms of bidules can be added, and homomorphisms of doohickeys can be added, then a functor F from modules to doohickeys is *additive* when $F(\psi + \varphi) = F(\varphi) + F(\psi)$ for all $\varphi, \psi : B \longrightarrow C$.

Propositions 1.1 and 1.2 now read as follows:

Proposition **1.1** (revisited). *For every left R-module M, $\mathrm{Hom}_R(M, -)$ is an additive functor from left R-modules to abelian groups.*

Proposition **1.2** (revisited). *For every left R-module A, $\mathrm{Hom}_R(-, A)$ is a contravariant additive functor from left R-modules to abelian groups.*

If F and G are functors from bidules to doohickeys, then a *natural transformation* τ from F to G is a construction that assigns to every bidule B a homomorphism $\tau_B : F(B) \longrightarrow G(B)$ of doohickeys, so that the square

$$
\begin{array}{ccc}
F(B) & \xrightarrow{\;\tau_B\;} & G(B) \\
{\scriptstyle F(\varphi)}\Big\downarrow & & \Big\downarrow{\scriptstyle G(\varphi)} \\
F(C) & \xrightarrow[\;\tau_C\;]{} & G(C)
\end{array}
$$

commutes for every homomorphism $\varphi : B \longrightarrow C$ of bidules. One also says that the homomorphism τ_B is *natural in B*.

Natural transformations of contravariant functors are defined similarly. As we shall see, homomorphisms that are "natural" in the ordinary sense, or are constructed in some canonical fashion, tend to be natural as defined above.

Proposition **1.3** (revisited). *If φ is a homomorphism of left R-modules, then $\mathrm{Hom}_R(M, \varphi)$ is natural in M. If ψ is a homomorphism of left R-modules, then $\mathrm{Hom}_R(\psi, A)$ is natural in A.*

Bimodules. The abelian group $\mathrm{Hom}_R(M, A)$ now wants to catch up with his neighbors and acquire module structures. Propositions 1.1 and 1.2 show that endomorphisms of $\mathrm{Hom}_R(M, A)$ arise readily from module endomorphisms of M and A; this suggests additional module structures on A and M.

Definition. *A left R-, right S-bimodule or just R-S-bimodule is an abelian group M with a left R-module structure and a right S-module structure such that $r(xs) = (rx)s$ for all $r \in R$, $x \in M$, and $s \in S$.*

The notation $_R M_S$ indicates that M is an R-S-bimodule. Examples abound: R is an R-R-bimodule; every left R-module is an R-$\mathbb{Z}$-bimodule; every left R-module M is a left R-, right $\mathrm{End}_R^{\mathrm{op}}(M)$-bimodule; the right R-module structure on a free left R-module, which depends on the choice of a basis, makes it an R-R-bimodule; if R is commutative, then every R-module is an R-R-bimodule.

R-S-bimodule structures on an abelian group M can be viewed as pairs of ring homomorphisms $R \longrightarrow \mathrm{End}_{\mathbb{Z}}(M)$, $S \longrightarrow \mathrm{End}_R^{\mathrm{op}}(M)$, or as pairs of ring homomorphisms $S \longrightarrow \mathrm{End}_{\mathbb{Z}}^{\mathrm{op}}(M)$, $R \longrightarrow \mathrm{End}_S(M)$ (see the exercises).

Proposition **1.4.** *If* M *is an* R-S-*bimodule and* A *is an* R-T-*bimodule, then* $\mathrm{Hom}_R(M, A)$ *is an* S-T-*bimodule, in which*

$$(s\alpha)(x) = \alpha(xs) \text{ and } (\alpha t)(x) = \alpha(x)t$$

for all $s \in S$, $x \in M$, $t \in T$, *and* $\alpha \in \mathrm{Hom}_R(M, A)$.

Proof. In the above, $s\alpha$ and αt are homomorphisms of left R-modules, since M and A are bimodules. Moreover, $s(\alpha + \beta) = s\alpha + s\beta$, and

$$s(s'\alpha)(x) = (s'\alpha)(xs) = \alpha((xs)s') = \alpha(x(ss')) = ((ss')\alpha)(x),$$

so that $s(s'\alpha) = (ss')\alpha$; thus $\mathrm{Hom}_R(M, A)$ is a left S-module. Similarly, $\mathrm{Hom}_R(M, A)$ is a right T-module, and

$$(s(\alpha t))(x) = (\alpha t)(xs) = \alpha(xs)t = ((s\alpha)(x))t = ((s\alpha)t)(x),$$

so that $s(\alpha t) = (s\alpha)t$ and $\mathrm{Hom}_R(M, A)$ is a bimodule. $\square$

Thus, $_RM_S$ and $_RA_T$ imply $_S\mathrm{Hom}_R(M, A)_T$. In particular, for every left R-module A, $\mathrm{Hom}_R(_RR, A)$ is an R-$\mathbb{Z}$-bimodule, that is, a left R-module, since $_RR$ is an R-R-bimodule; readers will verify that $\mathrm{Hom}_R(_RR, A) \cong A$.

Proposition 1.4 becomes simpler if R is commutative:

Corollary **1.5.** *If* R *is commutative, then* $\mathrm{Hom}_R(M, A)$ *is an* R-*module, for all* R-*modules* M *and* A.

For instance, if $R = K$ is a field and V, W are vector spaces over K, then $\mathrm{Hom}_K(V, W)$ is a vector space over K; the dual space $\mathrm{Hom}_K(V, K)$ is a vector space over K.

A *homomorphism* of R-S-bimodules is a homomorphism of left R-modules that is also a homomorphism of right S-modules.

Proposition **1.6.** *If* M *is an* R-S-*bimodule and* φ *is a homomorphism of* R-T-*bimodules, then* $\mathrm{Hom}_R(M, \varphi)$ *is a homomorphism of* S-T-*bimodules. If* A *is an* R-T-*bimodule and* ψ *is a homomorphism of* R-S-*bimodules, then* $\mathrm{Hom}_R(\psi, A)$ *is a homomorphism of* S-T-*bimodules.*

The proof is an exercise. In Proposition 1.6, $\mathrm{Hom}_R(M, -)$ is now a functor from R-T-bimodules to S-T-bimodules, and $\mathrm{Hom}_R(-, A)$ is a contravariant functor from R-S-bimodules to S-T-bimodules.

Exercises

1. Produce a one-to-one correspondence between R-S-bimodule structures on an abelian group M and pairs of ring homomorphisms $R \longrightarrow \mathrm{End}_{\mathbb{Z}}(M)$, $S \longrightarrow \mathrm{End}_R^{\mathrm{op}}(M)$.

2. Explain how $\mathrm{Hom}_R(M, A)$ is a T-S-bimodule when M is an S-R-bimodule and A is a T-R-bimodule.

3. Show that $\mathrm{Hom}_R(_RR, A) \cong A$ for every left R-module A, by an isomorphism that is natural in A.

4. Show that $\text{Hom}_R(M, \varphi)$ is a homomorphism of S-T-bimodules when M is an R-S-bimodule and φ is a homomorphism of R-T-bimodules.

5. Show that $\text{Hom}_R(\varphi, A)$ is a homomorphism of S-T-bimodules when A is an R-T-bimodule and φ is a homomorphism of R-S-bimodules.

The *dual* of a left R-module M is the right R-module $M^* = \text{Hom}_R(M, R)$.

6. Let M be a left R-module. Give details of the right action of R on M^* and of the left action of R on the double dual module M^{**}.

7. Produce a canonical homomorphism $M \longrightarrow M^{**}$ and show that it is natural in M.

2. Properties of Hom

This section brings basic properties of the Hom functors in Section 1, with applications to injective modules. These properties apply equally to left R-modules, right R-modules, and bimodules of various stripes.

Exactness. We begin with preservation of exact sequences.

Proposition **2.1** (Left Exactness). *If* $0 \longrightarrow A \xrightarrow{\mu} B \xrightarrow{\rho} C$ *is exact, then*

$$0 \longrightarrow \text{Hom}_R(M, A) \xrightarrow{\text{Hom}_R(M,\mu)} \text{Hom}_R(M, B) \xrightarrow{\text{Hom}_R(M,\rho)} \text{Hom}_R(M, C)$$

is exact.

Proof. If $\mu_*(\alpha) = 0$, then $\mu \circ \alpha = 0$, $\text{Im}\ \alpha \subseteq \text{Ker}\ \mu = 0$, and $\alpha = 0$. Similarly, $\rho_*(\mu_*(\alpha)) = \rho \circ \mu \circ \alpha = 0$. Conversely, if $\rho_*(\beta) = 0$, where $\beta \in \text{Hom}_R(M, B)$, then $\rho \circ \beta = 0$, $\text{Im}\ \beta \subseteq \text{Ker}\ \rho = \text{Im}\ \mu$, and β factors through μ: $\beta = \mu \circ \alpha = \mu_*(\alpha)$ for some $\alpha : M \longrightarrow A$. $\square$

$$0 \longrightarrow A \xrightarrow{\mu} B \xrightarrow{\rho} C$$
$$\alpha \nwarrow \quad \uparrow \beta$$
$$M$$

A covariant functor is *left exact* when it transforms left exact sequences into left exact sequences, and *exact* when it transforms short exact sequences into short exact sequences. Readers will show that an exact functor transforms all exact sequences into exact sequences. Proposition 2.1 now reads as follows:

Proposition **2.1** (revisited). $\text{Hom}_R(M, -)$ *is left exact.*

$\text{Hom}_R(M, -)$ does not preserve all exact sequences; the next result easily yields counterexamples.

Proposition **2.2.** *For a left R-module P the following conditions are equivalent:* (i) *P is projective;* (ii) $\text{Hom}_R(P, -)$ *preserves epimorphisms;* (iii) $\text{Hom}_R(P, -)$ *is exact.*

$\text{Hom}_R(-, A)$ has similar properties, which, like 2.2, make fine exercises. A contravariant functor is *left exact* when it transforms right exact sequences into left exact sequences, and *exact* when it transforms short exact sequences into short exact sequences. Readers may prove the following:

Proposition **2.3** (Left Exactness). $\operatorname{Hom}_R(-, M)$ *is left exact: if*
$$A \xrightarrow{\varphi} B \xrightarrow{\rho} C \longrightarrow 0 \text{ is exact, then so is}$$

$$0 \longrightarrow \operatorname{Hom}_R(C, M) \xrightarrow{\operatorname{Hom}_R(\rho, M)} \operatorname{Hom}_R(B, M) \xrightarrow{\operatorname{Hom}_R(\varphi, M)} \operatorname{Hom}_R(A, M).$$

Proposition **2.4.** *For a left R-module J the following conditions are equivalent:* (i) *J is injective;* (ii) $\operatorname{Hom}_R(\mu, J)$ *is an epimorphism for every monomorphism* μ; (iii) $\operatorname{Hom}_R(-, J)$ *is exact.*

Direct sums and products. $\operatorname{Hom}_R(M, -)$ preserves direct products:

Proposition **2.5.** *There is an isomorphism*

$$\operatorname{Hom}_R\big(M, \textstyle\prod_{i \in I} A_i\big) \cong \textstyle\prod_{i \in I} \operatorname{Hom}_R(M, A_i),$$

which is natural in M and in $(A_i)_{i \in I}$.

Proof. The projections $\pi_i : \prod_{i \in I} A_i \longrightarrow A_i$ induce homomorphisms $\operatorname{Hom}_R(M, \pi_i)$: $\operatorname{Hom}_R(M, \prod_{i \in I} A_i) \longrightarrow \operatorname{Hom}_R(M, A_i)$ and a homomorphism $\theta : \operatorname{Hom}_R(M, \prod_{i \in I} A_i) \longrightarrow \prod_{i \in I} \operatorname{Hom}_R(M, A_i)$ that sends $\alpha : M \longrightarrow \prod_{i \in I} A_i$ to $(\pi_i \circ \alpha)_{i \in I}$. The universal property of $\prod_{i \in I} A_i$ states precisely that θ is bijective.

$\operatorname{Hom}_R(-, \prod_{i \in I} A_i)$ and $\prod_{i \in I} \operatorname{Hom}_R(-, A_i)$ are contravariant functors that assign to every homomorphism $\varphi : M \longrightarrow N$ a homomorphism $\varphi^* = \operatorname{Hom}_R(\varphi, \prod_{i \in I} A_i)$, and a componentwise homomorphism $\overline{\varphi}$ that sends $(\alpha_i)_{i \in I}$ to $(\alpha_i \circ \varphi)_{i \in I}$. Naturality in M means that the square

$$
\begin{array}{ccc}
\operatorname{Hom}_R\big(M, \prod_{i \in I} A_i\big) & \xrightarrow{\;\theta_M\;} & \prod_{i \in I} \operatorname{Hom}_R(M, A_i) \\[4pt]
{\scriptstyle\varphi^*}\big\uparrow & & \big\uparrow{\scriptstyle\overline{\varphi}} \\[4pt]
\operatorname{Hom}_R\big(N, \prod_{i \in I} A_i\big) & \xrightarrow[\;\theta_N\;]{} & \prod_{i \in I} \operatorname{Hom}_R(N, A_i)
\end{array}
$$

commutes for every φ. Happily, for every $\alpha : N \longrightarrow \prod_{i \in I} A_i$, $\theta\big(\varphi^*(\alpha)\big) = (\pi_i \circ \alpha \circ \varphi)_{i \in I} = \overline{\varphi}\big(\theta(\alpha)\big)$. Thus θ is natural in M.

$\operatorname{Hom}_R(M, \prod_{i \in I} -_i)$ and $\prod_{i \in I} \operatorname{Hom}_R(M, -_i)$ are covariant functors that assign to every family $(\varphi_i)_{i \in I}$ of homomorphisms $\varphi_i : A_i \longrightarrow B_i$ a homomorphism $\varphi_* = \operatorname{Hom}_R(M, \prod_{i \in I} \varphi_i)$, $\alpha \longmapsto (\prod_{i \in I} \varphi_i) \circ \alpha$, and a componentwise homomorphism $\overline{\varphi} = \prod_{i \in I} \varphi_{i*}$ that sends $(\alpha_i)_{i \in I}$ to $(\varphi_i \circ \alpha_i)_{i \in I}$. Naturality in $(A_i)_{i \in I}$ means that the following square

$$
\begin{array}{ccc}
\operatorname{Hom}_R\big(M, \prod_{i \in I} A_i\big) & \xrightarrow{\;\theta_A\;} & \prod_{i \in I} \operatorname{Hom}_R(M, A_i) \\[4pt]
{\scriptstyle\varphi_*}\big\downarrow & & \big\downarrow{\scriptstyle\overline{\varphi}} \\[4pt]
\operatorname{Hom}_R\big(M, \prod_{i \in I} B_i\big) & \xrightarrow[\;\theta_B\;]{} & \prod_{i \in I} \operatorname{Hom}_R(M, B_i)
\end{array}
$$

commutes for every φ_i. Let $\rho_i : \prod_{i \in I} B_i \longrightarrow B_i$ be the projection. Then $\rho_i \circ \prod_{i \in I} \varphi_i = \varphi_i \circ \pi_i$ for all i. For every $\alpha : M \longrightarrow \prod_{i \in I} A_i$,

$$\theta_B(\varphi_*(\alpha)) = (\rho_i \circ (\prod_{i \in I} \varphi_i) \circ \alpha)_{i \in I} = (\varphi_i \circ \pi_i \circ \alpha)_{i \in I}$$
$$= \overline{\varphi}((\pi_i \circ \alpha)_{i \in I}) = \overline{\varphi}(\theta_A(\alpha)).$$

Thus θ is natural in $(A_i)_{i \in I}$. In other words, naturality is straightforward. $\square$

Readers will concoct an entirely similar proof for the next result:

Proposition 2.6. *There is an isomorphism*

$$\mathrm{Hom}_R(\bigoplus_{i \in I} M_i , A) \cong \prod_{i \in I} \mathrm{Hom}_R(M_i, A),$$

which is natural in $(M_i)_{i \in I}$ *and in* A.

Injective modules. We complete this section with the proof of Theorem X.4.11. The main lemma for this will seem more natural after Proposition 6.6.

Lemma 2.7. *Let* A *be an abelian group and let* M *be a left* R-module. *If* $\varphi \in \mathrm{Hom}_{\mathbb{Z}}(M, A)$, *then the mapping* ξ *that sends* $x \in M$ *to* $\xi(x): r \longmapsto \varphi(rx)$ *is a module homomorphism of* M *into* $\mathrm{Hom}_{\mathbb{Z}}(R_R, A)$. *This defines an isomorphism* $\mathrm{Hom}_{\mathbb{Z}}(M, A) \cong \mathrm{Hom}_R(M, \mathrm{Hom}_{\mathbb{Z}}(R_R, A))$ *of abelian groups, which is natural in* M *and* A.

Proof. $\mathrm{Hom}_{\mathbb{Z}}(R_R, A)$ is a left R-module, in which $(r\alpha)s = \alpha(sr)$ for all $r, s \in R$. If $x \in M$, then $\xi(x): r \longmapsto \varphi(rx)$ is in $\mathrm{Hom}_{\mathbb{Z}}(R_R, A)$, since φ is a homomorphism; similarly, $\xi(x + y) = \xi(x) + \xi(y)$, and

$$(r\xi(x))(s) = \xi(x)(sr) = \varphi(srx) = \xi(rx)(s),$$

so that $r\xi(x) = \xi(rx)$. Hence $\xi \in \mathrm{Hom}_R(M, \mathrm{Hom}_{\mathbb{Z}}(R_R, A))$.

To show that $\varphi \longmapsto \xi$ is an isomorphism we construct the inverse isomorphism. Let $\xi : M \longrightarrow \mathrm{Hom}_{\mathbb{Z}}(R_R, A)$ be any module homomorphism. Then $\xi(rx)(s) = (r\xi(x))(s) = \xi(x)(sr)$ for all $x \in M$ and $r, s \in R$. With $s = 1$ this reads $\xi(x)(r) = \xi(rx)(1)$, so that ξ is determined by the additive homomorphism $\overline{\xi} : x \longmapsto \xi(x)(1)$ of M into A. Hence $\Theta : \xi \longmapsto \overline{\xi}$ is an injection of $\mathrm{Hom}_R(M, \mathrm{Hom}_{\mathbb{Z}}(R_R, A))$ into $\mathrm{Hom}_{\mathbb{Z}}(M, A)$; Θ is an additive homomorphism, since $(\xi + \zeta)(x) = \xi(x) + \zeta(x)$. If $\varphi \in \mathrm{Hom}_{\mathbb{Z}}(M, A)$, and $\xi : M \longrightarrow \mathrm{Hom}_{\mathbb{Z}}(R_R, A)$ is the homomorphism constructed above, then $\varphi = \Theta(\xi)$. Hence Θ is an isomorphism.

Naturality in M means that the square

$$
\begin{array}{ccc}
\mathrm{Hom}_R(M, \mathrm{Hom}_{\mathbb{Z}}(R_R, A)) & \xrightarrow{\ \Theta\ } & \mathrm{Hom}_{\mathbb{Z}}(M, A) \\
\psi^* \uparrow & & \uparrow \psi^* \\
\mathrm{Hom}_R(N, \mathrm{Hom}_{\mathbb{Z}}(R_R, A)) & \xrightarrow{\ \Theta\ } & \mathrm{Hom}_{\mathbb{Z}}(N, A)
\end{array}
$$

commutes for every module homomorphism $\psi : M \longrightarrow N$. Let $\xi : N \longrightarrow \mathrm{Hom}_{\mathbb{Z}}(R_R, A)$. Then $\Theta(\xi) = \overline{\xi}$ sends $y \in N$ to $\xi(y)(1)$, and $\psi^*(\Theta(\xi)) =$

$\Theta(\xi) \circ \psi$ sends $x \in M$ to $\overline{\xi}(\psi(x)) = \xi(\psi(x))(1)$, which is exactly where $\Theta(\psi^*(\xi)) = \Theta(\xi \circ \psi)$ sends x. Hence $\Theta \circ \psi^* = \psi^* \circ \Theta$ and Θ is natural in M. Naturality in A is not used and is left to our readers. $\square$

Proposition **2.8.** *If A is a divisible abelian group, then* $\operatorname{Hom}_{\mathbb{Z}}(R_R, A)$ *is an injective left R-module.*

Proof. Let $\psi : M \longrightarrow N$ be any monomorphism of left R-modules. Since D is injective as a $\mathbb{Z}$-module, $\psi^* : \operatorname{Hom}_{\mathbb{Z}}(N, A) \longrightarrow \operatorname{Hom}_{\mathbb{Z}}(M, A)$ is surjective, by 2.4. Since the diagram above commutes,

$$\psi^* : \operatorname{Hom}_R(N, \operatorname{Hom}_{\mathbb{Z}}(R_R, A)) \longrightarrow \operatorname{Hom}_R(M, \operatorname{Hom}_{\mathbb{Z}}(R_R, A))$$

is surjective. Hence $\operatorname{Hom}_{\mathbb{Z}}(R_R, A)$ is injective, by 2.4. $\square$

Theorem X.4.11 can now be proved as follows. Let M be a left R-module. By X.4.10 there is a monomorphism $\mu \in \operatorname{Hom}_{\mathbb{Z}}(M, D)$ into a divisible abelian group D. Then $\operatorname{Hom}_{\mathbb{Z}}(R_R, D)$ is an injective left R-module, by 2.8. By 2.7, the mapping ξ that sends $x \in M$ to $\xi(x): r \longmapsto \mu(rx)$ is a module homomorphism of M into $\operatorname{Hom}_{\mathbb{Z}}(R_R, D)$. If $\xi(x) = 0$, then $\mu(x) = \xi(x)(1) = 0$ and $x = 0$. Hence ξ is injective. $\square$

Exercises

1. Prove that the following conditions are equivalent for a left R-module P: (i) P is projective; (ii) $\operatorname{Hom}_R(P, -)$ preserves epimorphisms; (iii) $\operatorname{Hom}_R(P, -)$ is exact.

2. Give an example of a module M and a short exact sequence $0 \longrightarrow A \longrightarrow B \longrightarrow C \longrightarrow 0$ such that $0 \longrightarrow \operatorname{Hom}_R(M, A) \longrightarrow \operatorname{Hom}_R(M, B) \longrightarrow \operatorname{Hom}_R(M, C) \longrightarrow 0$ is not exact.

3. Prove that $\operatorname{Hom}_R(-, M)$ is left exact.

4. Prove that the following conditions are equivalent for a left R-module J: (i) J is injective; (ii) $\operatorname{Hom}_R(\mu, J)$ is an epimorphism for every monomorphism μ; (iii) $\operatorname{Hom}_R(-, J)$ is exact.

5. Show that a functor that transforms short exact sequences into short exact sequences, also transforms every exact sequence into an exact sequence.

6. Prove that $0 \longrightarrow A \longrightarrow B \longrightarrow C$ is exact if and only if $0 \longrightarrow \operatorname{Hom}_R(M, A) \longrightarrow \operatorname{Hom}_R(M, B) \longrightarrow \operatorname{Hom}_R(M, C)$ is exact for every module M.

7. Prove that $A \longrightarrow B \longrightarrow C \longrightarrow 0$ is exact if and only if $0 \longrightarrow \operatorname{Hom}_R(C, M) \longrightarrow \operatorname{Hom}_R(B, M) \longrightarrow \operatorname{Hom}_R(A, M)$ is exact for every module M.

8. Define direct products of R-S-bimodules and prove their universal property.

9. Show that $\operatorname{Hom}_R(M, \prod_{i \in I} A_i) \cong \prod_{i \in I} \operatorname{Hom}_R(M, A_i)$ is an isomorphism of S-T-bimodule s when M is a R-S-bimodule and $(A_i)_{i \in I}$ are R-T-bimodules.

10. Show that $\operatorname{Hom}_R(M, -)$ preserves pullbacks of left R-modules (if $\varphi \circ \psi' = \psi \circ \varphi'$ is a pullback, then $\varphi_* \circ \psi'_* = \psi_* \circ \varphi'_*$ is a pullback).

11. Show that $\operatorname{Hom}_R(-, A)$ sends pushouts of left R-modules to pullbacks (if $\varphi_1 \circ \psi = \psi_1 \circ \varphi$ is a pushout, then $\psi^* \circ \varphi_1^* = \varphi^* \circ \psi_1^*$ is a pullback).

3. Direct Limits

Direct limits generalize directed unions. We study them for left R-modules, but, unlike most other constructions in this chapter, they are not specific to modules but apply to most algebraic objects.

Direct systems. Direct systems generalize directed families.

Definition. A preordered set or quasiordered set is an ordered pair $(I, \leqq)$ of a set I and a binary relation $\leqq$ on I that is reflexive ($i \leqq i$ for all $i \in I$) and transitive (if $i \leqq j$ and $j \leqq k$, then $i \leqq k$).

For example, every partially ordered set is a preordered set. If $(S_i)_{i \in I}$ is a family of subsets of a set, then I is a preordered set when $i \leqq j$ if and only if $S_i \subseteq S_j$. If $(X, \leqq)$ is a partially ordered set and $\pi : I \longrightarrow X$ is a surjection, then I is a preordered set when $i \leqq j$ if and only if $\pi(i) \leqq \pi(j)$; in fact, this construction yields all preordered sets (see the exercises).

Definition. A preordered set I is directed upward, or just directed, when for every $i, j \in I$ there exists $k \in I$ such that $i \leqq k$ and $j \leqq k$.

If I is directed, then every finite subset of I has an upper bound: for every $i_1, \ldots, i_n \in I$ there exists $k \in I$ such that $i_t \leqq k$ for all t. For example, every totally ordered set is directed; when $(S_i)_{i \in I}$ is a directed family of subsets of a set, then the preordered set I is directed.

Direct systems of sets and direct systems of modules are defined as follows.

Definition. Let I be a preordered set that is directed upward. A direct system of sets (of left R-modules) over I is an ordered pair $\mathcal{A} = (A, \alpha)$ of a family $A = (A_i)_{i \in I}$ of sets (left R-modules) and a family $\alpha = (\alpha_{ij})_{i, j \in I, i \leqq j}$ of mappings (module homomorphisms) $\alpha_{ij} : A_i \longrightarrow A_j (i \leqq j)$ such that α_{ii} is the identity on A_i, and $\alpha_{jk} \circ \alpha_{ij} = \alpha_{ik}$ for all $i \leqq j \leqq k$:

$$\begin{array}{ccc} & A_j \xrightarrow{\ \alpha_{jk}\ } A_k & \\ \alpha_{ij} \Big\uparrow & \nearrow \alpha_{ik} & \\ A_i & & \end{array}$$

Thus, direct systems are large commutative diagrams with arrows running upward. For example, a directed family $(A_i)_{i \in I}$ of subsets of a set can be viewed as a direct system of sets over I (preordered so that $i \leqq j$ if and only if $A_i \subseteq A_j$) in which $\alpha_{ij} : A_i \longrightarrow A_j$ is the inclusion mapping when $i \leqq j$; a directed family of submodules can be viewed as a similar direct system.

Definition. Let $\mathcal{A} = (A, \alpha)$, $\mathcal{B} = (B, \beta)$ be direct systems of sets (of left R-modules) over the same directed preordered set I. A homomorphism φ of $\mathcal{A}$ into $\mathcal{B}$ is a family $\varphi = (\varphi_i)_{i \in I}$ of mappings (module homomorphisms) $\varphi_i : A_i \longrightarrow B_i$ such that $\varphi_j \circ \alpha_{ij} = \beta_{ij} \circ \varphi_i$ whenever $i \leqq j$:

$$A_j \xrightarrow{\varphi_j} B_j$$
$$\alpha_{ij} \Big\uparrow \qquad \Big\uparrow \beta_{ij}$$
$$A_i \xrightarrow{\varphi_i} B_i$$

Homomorphisms of direct systems behave like other homomorphisms. There is an identity homomorphism $1_{\mathcal{A}} = (1_{A_i})_{i \in I} : \mathcal{A} \longrightarrow \mathcal{A}$. If $\varphi = (\varphi_i)_{i \in I} : \mathcal{A} \longrightarrow \mathcal{B}$ and $\psi = (\psi_i)_{i \in I} : \mathcal{B} \longrightarrow \mathcal{C}$ are homomorphisms of direct systems over I, then so is $\psi \circ \varphi = (\psi_i \circ \varphi_i)_{i \in I} : \mathcal{A} \longrightarrow \mathcal{C}$. Homomorphisms of direct systems of left R-modules can be added: if $\varphi = (\varphi_i)_{i \in I}$ and $\psi = (\psi_i)_{i \in I}$ are homomorphisms of $\mathcal{A}$ into $\mathcal{B}$, then so is $\psi + \varphi = (\psi_i + \varphi_i)_{i \in I}$.

Cones. Direct limits arise from cones with a universal property.

Definition. Let $\mathcal{A} = (A, \alpha)$ be a direct system of sets (of left R-modules) over a directed preordered set I. A cone $\varphi : \mathcal{A} \longrightarrow B$ from $\mathcal{A}$ to a set (left R-module) B is a family $\varphi = (\varphi_i)_{i \in I}$ of mappings (module homomorphisms) $\varphi_i : A_i \longrightarrow B$ such that $\varphi_j \circ \alpha_{ij} = \varphi_i$ whenever $i \leqq j$:

$$A_j \xrightarrow{\varphi_j} M$$
$$\alpha_{ij} \Big\uparrow \quad \nearrow \varphi_i$$
$$A_i$$

Equivalently, a cone from $\mathcal{A}$ to B is a homomorphism of $\mathcal{A}$ into the *constant* direct system (B, β) in which $B_i = B$ for all $i \in I$ and $\beta_{ij} = 1_B$ for all $i \leqq j$. For example, if $\mathcal{A} = (A_i)_{i \in I}$ is a directed family of submodules of a module M, and B is a submodule of M that contains $\bigcup_{i \in I} A_i$, then the inclusion homomorphisms $A_i \longrightarrow B$ constitute a cone from $\mathcal{A}$ to B.

If $\varphi = (\varphi_i)_{i \in I}$ is a cone from $\mathcal{A}$ to B and $\psi : B \longrightarrow C$ is a mapping (or a module homomorphism), then $\psi \circ \varphi = (\psi \circ \varphi_i)_{i \in I}$ is a cone from $\mathcal{A}$ to C. A *limit cone* of $\mathcal{A}$ is a cone that yields every cone uniquely in this fashion.

Definitions. Let $\mathcal{A} = (A, \alpha)$ be a direct system of sets (of left R-modules) over a directed preordered set I. A cone $\lambda = (\lambda_i)_{i \in I} : \mathcal{A} \longrightarrow L$ is a limit cone of $\mathcal{A}$, and L is a direct limit of $\mathcal{A}$, $L = \varinjlim \mathcal{A} = \varinjlim_{i \in I} A_i$, when, for every cone $\varphi = (\varphi_i)_{i \in I} : \mathcal{A} \longrightarrow B$, there exists a unique mapping (module homomorphism) $\overline{\varphi} : L \longrightarrow B$ such that $\varphi = \overline{\varphi} \circ \lambda$ ($\varphi_i = \overline{\varphi} \circ \lambda_i$ for all i):

$$L \xdashrightarrow{\overline{\varphi}} B$$
$$\lambda_i \Big\uparrow \quad \nearrow \varphi_i$$
$$A_i$$

Direct limits are also called *inductive limits* and *directed colimits*. Direct limits of groups, rings, fields, and bidules are defined similarly.

If $\lambda : \mathcal{A} \longrightarrow L$ is a limit cone and $\theta : L \longrightarrow M$ is a bijection (or an isomorphism), then $\theta \circ \lambda : \mathcal{A} \longrightarrow M$ is a limit cone. It follows in the usual way from the universal property of λ that this construction yields all limit cones:

Proposition 3.1. *The direct limit and limit cone of a direct system are unique up to isomorphism.*

Readers will verify that directed unions are direct limits:

Proposition 3.2. *If $(A_i)_{i \in I}$ is a directed family of subsets of a set (of submodules of a left R-module), then the inclusion mappings (homomorphisms) $A_i \longrightarrow \bigcup_{i \in I} A_i$ constitute a limit cone, and $\bigcup_{i \in I} A_i = \varinjlim_{i \in I} A_i$.*

For instance, $\mathbb{Z}_{p^\infty} \cong \varinjlim_{n > 0} \mathbb{Z}_{p^n}$. Similarly, the finitely generated submodules of a module M constitute a directed family whose union is M:

Proposition 3.3. *Every left R-module is a direct limit of finitely generated left R-modules.*

A different direct limit arises from any directed family $(A_i)_{i \in I}$ of submodules of a module M, namely, $M/(\bigcup_{i \in I} A_i) = \varinjlim_{i \in I} M/A_i$ (see the exercises).

Construction. We now prove the existence of direct limits.

Proposition 3.4. *Every direct system of sets has a direct limit.*

Proof. Let $\mathcal{A} = (A, \alpha)$ be a direct system of sets over a directed preordered set I. Every cone $\varphi : \mathcal{A} \longrightarrow M$ sends $x \in A_i$ and $\alpha_{ij}(x) \in A_j$ to the same element $\varphi_i(x) = \varphi_j(\alpha_{ij}(x))$ of M. More generally, if $x \in A_i$ and $y \in A_j$ have a common higher image in $\mathcal{A}$ (if $\alpha_{ik}(x) = \alpha_{jk}(y)$ for some $k \geq i, j$), then φ sends x and y to the same element $\varphi_i(x) = \varphi_k(\alpha_{ik}(x)) = \varphi_k(\alpha_{jk}(y)) = \varphi_j(y)$. We construct a limit cone that sends $x \in A_i$ and $y \in A_j$ to the same element if and only if x and y have a common higher image in $\mathcal{A}$.

Let $S = \{ (x, i) \mid i \in I, \ x \in A_i \}$. ($S$ is, up to bijections, the disjoint union of all A_i.) Let $\sim$ be the binary relation on S defined by

$$(x, i) \sim (y, j) \text{ if and only if } \alpha_{ik}(x) = \alpha_{jk}(y) \text{ for some } k \geq i, j$$

(and then $\alpha_{i\ell}(x) = \alpha_{j\ell}(y)$ for all $\ell \geq k$). For instance, $(x, i) \sim (\alpha_{ij}(x), j)$ when $j \geq i$. The relation $\sim$ is an equivalence relation on S: it is symmetric, reflexive since α_{ii} is the identity on A_i, and transitive: if $\alpha_{i\ell}(x) = \alpha_{j\ell}(y)$ for some $\ell \geq i, j$, and $\alpha_{jm}(y) = \alpha_{km}(z)$ for some $m \geq j, k$, then $n \geq \ell, m$ for some $n \in I$, since I is directed, and then $n \geq i, j, k$ and $\alpha_{in}(x) = \alpha_{jn}(y) = \alpha_{kn}(z)$. We denote by cls (x, i) the $\sim$-class of (x, i).

Let $L = S/\sim$ and let $\lambda_i : A_i \longrightarrow L$, $x \longmapsto$ cls (x, i). Then $\lambda_j \circ \alpha_{ij} = \lambda_i$ whenever $j \geq i$, since $(x, i) \sim (\alpha_{ij}(x), j)$; thus $\lambda = (\lambda_i)_{i \in I}$ is a cone from $\mathcal{A}$ to L. Let $\varphi_i : A_i \longrightarrow M$ be mappings such that $\varphi_j \circ \alpha_{ij} = \varphi_i$ whenever $j \geq i$. If $(x, i) \sim (y, j)$, then $\alpha_{ik}(x) = \alpha_{jk}(y)$ for some $k \geq i, j$ and $\varphi_i(x) =$

$\varphi_k(\alpha_{ik}(x)) = \varphi_k(\alpha_{jk}(y)) = \varphi_j(y)$. Therefore a mapping $\overline{\varphi} : L \longrightarrow M$ is well defined by $\overline{\varphi}(\text{cls}\,(x, i)) = \varphi_i(x)$. Then $\varphi_i = \overline{\varphi} \circ \lambda_i$ for all i, and $\overline{\varphi}$ is the only mapping of L into M with this property; so λ is a limit cone of $\mathcal{A}$. $\square$

Proposition **3.5.** *Every direct system of left R-modules has a direct limit.*

Proof. Let $\mathcal{A} = (A, \alpha)$ be a direct system of left R-modules over a directed preordered set I. As a direct system of sets, $\mathcal{A}$ has a limit cone $\lambda : \mathcal{A} \longrightarrow L$, in which $L = S/\!\sim$ and $\lambda_i(x) = \text{cls}\,(x, i)$ for all $x \in A_i$. We show that there is a unique left R-module structure on L such that every λ_i is a module homomorphism, and then λ is a limit cone of $\mathcal{A}$ as a direct system of left R-modules. A similar argument can be used for groups, rings, fields, etc.

We have $L = \{\, \text{cls}\,(x, i) \mid i \in I,\ x \in A_i \,\} = \bigcup_{i \in I} \lambda_i(A_i)$, and this is a directed union, since $i \leqq j$ implies $\lambda_i = \lambda_j \circ \alpha_{ij}$ and $\lambda_i(A_i) \subseteq \lambda_j(A_j)$. Hence every $a, b \in L$ can be written in the form $a = \lambda_i(x)$, $b = \lambda_i(y)$ for some $i \in I$ and $x, y \in A_i$. Then an addition on L is well defined by $a + b = \lambda_i(x + y)$ whenever $a = \lambda_i(x)$ and $b = \lambda_i(y)$: if $\lambda_i(x) = \lambda_j(u)$, $\lambda_i(y) = \lambda_j(v)$, then $\alpha_{ik}(x) = \alpha_{jk}(u)$, $\alpha_{ik}(y) = \alpha_{jk}(v)$ for some $k \geqq i, j$,

$$(x + y,\, i) \ \sim\ (\alpha_{ik}(x + y),\, k) = (\alpha_{ik}(x) + \alpha_{ik}(y),\, k)$$
$$= (\alpha_{jk}(u) + \alpha_{jk}(v),\, k) = (\alpha_{jk}(u + v),\, k) \ \sim\ (u + v,\, j),$$

and $\lambda_i(x + y) = \lambda_j(u + v)$. This addition makes L an abelian group, in which $0 = \text{cls}\,(0, i)$ (for any i) and $- \text{cls}\,(x, i) = \text{cls}\,(-x, i)$.

Similarly, a left action of R on L is well defined by $ra = \lambda_i(rx)$ whenever $a = \lambda_i(x)$: by the above, $\lambda_i(x)$ is any element of L, and $\lambda_i(x) = \lambda_j(y)$ implies $\alpha_{ik}(x) = \alpha_{jk}(y)$ for some $k \geqq i, j$,

$$(rx,\, i) \ \sim\ (\alpha_{ik}(rx),\, k) = (r\,\alpha_{ik}(x),\, k)$$
$$= (r\,\alpha_{jk}(y),\, k) = (\alpha_{jk}(ry),\, k) \ \sim\ (ry,\, j),$$

since α_{ik} and α_{jk} are homomorphisms, and $\lambda_i(x + y) = \lambda_j(u + v)$. It is immediate that this action makes L a left R-module. (Moreover, lucky L received the only module structure such that every λ_i is a module homomorphism.) Now, $\lambda : \mathcal{A} \longrightarrow L$ consists of module homomorphisms and is a cone of $\mathcal{A}$ as a direct system of left R-modules.

Let M be a left R-module and let $\varphi : \mathcal{A} \longrightarrow M$ be a cone. By 3.4 there is a unique mapping $\overline{\varphi} : L \longrightarrow M$ such that $\overline{\varphi} \circ \lambda_i = \varphi_i$ for every $i \in I$. This $\overline{\varphi}$ is a module homomorphism: if $a = \lambda_i(x)$, $b = \lambda_i(y) \in L$, then

$$\overline{\varphi}(a + b) \ =\ \overline{\varphi}(\lambda_i(x + y)) \ =\ \varphi_i(x + y)$$
$$=\ \varphi_i(x) + \varphi_i(y) \ =\ \overline{\varphi}(\lambda_i(x)) + \overline{\varphi}(\lambda_i(y)) \ =\ \overline{\varphi}(a) + \overline{\varphi}(b),$$

since φ_i is a homomorphism; similarly, $\overline{\varphi}(ra) = r\,\overline{\varphi}(a)$ for all $r \in R$ and $a \in A$. Hence $\overline{\varphi}$ is the only homomorphism such that $\overline{\varphi} \circ \lambda_i = \varphi_i$ for all i. $\square$

Readers will be happy to hear that direct limits of modules have a simpler characterization, which makes the construction above superfluous.

Proposition 3.6. *Let* $\mathcal{A} = (A, \alpha)$ *be a direct system of left R-modules over a directed preordered set* I. *A cone* $\varphi : \mathcal{A} \longrightarrow M$ *is a limit cone of* $\mathcal{A}$ *if and only if* (i) $M = \bigcup_{i \in I} \operatorname{Im} \varphi_i$ *and* (ii) $\operatorname{Ker} \varphi_i = \bigcup_{j \geq i} \operatorname{Ker} \alpha_{ij}$ *for every* $i \in I$. *Then* (iii) $\varphi_i(x) = \varphi_j(y)$ *if and only if* $\alpha_{ik}(x) = \alpha_{jk}(y)$ *for some* $k \geq i, j$.

Proof. The limit cone $\lambda : \mathcal{A} \longrightarrow L$ constructed above, $\lambda_i(x) = \operatorname{cls}(x, i)$, has properties (i) $L = \bigcup_{i \in I} \lambda_i(A_i)$ and (iii) $\lambda_i(x) = \lambda_j(y)$ if and only if $\alpha_{ik}(x) = \alpha_{jk}(y)$ for some $k \geq i, j$; hence (ii) $\lambda_i(x) = 0 = \lambda_i(0)$ if and only if $\alpha_{ij}(x) = \alpha_{ij}(0) = 0$ for some $j \geq i$. If $\varphi : \mathcal{A} \longrightarrow M$ is another limit cone, then $\varphi = \theta \circ \lambda$ for some isomorphism $\theta : L \longrightarrow M$, by 3.1; therefore φ also has properties (i), (ii), and (iii).

Conversely, let $\varphi : \mathcal{A} \longrightarrow M$ be a cone with properties (i) and (ii). There is a homomorphism $\overline{\varphi} : L \longrightarrow M$ such that $\varphi_i = \overline{\varphi} \circ \lambda_i$ for all i. Then $\operatorname{Im} \overline{\varphi} = \overline{\varphi}\big(\bigcup_{i \in I} \lambda_i(A_i)\big) = \bigcup_{i \in I} \varphi_i(A_i) = M$ by (i), and $\overline{\varphi}$ is surjective. If $a \in L$ and $\overline{\varphi}(a) = 0$, then $a = \lambda_i(x)$ for some $x \in A_i$, $\varphi_i(x) = 0$, $\alpha_{ij}(x) = 0$ for some $j \geq i$ by (ii), and $a = \lambda_j\big(\alpha_{ij}(x)\big) = 0$; hence $\overline{\varphi}$ is injective. Thus $\overline{\varphi}$ is an isomorphism, and $\varphi = \overline{\varphi} \circ \lambda$ is a limit cone. $\square$

Properties (i) and (ii) show that every direct limit is compounded from two simpler types of direct limits: when $\lambda : \mathcal{A} \longrightarrow L$ is a limit cone, then $L = \bigcup_{i \in I} \operatorname{Im} \lambda_i$ is a directed union; $\bigcup_{j \geq i} \operatorname{Ker} \alpha_{ij}$ is also a directed union, so that $\operatorname{Im} \lambda_i \cong A_i/\operatorname{Ker} \lambda_i \cong \varinjlim_{j \geq i} A_i/\operatorname{Ker} \alpha_{ij}$.

Properties. First, $\varinjlim_{i \in I}$ is an additive functor from direct systems of left R-modules (over I) to left R-modules:

Proposition 3.7. *Let* $\mathcal{A} = (A, \alpha)$ *and* $\mathcal{B} = (B, \beta)$ *be direct systems of left R-modules over the same directed preordered set* I, *with limit cones* $\lambda : \mathcal{A} \longrightarrow L$ *and* $\mu : \mathcal{B} \longrightarrow M$. *Every homomorphism* $\varphi = (\varphi_i)_{i \in I} : \mathcal{A} \longrightarrow \mathcal{B}$ *induces a homomorphism* $\overline{\varphi} = \varinjlim \varphi_i : \varinjlim A_i \longrightarrow \varinjlim B_i$ *unique such that* $\overline{\varphi} \circ \lambda = \mu \circ \varphi$:

$$
\begin{array}{ccc}
L & \overset{\overline{\varphi}}{\dashrightarrow} & M \\
\lambda_i \uparrow & & \uparrow \mu_i \\
A_i & \underset{\varphi_i}{\longrightarrow} & B_i
\end{array}
$$

Moreover, if φ *is the identity on* $\mathcal{A}$, *then* $\overline{\varphi}$ *is the identity on* $\varinjlim A_i$; *if* $\psi : \mathcal{B} \longrightarrow \mathcal{C}$ *is another homomorphism, then* $\overline{\psi \circ \varphi} = \overline{\psi} \circ \overline{\varphi}$; *if* $\chi : \mathcal{A} \longrightarrow \mathcal{B}$ *is another homomorphism, then* $\overline{\chi + \varphi} = \overline{\chi} + \overline{\varphi}$.

Proof. In the statement, $\mu \circ \varphi : \mathcal{A} \longrightarrow M$ is a cone and factors uniquely through λ. The last parts of the statement follow from this uniqueness. $\square$

Proposition 3.8. *Let* $\varphi : \mathcal{A} \longrightarrow \mathcal{B}$ *and* $\psi : \mathcal{B} \longrightarrow \mathcal{C}$ *be homomorphisms*

of direct systems of left R-modules over the same directed preordered set I. If
$A_i \xrightarrow{\varphi_i} B_i \xrightarrow{\psi_i} C_i$ *is exact for every i, then $\varinjlim A_i \longrightarrow \varinjlim B_i \longrightarrow \varinjlim C_i$ is exact.*

Proof. Let $\lambda : \mathcal{A} \longrightarrow L$, $\mu : \mathcal{B} \longrightarrow M$, and $\nu : \mathcal{C} \longrightarrow N$ be limit cones; let $\overline{\varphi} = \varinjlim \varphi_i$, $\overline{\psi} = \varinjlim \psi_i$, so that the diagram below commutes for every $i \in I$:

$$
\begin{array}{ccccc}
L & \xrightarrow{\overline{\varphi}} & M & \xrightarrow{\overline{\psi}} & N \\
\lambda_i \uparrow & & \mu_i \uparrow & & \uparrow \nu_i \\
A_i & \xrightarrow{\varphi_i} & B_i & \xrightarrow{\psi_i} & C_i
\end{array}
$$

Exactness of the top row is proved by diagram chasing, using properties (i) and (ii) in 3.6. First, $\overline{\psi} \circ \overline{\varphi} = 0$, since $\psi_i \circ \varphi_i = 0$ for every i, and $\operatorname{Im} \overline{\varphi} \subseteq \operatorname{Ker} \overline{\psi}$. Let $m \in \operatorname{Ker} \overline{\psi}$. Then $m = \mu_i(b)$ for some $i \in I$ and $b \in B_i$, by (i). Then $\nu_i(\psi_i(b)) = \overline{\psi}(\mu_i(b)) = 0$. Hence $\gamma_{ij}(\psi_i(b)) = 0$ for some $j \geqq i$, by (ii). Then $\psi_j(\beta_{ij}(b)) = \gamma_{ij}(\psi_i(b)) = 0$. By exactness, $\beta_{ij}(b) = \varphi_j(a)$ for some $a \in A_j$. Then $b = \mu_j(\beta_{ij}(b)) = mu_j(\varphi_j(a)) = \overline{\varphi}(\lambda_j(a)) \in \operatorname{Im} \overline{\varphi}$. $\square$

Corollary **3.9.** *If every φ_i is a monomorphism (an epimorphism), then $\varinjlim \varphi_i$ is a monomorphism (an epimorphism).*

The exercises give additional properties.

Exercises

1. Let I be a preordered set. Show that an equivalence relation on I is defined by $i \sim j$ if and only if $i \leqq j$ and $j \leqq i$. Prove that I is a preordered set if and only if there exists a surjection π of I to a partially ordered set J such that $i \leqq j$ in I if and only if $\pi(i) \leqq \pi(j)$ in J.

2. Let I be a preordered set with a greatest element. Show that I is directed. How do you find direct limits over I?

3. Show that the direct limit and limit cone of a direct system of left R-modules are unique up to isomorphism.

4. Let $(A_i)_{i \in I}$ be a directed family of submodules of a left R-module. Show that the inclusion homomorphisms $A_i \longrightarrow \bigcup_{i \in I} A_i$ constitute a limit cone.

5. Explain how $\mathbb{Z}_{p^\infty} \cong \varinjlim_{n>0} \mathbb{Z}_{p^n}$.

6. Show that every direct sum of left R-modules is a direct limit of finite direct sums: $\bigoplus_{i \in I} A_i = \varinjlim_{J \subseteq I, \, J \text{ finite}} \bigoplus_{j \in J} A_j$.

7. Let $(A_i)_{i \in I}$ be a directed family of submodules of a left R-module M. Arrange the quotients M/A_i into a direct system over I; show that $M/\left(\bigcup_{i \in I} A_i\right) = \varinjlim_{i \in I} M/A_i$.

8. For direct systems of left R-modules over a given directed preordered set I, what is the analogue of submodules? of quotient modules?

9. A subset J of a preordered set I is *cofinal* when, for every $i \in I$, J contains some $j \geqq i$. If I is directed, then so is J. Show that a direct system over I, and its restriction to J, have the same direct limit.

10. Define and construct direct limits of (not necessarily abelian) groups.

11. Define and construct direct limits of rings. Show that a direct limit of fields is a field.

12. Let φ be a homomorphism of direct systems of left R-modules. If every φ_i is an epimorphism, then prove, without using exactness, that $\varinjlim \varphi_i$ is an epimorphism.

13. Let φ be a homomorphism of direct systems of left R-modules. If every φ_i is a monomorphism, then prove, without using exactness, that $\varinjlim \varphi_i$ is a monomorphism.

14. Show that direct limits of left R-modules preserve finite direct sums ($\varinjlim (\mathcal{A} \oplus \mathcal{B}) \cong (\varinjlim \mathcal{A}) \oplus (\varinjlim \mathcal{B})$, whenever $\mathcal{A}$ and $\mathcal{B}$ are over the same directed preordered set).

15. Show that direct limits of left R-modules preserve pullbacks (if $\varphi, \varphi', \psi, \psi'$ are homomorphisms of direct systems over the same directed preordered set I, and $\varphi_i \circ \psi_i' = \psi_i \circ \varphi_i'$ is a pullback for every $i \in I$, then $(\varinjlim \varphi_i) \circ (\varinjlim \psi_i') = (\varinjlim \psi_i) \circ (\varinjlim \varphi_i')$ is a pullback).

16. Show that direct limits of left R-modules preserve pushouts (if $\varphi, \varphi', \psi, \psi'$ are homomorphisms of direct systems over the same directed preordered set I, and $\varphi_i' \circ \psi_i = \psi_i' \circ \varphi_i$ is a pushout for every $i \in I$, then $(\varinjlim \varphi_i') \circ (\varinjlim \psi_i) = (\varinjlim \psi_i') \circ (\varinjlim \varphi_i)$ is a pushout).

17. Prove that a ring R is left Noetherian if and only if every direct limit of injective left R-modules is an injective left R-module.

4. Inverse Limits

Inverse limits are obtained from direct limits by reversing all arrows. We study them for left R-modules but, like direct limits, they are not specific to modules but apply to most algebraic objects.

Inverse systems. Inverse systems are direct systems with arrows reversed.

Definition. Let I be a directed preordered set. An inverse system of left R-modules over I is an ordered pair $\mathcal{A} = (A, \alpha)$ of a family $A = (A_i)_{i \in I}$ of left R-modules and a family $\alpha = (\alpha_{ij})_{i,j \in I, i \geq j}$ of module homomorphisms $\alpha_{ij} : A_i \longrightarrow A_j$ ($i \geqq j$) such that α_{ii} is the identity on A_i, and $\alpha_{jk} \circ \alpha_{ij} = \alpha_{ik}$ for all $i \geqq j \geqq k$:

$$
\begin{array}{ccc}
& A_i & \\
{\scriptstyle \alpha_{ij}} \downarrow & & \searrow {\scriptstyle \alpha_{ik}} \\
A_j & \xrightarrow[\alpha_{jk}]{} & A_k
\end{array}
$$

Thus, inverse systems are commutative diagrams with arrows running downward. For example, a descending sequence $A_1 \supseteqq \cdots \supseteqq A_n \supseteqq A_{n+1} \supseteqq \cdots$ of submodules can be viewed as an inverse system over $\mathbb{N}$, in which $\alpha_{nm} : A_n \longrightarrow A_m$ is the inclusion mapping when $n \geqq m$.

Definition. Let $\mathcal{A} = (A, \alpha)$ and $\mathcal{B} = (B, \beta)$ be inverse systems of left R-modules over the same directed preordered set I. A homomorphism $\varphi : \mathcal{A} \longrightarrow \mathcal{B}$ of $\mathcal{A}$ into $\mathcal{B}$ is a family $\varphi = (\varphi_i)_{i \in I}$ of module homomorphisms $\varphi_i : A_i \longrightarrow B_i$ such that $\varphi_j \circ \alpha_{ij} = \beta_{ij} \circ \varphi_i$ whenever $i \geqq j$:

$$
\begin{array}{ccc}
A_i & \xrightarrow{\varphi_i} & B_i \\
{\scriptstyle\alpha_{ij}}\downarrow & & \downarrow{\scriptstyle\beta_{ij}} \\
A_j & \xrightarrow{\varphi_j} & B_j
\end{array}
$$

As in the case of direct systems, there is an identity homomorphism $1_{\mathcal{A}} = (1_{A_i})_{i \in I} : \mathcal{A} \longrightarrow \mathcal{A}$. If $\varphi = (\varphi_i)_{i \in I} : \mathcal{A} \longrightarrow \mathcal{B}$ and $\psi = (\psi_i)_{i \in I} : \mathcal{B} \longrightarrow \mathcal{C}$ are homomorphisms of inverse systems over I, then so is $\psi \circ \varphi = (\psi_i \circ \varphi_i)_{i \in I} : \mathcal{A} \longrightarrow \mathcal{C}$. If $\varphi = (\varphi_i)_{i \in I}$ and $\psi = (\psi_i)_{i \in I}$ are homomorphisms of $\mathcal{A}$ into $\mathcal{B}$, then so is $\psi + \varphi = (\psi_i + \varphi_i)_{i \in I}$.

Inverse limits come from cones with a universal property.

Definition. Let $\mathcal{A} = (A, \alpha)$ be an inverse system of left R-modules over a directed preordered set I. A cone $\varphi : M \longrightarrow \mathcal{A}$ from a left R-module M to $\mathcal{A}$ is a family $\varphi = (\varphi_i)_{i \in I}$ of module homomorphisms $\varphi_i : M \longrightarrow A_i$ such that $\alpha_{ij} \circ \varphi_i = \varphi_j$ whenever $i \geqq j$:

$$
\begin{array}{ccc}
M & \xrightarrow{\varphi_i} & A_i \\
& {\scriptstyle\varphi_j}\searrow & \downarrow{\scriptstyle\alpha_{ij}} \\
& & A_j
\end{array}
$$

Equivalently, a cone from M to $\mathcal{A}$ is a homomorphism of the *constant* inverse system (M, μ) (in which $M_i = M$ for all $i \in I$ and $\mu_{ij} = 1_M$ for all $i \leqq j$) into $\mathcal{A}$. For example, if $\mathcal{A}$ is a descending sequence $A_1 \supseteq \cdots \supseteq A_n \supseteq \cdots$ of submodules, and N is a submodule contained in $\bigcap_{n>0} A_n$, then the inclusion homomorphisms $N \longrightarrow A_n$ constitute a cone from N to $\mathcal{A}$.

If $\varphi = (\varphi_i)_{i \in I}$ is a cone from M to $\mathcal{A}$ and $\psi : N \longrightarrow M$ is a module homomorphism, then $\varphi \circ \psi = (\varphi_i \circ \psi)_{i \in I}$ is a cone from N to $\mathcal{A}$. A *limit cone* of $\mathcal{A}$ is a cone λ that yields every cone φ uniquely in this fashion:

$$
\begin{array}{ccc}
M & \longrightarrow & L \\
& {\scriptstyle\varphi_i}\searrow & \downarrow{\scriptstyle\lambda_i} \\
& & A_i
\end{array}
$$

Definitions. Let $\mathcal{A} = (A, \alpha)$ be an inverse system of left R-modules over a directed preordered set I. A cone $\lambda = (\lambda_i)_{i \in I}: L \longrightarrow \mathcal{A}$ is a limit cone of

$\mathcal{A}$, *and* L *is an* inverse limit *of* $\mathcal{A}$, $L = \varprojlim \mathcal{A} = \varprojlim_{i \in I} A_i$, *when, for every* cone $\varphi = (\varphi_i)_{i \in I}: M \longrightarrow \mathcal{A}$, *there exists a unique module homomorphism* $\overline{\varphi} : M \longrightarrow L$ *such that* $\varphi = \lambda \circ \overline{\varphi}$ ($\varphi_i = \lambda_i \circ \overline{\varphi}$ *for all* i).

Inverse limits are also called *projective limits* and *directed limits*. Inverse limits of sets, groups, rings, and bidules are defined similarly.

If $\lambda : L \longrightarrow \mathcal{A}$ is a limit cone and $\theta : M \longrightarrow N$ is an isomorphism, then $\lambda \circ \theta : M \longrightarrow \mathcal{A}$ is a limit cone. The universal property of λ implies in the usual way that this construction yields all limit cones.

Proposition **4.1.** *The inverse limit and limit cone of an inverse system are unique up to isomorphism.*

The exercises give some examples of inverse limits. For instance, readers will verify that intersections of decreasing sequences are inverse limits:

Proposition **4.2.** *If* $\mathcal{A}$ *is a descending sequence* $A_1 \supseteq \cdots \supseteq A_n \supseteq \cdots$ *of submodules, then the inclusion homomorphisms* $\bigcap_{n>0} A_n \longrightarrow A_n$ *constitute a limit cone of* $\mathcal{A}$.

Construction. The existence of inverse limits is proved as follows.

Proposition **4.3.** *Every inverse system of left R-modules has an inverse limit.*

Proof. Let $\mathcal{A} = (A, \alpha)$ be an inverse system of left R-modules over a directed preordered set I. We retrieve an inverse limit of $\mathcal{A}$ from the direct product $P = \prod_{i \in I} A_i$ and its projections $\pi_i : P \longrightarrow A_i$. Let

$$L = \{ (x_i)_{i \in I} \in P \mid x_j = \alpha_{ij}(x_i) \text{ whenever } i \geqq j \}.$$

Equivalently, $L = \bigcap_{i,j \in I, \, i \geqq j} \mathrm{Ker}\,(\pi_j - \alpha_{ij} \circ \pi_i)$; hence L is a submodule of P. Let $\lambda_i = \pi_{i|L} : (x_i)_{i \in I} \longmapsto x_i$. The definition of L shows that λ is a cone from L to $\mathcal{A}$. Let $\varphi : M \longrightarrow \mathcal{A}$ be any cone. In the diagram:

$$
\begin{array}{ccc}
 & M & \\
\overline{\varphi} \swarrow \quad \overline{\varphi} \downarrow & & \searrow \varphi_i \\
L \xrightarrow[\subseteq]{} P & \xrightarrow[\pi_i]{} & A_i
\end{array}
$$

the homomorphisms $\varphi_i : M \longrightarrow A_i$ induce a unique homomorphism $\overline{\varphi} : M \longrightarrow P$ such that $\pi_i \circ \overline{\varphi} = \varphi_i$ for all i, namely $\overline{\varphi}(x) = (\varphi_i(x))_{i \in I}$ for all $x \in M$. Since φ is a cone, $\varphi_j(x) = \alpha_{ij}(\varphi_i(x))$ whenever $i \geqq j$, and $\overline{\varphi}(x) \in L$ for all $x \in M$. Hence $\overline{\varphi}$ may be viewed as a homomorphism of M into L, and is then the only homomorphism such that $\lambda_i \circ \overline{\varphi} = \varphi_i$ for all i. $\square$

Inverse limits of sets, groups, rings, etc., are constructed similarly. The following properties may be used as a substitute for this construction.

Proposition **4.4.** *Let* $\mathcal{A} = (A, \alpha)$ *be an inverse system of left R-modules over a directed preordered set* I. *A cone* $\varphi : M \longrightarrow \mathcal{A}$ *is a limit cone of* $\mathcal{A}$ *if and* *only if* (i) $\bigcap_{i \in I} \operatorname{Ker} \varphi_i = 0$ *and* (ii) *if* $x_i \in A_i$ *and* $\alpha_{ij}(x_i) = x_j$ *whenever* $i \geqq j$, *then there exists* $x \in M$ *such that* $\varphi_i(x) = x_i$ *for all* i.

Proof. The limit cone $\lambda : L \longrightarrow \mathcal{A}$ constructed in the proof of 4.3 has properties (i) and (ii): if $x \in L$ and $\lambda_i(x) = 0$ for all i, then $x = 0$; if $x_i \in A_i$ and $\alpha_{ij}(x_i) = x_j$ whenever $i \geqq j$, then $x = (x_i)_{i \in I} \in L$ and $\lambda_i(x) = x_i$ for all i. If $\varphi : M \longrightarrow \mathcal{A}$ is another limit cone, then $\varphi = \lambda \circ \theta$ for some isomorphism $\theta : M \longrightarrow L$, by 4.1; therefore φ also has properties (i) and (ii).

Conversely, let $\varphi : M \longrightarrow \mathcal{A}$ be a cone with properties (i) and (ii). There is a homomorphism $\overline{\varphi} : M \longrightarrow L$ such that $\varphi_i = \lambda_i \circ \overline{\varphi}$ for all i. Then $\overline{\varphi}(x) = 0$ implies $\varphi_i(x) = 0$ for all i and $x = 0$, by (i); thus $\overline{\varphi}$ is injective. Let $x \in L$. Then $x = (x_i)_{i \in I} \in \prod_{i \in I} A_i$ and $\alpha_{ij}(x_i) = x_j$ whenever $i \geqq j$, by the construction of L. By (ii) there exists $y \in M$ such that $\varphi_i(y) = x_i$ for all i. Then $\lambda_i(\overline{\varphi}(y)) = \varphi_i(y) = \lambda_i(x)$ for all i and $\overline{\varphi}(y) = x$. Thus $\overline{\varphi}$ is surjective. Hence $\overline{\varphi}$ is an isomorphism, and $\varphi = \lambda \circ \overline{\varphi}$ is a limit cone. $\square$

Properties. First, $\varprojlim_{i \in I}$ is an additive functor from inverse systems of left R-modules (over I) to left R-modules. This is proved like Proposition 3.7:

Proposition 4.5. *Let* $\mathcal{A} = (A, \alpha)$ *and* $\mathcal{B} = (B, \beta)$ *be inverse systems of left R-modules over the same directed preordered set* I, *with limit cones* $\lambda : L \longrightarrow \mathcal{A}$ *and* $\mu : M \longrightarrow \mathcal{B}$. *Every homomorphism* $\varphi = (\varphi_i)_{i \in I}: \mathcal{A} \longrightarrow \mathcal{B}$ *induces a homomorphism* $\overline{\varphi} = \varprojlim \varphi_i : \varprojlim A_i \longrightarrow \varprojlim B_i$ *unique such that* $\mu \circ \overline{\varphi} = \varphi \circ \lambda$:

$$
\begin{array}{ccc}
L & \xrightarrow{\overline{\varphi}} & M \\
{\scriptstyle \lambda_i} \downarrow & & \downarrow {\scriptstyle \mu_i} \\
A_i & \xrightarrow{\varphi_i} & B_i
\end{array}
$$

Moreover: if φ *is the identity on* $\mathcal{A}$, *then* $\overline{\varphi}$ *is the identity on* $\varprojlim A_i$; *if* $\psi : \mathcal{B} \longrightarrow \mathcal{C}$ *is another homomorphism, then* $\overline{\psi \circ \varphi} = \overline{\psi} \circ \overline{\varphi}$; *if* $\psi : \mathcal{A} \longrightarrow \mathcal{B}$ *is another homomorphism, then* $\overline{\psi + \varphi} = \overline{\psi} + \overline{\varphi}$.

Proposition 4.6. *Let* $\varphi : \mathcal{A} \longrightarrow \mathcal{B}$ *and* $\psi : \mathcal{B} \longrightarrow \mathcal{C}$ *be homomorphism of inverse systems of left R-modules over the same directed preordered set* I. *If* $0 \longrightarrow A_i \xrightarrow{\varphi_i} B_i \xrightarrow{\psi_i} C_i$ *is exact for every* i, *then* $0 \longrightarrow \varprojlim A_i \longrightarrow \varprojlim B_i \longrightarrow \varprojlim C_i$ *is exact.*

Proof. Let $\lambda : \mathcal{A} \longrightarrow L$, $\mu : \mathcal{B} \longrightarrow M$, and $\nu : \mathcal{C} \longrightarrow N$ be limit cones; let $\overline{\varphi} = \varprojlim \varphi_i$, $\overline{\psi} = \varprojlim \psi_i$, so that the diagram below commutes for every $i \in I$:

$$
\begin{array}{ccccc}
0 \longrightarrow & L & \xrightarrow{\overline{\varphi}} & M & \xrightarrow{\overline{\psi}} & N \\
& {\scriptstyle \lambda_i} \downarrow & & {\scriptstyle \mu_i} \downarrow & & \downarrow {\scriptstyle \nu_i} \\
0 \longrightarrow & A_i & \xrightarrow{\varphi_i} & B_i & \xrightarrow{\psi_i} & C_i
\end{array}
$$

If $\overline{\varphi}(x) = 0$, then $\varphi_i\big(\lambda_i(x)\big) = \mu_i\big(\overline{\varphi}(x)\big) = 0$ for all i, $\lambda_i(x) = 0$ for all i, and $x = 0$ by (i); thus $\overline{\varphi}$ is injective. Next, $\overline{\psi} \circ \overline{\varphi} = 0$, since $\psi_i \circ \varphi_i = 0$ for every i, and $\operatorname{Im} \overline{\varphi} \subseteq \operatorname{Ker} \overline{\psi}$. Let $m \in \operatorname{Ker} \overline{\psi}$. Then $\psi_i\big(\mu_i(m)\big) = \nu_i\big(\overline{\psi}(m)\big) = 0$ and $\mu_i(m) = \varphi_i(a_i)$ for some $a_i \in A_i$. If $i \geqq j$, then

$$\varphi_j\big(\alpha_{ij}(a_i)\big) = \beta_{ij}\big(\varphi_i(a_i)\big) = \beta_{ij}\big(\mu_i(m)\big) = \mu_j(m) = \varphi_j(a_j)$$

and $\alpha_{ij}(a_i) = a_j$, since φ_j is injective. By (ii) there exists $\ell \in L$ such that $a_i = \lambda_i(\ell)$ for all i. Then $\mu_i\big(\overline{\varphi}(\ell)\big) = \varphi_i\big(\lambda_i(\ell)\big) = \varphi_i(a_i) = \mu_i(m)$ for all i, and $\overline{\varphi}(\ell) = m$ by (i). Thus $m \in \operatorname{Im} \overline{\varphi}$. $\square$

Corollary **4.7.** *An inverse limit of monomorphisms is a monomorphism.*

Corollary 4.7 does not extend to epimorphisms (readers will eagerly pursue counterexamples). This sad shortcoming keeps inverse limits from preserving exact sequences in general. But the exercises give additional properties.

Exercises

1. Show that the inverse limit and limit cone of an inverse system of left R-modules are unique up to isomorphism.

2. Let $\mathcal{A}$ be a descending sequence $A_1 \supseteq \cdots \supseteq A_n \supseteq \cdots$ of submodules. Show that the inclusion homomorphisms $\bigcap_{n>0} A_n \longrightarrow A_n$ constitute a limit cone of $\mathcal{A}$.

3. Define directed intersections of submodules and show that they are inverse limits.

4. Show that every direct product of left R-modules is an inverse limit of finite direct products: $\prod_{i \in I} A_i = \varprojlim_{J \subseteq I, \, J \text{ finite}} \prod_{j \in J} A_j$.

5. Define inverse limits of sets. Show that the inverse limit and limit cone of an inverse system of sets exist and are unique up to isomorphism.

6. Let $\mathcal{A} = (A, \alpha)$ be an inverse system of left R-modules over a directed preordered set I. Let $\lambda : L \longrightarrow A$ be a limit cone of $\mathcal{A}$ regarded as an inverse system of sets. Show that there exists a unique left R-module structure on L such that every λ_i is a module homomorphism, and then λ is a limit cone of $\mathcal{A}$ regarded as an inverse system of left R-modules.

7. Find an inverse limit of epimorphisms that is not an epimorphism.

8. Given a module M and an inverse system $\mathcal{A} = (A, \alpha)$ of modules, find an isomorphism $\operatorname{Hom}_R(M, \varprojlim A_i) \cong \varprojlim \operatorname{Hom}_R(M, A_i)$ that is natural in M and $\mathcal{A}$.

9. Given a module M and a direct system $\mathcal{A} = (A, \alpha)$ of modules, find an isomorphism $\operatorname{Hom}_R(\varinjlim A_i, M) \cong \varprojlim \operatorname{Hom}_R(A_i, M)$ that is natural in M and $\mathcal{A}$.

10. Show that inverse limits of modules preserve finite direct sums ($\varprojlim (\mathcal{A} \oplus \mathcal{B}) \cong (\varprojlim \mathcal{A}) \oplus (\varprojlim \mathcal{B})$, whenever $\mathcal{A}$ and $\mathcal{B}$ are over the same directed preordered set).

11. Show that inverse limits of modules preserve pullbacks (if $\varphi, \varphi', \psi, \psi'$ are homomorphisms of inverse systems over the same directed preordered set I, and $\varphi_i \circ \psi_i' = \psi_i \circ \varphi_i'$ is a pullback for every $i \in I$, then $(\varprojlim \varphi_i) \circ (\varprojlim \psi_i') = (\varprojlim \psi_i) \circ (\varprojlim \varphi_i')$ is a pullback).

12. Prove the following: a group G is an inverse limit of finite groups if and only if $\{1\}$ is the intersection of all normal subgroups of G of finite index. (These groups are called *profinite*; for example, all Galois groups are profinite.)

5. Tensor Products

This section gives the construction and first examples of tensor products. Properties are in the next section. As in the rest of this chapter, all rings have an identity element, and all modules are unital.

Bilinear mappings. Tensor products originate with vector spaces. Let V be a vector space over a field K with a finite basis $e_1, \ldots, e_m$, and let W be a vector space over K with a finite basis $f_1, \ldots, f_n$. The tensor product $V \otimes W$ is a vector space of dimension mn over K, with a basis $e_1 \otimes f_1, e_2 \otimes f_2, \ldots, e_m \otimes f_n$. Every $x = \sum_i x_i\, e_i \in V$ and $y = \sum_j x_j\, y_j \in W$ have a tensor product $x \otimes y = \sum_{i,j} x_i\, y_j\, e_i \otimes f_j$ in $V \otimes W$. The tensor map $\tau : (x, y) \longmapsto x \otimes y$ is bilinear: $(x + x') \otimes y = (x \otimes y) + (x' \otimes y)$, $x \otimes (y + y') = (x \otimes y) + (x \otimes y')$, and $(\lambda x) \otimes y = \lambda(x \otimes y) = x \otimes (\lambda y)$.

Conversely, if β is a bilinear mapping of $V \times W$ into another vector space W, then $\beta(x, y) = \beta\left(\sum_i x_i\, e_i,\ \sum_j y_j\, f_j\right) = \sum_{i,j} \left(x_i\, y_j\, \beta(e_i, f_j)\right)$ by bilinearity, and the linear transformation $T : V \otimes W \longrightarrow W$ that sends $e_i \otimes f_j$ to $\beta(e_i, f_j)$ also sends $x \otimes y$ to $\beta(x, y)$, for all $x \in V$ and $y \in W$. In fact, T is the only linear transformation such that $\beta = T \circ \tau$. In this sense every bilinear mapping of $V \times W$ factors uniquely through τ.

Tensor products of modules over a commutative ring R can be defined by the same universal property (which does not require bases). For the record:

Definition. Let R be a commutative ring and let A, B, C be R-modules. A mapping $\beta : A \times B \longrightarrow C$ is bilinear *when*

$$\beta(a + a',\, b) = \beta(a, b) + \beta(a', b),$$
$$\beta(a,\, b + b') = \beta(a, b) + \beta(a, b'),\quad and$$
$$\beta(ra, b) = r\,\beta(a, b) = \beta(a, rb),$$

for all $a, a' \in A$, $b, b' \in B$, and $r \in R$.

If $\beta : A \times B \longrightarrow C$ is bilinear, then the mappings $\beta(a, -): b \longmapsto \beta(a, b)$, $B \longrightarrow C$ and $\beta(-, b): a \longmapsto \beta(a, b)$, $A \longrightarrow C$ are module homomorphisms. This property characterizes bilinear mappings:

Proposition 5.1. Let R be a commutative ring and let A, B, C be R-modules. For a mapping $\beta : A \times B \longrightarrow C$ the following conditions are equivalent:

(1) *β is bilinear;*

(2) *$a \longmapsto \beta(a, -)$ is a module homomorphism of A into $\mathrm{Hom}_R(B, C)$;*

(3) *$b \longmapsto \beta(-, b)$ is a module homomorphism of B into $\mathrm{Hom}_R(A, C)$.*

Bihomomorphisms. Bilinear mappings can be defined in the same way for left R-modules over an arbitrary ring R, but then lose properties (2) and (3) above, if only because $\mathrm{Hom}_R(A, C)$ and $\mathrm{Hom}_R(B, C)$ are only abelian groups. It is more fruitful to keep properties (2) and (3), and to forgo bilinearity unless R is commutative. In the simplest form of (2) and (3), C is an abelian group; if B is a left R-module, then $\mathrm{Hom}_\mathbb{Z}(B, C)$ is a right R-module and A needs to be a right R-module; then B and $\mathrm{Hom}_\mathbb{Z}(A, C)$ are left R-modules.

Proposition 5.2. *Let R be a ring, let A be a right R-module, let B be a left R-module, and let C be an abelian group. For a mapping $\beta : A \times B \longrightarrow C$ the following conditions are equivalent:*

(1) *for all $a, a' \in A$, $b, b' \in B$, and $r \in R$,*
$$\beta(a + a', b) = \beta(a, b) + \beta(a', b),$$
$$\beta(a, b + b') = \beta(a, b) + \beta(a, b') \quad (\beta \text{ is biadditive}),$$
$$\beta(ar, b) = \beta(a, rb) \qquad (\beta \text{ is balanced});$$

(2) $a \longmapsto \beta(a, -)$ *is a module homomorphism of A into $\mathrm{Hom}_\mathbb{Z}(B, C)$;*

(3) $b \longmapsto \beta(-, b)$ *is a module homomorphism of B into $\mathrm{Hom}_\mathbb{Z}(A, C)$.*

Proof. Addition on $\mathrm{Hom}_\mathbb{Z}(B, C)$ is pointwise and that $r \in R$ acts on $\varphi \in \mathrm{Hom}_\mathbb{Z}(B, C)$ by $(\varphi r)(b) = \varphi(rb)$ for all $b \in B$; hence (1) states that $\beta(a + a', -) = \beta(a, -) + \beta(a', -)$, $\beta(a, -) \in \mathrm{Hom}_\mathbb{Z}(B, C)$, and $\beta(ar, -) = \beta(a, -) r$, and is equivalent to (2). The equivalence of (1) and (3) is similar. $\square$

Balanced biadditive mappings have been called *middle linear* mappings, *R-biadditive* mappings, and *balanced products* (a good name). The author prefers *bihomomorphisms:*

Definition. A bihomomorphism *of modules is a mapping that satisfies the equivalent conditions in Proposition* 5.2.

For example, the left action $(r, x) \longmapsto rx$ of R on any left R-module M is a bihomomorphism of $R_R \times M$ into the underlying abelian group M.

The tensor product. If $\beta : A \times B \longrightarrow C$ is a bihomomorphism and $\varphi : C \longrightarrow D$ is a homomorphism of abelian groups, then $\varphi \circ \beta : A \times B \longrightarrow D$ is a bihomomorphism. The *tensor product* of A and B is an abelian group $A \otimes_R B$ with a bihomomorphism τ of $A \times B$, from which every bihomomorphism of $A \times B$ can be recovered uniquely in this fashion:

$$
\begin{array}{ccc}
A \times B & & \\
\tau \downarrow & \searrow \beta & \\
A \otimes_R B & \dashrightarrow[\bar{\beta}] & C
\end{array}
$$

Definition. Let A be a right R-module and let B be a left R-module. A tensor product *of A and B is an abelian group $A \otimes_R B$ together with a bihomomorphism $\tau : A \times B \longrightarrow A \otimes_R B$, $(a, b) \longmapsto a \otimes b$, the* tensor map, *such that, for* *every abelian group* C *and bihomomorphism* $\beta : A \times B \longrightarrow C$ *there exists a unique homomorphism* $\overline{\beta} : A \otimes_R B$ *of abelian groups such that* $\beta = \overline{\beta} \circ \tau$.

Proposition **5.3.** *For every right R-module* A *and left R-module* B, $A \otimes_R B$ *and its tensor map exist, and they are unique up to isomorphism.*

Proof. Uniqueness follows from the universal property. Existence is proved by constructing a tensor product. Let $T = F/K$, where F is the free abelian group on the set $A \times B$, and K is the subgroup of F generated by all $(a + a', b) - (a, b) - (a', b)$, $(a, b + b') - (a, b) - \beta(a, b')$, and $(ar, b) - (a, rb) \in F$, where $a, a' \in A$, $b, b' \in B$, and $r \in R$. Let $\tau : A \times B \longrightarrow F \longrightarrow F/K$ be the canonical mapping. The definition of K shows that $\tau(a + a', b) = \tau(a, b) + \tau(a', b)$, $\tau(a, b + b') = \tau(a, b) + \tau(a, b')$, and $\tau(ar, b) = \tau(a, rb)$, for all $a, a' \in A$, $b, b' \in B$, and $r \in R$. Thus τ is a bihomomorphism.

Every mapping $\beta : A \times B \longrightarrow C$ extends to a homomorphism $\varphi : F \longrightarrow C$. If β is a bihomomorphism, then

$$\varphi\big((a + a', b) - (a, b) - (a', b)\big) = \beta(a + a', b) - \beta(a, b) - \beta(a', b) = 0$$

and $(a + a', b) - (a, b) - (a', b) \in \operatorname{Ker} \varphi$; similarly, $(a, b + b') - (a, b) - (a, b') \in \operatorname{Ker} \varphi$, and $(ar, b) - (a, rb) \in \operatorname{Ker} \varphi$. Thus, $\operatorname{Ker} \varphi$ contains every generator of K. Hence $\operatorname{Ker} \varphi$ contains K, and φ factors through the projection $\pi : F \longrightarrow F/K$: $\varphi = \overline{\beta} \circ \pi$ for some homomorphism $\overline{\beta} : F/K \longrightarrow C$:

$$
\begin{array}{ccc}
A \times B & \xrightarrow{\ \beta\ } & C \\
\subseteq \big\downarrow & \nearrow\varphi & \big\uparrow \overline{\beta} \\
F & \xrightarrow[\ \pi\ ]{} & F/K
\end{array}
$$

Then $\overline{\beta} \circ \tau = \beta$. Moreover, $\overline{\beta}$ is the only homomorphism $\overline{\beta} : F/K \longrightarrow C$ such that $\overline{\beta} \circ \tau = \beta$, since $F/K = \pi(F)$ is generated by $\pi(A \times B) = \tau(A \times B)$. Thus T is a tensor product of A and B, with tensor map τ. $\square$

Corollary **5.4.** (1) *Every element of* $A \otimes_R B$ *is a finite sum* $\sum_i (a_i \otimes b_i)$, *where* $a_i \in A$ *and* $b_i \in B$. (2) *If* $\sum_i (a_i \otimes b_i) = 0$ *in* $A \otimes_R B$, *then* $\sum_i (a_i \otimes b_i) = 0$ *in* $A' \otimes_R B'$ *for some finitely generated submodules* $A' \subseteq A$ *and* $B' \subseteq B$.

Proof. (1). Let $T = F/K$ as above. Every element of F is a finite linear combination $\sum_i n_i (a_i, b_i)$ with integer coefficients. Hence every element of $T = \pi(F)$ is a finite linear combination $\sum_i n_i (a_i \otimes b_i)$, and a finite sum $\sum_i (n_i a_i \otimes b_i)$ by bilinearity. Then (1) holds in every $A \otimes_R B \cong T$, by 5.3.

(2). If $\sum_i (a_i \otimes b_i) = 0$ in $A \otimes_R B$, then $\sum_i (a_i \otimes b_i) = 0$ in T and $\sum_i (a_i, b_i) \in K$ is a linear combination with integer coefficients of finitely many generators k_j of K. Let A' be the submodule of A generated by the finitely many elements of A that appear in $\sum_i (a_i, b_i)$ and in the generators k_j; let B' be the similar submodule of B. Let $T' = F'/K'$ be constructed as above from

A' and B'. Then K' contains all k_j, $\sum_i (a_i, b_i) \in K'$, $\sum_i (a_i \otimes b_i) = 0$ in T', and $\sum_i (a_i \otimes b_i) = 0$ in any $A' \otimes_R B' \cong T'$. $\square$

In general, not every element of $A \otimes_R B$ can be put in the form $a \otimes b$ (see the exercises). Corollary 5.4 excepted, the contruction above is not very helpful: only in a few cases will we know exactly what $A \otimes_R B$ looks like. Hence tensor products are usually manipulated through their properties, not their construction.

Homomorphisms. Tensor products yield additive functors, as follows.

Proposition 5.5. *If $\varphi : A \longrightarrow A'$ is a homomorphism of right R-modules, and $\psi : B \longrightarrow B'$ is a homomorphism of left R-modules, then there is a unique homomorphism $\varphi \otimes \psi : A \otimes_R B \longrightarrow A' \otimes_R B'$ such that*

$$(\varphi \otimes \psi)(a \otimes b) = \varphi(a) \otimes \psi(b)$$

for all $a \in A$ and $b \in B$:

$$
\begin{array}{ccc}
A \times B & \xrightarrow{\varphi \times \psi} & A' \times B' \\
{\scriptstyle \tau} \downarrow & & \downarrow {\scriptstyle \tau'} \\
A \otimes_R B & \dashrightarrow{\varphi \otimes \psi} & A' \otimes_R B'
\end{array}
$$

Moreover, $1_A \otimes 1_B = 1_{A \otimes_R B}$, $(\varphi \circ \varphi') \otimes (\psi \circ \psi') = (\varphi \otimes \psi) \circ (\varphi' \otimes \psi')$, $(\varphi + \varphi') \otimes \psi = (\varphi \otimes \psi) + (\varphi' \otimes \psi)$, and $\varphi \otimes (\psi + \psi') = (\varphi \otimes \psi) + (\varphi \otimes \psi')$, whenever defined. Thus, for every right R-module A, $A \otimes_R -$ is an additive functor from left R-modules to abelian groups; for every left R-module B, $- \otimes_R B$ is an additive functor from right R-modules to abelian groups.

Proof. Let $\tau : A \times B \longrightarrow A \otimes_R B$ and $\tau' : A' \times B' \longrightarrow A' \otimes_R B'$ be the tensor maps. Since τ' is a bihomomorphism, $\tau' \circ (\varphi \times \psi): (a, b) \longmapsto \varphi(a) \otimes \psi(b)$ is a bihomomorphism and factors uniquely through τ. Uniqueness in this factorization yields the properties in the second part of the statement. The last part follows, with $A \otimes \psi = 1_A \otimes \psi$ and $\varphi \otimes B = \varphi \otimes 1_B$. $\square$

If $\varphi : A \longrightarrow A'$ and $\psi : B \longrightarrow B'$ are homomorphisms, then the square below commutes, since $(\varphi \otimes 1_{B'}) \circ (1_A \otimes \psi) = \varphi \otimes \psi = (1_{A'} \otimes \psi) \circ (\varphi \otimes 1_B)$; hence $A \otimes \psi$ is natural in A and $\varphi \otimes B$ is natural in B.

$$
\begin{array}{ccc}
A \otimes_R B & \xrightarrow{A \otimes_R \psi} & A \otimes_R B' \\
{\scriptstyle \varphi \otimes B} \downarrow & & \downarrow {\scriptstyle \varphi \otimes B'} \\
A' \otimes_R B & \xrightarrow{A' \otimes_R \psi} & A' \otimes_R B'
\end{array}
$$

Bimodules. So far tensor products of modules are only abelian groups. As was the case for $\mathrm{Hom}_R(A, B)$, module structures on $A \otimes_R B$ arise from bimodule structures on A and B. Tensor products of bimodules can also be defined in terms of enhanced bihomomorphisms. These two approaches yield the same tensor products. First we define bihomomorphisms of bimodules.

Proposition **5.6.** *Let* R, S, T *be rings; let* A *be a left S-, right R-bimodule, let* B *be a left R-, right T-bimodule, and let* C *be a left S-, right T-bimodule. For a mapping* $\beta : A \times B \longrightarrow C$ *the following conditions are equivalent:*

(1) *for all* $a, a' \in A,\, b, b' \in B, r \in R, s \in S, and\ t \in T:$

$$\beta(a + a',\, b) = \beta(a, b) + \beta(a', b),$$
$$\beta(a,\, b + b') = \beta(a, b) + \beta(a, b'),$$
$$\beta(sa, b) = s\,\beta(a, b),\quad \beta(ar, b) = \beta(a, rb),\quad \beta(a, bt) = \beta(a, b)\,t;$$

(2) $a \longmapsto \beta(a, -)$ *is a bimodule homomorphism of* A *into* $\mathrm{Hom}_T (B, C);$

(3) $b \longmapsto \beta(b, -)$ *is a bimodule homomorphism of* B *into* $\mathrm{Hom}_S (A, C).$

In (2), A and $\mathrm{Hom}_T (B, C)$ are S-R-bimodule, by 6.4, and similarly for B and $\mathrm{Hom}_S (A, C)$ in (3). A *bihomomorphism* of bimodules is a mapping that satisfies the equivalent conditions in Proposition 5.6.

Proposition **5.7.** *Let* A *be a right R-module and let* B *be a left R-module.*

(1) *if* A *is an S-R-bimodule, then* $A \otimes_R B$ *is a left S-module, in which* $s\,(a \otimes b) = sa \otimes b$ *for all* $a \in A, b \in B, and\ s \in S;$

(2) *if* B *is an R-T-bimodule, then* $A \otimes_R B$ *is a right T-module, in which* $(a \otimes b)\,t = a \otimes bt$ *for all* $a \in A, b \in B, and\ t \in T.$

If A *is an S-R-bimodule and* B *is an R-T-bimodule, then:*

(3) $A \otimes_R B$ *is an S-T-bimodule;*

(4) *the tensor map* $A \times B \longrightarrow A \otimes_R B$ *is a bihomomorphism of bimodules;*

(5) *for every S-T-bimodule* C *and bihomomorphism* $\beta : A \times B \longrightarrow C$ *of bimodules, there exists a unique bimodule homomorphism* $\overline{\beta} : A \otimes_R B \longrightarrow C$ *such that* $\beta(a, b) = \overline{\beta}(a \otimes b)$ *for all* $a \in A$ *and* $b \in B;$

(6) *if* $\varphi : A \longrightarrow A'$ *and* $\psi : B \longrightarrow B'$ *are bimodule homomorphisms, then* $\varphi \otimes \psi$ *is a bimodule homomorphism;*

(7) $A \otimes_R -$ *is an additive functor from R-T-bimodules to S-T-bimodules, and* $- \otimes_R B$ *is an additive functor from S-R-bimodules to S-T-bimodules.*

Proof. We prove (1), (2), and (3), and leave the other parts to eager readers. Let A be an S-R-bimodule. If $s \in S$, then $\alpha_s : a \longmapsto sa$ is a right R-module endomorphism of A. By 5.5, there is a unique endomorphism $\overline{\alpha}_s = \alpha_s \otimes 1_B$ of $A \otimes_R B$ such that $\overline{\alpha}_s (a \otimes b) = \alpha_s(a) \otimes b = sa \otimes b$ for all $a \in A$ and $b \in B$. Since $s \longmapsto \alpha_s$ is a ring homomorphism, the addition and composition properties in 5.5 imply that $s \longmapsto \overline{\alpha}_s$ is a ring homomorphism of S into $\mathrm{End}_\mathbb{Z}(A \otimes_R B)$. Thus $A \otimes_R B$ is a left S-module, in which $s\,(a \otimes b) = sa \otimes b$ for all s, a, b.

Similarly, when B is an R-T-bimodule, then $A \otimes_R B$ is a right T-module, in which $(a \otimes b)\,t = a \otimes bt$ for all a, b, t. In (3), $s(xt) = (sx)t$ holds for all $s \in S$, $t \in T$, and $x \in A \otimes_R B$, since it holds for every generator $x = a \otimes b$:

$$s\big((a \otimes b)\,t\big) = s\,(a \otimes bt) = sa \otimes bt = \big(s\,(a \otimes b)\big)t. \quad \square$$

Thus $_S A_R$ and $_R B_T$ implies $_S (A \otimes_R B)_T$. If R is commutative, then R-modules are R-R-bimodules and Proposition 5.7 becomes nicer:

Corollary **5.8.** *If R is a commutative ring and A, B are R-modules, then:*

(1) *$A \otimes_R B$ is an R-module;*

(2) *the tensor map $(a, b) \longmapsto a \otimes b$ is bilinear;*

(3) *for every R-module C and bilinear mapping $\beta : A \times B \longrightarrow C$, there exists a unique module homomorphism $\overline{\beta} : A \otimes_R B \longrightarrow C$ such that $\beta(a, b) = \overline{\beta}(a \otimes b)$ for all $a \in A$ and $b \in B$;*

(4) *if $\varphi : A \longrightarrow A'$ and $\psi : B \longrightarrow B'$ are R-module homomorphisms, then $\varphi \otimes \psi$ is an R-module homomorphism;*

(5) *$A \otimes_R -$ and $- \otimes_R B$ are additive functors from R-modules to R-modules.*

Applications. We give two applications of tensor products of bimodules. First are tensor products by free modules, one of the rare cases in which we know what $A \otimes_R B$ looks like. A free right R-module F, with a basis $(e_i)_{i \in I}$, is an R-R-bimodule, in which $r\left(\sum_{i \in I} e_i x_i\right) = \sum_{i \in I} e_i (r x_i)$ for all $r, x_i \in R$. If R is commutative, this yields the usual R-R-bimodule on any right R-module. In general, the left R-module structure on F depends on the choice of a basis.

Similarly, a free left R-module F with a basis $(e_i)_{i \in I}$ is an R-R-bimodule, in which $\left(\sum_{i \in I} x_i e_i\right) r = \sum_{i \in I} (x_i r) e_i$ for all $r, x_i \in R$. In general, this bimodule structure depends on the choice of a basis. So do the isomorphisms in the next result.

Proposition **5.9.** *If F is a free right R-module with a basis $(e_i)_{i \in I}$, and B is a left R-module, then $F \otimes_R B \cong \bigoplus_{i \in I} B$; the isomorphism sends $\sum_{i \in I} (e_i \otimes b_i)$ to $(b_i)_{i \in I}$, and is natural in B. In particular, $R_R \otimes_R B \cong B$.*

If F is a free left R-module with a basis $(e_i)_{i \in I}$, and A is a right R-module, then $A \otimes_R F \cong \bigoplus_{i \in I} A$; the isomorphism sends $\sum_{i \in I} (a_i \otimes e_i)$ to $(a_i)_{i \in I}$, and is natural in A. In particular, $A \otimes_R {}_R R \cong A$.

Proof. First we show that $F \otimes_R B \cong \bigoplus_{i \in I} B$ as abelian groups. The equality $\tau\left(\sum_{i \in I} e_i x_i, b\right) = (x_i b)_{i \in I}$ defines a bihomomorphism $\tau : F \times B \longrightarrow \bigoplus_{i \in I} B$. We show that every bihomomorphism $\beta : F \times B \longrightarrow C$ factors uniquely through τ. Indeed, β induces an additive homomorphism $\overline{\beta} : (b_i)_{i \in I} \longmapsto \sum_{i \in I} \beta(e_i, b_i)$ of $\bigoplus_{i \in I} B$ into C. Then $\overline{\beta} \circ \tau = \beta$:

$$\beta\left(\sum_{i \in I} e_i x_i, b\right) = \sum_{i \in I} \beta(e_i x_i, b) = \sum_{i \in I} \beta(e_i, x_i b)$$
$$= \overline{\beta}\left((x_i b)_{i \in I}\right) = \overline{\beta}\left(\tau\left(\sum_{i \in I} e_i x_i, b\right)\right).$$

Now, $\tau(e_i, -)$ is the i injection $\iota_i : B \longrightarrow \bigoplus_{i \in I} B$. If $\beta = \varphi \circ \tau$ for some other additive homomorphism $\varphi : \bigoplus_{i \in I} B \longrightarrow C$, then $\varphi \circ \iota_i = \varphi(\tau(e_i, -)) = \beta(e_i, -) = \overline{\beta} \circ \iota_i$; hence $\varphi = \overline{\beta}$.

Since $F \otimes_R B$ and its tensor map are unique up to isomorphism, there is an additive isomorphism $\theta : F \otimes_R B \longrightarrow \bigoplus_{i \in I} B$ such that

$$\theta\big(\big(\textstyle\sum_{i \in I} e_i\, x_i\big) \otimes b\big) = \tau\big(\textstyle\sum_{i \in I} e_i\, x_i\, ,\, b\big) = (x_i\, b)_{i \in I}$$

for all $\sum_{i \in I} e_i\, x_i \in F$ and $b \in B$. Now, $F \otimes_R B$ is a left R-module by 5.7, since F is an R-R-bimodule. Then

$$\theta\big(r\big(\big(\textstyle\sum_{i \in I} e_i\, x_i\big) \otimes b\big)\big) = \theta\big(\big(\textstyle\sum_{i \in I} e_i\, (rx_i)\big) \otimes b = (rx_i\, b)_{i \in I}$$
$$= r\,(x_i\, b)_{i \in I} = r\,\theta\big(\big(\textstyle\sum_{i \in I} e_i\, x_i\big) \otimes b\big)$$

and $\theta(rt) = r\theta(t)$ for every generator $t = \big(\sum_{i \in I} e_i\, x_i\big) \otimes b$ of $F \otimes_R B$. Hence $\theta(rt) = r\theta(t)$ for all $r \in R$ and $t \in F \otimes_R B$, and θ is a module isomorphism. Readers will verify that θ is natural in B. The other isomorphism $A \otimes_R F \cong \bigoplus_{i \in I} A$ is similar (or follows from the first, by 6.1 below). $\square$

Change of rings. Readers know that a vector space over $\mathbb{R}$ can be enlarged to a vector space over $\mathbb{C}$: when $(e_i)_{i \in I}$ is a basis of V over $\mathbb{R}$, V consists of all linear combinations $\sum_{i \in I} x_i\, e_i$ with real coefficients, and can be enlarged to the vector space of all linear combinations $\sum_{i \in I} x_i\, e_i$ with complex coefficients. Similarly, a vector space over a field K can be enlarged to a vector space over any field extension of K. Tensor products provide a general construction that does not depend on bases and has a universal property.

Let $\rho : R \longrightarrow S$ be a ring homomorphism. Every left S-module M is also a left R-module, in which $r \cdot x = \rho(r)\, x$ for all $r \in R$ and $x \in M$; the R-module structure on M is the composition $R \longrightarrow S \longrightarrow \mathrm{End}_{\mathbb{Z}}(M)$. For example, vector spaces over $\mathbb{C}$ are also vector spaces over $\mathbb{R}$. The converse construction "enlarges" R-modules into S-modules.

Proposition **5.10.** *Let M be a left R-module and let $\rho : R \longrightarrow S$ be a homomorphism of rings [with identity].*

(1) *$S \otimes_R M$ is a left S-module, and $\iota : x \longmapsto 1 \otimes x$ is a homomorphism of left R-modules of M into $S \otimes_R M$;*

(2) *every R-module homomorphism of M into a left S-module factors uniquely through ι;*

(3) *if $(e_i)_{i \in I}$ is a basis of M, then $\big(\iota(e_i)\big)_{i \in I}$ is a basis of $S \otimes_R M$.*

Proof. (1). S is an S-R-bimodule, in which $s \cdot r = s\, \rho(r)$ for all $s \in S$ and $r \in R$. By 5.7, $S \otimes_R M$ is a left S-module. Hence $S \otimes_R M$ is also a left R-module, in which $r\,(s \otimes x) = \rho(r)\, s \otimes x$ for all r, s, x. Then $r\,\iota(x) = \rho(r) \otimes x = 1 \cdot r \otimes x = 1 \otimes rx = \iota(rx)$ for all r and x.

(2). Let N be an S-module and let $\varphi : M \longrightarrow N$ be a homomorphism of left R-modules. Then $s\, \varphi(rx) = s\, \rho(r)\, \varphi(x) = (s \cdot r)\, \varphi(x)$; hence $\beta : (s, x) \longmapsto s\, \varphi(x)$, $S \times M \longrightarrow N$ is a bihomomorphism of bimodules of $S \times M$ into the S-$\mathbb{Z}$-bimodule N and induces a unique S-module homomorphism $\psi : S \otimes_R M \longrightarrow N$ such that $\psi\,(s \otimes x) = s\, \varphi(x)$ for all s, x; ψ is the only S-module homomorphism such that $\psi \circ \iota = \varphi$, since this equality implies $\psi(s \otimes x) = s\, \psi(1 \otimes x) = s\, \varphi(x)$:

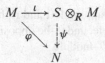

$$M \xrightarrow{\iota} S \otimes_R M$$

(3). If M has a basis $(e_i)_{i \in I}$, then 5.9 yields an isomorphism $\theta : S \otimes_R M \cong \bigoplus_{i \in I} S$, which sends $\sum_{i \in I} (s_i \otimes e_i)$ to $(s_i)_{i \in I}$, and every element of $S \otimes_R M$ can be written uniquely in the form $\sum_{i \in I} (s_i \otimes e_i) = \sum_{i \in I} s_i \, \iota(e_i)$. $\square$

Exercises

1. Prove the following: if A and B are finitely generated modules, then $A \otimes_R B$ is a finitely generated abelian group.

2. Give a direct proof that $R_R \otimes_R M \cong M$ for every left R-module M.

3. Let M be a left R-module and let I be a two-sided ideal of R. Show that there is an isomorphism $R/I \otimes_R M \cong M/IM$ that is natural in M.

4. Show that $\mathbb{Z}_m \otimes_{\mathbb{Z}} \mathbb{Z}_n \cong \mathbb{Z}_d$, where $d = \gcd(m, n)$.

5. Let R and S be rings. Make $R \otimes_{\mathbb{Z}} S$ a ring, in which $(r \otimes s)(r' \otimes s') = rr' \otimes ss'$.

6. For any abelian group M, produce a one-to-one correspondence bewteen R-S-bimodule structures on M and left $R \otimes_{\mathbb{Z}} S^{\mathrm{op}}$-module structures on M.

7. Let A be an S-R-bimodule, let B be an R-T-bimodule, and let C be an S-T-bimodule. Construct an abelian group $\mathrm{Bihom}\,(A \times B, C)$ of bihomomorphisms of bimodules of $A \times B$ into C.

8. Study the functorial properties of $\mathrm{Bihom}\,(A \times B, C)$ in the previous exercise.

In the next two exercises, $\rho : R \longrightarrow S$ is a ring homomorphism.

9. For every left S-modules A and B construct a monomorphism $\mathrm{Hom}_S\,({}_S A, {}_S B) \longrightarrow \mathrm{Hom}_R\,({}_R A, {}_R B)$ that is natural in A and B.

10. For every right S-module A and left S-module B construct an epimorphism $A_R \otimes_R {}_R B \longrightarrow A_S \otimes_S {}_S B$ that is natural in A and B.

In the next three exercises, R is commutative, M is an R-module, and S is a proper multiplicative subset of R.

11. Construct a *module of fractions* $S^{-1}M$.

12. State and prove a universal property of the canonical homomorphism $M \longrightarrow S^{-1}M$.

13. Prove that $S^{-1}R \otimes_R M \cong S^{-1}M$.

6. Properties of Tensor Products

This section contains basic properties of tensor products of modules, including commutativity, associativity, adjoint associativity, and right exactness.

First we look at the tensor product as a binary operation. The natural isomorphisms $A \otimes_R R \cong A$ and $R \otimes_R B \cong B$ in Proposition 5.9 provide an identity element of sorts, the R-R-bimodule R.

Commutativity requires swapping left and right modules. Recall that a right R-module A is a left R^{op}-module, under the opposite action $r * a = ar$; similarly, a left R-module B is a right R^{op}-module.

Proposition **6.1.** *For every right R-module A and left R-module B there is a commutativity isomorphism $B \otimes_{R^{\mathrm{op}}} A \cong A \otimes_R B$, which sends $b \otimes a$ to $a \otimes b$ and is natural in A and B. If A and B are bimodules, then the commutativity isomorphism is a bimodule isomorphism.*

Proof. If C is an abelian group, then β is a bihomomorphism of $A_R \times {}_R B$ into C if and only if $\beta^{\mathrm{op}} : (b, a) \longmapsto \beta(a, b)$ is a bihomomorphism of $B_{R^{\mathrm{op}}} \times {}_{R^{\mathrm{op}}} A$ into C. In particular, $\tau^{\mathrm{op}} : (b, a) \longmapsto \tau(a, b) = a \otimes b$ is a bihomomorphism of $B_{R^{\mathrm{op}}} \times {}_{R^{\mathrm{op}}} A$ into $A \otimes_R B$ and induces a unique homomorphism $\theta : B \otimes_{R^{\mathrm{op}}} A \longrightarrow A \otimes_R B$ such that $\theta (b \otimes a) = a \otimes b$ for all $a \in A$ and $b \in B$. Similarly, there is a unique homomorphism $\zeta : A \otimes_R B \longrightarrow B \otimes_{R^{\mathrm{op}}} A$ such that $\zeta (a \otimes b) = b \otimes a$ for all $a \in A$ and $b \in B$. Then θ and ζ are mutually inverse isomorphisms: $\theta\big(\zeta(a \otimes b)\big) = a \otimes b$ for all a, b; hence $\theta\big(\zeta(t)\big) = t$ for all $t \in A \otimes_R B$. Similarly, $\zeta\big(\theta(u)\big) = u$ for all $u \in B \otimes_{R^{\mathrm{op}}} A$. Naturality and the bimodule case are left to readers. $\square$

Associativity requires at least one bimodule B, so that $(A \otimes_R B) \otimes_S C$ and $A \otimes_R (B \otimes_S C)$ are defined.

Proposition **6.2.** *For every right R-module A, R-S-bimodule B, and left S-module C, there is an associativity isomorphism $(A \otimes_R B) \otimes_S C \cong A \otimes_R (B \otimes_S C)$, which sends $(a \otimes b) \otimes c$ to $a \otimes (b \otimes c)$ and is natural in A, B, and C. If A and C are bimodules, then the associativity isomorphism is a bimodule isomorphism.*

Proof. We prove the bimodule case and let A be a Q-R-bimodule and C be an S-T-bimodule, so that $(A \otimes_R B) \otimes_S C$ and $A \otimes_R (B \otimes_S C)$ are Q-T-bimodules; the first part of the statement is the case $Q = T = \mathbb{Z}$. For every $a \in A$ and $b \in B$, $\beta(a, b) : c \longmapsto a \otimes (b \otimes c)$ is a right T-module homomorphism of C into $A \otimes_R (B \otimes_S C)$. Moreover,

$$\beta(ar, b)(c) = ar \otimes (b \otimes c) = a \otimes r(b \otimes c) = a \otimes (rb \otimes c) = \beta(a, rb)(c)$$

for all a, b, c, r; similarly, $q\,\beta(a, b)(c) = \beta(qa, b)(c)$ and $\big(\beta(a, b)s\big)(c) = \beta(a, b)(sc) = \beta(a, bs)(c)$ for all a, b, c, q, s. Hence

$$\beta : A \times B \longrightarrow \mathrm{Hom}_T \big(C,\ A \otimes_R (B \otimes_S C)\big)$$

is a bihomomorphism of bimodules, and there is a unique homomorphism

$$\overline{\beta} : A \otimes_R B \longrightarrow \mathrm{Hom}_T \big(C,\ A \otimes_R (B \otimes_S C)\big)$$

of bimodules such that $\overline{\beta}(a \otimes b) = \beta(a, b)$ for all a, b. By 5.6, $(u, c) \longmapsto \overline{\beta}(u)(c)$ is a bihomomorphism of $(A \otimes_R B) \times C$ into $A \otimes_R (B \otimes_S C)$. Hence there is a bimodule homomorphism

$$\theta : (A \otimes_R B) \otimes_S C \longrightarrow A \otimes_R (B \otimes_S C)$$

such that $\theta \, (u \otimes c) = \overline{\beta}(u)(c)$ for all $u \in A \otimes_R B$ and $c \in C$; in particular, $\theta \left((a \otimes b) \otimes c \right) = \beta(a, b)(c) = a \otimes (b \otimes c)$ for all a, b, c. Similarly, there is a bimodule homomorphism $\zeta : A \otimes_R (B \otimes_S C) \longrightarrow (A \otimes_R B) \otimes_S C$ such that $\zeta \left(a \otimes (b \otimes c) \right) = (a \otimes b) \otimes c$ for all a, b, c. Then θ and ζ are mutually inverse isomorphisms. Readers will verify naturality. $\square$

If R is commutative, then Proposition 5.9, 6.1, and 6.2 yield module isomorphisms $R \otimes_R A \cong A$, $B \otimes_R A \cong A \otimes_R B$, and $(A \otimes_R B) \otimes_R C \cong A \otimes_R (B \otimes_R C)$, for all R-modules A, B, and C. Moreover, the tensor maps $(a, b, c) \longmapsto (a \otimes b) \otimes c$ and $(a, b, c) \longmapsto a \otimes (b \otimes c)$ are trilinear.

Longer tensor products. The above suggests that $A_1 \otimes \cdots \otimes A_n$ can be constructed directly. If R is commutative and $A_1, \ldots, A_n$, C are R-modules, then a mapping $\mu : A_1 \times \cdots \times A_n \longrightarrow C$ is *n-linear* or *multilinear* when

$$\mu \, (a_1, \ldots, a_{i-1}, -, a_{i+1}, \ldots, a_n) : a_i \longmapsto \mu \, (a_1, \ldots, a_i, \ldots, a_n)$$

is a module homomorphism of A_i into C, for every i. In fact, μ is multilinear if and only if, for any i, the mapping

$$(a_1, \ldots, a_{i-1}, a_{i+1}, \ldots, a_n) \longmapsto \mu \, (a_1, \ldots, a_{i-1}, -, a_{i+1}, \ldots, a_n)$$

of $A_1 \times \cdots \times A_{i-1} \times A_{i+1} \times \cdots \times A_n$ into $\mathrm{Hom}_R(A_i, C)$ is $(n-1)$-linear.

If R is commutative, then a *tensor product* of n R-modules $A_1, \ldots, A_n$ is an R-module $A_1 \otimes_R \cdots \otimes_R A_n$ with an n-linear mapping $\tau : (a_1, \ldots, a_n) \longmapsto a_1 \otimes \cdots \otimes a_n$ of $A_1 \times \cdots \times A_n$ into $A_1 \otimes_R \cdots \otimes_R A_n$, the *tensor map*, such that for every R-module C and n-linear mapping $\gamma : A_1 \times \cdots \times A_n \longrightarrow C$, there exists a unique homomorphism $\overline{\gamma} : A_1 \otimes_R \cdots \otimes_R A_n \longrightarrow C$ such that $\overline{\gamma} \circ \tau = \gamma$. Readers will easily prove the following.

Proposition 6.3. *Let R be a commutative ring and let $A_1, \ldots, A_n$ be R-modules. A tensor product $A_1 \otimes_R \cdots \otimes_R A_n$ and its tensor map exist, and they are unique up to isomorphism.*

For bimodules in general, we define *multihomomorphisms* $A_1 \times \cdots \times A_n \longrightarrow C$ of bimodules, also called *balanced products*. We do this when $n = 3$, and leave the general case and the proofs to interested readers.

Proposition 6.4. *Let Q, R, S, T be rings; let A be a left Q-, right R-bimodule, let B be a left R-, right S-bimodule, let C be a left S-, right T-bimodule, and let D be a left Q-, right T-bimodule. For a mapping $\gamma : A \times B \times C \longrightarrow D$ the following conditions are equivalent:*

(1) *for all* $a, a' \in A$, $b, b' \in B$, $c, c' \in C$, $q \in Q$, $r \in R$, $s \in S$, $t \in T$,

$$\gamma(a + a', b, c) = \gamma(a, b, c) + \gamma(a', b, c),$$
$$\gamma(a, b + b', c) = \gamma(a, b, c) + \gamma(a, b', c),$$
$$\gamma(a, b, c + c') = \gamma(a, b, c) + \gamma(a, b, c'),$$
$$\gamma(qa, b, c) = q \, \gamma(a, b, c),$$
$$\gamma(ar, b, c) = \gamma(a, rb, c),$$

$$\gamma(a, bs, c) = \gamma(a, b, sc),$$
$$\gamma(a, b, ct) = \gamma(a, b, c)\, t;$$

(2) $(a, b) \longmapsto \gamma(a, b, -)$ is a bihomomorphism of bimodules of $A \times B$ into $\mathrm{Hom}_T(C, D)$;

(3) $(b, c) \longmapsto \gamma(-, b, c)$ is a bihomomorphism of bimodules of $B \times C$ into $\mathrm{Hom}_Q(A, D)$.

A *trihomomorphism* of $A \times B \times C$ into D is a mapping that satisfies the equivalent conditions above. For example, the tensor maps $(a, b, c) \longmapsto (a \otimes b) \otimes c$ and $(a, b, c) \longmapsto a \otimes (b \otimes c)$ in Proposition 6.2 are trihomomorphisms.

Let A be a Q-R-bimodule, B be an R-S-bimodule, and C be an S-T-bimodule. A *tensor product* of A, B, and C is a Q-T-bimodule $A \otimes_R B \otimes_S C$ together with a trihomomorphism $\tau : A \times B \times C \longrightarrow A \otimes_R B \otimes_S C$, $(a, b, c) \longmapsto a \otimes b \otimes c$, the *tensor map*, such that for every Q-T-bimodule D and trihomomorphism $\gamma : A \times B \times C \longrightarrow D$ there exists a unique homomorphism $\overline{\gamma} : A \otimes_R B \otimes_S C \longrightarrow D$ of bimodules such that $\gamma = \overline{\gamma} \circ \tau$.

Proposition 6.5. *For every Q-R-bimodule A, R-S-bimodule B, and S-T-bimodule C, a tensor product $A \otimes_R B \otimes_S C$ and its tensor map exist, and they are unique up to isomorphism.*

To Proposition 6.2 can be added natural isomorphisms $(A \otimes_R B) \otimes_S C \cong A \otimes_R B \otimes_S C \cong A \otimes_R (B \otimes_S C)$, which send $(a \otimes b) \otimes c$ to $a \otimes b \otimes c$ and to $a \otimes (b \otimes c)$. These triple tensor products are normally written without parentheses.

Adjoint associativity is directly related to the definition of tensor products.

Proposition 6.6 (Adjoint Associativity). *Let A be a right R-module, let B be a left R-module, and let C be an abelian group. There are* adjoint associativity isomorphisms

$$\Theta : \mathrm{Hom}_{\mathbb{Z}}(A \otimes_R B, \ C) \cong \mathrm{Hom}_R\big(A, \ \mathrm{Hom}_{\mathbb{Z}}(B, C)\big),$$
$$\Xi : \mathrm{Hom}_{\mathbb{Z}}(A \otimes_R B, \ C) \cong \mathrm{Hom}_R\big(B, \ \mathrm{Hom}_{\mathbb{Z}}(A, C)\big),$$

which are natural in A, B, and C. For all $\varphi \in \mathrm{Hom}_{\mathbb{Z}}(A \otimes_R B, \ C)$, $\big(\Theta(\varphi)(a)\big)(b) = \big(\Xi(\varphi)(b)\big)(a) = \varphi(a \otimes b)$.

Proof. The set $\mathrm{Bihom}(A \times B, \ C)$ of all bihomomorphisms of $A \times B$ into C is an abelian group under pointwise addition. The universal property of the tensor map $\tau : (a, b) \longmapsto a \otimes b$ provides a bijection $\varphi \longmapsto \varphi \circ \tau$ of $\mathrm{Hom}_{\mathbb{Z}}(A \otimes_R B, \ C)$ onto $\mathrm{Bihom}(A \times B, \ C)$, which preserves pointwise addition. Proposition 5.2 provides two more bijections:

$$\mathrm{Bihom}(A \times B, C) \longrightarrow \mathrm{Hom}_R\big(A, \mathrm{Hom}_{\mathbb{Z}}(B, C)\big),$$
$$\mathrm{Bihom}(A \times B, C) \longrightarrow \mathrm{Hom}_R\big(B, \mathrm{Hom}_{\mathbb{Z}}(A, C)\big),$$

which send $\beta \in \mathrm{Bihom}(A \times B, \ C)$ to the homomorphisms $a \longmapsto \beta(a, -)$ in $\mathrm{Hom}_R\big(A, \ \mathrm{Hom}_{\mathbb{Z}}(B, C)\big)$ and $b \longmapsto \beta(-, b)$ in $\mathrm{Hom}_R\big(B, \ \mathrm{Hom}_{\mathbb{Z}}(A, C)\big)$, and preserve pointwise addition. Composing with $\mathrm{Hom}_{\mathbb{Z}}(A \otimes_R B, C) \longrightarrow$ $\mathrm{Bihom}(A \times B, C)$ yields Θ and Ξ. Readers will establish naturality. $\square$

There is a bimodule version of Proposition 6.6 (see the exercises). If R is commutative, then Θ and Ξ are module isomorphisms

$$\mathrm{Hom}_R(A \otimes_R B, C) \cong \mathrm{Hom}_R(A, \mathrm{Hom}_R(B, C)),$$
$$\mathrm{Hom}_R(A \otimes_R B, C) \cong \mathrm{Hom}_R(B, \mathrm{Hom}_R(A, C)).$$

Right exactness. Adjoint associativity yields other properties.

Proposition **6.7** (Right Exactness). *For every right R-module M and left R-module N, the functors $M \otimes_R -$ and $- \otimes_R N$ are right exact: if $A \longrightarrow B \longrightarrow C \longrightarrow 0$ is exact, then so is*

$$M \otimes_R A \longrightarrow M \otimes_R B \longrightarrow M \otimes_R C \longrightarrow 0;$$

if $A' \longrightarrow B' \longrightarrow C' \longrightarrow 0$ is exact, then so is

$$A' \otimes_R N \longrightarrow B' \otimes_R N \longrightarrow C' \otimes_R N \longrightarrow 0.$$

Proof. We prove the first half of the statement; then 6.1 yields the second half. Let $A \xrightarrow{\varphi} B \xrightarrow{\psi} C \longrightarrow 0$ be exact; let $\overline{\varphi} = 1_M \otimes \varphi$ and $\overline{\psi} = 1_M \otimes \psi$.

$$
\begin{array}{ccc}
0 & & 0 \\
\downarrow & & \downarrow \\
\mathrm{Hom}_{\mathbb{Z}}(M \otimes_R C, G) & \longrightarrow & \mathrm{Hom}_R(M, \mathrm{Hom}_{\mathbb{Z}}(C, G)) \\
\overline{\psi}^* \downarrow & & \downarrow \\
\mathrm{Hom}_{\mathbb{Z}}(M \otimes_R B, G) & \longrightarrow & \mathrm{Hom}_R(M, \mathrm{Hom}_{\mathbb{Z}}(B, G)) \\
\overline{\varphi}^* \downarrow & & \downarrow \\
\mathrm{Hom}_{\mathbb{Z}}(M \otimes_R A, G) & \longrightarrow & \mathrm{Hom}_R(M, \mathrm{Hom}_{\mathbb{Z}}(A, G))
\end{array}
$$

In the diagram above, G is any abelian group, $\overline{\varphi}^* = \mathrm{Hom}_{\mathbb{Z}}(M, \overline{\varphi})$, $\overline{\psi}^* = \mathrm{Hom}_{\mathbb{Z}}(M, \overline{\psi})$, and the horizontal arrows are adjoint associativity isomorphisms from 6.6. The diagram commutes, since the latter are natural, and the right column is exact by 2.3 and 2.1. Therefore, the left column is exact. Suitable choices of G now yield exactness of $M \otimes_R A \xrightarrow{\overline{\varphi}} M \otimes_R B \xrightarrow{\overline{\psi}} M \otimes_R C \longrightarrow 0$.

Let $\pi : M \otimes_R C \longrightarrow \mathrm{Coker}\, \overline{\psi} = (M \otimes_R C)/\mathrm{Im}\, \overline{\psi}$ be the projection. Then $\overline{\psi}^*(\pi) = \pi \circ \overline{\psi} = 0$. Hence $\pi = 0$, $\mathrm{Im}\, \overline{\psi} = M \otimes_R C$, and $\overline{\psi}$ is surjective.

Let $G = \mathrm{Coker}\, \overline{\varphi} = (M \otimes_R B)/\mathrm{Im}\, \overline{\varphi}$ and let $\rho : M \otimes_R B \longrightarrow G$ be the projection:

$$
\begin{array}{ccc}
M \otimes_R A & \xrightarrow{\overline{\varphi}} & M \otimes_R B & \xrightarrow{\overline{\psi}} & M \otimes_R C \\
 & & \rho \searrow & \downarrow \chi \\
 & & & G
\end{array}
$$

Then $\overline{\varphi}^*(\rho) = \rho \circ \overline{\varphi} = 0$. Therefore $\rho \in \mathrm{Im}\ \overline{\psi}^*$ and $\rho = \overline{\psi}^*(\chi) = \chi \circ \overline{\psi}$ for some homomorphism $\chi : M \otimes_R C \longrightarrow G$. Hence $\mathrm{Ker}\ \overline{\psi} \subseteq \mathrm{Ker}\ \rho = \mathrm{Im}\ \overline{\varphi}$. Conversely, $\mathrm{Im}\ \overline{\varphi} \subseteq \mathrm{Ker}\ \overline{\psi}$, since $\overline{\psi} \circ \overline{\psi} = 1_M \otimes (\psi \circ \varphi) = 1_M \otimes 0 = 0$. $\square$

The functors $M \otimes_R -$ and $- \otimes_R N$ are in general not exact: readers will show that they may fail to preserve monomorphisms. Modules M such that $- \otimes_R M$ is exact are studied in Section 8.

Readers may prove a two-variable analogue of Proposition 6.7:

Proposition 6.8. *If $A \longrightarrow B \longrightarrow C \longrightarrow 0$ and $A' \longrightarrow B' \longrightarrow C' \longrightarrow 0$ are exact, then so is*

$$(A \otimes_R B') \oplus (A' \otimes_R B) \longrightarrow B \otimes_R B' \longrightarrow C \otimes_R C' \longrightarrow 0.$$

Direct sums. The functors $M \otimes_R -$ and $- \otimes_R N$ preserve direct sums:

Proposition 6.9. *There are natural isomorphisms $M \otimes_R \left(\bigoplus_{i \in I} B_i \right) \cong \bigoplus_{i \in I} (M \otimes_R B_i)$ and $\left(\bigoplus_{i \in I} A_i \right) \otimes_R N \cong \bigoplus_{i \in I} (A_i \otimes_R N)$, which send $x \otimes (b_i)_{i \in I}$ to $(x \otimes b_i)_{i \in I}$ and $(a_i)_{i \in I} \otimes y$ to $(a_i \otimes y)_{i \in I}$.*

Proof. Let M be a right R-module and let $(A_i)_{i \in I}$ be a family of left R-modules. The injection $\iota_i : A_i \longrightarrow \bigoplus_{i \in I} A_i$ induces a homomorphism $\overline{\iota}_i = 1_M \otimes \iota_i : M \otimes_R A_i \longrightarrow M \otimes_R \left(\bigoplus_{i \in I} A_i \right)$. For every abelian group G there is a commutative diagram

$$
\begin{array}{ccc}
 & \prod_{i \in I} \mathrm{Hom}_R(M, \mathrm{Hom}_{\mathbb{Z}}(A_i, G)) & \\
\pi_i \swarrow & & \uparrow \cong \\
\mathrm{Hom}_R(M, \mathrm{Hom}_{\mathbb{Z}}(A_i, G)) \xleftarrow{\ \iota_i^*\ } & \mathrm{Hom}_R(M, \mathrm{Hom}_{\mathbb{Z}}(\bigoplus_{i \in I} A_i, G)) \\
\cong \uparrow & & \uparrow \cong \\
\mathrm{Hom}_{\mathbb{Z}}(M \otimes_R A_i, G) \xleftarrow{\ \overline{\iota}_i^*\ } & \mathrm{Hom}_{\mathbb{Z}}(M \otimes_R (\bigoplus_{i \in I} A_i), G)
\end{array}
$$

in which $\iota_i^* = \mathrm{Hom}_R\left(M, \mathrm{Hom}_{\mathbb{Z}}(\iota_i, G)\right)$ and $\overline{\iota}_i^* = \mathrm{Hom}_R(\overline{\iota}_i, G)$, 2.6 and 2.5 provide the top triangle, and the remaining two vertical arrows are adjoint associativity isomorphisms. Since the diagram commutes, there is for every family $(\varphi_i)_{i \in I}$ of additive homomorphisms $\varphi_i : M \otimes_R A_i \longrightarrow G$ a unique homomorphism $\varphi : M \otimes_R \left(\bigoplus_{i \in I} A_i \right) \longrightarrow G$ such that $\varphi_i = \overline{\iota}_i^*(\varphi)$ for all i, equivalently, $\varphi_i = \varphi \circ \iota_i$ for all i. This universal property characterizes the direct sum and its injections. Therefore there is an additive isomorphism $\theta : \bigoplus_{i \in I} (M \otimes_R A_i) \longrightarrow M \otimes_R \left(\bigoplus_{i \in I} A_i \right)$ such that $\theta \circ \kappa_i = \overline{\iota}_i$ for all i, where $\kappa_i : M \otimes_R A_i \longrightarrow \bigoplus_{i \in I} (M \otimes_R A_i)$ is the injection:

$$
\begin{array}{ccc}
M \otimes_R A_i & \xrightarrow{\ \kappa_i\ } & \bigoplus_{i \in I} (M \otimes_R A_i) \\
 & \overline{\iota}_i \searrow & \downarrow \theta \\
 & & M \otimes_R \left(\bigoplus_{i \in I} A_i \right)
\end{array}
$$

Then $\theta\big((m \otimes a_i)_{i \in I}\big) = m \otimes (a_i)_{i \in I}$. Naturality is straightforward. $\square$

If M, N, $(A_i)_{i \in I}$, and $(B_i)_{i \in I}$ are bimodules, then the isomorphisms in 6.8 are bimodule isomorphisms.

Direct limits. The functors $M \otimes_R -$ and $- \otimes_R N$ also preserve direct limits. The following result is proved like 6.8.

Proposition 6.10. *There are natural isomorphisms* $M \otimes_R \big(\varinjlim_{i \in I} B_i\big) \cong \varinjlim_{i \in I} (M \otimes_R B_i)$ *and* $\big(\varinjlim_{i \in I} A_i\big) \otimes_R N \cong \varinjlim_{i \in I} (A_i \otimes_R N)$.

Exercises

1. Show that the associativity isomorphism $(a \otimes b) \otimes c \longmapsto a \otimes (b \otimes c)$ is natural.

2. Find a definition of multihomomorphisms of bimodules by equivalent properties.

3. Prove the existence of triple tensor products $A \otimes_R B \otimes_S C$ of bimodules.

4. Let R be a commutative ring and let $A_1, \ldots, A_n$ be R-modules. Prove the existence and uniqueness of $A_1 \otimes_R \cdots \otimes_R A_n$ and its tensor map.

5. Choose one of the two adjoint associativity isomorphisms and prove that it is natural.

6. Let ${}_S A_R$, ${}_R B_T$, and ${}_S C_T$ be bimodules. Obtain natural isomorphisms

$$\mathrm{Hom}_{ST}\left(A \otimes_R B,\, C\right) \cong \mathrm{Hom}_{SR}\left(A,\, \mathrm{Hom}_T(B, C)\right),$$
$$\mathrm{Hom}_{ST}\left(A \otimes_R B,\, C\right) \cong \mathrm{Hom}_{RT}\left(B,\, \mathrm{Hom}_S(A, C)\right).$$

(Hom_{ST}, Hom_{SR}, and Hom_{RT} are abelian groups of bimodule homomorphisms.)

7. Choose one of the two adjoint associativity isomorphisms and prove that it is a module isomorphism when R is commutative.

8. Prove directly that $- \otimes_R N$ is right exact.

9. Prove the following: if M is a projective left R-module, then $- \otimes_R M$ is exact.

10. Find a monomorphism $A \longrightarrow B$ of abelian groups such that $\mathbb{Z}_2 \otimes_\mathbb{Z} A \longrightarrow \mathbb{Z}_2 \otimes_\mathbb{Z} B$ is not a monomorphism.

11. Prove the following: if ψ and ψ are epimorphisms, then $\varphi \otimes \psi$ is an epimorphism.

12. Use right exactness to give another proof that $R/I \otimes_R M \cong M/IM$ when I is a two-sided ideal of R.

13. Prove the following: if $A \longrightarrow B \longrightarrow C \longrightarrow 0$ is an exact sequence of right R-modules, and $A' \longrightarrow B' \longrightarrow C' \longrightarrow 0$ is an exact sequence of left R-modules, then

$$(A \otimes_R B') \oplus (A' \otimes_R B) \longrightarrow B \otimes_R B' \longrightarrow C \otimes_R C' \longrightarrow 0$$

is exact.

14. Prove the following: if A is a projective right R-module and B is a projective left R-module, then $A \otimes_R B$ is a projective (= free) abelian group.

15. Let M be a right R-module and let $\mathcal{A} = (A, \alpha)$ be a direct system of left R-modules over a directed preordered set I. Show that there is a natural isomorphism $M \otimes_R \big(\varinjlim_{i \in I} A_i\big) \cong \varinjlim_{i \in I} (M \otimes_R A_i)$.

16. Show that $M \otimes_R -$ preserves pushouts (if $\beta' \circ \alpha = \alpha' \circ \beta$ is a pushout, then so is $(M \otimes \beta') \circ (M \otimes \alpha) = (M \otimes \alpha') \circ (M \otimes \beta)$).

17. A submodule N of a left R-module M is *pure* when, for every $n > 0$ and $a_1, \ldots, a_n \in N$, every finite system of linear equations $\sum_j r_{ij} a_j = a_i$ with a solution in M has a solution in N. Show that a submodule N of a left R-module M is pure if and only if $A \otimes_R N \longrightarrow A \otimes_R M$ is injective for every right R-module A.

7. Dual Modules

This section extends to modules some familiar properties of vector spaces, and provides additional examples of tensor products, for use in the next section.

Definitions. The dual *of a left R-module M is the right R-module* $M^* = \mathrm{Hom}_R(M, {}_R R)$; *the* dual *of a right R-module N is the left R-module* $N^* = \mathrm{Hom}_R(N, R_R)$.

The R-R-bimodule structure on R provides the right action of R on M^*, $(\alpha r)(x) = \alpha(rx)$ for all $\alpha \in M^*$, $r \in R$, $x \in M$, and the left action of R on N^*, $(r\beta)(y) = \beta(yr)$, for all $\beta \in N^*$, $r \in R$, $y \in N$.

By Propositions 1.6, 2.1, 2.3, and 2.6, $\mathrm{Hom}_R(-, {}_R R)$ and $\mathrm{Hom}_R(-, R_R)$ are additive contravariant functors, from left R-modules to right R-modules and vice versa; both functors are left exact (if $A \longrightarrow B \longrightarrow C \longrightarrow 0$ is exact, then $0 \longrightarrow C^* \longrightarrow B^* \longrightarrow A^*$ is exact); $\left(\bigoplus_{i \in I} A_i \right)^* \cong \prod_{i \in I} A_i^*$; in particular, $(A_1 \oplus \cdots \oplus A_n)^* \cong A_1^* \oplus \cdots \oplus A_n^*$, and we have:

Proposition **7.1.** *If F is a free left (or right) R-module with a finite basis $(e_i)_{i \in I}$, then F^* is free with a finite basis $(e_i^*)_{i \in I}$, such that $e_i^*(e_i) = 1$, $e_i^*(e_j) = 0$ for all $j \neq i$.*

Then $(e_i^*)_{i \in I}$ is the *dual basis* of the given basis $(e_i)_{i \in I}$. Proposition 7.1 does not extend to all free modules: for instance, $\left(\bigoplus_{i \in I} \mathbb{Z} \right)^* \cong \prod_{i \in I} \mathbb{Z}$ is not free when I is infinite (see the exercises for Section X.3).

Corollary **7.2.** *If M is a finitely generated projective module, then so is M^*.*

Double dual. The *double dual* of a left (or right) R-module M is $M^{**} = (M^*)^*$. The following result is straightforward:

Proposition **7.3.** *For every left (or right) R-module M there is a canonical* evaluation homomorphism $\varepsilon_M : M \longrightarrow M^{**}$, *which is natural in M: namely,* $(\varepsilon_M(x))(\alpha) = \alpha(x)$ *for all $x \in M$ and $\alpha \in M^*$.*

Readers will recall that ε_V is an isomorphism when V is a finite-dimensional vector space over a field. In general we have:

Proposition **7.4.** *If M is a finitely generated projective module, then the* evaluation homomorphism $M \longrightarrow M^{**}$ *is an isomorphism.*

Proof. If F is free, with a basis $(e_i)_{i \in I}$, then applying 7.1 twice yields a basis $(e_i^{**})_{i \in I}$ of F^{**} such that $e_i^{**}(e_i^*) = 1$, $e_i^{**}(e_j^*) = 0$ for all $j \neq i$. We see that $e_i^{**} = \varepsilon_F(e_i)$. Therefore ε_F is an isomorphism.

If now P is finitely generated and projective, then there exist a finitely generated free module F and homomorphisms $\pi : F \longrightarrow P$, $\iota : P \longrightarrow F$ such that $\pi \circ \iota = 1_P$. Then 7.3 yields the two commutative squares below, in which $\pi^{**} \circ \iota^{**} = 1_{P^{**}}$ and ε_F is an isomorphism. Hence $\iota^{**} \circ \varepsilon_P = \varepsilon_F \circ \iota$ is injective, so that ε_P is injective, and $\varepsilon_P \circ \pi = \pi^{**} \circ \varepsilon_F$ is surjective, so that ε_P is surjective. $\square$

$$
\begin{array}{ccc}
P & \xrightarrow{\ \varepsilon_P\ } & P^{**} \\
\pi \big\Vert \iota & \quad \pi^{**} \big\Vert \iota^{**} & \\
F & \xrightarrow{\ \varepsilon_F\ } & F^{**}
\end{array}
$$

Some finiteness hypothesis is necessary in Proposition 7.4: when V is an infinite-dimensional vector space, then $\dim V^* > \dim V$, and $V^{**} \not\cong V$ (see the exercises).

Tensor products. The previous properties yield canonical homomorphisms and isomorphisms of tensor products.

Proposition **7.5.** *Let A and B be left R-modules. There is a homomorphism $\zeta : A^* \otimes_R B \longrightarrow \mathrm{Hom}_R(A, B)$, which is natural in A and B, such that $\zeta\,(\alpha \otimes b)(a) = \alpha(a)\,b$ for all $a \in A$, $b \in B$, and $\alpha \in A^*$. If A is finitely generated and projective, then ζ is an isomorphism.*

Proof. For every $\alpha \in A^*$ and $b \in B$, $\beta(\alpha, b) : a \longmapsto \alpha(a)\,b$ is a module homomorphism of A into B. We see that $\beta : A^* \times B \longrightarrow \mathrm{Hom}_R(A, B)$ is a bihomomorphism. Hence β induces an abelian group homomorphism $\zeta = \overline{\beta}$: $A^* \otimes_R B \longrightarrow \mathrm{Hom}_R(A, B)$ such that $\zeta\,(\alpha \otimes b)(a) = \beta(\alpha, b)(a) = \alpha(a)\,b$ for all a, b, α. Our tireless readers will prove naturality in A and B.

If A is free with a finite basis $(e_i)_{i \in I}$, then A^* is free, with the dual basis $(e_i^*)_{i \in I}$, and every element of $A^* \otimes_R B$ can be written in the form $\sum_{i \in I} e_i^* \otimes b_i$ for some unique $b_i \in B$, by 5.9. Then

$$
\zeta\left(\sum_{i \in I} e_i^* \otimes b_i\right)(e_j) = \sum_{i \in I} \left(e_i^*(e_j)\,b_i\right) = b_j,
$$

and $\zeta\left(\sum_{i \in I} e_i^* \otimes b_i\right)$ is the homomorphism that sends e_j to b_j for every j. Therefore ζ is bijective.

If now A is finitely generated and projective, then there exist a finitely generated free module F and homomorphisms $\pi : F \longrightarrow A$, $\iota : A \longrightarrow F$ such that $\pi \circ \iota = 1_F$. Naturality of ζ yields the two commutative squares below, in which ζ_F is an isomorphism and $\iota' = \iota^* \otimes B$, $\pi' = \pi^* \otimes B$, $\iota'' = \mathrm{Hom}_R(\iota, B)$, $\pi'' = \mathrm{Hom}_R(\pi, B)$, so that $\iota^* \circ \pi^*$, $\iota' \circ \pi'$, and $\iota'' \circ \pi''$ are identity mappings. Hence $\pi'' \circ \zeta_A = \zeta_F \circ \pi'$ is injective, so that ζ_A is injective, and $\zeta_A \circ \iota' = \iota'' \circ \zeta_F$ is surjective, so that ζ_A is surjective. $\square$

$$A^* \otimes_R B \xrightarrow{\ \zeta_A\ } \mathrm{Hom}_R(A, B)$$

$$\iota' \Big\Uparrow \pi' \qquad\qquad \iota'' \Big\Uparrow \pi''$$

$$F^* \otimes_R B \xrightarrow[\ \zeta_F\ ]{} \mathrm{Hom}_R(F, B)$$

Corollary 7.6. *Let A be a finitely generated projective right R-module. There is an isomorphism $A \otimes_R B \cong \mathrm{Hom}_R(A^*, B)$, which is natural in A and B.*

Proof. By 7.4, 7.5, $A \otimes_R B \cong A^{**} \otimes_R B \cong \mathrm{Hom}_R(A^*, B)$. $\square$

Corollary 7.7. *Let R be commutative and let A, B be finitely generated projective R-modules. There is an isomorphism $A^* \otimes_R B^* \cong (A \otimes_R B)^*$, which is natural in A and B.*

Proof. By 7.5, 6.6, $A^* \otimes_R B^* \cong \mathrm{Hom}_R(A, B^*) = \mathrm{Hom}_R\big(A, \mathrm{Hom}_R(B, R)\big)$ $\cong \mathrm{Hom}_R(A \otimes_R B, R) = (A \otimes_R B)^*$. $\square$

Exercises

1. Prove that $(R_R)^* \cong {}_R R$ and $({}_R R)^* \cong R_R$.

2. Let F be a free left R-module with a finite basis $(e_i)_{i \in I}$. Show that F^* is a free right R-module with a finite basis $(e_i^*)_{i \in I}$ such that $e_i^*(e_i) = 1$, $e_j^*(e_j) = 0$ for all $j \neq i$.

3. Prove the following: if M is a finitely generated projective left R-module, then M^* is a finitely generated projective right R-module.

4. Let $\varphi : E \longrightarrow F$ be a module homomorphism, where E and F are free left R-modules with given finite bases. Show that the matrix of $\varphi^* : F^* \longrightarrow E^*$ in the dual bases is the transpose of the matrix of φ.

5. Let N be a submodule of M. Let $N^\perp = \{\, \alpha \in M^* \mid \alpha(N) = 0 \,\}$. Show that $N^\perp \cong (M/N)^*$. Construct a monomorphism $M^*/N^\perp \longrightarrow N^*$.

6. Verify that the evaluation homomorphism $M \longrightarrow M^{**}$ is natural in M.

7. Let V be an infinite-dimensional vector space. Show that $\dim V^* > \dim V$. (Use results from Section A.5.)

8. Show that M^* is isomorphic to a direct summand of M^{***}.

9. Let R be commutative and let A, B be finitely generated projective R-modules. Let $\theta : A^* \otimes_R B^* \cong (A \otimes_R B)^*$ be the isomorphism in Corollary 7.7. Show that $\theta\,(\alpha \otimes \beta)$ sends $a \otimes b$ to $\alpha(a)\,\beta(b)$, for all $\alpha \in A^*$ and $\beta \in B^*$.

8. Flat Modules

This section gives basic properties of flat modules and proves Lazard's theorem [1969], which constructs flat modules as direct limits of free modules.

Definition. A left R-module M is flat *when the functor $- \otimes_R M$ is exact.*

Equivalently, M is flat when, for every short exact sequence $0 \longrightarrow A \longrightarrow$ $B \longrightarrow C \longrightarrow 0$ of right R-modules, the sequence

$$0 \longrightarrow A \otimes_R M \longrightarrow B \otimes_R M \longrightarrow C \otimes_R M \longrightarrow 0$$

is exact. By 6.7, M is flat if and only if, for every monomorphism $\mu : A \longrightarrow B$ of right R-modules, $\mu \otimes M : A \otimes_R M \longrightarrow B \otimes_R M$ is a monomorphism.

Proposition 8.1. *Every projective module is flat.*

Proof. Readers will verify that free modules are flat. Now, let P be a projective left R-module. There exist a free left R-module F and homomorphisms $\pi : F \longrightarrow P$, $\iota : P \longrightarrow F$ such that $\pi \circ \iota = 1_P$. Every monomorphism $\mu : A \longrightarrow B$ of right R-modules begets two commutative squares:

$$
\begin{array}{ccc}
A \otimes_R P & \xrightarrow{\mu \otimes P} & B \otimes_R P \\
{\scriptstyle \pi'} \big\uparrow \big\downarrow {\scriptstyle \iota'} & & {\scriptstyle \pi''} \big\uparrow \big\downarrow {\scriptstyle \iota''} \\
A \otimes_R F & \xrightarrow[\mu \otimes F]{} & B \otimes_R F
\end{array}
$$

in which $\pi' = A \otimes \pi$, $\iota' = A \otimes \iota$, $\pi'' = B \otimes \pi$, $\iota'' = B \otimes \iota$. Then $\mu \otimes F$ is injective, since F is flat, and ι' is injective, since $\pi' \circ \iota' = 1_{A \otimes P}$. Hence $\iota'' \circ (\mu \otimes P) = (\mu \otimes F) \circ \iota'$ is injective, and $\mu \otimes P$ is injective. $\square$

Properties. Readers will easily prove the following properties:

Proposition 8.2. *Every direct summand of a flat module is flat.*

Proposition 8.3. *Every direct sum of flat modules is flat.*

Proposition 8.4. *Every direct limit of flat modules is flat.*

Proposition 8.5. *A module M is flat if and only if $M \otimes \mu$ is a monomorphism whenever $\mu : A \longrightarrow B$ is a monomorphism and A, B are finitely generated.*

Proposition 8.6. *An abelian group is flat (as a $\mathbb{Z}$-module) if and only if it is torsion-free.*

Proposition 8.6 immediately yields examples, such as $\mathbb{Q}$, showing that flat modules need not be projective, even when the ring R has very nice properties.

Proof. Finitely generated torsion-free abelian groups are free, and are flat by 8.1. Now every torsion-free abelian group is the direct limit of its finitely generated subgroups, which are also torsion-free, and is flat by 8.4.

On the other hand, finite cyclic groups are not flat: if C is cyclic of order $m > 1$, then multiplication by m is a monomorphism $\mu(x) = mx$ of $\mathbb{Z}$ into $\mathbb{Z}$, but $\mu \otimes C$ is not a monomorphism since $(\mu \otimes c)(x \otimes c) = mx \otimes c = x \otimes mc = 0$ for all $x \in \mathbb{Z}$ and $c \in C$, whereas $\mathbb{Z} \otimes C \cong C \neq 0$.

If now the abelian group A is not torsion-free, then A contains a finite cyclic subgroup $C \neq 0$. The monomorphism $\mu : \mathbb{Z} \longrightarrow \mathbb{Z}$ above and the inclusion homomorphism $\iota : C \longrightarrow A$ yield the commutative square below, in which $\mathbb{Z} \otimes \iota$ is injective, but $\mu \otimes A$ is not injective: otherwise, $(\mathbb{Z} \otimes \iota) \circ (\mu \otimes C) = (\mu \otimes A) \circ (\mathbb{Z} \otimes \iota)$ and $\mu \otimes C$ would be injective. Therefore A is not flat. $\square$

$$
\begin{array}{ccc}
\mathbb{Z} \otimes C & \xrightarrow{\mu \otimes C} & \mathbb{Z} \otimes C \\
{\scriptstyle \mathbb{Z} \otimes \iota}\downarrow & & \downarrow{\scriptstyle \mathbb{Z} \otimes \iota} \\
\mathbb{Z} \otimes A & \xrightarrow[\mu \otimes A]{} & \mathbb{Z} \otimes A
\end{array}
$$

There is a duality of sorts between flat modules and injective modules.

Proposition **8.7.** *A left R-module M is flat if and only if the right R-module* $\mathrm{Hom}_{\mathbb{Z}} (M, \mathbb{Q}/\mathbb{Z})$ *is injective.*

Proof. First, $\mathrm{Hom}_{\mathbb{Z}} (-, \mathbb{Q}/\mathbb{Z})$ is exact, since the divisible abelian group $\mathbb{Q}/\mathbb{Z}$ is an injective $\mathbb{Z}$-module. If M is flat, then $- \otimes_R M$ is exact, $\mathrm{Hom}_{\mathbb{Z}} (- \otimes_R M, \mathbb{Q}/\mathbb{Z})$ is exact, $\mathrm{Hom}_R (-, \mathrm{Hom}_{\mathbb{Z}} (M, \mathbb{Q}/\mathbb{Z}))$ is exact by adjoint associativity, and $\mathrm{Hom}_{\mathbb{Z}} (M, \mathbb{Q}/\mathbb{Z})$ is injective.

For the converse we show that the homomorphisms of any abelian group G into $\mathbb{Q}/\mathbb{Z}$ *separate* the elements of G: if $g \neq 0$ in G, then $\varphi(g) \neq 0$ for some $\varphi : G \longrightarrow \mathbb{Q}/\mathbb{Z}$. Indeed, let $g \in G$, $g \neq 0$. Let $\psi : \langle g \rangle \longrightarrow \mathbb{Q}/\mathbb{Z}$ send g to $\frac{1}{n} + \mathbb{Z}$ if g has finite order $n > 1$, and to, say, $\frac{1}{2} + \mathbb{Z}$ if g has infinite order. Since $\mathbb{Q}/\mathbb{Z}$ is injective, ψ extends to a homomorphism $\varphi : G \longrightarrow \mathbb{Q}/\mathbb{Z}$, and then $\varphi(g) \neq 0$.

We show that $\mathrm{Hom}_{\mathbb{Z}} (C, \mathbb{Q}/\mathbb{Z}) \xrightarrow{\beta^*} \mathrm{Hom}_{\mathbb{Z}} (B, \mathbb{Q}/\mathbb{Z}) \xrightarrow{\alpha^*} \mathrm{Hom}_{\mathbb{Z}} (A, \mathbb{Q}/\mathbb{Z})$ exact implies $A \xrightarrow{\alpha} B \xrightarrow{\beta} C$ exact. If $c = \beta(\alpha(a)) \in \mathrm{Im}\,(\beta \circ \alpha)$, then $\varphi(c) = \varphi(\beta(\alpha(a))) = (\alpha^*(\beta^*(\varphi)))(c) = 0$ for all $\varphi : C \longrightarrow \mathbb{Q}/\mathbb{Z}$; therefore $c = 0$, and $\beta \circ \alpha = 0$. Conversely, let $b \in \mathrm{Ker}\,\beta$. Let $\pi : B \longrightarrow B/\mathrm{Im}\,\alpha = \mathrm{Coker}\,\alpha$ be the projection. For every homomorphism $\varphi : B/\mathrm{Im}\,\alpha \longrightarrow \mathbb{Q}/\mathbb{Z}$ we have $\alpha^*(\varphi \circ \pi) = \varphi \circ \pi \circ \alpha = 0$. Hence $\varphi \circ \pi \in \mathrm{Ker}\,\alpha^* = \mathrm{Im}\,\beta^*$ and $\varphi \circ \pi = \beta^*(\psi) = \psi \circ \beta$ for some homomorphism $\psi : C \longrightarrow \mathbb{Q}/\mathbb{Z}$:

$$
\begin{array}{ccc}
A \xrightarrow{\alpha} B & \xrightarrow{\beta} & C \\
{\scriptstyle \pi}\downarrow & & \downarrow{\scriptstyle \psi} \\
B/\mathrm{Im}\,\alpha & \xrightarrow[\varphi]{} & \mathbb{Q}/\mathbb{Z}
\end{array}
$$

and $\varphi(\pi(b)) = \psi(\beta(b)) = 0$. Therefore $\pi(b) = 0$ and $b \in \mathrm{Im}\,\alpha$.

If now $\mathrm{Hom}_{\mathbb{Z}} (M, \mathbb{Q}/\mathbb{Z})$ is injective, then $\mathrm{Hom}_R (-, \mathrm{Hom}_{\mathbb{Z}} (M, \mathbb{Q}/\mathbb{Z}))$ is exact, $\mathrm{Hom}_{\mathbb{Z}} (- \otimes_R M, \mathbb{Q}/\mathbb{Z})$ is exact by adjoint associativity, $- \otimes_R M$ is exact by the above, and M is flat. $\square$

The exercises list some consequences of these results.

Lazard's theorem. Lazard's theorem states that a module is flat if and only if it is a direct limit of free modules. A more detailed version is given below.

First we prove some preliminary results. A left R-module M is *finitely presented* when there is an exact sequence $F_1 \longrightarrow F_2 \longrightarrow M \longrightarrow 0$ in which F_1 and F_2 are finitely generated free left R-modules; equivalently, when M has a presentation $M \cong F_2/K$ with finitely many generators (F_2 is finitely generated) and finitely many defining relations (K is finitely generated).

Proposition 8.8. *Every module is a direct limit of finitely presented modules.*

Proof. Let $M \cong F/K$, where F is free with a basis X. We show that M is the direct limit of the finitely presented modules F_Y/S, where F_Y is the free submodule of F generated by a finite subset Y of X and S is a finitely generated submodule of $F_Y \cap K$.

Let $\mathcal{P}$ be the set of all ordered pairs $p = (Y_p, S_p)$ of a finite subset Y_p of X and a finitely generated submodule S_p of $F_{Y_p} \cap K$. Partially order $\mathcal{P}$ by $p \leq q$ if and only if $Y_p \subseteq Y_q$ and $S_p \subseteq S_q$. Then $\mathcal{P}$ is directed upward: for all $p, q \in \mathcal{P}$, $(Y_p, S_p), (Y_q, S_q) \leq (Y_p \cup Y_q, S_p + S_q) \in \mathcal{P}$.

Let $A_p = F_{Y_p}/S_p$. If $p \leq q$ in $\mathcal{P}$, then $S_p \subseteq S_q$ and there is a unique homomorphism $\alpha_{pq} : A_p \longrightarrow A_q$ such that the following square commutes:

$$
\begin{array}{ccc}
F_{Y_p} & \xrightarrow{\ \subseteq\ } & F_{Y_q} \\
\downarrow & & \downarrow \\
A_p = F_{Y_p}/S_p & \xrightarrow[\alpha_{pq}]{} & F_{Y_q}/S_q = A_q
\end{array}
$$

where the vertical arrows are projections; namely, $\alpha_{pq} : x + S_p \longmapsto x + S_q$. This constructs a direct system over $\mathcal{P}$. We show that $M = \varinjlim_{p \in \mathcal{P}} A_p$.

Since $S_p \subseteq K$, there is for every $p = (Y_p, S_p) \in \mathcal{P}$ a unique homomorphism $\lambda_p : A_p \longrightarrow M$ such that the square

$$
\begin{array}{ccc}
F_{Y_p} & \xrightarrow{\ \subseteq\ } & F \\
\downarrow & & \downarrow{\scriptstyle \pi} \\
A_p = F_{Y_p}/S_p & \xrightarrow[\lambda_p]{} & M
\end{array}
$$

commutes; namely, $\lambda_p : x + S_p \longmapsto \pi(x)$, where $\pi : F \longrightarrow M$ is the projection. This constructs a cone $\lambda = (\lambda_p)_{p \in \mathcal{P}}$, which we show is a limit cone. Every element of F belongs to a finitely generated submodule F_Y; therefore $M = \bigcup_{p \in \mathcal{P}} \operatorname{Im} \lambda_p$. If $\lambda_p(x + S_p) = 0$, where $x \in F_{Y_p}$, then $x \in F_{Y_p} \cap K$ belongs to a finitely generated submodule T of $F_{Y_p} \cap K$, $S_q = S_p + T \subseteq F_{Y_p} \cap K$ is finitely generated, $q = (Y_p, S_q) \in \mathcal{P}$, and $\alpha_{pq}(x + S_p) = x + S_q = 0$ in A_q. Hence $\operatorname{Ker} \lambda_p = \bigcup_{q \geq p} \operatorname{Ker} \alpha_{pq}$. Then λ is a limit cone, by 3.6. $\square$

Proposition 8.9. *Every homomorphism of a finitely presented module into a flat module factors through a finitely generated free module.*

Proof. Let $\varphi : M \longrightarrow A$ be a homomorphism of a flat left R-module M into a finitely presented left R-module A. There is an exact sequence $F_1 \xrightarrow{\tau} F_2 \xrightarrow{\sigma} A \longrightarrow 0$ in which F_1 and F_2 are free and finitely generated. Then $0 \longrightarrow A^* \xrightarrow{\sigma^*} F_2^* \xrightarrow{\tau^*} F_1^*$ is exact and 7.5 yields a commutative diagram

$$
\begin{array}{ccccc}
A^* \otimes_R M & \xrightarrow{\;\overline{\sigma}^*\;} & F_2^* \otimes_R M & \xrightarrow{\;\overline{\tau}^*\;} & F_1^* \otimes_R M \\
& & \zeta \downarrow & & \downarrow \zeta \\
& & \operatorname{Hom}_R(F_2, M) & \xrightarrow[\tau']{} & \operatorname{Hom}_R(F_1, M)
\end{array}
$$

in which $\overline{\sigma}^* = \sigma^* \otimes M$, $\overline{\tau}^* = \tau^* \otimes M$, $\tau' = \operatorname{Hom}_R(\tau, M)$, the vertical maps are isomorphisms, and the top row is exact since M is flat.

Since $\tau'(\varphi \circ \sigma) = \varphi \circ \sigma \circ \tau = 0$ we have $\zeta^{-1}(\varphi \circ \sigma) \in \operatorname{Ker} \overline{\tau}^*$ and $\varphi \circ \sigma = \zeta\big(\overline{\sigma}^*(t)\big)$ for some $t \in A^* \otimes_R M$. By 5.4, $t = \sum_i a_i^* \otimes x_i$ for some $a_1^*, \ldots, a_n^* \in A^*$ and $x_1, \ldots, x_n \in M$. Let F be a free right R-module with basis $e_1, \ldots, e_n$. Then F^{**} is free on $e_1^{**}, \ldots, e_n^{**}$, there is a homomorphism $\alpha : F^{**} \longrightarrow A^*$ such that $\alpha(e_i^{**}) = a_i^*$ for all i, and $t = \overline{\alpha}(u)$ for some $u \in F^{**} \otimes_R M$, where $\overline{\alpha} = \alpha \otimes M$. Since $F_2^{**} \cong F_2$ and $F^{***} \cong F^*$ we have $\sigma^* \circ \alpha = \xi^* : F^{**} \longrightarrow F_2^*$ for some homomorphism $\xi : F_2 \longrightarrow F^*$. Then $\tau^* \circ \xi^* = \tau^* \circ \sigma^* \circ \alpha = 0$, $\xi \circ \tau = 0$, $\operatorname{Ker} \sigma = \operatorname{Im} \tau \subseteq \operatorname{Ker} \xi$, and ξ factors through σ, $\xi = \chi \circ \sigma$ for some homomorphism $\chi : A \longrightarrow F^*$:

$$
\begin{array}{ccccccc}
F_1 & \xrightarrow{\tau} & F_2 & \xrightarrow{\sigma} & A & \longrightarrow & 0 \\
& & \xi \downarrow & \swarrow \chi & \downarrow \varphi & & \\
& & F^* & \xrightarrow[\psi]{\;-\,-\,\to\;} & M & &
\end{array}
$$

We now have a commutative square (below) in which $\xi' = \operatorname{Hom}_R(\xi, M)$ and the vertical arrows are isomorphisms. Then $\psi = \zeta(u) \in \operatorname{Hom}_R(F^*, M)$ satisfies $\psi \circ \chi \circ \sigma = \psi \circ \xi = \xi'(\psi) = \zeta\big(\overline{\xi}^*(u)\big) = \zeta\big(\overline{\sigma}^*(\overline{\alpha}(u))\big) = \zeta\big(\overline{\sigma}^*(t)\big) = \varphi \circ \sigma$; hence $\psi \circ \chi = \varphi$. $\square$

$$
\begin{array}{ccc}
F^{**} \otimes_R M & \xrightarrow{\;\overline{\xi}^*\;} & F_2^* \otimes_R M \\
\zeta \downarrow & & \zeta \downarrow \\
\operatorname{Hom}_R(F^*, M) & \xrightarrow[\xi']{} & \operatorname{Hom}_R(F_2, M)
\end{array}
$$

Corollary 8.10. *Every finitely presented flat module is projective.*

Proof. The identity on such a module factors through a free module. $\square$

In fact, Proposition 8.9 characterizes flat modules. This is part of the detailed version of Lazard's theorem:

Theorem **8.11** (Lazard [1969]). *For a left R-module* M *the following conditions are equivalent:*

(1) M *is flat;*

(2) *every homomorphism of a finitely presented free module into* M *factors through a finitely generated free module;*

(3) M *is a direct limit of finitely generated free modules;*

(4) M *is a direct limit of free modules.*

Proof. (3) implies (4); (4) implies (1), by 8.1 and 8.4; (1) implies (2), by 8.9. Now assume (2). Let $\pi : F \longrightarrow M$ be an epimorphism, where F is free with a basis X. Choose $\pi : F \longrightarrow M$ so that there are for every $m \in M$ infinitely many $x \in X$ such that $\pi(x) = m$; for instance, let F be free on $X = M \times \mathbb{N}$ and let π be the homomorphism such that $\pi(m, n) = m$. To prove (3) we use the construction in the proof of 8.8. Let $\mathcal{P}$ be the set of all ordered pairs $p = (Y_p, S_p)$ of a finite subset Y_p of X and a finitely generated submodule S_p of $F_{Y_p} \cap \operatorname{Ker} \pi$; then M is the direct limit of the finitely presented modules $A_p = F_{Y_p}/S_p$, with limit cone $\lambda_p : A_p \longrightarrow M$, $x + S_p \longmapsto \pi(x)$.

We show that $\mathcal{Q} = \{\, p \in \mathcal{P} \mid A_p \text{ is free} \,\}$ is cofinal in $\mathcal{P}$. Let $p = (Y_p, S_p) \in \mathcal{P}$. By (2), $\lambda_p : A_p \longrightarrow M$ factors through a finitely generated free module $F' : \lambda_p = \psi \circ \chi$, where $\chi : A_p \longrightarrow F'$, $\psi : F' \longrightarrow M$, and F' has a finite basis B. By the choice of π there is for each $b \in B$ at least one $z \in X \backslash Y_p$ such that $\pi(z) = \psi(b)$. Picking one for each b yields a finite subset Z of $X \backslash Y_p$ and an isomorphism $\theta : F_Z \longrightarrow F'$ such that $\psi(\theta(z)) = \pi(z)$ for all $z \in Z$. Then $F_{Y_p} \longrightarrow A_p \longrightarrow F'$ and $\theta : F_Z \longrightarrow F'$ induce an epimorphism $\rho : F_{Y_p \cup Z} \longrightarrow F'$ such that the diagram below commutes:

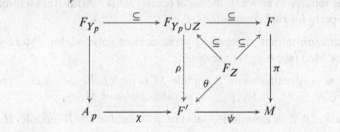

Now, $S_p \subseteq \operatorname{Ker} \rho \subseteq \operatorname{Ker} \pi$ and $\operatorname{Ker} \rho$ is finitely generated, since ρ splits; hence $q = (Y_p \cup Z, \operatorname{Ker} \rho) \in \mathcal{P}$. Then $p \leqq q$ and $q \in \mathcal{Q}$, since $A_q = F_{Y_p \cup Z}/\operatorname{Ker} \rho \cong F'$ is free. Thus $\mathcal{Q}$ is cofinal in $\mathcal{P}$, and $M = \varinjlim_{p \in \mathcal{P}} A_p = \varinjlim_{q \in \mathcal{Q}} A_q$ is a direct limit of finitely generated free modules. $\square$

Exercises

Prove the following:

1. Every direct summand of a flat module is flat.

2. Every direct sum of flat modules is flat.

3. Every direct limit of flat modules is flat.

4. Every abelian group can be embedded into a direct product of copies of $\mathbb{Q}/\mathbb{Z}$.

5. A right R-module M is flat if and only if $\mathrm{Hom}_R(M, \rho)$ is an epimorphism for every finitely presented module N and epimorphism $\rho : N \longrightarrow M$.

6. $I \otimes_R J \cong IJ$ when I and J are ideals of R and I is flat as a right R-module.

7. If R is commutative and S is a proper multiplicative subset of R, then $S^{-1}R$ is a flat R-module.

8. If M is generated by $(m_i)_{i \in I}$, then every element of $M \otimes_R A$ is a finite sum $\sum_{i \in I} m_i \otimes a_i$, and $\sum_{i \in I} m_i \otimes a_i = 0$ in $M \otimes_R A$ if and only if there exist finitely many $(b_j)_{j \in J} \in A$ and $r_{ij} \in R$ such that $a_i = \sum_{j \in J} r_{ij} b_j$ for all i and $\sum_{i \in I} m_i r_{ij} = 0$ for all j.

9. A right R-module M is flat if and only if $I \otimes_R M \longrightarrow R_R \otimes_R M$ is injective for every right ideal I of R, if and only if $I \otimes_R M \longrightarrow R_R \otimes_R M$ is injective for every finitely generated right ideal I of R. (You may want to use Proposition 8.5 and 8.7.)

10. A right R-module M is flat if and only if, for every finitely many $(r_i)_{i \in I} \in R$ and $(m_i)_{i \in I} \in M$ such that $\sum_{i \in I} m_i r_i = 0$, there exist finitely many $(n_j)_{j \in J} \in M$ and $t_{ij} \in R$ such that $m_i = \sum_{j \in J} n_j t_{ij}$ for all i and $\sum_{i \in I} t_{ij} r_i = 0$ for all j. (You may want to use the previous two exercises.)

9. Completions

Completions of modules are similar to the completions of rings in Section VI.9. They provide applications of inverse limits. In this section the emphasis is on completions relative to an ideal; the main results are the Artin-Rees lemma and a flatness property for ring completions.

Filtrations are infinite descending sequences of submodules. More general filters can be used (see the exercises).

Definition. A filtration *on an R-module M is an infinite descending sequence $M_1 \supseteq M_2 \supseteq \cdots \supseteq M_i \supseteq M_{i+1} \supseteq \cdots$ of submodules of M.*

For instance, if $\mathfrak{a}$ is a two-sided ideal of R, then, for any R-module M,

$$\mathfrak{a}M \supseteq \mathfrak{a}^2 M \supseteq \cdots \supseteq \mathfrak{a}^i M \supseteq \mathfrak{a}^{i+1} M \supseteq \cdots$$

is a filtration on M, the $\mathfrak{a}$-*adic filtration* on M, our main focus of interest in this section. Unfortunately, the $\mathfrak{a}$-adic filtration on a module M does not in general induce $\mathfrak{a}$-adic filtrations on its submodules ($N \cap \mathfrak{a}^i M$ need not equal $\mathfrak{a}^i N$). This leads to a larger class of filtrations that are more easily inherited, as shown by the Artin-Rees lemma, Lemma 9.2 below.

Definitions. Let $\mathfrak{a}$ be an ideal of R and let M be an R-module. An $\mathfrak{a}$-fil-tration on M is a filtration $M_1 \supseteq M_2 \supseteq \cdots$ on M such that $\mathfrak{a}M_i \subseteq M_{i+1}$

for all i. *An* $\mathfrak{a}$-stable filtration *on* M *is an* $\mathfrak{a}$-filtration $M_1 \supseteq M_2 \supseteq \cdots$ *such that* $\mathfrak{a}M_i = M_{i+1}$ *for all sufficiently large* i.

For instance, the $\mathfrak{a}$-adic filtration on M is $\mathfrak{a}$-stable.

Lemma **9.1.** *Let* $\mathfrak{a}$ *be an ideal of a commutative ring* R, *let* M *be an* R-module, *and let* $M_1 \supseteq M_2 \supseteq \cdots$ *be an* $\mathfrak{a}$-filtration *on* M.

(1) $R^+ = R \oplus \mathfrak{a} \oplus \mathfrak{a}^2 \oplus \cdots$ *is a ring* (*the* blown-up ring *of* R);

(2) $M^+ = M \oplus M_1 \oplus M_2 \oplus \cdots$ *is an* R^+-module (*the* blown-up module *of* M);

(3) *if* M *and all* M_i *are finitely generated* R-modules, *then the given filtration on* M *is* $\mathfrak{a}$-stable *if and only if* M^+ *is finitely generated as an* R^+-module.

Proof. (1). The elements of R^+ are infinite sequences $a = (a_0, a_1, \ldots, a_i, \ldots)$ in which $a_0 \in R$, $a_i \in \mathfrak{a}^i$ for all $i > 0$, and $a_i = 0$ for almost all i. Addition in R^+ is componentwise; multiplication is given by $ab = c$ when $c_k = \sum_{i+j=k} a_i b_j$ for all k. Then R^+ is a ring: R^+ is isomorphic to a subring of $R[X]$, $\{a(X) = a_0 + a_1 X + \cdots + a_n X^n \in R[X] \mid a_i \in \mathfrak{a}^i$ for all $i > 0\}$.

(2). The elements of M^+ are infinite sequences $x = (x_0, x_1, \ldots, x_i, \ldots)$ in which $x_0 \in M$, $x_i \in M_i$ for all $i > 0$, and $x_i = 0$ for almost all i. Now, R^+ acts on M^+ by $ax = y$ when $y_k = \sum_{i+j=k} a_i x_j$ for all k (note that $a_i x_j \in M_{i+j}$ when $a_i \in \mathfrak{a}^i$, $x_j \in M_j$, since $M_1 \supseteq M_2 \supseteq \cdots$ is an $\mathfrak{a}$-filtration). It is straightforward that M^+ is an R^+-module.

(3). If M^+ is a finitely generated R^+-module, then its generators are contained in $M \oplus M_1 \oplus \cdots \oplus M_n$ for some n; then M^+ is generated, as an R^+-module, by the generators of M, M_1, ..., M_n over R, and every element x of M^+ is a sum of $r_i x_j$'s in which r_i is a product of i elements of $\mathfrak{a}$ (or an element of R, if $i = 0$), $x_j \in M_j$, and $j \leqq n$. If $x \in M_k$ and $k \geqq n$, then either $j = n$ or $r_i x_j$ is the product of $i + j - n$ elements of $\mathfrak{a}$ and an element of $\mathfrak{a}^{n-j} M_j \subseteq M_n$. Therefore $M_k \subseteq \mathfrak{a}^{k-n} M_n$, and the filtration $M_1 \supseteq M_2 \supseteq \cdots$ is $\mathfrak{a}$-stable. Conversely, if the filtration $M_1 \supseteq M_2 \supseteq \cdots$ is $\mathfrak{a}$-stable and n is large enough, then $M_k = \mathfrak{a}^{k-n} M_n$ for all $k \geqq n$ and M^+ is generated, as an R^+-module, by $M \oplus M_1 \oplus \cdots \oplus M_n$; hence M^+ is finitely generated, like M, ..., M_n. $\square$

Lemma **9.2** (Artin-Rees). *Let* $\mathfrak{a}$ *be an ideal of a commutative Noetherian ring* R, *and let* M *be a finitely generated* R-module. *If* $M_1 \supseteq M_2 \supset \cdots$ *is an* $\mathfrak{a}$-stable filtration *on* M, *then* $N \cap M_1 \supseteq N \cap M_2 \supseteq \cdots$ *is an* $\mathfrak{a}$-stable filtration *on* N, *for every submodule* N *of* M.

Proof. First, $N \cap M_1 \supseteq N \cap M_2 \supseteq \cdots$ is an $\mathfrak{a}$-filtration on N; M is a Noetherian R-module, by VIII.8.5; N and all M_i, $N \cap M_i$ are finitely generated; and N^+ is an R^+-submodule of M^+. Since R is Noetherian, $\mathfrak{a}$ is a finitely generated ideal; R^+ is generated, as a ring, by R and the finitely many generators of $\mathfrak{a}$; and R^+ is Noetherian, by III.11.5. By 9.1, M^+ is a finitely generated R^+-module; hence M^+ is a Noetherian R^+-module, by VIII.11.5, N^+ is a finitely generated R^+-module, and $N \cap M_1 \supseteq N \cap M_2 \supseteq \cdots$ is $\mathfrak{a}$-stable, by 9.1. $\square$

Completions. We saw in Section VI.9 that the completion of R relative to a filtration $\mathcal{A} : \mathfrak{a}_1 \supseteq \mathfrak{a}_2 \supseteq \cdots$ on R is the ring $\widehat{R}_\mathcal{A} = \varprojlim_{i \to \infty} R/\mathfrak{a}_i$ of all infinite sequences $(x_1 + \mathfrak{a}_1, \ldots, x_i + \mathfrak{a}_i, \ldots)$ such that $x_i \in R$ and $x_i + \mathfrak{a}_j = x_j + \mathfrak{a}_j$ whenever $i \geq j$. Similarly, when $M_1 \supseteq M_2 \supseteq \cdots$ is a filtration on an R-module M, there is for every $i \geq j$ a canonical homomorphism $M/M_i \longrightarrow M/M_j$, $x + M_i \longmapsto x + M_j$; the modules M/M_i and homomorphisms $M/M_i \longrightarrow M/M_j$ constitute an inverse system.

Definition. The completion *of an R-module M relative to a filtration* $\mathcal{M}$: $M_1 \supseteq M_2 \supseteq \cdots$ *on M is* $\widehat{M}_\mathcal{M} = \varprojlim_{i \to \infty} M/M_i$. *If $\mathfrak{a}$ is an ideal of R, then the* $\mathfrak{a}$-adic completion *of M is its completion* $\widehat{M}_\mathfrak{a} = \varprojlim_{i \to \infty} M/\mathfrak{a}^i M$ *relative to the* $\mathfrak{a}$-adic filtration on M.

Thus, $\widehat{M}$ consists of all infinite sequences $(x_1 + M_1, \ldots, x_i + M_i, \ldots)$ such that $x_i \in M$ and $x_i + M_j = x_j + M_j$ whenever $i \geq j$; equivalently, $x_{i+1} \in x_i + M_i$ for all $i \geq 1$ ($x_{i+1} \in x_i + \mathfrak{a}^i M$, for the $\mathfrak{a}$-adic completion). The exercises give an alternate construction of $\widehat{M}$ by Cauchy sequences.

The projections $M \longrightarrow M/M_i$ constitute a cone from M and induce a homomorphism $M \longrightarrow \widehat{M}$.

Definition. If $\mathcal{M} : M_1 \supseteq M_2 \supseteq \cdots$ *is a filtration on M, then* $\iota : x \longmapsto (x + M_1, x + M_2, \ldots)$ *is the* canonical homomorphism *of M into* $\widehat{M}_\mathcal{M}$.

Readers will verify that ι is injective if and only if $\bigcap_{i > 0} M_i = 0$.

Definition. An R-module M is complete *relative to a filtration (or to an ideal of R) when the canonical homomorphism* $\iota : M \longrightarrow \widehat{M}$ *is an isomorphism.*

Properties. We show that $\widehat{M}_\mathcal{M}$ is always complete.

Proposition 9.3. If $\mathcal{M} : M_1 \supseteq M_2 \supseteq \cdots$ *is a filtration on an R-module M, then* $\widehat{M}_j = \{ (x_1 + M_1, \ldots, x_i + M_i, \ldots) \in \widehat{M}_\mathcal{M} \mid x_j \in M_j \}$ *is a submodule of* $\widehat{M}_\mathcal{M}$, $\widehat{\mathcal{M}} : \widehat{M}_1 \supseteq \widehat{M}_2 \supseteq \cdots$ *is a filtration on* $\widehat{M}_\mathcal{M}$, *and* $\widehat{M}_\mathcal{M}$ *is complete relative to* $\widehat{\mathcal{M}}$.

Note that $\widehat{M}_j = \{ (x_1 + M_1, x_2 + M_2, \ldots) \in \widehat{M}_\mathcal{M} \mid x_i \in M_i \text{ for all } i \leq j \}$; thus $(\overline{x}_1, \overline{x}_2, \ldots) \in \widehat{M}_j$ if and only if $\overline{x}_i = 0$ in M/M_i for all $i \leq j$.

Proof. Let $N = \widehat{M}$. The alternate description of $\widehat{M}_j$ shows that $\widehat{M}_1 \supseteq \widehat{M}_2 \supseteq \cdots$. Also, $\widehat{M}_j$ is a submodule of $\widehat{M}$, since it is the kernel of the homomorphism $\left(x_1 + M_1, x_2 + M_2, \ldots\right) \longmapsto x_j + M_j$ of $\widehat{M}$ into M/M_j. In particular, there is an isomorphism $\theta_i : N/\widehat{M}_j \longrightarrow M/M_j$, which sends $y + \widehat{M}_j = \left(x_1 + M_1, x_2 + M_2, \ldots\right) + \widehat{M}_j$ to the j component $x_j + M_j$ of y.

We see that $(\theta_i)_{i \in I}$ is an isomorphism of inverse systems. Therefore $(\theta_i)_{i \in I}$ induces an isomorphism $\theta : \widehat{N} \longrightarrow \widehat{M}$, which sends $\left(y_1 + \widehat{M}_1, y_2 + \widehat{M}_2, \ldots\right) \in \widehat{N}$ to $\left(x_1 + M_1, x_2 + M_2, \ldots\right)$, where $x_j + M_j$ is the j component of y_j. If $y \in N$, then $\iota(y) = \left(y + \widehat{M}_1, y + \widehat{M}_2, \ldots\right)$ and $\theta\left(\iota(y)\right) = \left(x_1 + M_1, x_2 + M_2, \ldots\right)$, where $x_j + M_j$ is the j component of y; in other words, $\theta\left(\iota(y)\right) = y$. Therefore $\iota = \theta^{-1} : N \longrightarrow \widehat{N}$ is an isomorphism. $\square$

Let $M_1 \supseteq M_2 \supseteq \cdots$ be a filtration of M and let $N_1 \supseteq N_2 \supseteq \cdots$ be a filtration of N. If $\varphi : M \longrightarrow N$ is a homomorphism of R-modules and $\varphi(M_i) \subseteq N_i$ for all i, then there is for every i a homomorphism $\varphi_i : M/M_i \longrightarrow N/N_i$, $x + M_i \longmapsto \varphi(x) + N_i$, and then $(\varphi_i)_{i \in I}$ is a homomorphism of inverse systems and induces a module homomorphism $\widehat{\varphi} : \widehat{M} \longrightarrow \widehat{N}$, which sends $\left(x_1 + M_1, x_2 + M_2, \ldots\right)$ to

$$\left(\varphi_1(x_1 + N_1), \varphi_2(x_2 + N_2), \ldots\right) = \left(\varphi(x_1) + N_1, \varphi(x_2) + N_2, \ldots\right).$$

In particular, when $\mathfrak{a}$ is an ideal of R, then $\varphi(\mathfrak{a}^i M) \subseteq \mathfrak{a}^i N$ for all i, so that every module homomorphism $\varphi : M \longrightarrow N$ induces a module homomorphism $\widehat{\varphi} : \widehat{M}_\mathfrak{a} \longrightarrow \widehat{N}_\mathfrak{a}$. This yields a completion functor:

Proposition 9.4. *For every ideal $\mathfrak{a}$ of R, $\widehat{}_\mathfrak{a}$ is an additive functor from R-modules to R-modules. In particular, $\widehat{}_\mathfrak{a}$ preserves finite direct sums.*

Proposition 9.5. *If $\varphi : M \longrightarrow N$ is surjective, then $\widehat{\varphi} : \widehat{M}_\mathfrak{a} \longrightarrow \widehat{N}_\mathfrak{a}$ is surjective.*

Proof. Since $\mathfrak{a}^i N$ is generated by all ry with $r \in \mathfrak{a}^i$ and $y \in N$ we have $\mathfrak{a}^i N = \varphi(\mathfrak{a}^i M)$. Let $\left(y_1 + \mathfrak{a} N, y_2 + \mathfrak{a}^2 N, \ldots\right) \in \widehat{N}_\mathfrak{a}$. Then $y_1 = \varphi(x_1)$ for some $x_1 \in M$. Construct $x_1, \ldots, x_i, \ldots \in M$ by induction so that $y_i = \varphi(x_i)$ and $x_{i+1} \in x_i + \mathfrak{a}^i M$ for all i. Given $\varphi(x_j) = y_j$, we have $y_{j+1} = \varphi(t)$ for some $t \in M$, $\varphi(t - x_j) = y_{j+1} - y_j \in \mathfrak{a}^j N$, and $\varphi(t - x_j) = \varphi(q)$ for some $q \in \mathfrak{a}^j M$. Then $x_{j+1} = x_j + q \in x_j + \mathfrak{a}^j M$ and $\varphi(x_j + q) = \varphi(t) = y_{j+1}$. Now $x = \left(x_1 + \mathfrak{a} M, x_2 + \mathfrak{a}^2 M, \ldots\right) \in \widehat{M}$ and $\widehat{\varphi}(x) = y$. $\square$

Proposition 9.6. *Let R be a commutative Noetherian ring and let $\mathfrak{a}$ be an ideal of R. If A, B, C are finitely generated R-modules and $0 \longrightarrow A \longrightarrow B \longrightarrow C \longrightarrow 0$ is exact, then $0 \longrightarrow \widehat{A}_\mathfrak{a} \longrightarrow \widehat{B}_\mathfrak{a} \longrightarrow \widehat{C}_\mathfrak{a} \longrightarrow 0$ is exact.*

Proof. Let $0 \longrightarrow A \overset{\mu}{\longrightarrow} B \overset{\sigma}{\longrightarrow} C \longrightarrow 0$ be exact. We may assume that $A = \operatorname{Ker} \sigma \subseteq B$ and that μ is the inclusion homomorphism. By 9.5, $\widehat{\sigma}$ is surjective.

Let $b \in B$. If $\sigma_i(b + \mathfrak{a}^i B) = 0$ in $C/\mathfrak{a}^i C$, then $\sigma(b) \in \mathfrak{a}^i C$, $\sigma(b) = \sigma(x)$

for some $x \in \mathfrak{a}^i B$, $b - x \in \text{Ker } \sigma = A$, and $b \in A + \mathfrak{a}^i B$. Conversely, if $b \in A + \mathfrak{a}^i B$, then $\sigma(b) \in \mathfrak{a}^i C$. Hence $\text{Ker } \sigma_i = (A + \mathfrak{a}^i B)/\mathfrak{a}^i B$. Then $A/(A \cap \mathfrak{a}^i B) \cong \text{Ker } \sigma_i$; the isomorphism sends $a + (A \cap \mathfrak{a}^i B)$ to $a + \mathfrak{a}^i B$. We now have an exact sequence

$$0 \longrightarrow A/(A \cap \mathfrak{a}^i B) \xrightarrow{\nu_i} B/\mathfrak{a}^i B \xrightarrow{\sigma_i} C/\mathfrak{a}^i C$$

where ν_i sends $a + (A \cap \mathfrak{a}^i B)$ to $a + \mathfrak{a}^i B$. Then the sequence

$$0 \longrightarrow \varprojlim_{i \to \infty} A/(A \cap \mathfrak{a}^i B) \xrightarrow{\widehat{\nu}} \widehat{B} \xrightarrow{\widehat{\sigma}} \widehat{C},$$

where $\widehat{\nu} = \varprojlim_{i \to \infty} \nu_i$, is exact, since inverse limits are left exact, by 4.6.

Since $\mathfrak{a}^i A \subseteq A \cap \mathfrak{a}^i B$ there is also a homomorphism $\rho_i : A/\mathfrak{a}^i A \longrightarrow A/(A \cap \mathfrak{a}^i B)$, which sends $a + \mathfrak{a}^i A$ to $a + (A \cap \mathfrak{a}^i B)$. Then $\nu_i \circ \rho_i = \mu_i$. Moreover, $(\rho_i)_{i \in I}$ is a homomorphism of inverse systems and induces a homomorphism $\widehat{\rho} : \widehat{A} = \varprojlim_{i \to \infty} A/\mathfrak{a}^i A \longrightarrow \varprojlim_{i \to \infty} A/(A \cap \mathfrak{a}^i B)$.

Since R is Noetherian and B is finitely generated, the $\mathfrak{a}$-stable $\mathfrak{a}$-adic filtration on B induces on A a filtration $A \cap \mathfrak{a}B \supseteq A \cap \mathfrak{a}^2 B \supseteq \cdots$, which is $\mathfrak{a}$-stable by the Artin-Rees lemma. Hence $\mathfrak{a}^i (A \cap \mathfrak{a}^j B) = A \cap \mathfrak{a}^{i+j} B$ for all i, whenever j is sufficiently large, $j \geqq k$. Then $A \cap \mathfrak{a}^i B \subseteq \mathfrak{a}^{i-j} A$ when $i > j \geqq k$. Since already $\mathfrak{a}^i A \subseteq A \cap \mathfrak{a}^i B$, Lemma 9.7 below, applied to $M = A$, $M_i = A \cap \mathfrak{a}^i B$, and $N_i = \mathfrak{a}^i A$, shows that $\widehat{\rho}$ is an isomorphism. Then $\widehat{\nu} \circ \widehat{\rho} = \widehat{\mu}$ implies that

$$0 \longrightarrow \widehat{A} \xrightarrow{\widehat{\mu}} \widehat{B} \xrightarrow{\widehat{\sigma}} \widehat{C} \text{ is exact. } \square$$

Lemma 9.7. Let $\mathcal{M} : M_1 \supseteq M_2 \supseteq \cdots$ and $\mathcal{N} : N_1 \supseteq N_2 \supseteq \cdots$ be filtrations on M such that $M_i \supseteq N_i$ and every N_i contains some M_j. The homomorphisms $\rho_i : M/N_i \longrightarrow M/M_i$, $x + N_i \longmapsto x + M_i$, induce an isomorphism $\widehat{\rho} : \widehat{M}_{\mathcal{N}} \longrightarrow \widehat{M}_{\mathcal{M}}$.

Proof. The homomorphism $\widehat{\rho} : \widehat{M}_{\mathcal{N}} \longrightarrow \widehat{M}_{\mathcal{M}}$ induced by $(\rho_i)_{i \in I}$ sends $x = (x_1 + N_1, x_2 + N_2, \dots) \in \widehat{M}_{\mathcal{N}}$ to $(x_1 + M_1, x_2 + M_2, \dots)$. Since every N_i contains some M_j we can choose $t(i)$ by induction so that $N_i \supseteq M_{t(i)}$, $t(i) \geqq i$, and $j \geqq i$ implies $t(j) \geqq t(i)$. If $x = (x_1 + N_1, x_2 + N_2, \dots) \in \widehat{M}_{\mathcal{N}}$ and $\rho(x) = 0$, then, for all i, $x_i \in M_i$, $x_{t(i)} \in M_{t(i)} \subseteq N_i$, and $x_i \in N_i$ since $x_{t(i)} + N_i = x_i + N_i$; hence $x = 0$. Thus $\widehat{\rho}$ is injective.

Let $x = (x_1 + M_1, x_2 + M_2, \dots) \in \widehat{M}_{\mathcal{M}}$. Let $y_i = x_{t(i)}$. If $j \geqq i$, then $t(j) \geqq t(i)$, $y_j + M_{t(i)} = y_i + M_{t(i)}$, and $y_j + N_i = y_i + N_i$. Hence $y = (y_1 + N_1, y_2 + N_2, \dots) \in \widehat{M}_{\mathcal{N}}$. For all i, $y_i + M_i = x_{t(i)} + M_i = x_i + M_i$, since $t(i) \geqq i$; hence $\widehat{\rho}(y) = x$. Thus $\widehat{\rho}$ is surjective. $\square$

Proposition 9.8. Let R be a commutative Noetherian ring, let $\mathfrak{a}$ be an ideal of R, and let M be a finitely generated R-module. There is an isomorphism $\widehat{M}_{\mathfrak{a}} \cong \widehat{R}_{\mathfrak{a}} \otimes_R M$, which is natural in M.

Proof. For every $i > 0$ there is an isomorphism $R/\mathfrak{a}^i \otimes_R M \cong M/\mathfrak{a}^i M$, which sends $(r + \mathfrak{a}^i) \otimes x$ to $rx + \mathfrak{a}^i M$ and is natural in M. Hence the projections $\widehat{R} \longrightarrow R/\mathfrak{a}^i$ induce homomorphisms $\widehat{R} \otimes_R M \longrightarrow R/\mathfrak{a}^i \otimes_R M \longrightarrow M/\mathfrak{a}^i M$ and a homomorphism $\alpha : \widehat{R} \otimes_R M \longrightarrow \widehat{M}$, which sends $(r_1 + \mathfrak{a}, r_2 + \mathfrak{a}^2, \ldots) \otimes x$ to $(r_1 x + \mathfrak{a}M, r_2 x + \mathfrak{a}^2 M, \ldots)$ and is natural in M.

If $M = {}_R R$, then $\widehat{R} \otimes M \cong \widehat{R} \cong \widehat{M}$ (as R-modules); the isomorphism sends $(r_1 + \mathfrak{a}, r_2 + \mathfrak{a}^2, \ldots) \otimes 1$ to $(r_1 + \mathfrak{a}, r_2 + \mathfrak{a}^2, \ldots)$ and coincides with α. By 9.4, $\widehat{}_\mathfrak{a}$ preserves finite direct sums; hence α is an isomorphism whenever M is free and finitely generated.

Let M be any finitely generated R-module. There is an exact sequence $0 \longrightarrow K \longrightarrow F \longrightarrow M \longrightarrow 0$ in which F is finitely generated free and K is a submodule of F. Since R is Noetherian, F is Noetherian by VIII.11.5, K is finitely generated, and there is an exact sequence $E \overset{\varphi}{\longrightarrow} F \overset{\sigma}{\longrightarrow} M \longrightarrow 0$ in which E and F are free and finitely generated. This yields a commutative diagram

$$
\begin{array}{ccccccc}
\widehat{R} \otimes_R E & \overset{\overline{\varphi}}{\longrightarrow} & \widehat{R} \otimes_R F & \overset{\overline{\sigma}}{\longrightarrow} & \widehat{R} \otimes_R M & \longrightarrow & 0 \\
{\scriptstyle \alpha_E} \downarrow & & {\scriptstyle \alpha_F} \updownarrow {\scriptstyle \alpha_F^{-1}} & & {\scriptstyle \alpha_M} \updownarrow {\scriptstyle \zeta} & & \\
\widehat{E} & \underset{\widehat{\varphi}}{\longrightarrow} & \widehat{F} & \underset{\widehat{\sigma}}{\longrightarrow} & \widehat{M} & \longrightarrow & 0
\end{array}
$$

in which $\overline{\varphi} = 1 \otimes \varphi$, $\overline{\sigma} = 1 \otimes \sigma$, α_E and α_F are isomorphisms by the above, and the top and bottom rows are exact by 6.7 and 9.6. Hence α_M is an isomorphism: indeed, $\operatorname{Ker} \widehat{\sigma} = \operatorname{Im} \widehat{\varphi} = \alpha_F(\operatorname{Im} \overline{\varphi}) = \operatorname{Ker}(\overline{\sigma} \circ \alpha_F^{-1})$; hence $\overline{\sigma} \circ \alpha_F^{-1} = \zeta \circ \widehat{\sigma}$ for some homomorphism $\zeta : \widehat{M} \longrightarrow \widehat{R} \otimes_R M$; since $\overline{\sigma}$ and $\widehat{\sigma}$ are epimorphisms, then ζ and α_M are mutually inverse isomorphisms. $\square$

Corollary **9.9.** *If R is Noetherian, then $\widehat{R}_\mathfrak{a}$ is a flat R-module, for every ideal $\mathfrak{a}$ of R.*

Proof. If $\mu : A \longrightarrow B$ is a monomorphism of R-modules, and A, B are finitely generated, then $\widehat{R}_\mathfrak{a} \otimes \mu$ is a monomorphism, by 9.6 and 9.8. $\square$

Exercises

1. Let $\mathcal{M} : M_1 \supseteq M_2 \supseteq \cdots$ be a filtration on an R-module M and let $\mathcal{N} : N_1 \supseteq N_2 \supseteq \cdots$ be a filtration on an R-module S. Let $\varphi : M \longrightarrow N$ be a module homomorphism such that $\varphi(M_i) \subseteq N_i$ for all i. Show that φ induces a homomorphism $\widehat{\varphi} : \widehat{M}_\mathcal{M} \longrightarrow \widehat{N}_\mathcal{N}$.

2. State and prove a uniqueness property in the previous exercise. Then state and prove a universal property of $\widehat{M}_\mathcal{M}$.

3. Let $\mathcal{M} : M_1 \supseteq M_2 \supseteq \cdots$ and $\mathcal{N} : N_1 \supseteq N_2 \supseteq \cdots$ be filtrations on an R-module M. Suppose that every M_i contains some N_j and that every N_j contains some M_i. Show that $\widehat{M}_\mathcal{M} \cong \widehat{M}_\mathcal{N}$.

In the following exercises, M is an R-module with a filtration $\mathcal{M} : M_1 \supseteq M_2 \supseteq \cdots ; x \in M$ is a *limit* of a sequence $(x_n)_{n>0}$ of elements of M when for every $i > 0$ there exists $N > 0$ such that $x - x_n \in M_i$ for all $n \geqq N$.

4. Prove the following limit laws: if $r \in R$, x is a limit of $(x_n)_{n>0}$, and y is a limit of $(y_n)_{n>0}$, then $x + y$ is a limit of $(x_n + y_n)_{n>0}$ and rx is a limit of $(rx_n)_{n>0}$; if $\mathcal{M}$ is the $\mathfrak{a}$-adic filtration and r is a limit of $(r_n)_{n>0}$ in R, then rx is a limit of $(r_n x_n)_{n>0}$.

5. A sequence $(x_n)_{n>0}$ of elements of M is *Cauchy* when for every $i > 0$ there exists $N > 0$ such that $x_m - x_n \in M_i$ for all $m, n \geqq N$. Show that every sequence of elements of R that has a limit is Cauchy. Show that M is complete if and only if every Cauchy sequence of elements of M has a unique limit in M.

6. Show that Cauchy sequences of elements of M constitute an R-module; show that sequences with limit zero constitute a submodule of that module; show that the quotient module is isomorphic to $\widehat{M}_{\mathcal{M}}$.

In the following exercises, a *filter* on a module M is a set $\mathcal{F}$ of submodules of M that is directed downward (for every $A, B \in \mathcal{F}$ there exists $C \in \mathcal{F}$ such that $A \cap B \supseteq C$).

7. Define a completion $\widehat{M}_{\mathcal{F}} = \varprojlim_{F \in \mathcal{F}} M/F$, and study its general properties (other than the following exercises). Don't expect to reach much depth.

8. Let R be commutative. Show that the submodules rM of an R-module M constitute a filter on M. State and prove a universal property of the corresponding completion.

9. Let $\mathcal{F}$ be a filter on an R-module M. Show that the cosets of the elements of $\mathcal{F}$ constitute a basis for a topology on R. (This is the *Krull topology* on R, similar to the topology on Galois groups.) Show that the operations on R are continuous for this topology.

10. Let M be complete relative to a filter. Show that the Krull topology on M is Hausdorff and totally disconnected.

XII
Ext and Tor

Homological algebra, the study of homology groups and related constructions, was a branch of algebraic topology until Eilenberg and MacLane [1942] devised a purely algebraic cohomology of groups, which shared many features with the cohomology of topological spaces. Recognition as a separate branch of algebra came with the book *Homological Algebra* [1956], by Cartan and Eilenberg. This chapter contains the basic properties of homology groups, resolutions, Ext, and Tor, with applications to groups and rings. As before, all rings have an identity element, and all modules are unital.

1. Complexes

This section covers basic properties of complexes of modules and their homology and cohomology. The results apply equally to left modules, right modules, and bimodules (over any given ring or rings).

Roots. Complexes and other concepts in this section arose from algebraic topology. *Standard simplexes* are generic points, straight line segments, triangles, tetrahedrons, etc., in Euclidean space; the standard simplex of dimension $n \geq 0$ may be defined as $\Delta_n = \{ (x_0, x_1, \ldots, x_n) \in \mathbb{R}^{n+1} \mid x_0, x_1, \ldots, x_n \geq 0, \ x_0 + x_1 + \cdots + x_n = 1 \}$. In a topological space X, a singular *simplex* of dimension n is a continuous mapping $\Delta_n \longrightarrow X$; singular *chains* of dimension n are formal linear combinations of simplexes with integer coefficients, and constitute a free abelian group $C_n(X)$. The *boundary* of a simplex of dimension $n \geq 1$ has a natural definition as a chain of dimension $n - 1$. This yields boundary homomorphisms $\partial_n : C_n(X) \longrightarrow C_{n-1}(X)$ such that $\partial_n \circ \partial_{n+1} = 0$. The nth singular *homology group* of X is $H_n(X) = \operatorname{Ker} \partial_n / \operatorname{Im} \partial_{n+1}$. These groups are determined by the singular *chain complex* of X, which is the null sequence $\mathcal{C}(X)$ of groups and homomorphisms $C_0(X) \longleftarrow C_1(X) \longleftarrow C_2(X) \longleftarrow \cdots$.

A *cochain* on X of dimension n, with coefficients in an abelian group G, is a homomorphism of $C_n(X)$ into G, such as might result from the assignment of an element of g to every singular simplex of dimension n. These homomorphisms constitute an abelian group $C^n(X, G) = \operatorname{Hom}_{\mathbb{Z}}(C_n(X), G)$. Every cochain $c : C_n(X) \longrightarrow G$ has a *coboundary* $\delta^n(c) = c \circ \partial_{n+1} : C_{n+1} \longrightarrow G$. Then

$\delta^{n+1} \circ \delta^n = 0$. The singular *cohomology groups* of X with coefficients in G are the groups $H^n(X, G) = \operatorname{Ker} \delta^{n+1}/\operatorname{Im} \delta^n$ determined by the singular *cochain complex* $C^0(X, G) \longrightarrow C^1(X, G) \longrightarrow C^2(X, G) \longrightarrow \cdots$. Homology groups with coefficients in G are similarly defined by $H_n(X, G) = \operatorname{Ker} \overline{\partial}_n/\operatorname{Im} \overline{\partial}_{n+1}$, where $\overline{\partial}_n : C_n(X) \otimes_{\mathbb{Z}} G \longrightarrow C_{n-1}(X) \otimes_{\mathbb{Z}} G$ is induced by ∂_n.

Homology. Homological algebra begins with the general concept of a complex and its homology groups or modules.

Definition. A chain complex *of modules is an infinite sequence*

$$\mathcal{C} : \cdots \longleftarrow C_{n-1} \overset{\partial_n}{\longleftarrow} C_n \overset{\partial_{n+1}}{\longleftarrow} C_{n+1} \longleftarrow \cdots$$

of modules and boundary *homomorphisms such that* $\partial_n \partial_{n+1} = 0$ *for all* n.

A *positive* complex $\mathcal{C}$ has $C_n = 0$ for all $n < 0$ and is usually written $C_0 \longleftarrow C_1 \longleftarrow \cdots$. The singular chain complex of a topological space is an example. A *negative* complex $\mathcal{C}$ has $C_n = 0$ for all $n > 0$, and is usually rewritten for convenience as a positive complex $C^0 \longrightarrow C^1 \longrightarrow \cdots$, with $C^n = C_{-n}$ and homomorphisms $\delta^n = \partial_{-n} : C^n \longrightarrow C^{n+1}$, so that $\delta_{n+1} \delta_n = 0$. The singular cochain complexes of a topological space are examples.

Definition. Let $\mathcal{C}$ *be a chain complex of modules. The* nth homology *module of* $\mathcal{C}$ *is* $H_n(\mathcal{C}) = \operatorname{Ker} \partial_n/\operatorname{Im} \partial_{n+1}$.

As in the case of singular chains in topology, the elements of C_n are *n-chains*; the elements of $\operatorname{Ker} \partial_n$ are *n-cycles*; the elements of $\operatorname{Im} \partial_{n+1}$ are *n-boundaries*. $\operatorname{Ker} \partial_n$ and $\operatorname{Im} \partial_{n+1}$ are often denoted by Z_n and B_n. We denote the *homology class* of $z \in \operatorname{Ker} \partial_n$ by $\operatorname{cls} z = z + \operatorname{Im} \partial_{n+1}$.

Definition. Let $\mathcal{A}$ *and* $\mathcal{B}$ *be chain complexes of modules. A* chain transformation $\varphi : \mathcal{A} \longrightarrow \mathcal{B}$ *is a family of module homomorphisms* $\varphi_n : A_n \longrightarrow B_n$ *such that* $\partial_n^{\mathcal{B}} \varphi_n = \varphi_{n-1} \partial_n^{\mathcal{A}}$ *for all* n:

$$
\begin{array}{ccccccccc}
\mathcal{A}: & \cdots \longleftarrow & A_{n-1} & \overset{\partial_n^{\mathcal{A}}}{\longleftarrow} & A_n & \overset{\partial_{n+1}^{\mathcal{A}}}{\longleftarrow} & A_{n+1} & \longleftarrow \cdots \\
\varphi \downarrow & & \varphi_{n-1} \downarrow & & \varphi_n \downarrow & & \downarrow \varphi_{n+1} & \\
\mathcal{B}: & \cdots \longleftarrow & B_{n-1} & \underset{\partial_n^{\mathcal{B}}}{\longleftarrow} & B_n & \underset{\partial_{n+1}^{\mathcal{B}}}{\longleftarrow} & B_{n+1} & \longleftarrow \cdots
\end{array}
$$

For example, every continuous mapping $f : X \longrightarrow Y$ of topological spaces induces a chain transformation $\mathcal{C}(f): \mathcal{C}(X) \longrightarrow \mathcal{C}(Y)$ of their singular chain complexes. In general, chain transformations are added and composed componentwise, the results being chain transformations.

Proposition 1.1. Every chain transformation $\varphi : \mathcal{A} \longrightarrow \mathcal{B}$ *induces a homomorphism* $H_n(\varphi): H_n(\mathcal{A}) \longrightarrow H_n(\mathcal{B})$, *which sends* $\operatorname{cls} z$ *to* $\operatorname{cls} \varphi_n(z)$ *for all* $z \in \operatorname{Ker} \partial_n^{\mathcal{A}}$. *Then* $H_n(-)$ *is an additive functor from chain complexes of modules to modules.*

Proof. Since $\partial_n \varphi_n = \varphi_{n-1} \partial_n$ for all n we have $\varphi_n(\operatorname{Im} \partial_{n+1}^{\mathcal{A}}) \subseteq \operatorname{Im} \partial_{n+1}^{\mathcal{B}}$ and $\varphi_n(\operatorname{Ker} \partial_n^{\mathcal{A}}) \subseteq \operatorname{Ker} \partial_n^{\mathcal{B}}$. By the factorization theorem there is a unique homomorphism $H_n(\varphi) \colon H_n(\mathcal{A}) \longrightarrow H_n(\mathcal{B})$ such that the diagram

$$
\begin{array}{ccccccc}
\operatorname{Im} \partial_{n+1}^{\mathcal{A}} & \overset{\subseteq}{\longrightarrow} & \operatorname{Ker} \partial_n^{\mathcal{A}} & \longrightarrow & H_n(\mathcal{A}) & \longrightarrow & 0 \\
\downarrow & & \downarrow & & \downarrow{\scriptstyle H_n(\varphi)} & & \\
\operatorname{Im} \partial_{n+1}^{\mathcal{B}} & \underset{\subseteq}{\longrightarrow} & \operatorname{Ker} \partial_n^{\mathcal{B}} & \longrightarrow & H_n(\mathcal{B}) & \longrightarrow & 0
\end{array}
$$

commutes, where the other vertical maps are restrictions of φ_n. By uniqueness in the above, $H_n(1_{\mathcal{A}})$ is the identity on $H_n(\mathcal{A})$; $H_n(\psi\,\varphi) = H_n(\psi)\,H_n(\varphi)$ when $\varphi \colon \mathcal{A} \longrightarrow \mathcal{B}$ and $\psi \colon \mathcal{B} \longrightarrow \mathcal{C}$; and $H_n(\psi + \varphi) = H_n(\psi) + H_n(\varphi)$ when $\varphi, \psi \colon \mathcal{A} \longrightarrow \mathcal{B}$. $\square$

We give a sufficient condition that $H_n(\varphi) = H_n(\psi)$ for all n.

Definitions. Let $\varphi, \psi \colon \mathcal{A} \longrightarrow \mathcal{B}$ be chain transformations. A chain homotopy $\sigma \colon \varphi \longrightarrow \psi$ *is a family of homomorphisms $\sigma_n \colon A_n \longrightarrow B_{n+1}$ such that $\varphi_n - \psi_n = \partial_{n+1}^{\mathcal{B}} \circ \sigma_n + \sigma_{n-1} \circ \partial_n^{\mathcal{A}}$ for all n:*

$$
\begin{array}{ccccc}
A_{n-1} & \overset{\partial_n^{\mathcal{A}}}{\longleftarrow} & A_n & \overset{\partial_{n+1}^{\mathcal{A}}}{\longleftarrow} & A_{n+1} \\
\| \| \quad {\scriptstyle \sigma_{n-1}}\!\!\searrow & & \| \| \quad {\scriptstyle \sigma_n}\!\!\searrow & & \| \| \\
B_{n-1} & \underset{\partial_n^{\mathcal{B}}}{\longleftarrow} & B_n & \underset{\partial_{n+1}^{\mathcal{B}}}{\longleftarrow} & B_{n+1}
\end{array}
$$

When there exists a chain homotopy $\sigma \colon \varphi \longrightarrow \psi$, φ and ψ are homotopic.

For example, if $\mathcal{C}(f), \mathcal{C}(g) \colon \mathcal{C}(X) \longrightarrow \mathcal{C}(Y)$ are induced by continuous transformations $f, g \colon X \longrightarrow Y$, then a homotopy (continuous deformation) of f into g induces just such a chain homotopy of $\mathcal{C}(f)$ into $\mathcal{C}(g)$.

Proposition 1.2. *If φ and ψ are homotopic, then $H_n(\varphi) = H_n(\psi)$ for all n.*

Proof. If $\varphi_n - \psi_n = \partial_{n+1}\sigma_n + \sigma_{n-1}\partial_n$ for all n, and $z \in \operatorname{Ker} \partial_n$, then $\varphi_n(z) - \psi_n(z) = (\partial_{n+1}\sigma_n + \sigma_{n-1}\partial_n)(z) = \partial_{n+1}(\sigma_n(z))$ and $H_n(\varphi)(\operatorname{cls} z) - H_n(\psi)(\operatorname{cls} z) = \operatorname{cls} \varphi_n(z) - \operatorname{cls} \psi_n(z) = 0$. $\square$

The exact homology sequence. A short exact sequence of complexes induces a long exact sequence that connects all their homology modules.

Definition. A sequence $\mathcal{A} \overset{\varphi}{\longrightarrow} \mathcal{B} \overset{\psi}{\longrightarrow} \mathcal{C}$ of chain complexes and transformations is exact *when the sequence $A_n \overset{\varphi_n}{\longrightarrow} B_n \overset{\psi_n}{\longrightarrow} C_n$ is exact for every n.*

Theorem 1.3 (Exact Homology Sequence). *Every short exact sequence $\mathcal{E} \colon 0 \longrightarrow \mathcal{A} \longrightarrow \mathcal{B} \longrightarrow \mathcal{C} \longrightarrow 0$ of chain complexes induces an exact sequence*

$$\cdots H_{n+1}(\mathcal{C}) \longrightarrow H_n(\mathcal{A}) \longrightarrow H_n(\mathcal{B}) \longrightarrow H_n(\mathcal{C}) \longrightarrow H_{n-1}(\mathcal{A}) \cdots,$$

which is natural in $\mathcal{E}$.

Proof. Exactness at $H_n(\mathcal{B})$ is proved by diagram chasing in:

$$
\begin{array}{ccccc}
A_{n+1} & \xrightarrow{\varphi_{n+1}} & B_{n+1} & \xrightarrow{\psi_{n+1}} & C_{n+1} \\
\Big\downarrow{\partial_{n+1}} & & \Big\downarrow{\partial_{n+1}} & & \Big\downarrow{\partial_{n+1}} \\
A_n & \xrightarrow{\varphi_n} & B_n & \xrightarrow{\psi_n} & C_n \\
\Big\downarrow{\partial_n} & & \Big\downarrow{\partial_n} & & \Big\downarrow{\partial_n} \\
A_{n-1} & \xrightarrow{\varphi_{n-1}} & B_{n-1} & \xrightarrow{\psi_{n-1}} & C_{n-1}
\end{array}
$$

First, $H_n(\psi)\big(H_n(\varphi)(\mathrm{cls}\,a)\big) = H_n(a)(\mathrm{cls}\,\varphi_n a) = \mathrm{cls}\,\psi_n\varphi_n a = 0$.

Conversely, if $b \in \mathrm{Ker}\,\partial_n^{\mathcal{B}}$ and $H_n(\psi)(\mathrm{cls}\,b) = \mathrm{cls}\,\psi_n b = 0$ in $H_n(\mathcal{C})$, then $\psi_n b = \partial_{n+1} c'$ for some $c' \in C_{n+1}$, $c' = \psi_{n+1} b'$ for some $b' \in B_{n+1}$, $\psi_n\,\partial_{n+1} b' = \partial_{n+1}\psi_{n+1} b' = \psi_n b$, $b - \partial_{n+1} b' \in \mathrm{Ker}\,\psi_n$, $b - \partial_{n+1} b' = \varphi_n a$ for some $a \in A_n$, and $a \in \mathrm{Ker}\,\partial_n$, since $\varphi_{n-1}\,\partial_n a = \partial_n\varphi_n a = \partial_n b = 0$; then $b = \varphi_n a + \partial_{n+1} b'$ yields $\mathrm{cls}\,b = \mathrm{cls}\,\varphi_n a = H_n(\varphi)(\mathrm{cls}\,a)$.

Next, construct the *connecting homomorphism* $\chi_{n+1} : H_{n+1}(\mathcal{C}) \longrightarrow H_n(\mathcal{A})$:

Lemma 1.4. *In Theorem 1.3, a connecting homomorphism $H_{n+1}(\mathcal{C}) \xrightarrow{\chi_{n+1}} H_n(\mathcal{A})$ is well defined by $\chi_{n+1}\mathrm{cls}\,c = \mathrm{cls}\,a$ whenever $c \in \mathrm{Ker}\,\partial_{n+1}$, $c = \psi_{n+1} b$, and $\partial_{n+1} b = \varphi_n a$, for some $b \in B_{n+1}$. This assigns a connecting homomorphism to every sequence $\mathcal{E}$ and integer $n+1$.*

Proof. First, $\partial_{n+1} b = \varphi_n a$ implies $\varphi_{n-1}\,\partial_n a = \partial_n\varphi_n a = \partial_n\,\partial_{n+1} b = 0$, $\partial_n a = 0$ since φ_{n-1} is injective, and $a \in \mathrm{Ker}\,\partial_n$. Assume that $c_1, c_2 \in \mathrm{Ker}\,\partial_{n+1}^{\mathcal{C}}$ and $\mathrm{cls}\,c_1 = \mathrm{cls}\,c_2$, $c_1 = \psi_{n+1} b_1$, $\partial_{n+1} b_1 = \varphi_n a_1$, $c_2 = \psi_{n+1} b_2$, $\partial_{n+1} b_2 = \varphi_n a_2$. Then $c_2 - c_1 = \partial_{n+2} c'$ for some $c' \in C_{n+2}$, $c' = \psi_{n+2} b'$ for some $b' \in B_{n+2}$, $\psi_{n+1}(b_2 - b_1) = c_2 - c_1 = \partial_{n+2}\psi_{n+2} b' = \psi_{n+1}\,\partial_{n+2} b'$, $b_2 - b_1 - \partial_{n+2} b' \in \mathrm{Ker}\,\psi_{n+1} = \mathrm{Im}\,\varphi_{n+1}$, $b_2 - b_1 - \partial_{n+2} b' = \varphi_{n+1} a$ for some $a \in A_{n+1}$, $\varphi_n(a_2 - a_1) = \partial_{n+1}(b_2 - b_1) = \partial_{n+1}\varphi_{n+1} a = \varphi_n\,\partial_{n+1} a$, $a_2 - a_1 = \partial_{n+1} a$ since φ_n is injective, and $\mathrm{cls}\,a_1 = \mathrm{cls}\,a_2$. Thus χ_{n+1} is well defined. It is immediate that χ_{n+1} is a homomorphism. $\square$

Exactness at $H_n(\mathcal{A})$ is proved as follows. If $c \in \mathrm{Ker}\,\partial_{n+1}$ and $\chi_{n+1}\mathrm{cls}\,c = \mathrm{cls}\,a$, then $c = \psi_{n+1} b$ and $\partial_{n+1} b = \varphi_n a$ for some $b \in B_{n+1}$, and $H_n(\varphi)(\mathrm{cls}\,a) = \mathrm{cls}\,\varphi_n a = 0$, since $\varphi_n a \in \mathrm{Im}\,\partial_{n+1}$. Conversely, if $H_n(\varphi)(\mathrm{cls}\,a) = 0$, where $a \in \mathrm{Ker}\,\partial_n$, then $\varphi_n a = \partial_{n+1} b$ for some $b \in B_{n+1}$ and $\partial_{n+1}\psi_{n+1} b = \psi_n\,\partial_{n+1} b = \psi_n\varphi_n a = 0$; hence $\psi_{n+1} b \in \mathrm{Ker}\,\partial_{n+1}$ and $\mathrm{cls}\,a = \chi_{n+1}\,\mathrm{cls}\,\psi_{n+1} b$.

Exactness at $H_{n+1}(\mathcal{C})$ is similar. If $\mathrm{cls}\,c = H_{n+1}(\psi)(\mathrm{cls}\,b)$, then in the computation of $\chi_{n+1}\mathrm{cls}\,c$ we may let $c = \psi_{n+1} b$, and then $\chi_{n+1}\mathrm{cls}\,c = \mathrm{cls}\,a$, where $\varphi_n a = \partial_{n+1} b = 0$, and $\chi_{n+1}\mathrm{cls}\,c = 0$. Conversely, if $\chi_{n+1}\mathrm{cls}\,c = 0$, then $c = \psi_{n+1} b$ and $\partial_{n+1} b = \varphi_n a$ for some $b \in B_{n+1}$ and $a \in \mathrm{Im}\,\partial_{n+1}$, $a = \partial_{n+1} a'$ for some $a' \in A_{n+1}$, $\partial_{n+1}\varphi_{n+1} a' = \varphi_n\,\partial_{n+1} a' = \varphi_n a = \partial_{n+1} b$,

$b' = b - \varphi_{n+1}a' \in \text{Ker } \partial_{n+1}$, $\psi_{n+1}b' = \psi_{n+1}b$, and $\text{cls } c = H_{n+1}(\psi)(\text{cls } b')$.

We show naturality in $\mathcal{E}$: if α, β, γ are chain transformations and the diagram

$$
\begin{array}{ccccccccc}
0 & \longrightarrow & \mathcal{A} & \xrightarrow{\varphi} & \mathcal{B} & \xrightarrow{\psi} & \mathcal{C} & \longrightarrow & 0 \\
& & \downarrow{\alpha} & & \downarrow{\beta} & & \downarrow{\gamma} & & \\
0 & \longrightarrow & \mathcal{A}' & \xrightarrow{\varphi'} & \mathcal{B}' & \xrightarrow{\psi'} & \mathcal{C}' & \longrightarrow & 0
\end{array}
$$

has exact rows and commutes, then the diagram

$$
\begin{array}{ccccccccc}
\cdots & H_{n+1}(\mathcal{C}) & \xrightarrow{\chi} & H_n(\mathcal{A}) & \longrightarrow & H_n(\mathcal{B}) & \longrightarrow & H_n(\mathcal{C}) & \xrightarrow{\chi} & H_{n-1}(\mathcal{A}) & \cdots \\
& \downarrow & & \downarrow & & \downarrow & & \downarrow & & \downarrow & \\
\cdots & H_{n+1}(\mathcal{C}') & \xrightarrow{\chi'} & H_n(\mathcal{A}') & \longrightarrow & H_n(\mathcal{B}') & \longrightarrow & H_n(\mathcal{C}') & \xrightarrow{\chi'} & H_{n-1}(\mathcal{A}') & \cdots
\end{array}
$$

where the vertical maps are induced by α, β, γ, also commutes. The middle squares commute since $\beta\varphi = \varphi'\alpha$, $\gamma\psi = \psi'\beta$ implies $H_n(\beta)\,H_n(\varphi) = H_n(\varphi')\,H_n(\alpha)$, $H_n(\gamma)\,H_n(\psi) = H_n(\psi')\,H_n(\beta)$.

For the left square (hence also for the right square) let $\chi_{n+1}\text{cls } c = \text{cls } a$, so that $c = \psi_{n+1}b$ and $\partial_{n+1}b = \varphi_n a$ for some $b \in B_{n+1}$. Then $H_n(\alpha)(\chi_{n+1}\text{cls } c) = \text{cls } \alpha_n a$; now $\gamma_{n+1}c = \gamma_{n+1}\psi_{n+1}b = \psi'_{n+1}\beta_{n+1}b$ and $\partial'_{n+1}\beta_{n+1}b = \beta_n\partial_{n+1}b = \beta_n\varphi_n a = \varphi'_n\alpha_n a$; hence $\chi'_{n+1}\text{cls } \gamma_{n+1}c = \text{cls } \alpha_n a$ and $\chi'_{n+1}H_n(\gamma)(\text{cls } c) = \chi'_{n+1}\text{cls } \gamma_{n+1}c = \text{cls } \alpha_n a = H_n(\alpha)(\chi_{n+1}\text{cls } c)$. $\square$

Theorem 1.3 also follows from the diagram lemma below (see the exercises):

Lemma **1.5** (Ker-Coker Sequence). *Every commutative diagram* $\mathcal{D}$

$$
\begin{array}{ccccccc}
A & \xrightarrow{\mu} & B & \xrightarrow{\sigma} & C & \longrightarrow & 0 \\
\downarrow{\alpha} & & \downarrow{\beta} & & \downarrow{\gamma} & & \\
0 & \longrightarrow & A' & \xrightarrow{\mu'} & B' & \xrightarrow{\sigma'} & C'
\end{array}
$$

with exact rows induces an exact sequence, which is natural in $\mathcal{D}$,

$$\text{Ker } \alpha \longrightarrow \text{Ker } \beta \longrightarrow \text{Ker } \gamma \longrightarrow \text{Coker } \alpha \longrightarrow \text{Coker } \beta \longrightarrow \text{Coker } \gamma.$$

Cohomology. Complexes of modules have cohomology groups with coefficients in modules:

Definitions. Let $\mathcal{A}$ *be a chain complex of left R-modules,*

$$\mathcal{A} : \cdots \longleftarrow A_{n-1} \xleftarrow{\partial_n} A_n \xleftarrow{\partial_{n+1}} A_{n+1} \longleftarrow \cdots .$$

An n-cochain of $\mathcal{A}$ *with coefficients in a left R-module G is a module homomorphism u of* A_n *into G; its cochain is the (n+1)-cochain* $\delta^n(u) = u \circ \partial_{n+1}$.

Under pointwise addition, the n-cochains of $\mathcal{A}$ with coefficients in G constitute an abelian group $C_R^n(\mathcal{A}, G) = \mathrm{Hom}_R(A_n, G)$. If G is a bimodule, or if $\mathcal{A}$ is a complex of bimodules, or both, then $C_R^n(\mathcal{A}, G)$ is a module or bimodule.

Definition. Let $\mathcal{A}$ be a chain complex of left R-modules and let G be a left R-module. The cohomology groups *of $\mathcal{A}$ with coefficients in G are the homology groups $H_R^n(\mathcal{A}, G) = \mathrm{Ker}\, \delta^n / \mathrm{Im}\, \delta^{n-1}$ of the cochain complex*

$$\mathrm{Hom}_R(\mathcal{A}, G): \ \cdots \ C_R^{n-1}(\mathcal{A}, G) \xrightarrow{\delta^{n-1}} C_R^n(\mathcal{A}, G) \xrightarrow{\delta^n} C_R^{n+1}(A_n, G) \ \cdots .$$

As in case of cochains in topology, the elements of $\mathrm{Ker}\, \delta^n$ are *n-cocycles*; the elements of $\mathrm{Im}\, \delta^{n+1}$ are *n-coboundaries*. $\mathrm{Ker}\, \delta^n$ and $\mathrm{Im}\, \delta^{n+1}$ are often denoted by Z^n and B^n. We denote the *cohomology class* of $z \in \mathrm{Ker}\, \delta^n$ by $\mathrm{cls}\, z = z + \mathrm{Im}\, \delta^{n-1}$.

One of our goals is to compute the cohomology groups of a complex from its homology groups or modules. Some results to that effect are given in Section 6. For now, Theorem 1.3 yields exact sequences for cohomology groups. This requires some restrictions, since short exact sequences of modules or chain complexes do not always induce short exact sequences of cochain complexes.

Theorem 1.6 (Exact Cohomology Sequence). *Let G be a left R-module and let $\mathcal{E} : 0 \longrightarrow \mathcal{A} \longrightarrow \mathcal{B} \longrightarrow \mathcal{C} \longrightarrow 0$ be a short exact sequence of chain complexes of left R-modules. If every A_n is injective, or if every C_n is projective, then $\mathcal{E}$ induces an exact sequence, which is natural in $\mathcal{E}$ and G,*

$$\cdots \longrightarrow H_R^n(\mathcal{C}, G) \longrightarrow H_R^n(\mathcal{B}, G) \longrightarrow H_R^n(\mathcal{A}, G) \longrightarrow H_R^{n+1}(\mathcal{C}, G) \longrightarrow \cdots .$$

Proof. If every A_n is injective, or if every C_n is projective, then the sequence $0 \longrightarrow A_n \longrightarrow B_n \longrightarrow C_n \longrightarrow 0$ splits; hence the sequence

$$0 \longrightarrow \mathrm{Hom}_R(C_n, G) \longrightarrow \mathrm{Hom}_R(B_n, G) \longrightarrow \mathrm{Hom}_R(A_n, G) \longrightarrow 0$$

splits, in particular is short exact; then

$$0 \longrightarrow \mathrm{Hom}_R(\mathcal{C}, G) \longrightarrow \mathrm{Hom}_R(\mathcal{B}, G) \longrightarrow \mathrm{Hom}_R(\mathcal{A}, G) \longrightarrow 0$$

is a short exact sequence of complexes (of abelian groups), which is natural in $\mathcal{E}$ and G, and the result follows from Theorem 1.3. $\square$

Theorem 1.7 (Exact Cohomology Sequence). *Let $\mathcal{A}$ be a chain complex of left R-modules and let $\mathcal{E} : 0 \longrightarrow G \longrightarrow G' \longrightarrow G'' \longrightarrow 0$ be a short exact sequence of left R-modules. If every A_n is projective, then $\mathcal{E}$ induces an exact sequence, which is natural in $\mathcal{E}$ and $\mathcal{A}$,*

$$\cdots \longrightarrow H_R^n(\mathcal{A}, G) \longrightarrow H_R^n(\mathcal{A}, G') \longrightarrow H_R^n(\mathcal{A}, G'') \longrightarrow H_R^{n+1}(\mathcal{A}, G) \longrightarrow \cdots .$$

Proof. If every A_n is projective, then the sequence

$$0 \longrightarrow \mathrm{Hom}_R(A_n, G) \longrightarrow \mathrm{Hom}_R(A_n, G') \longrightarrow \mathrm{Hom}_R(A_n, G'') \longrightarrow 0$$

is exact, by XI.2.2; hence

$$0 \longrightarrow \mathrm{Hom}_R(\mathcal{A}, G) \longrightarrow \mathrm{Hom}_R(\mathcal{A}, G') \longrightarrow \mathrm{Hom}_R(\mathcal{A}, G'') \longrightarrow 0$$

is a short exact sequence of complexes (of abelian groups), which is natural in $\mathcal{E}$ and $\mathcal{A}$, and the result follows from Theorem 1.3. $\square$

In Theorems 1.6 and 1.7, connecting homomorphisms can be assigned to every short exact sequence $\mathcal{E}$ and integer n. This follows from Lemma 1.4 (or from Lemma 1.5). Lemma 1.4 yields constructions of these homomorphisms.

Lemma 1.8. *If* $\mathcal{A} \xrightarrow{\varphi} \mathcal{B} \xrightarrow{\psi} \mathcal{C}'$ *in Theorem 1.6, then the connecting homomorphism* $H^n(\mathcal{A}, G) \xrightarrow{\chi^n} H^{n+1}(\mathcal{C}, G)$ *is well defined by* $\chi^n \mathrm{cls}\, \alpha = \mathrm{cls}\, \gamma$ *whenever* $\alpha \partial_{n+1} = 0$, $\alpha = \beta \varphi_n$, *and* $\beta \partial_{n+1} = \gamma \psi_{n+1}$ *for some* $\beta : B_n \longrightarrow G$:

$$
\begin{array}{ccccc}
A_{n+1} & \xrightarrow{\partial_{n+1}} & A_n & \xrightarrow{\alpha} & G \\
& {\scriptstyle\varphi_n}\downarrow & \swarrow{\scriptstyle\beta} & & \searrow{\scriptstyle\gamma} \\
& B_n & \xleftarrow[\partial_{n+1}]{} & B_{n+1} & \xrightarrow[\psi_{n+1}]{} & C_{n+1}
\end{array}
$$

Proof. The map χ^n is induced by $\mathrm{Hom}_R(\mathcal{C}, G) \xrightarrow{\psi^*} \mathrm{Hom}_R(\mathcal{B}, G) \xrightarrow{\varphi^*} \mathrm{Hom}_R(\mathcal{A}, G)$, in which $\mathrm{Hom}_R(\mathcal{C}, G)$ has coboundary $\delta^n = \partial_{n+1}^* = \mathrm{Hom}_R(\partial_{n+1}, G)$, and similarly for $\mathrm{Hom}_R(\mathcal{B}, G)$ and $\mathrm{Hom}_R(\mathcal{A}, G)$. By 1.4, $\chi^n \mathrm{cls}\, \alpha = \mathrm{cls}\, \gamma$ whenever $\alpha \in \mathrm{Ker}\, \delta^n$, $\alpha = \varphi_n^*(\beta)$, and $\delta^n(\beta) = \psi_{n+1}^*(\gamma)$ for some $\beta \in \mathrm{Hom}_R(B_n, G)$; equivalently, $\alpha \partial_{n+1}^{\mathcal{A}} = 0$, $\alpha = \beta \varphi_n$, and $\beta \partial_{n+1}^{\mathcal{B}} = \gamma \psi_{n+1}$ for some $\beta : B_n \longrightarrow G$. $\square$

Lemma 1.9. *If* $G \xrightarrow{\varphi} G' \xrightarrow{\psi} G''$ *in Theorem 1.7, then the connecting homomorphism* $H^n(\mathcal{A}, G'') \xrightarrow{\chi^n} H^{n+1}(\mathcal{A}, G)$ *is well defined by* $\chi^n \mathrm{cls}\, \alpha'' = \mathrm{cls}\, \alpha$ *whenever* $\alpha'' \partial_{n+1} = 0$, $\alpha'' = \psi \alpha'$, *and* $\alpha' \partial_{n+1} = \varphi \alpha$ *for some* $\alpha' : A_n \longrightarrow G'$:

$$
\begin{array}{ccc}
A_{n+1} & \xrightarrow{\partial_{n+1}} & A_n \\
{\scriptstyle\alpha}\downarrow & {\scriptstyle\alpha'}\downarrow & \searrow{\scriptstyle\alpha''} \\
G & \xrightarrow[\varphi]{} & G' & \xrightarrow[\psi]{} & G''
\end{array}
$$

Proof. The map χ^n is induced by $\mathrm{Hom}_R(\mathcal{A}, G) \xrightarrow{\varphi_*} \mathrm{Hom}_R(\mathcal{A}, G') \xrightarrow{\psi_*} \mathrm{Hom}_R(\mathcal{A}, G'')$, in which $\mathrm{Hom}_R(\mathcal{A}, G)$ has coboundary $\delta^n = \partial_{n+1}^* = \mathrm{Hom}_R(\partial_{n+1}, G)$ and similarly for $\mathrm{Hom}_R(\mathcal{A}, G')$ and $\mathrm{Hom}_R(\mathcal{A}, G'')$. By 1.4, $\chi^n \mathrm{cls}\, \alpha'' = \mathrm{cls}\, \alpha$ whenever $\alpha'' \in \mathrm{Ker}\, \delta^n$, $\alpha'' = \psi_*(\alpha')$, and $\delta^n(\alpha') = \psi_*(\alpha)$ for some $\alpha' \in \mathrm{Hom}_R(A_n, G')$; equivalently, $\alpha'' \partial_{n+1} = 0$, $\alpha'' = \psi \alpha'$, and $\alpha' \partial_{n+1} = \varphi \alpha$ for some $\alpha' : A_n \longrightarrow G'$. $\square$

Exercises

All complexes in the following exercises are chain complexes of left R-modules.

1. Define direct products of complexes and prove that H_n preserves direct products.

2. Define direct sums of complexes and prove that H_n preserves direct sums.

3. Define direct limits of complexes and prove that H_n preserves direct limits.

4. State and prove a homomorphism theorem for complexes.

5. Let $\varphi, \psi : A \longrightarrow B$ and $\chi, \omega : B \longrightarrow C$ be chain transformations. Prove the following: if there are chain homotopies $\varphi \longrightarrow \psi$ and $\chi \longrightarrow \omega$, then there is a chain homotopy $\chi \circ \varphi \longrightarrow \omega \circ \psi$.

6. Prove the following: if $0 \longrightarrow A \longrightarrow B \longrightarrow C \longrightarrow 0$ is a short exact sequence of complexes, and $H_n(A) = H_n(C) = 0$ for all n, then $H_n(B) = 0$ for all n.

7. Give a short proof of the nine lemma.

8. The *mapping cone* (also called *mapping cylinder*) of a chain transformation $\varphi : A \longrightarrow B$ is the complex M in which $M_n = A_{n-1} \oplus B_n$ and $\partial(a, b) = (-\partial a, \; \partial b + \varphi a)$. Verify that M is a complex and show that φ induces a sequence, which is natural in φ,

$$\cdots \longrightarrow H_n(B) \longrightarrow H_n(M) \longrightarrow H_{n-1}(A) \longrightarrow H_{n-1}(B) \longrightarrow \cdots .$$

9. Given a commutative diagram with exact rows

$$
\begin{array}{ccccccccc}
\cdots A_n & \xrightarrow{\varphi_n} & B_n & \xrightarrow{\psi_n} & C_n & \xrightarrow{\xi_n} & A_{n-1} & \xrightarrow{\varphi_{n-1}} & B_{n-1} \cdots \\
\alpha_n \downarrow & & \beta_n \downarrow & & \gamma_n \downarrow & & \alpha_{n-1} \downarrow & & \beta_{n-1} \downarrow \\
\cdots A'_n & \xrightarrow{\varphi'_n} & B'_n & \xrightarrow{\psi'_n} & C'_n & \xrightarrow{\xi'_n} & A'_{n-1} & \xrightarrow{\varphi'_{n-1}} & B'_{n-1} \cdots
\end{array}
$$

in which every γ_n is an isomorphism, show that there is an exact sequence

$$\cdots A_n \xrightarrow{\zeta_n} A'_n \oplus B_n \xrightarrow{\eta_n} B'_n \xrightarrow{\lambda_n} A_{n-1} \xrightarrow{\zeta_{n-1}} A'_{n-1} \oplus B_{n-1} \cdots$$

in which $\zeta_n a = (\alpha_n a, \; \varphi_b a)$, $\eta_n(a', b) = \varphi'_n a' - \beta_n b$, and $\lambda_n = \xi_n \gamma_n^{-1} \psi'_n$. (This is the purely algebraic version of the Mayer-Vietoris sequence in topology.)

10. Show that $\partial_n : A_n \longrightarrow A_{n-1}$ induces a homomorphism $\bar{\partial}_n : A_n / \mathrm{Im} \; \partial_{n+1} \longrightarrow \mathrm{Ker} \; \partial_{n-1}$, which is natural in A, such that $\mathrm{Ker} \; \bar{\partial}_n = H_n(A)$ and $\mathrm{Coker} \; \bar{\partial}_n = H_{n-1}(A)$.

11. Show that a commutative diagram D

$$
\begin{array}{ccccccc}
A & \xrightarrow{\mu} & B & \xrightarrow{\sigma} & C & \longrightarrow & 0 \\
\alpha \downarrow & & \beta \downarrow & & \gamma \downarrow & & \\
0 & \longrightarrow & A' & \xrightarrow{\mu'} & B' & \xrightarrow{\sigma'} & C'
\end{array}
$$

with exact rows induces an exact sequence, the *Ker-Coker sequence*, which is natural in D,

$$\mathrm{Ker} \; \alpha \longrightarrow \mathrm{Ker} \; \beta \longrightarrow \mathrm{Ker} \; \gamma \longrightarrow \mathrm{Coker} \; \alpha \longrightarrow \mathrm{Coker} \; \beta \longrightarrow \mathrm{Coker} \; \gamma.$$

12. Derive the exact homology sequence from the previous two exercises.

2. Resolutions

Projective resolutions and injective resolutions are exact sequences of projective modules, or of injective modules, used to analyze modules.

Definition. A module A can be analyzed by what we call *projective presentations*, which are short exact sequences $0 \longrightarrow K \longrightarrow P \longrightarrow A \longrightarrow 0$, where P is projective. A projective presentation can be assigned to every module A: e.g., let P be the free module $\bigoplus_A {}_R R$ generated by the set A, let $P \longrightarrow A$ be the module homomorphism induced by the identity on A, and let K be its kernel.

Module homomorphisms lift to projective presentations:

$$
\begin{array}{ccccccccc}
0 & \longrightarrow & K & \overset{\mu}{\longrightarrow} & P & \overset{\sigma}{\longrightarrow} & C & \longrightarrow & 0 \\
& & \alpha \downarrow & & \beta \downarrow & & \downarrow \gamma & & \\
0 & \longrightarrow & A' & \underset{\mu'}{\longrightarrow} & B' & \underset{\sigma'}{\longrightarrow} & C' & \longrightarrow & 0
\end{array}
$$

Lemma 2.1. *Given a diagram with exact rows (solid arrows) in which P is projective, there exist homomorphisms α and β (dotted arrows) that make the diagram commutative.*

Proof. Since P is projective, $\gamma \sigma$ factors through the epimorphism σ': $\gamma \sigma = \sigma' \beta$ for some homomorphism $\beta : P \longrightarrow B'$. Then $\sigma' \beta \mu = \gamma \sigma \mu = 0$; hence $\beta \mu$ factors uniquely through μ'. $\square$

Projective resolutions give a more extensive analysis by composting projective presentations into a long exact sequence.

Definition. A projective resolution *of a module A is an exact sequence*

$$
\cdots \longrightarrow P_n \overset{\partial_n}{\longrightarrow} P_{n-1} \longrightarrow \cdots \overset{\partial_1}{\longrightarrow} P_0 \overset{\varepsilon}{\longrightarrow} A \longrightarrow 0
$$

of modules and homomorphisms in which P_0, P_1, ..., P_n, ... are projective. A free resolution *is a projective resolution in which P_0, ..., P_n, ... are free.*

Thus, a projective resolution $\mathcal{P} \overset{\varepsilon}{\longrightarrow} A$ of A consists of a positive complex

$$
\mathcal{P}: \cdots \longrightarrow P_n \overset{\partial_n}{\longrightarrow} P_{n-1} \longrightarrow \cdots \overset{\partial_1}{\longrightarrow} P_0 \longrightarrow 0
$$

of projective modules and an epimorphism $\varepsilon : P_0 \longrightarrow A$, such that $H_n(\mathcal{P}) = 0$ for all $n > 0$ and $\operatorname{Im} \partial_1 = \operatorname{Ker} \varepsilon$. Then $H_0(\mathcal{P}) = P_0/\operatorname{Im} \partial_1 \cong A$; in particular, A is determined by $\mathcal{P}$, up to isomorphism. Moreover, $\mathcal{P} \overset{\varepsilon}{\longrightarrow} A$ is a composition of projective presentations

$$
0 \longrightarrow K_0 \longrightarrow P_0 \longrightarrow A \longrightarrow 0, \ ..., \ 0 \longrightarrow K_n \longrightarrow P_n \longrightarrow K_{n-1}, ...,
$$

where $K_0 = \operatorname{Ker} \varepsilon = \operatorname{Im} \partial_0$, $K_n = \operatorname{Ker} \partial_n = \operatorname{Im} \partial_{n+1}$ for all $n > 0$; K_n is the nth *syzygy* of A in the given projective resolution.

Every module has a projective resolution. In fact, a free resolution can be assigned to every module A: we saw that a presentation $0 \longrightarrow K_0 \longrightarrow F_0 \longrightarrow$

$A \longrightarrow 0$ (with F_0 free) can be assigned to A; continuing with the presentation $0 \longrightarrow K_1 \longrightarrow F_1 \longrightarrow K_0 \longrightarrow 0$ assigned to K_0, the presentation assigned to K_1, and so forth, assigns to A a free resolution $\cdots \longrightarrow F_1 \longrightarrow F_0 \longrightarrow A \longrightarrow 0$. We saw that in Section VIII.9 that free resolutions can be effectively computed for all finitely generated free R-modules if $R = K[X_1, ..., X_n]$.

Properties. Module homomorphisms *lift* to chain transformation of projective resolutions, and do so with considerable zest and some uniqueness.

Theorem **2.2** (Comparison Theorem). *Let* $\mathcal{P} \xrightarrow{\varepsilon} A$ *and* $\mathcal{Q} \xrightarrow{\zeta} B$ *be projective resolutions, and let* $\varphi : A \longrightarrow B$ *be a homomorphism. There is a chain transformation* $\overline{\varphi} = (\varphi_n)_{n \geq 0} \colon \mathcal{P} \longrightarrow \mathcal{Q}$ *such that* $\zeta \varphi_0 = \varphi \varepsilon$:

$$\begin{array}{ccc} \mathcal{P} & \dashrightarrow & \mathcal{Q} \\ \downarrow & & \downarrow \\ A & \longrightarrow & B \end{array}$$

Moreover, any two such chain transformations are homotopic.

Proof. Since P_0 is projective, and $\zeta : Q_0 \longrightarrow B$ is surjective, $\varphi \varepsilon$ factors through ζ, and $\varphi \varepsilon = \zeta \varphi_0$ for some $\varphi_0 : P_0 \longrightarrow Q_0$:

$$\begin{array}{ccc} P_0 & \xrightarrow{\varepsilon} & A \\ {\scriptstyle \varphi_0} \downarrow & & \downarrow {\scriptstyle \varphi} \\ Q_0 & \xrightarrow{\zeta} & B \end{array}$$

From this auspicious start φ_n is constructed recursively. Assume that $\varphi_0, ..., \varphi_n$ have been constructed, so that $\partial_n \varphi_n = \varphi_{n-1} \partial_n$. Then $\partial_n \varphi_n \partial_{n+1} = \varphi_{n-1} \partial_n \partial_{n+1} = 0$ (or $\zeta \varphi_0 \partial_1 = \varphi \partial_0 \partial_1 = 0$, if $n = 0$). Since $\mathcal{Q}$ is exact this implies Im $\varphi_n \partial_{n+1} \subseteq$ Im ∂_{n+1} (or Im $\varphi_0 \partial_1 \subseteq$ Im ∂_1, if $n = 0$). Hence $\varphi_n \partial_{n+1}$ and ∂_{n+1} induce a homomorphism $P_{n+1} \longrightarrow$ Im ∂_{n+1} and an epimorphism $Q_{n+1} \longrightarrow$ Im ∂_{n+1}. Since P_{n+1} is projective, $\varphi_n \partial_{n+1}$ factors through ∂_{n+1}, and $\varphi_n \partial_{n+1} = \varphi_{n+1} \partial_{n+1}$ for some $\varphi_{n+1} : P_{n+1} \longrightarrow Q_{n+1}$:

$$\begin{array}{ccccc} & & P_{n+1} & \xrightarrow{\partial_{n+1}} P_n \xrightarrow{\partial_n} P_{n-1} \\ {\scriptstyle \varphi_{n+1}} \swarrow & & \downarrow & {\scriptstyle \varphi_n} \downarrow \qquad \downarrow {\scriptstyle \varphi_{n-1}} \\ Q_{n+1} & \xrightarrow{\partial_{n+1}} & \text{Im } \partial_{n+1} \underset{\subseteq}{\longrightarrow} Q_n \xrightarrow{\partial_n} Q_{n-1} \end{array}$$

Let $(\psi_n)_{n \geq 0} \colon \mathcal{P} \longrightarrow \mathcal{Q}$ be another chain transformation that lifts φ ($\zeta \psi_0 = \varphi \varepsilon$). Then $(\psi_n - \varphi_n)_{n \geq 0}$ is a chain transformation that lifts 0. To complete the proof we show that if $\varphi = 0$ in the above (if $\zeta \varphi_0 = 0$), then $(\varphi_n)_{n \geq 0}$ is homotopic to 0: $\varphi_n = \partial_{n+1} \sigma_n + \sigma_{n-1} \partial_n$ for some $\sigma_n : P_n \longrightarrow Q_{n+1}$.

Since $\mathcal{P}$ and $\mathcal{Q}$ are positive complexes we start with $\sigma_n = 0$ for all $n < 0$. Since $\zeta \varphi_0 = 0$, φ_0 induces a homomorphism $P_0 \longrightarrow$ Ker $\zeta =$ Im ∂_1, which factors through the epimorphism $Q_1 \longrightarrow \mathrm{Im}\, \partial_1$ since P_0 is projective; then $\varphi_0 = \partial_1 \sigma_0 = \partial_1 \sigma_0 + \sigma_{-1} \partial_0$:

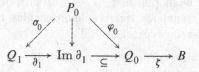

The remaining σ_n are constructed recursively. Assume that $\sigma_0, \ldots, \sigma_n$ have been constructed, so that $\varphi_n = \partial_{n+1} \sigma_n + \sigma_{n-1} \partial_n$. Then $\partial_{n+1} \varphi_{n+1} = \varphi_n \partial_{n+1} = \partial_{n+1} \sigma_n \partial_{n+1}$. Hence $\mathrm{Im}\, (\varphi_{n+1} - \sigma_n \partial_{n+1}) \subseteq \mathrm{Ker}\, \partial_{n+1} = \mathrm{Im}\, \partial_{n+2}$ and $\varphi_{n+1} - \sigma_n \partial_{n+1}$ induces a homomorphism $P_{n+1} \longrightarrow \mathrm{Im}\, \partial_{n+2}$, which factors through the epimorphism $Q_{n+2} \longrightarrow \mathrm{Im}\, \partial_{n+2}$ induced by ∂_{n+2}, since Q_{n+2} is projective; thus $\varphi_{n+1} - \sigma_n \partial_{n+1} = \partial_{n+2} \sigma_{n+1}$ for some $\sigma_{n+1} : P_{n+1} \longrightarrow Q_{n+2} : \square$

Lifting. We complete Theorem 2.2 by lifting short exact sequences and certain commutative diagrams to projective resolutions. This follows from the corresponding properties of projective presentations.

Lemma **2.3.** *The diagram below (solid arrows) with exact row and columns, in which P and R are projective, can be completed to a commutative 3×3 diagram (all arrows) with exact rows and columns, in which Q is projective.*

$$
\begin{array}{ccccccccc}
 & & 0 & & 0 & & 0 & & \\
 & & \downarrow & & \downarrow & & \downarrow & & \\
0 & \dashrightarrow & K & \overset{\nu}{\dashrightarrow} & L & \overset{\tau}{\dashrightarrow} & M & \dashrightarrow & 0 \\
 & & \downarrow{\scriptstyle i} & & \downarrow{\scriptstyle j} & & \downarrow{\scriptstyle k} & & \\
0 & \dashrightarrow & P & \overset{\kappa}{\dashrightarrow} & Q & \overset{\pi}{\dashrightarrow} & R & \dashrightarrow & 0 \\
 & & \downarrow{\scriptstyle \varepsilon} & & \downarrow{\scriptstyle \zeta} & & \downarrow{\scriptstyle \eta} & & \\
0 & \longrightarrow & A & \overset{\mu}{\longrightarrow} & B & \overset{\sigma}{\longrightarrow} & C & \longrightarrow & 0 \\
 & & \downarrow & & \downarrow & & \downarrow & & \\
 & & 0 & & 0 & & 0 & &
\end{array}
$$

Proof. Since the middle row must split we may as well let $Q = P \oplus R$, with $\kappa : p \longmapsto (p, 0)$ and $\pi : (p, r) \longmapsto r$. Then Q is projective and the middle row is exact. Maps $Q \longrightarrow B$ are induced by homomorphisms of P and R into B. Already $\mu \varepsilon : P \longrightarrow B$. Since R is projective, $\eta = \sigma \lambda$ for some $\lambda : R \longrightarrow B$. The resulting $\zeta : Q \longrightarrow B$ sends (p, r) to $\mu \varepsilon p + \lambda r$. Then $\zeta \kappa = \mu \varepsilon$, $\eta \pi = \sigma \zeta$, and ζ is surjective: if $b \in B$, then $\sigma b = \eta r$ for some $r \in R$, $\sigma (b - \lambda r) = 0$, $b - \lambda r = \mu a = \mu \varepsilon p$ for some $a \in A$, $p \in P$, and $b = \zeta (p, r)$.

Let $L = \mathrm{Ker}\,\zeta$ and let $j : L \longrightarrow Q$ be the inclusion homomorphism. Since $\zeta \kappa i = \mu \varepsilon i = 0$ there is a homomorphism $\nu : K \longrightarrow L$ such that $\kappa i = j \nu$. Similarly, there is a homomorphism $\tau : L \longrightarrow M$ such that $\pi j = k \tau$. Now, the whole 3×3 diagram commutes, and its columns and last two rows are exact. By the nine lemma, the top row is exact, too. $\square$

Proposition **2.4.** *For every short exact sequence* $0 \longrightarrow A \longrightarrow B \longrightarrow C \longrightarrow 0$ *of modules, and projective resolutions* $\mathcal{P} \longrightarrow A$, $\mathcal{R} \longrightarrow C$, *there exist a projective resolution* $\mathcal{Q} \longrightarrow B$ *and a short exact sequence* $0 \longrightarrow \mathcal{P} \longrightarrow \mathcal{Q} \longrightarrow \mathcal{R} \longrightarrow 0$ *such that the following diagram commutes:*

$$
\begin{array}{ccccc}
\mathcal{P} & \dashrightarrow & \mathcal{Q} & \dashrightarrow & \mathcal{R} \\
\downarrow & & \downarrow & & \downarrow \\
A & \longrightarrow & B & \longrightarrow & C
\end{array}
$$

Proof. By 2.3, applied to the given sequence and to $0 \longrightarrow K_0 = \mathrm{Ker}\,\varepsilon \longrightarrow P_0 \overset{\varepsilon}{\longrightarrow} A \longrightarrow 0$, $0 \longrightarrow M_0 = \mathrm{Ker}\,\eta \longrightarrow R_0 \overset{\eta}{\longrightarrow} C \longrightarrow 0$, there is a commutative 3×3 diagram with short exact rows and columns:

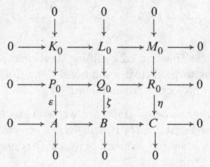

in which Q_0 is projective. Applying 2.3 again, to the top row and exact sequences $0 \longrightarrow K_1 = \mathrm{Ker}\,\partial_1^{\mathcal{P}} \longrightarrow P_1 \longrightarrow K_0 \longrightarrow 0$, $0 \longrightarrow M_1 = \mathrm{Ker}\,\partial_1^{\mathcal{Q}} \longrightarrow R_1 \longrightarrow M_0 \longrightarrow 0$, yields two more exact sequences, $0 \longrightarrow P_1 \longrightarrow Q_1 \longrightarrow R_1 \longrightarrow 0$ and $0 \longrightarrow K_1 \longrightarrow L_1 \longrightarrow M_1 \longrightarrow 0$. Repetition yields the required sequence $\mathcal{P} \longrightarrow \mathcal{Q} \longrightarrow \mathcal{R}$. $\square$

Next, we lift two short exact sequences together. (But we haven't tried three.)

Lemma **2.5.** *The commutative diagram next page (solid arrows) with short exact rows and columns, in which P, P', R, and R' are projective, can be completed to a comutative diagram (all arrows) with short exact rows and columns, in which Q and Q' are projective.*

Proof. By 2.3, the front and back faces can be filled in so that they commute, their rows and columns are short exact, and Q, Q' are projective. As in the proof of 2.3, we can let $Q = P \oplus R$, $\kappa : p \longmapsto (p,0)$, $\pi : (p,r) \longmapsto r$ and $Q' = P' \oplus R'$, $\kappa' : p' \longmapsto (p',0)$, $\pi' : (p',r') \longmapsto r'$; let $\zeta\,(p,r) = \mu\,\varepsilon p + \lambda r$, where $\lambda : R \longrightarrow B$ and $\sigma \lambda = \eta$, and $\zeta'\,(p',r') = \mu'\,\varepsilon' p' + \lambda' r'$,

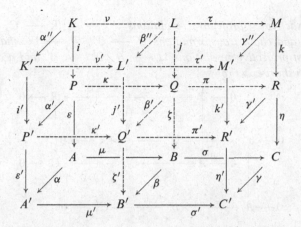

where $\lambda' : R' \longrightarrow B'$ and $\sigma' \lambda' = \eta'$; and let $L = \mathrm{Ker}\ \zeta$, $L' = \mathrm{Ker}\ \zeta'$ and $j : L \longrightarrow Q$, $j' : L' \longrightarrow Q'$ be the inclusion homomorphisms.

Maps $Q \longrightarrow Q'$ are induced by homomorphisms of P and R into P' and R'. We use α', γ', and $0 : P \longrightarrow R'$, and choose $\xi : R \longrightarrow P'$ so that the resulting homomorphism $\beta' : Q \longrightarrow Q'$, $\beta'(p, r) = (\alpha' p + \xi r,\ \gamma' r)$ makes the two lower cubes commute. Their upper faces commute for any ξ: $\beta' \mu' p = \beta'(p, 0) = (\alpha_0 p, 0) = \kappa' \alpha' p$ and $\pi' \beta'(p, r) = \pi'(\alpha' p + \xi r,\ \gamma' r) = \gamma' r = \gamma' \pi'(p, r)$. This leaves the face ζ', β', β, ζ. Since $\beta \mu \varepsilon = \mu' \alpha \varepsilon = \mu' \varepsilon' \alpha'$:

$$\beta \zeta(p, r) = \beta \mu \varepsilon p + \beta \lambda r = \mu' \varepsilon' \alpha' p + \beta \lambda r,$$
$$\zeta' \beta'(p, r) = \zeta'(\alpha' p + \xi r,\ \gamma' r) = \mu' \varepsilon' \alpha' p + \mu' \varepsilon' \xi r + \lambda' \gamma' r.$$

Hence $\zeta' \beta' = \beta \zeta$ if and only if $\mu' \varepsilon \xi = \beta \lambda - \lambda' \gamma'$. Now,

$$\sigma'(\beta \lambda - \lambda' \gamma') = \gamma \sigma \lambda - \sigma' \lambda' \gamma' = \gamma \eta - \eta' \gamma' = 0$$

by the hypothesis; hence $\beta \lambda - \lambda' \gamma'$ induces a homomorphism of R into $\mathrm{Ker}\ \sigma' = \mathrm{Im}\ \mu' = \mathrm{Im}\ \mu' \varepsilon'$, which factors through the epimorphism $P' \longrightarrow \mathrm{Im}\ \mu' \varepsilon'$ induced by $\mu' \varepsilon'$, since R is projective: thus, $\beta \lambda - \lambda' \gamma' = \mu' \varepsilon' \xi$ for some homomorphism $\xi : R \longrightarrow P'$. Then the two lower cubes commute.

Since the two lower cubes commute, β' induces a homomorphism $\beta'' : L \longrightarrow L'$: since $\zeta' \beta' j = \beta \zeta j = 0$, we have $\beta' j = j' \beta''$ for some $\beta'' : L \longrightarrow L'$. Then all faces of the two upper cubes commute, except perhaps the two upper faces. Since j', k' are monomorphisms, these also commute: indeed, $k' \tau' \beta'' = \pi' j' \beta'' = \pi' \beta' j = \gamma' \pi j = \gamma' k \tau = k' \gamma'' \tau$, whence $\tau' \beta'' = \gamma'' \tau$; similarly, $v' \alpha'' = \beta'' v$, and the entire diagram commutes. $\square$

Proposition 2.6. *For every commutative diagram*

$$\begin{array}{ccccccccc}
0 & \longrightarrow & A & \longrightarrow & B & \longrightarrow & C & \longrightarrow & 0 \\
& & \downarrow \alpha & & \downarrow & & \downarrow \gamma & & \\
0 & \longrightarrow & A' & \longrightarrow & B' & \longrightarrow & C' & \longrightarrow & 0
\end{array}$$

with short exact rows, projective resolutions $\mathcal{P} \longrightarrow A$, $\mathcal{R} \longrightarrow C$, $\mathcal{P}' \longrightarrow A'$, $\mathcal{R}' \longrightarrow C'$, and chain transformations $\mathcal{P} \longrightarrow \mathcal{P}'$, $\mathcal{R} \longrightarrow \mathcal{R}'$ that lift α and γ, there exist projective resolutions $\mathcal{Q} \longrightarrow B$, $\mathcal{Q}' \longrightarrow B'$ and a commutative diagram with short exact rows:

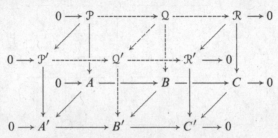

Proof. This follows from repeated applications of Lemma 2.5, just as Proposition 2.4 follows from repeated applications of Lemma 2.3. $\square$

Injective resolutions. A module A can also be analyzed by short exact sequences $0 \longrightarrow A \longrightarrow J \longrightarrow L \longrightarrow 0$, where J is injective. In fact, the construction in Section XI.2 assigns to every module A an embedding $\mu : A \longrightarrow J$ into an injective module J; the short exact sequence $0 \longrightarrow A \longrightarrow J \longrightarrow$ Coker $\mu \longrightarrow 0$ may then be assigned to A.

Injective resolutions are compositions of these injective "copresentations":

Definition. An injective resolution *of a module A is an exact sequence*

$$0 \longrightarrow A \longrightarrow J^0 \longrightarrow J^1 \longrightarrow \cdots \longrightarrow J^n \longrightarrow J^{n+1} \longrightarrow \cdots$$

of modules and homomorphisms in which J_0, J_1, ..., J_n, ... are injective.

Thus, an injective resolution $A \xrightarrow{\eta} \mathcal{J}$ of A consists of a negative complex

$$\mathcal{J} : 0 \longrightarrow J^0 \xrightarrow{\delta^0} J^1 \longrightarrow \cdots \longrightarrow J^n \xrightarrow{\delta^n} J^{n+1} \longrightarrow \cdots$$

of injective modules and a monomorphism $\eta : A \longrightarrow J^0$, such that $H^n(\mathcal{J}) = 0$ for all $n > 0$ and Ker $\delta^0 = \operatorname{Im} \eta$. Then $H^0(\mathcal{J}) = \operatorname{Ker} \delta^0 \cong A$. Also, $A \xrightarrow{\eta} \mathcal{J}$ is a composition of injective "copresentations"

$$0 \longrightarrow A \longrightarrow J_0 \longrightarrow L_0 \longrightarrow 0, ..., 0 \longrightarrow L_{n-1} \longrightarrow J_n \longrightarrow L_n, ...,$$

where $L_0 = \operatorname{Im} \delta^0 \cong \operatorname{Coker} \eta$, $L_n = \operatorname{Im} \delta^n \cong \operatorname{Coker} \partial_{n-1}$ for all $n > 0$; L_n is the nth *cosyzygy* of A (in the given injective resolution).

Every module has an injective resolution. In fact, an injective resolution can be assigned to every module A: we saw that an injective module J^0 and monomorphism $\eta : A \longrightarrow J^0$ can be assigned to A; an injective module J^1 and monomorphism Coker $\eta \longrightarrow J^1$ can then be assigned to Coker η, so that $0 \longrightarrow A \longrightarrow J^0 \longrightarrow J^1$ is exact; and so forth.

Module homomorphisms *lift*, uniquely up to homotopy, to chain transformations of injective resolutions.

$$0 \longrightarrow A' \longrightarrow B' \longrightarrow C' \longrightarrow 0$$
$$\quad\quad\quad \alpha\big\downarrow \quad\quad \beta\big\downarrow \quad\quad \gamma\big\downarrow$$
$$0 \longrightarrow A \longrightarrow J \longrightarrow L \longrightarrow 0$$

Lemma 2.7. *In a diagram with exact rows (solid arrows) in which J is injective, there exist homomorphisms β and γ (dotted arrows) that make the diagram commutative.*

Theorem 2.8 (Comparison Theorem). *Let $A \xrightarrow{\eta} \mathcal{J}$ and $B \xrightarrow{\zeta} \mathcal{K}$ be injective resolutions, and let $\varphi : A \longrightarrow B$ be a homomorphism. There is a chain transformation $(\varphi^n)_{n \geq 0} : \mathcal{J} \longrightarrow \mathcal{K}$ such that $\varphi^0 \eta = \zeta \varphi$:*

$$
\begin{array}{ccc}
\mathcal{J} & \dashrightarrow & \mathcal{K} \\
\big\uparrow & & \big\uparrow \\
A & \longrightarrow & B
\end{array}
$$

Moreover, any two such chain transformations are homotopic.

Proposition 2.9. *For every short exact sequence $0 \longrightarrow A \longrightarrow B \longrightarrow C \longrightarrow 0$ of modules, and injective resolutions $A \longrightarrow \mathcal{J}$, $C \longrightarrow \mathcal{L}$, there exist an injective resolution $B \longrightarrow \mathcal{K}$ and a short exact sequence $0 \longrightarrow \mathcal{J} \longrightarrow \mathcal{K} \longrightarrow \mathcal{K} \longrightarrow 0$ such that the following diagram commutes:*

$$
\begin{array}{ccccc}
\mathcal{J} & \dashrightarrow & \mathcal{K} & \dashrightarrow & \mathcal{L} \\
\big\uparrow & & \big\uparrow & & \big\uparrow \\
A & \longrightarrow & B & \longrightarrow & C
\end{array}
$$

Proposition 2.10. *For every commutative diagram*

$$
\begin{array}{ccccccccc}
0 & \longrightarrow & A & \longrightarrow & B & \longrightarrow & C & \longrightarrow & 0 \\
& & \big\downarrow & & \big\downarrow & & \big\downarrow & & \\
0 & \longrightarrow & A' & \longrightarrow & B' & \longrightarrow & C' & \longrightarrow & 0
\end{array}
$$

with short exact rows, injective resolutions $A \longrightarrow \mathcal{J}$, $C \longrightarrow \mathcal{L}$, $A' \longrightarrow \mathcal{J}'$, $C' \longrightarrow \mathcal{L}'$, and chain transformations $\mathcal{J} \longrightarrow \mathcal{J}'$, $\mathcal{L} \longrightarrow \mathcal{L}'$ that lift $A \longrightarrow A'$ and $C \longrightarrow C'$, there exist injective resolutions $B \longrightarrow \mathcal{K}$, $B' \longrightarrow \mathcal{K}'$ and a commutative diagram with short exact rows:

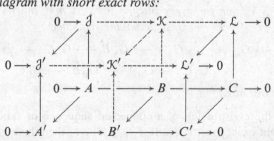

These results are obtained from 2.1, 2.2, 2.4, 2.6 by reversing all arrows. They have largely similar proofs, which may safely be left to readers as exercises.

Exercises

1. Given a diagram with exact rows (solid arrows)

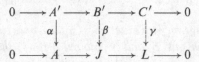

in which J is injective, show that there exist homomorphisms β and γ (dotted arrows) that make the diagram commutative.

2. Prove the comparison theorem for injective resolutions.

3. Show that a short exact sequence of modules lifts to a short exact sequence of their injective resolutions, as in Proposition 2.9.

4. Prove Proposition 2.10, in which a diagram of short exact sequence of modules lifts to a diagram of short exact sequences of their injective resolutions.

3. Derived Functors

The derived functors of a functor F are constructed by applying F to projective or injective resolutions. These functors repair the lack of exactness of F, when F is only right exact or left exact. Ext and Tor will be our main examples.

Left derived functors. Let F be a covariant functor from modules to modules, such as $M \otimes_R -$ or $- \otimes_R M$, that is right exact but not (in general) left exact: when $0 \longrightarrow A \longrightarrow B \longrightarrow C \longrightarrow 0$ is short exact, then $FA \longrightarrow FB \longrightarrow FC \longrightarrow 0$ is only right exact. The lack of left exactness in this last sequence can be measured by an exact sequence $F_1 A \longrightarrow F_1 B \longrightarrow F_1 C \longrightarrow FA \longrightarrow FB$, where F_1 is another functor. In turn, the lack of left exactness of $F_1 A \longrightarrow F_1 B \longrightarrow F_1 C$ can be measured by an exact sequence $F_2 A \longrightarrow F_2 B \longrightarrow F_2 C \longrightarrow F_1 A \longrightarrow F_1 B$, where F_2 is another functor. Then F, F_1, F_2, ... constitute a *connected sequence*:

Definition. A positive connected sequence of covariant functors consists of covariant functors G_0, G_1, ..., G_n, ... and of connecting homomorphisms $G_n C \longrightarrow G_{n-1} A$, one for every integer $n > 0$ and short exact sequence $\mathcal{E} : 0 \longrightarrow A \longrightarrow B \longrightarrow C \longrightarrow 0$, such that the sequence

$$\cdots \longrightarrow G_{n+1} C \longrightarrow G_n A \longrightarrow G_n B \longrightarrow G_n C \longrightarrow \cdots$$
$$\cdots G_1 C \longrightarrow G_0 A \longrightarrow G_0 B \longrightarrow G_0 C \longrightarrow 0$$

is exact and natural in $\mathcal{E}$.

Our goal is the construction of a connected sequence of functors that ends with a given right exact functor F. If F is additive, then there is a "best" such sequence, which is constructed as follows. We saw in Section 2 that to every module A can be assigned a projective resolution $\mathcal{P}^A \longrightarrow A$. Then $F\mathcal{P}^A$ is a positive complex of modules $\cdots \longrightarrow FP_2^A \longrightarrow FP_1^A \longrightarrow FP_0^A \longrightarrow 0$.

Proposition **3.1.** *Let F be a covariant additive functor. Assign a projective resolution $\mathcal{P}^A \longrightarrow A$ to every module A. For every $n \geqq 0$, let $L_n A = H_n(F\mathcal{P}^A)$; for every module homomorphism $\varphi : A \longrightarrow B$, let $L_n \varphi = H_n(F\overline{\varphi})$, where $\overline{\varphi} : \mathcal{P}^A \longrightarrow \mathcal{P}^B$ is a chain transformation that lifts φ. Then $L_n F = L_n$ is a well defined additive functor, which, up to natural isomorphisms, does not depend on the initial assignment of $\mathcal{P}^A$ to A.*

Proof. If $\overline{\varphi}$ and $\overline{\psi}$ both lift φ, then $\overline{\varphi}$ and $\overline{\psi}$ are homotopic, by 2.2; since F is additive, $F\overline{\varphi}$ and $F\overline{\psi}$ are homotopic, and $H_n(F\overline{\varphi}) = H_n(F\overline{\psi})$, by 1.2. Hence $L_n \varphi$ does not depend on the choice of $\overline{\varphi}$, and is well defined.

If φ is the identity on A, then $1_{\mathcal{P}^A}$ lifts φ to $\mathcal{P}^A$; hence $L_n 1_A = H_n(F 1_{\mathcal{P}^A})$ is the identity on $L_n A$. If $\varphi : A \longrightarrow B$ and $\psi : B \longrightarrow C$ are module homomorphisms, and $\overline{\varphi}$, $\overline{\psi}$ lift φ, ψ, then $\overline{\psi}\,\overline{\varphi}$ lifts $\psi\,\varphi$; hence $L_n(\psi\,\varphi) = (L_n \psi)(L_n \varphi)$. Similarly, if φ, $\psi : A \longrightarrow B$ and $\overline{\varphi}$, $\overline{\psi}$ lift φ, ψ, then $\overline{\varphi} + \overline{\psi}$ lifts $\varphi + \psi$; hence $L_n(\psi + \varphi) = L_n \psi + L_n \varphi$. Thus $L_n F$ is an additive functor.

Finally, let L_n, L_n' be constructed from two choices $\mathcal{P} \longrightarrow A$, $\mathcal{P}' \longrightarrow A$ of projective resolutions. By 2.2, $1_A : A \longrightarrow A$ lifts to $\overline{\theta} : \mathcal{P} \longrightarrow \mathcal{P}'$ and to $\overline{\zeta} : \mathcal{P}' \longrightarrow \mathcal{P}$. Then 1_A also lifts to $\overline{\zeta}\,\overline{\theta} : \mathcal{P} \longrightarrow \mathcal{P}$ and to $\overline{\theta}\,\overline{\zeta} : \mathcal{P}' \longrightarrow \mathcal{P}'$. Since $1_{\mathcal{P}}$ and $1_{\mathcal{P}'}$ also lift 1_A, $\overline{\zeta}\,\overline{\theta}$ is homotopic to $1_{\mathcal{P}}$, and $\overline{\theta}\,\overline{\zeta}$ is homotopic to $1_{\mathcal{P}'}$. Hence $(F\overline{\zeta})(F\overline{\theta})$ is homotopic to $1_{F(\mathcal{P})}$, $(F\overline{\theta})(F\overline{\zeta})$ is homotopic to $1_{F(\mathcal{P}')}$, and $H_n(F\overline{\theta}))$: $L_n A \longrightarrow L_n' A$, $H_n(F\overline{\zeta}))$: $L_n' A \longrightarrow L_n A$ are mutually inverse isomorphisms. That $H_n(F\overline{\theta})$ is natural in A is proved similarly. $\square$

Definition. In Proposition 3.1, $L_n F$ *is the nth* left derived functor *of F.*

Theorem **3.2.** *Let F be a covariant additive functor.*

(1) *If P is projective, then $(L_0 F)P \cong FP$ and $(L_n F)P = 0$ for all $n > 0$.*

(2) *If F is right exact, then there is for every module A an isomorphism $(L_0 F)A \cong FA$, which is natural in A.*

(3) $L_0 F$, $L_1 F$, ..., $L_n F$, ... *is a positive connected sequence of functors.*

(4) *If F is right exact, then F, $L_1 F$, ..., $L_n F$, ... is a positive connected sequence of functors.*

Proof. (1). If P is projective, then P has a projective resolution

$$\mathcal{P} \xrightarrow{\varepsilon} P : \quad \cdots \longrightarrow 0 \longrightarrow 0 \longrightarrow P \xrightarrow{\varepsilon} P \longrightarrow 0$$

in which $\varepsilon = 1_P$. Then $F\mathcal{P}$: $\cdots \longrightarrow 0 \longrightarrow 0 \longrightarrow FP \longrightarrow 0 \longrightarrow \cdots$, $H_0(F\mathcal{P}) \cong FP$, and $H_n(F\mathcal{P}) = 0$ for all $n \neq 0$.

(2). If F is right exact, then $FP_1 \xrightarrow{F\partial_1} FP_0 \xrightarrow{F\varepsilon} FA \longrightarrow 0$ is right exact, like $P_1 \xrightarrow{\partial_1} P_0 \xrightarrow{\varepsilon} A \longrightarrow 0$. Hence

$$L_0 A = H_0(F\mathcal{P}) = FP_0/\operatorname{Im} F\partial_1 = FP_0/\operatorname{Ker} F\varepsilon \cong FA.$$

The isomorphism $\theta^A : L_0 A \longrightarrow FA$ sends cls $t = t + \operatorname{Im} F\partial_1 = t + \operatorname{Ker} F\varepsilon$ to $(F\varepsilon)(t)$. If $\varphi : A \longrightarrow B$ is a homomorphism and $\overline{\varphi} = (\varphi_n)_{n \geq 0} : \mathcal{P}^A \longrightarrow \mathcal{P}^B$ lifts φ, then $\varepsilon^B \varphi_0 = \varphi \varepsilon^A$, $L_0\varphi = H_0(F\overline{\varphi})$ sends cls $t = t + \operatorname{Im} F\partial_1$ to cls $(F\varphi_0)t = (F\varphi_0)t + \operatorname{Im} F\partial_1$, and $\theta^B((L_0\varphi) \text{ cls } t) = \theta^B$ cls $(F\varphi_0)t = F(\varepsilon^B \varphi_0)t = F(\varphi \varepsilon^A)t = (F\varphi)\theta^A$cls t. Thus θ^A is natural in A.

(3). Let $\mathcal{E} : 0 \longrightarrow A \longrightarrow B \longrightarrow C \longrightarrow 0$ be a short exact sequence. By 2.3, $\mathcal{E}$ lifts to a short exact sequence $0 \longrightarrow \mathcal{P}^A \longrightarrow \mathcal{P}^B \longrightarrow \mathcal{P}^C \longrightarrow 0$. Since P_n^C is projective, $0 \longrightarrow P_n^A \longrightarrow P_n^B \longrightarrow P_n^C \longrightarrow 0$ splits for every n; hence $0 \longrightarrow FP_n^A \longrightarrow FP_n^B \longrightarrow FP_n^C \longrightarrow 0$ splits for every n and $0 \longrightarrow F\mathcal{P}^A \longrightarrow F\mathcal{P}^B \longrightarrow F\mathcal{P}^C \longrightarrow 0$ is short exact. Then 1.3 yields an exact homology sequence and its connecting homomorphisms

$$\cdots \longrightarrow H_{n+1}(F\mathcal{P}^C) \longrightarrow H_n(F\mathcal{P}^A) \longrightarrow H_n(F\mathcal{P}^B) \longrightarrow H_n(F\mathcal{P}^C) \longrightarrow$$
$$\cdots H_1(F\mathcal{P}^C) \longrightarrow H_0(F\mathcal{P}^A) \longrightarrow H_0(F\mathcal{P}^B) \longrightarrow H_0(F\mathcal{P}^C) \longrightarrow 0,$$

ending at $H_0(F\mathcal{P}^C)$, since all subsequent modules are null; equivalently,

$$\cdots \longrightarrow L_{n+1}C \longrightarrow L_n A \longrightarrow L_n B \longrightarrow L_n C \longrightarrow \cdots$$
$$\cdots L_1 C \longrightarrow L_0 A \longrightarrow L_0 B \longrightarrow L_0 C \longrightarrow 0.$$

We prove naturality. Given a commutative diagram with short exact rows:

$$\begin{array}{ccccccccc}
0 & \longrightarrow & A & \longrightarrow & B & \longrightarrow & C & \longrightarrow & 0 \\
 & & \downarrow & & \downarrow & & \downarrow & & \\
0 & \longrightarrow & A' & \longrightarrow & B' & \longrightarrow & C' & \longrightarrow & 0
\end{array}$$

there is, by 2.6, a commutative diagram with exact rows:

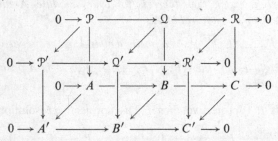

in which $\mathcal{P} = \mathcal{P}^A$, etc. Since the exact homology sequence in 1.3 is natural, the top two squares induce a commutative diagram:

$$\cdots\, H_{n+1}(\mathcal{R}) \longrightarrow H_n(\mathcal{P}) \longrightarrow H_n(\mathcal{Q}) \longrightarrow H_n(\mathcal{R}) \longrightarrow H_{n-1}(\mathcal{P}) \cdots$$

$$\cdots\, H_{n+1}(\mathcal{R}') \longrightarrow H_n(\mathcal{P}') \longrightarrow H_n(\mathcal{Q}') \longrightarrow H_n(\mathcal{R}') \longrightarrow H_{n-1}(\mathcal{R}') \cdots$$

which, up to natural isomorphisms, is none other than

$$\cdots\, L_{n+1} C \longrightarrow L_n A \longrightarrow L_n B \longrightarrow L_n C \longrightarrow L_{n-1} A \cdots$$

$$\cdots\, L_{n+1} C' \longrightarrow L_n A' \longrightarrow L_n B' \longrightarrow L_n C' \longrightarrow L_{n-1} A' \cdots$$

(4) follows from (2) and (3). $\square$

If F is right exact in Theorem 3.2, then the left derived functors of F constitute the "best" connected sequence that ends with F:

Theorem **3.3.** *Let F be a right exact, covariant additive functor, and let $G_0, G_1, \ldots, G_n, \ldots$ be a positive connected sequence of covariant functors. For every natural transformation $\varphi_0 : G_0 \longrightarrow F$ there exist unique natural transformations $\varphi_n : G_n \longrightarrow L_n F$ such that the square*

$$
\begin{array}{ccc}
G_{n+1}C & \xrightarrow{\;\xi^{\mathcal{E}}_{n+1}\;} & G_n A \\[2pt]
\varphi^C_{n+1} \downarrow & & \downarrow \varphi^A_n \\[2pt]
(L_{n+1}F)C & \xrightarrow[\;\chi^{\mathcal{E}}_{n+1}\;]{} & (L_n F)A
\end{array}
$$

commutes for every short exact sequence $\mathcal{E} : 0 \longrightarrow A \longrightarrow B \longrightarrow C \longrightarrow 0$ and every $n \geq 0$, where χ and ξ are the connecting homomorphisms.

Proof. We construct φ_n recursively. For every module A choose a projective presentation $\mathcal{E}^A : 0 \longrightarrow K \xrightarrow{\;\mu\;} P \longrightarrow A \longrightarrow 0$ (with P projective). Since φ_0 is natural, and $F, L_1, \ldots, L_n, \ldots$ and $G_0, G_1, \ldots, G_n, \ldots$ are connected sequences, there is a commutative diagram

$$
\begin{array}{ccccc}
G_1 A & \xrightarrow{\;\xi^A\;} & G_0 K & \xrightarrow{\;G_0\mu\;} & G_0 P \\[2pt]
\varphi^A_1 \downarrow & & \downarrow \varphi^K_0 & & \downarrow \varphi^P_0 \\[2pt]
L_1 P = 0 \longrightarrow & L_1 A & \xrightarrow[\;\chi^A\;]{} & FK & \xrightarrow[\;F\mu\;]{} FP
\end{array}
$$

with exact rows, in which $\xi^A = \xi_1^{\mathcal{E}^A}$, $\chi^A = \chi_1^{\mathcal{E}^A}$ and $L_1 P = 0$ by 3.2. We want φ_1^A to make the left square commute. Happily, $F\mu\, \varphi^K_0 \, \xi^A = \varphi^P_0 \, G_0\mu\, \xi^A = 0$, so that $\varphi^K_0 \, \xi^A$ factors through χ^A:

$$\varphi^K_0 \, \xi^A = \chi^A \, \varphi_1^A \qquad\qquad (1), \text{ case } n = 0$$

for some unique $\varphi_1^A : G_1 A \longrightarrow L_1 A$. In particular, φ_1 is unique, if it exists.

The construction of φ_{n+1} from φ_n when $n > 0$ is similar but simpler: now $\mathcal{E}^A : 0 \longrightarrow K \xrightarrow{\mu} P \longrightarrow A \longrightarrow 0$ induces a commutative diagram

$$
\begin{array}{ccc}
G_{n+1}A & \xrightarrow{\;\xi^A\;} & G_n K \\
{\scriptstyle\varphi^A_{n+1}}\downarrow & & \downarrow{\scriptstyle\varphi^K_n} \\
L_{n+1}P = 0 \longrightarrow L_{n+1}A & \xrightarrow[\chi^A]{} & L_n K \longrightarrow L_n P = 0
\end{array}
$$

with exact rows, in which $\xi^A = \xi^{\mathcal{E}^A}_{n+1}$, $\chi^A = \chi^{\mathcal{E}^A}_{n+1}$ and $L_{n+1}P = 0$, $L_n P = 0$ by 3.2. We want φ^A_{n+1} to make the square commute. Since χ^A is an isomorphism,

$$\varphi^K_n \, \xi^A \; = \; \chi^A \, \varphi^A_{n+1} \qquad\qquad (1),\ \text{case } n > 0$$

for some unique homomorphism $\varphi^A_{n+1} : G_{n+1}A \longrightarrow L_{n+1}A$. In particular, φ_{n+1} is unique, if it exists.

To show that φ_{n+1} has all required properties, consider a diagram

$$
\begin{array}{ccccccccc}
\mathcal{E}^C : & 0 \longrightarrow & L & \longrightarrow & Q & \longrightarrow & C & \longrightarrow & 0 \\
& & {\scriptstyle\alpha}\downarrow & & \downarrow & & \downarrow{\scriptstyle\gamma} & & \\
\mathcal{E}' : & 0 \longrightarrow & A' & \longrightarrow & B' & \longrightarrow & C' & \longrightarrow & 0
\end{array}
$$

with exact rows (solid arrows). By 2.1, there are dotted arrows such that the diagram commutes. Then the squares

$$
\begin{array}{ccc}
G_0 L & \xrightarrow{\;G_0\alpha\;} & G_0 A \\
{\scriptstyle\varphi^L_0}\downarrow & & \downarrow{\scriptstyle\varphi^A_0} \\
FL & \xrightarrow[F\alpha]{} & FA
\end{array}
\qquad
\begin{array}{ccc}
L_1 C & \xrightarrow{\;\chi^C\;} & FL \\
{\scriptstyle L_1\gamma}\downarrow & & \downarrow{\scriptstyle F\alpha} \\
L_1 C' & \xrightarrow[\chi^{\mathcal{E}'}]{} & FA'
\end{array}
\qquad
\begin{array}{ccc}
G_1 C & \xrightarrow{\;\xi^C\;} & G_0 L \\
{\scriptstyle G_1\gamma}\downarrow & & \downarrow{\scriptstyle G_0\alpha} \\
G_1 C' & \xrightarrow[\xi^{\mathcal{E}'}]{} & G_0 A'
\end{array}
$$

commute, since φ_0 and χ_1, ξ_1 are natural, and $\varphi^L_0 \, \xi^C = \chi^C \, \varphi^C_1$, by (1). Hence $\varphi^{A'}_0 \, \xi^{\mathcal{E}'} \, G_1\gamma = \varphi^{A'}_0 \, G_0\alpha \, \xi^C = F\alpha \, \varphi^L_0 \, \xi^C = F\alpha \, \chi^C \, \varphi^C_1 = \chi^{\mathcal{E}'} \, L_1\gamma \, \varphi^C_1$ and

$$\varphi^{A'}_0 \, \xi^{\mathcal{E}'} \, G_1\gamma = \chi^{\mathcal{E}'} \, L_1\gamma \, \varphi^C_1 . \qquad\qquad (2),\ \text{case } n = 0$$

If $n > 0$, then, similarly,

$$\varphi^{A'}_n \, \xi^{\mathcal{E}'} \, G_{n+1}\gamma = \chi^{\mathcal{E}'} \, L_{n+1}\gamma \, \varphi^C_{n+1} . \qquad\qquad (2),\ \text{case } n > 0$$

The required properties of φ_{n+1} follow from (2):

Let $\gamma = 1_C$ and let $\mathcal{E}' : 0 \longrightarrow M \longrightarrow R \longrightarrow C \longrightarrow 0$ be another projective presentation of C. By (2), $\varphi^M_n \, \xi^{\mathcal{E}'} = \chi^{\mathcal{E}'} \, \varphi^C_{n+1}$. Hence φ^C_{n+1} does not depend on the choice of $\mathcal{E}^C$, and φ_{n+1} is well defined by (1) (given φ_n).

Next, let $\gamma : C \longrightarrow C'$ be any homomorphism and $\mathcal{E}' = \mathcal{E}^{C'} : 0 \longrightarrow L' \longrightarrow Q' \longrightarrow C' \longrightarrow 0$. By (1) and (2), $\chi^{\mathcal{E}'} \, \varphi^{C'}_{n+1} \, G_{n+1}\gamma = \varphi^{C'}_n \, \xi^{\mathcal{E}'} \, G_{n+1}\gamma =$

$\chi^{\mathcal{E}'} L_{n+1} \gamma \, \varphi_{n+1}^C$. But $\chi^{\mathcal{E}'}$ is (at least) a monomorphism, since $L_{n+1} Q' = 0$ by 3.2. Hence $\varphi_{n+1}^{C'} G_{n+1} \gamma = L_{n+1} \gamma \, \varphi_{n+1}^C$; thus φ_{n+1}^C is natural in C.

Finally, if $\mathcal{E}' = \mathcal{E} : 0 \longrightarrow A \longrightarrow B \longrightarrow C \longrightarrow 0$ is any short exact sequence and $\gamma = 1$, then (2) reads $\varphi_n^A \, \xi^{\mathcal{E}} = \chi^{\mathcal{E}} \, \varphi_{n+1}^C$. $\square$

Readers who are still awake will verify that the universal property in Theorem 3.3 determines the left derived functors of F, up to natural isomorphisms.

Theorem 3.4. *Let $G_0, G_1, \ldots, G_n, \ldots$ be a positive connected sequence of covariant functors. If $G_n P = 0$ whenever P is projective and $n > 0$, then, up to natural isomorphisms, $G_1, \ldots, G_n, \ldots$ are the left derived functors of G_0.*

Proof. First, G_0 is right exact, by definition. By 3.3, the identity $G_0 \longrightarrow G_0$ induces natural transformations $\varphi_n : G_n \longrightarrow L_n$ to the derived functors $L_0 = G_0, L_1, \ldots, L_n, \ldots$ of G_0, which form a commutative square

$$
\begin{array}{ccc}
G_{n+1} C & \xrightarrow{\;\xi_{n+1}^{\mathcal{E}}\;} & G_n A \\
{\scriptstyle \varphi_{n+1}^C} \downarrow & & \downarrow {\scriptstyle \varphi_n^A} \\
L_{n+1} C & \xrightarrow[\;\chi_{n+1}^{\mathcal{E}}\;]{} & L_n A
\end{array}
$$

with the connecting homomorphisms, for every $n \geq 0$ and short exact sequence $\mathcal{E} : 0 \longrightarrow A \longrightarrow B \longrightarrow C \longrightarrow 0$. We prove by induction on n that φ_n^A is an isomorphism for every module A. Let $\mathcal{E}^A : 0 \longrightarrow K \longrightarrow P \longrightarrow A \longrightarrow 0$ be a projective presentation. First,

$$
\begin{array}{ccccccc}
G_1 P = 0 & \longrightarrow & G_1 A & \xrightarrow{\;\xi\;} & G_0 K & \longrightarrow & G_0 P \\
& & {\scriptstyle \varphi_1^A} \downarrow & & \| & & \| \\
L_1 P = 0 & \longrightarrow & L_1 A & \xrightarrow[\;\chi\;]{} & G_0 K & \longrightarrow & G_0 P
\end{array}
$$

is a commutative diagram with exact rows, in which $\xi = \xi_1^{\mathcal{E}^A}$ and $\chi = \chi_1^{\mathcal{E}^A}$ are the connecting homomorphisms and $L_1 P = 0$, $G_1 P = 0$ by 3.2 and the hypothesis. This readily implies that φ_1^A is an isomorphism. If $n > 0$, then

$$
\begin{array}{ccccccc}
G_{n+1} P = 0 & \longrightarrow & G_{n+1} A & \xrightarrow{\;\xi\;} & G_n K & \longrightarrow & G_n P = 0 \\
& & {\scriptstyle \varphi_{n+1}^A} \downarrow & & \downarrow {\scriptstyle \varphi_n^K} & & \\
L_{n+1} P = 0 & \longrightarrow & L_{n+1} A & \xrightarrow[\;\chi\;]{} & L_n K & \longrightarrow & L_n P = 0
\end{array}
$$

is a commutative diagram with exact rows, in which $\xi = \xi_{n+1}^{\mathcal{E}^A}$ and $\chi = \chi_{n+1}^{\mathcal{E}^A}$ are the connecting homomorphisms and $L_{n+1} P = 0$, $L_n P = 0$, $G_{n+1} P = 0$, $G_n P = 0$, by 3.2 and the hypothesis. Therefore ξ and χ are isomorphisms; if φ_n^K is an isomorphism, then so is φ_{n+1}^A. $\square$

Left exact functors. Covariant left exact functors from modules to modules, such as $\mathrm{Hom}_R(M, -)$, give rise to another kind of connected sequence:

Definition. A negative connected sequence of covariant functors *consists of covariant functors* G^0, G^1, ..., G^n, ... *and of* connecting homomorphisms $G^n C \longrightarrow G^{n+1} A$, *one for every integer* $n \geqq 0$ *and short exact sequence* $\mathcal{E} : 0 \longrightarrow A \longrightarrow B \longrightarrow C \longrightarrow 0$, *such that the sequence*

$$0 \longrightarrow G^0 A \longrightarrow G^0 B \longrightarrow G^0 C \longrightarrow G^1 A \longrightarrow G^1 B \longrightarrow \cdots$$
$$\cdots \longrightarrow G^n A \longrightarrow G^n B \longrightarrow G^n C \longrightarrow G^{n+1} A \longrightarrow \cdots$$

is exact and natural in $\mathcal{E}$.

We seek a connected sequence of functors that begins with a given left exact functor F. If F is additive, then there is again a "best" such sequence. We saw in Section 2 that to every module A can be assigned an injective resolution $A \longrightarrow \mathcal{J}_A$. Then $F\mathcal{J}_A$ is a negative complex of modules. The following result is proved like Proposition 3.1:

Proposition **3.5.** *Let* F *be a covariant additive functor. Assign an injective resolution* $A \longrightarrow \mathcal{J}_A$ *to every module* A. *For every* $n \geqq 0$, *let* $R^n A = H^n(F\mathcal{J}_A)$; *for every module homomorphism* $\varphi : A \longrightarrow B$, *let* $R^n \varphi = H^n(F\overline{\varphi})$, *where* $\overline{\varphi} : \mathcal{J}_A \longrightarrow \mathcal{J}_B$ *is a chain transformation that lifts* φ. *Then* $R^n F = R^n$ *is a well defined additive functor, which, up to natural isomorphisms, does not depend on the initial assignment of* $\mathcal{J}_A$ *to* A.

Definition. In Proposition 3.5, $R^n F$ is the *n*th *right derived functor of* F.

The following properties are proved much like Theorems 3.2, 3.3, 3.4:

Theorem **3.6.** *Let* F *be a covariant additive functor.*

(1) *If* J *is injective, then* $(R^0 F)J \cong FJ$ *and* $(R^n F)J = 0$ *for all* $n > 0$.

(2) *If* F *is left exact, then there is for every module* A *an isomorphism* $(R^0 F)A \cong FA$, *which is natural in* A.

(3) $R^0 F$, $R^1 F$, ..., $R^n F$, ... *is a negative connected sequence of functors.*

(4) *If* F *is left exact, then* F, $R^1 F$, ..., $R^n F$, ... *is a negative connected sequence of functors.*

Theorem **3.7.** *Let* F *be a left exact, covariant additive functor, and let* G^0, G^1, ..., G^n, ... *be a positive connected sequence of covariant functors. For every natural transformation* $\varphi_0 : F \longrightarrow G^0$ *there exist unique natural transformations* $\varphi_n : R^n F \longrightarrow G^n$ *such that the square*

$$
\begin{array}{ccc}
G^n C & \xrightarrow{\;\xi_{\mathcal{E}}^n\;} & G^{n+1} A \\
\varphi_C^n \uparrow & & \uparrow \varphi_A^{n+1} \\
(R^n F)C & \xrightarrow[\;\chi_{\mathcal{E}}^n\;]{} & (R^{n+1} F)A
\end{array}
$$

commutes for every short exact sequence $\mathcal{E} : 0 \longrightarrow A \longrightarrow B \longrightarrow C \longrightarrow 0$ *and every* $n \geqq 0$, *where* χ *and* ξ *are the connecting homomorphisms.*

Theorem **3.8.** *Let* G^0, G^1, ..., G^n, ... *be a negative connected sequence of covariant functors. If* $G^n J = 0$ *whenever* J *is injective and* $n > 0$, *then, up to natural isomorphisms,* G^1, ..., G^n, ... *are the right derived functors of* G^0.

Contravariant functors. Contravariant left exact functors from modules to modules, such as $\mathrm{Hom}_R(-, M)$, give rise to a third kind of connected sequence:

Definition. A negative connected sequence of contravariant functors *consists of contravariant functors* G^0, G^1, ..., G^n, ... *and of* connecting homomorphisms $G^n A \longrightarrow G^{n+1} C$, *one for every integer* $n \geq 0$ *and short exact sequence* $\mathcal{E} : 0 \longrightarrow A \longrightarrow B \longrightarrow C \longrightarrow 0$, *such that the sequence*

$$0 \longrightarrow G^0 C \longrightarrow G^0 B \longrightarrow G^0 A \longrightarrow G^1 C \longrightarrow G^1 B \longrightarrow \cdots$$
$$\cdots \longrightarrow G^n C \longrightarrow G^n B \longrightarrow G^n A \longrightarrow G^{n+1} C \longrightarrow \cdots$$

is exact and natural in $\mathcal{E}$.

If F is an additive, contravariant left exact functor, then a "best" connected sequence that begins with F is constructed as follows. Assign to every module A a projective resolution $\mathcal{P}^A \longrightarrow A$. Applying F to $\mathcal{P}^A$ yields a negative complex. Readers will easily establish the following properties.

Proposition **3.9.** *Let* F *be a contravariant additive functor. Assign a projective resolution* $\mathcal{P}^A \longrightarrow A$ *to every module* A. *For every* $n \geq 0$, *let* $R^n A = H^n(F\mathcal{P}^A)$; *for every module homomorphism* $\varphi : A \longrightarrow B$, *let* $R^n \varphi = H^n(F\overline{\varphi})$, *where* $\overline{\varphi} : \mathcal{P}^A \longrightarrow \mathcal{P}^B$ *is a chain transformation that lifts* φ. *Then* $R^n F = R^n$ *is a well defined additive contravariant functor, which, up to natural isomorphisms, does not depend on the initial assignment of* $\mathcal{P}^A$ *to* A.

Definition. In Proposition 3.9, $R^n F$ is the *n*th right derived functor *of* F.

Theorem **3.10.** *Let* F *be a contravariant additive functor.*

(1) *If* P *is projective, then* $(R^0 F)P \cong FP$ *and* $(R^n F)P = 0$ *for all* $n > 0$.

(2) *If* F *is left exact, then there is for every module* A *an isomorphism* $(R^0 F)A \cong FA$, *which is natural in* A.

(3) $R^0 F$, $R^1 F$, ..., $R^n F$, ... *is a negative connected sequence of contravariant functors.*

(4) *If* F *is left exact, then* F, $R^1 F$, ..., $R^n F$, ... *is a negative connected sequence of contravariant functors.*

Theorem **3.11.** *Let* F *be left exact, contravariant additive functor, and let* G^0, G^1, ..., G^n, ... *be a negative connected sequence of contravariant functors. For every natural transformation* $\varphi_0 : F \longrightarrow G_0$ *there exist unique natural transformations* $\varphi_n : R^n F \longrightarrow G^n$ *such that the square below, where* χ *and* ξ *are the connecting homomorphisms, commutes for every short exact sequence* $\mathcal{E} : 0 \longrightarrow A \longrightarrow B \longrightarrow C \longrightarrow 0$ *and every* $n \geq 0$.

$$G^n A \xrightarrow{\xi^n_{\mathcal{E}}} G^{n+1} C$$

$$\varphi^n_A \uparrow \qquad\qquad \uparrow \varphi^{n+1}_C$$

$$(R^n F)A \xrightarrow{\chi^n_{\mathcal{E}}} (R^{n+1} F)C$$

Theorem 3.12. *Let* G^0, G^1, ..., G^n, ... *be a negative connected sequence of contravariant functors. If* $G^n P = 0$ *whenever* P *is projective and* $n > 0$, *then, up to natural isomorphisms,* G^1, ..., G^n, ... *are the right derived functors of* G^0.

Contravariant right exact functors. A fourth construction of derived functors applies to contravariant right exact functors, where connected sequences of left derived contravariant functors are constructed from injective resolutions. But this construction has no applications here, due a serious shortage of good contravariant right exact functors. Details are left to interested readers.

Exercises

1. Let F be a covariant additive functor. Assign an injective resolution $A \longrightarrow \mathcal{J}_A$ to every module A. Let $R^n A = H^n(F\mathcal{J}_A)$; when $\varphi : A \longrightarrow B$, let $R^n \varphi = H^n(F\overline{\varphi})$, where $\overline{\varphi} : \mathcal{J}_A \longrightarrow \mathcal{J}_B$ is a chain transformation that lifts φ. Show that R^n is a well defined additive functor and, up to natural isomorphisms, does not depend on the initial assignment of $\mathcal{J}_A$ to A.

2. Let F be a covariant additive functor. Show that $(R^0 F)J \cong FJ$ and $(R^n F)J = 0$ for all $n > 0$, whenever J is injective.

3. Let F be a covariant additive functor. If F is left exact, then show that there is for every module A an isomorphism $(R^0 F)A \cong FA$, which is natural in A.

4. Let F be a covariant additive functor. Show that $R^0 F$, $R^1 F$, ..., $R^n F$, ... is a negative connected sequence of functors; if F is left exact, then F, $R^1 F$, ..., $R^n F$, ... is a negative connected sequence of functors.

5. Let F be a left exact, covariant additive functor, G^0, G^1, ..., G^n, ... be a positive connected sequence of covariant functors, and $\varphi_0 : F \longrightarrow G^0$ be a natural transformation. Show that there exist unique natural transformations $\varphi_n : R^n F \longrightarrow G^n$ such that the square

$$G^n C \xrightarrow{\xi^n_{\mathcal{E}}} G^{n+1} A$$

$$\varphi^n_C \uparrow \qquad\qquad \uparrow \varphi^{n+1}_A$$

$$(R^n F)C \xrightarrow{\chi^n_{\mathcal{E}}} (R^{n+1} F)A$$

commutes for every $n \geq 0$ and short exact sequence $\mathcal{E} : 0 \longrightarrow A \longrightarrow B \longrightarrow C \longrightarrow 0$, where χ and ξ are the connecting homomorphisms.

6. Let G^0, G^1, ..., G^n, ... be a negative connected sequence of covariant functors. Prove the following: if $G^n J = 0$ whenever J is injective and $n > 0$, then, up to natural isomorphisms, G^1, ..., G^n, ... are the right derived functors of G^0.

7. Let F be a contravariant additive functor. Assign a projective resolution $\mathcal{P}^A \longrightarrow A$ to every module A. Let $R^n A = H^n(F\mathcal{P}^A)$; when $\varphi : A \longrightarrow B$, let $R^n \varphi = H^n(F\overline{\varphi})$, where

$\overline{\varphi} : \mathcal{P}^A \longrightarrow \mathcal{P}^B$ is a chain transformation that lifts φ. Show that R^n is well defined and, up to natural isomorphisms, does not depend on the initial assignment of $\mathcal{P}^A$ to A.

8. Let F be a contravariant additive functor. Show that $(R^0 F) P \cong FP$ and $(R^n F) P = 0$ for all $n > 0$, whenever P is projective.

9. Let F be a contravariant additive functor. If F is left exact, then show that there is for every module A an isomorphism $(R^0 F) A \cong FA$, which is natural in A.

10. Let F be a contravariant additive functor. Show that $R^0 F$, $R^1 F$, ..., $R^n F$, ... is a negative connected sequence of contravariant functors; if F is left exact, then F, $R^1 F$, ..., $R^n F$, ... is a negative connected sequence of contravariant functors.

11. Let F be a left exact, contravariant additive functor; let G^0, G^1, ..., G^n, ... be a negative connected sequence of contravariant functors, and let $\varphi_0 : F \longrightarrow G^0$ be a natural transformation. Show that there exist unique natural transformations $\varphi_n : R^n F \longrightarrow G^n$ such that the square

$$
\begin{array}{ccc}
G^n A & \xrightarrow{\ \xi_{\mathcal{E}}^n\ } & G^{n+1} C \\
\varphi_A^n \big\uparrow & & \big\uparrow \varphi_C^{n+1} \\
(R^n F) A & \xrightarrow[\ \chi_{\mathcal{E}}^n\ ]{} & (R^{n+1} F) C
\end{array}
$$

commutes for every $n \geq 0$ and short exact sequence $\mathcal{E} : 0 \longrightarrow A \longrightarrow B \longrightarrow C \longrightarrow 0$, where χ and ξ are the connecting homomorphisms.

12. Let G^0, G^1, ..., G^n, ... be a negative connected sequence of contravariant functors. Prove the following: if $G^n P = 0$ whenever P is projective and $n > 0$, then, up to natural isomorphisms, G^1, ..., G^n, ... are the right derived functors of G^0.

*13. Define the left derived functors of a contravariant additive functor; then state and prove their basic properties.

4. Ext

This section constructs the functors Ext^n and gives their basic properties. All results are stated for left R-modules but apply equally to right R-modules.

Definition. For every left R-module A, $\mathrm{Hom}_R(A, -)$ is a covariant, left exact, additive functor from left R-modules to abelian groups, and has right derived functors, temporarily denoted by $\mathrm{RExt}_R^n(A, -)$. As in Section 3, $\mathrm{Hom}_R(A, -) = \mathrm{RExt}_R^0(A, -)$, $\mathrm{RExt}_R^1(A, -)$, $\mathrm{RExt}_R^2(A, -)$, , is a negative connected sequence of covariant functors from left R-modules to abelian groups, and $\mathrm{RExt}_R^n(A, J) = 0$ whenever J is injective and $n > 0$.

We find how $\mathrm{RExt}_R^n(A, -)$ depends on A. Every module homomorphism $\varphi : A \longrightarrow A'$ induces a natural transformation $\mathrm{Hom}_R(\varphi, -) : \mathrm{Hom}_R(A', -) \longrightarrow \mathrm{Hom}_R(A, -)$, which by Theorem 3.7 induces unique natural transformations $\mathrm{RExt}_R^n(\varphi, -) : \mathrm{RExt}_R^n(A', -) \longrightarrow \mathrm{RExt}_R^n(A, -)$ such that the square

$$\begin{array}{ccc} \mathrm{RExt}_R^n(A, B'') & \xrightarrow{\ \chi\ } & \mathrm{RExt}_R^{n+1}(A, B) \\ {\scriptstyle \mathrm{RExt}_R^n(\varphi, B'')}\Big\uparrow & & \Big\uparrow {\scriptstyle \mathrm{RExt}_R^{n+1}(\varphi, B)} \\ \mathrm{RExt}_R^n(A', B'') & \xrightarrow[\ \chi'\ ]{} & \mathrm{RExt}_R^{n+1}(A', B) \end{array}$$

commutes for every $n \geqq 0$ and short exact sequence $0 \longrightarrow B \longrightarrow B' \longrightarrow B'' \longrightarrow 0$, where χ and χ' are the connecting homomorphisms. Uniqueness in Theorem 3.7 implies that $\mathrm{RExt}_R^n(1_A, B)$ is the identity on $\mathrm{RExt}_R^n(A, B)$ and that $\mathrm{RExt}_R^n(\varphi \circ \psi, -) = \mathrm{RExt}_R^n(\psi, -) \circ \mathrm{RExt}_R^n(\varphi, -)$, $\mathrm{RExt}_R^n(\varphi + \psi, -) = \mathrm{RExt}_R^n(\varphi, -) + \mathrm{RExt}_R^n(\psi, -)$ whenever these operations are defined; moreover,

$$\begin{array}{ccc} \mathrm{RExt}_R^n(A, B) & \xrightarrow{\ \mathrm{RExt}_R^n(A, \psi)\ } & \mathrm{RExt}_R^n(A, B') \\ {\scriptstyle \mathrm{RExt}_R^n(\varphi, B)}\Big\uparrow & & \Big\uparrow {\scriptstyle \mathrm{RExt}_R^n(\varphi, B')} \\ \mathrm{RExt}_R^n(A', B) & \xrightarrow[\ \mathrm{RExt}_R^n(A', \psi)\ ]{} & \mathrm{RExt}_R^n(A', B') \end{array}$$

commutes for every homomorphism $\psi : B \longrightarrow B'$, since $\mathrm{RExt}_R^n(\varphi, B)$ is natural in B. Thus $\mathrm{RExt}_R^n(-, -)$ is, like $\mathrm{Hom}_R(-, -)$, an additive bifunctor from left R-modules to abelian groups, contravariant in the first variable and covariant in the second variable.

For every left R-module B, the contravariant functor $\mathrm{Hom}_R(-, B)$ also has right derived functors, temporarily denoted by $\mathrm{LExt}_R^n(-, B)$. By Theorem 3.10, $\mathrm{Hom}_R(-, B) = \mathrm{LExt}_R^0(-, B)$, $\mathrm{LExt}_R^1(-, B)$, $\mathrm{LExt}_R^2(-, B)$, ..., is a negative connected sequence of contravariant functors from left R-modules to abelian groups, and $\mathrm{LExt}_R^n(P, B) = 0$ whenever P is projective and $n > 0$.

Like RExt_R^n, LExt_R^n is a bifunctor. Every module homomorphism $\psi : B \longrightarrow B'$ induces a natural transformation $\mathrm{Hom}_R(-, \psi) \colon \mathrm{Hom}_R(-, B) \longrightarrow \mathrm{Hom}_R(-, B')$, which by Theorem 3.11 induces natural transformations $\mathrm{LExt}_R^n(-, \varphi) \colon \mathrm{LExt}_R^n(-, B) \longrightarrow \mathrm{LExt}_R^n(-, B')$ such that the square

$$\begin{array}{ccc} \mathrm{LExt}_R^n(A, B) & \xrightarrow{\ \chi\ } & \mathrm{LExt}_R^{n+1}(A'', B) \\ {\scriptstyle \mathrm{LExt}_R^n(A, \psi)}\Big\downarrow & & \Big\downarrow {\scriptstyle \mathrm{LExt}_R^{n+1}(A'', \psi)} \\ \mathrm{LExt}_R^n(A, B') & \xrightarrow[\ \chi'\ ]{} & \mathrm{LExt}_R^{n+1}(A'', B') \end{array}$$

commutes for every $n \geqq 0$ and short exact sequence $0 \longrightarrow A \longrightarrow A' \longrightarrow A'' \longrightarrow 0$, where χ and χ' are the connecting homomorphisms. As above, this makes $\mathrm{LExt}_R^n(-, -)$ an additive bifunctor from left R-modules to abelian groups, contravariant in the first variable and covariant in the second variable.

We show that $\mathrm{LExt}_R^n(A, B)$ and $\mathrm{RExt}_R^n(A, B)$ are naturally isomorphic; hence there is, up to this natural isomorphism, a single bifunctor Ext_R^n.

Theorem **4.1.** *For every* $n > 0$ *and left R-modules* A *and* B *there is an isomorphism* $\mathrm{LExt}^n_R(A, B) \cong \mathrm{RExt}^n_R(A, B)$, *which is natural in* A *and* B.

Proof. Let $\mathcal{A} : 0 \longrightarrow K \longrightarrow P \longrightarrow A \longrightarrow 0$ and $\mathcal{B} : 0 \longrightarrow B \longrightarrow J \longrightarrow L \longrightarrow 0$ be short exact sequences, with P projective and J injective. We have a commutative diagram, where $RA = \mathrm{RExt}^1_R(A, B)$ and $RK = \mathrm{RExt}^1_R(K, B)$,

$$
\begin{array}{ccccccccc}
& & 0 & & 0 & & 0 & & \\
& & \downarrow & & \downarrow & & \downarrow & & \\
0 \to & \mathrm{Hom}_R(A, B) & \to & \mathrm{Hom}_R(A, J) & \to & \mathrm{Hom}_R(A, L) & \to & RA \to 0 \\
& \downarrow & & \downarrow & & \downarrow & & \\
0 \to & \mathrm{Hom}_R(P, B) & \to & \mathrm{Hom}_R(P, J) & \xrightarrow{\sigma} & \mathrm{Hom}_R(P, L) & \to & 0 \\
& {\scriptstyle\alpha}\downarrow & & {\scriptstyle\beta}\downarrow & & {\scriptstyle\gamma}\downarrow & & \\
0 \to & \mathrm{Hom}_R(K, B) & \to & \mathrm{Hom}_R(K, J) & \xrightarrow[\tau]{} & \mathrm{Hom}_R(K, L) & \to & RK \to 0 \\
& \downarrow & & \downarrow & & \downarrow & & \\
& \mathrm{LExt}^1_R(A, B) & & 0 & & \mathrm{LExt}^1_R(A, L) & & \\
& \downarrow & & & & \downarrow & & \\
& 0 & & & & 0 & &
\end{array}
$$

which is natural in $\mathcal{A}$ and $\mathcal{B}$. The first and third rows are exact by Theorem 3.6 and end with zeros, since $\mathrm{RExt}^1_R(M, J) = 0$ for all M when J is injective. The first and third columns are exact by Theorem 3.10 and end with zeros, since $\mathrm{LExt}^1_R(P, M) = 0$ for all M when P is projective. The second row is exact since P is projective, and the second column is exact since J is injective.

Lemma 1.5, applied to α, β, γ, yields an exact sequence

$$\mathrm{Hom}_R(A, J) \longrightarrow \mathrm{Hom}_R(A, L) \longrightarrow \mathrm{LExt}^1_R(A, B) \longrightarrow 0.$$

So does the first row of the diagram:

$$\mathrm{Hom}_R(A, J) \longrightarrow \mathrm{Hom}_R(A, L) \longrightarrow \mathrm{RExt}^1_R(A, B) \longrightarrow 0.$$

By Lemma X.1.2, $\mathrm{Hom}_R(A, L) \longrightarrow \mathrm{LExt}^1_R(A, B)$ and $\mathrm{Hom}_R(A, L) \longrightarrow \mathrm{RExt}^1_R(A, B)$ factor uniquely through each other, which provides mutually inverse isomorphisms $\mathrm{LExt}^1_R(A, B) \cong \mathrm{RExt}^1_R(A, B)$. These isomorphisms are natural in $\mathcal{A}$ and $\mathcal{B}$; hence they are natural in A and B, since homomorphisms of A and B lift to homomorphisms of $\mathcal{A}$ and $\mathcal{B}$, by 2.1 and 2.7. This proves 4.1 when $n = 1$. Since σ and β are epimorphisms we also have isomorphisms

$$\mathrm{LExt}^1_R(A, L) \cong \mathrm{Coker}\, \gamma = \mathrm{Coker}\, \gamma\, \sigma$$

$$= \mathrm{Coker}\, \tau\, \beta = \mathrm{Coker}\, \tau \cong \mathrm{RExt}^1_R(K, B)$$

that are natural in $\mathcal{A}$ and $\mathcal{B}$.

For $n \geq 1$ we prove by induction on n that there are isomorphisms

$$\mathrm{LExt}^n_R(A, B) \cong \mathrm{RExt}^n_R(A, B) \quad \text{and} \quad \mathrm{LExt}^n_R(A, L) \cong \mathrm{RExt}^n_R(K, B)$$

that are natural in $\mathcal{A}$ and $\mathcal{B}$. We just proved this when $n = 1$. In general, Theorem 3.6, applied to $\mathcal{B}$, yields for every module M an exact sequence

$$\mathrm{RExt}_R^n(M, J) \longrightarrow \mathrm{RExt}_R^n(M, L) \longrightarrow \mathrm{RExt}_R^{n+1}(M, B) \longrightarrow \mathrm{RExt}_R^{n+1}(M, J),$$

which begins and ends with 0 since J is injective, and an isomorphism

$$\mathrm{RExt}_R^n(M, L) \cong \mathrm{RExt}_R^{n+1}(M, B) \tag{1}$$

that is natural in M and $\mathcal{B}$. Similarly, Theorem 3.10, applied to $\mathcal{A}$, yields for every module M an exact sequence

$$\mathrm{LExt}_R^n(P, M) \longrightarrow \mathrm{LExt}_R^n(K, M) \longrightarrow \mathrm{LExt}_R^{n+1}(A, M) \longrightarrow \mathrm{LExt}_R^{n+1}(P, M),$$

which begins and ends with 0 since P is projective, and an isomorphism

$$\mathrm{LExt}_R^n(K, M) \cong \mathrm{LExt}_R^{n+1}(A, M) \tag{2}$$

that is natural in $\mathcal{A}$ and M. Then (2) with $M = B$, (1) with $M = A$, and the induction hypothesis, yield isomorphisms

$$\mathrm{LExt}_R^{n+1}(A, B) \cong \mathrm{LExt}_R^n(K, B) \cong \mathrm{RExt}_R^n(K, B)$$
$$\cong \mathrm{LExt}_R^n(A, L) \cong \mathrm{RExt}_R^n(A, L) \cong \mathrm{RExt}_R^{n+1}(A, B)$$

that are natural in $\mathcal{A}$ and $\mathcal{B}$; and (2) with $M = L$, (1) with $M = K$, and the induction hypothesis, yield isomorphisms

$$\mathrm{LExt}_R^{n+1}(A, L) \cong \mathrm{LExt}_R^n(K, L) \cong \mathrm{RExt}_R^n(K, L) \cong \mathrm{RExt}_R^{n+1}(K, B)$$

that are natural in $\mathcal{A}$ and $\mathcal{B}$. This completes the induction. As before, the isomorphism $\mathrm{LExt}_R^n(A, B) \cong \mathrm{RExt}_R^n(A, B)$ is then natural in A and B. $\square$

Not surprisingly, $\mathrm{Ext}_R^n(A, B)$ is now defined as follows, up to natural isomorphisms:

Definition. Up to isomorphisms that are natural in A and B, $\mathrm{Ext}_R^n(A, B) \cong \mathrm{LExt}_R^n(A, B) \cong \mathrm{RExt}_R^n(A, B)$, for every left R-modules A and B.

The name Ext comes from a one-to-one correspondence between elements of $\mathrm{Ext}_R^1(C, A)$ and equivalence classes of extensions of A by C (short exact sequences $0 \longrightarrow A \longrightarrow B \longrightarrow C \longrightarrow 0$). See, for instance, MacLane, *Homology* [1963], for the details of this relationship and its generalization to every Ext^n.

Properties. First, Ext_R^n enjoys all the properties of LExt_R^n and RExt_R^n.

Proposition 4.2. The functor $\mathrm{Ext}_R^n(-, -)$ is an additive bifunctor from left R-modules to abelian groups, contravariant in the first variable and covariant in the second variable.

Proposition 4.3. For every projective resolution $\mathcal{P} \longrightarrow A$ and injective resolution $B \longrightarrow \mathcal{J}$ there are natural isomorphisms

$$\mathrm{Ext}_R^n(A, B) \cong H^n\big(\mathrm{Hom}_R(\mathcal{P}, B)\big) \cong H^n\big(\mathrm{Hom}_R(A, \mathcal{J})\big).$$

This follows from the definition of Ext and the definitions of derived functors.

Theorem **4.4.** (1) *If A is projective, then $\operatorname{Ext}^n_R(A, B) = 0$ for all $n \geqq 1$.*

(2) *If B is injective, then $\operatorname{Ext}^n_R(A, B) = 0$ for all $n \geqq 1$.*

(3) *For every exact sequence $\mathcal{A} : 0 \longrightarrow A \longrightarrow A' \longrightarrow A'' \longrightarrow 0$ and module B, there is an exact sequence, which is natural in $\mathcal{A}$ and B,*

$$0 \longrightarrow \operatorname{Hom}_R(A'', B) \longrightarrow \operatorname{Hom}_R(A', B) \longrightarrow \operatorname{Hom}_R(A, B)$$
$$\longrightarrow \operatorname{Ext}^1_R(A'', B) \longrightarrow \operatorname{Ext}^1_R(A', B) \longrightarrow \operatorname{Ext}^1_R(A, B) \longrightarrow \cdots$$
$$\longrightarrow \operatorname{Ext}^n_R(A'', B) \longrightarrow \operatorname{Ext}^n_R(A', B) \longrightarrow \operatorname{Ext}^n_R(A, B) \longrightarrow \cdots .$$

(4) *For every exact sequence $\mathcal{B} : 0 \longrightarrow B \longrightarrow B' \longrightarrow B'' \longrightarrow 0$ and module A, there is an exact sequence, which is natural in $\mathcal{A}$ and B,*

$$0 \longrightarrow \operatorname{Hom}_R(A, B) \longrightarrow \operatorname{Hom}_R(A, B') \longrightarrow \operatorname{Hom}_R(A, B'')$$
$$\longrightarrow \operatorname{Ext}^1_R(A, B) \longrightarrow \operatorname{Ext}^1_R(A, B') \longrightarrow \operatorname{Ext}^1_R(A, B'') \longrightarrow \cdots$$
$$\longrightarrow \operatorname{Ext}^n_R(A, B) \longrightarrow \operatorname{Ext}^n_R(A, B') \longrightarrow \operatorname{Ext}^n_R(A, B'') \longrightarrow \cdots .$$

This follows from Theorems 3.6 and 3.10. In fact, up to natural isomorphisms, $\operatorname{Ext}^n_R(A, B)$ is the only bifunctor with properties (1) and (3), and the only bifunctor with properties (2) and (4), by Theorems 3.8 and 3.12.

In addition, Ext_R inherits properties from Hom_R. Readers may establish the following properties:

Proposition **4.5.** $\operatorname{Ext}^1_R(C, A) = 0$ *if and only if every short exact sequence $0 \longrightarrow A \longrightarrow B \longrightarrow C \longrightarrow 0$ splits; a module M is projective if and only if $\operatorname{Ext}^1_R(M, B) = 0$ for all B; a module M is injective if and only if $\operatorname{Ext}^1_R(A, M) = 0$ for all A.*

Proposition **4.6.** *If A is a left R-, right S-bimodule and B is a left R-, right T-bimodule, then $\operatorname{Ext}^n_R(A, B)$ is a left S-, right T-bimodule. In particular, if R is commutative and A, B are R-modules, then $\operatorname{Ext}^n_R(A, B)$ is an R-module.*

Proposition **4.7.** *For every family $(B_i)_{i \in I}$ of left R-modules there is an isomorphism $\operatorname{Ext}^n_R\left(A, \prod_{i \in I} B_i\right) \cong \prod_{i \in I} \operatorname{Ext}^n_R(A, B_i)$, which is natural in A and $(B_i)_{i \in I}$.*

Proposition **4.8.** *For every family $(A_i)_{i \in I}$ of left R-modules there is an isomorphism $\operatorname{Ext}^n_R\left(\bigoplus_{i \in I} A_i, B\right) \cong \prod_{i \in I} \operatorname{Ext}^n_R(A_i, B)$, which is natural in $(A_i)_{i \in I}$ and B.*

Abelian groups. The case $R = \mathbb{Z}$ provides some examples of Ext groups.

Proposition **4.9.** $\operatorname{Ext}^n_{\mathbb{Z}}(A, B) = 0$ *for all $n \geqq 2$ and abelian groups A and B.*

Proof. Let R be a PID. Every submodule of a free R-module is free. Hence every R-module A has a free resolution $\mathcal{F} : 0 \longrightarrow F_1 \longrightarrow F_0 \longrightarrow A$ in which $F_n = 0$ for all $n \geqq 2$. Then $\mathrm{Hom}_R(F_n, B) = 0$ and $\mathrm{Ext}_R^n(A, B) \cong H^n\big(\mathrm{Hom}_R(\mathcal{F}, B)\big) = 0$ when $n \geqq 2$, for every R-module B. $\square$

If $A \cong \mathbb{Z}$ is infinite cyclic, then $\mathrm{Ext}_{\mathbb{Z}}^1(A, B) = 0$ for all B, since A is projective. The case of $A \cong \mathbb{Z}_m$ finite cyclic is an exercise:

Proposition 4.10. *For every abelian group B there is an isomorphism $\mathrm{Ext}_{\mathbb{Z}}^1(\mathbb{Z}_m, B) \cong B/mB$, which is natural in B.*

Combining Propositions 4.8 and 4.10 yields $\mathrm{Ext}_{\mathbb{Z}}^1(A, B)$ when A is finitely generated.

Exercises

1. Show that a left R-module P is projective if and only if $\mathrm{Ext}_R^1(P, B) = 0$ for every left R-module B. (In particular, the ring R is semisimple if and only if $\mathrm{Ext}_R^1(A, B) = 0$ for all left R-modules A and B.)

2. Show that a left R-module J is injective if and only if $\mathrm{Ext}_R^1(A, J) = 0$ for every left R-module A.

3. Prove that $\mathrm{Ext}_R^1(C, A) = 0$ if and only if every short exact sequence $0 \longrightarrow A \longrightarrow B \longrightarrow C \longrightarrow 0$ splits.

4. Show that a left R-module J is injective if and only if $\mathrm{Ext}_R^1(R/L, J) = 0$ for every left ideal L of R.

5. Explain how $\mathrm{Ext}_R^n(A, B)$ becomes a left S-, right T-bimodule when A is a left R-, right S-bimodule and B is a left R-, right T-bimodule.

6. Prove the following: for every family $(B_i)_{i \in I}$ of left R-modules there is an isomorphism $\mathrm{Ext}_R^n\big(A, \prod_{i \in I} B_i\big) \cong \prod_{i \in I} \mathrm{Ext}_R^n(A, B_i)$, which is natural in A and $(B_i)_{i \in I}$. (You may want to follow the proof of Theorem 4.1.)

7. Prove the following: for every family $(A_i)_{i \in I}$ of left R-modules there is an isomorphism $\mathrm{Ext}_R^n\big(\bigoplus_{i \in I} A_i, B\big) \cong \prod_{i \in I} \mathrm{Ext}_R^n(A_i, B)$, which is natural in $(A_i)_{i \in I}$ and B. (You may want to follqw the proof of Theorem 4.1.)

8. Prove that a ring R is left hereditary if and only if $\mathrm{Ext}_R^n(A, B) = 0$ for all left R-modules A and B.

9. Prove the following: for every abelian group B there is an isomorphism $\mathrm{Ext}_{\mathbb{Z}}^1(\mathbb{Z}_m, B) \cong B/mB$, which is natural in B.

10. Show that, if the abelian group A is divisible by m (if $mA = A$), then every short exact sequence $0 \longrightarrow A \longrightarrow B \longrightarrow \mathbb{Z}_m \longrightarrow 0$ splits.

11. Give another proof of Schur's theorem for abelian groups: if m and n are relatively prime, and $mA = nC = 0$, then every short exact sequence $0 \longrightarrow A \longrightarrow B \longrightarrow C \longrightarrow 0$ splits.

12. Show that $\mathrm{Ext}_{\mathbb{Z}}^1(A, \mathbb{Z}) \cong \mathrm{Hom}_{\mathbb{Z}}(A, \mathbb{Q}/\mathbb{Z})$ for every torsion abelian group A.

13. Let A be an abelian group such that $pA = 0$, where p is prime. Show that $\operatorname{Ext}_{\mathbb{Z}}^1(A, B) \cong \operatorname{Hom}_{\mathbb{Z}}(A, B/pB)$, for every abelian group B.

5. Tor

Tor is to tensor products as Ext is to Hom.

Definition. For every right R-module A and left R-module B, the right exact functors $A \otimes_R -$ and $- \otimes_R B$ have left derived functors, temporarily denoted by $\operatorname{RTor}_n^R(A, -)$ and $\operatorname{LTor}_n^R(-, B)$, which with $A \otimes_R B = \operatorname{RTor}_0^R(A, B) = \operatorname{LTor}_0^R(A, B)$ constitute positive connected sequences, such that $\operatorname{RTor}_n^R(A, P) = \operatorname{LTor}_n^R(Q, B) = 0$ whenever P, Q are projective and $n > 0$.

Module homomorphisms $\varphi : A \longrightarrow A'$ and $\psi : B \longrightarrow B'$ induce natural transformations $\varphi \otimes_R -$ from $A \otimes_R -$ to $A' \otimes_R -$ and $- \otimes_R \psi$ from $- \otimes_R B$ to $- \otimes_R B'$, which by Theorem 3.3 induce natural transformations $\operatorname{RTor}_n^R(\varphi, -)$ from $\operatorname{RTor}_n^R(A, -)$ to $\operatorname{RTor}_n^R(A', -)$ and $\operatorname{LTor}_n^R(-, \psi)$ from $\operatorname{LTor}_n^R(-, B)$ to $\operatorname{LTor}_n^R(-, B')$. As in Section 4, this makes RTor_n^R and LTor_n^R additive bifunctors from right and left R-modules to abelian groups, covariant in both variables.

Theorem 5.1. *For every right R-module A and left R-module B and every $n > 0$ there is an isomorphism $\operatorname{LTor}_n^R(A, B) \cong \operatorname{RTor}_n^R(A, B)$, which is natural in A and B.*

The proof of Theorem 5.1 is similar to that of Theorem 4.1, and may be entrusted to readers.

Definition. Up to isomorphisms that are natural in A and B, $\operatorname{Tor}_n^R(A, B) \cong \operatorname{LTor}_n^R(A, B) \cong \operatorname{RTor}_n^R(A, B)$, for every right R-module A and left R-module B.

The abelian groups $\operatorname{Tor}_n^R(A, B)$ are *torsion products* of A and B, after the case of abelian groups, where $\operatorname{Tor}_1^{\mathbb{Z}}(A, B)$ is determined by the torsion parts of A and B (see Proposition 5.9 below and the exercises).

Properties. First, Tor_n^R enjoys all the properties of LTor_n^R and RTor_n^R.

Proposition 5.2. $\operatorname{Tor}_n^R(\ ,-)$ is an additive bifunctor from right and left R-modules to abelian groups, covariant in both variables.

Proposition 5.3. For every projective resolution $\mathcal{P} \longrightarrow A$ and $\mathcal{Q} \longrightarrow B$ there are natural isomorphisms

$$\operatorname{Tor}_n^R(A, B) \cong H_n(\mathcal{P} \otimes_R B) \cong H_n(A \otimes_R \mathcal{Q}).$$

This follows from the definition of Tor and the definition of left derived functors.

Theorem **5.4.** (1) *If* A *is projective, then* $\text{Tor}_n^R(A, B) = 0$ *for all* $n \geq 1$.

(2) *If* B *is projective, then* $\text{Tor}_n^R(A, B) = 0$ *for all* $n \geq 1$.

(3) *For every exact sequence* $\mathcal{A} : 0 \longrightarrow A \longrightarrow A' \longrightarrow A'' \longrightarrow 0$ *and left* R-*module* B, *there is an exact sequence, which is natural in* $\mathcal{A}$ *and* B,

$$\cdots \longrightarrow \text{Tor}_n^R(A, B) \longrightarrow \text{Tor}_n^R(A', B) \longrightarrow \text{Tor}_n^R(A'', B) \longrightarrow \cdots$$

$$\longrightarrow \text{Tor}_1^R(A'', B) \longrightarrow A \otimes_R B \longrightarrow A' \otimes_R B \longrightarrow A'' \otimes_R B \longrightarrow 0$$

(4) *For every exact sequence* $\mathcal{B} : 0 \longrightarrow B \longrightarrow B' \longrightarrow B'' \longrightarrow 0$ *and right* R-*module* A, *there is an exact sequence, which is natural in* A *and* $\mathcal{B}$,

$$\cdots \longrightarrow \text{Tor}_n^R(A, B) \longrightarrow \text{Tor}_n^R(A, B') \longrightarrow \text{Tor}_n^R(A, B'') \longrightarrow \cdots$$

$$\longrightarrow \text{Tor}_1^R(A, B'') \longrightarrow A \otimes_R B \longrightarrow A \otimes_R B' \longrightarrow A \otimes_R B'' \longrightarrow 0$$

This follows from Theorem 3.2. In fact, up to natural isomorphisms, Tor_n^R is the only bifunctor with properties (1) and (3), and the only bifunctor with properties (2) and (4), by Theorem 3.4.

Moreover, Tor inherits a number of properties from tensor products.

Proposition **5.5.** *For every right* R-*module* A *and left* R-*module* B *there is an isomorphism* $\text{Tor}_n^R(A, B) \cong \text{Tor}_n^{R^{\text{op}}}(B, A)$, *which is natural in* A *and* B.

Proof. There is an isomorphism $A \otimes_R B \cong B \otimes_{R^{\text{op}}} A$, $a \otimes b \longmapsto b \otimes a$, which is natural in A and B. If $\mathcal{P} \longrightarrow A$ is a projective resolution of A as a right R-module, then $\mathcal{P} \longrightarrow A$ is a projective resolution of A as a left R^{op}-module, and the isomorphisms $P_n \otimes_R B \cong B \otimes_{R^{\text{op}}} P_n$ constitute a chain transformation $\mathcal{P} \otimes_R B \cong B \otimes_{R^{\text{op}}} \mathcal{P}$. Hence 5.3 yields isomorphisms

$$\text{Tor}_n^R(A, B) \cong H_n(\mathcal{P} \otimes_R B) \cong H_n(B \otimes_{R^{\text{op}}} \mathcal{P}) \cong \text{Tor}_n^{R^{\text{op}}}(B, A)$$

that are natural in $\mathcal{P}$ and B. Then $\text{Tor}_n^R(A, B) \cong \text{Tor}_n^{R^{\text{op}}}(B, A)$ is natural in A, since homomorphisms of A lift to its projective resolutions. $\square$

Hence $\text{Tor}_n^R(B, A) \cong \text{Tor}_n^R(A, B)$ if R is commutative.

The next two results are exercises.

Proposition **5.6.** *If* A *is a left* S-, *right* R-*bimodule and* B *is a left* R-, *right* T-*bimodule, then* $\text{Tor}_n^R(A, B)$ *is a left* S-, *right* T-*bimodule. In particular, if* R *is commutative and* A, B *are* R-*modules, then* $\text{Tor}_n^R(A, B)$ *is an* R-*module.*

Proposition **5.7.** *For every family* $(B_i)_{i \in I}$ *of left* R-*modules there is an isomorphism* $\text{Tor}_n^R(A, \bigoplus_{i \in I} B_i) \cong \bigoplus_{i \in I} \text{Tor}_n^R(A, B_i)$, *which is natural in* A *and* $(B_i)_{i \in I}$.

Similarly, $\mathrm{Tor}_n^R(\bigoplus_{i \in I} A_i, \, B) \cong \bigoplus_{i \in I} \mathrm{Tor}_n^R(A_i, B)$.

Abelian groups. The case $R = \mathbb{Z}$ provides examples of Tor groups.

Proposition **5.8.** $\mathrm{Tor}_n^{\mathbb{Z}}(A, B) = 0$ *for all* $n \geqq 2$ *and abelian groups* A *and* B.

Proof. More generally, let R be a PID. Every submodule of a free R-module is free. Hence every R-module A has a free resolution $\mathcal{F} : 0 \longrightarrow F_1 \longrightarrow F_0 \longrightarrow A$ in which $F_n = 0$ for all $n \geqq 2$. Then $F_n \otimes_R B = 0$ and $\mathrm{Tor}_n^{\mathbb{Z}}(A, B) \cong H_n(\mathcal{F} \otimes_R B) = 0$ when $n \geqq 2$, for every R-module B. $\square$

If $A \cong \mathbb{Z}$ is infinite cyclic, then $\mathrm{Tor}_1^{\mathbb{Z}}(A, B) = 0$ for all B, since A is projective. The case of $A \cong \mathbb{Z}_m$ finite cyclic is an exercise:

Proposition **5.9.** *For every abelian group* B *there is an isomorphism* $\mathrm{Tor}_1^{\mathbb{Z}}(\mathbb{Z}_m, \, B) \cong \{ b \in B \mid mb = 0 \}$, *which is natural in* B.

Propositions 5.7 and 5.9 yield $\mathrm{Tor}_1^{\mathbb{Z}}(A, B)$ whenever A is finitely generated.

Flat modules. Tor brings additional characterizations of flat modules.

Proposition **5.10.** *For a right* R-module A *the following properties are equivalent:* (1) A *is flat;* (2) $\mathrm{Tor}_1^R(A, B) = 0$ *for every left* R-module B; (3) $\mathrm{Tor}_n^R(A, B) = 0$ *for every left* R-module B *and* $n \geqq 1$.

Proof. Let $\mathcal{Q} \longrightarrow B$ be a projective resolution. If A is flat, then $A \otimes_R -$ is exact, the sequence $\cdots \longrightarrow A \otimes_R Q_1 \longrightarrow A \otimes_R Q_0 \longrightarrow A \otimes_R B \longrightarrow 0$ is exact, and $\mathrm{Tor}_n^R(A, B) \cong H_n(A \otimes_R \mathcal{Q}) = 0$ for all $n \geqq 1$.

Conversely, if $\mathrm{Tor}_1^R(A, B) = 0$ for every left R-module B, and $0 \longrightarrow B \longrightarrow B' \longrightarrow B'' \longrightarrow 0$ is a short exact sequence, then

$$0 = \mathrm{Tor}_1^R(A, B'') \longrightarrow A \otimes_R B \longrightarrow A \otimes_R B' \longrightarrow A \otimes_R B'' \longrightarrow 0$$

is short exact; hence A is flat. $\square$

In particular, an abelian group A is torsion-free if and only if $\mathrm{Tor}_1^{\mathbb{Z}}(A, B) = 0$ for every abelian group B.

Proposition **5.11.** *A right* R-module A *is flat if and only if* $A \otimes_R L \longrightarrow A \otimes_R {}_R R$ *is injective for every left ideal* L *of* R.

Proof. Assume that $A \otimes_R L \longrightarrow A \otimes_R R$ is injective for every left ideal L of R. Then $0 \longrightarrow L \longrightarrow {}_R R \longrightarrow R/L \longrightarrow 0$ induces an exact sequence

$$\mathrm{Tor}_1^R(A, {}_R R) \longrightarrow \mathrm{Tor}_1^R(A, R/L) \longrightarrow A \otimes_R L \longrightarrow A \otimes_R {}_R R$$

in which $\mathrm{Tor}_1^R(A, {}_R R) = 0$ by Theorem 5.4 and $A \otimes_R L \longrightarrow A \otimes_R {}_R R$ is injective. Therefore $\mathrm{Tor}_1^R(A, R/L) = 0$.

Now, let B be a submodule of a finitely generated left R-module C. Since C is finitely generated there is a tower

$$B = C_0 \subseteq C_1 \subseteq C_2 \subseteq \cdots \subseteq C_n = C$$

in which every C_{i+1} is generated by C_i and one of the generators of C. Then C_{i+1}/C_i is cyclic and $C_{i+1}/C_i \cong R/L$ for some left ideal L of R. Hence $\mathrm{Tor}_1^R(A, C_{i+1}/C_i) = 0$ and the exact sequence

$$\mathrm{Tor}_1^R(A, C_{i+1}/C_i) \longrightarrow A \otimes_R C_i \longrightarrow A \otimes_R C_{i+1}$$

shows that $A \otimes_R C_i \longrightarrow A \otimes_R C_{i+1}$ is injective. Therefore $A \otimes_R B \longrightarrow A \otimes_R C$ is injective. Since this holds whenever C is finitely generated, A is flat, by Proposition XI.8.5. $\square$

Exercises

1. Adjust the proof of Theorem 4.1 to show that the two definitions of Tor_n^R are equivalent (Theorem 5.1).

2. Explain how $\mathrm{Tor}_n^R(A, B)$ becomes a left S-, right T-bimodule when A is a left S-, right R-bimodule and B is a left R-, right T-bimodule.

3. Prove the following: for every family $(B_i)_{i \in I}$ of left R-modules there is an isomorphism $\mathrm{Tor}_n^R\big(A, \bigoplus_{i \in I} B_i\big) \cong \bigoplus_{i \in I} \mathrm{Tor}_n^R(A, B_i)$, which is natural in A and $(B_i)_{i \in I}$.

4. Let $(B_i)_{i \in I}$ be a direct system of left R-modules. Show that there is an isomorphism $\mathrm{Tor}_n^R\big(A, \varinjlim_{i \in I} B_i\big) \cong \varinjlim_{i \in I} \mathrm{Tor}_n^R(A, B_i)$, which is natural in A and $(B_i)_{i \in I}$.

5. Let R be left hereditary. Show that $\mathrm{Tor}_n^R(A, B) = 0$ for all A and B.

6. Prove that $\mathrm{Tor}_1^{\mathbb{Z}}(A, B)$ is torsion, for all abelian groups A and B.

7. Prove the following: for every abelian group B there is an isomorphism $\mathrm{Tor}_1^{\mathbb{Z}}(\mathbb{Z}_m, B) \cong \{ b \in B \mid mb = 0 \}$, which is natural in B.

8. Prove that $\mathrm{Tor}_1^{\mathbb{Z}}(A, B) \cong A \otimes_{\mathbb{Z}} B$ for all finite abelian groups A and B (the isomorphism is not natural or canonical).

9. Let m and n be relatively prime, and let A, B be abelian groups such that $mA = nB = 0$. Show that $\mathrm{Tor}_1^{\mathbb{Z}}(A, B) = 0$.

10. Show that there is a natural isomorphism $\mathrm{Tor}_1^{\mathbb{Z}}(\mathbb{Q}/\mathbb{Z}, B) \cong B$ for every torsion abelian group B.

11. Let $T(B) = \{ b \in B \mid nb = 0$ for some $n \neq 0 \}$ be the torsion part of B. Use the previous exercise to show that there is a natural isomorphism $\mathrm{Tor}_1^{\mathbb{Z}}(\mathbb{Q}/\mathbb{Z}, B) \cong T(B)$.

12. Show that a right R-module A is flat if and only if $\mathrm{Tor}_1^R(A, R/L) = 0$ for every finitely generated left ideal L of R.

13. A *flat resolution* of a right R-module A is an exact sequence $\mathcal{F} \longrightarrow A$ in which F_0, F_1, F_2, ... are flat. Show that $\mathrm{Tor}_n^R(A, B) \cong H_n(\mathcal{F} \otimes_R B)$, by an isomorphism that is natural in $\mathcal{F}$ and B.

6. Universal Coefficient Theorems

For suitable complexes $\mathcal{C}$, universal coefficient theorems compute the homology groups of $\mathrm{Hom}_R(\mathcal{C}, A)$ and $\mathcal{C} \otimes_R A$ from the homology of $\mathcal{C}$. In particular, the homology and cohomology groups of a topological space (with coefficients in any abelian group) are determined by its singular homology groups.

Both universal coefficient theorems require projective modules of cycles and boundaries. This occurs most naturally when every C_n is projective and R is *hereditary*, meaning that submodules of projective modules are projective. For example, R may be any Dedekind domain or PID, such as $\mathbb{Z}$.

Theorem **6.1** (Universal Coefficient Theorem for Cohomology). *Let R be a left hereditary ring; let $\mathcal{C}$ be a complex of projective left R-modules, and let M be any left R-module. For every $n \in \mathbb{Z}$ there is an exact sequence*

$$0 \longrightarrow \mathrm{Ext}^1_R\big(H_{n-1}(\mathcal{C}), M\big) \longrightarrow H^n(\mathcal{C}, M) \longrightarrow \mathrm{Hom}_R\big(H_n(\mathcal{C}), M\big) \longrightarrow 0,$$

which is natural in $\mathcal{C}$ and M and splits, by a homomorphism that is natural in M.

In particular, the singular cohomology of a topological space X is determined by its homology: $H^n(X, G) \cong \mathrm{Hom}_{\mathbb{Z}}\big(H_n(X), G\big) \oplus \mathrm{Ext}^1_{\mathbb{Z}}\big(H_{n-1}(X), G\big)$.

Proof. Every $\partial_n : C_n \longrightarrow C_{n-1}$ induces a commutative square

$$
\begin{array}{ccc}
C_n & \xrightarrow{\ \partial_n\ } & C_{n-1} \\
{\scriptstyle \pi_n}\big\downarrow & & \big\uparrow{\scriptstyle \iota_{n-1}} \\
B_{n-1} & \xrightarrow[\ \kappa_{n-1}\ ]{} & Z_{n-1}
\end{array}
$$

where $Z_{n-1} = \mathrm{Ker}\, \partial_{n-1}$, ι_{n-1} and κ_{n-1} are inclusion homomorphisms, $B_{n-1} = \mathrm{Im}\, \partial_n$, and $\pi_n x = \partial_n x$ for all $x \in C_n$. Then the sequence

$$0 \longrightarrow Z_n \xrightarrow{\ \iota_n\ } C_n \xrightarrow{\ \pi_n\ } B_{n-1} \longrightarrow 0$$

is exact. By definition of $H_n = H_n(\mathcal{C})$ there is also a short exact sequence

$$0 \longrightarrow B_n \xrightarrow{\ \kappa_n\ } Z_n \xrightarrow{\ \rho_n\ } H_n \longrightarrow 0.$$

These two sequences induce a commutative diagram $\mathcal{D}$ (top of next page) which is natural in $\mathcal{C}$ and M; the bottom row and first two columns are parts of the long exact sequences in Theorem 4.4, and end with 0 since B_n, $Z_n \subseteq C_n$ are projective, so that $\mathrm{Ext}^1_R(B_n, M) = 0$, $\mathrm{Ext}^1_R(Z_n, M) = 0$ for all n. Then Lemma 6.2 below yields the universal coefficients sequence

$$0 \longrightarrow \mathrm{Ext}^1_R\big(H_{n-1}(\mathcal{C}), M\big) \xrightarrow{\ \mu\ } H^n(\mathcal{C}, M) \xrightarrow{\ \sigma\ } \mathrm{Hom}_R\big(H_n(\mathcal{C}), M\big) \longrightarrow 0,$$

which is natural in $\mathcal{D}$ and therefore in $\mathcal{C}$ and M.

$$
\begin{array}{ccc}
0 & & 0 \\
\uparrow & & \downarrow \\
0 \longrightarrow \operatorname{Hom}_R(H_n, M) \xrightarrow{\rho_n^*} \operatorname{Hom}_R(Z_n, M) \xrightarrow{\kappa_n^*} \operatorname{Hom}_R(B_n, M) \\
\iota_n^* \uparrow \qquad\qquad \downarrow \pi_{n+1}^* \\
\operatorname{Hom}_R(C_{n-1}, M) \xrightarrow{\partial_n^*} \operatorname{Hom}_R(C_n, M) \xrightarrow{\partial_{n+1}^*} \operatorname{Hom}_R(C_{n+1}, M) \\
\iota_{n-1}^* \downarrow \qquad\qquad \uparrow \pi_n^* \\
\operatorname{Hom}_R(Z_{n-1}, M) \underset{\kappa_{n-1}^*}{\longrightarrow} \operatorname{Hom}_R(B_{n-1}, M) \underset{\chi}{\longrightarrow} \operatorname{Ext}_R^1(H_{n-1}, M) \longrightarrow 0 \\
\downarrow \qquad\qquad \uparrow \\
0 \qquad\qquad 0
\end{array}
$$

Moreover, the exact sequence

$$
0 \longrightarrow Z_n \xrightarrow{\iota_n} C_n \xrightarrow{\pi_n} B_{n-1} \longrightarrow 0
$$

splits, since B_{n-1} is projective, so that $\xi_n \iota_n = 1_{Z_n}$ for some $\xi_n : C_n \longrightarrow Z_n$. By Lemma 6.2 below, $\sigma \tau \xi_n^* \rho_n^*$ is the identity on $\operatorname{Hom}_R(H_n, M)$, where $\tau :$ $\operatorname{Ker} \partial_{n+1}^* \longrightarrow \operatorname{Ker} \partial_{n+1}^* / \operatorname{Im} \partial_n^*$ is the projection; hence the universal coefficients sequence splits. The homomorphism $\tau \xi_n^* \rho_n^*$ is natural in M (when $\mathcal{C}$ is fixed) but not in $\mathcal{C}$, since there is no natural choice for ξ_n. $\square$

Lemma **6.2.** *Every commutative diagram $\mathcal{D}$ in which the middle row is null*

$$
\begin{array}{ccc}
 & 0 & \quad 0 \\
 & \uparrow & \quad \downarrow \\
0 \longrightarrow A' \xrightarrow{\varphi'} B' \xrightarrow{\psi'} C' \\
 & \beta' \uparrow & \downarrow \gamma \\
 A \xrightarrow{\varphi} B \xrightarrow{\psi} C \\
\alpha \downarrow & \uparrow \beta'' \\
A'' \xrightarrow{\varphi''} B'' \xrightarrow{\psi''} C'' \longrightarrow 0 \\
\downarrow & \uparrow \\
0 & 0
\end{array}
$$

and the other rows and columns are exact induces an exact sequence

$$
0 \longrightarrow C'' \xrightarrow{\mu} \operatorname{Ker} \psi / \operatorname{Im} \varphi \xrightarrow{\sigma} A' \longrightarrow 0,
$$

which is natural in $\mathcal{D}$. Moreover, $\beta' \xi = 1_{B'}$ implies $\sigma \pi \xi \varphi' = 1_{A'}$, where $\pi : \operatorname{Ker} \psi \longrightarrow \operatorname{Ker} \psi / \operatorname{Im} \varphi$ is the projection.

Proof. Since γ is injective and α is surjective,

$$
\operatorname{Ker} \psi / \operatorname{Im} \varphi = \operatorname{Ker} \gamma \psi' \beta' / \operatorname{Im} \beta'' \varphi'' \alpha = \operatorname{Ker} \psi' \beta' / \operatorname{Im} \beta'' \varphi''.
$$

Now, β' sends $\operatorname{Ker} \psi' \beta' = \beta'^{-1} \operatorname{Ker} \psi'$ onto $\operatorname{Ker} \psi' = \operatorname{Im} \varphi' \cong A'$, and $\operatorname{Im} \beta'' \varphi''$ onto 0. Therefore β' induces a homomorphism $\sigma : \operatorname{Ker} \psi / \operatorname{Im} \varphi \longrightarrow$

A', which sends $\pi b = b + \operatorname{Im} \varphi$ to $\varphi'^{-1} \beta' b$, for all $b \in \operatorname{Ker} \psi$.

Next, β'' sends B'' onto $\operatorname{Ker} \beta' \subseteq \operatorname{Ker} \psi$ and sends $\operatorname{Im} \varphi''$ onto $\operatorname{Im} \beta'' \varphi'' = \operatorname{Im} \varphi$, so that $\pi \beta'' = 0$. By Lemma X.1.2, $\pi \beta'' = \mu \psi''$ for some homomorphism $\mu : C'' \longrightarrow \operatorname{Ker} \psi / \operatorname{Im} \varphi$; μ sends $\psi'' b''$ to $\pi \beta'' b''$, for all $b'' \in C''$:

$$
\begin{array}{ccccccc}
\operatorname{Im} \varphi & \xrightarrow{\subseteq} & \operatorname{Ker} \psi & \xrightarrow{\ \pi\ } & \operatorname{Ker} \psi / \operatorname{Im} \varphi & \longrightarrow & 0 \\
& & {\scriptstyle \beta''}\uparrow & & \uparrow{\scriptstyle \mu} & & \\
A'' & \xrightarrow{\ \varphi''\ } & B'' & \xrightarrow{\ \psi''\ } & C'' & \longrightarrow & 0
\end{array}
$$

The construction of μ and σ shows that they are natural in $\mathcal{D}$. Exactness of

$$
0 \longrightarrow C'' \xrightarrow{\ \mu\ } \operatorname{Ker} \psi / \operatorname{Im} \varphi \xrightarrow{\ \sigma\ } A' \longrightarrow 0
$$

is easy. If $\mu \psi'' b'' = 0$, then $\beta'' b'' \in \operatorname{Im} \varphi = \operatorname{Im} \beta'' \varphi''$, $\beta'' b'' = \beta'' \varphi'' a''$ for some $a'' \in A''$, $b'' = \varphi'' a''$, and $\psi'' b'' = 0$; hence μ is injective. Next, $\varphi' \sigma \mu \psi'' = \varphi' \sigma \pi \beta'' = \beta' \beta'' = 0$; hence $\sigma \mu = 0$. Conversely, if $\sigma \pi b = 0$, then $\beta' b = 0$, $b = \beta'' b''$ for some $b'' \in B''$, and $\pi b = \pi \beta'' b'' = \mu \psi'' b'' \in \operatorname{Im} \mu$. If $a' \in A'$, then $\varphi' a' = \beta' b$ for some $b \in B$, $b \in \operatorname{Ker} \psi$ since $\psi b = \gamma \psi' \beta' b = \gamma \psi' \varphi' a' = 0$, $\varphi' \sigma \pi b = \beta' b = \varphi' a'$, and $a' = \sigma \pi b$; hence σ is surjective.

Finally, $\beta' \xi = 1_{B'}$ implies $\psi \xi \varphi' = \gamma \psi' \beta' \xi \varphi' = \gamma \psi' \varphi' = 0$, $\operatorname{Im} \xi \varphi' \subseteq \operatorname{Ker} \psi$, $\varphi' \sigma \pi \xi \varphi' = \beta' \xi \varphi' = \varphi'$, and $\sigma \pi \xi \varphi' = 1_{A'}$. $\square$

Theorem 6.3 (Universal Coefficient Theorem for Homology). *Let R be a right hereditary ring; let $\mathcal{C}$ be a complex of projective right R-modules, and let M be any left R-module. For every $n \in \mathbb{Z}$ there is an exact sequence*

$$
0 \longrightarrow H_n(\mathcal{C}) \otimes_R M \longrightarrow H_n(\mathcal{C} \otimes_R M) \longrightarrow \operatorname{Tor}_1^R \big(H_{n-1}(\mathcal{C}), M \big) \longrightarrow 0,
$$

which is natural in $\mathcal{C}$ and M and splits, by a homomorphism that is natural in M.

In particular, the singular homology of a topological space X with coefficients in an abelian group G is determined by its plain homology:

$$
H_n(X, G) \cong \big(H_n(X) \otimes_{\mathbb{Z}} G \big) \oplus \operatorname{Tor}_1^{\mathbb{Z}} \big(H_{n-1}(X), G \big).
$$

Proof. The commutative square and exact sequences

$$
\begin{array}{ccc}
C_n & \xrightarrow{\ \partial_n\ } & C_{n-1} \\
{\scriptstyle \pi_n}\downarrow & & \uparrow{\scriptstyle \iota_{n-1}} \\
B_{n-1} & \xrightarrow{\ \kappa_{n-1}\ } & Z_{n-1}
\end{array}
$$

$$
0 \longrightarrow Z_n \xrightarrow{\ \iota_n\ } C_n \xrightarrow{\ \pi_n\ } B_{n-1} \longrightarrow 0, \qquad 0 \longrightarrow B_n \xrightarrow{\ \kappa_n\ } Z_n \xrightarrow{\ \rho_n\ } H_n \longrightarrow 0
$$

in the proof of Theorem 6.1 induce a commutative diagram $\mathcal{D}$ (top of next page) which is natural in $\mathcal{C}$ and M, whose bottom row and first two columns are parts

$$
\begin{array}{ccccc}
 & & 0 & & 0 \\
 & & \uparrow & & \downarrow \\
0 \longrightarrow \operatorname{Tor}_1^R(H_{n-1}, M) \xrightarrow{\ \chi\ } & B_{n-1} \otimes_R M & \xrightarrow{\ \overline{\kappa}_{n-1}\ } & Z_{n-1} \otimes_R M \\
 & & \overline{\pi}_n \uparrow & & \downarrow \overline{\iota}_{n-1} \\
C_{n+1} \otimes_R M \xrightarrow{\ \overline{\partial}_{n+1}\ } & C_n \otimes_R M & \xrightarrow{\ \overline{\partial}_n\ } & C_{n-1} \otimes_R M \\
\overline{\pi}_{n+1} \downarrow & & \overline{\iota}_n \uparrow & & \\
B_n \otimes_R M \xrightarrow[\ \overline{\kappa}_n\ ]{} & Z_n \otimes_R M & \xrightarrow[\ \overline{\rho}_n\ ]{} & H_n \otimes_R M \longrightarrow 0 \\
\downarrow & & \uparrow & & \\
0 & & 0 & &
\end{array}
$$

of the long exact sequences in Theorem 5.4, and end with 0 since $B_n, Z_n \subseteq C_n$ are projective, so that $\operatorname{Tor}_1^R(B_n, M) = 0$, $\operatorname{Tor}_1^R(Z_n, M) = 0$ for all n. Lemma 6.2 then yields the universal coefficients sequence

$$
0 \longrightarrow H_n(\mathcal{C}) \otimes_R M \xrightarrow{\ \mu\ } H_n(\mathcal{C} \otimes_R M) \xrightarrow{\ \sigma\ } \operatorname{Tor}_1^R(H_{n-1}(\mathcal{C}), M) \longrightarrow 0,
$$

which is natural in $\mathcal{D}$ and therefore in $\mathcal{C}$ and M. Moreover, the exact sequence

$$
0 \longrightarrow Z_n \xrightarrow{\ \iota_n\ } C_n \xrightarrow{\ \pi_n\ } B_{n-1} \longrightarrow 0
$$

splits, since B_{n-1} is projective, and $\pi_n \xi_n = 1_{B_{n-1}}$ for some $\xi_n : B_{n-1} \longrightarrow C_n$. By Lemma 6.2, $\sigma \tau \overline{\xi}_n \chi$ is the identity on $\operatorname{Tor}_1^R(H_n, M)$, where $\tau :$ $\operatorname{Ker} \overline{\partial}_{n+1} \longrightarrow \operatorname{Ker} \overline{\partial}_{n+1}/\operatorname{Im} \overline{\partial}_n$ is the projection. Thus, the universal coefficients sequence splits like a dry log. As before, $\tau \overline{\xi}_n \chi$ is natural in M (when $\mathcal{C}$ is fixed) but not in $\mathcal{C}$. $\square$

7. Cohomology of Groups

The cohomology of groups was discovered by Eilenberg and MacLane [1942] and has become an essential tool of group theory. This section contains a few definitions and early facts.

Cochains. In what follows, G is an arbitrary group, written multiplicatively. By Corollary II.12.7, a group extension $1 \longrightarrow A \longrightarrow E \longrightarrow G \longrightarrow 1$ of an abelian group A (written additively) by G is determined by a group action of G on A by automorphisms, equivalently, a homomorphism $G \longrightarrow \operatorname{Aut}(A)$, and a factor set s, which is a mapping $s : G \times G \longrightarrow A$ such that $s_{x,1} = 0 = s_{1,y}$ and

$$
s_{x,y} + s_{xy,z} = x s_{y,z} + s_{x,yz}
$$

for all $x, y, z \in G$. By Corollary II.12.8, two extensions of A by G are equivalent if and only if they share the same action of G on A and their factor sets s, t satisfy

$$s_{x,y} - t_{x,y} = u_x + xu_y - u_{xy}$$

for all $x, y \in G$, where $u_x \in A$ and $u_1 = 0$, that is, $s - t$ is a split factor set.

Cochains and their coboundaries are defined in every dimension $n \geqq 0$ so that factor sets are 2-cocycles and split factor sets are 2-coboundaries.

Definitions. Let A be an abelian group on which the group G acts by automorphisms. For every $n \geqq 0$, an n-cochain on G with values in A is a mapping $u : G^n \longrightarrow A$ such that $u_{x_1, \ldots, x_n} = 0$ whenever $x_i = 1$ for some i. The coboundary of an n-cochain u is the $(n + 1)$-cochain

$$(\delta^n u)_{x_1, \ldots, x_{n+1}} = x_1 u_{x_2, \ldots, x_{n+1}}$$
$$+ \sum_{1 \leq i \leq n} (-1)^i u_{x_1, \ldots, x_i x_{i+1}, \ldots, x_{n+1}} + (-1)^{n+1} u_{x_1, \ldots, x_n}.$$

Readers will verify that $(\delta u)_{x_1, \ldots, x_{n+1}} = 0$ whenever $x_i = 1$ for some i. The equality $\delta^{n+1} \delta^n u = 0$ follows from Proposition 7.5 below but may be verified directly.

Definitions. Let A be an abelian group on which the group G acts by automorphisms. Under pointwise addition, n-cochains, n-cocycles, and n-coboundaries constitute abelian groups $C^n(G, A)$, $Z^n(G, A) = \mathrm{Ker}\ \delta^n$, and $B^n(G, A) = \mathrm{Im}\ \delta^{n-1} \subseteq Z^n(G, A)$ (with $B^0(G, A) = 0$). The nth cohomology group of G with coefficients in A is $H^n(G, A) = Z^n(G, A)/B^n(G, A)$.

In particular, a 0-cochain $u : G^0 = \{\emptyset\} \longrightarrow A$ is an element of A; its coboundary is $(\delta u)_x = xu - u$. A 1-cochain is a mapping $u : G \longrightarrow A$ such that $u_1 = 0$. A 2-coboundary $(\delta u)_{x,y} = xu_y - u_{xy} + u_x$ is a split factor set. A 2-cochain is a mapping $u : G \times G \longrightarrow A$ such that $u_{x,1} = 0 = u_{1,y}$ for all $x, y \in G$; its coboundary is $(\delta u)_{x,y,z} = xu_{y,z} - u_{xy,z} + u_{x,yz} - u_{x,y}$. Hence factor sets are 2-cocycles. By the above, there is a one-to-one correspondence between the elements of $H^2(G, A)$ and the equivalence classes of group extensions of G by A with the given action: $H^2(G, A)$ *classifies* these extensions.

Readers will verify the following simpler examples:

Proposition **7.1.** (1) $H^0(G, A) \cong \{ a \in A \mid xa = a \text{ for all } x \in G \}$.

(2) *If G acts trivially on A (if $xa = a$ for all x, a), then $H^0(G, A) \cong A$ and $H^1(G, A) \cong \mathrm{Hom}(G, A)$.*

In (2), group homomorphisms of G into A can be added pointwise, since A is abelian; this makes the set $\mathrm{Hom}(G, A)$ an abelian group.

Proposition **7.2.** *If G is finite and $n \geqq 1$, then $H^n(G, A)$ is torsion, and the order of every element of $H^n(G, A)$ divides the order of G; if A is divisible and torsion-free, then $H^n(G, A) = 0$.*

Proof. Let u be an n-cochain. Define an $(n-1)$-cochain $v_{x_1, \ldots, x_{n-1}} = \sum_{x \in G} u_{x_1, \ldots, x_{n-1}, x}$. Then

$$\sum_{x \in G} (\delta^n u)_{x_1, \ldots, x_n, x} = x_1 \sum_{x \in G} u_{x_2, \ldots, x_n, x}$$
$$+ \sum_{1 \le i < n} (-1)^i \sum_{x \in G} u_{x_1, \ldots, x_i x_{i+1}, \ldots, x}$$
$$+ (-1)^n \sum_{x \in G} u_{x_1, \ldots, x_{n-1}, x_n x} + (-1)^{n+1} \sum_{x \in G} u_{x_1, \ldots, x_n}$$
$$= x_1 v_{x_2, \ldots, x_n} + \sum_{1 \le i < n} (-1)^i v_{x_1, \ldots, x_i x_{i+1}, \ldots, x_n}$$
$$+ (-1)^n v_{x_1, \ldots, x_{n-1}} + (-1)^{n+1} m u_{x_1, \ldots, x_n}$$
$$= (\delta^{n-1} v)_{x_1, \ldots, x_n} + (-1)^{n+1} m u_{x_1, \ldots, x_n},$$

where $m = |G|$. If u is an n-cocycle, then $m u \in \operatorname{Im} \delta^{n-1}$ and $m \operatorname{cls} u = 0$.

Let u be an n-cocycle and let v be as above. If A is divisible and torsion-free, then there is for every $a \in A$ a unique $b \in A$ such that $mb = a$. Therefore there is a unique n-cochain w such that $mw = v$. Then $m \delta w = \delta v = (-1)^n m u$; since A is torsion-free this implies $u = (-1)^n \delta w$ and $\operatorname{cls} u = 0$. $\square$

G-modules. We now construct a complex of modules whose cohomology groups are the cohomology groups of G.

With group algebras (just escaped from Section IX.7), abelian groups on which G acts by automorphisms acquire module status. Let R be a commutative ring [with identity]. There is a free R-module $R[G]$ with basis G; the elements of $R[G]$ are linear combinations $k = \sum_{x \in G} k_x x$ of elements of G with coefficients $k_x \in R$ such that $k_x = 0$ for almost all $x \in G$. On $R[G]$ there is a bilinear multiplication that extends the multiplication of G: when $k, \ell \in R[G]$, then $m = k\ell$ has

$$m_x = \sum_{y, z \in G, \ yz = x} k_y \ell_z$$

for all $x \in G$. This makes $R[G]$ a ring, the *group ring* or *group algebra* of G.

Proposition **7.3.** *For every abelian group A, there is a one-to-one correspondence between group actions of G on A by automorphisms, and [unital] $\mathbb{Z}[G]$-module structures on A.*

This is similar to Proposition IX.7.3. A group action $(g, a) \longmapsto ga$ of G extends to a module action $\left(\sum_{x \in G} k_x x\right) a = \sum_{x \in G} k_x xa$ of $\mathbb{Z}[G]$; readers will verify that this defines the one-to-one correspondence in Proposition 7.3.

In what follows, a *G-module* is a $\mathbb{Z}[G]$-module; equivalently, an abelian group on which G acts by automorphisms.

The bar resolution. The bar resolution derives its name from the original notation $[x_1 | x_2 | \ldots | x_n]$ for the generators $[x_1, \ldots, x_n]$ of B_n below.

For every $n \ge 0$ let B_n be the free G-module generated by all sequences

$[x_1, \ldots, x_n]$ of n elements $x_i \neq 1$ of G. It is convenient to let

$$[x_1, \ldots, x_n] = 0 \quad \text{whenever } x_i = 1 \text{ for some } i .$$

In particular, B_0 is the free G-module $\mathbb{Z}[G]$ with one generator $[\,]$.

Lemma 7.4. *For every $n \geq 1$ there is a unique module homomorphism $\partial_n :$ $B_n \longrightarrow B_{n-1}$ such that*

$$\partial_n [x_1, \ldots, x_n] = x_1 [x_2, \ldots, x_n]$$
$$+ \sum_{1 \leq i < n} (-1)^i [x_1, \ldots, x_i x_{i+1}, \ldots, x_n] + (-1)^n [x_1, \ldots, x_{n-1}]$$

for all $x_1, \ldots, x_n \in G$. Moreover, $\partial_{n+1} \partial_n = 0$.

The proof is an exercise. Another proof that $\partial_{n+1} \partial_n = 0$ is given below.

Definition. *The positive complex* $\cdots \longrightarrow B_2 \xrightarrow{\partial_2} B_1 \xrightarrow{\partial_1} B_0 \longrightarrow 0 \cdots$ *of G-modules is the* bar complex *of G.*

Proposition 7.5. *For every G-module A, there is an isomorphism $C^n(G, A)$ $\cong \mathrm{Hom}_{\mathbb{Z}[G]}(B_n, A)$, which is natural in A, such that the square*

$$
\begin{array}{ccc}
C^n(G, A) & \xrightarrow{\cong} & \mathrm{Hom}_{\mathbb{Z}[G]}(B_n, A) \\
\delta^n \downarrow & & \downarrow \partial^*_{n+1} \\
C^{n+1}(G, A) & \xrightarrow{\cong} & \mathrm{Hom}_{\mathbb{Z}[G]}(B_{n+1}, A)
\end{array}
$$

commutes for every $n \geq 0$. Hence the cohomology groups of G are those of its bar complex, up to isomorphisms that are natural in A.

Proof. Since B_n is free on all $[x_1, \ldots, x_n]$ such that $x_i \neq 1$ for all i, every n-cochain u induces a unique homomorphism $\theta^n u : B_n \longrightarrow A$ such that

$$(\theta^n u) [x_1, \ldots, x_n] = u_{x_1, \ldots, x_n}$$

whenever $x_i \neq 1$ for all i. Then $\theta^n u [x_1, \ldots, x_n] = u_{x_1, \ldots, x_n}$ for all $x_1, \ldots,$ $x_n \in G$, since both sides are 0 if any $x_i = 1$. It is immediate that $\theta^n :$ $C^n(G, A) \longrightarrow \mathrm{Hom}_{\mathbb{Z}[G]}(B_n, A)$ is an isomorphism and is natural in A.

The definition of ∂_n was tailored to ensure that $\partial^*_{n+1} \theta^n = \theta^{n+1} \delta^n$: for every $u \in C^n(G, A)$, $\partial^*_{n+1}(\theta^n u) = (\theta^n u) \circ \partial_{n+1}$; hence, for all $x_1, \ldots, x_{n+1} \in G$,

$$(\partial^*_{n+1} \theta^n u)[x_1, \ldots, x_{n+1}] = (\theta^n u) \partial_{n+1} [x_1, \ldots, x_{n+1}]$$
$$= (\theta^n u) x_1 [x_2, \ldots, x_{n+1}]$$
$$+ \sum_{1 \leq i \leq n} (-1)^i (\theta^n u) [x_1, \ldots, x_i x_{i+1}, \ldots, x_{n+1}]$$
$$+ (-1)^{n+1} (\theta^n u) [x_1, \ldots, x_n]$$
$$= x_1 u_{x_2, \ldots, x_{n+1}}$$
$$+ \sum_{1 \leq i \leq n} (-1)^i u_{x_1, \ldots, x_i x_{i+1}, \ldots, x_{n+1}} + (-1)^{n+1} u_{x_1, \ldots, x_n}$$
$$= (\delta^n u)_{x_1, \ldots, x_{n+1}} = (\theta^{n+1}(\delta^n u))[x_1, \ldots, x_{n+1}] . \quad \square$$

Since every B_n is a projective G-module, Theorem 1.7 yields:

Theorem **7.6** (Exact Cohomology Sequence). *For every short exact sequence* $\mathcal{E} : 0 \longrightarrow A \longrightarrow A' \longrightarrow A'' \longrightarrow 0$ *of G-modules there is an exact sequence*

$$\cdots \longrightarrow H^n(G, A) \longrightarrow H^n(G, A') \longrightarrow H^n(G, A'') \longrightarrow H^{n+1}(G, A) \longrightarrow \cdots,$$

which is natural in $\mathcal{E}$.

The bar complex has another property. Let $\mathbb{Z}$ be the trivial G-module. There is a unique module homomorphism $\varepsilon : B_0 \longrightarrow \mathbb{Z}$ such that $\varepsilon[\,] = 1$. Since G acts trivially on $\mathbb{Z}$ we have $\varepsilon x [\,] = x1 = 1$ and

$$\varepsilon\left(\sum_{x \in G} k_x\, x\, [\,]\right) = \sum_{x \in G} k_x \quad \text{for all } k = \sum_{x \in G} k_x\, x \in \mathbb{Z}[G].$$

Proposition **7.7.** $\mathcal{B} : \cdots \longrightarrow B_2 \xrightarrow{\partial_2} B_1 \xrightarrow{\partial_1} B_0 \xrightarrow{\varepsilon} \mathbb{Z}$ *is a free resolution of the trivial G-module $\mathbb{Z}$.*

Proof. Without its G-module structure, B_n is a free abelian group: every element of B_n is uniquely a linear combination with coefficients in $\mathbb{Z}[G]$ of generators $[x_1, \ldots, x_n]$ with $x_i \in G$ and $x_i \neq 1$ for all i; therefore every element of B_n is uniquely a linear combination with coefficients in $\mathbb{Z}$ of elements of the form $x[x_1, \ldots, x_n]$ with $x \in G$ and $x_i \in G$, $x_i \neq 1$ for all i.

By the above there is a unique homomorphism $s_n : B_n \longrightarrow B_{n+1}$ of abelian groups (which will yield a chain homotopy from $1_{\mathcal{B}}$ to 0) such that

$$s_n\, x\, [x_1, \ldots, x_n] = [x, x_1, \ldots, x_n]$$

whenever $x_i \neq 1$ for all i. The equality $s_n\, x\, [x_1, \ldots, x_n] = [x, x_1, \ldots, x_n]$ also holds if some $x_i = 1$. Let $s_{-1} : \mathbb{Z} \longrightarrow B_0$, $m \longrightarrow m[\,]$; let $s_n = 0$ for all $n < -1$. Then $\varepsilon s_{-1} = 1_{\mathbb{Z}}$. Since $\varepsilon x[\,] = 1$,

$$\partial^1 s_0 x[\,] = \partial^1[x] = x[\,] - [\,] = x[\,] - s_{-1}\varepsilon x[\,]$$

for all $x \in G$, and $\partial^1 s_0 + s_{-1}\varepsilon = 1_{B_0}$. Similarly,

$$\partial^{n+1} s_n\, x\, [x_1, \ldots, x_n] = \partial^{n+1}[x, x_1, \ldots, x_n]$$
$$= x[x_1, \ldots, x_n] - [xx_1, \ldots, x_n]$$
$$+ \sum_{2 \leq i \leq n} (-1)^i [x, \ldots, x_i x_{i+1}, \ldots, x_n] + (-1)^{n+1} [x, x_1, \ldots, x_{n-1}],$$
$$s_{n-1}\partial_n\, x\, [x_1, \ldots, x_n] = s_{n-1} x x_1 [x_2, \ldots, x_n]$$
$$+ \sum_{1 \leq i < n} (-1)^i s_{n-1} x [x_1, \ldots, x_i x_{i+1}, \ldots, x_n]$$
$$+ (-1)^n s_{n-1} x [x_1, \ldots, x_{n-1}];$$

hence $\partial^{n+1} s_n\, x\, [x_1, \ldots, x_n] = x[x_1, \ldots, x_n] - s_{n-1}\partial_n\, x\, [x_1, \ldots, x_n]$ and $\partial^{n+1} s_n + s_{n-1}\partial_n = 1_{B_n}$.

The equality $\varepsilon \partial^1 = 0$ then yields $\partial^1 \partial^2 s_1 = \partial^1 (1 - s_0 \partial_1) = \partial^1 - \partial^1 s_0 \partial^1 = \partial^1 - (1 - s_{-1}\varepsilon) \partial^1 = 0$, whence $\partial^1 \partial^2 = 0$, since B_2 is generated by $\operatorname{Im} s_1$.

Similarly, $\partial^{n-1} \partial^n = 0$ implies $\partial^n \partial^{n+1} s_n = \partial^n (1 - s_{n-1} \partial_n) = \partial^n - \partial^n s_{n-1} \partial^n = \partial^n - (1 - s_{n-2} \partial^{n-1}) \partial^n = 0$ and $\partial^n \partial^{n+1} = 0$, since Im s_n generates B_{n+1}. Hence $\partial^n \partial^{n+1} = 0$ for all n.

Now, s is a chain homotopy from the identity on $\mathcal{B}$ to $0 : \mathcal{B} \longrightarrow \mathcal{B}$. By 1.2, $H_n(1_{\mathcal{B}}) = H_n(0)$; hence $H_n(\mathcal{B}) = 0$ for all n and $\mathcal{B}$ is exact. $\square$

Definition. $\mathcal{B} : \cdots \longrightarrow B_2 \xrightarrow{\partial_2} B_1 \xrightarrow{\partial_1} B_0 \xrightarrow{\varepsilon} \mathbb{Z}$ *is the* bar resolution *of G.*

Propositions 7.7 and 4.3 now team up and produce the next result:

Theorem 7.8. *There is an isomorphism $H^n(G, A) \cong \operatorname{Ext}^n_{\mathbb{Z}[G]}(\mathbb{Z}, A)$ that is natural in A, where $\mathbb{Z}$ is the trivial G-module.*

Free groups. With Theorem 7.8, the cohomology of some groups can be calculated from other projective resolutions of $\mathbb{Z}$. We do this for free groups and leave cyclic groups as exercises. Let

$$I(G) = \{ \textstyle\sum_{x \in G} k_x \, x \in \mathbb{Z}[G] \mid \textstyle\sum_{x \in G} k_x = 0 \};$$

equivalently, $I(G) = \{ k \in \mathbb{Z}[G] \mid \varepsilon(k[\,]) = 0 \}$, so $I(G)$ is an ideal of $\mathbb{Z}[G]$.

Lemma 7.9. *If G is the free group on $(x_i)_{i \in I}$, then $I(G)$ is a free G-module, with basis $(x_i - 1)_{i \in I}$.*

Proof. Let M be the submodule of $I(G)$ generated by all $x_i - 1$. We show by induction on the length of the reduced word $x \in G$ that $x - 1 \in M$ for all $x \in G$: indeed, if $x \neq 1$, then either $x = x_i y$ or $x = x_i^{-1} y$, where y is shorter than x, and the induction hypothesis yields either $x_i y - 1 = x_i(y - 1) + (x_i - 1) \in M$ or $x_i^{-1} y - 1 = x_i^{-1}(y - 1) - x_i^{-1}(x_i - 1) \in M$. Then $k = \sum_{x \in G} k_x \, x \in I(G)$ implies $k = \sum_{x \in G} k_x \, (x - 1) \in M$, since $\sum_{x \in G} k_x = 0$. Thus $I(G) = M$ is generated by all $x_i - 1$.

To prove that $(x_i - 1)_{i \in I}$ is linearly independent in $I(G)$ we show that, for every G-module A and $a_i \in A$, there is a module homomorphism $\varphi : I(G) \longrightarrow A$ such that $\varphi(x_i - 1) = a_i$ for all i. In particular, there is a homomorphism $\varphi_j : I(G) \longrightarrow \mathbb{Z}[G]$ such that $\varphi_j(x_j - 1) = 1$ and $\varphi_j(x_i - 1) = 0$ for all $i \neq j$; hence $\sum_{i \in I} k_i (x_i - 1) = 0$ implies $k_j = \varphi_j(\sum_{i \in I} k_i (x_i - 1)) = 0$ for all j. (Alternately, $(x_i - 1)_{i \in I}$ has the universal property that characterizes bases.)

Given $a_i \in A$, define $u(x) \in A$ as follows for all $x \in G$ (so that $u(x) = \varphi(x - 1)$, once we construct φ). If $x = 1$, then $u(x) = 0$. If $x \neq 1$, then either $x = x_i y$ or $x = x_i^{-1} y$, where y is shorter than x; let

$$u(x_i y) = x_i u(y) + a_i, \quad u(x_i^{-1} y) = x_i^{-1} u(y) - x_i^{-1} a_i, \tag{1}$$

when $x_i y$ or $x_i^{-1} y$ is reduced. In particular, $u(x_i) = a_i$. Then (1) holds for all x_i and y: if $x_i y$ is not reduced, then $y = x_i^{-1} z$ and $x_i u(x_i^{-1} z) + a_i =$

$x_i\left(x_i^{-1}u(z) - x_i^{-1}a_i\right) + a_i = u(z) = u(x_i y)$; if $x_i^{-1}y$ is not reduced, then $y = x_i z$
and $x_i^{-1}u(x_i z) - x_i^{-1}a_i = x_i^{-1}\left(x_i u(z) + a_i\right) - x_i^{-1}a_i = u(z) = u(x_i^{-1}y)$.

We show, by induction on the length of $x \in G$, that

$$u(xy) = x\,u(y) + u(x) \tag{2}$$

for all $x, y \in G$ (so that u is a 1-cocycle). Indeed, (2) holds when $x = 1$. If
$x \neq 1$, then either $x = x_i z$ or $x = x_i^{-1}z$, where z is shorter than x; by (1) and the
induction hypothesis, either

$$
\begin{aligned}
u(xy) &= u(x_i z y) = x_i\, u(zy) + a_i \\
&= x_i z\, u(y) + x_i\, u(z) + a_i = x\, u(y) + u(x),
\end{aligned}
$$

or

$$
\begin{aligned}
u(xy) &= u(x_i^{-1} z y) = x_i^{-1} u(zy) - x_i^{-1} a_i \\
&= x_i^{-1} z\, u(y) + x_i^{-1} u(z) + x_i^{-1} a_i = x\, u(y) + u(x).
\end{aligned}
$$

Now, define, for every $k = \sum_{x \in G} k_x\, x = \sum_{x \in G} k_x\, (x - 1) \in I(G)$,

$$\varphi(k) = \sum_{x \in G} k_x\, u(x) \in A.$$

Then φ is additive and $\varphi(x_i - 1) = u(x_i) - u(1) = a_i$. For all $x \in G$ and
$k = \sum_{y \in G} k_y\, y \in I(G)$, $xk = \sum_{y \in G} k_y\, xy$ and (2) yields

$$
\begin{aligned}
\varphi(xk) &= \sum_{y \in G} k_y\, u(xy) = \sum_{y \in G} k_y\, xu(y) + \sum_{y \in G} k_y\, u(x) \\
&= \sum_{y \in G} k_y\, xu(y) = x\, \varphi(k),
\end{aligned}
$$

since $\sum_{y \in G} k_y = 0$. Hence φ is a module homomorphism. $\square$

Proposition **7.10.** $H^n(G, A) = 0$ *for all* $n \geq 2$ *when G is a free group.*

Proof. By Lemma 7.9, $\cdots \longrightarrow 0 \longrightarrow 0 \longrightarrow I(G) \longrightarrow \mathbb{Z}[G] \longrightarrow \mathbb{Z}$ is a
projective resolution of $\mathbb{Z}$; hence $\mathrm{Ext}^n_{\mathbb{Z}[G]}(\mathbb{Z}, A) = 0$ when $n \geq 2$. $\square$

Exercises

1. Let A be an abelian group. Show that there is a one-to-one correspondence between
group actions of G on A by automorphisms, and [unital] $\mathbb{Z}[G]$-module structures on A.

2. Let $u \in C^n(G, A)$. Verify directly that $\delta^n u \in C^{n+1}(G, A)$ and that $\delta^{n+1}\delta^n u = 0$.

3. Give a direct proof that $\partial_{n+1}\,\partial_n = 0$ in the bar resolution.

4. Show that $H^0(G, A) \cong \{\, a \in A \mid xa = a \text{ for all } x \in G \,\}$.

5. Show that $H^1(G, A) \cong \mathrm{Hom}(G, A)$ when G acts trivially on A.

6. Show that $H^2(G, \mathbb{Z}) \cong \mathrm{Hom}(G, \mathbb{R}/\mathbb{Z})$ when $\mathbb{R}$ and $\mathbb{Z}$ are trivial G-modules.

7. Use Proposition 7.2 to give another proof of Maschke's theorem: if G is finite and
K is a field whose characteristic does not divide $|G|$, then $K[G]$ is semisimple. (Hint:
multiplication by $|G|$ is an automorphism of any $K[G]$-module.)

8. Show that $\mathrm{Hom}_{\mathbb{Z}[G]}(I(G), A) \cong Z^1(G, A)$ for every group G and G-module A. (Hint: the isomorphism sends $\varphi : I(G) \longrightarrow A$ to $x \longmapsto \varphi(x - 1)$.)

9. Prove the following: if G is finite and A is a finitely generated G-module, then A is finitely generated as an abelian group and $H^n(G, A)$ is finite for all $n \geqq 1$.

10. Let $G = \langle c \mid c^m = 1 \rangle$ be cyclic of order m and let $p = 1 + c + \cdots + c^{m-1}$, $q = c - 1 \in \mathbb{Z}[G]$. Show that $\mathbb{Z}$ has a free resolution

$$\cdots \mathbb{Z}[G] \xrightarrow{q^*} \mathbb{Z}[G] \xrightarrow{p^*} \mathbb{Z}[G] \xrightarrow{q^*} \mathbb{Z}[G] \xrightarrow{\varepsilon} \mathbb{Z},$$

in which multiplication by p and q induces p^* and q^*. Then find $H^n(G, A)$.

8. Projective Dimension

The projective dimension of a module M is the least integer n for which there exists a projective resolution $0 \longrightarrow P_n \longrightarrow \cdots \longrightarrow P_0 \longrightarrow A \longrightarrow 0$. This section contains general properties.

Syzygies. We begin with some properties of the syzygies of a module A, which, in a projective resolution $\cdots P_2 \xrightarrow{\partial_2} P_1 \xrightarrow{\partial_1} P_0 \xrightarrow{\varepsilon} A \longrightarrow 0$ of A, are the modules $K_0 = \mathrm{Ker}\, \varepsilon$, $K_n = \mathrm{Ker}\, \partial_n$. Syzygies come with short exact sequences $0 \longrightarrow K_0 \longrightarrow P_0 \longrightarrow A \longrightarrow 0$, $0 \longrightarrow K_n \longrightarrow P_n \longrightarrow K_{n-1} \longrightarrow 0$.

Proposition 8.1. *For every module B and $n \geqq 1$ there are isomorphisms*

$$\mathrm{Ext}^{n+1}(A, B) \cong \mathrm{Ext}^n(K_0, B) \cong \mathrm{Ext}^{n-1}(K_1, B) \cong \cdots \cong \mathrm{Ext}^1(K_{n-1}, B)$$

that are natural in B.

Proof. Since all P_m are projective, Theorem 4.4 and the exact sequences $0 \longrightarrow K_0 \longrightarrow P_0 \longrightarrow A \longrightarrow 0$, $0 \longrightarrow K_m \longrightarrow P_m \longrightarrow K_{m-1} \longrightarrow 0$ yield exact sequences that are natural in B, for every $k, m \geqq 1$:

$$0 \longrightarrow \mathrm{Ext}^{n+1}(A, B) \longrightarrow \mathrm{Ext}^n(K_0, B) \longrightarrow 0,$$

$$0 \longrightarrow \mathrm{Ext}^{k+1}(K_m, B) \longrightarrow \mathrm{Ext}^k(K_{m-1}, B) \longrightarrow 0. \square$$

There is also a uniqueness result for syzygies.

Definition. *Two modules A and B are* projectively equivalent *when $P \oplus A \cong Q \oplus B$ for some projective modules P and Q.*

Proposition 8.2. *If K_0, K_1, ... and L_0, L_1, ... are the syzygies of a module A in two projective resolutions $\mathcal{P} \longrightarrow A$, $\mathcal{Q} \longrightarrow A$ of A, then K_n and L_n are projectively equivalent for all $n \geqq 0$.*

Proof. First we prove *Schanuel's lemma*: in the diagram with exact rows (solid arrows, next page), if P and Q are projective and θ is an isomorphism, then $P \oplus L \cong Q \oplus K$:

$$0 \longrightarrow K \xrightarrow{\mu} P \xrightarrow{\sigma} A \longrightarrow 0$$
$$\downarrow \alpha \qquad \downarrow \beta \qquad \downarrow \theta$$
$$0 \longrightarrow L \xrightarrow{\nu} Q \xrightarrow{\tau} A \longrightarrow 0$$

By 2.1, the diagram can be completed to a commutative diagram (all arrows). Define $\kappa : K \longrightarrow P \oplus L$ and $\rho : P \oplus L \longrightarrow Q$ by $\kappa x = (\mu x, \alpha x)$ and $\rho(p, y) = \beta p - \nu y$. Then κ is injective like μ, ρ is surjective like β, and $\rho \mu = 0$; moreover, $\rho(p, y) = \beta p - \nu y = 0$ implies $\theta \sigma p = \tau \beta p = \tau \nu y = 0$, $\sigma p = 0$, $p = \mu x$ for some $x \in K$, $\nu \alpha x = \beta \mu x = \beta p = \nu y$, $y = \alpha x$, and $(p, y) = \kappa x$. Thus the sequence

$$0 \longrightarrow K \xrightarrow{\kappa} P \oplus L \xrightarrow{\rho} Q \longrightarrow 0$$

is exact. Since Q is projective, this sequence splits, and $P \oplus L \cong Q \oplus K$.

That K_n and L_n are projectively equivalent is now proved by induction on n. First, K_0 and L_0 are projectively equivalent, by Schanuel's lemma. In general, arbitrary modules P and Q beget an exact sequence

$$0 \longrightarrow K_{n+1} \xrightarrow{\mu} P \oplus P_{n+1} \xrightarrow{\sigma} P \oplus K_n \longrightarrow 0,$$

in which $\mu x = (0, x)$ and $\sigma(x, y) = (x, \partial y)$, and a similar sequence

$$0 \longrightarrow L_{n+1} \xrightarrow{\nu} Q \oplus Q_{n+1} \xrightarrow{\tau} Q \oplus L_n \longrightarrow 0.$$

If P and Q are projective and $P \oplus K_n \cong Q \oplus L_n$, then $P \oplus P_{n+1}$ and $Q \oplus Q_{n+1}$ are projective and Schanuel's lemma yields $P \oplus P_{n+1} \oplus L_{n+1} \cong Q \oplus Q_{n+1} \oplus K_{n+1}$. $\square$

The *cosyzygies* of a module A in an injective resolution $0 \longrightarrow A \xrightarrow{\eta} J^0 \xrightarrow{\delta^0} J^1 \xrightarrow{\delta^1} \cdots$ are the modules $L^0 = \operatorname{Coker} \eta$, $L^1 = \operatorname{Coker} \delta^0$, ... They have similar properties (see the exercises).

Modules. We now define the projective dimension of a module.

Proposition 8.3. *For a module A the following conditions are equivalent:*

(1) *A has a projective resolution $0 \longrightarrow P_n \longrightarrow \cdots \longrightarrow P_0 \longrightarrow A \longrightarrow 0$;*

(2) *$\operatorname{Ext}^m(A, B) = 0$ for all $m \geqq n + 1$ and all modules B;*

(3) *$\operatorname{Ext}^{n+1}(A, B) = 0$ for all modules B;*

(4) *the $(n-1)$th syzygy is projective in every projective resolution of A;*

(5) *the $(n-1)$th syzygy is projective in some projective resolution of A.*

Proof. (3) implies (4): by 8.1, $\operatorname{Ext}^1(K_{n-1}, B) = \operatorname{Ext}^{n+1}(A, B) = 0$ for all B; hence K_{n-1} is projective, by 4.5. Clearly (1) implies (2), (2) implies (3), (4) implies (5), and (5) implies (1). $\square$

Definition. If there is an integer $n \geqq 0$ such that the equivalent conditions

in Proposition 8.3 hold, then the least such integer is the projective dimension
$\mathrm{pd}\,A$ *or* $\mathrm{pd}_R\,A$ *of the module* A*; otherwise,* $\mathrm{pd}\,A = \infty$.

For example, $\mathrm{pd}\,A = 0$ if and only if A is projective, by 4.5; $\mathrm{pd}_{\mathbb{Z}}\,A \leq 1$ for
every abelian group A, by 4.9. The exercises give an example with $\mathrm{pd}\,A = \infty$.

Proposition 8.4. *If* $0 \longrightarrow A \longrightarrow B \longrightarrow C \longrightarrow 0$ *is exact, then:*

(1) $\mathrm{pd}\,B \leq \max(\mathrm{pd}\,A, \mathrm{pd}\,C)$;

(2) $\mathrm{pd}\,A \leq \max(\mathrm{pd}\,B, \mathrm{pd}\,C - 1)$;

(3) $\mathrm{pd}\,C \leq \max(\mathrm{pd}\,A + 1, \mathrm{pd}\,B)$;

(4) *if* B *is projective, then either* A *and* C *are projective or* $\mathrm{pd}\,C = 1 + \mathrm{pd}\,A$.

Proof. (1). If $\mathrm{pd}\,A$, $\mathrm{pd}\,C \leq n$, then $0 = \mathrm{Ext}^{n+1}(C, M) \longrightarrow \mathrm{Ext}^{n+1}(B, M)$
$\longrightarrow \mathrm{Ext}^{n+1}(A, M) = 0$ is exact for every module M; hence $\mathrm{Ext}^{n+1}(B, M) = 0$
for all M and $\mathrm{pd}\,B \leq n$.

(2). If $\mathrm{pd}\,B \leq n$, $\mathrm{pd}\,C \leq n + 1$, then $0 = \mathrm{Ext}^{n+1}(B, M) \longrightarrow \mathrm{Ext}^{n+1}(A, M)$
$\longrightarrow \mathrm{Ext}^{n+2}(C, M) = 0$ is exact for every module M; hence $\mathrm{Ext}^{n+1}(A, M) = 0$
for all M and $\mathrm{pd}\,A \leq n$.

(3). If $\mathrm{pd}\,A \leq n$, $\mathrm{pd}\,B \leq n + 1$, then $0 = \mathrm{Ext}^{n+1}(A, M) \longrightarrow \mathrm{Ext}^{n+2}(C, M)$
$\longrightarrow \mathrm{Ext}^{n+2}(B, M) = 0$ is exact for every module M; hence $\mathrm{Ext}^{n+2}(C, M) = 0$
for all M and $\mathrm{pd}\,C \leq n + 1$.

(4). If B and C are projective, then the sequence splits, $B \cong A \oplus C$, and
A is projective. If B is projective but not C, then $\mathrm{pd}\,B = 0$, $\mathrm{pd}\,C \geq 1$, and
$\mathrm{pd}\,A \leq \mathrm{pd}\,C - 1$, $\mathrm{pd}\,C \leq \mathrm{pd}\,A + 1$ by (2), (3); hence $\mathrm{pd}\,C = \mathrm{pd}\,A + 1$. $\square$

Proposition 8.5. $\mathrm{pd}\left(\bigoplus_{i \in I} A_i\right) = \mathrm{l.u.b.}_{\,i \in I}\,\mathrm{pd}\,A_i$.

The proof is an exercise.

Injective dimension is defined similarly:

Proposition 8.6. *For a module* B *the following conditions are equivalent:*

(1) B *has an injective resolution* $0 \longrightarrow B \longrightarrow J_0 \longrightarrow \cdots \longrightarrow J_n \longrightarrow 0$;

(2) $\mathrm{Ext}^m(A, B) = 0$ *for all* $m \geq n + 1$ *and all modules* A;

(3) $\mathrm{Ext}^{n+1}(A, B) = 0$ *for all modules* A.

Definition. If there is an integer $n \geq 0$ *such that the equivalent conditions in
Proposition 8.6 hold, then the least such integer is the* injective dimension $\mathrm{id}\,B$
or $\mathrm{id}_R\,B$ *of the module* B*; otherwise,* $\mathrm{id}\,B = \infty$.

The injective dimension has properties similar to Propositions 8.4 and 8.5 (see
the exercises).

Exercises

1. Let A and B be projectively equivalent. Show that $\mathrm{Ext}^n(A, C) \cong \mathrm{Ext}^n(B, C)$ for
every module C.

2. Let F be a covariant right exact functor with left derived functors $L_1, L_2, \ldots$. Let $K_0, K_1, \ldots$ be syzygies of A. Show that
$$L_{n+1}A \cong L_n K_0 \cong L_{n-1}K_1 \cong \cdots \cong L_1 K_{n-1}.$$

3. Let F be a contravariant left exact functor with right derived functors $R^1, R^2, \ldots$. Let $L^0, L^1, \ldots$ be cosyzygies of B. Show that
$$R^{n+1}B \cong R^n L^0 \cong R^{n-1}L^1 \cong \cdots \cong R^1 L^{n-1}.$$
Hence $\mathrm{Ext}^{n+1}(A, B) \cong \mathrm{Ext}^n(A, L^0) \cong \mathrm{Ext}^{n-1}(A, L^1) \cong \cdots \cong \mathrm{Ext}^1(A, L^{n-1})$ for every module A.

4. Define injectively equivalent modules. If $L^0, L^1, \ldots$ and $M^0, M^1, \ldots$ are the cosyzygies of a module A in two injective resolutions $A \longrightarrow \mathcal{J}$, $A \longrightarrow \mathcal{K}$ of A, show that L^n and M^n are injectively equivalent for all $n \geq 0$.

5. Let p be prime and let $R = \mathbb{Z}/p^2\mathbb{Z}$ and $A = p\mathbb{Z}/p^2\mathbb{Z}$. Show that $R \xrightarrow{p_*} R \xrightarrow{p_*} R \xrightarrow{\varepsilon} A \longrightarrow 0$ is exact, where p_* is multiplication by p and $\varepsilon \overline{1} = \overline{p}$; hence $\mathrm{pd}_R A = \infty$.

6. Prove that $\mathrm{pd} \left(\bigoplus_{i \in I} A_i \right) = \mathrm{l.u.b.}_{i \in I} \mathrm{pd}\, A_i$.

7. For every R-module A, show that $\mathrm{pd}\, A = n < \infty$ implies $\mathrm{Ext}^n_R(A, R) \neq 0$.

8. Prove that a module B has an injective resolution $0 \longrightarrow B \longrightarrow J_0 \longrightarrow \cdots \longrightarrow J_n \longrightarrow 0$ if and only if $\mathrm{Ext}^{n+1}(A, B) = 0$ for all A, and then $\mathrm{Ext}^m(A, B) = 0$ for all A and all $m \geq n + 1$.

9. Show that $\mathrm{id}\, B \leq \max(\mathrm{id}\, A, \mathrm{id}\, C)$ when $0 \longrightarrow A \longrightarrow B \longrightarrow C \longrightarrow 0$ is exact.

10. Prove that $\mathrm{id} \left(\prod_{i \in I} A_i \right) = \mathrm{l.u.b.}_{i \in I} \mathrm{id}\, A_i$.

In the following exercises, $\rho : S \longrightarrow R$ is a ring homomorphism, so that every left R-module A is also a left S-module. Use Propositions 8.4, 8.5 to prove:

11. $\mathrm{pd}_S A \leq \mathrm{pd}_R A + \mathrm{pd}_S R$ for every left R-module A.

12. If $\mathrm{pd}_S A = \mathrm{pd}_R A + \mathrm{pd}_S R$ whenever $A \neq 0$ and $\mathrm{pd}_R A \leq 1$, then $\mathrm{pd}_S A = \mathrm{pd}_R A + \mathrm{pd}_S R$ whenever $A \neq 0$ and $\mathrm{pd}_R A < \infty$.

9. Global Dimension

The left global dimension of a ring R is the upper bound of the projective dimensions of its left R-modules. We prove Hilbert's theorem on syzygies, which gives the global dimension of $K[X_1, \ldots, X_n]$ when K is a field.

By Propositions 8.3 and 8.6, the following properties are equivalent for any ring R: (1) $\mathrm{pd}\, A \leq n$ for every left R-module A; (2) $\mathrm{Ext}^{n+1}_R(A, B) = 0$ for all left R-modules A and B; (3) $\mathrm{id}\, B \leq n$ for every left R-module B.

Definition. If there is an integer $n \geq 0$ such that $\mathrm{Ext}^{n+1}_R(A, B) = 0$ for all left R-modules A and B, then the least such integer is the left global dimension *$\mathrm{lgld}\, R$ of the ring R; otherwise, $\mathrm{lgld}\, R = \infty$.*

For example, $\mathrm{lgld}\, R = 0$ if and only if every left R-module is projective, if and only if R is semisimple. By Proposition 4.9, $\mathrm{lgld}\, \mathbb{Z} = 1$. In fact, $\mathrm{lgld}\, R = 1$ if and only if R is left hereditary, but not semisimple (see the exercises).

A ring R also has a *right global dimension* rgld R, which is defined similarly, using right R-modules instead of left R-modules. The example in Section X.6 of a ring that is left hereditary but not right hereditary shows that the left and right global dimensions may be different. If R is commutative, however, then the two global dimensions of R are equal, and R has a *global dimension* gld R = lgld R = rgld R.

Our main result gives gld $K[X_1, ..., X_n]$ when K is a field:

Theorem **9.1** (Hilbert's Theorem on Syzygies). gld $K[X_1, ..., X_n] = n$, *for every field* K.

Hilbert's original statement applied to ideals of $K[X_1, ..., X_n]$ and stated that their syzygies vanish in dimensions above n in a suitable projective resolution. We deduce Theorem 9.1 from a more general statement:

Theorem **9.2.** *For any ring* R, lgld $R[X]$ = lgld $R + 1$.

The proof of Theorem 9.2 uses change of ring constructions. Since R is a subring of $R[X]$, every left $R[X]$-module is, in particular, a left R-module. Conversely, every left R-module A has a universal left $R[X]$-module $\overline{A} = R[X] \otimes_R A$ (Proposition XI.5.10). We note the following properties.

Lemma **9.3.** *Let* A *be a left* R-module. *Every element of* $\overline{A} = R[X] \otimes_R A$ *can be written uniquely as a sum* $\sum_k X^k \otimes a_k$, *where* $a_k \in A$ *for all* k *and* $a_k = 0$ *for almost all* k. *So* $\overline{A}$ *is, an an* R-module, *a direct sum of copies of* A.

Proof. As a right R-module, $R[X]$ is free with basis 1, X, X^2, ...; hence 9.3 follows from XI.5.9. $\square$

Lemma **9.4.** *If* P *is a projective* R-module, *then* $\overline{P}$ *is a projective* $R[X]$-module. *Conversely, a projective* $R[X]$-module *is also projective as an* R-module.

Proof. If $F \cong \bigoplus_{i \in I} {}_R R$ is a free left R-module, then

$$\overline{F} \cong \bigoplus_{i \in I} R[X] \otimes_R {}_R R \cong \bigoplus_{i \in I} R[X]$$

is a free left $R[X]$-module. If now P is a projective R-module, then some $P \oplus Q$ is a free R-module, $\overline{P \oplus Q} \cong \overline{P \oplus Q}$ is a free $R[X]$-module, and $\overline{P}$ is a projective $R[X]$-module. Conversely, a free $R[X]$-module $\bigoplus_{i \in I} R[X]$ is a direct sum of copies of ${}_R R$ and is free as an R-module; hence a projective $R[X]$-module is also projective as an R-module. $\square$

Lemma **9.5.** $\mathrm{pd}_{R[X]} \overline{A} = \mathrm{pd}_R A$ *for every left* R-module A.

Proof. If $\mathrm{pd}_R A \leqq n$, then A has a projective resolution $0 \longrightarrow P_n \longrightarrow \cdots \longrightarrow P_0 \longrightarrow A \longrightarrow 0$ (over R); since $R[X]$ is flat as a right R-module, $0 \longrightarrow \overline{P}_n \longrightarrow \cdots \longrightarrow \overline{P}_0 \longrightarrow \overline{A} \longrightarrow 0$ is exact, and is, by 9.4, a projective resolution over $R[X]$; hence $\mathrm{pd}_{R[X]} \overline{A} \leqq n$. Conversely, if $\mathrm{pd}_{R[X]} \overline{A} \leqq n$, then $\overline{A}$ has a projective resolution $0 \longrightarrow Q_n \longrightarrow \cdots \longrightarrow Q_0 \longrightarrow \overline{A} \longrightarrow 0$ over

$R[X]$, which, by 9.4, is also a projective resolution over R; hence $\text{pd}_R \overline{A} \leqq n$, and $\text{pd}_R A = \text{pd}_R \overline{A} \leqq n$ by 8.5, 9.3. $\square$

Lemma **9.6.** *For every $R[X]$-module A there is an exact sequence $0 \longrightarrow \overline{A} \longrightarrow$* $\overline{A} \longrightarrow A \longrightarrow 0$.

Proof. By 9.3, every element of $\overline{A}$ is a sum $\sum_{0 \leq k \leq m} X^k \otimes a_k$, where $a_k \in A$. Hence there is an $R[X]$-homomorphism $\sigma : \overline{A} \longrightarrow A$ such that $\sigma(X^k \otimes a) = X^k a$, and σ is surjective. Since A is an $R[X]$-module, there is also an $R[X]$-homomorphism $\mu : \overline{A} \longrightarrow \overline{A}$ such that $\mu(X^k \otimes a) = (X^k \otimes Xa) - (X^{k+1} \otimes a)$. Then

$$\mu\left(\sum_{0 \leq k \leq m} X^k \otimes a_k\right) = \sum_{0 \leq k \leq m} X^k \otimes Xa_k - \sum_{0 \leq k \leq m} X^{k+1} \otimes a_k$$
$$= 1 \otimes Xa_0 + \sum_{1 \leq k \leq m} X^k \otimes (Xa_k - a_{k-1}) - X^{m+1} \otimes a_m.$$

We show that $0 \longrightarrow \overline{A} \xrightarrow{\mu} \overline{A} \xrightarrow{\sigma} A \longrightarrow 0$ is exact.

If $\mu\left(\sum_{0 \leq k \leq m} X^k \otimes a_k\right) = 0$, then $Xa_0 = 0$, $Xa_k - a_{k-1} = 0$ for all $1 \leq k \leq m$, and $a_m = 0$; hence $a_k = 0$ for all $0 \leq k \leq m$. Thus μ is injective.

Next, $\sigma\mu = 0$, since $\sigma\mu(X^k \otimes a) = \sigma(X^k \otimes Xa) - \sigma(X^{k+1} \otimes a) = 0$ for all k and a.

Finally, let $\sigma\left(\sum_{0 \leq k \leq m+1} X^k \otimes c_k\right) = \sum_{0 \leq k \leq m+1} X^k c_k = 0$. Define a_m, ..., a_1, a_0 by

$$-a_m = c_{m+1}, \; a_{k-1} = Xa_k - c_k \text{ for all } 1 \leq k \leq m.$$

Then $Xa_0 = X^2 a_1 - Xc_1 = X^3 a_2 - X^2 c_2 - Xc_1 = \cdots = X^{m+1} a_m - X^m c_m -$ $\cdots - Xc_1 = -X^{m+1} c_{m+1} - X^m c_m - \cdots - Xc_1 = c_0$ and

$$\mu\left(\sum_{0 \leq k \leq m} X^k \otimes a_k\right) = \sum_{0 \leq k \leq m+1} X^k \otimes c_k . \; \square$$

Corollary **9.7.** $\text{lgld } R[X] \leqq \text{lgld } R + 1$.

Proof. If $\text{lgld } R = \infty$, then, for every $n < \infty$, some R-module A has $\text{pd}_R A \geqq n$, whence $\text{pd}_{R[X]} \overline{A} \geqq n$, by 9.5; thus $\text{lgld } R[X] = \infty$. If $\text{lgld } R = n < \infty$, and A is any $R[X]$-module, then $\text{pd}_{R[X]} \overline{A} = \text{pd}_R A \leqq n$, by 9.5; by 9.6 and 8.4, $\text{pd}_{R[X]} A \leqq n + 1$; hence $\text{lgld } R[X] \leqq n + 1$. $\square$

To prove the converse inequality we use another change of rings. As a ring, $R \cong R[X]/(X)$; the projection $R[X] \longrightarrow R$ sends $r_0 + r_1 X + r_2 X^2 + \cdots \in R[X]$ onto r_0. Hence every left R-module A is also a left $R[X]$-module, in which $(r_0 + r_1 X + \cdots)a = r_0 a$, in particular, $XA = 0$. Moreover, $\text{Hom}_R(A, B)$ $= \text{Hom}_{R[X]}(A, B)$ for all R-modules A and B.

If C is an $R[X]$-module, then XC is a submodule of C, since X is central in $R[X]$, and C/XC is an R-module, in which $r(c + XC) = rc + XC$ for all $r \in R$ and $c \in C$. Readers will verify that $C/XC \cong R \otimes_{R[X]} C$ is the universal R-module of C.

Lemma **9.8.** *For every left R-module A there is a natural isomorphism* $\operatorname{Hom}_R(A, R) \cong \operatorname{Ext}^1_{R[X]}(A, R[X])$.

Proof. Let $\mu : R[X] \longrightarrow R[X]$ be multiplication by X. The exact sequence

$$0 \longrightarrow R[X] \overset{\mu}{\longrightarrow} R[X] \longrightarrow R \longrightarrow 0$$

of $R[X]$-modules and homomorphisms induces an exact sequence

$$\operatorname{Hom}_R(A, R) = \operatorname{Hom}_{R[X]}(A, R[X]) \longrightarrow \operatorname{Hom}_{R[X]}(A, R)$$
$$\longrightarrow \operatorname{Ext}^1_{R[X]}(A, R[X]) \overset{\mu_*}{\longrightarrow} \operatorname{Ext}^1_{R[X]}(A, R[X]),$$

which is natural in A, where $\mu_* = \operatorname{Ext}^1_{R[X]}(A, \mu)$. If $\varphi : A \longrightarrow R[X]$ is a homomorphism of $R[X]$-modules, then $X\varphi a = \varphi(Xa) = 0$ for all $a \in A$, since A is an R-module; hence $\varphi = 0$. Thus, $\operatorname{Hom}_{R[X]}(A, R[X]) = 0$.

Next, $\operatorname{Ext}^1_{R[X]}(A, R[X]) = H^1\big(\operatorname{Hom}_{R[X]}(A, \jmath)\big)$ for any injective resolution $\jmath : J^0 \longrightarrow J^1 \cdots$ of $R[X]$ over $R[X]$. Multiplication by X is a homomorphism $\mu^n : J^n \longrightarrow J^n$; these homomorphisms constitute a chain transformation that lifts μ, hence induces all $\operatorname{Ext}^1_{R[X]}(A, \mu)$. In particular, $\mu_*(\operatorname{cls} \alpha) = \operatorname{cls} \mu^1 \alpha$ for every cocycle $\alpha : A \longrightarrow J^1$. Since A is an R-module, $\mu^1 \alpha a = X\alpha a = \alpha(Xa) = 0$ for all $a \in A$. Hence $\mu_* = 0$. Therefore $\operatorname{Hom}_{R[X]}(A, R) \longrightarrow \operatorname{Ext}^1_{R[X]}(A, R[X])$ is an isomorphism. $\square$

Lemma **9.9.** $\operatorname{Ext}^n_R(A, R) \cong \operatorname{Ext}^{n+1}_{R[X]}(A, R[X])$, *for every R-module A.*

Proof. First, we show that $\operatorname{Ext}^n_{R[X]}(P, R[X]) = 0$ for every projective R-module P and $n > 0$. By 9.4, 9.3, $\overline{P}$ is a projective $R[X]$-module and $\overline{P}/X\overline{P} \cong P$. Now, multiplication by X, $\mu : \overline{P} \longrightarrow \overline{P}$, is injective on $\overline{P}$ since it is injective on some free $R[X]$-module $F \supseteq \overline{P}$. Hence the sequence

$$0 \longrightarrow \overline{P} \overset{\mu}{\longrightarrow} \overline{P} \longrightarrow P \longrightarrow 0$$

of $R[X]$-modules and homomorphisms is exact, and induces an exact sequence

$$\operatorname{Ext}^n_{R[X]}(\overline{P}, R[X]) \longrightarrow \operatorname{Ext}^n_{R[X]}(P, R[X]) \longrightarrow \operatorname{Ext}^{n+1}_{R[X]}(\overline{P}, R[X])$$

in which $\operatorname{Ext}^n_{R[X]}(\overline{P}, R[X]) = \operatorname{Ext}^{n+1}_{R[X]}(\overline{P}, R[X]) = 0$ since $\overline{P}$ is projective. Hence $\operatorname{Ext}^n_{R[X]}(P, R[X]) = 0$.

We now have functors $G^0 = \operatorname{Hom}_R(-, R)$ and $G^n = \operatorname{Ext}^{n+1}_{R[X]}(-, R[X])$, $n > 0$, which by 9.8, 4.4 constitute a negative connected sequence of contravariant functors, from R-modules to abelian groups. Since $G^n P = 0$ whenever P is projective and $n > 0$, it follows from 3.12 that $G^1, \ldots, G^n, \ldots$ are, up to natural isomorphisms, the right derived functors of G^0; in other words, there are natural isomorphisms $\mathrm{Ext}_{R[X]}^{n+1}(A, R[X]) \cong \mathrm{Ext}_R^n(A, R)$. $\square$

We can now complete the proof of Theorem 9.2. By 9.7, lgld $R[X] \leqq$ lgld $R + 1$, with equality if lgld $R = \infty$. Assume that lgld $R = n < \infty$. Then $\mathrm{pd}_R A = n$ for some R-module A: $\mathrm{Ext}_R^{n+1}(A, B) = 0$ for all B but $\mathrm{Ext}_R^n(A, B) \neq 0$. There is a short exact sequence $0 \longrightarrow K \longrightarrow F \longrightarrow B \longrightarrow 0$ of R-modules in which F is free; then $\mathrm{Ext}_R^n(A, F) \longrightarrow \mathrm{Ext}_R^n(A, B) \longrightarrow \mathrm{Ext}_R^{n+1}(A, K) = 0$ is exact; since $\mathrm{Ext}_R^n(A, B) \neq 0$ it follows that $\mathrm{Ext}_R^n(A, F) \neq 0$. By 8.5, $\mathrm{Ext}_R^n(A, R) \neq 0$. Then $\mathrm{Ext}_{R[X]}^{n+1}(A, R[X]) \cong \mathrm{Ext}_R^n(A, R) \neq 0$, by 9.9. Hence lgld $R[X] \geqq n + 1$. $\square$

Exercises

1. Show that lgld $R = 1$ if and only if R is left hereditary, but not semisimple.

2. Show that lgld $R =$ l.u.b. $(\mathrm{pd}\ R/L \mid L$ is a left ideal of $R)$. (Use cosyzygies.)

3. Verify that $R \otimes_{R[X]} C \cong C/XC$ for every $R[X]$-module C, where R is the left R-, right $R[X]$-bimodule in which $rX = 0$ for all $r \in R$.

4. Let B be an $R[X]$-module on which the action of X is injective. Show that $\mathrm{Ext}_R^n(A, B/XB) \cong \mathrm{Ext}_{R[X]}^{n+1}(A, B)$ for every R-module A. (You may want to adapt the proof of Lemma 9.8.)

5. Let S be a ring. Let $R = S/Sc$, where c is a central element of S and is not a unit or a zero divisor. Let B be an S-module on which the action of c is injective. Show that $\mathrm{Ext}_R^n(A, B/cB) \cong \mathrm{Ext}_{R[X]}^{n+1}(A, B)$ for every R-module A.

XIII
Algebras

Algebras, the last of the major algebraic objects in this book, are rings with a compatible vector space or module structure. Interest in algebras began with Hamilton's construction of the quaternions [1843] and Benjamin Peirce's paper *Linear Associative Algebras* [1864]. Algebras are fundamental to algebraic geometry and to the study of group representations (see Chapters VII and IX).

This chapter proves some basic properties, constructs algebras with various universal properties, and concludes with Frobenius's theorem [1877], which finds all division algebras over $\mathbb{R}$. The results add to our understanding of rings and fields and provide additional applications of tensor products.

1. Algebras over a Ring

This section contains the definition and initial properties of algebras and graded algebras. In what follows, R is a commutative ring (with identity element), and all modules are unital.

Definition. An algebra over a commutative ring R, or R-algebra, is an R-module A with a multiplication that is bilinear ($a(b+c) = ab + ac$, $(a+b)c = ac + bc$, and $(ra)b = a(rb) = r(ab)$, for all $a, b, c \in A$ and $r \in R$), associative ($a(bc) = (ab)c$ for all $a, b, c \in A$), and has an identity element 1 ($1a = a = a1$ for all $a \in A$).

Equivalently, an R-algebra is a ring (with identity element) with an R-module structure such that $(ra)b = a(rb) = r(ab)$, for all $a, b \in A$ and $r \in R$. Algebras as defined above are also called *associative* algebras (with identity); *nonassociative* algebras are defined similarly, but without the requirements that the multiplication be associative or have an identity element.

The earliest examples of algebras are $\mathbb{C}$ and the quaternion algebra $\mathbb{H}$, which are algebras over $\mathbb{R}$. Every ring is an algebra over $\mathbb{Z}$. Examples of R-algebras also include polynomial rings $R[X]$, $R[(X_i)_{i \in I}]$ in one or several variables; power series rings $R[[X]]$, $R[[(X_i)_{i \in I}]]$; matrix rings $M_n(R)$; quotients $R/\mathfrak{a}$ of R by its ideals; commutative ring extensions of R; and group algebras.

In an R-algebra A, $r1 + s1 = (r+s)1$ and $(r1)(s1) = (rs)1$ for all $r, s \in R$, so that $\iota : r \longmapsto r1$ is a ring homomorphism of R into A. Moreover, ι is a *central* homomorphism, meaning that every $\iota(r) = r1$ is *central* in A $((r1)\,a = a\,(r1)$ for all $a \in A)$. If ι is injective, then $r1$ may be identified with r and R becomes a subring of A; for instance, this is the case when R is a field and $A \neq 0$. In general, the R-module structure on A is determined by ι, since the action of $r \in R$ is simply multiplication by $\iota(r)$. Readers will prove a converse:

Proposition **1.1.** *For every ring A, there is a one-to-one correspondence between central ring homomorphisms of R into A, and R-module structures on A that make A an R-algebra.*

Definition. A homomorphism *of R-algebras is a ring homomorphism that is also an R-module homomorphism.*

Equivalently, an R-algebra homomorphism $\varphi : A \longrightarrow B$ is a mapping $\varphi : A \longrightarrow B$ such that $\varphi\,(a+b) = \varphi(a) + \varphi(b)$, $\varphi\,(ab) = \varphi(a)\,\varphi(b)$, $\varphi\,(ra) = r\,\varphi(a)$, and $\varphi(1) = 1$, for all $a, b \in A$ and $r \in R$.

We state and prove a homomorphism theorem for algebras. This unsurprising result requires some equally unsurprising definitions.

Definition. A subalgebra *of an R-algebra A is a subset S of A that is both a subring of A and a submodule of A.*

Every subalgebra S of A inherits from A a ring structure and an R-module structure that make S an R-algebra in its own right; this R-algebra is also called a *subalgebra* of A.

Definitions. A two-sided ideal *of an R-algebra A is a subset I of A that is both a two-sided ideal of A and a submodule of A. Then A/I is an R-algebra, the* quotient algebra *of A by I.*

In the above, A/I is both a ring and an R-module; it is immediate that this makes A/I an R-algebra. Moreover, the projection $A \longrightarrow A/I$ is an algebra homomorphism.

Theorem **1.2** (Homomorphism Theorem). *If $\varphi : A \longrightarrow B$ is a homomorphism of R-algebras, then $\mathrm{Im}\ \varphi$ is a subalgebra of B, $\mathrm{Ker}\ \varphi$ is an ideal of A, and $A/\mathrm{Ker}\ \varphi \cong \mathrm{Im}\ \varphi$; in fact, there is a unique algebra isomorphism $\theta : A/\mathrm{Ker}\ \varphi \longrightarrow \mathrm{Im}\ \varphi$ such that the following square commutes:*

$$
\begin{array}{ccc}
A & \xrightarrow{\ \varphi\ } & B \\
\downarrow & & \uparrow{\scriptstyle\subseteq} \\
A/\mathrm{Ker}\ \varphi & \dashrightarrow[\theta] & \mathrm{Im}\ \varphi
\end{array}
$$

Proof. The homomorphism theorems for rings and modules both yield the diagram above, with the same unique isomorphism θ; hence θ is an algebra isomorphism. $\square$

Theorem **1.3** (Factorization Theorem). *Let I be a two-sided ideal of an R-algebra A. Every algebra homomorphism whose kernel contains I factors uniquely through the projection $A \longrightarrow A/I$:*

Proof. The factorization theorems for rings and modules both yield the diagram above, with the same unique homomorphism ψ; hence ψ is an algebra homomorphism. $\square$

Graded algebras. A much longer book would show the importance of graded rings and modules in commutative algebra and homological algebra. Here, our interest is limited to the algebras in the next three sections.

Definition. A graded R-algebra *is an R-algebra A with submodules $(A_n)_{n \geq 0}$ such that $A = \bigoplus_{n \geq 0} A_n$, $1 \in A_0$, and $A_m A_n \subseteq A_{m+n}$ for all $m, n \geq 0$.*

For example, the polynomial algebra $R[(X_i)_{i \in I}]$ is a graded R-algebra, in which the submodule A_n in the definition consists of all homogeneous polynomials of degree n. In any graded algebra $A = \bigoplus_{n \geq 0} A_n$, every element a is a sum $a = \sum_{n \geq 0} a_n$ for some unique $a_n \in A_n$ (such that $a_n = 0$ for almost all n); a_n is the nth *homogeneous component* of a.

Definitions. A graded submodule (subring, subalgebra, two-sided ideal) *of a graded R-algebra $A = \bigoplus_{n \geq 0} A_n$ is a submodule (subring, subalgebra, two-sided ideal) S of A such that $S = \bigoplus_{n \geq 0} (A_n \cap S)$.*

Readers will verify that a submodule S of a graded algebra is a graded submodule if and only if the homogeneous components of every $s \in S$ are all in S. For example, in $\mathbb{Z}[X]$, polynomials with even coefficients constitute a graded ideal; multiples of $X^2 + 1$ do not.

A graded subalgebra $S = \bigoplus_{n \geq 0} (A_n \cap S)$ of a graded algebra $A = \bigoplus_{n \geq 0} A_n$ is itself a graded algebra, since $1 \in A_0 \cap S$ and $(A_m \cap S)(A_n \cap S) \subseteq A_{m+n} \cap S$ for all $m, n \geq 0$. Readers will show that the quotient algebra A/I of a graded algebra $A = \bigoplus_{n \geq 0} A_n$ by a graded two-sided ideal $I = \bigoplus_{n \geq 0} (A_n \cap I)$ is a graded algebra $A/I = \bigoplus_{n \geq 0} (A_n + I)/I$.

Definition. A homomorphism $\varphi : A \longrightarrow B$ *of graded R-algebras $A = \bigoplus_{n \geq 0} A_n$, $B = \bigoplus_{n \geq 0} B_n$ is a homomorphism of R-algebras such that $\psi(A_n) \subseteq B_n$ for all $n \geq 0$.*

The next result is a (graded?) exercise:

Theorem **1.4** (Homomorphism Theorem). *If $\varphi : A \longrightarrow B$ is a homomorphism of graded R-algebras, then $\operatorname{Im} \varphi$ is a graded subalgebra of B, $\operatorname{Ker} \varphi$ is a graded ideal of A, and $A/\operatorname{Ker} \varphi \cong \operatorname{Im} \varphi$; in fact, there is a unique* *isomorphism* $\theta : A/\mathrm{Ker}\, \varphi \longrightarrow \mathrm{Im}\, \varphi$ *of graded algebras such that the following square commutes:*

$$
\begin{array}{ccc}
A & \xrightarrow{\;\varphi\;} & B \\
\downarrow & & \uparrow{\scriptstyle \subseteq} \\
A/\mathrm{Ker}\, \varphi & \xdashrightarrow[\theta]{} & \mathrm{Im}\, \varphi
\end{array}
$$

Exercises

In the following exercises, R is a commutative ring (with identity).

1. Verify that $R/\mathfrak{a}$ is an R-algebra for every ideal $\mathfrak{a}$ of R.

2. Verify that $\mathrm{End}_R(M)$ is an R-algebra for every R-module M.

3. Let A be a ring. Show that there is a one-to-one correspondence between central ring homomorphisms of R into A, and R-module structures on A that make A an R-algebra.

4. Let A be an R-module. Show that there is a one-to-one correspondence between multiplications on A that make A an R-algebra, and pairs of module homomorphisms $\iota : R \longrightarrow A$, $\mu : A \otimes A \longrightarrow A$ such that the following diagrams commute (tensor products are over R):

$$
\begin{array}{ccccc}
A \otimes A & \xrightarrow{\;\mu\;} A \xleftarrow{\;\mu\;} & A \otimes A & \qquad A \otimes A \otimes A & \xrightarrow{\;\mu\otimes 1\;} A \otimes A \\
{\scriptstyle \iota\otimes 1}\uparrow & \| & \uparrow{\scriptstyle 1\otimes\iota} & {\scriptstyle 1\otimes\mu}\downarrow & \downarrow{\scriptstyle \mu} \\
R \otimes A & \xrightarrow[\cong]{} A \xleftarrow[\cong]{} & A \otimes R & \qquad A \otimes A & \xrightarrow[\;\mu\;]{} A \otimes A
\end{array}
$$

5. Show that $R[X]$ serves a the "free R-algebra" with one generator.

6. Show that $R[(X_i)_{i\in I}]$ serves a the "free commutative R-algebra" on the set I.

7. Show that the quotient algebra A/I of a graded algebra $A = \bigoplus_{n\geq 0} A_n$ by a graded two-sided ideal $I = \bigoplus_{n\geq 0} (A_n \cap I)$ is a graded algebra $A/I = \bigoplus_{n\geq 0} (A_n + I)/I$.

8. Let A and B be graded R-algebras. Show that an algebra homomorphism $\varphi : A \longrightarrow B$ is a homomorphism of graded algebras if and only if φ preserves the decomposition of elements into sums of homogeneous components.

9. Prove the homomorphism theorem for graded algebras.

2. The Tensor Algebra

The tensor algebra of a module is an algebra that is "freely" generated by that module, as shown by a suitable universal property. It is named after the tensor products used in its construction.

In what follows, R is a commutative ring (with identity element); all modules are unital; all algebras and tensor products are over R. Since R is commutative, the tensor product of n R-modules is an R-module; its tensor map is universal for n-linear mappings.

Generation. Let A be an R-algebra. Every intersection of subalgebras of A is a subalgebra of A. Hence there is for every subset S of A a smallest subalgebra of A that contains S, the subalgebra $\langle S \rangle$ of A *generated by* S, which is the intersection of all subalgebras of A that contain S, and is described as follows when S is a submodule:

Proposition **2.1.** *The subalgebra $\langle M \rangle$ of an R-algebra A generated by a submodule M of A consists of all sums of a scalar multiple of the identity element and finitely many nonempty products of elements of M.*

The elements of M usually satisfy relations in A, meaning that there usually are equalities between the sums in 2.1. The tensor algebra of M is constructed so that it is generated by M with the fewest possible relations. It is the "simplest," and also the "largest," algebra generated by M, as specified by 2.4, 2.5 below.

Seeking clues to the construction of the tensor algebra, we detect that the product $a_1 \cdots a_n \in A$ of $n \geq 2$ elements of M is an n-linear function of $a_1, \ldots, a_n$ and induces a module homomorphism of $M \otimes \cdots \otimes M$ into A, which sends $a_1 \otimes \cdots \otimes a_n$ to $a_1 \cdots a_n$. These maps, together with the central homomorphism $R \longrightarrow A$ and the inclusion homomorphism $M \longrightarrow A$, induce a module homomorphism of $R \oplus M \oplus (M \otimes M) \oplus \cdots$ into A, which is surjective by 2.1 when A is generated by M. In this sense every R-algebra that is generated by M is a child of the direct sum $R \oplus M \oplus (M \otimes M) \oplus \cdots$. We build the tensor algebra of M from this direct sum.

Construction. Let M be an R-module. The nth *tensor power* $T^n(M)$ or $\otimes^n M$ of M is defined as follows: $T^0(M) = R$; $T^1(M) = M$; when $n \geq 2$,

$$T^n(M) = M \otimes \cdots \otimes M$$

is the tensor product of n copies of M: $T^n(M) = M_1 \otimes \cdots \otimes M_n$, where $M_1 = \cdots = M_n = M$. For every $m, n > 0$ there is an isomorphism $T^m(M) \otimes T^n(M) \cong T^{m+n}(M)$ and a bilinear multiplication $\otimes$,

$$T^m(M) \times T^n(M) \xrightarrow{\ \otimes\ } T^m(M) \otimes T^n(M) \xrightarrow{\ \cong\ } T^{m+n}(M),$$

which sends $(a_1 \otimes \cdots \otimes a_m, \ b_1 \otimes \cdots \otimes b_n)$ to $a_1 \otimes \cdots \otimes a_m \otimes b_1 \otimes \cdots \otimes b_n$. Similarly, for every $n > 0$, the left and right actions of R on $T^n(M)$, and the multiplication on R itself, are bilinear multiplications $T^0(M) \times T^n(M) \longrightarrow T^n(M)$, $T^n(M) \times T^0(M) \longrightarrow T^n(M)$, and $T^0(M) \times T^0(M) \longrightarrow T^0(M)$, which send (r, t), (t, r), and (r, s), respectively, to rt, rt, and rs, and which we also denote by $\otimes$. (This uses the isomorphisms $R \otimes T^n(M) \cong T^n(M)$, $T^n(M) \otimes R \cong T^n(M)$, $R \otimes R \cong R$, as a pretext to identify $r \otimes t$ and rt, $t \otimes r$ and $tr = rt$, $r \otimes s$ and rs, when $r, s \in R$ and $t \in T^n(M)$.)

Definition. The tensor algebra *of an R-module M is*

$$T(M) = \bigoplus_{n \geq 0} T^n(M)$$

with multiplication given by $\left(\sum_{m \geq 0} t_m \right) \left(\sum_{m \geq 0} u_n \right) = \sum_{m, n \geq 0} (t_m \otimes u_n)$.

Proposition **2.2.** *The tensor algebra of a [unital] R-module M is a graded R-algebra and is generated by M.*

Corollary **2.3.** *If* M *is the free* R-*module on a set* X, *then* $T(M)$ *is a free* R-*module, with a basis that consists of* $1 \in R$, *all* $x \in X$, *and all* $x_1 \otimes \cdots \otimes x_n$ *with* $n \geqq 2$ *and* $x_1, \ldots, x_n \in X$.

The proofs are exercises, to enliven readers' long winter evenings.

Proposition **2.4.** *Every module homomorphism of an* R-*module* M *into an* R-*algebra* A *extends uniquely to an algebra homomorphism of* $T(M)$ *into* A.

Proof. Let $\varphi : M \longrightarrow A$ be a module homomorphism. For every $n \geqq 2$, multiplication in A yields an n-linear mapping $(a_1, \ldots, a_n) \longmapsto \varphi(a_1) \cdots \varphi(a_n)$ of M^n into A, which induces a module homomorphism $\varphi_n : T^n(M) \longrightarrow A$ that sends $a_1 \otimes \cdots \otimes a_n$ to $\varphi(a_1) \cdots \varphi(a_n)$ for every $a_1, \ldots, a_n \in M$. The central homomorphism $\varphi_0 = \iota : R \longrightarrow A$, the given map $\varphi_1 = \varphi : M \longrightarrow A$, and the homomorphisms $\varphi_n : T^n(M) \longrightarrow A$ induce a module homomorphism $\overline{\varphi} : T(M) \longrightarrow A$, which sends $t = \sum_{n \geqq 0} t_n$ to $\sum_{n \geqq 0} \varphi_n(t_n)$.

The equality $\overline{\varphi}(t) \overline{\varphi}(u) = \overline{\varphi}(tu)$ holds whenever $t = a_1 \otimes \cdots \otimes a_m$ and $u = b_1 \otimes \cdots \otimes b_n$ are generators of $T^m(M)$ and $T^n(M)$ (since φ_n is a module homomorphism, in case $m = 0$ or $n = 0$); therefore it holds whenever $t \in T^m(M)$ and $u \in T^n(M)$; therefore it holds for all $t, u \in T(M)$. Also $\overline{\varphi}(1) = \varphi_0(1) = 1$. Hence $\overline{\varphi}$ is an algebra homomorphism. If ψ is another algebra homomorphism that extends φ, then $S = \{ t \in T(M) \mid \psi(t) = \overline{\varphi}(t) \}$ is a subalgebra of $T(M)$ that contains M; therefore $S = T(M)$ and $\psi = \overline{\varphi}$. $\square$

Corollary **2.5.** *Every* R-*algebra that is generated by a submodule* M *is isomorphic to a quotient algebra of* $T(M)$.

Proof. If $A = \langle \varphi(M) \rangle$ in the proof of 2.4, then $\overline{\varphi}$ is surjective, by 2.1. $\square$

Corollary **2.6.** *If* M *is the free* R-*module on a set* X, *then* $T(M)$ *is the free* R-*algebra on the set* X: *every mapping of* X *into an* R-*algebra* A *extends uniquely to an algebra homomorphism of* $T(M)$ *into* A.

Proof. This follows from Proposition 2.4, since every mapping of X into an R-algebra A extends uniquely to a module homomorphism of M into A. $\square$

Exercises

Prove the following:

1. The tensor algebra of an R-module M is a graded R-algebra and is generated by M.

2. $T(_R R) \cong R[X]$.

3. If M is the free R-module on a set X, then $T(M)$ is a free R-module, with a basis that consists of $1 \in R$, all $x \in X$, and all $x_1 \otimes \cdots \otimes x_n$ with $n \geqq 2$ and $x_1, \ldots, x_n \in X$.

4. If M is free with basis $(e_i)_{i \in I}$, then $T(M)$ is isomorphic to the polynomial ring $R[(X_i)_{i \in I}]$ with indeterminates X_i that commute with constants but not with each other.

5. Every homomorphism $M \longrightarrow N$ of R-modules extends uniquely to a homomorphism $T(M) \longrightarrow T(N)$ of graded R-algebras.

3. The Symmetric Algebra

The symmetric algebra of an R-module M is the commutative analogue of its tensor algebra: a commutative R-algebra that is "freely" generated by M, as shown by its universal property. As before, R is a commutative ring [with identity]; all modules are unital; all algebras and tensor products are over R.

Construction. By 2.5, the symmetric algebra must be, up to isomorphism, the quotient algebra of $T(M)$ by some two-sided ideal I. Since the symmetric algebra must be commutative, there is an obvious candidate for I:

Definition. The symmetric algebra *of an R-module M is the quotient algebra $S(M) = T(M)/I$, where I is the ideal of $T(M)$ generated by all $t \otimes u - u \otimes t$ with $t, u \in T(M)$.*

We show that M may be identified with a submodule of $S(M)$.

Lemma **3.1.** $I \subseteq \bigoplus_{n \geq 2} T^n(M)$.

Proof. Let $t \in T^m(M)$ and $u \in T^n(M)$. If $m + n \geq 2$, then $t \otimes u - u \otimes t \in \bigoplus_{n \geq 2} T^n(M)$. If $m + n < 2$, then $m = 0$ or $n = 0$, and $t \otimes u - u \otimes t = 0$ since $r \otimes t = rt = t \otimes r$ in $T(M)$, for all $r \in R$ and $t \in T(M)$. Now, $\bigoplus_{n \geq 2} T^n(M)$ is an ideal of $T(M)$; hence $I \subseteq \bigoplus_{n \geq 2} T^n(M)$. $\square$

By 3.1, the projection $T(M) \longrightarrow S(M) = T(M)/I$ is injective on $R \oplus M$; therefore we may identity R and M with their images in $S(M)$. The product $a_1 \cdots a_n$ of $a_1, \ldots, a_n \in M$ in $S(M)$ is the image in $S(M)$ of their product $a_1 \otimes \cdots \otimes a_n$ in $T(M)$. By 2.1, 2.2, $S(M)$ is generated by M:

Corollary **3.2.** *The symmetric algebra of a [unital] R-module M is a commutative R-algebra and is generated by M.*

The definition of $S(M)$ also yields a universal property:

Proposition **3.3.** *Every module homomorphism of an R-module M into a commutative R-algebra A extends uniquely to an algebra homomorphism of $S(M)$ into A.*

Proof. Let $\pi : T(M) \longrightarrow S(M)$ be the projection and let $\varphi : M \longrightarrow A$ be a module homomorphism. By 2.4, φ extends to an algebra homomorphism $\psi : T(M) \longrightarrow A$. Since A is commutative we have $\psi(t \otimes u) = \psi(u \otimes t)$ and $t \otimes u - u \otimes t \in \text{Ker } \psi$, for all $t, u \in T(M)$. Hence $I \subseteq \text{Ker } \psi$. By 1.3, $\psi = \chi \circ \pi$ for some algebra homomorphism χ . $S(M) \longrightarrow A$:

$$
\begin{array}{ccc}
M & \overset{\subseteq}{\longrightarrow} & T(M) \\
{\scriptstyle\varphi}\downarrow & \swarrow{\scriptstyle\psi} & \downarrow{\scriptstyle\pi} \\
A & \overset{}{\underset{\chi}{\longleftarrow\!-\!-}} & S(M)
\end{array}
$$

Then χ extends φ, since we identified $a \in M$ with $\pi(a) \in S(M)$ for every

$a \in M$; as in the proof of 2.4, χ is the only algebra homomorphism with this property, since $S(M)$ is generated by M. $\square$

If M is the free R-module on a set X, then $S(M)$ is the free commutative R-algebra on the set X: every mapping of X into a commutative R-algebra A extends uniquely to a module homomorphism of M into A and thence to an algebra homomorphism of $S(M)$ into A. Since polynomial rings already have this universal property, we obtain the following:

Corollary 3.4. *If M is the free R-module on a set $(x_i)_{i \in I}$, then $S(M)$ is isomorphic to the polynomial ring $R[(X_i)_{i \in I}]$.*

Corollary 3.5. *If M is the free R-module on a totally ordered set X, then $S(M)$ is a free R-module, with a basis that consists of all $x_1 \cdots x_n$ with $n \geqq 0$, $x_1, \ldots, x_n \in X$, and $x_1 \leqq \cdots \leqq x_n$.*

Proof. The monomials of $R[(X_i)_{i \in I}]$ constitute a basis of $R[(X_i)_{i \in I}]$ as an R-module. Hence $S(M)$ is a free R-module when M is a free R-module, by 3.4. If $X = (x_i)_{i \in I}$ is totally ordered, every monomial in $R[(X_i)_{i \in I}]$ can be rewritten uniquely in the form $X_1 \cdots X_n$ with $n \geqq 0$, $X_1, \ldots, X_n \in X$, and $X_1 \leqq \cdots \leqq X_n$ (with $X_1 \cdots X_n = 1 \in R$ if $n = 0$, as usual); this yields the basis in the statement. $\square$

Description. We now give a more precise description of the ideal I and of $S(M)$ when M is not free.

First we note that the construction of $T(M)$ works because multiplication in an R-algebra is n-linear and the tensor map $M^n \longrightarrow T^n(M) = M \otimes \cdots \otimes M$ is universal for n-linear mappings. Multiplication in a commutative R-algebra yields n-linear mappings f that are *symmetric* ($f(a_{\sigma 1}, \ldots, a_{\sigma n}) = f(a_1, \ldots, a_n)$ for every permutation σ).

Lemma 3.6. *Let $S^n(M) = T^n(M)/I_n$, where $n \geq 2$ and I_n is the submodule of $T^n(M) = M \otimes \cdots \otimes M$ generated by all $a_{\sigma 1} \otimes \cdots \otimes a_{\sigma n} - a_1 \otimes \cdots \otimes a_n$, where $a_1, \ldots, a_n \in M$ and σ is a permutation of $\{1, \ldots, n\}$. The mapping $\mu_n : M^n \longrightarrow T^n(M) \longrightarrow S^n(M)$ is symmetric and n-linear. For every symmetric n-linear mapping v of M^n into an R-module N there is a unique module homomorphism $\overline{v} : S^n(M) \longrightarrow N$ such that $v = \overline{v} \circ \mu_n$:*

$$
\begin{array}{ccc}
M^n & \xrightarrow{\otimes} & T^n(M) \\
{\scriptstyle v} \downarrow & \swarrow{\scriptstyle \xi} & \downarrow {\scriptstyle \pi} \\
N & \xleftarrow[\overline{v}]{} & S^n(M)
\end{array}
$$

Proof. Let $\pi : T^n(M) \longrightarrow S^n(M)$ be the projection. Then $\mu_n(a_1, \ldots, a_n) = \pi(a_1 \otimes \cdots \otimes a_n)$ is symmetric, by the choice of I_n, and n-linear.

Let $v : M^n \longrightarrow N$ be n-linear. There is a module homomorphism $\xi : M^n \longrightarrow T^n(M)$ such that $\xi(a_1 \otimes \cdots \otimes a_n) = v(a_1, \ldots, a_n)$ for all $a_1, \ldots, a_n \in M$. If v is symmetric, then $a_{\sigma 1} \otimes \cdots \otimes a_{\sigma n} - a_1 \otimes \cdots \otimes a_n \in \text{Ker } \xi$

for all $a_1, \ldots, a_n \in M$ and permutations σ; hence $I_n \subseteq \mathrm{Ker}\, \xi$, and ξ factors through π: there is a module homomorphism $\overline{\nu} : S^n(M) \longrightarrow N$ such that $\overline{\nu} \circ \pi = \xi$. Then $\overline{\nu} \circ \mu_n = \nu$; $\overline{\nu}$ is the only module homomorphism with this property since $S^n(M)$ is generated by all $\mu_n(a_1, \ldots, a_n)$. $\square$

The module $S^n(M)$ is the nth *symmetric power* of M. It is convenient to let $I_0 = 0$, $I_1 = 0$, so that $S^0(M) = R$ and $S^1(M) = M$.

Proposition 3.7. *For every R-module M, I is a graded ideal of $T(M)$, $I = \bigoplus_{n \geq 0} I_n$, and $S(M) = T(M)/I = \bigoplus_{n \geq 0} S^n(M)$ is a graded R-algebra.*

Proof. Let $a_1, \ldots, a_n \in M$, where $n \geq 2$. Since $S(M)$ is commutative, $a_{\sigma 1} \cdots a_{\sigma n} = a_1 \cdots a_n$ for every permutation σ, and $a_{\sigma 1} \otimes \cdots \otimes a_{\sigma n} - a_1 \otimes \cdots \otimes a_n \in I$ for every permutation σ. Hence $I_n \subseteq I$, and $\bigoplus_{n \geq 0} I_n \subseteq I$.

To prove the converse inclusion we note that

$$(a_{\sigma 1} \otimes \cdots \otimes a_{\sigma m} - a_1 \otimes \cdots \otimes a_m) \otimes b_{m+1} \otimes \cdots \otimes b_{m+n} \in I_{m+n}$$

for all $m, n > 0$, permutations σ, and $a_1, \ldots, a_m, b_{m+1}, \ldots, b_{m+n} \in M$, since $\sigma 1, \ldots, \sigma m, m+1, \ldots, m+n$ is a permutation of $1, \ldots, m+n$. Hence

$$a_{\sigma 1} \otimes \cdots \otimes a_{\sigma m} \otimes u - a_1 \otimes \cdots \otimes a_m \otimes u \in I_{m+n}$$

for all $u \in T^n(M)$. Therefore $t \otimes u \in I_{m+n}$ for all $t \in I_m$ and $u \in T^n(M)$. Similarly, $t \otimes u \in I_{m+n}$ for all $t \in T^m(M)$ and $u \in I_n$. This also holds if $m = 0$ or $n = 0$. Therefore $\bigoplus_{n \geq 0} I_n$ is an ideal of $T(M)$.

If $t = a_1 \otimes \cdots \otimes a_m$ and $u = b_{m+1} \otimes \cdots \otimes b_{m+n}$ are generators of $T^m(M)$ and $T^n(M)$, then $t \otimes u - u \otimes t \in I_{m+n}$, since $m+1, \ldots, m+n, 1, \ldots, m$ is a permutation of $1, \ldots, m+n$. Hence $t \otimes u - u \otimes t \in I_{m+n}$ for all $t \in T^m(M)$ and $u \in T^n(M)$; $t \otimes u - u \otimes t \in \bigoplus_{n \geq 0} I_n$ for all $t \in T^m(M)$ and $u \in T(M)$; and $t \otimes u - u \otimes t \in \bigoplus_{n \geq 0} I_n$ for all $t, u \in T(M)$. Since $\bigoplus_{n \geq 0} I_n$ is an ideal of $T(M)$, this implies $I \subseteq \bigoplus_{n \geq 0} I_n$. $\square$

4. The Exterior Algebra

The exterior algebra of an R-module M is the "greatest" algebra generated by M in that $ab = -ba$ (and $a^2 = 0$) for all $a, b \in M$, according to its universal property.

Exterior algebras originate in the calculus of differential forms (due to Grassmann [1844]). If $\omega = P\,dx + Q\,dy + R\,dz$ is a differential form in three variables, the "exterior" differential $d\omega = (R_y - Q_z)\,dy\,dz + (P_z - R_x)\,dz\,dx + (Q_x - P_y)\,dx\,dy$ can be found by substituting $dP = P_x\,dx + P_y\,dy + P_z\,dz$, $dQ = Q_x\,dx + Q_y\,dy + Q_z\,dz$, $dR = R_x\,dx + R_y\,dy + R_z\,dz$ into ω, and using the rules $dx\,dx = dy\,dy = dz\,dz = 0$, $dz\,dy = -dy\,dz$, $dx\,dz = -dz\,dx$, $dy\,dx = -dx\,dy$. Differential forms are multiplied using the same rules. Terms in these products are usually separated by wedges $\wedge$ to distinguish them from ordinary products. Then $\omega \wedge \omega = 0$ for every first order differential form.

A submodule M of an algebra is *anticommutative* in that algebra when $a^2 = 0$ for all $a \in M$; then $a^2 + ab + ba + b^2 = (a + b)^2 = 0$ and $ba = -ab$ for all $a, b \in M$. (The converse holds when 2 is a unit.) The exterior algebra of M (named after exterior differentiation) is the "greatest" algebra generated by M in which M is anticommutative; examples include algebras of differential forms.

Construction. As before, R is a commutative ring [with identity]; all modules are unital; all algebras and tensor products are over R. By Corollary 2.5, the exterior algebra must be, up to isomorphism, the quotient algebra of $T(M)$ by some two-sided ideal J. There is an obvious candidate for J:

Definition. The exterior algebra *of an R-module M is the quotient algebra* $\Lambda(M) = T(M)/J$, *where J is the ideal of* $T(M)$ *generated by all* $a \otimes a$ *with* $a \in M$.

We have $J \subseteq \bigoplus_{n \geqq 2} T^n(M)$, since the latter is an ideal of $T(M)$ and contains $a \otimes a$ for all $a \in M$. Hence the projection $T(M) \longrightarrow \Lambda(M) = T(M)/J$ is injective on $R \oplus M$; therefore we may identity R and M with their images in $\Lambda(M)$. The product $a_1 \wedge \cdots \wedge a_n$ of $a_1, \ldots, a_n \in M$ in $\Lambda(M)$ is the image in $\Lambda(M)$ of their product $a_1 \otimes \cdots \otimes a_n$ in $T(M)$.

Proposition **4.1.** *The exterior algebra of a* (unital) *R-module M is an R-algebra generated by M, in which M is anticommutative.*

The definition of $\Lambda(M)$ also yields a universal property:

Proposition **4.2.** *Every module homomorphism φ of an R-module M into an R-algebra A in which $\varphi(M)$ is anticommutative extends uniquely to an algebra homomorphism of $\Lambda(M)$ into A:*

$$
\begin{array}{ccc}
M & \xrightarrow{\subseteq} & T(M) \\
{\scriptstyle \varphi} \downarrow & {\scriptstyle \psi} \nearrow & \downarrow {\scriptstyle \pi} \\
A & \xleftarrow{\quad \chi \quad} & \Lambda(M)
\end{array}
$$

Proof. Let $\pi : T(M) \longrightarrow \Lambda(M)$ be the projection. By 2.4, φ extends to an algebra homomorphism $\psi : T(M) \longrightarrow A$. Since $\varphi(M)$ is anticommutative in A we have $\psi(a \otimes a) = \varphi(a)^2 = 0$ and $a \otimes a \in \mathrm{Ker}\,\psi$, for all $a \in M$. Therefore $J \subseteq \mathrm{Ker}\,\psi$, and $\psi = \chi \circ \pi$ for some algebra homomorphism $\chi : \Lambda(M) \longrightarrow A$, by 1.3. Then χ extends φ, since we identified $a \in M$ with $\pi(a) \in \Lambda(M)$ for every $a \in M$; χ is the only algebra homomorphism with this property, since $\Lambda(M)$ is generated by M. $\square$

Further results require a more precise description of the ideal I and of $\Lambda(M)$.

Alternating maps. In an R-algebra in which M is anticommutative, multiplication of elements of M yields n-linear mappings f that are *alternating*:

$f(a_1, \ldots, a_n) = 0$ whenever $a_i = a_j$ for some $i \neq j$, hence $f(a_{\tau 1}, \ldots, a_{\tau n}) = -f(a_1, \ldots, a_n)$ for every transposition τ.

Lemma **4.3.** *Let* $\Lambda^n(M) = T^n(M)/J_n$, *where* $n \geq 2$ *and* J_n *is the submodule of* $T^n(M) = M \otimes \cdots \otimes M$ *generated by all* $a_1 \otimes \cdots \otimes a_n$, *where* $a_1, \ldots, a_n \in M$ *and* $a_i = a_j$ *for some* $i \neq j$. *The mapping* $\mu_n : M^n \longrightarrow T^n(M) \longrightarrow \Lambda^n(M)$ *is n-linear and alternating, and for every n-linear alternating mapping* v *of* M^n *into an R-module* N *there is a unique module homomorphism* $\overline{v} : \Lambda^n(M) \longrightarrow N$ *such that* $v = \overline{v} \circ \mu_n$.

Proof. Let $\pi : T^n(M) \longrightarrow \Lambda^n(M)$ be the projection. Then $\mu_n(a_1, \ldots, a_n)$ $= \pi(a_1 \otimes \cdots \otimes a_n)$ is alternating, by the choice of J_n, and n-linear.

Let $v : M^n \longrightarrow N$ be n-linear. There is a module homomorphism $\xi :$ $T^n(M) \longrightarrow N$ such that $\xi(a_1 \otimes \cdots \otimes a_n) = v(a_1, \ldots, a_n)$ for all $a_1, \ldots, a_n$ $\in M$. If v is alternating, then $a_1 \otimes \cdots \otimes a_n \in \mathrm{Ker}\, \xi$ whenever $a_1, \ldots, a_n \in M$ and $a_i = a_j$ for some $i \neq j$; hence $J_n \subseteq \mathrm{Ker}\, \xi$ and there is a module homomorphism $\overline{v} : \Lambda^n(M) \longrightarrow N$ such that $\overline{v} \circ \pi = \xi$:

$$\begin{array}{ccc} M^n & \overset{\otimes}{\longrightarrow} & T^n(M) \\ {\scriptstyle v}\downarrow & \nearrow{\scriptstyle \xi} & \downarrow{\scriptstyle \pi} \\ N & \overset{}{\underset{\overline{v}}{\longleftarrow -\!-\!-}} & \Lambda^n(M) \end{array}$$

Then $\overline{v} \circ \mu_n = v$; $\overline{v}$ is the only module homomorphism with this property since $\Lambda^n(M)$ is generated by all $\mu_n(a_1, \ldots, a_n)$. $\square$

The module $\Lambda^n(M)$ is the *n*th *exterior power* of M. It is convenient to let $J_0 = 0$, $J_1 = 0$, so that $\Lambda^0(M) = R$ and $\Lambda^1(M) = M$.

Proposition **4.4.** *For every R-module* M, J *is a graded ideal of* $T(M)$, $J = \bigoplus_{n \geq 0} J_n$, *and* $\Lambda(M) = T(M)/J = \bigoplus_{n \geq 0} \Lambda^n(M)$ *is a graded R-algebra.*

Proof. Let $a_1, \ldots, a_m \in M$, where $m \geq 2$. Since M is anticommutative in $\Lambda(M)$, $a_i = a_j$ for some $i \neq j$ implies $a_1 \wedge \cdots \wedge a_m = \pm a_i \wedge a_j \wedge a_1 \wedge \cdots \wedge a_m = 0$ and $a_1 \otimes \cdots \otimes a_m \in J$. Hence $J_m \subseteq J$, and $\bigoplus_{n \geq 0} J_m \subseteq J$. The converse inclusion is proved as follows.

If $a_i = a_j$ for some $i \neq j$, then $a_1 \otimes \cdots \otimes a_m \otimes b_1 \otimes \cdots \otimes b_n \in J_{m+n}$ for all $b_1, \ldots, b_n \in M$; hence $a_1 \otimes \cdots \otimes a_m \otimes t \in J_{m+n}$ for all $t \in T^n(M)$ and $j \otimes t \in J_{m+n}$ for all $t \in T^n(M)$, and $j \in J_m$. Similarly, $t \otimes j \in J_{m+n}$ for all $t \in T^n(M)$ and $j \in J_m$. Hence $\bigoplus_{n \geq 0} J_n$ is an ideal of $T(M)$. Since $a \otimes a \in J_2 \subseteq \bigoplus_{n \geq 0} J_n$ for all $a \in M$, it follows that $J \subseteq \bigoplus_{n \geq 0} J_n$. $\square$

Free modules. We use the last two results to give a more precise description of $\Lambda(M)$ when M is free.

Proposition **4.5.** *If* $n \geq 2$ *and* M *is a free R-module with a totally ordered basis* X, *then* $\Lambda^n(M)$ *is a free R-module, with a basis that consists of all* $x_1 \wedge \cdots \wedge x_n$ *with* $x_1, \ldots, x_n \in X$ *and* $x_1 < x_2 < \cdots < x_n$.

Proof. Let $x_1, \ldots, x_n \in X$. If there is a *duplication* $x_i = x_j$ for some $i \neq j$, then $x_1 \otimes \cdots \otimes x_n \in J_n$. Since M is anticommutative in $\Lambda(M)$, $x_{\tau 1} \wedge \cdots \wedge x_{\tau n} = -x_1 \wedge \cdots \wedge x_n$ in $\Lambda(M)$ for every transposition τ; hence, for every permutation σ, $x_{\sigma 1} \wedge \cdots \wedge x_{\sigma n} = \mathrm{sgn}\, \sigma\, x_1 \wedge \cdots \wedge x_n$ and

$$x_{\sigma 1} \otimes \cdots \otimes x_{\sigma n} - \mathrm{sgn}\, \sigma\, x_1 \otimes \cdots \otimes x_n \in J \cap T^n(M) = J_n.$$

We show that J_n is the submodule of $T^n(M)$ generated by the set Y_n of all $x_1 \otimes \cdots \otimes x_n$ with a duplication and all $x_1 \otimes \cdots \otimes x_n - \mathrm{sgn}\, \sigma\, x_{\sigma 1} \otimes \cdots \otimes x_{\sigma n}$ where $x_1, \ldots, x_n$ are distinct. By the above, $Y_n \subseteq J_n$. We show that every generator $a_1 \otimes \cdots \otimes a_n$ of J_n (with $a_1, \ldots, a_n \in M$ and $a_j = a_k$ for some $j \neq k$) is a linear combination of elements of Y_n. Indeed, $a_j = \sum_{i \in I} r_{ij}\, x_i$ and

$$a_1 \otimes \cdots \otimes a_n = \sum_{i_1, \ldots, i_n \in I} r_{1 i_1} \cdots r_{n i_n}\, x_{i_1} \otimes \cdots \otimes x_{i_n} = S.$$

In the sum S, $x_{i_1} \otimes \cdots \otimes x_{i_n} \in Y_n$ whenever $x_{i_1}, \ldots, x_{i_n}$ are not all distinct. If $a_1 = a_2$, then $r_{1i} = r_{2i}$ for all i and the terms of S in which $x_{i_1}, \ldots, x_{i_n}$ are distinct can be grouped in pairs

$$r_{1 i_1} \cdots r_{n i_n}\, (x_{i_1} \otimes x_{i_2} \otimes x_{i_3} \otimes \cdots \otimes x_{i_n} + x_{i_2} \otimes x_{i_1} \otimes x_{i_3} \otimes \cdots \otimes x_{i_n}),$$

with $x_{i_1} \otimes x_{i_2} \otimes x_{i_3} \otimes \cdots \otimes x_{i_n} + x_{i_2} \otimes x_{i_1} \otimes x_{i_3} \otimes \cdots \otimes x_{i_n} \in Y_n$. Thus $a_1 \otimes \cdots \otimes a_n$ is a linear combination of elements of Y_n. This also holds whenever $a_j = a_k$ for some $j \neq k$; the proof is the same, except for the notation, which is much worse. Hence J_n is the submodule generated by Y_n.

To complete the proof, let B be the basis of $T^n(M)$ that consists of all $x_1 \otimes \cdots \otimes x_n$. The symmetric group S_n acts on B: if $b = x_1 \otimes \cdots \otimes x_n$, then $\sigma b = x_{\sigma 1} \otimes \cdots \otimes x_{\sigma n}$. If $b = x_1 \otimes \cdots \otimes x_n$ has no duplication (if $x_1, \ldots, x_n$ are all distinct), then the orbit of b contains exactly one *strictly ascending* element $x_{\sigma 1} \otimes \cdots \otimes x_{\sigma n}$ such that $x_{\sigma 1} < \cdots < x_{\sigma n}$. Let

$$C = \{\, c \in B \mid c \text{ is strictly ascending} \,\}, \quad D = \{\, d \in B \mid d \text{ has a duplication} \,\},$$

and let B' be the set of all $b' = b - \varepsilon c$ where $b \in B \backslash (C \cup D)$, $c = \sigma b$ is strictly ascending, and $\varepsilon = \mathrm{sgn}\, \sigma$. Then $C \cup D \cup B'$ is a basis of $T^n(M)$. An element of Y_n is either in D or is a difference of two elements of B'; hence J_n is the submodule generated by $D \cup B' \subseteq Y_n$. Therefore $T^n(M) = J_n \oplus K_n$, where K_n is the submodule generated by C, and there is an isomorphism $K_n \cong \Lambda^n(M)$, which sends the basis C of K_n to a basis of $\Lambda^n(M)$ that consists of all $x_1 \wedge \cdots \wedge x_n$ with $x_1, \ldots, x_n \in X$ and $x_1 < x_2 < \cdots < x_n$. $\square$

Corollary **4.6.** *If M is the free R-module on a totally ordered set X, then $\Lambda(M)$ is a free R-module, with a basis that consists of all $x_1 \wedge \cdots \wedge x_n$ with $n \geqq 2$, $x_1, \ldots, x_n \in X$, and $x_1 < \cdots < x_n$.*

Exterior algebras provide an alternative approach to determinants that does not rely on Gauss elimination. Let $M = (_R R)^n$ and let $e_1, \ldots, e_n$ be its standard basis, totally ordered by $e_1 < e_2 < \cdots < e_n$. Then $\Lambda^n(M)$ is free on $\{e_1 \wedge \cdots \wedge e_n\}$, by Proposition 4.5. Hence there is an n-linear alternating form

$\delta(a_1, \ldots, a_n)$ such that $a_1 \wedge \cdots \wedge a_n = \delta \, e_1 \wedge \cdots \wedge e_n$; δ is nontrivial, since $\delta(e_1, \ldots, e_n) = 1$, and every n-linear alternating form is a multiple of δ, by Lemma 4.3. Readers will verify that δ is the usual determinant.

Exercises

1. Show that $v \wedge u = (-1)^{mn} \, u \wedge v$ for all $u \in \Lambda^m(M)$ and $v \in \Lambda^n(M)$.

2. Let M be free of rank r. Show that $\Lambda^n(M)$ has rank $\binom{r}{n}$ and that $\Lambda(M)$ has rank 2^r.

3. Let M be free of rank n, with a basis $x_1 < x_2 < \cdots < x_n$. Let $a_1, \ldots, a_n \in M$, $a_i = \sum_j r_{ij} \, e_j$. Show that $a_1 \wedge \cdots \wedge a_n = \det(r_{ij}) \, x_1 \wedge \cdots \wedge x_n$.

4. Let M be free with a totally ordered basis $X = (x_j)_{j \in J}$. Let $a_1, \ldots, a_n \in M$, $a_i = \sum_j r_{ij} \, x_j$. Show that $a_1 \wedge \cdots \wedge a_n = \sum \det(r_{ij_i}) \, x_{j_1} \wedge \cdots \wedge x_{j_n}$, where the sum has one term for every strictly increasing sequence $x_{j_1} < x_{j_2} < \cdots < x_{j_n}$.

5. Show that every homomorphism $M \longrightarrow N$ of R-modules extends uniquely to a homomorphism $\Lambda(M) \longrightarrow \Lambda(N)$ of graded R-algebras.

5. Tensor Products of Algebras

Tensor products are a fundamental tool in the study of algebras. This section contains basic properties and (in the exercises) applications to tensor, symmetric, and exterior algebras; the next section has applications to field theory. In what follows, all algebras and tensor products are over a commutative ring R.

Proposition **5.1.** *If A and B are R-algebras, then $A \otimes B$ is an R-algebra, in which* $(a \otimes b)(a' \otimes b') = aa' \otimes bb'$ *for all $a, a' \in A$ and $b, b' \in B$.*

Proof. The mapping $(a, b, a', b') \longmapsto aa' \otimes bb'$ of $A \times B \times A \times B$ into $A \otimes B$ is multilinear, since $\otimes$ and the multiplications on A and B are bilinear. Hence there is a unique module homomorphism $\mu : A \otimes B \otimes A \otimes B \longrightarrow A \otimes B$ such that $\mu(a \otimes b \otimes a' \otimes b') = aa' \otimes bb'$ for all a, a', b, b'. A bilinear multiplication on $A \otimes B$ is then defined by $tt' = \mu(t \otimes t')$, for all $t, t' \in A \otimes B$; in particular, $(a \otimes b)(a' \otimes b') = aa' \otimes bb'$. This multiplication is associative: when $t = a \otimes b$, $t' = a' \otimes b'$, $t'' = a'' \otimes b''$, then

$$(tt') \, t'' = (aa') \, a'' \otimes (bb') \, b'' = a \, (a'a'') \otimes b \, (bb'') = t \, (t't'');$$

since every element t of $A \otimes B$ is a finite sum $t = \sum_i a_i \otimes b_i$, it follows that $(tt') \, t'' = t \, (t't'')$ for all $t, t', t'' \in A \otimes B$. Also, $1 \otimes 1$ is the identity element of $A \otimes B$: indeed, $(1 \otimes 1)(a \otimes b) = a \otimes b = (a \otimes b)(1 \otimes 1)$ for all a, b, whence $(1 \otimes 1) \, t = t = t \, (1 \otimes 1)$ for all $t \in A \otimes B$. $\square$

Definition. Let A and B be R-algebras. The R-algebra $A \otimes B$ in Proposition 5.1 is the tensor product of A and B.

Properties. Tensor products of algebras inherit properties from tensor products of modules.

Proposition 5.2. *If $\varphi : A \longrightarrow A'$ and $\psi : B \longrightarrow B'$ are homomorphisms of R-algebras, then so is $\varphi \otimes \psi : A \otimes B \longrightarrow A' \otimes B'$.*

Proof. For all $a, a' \in A$ and $b, b' \in B$,

$$(\varphi \otimes \psi)\big((a \otimes b)(a' \otimes b')\big) = \varphi(aa') \otimes \psi(bb')$$
$$= \varphi(a)\,\varphi(a') \otimes \psi(b)\,\psi(b') = \big((\varphi \otimes \psi)(a \otimes b)\big)\big((\varphi \otimes \psi)(a' \otimes b')\big).$$

Therefore $(\varphi \otimes \psi)(tt') = (\varphi \otimes \psi)(t)\,(\varphi \otimes \psi)(t')$ for all $t, t' \in A \otimes B$. Moreover, $(\varphi \otimes \psi)(1 \otimes 1) = 1 \otimes 1$. $\square$

For all R-algebras A and B, $A \otimes -$ and $- \otimes B$ are now covariant functors from R-algebras to R-algebras, and $- \otimes -$ is a bifunctor. Readers may show that the canonical isomorphisms $R \otimes A \cong A$, $A \otimes B \cong B \otimes A$, $(A \otimes B) \otimes C \cong A \otimes (B \otimes C)$ are algebra isomorphisms when A, B, C are R-algebras.

Next, we note that the algebra $A \otimes B$ has a bimodule structure: since the multiplications on A and B are bilinear, A is a left A-, right R-bimodule, B is a left R-, right B-bimodule, and $A \otimes B$ is a left A-, right B-bimodule, in which $a\,(a' \otimes b) = aa' \otimes b$ and $(a \otimes b)\,b' = a \otimes bb'$ for all $a, a' \in A$ and $b, b' \in B$.

Proposition 5.3. *If A and B are commutative R-algebras, then $A \otimes B$ is a commutative R-algebra, and is also an A-algebra and a B-algebra.*

Proof. The algebra $A \otimes B$ is commutative since its generators $a \otimes b$ commute with each other. Its multiplication is bilinear over A: $(at)\,t' = a\,(tt') = t\,(at')$ for all $a \in A$ and $t, t' \in A \otimes B$, since commutativity in A yields

$$\big(a\,(a' \otimes b')\big)(a'' \otimes b'') = aa'a'' \otimes b'b'' = (a' \otimes b')\big(a\,(a'' \otimes b'')\big);$$

similarly, $t\,(t'b) = (tt')\,b = (tb)\,t'$ for all $t, t' \in A \otimes B$ and $b \in B$. $\square$

Proposition 5.4. *If A is free as an R-module, with basis $(e_i)_{i \in I}$, then $A \otimes B$ is free as a right B-module, with basis $(e_i \otimes 1)_{i \in I}$.*

Proof. As R-modules, $A \cong \bigoplus_{i \in I} {}_R R$ and $A \otimes B \cong \big(\bigoplus_{i \in I} {}_R R\big) \otimes B \cong \bigoplus_{i \in I} B$; when $a = \sum_{i \in I} r_i\,e_i$, these isomorphisms send a to $(r_i)_{i \in I}$ and $a \otimes b$ to $(r_i\,b)_{i \in I}$. The latter isomorphism θ preserves the right action of B; hence $A \otimes B$ is a free right B-module. Next, $e_j = \sum_{i \in I} \delta_{ij}\,e_i$, where $\delta_{ij} = 1$ if $i = j$, $\delta_{ij} = 0$ if $i \neq j$; hence θ sends $e_j \otimes 1$ to $f_j = (\delta_{ij})_{i \in I}$. Now, $(f_i)_{i \in I}$ is the standard basis of the free right B-module $\bigoplus_{i \in I} B$; hence $(e_i \otimes 1)_{i \in I}$ is a basis of the right B-module $A \otimes B$. $\square$

Finally, the algebra $A \otimes B$ comes with canonical algebra homomorphisms $\iota : A \longrightarrow A \otimes B$, $\kappa : B \longrightarrow A \otimes B$, defined by $\iota(a) = a \otimes 1$, $\kappa(b) = 1 \otimes b$. We see that ι and κ agree on R, and that $\iota(a)$ and $\kappa(b)$ commute for all $a \in A$ and $b \in B$. Readers may show that ι and κ are injective in some cases:

Proposition 5.5. *If A is free as an R-module, then $\iota : A \longrightarrow A \otimes B$ is injective. If B is free as an R-module, then $\kappa : B \longrightarrow A \otimes B$ is injective. If R is a field, then ι and κ are always injective.*

The tensor product of algebras and its "injections" have a universal property:

Proposition 5.6. *Let A and B be R-algebras and let $\iota : A \longrightarrow A \otimes B$, $\kappa : B \longrightarrow A \otimes B$ be the canonical homomorphisms. For every R-algebra C and algebra homomorphisms $\varphi : A \longrightarrow C$, $\psi : B \longrightarrow C$ such that $\varphi(a)$ and $\psi(b)$ commute for all $a \in A$ and $b \in B$, there is a unique algebra homomorphism $\chi : A \otimes B \longrightarrow C$ such that $\chi \circ \iota = \varphi$ and $\chi \circ \kappa = \psi$, which sends $a \otimes b$ to $\varphi(a)\,\psi(b)$:*

$$A \overset{\iota}{\longrightarrow} A \otimes B \overset{\kappa}{\longleftarrow} B$$

$$\varphi \searrow \quad \downarrow \chi \quad \swarrow \psi$$

$$C$$

Proof. If χ is an algebra homomorphism and $\chi \circ \iota = \varphi$ and $\chi \circ \kappa = \psi$, then $\chi\,(a \otimes b) = \chi\,\big(\iota(a)\,\kappa(b)\big) = \varphi(a)\,\psi(b)$ for all a, b; therefore χ is unique.

Conversely, the bilinear mapping $(a, b) \longmapsto \varphi(a)\,\psi(b)$ of $A \times B$ into C induces a unique module homomorphism $\chi : A \otimes B \longrightarrow C$ such that $\chi\,(a \otimes b) = \varphi(a)\,\psi(b)$ for all $a \in A$ and $b \in B$. Then $\chi \circ \iota = \varphi$, $\chi \circ \kappa = \psi$, and $\chi(1) = \varphi(1)\,\psi(1) = 1$. For all a, a', b, b',

$$\chi\big((a \otimes b)(a' \otimes b')\big) = \varphi(a)\,\psi(b)\,\varphi(a')\,\psi(b')$$
$$= \varphi(a)\,\varphi(a')\,\psi(b)\,\psi(b') = \chi\,(a \otimes b)\,\chi\,(a' \otimes b'),$$

since $\varphi(a')$ and $\psi(b)$ commute; therefore $\chi(tt') = \chi(t)\,\chi(t')$ for all $t, t' \in A \otimes B$, and χ is an algebra homomorphism. $\square$

We note a first consequence of Proposition 5.6:

Proposition 5.7 (Noether [1929]). *Let A and B be R-algebras. For every abelian group M there is a one-to-one correspondence between the left A-, right B-bimodule structures on M (with the same actions of R) and the left $A \otimes B^{\mathrm{op}}$-module structures on M.*

Proof. Let M be an left A-, right B-bimodule, with the same actions of R on M, so that M is, in particular, an R-module. A left A-, right B-bimodule structure on M consists of ring homomorphisms $\alpha : A \longrightarrow \mathrm{End}_{\mathbb{Z}}(M)$ and $\beta : B \longrightarrow \mathrm{End}_{\mathbb{Z}}^{\mathrm{op}}(M)$, equivalently $\beta : B^{\mathrm{op}} \longrightarrow \mathrm{End}_{\mathbb{Z}}(M)$, such that $\alpha(a)$ and $\beta(b)$ commute for all $a \in A$ and $b \in B$. Since R is central in A and B, all $a \in A$ and $b \in B$ act on M by R-endomorphisms, and α, β are algebra homomorphisms A, $B^{\mathrm{op}} \longrightarrow \mathrm{End}_R(M)$. By 5.6 they induce a unique algebra homomorphism $\gamma : A \otimes B^{\mathrm{op}} \longrightarrow \mathrm{End}_R(M)$ such that $\gamma \circ \iota = \alpha$, $\gamma \circ \kappa = \beta$, which is, in particular, a left $A \otimes B^{\mathrm{op}}$ module structure on M.

Conversely, let $\gamma : A \otimes B^{\mathrm{op}} \longrightarrow \mathrm{End}_{\mathbb{Z}}(M)$ be a left $A \otimes B^{\mathrm{op}}$ module structure on M. Then $\alpha = \gamma \circ \iota : A \longrightarrow \mathrm{End}_{\mathbb{Z}}(M)$ is a left A-module structure on M and $\beta = \gamma \circ \kappa : B^{\mathrm{op}} \longrightarrow \mathrm{End}_{\mathbb{Z}}(M)$ is a right B-module structure on M, which induces the same R-module structure on M as α, since ι and κ agree on R. Moreover, $\alpha(a)$ and $\beta(b)$ commute for all $a \in A$ and $b \in B$, since $\iota(a)$ and $\kappa(b)$ commute for all $a \in A$ and $b \in B$. Hence α

and β constitute a left A-, right B-bimodule structure on M. That we have constructed a one-to-one correspondence now follows from uniqueness in 5.6. $\square$

Exercises

1. Let A, B, C be R-algebras. Show that the canonical isomorphisms $R \otimes A \cong A$, $A \otimes B \cong B \otimes A$, $(A \otimes B) \otimes C \cong A \otimes (B \otimes C)$ are algebra isomorphisms.

2. Let A and B be R-algebras. Show that $\iota : A \longrightarrow A \otimes B$ is injective when A is free as an R-module.

3. Give an example in which $\iota : A \longrightarrow A \otimes B$ and $\kappa : B \longrightarrow A \otimes B$ are not injective.

4. Show that $S(M \oplus N) \cong S(M) \otimes S(N)$, for all R-modules M and N.

In the next three exercises, $K \longrightarrow M \longrightarrow N \longrightarrow 0$ is an exact sequence of R-modules.

5. Construct an exact sequence $T(M) \otimes K \otimes T(M) \longrightarrow T(M) \longrightarrow T(N) \longrightarrow 0$.

6. Construct an exact sequence $K \otimes S(M) \longrightarrow S(M) \longrightarrow S(N) \longrightarrow 0$.

7. Construct an exact sequence $K \otimes \Lambda(M) \longrightarrow \Lambda(M) \longrightarrow \Lambda(N) \longrightarrow 0$.

In the next three exercises, M is an R-module and A is an R-algebra. Prove the following:

8. There is an isomorphism $T_A(A \otimes_R M) \cong A \otimes_R T_R(M)$ that is natural in M.

9. There is an isomorphism $S_A(A \otimes_R M) \cong A \otimes_R S_R(M)$ that is natural in M.

10. There is an isomorphism $\Lambda_A(A \otimes_R M) \cong A \otimes_R \Lambda_R(M)$ that is natural in M.

11. Let $A = \bigoplus_{n \geq 0} A_n$ and $B = \bigoplus_{n \geq 0} B_n$ be graded algebras. Construct a *skew tensor product* $A \otimes' B$, in which $(a \otimes b)(a' \otimes b') = (-1)^{\ell m} (aa' \otimes bb')$ when $a \in A_k$, $a' \in A_\ell$, $b \in B_m$, $b' \in B_n$.

12. Show that $\Lambda(M \oplus N) \cong \Lambda(M) \otimes' \Lambda(N)$ for all R-modules M and N, using the skew tensor product in the previous exercise.

6. Tensor Products of Fields

A field extension of a field K is, in particular, a K-algebra. Hence any two field extensions of K have a tensor product that is a K-algebra. This section gives a few basic properties and examples, with applications to separability.

In what follows, K is any given field; unless otherwise specified, all algebras and tensor products are over K. Since a ring extension of K is a K-algebra, it follows from 5.1, 5.3 that any two commutative ring extensions of K have a tensor product over K, which is a commutative K-algebra. In particular, any two field extensions E and F of K have a tensor product $E \otimes F$ over K, which is a commutative K-algebra (but not, in general, a field, or even a domain, as we shall see). The canonical homomorphisms $E \longrightarrow E \otimes F$, $F \longrightarrow E \otimes F$ are injective, by 5.5, so that $E \otimes F$ can be regarded as a ring extension of both E and F. Since a field extension of K is free as a K-module, hence flat, $E \otimes F \longrightarrow E' \otimes F$ is injective whenever E is a subfield of E'.

We give two examples.

Proposition **6.1.** *Let* $K \subseteq E$ *be any field extension. Let* α *be algebraic over* K *and let* $q = \mathrm{Irr}\,(\alpha : K) = q_1^{m_1} \cdots q_r^{m_r}$, *where* $q_1, \ldots, q_r \in E[X]$ *are distinct monic irreducible polynomials. Then*

$$E \otimes_K K(\alpha) \cong E[X]/(q_1^{m_1}) \times \cdots \times E[X]/(q_r^{m_r}).$$

Proof. Readers will verify that there is an isomorphism $E \otimes K[X] \cong E[X]$ that sends $\gamma \otimes \left(\sum_{n \geq 0} a_n X^n \right)$ to $\sum_{n \geq 0} \gamma a_n X^n$. Hence the inclusion $(q) \longrightarrow K[X]$ induces a homomorphism $\xi : E \otimes (q) \longrightarrow E \otimes K[X] \cong E[X]$, which sends $\gamma \otimes \left(\sum_{n \geq 0} a_n X^n \right)$ to $\sum_{n \geq 0} \gamma a_n X^n$ whenever $\sum_{n \geq 0} a_n X^n$ is a multiple of $q(X)$. Then ξ sends $\gamma \otimes X^m q$ to $\gamma X^m q$ and $\mathrm{Im}\,\xi$ consists of all multiples of q in $E[X]$. The exact sequence $(q) \longrightarrow K[X] \longrightarrow K(\alpha) \longrightarrow 0$ induces exact sequences $E \otimes (q) \longrightarrow E \otimes K[X] \longrightarrow E \otimes K(\alpha) \longrightarrow 0$ and

$$0 \longrightarrow \mathrm{Im}\,\xi \longrightarrow E[X] \longrightarrow E \otimes K(\alpha) \longrightarrow 0.$$

Hence $E \otimes K(\alpha) \cong E[X]/\mathrm{Im}\,\xi$. In $E[X]$, $\mathrm{Im}\,\xi = (q_1^{m_1}) \cap \cdots \cap (q_r^{m_r})$, with $(q_i^{m_i}) + (q_j^{m_j}) = E[X]$ when $i \neq j$, since $q_i^{m_i}$ and $q_j^{m_j}$ are relatively prime; by the Chinese remainder theorem, $E[X]/\mathrm{Im}\,\xi \cong E[X]/(q_1^{m_1}) \times \cdots \times E[X]/(q_r^{m_r})$. $\square$

Using 6.1, readers will easily construct a tensor product of fields that is not a field, in fact, contains zero divisors and nontrivial nilpotent elements.

Proposition **6.2.** *For every field extension* $K \subseteq E$, $E \otimes_K K((X_i)_{i \in I})$ *is a domain, whose field of fractions is isomorphic to* $E((X_i)_{i \in I})$.

Proof. Just this once, let $X = (X_i)_{i \in I}$. As before, $a_m X^m$ denotes the monomial $a_m \prod_{i \in I} X_i^{m_i}$ of $K[X]$, and similarly in $E[X]$. Readers will verify that there is an isomorphism $E \otimes K[X] \cong E[X]$ that sends $\sum_m (\gamma a_m \otimes X^m) = \gamma \otimes \left(\sum_m a_m X^m \right)$ to $\sum_m \gamma a_m X^m$. The inclusion $K[X] \longrightarrow K(X)$ induces a monomorphism $E[X] \cong E \otimes K[X] \longrightarrow E \otimes K(X)$, which sends $\sum_m \alpha_m X^m$ to $\sum_m \alpha_m \otimes X^m$. Identify $\sum_m \alpha_m X^m$ and $\sum_m \alpha_m \otimes X^m$, so that $E[X]$ becomes a subalgebra of $E \otimes K(X)$.

Every $t \in E \otimes K(X)$ is a finite sum $t = \sum_i \alpha_i \otimes (f_i/g_i)$. Rewriting all $f_i/g_i \in K(X)$ with a common denominator $g \in K[X]$, $g \neq 0$, yields $t = \sum_i \alpha_i \otimes (f_i/g) = (\sum_i \alpha_i \otimes f_i)(1/g)$ in the $K(X)$-module $E \otimes K(X)$, so that $t = f/g$ for some $f \in E[X]$ and $0 \neq g \in K[X]$. Moreover, in the $K(X)$-module $E \otimes K(X)$, $f/g = f'/g'$ if and only if $g'f = gf'$ in $E \otimes K[X]$, if and only if $g'f = gf'$ in $E[X]$. Hence $E \otimes K(X)$ is isomorphic to the ring of fractions $R = S^{-1} E[X]$, where S is the proper multiplicative subset of all nonzero $g \in K[X]$. Therefore $E \otimes K(X)$ is a domain, and its field of fractions is isomorphic to that of R, which is $E(X)$ since $E[X] \subseteq R \subseteq E(X)$. $\square$

Separability. By MacLane's theorem IV.9.7, E is separable over K if and only if E and K^{1/p^∞} are linearly disjoint over K.

Tensor products yield another definition of linear disjointness. If $K \subseteq E \subseteq L$ and $K \subseteq F \subseteq L$ are fields, then E and F are linearly disjoint over K if and only if there exists a basis of E over K that is linearly independent over F, if and only if there exists a basis of F over K that is linearly independent over E (by Propositions IV.9.1 and IV.9.2). By Proposition 5.6, the inclusion homomorphisms of E and F into their composite $EF \subseteq L$ induce a multiplication homomorphism μ of $E \otimes F$ into EF, which sends $\alpha \otimes \beta$ to $\alpha\beta$ and is a homomorphism of E-modules and of F-modules.

Proposition 6.3. If $K \subseteq E \subseteq L$ and $K \subseteq F \subseteq L$ are fields, then E and F are linearly disjoint over K if and only if the homomorphism $E \otimes F \longrightarrow EF \subseteq L$ is injective.

Proof. Let $(\alpha_i)_{i \in I}$ be a basis of E over K. By 5.4, $(\alpha_i \otimes 1)_{i \in I}$ is a basis of $E \otimes F$ over F, which μ sends back to $(\alpha_i)_{i \in I}$. If μ is injective, then $(\alpha_i)_{i \in I} = \mu\big((\alpha_i \otimes 1)_{i \in I}\big)$ is linearly independent over F, hence E and F are linearly disjoint over K. Conversely, if E and F are linearly disjoint over K, then μ sends the basis $(\alpha_i \otimes 1)_{i \in I}$ of $E \otimes F$ over F to a family $(\alpha_i)_{i \in I}$ that is linearly independent over F; therefore μ is injective. $\square$

Definition. A commutative ring R is reduced when it contains no nonzero nilpotent element; equivalently, when its nilradical is 0.

Corollary 6.4. Let $K \subseteq E$ and L be fields. If E and F are linearly disjoint over K and $E \otimes_K L$ is reduced, then $EF \otimes_F L$ is reduced.

Proof. By 6.3, $E \otimes_K F \longrightarrow EF$ is injective, so that $E \otimes_K F$ is a domain; since $E \otimes_K F$ contains both E and F, its field of fractions is EF. Now, every $t \in EF \otimes_F L$ is a finite sum $t = \sum_{1 \le i \le n} \alpha_i \otimes \gamma_i$ in which $\alpha_1, \ldots, \alpha_n \in EF$ can be rewritten with a common denominator $\alpha_i = \beta_i/\delta$, where $\beta_i, \delta \in E \otimes_K F$. If $t = (1/\delta)\big(\sum_i \beta_i \otimes \gamma_i\big)$ is nilpotent, then $u = \sum_i \beta_i \otimes \gamma_i \in (E \otimes_K F) \otimes_F L \cong E \otimes_K L$ is nilpotent, $u = 0$, and $t = 0$. $\square$

We can now prove an souped-up version of MacLane's theorem.

Theorem 6.5 (MacLane). Let K be a field of characteristic $p \ne 0$. For a field extension $K \subseteq E$ the following conditions are equivalent:

(1) *E is separable over K;*

(2) *for every field extension F of K the ring $E \otimes_K F$ is reduced;*

(3) *the ring $E \otimes_K K^{1/p^\infty}$ is reduced;*

(4) *E and K^{1/p^∞} are linearly disjoint over K.*

Proof. (2) implies (3) since K^{1/p^∞} is a field extension of K; that (4) implies (1) was proved in Section IV.9 (Theorem IV.9.7).

(1) implies (2). First let E be a finite separable extension of K. Then $E = K(\alpha)$ for some separable $\alpha \in E$. By 6.1, $E \otimes F \cong F[X]/(q_1^{m_1}) \times \cdots \times F[X]/(q_r^{m_r})$, where $q_1, \ldots, q_r \in F[X]$ are distinct monic irreducible polynomials and $q = q_1^{m_1} \cdots q_r^{m_r}$. Now, $m_1 = \cdots = m_r = 1$, since α is separable over K; hence $E \otimes F$ is isomorphic to a direct product of fields, and is reduced.

Next, let E be finitely generated and separable over K, so that E has a transcendence base B such that E is (algebraic and) finite separable over $K(B)$. Then $E \otimes_K F \cong (E \otimes_{K(B)} K(B)) \otimes_K F \cong E \otimes_{K(B)} (K(B) \otimes_K F)$, since tensor products are associative, and these isomorphisms are algebra isomorphisms. By 6.2, $K(B) \otimes_K F$ is a domain, with quotient field $Q \cong F(B)$. Since $E \otimes_{K(B)} Q$ is reduced (by the above), the injection $E \otimes_K F \cong E \otimes_{K(B)} (K(B) \otimes_K F) \longrightarrow E \otimes_{K(B)} Q$ shows that $E \otimes_K F$ is reduced.

Finally, let E be any separable extension of K. Every $t \in E \otimes F$ is a finite sum $\alpha_1 \otimes \beta_1 + \cdots + \alpha_n \otimes \beta_n$, where $\alpha_1, \ldots, \alpha_n \in E$ belong to a finitely generated extension $E' = K(\alpha_1, \ldots, \alpha_n) \subseteq E$. Since E is separable over K, E' has a transcendence base B such that E' is separable over $K(B)$. By the above, $\alpha_1 \otimes \beta_1 + \cdots + \alpha_n \otimes \beta_n \in E' \otimes F$ is either zero or not nilpotent in $E' \otimes F$. Since $E' \otimes F \longrightarrow E \otimes F$ is injective, it follows that $t = \alpha_1 \otimes \beta_1 + \cdots + \alpha_n \otimes \beta_n \in E \otimes F$ is either zero or not nilpotent in $E \otimes F$.

(3) implies (4). First we show that E contains no $\alpha \in K^{1/p}\backslash K$. Indeed, α has a pth root β in K^{1/p^∞}, so that $\mathrm{Irr}\,(\alpha : K) = X^p - \alpha = (X - \beta)^p \in K^{1/p^\infty}[X]$. By 6.1, $K(\alpha) \otimes K^{1/p^\infty} \cong K^{1/p^\infty}[X]/(X - \beta)^p$. Now, the coset of $X - \beta$ in $K^{1/p^\infty}[X]/(X - \beta)^p$ is nilpotent. Hence $K(\alpha) \otimes K^{1/p^\infty}$ has a nonzero nilpotent element. If $\alpha \in E$, then the injection $K(\alpha) \otimes K^{1/p^\infty} \longrightarrow E \otimes K^{1/p^\infty}$ puts a nonzero nilpotent element in $E \otimes K^{1/p^\infty}$, contradicting (3).

To establish (4), we prove by induction on $[F' : K]$ that E and F' are linearly disjoint over K, for every finitely generated extension $K \subseteq F' \subseteq K^{1/p^\infty}$ $(\subseteq \overline{E})$ of K; then E and K^{1/p^∞} are linearly disjoint over K, by IV.9.1 and IV.9.2. We may assume that $F' \supsetneqq K$. Readers will verify that $F' = F(\alpha)$ for some subfield $K \subseteq F \subsetneqq F'$ and some $\alpha \in F'$ such that $\alpha^p \in F$ but $\alpha \notin F$. Then $EF' = (EF)(\alpha)$. Now, E and F are linearly disjoint over K, by the induction hypothesis. By (3) and 6.4, $EF \otimes_F K^{1/p^\infty}$ is reduced. But $F^{1/p^\infty} = K^{1/p^\infty}$, so $EF \otimes_F F^{1/p^\infty}$ is reduced. By the first part of the proof, EF does not contain $\alpha \in F^{1/p}\backslash F$. Hence EF and F' are linearly disjoint over F: since $[EF' : EF] = p = [F' : F]$, F' has a basis $1, \alpha, \ldots, \alpha^{p-1}$ over F

that is also a basis of EF' over EF. Since E and F are linearly disjoint over K, it follows from IV.9.4 that E and F' are linearly disjoint over K. $\square$

Exercises

1. Let $K \subseteq E$ and let α is algebraic over K. Show that $E \otimes K(\alpha)$ is a domain if and only if $\mathrm{Irr}\,(\alpha : K)$ is irreducible over E.

2. Let A be a K-algebra. Show that there is an isomorphism $A \otimes K[(X_i)_{i \in I}] \cong A[(X_i)_{i \in I}]$ that sends $\gamma \otimes \left(\sum_m a_m X^m \right)$ to $\sum_m \gamma a_m X^m$.

3. Give an example of fields $K \subseteq E$, F such that $E \otimes_K F$ contains zero divisors. Give an example of fields $K \subseteq E$, F such that $E \otimes_K F$ contains nonzero nilpotent elements.

4. Let $K \subsetneq F' \subseteq K^{1/p^\infty}$ be a finite extension of K. Show that $F' = F(\alpha)$ for some subfield $K \subseteq F \subsetneq F'$ and some $\alpha \in F'$ such that $\alpha^p \in F$ and $\alpha \notin F$.

7. Simple Algebras over a Field

This section explores some of the remarkable properties of simple Artinian algebras and division algebras over a field, culminating with Frobenius's theorem that finds all finite-dimensional division algebras over $\mathbb{R}$.

The center. In what follows, K is any given field. In a K-algebra A with identity element 1, we saw that to every $x \in K$ corresponds a central element $x1$ of A (meaning that $(x1)\,a = a\,(x1)$ for all $a \in A$) and that $x \longmapsto x1$ is a homomorphism of K into A, injective since K is a field. We identify $x \in K$ and $x1 \in A$, so that K becomes a central subfield of A.

An algebra is *simple* (*left Artinian*, a *division algebra*) when its underlying ring is simple (left Artinian, a division ring). For example, a field extension of K is a division K-algebra; a finite field extension of K is an Artinian division K-algebra. The quaternion algebra $\mathbb{H}$ is an Artinian division algebra over $\mathbb{R}$.

Matrix rings provide other examples of simple Artinian algebras. Let D be a division algebra over K. For every $x \in K$, the scalar $n \times n$ matrix with x on its diagonal is central in $M_n(D)$; this provides a central homomorphism of K into $M_n(D)$, which by Proposition 1.1 makes $M_n(D)$ a K-algebra. Then $M_n(D)$ is a simple Artinian K-algebra.

The basic result about Artinian simple algebras is Wedderburn's theorem, Theorem IX.3.8, which, recast in terms of K-algebras, reads as follows:

Theorem **7.1** (Wedderburn). *Let K be a field. A K-algebra A is a simple left Artinian K-algebra if and only if it is isomorphic to $M_n(D)$ for some $n > 0$ and division K-algebra $D \cong \mathrm{End}_A^{\mathrm{op}}\,(S)$, where S is a simple left A-module.*

Proof. Let A be a simple Artinian K-algebra. By IX.3.8, $A \cong M_n(D)$ for some $n > 0$ and some division ring $D \cong \mathrm{End}_A^{\mathrm{op}}\,(S)$, where S is a simple left

A-module. Since K is central in A, S is, in particular, a K-module, D is a K-algebra, $M_n(D)$ is a K-algebra, and the isomorphism $A \cong M_n(D)$ is an isomorphism of K-algebras. We proved the converse above. $\square$

Definition. The center *of a K-algebra A is*

$$Z(A) = \{ z \in A \mid az = za \text{ for all } z \in A \}.$$

The following example is an exercise:

Proposition **7.2.** *Let D be a division ring. The center of $M_n(D)$ is isomorphic to the center of D; so the center of a simple Artinian K-algebra is a field.*

Central simple algebras. If A is a simple Artinian K-algebra, then its center $C = Z(A)$ is a central subfield, by 7.2, which contains K; by 1.1, A is a C-algebra, and is then a simple Artinian C-algebra with center C. The study of simple Artinian algebras may therefore be limited to the case $C = K$.

Definition. A K-algebra A is central *when $Z(A) = K$.*

Theorem **7.3** (Noether [1929]). *Over any field, the tensor product of two central simple algebras is a central simple algebra.*

Proof. Let A and B be K-algebras. Buoyed by 5.5 we identify a and $a \otimes 1$, b and $1 \otimes b$, for all $a \in A$ and $b \in B$, so that A and B become subalgebras of $A \otimes_K B$. Theorem 7.3 then follows from a more detailed result:

Lemma **7.4** (Noether [1929]). *Let B be central simple. If A is simple, then $A \otimes_K B$ is simple. If A is central, then $A \otimes_K B$ is central. In fact:*

(1) *if $J \neq 0$ is an ideal of $A \otimes_K B$, then $J \cap A \neq 0$;*

(2) *every ideal of $A \otimes_K B$ has the form $I \otimes_K B$ for some ideal I of A;*

(3) *$Z(A \otimes B) = Z(A)$.*

Proof. (1). Choose $t = a_1 \otimes b_1 + \cdots + a_m \otimes b_m \in J \backslash 0$ (where $a_i \in A$, $b_i \in B$) so that m is the least possible. Then $a_1, \ldots, a_m$ are linearly independent over K, and so are $b_1, \ldots, b_m$. In particular, $b_m \neq 0$.

Since B is simple there exist $x_j, y_j \in B$ such that $1 = \sum_j x_j b_m y_j$. Then

$$u = \sum_j (1 \otimes x_j) t (1 \otimes y_j) = \sum_{i,j} a_i \otimes x_j b_i y_j = \sum_i a_i \otimes c_i,$$

where $c_i = \sum_j x_j b_i y_j \in B$. Then $u \in J$. We show that $u \in A$ and $u \neq 0$.

Since $a_1, \ldots, a_m$ are linearly independent over K, they are part of a basis of A over K; hence $a_1 \otimes 1, \ldots, a_m \otimes 1$ are part of a basis of $A \otimes B$ over B, by 5.4, and are linearly independent over B. Hence $u = \sum_i (a_i \otimes 1) c_i \neq 0$, since $c_m \neq 0$. For every $b \in B$ we now have

$$v = (1 \otimes b) u - u (1 \otimes b) = \sum_{1 \leq i \leq m} a_i \otimes (bc_i - c_i b)$$
$$= \sum_{1 \leq i \leq m-1} a_i \otimes (bc_i - c_i b),$$

since $c_m = 1$. Since $v \in J$, it follows from the choice of t that $v = 0$. Since $a_1 \otimes 1$, ..., $a_m \otimes 1$ are linearly independent over B, this implies $bc_i - c_i b = 0$, for all i and all $b \in B$. Thus c_1, ..., $c_m \in Z(B) = K$ and $u = \sum_i a_i \otimes c_i = \sum_i a_i c_i \otimes 1 \in A$.

(2). Let J be an ideal of $A \otimes B$. Then $I = J \cap A$ is an ideal of A and $I \otimes B \subseteq J$. We show that $I \otimes B = J$. Since vector spaces are flat, there is an exact sequence of vector spaces and homomorphisms

$$0 \longrightarrow I \otimes B \longrightarrow A \otimes B \overset{\pi}{\longrightarrow} (A/I) \otimes B \longrightarrow 0,$$

in which π is induced by the projection $A \longrightarrow A/I$ and is, therefore, an algebra homomorphism. Hence $\pi(J)$ is an ideal of $(A/I) \otimes B$. If $t \in J$ and $\pi(t) \in A/I$, then $\pi(t) = \pi(a)$ for some $a \in A$, $t - a \in \mathrm{Ker}\, \pi = I \otimes B \subseteq J$, $a \in J \cap A = I$, and $\pi(t) = \pi(a) = 0$; thus $\pi(J) \cap (A/I) = 0$. By (1), applied to A/I and B, $\pi(J) = 0$; therefore $J \subseteq I \otimes B$.

(3). In $A \otimes B$, every $a \in A$ commutes with every $b \in B$; hence $Z(A) \subseteq Z(A \otimes B)$. Conversely, let $t \in Z(A \otimes B)$. As in the proof of (1), write $t = a_1 \otimes b_1 + \cdots + a_m \otimes b_m$, where $a_i \in A$, $b_i \in B$, and m is the least possible (given t). Then a_1, ..., a_m are linearly independent over K and $a_1 \otimes 1$, ..., $a_m \otimes 1$ are linearly independent over B. For all $b \in B$,

$$\sum_i a_i \otimes (bb_i - b_i b) = \sum_i \big((1 \otimes b)t - t(1 \otimes b)\big) = 0;$$

hence $bb_i - b_i b = 0$ for all i and all $b \in B$, $b_i \in Z(B) = K$ for all i, $t = \sum_i a_i \otimes b_i = \sum_i a_i b_i \otimes 1 \in A$, and $t \in Z(A)$. $\square$

Theorem 7.5 (Skolem-Noether [1929]). *Let A be a simple K-algebra and let B be a central simple K-algebra, both of finite dimension over K. Any two homomorphisms $\varphi, \psi : A \longrightarrow B$ are conjugate (there exists a unit u of B such that $\psi(a) = u\, \varphi(b)\, u^{-1}$ for all $a \in A$).*

Proof. Since $\dim_K B$ is finite, then B is left Artinian; by 7.1, $B \cong M_n(D) \cong \mathrm{End}_D(S)$ for some division K-algebra $D \cong \mathrm{End}_B^{\mathrm{op}}(S)$, where S is a simple left B-module. Then S is a faithful B-module (by IX.3.8), a finite-dimensional right D-module $S \cong D^n$, and a left B-, right D-bimodule. Then $\varphi : A \longrightarrow B$ makes S a left A-, right D-bimodule and a left $A \otimes_K D^{\mathrm{op}}$-module, in which $(a \otimes d)s = \varphi(a)sd$ (by 5.7). Similarly, ψ yields another left $A \otimes_K D^{\mathrm{op}}$-module structure on S. We now have two left $A \otimes_K D^{\mathrm{op}}$-modules S_1 and S_2 on S with the same finite dimension over K.

The ring $R = A \otimes_K D^{\mathrm{op}}$ is simple by 7.4, and left Artinian since it has finite dimension over K. Hence there is, up to isomorphism, only one simple left R-module T, and every left R-module is a direct sum of copies of T. Hence $S_1 \cong T^k$ and $S_2 \cong T^\ell$. Since S_1 and S_2 have the same finite dimension over K, this implies $k = \ell$. Hence $S_1 \cong S_2$ as R-modules, and as left A-, right D-bimodules. The isomorphism $\theta : S_1 \longrightarrow S_2$ is a D-automorphism of S such that $\theta\big(\varphi(a)s\big) = \theta(as) = a\theta(s) = \psi(a)\,\theta(s)$ for all $a \in A$ and

$s \in S$. The isomorphism $\text{End}_D (S) \cong B$ sends θ to a unit u of B such that $u \varphi(a) s = \psi(a) u s$ for all $a \in A$ and $s \in S$; since S is a faithful B-module this implies $u \varphi(a) = \psi(a) u$ for all $a \in A$. $\square$

In a K-algebra A, the *centralizer* of a subset B of A is

$$C = \{ a \in A \mid ab = ba \text{ for all } b \in B \}.$$

Theorem 7.6. *Let A be a simple K-algebra of finite dimension over K and let B be a simple subalgebra of A. The centralizer C of B is a simple subalgebra of A. Moreover, B is the centralizer of C and $\dim_K A = (\dim_K B)(\dim_K C)$.*

In fact, readers may show that $B \otimes_K C \cong A$.

Proof. As in the proof of 7.5, $A \cong M_n(D) \cong \text{End}_D (S)$ for some division K-algebra $D \cong \text{End}_A^{\text{op}} (S)$, where S is a simple left A-module. Then S is a left A-, right D-bimodule and a left $A \otimes_K D^{\text{op}}$ module, and the isomorphism $\theta : A \longrightarrow \text{End}_D (S)$ induces the given left A-module structure on S. We may also view S as a left B-, right D-bimodule and a left $B \otimes_K D^{\text{op}}$ module. Again $R = B \otimes_K D^{\text{op}}$ is simple, by 7.4, and left Artinian, since $\dim_K R = (\dim_K B)(\dim_K D)$ is finite; hence $R \cong M_m(E) \cong \text{End}_E (T)$ for some division K-algebra $E \cong \text{End}_R^{\text{op}} (T)$, where T is a simple left R-module.

Since $\theta : A \cong \text{End}_D (S)$ is an isomorphism, then $a \in A$ is in C if and only if $\theta(a)$ commutes with $\theta(b)$ for all $b \in B$, if and only if $\theta(a)$ is a left B-, right D-bimodule endomorphism of S. Hence θ induces an isomorphism $C \cong \text{End}_R (S)$. Since R is simple we have $S \cong T^r$ for some $r > 0$ and $C \cong \text{End}_R (S) \cong M_r(\text{End}_R (T)) \cong M_r(E)$. Hence C is simple.

Since $A \cong M_n(D)$ and $C \cong M_r(E)$ we have

$$\dim A = n^2 \dim D \quad \text{and} \quad \dim C = r^2 \dim E$$

(dimensions are over K). Similarly, $B \otimes_K D^{\text{op}} \cong R \cong M_m(E)$ and 5.4 yield

$$(\dim B)(\dim D) = m^2 \dim E.$$

Since $S \cong D^n$, $T \cong E^m$, and $S \cong T^r$, we also have

$$n \dim D = \dim S = r \dim T = rm \dim E.$$

Hence

$$(\dim B)(\dim C) = \frac{m^2 \dim E}{\dim D} r^2 \dim E = n^2 \dim D = \dim A.$$

Finally, if B' is the centralizer of C, then $B \subseteq B'$ and, by the above, $(\dim C)(\dim B') = \dim A$; hence $\dim B = \dim B'$ and $B = B'$. $\square$

Division algebras. We now apply the previous theorems to division rings and algebras. First we note that every division ring D has a center, which is a subfield of D. The following properties make a tasty exercise:

Proposition **7.7.** *Let D be a division ring with center K. If $\dim_K D$ is finite, then $\dim_K D$ is a square. In fact, every maximal subfield $F \supseteq K$ of D is its own centralizer in D, so that $\dim_K D = (\dim_K F)^2$.*

Theorem **7.8** (Frobenius [1877]). *A division $\mathbb{R}$-algebra that has finite dimension over $\mathbb{R}$ is isomorphic to $\mathbb{R}$, $\mathbb{C}$, or the quaternion algebra $\mathbb{H}$.*

Proof. Let D be a division $\mathbb{R}$-algebra with center K and let $F \supseteq K$ be a maximal subfield of D, so that $\mathbb{R} \subseteq K \subseteq F \subseteq D$. By 7.7, $\dim_K D = (\dim_K F)^2$. Now, $[F : \mathbb{R}]$ is finite; hence either $F = \mathbb{R}$ or $F \cong \mathbb{C}$.

If $F = \mathbb{R}$, then $K = \mathbb{R} = F$, $\dim_K D = (\dim_K F)^2 = 1$, and $D = \mathbb{R}$.

If $F \cong \mathbb{C}$ and $K = F$, then again $\dim_K D = 1$ and $D = K \cong \mathbb{C}$.

Now, let $F \cong \mathbb{C}$ and $K \subsetneqq F$. Then $K = \mathbb{R}$ and $\dim_\mathbb{R} D = 4$. We identify F and $\mathbb{C}$ in what follows. There are two algebra homomorphisms of F into D, the inclusion homomorphism and complex conjugation. By 7.5 there exists a unit u of D such that $\bar{z} = uzu^{-1}$ for all $z \in F$. Then $u^2 z u^{-2} = z$ for all $z \in F$ and u^2 is in the centralizer of F. By 7.7, $u^2 \in F$. In fact, $u^2 \in \mathbb{R}$, since $\bar{u}^2 = u u^2 u^{-1} = u^2$. However, $u \notin F$, since $uiu^{-1} = \bar{i} \neq i$. Hence $u^2 < 0$, since $u^2 \geq 0$ would imply $u^2 = r^2$ for some $r \in \mathbb{R}$ and $u = \pm r \in F$; thus, $u^2 = -r^2$ for some $r \in \mathbb{R}$, $r \neq 0$.

Let $j = u/r$ and $k = ij$. Then $j \notin F$; hence $\{1, j\}$ is a basis of D over F, and $\{1, i, j, k\}$ is a basis of D over $\mathbb{R}$. Also, $j^2 = -1$, and $jzj^{-1} = jrzr^{-1}j^{-1} = uzu^{-1} = \bar{z}$ for all $z \in F$. Hence $ji = -ij = -k$, $k^2 = ijij = -ijij^{-1} = -i\bar{i} = -1$, $ik = -j$, $ki = iji = -i^2 j = j$, $kj = -i$, and $jk = jij = -jij^{-1} = -\bar{i} = i$. Thus $D \cong \mathbb{H}$. $\square$

Exercises

1. Let D be a division ring. Show that the center of $M_n(D)$ consists of scalar matrices and therefore is isomorphic to the center of D.

2. Let D be a division ring with center K. Show that every maximal subfield $F \supseteq K$ of D is its own centralizer in D, so that $\dim_K D = (\dim_K F)^2$. (Use Theorem 7.6.)

3. Show that a finite group is not the union of conjugates of any proper subgroup.

4. Use the previous exercise to prove that every finite division ring is a field.

In the following exercises, K is an arbitrary field, and A is a central simple K-algebra of finite dimension n over K. Prove the following:

5. $\dim_K A$ is a square. (You may want to use Proposition 7.7.)

6. $A \otimes_K A^{\mathrm{op}} \cong M_n(K)$.

7. $B \otimes_K C \cong A$ when B is a simple subalgebra of A and C is the centralizer of B.

XIV
Lattices

Lattices abound throughout many branches of Mathematics. This seems to have first been noticed by Dedekind [1897]. They are also of interest as algebraic systems. Systematic study began in the 1930s with the work of Birkhoff and Stone, and with Birkhoff's *Lattice Theory* [1940]. Lattice theory unifies various parts of algebra, though perhaps less successfully than the theories in the next two chapters.

This chapter draws examples from Chapters I, III, V, and VIII, and is otherwise independent of other chapters.

1. Definitions

This section introduces two kinds of partially ordered sets: semilattices and lattices.

Semilattices.

Definitions. In a partially ordered set $(S, \leqq)$, a lower bound *of a subset T of S is an element ℓ of S such that $\ell \leqq t$ for all $t \in T$; a* greatest lower bound *or* g.l.b. *of T, also called a* meet *or* infimum *of T, is a lower bound g of T such that $\ell \leqq g$ for every lower bound ℓ of T.*

A subset T that has a g.l.b. has only one g.l.b., which is denoted by $\bigwedge_{t \in T} t$; by $\bigwedge_{i \in I} t_i$, if T is written as a family $T = (t_i)_{i \in I}$; or by $t_1 \wedge \cdots \wedge t_n$, if $T = \{t_1, \ldots, t_n\}$ (by virtue of Proposition 1.1 below).

Definition. A lower semilattice *is a partially ordered set in which every two elements a and b have a greatest lower bound $a \wedge b$.*

By definition, $a \wedge b \leqq a$, $a \wedge b \leqq b$, and $x \leqq a$, $x \leqq b$ implies $x \leqq a \wedge b$. For example, the set 2^X of all subsets of a set X, partially ordered by inclusion, is a lower semilattice, in which the g.l.b. of two subsets is their intersection. This extends to any set of subsets of X that is closed under intersections: thus, the subgroups of a group, the subrings of a ring, the ideals of a ring, the submodules of a module, all constitute lower semilattices.

Proposition **1.1.** *The binary operation $\wedge$ on a lower semilattice $(S, \leqq)$ is* *idempotent* ($x \wedge x = x$ *for all* $x \in S$), *commutative, associative, and order preserving* ($x \leqq y$ *implies* $x \wedge z \leqq y \wedge z$ *for all* z). *Moreover, every finite subset* $\{ a_1, \ldots, a_n \}$ *of* S *has a greatest lower bound, namely* $a_1 \wedge \cdots \wedge a_n$.

The proof is an exercise.

Definitions. In a partially ordered set $(S, \leqq)$, *an* upper bound *of a subset* T *of* S *is an element* u *of* S *such that* $u \geqq t$ *for all* $t \in T$; *a* least upper bound *or* l.u.b. *of* T , *also called a* join *or* supremum *of* T, *is an upper bound* ℓ *of* T *such that* $\ell \leqq u$ *for every upper bound* u *of* T .

A subset T that has an l.u.b. has only one l.u.b., which is denoted by $\bigvee_{t \in T} t$; by $\bigvee_{i \in I} t_i$, if T is written as a family $T = (t_i)_{i \in I}$; or by $t_1 \vee \cdots \vee t_n$, if $T = \{ t_1, \ldots, t_n \}$ (by virtue of Proposition 1.2 below).

Definition. An upper semilattice *is a partially ordered set in which every two elements* a *and* b *have a least upper bound* $a \vee b$.

By definition, $a \leqq a \vee b$, $b \leqq a \vee b$, and $a \leqq x$, $b \leqq x$ implies $a \vee b \leqq x$. For example, the set 2^X of all subsets of a set X, partially ordered by inclusion, is an upper semilattice, in which the l.u.b. of two subsets is their union. The subgroups of a group constitute an upper semilattice, in which the supremum of two subgroups is the subgroup generated by their union; the ideals of a ring and the submodules of a module also constitute upper semilattices, in which the supremum of two ideals or submodules is their sum.

Least upper bounds and greatest lower bounds are related as follows.

Definition. The dual *or* opposite *of a partially ordered set* $(S, \leqq)$ *is the partially ordered set* $(S, \leqq)^{op} = (S, \leqq^{op})$ *on the same set with the opposite order relation,* $x \leqq^{op} y$ *if and only if* $y \leqq x$.

An l.u.b. in $(S, \leqq)$ is a g.l.b. in $(S, \leqq)^{op}$, and vice versa. Hence S is an upper semilattice if and only if S^{op} is a lower semilattice. Therefore the following statement follows from Proposition 1.1:

Proposition **1.2.** *The binary operation* $\vee$ *on an upper semilattice* $(S, \leqq)$ *is idempotent* ($x \vee x = x$ *for all* $x \in S$), *commutative, associative, and order preserving* ($x \leqq y$ *implies* $x \vee z \leqq y \vee z$ *for all* z). *Moreover, every finite subset* $\{ a_1, \ldots, a_n \}$ *of* S *has a least upper bound, namely* $a_1 \vee \cdots \vee a_n$.

There is a more general principle:

Metatheorem **1.3** (Duality Principle). *A theorem that holds in every partially ordered set remains true when the order relation is reversed.*

The duality principle is our first example of a *metatheorem*, which does not prove any specific statement, but is applied to existing theorems to yield new results. Like nitroglycerine, it must be handled with care. Order reversal in 1.3 applies to hypotheses as well as conclusions. Hence the duality principle does not apply to *specific* partially ordered sets: if Theorem T is true in S, then Theorem T^{op} is true in S^{op}, but it does not follow that T^{op} is true in S (unless T is also true in S^{op}). For example, every subset of $\mathbb{N}$ has a g.l.b., but it is deranged logic to deduce from the duality principle that every subset of $\mathbb{N}$ has a l.u.b.

Lattices.

Definition. A lattice *is a partially ordered set that is both a lower semilattice and an upper semilattice.*

Equivalently, a lattice is a partially ordered set in which every two elements a and b have a g.l.b. $a \wedge b$ and an l.u.b. $a \vee b$.

Every totally ordered set is a lattice: if, say, $a \leqq b$, then $a \wedge b = a$ and $a \vee b = b$; thus $\mathbb{N}$, $\mathbb{Q}$, and $\mathbb{R}$, with their usual order relations, are lattices. The subsets of a set, the subgroups of a group, the ideals of a ring, the submodules of a module, all constitute lattices.

Proposition **1.4.** *If X is a set, and L is a set of subsets of X that is closed under intersections and contains X, then L, partially ordered by inclusion, is a lattice.*

Proof. Let $A, B \in L$. Then $A \cap B \in L$ is the g.l.b. of A and B. The l.u.b. of A and B is the intersection of all $C \in L$ that contain $A \cup B$ (including X), which belongs to L by the hypothesis. $\square$

Finite partially ordered sets and lattices can be specified by directed graphs in which the elements are vertices and $x < y$ if and only if there is a path from x to y. Arrow tips are omitted; it is understood that all arrows point upward. For example, in the graph

$0 < 1$, but $a \nleqq b$; the graph shows that $a \wedge b = a \wedge c = b \wedge c = 0$ and $a \vee b = a \vee c = b \vee c = 1$, and represents a lattice.

Properties. If L is a lattice, then so is the opposite partially ordered set L^{op}; hence there is a duality principle for lattices:

Metatheorem **1.5** (Duality Principle). *A theorem that holds in every lattice remains true when the order relation is reversed.*

As was the case with 1.3, order reversal in Metatheorem 1.5 applies to hypotheses as well as conclusions; the duality principle does not apply to *specific* lattices.

Readers will verify that lattices can be defined as sets equipped with two operations that satisfy certain identities:

Proposition **1.6.** *Two binary operations* $\wedge$ *and* $\vee$ *on a set* S *are the infimum and supremum operations of a lattice if and only if the identities*

(1) (idempotence) $x \wedge x = x$, $x \vee x = x$,

(2) (commutativity) $x \wedge y = y \wedge x$, $x \vee y = y \vee x$,

(3) (associativity) $(x \wedge y) \wedge z = x \wedge (y \wedge z)$, $(x \vee y) \vee z = x \vee (y \vee z)$,

(4) (absorption laws) $x \wedge (x \vee y) = x$, $x \vee (x \wedge y) = x$

hold for all $x, y, z \in S$; *and then* $\wedge$ *and* $\vee$ *are the infimum and supremum operations of a unique lattice, in which* $x \leqq y$ *if and only if* $x \wedge y = x$, *if and only if* $x \vee y = y$.

Definition. A *sublattice of a lattice* L *is a subset of* L *that is closed under infimums and supremums.*

Equivalently, $S \subseteq L$ is a sublattice of L if and only if $x, y \in S$ implies $x \wedge y \in S$ and $x \vee y \in S$. Then it follows from Proposition 1.6 that S is a lattice in its own right, in which $x \leqq y$ if and only if $x \leqq y$ in L; this lattice is also called a *sublattice* of L.

Definitions. A homomorphism *of a lattice* A *into a lattice* B *is a mapping* $\varphi : A \longrightarrow B$ *such that* $\varphi(x \wedge y) = \varphi(x) \wedge \varphi(y)$ *and* $\varphi(x \vee y) = \varphi(x) \vee \varphi(y)$ *for all* $x, y \in A$. *An* isomorphism *of lattices is a bijective homomorphism.*

By Proposition 1.6, a lattice homomorphism φ is *order preserving*: $x \leqq y$ implies $\varphi(x) \leqq \varphi(y)$. But order preserving mappings between lattices are not necessarily homomorphisms; readers will easily find counterexamples. However, a bijection θ between lattices is an isomorphism if and only if both θ and θ^{-1} are order preserving (see the exercises); then θ^{-1} is also an isomorphism. More generally, an *isomorphism* of partially ordered sets is a bijection θ such that θ and θ^{-1} are order preserving.

There is a homomorphism theorem for lattices, for which readers are referred to the very similar result in Section XV.1.

Exercises

1. Show that the binary operation $\wedge$ on a lower semilattice $(S, \leqq)$ is idempotent, commutative, associative, and order preserving.

2. Prove the following: in a lower semilattice, every finite subset $\{ a_1, \ldots, a_n \}$ of S has a g.l.b., namely $a_1 \wedge \cdots \wedge a_n$.

3. Show that a binary operation that is idempotent, commutative, and associative is the infimum operation of a unique lower semilattice, in which $x \leqq y$ if and only if $xy = x$.

4. Show that two binary operations on the same set are the infimum and supremum operations of a lattice if and only if they are idempotent, commutative, associative, and satisfy the absorption laws; and then they are the infimum and supremum operations of a unique lattice.

5. Prove that every intersection of sublattices of a lattice L is a sublattice of L.

6. Prove that every directed union of sublattices of a lattice L is a sublattice of L.

7. Find an order preserving bijection between lattices that is not a lattice homomorphism.

8. Show that a bijection θ between lattices is an isomorphism if and only if both θ and θ^{-1} are order preserving.

2. Complete Lattices

A lattice is complete when every subset has an l.u.b. and a g.l.b. Most of the examples in the previous section have this property. This section contains basic examples and properties, and MacNeille's completion theorem.

Definition. A complete *lattice is a partially ordered set L (necessarily a lattice) in which every subset has an infimum and a supremum.*

In particular, a complete lattice L has a *least* element 0 (such that $0 \leqq x$ for all $x \in L$), which is the g.l.b. of L (and the l.u.b. of the empty subset of L). A complete lattice L also has a *greateast* element 1 (such that $x \leqq 1$ for all $x \in L$), which is the l.u.b. of L (and the g.l.b. of the empty subset of L).

The opposite of a complete lattice is a complete lattice. Hence there is a duality principle for complete lattices: a theorem that holds in every complete lattice remains true when the order relation is reversed.

Examples. Every finite lattice is complete (by Propositions 1.1 and 1.2). On the other hand, the lattices $\mathbb{N}$, $\mathbb{Q}$, $\mathbb{R}$ are not complete; but every closed interval of $\mathbb{R}$ is complete.

The following results are proved like Proposition 1.4:

Proposition 2.1. If X is a set, and L is a set of subsets of X that is closed under intersections and contains X, then L, partially ordered by inclusion, is a complete lattice.

Proposition 2.2. A partially ordered set S is a complete lattice if and only if it has a greatest element and every nonempty subset of S has an infimum.

In particular, the subsets of a set, the subgroups of a group, the ideals of a ring, the submodules of a module, all constitute complete lattices. We mention two more classes of examples.

Definitions (Moore [1910]). *A closure map on a partially ordered set S is a mapping $\Gamma : S \longrightarrow S$ that is order preserving (if $x \leqq y$, then $\Gamma x \leqq \Gamma y$), idempotent ($\Gamma\Gamma x = \Gamma x$ for all $x \in S$), and expanding ($\Gamma x \geqq x$ for all $x \in S$). Then $x \in S$ is* closed *relative to Γ when $\Gamma x = x$.*

The closure mapping $A \longmapsto \overline{A}$ on a topological space is a quintessential closure map. Algebraic closure maps include the subgroup generated by a subset, the ideal generated by a subset, and so forth.

Proposition **2.3.** *Relative to a closure map on a complete lattice* L, *the set of all closed elements of* L *is closed under infimums and is a complete lattice.*

The proof is an exercise.

Definition. A Galois connection *between two partially ordered sets* X *and* Y *is an ordered pair* (α, β) *of mappings* $\alpha : X \longrightarrow Y$ *and* $\beta : Y \longrightarrow X$ *that are order reversing (if* $x' \leqq x''$, *then* $\alpha x' \geqq \alpha x''$; *if* $y' \leqq y''$, *then* $\beta y' \geqq \beta y''$) *and satisfy* $\beta \alpha x \geqq x$, $\alpha \beta y \geqq y$ *for all* $x \in X$ *and* $y \in Y$.

The fixed field and Galois group constructions in Chapter V constitute a Galois connection; the exercises give other examples.

Proposition **2.4.** *If* (α, β) *is a Galois connection between two partially ordered sets* X *and* Y, *then* α *and* β *induce mutually inverse, order reversing bijections between* $\mathrm{Im}\ \alpha$ *and* $\mathrm{Im}\ \beta$; $\alpha \circ \beta$ *and* $\beta \circ \alpha$ *are closure maps; if* X *and* Y *are complete lattices, then* $\mathrm{Im}\ \alpha$ *and* $\mathrm{Im}\ \beta$ *are complete lattices.*

The proof is an exercise. In a Galois extension, Proposition 2.4 provides the usual bijections between intermediate fields and (closed) subgroups of the Galois group.

MacNeille's theorem is the following result.

Theorem **2.5** (MacNeille [1935]). *Every partially ordered set can be embedded into a complete lattice so that all existing infimums and supremums are preserved.*

Proof. Let $(X, \leqq)$ be a partially ordered set. For every subset S of X let

$$L(S) = \{\, x \in X \mid x \leqq s \text{ for all } s \in S \,\},$$
$$U(S) = \{\, x \in X \mid x \geqq s \text{ for all } s \in S \,\}$$

be the sets of all lower and upper bounds of S. Then $S \subseteq T$ implies $L(S) \supseteq L(T)$ and $U(S) \supseteq U(T)$; also, $S \subseteq U(L(S))$ and $S \subseteq L(U(S))$ for every $S \subseteq X$. Hence (L, U) is a Galois connection between 2^X and 2^X. By 2.4, $L \circ U$ is a closure map on 2^X; by 2.3, the set $\widehat{X}$ of all closed subsets of X is closed under intersections and is a complete lattice, in which infimums are intersections. For every $t \in X$ let

$$\lambda(t) = L(\{t\}) = \{\, x \in X \mid x \leqq t \,\}, \quad \upsilon(t) = U(\{t\}) = \{\, x \in X \mid x \geqq t \,\}.$$

Then $U(\lambda(t)) = \upsilon(t)$ and $L(\upsilon(t)) = \lambda(t)$. Therefore $\lambda(t)$ is closed. Hence λ is a mapping of X into $\widehat{X}$. We see that λ is injective, and that $x \leqq y$ in X if and only if $\lambda(x) \subseteq \lambda(y)$ in $\widehat{X}$.

For every subset S of X we have $L(S) = \bigcap_{s \in S} \lambda(s)$. If S has a g.l.b. t in X, then $L(S) = \lambda(t)$ and $\lambda(t) = \bigcap_{s \in S} \lambda(s)$ is the g.l.b. of $\lambda(S)$ in $\widehat{X}$.

Similarly, assume that S has an l.u.b. u in X. Then $U(S) = \upsilon(u)$, $L(U(S)) = \lambda(u)$, $\lambda(u)$ is closed, and $\lambda(u) \supseteq \lambda(s)$ for every $s \in S$. Conversely, if C is a closed subset of X that contains $\lambda(s)$ for every $s \in S$, then

$S \subseteq C$ and $\lambda(u) = L\big(U(S)\big) \subseteq L\big(U(C)\big) = C$. Hence $\lambda(u)$ is the l.u.b. of $\lambda(S)$ in $\widehat{X}$. $\square$

The complete lattice $\widehat{X}$ in the proof of Theorem 2.5 is the *MacNeille completion* of X. Examples include Dedekind's purely algebraic construction of $\mathbb{R}$ (see the exercises).

Exercises

1. Show that a partially ordered set S is a complete lattice if and only if it has a greatest element and every nonempty subset of S has a greatest lower bound.

2. Prove the following: when Γ is a closure map on a complete lattice L, then the set of all closed elements of L is closed under infimums and is a complete lattice.

3. Let L be a complete lattice and let C be a subset of L that is closed under infimums and contains the greatest element of L. Show that there is a closure map on L relative to which C is the set of all closed elements of L.

4. Let X be a set. When is a closure map on 2^X the closure mapping of a topology on X?

5. Prove the following: when (α, β) is a Galois connection between two partially ordered sets X and Y, then α and β induce mutually inverse, order reversing bijections between $\operatorname{Im} \alpha$ and $\operatorname{Im} \beta$; moreover, if X and Y are complete lattices, then $\operatorname{Im} \alpha$ and $\operatorname{Im} \beta$ are complete lattices.

6. Let R be a ring. The *annihilator* of a left ideal L of R is the right ideal $\operatorname{Ann}(L) = \{ x \in R \mid Lx = 0 \}$ of R. The *annihilator* of a right ideal T of R is the left ideal $\operatorname{Ann}(T) = \{ x \in R \mid xT = 0 \}$ of R. Show that these constructions constitute a Galois connection between left ideals and right ideals of R.

7. Let G be a group. The *centralizer* of a subgroup H of G is $C(H) = \{ x \in G \mid xh = hx$ for all $h \in H \}$. Show that (C, C) is a Galois connection between subgroups of G and subgroups of G.

8. Inspired by the previous exercise, construct a "centralizer" Galois connnection for subrings of a ring R.

9. Find the MacNeille completion of $\mathbb{N}$.

10. Show that the MacNeille completion of $\mathbb{Q}$ is isomorphic ro $\mathbb{R} \cup \{\infty\} \cup \{-\infty\}$.

3. Modular Lattices

Modular lattices have interesting chain properties. In this section we show that the length properties of abelian groups and modules are in fact properties of modular lattices.

Definition. A lattice L is modular *when $x \leqq z$ implies $x \vee (y \wedge z) = (x \vee y) \wedge z$, for all $x, y, z \in L$.*

In any lattice, $x \leq z$ implies that x and $y \wedge z$ are lower bounds of $\{ x \vee y, z \}$, whence $x \vee (y \wedge z) \leq (x \vee y) \wedge z$. Thus, a lattice L is modular if and only if $x \leq z$ implies $x \vee (y \wedge z) \geq (x \vee y) \wedge z$.

The opposite of a modular lattice is a modular lattice. Hence there is a duality principle for modular lattices: a theorem that holds in every modular lattice remains true when the order relation is reversed.

Examples. Modular lattices are named after the following example:

Proposition **3.1.** *The lattice of submodules of any module is modular.*

Proof. We saw that the submodules of a module constitute a lattice, in which $A \wedge B = A \cap B$ and $A \vee B = A + B$. Let A, B, C be submodules such that $A \subseteq C$. Then $A + (B \cap C) \subseteq (A + B) \cap C$ (this holds in every lattice). Conversely, if $x \in (A + B) \cap C$, then $x = a + b \in C$ for some $a \in A \subseteq C$ and $b \in B$, whence $b = x - a \in C$ and $x = a + b \in A + (B \cap C)$. $\square$

Readers will also verify that every totally ordered set is a modular lattice, and that the following lattice, which we call M_5, is modular:

But the following unfortunate lattice, which we call N_5, is not modular:

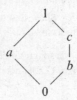

since $b \leq c$, $b \vee (a \wedge c) = b \vee 0 = b$, and $(b \vee a) \wedge c = 1 \wedge c = c$. The next result shows that N_5 is the quintessential nonmodular lattice.

Theorem **3.2.** *A lattice is modular if and only if it contains no sublattice that is isomorphic to N_5.*

Proof. A sublattice of a modular lattice is modular and not isomorphic to N_5.

Conversely, a lattice L that is not modular contains elements a, b, c such that $b \leq c$ and $u = b \vee (a \wedge c) < (b \vee a) \wedge c = v$. Then $v \leq b \vee a$, $a \leq u < v \leq c$, and $b \vee a \leq b \vee u \leq b \vee v \leq b \vee a$, so that $b \vee u = b \vee v = b \vee a$. Similarly, $b \wedge c \leq u$, $v \leq b \wedge c$, and $b \wedge c \leq b \wedge u \leq b \wedge v \leq b \wedge c$, so that $b \wedge u = b \wedge v = b \wedge c$. Thus b, u, v, $b \wedge u = b \wedge v$, and $b \vee u = b \vee v$ constitute a sublattice of L. We show that these five elements are distinct, so that our sublattice is isomorphic to N_5:

Already $u < v$. Moreover, $b \not\leqq c$: otherwise, $b = b \wedge c \leqq u < v$ and $u = b \vee u = b \vee v = v$; and $a \not\leqq b$: otherwise, $b = b \vee a \geqq v > u$ and $u = b \wedge u = b \wedge v = v$. Hence $v < b \vee v$ (otherwise, $b \leqq v \leqq c$), $b \wedge u < u$ (otherwise, $a \leqq u \leqq b$), $b \wedge u < b$ (otherwise, $b = b \wedge u = b \wedge c \leqq c$), $b < b \vee v$ (otherwise, $b = b \vee v = b \vee a \geqq a$). $\square$

Chains. A minimal element m of a lattice L is necessarily a least element, since $m \wedge x \leqq m$ implies $m = m \wedge x \leqq x$, for all $x \in L$. Dually, a maximal element of a lattice is necessarily a greatest element. If a finite chain $x_0 < x_1 < \cdots < x_n$ is a maximal chain of L, then x_0 is a minimal element of L and x_n is a maximal element of L; hence a lattice that has a finite maximal chain has a least element 0 and a greatest element 1, and every finite maximal chain $x_0 < x_1 < \cdots < x_n$ has $x_0 = 0$ and $x_n = 1$.

Theorem 3.3. In a modular lattice, any two finite maximal chains have the same length.

For example, a finite maximal chain of submodules of a module is a composition series; by Theorem 3.3, all composition series of that module have the same length.

Proof. In a partially ordered set X, an element b *covers* an element a when $a < b$ and there is no $x \in X$ such that $a < x < b$. We denote this relation by $b \succ a$, or by $a \prec b$. In a lattice, a finite chain $x_0 < x_1 < \cdots < x_n$ is maximal if and only if $x_0 = 0$, $x_n = 1$, and $x_i \prec x_{i+1}$ for all $i < n$.

Lemma 3.4. In a modular lattice, $x \wedge y \prec x$ if and only if $y \prec x \vee y$.

Proof. Suppose that $x \wedge y \prec x$ but $y \not\prec x \vee y$. Then $x \wedge y < x$, $x \not\leqq y$, $y < x \vee y$, and $y < z < x \vee y$ for some z. Then $x \vee y \leqq x \vee z \leqq x \vee y$ and $x \vee z = x \vee y$. Also $x \wedge y \leqq x \wedge z \leqq x$, and $x \wedge z \prec x$: otherwise, $x \leqq z$ and $x \vee z = z < x \vee y$. Hence $x \wedge z < x$, and $x \wedge z = x \wedge y$. Then $y \vee (x \wedge z) = y \vee (x \wedge y) = y < z = (y \vee x) \wedge z$, contradicting modularity. Therefore $x \wedge y \prec x$ implies $y \prec x \vee y$. The converse implication is dual. $\square$

With Lemma 3.4 we can "pull down" maximal chains as follows.

Lemma 3.5. In a modular lattice, if $n \geqq 1$, $0 \prec y_1 \prec \cdots \prec y_n$, and $x \prec y_n$, then $0 \prec x_1 \prec \cdots \prec x_{n-1} = x$ for some $x_1, \ldots, x_{n-1}$.

Proof. By induction on n. There is nothing to prove if $n = 1$. Let $n > 1$. If $x = y_{n-1}$, then $y_1, \ldots, y_{n-1}$ serve. Now, assume that $x \neq y_{n-1}$. Then $x \not\leqq y_{n-1}$, $y_{n-1} < y_{n-1} \vee x \leqq y_n$, and $y_{n-1} \vee x = y_n \succ y_{n-1}$. By 3.4, $t = x \wedge y_{n-1} \prec x$ and $t \prec y_{n-1}$. Hence the induction hypothesis yields $0 \prec x_1 \prec \cdots \prec x_{n-2} = t \prec x$ for some $x_1, \ldots, x_{n-2}$. $\square$

Armed with this property we assail Theorem 3.3. Let L be a modular lattice with a finite maximal chain $0 = x_0 \prec x_1 \prec \cdots \prec x_m = 1$. We prove by induction on m that in every such lattice all finite maximal chains have length m. This is clear if $m = 0$ (then $L = \{0\}$) or $m = 1$ (then $L = \{0, 1\}$). If $m > 1$ and $0 = y_0 \prec y_1 \prec \cdots \prec y_n = 1$ is another finite maximal chain, then, by 3.5, $0 \prec z_1 \prec \cdots \prec z_{n-1} = x_{m-1}$ for some $z_1, \ldots, z_{n-2}$; then $0 = x_0 \prec x_1 \prec \cdots \prec x_{m-1}$ and $0 = z_0 \prec z_1 \prec \cdots \prec z_{n-1}$ are finite maximal chains of the modular lattice $L(x_{m-1}) = \{ x \in L \mid x \leqq x_{m-1} \}$, and the induction hypothesis yields $m - 1 = n - 1$ and $m = n$. $\square$

Exercises

1. Show that a lattice L is modular if and only if the equality $x \vee \left(y \wedge (x \vee t) \right) = (x \vee y) \wedge (x \vee t)$ holds for all $x, y, t \in L$.

2. Show that a lattice is modular if and only if $x \leqq t$ and $z \leqq y$ implies $x \vee \left(y \wedge (z \vee t) \right) = \left((x \vee y) \wedge z \right) \vee t$.

3. Show that a lattice is modular if and only if $a \wedge b = a \wedge c$, $a \vee b = a \vee c$, $b \leqq c$ implies $b = c$, when $a, b, c \in L$.

4. Show that every totally ordered set is a modular lattice.

5. Show that the normal subgroups of any group constitute a modular lattice.

6. Show that the lattice of all subgroups of a group need not be modular.

7. Verify directly that M_5 is modular.

In the following two exercises, a *closed interval* of a lattice L is a sublattice $[a, b] = \{ x \in L \mid a \leqq x \leqq b \}$ of L, where $a \leqq b$ in L.

8. In a modular lattice, show that $[a \wedge b, a]$ and $[b, a \vee b]$ are isomorphic lattices, for all a and b.

9. Let L be a lattice in which all maximal chains are finite and have the same length (the *length* of L). Further assume that $[a \wedge b, a]$ and $[b, a \vee b]$ have the same length, for all $a, b \in L$. Prove that L is modular.

10. Prove the following: in a modular lattice that has a finite maximal chain of length n, every chain is finite, of length at most n.

11. Let L be a lattice in which $x \prec x \vee y$ and $y \prec x \vee y$ implies $x \wedge y \prec x$ and $x \wedge y \prec y$. Prove that any two finite maximal chains of L have the same length.

4. Distributive Lattices

Distributive lattices are less general than modular lattices but still include some important examples. This section contains some structure results.

Distributive lattices are defined by the following equivalent properties.

Proposition **4.1.** *In a lattice L, the distributivity conditions*

(1) $x \vee (y \wedge z) = (x \vee y) \wedge (x \vee z)$ *for all $x, y, z \in L$,*

(2) $x \wedge (y \vee z) = (x \wedge y) \vee (x \wedge z)$ *for all $x, y, z \in L$,*

are equivalent, and imply modularity.

Proof. Assume (1). Then $x \leqq z$ implies $x \vee (y \wedge z) = (x \vee y) \wedge (x \vee z) = (x \vee y) \wedge z$. Hence L is modular. Then $x \wedge z \leqq x$ yields $(x \wedge z) \vee (y \wedge x) = x \wedge (z \vee (y \wedge x)) = x \wedge (z \vee y) \wedge (z \vee x) = x \wedge (z \vee y)$ and (2) holds. Dually, (2) implies (1). $\square$

Definition. A lattice is distributive *when it satisfies the equivalent conditions in Proposition 4.1.*

For example, the lattice 2^X of all subsets of a set X is distributive; so is every sublattice of 2^X. Other examples are given below and in the next section.

The opposite of a distributive lattice is a distributive lattice. Hence there is a duality principle for distributive lattices: a theorem that holds in every distributive lattice remains true when the order relation is reversed.

By Proposition 4.1, a distributive lattice is modular. The lattice M_5:

is modular but not distributive, since $a \vee (b \wedge c) = a$ but $(a \vee b) \wedge (a \vee c) = 1$. In fact, M_5 and N_5 are the quintessential nondistributive lattices:

Theorem **4.2** (Birkhoff [1934]). *A lattice is distributive if and only if it contains no sublattice that is isomorphic to M_5 or to N_5.*

Proof. A sublattice of a distributive lattice is distributive and is not, therefore, isomorphic to M_5 or N_5.

Conversely, assume that the lattice L is not distributive. We may assume that L is modular: otherwise, L contains a sublattice that is isomorphic to N_5, by 3.2, and the theorem is proved. Since L is not distributive, $a \wedge (b \vee c) \neq (a \wedge b) \vee (a \wedge c)$ for some $a, b, c \in L$. Let

$$u = (a \wedge b) \vee (b \wedge c) \vee (c \wedge a) \text{ and } v = (a \vee b) \wedge (b \vee c) \wedge (c \vee a).$$

Then $u \leqq v$, since $a \wedge b \leqq a \vee b$, etc. Let

$$
\begin{aligned}
x &= u \vee (a \wedge v) = (u \vee a) \wedge v, \\
y &= u \vee (b \wedge v) = (u \vee b) \wedge v, \\
z &= u \vee (c \wedge v) = (u \vee c) \wedge v.
\end{aligned}
$$

Since L is modular and $a \wedge v = a \wedge (b \vee c)$, $b \vee u = b \vee (c \wedge a)$,

$$
\begin{aligned}
x \wedge y &= \big(u \vee (a \wedge v)\big) \wedge \big(u \vee (b \wedge v)\big) \\
&= u \vee \big((a \wedge v) \wedge \big((u \vee b) \wedge v\big)\big) = u \vee \big((a \wedge v) \wedge (u \vee b)\big) \\
&= u \vee \big((a \wedge (b \vee c)) \wedge (b \vee (c \wedge a))\big) \\
&= u \vee \big(((a \wedge (b \vee c)) \wedge b) \vee (c \wedge a)\big) \\
&= u \vee (a \wedge b) \vee (c \wedge a) = u.
\end{aligned}
$$

Permuting a, b, and c yields $y \wedge z = z \wedge x = u$. Dually, $x \vee y = y \vee z = z \vee x = v$. Thus $\{u, v, x, y, z\}$ is a sublattice of L. We show that u, v, x, y, z are distinct, so that $\{u, v, x, y, z\} \cong M_5$:

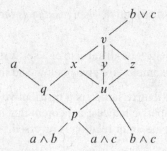

Since $a \wedge (b \vee c) \neq (a \wedge b) \vee (a \wedge c)$, but $a \wedge b \leqq a \wedge (b \vee c)$ and $a \wedge c \leqq a \wedge (b \vee c)$, we have

$$
p = (a \wedge b) \vee (a \wedge c) < a \wedge (b \vee c) = q.
$$

Now, $a \wedge v = a \wedge (b \vee c) = q$ and modularity yields

$$
\begin{aligned}
u \wedge a &= \big(((a \wedge b) \vee (a \wedge c)) \vee (b \wedge c)\big) \wedge a \\
&= \big((a \wedge b) \vee (a \wedge c)\big) \vee \big((b \wedge c) \wedge a\big) = (a \wedge b) \vee (a \wedge c) = p.
\end{aligned}
$$

Therefore $u < v$. Hence x, y, z are distinct (if, say, $x = y$, then $u = x \wedge y = x \vee y = v$) and distinct from u and v (if, say, $x = u$, then $y = x \vee y = v$ and $z = x \vee z = v = y$). $\square$

Irreducible elements. We now turn to structure theorems, the first of which uses order ideals and irreducible elements. An element i of a lattice L is *irreducible* (short for sup irreducible) when i is not a minimal element of L and $x \vee y = i$ implies $x = i$ or $y = i$. For example, the irreducible elements of 2^X are the one element subsets of X.

Lemma **4.3.** *In a lattice L that satisfies the descending chain condition, every element of L is the supremum of a set of irreducible elements of L.*

Proof. Assume that there is an element of L that is not the supremum of a set of irreducible elements of L. By the d.c.c., there is an element m of L that is minimal with this unsavory property. Then m is not a minimal element of L: otherwise, m is the least element of L and is the supremum of an empty set of irreducible elements. Also, m is not irreducible: otherwise, m is the supremum of the set $\{m\}$ of irreducible elements. Therefore $m = x \vee y$ for some $x, y \neq m$. Then $x < m$ and $y < m$; by the choice of m, x and y are supremums of sets of irreducible elements. But then so is $m = x \vee y$. $\square$

We denote by $\mathrm{Irr}\,(L)$ the set of all irreducible elements of L; $\mathrm{Irr}\,(L) \subseteq L$ is partially ordered, with $i \leqq j$ in $\mathrm{Irr}\,(L)$ if and only if $i \leqq j$ in L.

Definition. An order ideal *of a partially ordered set* S *is a subset* I *of* S *such that* $x \leqq y \in I$ *implies* $x \in I$.

Order ideals have been called a variety of other names.

Proposition **4.4.** *The order ideals of a partially ordered set* S, *partially ordered by inclusion, constitute a distributive lattice* $\mathrm{Id}\,(S)$.

Proof. First, S is an order ideal of itself, and every intersection of order ideals of S is an order ideal of S. By 2.1, $\mathrm{Id}\,(S)$ is a complete lattice, in which infimums are intersections. Moreover, every union of order ideals of S is an order ideal of S, so that supremums in $\mathrm{Id}\,(S)$ are unions. Hence $\mathrm{Id}\,(S)$ is a sublattice of 2^S and is distributive. $\square$

We show that Proposition 4.4 yields all finite distributive lattices.

Theorem **4.5.** *A finite lattice* L *is distributive if and only if* $L \cong \mathrm{Id}\,(S)$ *for some finite partially ordered set* S, *namely* $S = \mathrm{Irr}\,(L)$.

In Theorem 4.5, S is unique up to isomorphism: readers will show that $\mathrm{Irr}\,\big(\mathrm{Id}\,(S)\big) \cong S$ when S is finite. hence $L \cong \mathrm{Id}\,(S)$ implies $S \cong \mathrm{Irr}\,\big(\mathrm{Id}\,(S)\big)$ $\cong \mathrm{Irr}\,(L)$.

Proof. Let L be distributive. For every $x \in L$,

$$\theta(x) = \{\, i \in \mathrm{Irr}\,(L) \mid i \leqq x \,\}$$

is an order ideal of $\mathrm{Irr}\,(L)$. By 4.3, x is the supremum of a set J of irreducible elements of L; then $J \subseteq \theta(x)$; therefore $x = \bigvee_{i \in \theta(x)} i$. Hence θ is injective.

Let I be an order ideal of $\mathrm{Irr}\,(L)$. Let $x = \bigvee_{i \in I} i$. Then $I \subseteq \theta(x)$. Conversely, if $j \in \theta(x)$, then, since L is distributive,

$$j = j \wedge \big(\textstyle\bigvee_{i \in I} i\big) = \bigvee_{i \in I} (j \wedge i);$$

since j is irreducible, $j = j \wedge i$ for some $i \in I$, so that $j \leqq i \in I$ and $j \in I$. Hence $I = \theta(x)$. Thus θ is a bijection of L onto $\mathrm{Id}\,\big(\mathrm{Irr}\,(L)\big)$. The inverse bijection sends I to $\bigvee_{i \in I} i$; both θ and its inverse are order preserving. $\square$

Birkhoff's theorem. Theorem 4.5 implies that a finite distributive lattice is isomorphic to a sublattice of 2^X for some set X. Birkhoff's theorem extends this property to every distributive lattice. The proof uses ideals of lattices.

Definitions. An ideal *of a lattice L is an order ideal I of L such that* $x, y \in I$ *implies* $x \vee y \in I$. *A principal ideal of L is an ideal of the form* $L(a) = \{ x \in L \mid x \leqq a \}$ *for some* $a \in L$.

The name "ideal" was bestowed in earlier, more innocent times, when $x \wedge y$ was denoted by xy and $x \vee y$ was denoted by $x + y$, and the definition of ideals of a lattice resembled the definition of ideals of a commutative ring. Ideals of distributive lattices also share properties with ideals of rings, such as Proposition 4.6 and Lemma 4.7 below.

Every lattice L is an ideal of itself, and every intersection of ideals of L is an ideal of L; by Proposition 2.1, the ideals of L constitute a complete lattice, in which infimums are intersections. Supremums are more complex; however, we have:

Proposition **4.6.** *Let A and B be ideals of a distributive lattice L. In the lattice of ideals of L,* $A \vee B = \{ a \vee b \mid a \in A, b \in B \}$.

Proof. Let $C = \{ a \vee b \mid a \in A, b \in B \}$. An ideal of L that contains both A and B also contains C. Hence it suffices to show that C is an ideal. If $x \leqq y \in C$, then $y = a \vee b$ for some $a \in A$ and $b \in B$, and $x = x \wedge (a \vee b) = (x \wedge a) \vee (x \wedge b) \in C$, since $x \wedge a \in A$ and $x \wedge b \in B$. Moreover, $x, y \in C$ implies $x \vee y \in C$, since both A and B have this property. $\square$

Definition. A prime *ideal of a lattice L is an ideal* $P \neq \emptyset, L$ *such that* $x \wedge y \in P$ *implies* $x \in P$ *or* $y \in P$.

Lemma **4.7.** *Let I be an ideal of a distributive lattice L. For every* $a \in L$, $a \notin I$ *there exists a prime ideal P of L that contains I but not a.*

Proof. The union of a chain of ideals of L is an ideal of L; hence the union of a chain of ideals of L that contain I but not a is an ideal of L that contains I but not a. By Zorn's lemma, there is an ideal P of L that contains I but not a and is maximal with this property. Then P is prime: Let $x, y \in L, x, y \notin P$. In the lattice of ideals of L, $P \vee L(x)$ and $P \vee L(y)$ properly contain P. By the choice of P, both $P \vee L(x)$ and $P \vee L(y)$ contain a. By 4.6, $a = p \vee z = q \vee t$ for some $p, q \in P$ and $z \leqq x, t \leqq y$. Hence

$$a = (p \vee z) \wedge (q \vee t) = (p \wedge q) \wedge (p \wedge t) \wedge (z \wedge q) \wedge (z \wedge t).$$

Now, $p \wedge q, p \wedge t, z \wedge q \in P$ but $a \notin P$; therefore $z \wedge t \notin P$ and $x \wedge y \notin P$. $\square$

We can now prove Birkhoff's theorem:

Theorem **4.8** (Birkhoff [1933]). *A lattice L is distributive if and only if it is isomorphic to a sublattice of* 2^X *for some set X.*

Proof. Let X be the set of all prime ideals of L. Define $V : L \longrightarrow 2^X$ by

$$V(x) = \{ P \in X \mid x \notin P \}.$$

By 4.9 below (stated separately for future reference), V is a lattice homomorphism, so that Im V is a sublattice of 2^X. Since $x \leqq y$ if and only if $V(x) \subseteq V(y)$, V is an isomorphism of L onto Im V. $\square$

Lemma 4.9. *Let L be a distributive lattice. Define*

$$V(a) = \{ P \in X \mid a \notin P \}, \text{ where } a \in L.$$

Then V is injective; $a \leqq b$ if and only if $V(a) \subseteq V(b)$; $V(a \wedge b) = V(a) \cap V(b)$; and $V(a \vee b) = V(a) \cup V(b)$, for all $a, b \in L$.

Proof. If P is a prime ideal, then $a \wedge b \notin P$ if and only if $a \notin P$ and $b \notin P$; therefore $V(a \wedge b) = V(a) \cap V(b)$. If P is any ideal, then $a \vee b \notin P$ if and only if $a \notin P$ or $b \notin P$; therefore $V(a \vee b) = V(a) \cup V(b)$.

If $a \not\leqq b$, then 4.7 provides a prime ideal P that contains $L(b)$ but not a, so that $P \in V(a) \setminus V(b)$. If $b \neq a$, then either $a \not\leqq b$ or $b \not\leqq a$; in either case, $V(b) \neq V(a)$. Hence V is injective. Then $V(a) \subseteq V(b)$ implies $V(a) = V(a) \cap V(b) = V(a \wedge b)$ and $a = a \wedge b \leqq b$. Conversely, $a \leqq b$ implies $V(a) \subseteq V(b)$. $\square$

Exercises

1. Show that a lattice L is distributive if and only if $(x \wedge y) \vee (y \wedge z) \vee (z \wedge x) = (x \vee y) \wedge (y \vee z) \wedge (z \vee x)$ for all $x, y, z \in L$.

2. Show that a lattice L is distributive if and only if $a \wedge b = a \wedge c$, $a \vee b = a \vee c$ implies $b = c$, when $a, b, c \in L$.

3. Prove or disprove: the lattice of all subgroups of $\mathbb{Z}$ is distributive.

4. Find an abelian group G such that the lattice of all subgroups of G is not distributive.

5. Show that $\mathrm{Irr}\left(\mathrm{Id}\,(S)\right) \cong S$ for every finite partially ordered set S.

6. Say that $x = x_1 \vee x_2 \vee \cdots \vee x_n$ is an irredundant supremum when $x_1 \vee \cdots \vee x_{i-1} \vee x_{i+1} \vee \cdots \vee x_n < x$ for all i. Show that every element of a finite distributive lattice can be written uniquely as an irredundant supremum of irreducible elements.

7. Prove that every maximal chain of a finite distributive lattice L has length $\left|\mathrm{Irr}\,(L)\right|$.

8. Show that the lattice of subgroups of a group G is distributive if and only if every finitely generated subgroup of G is cyclic.

5. Boolean Lattices

Boolean lattices generalize the lattice of subsets of a set. They were introduced by Boole [1847] for use in mathematical logic, as formal algebraic systems in which the properties of infimums, supremums, and complements match those of conjunctions, disjunctions, and negations. Boolean lattices are still in use today, as a source of models of set theory, and in the design of electronic logic circuits.

Definition. Let L be a lattice with a least element 0 and a greatest element 1. A *complement* of an element a of L is an element a' of L such that $a \wedge a' = 0$ and $a \vee a' = 1$. For example, the usual complement $X \backslash Y$ of a subset Y of a set X is a complement in the lattice 2^X.

Proposition **5.1.** *In a distributive lattice with a least element and a greatest element:* (1) *an element has at most one complement;* (2) *if a' is the complement of a and b' is the complement of b, then $a' \vee b'$ is the complement of $a \wedge b$, and $a' \wedge b'$ is the complement of $a \vee b$.*

Proof. (1). If b and c are complements of a, then

$$b = b \wedge (a \vee c) = (b \wedge a) \vee (b \wedge c) = b \wedge c \leqq c;$$

exchanging b and c then yields $c \leqq b$. (2). By distributivity,

$$(a \wedge b) \wedge (a' \vee b') = (a \wedge b \wedge a') \vee (a \wedge b \wedge b') = 0 \vee 0 = 0.$$

Dually, $(a' \vee b') \vee (a \wedge b) = 1$. Hence $a' \vee b'$ is a complement of $a \wedge b$. Dually, $a' \wedge b'$ is a complement of $a \vee b$. $\square$

Definition. A Boolean lattice, *also called a* Boolean algebra, *is a distributive lattice with a least element and a greatest element, in which every element has a complement.*

For example, the lattice 2^X of all subsets of a set X is a Boolean lattice.

The opposite of a Boolean lattice L is a Boolean lattice; in fact, $L^{\mathrm{op}} \cong L$, by Proposition 5.1. Hence there is a duality principle for Boolean lattices: a theorem that holds in every Boolean lattice remains true when the order relation is reversed.

Boolean rings. The next examples come from rings.

Definition. A Boolean ring *is a ring R [with an identity element] in which $x^2 = x$ for all $x \in R$.*

The name "Boolean" is justified by Proposition 5.3, 5.4 below.

Lemma **5.2.** *A Boolean ring is commutative and has characteristic 2.*

Proof. If R is Boolean, then

$$x + x = (x + x)(x + x) = x^2 + x^2 + x^2 + x^2 = x + x + x + x$$

and $x + x = 0$, for all $x \in R$. Then

$$x + y = (x + y)(x + y) = x^2 + xy + yx + y^2 = x + xy + yx + y,$$

whence $yx = -xy = xy$, for all $x, y \in R$. $\square$

Proposition **5.3.** *If R is a Boolean ring, then R, partially ordered by $x \leqq y$ if and only if $xy = x$, is a Boolean lattice $\mathrm{L}(R)$, in which $x \wedge y = xy$, $x \vee y = x + y + xy$, and $x' = 1 - x$.*

The proof is an exercise.

Proposition **5.4** (Stone [1936]). *If L is a Boolean lattice, then L, with addition and multiplication*

$$x + y = (x' \wedge y) \vee (x \wedge y'), \quad xy = x \wedge y,$$

is a Boolean ring $R(L)$. *Moreover,* $L(R(L)) = L$ *and* $R(L(R)) = R$ *for every Boolean ring* R.

Proof. The addition on L is commutative; readers who love computation will delight in showing that it is associative. Moreover, $x + 0 = x$ and $x + x = 0$ for all $x \in L$; hence $(L, +)$ is an abelian group.

The multiplication on L is commutative, associative, and idempotent, by 1.2. Moreover, $1x = x$ for all $x \in L$, and 5.1 yields

$$
\begin{aligned}
xz + yz &= \left((x' \vee z') \wedge (y \wedge z)\right) \vee \left((x \wedge z) \wedge (y' \vee z')\right) \\
&= (x' \wedge y \wedge z) \vee (z' \wedge y \wedge z) \vee (x \wedge z \wedge y') \vee (x \wedge z \wedge z') \\
&= (x' \wedge y \wedge z) \vee (x \wedge z \wedge y') \\
&= \left((x' \wedge y) \vee (x \wedge y')\right) \wedge z = (x + y)z
\end{aligned}
$$

for all $x, y, z \in L$. Thus $R(L)$ is a Boolean ring. Readers will verify that $L(R(L)) = L$ and $R(L(R)) = R$ for every Boolean ring R : $\square$

Finite Boolean lattices. We now apply Theorems 4.5 and 4.8 to Boolean lattices. An *atom* of a Boolean lattice is a minimal nonzero element (an element $a > 0$ with no $a > b > 0$). For example, the atoms of 2^X are the one element subsets of X. Readers will verify that the atoms of a Boolean lattice are precisely its irreducible elements.

Theorem **5.5.** *A finite lattice L is Boolean if and only if $L \cong 2^X$ for some finite set X.*

Proof. The lattice 2^X is always Boolean. Conversely, let L be Boolean. By 4.5, $L \cong \mathrm{Id}(S)$, where $S = \mathrm{Irr}(L)$ is the partially ordered set of all atoms (irreducible elements) of L. Since the atoms of L satisfy no strict inequality $i < j$, every subset of S is an order ideal of S, and $L \cong \mathrm{Id}(S) = 2^S$. $\square$

A *Boolean sublattice* of a Boolean lattice L is a sublattice S such that $0 \in S$, $1 \in S$, and $x \in S$ implies $x' \in S$. A Boolean sublattice of L is a Boolean lattice in its own right; this lattice is also called a Boolean sublattice of L.

We saw that groups, defined as sets with a suitable binary operation, also enjoy a constant "identity element" operation and a unary $x \longmapsto x^{-1}$ operation; and that a subgroup is a subset that is closed under all three operations. Similarly, Boolean lattices have, besides their two binary operations $\wedge$ and $\vee$, two constant 0 and 1 operations and a unary $x \longmapsto x'$ operation; a Boolean sublattice is a subset that is closed under all five operations.

Theorem **5.6** (Birkhoff [1933]). *A lattice L is Boolean if and only if it is isomorphic to a Boolean sublattice of 2^X for some set X.*

Proof. Let L be Boolean and let X be the set of all prime ideals of L. Define $V : L \longrightarrow 2^X$ by $V(x) = \{\, P \in X \mid x \notin P \,\}$. By 4.9 and 5.7 below, V is a homomorphism of Boolean lattices, so that Im V is a Boolean sublattice of 2^X, and V is an isomorphism of L onto Im V. $\square$

Stone's theorem. Stone [1934] used topology to sharpen Theorem 5.6.

Lemma **5.7.** *Let L be a Boolean lattice. The sets*

$$V(a) = \{\, P \in X \mid a \notin P \,\}, \text{ where } a \in L,$$

constitute a basis for a topology on the set X of all prime ideals of L. Moreover, $V(0) = \emptyset$, $V(1) = X$, and $V(a') = V(a)'$, for all $a \in L$.

Proof. By 4.9, $V(a \wedge b) = V(a) \cap V(b)$ for all $a, b \in L$; hence the sets $V(a)$ with $a \in L$ constitute a basis of open sets for a topology on X. If P is a prime ideal of L, then $0 \in P$, since $P \neq \emptyset$, and $1 \notin P$, since $P \neq L$; therefore $V(0) = \emptyset$ and $V(1) = X$. By 4.9, $V(a \vee a') = V(a) \cup V(a') = X$, $V(a) \cap V(a') = V(a \wedge a') = \emptyset$, and $V(a') = V(a)'$. $\square$

The *Stone space* of a Boolean lattice L is set of all its prime ideals, with the topology specified by Lemma 5.7.

Proposition **5.8.** *The Stone space of a Boolean lattice is compact Hausdorff and totally disconnected.*

Proof. Let X be the Stone space of a Boolean lattice L. If $P \neq Q$ in X, then, say, $a \in P$, $a \notin Q$ for some $a \in L$, and then $Q \in V(a)$ and $P \in V(a') = V(a)'$. Therefore X is Hausdorff. Moreover, every $V(a)$ is open and closed (since $V(a)' = V(a')$ is open); hence X is totally disconnected.

To prove that X is compact we show that every ultrafilter $\mathcal{U}$ on X converges. Since $\mathcal{U}$ is a ultrafilter, $V(a) \notin \mathcal{U}$ implies $V(a') = V(a)' \in \mathcal{U}$; conversely, $V(a') \in \mathcal{U}$ implies $V(a) \notin \mathcal{U}$: otherwise, $\emptyset = V(a) \cap V(a') \in \mathcal{U}$. Let

$$P = \{\, a \in L \mid V(a) \notin \mathcal{U} \,\}.$$

If $a \leqq b \in P$, then $V(a) \subseteq V(b) \notin \mathcal{U}$; hence $V(a) \notin \mathcal{U}$ and $a \in P$. If $a, b \in P$, then $V(a), V(b) \notin \mathcal{U}$, $V(a'), V(b') \in \mathcal{U}$, $V(a' \wedge b') = V(a') \cap V(b') \in \mathcal{U}$, $V(a \vee b) \notin \mathcal{U}$, and $a \vee b \in P$. If $a, b \notin P$, then $V(a), V(b) \in \mathcal{U}$, $V(a \wedge b) = V(a) \cap V(b) \in \mathcal{U}$, and $a \wedge b \notin P$. Thus $P \in X$. Then $\mathcal{U}$ converges to P: when $V(a)$ is a neighborhood of P, then $P \in V(a)$, $a \notin P$, and $V(a) \in \mathcal{U}$. $\square$

A *Stone space* is a topological space that is compact Hausdorff and totally disconnected. The Stone space of a Boolean lattice L has these properties. Conversely, in any topological space X, every finite union, finite intersection, or complement of closed and open subsets is closed and open; hence the closed and open subsets of X constitute a Boolean sublattice $L(X)$ of 2^X.

Theorem **5.9** (Stone [1934]). *Every Boolean lattice is isomorphic to the lattice of closed and open subsets of its Stone space.*

Readers may prove a converse: every Stone space is homeomorphic to the Stone space of its lattice of closed and open subsets.

Proof. Let L be a Boolean lattice and let X be its Stone space. For every $a \in L$, $V(a)$ is open in X, and is closed in X since $X \backslash V(a) = V(a')$ is open. Conversely, if $U \in L(X)$ is a closed and open subset of X, then U is a union of basic open sets $V(a) \subseteq U$; since U is closed, U is compact, U is a finite union $V(a_1) \cup \cdots \cup V(a_n) = V(a_1 \vee \cdots \vee a_n)$, and $U = V(a)$ for some $a \in L$. Thus V is a mapping of L onto $L(X)$. By 4.9 and 5.7, $V : L \longrightarrow L(X)$ is a lattice isomorphism. $\square$

Exercises

1. Let D be the set of all positive divisors of some $n \in \mathbb{N}$, partially ordered by $x \leqq y$ if and only if x divides y. Show that D is a distributive lattice. When is D a Boolean lattice?

2. A *cofinite* subset of a set X is a subset S of X whose complement $X \backslash S$ is finite. Show that the subsets of X that are either finite or cofinite constitute a Boolean lattice.

3. Show that a direct product of Boolean lattices is a Boolean lattice, when ordered componentwise.

4. A *central idempotent* of a ring R [with an identity element] is an element e of R such that $e^2 = e$ and $ex = xe$ for all $x \in R$. Show that the central idempotents of R constitute a Boolean lattice when ordered by $e \leqq f$ if and only if $ef = e$. (Hint: $e \vee f = e + f - ef$.)

5. Verify that a Boolean ring, partially ordered by $x \leqq y$ if and only if $xy = x$, is a Boolean lattice, in which $x \wedge y = xy$, $x \vee y = x + y + xy$, and $x' = 1 - x$.

6. Verify that the addition $x + y = (x' \wedge y) \vee (x \wedge y')$ on a Boolean lattice is associative.

7. Verify that $L(R(L)) = L$ for every Boolean lattice L, and that $R(L(R)) = R$ for every Boolean ring R.

8. Show that a Boolean lattice L and its Boolean ring $R(L)$ have the same ideals.

9. Construct a purely lattice-theoretic quotient L/I of a Boolean lattice L by a lattice ideal I of L.

10. Verify that $R\left(\prod_{i \in I} L_i\right) \cong \prod_{i \in I} R(L_i)$ for all Boolean lattices $(L_i)_{i \in I}$.

11. Recall that a *closed interval* of a lattice L is a sublattice $[a, b] = \{ x \in L \mid a \leqq x \leqq b \}$ of L, where $a \leqq b$ in L. Show that every closed interval of a Boolean lattice L is a Boolean lattice (though not a Boolean sublattice of L).

12. Show that the irreducible elements of a Boolean lattice are its atoms.

A *generalized Boolean lattice* is a lattice L with a least element 0 such that every interval $[0, x]$ is a Boolean lattice.

13. Show that the finite subsets of any set X constitute a generalized Boolean lattice.

14. Show that Propositions 5.3 and 5.4 extend to generalized Boolean lattices, if rings are not required to have an identity element.

15. Show that the identities $\left(\bigvee_{i \in I} x \right) \wedge y = \bigvee_{i \in I} (x_i \wedge y)$ and $\left(\bigwedge_{i \in I} x \right) \vee y = \bigwedge_{i \in I} (x_i \vee y)$ hold in every Boolean lattice that is complete.

16. Show that the identities $\left(\bigvee_{i \in I} x \right)' = \bigwedge_{i \in I} x_i'$ and $\left(\bigwedge_{i \in I} x \right)' = \bigvee_{i \in I} x_i'$ hold in every Boolean lattice that is complete.

17. Show that a complete Boolean lattice L is isomorphic to the lattice of all subsets of a set if and only if every element of L is the supremum of a set of atoms of L.

18. Show that every compact Hausdorff and totally disconnected topological space is homeomorphic to the Stone space of its lattice of closed and open subsets.

XV
Universal Algebra

Universal algebra is the study of algebraic objects in general, also called *universal algebras*. These general objects were first considered by Whitehead [1898]. Birkhoff [1935], [1944] initiated their systematic study.

Varieties are classes of universal algebras defined by identities. Groups, rings, left *R*-modules, etc., constitute varieties, and many of their properties are in fact properties of varieties. The main results in this chapter are two theorems of Birkhoff, one that characterizes varieties, one about subdirect decompositions. The chapter draws examples from Chapters I, III, V, VIII, and XIV, and is otherwise independent of previous chapters.

1. Universal Algebras

A universal algebra is a set with any number of operations. This section gives basic properties, such as the homomorphism and factorization theorems.

Definitions. Let $n \geq 0$ be a nonnegative integer. An n-ary operation ω on a set X is a mapping of X^n into X, where X^n is the Cartesian product of n copies of X; the number n is the arity *of ω.*

An operation of arity 2 is a *binary operation*. An operation of arity 1 or *unary* operation on a set X is simply a mapping of X into X. By convention, the empty cardinal product X^0 is your favorite one element set, for instance, $\{\emptyset\}$; hence an operation of arity 0 or *constant* operation on a set X merely selects one element of X. Binary operations predominate in previous chapters, but constant and unary operations were encountered occasionally.

There are operations of infinite arity (for instance, infimums and supremums in complete lattices), but many properties in this chapter require finite arity. Order relations and partial operations are excluded for the same reason (a *partial operation* on a set X is a mapping of a subset of X^n into X and need not be defined for all $(x_1, \ldots, x_n) \in X^n$).

Universal algebras are classified by their *type*, which specifies number of operations and arities:

Definitions. A type *of universal algebras is an ordered pair of a set* T *and a mapping* $\omega \longmapsto n_\omega$ *that assigns to each* $\omega \in T$ *a nonnegative integer* n_ω, *the formal* arity *of* ω. A universal algebra, *or just* algebra, *of type* T *is an ordered pair of a set* A *and a mapping, the* type-T algebra structure *on* A, *that assigns to each* $\omega \in T$ *an operation* ω_A *on* A *of arity* n_ω.

For clarity ω_A is often denoted by just ω. For example, rings and lattices are of the same type, which has two elements of arity 2. Sets are universal algebras of type $T = \emptyset$. Groups and semigroups are of the same type, which has one element of arity 2. Groups may also be viewed as universal algebras with one binary operation, one constant operation that selects the identity element, and one unary operation $x \longmapsto x^{-1}$; the corresponding type has one element of arity 0, one element of arity 1, and one element of arity 2. Left R-modules are universal algebras with one binary operation (addition) and one unary operation $x \longmapsto rx$ for every $r \in R$. These descriptions will be refined in Section 2 when we formally define identities.

On the other hand, partially ordered sets and topological spaces are not readily described as universal algebras. Section XVI.10 explains why, to some extent.

Subalgebras of an algebra are subsets that are closed under all operations:

Definition. A subalgebra *of a universal algebra* A *of type* T *is a subset* S *of* A *such that* $\omega\,(x_1, \,\ldots,\, x_n) \in S$ *for all* $\omega \in T$ *of arity* n *and* $x_1, \,\ldots,\, x_n \in S$.

Let S be a subalgebra of A. Every operation ω_A on A has a restriction ω_S to S (sends S^n into S, if ω has arity n). This makes S an algebra of the same type as A, which is also called a *subalgebra* of A.

Readers will verify that the definition of subalgebras encompasses subgroups, subrings, submodules, etc., provided that groups, rings, modules, etc. are defined as algebras of suitable types. Once started, they may as well prove the following:

Proposition **1.1.** *The intersection of subalgebras of a universal algebra* A *is a subalgebra of* A.

Proposition **1.2.** *The union of a nonempty directed family of subalgebras of a universal algebra* A *is a subalgebra of* A. *In particular, the union of a nonempty chain of subalgebras of a universal algebra* A *is a subalgebra of* A.

Proposition 1.2 becomes false if infinitary operations are allowed.

Homomorphisms are mappings that preserve all operations.

Definition. Let A *and* B *be universal algebras of the same type* T. *A homomorphism of* A *into* B *is a mapping* $\varphi : A \longrightarrow B$ *such that*

$$\varphi\left(\omega_A\,(x_1, \,\ldots,\, x_n)\right) = \omega_B\left(\varphi(x_1), \,\ldots,\, \varphi(x_n)\right)$$

for all $n \geqq 0$, *all* $\omega \in T$ *of arity* n, *and all* $x_1, \,\ldots,\, x_n \in A$.

Readers will see that this definition yields homomorphisms of groups, rings, R-modules, lattices, and so forth. In general, the identity mapping 1_A on a universal algebra A is a homomorphism. If $\varphi : A \longrightarrow B$ and $\psi : B \longrightarrow C$ are homomorphisms of algebras of the same type, then so is $\psi \circ \varphi : A \longrightarrow C$.

An *isomorphism* of universal algebras of the same type is a bijective homomorphism; then the inverse bijection is also an isomorphism. If S is a subalgebra of A, then the inclusion mapping $S \longrightarrow A$ is a homomorphism, the *inclusion homomorphism* of S into A.

Quotient algebras. Universal algebras differ from groups, and from group based structures like rings and modules, in that quotient algebras must in general be constructed from equivalence relations, rather than from subalgebras. For example, this is the case with sets, semigroups, and lattices.

In the case of sets, every mapping $f : X \longrightarrow Y$ induces an equivalence relation $f(x) = f(y)$ on X, which we denote by $\ker f$. Conversely, when $\mathcal{E}$ is an equivalence relation on a set X, there is a *quotient set* $X/\mathcal{E}$, which is the set of all equivalence classes, and a *canonical projection* $\pi : X \longrightarrow X/\mathcal{E}$, which assigns to each $x \in X$ its equivalence class; and then $\mathcal{E} = \ker \pi$.

Algebra structures are inherited by quotient sets as follows.

Proposition **1.3.** *Let A be a universal algebra of type T. For an equivalence relation $\mathcal{E}$ on A the following conditions are equivalent:*

(1) *there exists a type-T algebra structure on $A/\mathcal{E}$ such that the canonical projection $\pi : A \longrightarrow A/\mathcal{E}$ is a homomorphism;*

(2) *there exists a homomorphism $\varphi : A \longrightarrow B$ of universal algebras of type T such that $\ker \varphi = \mathcal{E}$;*

(3) *$x_1 \mathcal{E} y_1, ..., x_n \mathcal{E} y_n$ implies $\omega(x_1, ..., x_n \mathcal{E} \omega(y_1, ..., y_n)$, for all $n \geqq 0$, all $\omega \in T$ of arity n, and all $x_1, ..., x_n, y_1, ..., y_n \in A$.*

Then the algebra structure in (1) *is unique.*

Proof. (1) implies (2); that (2) implies (3) follows from the definitions.

(3) implies (1). Let $Q = A/\mathcal{E}$ and let $\pi : A \longrightarrow Q$ be the projection. For every $\omega \in T$ of arity n and every equivalence classes $E_1, ..., E_n$, the set

$$\omega_A(E_1, ..., E_n) = \{ \omega_A(x_1, ..., x_n) \mid x_1 \in E_1, ..., x_n \in E_n \}$$

is contained in a single equivalence class, by (3). This yields a mapping $\omega_Q : Q^n \longrightarrow Q$, which assigns to $(E_1, ..., E_n) \in Q^n$ the equivalence class $\omega_Q(E_1, ..., E_n)$ that contains $\omega_A(E_1, ..., E_n)$. Then

$$\pi\left(\omega_A(x_1, ..., x_n)\right) = \omega_Q\left(\pi(x_1), ..., \pi(x_n)\right)$$

for all $x_1, ..., x_n \in A$, by definition of $\omega_A(E_1, ..., E_n)$; equivalently, $\pi \circ \omega_A = \omega_Q \circ \pi^n$. Moreover, ω_Q is the only mapping with this property, since π^n is surjective. This constructs a type-T algebra structure on Q, which is the only structure such that π is a homomorphism. $\square$

Definitions. A congruence *on a universal algebra A is an equivalence relation* $\mathcal{E}$ *on A that satisfies the equivalent conditions in Proposition* 1.3; *then the universal algebra* $A/\mathcal{E}$ *is the* quotient *of A by* $\mathcal{E}$.

The quotient of a group G by a normal subgroup N is really the quotient of G by a congruence on G, namely, the partition of G into cosets of N, which is a congruence since $xN\,yN \subseteq xyN$ for all $x, y \in G$. In fact, all congruences on a group arise from normal subgroups (see the exercises). Readers will easily establish the following properties:

Proposition **1.4.** *The intersection of congruences on a universal algebra A is a congruence on A.*

Proposition **1.5.** *The union of a nonempty directed family of congruences on a universal algebra A is a congruence on A. In particular, the union of a nonempty chain of congruences on A is a congruence on A.*

Quotient algebras have a universal property:

Theorem **1.6** (Factorization Theorem). *Let A be a universal algebra and let* $\mathcal{E}$ *be a congruence on A. Every homomorphism of universal algebras* $\varphi : A \longrightarrow B$ *such that* ker φ *contains* $\mathcal{E}$ *factors uniquely through the canonical projection* $\pi : A \longrightarrow A/\mathcal{E}$ *(there exists a homomorphism* $\psi : A/\mathcal{E} \longrightarrow B$ *unique such that* $\varphi = \psi \circ \pi$):

$$A \xrightarrow{\ \pi\ } A/\mathcal{E}$$
$$\varphi \searrow \quad \downarrow \psi$$
$$B$$

Readers will prove a more general property:

Theorem **1.7** (Factorization Theorem). *Let* $\varphi : A \longrightarrow B$ *be a homomorphism of universal algebras. If* φ *is surjective, then every homomorphism* $\psi : A \longrightarrow C$ *of universal algebras such that* ker ψ *contains* ker φ *factors uniquely through* φ *(there exists a homomorphism* $\chi : B \longrightarrow C$ *unique such that* $\psi = \chi \circ \varphi$):

$$A \xrightarrow{\ \varphi\ } B$$
$$\psi \searrow \quad \downarrow \chi$$
$$C$$

The homomorphism theorem for universal algebras reads as follows:

Theorem **1.8** (Homomorphism Theorem). *If* $\varphi : A \longrightarrow B$ *is a homomorphism of universal algebras, then* ker φ *is a congruence on A,* Im φ *is a subalgebra of B, and*

$$A/\mathrm{ker}\,\varphi \cong \mathrm{Im}\,\varphi;$$

in fact, there is an isomorphism $\theta : A/\mathrm{ker}\,f \longrightarrow$ Im f *unique such that* $\varphi = \iota \circ \theta \circ \pi$, *where* $\iota :$ Im $f \longrightarrow B$ *is the inclusion homomorphism and* $\pi : A \longrightarrow A/\mathrm{ker}\,f$ *is the canonical projection:*

$$A \xrightarrow{\;\varphi\;} B$$

$$\pi \downarrow \qquad \uparrow \iota$$

$$A/\ker \varphi \dashrightarrow_{\theta} \operatorname{Im} \varphi$$

Proof. First, $\ker \varphi$ is a congruence on A by definition, and it is clear that $\operatorname{Im} \varphi$ is a subalgebra of B. Let $\theta : A/\ker \varphi \longrightarrow \operatorname{Im} \varphi$ be the bijection that sends an equivalence class E of $\ker \varphi$ to the sole element of $\varphi(E)$. Then $\iota \circ \theta \circ \pi = \varphi$, and θ is the only mapping of $A/\ker \varphi$ into $\operatorname{Im} \varphi$ with this property. We show that θ is a homomorphism. If $\omega \in T$ has arity n and $x_1, \ldots, x_n \in A$, then

$$\iota\big(\theta\big(\omega\left(\pi(x_1), \ldots, \pi(x_n)\right)\big)\big) = \iota\big(\theta\big(\pi\left(\omega(x_1, \ldots, x_n)\right)\big)\big)$$
$$= \omega\big(\iota\big(\theta\left(\pi(x_1)\right)\big), \ldots, \iota\big(\theta\left(\pi(x_n)\right)\big)\big) = \iota\big(\omega\big(\theta\left(\pi(x_1)\right), \ldots, \theta\left(\pi(x_n)\right)\big)\big),$$

since π, $\varphi = \iota \circ \theta \circ \pi$, and ι are homomorphisms. Hence

$$\theta\left(\omega\left(y_1, \ldots, y_n\right)\right) = \omega\left(\theta(y_1), \ldots, \varphi(y_n)\right)$$

for all $y_1, \ldots, y_n \in A/\ker \varphi$, since ι is injective and π is surjective. $\square$

The isomorphism theorems extend to universal algebras.

Proposition **1.9.** *Let* $\varphi : A \longrightarrow B$ *be a homomorphism of universal algebras. If* $\mathcal{E}$ *is a congruence on* B, *then* $\varphi^{-1}(\mathcal{E})$, *defined by*

$$x \, \varphi^{-1}(\mathcal{E}) \, y \text{ if and only if } \varphi(x) \, \mathcal{E} \, \varphi(y),$$

is a congruence on A. *If* φ *is surjective, then* $A/\varphi^{-1}(\mathcal{E}) \cong B/\mathcal{E}$, *and the above defines a one-to-one correspondence between congruences on* B *and congruences on* A *that contain* $\ker \varphi$.

The proof is an exercise; so is the second isomorphism theorem.

Exercises

1. Show that the intersection of subalgebras of an algebra A is a subalgebra of A.

2. Show that the union of a nonempty directed family of subalgebras of an algebra A is a subalgebra of A.

3. Let $A = \mathbb{R} \cup \{\infty\}$ be the algebra with one infinitary operation that assigns to each infinite sequence its least upper bound in A. Show that a directed union of subalgebras of A need not be a subalgebra of A.

4. Let $\varphi : A \longrightarrow B$ be a homomorphism of universal algebras, and let S be a subalgebra of A. Show that $\varphi(S) = \{ \varphi(x) \mid x \in S \}$ is a subalgebra of B.

5. Let $\varphi : A \longrightarrow B$ be a homomorphism of universal algebras, and let T be a subalgebra of B. Show that $\varphi^{-1}(T) = \{ x \in A \mid \varphi(x) \in T \}$ is a subalgebra of A.

6. Use the previous two exercises to produce a one-to-one correspondence between certain subalgebras of A and certain subalgebras of B.

7. Show that every congruence on a group is the partition into cosets of a unique normal subgroup; this defines a one-to-one correspondence between normal subgroups and congruences.

8. Produce a one-to-one correspondence between the ideals of a ring and its congruences.

9. Let S be a semigroup in which $xy = x$ for all $x, y \in S$. Show that every equivalence relation on S is a congruence. If S has five elements, then show that S has more congruences than subsets (hence there cannot be a one-to-one correspondence between suitable subsets of S and congruences on S).

10. Show that an equivalence relation on an algebra A is a congruence on A if and only if it is a subalgebra of $A \times A$.

11. Show that the intersection of congruences on an algebra A is a congruence on A.

12. Show that the union of a nonempty directed family of congruences on an algebra A is a congruence on A.

13. Let $\varphi : A \longrightarrow B$ be a surjective homomorphism. Show that every homomorphism $\psi : A \longrightarrow C$ such that ker ψ contains ker φ factors uniquely through φ.

14. Let $\varphi : A \longrightarrow B$ be a homomorphism and let $\mathcal{E}$ be a congruence on B. Show that $\varphi^{-1}(\mathcal{E})$ is a congruence on A.

15. If φ is surjective, then show that the previous exercise defines a one-to-one correspondence between congruences on B, and congruences on A that contain ker φ; and that $A/\varphi^{-1}(\mathcal{E}) \cong B/\mathcal{E}$.

16. Let A be a universal algebra, let S be a subalgebra of A, and let $\mathcal{E}$ be a congruence on A. Show that $T = \{ x \in A \mid x \, \mathcal{E} \, s \text{ for some } s \in S \}$ is a subalgebra of A. Show that $\mathcal{E}$ induces congruences $\mathcal{A}$ on S and $\mathcal{B}$ on T, and that $T/\mathcal{B} \cong S/\mathcal{A}$.

2. Word Algebras

Word algebras are free universal algebras of a given type, and lead to a formal definition of identities.

Generators. Since every intersection of subalgebras of a universal algebra A is a subalgebra of A, there is, for every subset X of A, a subalgebra of A *generated by* X, which is the least subalgebra of A that contains X, and is the intersection of all subalgebras of A that contain X. The following is an exercise:

Proposition **2.1.** *Let X be a subset of a universal algebra A of type T. Define $S_k \subseteq A$ for every integer $k \geqq 0$ by $S_0 = X$; if $k > 0$, then S_k is the set of all $\omega(w_1, \ldots, w_n)$ in which $\omega \in T$ has arity n and $w_1 \in S_{k_1}, \ldots, w_n \in S_{k_n}$, with $k_1, \ldots, k_n \geqq 0$ and $1 + k_1 + \cdots + k_n = k$. The subalgebra $\langle X \rangle$ of A generated by X is $\langle X \rangle = \bigcup_{k \geqq 0} S_k$.*

By 2.1, every element of $\langle X \rangle$ can be calculated in finitely many steps from elements of X and operations on A (using k operations when $x \in S_k$). In general, this calculation can be performed in several different ways. The simplest way to construct an algebra of type T that is generated by X is to ensure that different calculations yield different results. This is precisely what happens in the word algebra. Thus, word algebras are similar to free groups, except that, in word algebras, words like $x\,(yz)$ and $(xy)\,z$ are distinct, and words like xx^{-1} need not be omitted. Indeed, free groups must satisfy certain identities; word algebras are exempt from this requirement.

Construction. Given a type T of universal algebras and a set X, define a set W_k as follows: let $W_0 = X$; if $k > 0$, then W_k is the set of all sequences $(\omega, w_1, \ldots, w_n)$ in which $\omega \in T$ has arity n and $w_1 \in W_{k_1}, \ldots, w_n \in W_{k_n}$, where $k_1, \ldots, k_n \geqq 0$ and $1 + k_1 + \cdots + k_n = k$. This classifies words by the number k of operations $\omega \in T$ that appear in them.

For instance, if T consists of a single element μ of arity 2, then $W_0 = X$; the elements of W_1 are all (μ, x, y) with $x, y \in X$; the elements of W_2 are all $(\mu, x, \mu(y, z))$ and $(\mu, \mu(x, y), z)$ with $x, y, z, t \in X$; and so forth.

Definition. The word algebra *of type T on the set X is the union $W = W_X^T = \bigcup_{k \geqq 0} W_k$, with operations defined as follows: if $\omega \in T$ has arity n and $w_1 \in W_{k_1}, \ldots, w_n \in W_{k_n}$, then $\omega_W(w_1, \ldots, w_n) = (\omega, w_1, \ldots, w_n) \in W_k$, where $k = 1 + k_1 + \cdots + k_n$.*

Proposition 2.2. *If $w \in W_X^T$, then $w \in W_Y^T$ for some finite subset Y of X.*

Proof. We have $w \in W_k$ for some k and prove the result by induction on k. If $w \in W_0$, then $w \in X$ and $Y = \{w\}$ serves. If $k > 0$ and $w \in W_k$, then $w = (\omega, w_1, \ldots, w_n)$, where $\omega \in T$ has arity n and $w_1 \in W_{k_1}, \ldots, w_n \in W_{k_n}$ for some $k_1, \ldots, k_n < k$. By the induction hypothesis, $w_i \in W_{Y_i}^T$ for some finite subset Y_i of X. Then $w_1, \ldots, w_n \in W_Y^T$, where $Y = Y_1 \cup \cdots \cup Y_n$ is a finite subset of X, and $w = (\omega, w_1, \ldots, w_n) \in W_Y^T$. $\square$

Word algebras are blessed with a universal property:

Proposition 2.3. *The word algebra W_X^T of type T on a set X is generated by X. Moreover, every mapping of X into a universal algebra of type T extends uniquely to a homomorphism of W_X^T into A.*

Proof. $W = W_X^T$ is generated by X, by 2.1. Let f be a mapping of X into a universal algebra A of type T. If $\varphi : W \longrightarrow A$ is a homomorphism that extends f, then necessarily $\varphi(x) = f(x)$ for all $x \in X$ and $\varphi(\omega, w_1, \ldots, w_n) = \omega_A(\varphi(w_1), \ldots, \varphi(x_n))$ for all $(\omega, w_1, \ldots, w_n) \in W_k$. These conditions define (recursively) a unique mapping of W into A; therefore φ is unique; and we see that our mapping is a homomorphism. $\square$

Identities. Word algebras yield precise definitions of relations and identities, which resemble the definition of group relations in Section I.7, except that an identity that holds in an algebra must hold for all elements of that algebra.

Formally, a *relation* of type T between the elements of a set X is a pair

(u, v), often written as an equality $u = v$, of elements of the word algebra W_X^T of type T; the relation (u, v) *holds* in a universal algebra A of type T *via* a mapping $f : X \longrightarrow A$ when $\varphi(u) = \varphi(v)$, where $\varphi : W_X^T \longrightarrow A$ is the homomorphism that extends f.

An *identity* is a relation that holds via every mapping. Since identities involve only finitely many elements at a time, the set X needs only arbitrarily large finite subsets and could be any infinite set. In the formal definition, X is your favorite countable infinite set (for instance, $\mathbb{N}$).

Definitions. Let X be a countable infinite set. An identity *of type T is a pair (u, v), often written as an equality $u = v$, of elements of the word algebra W_X^T of type T on the set X. An identity (u, v) holds in a universal algebra A of type T when $\varphi(u) = \varphi(v)$ for every homomorphism $\varphi : W_X^T \longrightarrow A$; then A satisfies the identity (u, v).*

In this definition, the choice of X is irrelevant in the following sense. Between any two countable infinite sets X and Y, there is a bijection $X \longrightarrow Y$, which induces an isomorphism $\theta : W_X^T \cong W_Y^T$. If $u, v \in W_X^T$, then the identity (u, v) holds in A if and only if the identity $(\theta(u), \theta(v))$ holds in A. In this sense the identities that hold in A do not depend on the choice of X.

For example, *associativity* for a binary operation μ is the identity

$$((\mu, x, \mu(y, z)), (\mu, \mu(x, y), z)),$$

where x, y, z are any three distinct elements of X. This identity holds in a universal algebra A if and only if

$$\mu_A(\varphi(x), \mu_A(\varphi(y), \varphi(z))) = \varphi((\mu, x, \mu(y, z))$$
$$= \varphi(\mu, \mu(x, y), z) = \mu_A(\mu_A(\varphi(x), \varphi(y)), \varphi(z))$$

for every homomorphism $\varphi : W_X^T \longrightarrow A$. By 2.3, there is for every $a, b, c \in A$ a homomorphism $\varphi : W_X^T \longrightarrow A$ that sends x, y, z to a, b, c; hence the associativity identity holds in A if and only if $\mu_A(a, \mu_A(b, c)) = \mu_A(\mu_A(a, b), c)$ for all $a, b, c \in A$, if and only if μ_A is associative in the usual sense.

Exercises

1. Let X be a subset of a universal algebra A of type T. Show that the subalgebra $\langle X \rangle$ of A generated by X is $\langle X \rangle = \bigcup_{k \geq 0} S_k$, where $S_k \subseteq A$ is defined by: $S_0 = X$; if $k > 0$, then S_k is the set of all $\omega(w_1, \ldots, w_n)$ in which $\omega \in T$ has arity n and $w_1 \in S_{k_1}, \ldots, w_n \in S_{k_n}$, with $k_1, \ldots, k_n \geqq 0$ and $1 + k_1 + \cdots + k_n = k$.

2. Show that every mapping $f : X \longrightarrow Y$ induces a homomorphism $W_f^T : W_X^T \longrightarrow W_Y^T$ of word algebras of type T, so that W_-^T becomes a functor from sets to universal algebras of type T.

3. Show that every universal algebra of type T is a homomorphic image of a word algebra of type T.

4. Let $T = \{\varepsilon, \iota, \mu\}$, where ε has arity 0, ι has arity 1, and μ has arity 2. Describe all elements of $W_0 \cup W_1 \cup W_2 \subseteq W_X^T$.

5. Let $T = \{\alpha\} \cup R$, where α has arity 2 and every $r \in R$ has arity 1. Describe all elements of $W_0 \cup W_1 \cup W_2 \subseteq W_X^T$.

6. Write commutativity as a formal identity.

7. Write distributivity in a ring as a formal identity.

8. Given a countable infinite set X, show that the set of all identities that hold in a universal algebra A of type T is a congruence on W_X^T.

3. Varieties

A variety consists of all algebras of the same type that satisfy a given set of identities. Most of the algebraic objects in this book (groups, rings, modules, etc.) constitute varieties. Many of their properties extend to all varieties. This section contains general characterizations and properties of varieties. Additional properties will be found in Section XVI.10.

Definition. Let T be a type of universal algebras and let X be a given countable infinite set. Every set $\mathfrak{I} \subseteq W_X^T \times W_X^T$ of identities of type T defines a class $V(\mathfrak{I})$, which consists of all universal algebras of type T that satisfy every identity $(u, v) \in \mathfrak{I}$.

Definition. Let X be a given countable infinite set. A variety of type T is a class $\mathcal{V} = V(\mathfrak{I})$, which consists of all universal algebras of type T that satisfy some set $\mathfrak{I} \subseteq W_X^T \times W_X^T$ of identities of type T.

The class of all universal algebras of type T is a variety, namely $V(\emptyset)$. At the other extreme is the *trivial* variety $\mathcal{T}$ of type T, which consists of all universal algebras of type T with at most one element, and is characterized by the single identity $x = y$, where $x \neq y$; $\mathcal{T}$ is contained in every variety of type T.

Groups constitute a variety. The definition of groups as algebras with one binary operation is not suitable for this, since the existence of an identity element, or the existence of inverses, is not an identity. But we may regard groups as algebras with one binary operation, one constant "identity element" operation 1, and one unary operation $x \longmapsto x^{-1}$. An algebra of this type is a group if and only if $1x = x$ for all $x \in G$, $x1 = x$ for all $x \in G$, $xx^{-1} = 1$ for all $x \in G$, $x^{-1}x = 1$ for all $x \in G$, and $x(yz) = (xy)z$ for all $x, y, z \in G$; these five conditions are identities. (Dedicated readers will write them as formal identities.)

Abelian groups constitute a variety (of algebras with one binary operation, one constant "identity element" operation 0, and one unary operation $x \longmapsto -x$) defined by the five identities above and one additional commutativity identity $x + y = y + x$. Readers will verify that rings, R-modules, R-algebras, lattices, etc., constitute varieties, when suitably defined. But fields do not constitute a variety; this follows from Proposition 3.1 below.

Properties. Every variety is closed under certain constructions.

A *homomorphic image* of a universal algebra A is a universal algebra B of the same type such that there exists a surjective homomorphism of A onto B; equivalently, that is isomorphic to the quotient of A by a congruence on A.

The *direct product* of a family $(A_i)_{i \in I}$ of algebras of the same type T is the Cartesian product $\prod_{i \in I} A_i$, equipped with componentwise operations,

$$\omega \left((x_{1i})_{i \in I}, \ldots, (x_{ni})_{i \in I} \right), = \left(\omega (x_{1i}, \ldots, x_{ni}) \right)_{i \in I}$$

for all $(x_{1i})_{i \in I}, \ldots, (x_{ni})_{i \in I} \in \prod_{i \in I} A_i$ and $\omega \in T$ of arity n. The direct product comes with a *projection* $\pi_j : \prod_{i \in I} A_i \longrightarrow A_j$ for each $j \in J$, which sends $(x_i)_{i \in I} \in \prod_{i \in I} A_i$ to its j component x_j. The operations on $\prod_{i \in I} A_i$ are the only operations such that every projection is a homomorphism.

A *directed family* of algebras is a family $(A_i)_{i \in I}$ of algebras of the same type T, such that for every $i, j \in I$ there exists $k \in I$ such that A_i and A_j are subalgebras of A_k. A *directed union* of algebras of the same type T is the union $A = \bigcup_{i \in I} A_i$ of a directed family $(A_i)_{i \in I}$ of algebras of type T. Readers will verify that there is unique type T algebra structure on A such that every A_i is a subalgebra of A. Directed unions are particular cases of direct limits.

Proposition 3.1. *Every variety is closed under subalgebras, homomorphic images, direct products, and directed unions.*

Proof. Let $\mathcal{V} = \mathrm{V}(\mathcal{I})$ be the variety of all universal algebras A of type T that satisfy a set $\mathcal{I} \subseteq W_X^T \times W_X^T$ of identities. An algebra A of type T belongs to $\mathcal{V}$ if and only if $\varphi(u) = \varphi(v)$ for every $(u, v) \in \mathcal{I}$ and homomorphism $\varphi : W_X^T \longrightarrow A$. Readers will verify that $\mathcal{V}$ contains every subalgebra of every $A \in \mathcal{V}$, and every direct product of algebras $A_i \in \mathcal{V}$.

Let $A \in \mathcal{V}$ and let $\sigma : A \longrightarrow B$ be a surjective homomorphism. Let $\psi : W_X^T \longrightarrow B$ be a homomorphism. Since σ is surjective one can choose for each $x \in X$ one $f(x) \in A$ such that $\sigma(f(x)) = \psi(x)$. By 2.3, f extends to a homomorphism $\varphi : W_X^T \longrightarrow A$:

$$\begin{array}{ccc} X & \xrightarrow{\ f\ } & A \\ {\scriptstyle \subseteq} \downarrow & {\scriptstyle \varphi} \nearrow & \downarrow {\scriptstyle \sigma} \\ W_X^T & \xrightarrow[\ \psi\ ]{} & B \end{array}$$

Then $\psi = \sigma \circ \varphi$, since both agree on X. If now $(u, v) \in \mathcal{I}$, then $\varphi(u) = \varphi(v)$ and $\psi(u) = \sigma(\varphi(u)) = \sigma(\varphi(v)) = \psi(v)$. Therefore $B \in \mathcal{V}$.

Let $A = \bigcup_{i \in I} A_i$ be a directed union of universal algebras $A_i \in \mathcal{V}$. Let $(u, v) \in \mathcal{I}$ and let $\psi : W_X^T \longrightarrow A$ be a homomorphism. By 2.2, $u, v \in$

W_Y^T for some finite subset Y of X. Since Y is finite, some A_i contains all $\psi(y) \in \psi(Y)$. By 2.3, the restriction of ψ to Y extends to a homomorphism $\varphi : W_Y^T \longrightarrow A_i$:

$$
\begin{array}{ccc}
Y & \xrightarrow{\;\;f\;\;} & A_i \\
{\scriptstyle \subseteq}\big\downarrow & {\scriptstyle \varphi}\nearrow & \big\uparrow{\scriptstyle \subseteq} \\
W_X^T & \xrightarrow[\;\;\psi\;\;]{} & A
\end{array}
$$

Then $\varphi(w) = \psi(w)$ for all $w \in W_Y^T$, since φ and ψ agree on Y. Hence $\psi(u) = \varphi(u) = \varphi(v) = \psi(v)$. Therefore $A \in \mathcal{V}$. $\square$

By 3.1, fields do not constitute a variety (of any type), since, say, the direct product of two fields is not a field.

Free algebras. Free algebras are defined by their universal property:

Definition. Let X be a set and let $\mathcal{C}$ be a class of universal algebras of type T. A universal algebra F is free on the set X in the class $\mathcal{C}$ when $F \in \mathcal{C}$ and there exists a mapping $\eta : X \longrightarrow F$ such that, for every mapping f of X into a universal algebra $A \in \mathcal{C}$, there exists a unique homomorphism $\varphi : F \longrightarrow A$ such that $\varphi \circ \eta = f$.

$$
\begin{array}{ccc}
X & \xrightarrow{\;\;\eta\;\;} & F \\
& {\scriptstyle f}\searrow & \big\downarrow{\scriptstyle \varphi} \\
& & A
\end{array}
$$

For example, free groups are free in this sense in the class of all groups; W_X^T is free on X in the class of all universal algebras of type T, by Proposition 2.3. Some definitions of free algebras require the mapping η to be injective; readers will verify that this property holds when $\mathcal{C}$ is not trivial (when some $A \in \mathcal{C}$ has at least two elements). Readers will also prove the following:

Proposition 3.2. Let X be a set and let $\mathcal{C}$ be a class of universal algebras of the same type. If there exists a universal algebra F that is free on X in the class $\mathcal{C}$, then F and the mapping $\eta : X \longrightarrow F$ are unique up to isomorphism; moreover, F is generated by $\eta(X)$.

Existence of free algebras is a main property of varieties. More generally:

Theorem 3.3. Let $\mathcal{C}$ be a class of universal algebras of the same type, that is closed under isomorphisms, direct products, and subalgebras (for instance, a variety). For every set X there exists a universal algebra that is free on X in the class $\mathcal{C}$.

Proof. We give a direct proof; a better proof will be found in Section XVI.10. Given a set X, let $(\mathcal{E}_i)_{i \in I}$ be the set of all congruences $\mathcal{E}_i$ on W_X^T such that $W_X^T / \mathcal{E}_i \in \mathcal{C}$; let $C_i = W_X^T / \mathcal{E}_i$ and $\pi_i : W_X^T \longrightarrow C_i$ be the projection. Then $P = \prod_{i \in I} C_i \in \mathcal{C}$. Define a mapping $\eta : X \longrightarrow P$ by $\eta(x) = \big(\pi_i(x)\big)_{i \in I}$.

If $C \in \mathcal{C}$, then every mapping $f : X \longrightarrow C$ extends to a homomorphism φ of W_X^T into C. Then $\operatorname{Im} \varphi$ is a subalgebra of $C \in \mathcal{C}$, $W_X^T / \ker \varphi \cong \operatorname{Im} \varphi \in \mathcal{C}$, and $\ker \varphi = \mathcal{E}_i$ for some i. Composing $\pi_i : W_X^T \longrightarrow C_i = W_X^T / \ker \varphi$ and $W_X^T / \ker \varphi \cong \operatorname{Im} \varphi \subseteq C$ yields a homomorphism $\psi : P \longrightarrow C$ such that $\psi \circ \eta = f$. But ψ need not be unique with this property.

Let F be the set of all $p \in P$ such that $\zeta(p) = p$ for every endomorphism ζ of P such that $\zeta(\eta(x)) = \eta(x)$ for all $x \in X$. Then $\eta(X) \subseteq F$, F is a subalgebra of P, and $F \in \mathcal{C}$. If $C \in \mathcal{C}$ and $f : X \longrightarrow C$ is a mapping, then the above yields a homomorphism ψ of $F \subseteq P$ into C such that $\psi \circ \eta = f$. We show that ψ is unique, so that F is free on X in the class $\mathcal{C}$.

Let $\varphi, \psi : F \longrightarrow C$ be homomorphisms such that $\varphi \circ \eta = \psi \circ \eta$. Then $E = \{ p \in F \mid \varphi(p) = \psi(p) \}$ contains $\eta(X)$ and is a subalgebra of F. Since $\eta : X \longrightarrow E$ and $E \in \mathcal{C}$, there is a homomorphism $\zeta : P \longrightarrow E$ such that $\zeta \circ \eta = \eta$. Then ζ is an endomorphism of P, $\zeta(\eta(x)) = \eta(x)$ for all $x \in X$, $p = \zeta(p) \in E$ for every $p \in F$, $\varphi(p) = \psi(p)$ for every $p \in F$, and $\varphi = \psi$. $\square$

Birkhoff's theorem on varieties is the converse of Proposition 3.1:

Theorem **3.4** (Birkhoff [1935]). *A nonempty class of universal algebras of the same type is a variety if and only if it is closed under direct products, subalgebras, and homomorphic images.*

Proof. First we prove the following: when F is free in a class $\mathcal{C}$ on an infinite set, relations that hold in F yield identities that hold in every $C \in \mathcal{C}$:

Lemma **3.5.** *Let X be a given infinite countable set, let Y be an infinite set, and let $p, q \in W_Y^T$.*

(1) *There exist homomorphisms $\sigma : W_Y^T \longrightarrow W_X^T$ and $\mu : W_X^T \longrightarrow W_Y^T$ such that $\sigma \circ \mu$ is the identity on W_X^T and $\mu(\sigma(p)) = p$, $\mu(\sigma(q)) = q$.*

(2) *Let F be free on Y in a class $\mathcal{C}$ of universal algebras of type T and let $\varphi : W_Y^T \longrightarrow F$ be the homomorphism that extends $\eta : Y \longrightarrow F$. If $\varphi(p) = \varphi(q)$, then the identity $\sigma(p) = \sigma(q)$ holds in every algebra $C \in \mathcal{C}$.*

Proof. (1). By 2.2, $p, q \in W_Z^T$ for some finite subset Z of Y. There is an injection $h : X \longrightarrow Y$ such that $h(X)$ contains Z. The inverse bijection $h(X) \longrightarrow X$ can be extended to a surjection $g : Y \longrightarrow X$; then $g \circ h$ is the identity on X and $h(g(z)) = z$ for all $z \in Z$. By 2.3, $g : Y \longrightarrow W_X^T$ and $h : X \longrightarrow W_Y^T$ extend to homomorphisms σ and μ such that $\sigma \circ \mu$ is the identity on W_X^T and $\mu(\sigma(z)) = z$ for all $z \in Z$:

$$
\begin{array}{ccc}
X & \overset{\subseteq}{\longrightarrow} & W_X^T \\
{\scriptstyle g}\big\updownarrow{\scriptstyle h} & & {\scriptstyle \sigma}\big\updownarrow{\scriptstyle \mu} \\
Y & \underset{\subseteq}{\longrightarrow} & W_Y^T
\end{array}
$$

Then $\mu\big(\sigma(w)\big) = w$ for all $w \in W_Z^T$, and $\mu\big(\sigma(p)\big) = p$, $\mu\big(\sigma(q)\big) = q$.

(2). Let $\xi : W_X^T \longrightarrow C$ be any homomorphism. Since $C \in \mathcal{C}$, the restriction of $\xi \circ \sigma$ to Y factors through η: there is a homomorphism $\chi : F \longrightarrow A$ such that $\chi\big(\eta(y)\big) = \xi\big(\sigma(y)\big)$ for all $y \in Y$:

$$Y \xrightarrow{\subseteq} W_Y^T \underset{\mu}{\overset{\sigma}{\rightleftarrows}} W_X^T$$

(diagram with η, φ, ξ, χ: $Y \xrightarrow{\subseteq} W_Y^T \rightleftarrows W_X^T$, η and φ down to F, ξ down to C, $F \dashrightarrow{\chi} C$)

Then uniqueness in Proposition 2.3 yields $\chi \circ \varphi = \xi \circ \sigma$. Hence $\varphi(p) = \varphi(q)$ implies $\xi\big(\sigma(p)\big) = \xi\big(\sigma(q)\big)$. Thus, the identity $\sigma(p) = \sigma(q)$ holds in C. $\square$

Lemma 3.6. *Let $\mathcal{C}$ be a class of universal algebras of the same type T, that is closed under isomorphisms, direct products, and subalgebras. Let A be a nonempty universal algebra of type T such that every identity that holds in every $C \in \mathcal{C}$ also holds in A. Then A is a homomorphic image of some $C \in \mathcal{C}$.*

Proof. There is an infinite set Y and a mapping f of Y into A such that A is generated by $f(Y)$: indeed, A is generated by some subset S; if S is infinite, then $Y = S$ serves; otherwise, construct Y by adding new elements to S, which f sends anywhere in A. Then 2.3 yields a homomorphism $\psi : W_Y^T \longrightarrow A$ that extends f. Since W_Y^T is generated by Y, Im ψ is generated by $\psi(Y)$, Im $\psi = A$, and ψ is surjective. By 3.3 there exists an algebra F that is free on Y in $\mathcal{C}$. The homomorphism $\varphi : W_Y^T \longrightarrow F$ that extends $\eta : Y \longrightarrow F$ is surjective: since W_Y^T is generated by Y, Im φ is generated by $\varphi(Y) = \eta(Y)$ and Im $\varphi = F$ by 3.2. We show that ker $\varphi \subseteq$ ker ψ: if $\varphi(p) = \varphi(q)$, then 3.5 yields homomorphisms μ and σ such that $\sigma \circ \mu$ is the identity on W_X^T and the identity $\sigma(p) = \sigma(q)$ holds in every $C \in \mathcal{C}$; then the identity $\sigma(p) = \sigma(q)$ holds in A, $\psi\big(\mu(\sigma(p))\big) = \psi\big(\mu(\sigma(q))\big)$, and $\psi(p) = \psi(q)$. Therefore $\psi = \chi \circ \varphi$ for some homomorphism $\chi : F \longrightarrow A$; then χ is surjective, like ψ: $\square$

$$Y \xrightarrow{\subseteq} W_Y^T \underset{\mu}{\overset{\sigma}{\rightleftarrows}} W_X^T$$

(diagram with η, φ, ψ: $F \dashrightarrow{\chi} A$)

Now, let $\mathcal{C}$ be a class of universal algebras of the same type T, that is closed under direct products, subalgebras, and homomorphic images (hence, closed under isomorphisms). Let X be any given countable infinite set and let $\mathfrak{I} \subseteq W_X^T \times W_X^T$ be the set of all identities that hold in every algebra $C \in \mathcal{C}$. Then $\mathcal{C} \subseteq V(\mathfrak{I})$. Conversely, let $A \in V(\mathfrak{I})$. If $A = \emptyset$, then A is isomorphic to the empty subalgebra of some $C \in \mathcal{C}$ and $A \in \mathcal{C}$. If $A \neq \emptyset$, then A is a homomorphic image of some $C \in \mathcal{C}$, by 3.6, and again $A \in \mathcal{C}$. Thus $\mathcal{C} = V(\mathfrak{I})$ is a variety. $\square$

We note some consequences of Birkhoff's theorem and its proof. First, every intersection of varieties is a variety: indeed, let $(\mathcal{V}_i)_{i \in I}$ be varieties of universal algebras of type T; then $\bigcap_{i \in I} \mathcal{V}_i$ is, like every $\mathcal{V}_i$, closed under direct products, subalgebras, and homomorphic images, and is therefore a variety. Consequently, every class $\mathcal{C}$ of algebras of type T *generates* a variety, which is the smallest variety of type T that contains $\mathcal{C}$.

Proposition 3.7. *Let $\mathcal{C}$ be a class of universal algebras of type T. The variety generated by $\mathcal{C}$ consists of all homomorphic images of subalgebras of direct products of members of $\mathcal{C}$.*

Proof. For any class $\mathcal{C}$ of universal algebras of type T:

(1) a homomorphic image of a homomorphic image of a member of $\mathcal{C}$ is a homomorphic image of a member of $\mathcal{C}$; symbolically, $\text{HH}\mathcal{C} \subseteq \text{H}\mathcal{C}$;

(2) a subalgebra of a subalgebra of a member of $\mathcal{C}$ is a subalgebra of a member of $\mathcal{C}$; symbolically, $\text{SS}\mathcal{C} \subseteq \text{S}\mathcal{C}$;

(3) a direct product of direct products of members of $\mathcal{C}$ is a direct product of members of $\mathcal{C}$; symbolically, $\text{PP}\mathcal{C} \subseteq \text{P}\mathcal{C}$;

(4) a subalgebra of a homomorphic image of a member of $\mathcal{C}$ is a homomorphic image of a subalgebra of a member of $\mathcal{C}$; symbolically, $\text{SH}\mathcal{C} \subseteq \text{HS}\mathcal{C}$;

(5) a direct product of subalgebras of members of $\mathcal{C}$ is a subalgebra of a direct product of members of $\mathcal{C}$; symbolically, $\text{PS}\mathcal{C} \subseteq \text{SP}\mathcal{C}$;

(6) a direct product of homomorphic images of members of $\mathcal{C}$ is a homomorphic image of a direct product of members of $\mathcal{C}$; symbolically, $\text{PH}\mathcal{C} \subseteq \text{HP}\mathcal{C}$.

Now, every variety $\mathcal{V}$ that contains $\mathcal{C}$ also contains all homomorphic images of subalgebras of direct products of members of $\mathcal{C}$: symbolically, $\text{HSP}\mathcal{C} \subseteq \mathcal{V}$. Conversely, $\text{HSP}\mathcal{C}$ is closed under homomorphic images, subalgebras, and direct products: by the above, $\text{HHSP}\mathcal{C} \subseteq \text{HSP}\mathcal{C}$, $\text{SHSP}\mathcal{C} \subseteq \text{HSSP}\mathcal{C} \subseteq \text{HSP}\mathcal{C}$, and $\text{PHSP}\mathcal{C} \subseteq \text{HPSP}\mathcal{C} \subseteq \text{HSPP}\mathcal{C} \subseteq \text{HSP}\mathcal{C}$; therefore $\text{HSP}\mathcal{C}$ is a variety. $\square$

Once generators are found for a variety, Proposition 3.7 provides very loose descriptions of all members of that variety. This is useful for structures like semigroups or lattices, that are difficult to describe precisely.

Another consequence of the above is a one-to-one correspondence between varieties of type T and certain congruences on W_X^T. A congruence $\mathcal{E}$ on a universal algebra A is *fully invariant* when $a \, \mathcal{E} \, b$ implies $\zeta(a) \, \mathcal{E} \, \zeta(b)$, for every $a, b \in A$ and endomorphism ζ of A.

Proposition 3.8. *Let X be a given infinite countable set. There is a one-to-one, order reversing correspondence between varieties of type T and fully invariant congruences on W_X^T.*

Proof. For each variety $\mathcal{V}$, let $\text{I}(\mathcal{V}) \subseteq W_X^T \times W_X^T$ be the set of all identities that hold in every $A \in \mathcal{V}$: the set of all $(u, v) \in W_X^T \times W_X^T$ such that $\xi(u) = \xi(v)$ for every homomorphism $\xi : W_X^T \longrightarrow A$ such that $A \in \mathcal{V}$. If $(u, v) \in \text{I}(\mathcal{V})$ and ζ is an endomorphism of W_X^T, then $\xi\big(\zeta(u)\big) = \xi\big(\zeta(v)\big)$ for every homomorphism $\xi : W_X^T \longrightarrow A$ such that $A \in \mathcal{V}$, since $\xi \circ \zeta$ is another such homomorphism, and $(\zeta(u),\ \zeta(v)) \in I(\mathcal{V})$. Moreover, $I(\mathcal{V})$ is the intersection of congruences ker ξ and is a congruence on W_X^T. Hence $I(\mathcal{V})$ is a fully invariant congruence on W_X^T.

Conversely, a fully invariant congruence $\mathcal{E}$ on W_X^T is a set of identities and determines a variety $V(\mathcal{E})$ of type T. The constructions I and V are order reversing; we show that $I(V(\mathcal{E})) = \mathcal{E}$ and $V(I(\mathcal{V})) = \mathcal{V}$ for every fully invariant congruence $\mathcal{E}$ and variety $\mathcal{V}$, so that I and V induce the one-to-one correspondence in the statement.

First we show that $F = W_X^T/\mathcal{E} \cdot \in V(\mathcal{E})$ when $\mathcal{E}$ is fully invariant. Let $\pi : W_X^T \longrightarrow F$ be the projection and let $\xi : W_X^T \longrightarrow F$ be any homomorphism. For every $x \in X$ choose $g(x) \in W_X^T$ such that $\pi(g(x)) = \xi(x)$. By 2.3, $g : X \longrightarrow W_X^T$ extends to an endomorphism ζ of W_X^T; then $\pi \circ \zeta = \xi$, since they agree on X. Hence $(u, v) \in \mathcal{E}$ implies $(\zeta(u),\ \zeta(v)) \in \mathcal{E}$ and $\xi(u) = \pi(\zeta(u)) = \pi(\zeta(v)) = \xi(v)$. Thus $F \in V(\mathcal{E})$. (Readers will verify that F is the free algebra on X in $V(\mathcal{E})$, and generates $V(\mathcal{E})$.)

Now, $\mathcal{E} \subseteq I(V(\mathcal{E}))$, since every $(u, v) \in \mathcal{E}$ holds in every $A \in V(\mathcal{E})$. Conversely, if $(u, v) \in I(V(\mathcal{E}))$ holds in every $A \in V(\mathcal{E})$, then (u, v) holds in $F \in V(\mathcal{E})$, $\pi(u) = \pi(v)$, and $(u, v) \in \mathcal{E}$. Thus $I(V(\mathcal{E})) = \mathcal{E}$.

Conversely, let $\mathcal{V}$ be a variety of type T. Then $\mathcal{V} \subseteq V(I(\mathcal{V}))$, since every member of $\mathcal{V}$ satisfies every identity in $I(\mathcal{V})$. Conversely, let $A \in V(I(\mathcal{V}))$. If $A = \varnothing$, then A is isomorphic to the empty subalgebra of some $C \in \mathcal{V}$ and $A \in \mathcal{V}$. If $A \neq \varnothing$, then A is a homomorphic image of some $C \in \mathcal{V}$, by 3.6, and again $A \in \mathcal{V}$. Thus $\mathcal{V} = V(I(\mathcal{V}))$. $\square$

Exercises

1. Write a set of formal identities that characterize groups.

2. Write a set of formal identities that characterizes rings [with identity elements].

3. Show that lattices constitute a variety (of universal algebras with two binary operations).

4. Show that modular lattices constitute a variety (of universal algebras with two binary operations).

5. Show that Boolean lattices constitute a variety.

6. Show that the direct product $\prod_{i \in I} A_i$ of universal algebras $(A_i)_{i \in I}$ of type T, and its projections $\pi_j : \prod_{i \in I} A_i \longrightarrow A_j$, have the following universal property: for every universal algebra A of type T and homomorphisms $\varphi_i : A \longrightarrow A_i$, there is a unique homomorphism $\varphi : A \longrightarrow \prod_{i \in I} A_i$ such that $\pi_i \circ \varphi = \varphi_i$ for all $i \in I$.

7. Let $A = \bigcup_{i \in I} A_i$ be the directed union $A = \bigcup_{i \in I} A_i$ of a directed family $(A_i)_{i \in I}$ of universal algebras of the same type T. Show that there is unique type-T algebra structure on A such that every A_i is a subalgebra of A.

8. Define and construct direct limits of universal algebras of type T; verify that the direct limit is a directed union of homomorphic images.

9. Define and construct inverse limits of universal algebras of type T.

Prove the following:

10. Every variety is closed under subalgebras and closed under direct products.

11. If F is free on a set X in a class $\mathcal{C}$ that contains an algebra with more than one element, then $\eta : X \longrightarrow F$ is injective.

12. Let $\mathcal{C}$ be a class of universal algebras of the same type. The free algebra F on a set X in the class $\mathcal{C}$ and the corresponding mapping $\eta : X \longrightarrow F$, if they exist, are unique up to isomorphism.

13. Let $\mathcal{C}$ be a class of universal algebras of the same type. If F is free on a set X in the class $\mathcal{C}$, then F is generated by $\eta(X)$.

14. Let $\mathcal{C}$ be a class of universal algebras of the same type. If F is free in $\mathcal{C}$ on an infinite set X, then every identity that holds in F holds in every member of $\mathcal{C}$.

15. If $\mathcal{V}$ is a variety, and F is free in $\mathcal{V}$ on an infinite set, then $\mathcal{V}$ is generated by F (so that every member of $\mathcal{V}$ is a homomorphic image of a subalgebra of a direct product of copies of F). (Use the previous exercice.)

16. The variety of all abelian groups is generated by $\mathbb{Z}$.

17. The variety of all commutative semigroups is generated by $\mathbb{N}$.

18. Given a countable infinite set X, when $\mathcal{E}$ is a fully invariant congruence on W_X^T, then $W_X^T / \mathcal{E}$ is free on X in the variety $V(\mathcal{E})$.

19. There are no more than $|\mathbb{R}|$ varieties of groups.

4. Subdirect Products

Subdirect products were introduced by Birkhoff [1944]. They provide loose but useful descriptions of structures that are difficult to describe more precisely; examples in this section include distributive lattices and commutative semigroups.

Definition. A subdirect product of a family $(A_i)_{i \in I}$ of universal algebras of the same type T is a subalgebra P of the Cartesian product $\prod_{i \in I} A_i$ such that $\pi_i(P) = A_i$ for all $i \in I$, where $\pi_j : \prod_{i \in I} A_i \longrightarrow A_j$ is the projection.

For example, in the vector space $\mathbb{R}^3 = \mathbb{R} \times \mathbb{R} \times \mathbb{R}$, a straight line $x = at$, $y = bt$, $z = ct$ is a subdirect product of $\mathbb{R}$, $\mathbb{R}$, and $\mathbb{R}$ if and only if $a, b, c \neq 0$. Thus, a subdirect product of algebras may be very thinly spread in their direct product. Only the conditions $\pi_i(P) = A_i$ prevent subdirect products from being too dangerously thin.

Proposition 4.1. Let $(A_i)_{i \in I}$ be universal algebras of type T. A universal algebra A of type T is isomorphic to a subdirect product of $(A_i)_{i \in I}$ if and only *if there exist surjective homomorphisms* $\varphi_i : A \longrightarrow A_i$ *such that* $\bigcap_{i \in I} \ker \varphi_i$ *is the equality on* A.

Here, $\bigcap_{i \in I} \ker \varphi_i$ is the equality on A if and only if $\varphi_i(x) = \varphi_i(y)$ for all $i \in I$ implies $x = y$, if and only if $x \neq y$ in A implies $\varphi_i(x) \neq \varphi_i(y)$ for some $i \in I$. Homomorphisms with this property are said to *separate* the elements of A.

Proof. Let P be a subdirect product of $(A_i)_{i \in I}$. The inclusion homomorphism $\iota : P \longrightarrow \prod_{i \in I} A_i$ and projections $\pi_j : \prod_{i \in I} A_i \longrightarrow A_j$ yield surjective homomorphisms $\rho_i = \pi_i \circ \iota : P \longrightarrow A_i$ that separate the elements of P, since elements of the product that have the same components must be equal. If now $\theta : A \longrightarrow P$ is an isomorphism, then the homomorphisms $\varphi_i = \rho_i \circ \theta$ are surjective and separate the elements of A.

Conversely, assume that there exist surjective homomorphisms $\varphi_i : A \longrightarrow A_i$ that separate the elements of A. Then $\varphi : x \longmapsto (\varphi_i(x))_{i \in I}$ is an injective homomorphism of A into $\prod_{i \in I} A_i$. Hence $A \cong \operatorname{Im} \varphi$; moreover, $\operatorname{Im} \varphi$ is a subdirect product of $(A_i)_{i \in I}$, since $\pi_i(\operatorname{Im} \varphi) = \varphi_i(A) = A_i$ for all i. $\square$

Direct products are associative: if $I = \bigcup_{j \in J} I_j$ is a partition of I, then $\prod_{i \in I} A_i \cong \prod_{j \in J} (\prod_{i \in I_j} A_i)$. So are subdirect products, as readers will deduce from Proposition 4.1:

Proposition 4.2. *Let* $(A_i)_{i \in I}$ *be universal algebras of type* T *and let* $I = \bigcup_{j \in J} I_j$ *be a partition of* I. *An algebra* A *of type* T *is isomorphic to a subdirect product of* $(A_i)_{i \in I}$ *if and only if* A *is isomorphic to a subdirect product of algebras* $(P_j)_{j \in J}$ *in which each* P_j *is a subdirect product of* $(A_i)_{i \in I_j}$.

Subdirect decompositions. A *subdirect decomposition* of A into algebras $(A_i)_{i \in I}$ of the same type is an isomorphism of A onto a subdirect product of $(A_i)_{i \in I}$. By 4.1, subdirect decompositions of A can be set up from within A from suitable families of congruences on A. They are inherited by every variety $\mathcal{V}$: when A has a subdirect decomposition into algebras $(A_i)_{i \in I}$, then $A \in \mathcal{V}$ if and only if $A_i \in \mathcal{V}$ for all i, by 3.1.

Subdirect decompositions of A give loose descriptions of A in terms of presumably simpler components $(A_i)_{i \in I}$. The simplest possible components are called *subdirectly irreducible*:

Definition. A universal algebra A *is* subdirectly irreducible *when* A *has more than one element and, whenever* A *is isomorphic to a subdirect product of* $(A_i)_{i \in I}$, *at least one of the projections* $A \longrightarrow A_i$ *is an isomorphism.*

Proposition 4.3. *A universal algebra* A *is subdirectly irreducible if and only if* A *has more than one element and the equality on* A *is not the intersection of congruences on* A *that are different from the equality.*

The proof is an exercise in further deduction from Proposition 4.1.

Theorem **4.4** (Birkhoff [1944]). *Every nonempty universal algebra is isomorphic to a subdirect product of subdirectly irreducible universal algebras. In any variety* $\mathcal{V}$, *every nonempty universal algebra* $A \in \mathcal{V}$ *is isomorphic to a subdirect product of subdirectly irreducible universal algebras* $A_i \in \mathcal{V}$.

Proof. Let A be a nonempty algebra of type T. By 1.5, the union of a chain of congruences on A is a congruence on A. Let $a, b \in A$, $a \neq b$ of A. If $(\mathcal{C}_i)_{i \in I}$ is a chain of congruences on A, none of which contains the pair (a, b), then the union $\mathcal{C} = \bigcup_{i \in I} \mathcal{C}_i$ is a congruence on A that does not contain the pair (a, b). By Zorn's lemma, there is a congruence $\mathcal{M}_{a,b}$ on A that is maximal such that $(a, b) \notin \mathcal{M}_{a,b}$. The intersection $\bigcap_{a,b \in A, \, a \neq b} \mathcal{M}_{a,b}$ cannot contain any pair (a, b) with $a \neq b$ and is the equality on A. By 4.1, A is isomorphic to a subdirect product of the quotient algebras $A/\mathcal{M}_{a,b}$.

The algebra $A/\mathcal{M}_{a,b}$ has at least two elements, since $\mathcal{M}_{a,b}$ does not contain the pair (a, b). Let $(\mathcal{C}_i)_{i \in I}$ be congruences on $A/\mathcal{M}_{a,b}$, none of which is the equality. Under the projection $\pi : A \longrightarrow A/\mathcal{M}_{a,b}$, the inverse image $\pi^{-1}(\mathcal{C}_i)$ is, by 1.9, a congruence on A, which properly contains $\ker \pi = \mathcal{M}_{a,b}$, hence contains the pair (a, b), by the maximality of $\mathcal{M}_{a,b}$. Hence $(\pi(a), \pi(b)) \in \mathcal{C}_i$ for every i, and $\bigcap_{i \in I} \mathcal{C}_i$ is not the equality on $A/\mathcal{M}_{a,b}$. Thus $A/\mathcal{M}_{a,b}$ is subdirectly irreducible, by 4.3. $\square$

Abelian groups. Abelian groups can be used to illustrate these results.

Congruences on an abelian group are induced by its subgroups. Hence an abelian group A (written additively) is isomorphic to a subdirect product of abelian groups $(A_i)_{i \in I}$ if and only if there exist surjective homomorphisms $\varphi_i : A \longrightarrow A_i$ such that $\bigcap_{i \in I} \operatorname{Ker} \varphi_i = 0$; an abelian group A is subdirectly irreducible if and only if A has more than one element and 0 is not the intersection of nonzero subgroups of A.

By Theorem 4.4, every abelian group is isomorphic to a subdirect product of subdirectly irreducible abelian groups. The latter are readily determined.

Proposition **4.5.** *An abelian group is subdirectly irreducible if and only if it is isomorphic to* $\mathbb{Z}_{p^\infty}$ *or to* $\mathbb{Z}_{p^n}$ *for some* $n > 0$.

Proof. Readers will verify that $\mathbb{Z}_{p^\infty}$ and $\mathbb{Z}_{p^n}$ (where $n > 0$) are subdirectly irreducible. Conversely, every abelian group A can, by X.4.9 and X.4.10, be embedded into a direct product of copies of $\mathbb{Q}$ and $\mathbb{Z}_{p^\infty}$ for various primes p. Hence A is isomorphic to a subdirect product of subgroups of $\mathbb{Q}$ and $\mathbb{Z}_{p^\infty}$.

Now, $\mathbb{Q}$ has subgroups $\mathbb{Z}$, $2\mathbb{Z}$, ..., $2^k \mathbb{Z}$, ..., whose intersection is 0; since $\mathbb{Q}/ 2^k \mathbb{Z} \cong \mathbb{Q}/\mathbb{Z}$, $\mathbb{Q}$ is isomorphic to a subdirect product of subgroups of $\mathbb{Q}/\mathbb{Z}$. Readers will verify that $\mathbb{Q}/\mathbb{Z}$ is isomorphic to a direct sum of $\mathbb{Z}_{p^\infty}$'s (for various primes p). By 4.2, $\mathbb{Q}$ is isomorphic to a subdirect product of subgroups of $\mathbb{Z}_{p^\infty}$ (for various primes p). Then every abelian group A is isomorphic to a subdirect product of subgroups of $\mathbb{Z}_{p^\infty}$ (for various primes p). If A is subdirectly irreducible, then A is isomorphic to a subgroup of some $\mathbb{Z}_{p^\infty}$. $\square$

Distributive lattices. Birkhoff's earlier theorem, XIV.4.8, states that every distributive lattice is isomorphic to a sublattice of the lattice of all subsets 2^X of some set X. We give another proof of this result, using subdirect products.

Since distributive lattices constitute a variety, every distributive lattice is isomorphic to a subdirect product of subdirectly irreducible distributive lattices, by 4.4. One such lattice is the two-element lattice $L_2 = \{0, 1\}$, which has only two congruences and is subdirectly irreducible by 4.3.

Proposition **4.6.** *Every distributive lattice is isomorphic to a subdirect product of two-element lattices. A distributive lattice is subdirectly irreducible if and only if it has just two elements.*

Proof. To each prime ideal $P \ne \emptyset, L$ of a distributive lattice L there corresponds a lattice homomorphism φ_P of L onto L_2, defined by $\varphi_P(x) = 0$ if $x \in P$, $\varphi_P(x) = 1$ if $x \notin P$. The homomorphisms φ_P separate the elements of L: if $a, b \in L$ and $a \ne b$, then, say, $a \nleq b$, and Lemma XIV.4.7 provides a prime ideal P of L that contains the ideal $I = \{\, x \in L \mid x \leqq b \,\}$ but does not contain $a \notin I$, so that $\varphi_P(b) \ne \varphi_P(a)$. By 4.1, L is isomorphic to a subdirect product of copies of L_2. If L is subdirectly irreducible, then some φ_P is an isomorphism and $L \cong L_2$ has just two elements. $\square$

A direct product $\prod_{i \in I} L_2$ of copies of L_2 is isomorphic to the lattice 2^I of all subsets of the index set I; the isomorphism sends $(x_i)_{i \in I} \in \prod_{i \in I} L_2$ to $\{\, i \in I \mid x_i = 1 \,\}$. Hence a subdirect product of copies of L_2 is, in particular, isomorphic to a sublattice of some 2^I; thus, Theorem XIV.4.8 follows from 4.6.

Commutative semigroups include abelian groups but can be much more complex; for instance, there are about 11.5 million nonisomorphic commutative semigroups of order 9. We use subdirect products to assemble finitely generated commutative semigroups from the following kinds of semigroups.

Definitions. A semigroup S is cancellative *when $ac = bc$ implies $b = c$, and $ca = cb$ implies $a = b$, for all $a, b, c \in S$. A* nilsemigroup *is a semigroup S with a zero element z (such that $sz = z = zs$ for all $s \in S$) in which every element is nilpotent ($s^m = z$ for some $m > 0$).*

Finitely generated commutative semigroups are related to ideals of polynomial rings. By Proposition 1.3, a congruence $\mathcal{E}$ on a commutative semigroup S is an equivalence relation $\mathcal{E}$ on S such that $a \mathrel{\mathcal{E}} b$ and $c \mathrel{\mathcal{E}} d$ implies $ac \mathrel{\mathcal{E}} bd$. We saw in Section I.1 that the free commutative semigroup with generators $x_1, \ldots, x_n$ consists of all nonconstant monomials $X^a = X_1^{a_1} X_2^{a_2} \cdots X_n^{a_n} \in R[X_1, \ldots, X_n]$, where R is any commutative ring [with identity]. Every ideal $\mathfrak{E}$ of $R[X_1, \ldots, X_n]$ induces a congruence $\mathcal{E}$ on F, in which $X^a \mathrel{\mathcal{E}} X^b$ if and only if $X^a - X^b \in \mathfrak{E}$; then $\mathfrak{E}$ determines a commutative semigroup $F/\mathcal{E}$ and, by extension, every commutative semigroup $S \cong F/\mathcal{E}$.

Lemma **4.7.** *Let F be the free commutative semigroup on $X_1, \ldots, X_n$. Every congruence $\mathcal{E}$ on F is induced by an ideal of $\mathbb{Z}[X_1, \ldots, X_n]$. Every commutative semigroup with n generators is determined by an ideal of $\mathbb{Z}[X_1, \ldots, X_n]$.*

Proof. Let $\mathfrak{E}$ be the ideal of $\mathbb{Z}[X_1, \ldots, X_n]$ generated by all binomials $X^a - X^b$ such that $X^a \mathcal{E} X^b$ in F. The ideal $\mathfrak{E}$ induces a congruence $\overline{\mathcal{E}}$ on F, in which $X^a \overline{\mathcal{E}} X^b$ if and only if $X^a - X^b \in \mathfrak{E}$. Then $\mathcal{E} \subseteq \overline{\mathcal{E}}$. We prove the converse inclusion. Since $X^a \mathcal{E} X^b$ implies $X^a X^c \mathcal{E} X^b X^c$, the ideal $\mathfrak{E}$ consists of all finite sums $\sum_i n_i (X^{a_i} - X^{b_i})$ in which $n_i \in \mathbb{Z}$ and $X^{a_i} \mathcal{E} X^{b_i}$ for all i. Since $\mathcal{E}$ is symmetric we may further assume that $n_i > 0$ for all i. Hence $X^a \overline{\mathcal{E}} X^b$ if and only if there is an equality

$$X^a - X^b = \sum_i n_i (X^{a_i} - X^{b_i}) \tag{1}$$

in which $n_i > 0$ and $X^{a_i} \mathcal{E} X^{b_i}$ for all i.

If $\mathcal{E} \subsetneqq \overline{\mathcal{E}}$, then there is an equality (1) in which $X^a \mathcal{E} X^b$ does not hold, and in which $\sum_i n_i$ is as small as possible. Since X^a appears in the left hand side of (1), it must also appear in the right hand side, and $X^{a_k} = X^a$ for some k. Subtracting $X^{a_k} - X^{b_k}$ from both sides of (1) then yields an equality

$$X^{b_k} - X^b = \sum_i m_i (X^{a_i} - X^{b_i})$$

in which $X^b \mathcal{E} X^{b_k}$ does not hold (otherwise, $X^a \mathcal{E} X^b$) and $\sum_i m_i = \left(\sum_i n_i\right) - 1$, an intolerable contradiction. Therefore $\mathcal{E} = \overline{\mathcal{E}}$ is induced by $\mathfrak{E}$.

Now, let S be a commutative semigroup with n generators $x_1, \ldots, x_n$. There is a homomorphism $\pi : F \longrightarrow S$ that sends X_i to x_i, defined by $\pi \left(X_1^{a_1} \cdots X_n^{a_n} \right) = x_1^{a_1} \cdots x_n^{a_n}$; π is surjective, since S is generated by $x_1, \ldots, x_n$. The congruence $\mathcal{E} = \ker \pi$ on F is induced by an ideal $\mathfrak{E}$ of $\mathbb{Z}[X_1, \ldots, X_n]$; hence $S \cong F/\mathcal{E}$ is determined by $\mathfrak{E}$. $\square$

Proposition **4.8.** *Every commutative semigroup with n generators has a subdirect decomposition into finitely many commutative semigroups determined by primary ideals of $\mathbb{Z}[X_1, \ldots, X_n]$.*

Proof. Let S be a commutative semigroup with n generators $x_1, \ldots, x_n$. By 4.7, $S \cong F/\mathcal{E}$, where $\mathcal{E}$ is induced by an ideal $\mathfrak{E}$ of $\mathbb{Z}[X_1, \ldots, X_n]$. In the Noetherian ring $\mathbb{Z}[X_1, \ldots, X_n]$, the ideal $\mathfrak{E}$ is the intersection of finitely many primary ideals $\mathfrak{Q}_1, \ldots, \mathfrak{Q}_r$. Hence $\mathcal{E}$ is the intersection of the congruences $\mathcal{Q}_1, \ldots, \mathcal{Q}_r$ induced by $\mathfrak{Q}_1, \ldots, \mathfrak{Q}_r$. By 1.9, $\mathcal{Q}_1, \ldots, \mathcal{Q}_r$ are the inverse images under $\pi : F \longrightarrow S$ of congruences $\mathcal{C}_1, \ldots, \mathcal{C}_r$ on S such that $S/\mathcal{C}_i \cong F/\mathcal{Q}_i$. Since $\mathcal{E} = \mathcal{Q}_1 \cap \cdots \cap \mathcal{Q}_r$, the equality on S is the intersection of $\mathcal{C}_1, \ldots, \mathcal{C}_r$, and S is isomorphic to a subdirect product of the semigroups $S/\mathcal{C}_1, \ldots, S/\mathcal{C}_r$ determined by $\mathfrak{Q}_1, \ldots, \mathfrak{Q}_r$. $\square$

Now, let S be determined by a primary ideal $\mathfrak{Q}$ of $\mathbb{Z}[X_1, \ldots, X_n]$, so that

$\pi(X^a) = \pi(X^b)$ if and only if $X^a - X^b \in \mathfrak{Q}$, where $\pi : F \longrightarrow S$ is the projection. The radical $\mathfrak{P}$ of $\mathfrak{Q}$ is a prime ideal of $\mathbb{Z}[X_1, ..., X_n]$. Moreover:

(1) If $X^c \in \mathfrak{Q}$, then $z = \pi(X^c)$ is a zero element of S: indeed, $X^a X^c - X^c \in \mathfrak{Q}$ for all $X^a \in F$, hence $sz = z$ for all $s = \pi(X^a) \in S$.

(2) If $X^c \in \mathfrak{P}$, then $(X^c)^m \in \mathfrak{Q}$ for some $m > 0$; hence S has a zero element z, and $s = \pi(X^c) \in S$ is nilpotent ($s^m = z$). Since $\mathfrak{P}$ is an ideal of $\mathbb{Z}[X_1, ..., X_n]$, the elements $s = \pi(X^c)$ such that $X^c \in \mathfrak{P}$ constitute an *ideal* N of S ($s \in N$ implies $st \in N$ for all $t \in S$).

(3) If $X^c \notin \mathfrak{P}$, then $X^c(X^a - X^b) \in \mathfrak{Q}$ implies $X^a - X^b \in \mathfrak{Q}$; hence $s = \pi(X^c) \in S$ is *cancellative in* S ($st = su$ implies $t = u$, when $t, u \in S$). Since $\mathfrak{P}$ is a prime ideal of $\mathbb{Z}[X_1, ..., X_n]$, X^c, $X^d \notin \mathfrak{P}$ implies $X^c X^d \notin \mathfrak{P}$; hence the elements $s = \pi(X^c)$ such that $X^c \notin \mathfrak{P}$ constitute a *subsemigroup* C of S ($s, t \in C$ implies $st \in C$).

By (2) and (3), a semigroup that is determined by a primary ideal $\mathfrak{Q}$ of $\mathbb{Z}[X_1, ..., X_n]$ is either a nilsemigroup (if $\mathfrak{P}$ contains every $X^c \in F$), or cancellative (if $\mathfrak{P}$ contains no $X^c \in F$), or *subelementary* in the following sense:

Definition. A subelementary *semigroup is a commutative semigroup that is the disjoint union $S = N \cup C$ of an ideal N and a nonempty subsemigroup C, such that N is a nilsemigroup and every element of C is cancellative in S.*

Subelementary semigroups are named for their relationship, detailed in the exercises, to previously defined "elementary" semigroups.

Every finitely generated commutative semigroup is now a subdirect product of finitely many nilsemigroups, cancellative semigroups, and subelementary semigroups. Readers will verify that a subdirect product of finitely many nilsemigroups is a nilsemigroup, and that a subdirect product of cancellative semigroups is cancellative. Subdirect decompositions need only one of each; hence we have:

Theorem **4.9** (Grillet [1975]). *Every finitely generated commutative semigroup is isomorphic to a subdirect product of a nilsemigroup, a cancellative semigroup, and finitely many subelementary semigroups.*

Exercises

1. Let $(A_i)_{i \in I}$ be universal algebras of type T and let $I = \bigcup_{j \in J} I_j$ be a partition of I. Show that a universal algebra A of type T is isomorphic to a subdirect product of $(A_i)_{i \in I}$ if and only if A is isomorphic to a subdirect product of algebras $(P_j)_{j \in J}$ in which each P_j is a subdirect product of $(A_i)_{i \in I_j}$.

2. Let $\mathcal{C}$ be a class of universal algebras of type T. If every A_i is a subdirect product of members of $\mathcal{C}$, then show that every subdirect product of $(A_i)_{i \in I}$ is a subdirect product of members of $\mathcal{C}$.

3. Let $\mathcal{C}$ be a class of universal algebras of type T. Show that every subalgebra of a subdirect product of members of $\mathcal{C}$ is a subdirect product of subalgebras of members of $\mathcal{C}$.

4. Prove that an algebra A is subdirectly irreducible if and only if A has more than one element and the equality on A is not the intersection of congruences that are different from the equality.

5. Show that $\mathbb{Z}_{p^\infty}$ and $\mathbb{Z}_{p^n}$ are subdirectly irreducible (when $n > 0$).

6. Show that $\mathbb{Z}$ is isomorphic to a subdirect product of cyclic groups of prime order p, one for each prime p.

Readers who are allergic to semigroups should avoid the remaining exercises.

7. Show that the zero element of a semigroup, if it exists, is unique.

8. Show that a finitely generated commutative nilsemigroup is finite.

9. Prove *Rédei's theorem* [1956]: the congruences on a finitely generated commutative semigroup satisfy the ascending chain condition.

10. Verify that a subdirect product of finitely many nilsemigroups is a nilsemigroup.

11. Verify that a subdirect product of cancellative semigroups is cancellative.

12. What can you say of a commutative semigroup that is determined by a prime ideal of $\mathbb{Z}[X_1, ..., X_n]$? by a semiprime ideal of $\mathbb{Z}[X_1, ..., X_n]$?

13. Show that a finite cancellative semigroup is a group.

14. Show that a cancellative commutative semigroup has a group of fractions, in which it can be embedded.

15. Show that a cancellative commutative semigroup is subdirectly irreducible if and only if its group of fractions is subdirectly irreducible.

16. Prove the following: if a subelementary semigroup $S = N \cup C$ is subdirectly irreducible, then its cancellative part C is subdirectly irreducible, or has just one element.

17. Prove *Malcev's theorem* [1958]: every subdirectly irreducible, finitely generated commutative semigroup is finite; hence every finitely generated commutative semigroup is isomorphic to a subdirect product of of finite semigroups. (Use the previous two exercises.)

18. A commutative semigroup S is *elementary* when it is the disjoint union $S = N \cup G$ of an ideal N and an abelian group G, such that N is a nilsemigroup, G is a group, and every element of G is cancellative in S. Show that a subelementary semigroup $S = N \cup C$ can be embedded into an elementary semigroup (e.g., a semigroup of fractions s/c).

19. Show that every finite semigroup is isomorphic to a subdirect product of a nilsemigroup, a group, and finitely many elementary semigroups.

XVI
Categories

A characteristic feature of abstract algebra is that it ignores what groups, rings, modules, etc., are made of, and studies only how their elements relate to each other (by means of operations, subgroups, etc.).

Category theory is the next step in abstraction: it ignores elements and studies only how groups, rings, modules, etc., relate to each other (by means of homomorphisms). This fruitful idea was introduced by Eilenberg and MacLane [1945]. It unifies concepts from many parts of mathematics and provides essential conceptual understanding. It also gives quick access to a number of useful properties. This chapter is a short introduction to the subject, including functors, limits, abelian categories, adjoint functors, tripleability, and properties of varieties.

1. Definition and Examples

Category theory challenges the foundations of mathematics in that it applies to collections called *proper classes*, such that the class of all sets, the class of all groups, etc., that are too large to be sets and are banned from standard Zermelo-Fraenkel set theory because their very consideration leads to paradoxes (see Section A.3).

This difficulty can be finessed in at least three ways:

(1) Use the Gödel-Bernays axioms of set theory. These axioms are essentially equivalent to Zermelo-Fraenkel's (the same results can be proved from both), but they recognize classes from the start and include an axiom of choice that applies to classes as well as sets. But they are not the generally accepted standard axioms of set theory.

(2) Like MacLane [1971], assume at the start that there exists a set model U of Zermelo-Fraenkel set theory (a *universe*). This hypothesis is not part of the Zermelo-Fraenkel axioms, but it is consistent with them, and allows all necessary business to be conducted within the set U, avoiding proper classes.

(3) Sneak classes in through the back door, like Jech [1978]. Zermelo-Fraenkel set theory does not allow classes but it allows statements of *membership* in

classes, such as "G is a group." This loophole allows limited use of "classes," not as actual collections, but as convenient abbreviations of membership statements. One must then carefully avoid forbidden practices, such as applying the axiom of choice to a proper class, or having it belong to another class.

The author thinks that the third approach is more natural for graduate students (though not necessarily better for working mathematicians). It leads to a more detailed if occasionally more awkward exposition.

Besides, life with classes is not all bleak. Some definitions extend from sets to classes: for instance, inclusion, $\mathcal{A} \subseteq \mathcal{B}$ means "$A \in \mathcal{A}$ implies $A \in \mathcal{B}$", and Cartesian products, $X \in \mathcal{A} \times \mathcal{B}$ means "X is an ordered pair $X = (A, B)$ in which $A \in \mathcal{A}$ and $B \in \mathcal{B}$." A *class function* of $\mathcal{A}$ into $\mathcal{B}$ is a class $\mathcal{F} \subseteq \mathcal{A} \times \mathcal{B}$, such that $(A, B) \in \mathcal{F}$, $(A, C) \in \mathcal{F}$ implies $B = C$; then $(A, B) \in \mathcal{F}$ is abbreviated as $B = \mathcal{F}(A)$, and $\mathcal{F}$ *assigns* B to A. A *small* class is a set.

Definition. Categories are defined so that sets and mappings constitute a category, groups and their homomorphisms constitute a category, topological spaces and continuous mappings constitute a category, and so forth.

Definition. A category $\mathcal{C}$ *has*

a class whose elements are the objects *of* $\mathcal{C}$;

a class whose elements are the morphisms *or* arrows *of* $\mathcal{C}$;

two class functions that assign to every morphism of $\mathcal{C}$ *a* domain *and a* codomain, *which are objects of* $\mathcal{C}$;

a class function that assigns to certain pairs (α, β) *of morphisms of* $\mathcal{C}$ *their* composition *or* product $\alpha\beta$, *which is a morphism of* $\mathcal{C}$, *such that:*

(1) $\alpha\beta$ *is defined if and only if the domain of* α *is the codomain of* β, *and then the domain of* $\alpha\beta$ *is the domain of* β *and the codomain of* $\alpha\beta$ *is the codomain of* α;

(2) *for every object* A *of* $\mathcal{C}$ *there exists an* identity morphism 1_A *whose domain and codomain are* A, *such that* $\alpha 1_A = \alpha$ *whenever* A *is the domain of* α *and* $1_A \beta = \beta$ *whenever* A *is the codomain of* α;

(3) *if* $\alpha\beta$ *and* $\beta\gamma$ *are defined, then* $\alpha(\beta\gamma) = (\alpha\beta)\gamma$.

Axiom (1) models morphisms on maps that are written on the left. A morphism α with domain A and codomain B is *from A to B* and is denoted by an arrow, $\alpha : A \longrightarrow B$ or $A \xrightarrow{\alpha} B$. Then $\alpha\beta$ is defined if and only if $A \xrightarrow{\beta} B \xrightarrow{\alpha} C$ for some objects A, B, C. Readers will show that the identity morphism 1_A in Axiom (2) is unique, for every object A. In Axiom (3), if $\alpha\beta$ and $\beta\gamma$ are defined, then $\alpha(\beta\gamma)$ and $(\alpha\beta)\gamma$ are defined, by (1).

Examples. Sets and mappings become the objects and morphisms of a category once a small adjustment is made in the definition of mappings: one must regard as different morphisms a mapping $A \longrightarrow B$ and its composition

$A \longrightarrow B \overset{\subset}{\longrightarrow} C$ with a strict inclusion, which consists of the same ordered pairs but has a different codomain. This can be achieved by defining a mapping of A into B as an ordered triple (A, f, B) in which f is the usual set of ordered pairs. Then sets and mappings (written on the left) become the objects and morphisms of a category *Sets*; the identity morphism of a set is its usual identity mapping; composition in *Sets* is the usual composition of mappings.

With the same definition of mappings, groups and their homomorphisms are the objects and morphisms of a category *Grps*; abelian groups and their homomorphisms are the objects and morphisms of a category *Abs*; left R-modules and their homomorphisms are the objects and morphisms of a category $_R$*Mods*; etc.

A category is *small* when its class of objects and its class of morphisms are sets; its components can be bundled into a single set. The categories *Sets*, *Grps*, $_R$*Mods*, etc., are not small, but there are useful examples of small categories.

Recall that a *preordered set* is a set with a binary relation $\leq$ that is reflexive and transitive; partially ordered sets are an example. Every preordered set I can be viewed as a small category, whose objects are the elements of I and whose morphisms are the elements of $\leq$ (namely, ordered pairs (i, j) of elements of I such that $i \leq j$). Domain and codomain are given by $(i, j): i \longrightarrow j$, so that "arrows point upward;" composition is given by $(j, k)(i, j) = (i, k)$.

Graphs. Graphs are like small categories without composition and give rise to additional examples of small categories.

Definition. A small directed graph or just graph $\mathcal{G}$ consists of

a set whose elements are the vertices *or* nodes *of $\mathcal{G}$;*

a set whose elements are the edges *or* arrows *of $\mathcal{G}$;*

two mappings, that assign to every edge of $\mathcal{G}$ an origin *and a* destination, *which are vertices of $\mathcal{G}$.*

An edge a with origin i and destination j is *from i to j* and is denoted by an arrow, $a : i \longrightarrow j$ or $i \overset{a}{\longrightarrow} j$. Graphs will be used as abstract diagrams, such as the *square* and *triangle* graphs:

Every small category is a graph. Conversely, every graph $\mathcal{G}$ generates a small category, as follows. In a graph $\mathcal{G}$, a nonempty *path*

$$i \overset{a_1}{\longrightarrow} \bullet \overset{a_2}{\longrightarrow} \bullet \cdots \bullet \overset{a_n}{\longrightarrow} j$$

from a vertex i to a vertex j is a sequence $(i, a_1, \ldots, a_n, j)$ in which $n > 0$, i is the origin of a_1, the destination of a_k is the origin of a_{k+1} for all $k < n$, and the destination of a_n is j. For every vertex i there is also an *empty path*

(i, i) from i to i. Paths are composed by concatenation:

$$(j, b_1, \ldots, b_n, k)(i, a_1, \ldots, a_m, j) = (i, a_1, \ldots, a_m, b_1, \ldots, b_n, k).$$

Proposition **1.1.** *The vertices and paths of a graph $\mathcal{G}$ are the objects and morphisms of a small category $\widehat{\mathcal{G}}$, the* free category *or* category of paths *of $\mathcal{G}$.*

This is an exercise. A universal property of $\widehat{\mathcal{G}}$ is stated in Section 3.

Monomorphisms and epimorphisms. Among mappings one distinguishes injections, surjections, and bijections. The composition properties of these special mappings are expressed by the following definitions.

Definitions. In a category, a monomorphism *is a morphism μ such that $\mu\alpha = \mu\beta$ implies $\alpha = \beta$; an* epimorphism *is a morphism σ such that $\alpha\sigma = \beta\sigma$ implies $\alpha = \beta$; an* isomorphism *is a morphism $\alpha : A \longrightarrow B$ that has an* inverse $\beta : B \longrightarrow A$ *such that $\alpha\beta = 1_B$ and $\beta\alpha = 1_A$.*

In many categories, monomorphisms coincide with injective morphisms, and the two terms are often used interchangeably.

Proposition **1.2.** *In the category Grps of groups and homomorphisms, a morphism is a monomorphism if and only if it is injective.*

Proof. Let $\mu : A \longrightarrow B$ be a monomorphism in *Grps*. Assume that $\mu(x) = \mu(y)$. Since $\mathbb{Z}$ is free on $\{1\}$ there exist homomorphisms $\alpha, \beta : \mathbb{Z} \longrightarrow A$ such that $\alpha(1) = x$ and $\beta(1) = y$. Then $\mu(\alpha(1)) = \mu(\beta(1))$, $\mu \circ \alpha = \mu \circ \beta$, $\alpha = \beta$, and $x = y$. Thus μ is injective. The converse is clear. $\square$

Readers will prove similar results for *Sets*, *Rings*, $_R$*Mods*, and so forth.

Identification of epimorphisms with surjective morphisms is more difficult, and less successful. For *Grps*, it requires free products with amalgamation.

Proposition **1.3.** *In the category Grps of groups and homomorphisms, a morphism is an epimorphism if and only if it is surjective.*

Proof. Surjective homomorphisms are epimorphisms. Conversely, let $\varphi : G \longrightarrow H$ be a homomorphism of groups that is not surjective. Construct isomorphic copies H_1, H_2 of H that contain Im φ as a subgroup and satisfy $H_1 \cap H_2 = \text{Im } \varphi$; embed the group amalgam $H_1 \cup H_2$ into its free product with amalgamation P. The isomorphisms $H \cong H_i$ and injections $H_i \longrightarrow P$ yields homomorphisms $\alpha_1 \neq \alpha_2 : H \longrightarrow P$ such that $\alpha_1 \circ \varphi = \alpha_2 \circ \varphi$. $\square$

A similar result holds in *Sets* and $_R$*Mods* but not in the categories *Rings* and *R-Algs* (see the exercises). In these categories, using "epimorphism" to mean "surjective homomorphism" should be discouraged, or outright forbidden.

Duality. Every category $\mathcal{C}$ has an opposite category $\mathcal{C}^{\text{op}}$, which is constructed as follows. The objects and morphisms of $\mathcal{C}^{\text{op}}$ are those of $\mathcal{C}$. The domain and codomain functions of $\mathcal{C}^{\text{op}}$ are the codomain and domain functions of $\mathcal{C}$, so that $\alpha : A \longrightarrow B$ in $\mathcal{C}$ if and only if $\alpha : B \longrightarrow A$ in $\mathcal{C}^{\text{op}}$. The composition $\alpha * \beta$

of α and β in $\mathcal{C}^{op}$ is the composition $\beta\alpha$ of β and α in $\mathcal{C}$. Thus, $\mathcal{C}^{op}$ is the same as $\mathcal{C}$ but with all arrows and compositions reversed. The next result is an easy exercise:

Proposition **1.4.** *If $\mathcal{C}$ is a category, then $\mathcal{C}^{op}$ is a category, the* dual *or* opposite *of $\mathcal{C}$.*

For example, a preordered set I has an opposite preordered set I^{op}, which is the same set but preordered by $i \leqq j$ in I^{op} if and only if $i \geqq j$ in I; then I^{op} is the opposite category of I. Similarly, every graph G has an opposite graph G^{op}, and then $\widehat{G^{op}} = \widehat{G}^{op}$. Proposition 1.4 implies a duality principle:

Metatheorem **1.5** (Duality Principle). *A theorem that applies to all categories remains true when all arrows and compositions are reversed.*

Like Metatheorem XIV.1.3, the duality principle does not prove any specific statement, but is applied to existing theorems to yield new results. One should keep in mind that reversal of arrows and compositions in 1.5 applies to hypotheses as well as conclusions. The duality principle does not apply to *specific* categories: if Theorem T is true in $\mathcal{C}$, then Theorem T^{op} is true in $\mathcal{C}^{op}$, but it does not follow that T^{op} is true in $\mathcal{C}$ (unless T is also true in $\mathcal{C}^{op}$). For example, in the category of fields and homomorphisms, every morphism is a monomorphism; it would be a gross misuse of the duality principle to conclude that every homomorphism of fields is an epimorphism; Artin would turn in his grave.

We illustrate proper use of duality with the following result (left to readers):

Proposition **1.6.** *Let α and β be morphisms in a category and let $\alpha\beta$ be defined. If α and β are monomorphisms, then $\alpha\beta$ is a monomorphism. If $\alpha\beta$ is a monomorphism, then β is a monomorphism.*

Reversing arrows and compositions transforms monomorphisms into epimorphisms, and vice versa. Hence 1.5, 1.6 yield the following result:

Proposition **1.7.** *Let α and β be morphisms in a category and let $\alpha\beta$ be defined. If α and β are epimorphisms, then $\alpha\beta$ is an epimorphism. If $\alpha\beta$ is an epimorphism, then α is an epimorphism.*

Exercises

1. Show that a small category is [the category that arises from] a preordered set if and only if for every objects i and j there is at most one morphism from i to j.

2. Let I be a preordered set (viewed as a category). What are the monomorphisms of I? its epimorphisms? its isomorphisms?

3. Show that a mapping is a monomorphism in *Sets* if and only if it is injective.

4. Show that a homomorphism of rings [with identity elements] is a monomorphism in *Rings* if and only if it is injective.

5. Show that a homomorphism of left R-modules is a monomorphism in $_R$*Mods* if and only if it is injective.

6. Show that a mapping is an epimorphism in *Sets* if and only if it is surjective.

7. Show that a homomorphism of left R-modules is an epimorphism in $_R$*Mods* if and only if it is surjective.

8. Show that the inclusion homomorphism $\mathbb{Z} \longrightarrow \mathbb{Q}$ is an epimorphism in *Rings*.

9. Let α and β be morphisms in a category and let $\alpha\beta$ be defined. Prove the following: if α and β are monomorphisms, then $\alpha\beta$ is a monomorphism; if $\alpha\beta$ is a monomorphism, then β is a monomorphism.

10. Let α and β be morphisms in a category and let $\alpha\beta$ be defined. Give a direct proof of the following: if α and β are epimorphisms, then $\alpha\beta$ is an epimorphism; if $\alpha\beta$ is an epimorphism, then α is an epimorphism.

11. Find an "objectless" set of axioms for categories, using only morphisms and composition. (Hint: replace objects by identity morphisms ε, which can be defined by the conditions: $\alpha\varepsilon = \alpha$, $\varepsilon\beta = \beta$, whenever $\alpha\varepsilon$ and $\varepsilon\beta$ are defined.)

2. Functors

Objects relate to each other by means of morphisms. Categories relate to each other by means of functors; and functors, by means of natural transformations.

Definition. A functor or covariant functor F from a category $\mathcal{A}$ to a category $\mathcal{C}$ assigns to each object A of $\mathcal{A}$ an object $F(A)$ of $\mathcal{C}$, and assigns to each morphism α of $\mathcal{A}$ a morphism $F(\alpha)$ of $\mathcal{C}$, so that

(1) *if $\alpha : A \longrightarrow B$, then $F(\alpha): F(A) \longrightarrow F(B)$;*

(2) *$F(1_A) = 1_{F(A)}$ for every object A of $\mathcal{A}$;*

(3) *$F(\alpha\beta) = F(\alpha)\,F(\beta)$ whenever $\alpha\beta$ is defined.*

Examples. Our first official examples of functors (in Chapter XI) were the functors $\mathrm{Hom}_R(A, -)$ from $_R$*Mods* to the category *Abs* of abelian groups and homomorphisms, and the functors $A \otimes_R -$, also from $_R$*Mods* to *Abs*.

Other examples have been hiding as far back as Chapter I. The *forgetful functor* from *Grps* to *Sets* assigns to every group G its underlying set, and to every homomorphism of groups its underlying mapping. (The only difference between a homomorphism of groups and its underlying mapping is that the domain and codomain of the former are groups, whereas the domain and codomain of the latter are sets.) There are similar forgetful functors from *Rings* to *Abs*, from $_R$*Mods* to *Abs*, from *R-Algs* to *Rings*, from *R-Algs* to $_R$*Mods*, and so forth.

The *free group functor* F assigns to each set X the free group F_X on X constructed in Section I.6, which comes with an injection $\iota_X : X \longrightarrow F_X$; to a mapping $f : X \longrightarrow Y$ the free group functor assigns the unique homomorphism $F_f : F_X \longrightarrow F_Y$ such that $F_f \circ \iota_X = \iota_Y \circ f$. Curious readers will verify that F is indeed a functor, and detect other free functors in previous chapters.

Properties. Functors compose: if F is a functor from $\mathcal{A}$ to $\mathcal{B}$, and G is a functor from $\mathcal{B}$ to $\mathcal{C}$, then $G \circ F$ is a functor from $\mathcal{A}$ to $\mathcal{C}$. This composition is associative. Moreover, there is for each category $\mathcal{C}$ an *identity functor* $1_{\mathcal{C}}$ on $\mathcal{C}$, which is the identity on objects and morphisms of $\mathcal{C}$.

Categories and functors now look like the objects and morphisms of an enormous category ... except that we may not collect proper classes into a class, or collect all categories into a class. This restriction does not, however, apply to *small* categories:

Proposition **2.1.** *Small categories and their functors are the objects and morphisms of a category Cats.*

Definition. Let F and G be functors from $\mathcal{A}$ to $\mathcal{B}$. A natural transformation $\tau : F \longrightarrow G$ assigns to each object A of $\mathcal{A}$ a morphism $\tau_A : F(A) \longrightarrow G(A)$, so that $G(\alpha)\, \tau_A = \tau_B\, F(\alpha)$ for every morphism $\alpha : A \longrightarrow B$ of $\mathcal{A}$:

$$
\begin{array}{ccc}
F(A) & \xrightarrow{\ \tau_A\ } & G(A) \\
{\scriptstyle F(\alpha)}\Big\downarrow & & \Big\downarrow{\scriptstyle G(\alpha)} \\
F(B) & \xrightarrow[\ \tau_B\ ]{} & G(B)
\end{array}
$$

One also says that that the morphism τ_A (that depends on A) is *natural* in A. If every τ_A is an isomorphism, then the inverse isomorphism τ_A^{-1} is natural in A, and τ is a *natural isomorphism*. Examples of natural transformations have been seen in previous chapters.

Natural transformations compose: if $F, G, H : \mathcal{A} \longrightarrow \mathcal{B}$ are functors and $\tau : F \longrightarrow G$, $\upsilon : G \longrightarrow H$ are natural transformations, then $\upsilon\tau : F \longrightarrow H$ is a natural transformation, which assigns $\upsilon_A\, \tau_A$ to A. This composition is associative. Moreover, there is for each functor $F : \mathcal{A} \longrightarrow \mathcal{B}$ an identity natural transformation 1_F on F, which assigns $1_{F(A)}$ to each object A of $\mathcal{A}$. Functors from $\mathcal{A}$ to $\mathcal{B}$ and their natural transformations now look like the objects and morphisms of a category ... but, in general, functors are proper classes and may not be collected into a class. If $\mathcal{A}$ is small, however, then functors from $\mathcal{A}$ to $\mathcal{B}$ and their natural transformations are sets, and we obtain a *functor category*:

Proposition **2.2.** *Let $\mathcal{A}$ be a small category and let $\mathcal{B}$ be a category. Functors from $\mathcal{A}$ to $\mathcal{B}$ and their natural transformations are the objects and morphisms of a category Func $(\mathcal{A}, \mathcal{B})$.*

Natural transformations also compose with functors, as readers will show:

Proposition **2.3.** *If $F, G : \mathcal{A} \longrightarrow \mathcal{B}$ and $H : \mathcal{B} \longrightarrow \mathcal{C}$ are functors, and $\tau : F \longrightarrow G$ is a natural transformation, then $H \circ \tau : H \circ F \longrightarrow H \circ G$ is a natural transformation, which assigns $H(\tau_A)$ to A.*

If $F, G : \mathcal{A} \longrightarrow \mathcal{B}$ and $K : \mathcal{C} \longrightarrow \mathcal{A}$ are functors, and $\tau : F \longrightarrow G$ is a natural transformation, then $\tau \circ K : F \circ K \longrightarrow G \circ K$ is a natural transformation, which assigns $\tau_{K(C)}$ to C.

With functors and natural transformations we can define when two categories are essentially identical.

Definition. Two categories $\mathcal{A}$ and $\mathcal{B}$ are isomorphic *when there exist functors $F : \mathcal{A} \longrightarrow \mathcal{B}$ and $G : \mathcal{B} \longrightarrow \mathcal{A}$ such that $F \circ G = 1_{\mathcal{B}}$ and $G \circ F = 1_{\mathcal{A}}$.*

For example, an isomorphism of preordered sets is also an isomorphism of categories; the category of Boolean lattices and the category of Boolean rings are isomorphic. But isomorphisms of categories are somewhat rare. The exercises explore a less restrictive definition:

Definition. Two categories $\mathcal{A}$ and $\mathcal{B}$ are equivalent *when there exist functors $F : \mathcal{A} \longrightarrow \mathcal{B}$ and $G : \mathcal{B} \longrightarrow \mathcal{A}$ and natural isomorphisms $F \circ G \cong 1_{\mathcal{B}}$ and $G \circ F \cong 1_{\mathcal{A}}$.*

Contravariant functors. Some functors, like $\mathrm{Hom}_R(-, A)$, reverse arrows:

Definition. A contravariant functor *F from a category $\mathcal{A}$ to a category $\mathcal{C}$ assigns to each object A of $\mathcal{A}$ an object $F(A)$ of $\mathcal{C}$, and assigns to each morphism α of $\mathcal{A}$ a morphism $F(\alpha)$ of $\mathcal{C}$, so that*

(1) *if $\alpha : A \longrightarrow B$, then $F(\alpha): F(B) \longrightarrow F(A)$;*

(2) *$F(1_A) = 1_{F(A)}$ for every object A of $\mathcal{A}$;*

(3) *$F(\alpha\beta) = F(\beta)\, F(\alpha)$ whenever $\alpha\beta$ is defined.*

Equivalently, a contravariant functor from $\mathcal{A}$ to $\mathcal{C}$ is a covariant functor from $\mathcal{A}$ to $\mathcal{C}^{\mathrm{op}}$; or a covariant functor from $\mathcal{A}^{\mathrm{op}}$ to $\mathcal{C}$.

Bifunctors. Bifunctors are functors with two variables, like $\mathrm{Hom}_R(-, -)$ and $- \otimes_R -$. For a general definition, construct the following:

Definition. The Cartesian product *of two categories $\mathcal{A}$ and $\mathcal{B}$ is the category $\mathcal{A} \times \mathcal{B}$ defined as follows: an object of $\mathcal{A} \times \mathcal{B}$ is an ordered pair (A, B) of an object A of $\mathcal{A}$ and an object B of $\mathcal{B}$; a morphism of $\mathcal{A} \times \mathcal{B}$ is an ordered pair (α, β) of a morphism α of $\mathcal{A}$ and a morphism β of $\mathcal{B}$; domain, codomain, and composition are componentwise.*

Thus, if $\alpha : A \longrightarrow A'$ in $\mathcal{A}$ and $\beta : B \longrightarrow B'$ in $\mathcal{B}$, then $(\alpha, \beta): A \times B \longrightarrow A' \times B'$ in $\mathcal{A} \times \mathcal{B}$. Also, $1_{(A,B)} = (1_A, 1_B)$.

Definition. A bifunctor *from categories $\mathcal{A}$ and $\mathcal{B}$ to a category $\mathcal{C}$ is a functor from $\mathcal{A} \times \mathcal{B}$ to $\mathcal{C}$.*

Strictly speaking, this defines bifunctors that are covariant in both variables. For example, $- \otimes_R -$ is a bifunctor from $Mods_R$ and $_RMods$ to Abs; $- \otimes_R -$ also denotes a bifunctor from $_SBims_R$ and $_RBims_T$ to $_SBims_T$. ($_PBims_Q$ is the category of left P-, right Q-bimodules and homomorphisms.) A bifunctor from $\mathcal{A}$ and $\mathcal{B}$ into $\mathcal{C}$ that is contravariant in the first variable and covariant in the second variable is a functor from $\mathcal{A}^{\mathrm{op}} \times \mathcal{B}$ to $\mathcal{C}$. For example, $\mathrm{Hom}_R(-, -)$ is a bifunctor that is contravariant in the first variable and covariant in the second variable, from $Mods_R$ and $_RMods$ to Abs; $\text{Hom}_R(-, -)$ also denotes the similar bifunctors for bimodules.

Bifunctors relate to functors as functions of two variables everywhere relate to functions of one variable, which readers will show:

Proposition **2.4.** *Let* F *be a bifunctor from* $\mathcal{A}$ *and* $\mathcal{B}$ *to* $\mathcal{C}$. *For each object* A *of* $\mathcal{A}$, $F(A, -)$ *is a functor from* $\mathcal{B}$ *to* C. *For each morphism* $\alpha : A \longrightarrow A'$ *of* $\mathcal{A}$, $F(\alpha, -)$ *is a natural transformation from* $F(A, -)$ *to* $F(A', -)$. *Moreover,* $F(1_A, -) = 1_{F(A,-)}$ *and* $F(\alpha\alpha', -) = F(\alpha, -) \circ F(\alpha', -)$ *whenever* $\alpha\alpha'$ *is defined.*

Conversely, assign to each object A *of* $\mathcal{C}$ *a functor* F_A *from* $\mathcal{B}$ *to* $\mathcal{C}$; *to each morphism* $\alpha : A \longrightarrow A'$ *of* $\mathcal{A}$, *assign a natural transformation* F_α *from* F_A *to* $F_{A'}$; *assume that* $F_{1_A} = 1_{F_A}$ *and* $F_{\alpha\alpha'} = F_\alpha \circ F_{\alpha'}$ *whenever* $\alpha\alpha'$ *is defined. Then there exists a unique bifunctor* F *from* $\mathcal{A}$ *and* $\mathcal{B}$ *to* $\mathcal{C}$ *such that* $F_A = F(A, -)$ *and* $F_\alpha = F(\alpha, -)$ *for all* A *and* α.

In particular, if $\mathcal{B}$ is small, then Proposition 2.4 provides a one-to-one correspondence between bifunctors from $\mathcal{A}$ and $\mathcal{B}$ to $\mathcal{C}$, and functors from $\mathcal{A}$ to $Func\,(\mathcal{B}, \mathcal{C})$. If $\mathcal{A}$ is small, there is a similar one-to-one correspondence between bifunctors from $\mathcal{A}$ and $\mathcal{B}$ to $\mathcal{C}$, and functors from $\mathcal{B}$ to $Func\,(\mathcal{A}, \mathcal{C})$.

Hom. A bifunctor $\text{Hom}(-, -)$ can be defined in many categories. If A and B are objects of a category $\mathcal{C}$, then $\text{Hom}_\mathcal{C}(A, B)$ denotes the class of all morphisms of $\mathcal{C}$ from A to B (also denoted by $\mathcal{C}(A, B)$).

Definition. A category $\mathcal{C}$ *is* locally small *when* $\text{Hom}_\mathcal{C}(A, B)$ *is a set for all objects* A *and* B *of* $\mathcal{C}$.

For example, $Sets$, $Grps$, $Rings$, $_RMods$, are locally small categories. Every small category is locally small.

If $\mathcal{C}$ is locally small, there is a bifunctor $\text{Hom}_\mathcal{C}(-, -)$ from $\mathcal{C}^{op}$ and $\mathcal{C}$ to $Sets$. For each pair of objects A, C of $\mathcal{C}$, $\text{Hom}_\mathcal{C}(A, C)$ is the set of all morphisms $A \longrightarrow C$ of $\mathcal{C}$. For each pair of morphisms $\alpha : A \longrightarrow B$, $\gamma : C \longrightarrow D$, $\text{Hom}_\mathcal{C}(\alpha, \gamma) : \text{Hom}_\mathcal{C}(B, C) \longrightarrow \text{Hom}_\mathcal{C}(A, D)$ sends $\beta : B \longrightarrow C$ to $\gamma\beta\alpha$:

$$
\begin{array}{ccc}
B & \xrightarrow{\ \beta\ } & C \\
{\scriptstyle \alpha}\uparrow & & \downarrow{\scriptstyle \gamma} \\
A & \dashrightarrow[\gamma\beta\alpha] & D
\end{array}
$$

Readers will verify that $\text{Hom}_\mathcal{C}(-, -)$ is indeed a bifunctor.

If $\mathcal{C}$ is locally small, then Proposition 2.4 yields for every object A of $\mathcal{C}$ a functor $\text{Hom}_\mathcal{C}(A, -)$ from $\mathcal{C}$ to $Sets$ and for every morphism $\alpha : A \longrightarrow B$ of $\mathcal{C}$ a natural transformation $\text{Hom}_\mathcal{C}(\alpha, -) : \text{Hom}_\mathcal{C}(B, -) \longrightarrow \text{Hom}_\mathcal{C}(A, -)$. Proposition 2.4 also yields for every object C of $\mathcal{C}$ a contravariant functor

$\operatorname{Hom}_{\mathcal{C}}(-,C)$ and for every morphism $\gamma : C \longrightarrow D$ of $\mathcal{C}$ a natural transformation $\operatorname{Hom}_{\mathcal{C}}(-,\gamma) \colon \operatorname{Hom}_{\mathcal{C}}(-,C) \longrightarrow \operatorname{Hom}_{\mathcal{C}}(-,D)$.

Exercises

1. Show that equivalence of categories is reflexive, symmetric, and transitive.

In the following three problems, a *skeleton* of a category $\mathcal{C}$ is a category $\mathcal{S}$ such that: (i) every object of $\mathcal{S}$ is an object of $\mathcal{C}$; (ii) every object of $\mathcal{C}$ is isomorphic to a unique object of $\mathcal{S}$; (iii) when S and T are objects of $\mathcal{S}$, the morphisms from S to T are the same in $\mathcal{S}$ and $\mathcal{C}$. (And now it is out of the closet.)

2. Show that a category is equivalent to any of its skeletons.

3. Let $\mathcal{S}$ be a skeleton of $\mathcal{A}$ and let $\mathcal{T}$ be a skeleton of $\mathcal{B}$. Show that $\mathcal{A}$ and $\mathcal{B}$ are equivalent if and only if $\mathcal{S}$ and $\mathcal{T}$ are isomorphic.

4. Show that every preordered set is equivalent (as a category) to a partially ordered set.

5. Let $\mathcal{B}$ be a category and let $\mathcal{A}$ be a small category. Show that evaluation $(A,F) \longmapsto F(A)$ is a bifunctor from $\mathcal{A}$ and $\mathit{Func}\,(\mathcal{A},\mathcal{B})$ to $\mathcal{B}$.

6. Show that there is a one-to-one correspondence between bifunctors from $\mathcal{A}$ and $\mathcal{B}$ to $\mathcal{C}$, and "functorial" assignments of functors from $\mathcal{B}$ to $\mathcal{C}$ to objects of $\mathcal{A}$, as in 2.4.

7. Give a direct proof that there is a one-to-one correspondence between bifunctors from $\mathcal{A}$ and $\mathcal{B}$ to $\mathcal{C}$, and "functorial" assignments of functors from $\mathcal{A}$ to $\mathcal{C}$ to objects and morphisms of $\mathcal{B}$.

8. Prove *Yoneda's lemma*: when $\mathcal{C}$ is locally small and F is a functor from $\mathcal{C}$ to *Sets*, there is for each object C of $\mathcal{C}$ a one-to-one correspondence between elements of $F(C)$ and natural transformations from $\operatorname{Hom}_{\mathcal{C}}(C,-)$ to F. (You may use $\tau_C(1_C) \in F(C)$ when $\tau : \operatorname{Hom}_{\mathcal{C}}(C,-) \longrightarrow F$ is a natural transformation.)

3. Limits and Colimits

Limits and colimits generalize many of the constructions seen in previous chapters: direct products, direct sums, pullbacks, kernels, direct limits, and others.

Diagrams. Limits apply to diagrams, which are formally defined as follows.

Definition. A diagram *in a category* $\mathcal{C}$ *over a [small] graph* $\mathcal{G}$ *is an ordered pair of mappings, one that assigns to each vertex i of $\mathcal{G}$ an object D_i of $\mathcal{C}$, and one that assigns to each edge $a : i \longrightarrow j$ of $\mathcal{G}$ a morphism $D_a : D_i \longrightarrow D_j$ of $\mathcal{C}$.*

For example, a direct system of left R-modules is, in particular, a diagram in $_R\mathit{Mods}$ over a preordered set, viewed as a graph. A *square* or a *triangle* in a category $\mathcal{C}$ is a diagram in $\mathcal{C}$ over the square or triangle graph:

Diagrams are objects of suitable categories:

Definition. Let D and E be diagrams in a category $\mathcal{C}$ over a graph $\mathcal{G}$. A morphism from D to E is a mapping φ that assigns to each vertex i of $\mathcal{G}$ a morphism $\varphi_i : D_i \longrightarrow E_i$ of $\mathcal{C}$, so that $E_a \varphi_i = \varphi_j D_a$ for every edge $a : i \longrightarrow j$ of $\mathcal{G}$:

$$
\begin{array}{ccc}
D_i & \xrightarrow{\varphi_i} & E_i \\
D_a \downarrow & & \downarrow E_a \\
D_j & \xrightarrow{\varphi_j} & E_j
\end{array}
$$

Morphisms of diagrams compose: if $\varphi : D \longrightarrow E$ and $\psi : E \longrightarrow F$ are morphisms of diagrams over $\mathcal{G}$ (in the same category), then so is $\psi\varphi : D \longrightarrow F$, which assigns $\psi_i \varphi_i : D_i \longrightarrow F_i$ to each vertex i of $\mathcal{G}$. Moreover, there is for every diagram D an identity morphism $1_D : D \longrightarrow D$, which assigns $1_{D_i} : D_i \longrightarrow D_i$ to each vertex i of $\mathcal{G}$. We now have a category:

Proposition **3.1.** *Let $\mathcal{C}$ is a category and let $\mathcal{G}$ be a [small] graph. Diagrams in $\mathcal{C}$ over $\mathcal{G}$ and their morphisms are the objects and morphisms of a category Diag $(\mathcal{G}, \mathcal{C})$.*

The category *Diag* $(\mathcal{G}, \mathcal{C})$ resembles a functor category and is in fact isomorphic to one:

Proposition **3.2.** *Let $\mathcal{C}$ be a category and let $\mathcal{G}$ be a [small] graph. Every diagram $D : \mathcal{G} \longrightarrow \mathcal{C}$ extends uniquely to a functor $\widehat{D} : \widehat{\mathcal{G}} \longrightarrow \mathcal{C}$; every morphism $D \longrightarrow E$ of diagrams over $\mathcal{G}$ is a natural transformation $\widehat{D} \longrightarrow \widehat{E}$; hence the categories Diag $(\mathcal{G}, \mathcal{C})$ and Func $(\widehat{\mathcal{G}}, \mathcal{C})$ are isomorphic.*

Proof. The objects of $\widehat{\mathcal{G}}$ are the vertices of $\mathcal{G}$, and the morphisms of $\widehat{\mathcal{G}}$ are the empty paths (i, i) and all nonempty paths $(i, a_1, \ldots, a_n, j)$. The latter are compositions in $\widehat{\mathcal{G}}$: $(i, a_1, \ldots, a_n, j) = (\bullet, a_n, j) \cdots (\bullet, a_2, \bullet)(i, a_1, \bullet)$. The unique functor $\widehat{D}$ that extends D is given by $\widehat{D}(i) = D_i$, $\widehat{D}(i, i) = 1_{D_i}$, and $\widehat{D}(i, a_1, \ldots, a_n, j) = D_{a_n} \cdots D_{a_2} D_{a_1}$. The remaining parts of the statement are equally immediate. $\square$

The construction of $\widehat{D}$ yields a precise definition of commutative diagrams: D is *commutative* when $\widehat{D}_p = \widehat{D}_q$ whenever $p, q : i \longrightarrow j$ are nonempty paths with the same domain and codomain.

Limits are generalizations of direct products, pullbacks, and inverse limits.

Definition. Let A be an object of a category $\mathcal{C}$ and let D be a diagram in $\mathcal{C}$ over a graph $\mathcal{G}$. A cone φ from A to D assigns to each vertex i of $\mathcal{G}$ a morphism $\varphi_i : A \longrightarrow D_i$, so that $D_a \varphi_i = \varphi_j$ for every edge $a : i \longrightarrow j$ of $\mathcal{G}$:

$$A \xrightarrow{\varphi_i} D_i$$
$$\varphi_j \searrow \quad \downarrow D_a$$
$$D_j$$

Equivalently, a cone from A to D is a morphism from $C(A)$ to D, where $C(A)$ is the *constant* diagram, which assigns A to every vertex and 1_A to every edge. Since morphisms of diagrams compose, the following hold: if $\varphi : A \longrightarrow D$ is a cone and $\psi : D \longrightarrow E$ is a morphism of diagrams, then $\psi\varphi : A \longrightarrow E$ is a cone; if $\varphi : B \longrightarrow D$ is a cone and $\psi : A \longrightarrow B$ is a morphism, then $\varphi\psi : A \longrightarrow D$ is a cone.

Definitions. Let D be a diagram in a category $\mathcal{C}$ over a [small] graph $\mathcal{G}$. A limit cone of D is a cone $\lambda : L \longrightarrow D$ such that, for every cone $\varphi : A \longrightarrow D$ there is a unique morphism $\overline{\varphi} : A \longrightarrow L$ such that $\varphi = \lambda\overline{\varphi}$ ($\varphi_i = \lambda_i \overline{\varphi}$ for every vertex i of $\mathcal{G}$):

$$L \xrightarrow{\lambda_i} D_i$$
$$\overline{\varphi} \uparrow \quad \nearrow \varphi_i$$
$$A$$

Then the object L is a limit *of the diagram D.*

Readers will easily establish the following properties:

Proposition 3.3. If $\lambda : L \longrightarrow D$ and $\lambda' : L' \longrightarrow D$ are limit cones of D, then there is an isomorphism $\theta : L' \longrightarrow L$ such that $\lambda' = \lambda\theta$. Conversely, if $\lambda : L \longrightarrow D$ is a limit cone of D and $\theta : L' \longrightarrow L$ is an isomorphism, then $\lambda\theta : L' \longrightarrow D$ is a limit cone of D.

Examples. Limits include a number of constructions from previous chapters. Inverse limits are the most obvious example. We define some general types.

Definition. In a category $\mathcal{C}$, the product *of a family $(D_i)_{i \in I}$ of objects of $\mathcal{C}$ consists of an object $P = \prod_{i \in I} D_i$ and a family $(\pi_i)_{i \in I}$ of projections $\pi_i : P \longrightarrow D_i$ such that, for every object A and family $(\varphi_i)_{i \in I}$ of morphisms $\varphi_i : A \longrightarrow D_i$ there is a unique morphism $\overline{\varphi} : A \longrightarrow P$ such that $\varphi_i = \pi_i \overline{\varphi}$ for every $i \in I$:*

$$P \xrightarrow{\pi_i} D_i$$
$$\overline{\varphi} \uparrow \quad \nearrow \varphi_i$$
$$A$$

The object P is also called a product *of the objects $(D_i)_{i \in I}$.*

For example, every family of sets has a product in *Sets*, which is their Cartesian product with the usual projections; and similarly for *Grps* , $_R Mods$, etc. The product of a finite family $D_1, \ldots, D_n$ is denoted by $D_1 \times \cdots \times D_n$.

Products are limits of diagrams over certain graphs. A *discrete* graph is a graph without edges, and may be identified with its set I of vertices. A diagram

D over a discrete graph I is a family $(D_i)_{i \in I}$ of objects; a cone $\varphi : A \longrightarrow D$ is a family $(\varphi_i)_{i \in I}$ of morphisms $\varphi_i : A \longrightarrow D_i$. Thus, the object P and projections $\pi_i : P \longrightarrow D_i$ constitute a product of $(D_i)_{i \in I}$ if and only if $\pi : P \longrightarrow D$ is a limit cone of D.

Readers will verify that products are associative: if $I = \bigcup_{j \in J} I_j$ is a partition of I, then there is a natural isomorphism $\prod_{i \in I} D_i \cong \prod_{j \in J} \left(\prod_{i \in I_j} D_j \right)$ (when these products exist). For instance, $A \times B \times C$ is naturally isomorphic to $(A \times B) \times C$ and to $A \times (B \times C)$.

Definition. In a category $\mathcal{C}$, *the* equalizer *of two morphisms* $\alpha, \beta : A \longrightarrow B$ *is a morphism* $\varepsilon : E \longrightarrow A$ *such that* (i) $\alpha \varepsilon = \beta \varepsilon$ *and* (ii) *every morphism* φ *such that* $\alpha \varphi = \beta \varphi$ *factors uniquely through* ε:

$$E \xrightarrow{\ \varepsilon\ } A \underset{\beta}{\overset{\alpha}{\rightrightarrows}} B$$
$$\begin{array}{c} \uparrow \nearrow^{\varphi} \\ C \end{array}$$

For example, in $_R Mods$, the equalizer of $\alpha : M \longrightarrow N$ and 0 is the inclusion homomorphism $\mathrm{Ker}\,\alpha \longrightarrow M$. The exercises give other examples.

In general, $A \underset{\beta}{\overset{\alpha}{\rightrightarrows}} B$ is a diagram D over the graph $\bullet \rightrightarrows \bullet$. A cone into D consists of morphisms $\varphi : C \longrightarrow A$, $\psi : C \longrightarrow B$ such that $\psi = \alpha \varphi = \beta \varphi$, and is uniquely determined by φ. Hence (ε, η) is a limit cone of D if and only if ε is an equalizer of α and β and $\eta = \alpha \varepsilon = \beta \varepsilon$.

More generally, the *equalizer* of a set S of morphisms from A to B is a morphism $\varepsilon : E \longrightarrow A$ such that (i) $\sigma \varepsilon = \tau \varepsilon$ for all $\sigma, \tau \in S$, and (ii) every morphism φ with this property factors uniquely through ε. Every equalizer ε is a monomorphism, since factorization through ε is unique.

A *pullback* in a category $\mathcal{C}$ is a commutative square $\alpha \beta' = \beta \alpha'$ such that for every commutative square $\alpha \psi = \beta \varphi$ there exists a unique morphism χ such that $\varphi = \alpha' \chi$ and $\psi = \beta' \chi$; pullbacks in $_R Mods$ are an example. A pullback $\alpha \beta' = \beta \alpha'$ consists of a limit and part of the limit cone of the following diagram:

$$\begin{array}{ccc} & A & \\ & \downarrow{\alpha} & \\ B & \xrightarrow[\beta]{} & C \end{array}$$

Indeed, a cone into D consists of morphisms $\varphi : X \longrightarrow A$, $\psi : X \longrightarrow B$, and $\xi : X \longrightarrow C$ such that $\alpha \varphi = \beta \psi = \xi$, and is uniquely determined by φ and ψ. Hence (φ, ψ, ξ) is a limit cone of D if and only if $\alpha \varphi = \beta \psi$ is a pullback.

Colimits. A diagram in $\mathcal{C}$ over $\mathcal{G}$ is also a diagram in $\mathcal{C}^{\mathrm{op}}$ over $\mathcal{G}^{\mathrm{op}}$.

Definition. The colimit *and* colimit cone *of a diagram* D *in a category* $\mathcal{C}$ *are the limit and limit cone of* D *in* $\mathcal{C}^{\mathrm{op}}$.

Thus, limits and colimits are dual concepts. We give more detailed definitions.

Definition. Let A be an object of a category $\mathcal{C}$ and let D be a diagram in $\mathcal{C}$ over a graph $\mathcal{G}$. A cone φ from D to A assigns to each vertex i of $\mathcal{G}$ a morphism $\varphi_i : D_i \longrightarrow A$, so that $\varphi_j D_a = \varphi_i$ for every edge $a : i \longrightarrow j$ of $\mathcal{G}$:

$$
\begin{array}{ccc}
D_i & \xrightarrow{\ \varphi_i\ } & A \\
{\scriptstyle D_a}\downarrow & \nearrow{\scriptstyle \varphi_j} & \\
D_j & &
\end{array}
$$

Equivalently, a cone from D to A is a morphism of diagrams from D to the constant diagram $C(A)$. (No one could muster the nerve to use the rightful name, *cocone*.) Thus, a *colimit cone* of D is a cone $\lambda : D \longrightarrow L$ such that, for every cone $\varphi : D \longrightarrow A$ there is a unique morphism $\overline{\varphi} : L \longrightarrow A$ such that $\varphi = \overline{\varphi}\lambda$ ($\varphi_i = \overline{\varphi}\lambda_i$ for every vertex i of $\mathcal{G}$):

$$
\begin{array}{ccc}
D_i & \xrightarrow{\ \lambda_i\ } & L \\
 & {\scriptstyle \varphi_i}\searrow & \downarrow{\scriptstyle \overline{\varphi}} \\
 & & A
\end{array}
$$

Then the object L and cone λ constitute a *colimit* of D; the object L is also called a *colimit* of D.

Proposition 3.4. *If $\lambda : D \longrightarrow L$ and $\lambda' : D \longrightarrow L'$ are colimit cones of D, then there is an isomorphism $\theta : L \longrightarrow L'$ such that $\lambda' = \theta\lambda$. Conversely, if $\lambda : D \longrightarrow L$ is a colimit cone of D and $\theta : L \longrightarrow L'$ is an isomorphism, then $\theta\lambda : D \longrightarrow L'$ is a colimit cone of D.*

Examples. Direct limits and direct sums of modules are examples of colimits. In general, the *coproduct* of a family of objects of $\mathcal{C}$ is their product in $\mathcal{C}^{\mathrm{op}}$:

Definition. In a category $\mathcal{C}$, the coproduct of a family $(D_i)_{i \in I}$ of objects of $\mathcal{C}$ consists of an object $P = \coprod_{i \in I} D_i$ and a family $(\iota_i)_{i \in I}$ of injections $\iota_i : D_i \longrightarrow P$ such that, for every object A and family $(\varphi_i)_{i \in I}$ of morphisms $\varphi_i : D_i \longrightarrow A$ there is a unique morphism $\overline{\varphi} : P \longrightarrow A$ such that $\varphi_i = \overline{\varphi}\iota_i$ for every $i \in I$:

$$
\begin{array}{ccc}
D_i & \xrightarrow{\ \iota_i\ } & P \\
 & {\scriptstyle \varphi_i}\searrow & \downarrow{\scriptstyle \overline{\varphi}} \\
 & & A
\end{array}
$$

The object P is also called a coproduct *of $(D_i)_{i \in I}$.*

For example, the free product of a family of groups is its coproduct in *Grps*; the direct sum of a family of left R-modules is its coproduct in $_R$*Mods*. In the category of commutative R-algebras, tensor products are coproducts.

We denote the coproduct of a finite family $D_1, \ldots, D_n$ by $D_1 \amalg \cdots \amalg D_n$ (a number of other symbols are in use for coproducts). Coproducts are associative: if $I = \bigcup_{j \in J} I_j$ is a partition of I, then there is a natural isomorphism

$\coprod_{i \in I} D_i \cong \coprod_{j \in J} \left(\coprod_{i \in I_j} D_j \right)$ (when these coproducts exist). For instance, $A \amalg B \amalg C$ is naturally isomorphic to $(A \amalg B) \amalg C$ and to $A \amalg (B \amalg C)$.

The *coequalizer* of two coterminal morphisms of $\mathcal{C}$ is their equalizer in $\mathcal{C}^{op}$, necessarily an epimorphism of $\mathcal{C}$. In detail:

Definition. In a category $\mathcal{C}$, the coequalizer *of two morphisms $\alpha, \beta : A \longrightarrow B$ is a morphism $\gamma : B \longrightarrow C$ such that* (i) $\gamma\alpha = \gamma\beta$ *and* (ii) *every morphism φ such that $\varphi\alpha = \varphi\beta$ factors uniquely through γ :*

$$A \underset{\beta}{\overset{\alpha}{\rightrightarrows}} B \overset{\gamma}{\longrightarrow} C$$
$$\varphi \searrow \quad \vdots$$
$$X$$

For example, in $_R Mods$, the coequalizer of $\alpha : M \longrightarrow N$ and 0 is the projection $N \longrightarrow \mathrm{Coker}\,\alpha$. The exercises give other examples.

A *pushout* in a category $\mathcal{C}$ is a commutative square $\alpha'\beta = \beta'\alpha$ such that for every commutative square $\psi\alpha = \varphi\beta$ there exists a unique morphism χ such that $\varphi = \chi\alpha'$ and $\psi = \chi\beta'$. Pushouts in $_R Mods$ are an example. A pushout $\alpha'\beta = \beta'\alpha$ consists of a colimit and part of the colimit cone of a diagram:

$$\begin{array}{ccc} C & \overset{\alpha}{\longrightarrow} & A \\ \beta \downarrow & & \\ B & & \end{array}$$

Exercises.

1. Show that every diagram in a preordered set is commutative, and that preordered sets are the only categories with this property.

2. Let I be a preordered set. What is the product of a family of elements of I ?

3. Let I be a preordered set. Show that the limit of a diagram in I is the product of the objects in the diagram.

4. Prove the following associativity property of products: if $I = \bigcup_{j \in J} I_j$ is a partition of I, then there is a natural isomorphism $\prod_{i \in I} D_i \cong \prod_{j \in J} \left(\prod_{i \in I_j} D_j \right)$ (when these products exist).

5. Describe equalizers in (i) *Sets* ; (ii) *Grps* ; (iii) *Rings* ; (iv) $_R Mods$.

6. Show that every monomorphism of *Grps* is the equalizer of two homomorphisms.

7. Show that every monomorphism of $_R Mods$ is the equalizer of two homomorphisms.

8. Find a monomorphism of *Rings* that is not the equalizer of two homomorphisms.

9. Describe coproducts in *Sets* .

10. Let I be a preordered set. What is the coproduct of a family of elements of I ? Show that the colimit of a diagram in I is the coproduct of the objects in the diagram.

11. Construct coequalizers in *Sets* .

12. Construct coequalizers in *Grps*.

13. Show that every epimorphism of *Grps* is the coequalizer of two homomorphisms.

14. Describe coequalizers in $_R$*Mods*. Is every epimorphism of $_R$*Mods* the coequalizer of two homomorphisms?

15. Construct coequalizers in *Rings*. Is every epimorphism of *Rings* the coequalizer of two homomorphisms?

4. Completeness

This section contains constructions of limits and colimits, and results pertaining to their existence. In particular, in *Sets*, *Grps*, $_R$*Mods*, and so on, every diagram has a limit and a colimit.

Definition. A category $\mathcal{C}$ *is* complete, *or* has limits, *when every* [small] *diagram in* $\mathcal{C}$ *has a limit.*

Many applications require the stronger property, which holds in many categories, that *a limit can be assigned to every diagram* in $\mathcal{C}$; more precisely, that there is for every [small] graph $\mathcal{G}$ a class function that assigns a limit cone to every diagram in $\mathcal{C}$ over $\mathcal{G}$. If $\mathcal{C}$ is small, then this property is equivalent to completeness, by the axiom of choice.

Proposition **4.1.** *The category Sets is complete; in fact, in Sets, a limit can be assigned to every diagram.*

Proof. Let D be a diagram in *Sets* over a graph $\mathcal{G}$. Let $P = \prod_{i \in \mathcal{G}} D_i$ be the Cartesian product (where "$i \in \mathcal{G}$" is short for "i is a vertex of $\mathcal{G}$"), with projections $\pi_i : P \longrightarrow D_i$. We show that

$$L = \{ (x_i)_{i \in \mathcal{G}} \in P \mid D_a(x_i) = x_j \text{ whenever } a : i \longrightarrow j \}$$

is a limit of D, with limit cone $\lambda_i = \pi_{i|D_i} : L \longrightarrow D_i$. Indeed, $\lambda : L \longrightarrow D$ is a cone, by definition of L. If $\varphi : A \longrightarrow D$ is a cone, then $D_a\big(\varphi_i(x)\big) = \varphi_j(x)$, for every edge $a : i \longrightarrow j$ and $x \in A$. Hence $\overline{\varphi}(x) = \big(\varphi_i(x)\big)_{i \in \mathcal{G}} \in L$ for all $x \in A$. This defines a mapping $\overline{\varphi} : A \longrightarrow L$ such that $\lambda_i \circ \overline{\varphi} = \varphi_i$ for all $i \in \mathcal{G}$, and $\overline{\varphi}$ is the only such mapping. $\square$

We call the limit constructed above the *standard limit* of D.

Standard limits spill into neighboring categories. Let D be a diagram in *Grps*, over a graph $\mathcal{G}$. Let $P = \prod_{i \in \mathcal{G}} D_i$ be the Cartesian product and let

$$L = \{ (x_i)_{i \in \mathcal{G}} \in P \mid D_a(x_i) = x_j \text{ whenever } a : i \longrightarrow j \}$$

be the standard limit of D in *Sets*, with limit cone $\lambda : L \longrightarrow D$. Since every D_a is a homomorphism, L is a subgroup of P; then every λ_i is a homomorphism. If $\varphi : A \longrightarrow D$ is a cone in *Grps*, then every φ_i is a homomorphism, and so is the unique mapping $\overline{\varphi}(x) = \big(\varphi_i(x)\big)_{i \in \mathcal{G}}$ such that $\lambda_i \circ \overline{\varphi} = \varphi_i$ for all

$i \in \mathcal{G}$; hence $\overline{\varphi}$ is the only homomorphism such that $\lambda_i \circ \overline{\varphi} = \varphi_i$ for all $i \in \mathcal{G}$. Thus, the standard limit in *Sets* yields a limit in *Grps*.

Proposition 3.3 yields an additional property: if $\lambda' : L' \longrightarrow D$ is another limit cone in *Grps*, then there is an isomorphism $\theta : L' \longrightarrow L$ such that $\lambda' = \lambda\theta$, whence $\lambda' : L' \longrightarrow D$ is a limit cone of D in *Sets*.

Definition. A functor $F : \mathcal{A} \longrightarrow \mathcal{B}$ preserves limits of diagrams over a graph $\mathcal{G}$ when, for every diagram D over $\mathcal{G}$ with a limit cone $\lambda = (\lambda_i)_{i \in \mathcal{G}}$ in $\mathcal{A}$, $F(\lambda) = \big(F(\lambda_i)\big)_{i \in \mathcal{G}}$ is a limit cone of $F(D)$ in $\mathcal{B}$.

We have proved the following:

Proposition 4.2. *The category Grps is complete; in fact, in Grps, a limit can be assigned to every diagram. Moreover, the forgetful functor from Grps to Sets preserves all limits.*

The categories *Rings*, $_R$*Mods*, *R-Algs*, etc., have similar properties. A general result to that effect is proved in Section 10.

Readers may show that Hom functors preserve limits:

Proposition 4.3. *Let $\mathcal{C}$ be a locally small category. For every object A of $\mathcal{C}$, the functor $\mathrm{Hom}_{\mathcal{C}}(A, -)$ preserves all existing limits. In fact, λ is a limit cone of a diagram D in $\mathcal{C}$ if and only if $\mathrm{Hom}_{\mathcal{C}}(A, \lambda)$ is a limit cone of the diagram $\mathrm{Hom}_{\mathcal{C}}(A, D)$ for every object A of $\mathcal{C}$.*

Dually, $\mathrm{Hom}_{\mathcal{C}}(-A)$ changes colimit cones to limit cones and colimits to limits. For R-modules, the properties $\mathrm{Hom}_R(A, \prod_{i \in I} B_i) \cong \prod_{i \in I} \mathrm{Hom}_R(A, B_i)$ and $\mathrm{Hom}_R(\bigoplus_{i \in I} A_i, B) \cong \prod_{i \in I} \mathrm{Hom}_R(A_i, B)$ are particular cases of Proposition 4.3 and its dual.

Limits by products and equalizers. Our next result is based on a general construction of limits.

Proposition 4.4. *A category that has products and equalizers is complete. If in a category $\mathcal{C}$ a product can be assigned to every family of objects of $\mathcal{C}$, and an equalizer can be assigned to every pair of coterminal morphisms of $\mathcal{C}$, then a limit can be assigned to every diagram in $\mathcal{C}$.*

Proof. Let $\mathcal{G}$ be a graph, let E be the set of its edges, and let o, d be the origin and destination mappings of $\mathcal{G}$, so that $a : o(a) \longrightarrow d(a)$ for every edge a. Let D be a diagram over $\mathcal{G}$. Let $P = \prod_{i \in \mathcal{G}} D_i$ and $Q = \prod_{a \in E} D_{d(a)}$, with projections $\pi_i : P \longrightarrow D_i$ and $\rho_a : Q \longrightarrow D_{d(a)}$, be products in $\mathcal{C}$ (or be the assigned products and projections in $\mathcal{C}$). The universal property of Q yields unique morphisms $\alpha, \beta : P \longrightarrow Q$ such that $\rho_a \alpha = \pi_{d(a)}$ and $\rho_a \beta = D_a \pi_{o(a)}$ for all $a \in E$. Let $\varepsilon : L \longrightarrow P$ be an equalizer of α and β in $\mathcal{C}$ (or their assigned equalizer) and let $\lambda_i = \pi_i \varepsilon$:

$$A \xrightarrow{\varphi_i} D_i \qquad D_{d(a)}$$

We show that $\lambda = (\lambda_i)_{i \in \mathcal{G}}$ is a limit cone of D. (The standard limit in Proposition 4.1 and its limit cone are constructed in just this way.) If $a : i \longrightarrow j$, then $o(a) = i$, $d(a) = j$, $\rho_a \alpha = \pi_j$, $\rho_a \beta = D_a \pi_i$, and $D_a \lambda_i = D_a \pi_i \varepsilon = \rho_a \beta \varepsilon = \rho_a \alpha \varepsilon = \pi_j \varepsilon = \lambda_j$. Thus λ is a cone from L to D. Now, let $\varphi : A \longrightarrow D$ be any cone. Let $\psi : A \longrightarrow P$ be the unique morphism such that $\pi_i \psi = \varphi_i$ for all i. Since φ is a cone, $\rho_a \alpha \psi = \pi_j \psi = \varphi_j = D_a \varphi_i = D_a \pi_i \psi = \rho_a \beta \psi$ for every edge $a : i \longrightarrow j$. Therefore $\alpha \psi = \beta \psi$ and there is a unique morphism $\overline{\varphi} : A \longrightarrow L$ such that $\psi = \varepsilon \overline{\varphi}$. Then $\overline{\varphi}$ is unique such that $\varphi_i = \lambda_i \overline{\varphi}$. $\square$

A *finite* graph is a graph with finitely many vertices and finitely many edges. A *finite* limit is a limit of a diagram over a finite graph.

Corollary **4.5.** *A category that has equalizers and finite products has finite limits.*

Cocompleteness. A category is *cocomplete* when its opposite is complete:

Definition. A category $\mathcal{C}$ *is* cocomplete, *or has* colimits, *when every [small] diagram in* $\mathcal{C}$ *has a colimit.*

Many applications require the stronger property that *a colimit can be assigned to every diagram* in $\mathcal{C}$; more precisely, that there is for every [small] graph $\mathcal{G}$ a class function that assigns a colimit cone to every diagram in $\mathcal{C}$ over $\mathcal{G}$. By the axiom of choice, this property is equivalent to cocompleteness if $\mathcal{C}$ is small.

Proposition **4.6.** *The category* $_R$*Mods is cocomplete; in fact, in* $_R$*Mods, a colimit can be assigned to every diagram.*

Proof. Let D be a diagram in $_R$*Mods* over a graph $\mathcal{G}$. Let $S = \bigoplus_{i \in \mathcal{G}} D_i$ be the direct sum and let $\iota_i : D_i \longrightarrow S$ be the injections. Let K be the submodule of S generated by all $\iota_j (D_a(x)) - \iota_i(x)$ in which $a : i \longrightarrow j$ and $x \in D_i$. Let $L = S/K$, let $\pi : S \longrightarrow L$ be the projection, and let $\lambda_i = \pi \circ \iota_i$. We show that $\lambda = (\lambda_i)_{i \in \mathcal{G}}$ is a colimit cone of D.

A cone $\varphi : D \longrightarrow A$ induces a unique homomorphism $\psi : S \longrightarrow A$ such that $\psi \circ \iota_i = \varphi_i$ for all i. Since φ is a cone, $\varphi_j (D_a(x)) - \varphi_i(x) = 0$ for all $a : i \longrightarrow j$ and $x \in D_i$; hence Ker ψ contains every generator of K, Ker ψ contains K, and there is a unique homomorphism $\overline{\varphi}$ such that $\psi = \overline{\varphi} \circ \pi$ (see the diagram next page). Then $\overline{\varphi}$ is unique such that $\overline{\varphi} \circ \pi \circ \iota_i = \psi \circ \iota_i = \varphi_i$ for all i. $\square$

$$D_i \xrightarrow{\varphi_i} A$$

$$\iota_i \downarrow \quad \overset{\psi}{\nearrow} \quad \uparrow \bar{\varphi}$$

$$S \xrightarrow{\pi} L$$

Readers will set up similar arguments, using coproducts, to show that *Sets*, *Grps*, etc., are cocomplete. The case of coproducts shows that forgetful functors to *Sets*, though otherwise virtuous, do not in general preserve colimits.

Limit functors. Let $\lambda : L \longrightarrow D$ and $\lambda' : L' \longrightarrow D'$ be limit cones of diagrams over the same graph $\mathcal{G}$. If $\alpha : D \longrightarrow D'$ is a morphism of diagrams, then $\alpha\lambda$ is a cone to D' and there is a unique morphism $\lim \alpha : L \longrightarrow L'$ such that $\alpha_i \, \lambda_i = \lambda'_i (\lim \alpha)$ for all i :

$$
\begin{array}{ccc}
L & \xrightarrow{\lambda_i} & D_i \\
\lim \alpha \downarrow & & \downarrow \alpha_i \\
L' & \xrightarrow{\lambda'_i} & D'_i
\end{array}
$$

For instance, if $\prod_{i \in I} A_i$ and $\prod_{i \in I} B_i$ exist, every family of morphisms $\alpha_i : A_i \longrightarrow B_i$ induces a unique morphism $\bar{\alpha} = \prod_{i \in I} \alpha_i : \prod_{i \in I} A_i \longrightarrow \prod_{i \in I} B_i$ such that $\rho_i \, \bar{\alpha} = \alpha_i \, \pi_i$ for all i, where $\pi_i : \prod_{i \in I} A_i \longrightarrow A_i$ and $\rho_i : \prod_{i \in I} B_i \longrightarrow B_i$ are the projections.

In general, it is immediate that $\lim 1_D = 1_{\lim D}$ and that $\lim (\alpha\beta) = (\lim \alpha)(\lim \beta)$. Hence we now have a functor:

Proposition **4.7.** *Let* $\mathcal{G}$ *be a graph and let* $\mathcal{C}$ *be a category in which a limit can be assigned to every diagram over* $\mathcal{G}$. *There is a* limit functor *from* $Diag\,(\mathcal{G}, \mathcal{C})$ *to* $\mathcal{C}$ *that assigns to each diagram* D *over* $\mathcal{G}$ *its assigned limit* $\lim D$, *and to each morphism* $\alpha : D \longrightarrow D'$ *of diagrams over* $\mathcal{G}$ *the induced morphism* $\lim \alpha : \lim D \longrightarrow \lim D'$.

Dually, if a colimit can be assigned to every diagram over $\mathcal{G}$, then there is a *colimit functor* $Diag\,(\mathcal{G}, \mathcal{C}) \longrightarrow \mathcal{C}$.

Exercises

1. Show that $_R Mods$ is complete; in fact, a limit can be assigned to every diagram in $_R Mods$, and the forgetful functor from $_R Mods$ to *Sets* preserves limits.

2. Show that the forgetful functor from $_R Mods$ to *Abs* preserves limits.

3. Show that the forgetful functor F from *Grps* to *Sets creates limits*: if $D : \mathcal{G} \longrightarrow Grps$ is a diagram and $\mu : M \longrightarrow F(D)$ is a limit cone of $F(D)$ in *Sets*, then there is a unique cone $\lambda : L \longrightarrow D$ such that $F(\lambda) = \mu$, and it is a limit cone of D in *Grps*.

4. Show that the forgetful functor from $_R Mods$ to *Sets* creates limits.

5. Show that a category is complete if and only if it has products and pullbacks.

6. Prove that *Sets* is cocomplete; in fact, a colimit can be assigned to every diagram in *Sets*.

7. Prove that *Grps* is cocomplete; in fact, a colimit can be assigned to every diagram in *Grps*.

8. Let $\mathcal{A}$ be a small category and let $\mathcal{C}$ be a category in which a limit can be assigned to every diagram. Prove that *Func* $(\mathcal{A}, \mathcal{C})$ is complete; in fact, a "pointwise" limit can be assigned to every diagram in *Func* $(\mathcal{A}, \mathcal{C})$.

9. Let $\mathcal{C}$ be a locally small category. Show that the functor $\text{Hom}_{\mathcal{C}}(A, -)$ preserves all existing limits, for every object A of $\mathcal{C}$; in fact, λ is a limit cone of a diagram D in $\mathcal{C}$ if and only if $\text{Hom}_{\mathcal{C}}(A, \lambda)$ is a limit cone of the diagram $\text{Hom}_{\mathcal{C}}(A, D)$ for every object A of $\mathcal{C}$.

10. Let D be a diagram in a locally small category $\mathcal{C}$. Prove that $\lambda : D \longrightarrow L$ is a colimit cone of D if and only if $\text{Hom}_{\mathcal{C}}(\lambda, A)$ is a limit cone of $\text{Hom}_{\mathcal{C}}(D, A)$ for every object A of $\mathcal{C}$.

11. Prove that every limit functor preserves products.

12. Prove that every limit functor preserves equalizers.

5. Additive Categories

Additive categories and abelian categories share some of the special properties of *Abs* and $_R$*Mods*. This section contains definitions and elementary properties; it can be skipped at first reading.

Definition. An additive category *is a locally small category* $\mathcal{A}$ *with an abelian group operation on each* $\text{Hom}_A(A, B)$ *such that composition is biadditive:* $(\alpha + \beta)\gamma = \alpha\gamma + \beta\gamma$ *and* $\alpha(\gamma + \delta) = \alpha\gamma + \alpha\delta$ *whenever the sums and compositions are defined.*

Some definitions of additive categories also require finite products; others omit the locally small requirement. As defined above, *Abs*, $_R$*Mods*, and $_R$*Bims$_S$* are additive categories; readers will verify that a ring [with an identity element] is precisely an additive category with one object.

In an additive category $\mathcal{A}$, the *zero morphism* $0 : A \longrightarrow B$ is the zero element of $\text{Hom}_A(A, B)$ (the morphism 0 such that $0 + \alpha = \alpha = \alpha + 0$ for every $\alpha : A \longrightarrow B$). If 0β is defined, then $0\beta = (0 + 0)\beta = 0\beta + 0\beta$ and $0\beta = 0$ (0β is a zero morphism). Similarly, $\gamma 0 = 0$ whenever $\gamma 0$ is defined.

In an additive category $\mathcal{A}$, a *zero object* is an object Z of A such that $\text{Hom}_A(A, Z) = \{0\}$ and $\text{Hom}_A(Z, A) = \{0\}$ for every object A; equivalently, such that there is one morphism $A \longrightarrow Z$ and one morphism $Z \longrightarrow A$ for each object A. Then $\alpha : A \longrightarrow B$ is a zero morphism if and only if α factors through Z. A zero object, if it exists, is unique up to isomorphism, and is normally denoted by 0.

The definition of additive categories is self-dual: if $\mathcal{A}$ is an additive category, then $\mathcal{A}^{\text{op}}$, with the same addition on each $\text{Hom}_{A^{\text{op}}}(A, B) = \text{Hom}_A(B, A)$, is an additive category. We see that $\mathcal{A}$ and $\mathcal{A}^{\mathrm{op}}$ have the same zero objects and the same zero morphisms.

Biproducts. In an additive category, finite products (products of finitely many objects) coincide with finite coproducts.

Proposition **5.1.** *For objects* A, B, P *of an additive category the following conditions are equivalent:*

(1) *There exist morphisms* $\pi : P \longrightarrow A$ *and* $\rho : P \longrightarrow A$ *such that* P *is a product of* A *and* B *with projections* π *and* ρ.

(2) *There exist morphisms* $\iota : A \longrightarrow P$ *and* $\kappa : B \longrightarrow P$ *such that* P *is a coproduct of* A *and* B *with injections* ι *and* κ.

(3) *There exist morphisms* $\pi : P \longrightarrow A$, $\rho : P \longrightarrow A$, $\iota : A \longrightarrow P$, *and* $\kappa : B \longrightarrow P$ *such that* $\pi\iota = 1_A$, $\rho\kappa = 1_B$, $\pi\kappa = 0$, $\rho\iota = 0$, *and* $\iota\pi + \kappa\rho = 1_P$.

Proof. (1) implies (3). The universal property of products yields morphisms $\iota : A \longrightarrow P$, $\kappa : B \longrightarrow P$ such that $\pi\iota = 1_A$, $\rho\iota = 0$, $\pi\kappa = 0$, and $\rho\kappa = 1_B$. Then $\pi\,(\iota\pi + \kappa\rho) = \pi$, $\rho\,(\iota\pi + \kappa\rho) = \rho$, and $\iota\pi + \kappa\rho = 1_P$.

(3) implies (1). Let $\alpha : C \longrightarrow A$ and $\beta : C \longrightarrow B$ be morphisms. Then $\gamma = \iota\alpha + \kappa\beta : C \longrightarrow P$ satisfies $\pi\gamma = \alpha$ and $\rho\gamma = \beta$. Conversely, if $\pi\delta = \alpha$ and $\rho\delta = \beta$, then $\delta = (\iota\pi + \kappa\rho)\,\delta = \iota\alpha + \kappa\beta = \gamma$.

Dually, (2) and (3) are equivalent. $\square$

Definitions. A biproduct *of two objects* A *and* B *is an object* $A \oplus B$ *that is both a product of* A *and* B *and a coproduct of* A *and* B. *A category* $\mathcal{C}$ *has biproducts* when every two objects of $\mathcal{C}$ have a biproduct in $\mathcal{C}$.

For example, the (external) direct sum of two left R-modules is a biproduct in $_R Mods$ (and also a byproduct of Proposition 5.1).

We saw that morphisms $\alpha : A \longrightarrow C$ and $\beta : B \longrightarrow D$ induce morphisms $\alpha \times \beta : A \times B \longrightarrow C \times D$ and $\alpha \amalg \beta : A \amalg B \longrightarrow C \amalg D$ (if the products and coproducts exist). If the biproducts $A \oplus B$ and $C \oplus D$ exist, then $\alpha \times \beta$ and $\alpha \amalg \beta$ coincide (see the exercises) and $\alpha \times \beta = \alpha \amalg \beta$ is denoted by $\alpha \oplus \beta$.

A biproduct $A \oplus A$ with projections π, ρ and injections ι, κ has a *diagonal* morphism $\Delta_A : A \longrightarrow A \oplus A$ such that $\pi\,\Delta_A = \rho\,\Delta_A = 1_A$, and a *codiagonal* morphism $\nabla_A : A \oplus A \longrightarrow A$ such that $\nabla_A\,\iota = \nabla_A\,\kappa = 1_A$. If biproducts exist, then these maps determine the addition on $\mathrm{Hom}_{\mathcal{A}}\,(A,\,B)$.

Proposition **5.2.** *If* α, $\beta : A \longrightarrow B$ *are morphisms of an additive category, and the biproducts* $A \oplus A$ *and* $B \oplus B$ *exist, then* $\alpha + \beta = \nabla_B\,(\alpha \oplus \beta)\,\Delta_A$.

Proof. Let π, $\rho : A \oplus A \longrightarrow A$ and π', $\rho' : B \oplus B \longrightarrow B$ be the projections, and let ι', $\kappa' : B \longrightarrow B \oplus B$ be the injections. By 5.1, $\iota'\pi' + \kappa'\rho' = 1_{B \oplus B}$. Hence $\nabla_B\,(\alpha \oplus \beta)\,\Delta_A = \nabla_B\,(\iota'\pi' + \kappa'\rho')(\alpha \times \beta)\,\Delta_A = \pi'\,(\alpha \times \beta)\,\Delta_A + \rho'\,(\alpha \times \beta)\,\Delta_A = \alpha\pi\,\Delta_A + \beta\rho\Delta_A = \alpha + \beta$. $\square$

Kernels and cokernels are defined as follows in any additive category.

Definition. In an additive category, a kernel *of a morphism* α *is an equalizer of* α *and* 0; *a* cokernel *of a morphism* α *is a coequalizer of* α *and* 0.

Thus, $\kappa : K \longrightarrow A$ is a kernel of $\alpha : A \longrightarrow B$ if and only if $\alpha\kappa = 0$, and every morphism φ such that $\alpha\varphi = 0$ factors uniquely through κ:

$$K \xrightarrow{\kappa} A \xrightarrow{\alpha} B$$
$$\Big\uparrow \quad \nearrow_{\varphi}$$
$$X$$

Then κ is a monomorphism and is unique up to isomorphism. The object K is also called a *kernel* of α. Dually, $\gamma : B \longrightarrow C$ is a cokernel of $\alpha : A \longrightarrow B$ if and only if $\gamma\alpha = 0$, and every morphism φ such that $\varphi\alpha = 0$ factors uniquely through γ:

$$A \xrightarrow{\alpha} B \xrightarrow{\gamma} C$$
$$\searrow_{\varphi} \quad \Big\downarrow$$
$$X$$

Then γ is an epimorphism and is unique up to isomorphism. The object C is also called a *cokernel* of α.

Abelian categories. Abelian categories are blessed with additional properties that hold in Abs, $_R Mods$, $_R Bims_S$ but not in every additive category.

Definition. An abelian category *is an additive category that has a zero object, biproducts, kernels, and cokernels, in which every monomorphism is a kernel and every epimorphism is a cokernel.*

Some applications (not included here) require the stronger property that a biproduct can be assigned to every pair of objects, and a kernel and cokernel can be assigned to every morphism. Abelian categories with this stronger property include Abs, $_R Mods$, and $_R Bims_S$. Conversely, some elementary properties of Abs, $_R Mods$, and $_R Bims_S$ hold in every abelian category.

Proposition 5.3. An abelian category has finite limits and colimits.

Proof. In an abelian category, every nonempty finite family A_1, ..., A_n has a product $(...((A_1 \times A_2) \times A_3) \times ...) \times A_n$. The empty sequence also has a product, the zero object. Hence an abelian category has finite products. Any two coterminal morphisms α and β also have an equalizer, namely, any kernel of $\alpha - \beta$. By 4.5, every finite diagram has a limit. Dually, an abelian category has finite coproducts, coequalizers, and finite colimits. $\square$

Lemma 5.4. Let μ be a monomorphism and let σ be an epimorphism. In an abelian category, μ is a kernel of σ if and only if σ is a cokernel of μ.

Proof. First, σ is a cokernel of some α. Assume that μ is a kernel of σ. Then $\sigma\mu = 0$, and α factors through μ: $\alpha = \mu\xi$, since $\sigma\alpha = 0$. If $\varphi\mu = 0$, then $\varphi\alpha = 0$ and φ factors uniquely through σ. Thus σ is a cokernel of μ. The converse implication is dual. $\square$

Proposition **5.5.** *A morphism of an abelian category is a monomorphism if and only if it has a kernel that is a zero morphism; an epimorphism if and only if it has a cokernel that is a zero morphism; an isomorphism if and only if it is both a monomorphism and an epimorphism.*

Proof. We prove the last part and leave the first two parts as exercises. In any category, an isomorphism is both a monomorphism and an epimorphism. Conversely, let $\alpha : A \longrightarrow B$ be both a monomorphism and an epimorphism of an abelian category. Let γ be a cokernel of α. Then $\gamma\alpha = 0 = 0\alpha$ and $\gamma = 0$, since α is an epimorphism. By 5.4, α is a kernel of γ; hence $\gamma\, 1_B = 0$ implies $1_B = \alpha\beta$ for some $\beta : B \longrightarrow A$. Then $\alpha\beta\alpha = \alpha = \alpha\, 1_A$ and $\beta\alpha = 1_A$. $\square$

Definition. Let α be a morphism of an abelian category. An *image* of α is a *kernel of a cokernel of α. A coimage of α is a cokernel of a kernel of α.*

The image and coimage of a morphism are, like kernels and cokernels, unique up to isomorphism.

If $\varphi : A \longrightarrow B$ is a homomorphism of abelian groups or modules, then the projection $B \longrightarrow B/\mathrm{Im}\,\varphi$ is a cokernel of φ; hence the inclusion homomorphism $\mathrm{Im}\,\varphi \longrightarrow B$ and its domain $\mathrm{Im}\,\varphi$ are images of φ as defined above; the inclusion homomorphism $\mathrm{Ker}\,\varphi \longrightarrow B$ is a kernel of φ; hence the projection $A \longrightarrow A/\mathrm{Ker}\,\varphi$ and its codomain $A/\mathrm{Ker}\,\varphi$ are coimages of φ. In *Abs* and $_R Mods$, the homomorphism theorem implies that the image and coimage of a morphism are isomorphic. This holds in every abelian category.

Proposition **5.6** (Homomorphism Theorem). *Let α be a morphism of an abelian category; let ι be an image of α, and let ρ be a coimage of α. There exists a unique isomorphism θ such that $\alpha = \iota\theta\rho$.*

If no specific image has been assigned to α, then $\iota\theta$ is as good an image as ι, and α is the composition of an image and a coimage.

Proof. Construct the following diagram:

$$
\begin{array}{ccccccc}
 & & & & D & & \\
 & & & \delta\uparrow & \;\;\nwarrow\xi & & \\
K & \xrightarrow{\;\kappa\;} & A & \xrightarrow{\;\alpha\;} & B & \xrightarrow{\;\gamma\;} & C \\
 & & \rho\downarrow & \searrow\beta & \uparrow\iota & & \\
 & & & Q & \xrightarrow{\;\theta\;} & I & \xrightarrow{\;\varphi\;} & X \\
 & & & & \searrow\zeta \;\; \eta\uparrow\downarrow\lambda & & \\
 & & & & L & &
\end{array}
$$

Let $\kappa : K \longrightarrow A$ and $\gamma : B \longrightarrow C$ be a kernel and cokernel of $\alpha : A \longrightarrow B$. Let $\iota : I \longrightarrow B$ be a kernel of γ and let $\rho : A \longrightarrow Q$ be a cokernel of κ. Since

$\gamma\alpha = 0$, $\alpha = \iota\beta$ for some $\beta : A \longrightarrow I$. Then $\beta\kappa = 0$, since $\iota\beta\kappa = \alpha\kappa = 0$, and $\beta = \theta\rho$ for some $\theta : Q \longrightarrow I$. Now, $\alpha = \iota\theta\rho$; since ι is a monomorphism and ρ is an epimorphism, θ is the only morphism with this property. It remains to show that θ is an isomorphism.

Suppose that $\varphi\theta = 0$. Let λ be a kernel of φ and let δ be a cokernel of $\iota\lambda$. Since $\varphi\theta = 0$ we have $\theta = \lambda\zeta$ for some ζ. Then $\delta\alpha = \delta\iota\theta\rho = \delta\iota\lambda\zeta\rho = 0$ and $\delta = \xi\gamma$ for some ξ. Hence $\delta\iota = \xi\gamma\iota = 0$. By 5.4, $\iota\lambda$ is a kernel of δ; hence $\iota = \iota\lambda\eta$ for some η. Then $\lambda\eta = 1_I$ and $\varphi = \varphi\lambda\eta = 0$. By 5.5, θ is an epimorphism. Dually, θ is a monomorphism, hence an isomorphism, by 5.5. $\square$

Exercises

1. Show that an object Z of an additive category $\mathcal{A}$ is a zero object if and only if $1_Z = 0$, if and only if $\mathrm{Hom}_{\mathcal{A}}(Z, Z) = 0$.

2. Let $\mathcal{A}$ and $\mathcal{B}$ be additive categories. A functor $F : \mathcal{A} \longrightarrow \mathcal{B}$ is *additive* when $F(\alpha + \beta) = F(\alpha) + F(\beta)$ whenever $\alpha + \beta$ is defined. If $\mathcal{A}$ has biproducts, show that a functor from $\mathcal{A}$ to $\mathcal{B}$ is additive if and only if it preserves biproducts.

3. Extend Proposition 5.1 to all nonempty finite families of objects.

4. Assume that the biproducts $A \oplus B$ and $C \oplus D$ exist. Show that $\alpha \times \beta$ and $\alpha \amalg \beta$ coincide for all $\alpha : A \longrightarrow C$ and $\beta : B \longrightarrow D$.

5. Show that a morphism in an abelian category is a monomorphism if and only if it has a kernel that is a zero morphism.

6. Let α be a morphism in an abelian category. Prove the following: if $\alpha = \mu\sigma$, where μ is a monomorphism and σ is an epimorphism, then μ is an image of α and σ is a coimage of α.

7. Let $\alpha\beta' = \beta\alpha'$ be a pullback in an abelian category. Prove the following: if α is a monomorphism, then α' is a monomorphism.

8. Extend the construction of pullbacks in $_R\mathit{Mods}$ to all abelian categories.

9. Define exact sequences and split exact sequences in any abelian category.

10. Let $\mathcal{A}$ be a small category and let $\mathcal{B}$ be an abelian category in which a limit and colimit can be assigned to every finite diagram. Show that $\mathit{Func}\,(\mathcal{A}, \mathcal{B})$ is abelian.

11. Let $\mathcal{A}$ be a small category and let $\mathcal{B}$ be an abelian category in which a limit and colimit can be assigned to every finite diagram. When is a sequence in $\mathit{Func}\,(\mathcal{A}, \mathcal{B})$ exact?

6. Adjoint Functors

Limits and colimits do not account for all universal properties encountered in previous chapters. Free groups, free modules, tensor products, have universal properties of a different kind; they are particular cases of adjoint functors.

Definition. A precise statement of the universal property of free groups requires the two categories *Grps* and *Sets*, the free group functor F from *Sets* to *Grps*, and the forgetful functor S from *Grps* to *Sets*. The injection

$\iota_X : X \longrightarrow F_X$ is a morphism from X to $S(F_X)$ and is natural in X by definition of F. The universal property of F_X reads: for every morphism $f : X \longrightarrow S(G)$ (in *Sets*) there is a unique morphism $\varphi : F_X \longrightarrow G$ (in *Grps*) such that $S(\varphi) \circ \iota_X = f$. Adjoint functors are defined by a very similar property and its dual.

Proposition **6.1.** *For two functors* $F : \mathcal{A} \longrightarrow \mathcal{C}$ *and* $G : \mathcal{C} \longrightarrow \mathcal{A}$ *the following conditions are equivalent:*

(1) *there exists a natural transformation* $\eta : 1_{\mathcal{A}} \longrightarrow G \circ F$ *such that for every morphism* $\alpha : A \longrightarrow G(C)$ *of* $\mathcal{A}$ *there exists a unique morphism* $\gamma : F(A) \longrightarrow C$ *of* $\mathcal{C}$ *such that* $\alpha = G(\gamma) \eta_A$:

$$
\begin{array}{ccc}
A \xrightarrow{\ \eta_A\ } G(F(A)) & \qquad & F(A) \\
\ \ \searrow{\scriptstyle \alpha} \quad \downarrow{\scriptstyle G(\gamma)} & & \ \ \downarrow{\scriptstyle \gamma} \\
G(C) & & C
\end{array}
$$

(2) *there exists a natural transformation* $\varepsilon : F \circ G \longrightarrow 1_{\mathcal{C}}$ *such that for every morphism* $\gamma : F(A) \longrightarrow C$ *of* $\mathcal{C}$ *there exists a unique morphism* $\alpha : A \longrightarrow G(C)$ *of* $\mathcal{A}$ *such that* $\gamma = \varepsilon_C F(\alpha)$:

$$
\begin{array}{ccc}
F(G(C)) \xrightarrow{\ \varepsilon_C\ } C & \qquad & G(C) \\
\ \ \uparrow{\scriptstyle F(\alpha)} \quad \nearrow{\scriptstyle \gamma} & & \ \ \uparrow{\scriptstyle \alpha} \\
F(A) & & A
\end{array}
$$

Proof. (1) implies (2). Applying (1) to $1_{G(C)} : G(C) \longrightarrow G(C)$ yields a morphism $\varepsilon_C : F(G(C)) \longrightarrow C$, unique such that

$$G(\varepsilon_C) \, \eta_{G(C)} = 1_{G(C)}.$$

We show that ε_C is natural in C. Every $\gamma : C \longrightarrow D$ induces a diagram

$$
\begin{array}{ccc}
G(C) \xrightarrow{\qquad G(\gamma) \qquad} & & G(D) \\
\Big\| \quad \searrow{\scriptstyle \eta_{G(C)}} & & \Big\| \quad \searrow{\scriptstyle \eta_{G(D)}} \\
\quad G(F(G(C))) \xrightarrow{\ G(F(G(\gamma)))\ } & & G(F(G(D))) \\
\Big\| \quad \swarrow{\scriptstyle G(\varepsilon_C)} & & \Big\| \quad \swarrow{\scriptstyle G(\varepsilon_D)} \\
G(C) \xrightarrow{\qquad G(\gamma) \qquad} & & G(D)
\end{array}
$$

in which the side triangles commute by definition of ε, the back face commutes, and the upper face commutes since η is a natural transformation. Hence

$$G(\varepsilon_D) \, G(F(G(\gamma))) \, \eta_{G(C)} = G(\gamma) = G(\gamma) \, G(\varepsilon_C) \, \eta_{G(C)}$$

and uniqueness in (1) yields $\varepsilon_D \, F(G(\gamma)) = \gamma \, \varepsilon_C$. Thus $\varepsilon : F \circ G \longrightarrow 1_{\mathcal{C}}$ is a natural transformation.

Now, let $\gamma : F(A) \longrightarrow C$ be a morphism in $\mathcal{C}$. If $\alpha : A \longrightarrow G(C)$ and $\varepsilon_C F(\alpha) = \gamma$, then $\alpha = G(\varepsilon_C) \eta_{G(C)} \alpha = G(\varepsilon_C) G(F(\alpha)) \eta_A = G(\gamma) \eta_A$, since η is a natural transformation. Hence α is unique. If, conversely, $\alpha = G(\gamma) \eta_A : A \longrightarrow G(C)$, then $G(\varepsilon_C) G(F(\alpha)) \eta_A = G(\varepsilon_C) \eta_{G(C)} \alpha = \alpha = G(\gamma) \eta_A$, and uniqueness in (1) yields $\varepsilon_C F(\alpha) = \gamma$. Thus (2) holds.

In particular, if $\gamma = 1_{F(A)}$, then $\alpha = \eta_A$ and $\varepsilon_C F(\alpha) = \gamma$ becomes: $\varepsilon_{F(A)} F(\eta_A) = 1_{F(A)}$.

(2) implies (1). Reversing all arrows and compositions, and exchanging $\mathcal{A}$ and $\mathcal{C}$, F and G, ε and η, also exchanges (1) and (2). Therefore, that (2) implies (1) follows from (1) implies (2). $\square$

Definitions. If the equivalent conditions in Proposition 6.1 hold, then F and G are a pair of mutually adjoint functors*; F is a* left adjoint *of G; G is a* right adjoint *of F; and $(F, G, \eta, \varepsilon)$ is an* adjunction *from $\mathcal{A}$ to $\mathcal{C}$.*

The terminology "left adjoint" and "right adjoint" comes from the isomorphism $\mathrm{Hom}_{\mathcal{C}}(F(A), C) \cong \mathrm{Hom}_{\mathcal{A}}(A, G(C))$ in Proposition 6.4 below, which resembles the definition $\langle Fx, y \rangle = \langle x, Gy \rangle$ of adjoint linear transformations.

As expected, the free group functor from *Sets* to *Grps* is a left adjoint of the forgetful functor from *Grps* to *Sets*. The free left R-module functor from *Sets* to $_R$*Mods* is a left adjoint of the forgetful functor from $_R$*Mods* to *Sets*.

Limit functors are right adjoints. Let $\mathcal{C}$ be a category in which a limit can be assigned to every diagram over some graph $\mathcal{G}$, so that there is a limit functor from $Diag(\mathcal{G}, \mathcal{C})$ to $\mathcal{C}$. Let $C : \mathcal{C} \longrightarrow Diag(\mathcal{G}, \mathcal{C})$ be the constant diagram functor. The limit cone $\lambda_D : C(\lim D) \longrightarrow D$ of D is natural in D, and the universal property of limits states that for every morphism $\varphi : C(A) \longrightarrow D$ of $Diag(\mathcal{G}, \mathcal{C})$ there exists a unique morphism $\overline{\varphi} : A \longrightarrow \lim D$ such that $\varphi = \lambda_D C(\overline{\varphi})$. Thus $\lim$ is a right adjoint of C. Dually, a colimit functor is a left adjoint of a constant diagram functor.

Properties. We note two easy consequences of the definition.

Proposition 6.2. Any two left adjoints of the same functor are naturally isomorphic. Any two right adjoints of the same functor are naturally isomorphic.

This follows from the universal properties. There is also an easy condition for existence (used in the next section for deeper existence results):

Proposition 6.3. A functor $G : \mathcal{C} \longrightarrow \mathcal{A}$ has a left adjoint if and only if to each object A of $\mathcal{A}$ can be assigned an object $F(A)$ of $\mathcal{C}$ and a morphism $\eta_A : A \longrightarrow G(F(A))$ of $\mathcal{A}$, so that for every morphism $\alpha : A \longrightarrow G(C)$ of $\mathcal{A}$ there exists a unique morphism $\gamma : F(A) \longrightarrow C$ of $\mathcal{C}$ such that $\alpha = G(\gamma) \eta_A$.

In locally small categories, Proposition 6.1 can be expanded as follows:

Proposition 6.4. If $\mathcal{A}$ and $\mathcal{C}$ are locally small, then $F : \mathcal{A} \longrightarrow \mathcal{C}$ and

$G : \mathcal{C} \longrightarrow \mathcal{A}$ *are mutually adjoint functors if and only if there is a bijection*

$$\theta_{A,C} : \mathrm{Hom}_{\mathcal{C}}\,(F(A), C) \longrightarrow \mathrm{Hom}_{\mathcal{A}}\,(A, G(C))$$

that is natural in A and C.

In case $\mathcal{A}$ and $\mathcal{C}$ are locally small, an *adjunction* from $\mathcal{A}$ to $\mathcal{C}$ generally consists of F, G, and at least one of η, ε, and θ.

Proof. Let F and G be mutually adjoint functors. By (1) in Proposition 6.1, $\theta_{A,C}(\gamma) = G(\gamma)\,\eta_A$ is a bijection of $\mathrm{Hom}_{\mathcal{C}}\,(F(A), C)$ onto $\mathrm{Hom}_{\mathcal{A}}\,(A, G(C))$. Let $\beta : A \longrightarrow B$ and $\delta : C \longrightarrow D$. Since η is a natural transformation, $G(F(\beta))\,\eta_A = \eta_B\,\beta$. Hence, for every $\gamma : F(B) \longrightarrow C$,

$$\theta_{A,D}\big(\mathrm{Hom}_{\mathcal{C}}\,(F(\beta),\delta)(\gamma)\big) = \theta_{A,D}\big(\delta\gamma\,F(\beta)\big) = G(\delta)\,G(\gamma)\,G(F(\beta))\,\eta_A$$
$$= G(\delta)\,G(\gamma)\,\eta_B\,\beta = \mathrm{Hom}_{\mathcal{A}}\,(\beta, G(\delta))\,(\theta_{B,C}(\gamma)).$$

Thus the square below commutes and θ is a natural transformation.

$$
\begin{array}{ccc}
\mathrm{Hom}_{\mathcal{C}}\,(F(B), C) & \xrightarrow{\ \theta_{B,C}\ } & \mathrm{Hom}_{\mathcal{A}}\,(B, G(C)) \\
{\scriptstyle \mathrm{Hom}_{\mathcal{C}}\,(F(\beta),\delta)}\Big\downarrow & & \Big\downarrow{\scriptstyle \mathrm{Hom}_{\mathcal{A}}\,(\beta, G(\delta))} \\
\mathrm{Hom}_{\mathcal{C}}\,(F(A), D) & \xrightarrow[\ \theta_{A,D}\ ]{} & \mathrm{Hom}_{\mathcal{A}}\,(A, G(D))
\end{array}
$$

Conversely, assume that $\theta : \mathrm{Hom}_{\mathcal{C}}\,(F(-), -) \longrightarrow \mathrm{Hom}_{\mathcal{A}}\,(-, G(-))$ is a natural bijection. For all $\beta : A \longrightarrow B$, $\delta : C \longrightarrow D$, and $\gamma : F(B) \longrightarrow C$,

$$\theta_{A,D}\big(\delta\gamma\,F(\beta)\big) = \theta_{A,D}\big(\mathrm{Hom}_{\mathcal{C}}\,(F(\beta),\delta)(\gamma)\big)$$
$$= \mathrm{Hom}_{\mathcal{A}}\,(\beta, G(\delta))\,(\theta_{B,C}(\gamma)) = G(\delta)\,\theta_{B,C}(\gamma)\,\beta. \qquad (*)$$

Let $\eta_A = \theta_{A,\,F(A)}(1_{F(A)}) : A \longrightarrow G(F(A))$. With $A = B$, $C = F(A) = F(B)$, $\beta = 1_A$, and $\gamma = 1_{F(A)}$, $(*)$ reads:

$$\theta_{A,D}(\delta) = G(\delta)\,\eta_A, \text{ for all } \delta : F(A) \longrightarrow D.$$

With $C = D = F(B)$ and $\gamma = \delta = 1_{F(B)}$, $(*)$ then yields:

$$G(F(\beta))\,\eta_A = \theta_{A,F(B)}(\beta) = \eta_B\,\beta, \text{ for all } \beta : A \longrightarrow B;$$

hence η is a natural transformation. Then (1) in Proposition 6.1 holds, since $\theta_{A,C}$ is bijective. $\square$

For example, the adjoint associativity of tensor products $\mathrm{Hom}_{\mathbb{Z}}\,(A \otimes_R B, C)$ $\cong \mathrm{Hom}_R(A,\ \mathrm{Hom}_{\mathbb{Z}}\,(B, C))$ shows that, for every left R-module B, $- \otimes_R B$ is a left adjoint of the functor $\mathrm{Hom}_{\mathbb{Z}}\,(B, -)$ from *Abs* to *Mods*$_R$.

If $\alpha : A \longrightarrow G(C)$ and $\gamma : F(A) \longrightarrow C$, then we saw in the proof of Proposition 6.1 that $\alpha = G(\gamma)\,\eta_A$ implies $\varepsilon_C\,F(\alpha) = \gamma$. Hence the equality $\theta_{A,C}(\gamma) = G(\gamma)\,\eta_A$ above implies $\theta_{A,C}^{-1}(\alpha) = \varepsilon_C\,F(\alpha)$. The remaining properties in the next lemma were proved incidentally with Proposition 6.1:

Lemma **6.5.** *If* $(F, G, \eta, \varepsilon)$ *is an adjunction from* $\mathcal{A}$ *to* $\mathcal{C}$, *then*

$$G(\varepsilon_C)\,\eta_{G(C)} = 1_{G(C)} \ and \ \varepsilon_{F(A)}\,F(\eta_A) = 1_{F(A)},$$

for all objects A *of* $\mathcal{A}$ *and* C *of* $\mathcal{C}$. *If* $\mathcal{A}$ *and* $\mathcal{C}$ *are locally small and* $(F, G, \eta, \varepsilon, \theta)$ *is an adjunction from* $\mathcal{A}$ *to* $\mathcal{C}$, *then*

$$\theta_{A,C}(\gamma) = G(\gamma)\,\eta_A \ and \ \theta_{A,C}^{-1}(\alpha) = \varepsilon_C\,F(\alpha),$$

for all objects A *of* $\mathcal{A}$ *and* C *of* $\mathcal{C}$.

Proposition **6.6.** *Every right adjoint functor preserves limits. Every left adjoint functor preserves colimits.*

Proof. Let $F : \mathcal{A} \longrightarrow \mathcal{C}$ be a left adjoint of $G : \mathcal{C} \longrightarrow \mathcal{A}$, and let $\lambda : L \longrightarrow D$ be a limit cone of a diagram D in $\mathcal{C}$ over some graph $\mathcal{G}$. We want to show that $G(\lambda)\colon G(L) \longrightarrow G(D)$ is a limit cone of $G(D)$. First, $G(\lambda)$ is a cone to $G(D)$. Let $\alpha : A \longrightarrow G(D)$ be any cone to $G(D)$. For every vertex i of $\mathcal{G}$ there is a unique morphism $\gamma_i : F(A) \longrightarrow D_i$ such that $\alpha_i = G(\gamma_i)\,\eta_A$, namely, $\gamma_i = \varepsilon_{D_i}\,F(\alpha_i)$. If $a : i \longrightarrow j$ is an edge of $\mathcal{G}$, then $\alpha_j = G(D_a)\,\alpha_i$ and

$$G(\gamma_j)\,\eta_A \;=\; \alpha_j \;=\; G(D_a)\,\alpha_i \;=\; G(D_a\,\gamma_i)\,\eta_A\,;$$

therefore $\gamma_j = D_a\,\gamma_i$. Thus γ is a cone to D. Hence there is a unique morphism $\overline{\gamma} : F(A) \longrightarrow L$ such that $\gamma_i = \lambda_i\,\overline{\gamma}$ for all i. Then $\overline{\alpha} = G(\overline{\gamma})\,\eta_A$ satisfies

$$\alpha_i \;=\; G(\gamma_i)\,\eta_A \;=\; G(\lambda_i)\,G(\overline{\gamma})\,\eta_A \;=\; G(\lambda_i)\,\overline{\alpha}$$

for all i. Also, $\overline{\alpha}$ is unique with this property: if $\alpha_i = G(\lambda_i)\,\beta$ for all i, then

$$\lambda_i\,\varepsilon_L\,F(\beta) = \varepsilon_{D_i}\,F(G(\lambda_i))\,F(\beta) = \varepsilon_{D_i}\,F(\alpha_i) = \gamma_i = \lambda_i\,\overline{\gamma}$$

for all i; this implies $\varepsilon_L\,F(\beta) = \overline{\gamma}$ and $\beta = G(\overline{\gamma})\,\eta_A = \overline{\alpha}$. $\square$

Proposition 6.6 implies, in one fell swoop, that limit functors preserve limits, that the forgetful functors from *Grps* to *Sets* and from $_R$*Mods* to *Sets* preserve limits, and that $- \otimes_R B$ preserves colimits.

Exercises

1. Show that $R \otimes_{\mathbb{Z}} -$ is a left adjoint of the forgetful functor from $_R$*Mods* to *Abs*.

2. Show that, for every set B, $- \times B$ is a left adjoint of the functor $\mathrm{Hom}(B, -)$ from *Sets* to *Sets*.

3. Show that an equivalence of categories is an adjunction.

4. Give examples of adjoint functors that have not been mentioned in this section, or in the previous exercises.

5. Prove that any two left adjoints of a functor are naturally isomorphic.

6. Prove that $(F, G, \eta, \varepsilon)$ is an adjunction from $\mathcal{A}$ to $\mathcal{C}$ if and only if $G(\varepsilon_C)\,\eta_{G(C)} = 1_{G(C)}$ and $\varepsilon_{F(A)}\,F(\eta_A) = 1_{F(A)}$ for all objects A of $\mathcal{A}$ and C of $\mathcal{C}$.

7. Let $\mathcal{A}$ and $\mathcal{C}$ be locally small. Deduce Proposition 6.6 from 6.5 and 4.3.

8. Let F and G be mutually adjoint functors between abelian categories. Show that F and G are additive.

9. Let (F, G, θ) be an adjunction between abelian categories. Show that θ is an isomorphism of abelian groups.

7. The Adjoint Functor Theorem

The adjoint functor theorem gives sufficient conditions for the existence of a left adjoint functor. Its present formulation is essentially due to Freyd [1964]. Following MacLane [1971] we deduce it from more general results, and from a most general view of universal properties.

Initial and terminal objects. First, a definition:

Definitions. An initial object *of a category* $\mathcal{C}$ *is an object* C *such that there is for every object* A *of* $\mathcal{C}$ *exactly one morphism from* C *to* A. *A* terminal object *of a category* $\mathcal{C}$ *is an object* C *such that there is for every object* A *of* $\mathcal{C}$ *exactly one morphism from* A *to* C.

For example, in the category *Sets*, $\emptyset$ is an initial object, and every one element set is a terminal object. In an additive category, a zero object is both an initial and a terminal object. The exercises give other examples.

The product of an empty family of objects, if it exists, is a terminal object. Therefore every complete category has a terminal object. Dually, every cocomplete category has an initial object.

Proposition **7.1.** *In every category, any two initial objects are isomorphic, and any two terminal objects are isomorphic.*

Proof. Let A and B be initial objects. There exists a morphism $\alpha : A \longrightarrow B$ and a morphism $\beta : B \longrightarrow A$. Then 1_A and $\beta\alpha$ are morphisms from A to A; since A is an initial object, $\beta\alpha = 1_A$. Similarly, $\alpha\beta = 1_B$. Thus A and B are isomorphic. Dually, any two terminal objects are isomorphic. $\square$

Readers will recognize this proof as a standard uniqueness argument for objects with universal properties. In fact, *every* universal property we have encountered is equivalent to the existence of an initial object in a suitable category:

Let D be a diagram over a graph $\mathcal{G}$ in a category $\mathcal{A}$. The cones from D are the objects of a cone category $\mathcal{C}$, in which a morphism from $\varphi : D \longrightarrow A$ to $\psi : D \longrightarrow B$ is a morphism $\alpha : A \longrightarrow B$ such that $\alpha\varphi = \psi$ ($\alpha \varphi_i = \psi_i$ for every vertex i of $\mathcal{G}$). A colimit cone of D is precisely an initial object of $\mathcal{C}$. Dually, a limit cone of D is a terminal object in a category of cones to D.

Let (F, G, η) be an adjunction from $\mathcal{A}$ to $\mathcal{C}$. The universal property of $\eta_A : A \longrightarrow G(F(A))$ states that, for every $\alpha : A \longrightarrow G(C)$, there is a unique $\gamma : F(A) \longrightarrow C$ such that $\alpha = G(\gamma)\eta_A$. Let $\mathcal{C}_A$ be the following category. An object of $\mathcal{C}_A$ is an ordered pair (α, C) in which C is an object of $\mathcal{C}$ and

α is a morphism of $\mathcal{A}$ from A to $G(C)$. A morphism of $\mathcal{C}_A$ from (α, C) to (β, D) is a morphism $\gamma : C \longrightarrow D$ of $\mathcal{C}$ such that $\beta = G(\gamma)\,\alpha$. The universal property of η_A states precisely that $(\eta_A,\, F(A))$ is an initial object of $\mathcal{C}_A$.

Existence. Due to this last example, existence results for initial objects yield existence results for adjoint functors.

Proposition **7.2.** *A locally small, complete category $\mathcal{C}$ has an initial object if and only if it satisfies the* solution set condition*:*

there exists a set $\mathcal{S}$ of objects of $\mathcal{C}$ such that there is for every object A of $\mathcal{C}$ at least one morphism from some $S \in \mathcal{S}$ to A.

Proof. If $\mathcal{C}$ has an initial object C, then $\{C\}$ is a solution set. Conversely, let $\mathcal{C}$ have a solution set $\mathcal{S}$. Since $\mathcal{C}$ is complete, $\mathcal{S}$ has a product P in $\mathcal{C}$. Then $\{P\}$ is a solution set: for every object A of $\mathcal{C}$ there is at least one morphism $P \longrightarrow S \longrightarrow A$. Since $\mathcal{C}$ is locally small, $\mathrm{Hom}_{\mathcal{C}}(P, P)$ is a set of morphisms from P to P, and has an equalizer $\varepsilon : E \longrightarrow P$. For every object A of $\mathcal{C}$ there is at least one morphism $\alpha : E \longrightarrow P \longrightarrow A$. We show that α is unique.

Let $\alpha,\ \beta : E \longrightarrow A$. Then α and β have an equalizer $\zeta : F \longrightarrow E$. By the above there exists a morphism $\varphi : P \longrightarrow F$:

$$F \xrightarrow{\ \zeta\ } E \underset{\beta}{\overset{\alpha}{\rightrightarrows}} A$$
$$\varphi \searrow \quad \downarrow \varepsilon$$
$$P$$

Then $\varepsilon\zeta\varphi : P \longrightarrow P$, $\varepsilon\zeta\varphi\varepsilon = 1_P\,\varepsilon = \varepsilon\,1_E$, $\zeta\varphi\varepsilon = 1_E$, and $\alpha\zeta = \beta\zeta$ yields $\alpha = \alpha\zeta\varphi\varepsilon = \beta\zeta\varphi\varepsilon = \beta$. $\square$

In Proposition 7.2, if every diagram can be assigned a limit, then the proof constructs a specific initial object.

The adjoint functor theorem. As expected, Proposition 7.2 yields a sufficient condition for the existence of adjoints:

Theorem **7.3** (Adjoint Functor Theorem). *Let $\mathcal{A}$ be a category and let $\mathcal{C}$ be a locally small category in which a limit can be assigned to every diagram. A functor $G : \mathcal{C} \longrightarrow \mathcal{A}$ has a left adjoint if and only if it preserves limits and satisfies the* solution set condition*:*

to every object A of $\mathcal{A}$ can be assigned a set $\mathcal{S}$ of morphisms $\sigma : A \longrightarrow G(C_\sigma)$ of $\mathcal{C}$ such that every morphism $\alpha : A \longrightarrow G(C)$ is a composition $\alpha = G(\gamma)\,\sigma$ for some $\sigma \in \mathcal{S}$ and homomorphism $\gamma : C_\sigma \longrightarrow C$.

Before proving Theorem 7.3, we show how it implies the existence of free groups. We know that *Grps* is locally small, that a limit can be assigned to every diagram of groups and homomorphisms, and that the forgetful functor $G : Grps \longrightarrow Sets$ preserves limits. The existence of free groups then follows from the assignment of a solution set to every set X. We construct one as follows.

Let X be a set. Every mapping f of X into a group G factors through the subgroup H of G generated by $Y = f(X)$. Now, every element of H is a product of finitely many elements of $Z = Y \cup Y^{-1}$. If Z is finite, then H is finite or countable. If Z is infinite, then, by A.5.10, there are $|Z|$ finite sequences of elements of Z and $|H| \leq |Z|$. In either case, $|Z| \leq |X| + |X|$ and $|H| \leq \lambda$, where $\lambda = \max(|X|, \aleph_0)$ depends only on X.

For every cardinal number $\kappa \leq \lambda$ we can choose one set T of cardinality κ (for instance, κ itself) and place all possible group operations $T \times T \longrightarrow T$ on T. The result is a set $\mathcal{T}$ of groups such that every group H of cardinality $|H| \leq \lambda$ is isomorphic to some $T \in \mathcal{T}$. Then every mapping f of X into a group factors through some $T \in \mathcal{T}$. There is only a set $\mathcal{S}$ of mappings $X \longrightarrow T \in \mathcal{T}$; thus, $\mathcal{S}$ is a solution set that can be assigned to X.

Comma categories. To prove the adjoint functor theorem we apply Proposition 7.2 to the following *comma category* $(A \downarrow G)$ (denoted by $\mathcal{C}_A$ earlier): given a functor $G : \mathcal{C} \longrightarrow \mathcal{A}$ and an object A of $\mathcal{A}$, an object of $(A \downarrow G)$ is an ordered pair (α, C) in which C is an object of $\mathcal{C}$ and α is a morphism of $\mathcal{A}$ from A to $G(C)$; a morphism of $(A \downarrow G)$ from (α, C) to (β, D) is a morphism $\gamma : C \longrightarrow D$ of $\mathcal{C}$ such that $\beta = G(\gamma)\alpha$.

Composition in $(A \downarrow G)$ is composition in $\mathcal{C}$. It is immediate that $(A \downarrow G)$ is a category. There is a *projection* functor from $(A \downarrow G)$ to $\mathcal{C}$, which assigns to each object (α, C) of $(A \downarrow G)$ the object C of $\mathcal{C}$, and to each morphism $\gamma : (\alpha, C) \longrightarrow (\beta, D)$ of $(A \downarrow G)$ the morphism $\gamma : C \longrightarrow D$ of $\mathcal{C}$.

Lemma **7.4.** *Let $\mathcal{C}$ be a locally small category in which a limit can be assigned to every diagram, let $G : \mathcal{C} \longrightarrow \mathcal{A}$ be a functor, and let A be an object of $\mathcal{A}$. If G preserves limits, then a limit can be assigned to every diagram in $(A \downarrow G)$, and the projection functor $(A \downarrow G) \longrightarrow \mathcal{C}$ preserves limits.*

Proof. Let Δ be a diagram in $(A \downarrow G)$ over a graph $\mathcal{G}$. For every vertex i of $\mathcal{G}$, $\Delta_i = (\delta_i, D_i)$, where $\delta_i : A \longrightarrow G(D_i)$; for every edge $a : i \longrightarrow j$, $\Delta_a = D_a : D_i \longrightarrow D_j$ and $\delta_j = G(D_a)\delta_i$. Then D is a diagram in $\mathcal{C}$ and δ is a cone from A to $G(D)$. Let $\lambda : L \longrightarrow D$ be the limit cone assigned to D. Then $G(\lambda): G(L) \longrightarrow G(D)$ is a limit cone of $G(D)$, and there is a morphism $\overline{\delta} : A \longrightarrow G(L)$ unique such that $\delta_i = G(\lambda_i)\overline{\delta}$ for all i. We show that $(\overline{\delta}, L)$ is a limit of Δ, with limit cone λ.

First, $\lambda_i : (\overline{\delta}, L) \longrightarrow (\delta_i, D_i)$ is a morphism of $(A \downarrow G)$ and $\Delta_a \lambda_i = D_a \lambda_i = \lambda_j$ when $a : i \longrightarrow j$. Thus $\lambda : (\overline{\delta}, L) \longrightarrow \Delta$ is a cone into Δ. Let $\varphi : (\alpha, C) \longrightarrow \Delta$ be any cone to Δ. Then [the projection of] φ is a cone from C to D, and there is a unique morphism $\overline{\varphi}$ such that $\varphi_i = \lambda_i \overline{\varphi}$ for all i:

$$
\begin{array}{ccc}
A \xrightarrow{\;\alpha\;} G(C) & & C \\
{\scriptstyle\overline{\delta}}\downarrow \; {\scriptstyle G(\overline{\varphi})}\swarrow \;\; \downarrow{\scriptstyle G(\varphi_i)} & & {\scriptstyle\overline{\varphi}}\swarrow \;\; \downarrow{\scriptstyle\varphi_i} \\
G(L) \xrightarrow[G(\lambda_i)]{} G(D_i) & & L \xrightarrow[\lambda_i]{} D_i
\end{array}
$$

Since λ_i and $\varphi_i : (\alpha, C) \longrightarrow (\delta_i, D_i)$ are morphisms of $(A \downarrow G)$ we have $G(\lambda_i) \, G(\overline{\varphi}) \, \alpha = G(\varphi_i) \, \alpha = \delta_i = G(\lambda_i) \, \overline{\delta}$ for all i; since $G(\lambda)$ is a limit cone of $G(D)$ this implies $G(\overline{\varphi}) \, \alpha = \overline{\delta}$. Thus $\overline{\varphi} : (\alpha, C) \longrightarrow (\overline{\delta}, L)$ is a morphism of $(A \downarrow G)$, and is the only morphism of $(A \downarrow G)$ such that $\varphi_i = \lambda_i \, \overline{\varphi}$ for all i. $\square$

We now prove Theorem 7.3. Let $\mathcal{C}$ be a locally small category in which a limit can be assigned to every diagram, and let $G : \mathcal{C} \longrightarrow \mathcal{A}$ be a functor. If there is an adjunction (F, G, η), then G preserves limits by 6.6, and every object A of $\mathcal{A}$ has a solution set $\{\eta_A\}$. Conversely, assume that G preserves limits and that a solution set $\mathcal{S}$ of morphisms $\sigma : A \longrightarrow G(C_\sigma)$ of $\mathcal{C}$ can be assigned to every object A of $\mathcal{A}$, so that every morphism $\alpha : A \longrightarrow G(C)$ is a composition $\alpha = G(\gamma) \, \sigma$ from some $\sigma \in \mathcal{S}$ and $\gamma : C_\sigma \longrightarrow C$. Then $(A \downarrow G)$ is complete, by 7.4, and $\{(\sigma, C_\sigma) \mid \sigma \in \mathcal{S}\}$ is a solution set in $(A \downarrow G)$; therefore $(A \downarrow G)$ has an initial object, by 7.2. In fact, since every diagram in $(A \downarrow G)$ can be assigned a limit, a specific initial object $(\eta_A, F(A))$ can be selected in $(A \downarrow G)$. Then G has a left adjoint, by 6.3. $\square$

Exercises

1. Find all initial and terminal objects in the following categories: *Grps*; *Rings*; $_R$*Mods*; a partially ordered set I.

2. Prove directly that a limit cone of a diagram is a terminal object in a suitable cone category.

3. Use the adjoint functor theorem to show that the forgetful functor from $_R$*Mods* to *Sets* has a left adjoint.

4. Use the adjoint functor theorem to show that the forgetful functor from *Rings* to *Sets* has a left adjoint.

5. Let $G : \mathcal{C} \longrightarrow \mathcal{A}$ preserve limits. Show that the projection functor $(A \downarrow G) \longrightarrow \mathcal{C}$ creates limits.

In the following exercises, $\mathcal{C}$ is a locally small category. A functor $F : \mathcal{C} \longrightarrow$ *Sets* is *representable* when it is naturally isomorphic to $\mathrm{Hom}_\mathcal{C}(C, -)$ for some object C of $\mathcal{C}$. Representable functors provide one more approach to universal properties.

6. Let $F \cong \mathrm{Hom}_\mathcal{C}(C, -)$ be representable. Formulate a universal property for the object C, and show that C is unique up to isomorphism.

7. Show that $\mathcal{C}$ has an initial object if and only if the constant functor $X \longmapsto \{1\}$ is representable.

8. Let D be a diagram in $\mathcal{C}$. Define a functor $F : \mathcal{C} \longrightarrow$ *Sets* such that $F(A)$ is the set of all cones from D to A. Show that D has a colimit in $\mathcal{C}$ if and only if F is representable.

9. Let $G : \mathcal{C} \longrightarrow \mathcal{A}$ be a functor, where $\mathcal{A}$ is locally small. Show that G has a left adjoint if and only if $\mathrm{Hom}_\mathcal{A}(A, G(-))$ is representable for every object A of $\mathcal{A}$.

10. Show that $F : \mathcal{C} \longrightarrow$ *Sets* is representable if and only if it preserves limits and satisfies the solution set condition: there exists a set $\mathcal{S}$ of objects such that, for every object A of D and every $x \in F(A)$, there exists $S \in \mathcal{S}$, $y \in F(S)$, and $\alpha : S \longrightarrow A$ such that $F(\alpha)(y) = x$.

8. Triples

Triples provide a unified description of most algebraic systems, using one functor and two natural transformations. This construction took its final shape in a paper by Eilenberg and Moore [1965].

Definition. Our examples of adjunctions $(F, G, \eta, \varepsilon)$ have emphasized η at the expense of ε. To redress this injustice we look at ε when F is the free left R-module functor and G is the forgetful functor from left R-modules to sets. For every set X and module M, the natural bijection

$$\theta_{X,M} : \operatorname{Hom}_{RMods}(F(X), M) \longrightarrow \operatorname{Hom}_{Sets}(X, G(M))$$

sends a homomorphism $F(X) \longrightarrow M$ to its restriction to X. By 6.5, $\varepsilon_M = \theta^{-1}_{G(M),M}(1_{G(M)})$ is the homomorphism $F(G(M)) \longrightarrow M$ that extends the identity on M. Hence ε_M takes an element of $F(G(M))$, written uniquely as a linear combination $\sum_i r_i\, m_i$ of elements of M with coefficients in R, and sends it to $\sum_i r_i\, m_i \in M$ as calculated in M. In particular, the addition and action of R on M are completely determined, within the category $Sets$, by the set $G(M)$ and the mapping $G(\varepsilon_M) : G(F(G(M))) \longrightarrow G(M)$.

To avoid unsightly pile-ups of parentheses, we write functors as left operators in what follows (FA instead of $F(A)$). Every adjunction $(F, G, \eta, \varepsilon)$ from $\mathcal{A}$ to $\mathcal{C}$ determines a functor $GF : \mathcal{A} \longrightarrow \mathcal{A}$ and natural transformations $\eta : 1_{\mathcal{A}} \longrightarrow GF$ and $G\varepsilon F : GFGF \longrightarrow GF$, whose basic properties are as follows.

Proposition 8.1 Let $(F, G, \eta, \varepsilon)$ be an adjunction from $\mathcal{A}$ to $\mathcal{C}$. Let $T = GF$ and $\mu = G\varepsilon F : TT \longrightarrow T$. The following diagrams commute:

$$
\begin{array}{ccc}
TTT & \xrightarrow{\ \mu T\ } & TT \\
{\scriptstyle T\mu}\downarrow & & \downarrow{\scriptstyle \mu} \\
TT & \xrightarrow{\ \mu\ } & T
\end{array}
\qquad\qquad
\begin{array}{ccccc}
T & \xrightarrow{\ \eta T\ } & TT & \xleftarrow{\ T\eta\ } & T \\
 & \searrow_{1} & \downarrow{\scriptstyle \mu} & \swarrow_{1} & \\
 & & T & &
\end{array}
$$

Proof. Since ε is a natural transformation, the square below left commutes

$$
\begin{array}{ccc}
FGFGC & \xrightarrow{\ \varepsilon_{FGC}\ } & FGC \\
{\scriptstyle FG\varepsilon_C}\downarrow & & \downarrow{\scriptstyle \varepsilon_C} \\
FGC & \xrightarrow{\ \varepsilon_C\ } & C
\end{array}
\qquad\qquad
\begin{array}{ccc}
GFGFGFA & \xrightarrow{\ G\varepsilon_{FGFA}\ } & GFGFA \\
{\scriptstyle GFG\varepsilon_{FA}}\downarrow & & \downarrow{\scriptstyle G\varepsilon_{FA}} \\
GFGFA & \xrightarrow{\ G\varepsilon_{FA}\ } & GFA
\end{array}
$$

for every object C of $\mathcal{C}$. If $C = FA$, applying G yields the square above right; hence $\mu_A\, \mu_{TA} = \mu_A\,(T\mu_A)$ for every object A of $\mathcal{A}$. By 6.5, $\mu_A\, \eta_{TA} = (G\varepsilon_{FA})\,\eta_{GFA} = 1_{GFA}$ and $\mu_A\,(T\eta_A) = (G\varepsilon_{FA})(GF\eta_A) = G1_{FA} = 1_{GFA}$. $\square$

Definition. A triple (T, η, μ) on a category $\mathcal{A}$ is a functor $T : \mathcal{A} \longrightarrow \mathcal{A}$ with two natural transformations $\eta : 1_{\mathcal{A}} \longrightarrow T$ and $\mu : TT \longrightarrow T$ such that the following diagrams commute:

$$TTT \xrightarrow{\mu T} TT \qquad\qquad T \xrightarrow{\eta T} TT \xleftarrow{T\eta} T$$
$$T\mu \downarrow \qquad\qquad \downarrow \mu \qquad\qquad {}_1 \searrow \quad \downarrow \mu \quad \swarrow {}_1$$
$$TT \xrightarrow[\mu]{} T \qquad\qquad\qquad T$$

Triples are also called *monads* and (in older literature) *standard constructions*. Monads have a formal similarity with monoids, which also have a multiplication $M \times M \longrightarrow M$ and identity element $\{1\} \longrightarrow M$ with similar properties.

T-algebras. By 8.1, every adjunction from $\mathcal{A}$ to $\mathcal{C}$ induces a triple on $\mathcal{A}$. We show that, conversely, every triple is induced by an adjunction.

Definitions. Let (T, η, μ) be a triple on a category $\mathcal{A}$. A T-algebra is a pair (A, φ) of an object A of $\mathcal{A}$ and a morphism $\varphi : TA \longrightarrow A$ such that the following diagrams commute:

$$TTA \xrightarrow{T\varphi} TA \qquad\qquad A \xrightarrow{\eta_A} TA$$
$$\mu_A \downarrow \qquad\quad \downarrow \varphi \qquad\qquad {}_1 \searrow \quad \downarrow \varphi$$
$$TA \xrightarrow[\varphi]{} A \qquad\qquad\qquad A$$

A morphism *or* homomorphism $\alpha : (A, \varphi) \longrightarrow (B, \psi)$ *of T-algebras is a morphism $\alpha : A \longrightarrow B$ such that $\alpha\varphi = \psi\,(T\alpha)$:*

$$TA \xrightarrow{T\alpha} TB$$
$$\varphi \downarrow \qquad\quad \downarrow \psi$$
$$A \xrightarrow[\alpha]{} B$$

For example, let (T, η, μ) be the triple on *Sets* induced by the adjunction $(F, G, \eta, \varepsilon)$ from *Sets* to ${}_R Mods$. If M is a left R-module, then $(GM, G\varepsilon_M)$ is a T-algebra, by 8.2 below. We saw that $G\varepsilon_M : GFGM \longrightarrow GM$ determines the operations on M; thus, every left R-module "is" a T-algebra. A converse is proved in the next section.

Proposition **8.2.** *Let $(F, G, \eta, \varepsilon)$ be an adjunction from $\mathcal{A}$ to $\mathcal{C}$ and let $(T, \eta, \mu) = (GF, \eta, G\varepsilon F)$ be the triple it induces on $\mathcal{A}$. For every object C of $\mathcal{C}$, $(GC, G\varepsilon_C)$ is a T-algebra. If $\gamma : C \longrightarrow D$ is a morphism of $\mathcal{C}$, then $G\gamma : (GC, G\varepsilon_C) \longrightarrow (GD, G\varepsilon_D)$ is a homomorphism of T-algebras.*

Proof. The diagrams below commute, the first since ε is a natural transformation, the second by 6.5; hence $(GC, G\varepsilon_C)$ is a T-algebra.

$$GFGFGC \xrightarrow{GFG\varepsilon_C} GFGC \qquad\qquad GC \xrightarrow{\eta GC} GFGC$$
$$G\varepsilon_{FGC} \downarrow \qquad\qquad \downarrow G\varepsilon_C \qquad\qquad {}_1 \searrow \quad \downarrow G\varepsilon$$
$$GFGC \xrightarrow[G\varepsilon_C]{} GC \qquad\qquad\qquad GC$$

If $\gamma : C \longrightarrow D$ is a morphism of $\mathcal{C}$, then the square below commutes, since ε is a natural transformation; hence γ is a homomorphism of T-algebras. $\square$

$$GFGC \xrightarrow{GF\gamma} GFGD$$

$$G\varepsilon_C \downarrow \qquad\qquad \downarrow G\varepsilon_D$$

$$C \xrightarrow{\qquad \gamma \qquad} D$$

Proposition **8.3.** *If* (T, η, μ) *is a triple on a category* $\mathcal{A}$, *then* T-*algebras and their homomorphisms are the objects and morphisms of a category* $\mathcal{A}^T$; *moreover, there is an adjunction* $(F^T, G^T, \eta^T, \varepsilon^T)$ *from* $\mathcal{A}$ *to* $\mathcal{A}^T$ *that induces on* $\mathcal{A}$ *the given triple, given by*

$$G^T(A, \varphi) = A, \quad F^T A = (TA, \mu_A), \quad \eta_A^T = \eta_A, \quad \varepsilon_{(A,\varphi)}^T = \varphi.$$

Proof. It is immediate that T-algebras and their homomorphisms constitute a category $\mathcal{A}^T$ with a forgetful functor $G^T : \mathcal{A}^T \longrightarrow \mathcal{A}$. The definition of triples shows that $F^T A = (TA, \mu_A)$ is a T-algebra. If $\alpha : A \longrightarrow B$ is a morphism of $\mathcal{A}$, then $F^T \alpha = T\alpha : (TA, \mu_A) \longrightarrow (TB, \mu_B)$ is a homomorphism of T-algebras, since μ is a natural transformation. This constructs a functor $F^T : \mathcal{A} \longrightarrow \mathcal{A}^T$. We have $G^T F^T = T$, so that $\eta \stackrel{.}{:} 1_{\mathcal{A}} \longrightarrow G^T F^T$.

We show that, for every morphism $\alpha : A \longrightarrow C = G^T(C, \psi)$ of $\mathcal{A}$, there is a unique homomorphism $\gamma : F^T A = (TA, \mu_A) \longrightarrow (C, \psi)$ such that $\gamma \eta_A = (G^T \gamma) \eta_A = \alpha$. Necessarily $\gamma \mu_A = \psi(T\gamma)$ and $\gamma = \gamma \mu_A(T\eta_A) = \psi(T\gamma)(T\eta_A) = \psi(T\alpha)$; hence γ is unique. Conversely, let $\gamma = \psi(T\alpha)$. Since μ is a natural transformation and (C, ψ) is a T-algebra,

$$\gamma \eta_A = \psi(T\alpha)\mu_A = \psi\mu_C(TT\alpha) = \psi(T\psi)(TT\alpha) = \psi(T\gamma);$$

hence γ is a homomorphism from (TA, μ_A) to (C, ψ). Moreover, $\gamma \eta_A = \psi (T\alpha) \eta_A = \psi \eta_C \alpha = \alpha$, since η is natural and (C, ψ) is a T-algebra.

We now have an adjunction $(F^T, G^T, \eta^T, \varepsilon^T)$, in which $\eta^T = \eta$. By 6.5, $\varepsilon_{(A,\varphi)}^T$ is the homomorphism from $F^T G^T(A, \varphi) = F^T A = (TA, \mu_A)$ to (A, φ) such that $(G^T \varepsilon_{(A,\varphi)}^T) \eta_{G^T(A,\varphi)}^T = 1_{G^T(A,\varphi)}$, equivalently $\varepsilon_{(A,\varphi)}^T \eta_A = 1_A$. But φ has these properties, since (A, φ) is a T-algebra. Hence $\varepsilon_{(A,\varphi)}^T = \varphi$. $\square$

In the following sense, the category $\mathcal{A}^T$ of T-algebras is the "greatest" category with an adjunction from $\mathcal{A}$ that induces T.

Proposition **8.4.** *Let* $(F, G, \eta, \varepsilon)$ *be an adjunction from* $\mathcal{A}$ *to* $\mathcal{C}$ *and let* $(T, \eta, \mu) = (GF, \eta, G\varepsilon F)$ *be the triple it induces on* $\mathcal{A}$. *There is a unique functor* $Q : \mathcal{C} \longrightarrow \mathcal{A}^T$ *such that* $F^T = QF$ *and* $G^T Q = G$, *given by*

$$QC = (GC, G\varepsilon_C), \quad Q\gamma = G\gamma.$$

Proof. By 8.2, $(GC, G\varepsilon_C)$ is a T-algebra for every object C of $\mathcal{C}$, and, if $\gamma : C \longrightarrow D$ is a morphism of $\mathcal{C}$, then $G\gamma : (GC, G\varepsilon_C) \longrightarrow (GD, G\varepsilon_D)$ is a homomorphism of T-algebras. Hence the equalities $QC = (GC, G\varepsilon_C)$, $Q\gamma = G\gamma$ define a functor Q from $\mathcal{C}$ to $\mathcal{A}^T$. Moreover, $QF = F^T$, $G^T Q = G$.

Conversely, let $Q : \mathcal{C} \longrightarrow \mathcal{A}^T$ be a functor such that $F^T = QF$ and $G^T Q = G$. We have $Q \varepsilon_C = \varepsilon_{QC}^T$ for every object C of $\mathcal{C}$, since both $Q \varepsilon_C$ and ε_{QC}^T satisfy $\left(G^T \varepsilon_{QC}^T \right) \eta_{G^T QC}^T = 1_{G^T QC}$ and $\left(G^T Q \varepsilon_C \right) \eta_{G^T QC}^T = \left(G \varepsilon_C \right) \eta_{GC} = 1_{GC} = 1_{G^T QC}$. Since $G^T Q = G$ we have $QC = (GC, \varphi)$ for some $\varphi : GFGC \longrightarrow GC$. By 8.3, $\varphi = \varepsilon_{QC}^T = Q \varepsilon_C = G^T Q \varepsilon_C = G \varepsilon_C$. Thus $QC = (GC, G \varepsilon_C)$, for every object C of $\mathcal{C}$. Moreover, $G \gamma = G^T Q \gamma = Q \gamma$ for every morphism γ of $\mathcal{C}$. Thus Q is unique. $\square$

Exercises

1. Describe the triple on *Sets* induced by the free group functor from *Sets* to *Grps* and its forgetful right adjoint.

2. Describe the triple on *Abs* induced by the forgetful functor from $_R$*Mods* to *Abs* and its left adjoint.

3. Describe the triple on *Abs* induced by the forgetful functor from commutative rings to *Abs* and its left adjoint.

4. Let I be a preordered set, viewed as a category. What is a triple on I? What is a T-algebra?

5. Let G be a group. Show that a triple (T, η, μ) on *Sets* is defined by $TX = G \times X$, $\eta_X : x \longmapsto (1, x)$, and $\mu_X : \big(g, (h, x) \big) \longmapsto (gh, x)$. Show that T-algebras coincide with group actions of G.

6. Let F and F' be two left adjoints of G. What can you say about the triples induced by (F, G) and (F', G)?

7. Show that G^T creates limits.

9. Tripleability

In this section we prove Beck's theorem [1966], which characterizes categories of T-algebras and implies that *Grps*, *Rings*, $_R$*Mods*, etc., are isomorphic to categories of T-algebras, for suitable triples T.

Definition. A functor $G : \mathcal{C} \longrightarrow \mathcal{A}$ is tripleable *when it has a left adjoint and the functor Q in Proposition 8.4 is an isomorphism of categories.*

Similarly, $\mathcal{C}$ is *tripleable over* $\mathcal{A}$ when there is a canonical functor $G : \mathcal{C} \longrightarrow \mathcal{A}$ (usually a forgetful functor) that is tripleable.

Beck's theorem implies that *Grps*, *Rings*, $_R$*Mods*, etc. are tripleable over *Sets*. But not every category is tripleable over *Sets*. For example, let *POS* be the category of partially ordered sets and order preserving mappings. The forgetful functor G from *POS* to *Sets* has a left adjoint, which assigns to each set X the *discrete* partially ordered set FX on X, in which $x \leq y$ if and only if $x = y$: indeed, every mapping of X into a partially ordered set S "extends" uniquely to an order preserving mapping of FX into S. The induced triple

(T, η, μ) has $TX = X$ and $\eta_X = \mu_X = 1_X$ for every set X. Hence T-algebras "are" sets, that is, G^T is an isomorphism of categories. But $G^T Q = G$ is not an isomorphism, so Q is not an isomorphism. Readers will set up a similar argument for the category $Tops$ of topological spaces and continuous mappings. Thus, tripleability is a property of sets with operations and is not generally shared by other mathematical structures.

Split coequalizers. Beck's theorem requires properties that are specific to T-algebras and their forgetful functors G^T. Beck observed that, in a T-algebra (A, φ), the morphism $\varphi : TA \longrightarrow A$ is a coequalizer of μ_A and $T\varphi$. Indeed, $\varphi \mu_A = \varphi (T\varphi)$ and $\varphi \eta_A = 1_A$, since (A, φ) is a T-algebra, $\mu_A \eta_{TA} = 1_{TA}$ since (T, η, μ) is a triple, and $(T\varphi) \eta_{TA} = \eta_A \varphi$, since η is a natural transformation. If now $\psi \mu_A = \psi (T\varphi)$, then $\psi = \psi \mu_A \eta_{TA} = \psi (T\varphi) \eta_{TA} = \psi \eta_A \varphi$ factors through φ; this factorization is unique, since $\varphi \eta_A = 1_A$. Thus, φ is a coequalizer because η_A and η_{TA} have certain properties.

Definition. Let $\alpha, \beta : A \longrightarrow B$. A split coequalizer *of* α *and* β *is a morphism* $\sigma : B \longrightarrow C$ *such that* $\sigma \alpha = \sigma \beta$ *and there exist morphisms* $\kappa : B \longrightarrow A$ *and* $\nu : C \longrightarrow B$ *such that* $\alpha \kappa = 1_B$, $\sigma \nu = 1_C$, *and* $\beta \kappa = \nu \sigma$.

Proposition **9.1.** (1) *If* (A, φ) *is a* T-algebra, then φ *is a split coequalizer of* μ_A *and* $T\varphi$.

(2) *A split coequalizer of* α *and* β *is a coequalizer of* α *and* β.

(3) *Every functor preserves split coequalizers.*

Proof. (1). We saw that φ is a split coequalizer of $\alpha = \mu_A$ and $\beta = T\varphi$, with $\nu = \eta_A$ and $\kappa = \eta_{TA}$. (2). Assume $\sigma \alpha = \sigma \beta$ and $\alpha \kappa = 1$, $\sigma \nu = 1$, $\beta \kappa = \nu \sigma$. If $\varphi \alpha = \varphi \beta$, then $\varphi = \varphi \alpha \kappa = \varphi \beta \kappa = \varphi \nu \sigma$ factors through σ; this factorization is unique, since $\sigma \nu = 1$. (3) is clear. $\square$

Definition. Let $\gamma, \delta : C \longrightarrow D$ be morphisms of $\mathcal{C}$. A functor $G : \mathcal{C} \longrightarrow \mathcal{A}$ creates coequalizers *of* γ *and* δ *when every coequalizer of* $G\gamma$ *and* $G\delta$ *in* $\mathcal{A}$ *is the image under* G *of a unique morphism* $\sigma : D \longrightarrow K$ *of* $\mathcal{C}$, *and* σ *is a coequalizer of* γ *and* δ.

Proposition **9.2.** *The functor* G^T *creates coequalizers of pairs* α, β *with a split coequalizer in* $\mathcal{A}$.

Proof. Let $\alpha, \beta : (A, \varphi) \longrightarrow (B, \psi)$ be homomorphisms of T-algebras that have a split coequalizer $\sigma : B \longrightarrow C$ in $\mathcal{A}$. We need to show that there is a unique T-algebra (C, χ) such that σ is a homomorphism:

$$
\begin{array}{ccccc}
TA & \underset{T\beta}{\overset{T\alpha}{\rightrightarrows}} & TB & \overset{T\sigma}{\longrightarrow} & TC \\
{\scriptstyle\varphi}\downarrow & & \downarrow{\scriptstyle\psi} & & \downarrow{\scriptstyle\chi} \\
A & \underset{\beta}{\overset{\alpha}{\rightrightarrows}} & B & \underset{\sigma}{\longrightarrow} & C
\end{array}
$$

for then, σ is a split coequalizer in $\mathcal{A}^T$. By 9.1, $T\sigma$ is a split coequalizer of $T\alpha$ and $T\beta$. Since $\sigma\,\psi\,(T\alpha) = \sigma\alpha\varphi = \sigma\beta\varphi = \sigma\,\psi\,(T\beta)$ there is a unique morphism $\chi : TC \longrightarrow C$ such that $\sigma\psi = \chi\,(T\sigma)$. We show that (C, χ) is a T-algebra.

Since η is a natural transformation there is a commutative diagram:

$$
\begin{array}{ccccc}
B & \xrightarrow{\;\eta_B\;} & TB & \xrightarrow{\;\psi\;} & B \\
{\scriptstyle\sigma}\downarrow & & {\scriptstyle T\sigma}\downarrow & & \downarrow{\scriptstyle\sigma} \\
C & \xrightarrow{\;\eta_C\;} & TC & \xrightarrow{\;\chi\;} & C
\end{array}
$$

Since (B, φ) is a T-algebra, we have $\chi\,\eta_C\,\sigma = \sigma\psi\,\eta_B = \sigma = 1_C\,\sigma$; hence $\chi\,\eta_C = 1_C$. Similarly, in the diagram

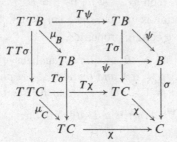

the top face commutes since (B, ψ) is a T-algebra, and the four side faces commute, by definition of χ and naturality of μ. Hence the bottom face commutes: using the other faces, $\chi\,(T\chi)\,(TT\sigma) = \chi\,\mu_C\,(TT\sigma)$, whence $\chi\,(T\chi) = \chi\,\mu_C$, since $TT\sigma$ is a split coequalizer, hence an epimorphism. $\square$

Beck's theorem is the converse of 9.2:

Theorem **9.3** (Beck). *A functor* $G : \mathcal{C} \longrightarrow \mathcal{A}$ *with a left adjoint is tripleable if and only if it creates coequalizers of pairs* α, β *such that* $G\alpha$, $G\beta$ *have a split coequalizer in* $\mathcal{A}$.

Theorem 9.3 follows from Proposition 9.2 and from a result that is similar to Proposition 8.4:

Lemma **9.4.** *Let* $(F, G, \eta, \varepsilon): \mathcal{A} \longrightarrow \mathcal{C}$ *and* $(F', G', \eta', \varepsilon'): \mathcal{A}' \longrightarrow \mathcal{C}'$ *be adjunctions that induce on* $\mathcal{A}$ *the same triple* $(T, \eta, \mu) = (GF, \eta, G\varepsilon F) = (G'F', \eta', G'\varepsilon'F')$. *If* G' *creates coequalizers of pairs* α, β *such that* $G'\alpha$, $G'\beta$ *have a split coequalizer in* $\mathcal{A}$, *then there is a unique functor* $R : \mathcal{C} \longrightarrow \mathcal{C}'$ *such that* $F' = RF$ *and* $G'R = G$.

Proof. Existence. Let C be an object of $\mathcal{C}$. By 8.2, 9.1, $(GC, G\varepsilon_C)$ is a T-algebra and $G\varepsilon_C$ is a split coequalizer of $G'\varepsilon'_{F'GC} = \mu_{GC}$ and $G'F'G\varepsilon_C = TG\varepsilon_C$. Then $G\varepsilon_C$ is the image under G' of a unique morphism $\rho'_C : F'GC \longrightarrow RC$, which is a coequalizer of $\varepsilon'_{F'GC}$ and $F'G\varepsilon_C : F'G'F'GC \longrightarrow F'GC$. This defines R on objects. In particular, $G'\rho'_C = G\varepsilon_C$ and $G'RC = GC$.

For every morphism $\gamma : C \longrightarrow D$ in $\mathcal{C}$ we have a diagram

$$
\begin{array}{ccc}
F'G'F'GC \underset{F'G\varepsilon_C}{\overset{\varepsilon'_{F'GC}}{\rightrightarrows}} F'GC & \overset{\rho'_C}{\longrightarrow} & RC \\
{\scriptstyle F'G'F'G\gamma}\downarrow \qquad\qquad \downarrow{\scriptstyle F'G\gamma} & & \downarrow{\scriptstyle R\gamma} \\
F'G'F'GD \underset{F'G\varepsilon_D}{\overset{\varepsilon'_{F'GD}}{\rightrightarrows}} F'GD & \underset{\rho'_D}{\longrightarrow} & RD
\end{array}
$$

in which

$$
\rho'_D \, (F'G\gamma) \, \varepsilon'_{F'GC} = \rho'_D \, \varepsilon'_{F'GD} \, (F'G'F'G\gamma)
$$
$$
= \rho'_D \, (F'G\varepsilon_D)(F'G'F'G\gamma) = \rho'_D \, (F'G\gamma) \, F'G\varepsilon_C,
$$

since ε' is a natural transformation. Therefore $\rho'_D \, (F'G\gamma) = (R\gamma) \, \rho'_C$ for some unique $R\gamma : RC \longrightarrow RD$. Uniqueness in this factorization readily implies that R is a functor.

We saw that $GC = G'RC$. In the above, $(G'R\gamma)(G\varepsilon_C) = (G'R\gamma)(G'\rho'_C) = (G'\rho'_D)(G'F'G\gamma) = (G\varepsilon_D)(GFG\gamma) = (G\gamma)(G\varepsilon_C)$, since ε is a natural transformation. Since $(G\varepsilon_C) \, \eta_{GC} = 1_{GC}$ this implies $G'R\gamma = G\gamma$. Thus $G'R = G$.

If A is an object of $\mathcal{A}$, then $G'\varepsilon'_{F'A} = G\varepsilon_{FA}$; therefore $\rho'_{FA} = \varepsilon'_{F'A}$, and $RFA = F'A$. If $\alpha : A \longrightarrow B$ is a morphism of $\mathcal{A}$, then

$$
(RF\alpha) \, \rho'_{FA} = \rho'_{FB} \, (F'GF\alpha) = \varepsilon'_{F'B} \, (F'GF\alpha) = (F'\alpha)(\varepsilon'_{FA}) = (F'\alpha) \, \rho'_{FA},
$$

since ε' is natural; therefore $RF\alpha = F'\alpha$. Thus $RF = F'$.

Uniqueness. Let $S : \mathcal{C} \longrightarrow \mathcal{C}'$ be a functor such that $F' = SF'$ and $G'S = G$. We have $S\varepsilon_C = \varepsilon'_{SC}$ for every object C of $\mathcal{C}$, since $(G'\varepsilon'_{SC}) \, \eta'_{G'SC} = 1_{G'SC}$ and $(G'S\varepsilon_C) \, \eta'_{G'SC} = (G\varepsilon_C) \, \eta_{GC} = 1_{GC} = 1_{G'SC}$.

Let C be an object of $\mathcal{C}$. As above, $G\varepsilon_C$ is a split coequalizer of $G'\varepsilon'_{F'GC} = \mu_{GC}$ and $G'F'G\varepsilon_C = TG\varepsilon_C$, so that $G\varepsilon_C$ is the image under G' of a unique morphism, namely the epimorphism $\rho'_C : F'GC \longrightarrow RC$. Since $G'\varepsilon'_{SC} = G'S\varepsilon_C = G\varepsilon_C$, it follows that $\varepsilon'_{SC} = \rho'_C$. In particular, $SC = RC$.

Let $\gamma : C \longrightarrow D$ be a morphism of $\mathcal{C}$. Since ε' is a natural transformation, we have $(S\gamma) \, \varepsilon'_{SC} = \varepsilon'_{SD} \, (F'G'S\gamma)$, equivalently $(S\gamma) \, \rho'_C = \rho'_D \, (F'G\gamma)$. Since $(R\gamma) \, \rho'_C = \rho'_D \, (F'G\gamma)$, it follows that $S\gamma = R\gamma$. Thus $S = R$. $\square$

We now prove Theorem 9.3. By 8.3, 9.2, G^T has a left adjoint and creates coequalizers of pairs α, β such that $G^T\alpha$, $G^T\beta$ have a split coequalizer in $\mathcal{A}$. These properties are inherited by $G^T Q$ whenever Q is an isomorphism of categories, and by every tripleable functor G.

Conversely, let $G : \mathcal{C} \longrightarrow \mathcal{A}$ have a left adjoint and create coequalizers of pairs α, β such that $G\alpha$, $G\beta$ have a split coequalizer in $\mathcal{A}$. Let $(F, G, \eta, \varepsilon)$ be an adjunction and let $(T, \eta, \mu) = (GF, \eta, G\varepsilon F)$ be the triple it induces on

$\mathcal{A}$. Let $Q : \mathcal{C} \longrightarrow \mathcal{A}^T$ be the unique functor in 8.4, such that $F^T = QF$ and $G^T Q = G : QC = (GC, G\varepsilon_C)$ and $Q\gamma = G\gamma$ for every C and γ in $\mathcal{C}$.

Since G creates coequalizers of pairs α, β such that $G\alpha$, $G\beta$ have a split coequalizer in $\mathcal{A}$, there is by 9.4 a functor $R : \mathcal{A}^T \longrightarrow \mathcal{C}$ such that $F = RF^T$ and $GR = G^T$. Then $F^T = QRF^T$ and $G^T QR = G^T$; by the uniqueness in 9.4, $QR = 1_{\mathcal{A}^T}$. Similarly, $F = RQF$ and $GRQ = G$; by the uniqueness in 9.4, $RQ = 1_{\mathcal{C}}$. Thus Q is an isomorphism. $\square$

Examples. We use Beck's theorem to prove the following:

Proposition **9.5.** *The forgetful functor from* Grps *to* Sets *is tripleable.*

Proof. We show that this functor creates coequalizers of pairs α, $\beta : G \longrightarrow H$ of group homomorphisms that have a split coequalizer $\sigma : H \longrightarrow K$ in *Sets*.

Let $m_G : G \times G \longrightarrow G$ and $m_H : H \times H \longrightarrow H$ be the group operations on G and H. Since α and β are homomorphisms we have $\alpha \circ m_G = m_H \circ (\alpha \times \alpha)$ and $\beta \circ m_G = m_H \circ (\beta \times \beta)$. Hence $\sigma \circ m_H \circ (\alpha \times \alpha) = \sigma \circ m_H \circ (\beta \times \beta)$. By 9.1, $\sigma \times \sigma$ is a split coequalizer of $\alpha \times \alpha$ and $\beta \times \beta$; therefore there is a unique mapping $m_K : K \times K \longrightarrow K$ such that $\sigma \circ m_H = m_K \circ (\sigma \times \sigma)$:

$$
\begin{array}{ccccc}
G \times G & \underset{\beta \times \beta}{\overset{\alpha \times \alpha}{\rightrightarrows}} & H \times H & \overset{\sigma \times \sigma}{\longrightarrow} & K \times K \\
\downarrow{\scriptstyle m_G} & & \downarrow{\scriptstyle m_H} & & \downarrow{\scriptstyle m_K} \\
G & \underset{\beta}{\overset{\alpha}{\rightrightarrows}} & H & \underset{\sigma}{\longrightarrow} & K
\end{array}
$$

Since σ is surjective, m_K inherits associativity, identity element, and inverses from m_H. Thus, there is a unique group operation on the set K such that σ is a homomorphism. It remains to show that σ is a coequalizer in *Grps*.

Let $\varphi : H \longrightarrow L$ be a group homomorphism such that $\varphi \circ \alpha = \varphi \circ \beta$. Since σ is a coequalizer in *Sets* there is a unique mapping $\psi : K \longrightarrow L$ such that $\varphi = \psi \circ \sigma$. Since φ and σ are homomorphisms, $\psi\big(\sigma(x)\sigma(y)\big) = \psi\big(\sigma(xy)\big) = \varphi(xy) = \varphi(x)\varphi(y) = \psi\big(\sigma(x)\big)\psi\big(\sigma(y)\big)$; hence ψ is a homomorphism. $\square$

Readers will prove similar results for *Rings*, $_R$*Mods*, etc. These also follow from a theorem in the next section.

Exercises

1. Topological spaces and continuous mappings are the objects and the morphisms of a category *Tops*. Show that the forgetful functor from *Tops* to *Sets* has a left adjoint but is not tripleable. (But see Exercise 6 below.)

2. Prove that the forgetful functor from *Rings* to *Sets* is tripleable.

3. Prove that the forgetful functor from $_R$*Mods* to *Sets* is tripleable.

4. Prove that the forgetful functor from *Rings* to *Abs* is tripleable.

5. Prove that the forgetful functor from $_R$*Mods* to *Abs* is tripleable.

*6. For readers who know some topology: show that the forgetful functor from compact Hausdorff spaces to sets is tripleable.

*7. A coequalizer σ of two morphisms α and β is an *absolute coequalizer* when $F(\sigma)$ is a coequalizer of $F(\alpha)$ and $F(\beta)$ for every functor F. For example, split coequalizers are absolute coequalizers. Prove the following: a functor $G : \mathcal{C} \longrightarrow \mathcal{A}$ with a left adjoint is tripleable if and only if it creates coequalizers of pairs α, β such that $G\alpha$, $G\beta$ have an absolute coequalizer in $\mathcal{A}$.

10. Varieties

This section collects categorical properties of varieties: completeness, cocompleteness, existence of free algebras, and tripleability. The reader is referred to Section XV.3 for examples and basic properties.

Every variety, more generally, every class $\mathcal{C}$ of algebras of the same type, defines a category, whose class of objects is $\mathcal{C}$, and whose morphisms are all homomorphisms from a member of $\mathcal{C}$ to another. This category comes with a forgetful functor to *Sets*, which assigns to each member of $\mathcal{C}$ its underlying set.

Proposition **10.1.** *If $\mathcal{C}$ is a class of universal algebras of type T that is closed under products and subalgebras (for instance, a variety), then $\mathcal{C}$ is complete; in fact, a limit can be assigned to every diagram in $\mathcal{C}$, and the forgetful functor from $\mathcal{C}$ to Sets preserves limits.*

Proof. Let D be a diagram in $\mathcal{C}$ over a graph $\mathcal{G}$. Let $P = \prod_{i \in \mathcal{G}} D_i$ be the direct product, with projections $\pi_i : P \longrightarrow D_i$ and componentwise operations,

$$\omega\left((x_{1i})_{i \in \mathcal{G}}, \ldots, (x_{ni})_{i \in \mathcal{G}}\right) = \left(\omega(x_{1i}, \ldots, x_{ni})\right)_{i \in \mathcal{G}}$$

whenever $\omega \in T$ has arity n and $x_1, \ldots, x_n \in P$, $x_k = (x_{ki})_{i \in \mathcal{G}}$. In *Sets*, the [underlying] diagram D has a standard limit

$$L = \left\{ (x_i)_{i \in \mathcal{G}} \in P \mid D_a(x_i) = x_j \text{ whenever } a : i \longrightarrow j \right\}$$

with limit cone $\lambda = (\lambda_i)_{i \in \mathcal{G}}$, where $\lambda_i = \pi_{i|D_i} : L \longrightarrow D_i$. Then L is a subalgebra of P: if $\omega \in T$ has arity n, $x_1, \ldots, x_n \in L$, and $a : i \longrightarrow j$, then D_a sends the i component $\omega(x_{1i}, \ldots, x_{ni})$ of $\omega(x_1, \ldots, x_n)$ to

$$D_a\left(\omega(x_{1i}, \ldots, x_{ni})\right) = \omega\left(D_a(x_{1i}), \ldots, D_a(x_{ni})\right) = \omega(x_{1j}, \ldots, x_{nj}),$$

which is the j component of $\omega(x_1, \ldots, x_n)$; hence $\omega(x_1, \ldots, x_n) \in L$. By the hypothesis, $L \in \mathcal{C}$. Also, every λ_i is a homomorphism, and $\lambda : L \longrightarrow D$ is a cone in $\mathcal{C}$. Then λ is a limit cone in $\mathcal{C}$: if $\varphi = (\varphi_i)_{i \in \mathcal{G}} : A \longrightarrow D$ is another cone in $\mathcal{C}$, then φ is a cone in *Sets*, there is a unique mapping $\overline{\varphi} : A \longrightarrow L$ such that $\varphi_i = \lambda_i \circ \overline{\varphi}$ for all i, namely $\overline{\varphi}(x) = \left(\varphi_i(x)\right)_{i \in \mathcal{G}}$; $\overline{\varphi}$ is a homomorphism since the operations on L are componentwise, and is the only homomorphism such that $\varphi_i = \lambda_i \circ \overline{\varphi}$ for all i. $\square$

Proposition 10.1 and the adjoint functor theorem yield a better proof of Theorem XV.3.3:

Theorem 10.2. *Let $\mathcal{C}$ be a class of universal algebras of the same type, that is closed under isomorphisms, direct products, and subalgebras (for instance, a variety). The forgetful functor from $\mathcal{C}$ to Sets has a left adjoint. Hence there is for every set X a universal algebra that is free on X in the class $\mathcal{C}$; in fact, a universal algebra that is free on X in the class $\mathcal{C}$ can be assigned to every set X.*

Proof. We show that the forgetful functor $G : \mathcal{C} \longrightarrow Sets$ has a left adjoint F; then F assigns to each set X an algebra $F_X \in \mathcal{C}$ that is free on X in the class $\mathcal{C}$: for every mapping f of X into a universal algebra $A \in \mathcal{C}$, there exists a unique homomorphism $\varphi : F_X \longrightarrow A$ such that $\varphi \circ \eta = f$.

By 10.1, $\mathcal{C}$ is a locally small category in which a limit can be assigned to every diagram, and G preserves limits. This leaves the solution set condition:

to every set X can be assigned a set $\mathcal{S}$ of mappings $s : X \longrightarrow C_s$ of X into algebras $C_s \in \mathcal{C}$ such that every mapping f of X into an algebra $C \in \mathcal{C}$ is a composition $f = \gamma \circ s$ for some $s \in \mathcal{S}$ and $\gamma : C_s \longrightarrow C$.

Let $\mathcal{S}$ be the set of all mappings $s_{\mathcal{E}} : X \longrightarrow W_X^T \longrightarrow W_X^T/\mathcal{E}$, where $\mathcal{E}$ is a congruence on the word algebra W_X^T and $W_X^T/\mathcal{E} \in \mathcal{C}$. Every mapping f of X into an algebra $C \in \mathcal{C}$ extends to a homomorphism $\varphi : W_X^T \longrightarrow C$; then $\mathcal{E} = \ker \varphi$ is a congruence on W_X^T, and $W_X^T/\mathcal{E} \in \mathcal{C}$, since $W_X^T/\mathcal{E} \cong \operatorname{Im} \varphi \subseteq C \in \mathcal{C}$. Since $\ker \varphi = \mathcal{E}$, φ factors through the projection $\pi : W_X^T \longrightarrow W_X^T/\mathcal{E}$, $\varphi = \psi \circ \pi$ for some homomorphism $\psi : W_X^T/\mathcal{E} \longrightarrow C$:

$$X \xrightarrow{\subseteq} W_X^T \xrightarrow{\pi_{\mathcal{E}}} W_X^T/\mathcal{E}$$

$$f \searrow \quad \downarrow \varphi \quad \swarrow \psi$$

$$C$$

Then $f = \psi \circ s_{\mathcal{E}}$. Thus $\mathcal{S}$ is a solution set. $\square$

Proposition 10.3. *Every variety $\mathcal{V}$ is cocomplete; in fact, a colimit can be assigned to every diagram in $\mathcal{V}$.*

Proof. Let $\mathcal{V}$ be a variety of type T and let D be a diagram in $\mathcal{V}$ over a graph $\mathcal{G}$. Let $X = \bigcup_{i \in \mathcal{G}} (D_i \times \{i\})$ be the disjoint union of the underlying sets D_i. By 10.2 an algebra $F \in \mathcal{V}$ that is free on X can be assigned to X. Composing $\eta : X \longrightarrow F$ and the inclusion $\iota_i : D_i \longrightarrow X$ yields a mapping $\eta \circ \iota_i : D_i \longrightarrow F$. Since every intersection of congruences on F is a congruence on F, there is a least congruence $\mathcal{E}$ on F such that:

(1) $\eta\big(\iota_i\,(\omega\,(x_1,\ \ldots,\ x_n))\big) \; \mathcal{E} \; \omega\big(\eta(\iota_i(x_1)),\ \ldots,\ \eta(\iota_i(x_n))\big)$ for all $i \in \mathcal{G}$, $n \geqq 0$, $\omega \in T$ of arity n, and $x_1,\ \ldots,\ x_n \in D_i$; and

(2) $\eta\big(\iota_i(x)\big) \; \mathcal{E} \; \eta\big(\iota_j(D_a(x))\big)$ for all $a : i \longrightarrow j$ and $x \in D_i$.

Let $L = F/\mathcal{E}$ and let $\pi : F \longrightarrow L$ be the projection. We show that L is a colimit of D, with colimit cone $\lambda = (\lambda_i)_{i \in \mathcal{G}}$, $\lambda_i = \pi \circ \eta \circ \iota_i$.

First, λ is a cone from D to L in $\mathcal{V}$: $L = F/\mathcal{E} \in \mathcal{V}$ since $\mathcal{V}$ is a variety; every λ_i is a homomorphism, since π is a homomorphism and (1) holds; when $a : i \longrightarrow j$, then $\lambda_i = \lambda_j \circ D_a$, since (2) holds.

Let $\varphi : D \longrightarrow A$ be a cone in $\mathcal{V}$. The mappings $\varphi_i : D_i \longrightarrow A$ induce a mapping $g : X \longrightarrow A$ such that $g \circ \iota_i = \varphi_i$ for all i, and a homomorphism $\psi : F \longrightarrow A$ such that $\psi \circ \eta = g$:

$$
\begin{array}{ccccc}
X & \xrightarrow{\eta} & F & \xrightarrow{\pi} & L \\
\iota_i \uparrow & {\scriptstyle g} \searrow & \downarrow {\scriptstyle \psi} & \swarrow {\scriptstyle \overline{\varphi}} & \\
D_i & \xrightarrow{\varphi_i} & A & &
\end{array}
$$

Then $\psi \circ \eta \circ \iota_i = \varphi_i$ for all i, and ker ψ satisfies (1) and (2): for all $i \in \mathcal{G}$, $\omega \in T$ of arity n, and $x_1, \ldots, x_n \in D_i$,

$$
\begin{aligned}
\psi\big(\eta\big(\iota_i\big(\omega\,(x_1, \ldots, x_n)\big)\big)\big) &= \omega\big(\psi\big(\eta(\iota_i(x_1))\big), \ldots, \psi\big(\eta(\iota_i(x_n))\big)\big) \\
&= \psi\big(\omega\big(\eta(\iota_i(x_1)), \ldots, \eta(\iota_i(x_n))\big)\big)
\end{aligned}
$$

since φ_i and ψ are homomorphisms; for all $a : i \longrightarrow j$ and $x \in D_i$, $\psi\big(\eta(\iota_i(x))\big) = \psi\big(\eta(\iota_j(D_a(x)))\big)$, since $\varphi_i = \varphi_j \circ D_a$. Therefore ker $\psi \supseteq \mathcal{E} = $ ker π and ψ factors through π: $\psi = \overline{\varphi} \circ \pi$ for some homomorphism $\overline{\varphi} : L \longrightarrow A$. Then $\overline{\varphi} \circ \lambda_i = \varphi_i$ for all i; $\overline{\varphi}$ is unique with this property, since F is generated by $\eta(X)$ by XV.3.2, so that L is generated by $\bigcup_{i \in \mathcal{G}} \lambda_i (D_i)$. $\square$

Theorem 10.4. *For every variety $\mathcal{V}$, the forgetful functor from $\mathcal{V}$ to Sets is tripleable.*

Proof. We invoke Beck's theorem. Let $\mathcal{V}$ be a variety of type T. The forgetful functor from $\mathcal{V}$ to *Sets* has a left adjoint by 10.2; we show that it creates coequalizers of pairs $\alpha, \beta : A \longrightarrow B$ of homomorphisms of algebras $A, B \in \mathcal{V}$, which have a split coequalizer $\sigma : B \longrightarrow C$ in *Sets*.

Let $\alpha, \beta : A \longrightarrow B$ be homomorphisms of algebras $A, B \in \mathcal{V}$ that have a split coequalizer $\sigma : B \longrightarrow C$ in *Sets*. Let $\omega \in T$ have arity n. Since α, β are homomorphisms, $\alpha \circ \omega_A = \omega_B \circ \alpha^n$ and $\beta \circ \omega_B = \omega_B \circ \beta^n$; hence $\sigma \circ \omega_B \circ \alpha^n = \sigma \circ \omega_B \circ \beta^n$. By 9.1, σ^n is a split coequalizer of α^n and β^n; hence there is a unique operation $\omega_C : C^n \longrightarrow C$ such that $\sigma \circ \omega_B = \omega_C \circ \sigma^n$:

$$
\begin{array}{ccccc}
A^n & \underset{\beta^n}{\overset{\alpha^n}{\rightrightarrows}} & B^n & \xrightarrow{\sigma^n} & C^n \\
\omega_A \downarrow & & \downarrow \omega_B & & \downarrow \omega_C \\
A & \underset{\beta}{\overset{\alpha}{\rightrightarrows}} & B & \xrightarrow{\sigma} & C
\end{array}
$$

In this way C becomes an algebra of type T, and σ becomes a homomorphism. Then $C \in \mathcal{V}$, since σ is surjective, and C is the only algebra on the set C such that σ is a homomorphism.

It remains to show that σ is a coequalizer in $\mathcal{V}$. Let $X \in \mathcal{V}$ and let $\varphi : B \longrightarrow X$ be a homomorphism such that $\varphi \circ \alpha = \varphi \circ \beta$. Since σ is a coequalizer in *Sets* there is a unique mapping $\psi : C \longrightarrow X$ such that $\varphi = \psi \circ \sigma$. If $\omega \in T$ has arity n and $x_1, \ldots, x_n \in B$, then

$$\psi\big(\omega\left(\sigma(x_1), \ldots, \sigma(x_n)\right)\big) = \psi\big(\sigma\left(\omega\left(x_1, \ldots, x_n\right)\right)\big)$$
$$= \omega\left(\psi(\sigma(x_1)), \ldots, \psi(\sigma(x_n))\right),$$

since σ and φ are homomorphisms; hence ψ is a homomorphism. $\square$

Exercises

1. Let $\mathcal{C}$ be a class of universal algebras of type T that is closed under products and subalgebras (for instance, a variety). Show that the forgetful functor from $\mathcal{C}$ to *Sets* creates limits.

2. Let $\mathcal{V}$ be a variety. Show that $\mathcal{V}$ has direct limits; in fact, a direct limit can be assigned to every direct system in $\mathcal{V}$, and the forgetful functor from $\mathcal{V}$ to *Sets* preserves direct limits.

3. Let $\mathcal{V}$ be a variety. Show that the forgetful functor from $\mathcal{V}$ to *Sets* creates direct limits.

4. Give a direct proof that the forgetful functor from lattices to sets is tripleable.

5. Show that Boolean lattices constitute a variety (of algebras with two binary operations, two constant operations 0 and 1, and one unary complement operation). Describe the free Boolean lattice on finite set X: describe its elements and operations, and prove that your guess is correct.

6. Show that distributive lattices constitute a variety (of algebras with two binary operations). Describe the free distributive lattice on a three element set $\{a, b, c\}$: list its elements, draw a diagram, and prove that your guess is correct. (Hint: it has 18 elements.)

*7. Show that modular lattices constitute a variety (of algebras with two binary operations). Describe the free modular lattice on a three element set $\{a, b, c\}$: list its elements, draw a diagram, and prove that your guess is correct. (Hint: it has 28 elements.)

A
Appendix

This appendix collects various properties used throughout the text: several formulations of the ascending and descending chain conditions; several formulations of the axiom of choice; basic properties of ordinal and cardinal numbers.

1. Chain Conditions

The conditions in question are two useful finiteness conditions: the ascending chain condition and the descending chain condition.

The ascending chain condition is a property of some partially ordered sets. Recall that a *partially ordered set*, $(X, \leqq)$ or just X, is an ordered pair of a set X and a binary relation $\leqq$ on X, the *partial order relation* on X, that is reflexive ($x \leqq x$), transitive ($x \leqq y$, $y \leqq z$ implies $x \leqq z$), and antisymmetric ($x \leqq y$, $y \leqq x$ implies $x = y$). (A *total* order relation also has $x \leqq y$ or $y \leqq x$, for every $x, y \in X$; then X is *totally ordered*.)

Proposition **1.1.** *For a partially ordered set X the following conditions are equivalent:*

(1) *every infinite ascending sequence $x_1 \leqq x_2 \leqq \cdots \leqq x_n \leqq x_{n+1} \leqq \cdots$ of elements of X terminates (is eventually stationary): there exists $N > 0$ such that $x_n = x_N$ for all $n \geqq N$;*

(2) *there is no infinite strictly ascending sequence $x_1 < x_2 < \cdots < x_n < x_{n+1} < \cdots$ of elements of X;*

(3) *every nonempty subset S of X has a* maximal *element (an element s of S such that there is no $s < x \in S$).*

Proof. (1) implies (2), since a strictly ascending infinite sequence cannot terminate.

(2) implies (3). If the nonempty set S in (c) has no maximal element, then one can choose $x_1 \in S$; since x_1 is not maximal in S one can choose $x_1 < x_2 \in S$; since x_2 is not maximal in S one can choose $x_1 < x_2 < x_3 \in S$; this continues indefinitely and, before you know it, you are saddled with an infinite strictly ascending sequence. (This argument implicitly uses the axiom of choice.)

(3) implies (1): some x_N must be maximal in the sequence $x_1 \leqq x_2 \leqq \cdots$ (actually, in the set $\{\, x_n \mid n > 0 \,\}$), and then $x_N \leqq x_n$ implies $x_N = x_n$ when $n \geqq N$, since $x_N < x_n$ is impossible. $\square$

A *chain* of a partially ordered set $(X, \leqq)$ is a subset of X that is totally ordered by $\leqq$. The infinite ascending sequences in (1) and (2) are traditionally called chains; this has been known to lure unwary readers into a deranged belief that all chains are ascending sequences.

Definition. The ascending chain condition *or* a.c.c. *is condition* (2) *in Proposition* 1.1.

The a.c.c. is a *finiteness condition*, meaning that it holds in every finite partially ordered set. Partially ordered sets that satisfy the a.c.c. are sometimes called *Noetherian*; this terminology is more often applied to bidules such as rings and modules whose ideals or submodules satisfy the a.c.c., when partially ordered by inclusion. In these cases the a.c.c. usually holds if and only if the subbidules are finitely generated. We prove this in the case of groups.

Proposition 1.2. *The subgroups of a group G satisfy the ascending chain condition if and only if every subgroup of G is finitely generated.*

Proof. Assume that every subgroup of G is finitely generated, and let $H_1 \subseteq H_2 \subseteq \cdots \subseteq H_n \subseteq H_{n+1} \subseteq \cdots$ be an infinite ascending sequence of subgroups of G. The union $H = \bigcup_{n>0} H_n$ is a subgroup, by I.3.9, and is generated by finitely many elements $x_1, \ldots, x_k$ of H. Then every x_i belongs to some H_{n_i}. If $N \geqq \max(n_1, \ldots, n_k)$, then H_N contains every x_i, $H \subseteq H_N$, and $H_n = H_N$ for all $n \geqq N$, since $H_N \subseteq H_n \subseteq H \subseteq H_N$. Thus the subgroups of G satisfy the ascending chain condition.

Conversely, assume that the subgroups of G satisfy the a.c.c. Let H be a subgroup of G. The set $\mathcal{S}$ of finitely generated subgroups of H is not empty, since, for instance, $\{1\} \in \mathcal{S}$. Therefore $\mathcal{S}$ has a maximal element M, which is generated by some $x_1, \ldots, x_k \in M$. For every $h \in H$, the subgroup K of H generated by $x_1, \ldots, x_k$ and h is finitely generated and contains M; hence $K = M$ and $h \in M$. Thus $H = M$, and H is finitely generated. $\square$

Noetherian induction uses the a.c.c. to produce maximal elements, as in this last proof. Proposition 1.2 could be proved by ordinary induction: if H is not finitely generated, then H has a finitely generated subgroup $H_1 \subsetneqq H$; adding a generator $h \in H \backslash H_1$ to the generators of H_1 yields a finitely generated subgroup $H_1 \subsetneqq H_2 \subsetneqq H$, since H is not finitely generated; continuing thus contradicts the a.c.c. We recognize the proof that (2) implies (3) in Proposition 1.1. Noetherian induction is more elegant but not essentially different.

The descending chain condition is also a property of some partially ordered sets, which the next result shows is closely related to the a.c.c.

Proposition 1.3. *If $\leqq$ is a partial order relation on a set X, then so is the opposite relation, $x \leqq^{\mathrm{op}} y$ if and only if $y \leqq x$.*

We omit the proof, to avoid insulting our readers.

Definition. If $X = (X, \leqq)$ is a partially ordered set, then $X^{\mathrm{op}} = (X, \leqq^{\mathrm{op}})$ is its opposite *partially ordered set.*

By Proposition 1.3, a theorem that holds in *every* partially ordered set X also holds in its opposite, and remains true when all inequalities are reversed. Thus Proposition 1.1 yields:

Proposition **1.4.** *For a partially ordered set X the following conditions are equivalent:*

(1) *every infinite descending sequence $x_1 \geqq x_2 \geqq \cdots \geqq x_n \geqq x_{n+1} \geqq \cdots$ of elements of X terminates: there exists $N > 0$ such that $x_n = x_N$ for all $n \geqq N$;*

(2) *there is no infinite strictly descending sequence $x_1 > x_2 > \cdots > x_n > x_{n+1} > \cdots$ of elements of X;*

(3) *every nonempty subset S of X has a* minimal *element (an element s of S such that there is no $s > x \in S$).*

Definition. The descending chain condition *or* d.c.c. *is condition* (2) *in Proposition* 1.4.

Like the a.c.c., the d.c.c. is a finiteness condition. Partially ordered sets that satisfy the d.c.c. are sometimes called *Artinian*; this terminology is more often applied to bidules such as rings and modules whose ideals or submodules satisfy the d.c.c., when partially ordered by inclusion.

Artinian induction uses the d.c.c. to produce minimal elements. This includes *strong induction* on a natural number n, in which the induction hypothesis is that the desired result holds for all smaller values of n. This works because the natural numbers satisfy the d.c.c.: if the desired result was false for some n, then it would be false for some minimal n, and true for all smaller values of n, which is precisely the situation ruled out by strong induction.

Exercises

1. Show that the subgroups of the additive group $\mathbb{Z}$ do not satisfy the d.c.c.

2. Show that the a.c.c. does not imply the d.c.c., and that the d.c.c. does not imply the a.c.c. (Hence neither condition implies finiteness.)

3. Prove the following: when a partially ordered set X satisfies the a.c.c. and the d.c.c., then every chain of elements of X is finite.

4. Construct a partially ordered set X that satisfies the a.c.c. and the d.c.c., and contains a finite chain with n elements for every positive integer n.

5. Show that a partially ordered set X satisfies the a.c.c. if and only if every nonempty chain C of X has a *greatest* element (an element m of C such that $x \leqq m$ for all $x \in C$).

6. Greatest elements are sometimes inaccurately called *unique maximal* elements. Construct a partially ordered set X with just one maximal element and no greatest element.

7. Let the partially ordered set X satisfy the d.c.c. You have just devised a proof that if a certain property of elements of X is true for every $y < x$ in X, then it is true for x (where $x \in X$ is arbitrary). Can you conclude that your property is true for all $x \in X$?

2. The Axiom of Choice

This section contains the axiom of choice (first formulated by Zermelo [1904]) and some of its useful consequences, including the most useful, Zorn's lemma.

Axiom of Choice: *Every set has a choice function.*

A *choice function* on a set S is a mapping c that assigns to every nonempty subset T of S an element $c(T)$ of T. (Thus c chooses one element $c(T)$ in each nonempty T.) Though less "intuitively obvious" than other axioms, the axiom of choice became one of the generally accepted axioms of set theory after Gödel [1938] proved that it is consistent with the other generally accepted axioms, and may therefore be assumed without generating contradictions.

In this section we give a number of useful statements that are equivalent to the axiom of choice (assuming the other axioms of set theory).

Proposition **2.1.** *The axiom of choice is equivalent to the following statement: when I is a nonempty set, and $(S_i)_{i \in I}$ is a family of nonempty sets, then $\prod_{i \in I} S_i$ is nonempty.*

Proof. Recall that $\prod_{i \in I} S_i$ is the set of all mappings (usually written as families) that assign to each $i \in I$ some element of S_i. If $I \neq \emptyset$ and $S_i \neq \emptyset$ for all i, and the axiom of choice holds, then $\bigcup_{i \in I} S_i$ has a choice function c, and then $(c(S_i))_{i \in I} \in \prod_{i \in I} S_i$, so that $\prod_{i \in I} S_i \neq \emptyset$.

Conversely, assume that $\prod_{i \in I} S_i$ is nonempty whenever I is nonempty and $(S_i)_{i \in I}$ is a family of nonempty sets. Let S be any set. If $S = \emptyset$, then the empty mapping is a choice function on S. If $S \neq \emptyset$, then so is $\prod_{T \subseteq S, \, T \neq \emptyset} T$; an element of $\prod_{T \subseteq S, \, T \neq \emptyset} T$ is precisely a choice function on S. $\square$

Zorn's lemma is due to Zorn [1935], though Hausdorff [1914] and Kuratowski [1922] had published closely related statements. Recall that a *chain* of a partially ordered set X is a subset C of X such that at least one of the statements $x \leq y$, $y \leq x$ holds for every $x, y \in C$. An *upper bound* of C in X is an element b of X such that $x \leq b$ for all $x \in C$.

Zorn's Lemma: *when X is a nonempty partially ordered set, and every nonempty chain of X has an upper bound in X, then X has a maximal element.*

Theorem **2.2.** *The axiom of choice is equivalent to Zorn's lemma.*

We defer the proof. That Zorn's lemma implies the axiom of choice is shown later in this section, with Theorem 2.4. The converse is proved in Section 4.

Zorn's lemma provides a method of proof, *transfinite induction,* which is similar to integer induction and to Noetherian induction but is much more powerful. For instance, suppose that we want to prove that some nonempty partially ordered set X has a maximal element. Using ordinary induction we could argue as follows. Since X is not empty there exists $x_1 \in X$. If x_1 is not maximal, then $x_1 < x_2$ for some $x_2 \in X$. If x_2 is not maximal, then $x_2 < x_3$ for some $x_3 \in X$. Continuing in this fashion yields a strictly ascending sequence, which is sure to reach a maximal element only if X satisfies the ascending chain condition (equivalently, if every nonempty chain of X has a greatest element). Zorn's lemma yields a maximal element under the much weaker hypothesis that every nonempty chain of X has an upper bound. The proof in Section 4 reaches a maximal element by constructing a strictly ascending sequence that is indexed by ordinal numbers and can be as infinitely long as necessary.

Previous chapters contain numerous applications of Zorn's lemma. The author feels that this section should contain one; the exercises give more. Recall that a *cross section* of an equivalence relation on a set X is a subset S of X such that every equivalence class contains exactly one element of X.

Corollary **2.3.** *Every equivalence relation has a cross section.*

Proof. Let X be a set with an equivalence relation. Let $\mathcal{S}$ be the set of all subsets S of X such that every equivalence class contains at most one element of S. Then $\mathcal{S} \neq \emptyset$, since $\emptyset \in \mathcal{S}$. Partially order $\mathcal{S}$ by inclusion. We show that $S = \bigcup_{i \in I} S_i \in \mathcal{S}$ when $(S_i)_{i \in I}$ is a chain of elements of $\mathcal{S}$. If $x, y \in S$, then $x \in S_i$ and $y \in S_j$ for some $i, j \in I$, with $S_i \subseteq S_j$ or $S_j \subseteq S_i$, since $(S_i)_{i \in I}$ is a chain, so that, say, $x, y \in S_j$. If x and y are equivalent, then $x = y$, since $S_j \in \mathcal{S}$. Thus $S \in \mathcal{S}$: every chain of $\mathcal{S}$ has an upper bound in $\mathcal{S}$.

By Zorn's lemma, $\mathcal{S}$ has a maximal element S. Then every equivalence class C contains an element of S: otherwise, $S \cup \{c\} \in \mathcal{S}$ for any $c \in C$, in defiance of the maximality of S. Hence S is a cross section. $\square$

Readers can also derive Corollary 2.3 directly from the axiom of choice.

Well ordered sets. A *well ordered* set is a partially ordered set X in which every nonempty subset S has a *least* element (an element s of S such that $s \leq x$ for every $x \in S$).

For example, $\mathbb{N}$ is well ordered. A well ordered set is totally ordered (since every subset $\{x, y\}$ must have a least element) and satisfies the descending chain condition (since a least element of S is, in particular, a minimal element of S).

Theorem **2.4** (Zermelo [1904]). *The axiom of choice is equivalent to the* well-ordering principle: *every set can be well ordered.*

Proof. A well ordered set S has a choice function, which assigns to each nonempty subset of S its least element. Hence the well-ordering principle implies the axiom of choice. We show that Zorn's lemma implies the well-ordering principle (hence implies the axiom of choice). That the axiom of choice implies Zorn's lemma is proved in Section 4.

Given a set S, let $\mathcal{W}$ be the set of all ordered pairs $(X, \leqq_X)$ such that $X \subseteq S$ and X is well ordered by $\leqq_X$. Then $\mathcal{W} \neq \varnothing$, since $\varnothing \subseteq S$ is well ordered by the empty order relation. Let $(X, \leqq_X) \leqq (Y, \leqq_Y)$ in $\mathcal{W}$ when

(a) $X \subseteq Y$;

(b) when $x', x'' \in X$, then $x' \leqq_X x''$ if and only if $x' \leqq_Y x''$; and

(c) if $y \in Y$ and $y \leqq_Y x \in X$, then $y \in X$;

equivalently, when $(X, \leqq_X)$ is the lower part of $(Y, \leqq_Y)$ with the induced order relation. It is immediate that this defines an order relation on $\mathcal{W}$.

Let $(X_i, \leqq_i)_{i \in I}$ be a chain of $\mathcal{W}$. If $x, y \in X = \bigcup_{i \in I} X_i$, then let $x \leqq_X y$ if and only if $x, y \in X_i$ and $x \leqq_i y$, for some $i \in I$. If also $x, y \in X_j$ and $x \leqq_j y$ for some $j \in I$, then, say, $(X_i, \leqq_i) \leqq (X_j, \leqq_j)$, and $x \leqq_i y$ if and only if $x \leqq_j y$ by (b). Similarly, if $x \leqq_X y$ and $y \leqq_X x$, then $x, y \in X_i$, $x \leqq_i y$ and $x, y \in X_j$, $y \leqq_j x$ for some $i, j \in I$; if, say, $(X_i, \leqq_i) \leqq (X_j, \leqq_j)$, then $x \leqq_j y$ and $y \leqq_j x$, whence $x = y$. Thus $\leqq_X$ is antisymmetric. Similar arguments show that $\leqq_X$ is reflexive and transitive.

Let T be a nonempty subset of X. Then $T \cap X_i \neq \varnothing$ for some i, and $T \cap X_i$ has a least element t under $\leqq_i$. In fact, t is the least element of T. Indeed, let $u \in T$, $u \in X_j$ for some j. If $(X_i, \leqq_i) \leqq (X_j, \leqq_j)$, then $t \leqq_X u$, since $u <_j t \in X_i$ would imply $u \in X_i$ by (c) and $u <_i t$ by (b), so that t would not be the least element of $T \cap X_i$. If $(X_j, \leqq_j) \leqq (X_i, \leqq_i)$, then $u \in X_i$ and $t \leqq_X u$. Thus X is well ordered by $\leqq_X$, and $(X, \leqq_X) \in \mathcal{W}$.

If $x, y \in X_i$, then we saw at the beginning of the proof that $x \leqq_X y$ if and only if $x \leqq_i y$. Let $y \in X$ and $y \leqq_X x \in X_i$. Then $x, y \in X_j$ and $y \leqq_j x$ for some j. If $(X_i, \leqq_i) \leqq (X_j, \leqq_j)$, then $y \in X_i$ by (c). If $(X_j, \leqq_j) \leqq (X_i, \leqq_i)$, then again $y \in X_i$. Thus $(X_i, \leqq_i) \leqq (X, \leqq_X)$ for all i, and $(X, \leqq_X)$ is an upper bound of $(X_i, \leqq_i)_{i \in I}$ in $\mathcal{W}$.

At this point we invoke Zorn's lemma and are rewarded with a maximal element $(M, \leqq_M)$ of $\mathcal{W}$. We show that $M = S$. Suppose that $(X, \leqq_X) \in \mathcal{W}$ and $X \subsetneqq S$. Let $s \in S \backslash X$. Extend $\leqq_X$ to $Y = X \cup \{s\}$ so that s is the greatest element of Y. Then Y is well ordered: when $T \subseteq Y$, $T \neq \varnothing$, then s is the least element of T if $T = \{s\}$; otherwise, the least element of $T \cap X$ is also the least element of T. Hence $(X, \leqq_X) < (Y, \leqq_Y)$ and $(X, \leqq_X)$ is not maximal. Therefore $M = S$, and then S is well ordered by $\leqq_M$. $\square$

Exercises

1. Prove that every equivalence relation on a set has a cross section, using the axiom of choice but not Zorn's lemma.

2. The *domain* of a binary relation R is the set $\{\, x \mid (x, y) \in R \text{ for some } y \,\}$. Show that the axiom of choice is equivalent to the following statement: every binary relation contains a mapping that has the same domain.

3. Show that a partially ordered set is well ordered if and only if it is totally ordered and satisfies the descending chain condition.

4. Let G be a group and let $a \in G$, $a \neq 1$. Use Zorn's lemma to prove that there is a subgroup M of G that is maximal such that $a \notin M$ (that is, $a \notin M$, and $M < H \leqq G$ implies $a \in H$).

5. Let G be a group and let A be a subgroup of G. Use Zorn's lemma to prove that there is a subgroup M of G that is maximal such that $M \cap A = 1$ (that is, $M \cap A = 1$, and $M < H \leqq G$ implies $H \cap A \neq 1$).

6. Use Zorn's lemma to prove that every vector space has a maximal linearly independent subset; then show that the latter is a basis.

7. Use Zorn's lemma to prove that every order relation is an intersection of total order relations.

3. Ordinal Numbers

This section contains basic general properties of ordinal numbers.

Definition. Ordinal numbers are most naturally defined as isomorphy classes of well ordered sets. Unfortunately, isomorphy classes of well ordered sets are very large, and embarrassing contradictions arise when such large classes are collected into sets or classes. The most famous is *Russell's paradox*: Let R be the "set" of all sets X such that $X \notin X$. If $R \notin R$, then R is one of the sets X such that $X \notin X$; therefore $R \in R$. But then R is not one of the sets X such that $X \notin X$; therefore $R \notin R$.

Contradictions can be avoided if "large" collections like R are not allowed among sets, and are denied all rights and privileges enjoyed by sets: in this case, the right to belong to a set or collection. Modern ordinal numbers are well ordered sets, chosen so that there is only one in each isomorphy class (as, for instance, in Jech [1978]).

Definition. A set X is transitive *when $x \in X$ and $t \in x$ implies $t \in X$; equivalently, when every element of X is a subset of X.*

The empty set is transitive. Since X transitive implies $X \cup \{X\}$ transitive, the sets $\{\emptyset\}$, $\{\emptyset, \{\emptyset\}\}$, $\{\emptyset, \{\emptyset\}, \{\emptyset, \{\emptyset\}\}\}$, etc., are transitive.

Definition. An ordinal number is a well ordered transitive set in which $x < y$ if and only if $x \in y$.

The first ordinals are readily found. The empty set is an ordinal. A nonempty ordinal σ has a least element α, which must be empty since $t \in \alpha$ would imply $t \in \sigma$ and $t < \alpha$. If σ has no other element, then $\sigma = \{\emptyset\}$. Otherwise, there is a least $\beta > \alpha$ in σ. Then $\alpha \in \beta$; conversely, $x \in \beta$ implies $x \in \sigma$,

$\alpha \leq x < \beta$, and $x = \alpha$; hence $\beta = \{\alpha\} = \{\emptyset\}$. If σ has no other element, then $\sigma = \{\alpha, \beta\} = \{\emptyset, \{\emptyset\}\}$. Continuing this process yields the first ordinals, which are generally identified with nonnegative integers:

$$0 = \emptyset, \quad 1 = \{\emptyset\}, \quad 2 = \{\emptyset, \{\emptyset\}\}, \quad 3 = \{\emptyset, \{\emptyset\}, \{\emptyset, \{\emptyset\}\}\}, \quad \text{etc.}$$

We see that $1 = \{0\}$, $2 = \{0, 1\}$, $3 = \{0, 1, 2\}$, etc.

Readers may prove the following:

Proposition 3.1. *Every element of an ordinal number is an ordinal number.*

Ordering. Ordinal numbers are ordered by inclusion:

Proposition 3.2. *If α and β are ordinal numbers, then $\alpha \in \beta$ if and only $\alpha \subsetneqq \beta$. Hence the class Ord of all ordinal numbers is totally ordered, with $\alpha < \beta$ if and only if $\alpha \in \beta$.*

Proof. Since β is transitive, $\alpha \in \beta$ implies $\alpha \subseteq \beta$; moreover, $\alpha \notin \alpha$ (otherwise, $\alpha < \alpha$ in β), whence $\alpha \subsetneqq \beta$. Conversely, assume $\alpha \subsetneqq \beta$. Then $\beta \backslash \alpha$ has a least element γ. If $x \in \gamma$, then $x \in \alpha$: otherwise, γ would not be least. Conversely, $x \in \alpha$ implies $x \neq \gamma$, since $\gamma \notin \alpha$, and $\gamma \notin x$, otherwise, $\gamma \in \alpha$; in the totally ordered set β this implies $x \in \gamma$. Thus $\alpha = \gamma \in \beta$.

Now, let α and β be any ordinal numbers. Then $\delta = \alpha \cap \beta$ is transitive, since α and β are transitive, and is well ordered (as a subset of α) with $x < y$ in δ if and only if $x \in y$. In other words, δ is an ordinal. If $\delta \neq \alpha, \beta$, then $\delta \subsetneqq \alpha, \beta$, $\delta \in \alpha, \beta$ by the above, and $\delta \in \delta$, a contradiction. Therefore $\delta = \alpha$ or $\delta = \beta$; hence $\alpha \subseteq \beta$ or $\beta \subseteq \alpha$. Thus *Ord* is totally ordered by inclusion. $\square$

Propositions 3.1 and 3.2 imply $\alpha = \{\beta \in Ord \mid \beta < \alpha\}$ for every ordinal α.

Proposition 3.3. *Every nonempty class of ordinal numbers has a least element.*

Proof. Let $\mathcal{C}$ be a nonempty class of ordinals. Let $\alpha \in \mathcal{C}$. We may assume that α is not the least element of $\mathcal{C}$. Then $\mathcal{C} \cap \alpha \neq \emptyset$ and $\mathcal{C} \cap \alpha \subseteq \alpha$ has a least element γ. In fact, γ is the least element of $\mathcal{C}$: if $\beta \in \mathcal{C}$, then, by 3.2, either $\gamma < \alpha \leq \beta$, or $\beta \in \alpha \cap \mathcal{C}$ and $\beta \geq \gamma$. $\square$

Thus *Ord* is a well ordered class.

Proposition 3.4. *The union of a set of ordinal numbers is an ordinal number.*

Proof. Let S be a set of ordinal numbers. Then $\upsilon = \bigcup_{\sigma \in S} \sigma$ is a set. By 3.2, S is a chain, and any two elements of S are elements of some $\sigma \in S$. It follows that υ is transitive, and is totally ordered, with $x < y$ in υ if and only if $x < y$ in some $\sigma \in S$, if and only if $x \in y$.

Let T be a nonempty subset of υ. Then $T \cap \sigma \neq \emptyset$ for some $\sigma \in S$, and $T \cap \sigma$ has a least element γ. As in the proof of 3.3, γ is the least element of T: if $\tau \in T$, then either $\gamma < \sigma \leq \tau$, or $\tau \in \sigma \cap T$ and $\tau \geq \gamma$. $\square$

Corollary 3.5. *The class Ord of all ordinal numbers is not a set.*

Proof. Let α be an ordinal. The *successor* $\beta = \alpha \cup \{\alpha\}$ of α is also an ordinal, as readers will verify. The result follows from this and 3.4. If *Ord* were a set, then $\bigcup_{\alpha \in Ord} \alpha$ would be an ordinal number, and would be the greatest ordinal number, in particular, would be greater than his successor, a feat easily achieved by King Louis XIV of France but not possible for ordinal numbers. $\square$

Well ordered sets. Now, we show that there is one ordinal number in every isomorphy class of well ordered sets. First, a *lower section* (or *order ideal*) of a partially ordered set X is a subset S of X such that $x \leqq s \in S$ implies $x \in S$. Then X is a lower section of itself; for every $a \in X$ there is a lower section $X(a) = \{ x \in X \mid x < a \}$.

Lemma **3.6.** *A subset S of a well ordered set X is a lower section of X if and only if either $S = X$ or $S = X(a)$ for some $a \in X$.*

Proof. Let $S \neq X$ be a lower section. Then $X \backslash S$ has a least element a. If $x < a$, then $x \in S$: otherwise, a would not be the least element of $X \backslash S$. If $x \in S$, then $x < a$: otherwise, $a \leqq x \in S$ and $a \in S$. Thus $S = X(a)$. $\square$

In general, an *isomorphism* of a partially ordered set X onto a partially ordered set Y is a bijection $\theta : X \longrightarrow Y$ such that $x' \leqq x''$ in X if and only if $\theta(x') \leqq \theta(x'')$ in Y. If X and Y are totally ordered, one needs only the implication "$x' \leqq x''$ implies $\theta(x') \leqq \theta(x'')$": then $\theta(x') \leqq \theta(x'')$ implies $x' \leqq x''$, since $x' > x''$ would imply $\theta(x') > \theta(x'')$.

Lemma **3.7.** *Let S and T be lower sections of a well ordered set X. If $S \cong T$, then $S = T$.*

Proof. Let $S \neq T$ and let $\theta : S \longrightarrow T$ be an isomorphism. By 3.6, $S \subseteq T$ or $T \subseteq S$, and we may exchange S and T if necessary and assume that $S \nsubseteq T$. Then we cannot have $\theta(x) = x$ for all $x \in S$, and the set $\{ x \in S \mid \theta(x) \neq x \}$ has a least element a. Then $\theta(x) = x$ when $x \in S$ and $x < a$, but $\theta(a) \neq a$. If $a < \theta(a) \in T$, then $a \in T$, $a = \theta(x)$ for some $x \in S$, and $\theta(x) < \theta(a)$ implies $x < a$ and $\theta(x) = x < a$, a contradiction. Therefore $\theta(a) < a \in S$, but then $\theta(a) \in S$, $\theta(\theta(a)) = \theta(a)$, and $\theta(a) = a$, another contradiction. $\square$

Proposition **3.8.** *Every well ordered set is isomorphic to a unique ordinal number.*

Proof. Uniqueness follows from 3.7: if, say, $\alpha < \beta$ in *Ord*, then $\alpha = \{ \gamma \in \beta \mid \gamma < \alpha \}$ is a lower section of β, and $\alpha \ncong \beta$.

Now, let X be a well ordered set. Let φ be the set of all ordered pairs (a, α) such that $a \in X$, $\alpha \in Ord$, and $X(a) \cong \alpha$. Then φ is a mapping: if $(a, \alpha), (b, \beta) \in \varphi$ and $a = b$, then $\alpha \cong X(a) \cong \beta$ and $\alpha = \beta$. Similarly, φ is injective: if $(a, \alpha), (b, \beta) \in \varphi$ and $\alpha = \beta$, then $X(a) \cong X(b)$, $X(a) = X(b)$ by 3.7, and $a = b$ since a is the least element of $X \backslash X(a)$ and similarly for b.

Assume that $(a, \alpha), (b, \beta) \in \varphi$ and $a < b$. Let $\theta : X(b) \longrightarrow \beta$ be an isomorphism. Then $\theta(a) < \beta$ and θ induces an isomorphism of $X(a)$ onto

$\{\gamma \in \beta \mid \gamma < \theta(a)\} = \theta(a)$. (This argument also shows that the domain dom φ of φ is a lower section of X.) Therefore $(a, \theta(a)) \in \varphi$. Hence $\alpha = \theta(a) < \beta$.

Similarly, assume that $(a, \alpha), (b, \beta) \in \varphi$ and $\alpha < \beta$. Let $\zeta : \beta \longrightarrow X(b)$ be an isomorphism. Then $\zeta(\alpha) < b$ and ζ induces an isomorphism of α onto $X(\zeta(\alpha))$. Therefore $(\zeta(\alpha), \alpha) \in \varphi$. (This argument also shows that the range ran φ of φ is a lower section of Ord.) Hence $a = \zeta(\alpha) < b$, since φ is injective. Thus φ is an isomorphism of dom φ onto ran φ.

Since Ord is not a set, ran φ cannot be all of Ord, and there is a least ordinal $\gamma \notin \mathrm{ran}\,\varphi$. Then, as in the proof of 3.6, ran $\varphi = \{\alpha \in Ord \mid \alpha < \gamma\} = \gamma$. If dom φ is not all of X, then dom $\varphi = X(c)$ for some $c \in X$ by 3.6, φ is an isomorphism of $X(c)$ onto γ, $(c, \gamma) \in \varphi$, and $c \in \mathrm{dom}\,\varphi$, a contradiction; therefore dom $\varphi = X$ and $X \cong \gamma$. $\square$

Successor and limit ordinals. We show that all ordinals are generated by two constructions: unions from Proposition 3.4, and successors, whose definition follows.

Proposition **3.9.** *If α is an ordinal number, then so is $\alpha \cup \{\alpha\}$; in fact, $\alpha \cup \{\alpha\}$ is the least ordinal $\beta > \alpha$.*

The proof is an exercise for our avid readers.

Definition. The successor *of an ordinal number α is the ordinal number* $\alpha \cup \{\alpha\}$.

The successor $\alpha \cup \{\alpha\}$ of α is normally denoted by $\alpha + 1$. (The sum of any two ordinals is defined in the exercises.) It *covers* α: there is no ordinal $\alpha < \beta < \alpha + 1$, since there is no set $\alpha \subsetneqq S \subsetneqq \alpha \cup \{\alpha\}$.

Proposition **3.10.** *An ordinal number α is a successor if and only if it has a greatest element; then the greatest element of α is $\bigcup_{\gamma < \alpha} \gamma < \alpha$ and α is its successor. Otherwise, $\bigcup_{\gamma < \alpha} \gamma = \alpha$.*

Proof. A successor $\alpha = \beta \cup \{\beta\}$ has a greatest element β. Conversely, assume that α has a greatest element β. Then $\bigcup_{\gamma < \alpha} \gamma = \beta < \alpha$, and $\beta < \delta$ implies $\delta \geqq \alpha$, since $\delta < \alpha$ implies $\delta \leqq \beta$. Hence α is the successor of β.

The inclusion $\bigcup_{\gamma < \alpha} \gamma \subseteq \alpha$ holds for every ordinal α. If $\beta = \bigcup_{\gamma < \alpha} \gamma \subsetneqq \alpha$, then $\beta \in Ord$ by 3.4 and $\beta < \alpha$, so that β is the greatest element of α. Therefore $\bigcup_{\gamma < \alpha} \gamma = \alpha$ when α does not have a greatest element. $\square$

Definition. A limit ordinal *is an ordinal $\alpha \neq 0$ such that $\alpha = \bigcup_{\gamma < \alpha} \gamma$.*

Thus, a nonzero ordinal is either 0 or a successor or a limit ordinal (a union or "limit" of lesser ordinals). We can now form a clearer picture of Ord. An ordinal α and its successors constitute a sequence $\alpha < \alpha + 1 < \alpha + 2 < \cdots < \alpha + n < \cdots$ whose union is a limit ordinal. Thus Ord is made of sequences

$$0 < 1 < 2 < \cdots < \omega < \omega + 1 < \omega + 2 < \cdots < \omega + \omega < \omega + \omega + 1 < \cdots$$

that begin with 0 and with limit ordinals $\omega = \bigcup_{n<\omega} n$, $\omega + \omega = \bigcup_{n>0} (\omega + n)$, ... The limit ordinals themselves are arranged into similar sequences

$$\omega < \omega + \omega < \cdots < \omega^2 < \omega^2 + \omega < \cdots < \omega^2 + \omega^2 < \omega^2 + \omega^2 + \omega < \cdots$$

that begin with ω and with unions of lesser limit ordinals; these sequences extend indefinitely, with no end in sight.

Exercises

1. Prove the following: when the well ordered sets X and Y are isomorphic, then there is only one isomorphism of X onto Y.

2. Prove the following: when α is an ordinal number, then so is $\alpha \cup \{\alpha\}$; in fact, $\alpha \cup \{\alpha\}$ is the least ordinal $\beta > \alpha$.

3. Let X and Y be disjoint well ordered sets. Order $Z = X \cup Y$ so that $x \leq y$ in Z if and only if either $x \leq y$ in X, or $x \leq y$ in Y, or $x \in X$ and $y \in Y$. Show that Z is well ordered. Show that $X \cong X'$, $Y \cong Y'$ implies $Z \cong Z'$.

The *sum* of two ordinal numbers α and β is the ordinal number $\alpha + \beta \cong Z$, where Z is constructed as in the previous exercise from $X \cong \alpha$ and $Y \cong \beta$.

4. Show that ordinal addition is associative.

5. Show that $\omega + 1 \neq 1 + \omega$. ($\omega$ is the least infinite ordinal.)

6. Prove that every ordinal can be written uniquely as a sum $\alpha + n$, where α is 0 or a limit ordinal, and n is a finite ordinal.

7. Let X and Y be well ordered sets. Order $Z = X \times Y$ so that $(x', y') < (x'', y'')$ in Z if and only if either $y' < y''$, or $y' = y''$ and $x' < x''$. (Thus Z consists of $|Y|$ copies of X placed end to end.) Show that Z is well ordered. Show that $X \cong X'$, $Y \cong Y'$ implies $Z \cong Z'$.

The *product* $\alpha\beta$ of two ordinal numbers α and β is the ordinal number $\alpha\beta \cong Z$, where Z is constructed as in the previous exercise from $X \cong \alpha$ and $Y \cong \beta$.

8. Show that ordinal multiplication is associative.

9. Show that $\alpha (\beta + \gamma) = \alpha\beta + \alpha\gamma$ for all ordinals α, β, γ.

10. Show that $2\omega \neq \omega 2$.

11. Show that $(1 + 1) \omega \neq 1\omega + 1\omega$.

4. Ordinal Induction

Ordinal numbers can be used instead of integers in inductive proofs and constructions. This method of proof, known as ordinal induction, is as powerful as Zorn's lemma, and sometimes more convenient or more natural.

Ordinary induction is based on the following property of natural numbers: if S is a subset of $\mathbb{N} = \{1, 2, \ldots\}$ such that $1 \in S$ and that $n \in S$ implies $n + 1 \in S$, then $S = \mathbb{N}$. Ordinal numbers have a similar property:

Proposition **4.1.** *Let* $\mathcal{C}$ *be a class of ordinal numbers such that*

(1) $0 \in \mathcal{C}$;

(2) $\alpha \in \mathcal{C}$ *implies* $\alpha + 1 \in \mathcal{C}$;

(3) *if* α *is a limit ordinal and* $\beta \in \mathcal{C}$ *for all* $\beta < \alpha$, *then* $\alpha \in \mathcal{C}$.

Then $\mathcal{C} = Ord$.

Proof. If $\mathcal{C} \neq Ord$, then $Ord \setminus \mathcal{C}$ has a least element α, by 3.3. Then $\mathcal{C}$ contains every $\beta < \alpha$. But $\alpha \neq 0$, by (1); α is not a successor ordinal, by (2); and α is not a limit ordinal, by (3). Therefore $\mathcal{C} = Ord$. $\square$

Ordinal induction is a method of proof based on Proposition 4.1. It resembles ordinary induction, except for (3). There are some variants, which readers will easily establish. Induction can be limited to a given ordinal σ:

Proposition **4.2.** *Let* σ *be an ordinal number and let* $\mathcal{C}$ *be a class of ordinal numbers such that*

(1) $0 \in \mathcal{C}$;

(2) *if* $\alpha \in \mathcal{C}$ *and* $\alpha + 1 < \sigma$, *then* $\alpha + 1 \in \mathcal{C}$;

(3) *if* $\alpha < \sigma$ *is a limit ordinal and* $\beta \in \mathcal{C}$ *for all* $\beta < \alpha$, *then* $\alpha \in \mathcal{C}$.

Then $\mathcal{C}$ *contains every ordinal number* $\alpha < \sigma$.

There are also "strong" versions of Propositions 4.1 and 4.2, which follow from Proposition 3.3:

Proposition **4.3.** *Let* $\mathcal{C}$ *be a class of ordinal numbers such that* $\beta \in \mathcal{C}$ *for all* $\beta < \alpha$ *implies* $\alpha \in \mathcal{C}$. *Then* $\mathcal{C} = Ord$.

Let σ *be an ordinal number and let* $\mathcal{C}$ *be a class of ordinal numbers such that* $\beta \in \mathcal{C}$ *for all* $\beta < \alpha$ *implies* $\alpha \in \mathcal{C}$ *when* $\alpha < \sigma$. *Then* $\mathcal{C}$ *contains every ordinal number* $\alpha < \sigma$.

Recursion. A *transfinite sequence* is a family $(x_\alpha)_{\alpha \in Ord}$ or $(x_\alpha)_{\alpha < \sigma}$ indexed by Ord or indexed by an ordinal number σ. By the well ordering principle, the elements of every set can be arranged into a transfinite sequence: well order X, so that X is isomorphic to some ordinal σ, and let $x_\alpha = \theta(\alpha)$, where $\theta : \sigma \longrightarrow X$ is the isomorphism. On the other hand, we have:

Lemma **4.4.** *No set can contain a transfinite sequence* $(x_\alpha)_{\alpha \in Ord}$ *indexed by all ordinals, such that* $x_\alpha \neq x_\beta$ *whenever* $\alpha \neq \beta$.

Proof. In the next section we shall see that such a sequence would force the poor set to have entirely too many elements. For now we argue as follows. Let X be the subset of all x_α. Order X so that $x_\alpha < x_\beta$ if and only if $\alpha < \beta$. Then X is well ordered, by 3.3; in fact, X is isomorphic to Ord. By 3.9, X is isomorphic to an ordinal number σ. The isomorphism $Ord \cong X \cong \sigma$ sends the lower section σ of Ord to a lower section $\tau < \sigma$ of σ, contradicting 3.7. $\square$

As a rather complicated first example of ordinal induction we show that transfinite sequences can be constructed recursively.

Proposition **4.5** (Recursion). *Let S be a set and let $F : 2^S \longrightarrow S$ be a mapping, where 2^S is the set of all subsets of S. There exists a unique transfinite sequence $(x_\alpha)_{\alpha \in Ord}$ indexed by all ordinals, such that*

$$x_\alpha = F(\{ x_\gamma \mid \gamma < \alpha \}) \qquad (*)$$

holds for all $\alpha \in Ord$.

Informally we say that $(*)$ "defines x_α by induction". The exercises give more general forms of recursion.

Proof. First we show by induction on $\sigma \in Ord$ that there is at most one sequence $(x_\alpha)_{\alpha < \sigma}$ indexed by σ such that $(*)$ holds for all $\alpha < \sigma$. Assume this uniqueness for every $\tau < \sigma$. Let $(x'_\alpha)_{\alpha < \sigma}$ and $(x''_\alpha)_{\alpha < \sigma}$ satisfy $(*)$ for all $\alpha < \sigma$. If $\sigma = 0$, then $x'_\alpha = x''_\alpha$ for all $\alpha < \sigma$, vacuously. If $\sigma = \tau + 1$ is a successor ordinal, then $x'_\alpha = x''_\alpha$ for all $\alpha < \tau$ by the induction hypothesis and

$$x'_\tau = F(\{ x'_\gamma \mid \gamma < \tau \}) = F(\{ x''_\gamma \mid \gamma < \tau \}) = x''_\tau ;$$

hence $x'_\alpha = x''_\alpha$ for all $\alpha < \sigma$. If σ is a limit ordinal, then $x'_\alpha = x''_\alpha$ for every $\alpha < \tau < \sigma$ by the induction hypothesis, and for every $\alpha < \sigma = \bigcup_{\tau < \sigma} \tau$.

Next we show by induction on $\sigma \in Ord$ that there exists a sequence $(x_\alpha)_{\alpha < \sigma}$ indexed by σ such that $(*)$ holds for all $\alpha < \sigma$. Assume that such a sequence exists for all $\tau < \sigma$. The empty sequence serves if $\sigma = 0$. If $\sigma = \tau + 1$, then by the induction hypothesis there is a sequence $(x_\alpha)_{\alpha < \tau}$ indexed by τ such that $(*)$ holds for all $\alpha < \tau$; define $x_\tau = F(\{ x_\gamma \mid \gamma < \tau \})$; then $(*)$ holds for all $\alpha < \sigma$. Now, let σ be a limit ordinal. For every $\tau < \sigma$ the induction hypothesis provides a sequence $(x^\tau_\alpha)_{\alpha < \tau}$ indexed by τ such that $(*)$ holds for all $\alpha < \tau$. By the first part of the proof, $x^\tau_\alpha = x^\upsilon_\alpha$ whenever $\alpha < \tau, \upsilon < \sigma$. Since $\sigma = \bigcup_{\tau < \sigma} \tau$, a sequence $(x^\sigma_\alpha)_{\alpha < \sigma}$ indexed by σ is well defined by $x^\sigma_\alpha = x^\tau_\alpha$ whenever $\alpha < \tau < \sigma$. Then x^σ satisfies $(*)$ for all $\alpha < \sigma$.

We now have for every ordinal σ a sequence $(x^\sigma_\alpha)_{\alpha < \sigma}$ indexed by σ such that $(*)$ holds for all $\alpha < \sigma$. As above, the first part of the proof implies $x^\sigma_\alpha = x^\tau_\alpha$ whenever $\alpha < \sigma, \tau$. Hence a sequence $(x_\alpha)_{\alpha \in Ord}$ indexed by Ord is well defined by $x_\alpha = x^\sigma_\alpha$ whenever $\alpha < \sigma$; this sequence satisfies $(*)$ for all α, and is unique by the first part of the proof. $\square$

Zorn's lemma. We use recursion to show that the axiom of choice implies Zorn's lemma. This completes the proofs of Theorems 2.2 and 2.4. Readers will make sure that the author did not pull a fast one and invoked Zorn in proofs, in either this section or Section 3.

Let X be a nonempty partially ordered set in which every nonempty chain has an upper bound. Assume that X has a choice function c but no maximal element. For every subset S of X let

$$u(S) = \{ x \in X \mid s < x \text{ for all } s \in S \}$$

be the set of all (strict) upper bounds of S. Choose some $a \in X$ and let

$$F(S) = \begin{cases} c\big(u(S)\big) & \text{if } u(S) \neq \emptyset, \\ a & \text{if } u(S) = \emptyset. \end{cases}$$

By 4.5 there is a transfinite sequence $(x_\alpha)_{\alpha \in Ord}$ indexed by all ordinals, such that $x_\alpha = F(\{x_\gamma \mid \gamma < \alpha\})$ for all $\alpha \in Ord$. We prove by induction on α that $x_\gamma < x_\alpha$ for all $\gamma < \alpha$; this contradicts 4.4. There is nothing to prove if $\alpha = 0$. Let $\alpha > 0$; assume that $x_\gamma < x_\beta$ whenever $\gamma < \beta < \alpha$. Let

$$S_\alpha = \{x_\gamma \mid \gamma < \alpha\}.$$

If $\alpha = \beta + 1$ is a successor, then S_α is a chain. We have $x_\beta < t$ for some $t \in X$, since x_β is not a maximal element of X. Then $x_\gamma \leq x_\beta < t$ for all $\gamma < \alpha$ and $u(S_\alpha) \neq \emptyset$. Hence $x_\alpha \in u(S_\alpha)$ and $x_\gamma < x_\alpha$ for all $\gamma < \alpha$. If α is a limit ordinal, then the nonempty chain S_α has an upper bound t in X. Since α is a limit ordinal, $\gamma < \alpha$ implies $\gamma + 1 < \alpha$, so that $x_\gamma < x_{\gamma+1} \leq t$ for all $\gamma < \alpha$; hence again $u(S_\alpha) \neq \emptyset$, $x_\alpha \in u(S_\alpha)$, and $x_\gamma < x_\alpha$ for all $\gamma < \alpha$.

The sequence $(x_\alpha)_{\alpha \in Ord}$ in this proof is normally constructed more informally, as follows. Assume that x_γ has been constructed for all $\gamma < \alpha$, so that $x_\gamma < x_\beta$ for all $\gamma < \beta < \alpha$. Choose any $x_0 \in X$. If $\alpha = \beta + 1$ is a successor, we can choose some $x_\alpha > x_\beta$, since x_β is not a maximal element of X; then $x_\gamma \leq x_\beta < x_\alpha$ for all $\gamma < \alpha$. If α is a limit ordinal, we can choose an upper bound $x_\alpha \in X$ of the nonempty chain $\{x_\gamma \mid \gamma < \alpha\}$, and then $x_\gamma < x_{\gamma+1} \leq x_\alpha$ for all $\gamma < \alpha$. We now have $x_\beta < x_\alpha$ whenever $\beta < \alpha$, blatantly contradicting 4.4. In this argument it is understood that x_α is constructed by ordinal recursion, and that a choice function provides all required choices.

Exercises

1. Let σ be an ordinal number and let $\mathcal{C}$ be a class of ordinal numbers such that (1) $0 \in \mathcal{C}$; (2) if $\alpha \in \mathcal{C}$ and $\alpha + 1 < \sigma$, then $\alpha + 1 \in \mathcal{C}$; and (3) if $\alpha < \sigma$ is a limit ordinal and $\beta \in \mathcal{C}$ for all $\beta < \alpha$, then $\alpha \in \mathcal{C}$. Prove that $\mathcal{C}$ contains every ordinal number $\alpha < \sigma$.

2. Let S be a set and let $F : D \longrightarrow S$ be a mapping, where D is a set of subsets of S. Prove that there exists a unique transfinite sequence (x_α), indexed by all ordinals or by some ordinal σ, such that $x_\alpha = F(\{x_\gamma \mid \gamma < \alpha\})$ whenever $F(\{x_\gamma \mid \gamma < \alpha\})$ is defined (whenever x_γ is defined for all $\gamma < \alpha$, and $\{x_\gamma \mid \gamma < \alpha\} \in D$).

3. Let G be a group and let $a \in G$, $a \neq 1$. Use ordinal induction to prove that there is a subgroup M of G that is maximal such that $a \notin M$.

4. Let G be a group and let A be a subgroup of G. Use ordinal induction to prove that there is a subgroup M of G that is maximal such that $M \cap A = 1$.

5. Use ordinal induction to prove that every vector space has a maximal linearly independent subset.

6. Given a field K, use ordinal induction to construct a field $F \supseteq K$ in which every

irreducible polynomial $q \in K[X]$ has a root. (First, arrange these polynomials into a transfinite sequence.)

7. Let G be a group. Construct a transfinite sequence of subgroups of G (the transfinite ascending central series),

$$1 = Z_0(G) \trianglelefteq Z_1(G) \trianglelefteq \cdots \trianglelefteq Z_\alpha(G) \trianglelefteq Z_{\alpha+1}(G) \trianglelefteq \cdots,$$

in which $Z_{\alpha+1}(G) / Z_\alpha(G) = Z\big(G/Z_\alpha(G)\big)$ for every ordinal α.

5. Cardinal Numbers

Cardinal numbers, introduced by Cantor [1873], are used to assign a number of elements to every set. This section contains a number of notable properties.

Number of elements. Deciding when one set has fewer elements than another, or has as many elements, is easier than actually counting its elements.

Definitions. A set X has as many elements as *a set Y when there exists a bijection of X onto Y. A set X has at most as many elements as a set Y when there exists an injection of X into Y. A set X has fewer elements than a set Y when there exists an injection of X into Y but no bijection of X onto Y.*

Thus, X has at most as many elements as Y if and only if X has as many elements as a subset of Y. For example, we have:

Proposition 5.1. Let $I_n = \{1, 2, \ldots, n\}$. *If $m < n$, then I_m has fewer elements than I_n; in fact, there is no injection $I_n \longrightarrow I_m$.*

Proof. This is not obvious since we have not established that we can count elements as usual. What is obvious is that $I_m \subseteq I_n$ has at most as many elements as I_n. We prove by induction on m that there is no injection $f : I_n \longrightarrow I_m$.

If $m = 0 < n$, then there is no injection of I_n into $I_0 = \emptyset$. Let $m > 0$ and let $f : I_n \longrightarrow I_m$ be any mapping. If $f(n) = f(i)$ for some $i < n$, then f is not injective. Assume that $f(n) \neq f(i)$ for all $i < n$. Let σ be a permutation of I_m such that $\sigma(f(n)) = m$. Let $g = \sigma \circ f : I_n \longrightarrow I_m$. Then $g(n) = m$ and $g(i) \neq g(n) = m$ for all $i < n$. Hence $g(I_{n-1}) \subseteq I_{m-1}$. By the induction hypothesis, the restriction of g to I_{n-1} is not injective. Hence neither is g. $\square$

Proposition 5.2 (Cantor [1883]). *Every set X has fewer elements than the set 2^X of all its subsets.*

Proof. There is an injection $x \longmapsto \{x\}$ of X into 2^X. To show that X has fewer elements than 2^X we prove that there is no bijection of X onto 2^X. Let $f : X \longrightarrow 2^X$ be any mapping. Then $S = \{x \in X \mid x \notin f(x)\} \in 2^X$. But $x \in X$ implies either $x \in S$ and $x \notin f(x)$, or $x \notin S$ and $x \in f(x)$; therefore $S \neq f(x)$ for all $x \in X$, and f is not surjective. $\square$

The next result will fully establish that our terminology is sensible.

Theorem **5.3** (Cantor-Bernstein). *Let* X *and* Y *be sets. If there exist an injection of* X *into* Y *and an injection of* Y *into* X, *then there exists a bijection of* X *onto* Y.

Proof. We may assume that X and Y are disjoint. Let $f : X \longrightarrow Y$ and $g : Y \longrightarrow X$ be injections. Arrange $X \cup Y$ into disjoint families in which every element of one set begets (all by itself) one child in the other set. This imagery is due to Halmos. The *child* of $x \in X$ is $f(x) \in Y$, and x is the *parent* of $f(x)$ (the sole parent, since f is injective); similarly, the *child* of $y \in Y$ is $g(y) \in X$, and y is the sole *parent* of $g(y)$. The *descendants* of $x \in X$ are $f(x)$, $g(f(x))$, $f(g(f(x)))$, ...; the *descendants* of $y \in Y$ are $g(y)$, $f(g(y))$, $g(f(g(y)))$, ... The elements of $f(X)$ or $g(Y)$ have a parent, but the elements of $Y \backslash f(X)$, and the elements of $X \backslash g(Y)$, are *orphans*.

The ancestry of an element of $X \cup Y$ either extends indefinitely upward or ends, or rather begins, with an orphan. Thus, an element of X either descends from an orphan of X (or is an orphan itself), or descends from an orphan in Y, or has infinite ancestry; these constitute disjoint sets X_X, X_Y, X_∞ whose union is X. Similarly, an element of Y either descends from an orphan in X, or descends from an orphan of Y (or is an orphan itself), or has infinite ancestry; these constitute disjoint sets Y_X, Y_Y, Y_∞ whose union is Y.

We see that $f(X_X) = Y_X$, $g(Y_Y) = X_Y$, and $f(X_\infty) = Y_\infty$ (also, $g(Y_\infty) = X_\infty$). Hence f and g induce bijections $X_X \longrightarrow Y_X$, $X_Y \longrightarrow Y_Y$, and $X_\infty \longrightarrow Y_\infty$, which can be pasted together into a bijection $X \longrightarrow Y$. $\square$

Cardinal numbers. The *equipotence* relation "X has as many elements as Y" is reflexive, symmetric, and transitive. Cardinal numbers are most naturally defined as equivalence classes of equipotent sets. Unfortunately, as was the case with ordinals, equipotence classes are too large to be allowed membership in a set or collection. Modern cardinal numbers are sets, chosen so that there is only one in each equipotence class (as in Jech [1978], for instance).

Definition. A cardinal number *is an ordinal number* κ *such that every ordinal number* $\alpha < \kappa$ *has fewer elements than* κ. $\square$

For example, every finite ordinal number $0 = \emptyset$, $1 = \{0\}$, $2 = \{0, 1\}$, ... is a cardinal number, by Proposition 5.1. These are the *finite* cardinals; the remaining cardinals are *infinite*. The first limit ordinal ω is a cardinal: indeed, every finite ordinal $n < \omega$ has fewer elements than ω: if there were a bijection $\omega \longrightarrow n$, then there would be an injection $n + 1 \longrightarrow n$, in defiance of 5.1. Readers will verify that infinite cardinals are limit ordinals and can be arranged into a transfinite sequence, traditionally denoted by

$$\aleph_0 < \aleph_1 < \cdots < \aleph_\alpha < \aleph_{\alpha+1} < \cdots,$$

indexed by all ordinals. (Thus, $\aleph_0$ is another name for ω.)

We show that there is one cardinal number in every equipotence class of sets.

Proposition **5.4.** *For every set* X *there exists a unique cardinal number* $|X|$ *such that there is a bijection of* X *onto* $|X|$.

Proof. By the axiom of choice, every set X can be well ordered (Theorem 2.4) and has the same number of elements as an ordinal number, by 3.9. The least ordinal number κ with this property is a cardinal number (since all ordinals $\alpha < \kappa$ have fewer elements). Moreover, κ is the only cardinal number with a bijection $X \longrightarrow \kappa$: there is no bijection $\kappa \longrightarrow \lambda$ between cardinal numbers $\kappa < \lambda$, since κ has fewer elements than λ. $\square$

Definition. In Proposition 5.4, $|X|$ *is the* cardinality *or* number of elements *of* X.

Then X has as many elements as Y, as defined earlier, if and only if $|X| = |Y|$; readers will show that X has at most as many elements as Y if and only if $|X| \leqq |Y|$.

Countable sets. A set X is *finite* when its cardinality $|X|$ is finite; equivalently, when there is a bijection of X onto some $I_n = \{1, 2, \ldots, n\}$. Then $n = |X|$; in particular, n is unique (by Proposition 5.1 or 5.4). Otherwise, X is *infinite*.

Definition. A set X is countable *when* $|X| \leqq \aleph_0$.

Countable sets are often defined by the stricter condition $|X| = \aleph_0$; then a set X such that $|X| \leqq \aleph_0$ is *finite or countable*.

Readers will verify that $0 < |X| \leqq \aleph_0 = |\mathbb{N}|$ if and only if there is a surjection of $\mathbb{N}$ onto X; hence a nonempty set X is countable if and only if all the elements of X can be arranged into a finite or infinite sequence $x_1, \ldots, x_n, \ldots$ (indexed by natural numbers). For example, $\mathbb{N}$ and every $I_n = \{1, 2, \ldots, n\}$ are countable. The next result yields more examples.

Proposition **5.5.** *A direct product of finitely many countable sets is countable. A union of countably many countable sets is countable.*

Proof. The elements of $\mathbb{N} \times \mathbb{N}$ can be arranged by increasing sums into a sequence $(1, 1)$; $(1, 2), (2, 1)$; $(1, 3), (2, 2), (3, 1)$; ... Thus $\mathbb{N} \times \mathbb{N}$ is countable. If now X and Y are countable, there are injections $X \longrightarrow \mathbb{N}$, $Y \longrightarrow \mathbb{N}$, and $X \times Y \longrightarrow \mathbb{N} \times \mathbb{N}$, and $X \times Y$ is countable. It follows, by induction on n, that the direct product of n countable sets is countable (for every $n \in \mathbb{N}$).

A countable family of sets can be arranged into a finite or infinite sequence $X_1, \ldots, X_n, \ldots$. Its union $X = \bigcup_{n>0} X_n$ is also the disjoint union of the countable sets $X'_n = X_n \backslash (X_1 \cup \cdots \cup X_{n-1})$. Injections $X'_n \longrightarrow \mathbb{N} \longrightarrow \mathbb{N} \times \{n\}$ then combine into an injection $X \longrightarrow \mathbb{N} \times \mathbb{N}$. Hence X is countable. $\square$

By Proposition 5.5, $\mathbb{Z} = \{0\} \cup \mathbb{N} \cup -\mathbb{N}$ and $\mathbb{Q} = \bigcup_{n \in \mathbb{N}} \{a/n \mid a \in \mathbb{Z}\}$ are countable. But not every set is countable.

Proposition **5.6** (Cantor [1873]). $\mathbb{R}$ *is not countable.*

Proof. Let X be the set of all real numbers with a decimal expansion $0 . d_1 d_2 \ldots d_n \ldots$ in which every digit d_n is either 0 or 1. Every such $0 . d_1 d_2 \ldots d_n \ldots$ is determined by a subset $\{ n \in \mathbb{N} \mid d_n = 1 \}$ of $\mathbb{N}$. This constructs a bijection $X \longrightarrow 2^{\mathbb{N}}$. By 5.2, $|\mathbb{R}| \geqq |X| = |2^{\mathbb{N}}| > |\mathbb{N}| = \aleph_0 . \square$

Readers can now follow in Cantor's footsteps and show that there are only countably many real numbers that are algebraic over $\mathbb{Q}$; this leaves uncountably many real numbers that are transcendental over $\mathbb{Q}$.

Operations. Readers will verify that the union of a set of cardinals is a cardinal (not a pope). Hence the next result is proved like Corollary 3.5 (but using Proposition 5.2):

Proposition **5.7.** *The class Card of all cardinal numbers is not a set.*

In particular, sets do not constitute a set: if they did, then the smaller classes *Ord* and *Card* would be sets, contradicting 3.5 and 5.7.

Addition, multiplication, and *exponentiation* of cardinal numbers are defined as follows. If κ and λ are cardinals, then

$$\kappa + \lambda = |X \cup Y|, \; \kappa\lambda = |X \times Y|, \text{ and } \kappa^\lambda = |X^Y|,$$

where $|X| = \kappa$, $|Y| = \lambda$, $X \cap Y = \emptyset$, and X^Y denotes the set of all mappings of Y into X. The exercises also define infinite sums and products. Readers will verify that these operations are well defined and have good properties. They also have amusing little quirks.

Proposition **5.8.** *Let κ and λ be cardinal numbers. If κ or λ is infinite, then $\kappa + \lambda = \max(\kappa, \lambda)$.*

Proof. We show that $\kappa + \kappa = \kappa$ when κ is infinite. Then $\lambda \leqq \kappa$ implies $\kappa \leqq \kappa + \lambda \leqq \kappa + \kappa = \kappa$ and $\kappa + \lambda = \kappa$, and 5.8 holds.

For every set X, $|X| + |X| = |2 \times X|$, since $2 \times X = \{0, 1\} \times X$ is the disjoint union of $\{0\} \times X$ and $\{1\} \times X$. Let A be an infinite set. Let $\mathcal{S}$ be the set of all ordered pairs (X, f) such that $X \subseteq A$ and f is a bijection of X onto $2 \times X$. Since $|A| \geqq \aleph_0$ there exists an injection $\mathbb{N} \longrightarrow A$, A contains an infinite countable subset X, $2 \times X$ is countable by 5.5, and there is a bijection of X onto $2 \times X$; hence $\mathcal{S} \neq \emptyset$. Partially order $\mathcal{S}$ by $(X, f) \leqq (Y, g)$ if and only if $X \subseteq Y$ and $f = g_{|X}$. It is immediate that every nonempty chain of $\mathcal{S}$ has an upper bound in $\mathcal{S}$. By Zorn's lemma, $\mathcal{S}$ has a maximal element (M, m). Then $|M| + |M| = |M|$; we show that $|M| = |A|$.

If $A \backslash M$ is infinite, then $A \backslash M$ contains an infinite countable subset X, and any bijection $f : X \longrightarrow 2 \times X$ can be combined with $m : M \longrightarrow 2 \times M$ into a bijection $M \cup X \longrightarrow 2 \times (M \cup X)$ that extends m, contradicting the maximality of M. Therefore $A \backslash M$ is finite. Hence M is infinite and contains an infinite countable subset Y. Then $|Y \cup (A \backslash M)| = |Y|$ by 5.5 and

$$|A| = |Y \cup (A \backslash M)| + |M \backslash Y| = |Y| + |M \backslash Y| = |M| . \square$$

Proposition **5.9.** *Let κ and λ be nonzero cardinal numbers. If κ or λ is infinite, then $\kappa\lambda = \max(\kappa, \lambda)$.*

Proof. We show that $\kappa\kappa = \kappa$ when κ is infinite. Then $1 \leqq \lambda \leqq \kappa$ implies $\kappa \leqq \kappa\lambda \leqq \kappa\kappa = \kappa$ and $\kappa\lambda = \kappa$, and 5.9 holds.

Let A be an infinite set. As in the proof of 5.8, let $\mathcal{S}$ be the set of all ordered pairs (X, f) such that $X \subseteq A$ and f is a bijection of X onto $X \times X$. Since A is infinite, A contains an infinite countable subset X, $X \times X$ is countable by 5.5, and there is a bijection of X onto $X \times X$; hence $\mathcal{S} \neq \emptyset$. Partially order $\mathcal{S}$ by $(X, f) \leqq (Y, g)$ if and only if $X \subseteq Y$ and $f = g_{|X}$. It is immediate that every nonempty chain of $\mathcal{S}$ has an upper bound in $\mathcal{S}$. By Zorn's lemma, $\mathcal{S}$ has a maximal element (M, m). Then $|M|\,|M| = |M|$; we show that $|M| = |A|$.

Assume $|M| < |A|$. Then $|A \backslash M| = |A|$ by 5.8, since $|A| = |M| + |M \backslash A|$. Hence there is an injection $M \longrightarrow A \backslash M$ and $A \backslash M$ contains a subset X such that $|X| = |M|$. Then there is a bijection $f : X \longrightarrow M \longrightarrow M \times M \longrightarrow X \times X$, which can be combined with $m : M \longrightarrow M \times M$ into a bijection $M \cup X \longrightarrow (M \cup X) \times (M \cup X)$ that extends m, in utter disregard of the maximality of M. $\square$

Corollary **5.10.** *An infinite set X has $|X|$ finite subsets; moreover, there are $|X|$ finite sequences of elements of X.*

Proof. The set X has at least $|X|$ finite subsets, since it has $|X|$ subsets with one element. On the other hand, X has $1 \leq |X|$ empty subset, $|X|$ subsets with one element, at most $|X|\,|X| = |X|$ subsets with two elements, and generally at most $|X|\,|X| \cdots |X| = |X|^n = |X|$ subsets with n elements. Hence X has at most $|X| + |X| + \cdots = \aleph_0|X|$ finite subsets, and $\aleph_0|X| = |X|$ by 5.9. Subsets can be replaced by sequences in this argument. $\square$

Readers will use Propositions 5.8, 5.9 to show that there are groups of arbitrary cardinality, and that for every ring R there are R-modules of arbitrary infinite cardinality $\kappa \geq |R|$; hence groups do not constitute a set, and modules over a given ring R do not constitute a set.

Exercises

1. Given two sets X and Y, show that there exists an injection of X into Y if and only if there exists a surjection of Y onto X.

2. Show that the union of a set of cardinal numbers is a cardinal number.

3. Show that every infinite cardinal number is a limit ordinal.

4. Show that all infinite cardinals can be arranged into a transfinite sequence $\aleph_0 < \aleph_1 < \cdots < \aleph_\alpha < \aleph_{\alpha+1} < \cdots$ indexed by all ordinals.

5. Prove the following: there exists an injection $X \longrightarrow Y$ if and only if $|X| \leqq |Y|$.

6. Prove that there are countably many real numbers that are algebraic over $\mathbb{Q}$ and uncountably many real numbers that are transcendental over $\mathbb{Q}$.

7. Prove that there are uncountably many mappings of $\mathbb{N}$ into $\mathbb{N}$.

8. For every cardinal $\kappa > 0$, prove that sets of cardinality κ do not constitute a set.

Readers are warned that the sum and product of two cardinals are usually not the same as their sum and product as ordinals.

9. Show that the addition of cardinals is commutative and associative.

10. Show that the multiplication of cardinals is commutative and associative.

11. Show that $(\kappa + \lambda)\,\mu = \kappa\mu + \lambda\mu$ for all cardinals κ, λ, μ.

12. Show that $\kappa^{\lambda+\mu} = \kappa^\lambda\,\kappa^\mu$ for all cardinals κ, λ, μ.

13. Show that $\kappa^{\lambda\mu} = (\kappa^\lambda)^\mu$ for all cardinals κ, λ, μ.

14. Verify that infinite sums and products of cardinals $(\kappa_i)_{i\in I}$ are well defined by $\sum_{i\in I} \kappa_i = \left| \bigcup_{i\in I} X_i \right|$ and $\prod_{i\in I} \kappa_i = \left| \prod_{i\in I} X_i \right|$, where $|X_i| = \kappa_i$ for all i and the sets X_i are pairwise disjoint.

15. Show that $\sum_{i\in I} \kappa_i = \sum_{j\in J} \left(\sum_{i\in I_j} \kappa_i \right)$ when $I = \bigcup_{j\in J} I_j$ is a partition of I.

16. Show that $\prod_{i\in I} \kappa_i = \prod_{j\in J} \left(\prod_{i\in I_j} \kappa_i \right)$ when $I = \bigcup_{j\in J} I_j$ is a partition of I.

17. Show that $\prod_{i\in I} \kappa^{\lambda_i} = \kappa^{\sum_{i\in I} \lambda_i}$.

18. Show that $\left(\prod_{i\in I} \kappa_i \right)^\lambda = \prod_{i\in I} \kappa_i^\lambda$.

19. Show that every cardinal number is the cardinality of a group.

20. Let R be a ring. Show that every infinite cardinal number $\kappa \geq |R|$ is the cardinality of an R-module.

References

The following are books and papers cited in the text.

[1824] Abel, N., *Mémoire sur les équations algébriques, où l'on démontre l'impossibilité de la résolution de l'équation générale du cinquième degré*, Christiania, 1824; published in J. reine angew. Math. 1 (1826), and in Oeuvres complètes, vol. I, Christiania, 1881, 28–33, 66–87.

[1926] Artin, E. and Schreier, O., *Algebraische Konstruktion reeller Körper*, Abh. Math. Sem. Hamburg 5 (1926), 83–115.

[1927] Artin, E., *Zur Theorie der hyperkomplexen Zahlen*, Abh. Math. Sem. Hamburg 5 (1927), 251–260.

[1940] Baer, R., *Abelian groups that are direct summands of every containing abelian group*, Bull. Amer. Math. Soc. 46 (1940), 800–806.

[1966] Beck, J., According to MacLane [1971], Beck's theorem was "unpublished, but presented at a conference in 1966"; it is likely that MacLane was there.

[1933] Birkhoff, G., *On the combination of subalgebras*, Proc. Cambridge Philos. Soc. 29 (1933), 441–464.

[1934] Birkhoff, G., *On the lattice theory of ideals*, Bull. Amer. Math. Soc. 40 (1934), 613–619.

[1935] Birkhoff, G., *On the structure of abstract algebras*, Proc. Cambridge Philos. Soc. 31 (1935), 433–454.

[1940] Birkhoff, G., *Lattice Theory*, Amer. Math. Soc., 1940.

[1944] Birkhoff, G., *Subdirect unions in universal algebra*, Bull. Amer. Math. Soc. 50 (1944), 764–768.

[1847] Boole, R., *The Mathematical Analysis of Logic*, 1847.

[1965] Buchberger, B., Doctoral Dissertation, Innsbrück, 1965; see also *A theoretical basis for the reduction of polynomials to canonical forms*, ACM SIGSAM Bull. 39 (1976), 19–29.

[1897] Burnside, W., *Theory of groups of finite order*, Cambridge Univ. Press, 1897.

[1905] Burnside, W., *On the condition of reducibility of any group of linear substitutions*, Proc. London Math. Soc. (2) 3 (1905), 430–434.

[1873] Cantor, G., *Über eine Eigenschaft des Inbegriffs aller reellen algebraischen Zahlen*, J. reine angew. Math. 77 (1873), 258–263.

[1883] Cantor, G., *Fondements d'une théorie générale des ensembles*, Acta Math 2 (1883), 381–408.

[1545] Cardano, G., *Ars Magna*, 1545.

[1956] Cartan, H. and Eilenberg, S., *Homological Algebra*, Princeton, 1956.

[1815] Cauchy, A., *Mémoire sur le nombre des valeurs qu'une fonction peut acquérir, lorsqu'on y permute de toutes les manières possibles les quantités qu'elle renferme*, J. Ecole Polytechn. X (1815), 1–28.

[1961] Chase, S., *Direct products of modules*, Trans. Amer. Math. Soc. 97 (1961), 457–473.

[1871] Dedekind, R., Xth supplement to Dirichlet's *Vorlesungen über Zahlentheorie*, Braunschweig, 1871.

[1897] Dedekind, R., *Über Zerlegungen von Zahlen durch ihre grössten gemeinsamen Teiler*, Festschr. Techn. Hoch. Braunschweig (1897), 1–40.

[1900] Dedekind, R., *Über die von drei Moduln erzeugte Dualgruppe*, Math. Ann. 53 (1900), 371–403.

[1837] Dirichlet, G., *Beweis des Satzes, dass jede unbegrentze arithmetische Progression, deren erstes Glied und Differenz ganze Zahlen ohne gemeinschaftlichen Factor sind, unendlich viele Primzahlen enthält*, Abh. König. Akad. Wiss. Berlin, math. Abh. 1837, 45–71.

[1882] Dyck, W., *Gruppentheoretische Studien*, Math. Ann. 20 (1882), 1–45.

[1942] Eilenberg, S. and MacLane, S., *Group extensions and homology*, Ann. of Math. 43 (1942), 757–831.

[1945] Eilenberg, S. and MacLane, S., *General theory of natural equivalences*, Trans. Amer. Math. Soc. 58 (1945), 231–294.

[1965] Eilenberg, S. and Moore, J.C., *Adjoint functors and triples*, Illinois J. Math. 9 (1965), 381–398.

[1850] Eisenstein, G., *Über die Irreductibilität und einige andere Eigenschaften der Gleichung, von welcher die Theilung der ganzen Lemniscate abhängt*, J. reine angew. Math. 39 (1850), 160–179.

[1963] Feit, W. and Thompson, J.G., *Solvability of groups of odd order*, Pacific J. Math. 13 (1963), 775–1029.

[1964] Freyd, P., *Abelian Categories*, Harper and Row, New York, 1964.

[1877] Frobenius, G., *Über lineare Substitutionen und bilineare Formen*, J. reine angew. Math. 84 (1877), 1–63.

[1878] Frobenius, G. and Stickelberger, L., *Über Gruppen mit vertauschbaren Elementen*, J. reine angew. Math. 86 (1878), 217–262.

[1830] Galois, E., *Mémoire sur les conditions de résolubilité des équations par radicaux*, 1930; published in J. Math. Pures Appl. 11 (1846), 381–444, and in Oeuvres Mathématiques, Paris, 1897, 33–50.

[1799] Gauss, G., Doctoral dissertation, Helmstadt, 1799.

[1801] Gauss, G., *Disquisitiones arithmeticae*, Leipzig, 1801.

[1938] Gödel, K., *The consistency of the axiom of choice and of the generalized continuum hypothesis*, Proc. Nat. Acad. Sci. 24 (1938), 556–557.

[1994 up] Gorenstein, D., Lyons, R., Solomon, R., *The Classification of the Finite Simple Groups*, Amer. Math. Soc., Providence, RI, 1994, 1996, 1998, 1998, 2002, 2005.

[1844] Grassmann, H., *Grundzüge zu einer rein geometrischen Theorie der Curven, mit Anwendung einer rein geometrischen Analyse*, J. reine angew. Math. 31 (1844), 111–132.

[1927] Grell, H., *Beziehungen zwischen den Idealen verschiedener Ringe*, Math. Ann. 97 (1927), 490–523.

[1975] Grillet, P., *Primary semigroups*, Michigan Math. J. 22 (1975), 321–336.

[1939] Gröbner, W., *Über die algebraischen Eigenschaften der Integrale von linearen Differentialgleichungen mit konstanten Koeffizienten*, Monatsh. Math. Phys. 47 (1939), 247–284.

[1928] Hall, P., *A note on soluble groups*, J. London Math. Soc. 3 (1928), 98–105.

[1843] Hamilton, W.R., *On Quaternions; or on a new System of Imaginaries in Algebra* (letter to John T. Graves, dated October 17, 1843), Philos. Magazine 25, 489–495.

[1914] Hausdorff, F., *Grundzüge der Mengenlehre*, Leipzig, Veit & Co., 1914.

[1897] Hensel, K., *Über die Fundamentalgleichung und die ausserwesentlichen Diskriminantentheiler eines algebraischen Körpers*, Gött. Nachr. (1897), 254–260; see also *Theorie der algebraischen Zahlen*, Teubner, Lepzig, 1908.

[1904] Hensel, K., *Neue Grundlagen der Arithmetik*, J. Reine Angew. Math. 127 (1904), 51–84.

[1890] Hilbert, D., *Über die Theorie der algebraischen Formen*, Math. Ann. 36 (1890), 473–534.

[1893] Hilbert, D., *Über die vollen Invariantensysteme*, Math. Ann. 42 (1893), 313–373.

[1897] Hilbert, D., *Zahlbericht*, Jahresber. D.M.V., 4 (1897), 175–546.

[1889] Hölder, O., *Zurückführung einer beliebigen algebraischen Gleichung auf eine Kette von Gleichungen*, Math. Ann. 34 (1889), 26–56.

[1939] Hopkins, C., *Rings with minimal condition for left ideals*, Ann. Math. (2) 40 (1939), 712–730.

[1941] Hurewicz, W., *On duality theorems*, Abstract # 47-7-329, Bull. Amer. Math. Soc. 47 (1941), 562–563.

[1945a] Jacobson, N., *The radical and semi-simplicity for arbitrary rings*, Amer. J. Math. 67 (1945), 300–320.

[1945b] Jacobson, N., *Structure theory of simple rings without finiteness assumptions*, Trans. Amer. Math. Soc. 57 (1945), 228–245.

[1978] Jech, T., *Set Theory*, Academic Press, 1978.

[1869] Jordan, C., *Théorèmes sur les équations algébriques*, J. math. pures appl. (2) 14 (1869), 139–146; *Commentaires sur Galois*, Math. Ann. 1 (1869), 141–160.

[1870] Jordan, C., *Traité des substitutions et des équations algébriques*, Gauthier-Villars, Paris, 1870, 114–125.

[1872] Klein, F., Inaugural address, Erlangen Univ., 1872.

[1870] Kronecker, L., *Auseinandersetzungen einiger Eigenschaften der Klassenanzahl idealer complexer Zahlen*, Monatsh. Abh. Berlin (1870), 881.

[1887] Kronecker, L., *Ein Fundamentalsatz der allgemeinen Arithmetik*, J. reine angew. Math. 100 (1887), 490–510.

[1925] Krull, W., *Über verallgemeinerte endliche Abelsche Gruppen*, Math. Zeitschr. 23 (1925), 161–196.

[1928] Krull, W., *Primidealketten in allgemeinen Ringbereichen*, Sitz. Heidelberg Akad. Wiss. 1928, No. 7.

[1932] Krull, W., *Allgemeine Bewertungstheorie*, J. reine angew. Math. 167 (1932), 160–196.

[1922] Kuratowski, C., *Une méthode d'élimination des nombres transfinis des raisonnements mathématiques*, Fund. Math. 3 (1922), 76–108.

[1913] Kürschak, J., *Über Limesbildung und allgemeine Körpertheorie*, Proc. 5. Intern. Math. Congr. 1 (1913), 285–289.

[1770] Lagrange, J., *Réflexions sur la résolution algébrique des équations*, 1770; Oeuvres de Lagrange, 3, 205–421.

[1905] Lasker, E., *Zur Theorie der Moduln und Ideale*, Math. Ann. 60 (1905), 20–116.

[1969] Lazard, D., *Autour de la platitude*, Bull. Soc. Math. France 97 (1969), 81–128.

[1939] Levitzki, J., *On rings which satisfy the minimum condition for the right-hand ideals*, Compos. Math. 7 (1939), 214–222.

[1939] MacLane, S., *Modular fields (I)*, Duke Math. J. 5 (1939), 372–393.

[1963] MacLane, S., *Homology*, Springer, 1963.

[1971] MacLane, S., *Categories for the Working Mathematician*, Springer, 1971.

[1935] MacNeille, H., Doct. diss., Harvard, 1935; see also *Partially ordered sets*, Trans. Amer. Math. Soc. 42 (1937), 416–460.

[1958] Mal'cev, A.I., *On homomorphisms onto finite groups* [Russian], Uch. Zap. Ivanov. Gos. Ped. Inst. 18 (1958), 49–60.

[1898] Maschke, H., *Über den arithmetischen Charakter der Coefficienten der Substitutionen endlicher linearer Substitutionsgruppen*, Math. Ann. 50 (1898), 492–498.

[1966] McCarthy, P., *Algebraic Extensions of Fields*, Chelsea, 1966.

[1910] Moore, E., *Introduction to a Form of General Analysis*, The New Haven Math. Coll., Yale Univ. Press, New Haven, 1910.

[1962] Nagata, M., *Local Rings*, Wiley, 1962, p. 213.

[1924] Nielsen, J., *Die Isomorphismengruppe der freien Gruppen*, Math. Ann. 91 (1924), 169–209.

[1921] Noether, E., *Idealtheorie in Ringbereichen*, Math. Ann. 83 (1921), 24–66.

[1926] Noether, E., *Abstrakter Aufbau der Idealtheorie in algebraischen Zahl und Funktionenkörpern*, Math. Ann. 96 (1926), 26–61.

[1929] Noether, E., *Hyperkomplexe Grössen und Darstellungstheorie*, Math. Z. 30 (1929), 641–692.

[1918] Ostrowski, A., *Über einige Lösungen der Funktionalgleichung $\varphi(x)\,\varphi(y) = \varphi(xy)$*, Acta Math. 41 (1918), 271–284.

[1934] Ostrowski, A., *Untersuchungen zur arithmetischen Theorie der Körper. (Die Theorie der Teilbarkeit in allgemeinen Körpern. I-III.)*, Math. Z. 39 (1934), 296–404.

[1864] Peirce, B., *Linear Associative Algebra*, lecture to the American Association for the Advancement of Science; published in Amer. J. Math. 4 (1881), 97–229.

[1956] Rédei, L., *The Theory of Finitely Generated Commutative Semigroups*, Akad. Kiadó, Budapest, 1956; English translation, Pergamon Press, Oxford, 1963.

[1911] Remak, R., *Über die Zerlegung der endlichen Gruppen in direkte unzerlegbare Faktoren*, J. reine angew. Math. 139 (1911), 293–308.

[1930] Remak, R., *Über minimale invariante Untergruppen in der Theorie der endlichen Gruppen*, J. reine angew. Math. 162 (1930), 1–16.

[1868] Schering, E., *Der fundamental Classen der zusammengesetzbaren arithmetischen Formen*, Abh. Ges. Göttingen 14 (1868-69), 13.

[1912] Schmidt, O., *Über die Zerlegung endlicher Gruppen in direkte unzerlegbare Faktoren*, Izv. Kiev Univ. 1912, 1–6.

[1928] Schmidt, O., *Über unendlich Gruppen mit endlicher Kette*, Math. Z. 29 (1928), 34–41.

[1926] Schreier, O., *Über die Erweiterung von Gruppen, I*, Monatsh. Math. Phys. 34 (1926), 165–180; *Über die Erweiterung von Gruppen, II*, Abh. math. Sem. Hamburg 4 (1926), 321–346.

[1928] Schreier, O., *Über der Jordan-Hölderschen Satz*, Abh. math. Sem. Hamburg 6 (1928), 300–302.

[1980] Schreyer, F., *Die Berechnung von Syzygien mit dem verallgemeinerten Weierstrass's-chen Divisionsatz*, Diplom Thesis, Univ. of Hamburg, 1980; see also *A standard basis approach to syzygies of canonical curves*, J. Reine Angew. Math. 421 (1991), 83–123.

[1904] Schur, I., *Über die Darstellung der endlichen Gruppen durch gebrochene lineare Substitutionen*, J. reine angew. Math. 127 (1904), 20–50.

[1910] Steinitz, E., *Algebraische Theorie der Körper*, J. reine angew. Math. 137 (1910), 167–309.

[1934] Stone, M., *Boolean algebras and their applications to topology*, Proc. Nat. Acad. Sci. 20 (1934), 197–202.

[1935] Stone, M., *Subsumption of the theory of Boolean algebras under the theory of rings*, Proc. Nat. Acad. Sci. 21 (1935), 103–105.

[1872] Sylow, L., *Théorèmes sur les groupes de substitutions*, Math. Ann. 5 (1872), 584–594.

[1948] Uzkov, A.I., *On rings of quotients of commutative rings*, Mat. Sbornik (N.S.) 13 (1948), 71–78.

[1930] van der Waerden, B., *Moderne Algebra*, Springer, 1930 (vol. I), 1931 (vol. II).

[1905] Wedderburn, J., *A theorem on finite algebras*, Trans. Amer. Math. Soc. 6 (1905), 349–352.

[1907] Wedderburn, J., *Note on hypercomplex numbers*, Proc. Edinburgh Math. Soc. 25 (1907), 2–4.

[1952] Weil, A., *Fibre Spaces in Algebraic Geometry*, lectures at the University of Chicago, 1952.

[1898] Whitehead, A., *A Treatise on Universal Algebra*, Cambridge Univ. Press, 1898.

[1944] Zariski, O., *The compactness of the Riemann manifold of an abstract field of algebraic functions*, Bull. Amer. Math. Soc. 50 (1944), 683–691.

[1933] Zassenhaus, H., *Zum Satz von Jordan-Hölder-Schreier*, Abh. math. Sem. Hamburg 10 (1934), 106–108.

[1937] Zassenhaus, H., *Lehrbuch der Gruppentheorie*, vol. 1, Teubner, Lepizig, 1937.

[1904] Zermelo, E., *Beweis, dass jede Menge wohlgeordnet werden kann*, Math. Ann. 59 (1904), 514–516.

[1935] Zorn, M., *A remark on method in transfinite algebra*, Bull. Amer. Math. Soc. 41 (1935), 667–670.

Further Readings

The following are some historically important books, and some books I like. They do not constitute a complete bibliography.

General

van der Waerden, B.L., *Moderne Algebra, vol. I, vol. II*, Springer, 1930, 1931.

Lang, Serge, *Algebra*, Addison-Wesley, 1965.

Hungerford, Thomas W., *Algebra*, Springer, 1980.

Isaacs, I. Martin, *Algebra, a Graduate Course*, Brooks/Cole, 1994.

Kempf, George R., *Algebraic Structures*, Vieweg, 1995.

Groups

Burnside, W., *Theory of Groups of Finite Order*, Cambridge, 1897.

Suzuki, Michio, *Group Theory*, Springer, 1986.

Gorenstein, Daniel, *Finite Groups*, 2nd. ed., Chelsea, 1980.

Rings and modules

Jacobson, Nathan, *Structure of Rings*, American Mathematical Society, 1956.

Dauns, John, *Rings and Modules*, Cambridge, 1994.

Lam. T.Y., *A First Course in Noncommutative Rings*, Springer, 1991.

Commutative Algebra

Zariski, Oscar and Samuel, Pierre, *Commutative Algebra*, vol. I, vol. II, van Nostrand, 1958, 1960.

Eisenbud, David, *Commutative Algebra With a View to Algebraic Geometry*, Springer, 1994.

Homological Algebra

Cartan, Henri and Eilenberg, Samuel, *Homological Algebra*, Princeton, 1956.

MacLane, Saunders, *Homology*, Springer, 1963.

Rotman, Joseph J., *An Introduction to Homological Algebra*, Academic Press, 1979.

Lattices

Birkhoff, G., *Lattice Theory*, American Mathematical Society, 1940.

Grätzer, Goerge, *General Lattice Theory*, Birkhäuser, 1978.

Universal Algebra

Cohn, P.M., *Universal Algebra*, Harper and Row, 1965.

Grätzer, Goerge, *Universal Algebra*, van Nostrand, 1968.

Categories

MacLane, Saunders, *Categories for the Working Mathematician*, Springer, 1971.

Set Theory

Jech, T., *Set Theory*, Academic Press, 1978.

History

Lubos Nový, *Origins of Modern Algebra*, Noordhoff, Leyden 1973.

Index

Graduate Texts in Mathematics

(*continued from page ii*)